Customer Support Information

Plunkett's Biotech & Genetics Industry Almanac 2008

Please register your book immediately...

if you did not purchase it directly from Plunkett Research, Ltd. This will enable us to fulfill your replacement request if you have a damaged product, or your requests for assistance. Also it will enable us to notify you of future editions, so that you may purchase them from the source of your choice.

If you are an actual purchaser but did not receive a FREE CD-ROM version with your book...

you may request it by returning this form. (Proof of purchase may be required.)

_____ YES, please register me as a purchaser of the book.
I did not buy it directly from Plunkett Research, Ltd.

_____ YES, please send me a free CD-ROM version of the book.
I am an actual purchaser, but I did not receive one with my book.

Customer Name _____

Title_____

Organization _____

Address_____

City_____State_____Zip_____

Country (if other than USA) _____

Phone_____Fax _____

E-mail _____

Mail or Fax to: **Plunkett Research, Ltd.**
Attn: FREE CD-ROM and/or Registration
P.O. Drawer 541737, Houston, TX 77254-1737 USA
713.932.0000 · Fax 713.932.7080 · www.plunkettresearch.com

PLUNKETT'S BIOTECH & GENETICS INDUSTRY ALMANAC 2008

The only comprehensive
guide to biotechnology and genetics companies
and trends

Jack W. Plunkett

Published by:
Plunkett Research, Ltd., Houston, Texas
www.plunkettresearch.com

PLUNKETT'S
BIOTECH & GENETICS
INDUSTRY ALMANAC
2008

Editor and Publisher:
Jack W. Plunkett

Executive Editor and Database Manager:
Martha Burgher Plunkett

Senior Editors and Researchers:
Addie K. FryeWeaver
Christie Manck
John Peterson

Editors, Researchers and Assistants:
Andreea Balan
Brandon Brison
Daniel Jordan
Kara Jordan
Maria Kolber
Lindsey Meyn
Kristen Morrow
Kyle Wark
Suzanne Zarosky

E-Commerce Managers:
Heather M. Cook
Jared Deter
Emily Hurley
Ian Markham
Lynne Zarosky

Information Technology Managers:
Wenping Guo
Carl Thomsen

Cover Design:
Kim Paxson, Just Graphics
Junction, TX

Special Thanks to:
Burrill & Company
Centers for Medicare & Medicaid Services (CMS)
Ernst & Young
ISAAA
IMS Health
National Science Foundation
Pharmaceutical Research & Manufacturers
Association (PhRMA)
Tufts Center for the Study of Drug Development
U.S. Department of Labor, Bureau of Labor Statistics
U.S. Food & Drug Administration (FDA)
U.S. National Science Foundation
U.S. Patent & Trademark Office

Plunkett Research, Ltd.
P. O. Drawer 541737, Houston, Texas 77254 USA
Phone: 713.932.0000   Fax: 713.932.7080
www.plunkettresearch.com

Published by:

Plunkett Research, Ltd.

P. O. Drawer 541737

Houston, Texas 77254-1737

Phone: 713.932.0000

Fax: 713.932.7080

Internet: www.plunkettresearch.com

ISBN10 # 1-59392-087-3

ISBN13 # 978-1-59392-087-6

Disclaimer of liability
for use and results of use:

The editors and publishers assume no responsibility for your own success in making an investment or business decision, in seeking or keeping any job, in succeeding at any firm or in obtaining any amount or type of benefits or wages. Your own results and the job stability or financial stability of any company depend on influences outside of our control. All risks are assumed by the reader. Investigate any potential employer or business relationship carefully and carefully verify past and present finances, present business conditions and the level of compensation and benefits currently paid. Each company's details are taken from sources deemed reliable; however, their accuracy is not guaranteed. The editors and publishers assume no liability, beyond the actual payment received from a reader, for any direct, indirect, incidental or consequential, special or exemplary damages, and they do not guarantee, warrant nor make any representation regarding the use of this material. Trademarks or tradenames are used without symbols and only in a descriptive sense, and this use is not authorized by, associated with or sponsored by the trademarks' owners. Ranks and ratings are presented as an introductory and general glance at corporations, based on our research and our knowledge of businesses and the industries in which they operate. The reader should use caution.

PLUNKETT'S BIOTECH & GENETICS INDUSTRY ALMANAC 2008

CONTENTS

Continued on next page

A Short Biotech & Genetics Industry Glossary

10-K (10K): An annual report filed by publicly held companies. It provides a comprehensive overview of the company's business and its finances. By law, it must contain specific information and follow a given form, the "Annual Report on Form 10-K." The U.S. Securities and Exchange Commission requires that it be filed within 90 days after fiscal year end. However, these reports are often filed late due to extenuating circumstances. Variations of a 10-K are often filed to indicate amendments and changes. Most publicly held companies also publish an "annual report" that is not on Form 10-K. These annual reports are more informal and are frequently used by a company to enhance its image with customers, investors and industry peers.

510 K: An application filed with the FDA for a new medical device to show that the apparatus is "substantially equivalent" to one that is already marketed.

ADME (Absorption, Distribution, Metabolism and Excretion): In clinical trials, the bodily processes studied to determine the extent and duration of systemic exposure to a drug.

AE (Adverse Event): In clinical trials, a condition not observed at baseline or worsened if present at baseline. Sometimes called Treatment Emergent Signs and Symptoms (TESS).

Agricultural Biotechnology (AgriBio): The application of biotechnology methods to enhance agricultural plants and animals.

Amino Acid: Any of a class of 20 molecules that combine to form proteins.

ANDA (Abbreviated New Drug Application): An application filed with the FDA showing that a substance is the same as an existing, previously approved drug (i.e., a generic version).

Angiogenesis: Blood vessel formation, typically in the growth of malignant tissue.

Angioplasty: The re-opening of a blood vessel by non-surgical techniques such as balloon dilation or laser, or through surgery.

Antibody: A protein produced by white blood cells in response to a foreign substance (see "Antigen"). Each antibody can bind only to one specific antigen.

Antigen: A foreign substance that causes the immune system to create an antibody (see "Antibody").

Antisense Technology: The use of RNA-like oligonucleotides that bind to RNA and inhibit the expression of a gene.

Apoptosis: A normal cellular process leading to the termination of a cell's life.

Applied Research: The application of compounds, processes, materials or other items discovered during basic research to practical uses. The goal is to move discoveries along to the final development phase.

Array: An orderly arrangement, such as a rectangular matrix of data. In some laboratory systems, such as microarrays, multiple detectors (probes) are positioned in an array in order to best perform research. See "Microarray."

Artificial Life (AL): See "Synthetic Biology."

Assay: A laboratory test to identify and/or measure the amount of a particular substance in a sample. Types of assays include endpoint assays, in which a single measurement is made at a fixed time; kinetic assays, in which increasing amounts of a product are formed with time and are monitored at multiple points; microbiological assays, which measure the concentration of antimicrobials in biological material; and immunological assays, in which analysis or measurement is based on antigen-antibody reactions.

Baby Boomer: Generally refers to people born in the U.S. and Western Europe from 1946 to 1964. In the U.S., the total number of Baby Boomers is about 78 million--one of the largest and most affluent demographic groups. The term evolved to include the children of soldiers and war industry workers who were involved in World War II. When those veterans and workers returned to civilian life they started or added to families in large numbers. As a result, the baby boom generation is one of the largest demographic segments in the U.S. Some baby boomers have already started reaching early retirement age. By 2011, millions will begin turning traditional retirement age (65), resulting in extremely rapid growth in the senior portion of the population.

Baseline: A set of data used in clinical studies, or other types of research, for control or comparison.

Basic Research: Attempts to discover compounds, materials, processes or other items that may be largely or entirely new and/or unique. Basic research may start with a theoretical concept that has yet to be proven. The goal is to create discoveries that can be moved along to applied research. Basic research is sometimes referred to as "blue sky" research.

Big Pharma: The top tier of pharmaceutical companies in terms of sales and profits (e.g., Pfizer, Merck, Johnson & Johnson).

Bioaccumulation: A process in which chemicals are retained in fatty body tissue and increased in concentration over time.

Bioavailability: In pharmaceuticals, the rate and extent to which a drug is absorbed or is otherwise available to the treatment site in the body.

Biochemical Engineering: A sector of chemical engineering that deals with biological structures and processes. Biochemical engineers may be found in the pharmaceutical, biotechnology and environmental fields, among others.

Biochemicals: Chemicals that either naturally occur or are identical to naturally occurring substances. Examples include hormones, pheromones and enzymes. Biochemicals function as pesticides through non-toxic, non-lethal modes of action, disrupting insect mating patterns, regulating growth or acting as repellants. They tend to be environmentally desirable, and may be produced by industry from organic sources such as plant waste (biomass). Biochemicals also may be referred to as bio-based chemicals, green chemicals or plant-based chemicals.

Biodiesel: A fuel derived when glycerin is separated from vegetable oils or animal fats. The resulting byproducts are methyl esters (the chemical name for biodiesel) and glycerin which can be used in soaps and cleaning products. It has lower emissions than petroleum diesel and is currently used as an additive to that fuel since it helps with lubricity.

Bioengineering: Engineering principles applied when working in biology and pharmaceuticals.

Bioequivalence: In pharmaceuticals, the demonstration that a drug's rate and extent of absorption are not significantly different from those of an existing drug that is already approved by the FDA. This is the basis upon which generic and brand name drugs are compared.

Bioethanol: A fuel produced by the fermentation of plant matter such as corn. Fermentation is enhanced through the use of enzymes that are created through biotechnology. (Also, see "Ethanol").

Biogeneric: See "Follow-on Biologics."

Biogenerics: Genetic versions of drugs that have been created via biotechnology. Also, see "Follow-on Biologics".

Bioinformatics: Research, development or application of computational tools and approaches for expanding the use of biological, medical, behavioral or health data, including those to acquire, store, organize, archive, analyze or visualize such data. Bioinformatics is often applied to the study of genetic data. It applies principles of information sciences and technologies to make vast, diverse and complex life sciences data more understandable and useful.

Biologic: Drugs that are synthesized from living organisms. Specifically, biologics may be any virus, therapeutic serum, toxin, antitoxin, vaccine, blood, blood component or derivative, allergenic or analogous product, or arsphenamine or one of its derivatives used for the prevention, treatment or cure of disease.

Biomagnification: The increase of tissue accumulation in species higher in the natural food chain as contaminated food species are eaten.

Biomass: Organic, non-fossil material of biological origin constituting a renewable energy source. The biomass can be burnt as fuel in a system that creates steam to turn a turbine, generating electricity. For example, biomass can include wood chips and agricultural crops.

BioMEMS: MEMS used in medicine. See "MEMS (Micro Electro Mechanical Systems)."

Biomimetic: Mimicking, imitating, copying or learning from nature.

Biopharmaceutical: That portion of the pharmaceutical industry focused on the use of biotechnology to create new drugs. A biopharmaceutical can be any biological compound that is intended to be used as a therapeutic drug, including recombinant proteins, monoclonal and polyclonal antibodies, antisense oligonucleotides, therapeutic genes, and recombinant and DNA vaccines.

Biopolymeroptoelectromechanical Systems (BioPOEMS): Combination of MEMS and optics used in biological applications.

Biorefinery: A refinery that produces fuels from biomass. These fuels may include bioethanol (produced from corn or other plant matter) or biodiesel (produced from plant or animal matter).

Biosensor: A sensor based on the use of biological materials or that targets biological analytes.

Biosimilar: See "Follow-on Biologics."

Biotechnology: A set of powerful tools that employ living organisms (or parts of organisms) to make or modify products, improve plants or animals (including humans) or develop microorganisms for specific uses. Early uses of biotechnology included traditional animal and plant breeding techniques, based on improving genetic lineage in order to create a plant or animal with desirable characteristics. Early uses also included the use of yeast in making bread, beer, wine and cheese. Today, biotechnology is most commonly thought of to include the development of human medical therapies and processes using recombinant DNA, cell fusion, other genetic techniques and bioremediation. Modern biotechnology uses advanced technologies to modify the genes of cells so they will produce new substances or perform new functions. A good example is recombinant DNA technology, in which a copy of DNA containing one or more genes is transferred between organisms or recombined within an organism.

BLA (Biologics License Application): An application to be submitted to the FDA when a firm wants to obtain permission to market a biological product. It is typically submitted after completion of Phase III clinical trials. It was formerly known as Product License Application (PLA).

Blastocyst: A fertilized embryo, aged four to 11 days, which consists of multiplying cells both outside and inside a cavity. It is the blastocyst that embeds itself in the uterine wall and ultimately develops into a fetus. Blastocysts are utilized outside the womb during the process of stem cell cultivation.

Branding: A marketing strategy that places a focus on the brand name of a product, service or firm in order to increase the brand's market share, increase sales, establish credibility, improve satisfaction, raise the profile of the firm and increase profits.

B-to-B, or B2B: See "Business-to-Business."

B-to-C, or B2C: See "Business-to-Consumer."

Business Process Outsourcing (BPO): The outsourcing of non-mission-critical business processes that may include call centers, basic accounting or human resources management, depending on the industry involved. Also, see "ITES (IT-Enabled Services)."

Business-to-Business: An organization focused on selling products, services or data to commercial customers rather than individual consumers. Also known as B2B.

Business-to-Consumer: An organization focused on selling products, services or data to individual consumers rather than commercial customers. Also known as B2C.

CANDA (Computer-Assisted New Drug Application): An electronic submission of a new drug application (NDA) to the FDA.

Captive Offshoring: Used to describe a company-owned offshore operation. For example, Microsoft owns and operates significant captive offshore research and development centers in China and elsewhere that are offshore from Microsoft's U.S. home base. Also see "Offshoring."

Carcinogen: A substance capable of causing cancer. A suspected carcinogen is a substance that may cause cancer in humans or animals but for which the evidence is not conclusive.

CBER (Center for Biologics Evaluation and Research): The branch of the FDA responsible for the regulation of biological products, including blood, vaccines, therapeutics and related drugs and devices, to ensure purity, potency, safety, availability and effectiveness. www.fda.gov/cber

CDC (Centers for Disease Control and Prevention): The federal agency charged with protecting the public health of the nation by providing leadership and direction in the prevention and control of diseases and other preventable conditions and responding to public health emergencies. Headquartered in Atlanta, it was established as an operating health agency within the U.S. Public Health Service on July 1, 1973. See www.cdc.gov.

CDER (Center for Drug Evaluation and Research): The branch of the FDA responsible for the regulation of drug products. www.fda.gov/cder

CDRH (Center for Devices and Radiological Health): The branch of the FDA responsible for the regulation of medical devices. www.fda.gov/cdrh

Chromosome: A structure in the nucleus of a cell that contains genes. Chromosomes are found in pairs.

Class I Device: An FDA classification of medical devices for which general controls are sufficient to ensure safety and efficacy.

Class II Device: An FDA classification of medical devices for which performance standards and special controls are sufficient to ensure safety and efficacy.

Class III Device: An FDA classification of medical devices for which pre-market approval is required to ensure safety and efficacy, unless the device is substantially equivalent to a currently marketed device. See "510 K."

Clinical Trial: See "Phase I Clinical Trials," along with definitions for Phase II, Phase III and Phase IV.

Clone: A group of identical genes, cells or organisms derived from one ancestor. A clone is an identical copy. "Dolly" the sheep is a famous case of a clone of an animal. Also see "Cloning (Reproductive)" and "Cloning (Therapeutic)."

Cloning (Reproductive): A method of reproducing an exact copy of an animal or, potentially, an exact copy of a human being. A scientist removes the nucleus from a donor's unfertilized egg, inserts a

nucleus from the animal to be copied and then stimulates the nucleus to begin dividing to form an embryo. In the case of a mammal, such as a human, the embryo would then be implanted in the uterus of a host female. Also see "Cloning (Therapeutic)."

Cloning (Therapeutic): A method of reproducing exact copies of cells needed for research or for the development of replacement tissue or organs. A scientist removes the nucleus from a donor's unfertilized egg, inserts a nucleus from the animal whose cells are to be copied and then stimulates the nucleus to begin dividing to form an embryo. However, the embryo is never allowed to grow to any significant stage of development. Instead, it is allowed to grow for a few hours or days, and stem cells are then removed from it for use in regenerating tissue. Also see "Cloning (Reproductive)."

CMS (Centers for Medicare and Medicaid Services): A federal agency responsible for administering Medicare and monitoring the states' operations of Medicaid. See www.cms.hhs.gov.

Combinatorial Chemistry: A chemistry in which molecules are found that control a pre-determined protein. This advanced computer technique enables scientists to use automatic fluid handlers to mix chemicals under specific test conditions at extremely high speed. Combinatorial chemistry can generate thousands of chemical compound variations in a few hours. Previously, traditional chemistry methods could have required several weeks to do the same work.

Complementary-DNA (cDNA): A sequence acquired by copying a messenger RNA (mRNA) molecule back into DNA. In contrast to the original DNA, mRNA codes for an expressed protein without non-coding DNA sequences (introns). Therefore, a cDNA probe can also be used to find the specific gene in a complex DNA sample from another organism with different non-coding sequences. Also called Copy DNA.

Computational Biology: The development and application of data-analytical and theoretical methods, mathematical modeling and computational simulation techniques to the study of biological, behavioral, and social systems. Computational biology uses mathematical and computational approaches to address theoretical and experimental questions in biology.

Contract Manufacturer: A company that manufactures products that will be sold under the brand names of its client companies. For example, a large number of consumer electronics, such as laptop computers, are manufactured by contract manufacturers for leading brand-name computer companies such as Dell. Many other types of products are made under contract manufacturing, from apparel to pharmaceuticals. Also see "OEM (Original Equipment Manufacturer)" and "ODM (Original Design Manufacturer)."

Coordinator: In clinical trials, the person at an investigative site who handles the administrative responsibilities of the trial, acts as a liaison between the investigative site and the sponsor, and reviews data and records during a monitoring visit.

CPMP (Committee on Proprietary Medicinal Products): A committee, composed of two people from each EU Member State (see "EU (European Union)"), that is responsible for the scientific evaluation and assessment of marketing applications for medicinal products in the EU. The CPMP is the major body involved in the harmonization of pharmaceutical regulations within the EU and receives administrative support from the European Medicines Evaluation Agency. See "EMEA (European Medicines Evaluation Agency)."

CRA (Clinical Research Associate): An individual responsible for monitoring clinical trial data to ensure compliance with study protocol and FDA GCP regulations.

CRF (Case Report Form): In clinical trials, a standard document used by clinicians to record and report subject data pertinent to the study protocol.

CRO (Contract Research Organization): An independent organization that contracts with a client to conduct part of the work on a study or research project. For example, drug and medical device makers frequently outsource clinical trials and other research work to CROs.

CRT (Case Report Tabulation): In clinical trials, a tabular listing of all data collected on study case report forms.

DBA: Doing business as.

Development: The phase of research and development (R&D) in which researchers attempt to create new products from the results of discoveries and applications created during basic and applied research.

Device: In medical products, an instrument, apparatus, implement, machine, contrivance, implant, in vitro reagent or other similar or related article, including any component, part or accessory, that 1) is recognized in the official National Formulary or United States Pharmacopoeia or any supplement to them, 2) is intended for use in the diagnosis of disease or other conditions, or in the cure, mitigation, treatment or prevention of disease, in man or animals or 3) is intended to affect the structure of the body of man or animals and does not achieve any of its principal intended purposes through chemical action within or on the body of man or animals and is not dependent upon being metabolized for the achievement of any of its principal intended purposes.

Diagnostic Radioisotope Facility: A medical facility in which radioactive isotopes (radiopharmaceuticals) are used as tracers or indicators to detect an abnormal condition or disease in the body.

Distributor: An individual or business involved in marketing, warehousing and/or shipping of products manufactured by others to a specific group of end users. Distributors do not sell to the general public. In order to develop a competitive advantage, distributors often focus on serving one industry or one set of niche clients. For example, within the medical industry, there are major distributors that focus on providing pharmaceuticals, surgical supplies or dental supplies to clinics and hospitals.

DMB (Data Monitoring Board): A committee that monitors the progress of a clinical trial and carefully observes the safety data.

DNA (Deoxyribonucleic Acid): The carrier of the genetic information that cells need to replicate and to produce proteins.

DNA Chip: A revolutionary tool used to identify mutations in genes like BRCA1 and BRCA2. The chip, which consists of a small glass plate encased in plastic, is manufactured using a process similar to the one used to make computer microchips. On the surface, each chip contains synthetic single-stranded DNA sequences identical to a normal gene.

Drug Utilization Review: A quantitative assessment of patient drug use and physicians' patterns of prescribing drugs in an effort to determine the usefulness of drug therapy.

DSMB (Data and Safety Monitoring Board): See "DMB (Data Monitoring Board)."

EC (European Community): See "EU (European Union)."

Ecology: The study of relationships among all living organisms and the environment, especially the totality or pattern of interactions; a view that includes all plant and animal species and their unique contributions to a particular habitat.

EDI (Electronic Data Interchange): An accepted standard format for the exchange of data between various companies' networks. EDI allows for the transfer of e-mail as well as orders, invoices and other files from one company to another.

Efficacy: A drug or medical product's ability to effectively produce beneficial results within a patient.

EFGCP (European Forum for Good Clinical Practices): The organization dedicated to finding common ground in Europe on the implementation of Good Clinical Practices. See "GCP (Good Clinical Practices)." www.efgcp.org

ELA (Establishment License Application): Required for the approval of a biologic (see "Biologic"). It permits a specific facility to manufacture a biological product for commercial purposes. Compare to "PLA (Product License Agreement)."

Electroporation: A health care technology that uses short pulses of electric current (DC) to create openings (pores) in the membranes of cancerous cells, thus leading to death of the cells. It has potential as a treatment for prostate cancer. In the laboratory, it is a means of introducing foreign proteins or DNA into living cells.

EMEA (European Medicines Evaluation Agency): The European agency responsible for supervising and coordinating applications for marketing medicinal products in the European Union (see "EU (European Union)" and "CPMP (Committee on Proprietary Medicine)"). The EMEA is headquartered in the U.K. www.eudraportal.eudra.org

Endpoint: A clinical or laboratory measurement used to assess safety, efficacy or other trial objectives of a test article in a clinical trial.

Enterprise Resource Planning (ERP): An integrated information system that helps manage all aspects of a business, including accounting, ordering and human resources, typically across all locations of a major corporation or organization. ERP is considered to be a critical tool for management of large organizations. Suppliers of ERP tools include SAP and Oracle.

Enzyme: A protein that acts as a catalyst, affecting the chemical reactions in cells.

Epigenetics: A relatively new branch of biology focused on gene "silencers." Scientists involved in epigenetics are studying the function within a gene that regulates whether a gene is operating a full capacity or is toned down to a lower level. The level of operation of a given gene may lead to a higher risk of disease, such as certain types of cancer, within a patient.

EST (Expressed Sequence Tags): Small pieces of DNA sequence (usually 200 to 500 nucleotides long) that are generated by sequencing either one or both ends of an expressed gene. The idea is to sequence bits of DNA that represent genes expressed in certain cells, tissues or organs from different organisms and use these tags to fish a gene out of a portion of chromosomal DNA by matching base pairs. See "Gene Expression."

Ethanol: A clear, colorless, flammable, oxygenated hydrocarbon, also called ethyl alcohol. In the U.S., it is used as a gasoline octane enhancer and oxygenate in a 10% blend called E10. Ethanol can be used in higher concentrations (such as an 85% blend called E85) in vehicles designed for its use. It is typically produced chemically from ethylene or biologically from fermentation of various sugars from carbohydrates found in agricultural crops and cellulose residues from crops or wood. Grain ethanol production is typically based on corn or sugarcane. Cellulosic ethanol production is based on agricultural waste, such as wheat stalks, that has been treated with enzymes to break the waste down into component sugars.

Etiology: The study of the causes or origins of diseases.

EU (European Union): A consolidation of European countries (member states) functioning as one body to facilitate trade. Previously known as the European Community (EC), the EU expanded to include much of Eastern Europe in 2004, raising the total number of member states to 25. In 2002, the EU launched a unified currency, the Euro. See europa.eu.int.

EU Competence: The jurisdiction in which the EU can take legal action.

FD&C Act (Federal Food, Drug and Cosmetic Act): A set of laws passed by the U.S. Congress, which controls, among other things, residues in food and feed.

FDA (Food and Drug Administration): The U.S. government agency responsible for the enforcement of the Federal Food, Drug and Cosmetic Act, ensuring industry compliance with laws regulating products in commerce. The FDA's mission is to protect the public from harm and encourage technological advances that hold the promise of benefiting society. www.fda.gov

Follow-on Biologics: A term used to describe generic versions of drugs that have been created using biotechnology. Because biotech drugs ("biologics") are made from living cells, a generic version of a drug probably won't be biochemically identical to the original branded version of the drug. Consequently, they are described as "follow-on" drugs to set them apart. Since these drugs won't be biochemically the same as the original drugs, there are concerns that they may not be as safe or effective unless they go through clinical trials for proof of quality. In Europe, these drugs are referred to as "biosimilars".

Formulary: A preferred list of drug products that typically limits the number of drugs available within a therapeutic class for purposes of drug purchasing, dispensing and/or reimbursement. A government body, third-party insurer or health plan, or an institution may compile a formulary. Some institutions or health plans develop closed (i.e. restricted) formularies where only those drug products listed can be dispensed in that institution or reimbursed by the health plan. Other formularies may have no restrictions (open formulary) or may have certain restrictions such as higher patient cost-sharing requirements for off-formulary drugs.

Functional Foods: Food products that contain nutrients, such as vitamins, associated with certain health benefits. The nutrients may occur naturally, or the foods may have been enhanced with them.

Functional Genomics: The process of attempting to convert the molecular information represented by DNA into an understanding of gene functions and effects. To address gene function and expression specifically, the recovery and identification of mutant and over-expressed phenotypes can be employed. Functional genomics also entails research on the protein function (proteomics) or, even more broadly, the whole metabolism (metabolomics) of an organism.

Functional Imaging: The uses of PET scan, MRI and other advanced imaging technology to see how an area of the body is functioning and responding. For example, brain activity can be viewed, and the reaction of cancer tumors to therapies can be judged using functional imaging.

Functional Proteomics: The study of the function of all the proteins encoded by an organism's entire genome.

GCP (Good Clinical Practices): FDA regulations and guidelines that define the responsibilities of the key figures involved in a clinical trial, including the sponsor, the investigator, the monitor and the Institutional Review Board. See "IRB (Institutional Review Board)."

GDP (Gross Domestic Product): The total value of a nation's output, income and expenditures produced with a nation's physical borders.

Gene: A working subunit of DNA; the carrier of inheritable traits.

Gene Chip: See "DNA Chip."

Gene Expression: The term used to describe the transcription of the information contained within the DNA (the repository of genetic information) into messenger RNA (mRNA) molecules that are then translated into the proteins that perform most of the critical functions of cells. Scientists study the kinds and amounts of mRNA produced by a cell to learn which genes are expressed, which in turn provides insights into how the cell responds to its changing needs.

Gene Knock-Out: The inhibition of gene expression through various scientific methods.

Gene Therapy: Treatment based on the alteration or replacement of existing genes. Genetic therapy involves splicing desired genes insolated from one patient into a second patient's cells in order to compensate for that patient's inherited genetic defect, or to enable that patient's body to better fight a specific disease.

Genetic Code: The sequence of nucleotides, determining the sequence of amino acids in protein synthesis.

Genetically Modified (GM) Foods: Food crops that are bioengineered to resist herbicides, diseases or insects; have higher nutritional value than non-engineered plants; produce a higher yield per acre; and/or last longer on the shelf. Additional traits may include resistance to temperature and moisture extremes. Agricultural animals also may be genetically modified organisms.

Genetics: The study of the process of heredity.

Genome: The genetic material (composed of DNA) in the chromosomes of a living organism.

Genomics: The study of genes, their role in diseases and our ability to manipulate them.

Genotype: The genetic constitution of an organism.

Globalization: The increased mobility of goods, services, labor, technology and capital throughout the world. Although globalization is not a new development, its pace has increased with the advent of new technologies, especially in the areas of telecommunications, finance and shipping.

GLP (Good Laboratory Practices): A collection of regulations and guidelines to be used in laboratories where research is conducted on drugs, biologics or devices that are intended for submission to the FDA.

GM (Genetically-Modified): See "Genetically Modified (GM) Foods" and "GMO (Genetically Modified Organism)."

GMO (Genetically Modified Organism): An organism that has undergone genome modification by the insertion of a foreign gene. The genetic material of a GMO is not found through mating or natural recombination.

GMP (Good Manufacturing Practices): A collection of regulations and guidelines to be used in manufacturing drugs, biologics and medical devices.

GNP (Gross National Product): A country's total output of goods and services from all forms of economic activity measured at market prices for one calendar year. It differs from GDP (Gross Domestic Product) in that GNP includes income from investments made in foreign nations.

HESC (Human Embryonic Stem Cell): See "Stem Cells."

HHS (U.S. Department of Health and Human Services): This agency has more than 300 major programs related to human health and welfare, the largest of which is Medicare. See www.hhs.gov

High-Throughput Screening (HTP): Makes use of techniques that allow for a fast and simple test on the presence or absence of a desirable structure, such as a specific DNA sequence. HTP screening often uses DNA chips or microarrays and automated data processing for large-scale screening, for instance, to identify new targets for drug development.

IDE (Investigational New Device Exemption): A document that must be filed with the FDA prior to initiating clinical trials of medical devices considered to pose a significant risk to human subjects.

IEEE: The Institute of Electrical and Electronic Engineers. The IEEE sets global technical standards and acts as an authority in technical areas including computer engineering, biomedical technology, telecommunications, electric power, aerospace and consumer electronics, among others. www.ieee.org.

Imaging: In medicine, the viewing of the body's organs through external, high-tech means. This reduces the need for broad exploratory surgery. These advances, along with new types of surgical instruments, have made minimally invasive surgery possible. Imaging includes MRI (magnetic resonance imaging), CT (computed tomography or CAT scan), MEG (magnetoencephalography), improved x-ray technology, mammography, ultrasound and angiography.

Imaging Contrast Agent: A molecule or molecular complex that increases the intensity of the signal detected by an imaging technique, including MRI and ultrasound. An MRI contrast agent, for example, might contain gadolinium attached to a targeting antibody. The antibody would bind to a specific target, a metastatic melanoma cell for example, while the gadolinium would increase the magnetic signal detected by the MRI scanner.

Immunoassay: An immunological assay. Types include agglutination, complement-fixation, precipitation, immunodiffusion and electrophoretic assays. Each type of assay utilizes either a particular type of antibody or a specific support medium (such as a gel) to determine the amount of antigen present.

In Vitro: Laboratory experiments conducted in the test tube, or otherwise, without using live animals and/or humans.

In Vivo: Laboratory experiments conducted with live animals and/or humans.

IND (Investigational New Drug Application): A document that must be filed with the FDA prior to initiating clinical trials of drugs or biologics.

Indication: Refers to a specific disease, illness or condition for which a drug is approved as a treatment. Typically, a new drug is first approved for one indication. Then, an application to the FDA is later made for approval of additional indications.

Informatics: See "Bioinformatics."

Informed Consent: Must be obtained in writing from people who agree to be clinical trial subjects prior to their enrollment in the study. The document must explain the risks associated with the study and treatment and describe alternative therapy available to the patient. A copy of the document must also be provided to the patient.

Initial Public Offering (IPO): A company's first effort to sell its stock to investors (the public).

Investors in an up-trending market eagerly seek stocks offered in many IPOs because the stocks of newly public companies that seem to have great promise may appreciate very rapidly in price, reaping great profits for those who were able to get the stock at the first offering. In the United States, IPOs are regulated by the SEC (U.S. Securities Exchange Commission) and by the state-level regulatory agencies of the states in which the IPO shares are offered.

Insertion Mutants: Mutants of genes that are obtained by inserting DNA, for instance through mobile DNA sequences. In plant research, the capacity of the bacterium Agrobacterium to introduce DNA into the plant genome is employed to induce mutants. In both cases, mutations lead to lacking or changing gene functions that are revealed by aberrant phenotypes. Insertion mutant isolation, and subsequent identification and analysis are employed in model plants such as Arabiopsis and in crop plants such as maize and rice.

Interactomics (Interactome): The study of the interactions between genes, RNA, proteins and metabolites within the cell.

Interferon: A type of biological response modifier (a substance that can improve the body's natural response to disease).

Investigator: In clinical trials, a clinician who agrees to supervise the use of an investigational drug, device or biologic in humans. Responsibilities of the investigator, as defined in FDA regulations, include administering the drug, observing and testing the patient, collecting data and monitoring the care and welfare of the patient.

Iontophoresis: The transfer of ions of medicine through the skin using a local electric current.

IRB (Institutional Review Board): A group of individuals usually found in medical institutions that is responsible for reviewing protocols for ethical consideration (to ensure the rights of the patients). An IRB also evaluates the benefit-to-risk ratio of a new drug to see that the risk is acceptable for patient exposure. Responsibilities of an IRB are defined in FDA regulations.

ISO 9000, 9001, 9002, 9003: Standards set by the International Organization for Standardization. ISO

9000, 9001, 9002 and 9003 are the highest quality certifications awarded to organizations that meet exacting standards in their operating practices and procedures.

ITES (IT-Enabled Services): The portion of the Information Technology industry focused on providing business services, such as call centers, insurance claims processing and medical records transcription, by utilizing the power of IT, especially the Internet. Most ITES functions are considered to be back-office procedures. Also, see "Business Process Outsourcing (BPO)."

Kinase (Protein Kinase): Enzymes that influence certain basic functions within cells, such as cell division. Kinases catalyze the transfer of phosphates from ATP to proteins, thus causing changes in protein function. Defective kinases can lead to diseases such as cancer.

Lifestyle Drug: Lifestyle drugs target a variety of medical conditions ranging from the painful to the inconvenient, including obesity, impotence, memory loss and depression, rather than illness or disease. Drug companies continue to develop lifestyle treatments for hair loss and skin wrinkles in an effort to capture their share of the huge anti-aging market aimed at the baby-boomer generation.

Ligand: Any atom or molecule attached to a central atom in a complex compound.

Liposome: A micro or nanoscale lipid or phospholipid layer enclosing a liquid core used for transport for particular molecules or biological structures or as a model for membranes.

Marketing: Includes all planning and management activities and expenses associated with the promotion of a product or service. Marketing can encompass advertising, customer surveys, public relations and many other disciplines. Marketing is distinct from selling, which is the process of sell-through to the end user.

Mass Spectrometry: Usage of analytical devices that can determine the mass (or molecular weight) of proteins and nucleic acids, the sequence of protein molecules, the chemical organization of almost all substances and the identification of gram-negative and gram-positive microorganisms.

Medical Device: See "Device."

Metabolomics: The study of low-molecular-weight materials produced during genomic expression within a cell. Such studies can lead to a better understanding of how changes within genes and proteins affect the function of cells.

Metagenomics: An advanced form of genomics that increases the understanding of complex microbial systems.

Microarray: A DNA analysis tool consisting of a microscopic ordered array that enables parallel analysis of complex biochemical samples; used to analyze how large numbers of genes interact with each other; used for genotyping, mapping, sequencing, sequence detection; usually constructed by applying biomolecules onto a slide or chip and then scanning with microscope or other imaging equipment.

Microfluidics: Refers to the manipulation of microscopic amounts of fluid, generally for analysis in microarrays using high throughput screening. The volume of the liquid involved is on the nanolitre scale.

Molecular Imaging: An emerging field in which advanced biology on the molecular level is combined with noninvasive imaging to determine the presence of certain proteins and other important genetic material.

Monoclonal Antibodies (mAb, Human Monoclonal Antibody): Antibodies that have been cloned from a single antibody and massed produced as a therapy or diagnostic test. An example is an antibody specific to a certain protein found in cancer cells.

Nanocantilever: The simplest micro-electro-mechanical system (MEMS) that can be easily machined and mass-produced via the same techniques used to make computer chips. The ability to detect extremely small displacements make nanocantilever beams an ideal device for detecting extremely small forces, stresses and masses. Nanocantilevers coated with antibodies, for example, will bend from the mass added when substrate binds to its antibody, providing a detector capable of sensing the presence of single molecules of clinical importance.

Nanoparticle: A nanoscale spherical or capsule-shaped structure. Most, though not all, nanoparticles are hollow, which provides a central reservoir that can be filled with anticancer drugs, detection agents, or chemicals, known as reporters, that can signal if a drug is having a therapeutic effect. The surface of a nanoparticle can also be adorned with various targeting agents, such as antibodies, drugs, imaging agents, and reporters. Most nanoparticles are constructed to be small enough to pass through blood capillaries and enter cells.

Nanopharmaceuticals: Nanoscale particles that modulate drug transport in drug uptake and delivery applications.

Nanoshell: A nanoparticle composed of a metallic shell surrounding a semiconductor. When nanoshells reach a target cancer cell, they can be irradiated with near-infrared light or excited with a magnetic field, either of which will cause the nanoshell to become hot, killing the cancer cell.

Nanotechnology: The science of designing, building or utilizing unique structures that are smaller than 100 nanometers (a nanometer is one billionth of a meter). This involves microscopic structures that are no larger than the width of some cell membranes.

Nanowires: A nanometer-scale wire made of metal atoms, silicon, or other materials that conduct electricity. Nanowires are built atom by atom on a solid surface, often as part of a microfluidic device. They can be coated with molecules such as antibodies that will bind to proteins and other substances of interest to researchers and clinicians. By the very nature of their nanoscale size, nanowires are incredibly sensitive to such binding events and respond by altering the electrical current flowing through them, and thus can form the basis of ultra sensitive molecular detectors.

National Drug Code (NDC): An identifying drug number maintained by the FDA.

NCE (New Chemical Entity): A new molecular compound not previously approved for human use in the U.S., excluding diagnostic agents, vaccines and other biologic compounds not approved by the FDA's Center for Drug Evaluation and Research (CDER).

NDA (New Drug Application): An application requesting FDA approval, after completion of Phase III studies, to market a new drug for human use in interstate commerce. Clinical trial results generally account for approximately 80% of the NDA.

New Molecular Entity (NME): Defined by the FDA as a medication containing an active substance that has never before been approved for marketing in any form in the U.S.

NIH (National Institutes of Health): A branch of the U.S. Public Health Service that conducts biomedical research. www.nih.gov

NME: See "New Molecular Entity (NME)."

Nonclinical Studies: In vitro (laboratory) or in vivo (animal) pharmacology, toxicology and pharmacokinetic studies that support the testing of a product in humans. Usually at least two species are evaluated prior to Phase I clinical trials. Nonclinical studies continue throughout all phases of research to evaluate long-term safety issues.

Nucleic Acid: A large molecule composed of nucleotides. Nucleic acids include RNA, DNA and antisense oligonucleotides.

Nutraceutical: Nutrient + pharmaceutical – a food or part of a food that has been isolated and sold in a medicinal form and claims to offer benefits such as the treatment or prevention of disease.

Nutraceuticals: Food products and dietary supplements that may have certain health benefits. Nutraceuticals may offer specific vitamins or minerals. Also see "Functional Foods."

Nutrigenomics: The study of how food interacts with genes.

ODM (Original Design Manufacturer): A contract manufacturer that offers complete, end-to-end design, engineering and manufacturing services. ODMs design and build products, such as consumer electronics, that client companies can then brand and sell as their own. For example, a large percentage of laptop computers, cell phones and PDAs are made by ODMs. Also see "OEM (Original Equipment Manufacturer)" and "Contract Manufacturer."

OEM (Original Equipment Manufacturer): A company that manufactures a product or component for sale to a customer that will integrate the component into a final product or assembly. The OEM's customer will distribute the end product or resell it to an end user. For example, a personal computer made under a brand name by a given company may contain various components, such as hard drives, graphics cards or speakers, manufactured by several different OEM "vendors," but the firm doing the final assembly/manufacturing process is the final manufacturer. Also see "ODM (Original Design Manufacturer)" and "Contract Manufacturer."

Offshoring: The rapidly growing tendency among U.S., Japanese and Western European firms to send knowledge-based and manufacturing work overseas. The intent is to take advantage of lower wages and operating costs in such nations as China, India, Hungary and Russia. The choice of a nation for offshore work may be influenced by such factors as language and education of the local workforce, transportation systems or natural resources. For example, China and India are graduating high numbers of skilled engineers and scientists from their universities. Also, some nations are noted for large numbers of workers skilled in the English language, such as the Philippines and India. Also see "Captive Offshoring" and "Outsourcing."

Oncogene: A unit of DNA that normally directs cell growth, but which can also promote or allow the uncontrolled growth of cancer.

Oncology: The diagnosis, study and treatment of cancer.

Orphan Drug: A drug, biologic or antibiotic designated by the FDA as providing therapeutic benefit for an indication (disease or condition) affecting less than 200,000 people in the U.S. Companies that market orphan drugs are granted a period of market exclusivity in return for the limited commercial potential of the drug.

OTC (Over-the-Counter Drugs): FDA-regulated products that do not require a physician's prescription. Some examples are aspirin, sunscreen, nasal spray and sunglasses.

Outsourcing: The hiring of an outside company to perform a task otherwise performed internally by the company, generally with the goal of lowering costs and/or streamlining work flow. Outsourcing contracts are generally several years in length. Companies that hire outsourced services providers often prefer to focus on their core strengths while sending more routine tasks outside for others to perform. Typical outsourced services include the running of human resources departments, telephone call centers and computer departments. When outsourcing is performed overseas, it may be referred to as offshoring. Also see "Offshoring."

Patent: A property right granted by the U.S. government to an inventor to exclude others from making, using, offering for sale, or selling the invention throughout the U.S. or importing the invention into the U.S. for a limited time in exchange for public disclosure of the invention when the patent is granted.

Pathogen: Any microorganism (e.g., fungus, virus, bacteria or parasite) that causes a disease.

Peer Review: The process used by the scientific community, whereby review of a paper, project or report is obtained through comments of independent colleagues in the same field.

Pharmacodynamics (PD): The study of reactions between drugs and living systems. It can be thought of as the study of what a drug does to the body.

Pharmacoeconomics: The study of the costs and benefits associated with various drug treatments.

Pharmacogenetics: The investigation of the different reactions of human beings to drugs and the underlying genetic predispositions. The differences in reaction are mainly caused by mutations in certain enzymes responsible for drug metabolization. As a result, the degradation of the active substance can lead to harmful by-products, or the drug might have no effect at all.

Pharmacogenomics: The use of the knowledge of DNA sequences for the development of new drugs.

Pharmacokinetics (PK): The study of the processes of bodily absorption, distribution, metabolism and excretion of compounds and medicines. It can be thought of as the study of what the body does to a drug. See "ADME (Absorption, Distribution, Metabolism and Excretion)."

Pharmacology: The science of drugs, their characteristics and their interactions with living organisms.

Pharmacy Benefit Manager (PBM): An organization that provides administrative services in processing and analyzing prescription claims for pharmacy benefit and coverage programs. Many PBMs also operate mail order pharmacies or have arrangements to include prescription availability through mail order pharmacies.

Phase I Clinical Trials: Studies in this phase include initial introduction of an investigational drug into humans. These studies are closely monitored and are usually conducted in healthy volunteers. Phase I trials are conducted after the completion of extensive nonclinical or pre-clinical trials not involving humans. Phase I studies include the determination of clinical pharmacology, bioavailability, drug interactions and side effects associated with increasing doses of the drug.

Phase II Clinical Trials: Include randomized, masked, controlled clinical studies conducted to evaluate the effectiveness of a drug for a particular indication(s). During Phase II trials, the minimum effective dose and dosing intervals should be determined.

Phase III Clinical Trials: Consist of controlled and uncontrolled trials that are performed after preliminary evidence of effectiveness of a drug has been established. They are conducted to document the safety and efficacy of the drug, as well as to determine adequate directions (labeling) for use by the physician. A specific patient population needs to be clearly identified from the results of these studies. Trials during Phase III are conducted using a large number of patients to determine the frequency of adverse events and to obtain data regarding intolerance.

Phase IV Clinical Trials: Conducted after approval of a drug has been obtained to gather data supporting new or revised labeling, marketing or advertising claims.

Phenomics: The study of how an organism's structure responds to such things as toxins or drugs.

Phenotype: Observable characteristics of an organism produced by the organism's genotype interacting with the environment.

PHS (Public Health Service): May stand for the Public Health Service Act, a law passed by the U.S. Congress in 1944. PHS also may stand for the Public Health Service itself, a U.S. government agency established by an act of Congress in July 1798, originally authorizing hospitals for the care of American merchant seamen. Today, the Public Health Service sets national health policy; conducts medical and biomedical research; sponsors programs for disease control and mental health; and enforces laws to assure the safety and efficacy of drugs, foods, cosmetics and medical devices. The FDA (Food and Drug Administration) is part of the Public Health Service, as are the Centers for Disease Control and Prevention (CDC).

Phylogenetic Systematics: The field of biology that deals with identifying and understanding the evolutionary relationships among the many different kinds of life on earth, both living (extant) and dead (extinct).

Pivotal Studies: In clinical trials, a Phase III trial that is designed specifically to support approval of a product. These studies are well-controlled (usually by placebo) and are generally designed with input from the FDA so that they will provide data that is adequate to support approval of the product. Two pivotal studies are required for drug product approval, but usually only one study is required for biologics.

PLA (Product License Agreement): See "BLA (Biologics License Application)."

Plant Patent: A plant patent may be granted by the U.S. Patent and Trademark Office to anyone who invents or discovers and asexually reproduces any distinct and new variety of plant.

Platform or Technology Platform Companies: Firms hoping to profit by providing information systems, software, databases and related support to biopharmaceutical companies.

PMA (Pre-Market Approval): Required for the approval of a new medical device or a device that is to be used for life-sustaining or life-supporting purposes, is implanted in the human body or presents potential risk of illness or injury.

PMCs: Postmarketing study commitments. PMCs are clinical studies that are not required for FDA initial approval of a drug, but the FDA nonetheless feels these studies will provide important data on a newly marketed drug. Consequently, the drug firm makes a commitment for continuing studies.

Positional Cloning: The identification and cloning of a specific gene, with chromosomal location as the only source of information about the gene.

Post-Marketing Surveillance: The FDA's ongoing safety monitoring of marketed drugs.

Preclinical Studies: See "Nonclinical Studies."

Priority Drugs: NCEs (New Chemical Entities) that the FDA feels may offer high therapeutic value that are therefore marked for priority review.

Protein: A large, complex molecule made up of amino acids.

Proteome: The genetic material (composed of amino acids) in the chromosomes of a living organism.

Proteomics: The study of gene expression at the protein level, by the identification and characterization of proteins present in a biological sample.

Qdots: See "Quantum Dots (Qdots)."

QOL (Quality of Life): In medicine, an endpoint of therapeutic assessment used to adjust measures of effectiveness for clinical decision-making. Typically, QOL endpoints measure the improvement of a patient's day-to-day living as a result of specific therapy.

Quantum Dots (Qdots): Nanometer sized semiconductor particles, made of cadmium selenide (CdSe), cadmium sulfide (CdS) or cadmium telluride (CdTe) with an inert polymer coating. The semiconductor material used for the core is chosen based upon the emission wavelength range being targeted: CdS for UV-blue, CdSe for the bulk of the visible spectrum, CdTe for the far red and near-infrared, with the particle's size determining the exact color of a given quantum dot. The polymer

coating safeguards cells from cadmium toxicity but also affords the opportunity to attach any variety targeting molecules, including monoclonal antibodies directed to tumor-specific biomarkers. Because of their small size, quantum dots can function as cell- and even molecule-specific markers that will not interfere with the normal workings of a cell. In addition, the availability of quantum dots of different colors provides a powerful tool for following the actions of multiple cells and molecules simultaneously.

R&D: Research and development. Also see "Applied Research" and "Basic Research."

Radioisotope: An object that has varying properties that allows it to penetrate other objects at different rates. For example, a sheet of paper can stop an alpha particle, a beta particle can penetrate tissues in the body and a gamma ray can penetrate concrete. The varying penetration capabilities allow radioisotopes to be used in different ways. (Also called radioactive isotope or radionuclide.)

Receptor: Proteins in or on a cell that selectively bind a specific substance called a ligand. See "Ligand."

Recombination: The natural process of breaking and rejoining DNA strands to produce new combinations of genes.

Reporter Gene: A gene that is inserted into DNA by researchers in order to indicate when a linked gene is successfully expressed or when signal transduction has taken place in a cell.

Return on Investment (ROI): A measure of a company's profitability, expressed in percentage as net profit (after taxes) divided by total dollar investment.

RNA (Ribonucleic Acid): A macromolecule found in the nucleus and cytoplasm of cells; vital in protein synthesis.

RNAi (RNA interference): A biological occurrence where double-stranded RNA is used to silence genes.

Semiconductor: A generic term for a device that controls electrical signals. It specifically refers to a material (such as silicon, germanium or gallium arsenide) that can be altered either to conduct electrical current or to block its passage. Carbon nanotubes may eventually be used as semiconductors. Semiconductors are partly responsible for the miniaturization of modern electronic devices, as they are vital components in computer memory and processor chips. The manufacture of semiconductors is carried out by small firms, and by industry giants such as Intel and Advanced Micro Devices.

Single Nucleotide Polymorphisms (SNPs): Stable mutations consisting of a change at a single base in a DNA molecule. SNPs can be detected by HTP analyses, such as gene chips, and they are then mapped by DNA sequencing. They are the most common type of genetic variation.

SMDA (Safe Medical Devices Act): An act that amends the Food, Drug and Cosmetic Act to impose additional regulations on medical devices. The act became law in 1990.

SNP: See "Single-Nucleotide Polymorphisms (SNPs)."

Sponsor: The individual or company that assumes responsibility for the investigation of a new drug, including compliance with the FD&C Act and regulations. The sponsor may be an individual, partnership, corporation or governmental agency and may be a manufacturer, scientific institution or investigator regularly and lawfully engaged in the investigation of new drugs. The sponsor assumes most of the legal and financial responsibility of the clinical trial.

Standard Drugs: NCEs (New Chemical Entities) that the FDA feels offer few advantages over existing drugs that are therefore given lower status for review.

Stem Cells: Cells found in human bone marrow, the blood stream and the umbilical cord that can be replicated indefinitely and can turn into any type of mature blood cell, including platelets, white blood cells or red blood cells. Also referred to as pluripotent cells.

Study Coordinator: See "Coordinator."

Subsidiary, Wholly-Owned: A company that is wholly controlled by another company through stock ownership.

Supply Chain: The complete set of suppliers of goods and services required for a company to operate its business. For example, a manufacturer's supply chain may include providers of raw materials, components, custom-made parts and packaging materials.

Synthetic Biology: Synthetic biology can be defined as the design and construction of new entities, including enzymes and cells, or the reformatting of existing biological systems. This science capitalizes on previous advances in molecular biology and systems biology, by applying a focus on the design and construction of unique core components that can be integrated into larger systems in order to solve specific problems.

Systems Biology: The use of combinations of advanced computer hardware, software and database technologies to take a systemic approach to biological research. Advanced technologies will enable scientists to view genetic predisposition by integrating information about entire biological systems, from DNA to proteins to cells to tissues.

T Cell (T-Cell): A white blood cell that carries out immune system responses. The T cell originates in the bone marrow and matures in the thymus.

Targets: The proteins involved in a specific disease. Drug compounds concentrate on specific targets in order to have the greatest positive effect and cut down on the incidence of side effects.

Taste Masking: The creation of a barrier between a drug molecule and taste receptors so the drug is easier to take. It masks bitter or unpleasant tastes.

TESS: See "AE (Adverse Event)."

Toxicogenomics: The study of the relationship between responses to toxic substances and the resulting genetic changes.

Trial Coordinator: See "Coordinator."

Utility Patent: A utility patent may be granted by the U.S. Patent and Trademark Office to anyone who invents or discovers any new, useful, and non-obvious process, machine, article of manufacture, or composition of matter, or any new and useful improvement thereof.

Validation of Data: The procedure carried out to ensure that the data contained in a final clinical trial report match the original observations.

WHO (World Health Organization): A United Nations agency that assists governments in strengthening health services, furnishing technical assistance and aid in emergencies, working on the prevention and control of epidemics and promoting cooperation among different countries to improve nutrition, housing, sanitation, recreation and other aspects of environmental hygiene. Any country that is a member of the United Nations may become a member of the WHO by accepting its constitution. The WHO currently has 191 member states.

Xenotransplantation: The science of transplanting organs such as kidneys, hearts or livers into humans from other mammals, such as pigs or other agricultural animals grown with specific traits for this purpose.

Zoonosis: An animal disease that can be transferred to man.

INTRODUCTION

PLUNKETT'S BIOTECH & GENETICS INDUSTRY ALMANAC, the sixth edition of our guide to the biotech and genetics field, is designed to be used as a general source for researchers of all types.

The data and areas of interest covered are intentionally broad, ranging from the ethical questions facing biotechnology, to emerging technology, to an in-depth look at the 400 major for-profit firms (which we call "THE BIOTECH 400") within the many industry sectors that make up the biotechnology and genetics arena.

This reference book is designed to be a general source for researchers. It is especially intended to assist with market research, strategic planning, employment searches, contact or prospect list creation (be sure to see the export capabilities of the accompanying CD-ROM that is available to book and eBook buyers) and financial research, and as a data resource for executives and students of all types.

PLUNKETT'S BIOTECH & GENETICS INDUSTRY ALMANAC takes a rounded approach for the general reader. This book presents a complete overview of the entire biotechnology and genetics system (see "How To Use This Book"). For example, you will find trends in the biopharmaceuticals market, along with easy-to-use charts and tables on all facets of biotechnology in general: from the sales and profits of the major drug companies to the amounts of time required in the various stages of drug approval.

THE BIOTECH 400 is our unique grouping of the biggest, most successful corporations in all segments of the global biotechnology and genetics industry. Tens of thousands of pieces of information, gathered from a wide variety of sources, have been researched and are presented in a unique form that can be easily understood. This section includes thorough indexes to THE BIOTECH 400, by geography, industry, sales, brand names, subsidiary names and many other topics. (See Chapter 4.)

Especially helpful is the way in which PLUNKETT'S BIOTECH & GENETICS INDUSTRY ALMANAC enables readers who have no business or scientific background to readily compare the financial records and growth plans of large biotech companies and major industry groups. You'll see the mid-term financial record of each firm, along with the impact of earnings, sales and strategic plans on each company's potential to fuel growth, to serve new markets and to provide investment and employment opportunities.

No other source provides this book's easy-to-understand comparisons of growth, expenditures, technologies, corporations and many other items of great importance to people of all types who may be studying this, one of the most exciting industries in the world today.

By scanning the data groups and the unique indexes, you can find the best information to fit your personal research needs. The major growth companies in biotechnology and genetics are profiled and then ranked using several different groups of specific criteria. Which firms are the biggest employers? Which companies earn the most profits? These things and much more are easy to find.

In addition to individual company profiles, an overview of biotechnology markets and trends is provided. This book's job is to help you sort through easy-to-understand summaries of today's trends in a quick and effective manner.

Whatever your purpose for researching the biotechnology and genetics field, you'll find this book to be a valuable guide. Nonetheless, as is true with all resources, this volume has limitations that the reader should be aware of:

- Financial data and other corporate information can change quickly. A book of this type can be no more current than the data that was available as of the time of editing. Consequently, the financial picture, management and ownership of the firm(s) you are studying may have changed since the date of this book. For example, this almanac includes the most up-to-date sales figures and profits available to the editors as of mid 2006. That means that we have typically used corporate financial data as of late-2006.

- Corporate mergers, acquisitions and downsizing are occurring at a very rapid rate. Such events may have created significant change, subsequent to the publishing of this book, within a company you are studying.

- Some of the companies in THE BIOTECH 400 are so large in scope and in variety of business endeavors conducted within a parent organization, that we have been unable to completely list all subsidiaries, affiliations, divisions and activities within a firm's corporate structure.

- This volume is intended to be a general guide to a vast industry. That means that researchers should look to this book for an overview and, when

conducting in-depth research, should contact the specific corporations or industry associations in question for the very latest changes and data. Where possible, we have listed contact names, toll-free telephone numbers and World Wide Web site addresses for the companies, government agencies and industry associations involved so that the reader may get further details without unnecessary delay.

- Tables of industry data and statistics used in this book include the latest numbers available at the time of printing, generally through late-2005. In a few cases, the only complete data available was for earlier years.

- We have used exhaustive efforts to locate and fairly present accurate and complete data. However, when using this book or any other source for business and industry information, the reader should use caution and diligence by conducting further research where it seems appropriate. We wish you success in your endeavors, and we trust that your experience with this book will be both satisfactory and productive.

Jack W. Plunkett
Houston, Texas
August 2007

HOW TO USE THIS BOOK

The two primary sections of this book are devoted first to the biotechnology and genetics industry as a whole and then to the "Individual Data Listings" for THE BIOTECH 400. If time permits, you should begin your research in the front chapters of this book. Also, you will find lengthy indexes in Chapter 4 and in the back of the book.

THE BIOTECH AND GENETICS INDUSTRY

Glossary: A short list of biotech and genetics industry terms.

Chapter 1: Major Trends Affecting the Biotech & Genetics Industry. This chapter presents an encapsulated view of the major trends that are creating rapid changes in the biotech and genetics industry today.

Chapter 2: Biotech & Genetics Industry Statistics. This chapter presents in-depth statistics on spending, research, pharmaceuticals and more.

Chapter 3: Important Biotech & Genetics Industry Contacts – Addresses, Telephone Numbers and World Wide Web Sites. This chapter covers contacts for important government agencies, biotech organizations and trade groups. Included are numerous important World Wide Web sites.

THE BIOTECH 400

Chapter 4: THE BIOTECH 400: Who They Are and How They Were Chosen. The companies compared in this book were carefully selected from the biotech and genetics industry, largely in the United States. 75 of the firms are based outside the U.S. For a complete description, see THE BIOTECH 400 indexes in this chapter.

  Individual Data Listings:

  Look at one of the companies in THE BIOTECH 400's Individual Data Listings. You'll find the following information fields:

  Company Name:

  The company profiles are in alphabetical order by company name. If you don't find the company you are seeking, it may be a subsidiary or division of one of the firms covered in this book. Try looking it up in the Index by Subsidiaries, Brand Names and Selected Affiliations in the back of the book.

  Ranks:

  Industry Group Code: An NAIC code used to group companies within like segments. (See Chapter 4 for a list of codes.)

Ranks Within This Company's Industry Group: Ranks, within this firm's segment only, for annual sales and annual profits, with 1 being the highest rank.

Business Activities:

A grid arranged into six major industry categories and several sub-categories. A "Y" indicates that the firm operates within the sub-category. A complete Index by Industry is included in the beginning of Chapter 4.

Types of Business:

A listing of the primary types of business specialties conducted by the firm.

Brands/Divisions/Affiliations:

Major brand names, operating divisions or subsidiaries of the firm, as well as major corporate affiliations—such as another firm that owns a significant portion of the company's stock. A complete Index by Subsidiaries, Brand Names and Selected Affiliations is in the back of the book.

Contacts:

The names and titles up to 27 top officers of the company are listed, including human resources contacts.

Address:

The firm's full headquarters address, the headquarters telephone, plus toll-free and fax numbers where available. Also provided is the World Wide Web site address.

Financials:

Annual Sales (2006 or the latest fiscal year available to the editors, plus up to four previous years): These are stated in thousands of dollars (add three zeros if you want the full number). This figure represents consolidated worldwide sales from all operations. 2006 figures may be estimates or may be for only part of the year—partial year figures are appropriately footnoted.

Annual Profits (2006 or the latest fiscal year available to the editors, plus up to four previous years): These are stated in thousands of dollars (add three zeros if you want the full number). This figure represents consolidated, after-tax net profit from all operations. 2006 figures may be estimates or may be for only part of the year—partial year figures are appropriately footnoted.

Stock Ticker, International Exchange, Parent Company: When available, the unique stock market symbol used to identify this firm's common stock for trading and tracking purposes is indicated. Where appropriate, this field may contain "private" or "subsidiary" rather than a ticker symbol. If the firm is a publicly-held company headquartered outside of the

U.S., its international ticker and exchange are given. If the firm is a subsidiary, its parent company is listed.

Total Number of Employees: The approximate total number of employees, worldwide, as of the end of 2006 (or the latest data available to the editors).

Apparent Salaries/Benefits:

(The following descriptions generally apply to U.S. employers only.)

A "Y" in appropriate fields indicates "Yes."

Due to wide variations in the manner in which corporations report benefits to the U.S. Government's regulatory bodies, not all plans will have been uncovered or correctly evaluated during our effort to research this data. Also, the availability to employees of such plans will vary according to the qualifications that employees must meet to become eligible. For example, some benefit plans may be available only to salaried workers—others only to employees who work more than 1,000 hours yearly. Benefits that are available to employees of the main or parent company may not be available to employees of the subsidiaries. In addition, employers frequently alter the nature and terms of plans offered.

NOTE: Generally, employees covered by wealth-building benefit plans do not *fully* own ("vest in") funds contributed on their behalf by the employer until as many as five years of service with that employer have passed. All pension plans are voluntary—that is, employers are not obligated to offer pensions.

Pension Plan: The firm offers a pension plan to qualified employees. In this case, in order for a "Y" to appear, the editors believe that the employer offers a defined benefit or cash balance pension plan (see discussions below).The type and generosity of these plans vary widely from firm to firm. Caution: Some employers refer to plans as "pension" or "retirement" plans when they are actually 401(k) savings plans that require a contribution by the employee.

- Defined Benefit Pension Plans: Pension plans that do not require a contribution from the employee are infrequently offered. However, a few companies, particularly larger employers in high-profit-margin industries, offer defined benefit pension plans where the employee is guaranteed to receive a set pension benefit upon retirement. The amount of the benefit is determined by the years of service with the company and the employee's salary during the later years of employment. The longer a person works for the employer, the higher the retirement benefit. These defined benefit plans are funded entirely by the employer. The benefits, up to a reasonable limit, are guaranteed by the Federal Government's Pension Benefit Guaranty Corporation. These plans are not portable—if you leave the company, you cannot transfer your benefits into a different plan. Instead, upon retirement you will receive the benefits that vested during your service with the company. If your employer offers a pension plan, it must give you a summary plan description within 90 days of the date you join the plan. You can also request a summary annual report of the plan, and once every 12 months you may request an individual benefit statement accounting of your interest in the plan.

- Defined Contribution Plans: These are quite different. They do not guarantee a certain amount of pension benefit. Instead, they set out circumstances under which the employer will make a contribution to a plan on your behalf. The most common example is the 401(k) savings plan. Pension benefits are not guaranteed under these plans.

- Cash Balance Pension Plans: These plans were recently invented. These are hybrid plans—part defined benefit and part defined contribution. Many employers have converted their older defined benefit plans into cash balance plans. The employer makes deposits (or credits a given amount of money) on the employee's behalf, usually based on a percentage of pay. Employee accounts grow based on a predetermined interest benchmark, such as the interest rate on Treasury Bonds. There are some advantages to these plans, particularly for younger workers: a) The benefits, up to a reasonable limit, are guaranteed by the Pension Benefit Guaranty Corporation. b) Benefits are portable—they can be moved to another plan when the employee changes companies. c) Younger workers and those who spend a shorter number of years with an employer may receive higher benefits than they would under a traditional defined benefit plan.

ESOP Stock Plan (Employees' Stock Ownership Plan): This type of plan is in wide use. Typically, the plan borrows money from a bank and uses those funds to purchase a large block of the corporation's stock. The corporation makes contributions to the plan over a period of time, and the stock purchase loan is eventually paid off. The value of the plan grows significantly as long as the market price of the stock holds up. Qualified employees are allocated a share of the plan based on their length of service and their level of salary. Under federal regulations, participants in ESOPs are allowed to diversify their account holdings in set percentages that rise as the employee ages and gains years of service with the company. In this manner, not all of the employee's assets are tied up in the employer's stock.

Savings Plan, 401(k): Under this type of plan, employees make a tax-deferred deposit into an account. In the best plans, the company makes annual matching donations to the employees' accounts, typically in some proportion to deposits made by the employees themselves. A good plan will match one-half of employee deposits of up to 6% of wages. For example, an employee earning $30,000 yearly might deposit $1,800 (6%) into the plan. The company will match one-half of the employee's deposit, or $900. The plan grows on a tax-deferred basis, similar to an IRA. A very generous plan will match 100% of employee deposits. However, some plans do not call for the employer to make a matching deposit at all. Other plans call for a matching contribution to be made at the discretion of the firm's board of directors. Actual terms of these plans vary widely from firm to firm. Generally, these savings plans allow employees to deposit as much as 15% of salary into the plan on a tax-deferred basis. However, the portion that the company uses to calculate its matching deposit is generally limited to a maximum of 6%. Employees should take care to diversify the holdings in their 401(k) accounts, and most people should seek professional guidance or investment management for their accounts.

Stock Purchase Plan: Qualified employees may purchase the company's common stock at a price below its market value under a specific plan. Typically, the employee is limited to investing a small percentage of wages in this plan. The discount may range from 5 to 15%. Some of these plans allow for deposits to be made through regular monthly payroll deductions. However, new accounting rules for corporations, along with other factors, are leading many companies to curtail these plans—dropping the discount allowed, cutting the maximum yearly stock purchase or otherwise making the plans less generous or appealing.

Profit Sharing: Qualified employees are awarded an annual amount equal to some portion of a company's profits. In a very generous plan, the pool of money awarded to employees would be 15% of profits. Typically, this money is deposited into a long-term retirement account. Caution: Some employers refer to plans as "profit sharing" when they are actually 401(k) savings plans. True profit sharing plans are rarely offered.

Highest Executive Salary: The highest executive salary paid, typically a 2006 amount (or the latest year available to the editors) and typically paid to the Chief Executive Officer.

Highest Executive Bonus: The apparent bonus, if any, paid to the above person.

Second Highest Executive Salary: The next-highest executive salary paid, typically a 2006 amount (or the latest year available to the editors) and typically paid to the President or Chief Operating Officer.

Second Highest Executive Bonus: The apparent bonus, if any, paid to the above person.

Other Thoughts:

Apparent Women Officers or Directors: It is difficult to obtain this information on an exact basis, and employers generally do not disclose the data in a public way. However, we have indicated what our best efforts reveal to be the apparent number of women who either are in the posts of corporate officers or sit on the board of directors. There is a wide variance from company to company.

Hot Spot for Advancement for Women/Minorities: A "Y" in appropriate fields indicates "Yes." These are firms that appear either to have posted a substantial number of women and/or minorities to high posts or that appear to have a good record of going out of their way to recruit, train, promote and retain women or minorities. (See the Index of Hot Spots For Women and Minorities in the back of the book.) This information may change frequently and can be difficult to obtain and verify. Consequently, the reader should use caution and conduct further investigation where appropriate.

Growth Plans/ Special Features:

Listed here are observations regarding the firm's strategy, hiring plans, plans for growth and product development, along with general information regarding a company's business and prospects.

Locations:

A "Y" in the appropriate field indicates "Yes."

Primary locations outside of the headquarters, categorized by regions of the United States and by international locations. A complete index by locations is also in the front of this chapter.

Chapter 1

MAJOR TRENDS AFFECTING THE BIOTECH & GENETICS INDUSTRY

Trends Affecting the Biotechnology and Genetics Industry:

1) A Short History of Biotechnology
2) The State of the Biotechnology Industry Today
3) Ethanol Use Grows Quickly
4) New Money Pours into Biotech Firms
5) Major Drug Companies Bet on Partnerships With Smaller Biotech Research Firms
6) From India to Singapore to China, Nations Compete Fiercely in Biotech Development
7) Gene Therapies and Patients' Genetic Profiles Promise a Personalized Approach to Medicine
8) New Kinase Inhibitors Are Breakthrough Drugs for Cancer Treatment—Many More Will Follow
9) Pharmaceutical Costs Soar in U.S., Controversy over Drug Prices Rages On
10) Biotech Drugs Pick Up the Slack as Blockbuster Mainstream Drugs Age
11) Biogenerics are in Limbo in the U.S.
12) Breakthrough Drug Delivery Systems Evolve
13) Stem Cells—Controversy in the U.S. Threatens to Leave America Far Behind in the Research Race
14) Stem Cell Funding Trickles in the U.S. at the Federal Level While a Few States Create Funding of Their Own
15) Stem Cells—Therapeutic Cloning Techniques Advance

16) Stem Cells—A New Era of Regenerative Medicine Looms
17) Nanotechnology Converges with Biotech
18) Agricultural Biotechnology Scores Breakthroughs but Causes Controversy/Selective Breeding Offers a Compromise
19) Focus on Vaccines to Counter Potential Bioterror Attacks in the U.S.
20) Ethical Issues Abound
21) Technology Discussion—Genomics
22) Technology Discussion—Proteomics
23) Technology Discussion—Microarrays
24) Technology Discussion—DNA Chips
25) Technology Discussion—SNPs ("Snips")
26) Technology Discussion—Combinatorial Chemistry
27) Technology Discussion—Synthetic Biology
28) Technology Discussion—Recombinant DNA
29) Technology Discussion—Polymerase Chain Reaction (PCR)

1) A Short History of Biotechnology

While the 1900s will be remembered by industrial historians as the Information Technology Era and the Physics Era, the 2000s will most likely be marked as the Biotechnology Era because rapid advances in biotechnology will completely revolutionize many aspects of life in coming decades. However, the field of biotechnology can trace its true birth back to the dawn of civilization, when early man discovered the ability to ferment grains to make alcoholic beverages, and learned of the usefulness of cross-pollinating crops in order to create new hybrid strains—the earliest form of genetic engineering. In ancient China, people are thought to have harvested mold from soybean curd to use as an antibiotic as early as 500 B.C.

Robert Hooke first described cells as a concept in 1663 A.D., and in the late 1800s, Gregor Mendel conducted experiments that became the basis of modern theories about heredity. Alexander Fleming discovered the first commercial antibiotic, penicillin, in 1928.

The modern, more common concept of "biotech" could reasonably be said to have its beginnings shortly after World War II. In 1953, scientists James Watson and Francis Crick conceived the "double helix" model of DNA, and thus encouraged a spate of scientists to consider the further implications of human DNA. The Watson/Crick three-dimensional model began to unlock the mysteries of heredity and the methods by which replication of genetic material takes place within cells.

Significant steps toward biotech drugs occurred in the early 1970s. In 1973, Dr. Stanley N. Cohen, a Stanford University genetics professor, and Dr. Herb Boyer, a biochemist, genetic engineer and educator at UC-San Francisco, introduced the concept of gene-splicing and created the first form of recombinant DNA. In 1974, Cesar Milstein and Georges Kohler created monoclonal antibodies, cells that clone over and over again to create large quantities of a specific antibody. Many of today's top biotech drugs are monoclonal antibodies. These two discoveries (recombinant DNA and monoclonal antibodies) created the building blocks of the first modern commercial biotech drugs.

Boyer and Cohen's gene-splicing technique enabled scientists to cut genetic material from the cells of one organism and paste it into another organism. This was an important discovery because the genetic material they moved from one place to another instructs a cell as to how to make a particular protein. The organism on the receiving end of the gene-splicing technique is then able to make that protein. Over time, scientists have perfected the technique of splicing material that enables cells to create proteins that control the creation of insulin, the level of blood pressure and many other human functions. Such genetic engineering enabled, for the first time, the creation of massive vats of isolated proteins grown in bacteria or in cells harvested from mammals—in quantities large enough for the

commercial production of new drugs. (In fact, Boyer and Cohen's early experiments involved inserting a gene from an African clawed toad into bacterial DNA for duplication.)

In 1975, the first human gene was isolated, opening the door to gene therapy and creating the interest that led to the beginning of the massive, publicly funded Human Genome Project in 1990. In 1976, Bob Swanson of the now-famous Silicon Valley venture capital firm Kleiner Perkins formed a new business, Genentech, in conjunction with Herb Boyer (see above). Other early biotech firms arrived soon after, generally funded by venture capital firms, angel investors and corporate venture partners. These early biotech startups included many companies that grew into today's super-successful biopharma corporations: Amgen, Chiron, Biogen Idec and Genzyme. The creation of these startups, focused on the development of new drugs, was particularly noteworthy because it was the first time in decades that new drug companies were launched in significant numbers. In fact, most major drug companies in existence at the beginning of the 1970s could trace their histories back to the early 1900s or before.

2) The State of the Biotechnology Industry Today

Prescription drug purchases in the U.S. totaled about $229 billion during 2007, according to U.S. Government estimates, representing about $760 per capita. The total is up from a mere $40 billion in 1990. In 2016, government estimates show that American drug purchases may reach $497 billion, thanks to a rapidly aging U.S. population, inflation and the continued introduction of expensive new drugs. Biotech-related drugs accounted for perhaps $44 billion of the U.S. market in 2007, or 20%. Generic drugs account for about 58% of all drug expenditures in the U.S., up from only 33% in 1990.

Veterinary drugs are about a $2.5 billion market in the U.S., and about $5 billion globally. Agricultural biotechnology comprises an immense global sector, boosting production of food crops and enhancing nutritional qualities.

Ernst & Young estimates that there were 4,275 global biotech firms in 2006 (of which 1,452 were in the U.S.); including 710 that were publicly held (336 were in the U.S.). Global employment totals about 190,500, including 130,600 in the U.S.

Near the end of 2006, the Pharmaceutical Research and Manufacturers of America (PhRMA, www.phrma.org) estimated that $43.0 billion was invested in research and development that year by member biotechnology and pharmaceutical research member companies, about 17.5% of total global sales for these firms.

Advanced generations of drugs developed through biotechnology continue to enter the marketplace. The results promise to be spectacular for patients, as a technology tipping point of medical care is nearing, where drugs that target specific genes and proteins will become widespread. However, it continues to become more difficult and more expensive to introduce a new drug in the U.S. For example, during 2006, the FDA (Food and Drug Administration) approved only 22 new molecular entities or "NMEs" (medications containing active substances that have never before been approved for marketing in the U.S.). This is up from 20 during 2005, but down considerably from 36 in 2004. These are novel new drugs that are categorized differently from "NDAs" or New Drug Applications.

Conditions treated by these new NMEs ranged from gastrointestinal tumors to angina, Parkinson's disease, leukemia, hepatitis B and schizophrenia.

During 2006, 93 NDAs were approved by the FDA (up from 79 during 2005), while an additional two received tentative approval and five AIDS drugs received tentative "emergency" approval.

Developing a new drug is an excruciatingly slow and expensive endeavor. According to PhRMA, the average time required for the drug discovery, development and clinical trials process is 16 years. The good news is that the median FDA approval time for a "priority" NME was down to about 6.0 months during 2003 through 2006 from 16.3 in 2002, and for "standard" NMEs to 13.7 months in 2006 from 23.0 in 2005.

The promising era of personalized medicine is slowly, slowly moving closer to fruition. Dozens of exciting new drugs for the treatment of dire diseases such as cancer, AIDS, Parkinson's and Alzheimer's are either on the market or are very close to regulatory approval. PhRMA estimates that more than 250 million patients have already been treated with biotech drugs.

Stem cell research is moving ahead briskly on a global basis, despite the negative effect of restrictive research funding rules of the U.S. Federal Government. Stem cell breakthroughs are occurring rapidly. There is truly exciting evidence of the ability of stem cells to treat many problems, from cardiovascular disease to neurological disorders. Menlo Park, California-based Geron Corporation, for example, has published the results of its experiments that show that when certain cells (OPCs) derived from stem cells were injected in rats that had spinal cord injuries, the rats quickly recovered. According to the company, "Rats transplanted seven days after injury showed improved walking ability compared to animals receiving a control transplant. The OPC-treated animals showed improved hindlimb-forelimb coordination and weight bearing capacity, increased stride length, and better paw placement compared to control-treated animals."

Despite exponential advances in biopharmaceutical knowledge and technology, biotech companies enduring the task of getting new drugs to market continue to face long timeframes, daunting costs and immense risks.

Although the number of NDAs submitted to the FDA has grown dramatically since 1996, the number of new drugs receiving final approval remains relatively small. On average, of every 1,000 experimental drug compounds in some form of pre-clinical testing, only one actually makes it to clinical trials. Then, only one in five of those drugs makes it to market. Of the drugs that get to market, only one in three recover their costs. Meanwhile, the patent expiration clock is ticking—soon enough, manufacturers of generic alternatives steal market share from the firms that invested all that time and money in the development of the original drug.

Global Factors Boosting Biotech Today:

1) A rapid aging of the population base of industrialized nations such as Japan and the U.S., including the 78 million Baby Boomers in America who are entering senior years and needing a growing level of health care
2) A U.S. government focus on developing vast stores of highly-effective vaccines to combat bioterrorism strikes
3) Aggressive, global investment firms that are willing to risk their funds on biotech research and development
4) Vast research budgets at major pharmaceuticals firms
5) A growing global dependence on genetically-engineered agricultural seeds
6) Aggressive investment in biotechnology research in Singapore, China and India, often with government sponsorship
7) A quickly growing emphasis on bioethanol as a substitute for petroleum

---

Internet Research Tip:

You can review current and historical drug approval reports at the following page at the Center for Drug Evaluation and Research (CDER).
www.fda.gov/cder/rdmt

The Center for Biologics Evaluation and Research (CBER) regulates biologic products for use in humans. It is a source of a broad variety of data on drugs, including blood products, counterfeit drugs, exports, drug shortages, recalls and drug safety.
www.fda.gov/cber

According to a study released in 2001 by the Tufts Center for the Study of Drug Development, the cost of developing a new drug and getting it to market averaged $802 million, up from about $500 million in 1996. (Averaged into these figures are the costs of developing and testing drugs that never reach the market.) Expanding on the study to include post-approval research (Phase IV clinical studies), Tufts increased the number to $897 million. Tufts estimated the average cost to develop a new biologic at $1.2 billion in 2006. Even more pessimistic is research released in 2003 by Bain & Co., a consulting firm, which states that the cost is more on the order of $1.7 billion, including such factors as marketing and advertising expenses.

The typical time elapsed from the synthesis of a new chemical compound to its introduction to the market remains 12 to 20 years. Considering that the patent for a new compound only lasts about 20 years, a limited amount of time is available to reclaim the considerable investments in research, development, trials and marketing. As a result of these costs and the lengthy time-to-market, young biotech companies encounter a harsh financial reality: commercial profits take years and years to emerge from promising beginnings in the laboratory.

However, advances in systems biology (the use of a combination of state-of-the-art technologies, such as molecular diagnostics, advanced computers and extremely deep, efficient genetic databases) may eventually lead to more efficient, faster drug development at reduced costs. Much of this advance will stem from the use of technology to efficiently target the genetic causes of, and develop novel cures for, niche diseases.

The FDA is attempting to help the drug industry bring the most vital drugs to market in shorter time with three programs: Fast Track, Priority Review and Accelerated Approval. The benefits of Fast

Track include scheduled meetings to seek FDA input into development as well as the option of submitting a New Drug Application in sections rather than all components at once. The Fast Track designation is intended for drugs that address an unmet medical need, but is independent of Priority Review and Accelerated Approval. Priority drugs are those considered by the FDA to offer improvements over existing drugs or to offer high therapeutic value. The priority program, along with increased budget and staffing at the FDA, are having a positive effect on total approval times for new drugs.

For example, in May 2001 the FDA approved Novartis's new drug Gleevec (a revolutionary and highly effective treatment for patients suffering from chronic myeloid leukemia) after an astonishingly brief two and one-half months in the approval process (compared to a more typical six months). This rapid approval, which enabled the drug to promptly begin saving lives, was possible because of two factors aside from the FDA's cooperation. One, Novartis mounted a targeted approach to this niche disease. Its research determined that a specific genetic malfunction causes the disease, and its drug specifically blocks the protein that causes the genetic malfunction. Two, thanks to its use of advanced genetic research techniques, Novartis was so convinced of the effectiveness of this drug that it invested heavily and quickly in its development.

Generally, Fast Track approval is reserved for life-threatening diseases such as rare forms of cancer, but new policies are setting the stage for accelerated approval for less deadly but more pervasive conditions such as diabetes and obesity. Approval is also being made easier by the use of genetic testing to determine a drug's efficacy, as well as the practice of drug companies working closely with federal organizations. Examples of these new policies are exemplified in the approval of Iressa, which helps fight cancer in only 10% of patients but is associated with a genetic marker that can help predict a patient's receptivity; and VELCADE, a cancer drug that received initial approval in only four months because the company that makes it, Millennium Pharmaceuticals, worked closely with the National Cancer Institute to review trials.

Small- to mid-size biotech firms continue to look to mature, global pharmaceutical companies for cash, marketing muscle, distribution channels and regulatory expertise. Good examples are the agreement between Millennium Pharmaceuticals and Johnson & Johnson for the development of VELCADE, and Isis Pharmaceuticals' multiple deals with Novartis, Pfizer and other partners for research into new antisense drugs for inflammatory and metabolic diseases.

Meanwhile, major projects are underway, backed by diverse sponsors, to add to or build from scratch massive databases of genetic data on a scale not before imagined. In Iceland, for example, a group called DeCode Genetics has amassed a database of essentially all of the DNA in Iceland's unique, isolated population.

Internet Research Tip:
For extensive commentary and analysis on the development and approval of new drugs see:

Tufts Center for the Study of Drug Development
csdd.tufts.edu
Note: This website gives you the opportunity to download the latest annual edition of the "Outlook", an excellent summary review of trends in drug development.

With progress come setbacks, including a massive award for damages (more than $250 million) that occurred in a small-town Texas court in August 2005. The award was made to the widow of a patient who allegedly had a fatal reaction to Merck & Co.'s Vioxx pain medication (which had previously been removed from the market due to safety concerns). Texas laws capping medical case awards will reduce the damages significantly. Nonetheless, recent drug safety issues and a proliferation of lawsuits such as this may accelerate changes in the business models of drug development firms, discouraging them from risking funds on long-shot drugs intended to benefit the mass market. Meanwhile, drug makers will continue to alter marketing methods and greatly reduce consumer advertising. Virtually all drugs have significant side effect risks for certain types of patients. While drug makers have long practiced a high level of disclosure, those risks will be more clearly communicated in the future.

Global trends are affecting the biotech industry in a big way. Post 9/11, an emphasis is being placed by government agencies on the prevention of bioterror risks, such as attacks by the spread of anthrax. This factor, combined with global concern about the possible spread of avian flu, has been a significant boost to vaccine research and production. Meanwhile, the rapid rise of offshoring and globalization is contributing to the movement of research, development and clinical trials away from

the U.S., U.K. and France into lower cost technology centers in India and elsewhere. In fact, biotech firms are rising rapidly in India, China, Singapore and South Korea that will provide serious future competition to older companies in the West.

3) Ethanol Use Grows Quickly

Soaring gasoline prices, effective lobbying by agricultural and industrial interests, and a growing interest in cutting reliance on imported oil has put a high national focus on bioethanol in America. Corn and other organic materials, including agricultural waste, can be converted into ethanol through the use of engineered bacteria and superenzymes manufactured by biotechnology firms. This trend is giving a boost to the biotech, agriculture and alternative energy sectors.

In addition to the use of ethanol in cars and trucks, the chemicals industry, faced with daunting increases in petrochemicals costs, has a new appetite for bioethanol. In fact, bioethanol can be used to create plastics—an area that consumes vast quantities of oil in America and around the globe. Archer Daniels Midland is constructing a plant in Clinton, Iowa that will product 50,000 tons of plastic per year through the use of biotechnology to convert corn into polymers.

Ethanol is an alcohol produced by a distilling process similar to that used to produce liquors. A small amount of ethanol is added to about 30% of the gasoline sold in America, and most U.S. autos are capable of burning "E10," a gasoline blend that contains 10% ethanol. E85 is an 85% ethanol blend that may grow in popularity due to a shift in automotive manufacturing. Although only 800 of the 170,000 U.S. service stations sold E85 as of the middle of 2006, there may be an increase in demand for ethanol in the U.S. due to Detroit's increased production of vehicles than can run on a mixture of gasoline and ethanol. Ford offers E85-capable F-150 pickup trucks, Ford Crown Victorias, Mercury Grand Marquis and Lincoln Town Cars. At GM, 400,000 "flexfuel" vehicles were built in 2006, including the Chevy Avalanche, Chevy Monte Carlo, Chevy Tahoe, GMC Yukon and GMC Sierra.

Ethanol is a very popular fuel source in Brazil. In fact, Brazil is the world's largest producer of ethanol, which provides a significant amount of the fuel used in Brazil's cars. This is due to a concerted effort by the government to reduce dependency on petroleum product imports. After getting an initial boost due to government subsidies and fuel tax strategies beginning in 1975, Brazilian producers have developed plants (typically sugars) that enable them to produce ethanol at moderate cost. The fact that Brazil's climate is ideally suited for sugarcane is a great asset. Brazilian ethanol production is so successful that it is a major exporter of ethanol to the U.S. Brazilian automobiles are typically equipped with engines that can burn pure ethanol or a blend of the gasoline and ethanol. Brazilian car manufacturing plants operated by Ford, GM and Volkswagen all make such cars.

In America, partly in response to the energy crisis of the 1970s, Congress instituted federal ethanol production subsidies in 1979. Corn-based grain ethanol production picked up quickly, and federal subsidies have totaled about $10 billion through 2005. The size of these subsidies and environmental concerns about the production of grain ethanol produced a steady howl of protest from observers through the years. Nonetheless, the Clean Air Act of 1990 further boosted ethanol production by increasing the use of ethanol as an additive to gasoline. Meanwhile, the largest producers of ethanol, such as Archer Daniels Midland (ADM), have reaped significant amounts of subsidies from Washington for their output. The ethanol cause was further promoted by President Bush in his state of the union address in 2006, in which he chastised Americans for their "addiction to oil" and promoted the use of ethanol among other alternative fuels. The Bush administration claims that ethanol could provide more than a third of the U.S.'s gasoline needs by 2025.

The future of ethanol as a viable, environmentally friendly, cost-effective fuel that can be economically produced without government support may lie in new technologies based on production from cellulose rather than corn. The U.S. Energy Act of 2005 specifically requires immense increases in the amount of ethanol or biodiesel used as fuel.

Traditional grain ethanol is typically made from corn or sugarcane. In contrast to grain ethanol, "cellulosic" ethanol is made more efficiently out of more plentiful materials. This type of ethanol is made from agricultural waste like corncobs, wheat husks, stems, stalks and leaves, which are treated with specially engineered enzymes to break the waste down into its component sugars. The sugars (or sucrose) are used to make ethanol. Since agricultural waste is plentiful, turning it into energy is wonderful strategy.

The trick to cellulosic ethanol production is the creation of efficient enzymes to treat the agricultural waste. The U.S. Department of Energy is investing $20 million per year in funding along with major chemical companies such as Dow Chemical, DuPont and Cargill.

Iogen, a Canadian biotechnology company, makes just such an enzyme and is presently building production-size cellulosic ethanol facilities in the U.S., Canada and Germany, to start commercialization and determine how economical the process is. The company plans to construct a $300-million, large-scale biorefinery by 2008 with a potential output of 50 million gallons per year. Its pilot plant in Ottawa is capable of producing 260,000 gallons per year from 20 million tons of wheat straw and corn stalks.

Construction of new ethanol production plants pushed total production capacity in the U.S. to about 5.4 billion gallons by the end of 2006 (about 3.4% of total U.S. gasoline consumption), up from 3.9 billion as of June 2005. As of August 1, 2007, there were 124 ethanol biorefineries in the U.S. (totaling 6.5 billion barrels of capacity), with another 76 under construction (which will add another 6.4 billion barrels).

Other companies, such as Syngenta, DuPont and Ceres, are genetically engineering crops so that they can be more easily converted to ethanol or other energy producing products. Syngenta, for example, is testing an engineered corn that contains the enzyme amylase. Amylase breaks down the corn's starch into sugar, which is then fermented into ethanol. The process currently used with traditional corn crops adds amylase to begin the process.

Environmentalists are concerned that genetically engineering crops for use in energy-related yields will endanger the food supply through cross-pollination with traditional plants. Monsanto is focusing on conventional breeding of plants with naturally higher fermentable starch contents as an alternative to genetic engineering.

Another concern relating to ethanol use is that its production is not as energy efficient as that of biodiesel made from soybeans. According to a study at the University of Minnesota, the farming and processing of corn grain for ethanol yields approximately 25% more energy than it consumes, compared to 93% for biodiesel. Likewise, greenhouse gas emissions savings are greater for biodiesel. Producing and burning ethanol results in 12% less greenhouse gas emissions than burning gasoline, while producing and burning biodiesel results in a 41% reduction compared to making and burning regular diesel fuel.

4) New Money Pours Into Biotech Firms

Restructuring, mergers and bankruptcies are regular occurrences at smaller biotech firms, as their cash hoards from IPOs or initial rounds of venture capital dry up. Many of these companies are unable to achieve meaningful levels of revenues, and some of them incur immense annual losses due to the costs of research, development and regulatory requirements.

Much of the financial success that has been achieved by biotech firms lies with only a handful of companies such as Amgen and Genentech. These companies' initial investors have benefited greatly from growing sales and profits along with soaring stock market values, making some investors willing to take the enormous risk for the chance at a huge return.

New drugs and profitable drug companies take a long, long time to develop. The years of operation required to reach profitability at biopharmaceutical firms ranges from six years for Amgen to 10 years for Chiron and 15 years for Cephalon. These timeframes do not fit the requirements of many investors.

Despite this air of risk, investors seem to have regained their desire to bet on biotech, with an estimated $25 billion in new financings achieved by U.S. biotech companies in 2006 (up from $17.5 billion in 2005), according to corporate finance experts at Burrill & Co. In addition, about 18 firms achieved successful IPOs.

Key Food & Drug Administration (FDA) terms relating to human clinical trials and approval stages:

Phase I—Small-scale human trials to determine safety. Typically include 20 to 60 patients and are six months to one year in length.

Phase II—Preliminary trials on a drug's safety/efficacy. Typically include 100 to 500 patients and are one and a half to two years in length.

Phase III—Large-scale controlled trials for efficacy/safety; also the last stage before a request for approval for commercial distribution is made to the FDA. Typically include 1,000 to 7,500 patients and are three to five years in length.

Phase IV—Follow-up trials after a drug is released to the public.

NDA (New Drug Application)—A document submitted to the FDA that states the results of clinical trials and asks permission to market the drug to the public.

PMA (Pre-Market Approval)—An application to the FDA seeking permission to sell a medical device (rather than a drug) that is implanted, life-sustaining or life-supporting.

Source: Plunkett Research, Ltd.

---

5) Major Drug Companies Bet on Partnerships With Smaller Biotech Research Firms

At pharmaceutical firms both large and small, profits are under constant pressure because blockbuster drugs that have made immense profits for many years quickly lose their patent protection and face vast competition from generic versions. In the U.S., generic drugs now hold about a 58% market share by volume. This puts pressure on large research-based drug firms to develop new avenues for profits. One such avenue is partnerships with and investments in young biotech companies, but profits from such ventures will, in most cases, be slow to appear. Meanwhile, the major, global drug firms are investing billions in-house on biotech research and development projects, but new blockbusters are elusive.

For example, Pfizer invests more than $7.5 billion yearly on R&D. That money is invested in carefully designed research programs with specific goals. In 2007, its goal was to triple the number of late-stage compounds in its research portfolio by 2009. In 2007, it had 11 drug programs in final testing stages, up from eight the previous year, and 47 programs in mid-stage testing, up from 32.

Much of future success for the world's major drug companies will lie in harnessing their immense financial power along with their legions of salespeople and marketing specialists to license and sell innovative new drugs that are developed by smaller companies. There are dozens of exciting, smaller biotech companies that are focused on state-of-the-art research that lack the marketing muscle needed to effectively distribute new drugs in the global marketplace. To a large degree, these companies rely on contracts and partnerships with the world's largest drug manufacturers. In addition to money to finance research and salespeople to promote new drugs to doctors, the major drug makers can offer expertise in guiding new drugs through the intricacies of the regulatory process. While these arrangements may not lead to blockbuster drugs that will sell billions of pills yearly to treat mass market diseases, they can and often do lead to very exciting targeted drugs that can produce $300 million to $1 billion in yearly revenues once they are commercialized. A string of these mid-level revenue drugs can add up to a significant amount of yearly income. According to Burrill & Co., U.S. biotech firms reaped $15 billion in capital through partnering during 2006.

A good example is the relationship between Pfizer, one of the world's leading drug companies, and Medarex (www.medarex.com), a relative newcomer. In late 2004, Medarex announced a major partnership with Pfizer that included an initial cash infusion to Medarex of $80 million and a purchase of $30 million in Medarex stock by Pfizer. Their long-term goal is for Medarex to use its unique UltiMAb antibody technology to create up to 50 new drugs targeted at diseases specified by Pfizer. Eventually, Medarex may earn an additional $400 million in payments under the deal. Medarex has additional partnerships with other leading firms.

Another example is the $20 million invested by Colgate-Palmolive in late 2005 in Introgen Therapeutics, Inc., an Austin, Texas biotech firm that is studying new ways to treat oral cancer. Introgen created a mouthwash that contains a virus that carries known cancer-suppressing genes, which theoretically will treat precancerous growths in the mouth.

6) From India to Singapore to China, Nations Compete Fiercely in Biotech Development

While pharmaceutical companies based in the nations with the largest economies, such as the U.S., U.K. and Japan, struggle to discover the next important drug, companies and government agencies in many other countries are enhancing their positions on the biotech playing field, building their own educational and technological infrastructures. Not surprisingly, countries such as India, Singapore and China, which have already made deep inroads into the information technology industry, are making major efforts in biotechnology, which is very much an information-based science. Firms that manufacture generics and provide contract research, development and clinical trials services are already common in such nations (in India alone, clinical trials are expected to reach $1 billion annually by 2010). In most cases, this is just a beginning, with original drug and technology development the ultimate goal.

The government of Singapore, for example, has made biotechnology one of its top priorities for development, vowing to make it one of the staples of its economy. Its $286-million "Biopolis," a research and development center opened in 2004 that encompasses 2 million square feet of laboratories and offices, is only a small part of its $2.3-billion initiative to foster biotech in Singapore. Eventually, Biopolis will house more than 2,000 scientists. Even Dubai is planning a major biotech research center by investing in a new Dubai Biotechnology Initiative.

Outsourcing of biotech tasks to India is growing at a very high annual rate, reaching an estimated $120 million in 2004. (India has dozens of drug manufacturing plants that meet FDA specifications.) India's total pharmaceuticals industry revenues, largely from the manufacturing of generics, approached $10 billion in 2006. The president of India's Association of Biotechnology Led Enterprises recently forecast that biotech revenues alone could reach $5 billion in that nation by 2010.

India already has over 200 firms involved in biotechnology and related support services. In March 2005, the nation tightened its intellectual property laws in order to provide strong patent protection to the drug industry. This means that major drug firms will be opening facilities in India, and that startups will flourish. Drug sector firms to watch that are based in India include Nicholas Piramal, Ltd., Ranbaxy Laboratories Ltd. and Dr. Reddy's Laboratories Ltd. Services firms and research labs in India will continue to have a briskly growing business in outsourced services for the biotech sector. The costs of developing a new drug in India can be a small fraction of those in the U.S., although drugs developed in India still are required to go through the lengthy and expensive U.S. FDA approval process before they can be sold to American patients.

Stem cell (and cloning) research activity is brisk in a number of nations outside the U.S. as well. To begin with, certain institutions around the world have stem cell lines in place, and some make them available for purchase. Groups that own existing lines include the National University of Singapore, Monash University in Australia and Hadassah Medical Centre in Israel. Sweden has also stepped onto the stage as a major player in stem cell research, with 40 companies focused on the field, including rising stars such as Cellartis AB, which has one of the largest lines of stem cells in the world, and NeuroNova AB, which is focusing on regenerating nerve tissue. It is estimated that Sweden is home to 32% of all harvested stem cells worldwide.

More importantly, several Asian nations, including Singapore, South Korea, Japan and China, are investing intensely in biotech research centered on cloning and the development of stem cell therapies. The global lead in the development of stem cell therapies may eventually pass to China, where the Chinese Ministry of Science and Technology readily sees the commercial potential and is enthusiastically funding research. On top of funding from the Chinese government, investments in labs and research are being backed by Chinese universities, private companies, venture capitalists and Hong Kong-based investors.

China and India both have booming offshore business centers for clinical trials. The savings can be up to 50% to 80% over conducting trials in the U.S. This is a very large global business sector involving tens of billions of dollars in annual expenses.

Meanwhile, leading biotech firms, including Roche, Pfizer and Eli Lilly, are taking advantage of China's very high quality education systems and low operating costs to establish R&D centers there. This is a model that has already been used successfully by leading IT firms operating large research labs in China, including Microsoft. Clinical trials are underway in China for what could become one of the first vaccines against SARS, and at least four AIDS vaccines are under development or trials there.

Taiwan recently opened the second of four planned biotech research parks. Vietnam has plans to open six biotech research labs. Australia has a rapidly developing biotechnology industry.

Scientist Alan Colman, a leader of the team of geneticists that successfully cloned Dolly the sheep, relocated from the U.K. to Singapore to take advantage of that nation's immense focus on investing in and nurturing biotech research. Today,

he is CEO of Singapore-based ES Cell International, working on developing stem cells that will be capable of producing insulin, with the potential to cure patients suffering from diabetes.

South Korea is a world leader in research and development in a wide variety of technical sectors, and it is pushing ahead boldly into biotechnology. Initiatives include the Korea Research Institute of Bioscience and Biotechnology. The combination of government backing and extensive private capital in South Korea could make this nation a biotech powerhouse. One area of emphasis there is stem cell research.

In addition to fewer restrictions, many countries outside of the U.S. have cheaper labor costs, even for highly educated professionals such as doctors and scientists.

7) Gene Therapies and Patients' Genetic Profiles Promise a Personalized Approach to Medicine

Scientists now believe that almost all diseases have some genetic component. For example, some people have a genetic predisposition for breast cancer or heart disease. Understanding of human genetics will soon lead to breakthroughs in gene therapy for many ailments. Organizations ranging from the Mayo Clinic to drug giant GlaxoSmithKline are experimenting with personalized drugs that are designed to provide appropriate therapies based on a patient's personal genetic makeup or their lack of specific genes. Genetic therapy involves splicing desired genes taken from one patient into a second patient's cells in order to compensate for that patient's inherited genetic defect, or to enable that patient's body to better fight a specific disease.

For example, drugs that target the genetic origins of tumors may offer more effective, longer-lasting and far less toxic alternatives to conventional chemotherapy and radiation. One of the most noted drugs that target specific genetic action is Herceptin, a monoclonal antibody that was developed by Genentech. Approved by the FDA in 1998, Herceptin, when used in conjunction with chemotherapy, shows great promise in significantly reducing breast cancer for certain patients who are known to overproduce the HER2 gene. (A simple test is used to determine if this gene is present in the patient.) Herceptin, which works by blocking genetic signals, thus preventing the growth of cancerous cells, may show potential in treating other types of cancer, such as ovarian, pancreatic or prostate cancer.

More recent discoveries include tests that predict side effects from colon cancer drug Irinotecan and best dosage levels for patients taking the blood thinner Warfarin. The Mayo Clinic's Irinotecan test was released in late 2005 and is one of several genetic tests marketed by the clinic. It planned to release an additional 12 tests through 2006. Mayo is also marketing a genetic test that determines whether a patient will respond poorly to antidepressants.

One of the fastest-growing genetic tests is marketed by Genomic Health, based in Redwood City, California. Its Oncotype DX test provides breast cancer patients an assessment of the likelihood of the recurrence of their cancer based on the expression of 21 different genes in a tumor. The test enables patients to evaluate the results they may expect from post-operative therapies such as Tamoxifen or chemotherapy. The test costs about $3,500, and to-date has generally not been covered by health insurance, despite its potential to save treatment costs. The firm is also doing full-scale clinical development on a test to predict the likelihood of recurrence of colon cancer. Such tests will be standard preventative treatment in coming decades.

The scientific community's rapidly improving knowledge of genes and the role they play in disease is leading to several different tracks for improved treatment results. One track is to profile a patient's genetic makeup for a better understanding of a) which drugs a patient may respond to effectively, and b) whether certain defective genes reside in a patient and are causing a patient's disease or illness. Yet another application of gene testing is to study how a patient's liver is able to metabolize medication, which could help significantly when deciding upon proper dosage. Since today's widely used drugs often produce desired results in only about 50% of patients who receive them, the use of specific medications based on a patient's genetic profile could greatly boost treatment results while cutting costs. Each year, 2.2 million Americans suffer side effects from prescription drugs. Of those, more than 100,000 die, making adverse drug reaction a leading cause of death in the U.S. A Journal of the American Medical Association study states that the annual cost of treating these drug reactions totals $4 billion each year.

A second track for use of genetic knowledge is to attack, and attempt to alter, specific defective genes—this approach is sometimes referred to as "gene therapy." Generally, pure gene therapy attempts to target defective genes within a patient by introducing new copies of normal genes. These new genes may be introduced through the use of viruses or proteins that carry them into the patient's body.

China-based Shenzhen Sibiono GeneTech achieved the world's first pure gene therapy to be approved for wide commercial use in October 2003. The drug is sold under the brand name Gendicine as a treatment for a head and neck cancer known as squamous cell carcinoma (HNSC). The drug has proven highly effective in trials, and the company is testing its technology for several other types of cancer, including esophageal, gastric, colon, liver and rectum. As of mid-2006, the drug was available to Chinese patients at the Haidian Hospital in Beijing and to foreign patients at a facility in Shanghai.

Other applications of gene therapy are already in research in the U.S. and elsewhere for treatment of a wide variety of diseases. For example, gene therapy may be highly effective in the treatment of rare immune system disorders, melanoma (a malignant skin cancer for which there is currently no effective cure once the disease has spread to other organs) and cystic fibrosis. In 2006, two male patients suffering from a rare malady called chronic granulomatous disease (CGD), which makes patients terribly vulnerable to infections, were able to cease taking daily antibiotics due to a gene therapy that introduced healthy genes to replace defective ones in their bloodstreams.

In 2006, researchers at the National Cancer Institute in Maryland reported exciting results of a targeted gene therapy used on 17 cancer patients who had advanced melanoma to such a degree that they were not expected to live more than a few months. Two of the patients were completely cancer free 18 months after the start of the treatment, which succeeded in either eradicating or shrinking large tumors to the point that they could be surgically removed. This groundbreaking therapy alters immune system cells so that they target cancerous tumors. The technology has the potential to be adapted for use against other specific cancers. For this study, scientists led by Steven Rosenberg, MD, isolated a genetic component, a T-cell receptor, in a previous patient who had responded particularly well to a T-cell infusion. (T-cells are particular white blood cells that mature in the thymus and work as part of the immune system.) This patient's tumors had shrunk by 95%. Dr. Rosenberg's team spliced this receptor into T-cells removed from the 17 patients in the study, multiplied those cells into the billions, and then transfused them into the melanoma patients.

Great strides in potential gene therapy for muscular dystrophy are being made at two major universities. In 2002, scientists at the University of Washington found that, using a virus-delivery model, gene therapy might successfully be used to implant a healthy, protein-producing gene in patients suffering from the disease. Studies in 2005 at the University of Pittsburgh found that a miniaturized version of this gene could be successfully delivered throughout the bodies of mice that are afflicted with the disease, and that the mice showed some improvement. Success of this type could eventually lead to gene therapy programs for use in humans who suffer from muscular dystrophy.

Roche Pharmaceuticals and Affymetrix, Inc. have developed a lab-on-a-chip (the AmpliChip CYP450) that can detect more than two dozens variations of two different genes. These genes are important to a patient's reaction to and use of drug therapies because the genes regulate the way in which the liver metabolizes a large number of common pharmaceuticals, such as beta blockers and antidepressants. A quick analysis of a tiny bit of a patient's blood can lead to much more effective use of prescriptions. Additional chips for specific types of genetic analysis will follow, such as the 2006 launch of a chip that scans for malignant tumor-causing variations of the human p53 gene.

Another gene discovered by scientists appears to play a major role in widespread forms of the more than 30 different types of congenital heart defects, the most common of all human birth defects. Genes that are used by bacteria to trigger the infection process have also been identified, which could lead to powerful vaccines and antibiotics against life-threatening bacteria such as salmonella.

In the area of heart disease, about 300,000 patients yearly undergo heart bypass surgery in an effort to deliver increased blood flow to and from the heart. Clogged arteries are bypassed with arteries moved from the leg or elsewhere in the patient's body. Genetic experts have now developed the biobypass. That is, they have determined which genes and human proteins create a condition known as angiogenesis, which is the growth of new blood vessels that can increase blood flow without traumatic bypass surgery. This technique may become widespread in the near future.

Gene therapy is still in its infancy, and is not without its failures. An 18-year-old patient suffering from a disorder that builds toxic levels of ammonia in the bloodstream died during a clinical trial conducted by the University of Pennsylvania in 1999. The FDA subsequently shut down a number of drug trials. However, the potential for gene therapy and genetic testing used for the choice and dosage of medications is almost limitless. Watch for major investment by

pharmaceutical companies over the mid-term in further study and test development.

8) New Kinase Inhibitors Are Breakthrough Drugs for Cancer Treatment—Many More Will Follow

Multi-kinase inhibitors are now on the leading edge of new drug development. This exciting class of drugs has the potential to target a wide variety of defective proteins. Kinases are enzymes that influence certain basic functions within cells, such as cell division. (This enzyme catalyzes the transfer of phosphates from ATP to proteins, thus causing changes in protein function.) Defective kinases can lead to diseases such as cancer. Kinase-blocking drugs are able to inhibit cell activity that can cause both tumor growth and blood vessel creation—thus they are able to shut down the blood vessels that enable tumors to grow and thrive. A good example is PTK, the experimental drug developed by Novartis for the treatment of cancer of the colon, brain and other organs.

Earlier, in 2001, Novartis launched the drug Gleevec for the treatment of a blood cell cancer known as chronic myeloid leukemia. Gleevec operates by destroying diseased blood cells without harming normal cells. Patient results have been excellent. Additional projects include Genentech's highly effective cancer drug Avastin, and the lung cancer drug Tarceva, developed by OSI Pharmaceuticals, Genentech and Roche Group.

Dozens of new kinase-blocking drugs are in various stages of development and testing. This could become a tremendous breakthrough in treatment of cancer patients and other groups. From a business standpoint, the market for such drugs is growing rapidly. By 2010, the global market should top $11.8 billion.

9) Pharmaceutical Costs Soar in U.S., Controversy over Drug Prices Rages On

Historically, the drug industry has been one of the world's most profitable business sectors. Nearly 20% of pharmaceutical revenue is redirected toward discovering new medicines, well above the average of 4% reinvested in research and development in most industries. Advanced technology has allowed drug companies to saturate their development programs with smarter, more promising drugs. R&D (research and development) budgets are staggering. For example, Pfizer invests more than $7.5 billion yearly in R&D. Total spending on R&D by U.S.

drug firms $43 billion reached in 2007, according to PhRMA, an industry association.

However, both profits and research budgets are under pressure at pharmaceutical companies due to a wide range of challenges. Blockbuster drugs that might provide the high returns needed to bring a drug to market are increasingly difficult to develop. Patents that have protected today's blockbuster drugs from competition are rapidly expiring. Lawsuits against drug makers abound, and drug safety issues rank among the top challenges faced by the pharmaceutical industry today.

Meanwhile, payors are fighting to reduce the growing amounts that they spend on prescription drugs, and future revenues for drug makers will be impacted by cost-control measures at the individual level. In addition, the U.S. Government's Medicare drug benefit is putting even greater pressure on drug companies to sell drugs at lower prices. Drug makers are reacting to these pressures as best they can. For example, in mid-2007, Johnson & Johnson announced job cuts affecting 4,800 employees in an effort to reduce operating costs.

Highly effective new drug therapies can be incredibly expensive for patients. Take Iressa, a breakthrough treatment for lung cancer. It costs about $1,800 monthly per patient. Likewise, the groundbreaking colon cancer treatment Erbitux can cost as much a $30,000 for a seven-week treatment. A standard treatment of the latest drug regimens for some diseases can run up a $250,000 or higher bill over a couple of years.

The good news is that as the demand for new and improved treatments intensifies, so do the abilities of modern technology. In addition to expediting the process and lowering the costs of drug discovery and development, advanced pharmaceutical technology promises to increase the number of diseases that are treatable with drugs, enhance the effectiveness of those drugs and increase the ability to predict disease, not just the ability to react to it.

More good news is the FDA's announcement in July 2006 of plans to adapt the ways in which some experimental drugs are brought to clinical trial. In the past, the necessary effort and expense to prepare a request for human trials of a new drug amounted to between $500,000 and $1 million. Under the new proposal, the FDA would require less stringent testing (under limited circumstances) in the lab and on animals before being tried on humans. The FDA's willingness to consider adaptive designs is a major shift.

Consumers' voracious appetites for new drugs continue to grow. Insurance companies may raise co-payments for drugs, strike deals with drug companies and employ pharmacy benefit management tactics in an effort to fend off rising pharmaceutical costs. Also, new drugs have very high prices when they initially hit the market, since they have no generic equivalent.

Until recently, pharmaceutical research was focused primarily on curing life-threatening or severely debilitating illnesses. But a current generation of drugs, commonly referred to as "lifestyle" drugs, is transforming the pharmaceutical industry. Lifestyle drugs target a variety of medical conditions, ranging from the painful to the inconvenient, including obesity, impotence, memory loss and depression. Drug companies also continue to develop lifestyle treatments for hair loss and skin wrinkles in an effort to capture their share of the huge anti-aging market aimed at the Baby Boomer generation. The use of lifestyle drugs dramatically increases the total annual consumer intake of pharmaceuticals, and creates a great deal of controversy over which drugs should be covered by managed care and which should be paid for by the consumer alone.

In coming years, taming pharmaceutical costs will be one of the biggest challenges facing the health care system. Prescription drug costs already account for more than 10% of all health care expenditures in the U.S. Managed care must be able to determine which promising new drugs can deliver meaningful clinical benefits proportionate to their monetary costs.

Several developments are fueling the fire under the controversy over drug costs in the U.S. To begin with, it has become common knowledge that pharmaceutical firms tend to price their drugs at vastly lower prices outside the U.S. market. The result is that U.S. consumers and their managed care payors are bearing a disproportionate share of the costs of developing new drugs. Whereas prices in the U.S. are determined by a free market and are mostly limited only by competition, nations such as Australia and Canada put a cap on drug prices that cannot exceed a given amount. Taking advantage of this discrepancy, generally a 30% to 80% difference, border-crossing U.S. citizens have saved bundles of money on prescription drugs by importing them from Canada. Canadian Internet-based drug retailers have been selling to anyone with a credit card and a faxed prescription. Naturally, U.S. drug companies have raised a furor. Many have threatened to limit

Canada's supplies so it only meets the demands of Canadians, as well as putting a corporate embargo on any pharmacy or distributor suspected of selling to Internet companies. However, the U.S. Government ahs decided to stop interfering. In October 2006, U.S. Customs authorities announced that they would no longer seize individual shipments of drugs from Canada to the U.S. if they contained no more than a 90-day supply.

Meanwhile, drug makers are challenged to price their drugs in a way that will earn a good return on investment prior to the expiration of patent protection. The current U.S. patent policy grants drug manufacturers the normal 20 years protection from the date of the original patent (which is most likely filed very early in the research state), plus a period of 14 years after FDA approval. Since typical drugs take 10 to 15 years to research, prove in trials and bring to market, this effectively gives the patent holder only 19 to 24 years before low-price competition from generic manufacturers begins.

Branded drugs tend to lose 15% to 30% of market share when generic versions come on the market, eventually losing as much as 75% to 90% after a few years.

Then there is the issue of "follow-on" drugs. These are drugs that hold their own patent for therapies similar to or competing with the original breakthrough patent. Follow-on drugs tend to get through regulatory approval much faster, and often are brought to market at much lower prices. Both factors create a significant competitive advantage for follow-on patent owners.

Major drug companies are trying to get on the generic gravy train by quietly creating their own generic drug subsidiaries. Pfizer, for example, has a division called Greenstone, Ltd., which produces generic versions of its blockbuster drugs including Zoloft, an antidepressant that brought in $2.5 billion in 2005 sales. Likewise, Johnson & Johnson has a generic subsidiary called Patriot Pharmaceuticals and Schering-Plough created Warrick Pharmaceuticals. Naturally, independent generic drug companies oppose the practice.

Genetic drugs already account for about 58% of all drug purchases (by volume) in the U.S., up from 33% in 1990. Consumers (and their insurance companies) save from 30% to 75% over branded drug costs. From 2007 through 2012, patents will expire on branded drugs in the U.S. with about $60 billion in combined annual sales. For example, by 2011 a generic version of cholesterol-controlling Lipitor is expected to be on the market. Lipitor has

long been the world's best selling branded drug (selling as much as $13 billion yearly at one time), but sales were declining in mid-2007 thanks to a generic substitute for a competing drug named Zocor; the generic of Zocor costs about 60% less than Lipitor.

Factors leading to soaring drug costs in the American health care system:

- 78 million Baby Boomers are beginning to enter their senior years. The lifespan of Americans is increasing, and chronic illnesses are increasing as the population ages.
- Medicare Part D, the recently introduced component of Medicare that pays for prescription drugs for seniors, made affordable drugs available to millions of Americans, fueling overall drug sales.
- The drug industry intensified its sales effort over recent years. Direct-to-consumer advertising and legions of sales professionals calling on physicians increased demand for the newest, most expensive drugs.
- Convoluted and uncoordinated lists of drug "formularies" (available drugs, their uses and their interactions) require increased administrative work to sort through, thus forcing costs upward.
- Physicians often prescribe name-brand drugs when a generic equivalent may be available at a fraction of the cost.
- Research budgets are immense. Breakthroughs in research and development are creating significant new drug therapies, allowing a wide range of popular treatments that were not previously available. An excellent example is the rampant use of antidepressants such as Prozac and Zoloft. Meanwhile, major drug companies face the loss of patent protection on dozens of leading drugs. They are counting on expensive research, partnerships and acquisitions to replace those marquis drugs.
- "Lifestyle" drug use is increasing, as shown by the popularity of such drugs as Viagra (for the treatment of sexual dysfunction), Propecia (for the treatment of male baldness) and Botox (for the treatment of facial wrinkles).

Source: Plunkett Research, Ltd.

10) Biotech Drugs Pick Up the Slack as Blockbuster Mainstream Drugs Age

While Big Pharma has concentrated mainly on mass-market drugs that treat a broad spectrum of ailments, biotech companies have largely developed treatments for rare disorders or maladies such as certain cancers that only affect a small portion of the population. For example, biotech pioneers Genentech and Biogen Idec developed Rituxan for the treatment of non-Hodgkin's lymphoma, which had only $1.8 billion in sales in 2005.

Drugs such as Rituxan are commonly referred to as "orphan drugs," which means that they treat illnesses that no other drug on the market addresses. Technically, a drug designated by the FDA with orphan status provides therapeutic benefit for a disease or condition that affects less than 200,000 people in the U.S. Almost half of all drugs produced by biotech companies are for orphan diseases. These drugs enjoy a unique status due to the Orphan Drug Act of 1983, which gives pharmaceutical companies a seven-year monopoly on the drug without having to file for patent protection, plus a 50% tax credit for research and development costs. As of early-2007 there 1,679 drugs designated "orphan," including a few hundred on the market and the balance pending approval.

Meanwhile, blockbuster drugs are becoming more and more difficult to produce. Although drugs such as Zocor, Zoloft and Ambien faced generic competition for the first time in 2006 when their patents expire, new drugs to replace them are slow in coming.

As the wellspring of mainstream Big Pharma blockbusters begins to dry, many big pharmaceutical companies are rushing to develop their own orphan drugs. Pfizer released Sutent, a drug used to treat cancerous kidney and stomach-lining tumors in 2006, long after biotech firm Genentech had four drugs used for cancer on the market. Genentech's colon cancer treatment, Avastin, is expected to reach $6.2 billion in sales in 2009, while Pfizer's Sutent is projected to bring in only $570 million that same year.

Commentary: The challenges facing the biopharmaceuticals industry from 2004-2010

- Working with governments to develop methods to safely and effectively speed approval of new drugs.
- Working with the investment community to build confidence and foster patience in the investors for the lengthy timeframe required for commercialization of promising new compounds.
- Working with civic, government, religious and academic leaders to deal with ethical questions centered on stem cells and other new technologies in a manner that will enable research and development to move forward.
- Overcoming, through research, the technical obstacles to therapeutic cloning.
- Enhancing sales and distribution channels so that they educate patients, payors and physicians about new drugs in a cost-effective manner.
- Emphasizing fair and appropriate pricing models that will enable payors (both private and public) to afford new drugs and diagnostics while providing ample profit incentives to the industry.
- Developing appropriate standards that fully realize the potential of systems biology (that is, the use of advanced information technology and the resources of genetic databases) in a manner that will create the synergies necessary to accelerate and lower the total cost of new drug development.
- Fostering payor acceptance, diagnostic practices and physician practices that will harness the full potential of genetically targeted, personalized medicine when a large base of new biopharma drugs becomes available.

Source: Plunkett Research, Ltd.

---

11) Biogenerics are in Limbo in the U.S.

Patents on the first biotech-based drugs began expiring in 2001. Nonetheless, the FDA has no formal path for enabling drug makers to obtain approval of a generic version of biotechnology-based drugs without extensive clinical trials. Consequently, biotech drug makers as of yet haven't faced the type of generic competition that constantly challenges makers of traditional drugs, despite the fact that biotech-based drugs are about a $44 billion industry (2007) in the U.S. alone. However, this situation is likely to change in the near future as more and more biotech drugs go off-patent.

Because biotech drugs ("biologics") are made from living cells, a generic version of a drug probably won't be biochemically identical to the original branded version of the drug. Consequently, they are described as "follow-on biologics" to set them apart. There are concerns that follow-on biologics may not be as safe or effective as the originals unless they go through clinical trials for proof of quality.

In Europe, these drugs are referred to as "biosimilars". In the European Union, the first biosimilars were approved in April 2006 by the EMEA (European Agency for the Evaluation of Medicinal Products), the regulatory body responsible for new drugs. The first approved biosimilar was Omnitrope, a generic substitute for growth hormone Genotropin. The second was Valtropin.

FDA regulations for approval of generics, written in 1984, allow companies to achieve approval of generic versions of chemically-based, traditionally-manufactured pharmaceuticals with expired patents in a relatively short period of time. However, the law doesn't address drugs developed through biotechnology. The question remains whether the FDA will approve such generic biologics without lengthy and expensive clinical trials. It's a whole new game in the generics sector. Would-be generics makers also face the fact that biotech drugs can be vastly more complicated to manufacture than chemicals-based drugs.

Like the Europeans (EMEA), the FDA did approve a simple generic biotech drug, a human growth hormone, in May 2006. However, there are no published guidelines for getting more complex biogenerics to market. The FDA is under mounting pressure from the U.S. House of Representatives and from a number of state governors who are petitioning for these drugs.

In order to obtain approval of a biogeneric under existing regulations, a manufacturer would be forced to approve that the new drug contains the same ingredient as the original drug and that the two drugs are bioequivalent. This may be impossible without clinical trials.

In Europe, the EMEA has issued guidelines for approval of biosimilars. These include:

1)   Comparability studies are required. The proof or lack of proof of comparability to the original drug will dictate how many new clinical studies may be required.

2)   Clinical studies may be required to prove the biosimilar's safety and efficacy.

3)   Nonclinical studies may be required as well.

4)   After the biosimilar is approved and brought to market, continuing safety and efficacy study commitments will be required.

Watch for similar requirements to emerge in the U.S.

12) Breakthrough Drug Delivery Systems Evolve

Controlling how drugs are delivered is a huge business. Sales of drugs using new drug delivery systems was expected to balloon to $30 billion by 2007, up from $9.8 billion in 1998.

Until the biotech age, drugs were generally comprised of small chemical molecules capable of being absorbed by the stomach and passed into the blood stream—drugs that were swallowed as pills or liquids. However, many new biotech drugs require injection (or some other form of delivery) directly into the bloodstream, because they are based on larger molecules that cannot be absorbed by the stomach. Many new drug delivery techniques that provide an alternative to needles are in development.

In the near future, there may be an implantable microchip, controlled by a miniature computer, capable of releasing variable doses of multiple potent medications over an extended period, potentially up to one year. The miniscule silicon chips will bear a series of tiny wells, sealed with membranes that dissolve and release the contents when a command is received by the computer. Chips that can receive commands beamed through the skin are also theoretically possible. This technology would help treatment of conditions such as Parkinson's disease or cancer, where doctors need to vary medications and dosages. This technology must first be tested in animals and then in humans to ensure that the chips are biocompatible. The chips will more likely be used first in external applications, which may facilitate laboratory testing and drug development.

Other potential needle-free drug delivery systems include synthetic molecules attached to a drug, making it harder for the stomach to render the medicine useless before it reaches the blood. High-tech inhalers, which force medicine through the lungs, are also in the works. For the patient, this means less pain and the promise of better outcomes. Needle-free systems may also make toxic drugs safer and give older drugs new life. For example, the painkiller Fetanyl was recently converted into a lozenge.

Alkermes and Alza Corp. are developing techniques to encapsulate or rearrange drug molecules into more sturdy compounds that release steady, even doses over a prolonged period. A patch being developed by Alza Corp., a subsidiary of Johnson & Johnson, has a network of microscopic needles that penetrate painlessly into the first layer of the skin. In addition, Alza is in competition with Vyteris to develop a patch that delivers drugs via a small electric current, activated by a button on the patch itself. Another device, made by Sontra Medical Corp., uses ultrasound and gel to agitate and open temporary pores in the skin that can then receive a drug. The same technology can provide continuous, transdermal glucose monitoring. Yet another potential drug delivery system is edible film; quickly dissolving films treated with medication that melt on the tongue. Already in use as a breath freshener (Listerine brand PocketPaks, made by Pfizer), edible film is now the delivery method of choice for Novartis' Triaminic and Theraflu Thin Strips and Gas-X Thin Strips.

13) Stem Cells—Controversy in the U.S. Threatens to Leave America Far Behind in the Research Race

During the 1980s, a biologist at Stanford University, Irving L. Weissman, was the first to isolate the stem cell that builds human blood (the mammalian hematopoietic cell). Later, Weissman isolated a stem cell in a laboratory mouse and went on to co-found SysTemix, Inc. (now part of drug giant Novartis) and StemCells, Inc. to continue this work in a commercial manner.

In November 1998, two different university-based groups of researchers announced that they had accomplished the first isolation and characterization of the human embryonic stem cell (HESC). One group was led by James A. Thomson at the University of Wisconsin at Madison. The second was led by John D. Gearhart at the Johns Hopkins University School of Medicine at Baltimore. The HESC is among the most versatile basic building blocks in the human body. Embryos, when first conceived, begin creating small numbers of HESCs, and these cells eventually differentiate and develop into the more than 200 cell types that make up the distinct tissues and organs of the human body. If scientists can reproduce and then guide the development of these basic HESCs, then they could theoretically grow replacement organs and tissues in the laboratory—even such complicated tissue as brain cells or heart cells.

Ethical and regulatory difficulties have arisen from the fact that the only source for human "embryonic" stem cells is, logically enough, human embryos. A laboratory can obtain these cells in one of three ways: 1) by inserting a patient's DNA into an egg, thus producing a blastocyst that is a clone of the patient—which is then destroyed after only a few days of development; 2) by harvesting stem cells from aborted fetuses; or 3) by harvesting stem cells from embryos that are left over and unused after an in vitro fertilization of a hopeful mother. (Artificial in vitro fertilization requires the creation of a large number of test tube embryos per instance, but only one of these embryos is used in the final process.)

A rich source of similar but "non-embryonic" stem cells is bone marrow. Doctors have been performing bone marrow transplants in humans for years. This procedure essentially harnesses the healing power of stem cells, which proliferate to create healthy new blood cells in the recipient. Several other non-embryonic stem cell sources have great promise (see "Potential methods of developing 'post-embryonic' stem cells without the use of human embryos" below).

In the fall of 2001, a small biotech company called Advanced Cell Technology, Inc. announced the first cloning of a human embryo. The announcement set off yet another firestorm of rhetoric and debate on the scientific and ethical questions that cloning and related stem cell technology inspire. While medical researchers laud the seemingly infinite possibilities stem cells promise for fighting disease and the aging process, conservative theologians, many government policy makers, certain ethics organizations and pro-life groups decry the harvest of cells from aborted fetuses and the possibility of cloning as an ethical and moral abomination.

Stem cell research has been underway for years at biotech companies including Stem Cells, Inc., Geron and ViaCell. One company, Osiris Therapeutics, has been at work long enough to have several clinical trial programs in progress. Osiris derives its stem cells from the bone marrow of healthy adults between the ages of 18 and 30 who are volunteers. Prior to harvesting the stem cells, Osiris screens blood samples of the donors to make sure that they are free of diseases such as HIV and hepatitis. Osiris believes that this approach to gathering stem cells places them outside of the embryonic source controversy.

The potential benefits of stem cell-based therapies are staggering. Neurological disorders might be aided with the growth of healthy cells in the brain. Injured cells in the spinal column might be regenerated. Damaged organs such as hearts, livers and kidneys might be infused with healthy cells.

Potential methods of developing "post-embryonic" stem cells without the use of human embryos:

- Adult Skin Cells—Exposure of harvested adult skin cells to viruses that carry specific genes, capable of reprogramming the skin cells so that they act as stem cells
- Parthenogenesis—manipulation of unfertilized eggs.
- Other Adult Cells—Harvesting adult stem cells from bone marrow or brain tissue.
- Other Cells—harvesting of stem cells from human umbilical cords, placentas or other cells.
- De-Differentiation—use of the nucleus of an existing cell, such as a skin cell, that is altered by an egg that has had its own nucleus removed.
- Transdifferentiation—making a skin cell de-differentiate back to its primordial state so that it can then morph into a useable organ cell, such as heart tissue.

HESCs (typically harvested from five-day-old human embryos which are destroyed during the process) are used because of their ability to evolve into any cell or tissue in the body. Their versatility is undeniable, yet the implications of the death of the embryo are at the heart of the ethical and moral concerns.

Meanwhile, scientists have discovered that there are stem cells in existence in many diverse places in the adult human body, and they are thus succeeding in creating stem cells without embryos, by utilizing "post-embryonic" cells, such as cells from marrow. Such cells are already showing the ability to differentiate and function in animal and human recipients. Best of all, these types of stem cells may not be plagued by problems found in the use of HESCs, such as the tendency for HESCs to form tumors when they develop into differentiated cells. However, Thomas Okarma, CEO and President of Geron, argues that stem cells derived from bone marrow and other adult sources are fundamentally limited in their application, because it is nearly impossible to get them to produce anything besides blood cells. This makes it very difficult to harvest organ or nerve cells.

The biggest news lies in studies published in 2006 and 2007 regarding the reprogramming of adult mouse skin cells into stem cells. Initially, Shinya Yamanaka of Kyoto (Japan) University announced that he and coworkers had exposed skin cells harvested from adult mice to viruses carrying four specific genes. Yamanaka had determined that these four genes are apparently responsible for a stem cell's ability to develop into virtually any type of tissue. The technique is easy to replicate, and it was confirmed by additional studies in the U.S. published in 2007. This may be a tremendous breakthrough in stem cell research. However, scientists are still a long way from overcoming daunting technical challenges and adapting this technique for use in humans.

14) Stem Cell Funding Trickles in the U.S. at the Federal Level While a Few States Create Funding of Their Own

The U.S. Government, in August 2001, set strict limits on the use of federal funding for embryonic stem cell research. This action was taken in spite of impassioned testimony regarding the healing potential of stem cells by physicians, researchers and celebrities, such as actor Michael J. Fox on behalf of research for Parkinson's disease, the late actor Christopher Reeve for spinal injury study and former first lady Nancy Reagan on behalf of Alzheimer's research. The use of non-federal funding, however, is not currently restricted, although many groups would like to see further state or federal level restrictions on stem cell research or usage. A major confrontation is underway between American groups that advocate the potential health benefits of stem cell therapies and groups that decry the use of stem cells on ethical or religious terms. Meanwhile, stem cell development is forging ahead in other technologically advanced nations.

Under current U.S. regulations, federal research funds may be granted only for work with the 78 lines of stem cells that existed in 2001. Harvesting and developing new embryonic lines would not qualify. Once a stem cell starts to replicate, a large colony, or line, of self-replenishing cells can theoretically continue to reproduce forever. Unfortunately, only about a dozen of the stem cell lines existing at the time were considered to be useful, and some scientists believe that these lines are getting tired.

This is not to say that federal funds aren't being used in stem cell research at all. In fact, the Office of Management and Budget reports that total National

Institutes of Health is providing some funding for stem cell programs, obviously from "existing" lines.

In November 2004, voters in California approved a unique measure that provides $3 billion in state funding for stem cell research. Connecticut, Massachusetts and New Jersey have also passed legislation that permits embryonic-stem cell research. California already has a massive biotech industry, spread about San Diego and San Francisco in particular. As approved, California's Proposition 71 creates an oversight committee that will determine how and where grants will be made, and an organization, the California Institute for Regenerative Medicine (www.cirm.ca.gov), to issue bonds for funding and to manage the entire program. The money will be invested in research at a rate of about $295 million yearly over 10 years.

However, a pro-life, not-for-profit group called Life Legal Defense Foundation and a taxpayers group called People's Advocate held up the institute's ability to issue the bonds needed to fund the grants. A lawsuit was been filed in Superior Court in Alameda County, California asserting that the institute cannot legally issue state-backed obligations because the institute doesn't operate under direct state control. In April 2006, the court ruled that the law behind Proposition 71 and California's state-funded stem cell effort is constitutional in its entirety.

In June 2007, the California Institute for Regenerative Medicine approved grants totaling more than $50 million to finance construction of shared research laboratories at 17 academic and non-profit institutions. These facilities are scheduled to be complete and available to researchers within six months to two years of the grant awards. This pushed total funding granted to-date by the organization to more than $200 million. The grants will fund dedicated laboratory space for the culture of human embryonic stem cells (HESCs), particularly those that fall outside federal guidelines.

Meanwhile, many members of the U.S. Congress want to see much greater funding of stem cell research from federal coffers. In a dramatic move in the summer of 2005, Senate majority leader Bill Frist (a physician himself) made a statement backing deeper federal involvement in stem cell research, breaking with the well-known views of President Bush. In 2006, a new bill for the expansion of embryonic stem cell research funding was passed in both the U.S. Senate and the House of Representatives. However, President Bush vetoed the bill in July, the first veto of his administration. The House vote, which was 235 to 193, was not a

two-thirds majority, which would have been the necessary number to override a presidential veto. Reform of federally-funded stem cell research is dead for now.

In the private sector, funding for stem cell research has been generous. For example, the Juvenile Diabetes Research Foundation has an $8 million stem cell research program underway and Stanford University has used a $12 million donation to create a research initiative. Likewise, major, privately funded efforts have been launched at Harvard and at the University of California at San Francisco.

15) Stem Cells—Therapeutic Cloning Techniques Advance

For scientists, the biggest challenge at present may be to discover the exact process by which stem cells are signaled to differentiate. Another big challenge lies in the fact that broad use of therapeutic cloning may require immense numbers of human eggs in which to grow blastocysts.

The clearest path to "therapeutic" cloning may lie in "autologous transplantation." In this method, a tiny amount of a patient's muscle or other tissue would be harvested. This sample's genetic material would then be de-differentiated; that is, reduced to a simple, unprogrammed state. The patient's DNA sample would then be inserted into an egg to grow a blastocyst. The blastocyst would be manipulated so that its stem cells would differentiate into the desired type of tissue, such as heart tissue. That newly grown tissue would then be transplanted to the patient's body. Many obstacles must be overcome before such a transplant can become commonplace, but the potential is definitely there to completely revolutionize healing through such regenerative, stem cell-based processes. One type of bone marrow stem cell, recently discovered by scientists at the University of Minnesota, appears to have a wide range of differentiation capability.

It is instructive to note that there are two distinct types of embryonic cloning: "reproductive" cloning and "therapeutic" cloning. While they have similar beginnings, the desired end results are vastly different.

"Reproductive" cloning is a method of reproducing an exact copy of an animal—or potentially an exact copy of a human being. A scientist would remove the nucleus from a donor's unfertilized egg, insert a nucleus from the animal, or human, to be copied, and then stimulate the nucleus to begin dividing to form an embryo. In the case of a mammal, such as a human, the embryo would then be implanted in the uterus of a host female for gestation and birth. The successful birth of a cloned human baby doesn't necessarily mean that a healthy adult human will result. To date, cloned animals have tended to develop severe health problems. For example, a U.S. firm, Advanced Cell Technology, reports that it has engineered the birth of cloned cows that appeared healthy at first but developed severe health problems after a few years. Nonetheless, successful cloning of animals is progressing at labs in many nations.

On the other hand, "therapeutic" cloning is a method of reproducing exact copies of cells needed for research or for the development of replacement tissue. In this case, once again a scientist removes the nucleus from a donor's unfertilized egg, inserts a nucleus from the animal, or human, whose cells are to be copied, and then stimulates the nucleus to begin dividing to form an embryo. However, in therapeutic use, the embryo would never be allowed to grow to any significant stage of development. Instead, it would be allowed to grow for a few hours or days, and stem cells would then be removed from it for use in regenerating tissue.

Because it can provide a source of stem cells, cloning has uses in regenerative medicine that can be vital in treating many types of disease. The main differences between stem cells derived from clones and those derived from aborted fetuses or fertility specimens is that a) they are made from only one source of genes, rather than by mixing sperm and eggs; and b) they are made specifically for scientific purposes, rather than being existing specimens, putting them up to more intense ethical discussions. Cloned stem cells have the added advantage of being 100% compatible with their donors, because they share the same genes, and so would provide the best possible source for replacement organs and tissues. Although the use of cloning for regeneration has stirred heated debate as well, it has not resulted in universal rejection. Most of the industrialized countries, including Canada, Russia, most of Western Europe and most of Asia, have made some government-sanctioned allowances for research into this area.

As a result of government sanction of research and development, some countries have already made progress in the field of regenerative cloning. In an important development in August 2004, scientists at Newcastle University in the U.K. announced that they have been granted permission by the Human Fertilisation and Embryology Authority (HFEA), a unit of the British Government, to create human embryos as a source of stem cells for certain therapeutic purposes. Specifically, researchers will clone early-stage embryos in search of new treatments for such degenerative diseases as Parkinson's disease, Alzheimer's and diabetes. The embryos will be destroyed before they are two weeks old and will therefore not develop beyond a tiny cluster of cells.

The biggest news to hit therapeutic cloning was the claim by South Korean scientist Hwang Woo Suk of Seoul National University announced that he and his team of researchers had created patient-specific stem cells; that is, cells taken from adult patients and then cloned and further manipulated to become stem cells. Backed by the South Korean government, the breakthrough promised the dawn of a new era of stem cell research with Hwang as the star. However, by December 2005, it became clear that the Korean claims were fraudulent. There were, in fact, no cloned stem cell lines at all.

The good news is that several other cloning methods are on the horizon. Markus Grompe, director of the Oregon Stem Cell Center at Oregon Health and Science University in Portland is working on research similar to that at Newcastle University. Adult donor cells are forced to create a protein called nanog, which is only found in stem cells, yet the process is altered in a way that keeps the cells from forming into embryos. At the same time, researchers at MIT are experimenting with defusing CDX2, a gene in the nucleus of a cell taken from an adult, before transferring the nucleus into a donated human egg that has been stripped of its own DNA. The resulting egg could be developed into stem cells, but due to the lack of CDX2, would be unable to develop into an embryo.

16) Stem Cells—A New Era of Regenerative Medicine Looms

Many firms are conducting product development and research in the areas of skin replacement, vascular tissue replacement and bone grafting or regeneration. Stem cells, as well as transgenic organs harvested from pigs, are under study for use in humans. At its highest and most promising level, regenerative medicine may eventually utilize human stem cells to create virtually any type of replacement organ or tissue.

In one recent, exciting experiment, doctors took stem cells from bone marrow and injected them into the hearts of patients undergoing bypass surgery. The study showed that the bypass patients who received the stem cells were pumping blood 24% better than patients who had not received them.

In another experiment, conducted by Dr. Mark Keating at Harvard, the first evidence was shown that stem cells may be used for regenerating lost limbs and organs. The regenerative abilities of amphibians have long been known, but exactly how they do it, or how it could be applied to mammals, has been little understood. Much of the regenerative challenge lies in differentiation, or the development of stem cells into different types of adult tissue such as muscle and bone. Creatures such as amphibians have the ability to turn their complex cells back into stem cells in order to regenerate lost parts. In the experiment, Dr. Keating made a serum from the regenerating nub (stem cells) of a newt's leg and applied it to adult mouse cells in a petri dish. He observed the mouse cells to "de-differentiate," or turn into stem cells. In a later experiment, de-differentiated cells were turned back into muscle, bone and fat. These experiments could be the first steps to true human regeneration. Keating is continuing to make exciting breakthroughs in regenerative research.

The potential of the relatively young science of tissue engineering appears to be unlimited. Transgenics (the use of organs and tissues grown in laboratory animals for transplantation to humans) is considered by many to have great future potential, and improvements in immune system suppression will eventually make it possible for the human body to tolerate foreign tissue instead of rejecting it. There is also increasing theoretical evidence that malfunctioning or defective vital organs such as livers, bladders and kidneys could be replaced with perfectly functioning "neo-organs" (like spare parts) grown in the laboratory from the patient's own stem cells, with minimal risk of rejection.

The ability of most human tissue to repair itself is a result of the activity of these cells. The potential that cultured stem cells have for transplant medicine and basic developmental biology is enormous.

Diabetics who are forced to cope with daily insulin injection treatments could also benefit from engineered tissues. If they could receive a fully functioning replacement pancreas, diabetics might be able to throw away their hypodermic needles once and for all. This could also save the health care system immense sums, since diabetics tend to suffer from many ailments that require hospitalization and intensive treatment, including blindness, organ failure, diabetic coma and circulatory diseases.

Elsewhere, the harvesting of replacement cartilage, which does not require the growth of new blood vessels, is being used to repair damaged joints and treat urological disorders. Genzyme Corp. recently won FDA approval for its replacement cartilage product Carticel. Genzyme's process involves harvesting the patient's own cartilage-forming cells, and, from those cells, re-growing new cartilage in the laboratory. The physician then injects the new cartilage into the damaged area. Full regeneration of the replacement cartilage is expected to take up to 18 months. The Genzyme process can cost up to $30,000, compared to $10,000 for typical cartilage surgery.

In 1997, Carticel (see www.carticel.com) became the first biologic cell therapy licensed by the FDA under accelerated approval. Since then, more than 10,000 patients have been treated with the Carticel method for treating knee pain. Other companies are exploring alternative methods that may be less expensive and therefore more attractive to payors.

Companies to Watch: StemCells, Inc., in Palo Alto, California (www.stemcellsinc.com), is focusing on the use of stem cells to treat damage to major organs such as the liver, pancreas and central nervous system. ViaCell, Inc., in Boston, Massachusetts (www.viacellinc.com), develops therapies using umbilical cord stems. Also, their ViaCord product enables families to preserve their baby's umbilical cord at the time of birth for possible future use in treating over 40 diseases and genetic disorders.

Internet Research Tip:
For an excellent primer on genetics and basic biotechnology techniques, see:

National Center for Biotechnology Information
www.ncbi.nlm.nih.gov

---

17) Nanotechnology Converges with Biotech

Because of their small size, nanoscale devices can readily interact with biomolecules on both the surface and the inside of cells. By gaining access to so many areas of the body, they have the potential to detect disease and deliver treatment in unique ways. Nanotechnology will create "smart drugs" that are more targeted and have fewer side effects than traditional drugs.

Current applications of nanotechnology in health care include immunosuppressants, hormone therapies, drugs for cholesterol control, and drugs for appetite enhancement, as well as advances in imaging, diagnostics and bone replacement. For example, the NanoCrystal technology developed by

Elan, a major biotechnology company, enhances drug delivery in the form of tiny particles, typically less than 1,000 nanometers in diameter. The technology can be used to provide more effective delivery of drugs in tablet form, capsules, powders and liquid dispersions. Abbot Laboratories uses Elan's technology to improve results in its cholesterol drug Tricor. Par Pharmaceutical Companies uses NanoCrystal in its Megace ES drug for the improvement of appetite in people with anorexia.

Since biological processes, including events that lead to cancer, occur at the nanoscale at and inside cells, nanotechnology offers a wealth of tools that are providing cancer researchers with new and innovative ways to diagnose and treat cancer. In America, the National Cancer Institute has established the Alliance for Nanotechnology in Cancer in order to foster breakthrough research.

Nanoscale devices have the potential to radically change cancer therapy for the better and to dramatically increase the number of effective therapeutic agents. These devices can serve as customizable, targeted drug delivery vehicles capable of ferrying large doses of chemotherapeutic agents or therapeutic genes into malignant cells while sparing healthy cells, greatly reducing or eliminating the often unpalatable side effects that accompany many current cancer therapies.

At the University of Michigan at Ann Arbor, Dr. James Baker is working with molecules known as dendrimers to create new cancer diagnostics and therapies, thanks to grants from the National Institutes of Health and other funds. A dendrimer is a spherical molecule of uniform size (five to 100 nanometers) and well-defined chemical structure. Dr. Baker's lab is able to build a nanodevice with four or five attached dendrimers. To deliver cancer-fighting drugs directly to cancer cells, Dr. Baker loads some dendrimers on the device with folic acid, while loading others with drugs that fight cancer. Since folic acid is a vitamin, many proteins in the body will bind with it, including proteins on cancer cells. When a cancer cell binds to and absorbs the folic acid on the nanodevice, it also absorbs the anticancer drug. For use in diagnostics, Dr. Baker is able to load a dendrimer with molecules that are visible to an MRI. When the dendrimer, due to its folic acid, binds with a cancer cell, the location of that cancer cell is shown on the MRI. Each of these nanodevices may be developed to the point that they are able to perform several advanced functions at once, including cancer cell recognition, drug delivery, diagnosis of the cause of a cancer cell,

cancer cell location information and reporting of cancer cell death. Universities that are working on the leading edge of cancer drug delivery and diagnostics using nanotechnology include MIT and Harvard, as well as Rice University and the University of Michigan.

18) Agricultural Biotechnology Scores Breakthroughs but Causes Controversy/Selective Breeding Offers a Compromise

The biotech-era technology of "molecular farming" will soon lead to broad commercialization of human drug therapies that are grown via agricultural methods. For example, by inserting human genes into plants, scientists can manipulate them so they grow certain human proteins instead of natural plant proteins. The growth in plants of transgenic protein therapies for humans may become widespread. Such drug development methods may prove to be extremely cost-effective. At the same time, hundreds of antibodies produced in farm animals for use in human drug therapies are currently under development or in clinical trials.

Meanwhile, genetically modified foods (frequently referred to as "GM" for genetically modified, or "GMO" for genetically modified organisms) offer tremendous promise in agriculture—particularly in high-population nations like China and India. A study completed by the Agriculture Policy Research Center at the Chinese Academy of Sciences in Beijing in mid-2005 found immense potential in the use of genetically modified rice in China. The center's director estimates that Chinese farmers could increase their total annual income by $4 billion through planting GM rice, which would result in higher yields per acre. China has made massive investments in agricultural biotechnology research. However, the planting of GM rice seeds is currently authorized only in experimental plots within China.

Agricultural biotechnology became a significant commercial industry during the 1980s. It was fostered both by startups and by large chemical or seed companies. All of these players were focused on developing genetically modified seeds and plants that had higher yields, better nutritional qualities and/or resistance to diseases or insects. Additional traits of GM plants may include resistance to temperature and moisture extremes. By 2006, over 42 billion acres of GM crops were planted worldwide (a 14% increase over 2005)—mostly in the U.S., but large amounts were also planted in Argentina,

Canada, Mexico, Romania, Uruguay and South Africa. Meanwhile, GM seeds have the potential to create vast benefits in low-income nations where reliance on small farms or gardens is high and food is scarce.

At the same time, researchers are modifying the structural makeup of some plants in order to alter leaves, stems, branches, roots or seed structures. The ability to modify the nutritional makeup of plants can have highly desirable effects. For example, Mycogen, an affiliate of Dow AgroSciences (www.dowagro.com), has developed sunflower seeds with higher levels of oleic and linoleic acids—acids with exceptional nutritional value.

There are currently dozens of agribio food products on the market, including a range of fruits, vegetables and nuts. There is significant potential for rapid development of new products, thanks to the same technologies that are pushing development of human gene therapies in the pharmaceutical industry.

U.S. farmers have enjoyed greatly increased crop yields and crop quality thanks to GM seeds, and by some estimates as much as 70% of U.S. food may contain ingredients that have been grown with GM methods. In particular, U.S. farmers are reaping tremendous crops of GM soybeans (89% of the U.S. market), cotton (83%) and corn (61%). These crops eventually become ingredients in everything from baked goods to soft drinks to clothing.

Although scientists have been able to engineer highly desirable traits in GM seeds for crops (such as disease-resistance and insect-resistance), and the scientific community has given GM foods a clean bill of health for years, such modified foods face stiff resistance among many consumers, particularly in Europe. While many areas of biotechnology are controversial, agricultural biotech is currently the largest target for consumer backlash and government intervention into the marketplace. The U.S. market for genetically engineered seeds has already reached the $4-billion range, but consumer resistance to food products containing material grown in this manner is sometimes fierce.

In the fall of 2000, a relatively small quantity of taco shells were found to have been manufactured with cornmeal made from GM corn known as StarLink. Unfortunately, StarLink was intended for hog feed rather than human foods. Interestingly, the StarLink affair seems to have had little long-term effect on consumer preferences in the U.S., although there were loud cries for more government regulation at the time.

Consumers in Europe have a strong fear of GM foods. It may stem in part from a cultural preference for locally grown, natural foods. Basic European grocery shopping habits and food preparation habits vary from those of U.S. consumers. For example, Europeans tend to shop today for tonight's meal, rather than stocking up on several days' worth of food as many Americans do. Europeans also suffered mightily from the outbreak of mad cow disease in 1996, and in subsequent outbreaks such as the one in the U.K. in 2003, which contributed to their concerns about food sources.

In 2000, two of the world's largest biotech companies, AstraZeneca and Novartis, spun off and combined their agriculture divisions into one new company, Syngenta. The newly merged Syngenta cut employee count and closed overlapping business units. The goal was to create an entity focused on seeds, crop protection products, insecticides and other agricultural products. With this focus, Syngenta is in a position to make the best research, development and marketing decisions. The firm's annual investment in research and development is substantial, at about 10% of revenues. In 2006, its sales of seeds of all types totaled about $1.74 billion.

Meanwhile, Monsanto, a major competitor to Syngenta, has invested heavily in biotech seed research with terrific results. From a 2002 loss of $1.7 billion, Monsanto has evolved to a 2006 operating profit of about $689 million (on sales of about $7.3 billion).

A particular concern among farmers in many parts of the world is that GM crops may infest neighboring plants when they pollinate, thus triggering unintended modification of plant DNA. In any event, there is a vast distrust of GM foods in certain locales. U.S. food growers and processors face significant difficulty exporting to the European Union (EU) because of the reliance that American farmers place on GM seeds.

The European Union, as well as specific nations in Europe, has kept many regulations in place that make the use of GM seeds or the import of GM food products a difficult to impossible task. These restrictions remain a hot topic of debate at the World Trade Organization and elsewhere. Meanwhile, a handful of localities in the U.S. have banned or restricted the planting of GM seeds, hoping to protect traditional crops that local growers are widely-known for. A typical restriction is to require that GM seeds be planted at least a certain distance away from non-GM crops.

Some anti-GM activists have arguments with big business—particularly with the giant corporations like Monsanto that make GM seeds. Some people have accused Monsanto of persecuting farmers who appear to be using Monsanto-developed seeds without paying for them. The company has also received criticism for its history of manufacturing chemicals that have risen to varying levels of infamy, such as SBCs, DDT and Agent Orange. Unfortunately, protestors are sometimes violent or destructive. In 2000 alone, there were over 30 acts of anti-biotech terrorism around the world, including the firebombing of a biotech lab at the University of Michigan. In 2006, Greenpeace sent out 1,500 volunteers to seek out farms with GM crops and identify those farms on maps published for the anti-GM community.

While concern in many European countries continues, the number of acres sown with genetically modified corn (which is the only transgenic seed allowed by EU rules) is slowly growing due to the fact that some farmers can no longer ignore the cost savings and improved crop yields. In France, for example, the number of acres planted with GM corn in 2006 was 12,350, up from 1,215 in 2005 according to Agricultural Biotechnology in Europe (www.abeurope.info).

A landmark compromise may be on the horizon thanks to a new selective breeding technique that introduces no foreign DNA such as that used in GM seeds. The technology uses old-school practices in which plants with desirable characteristics such as longer shelf life or resistance to insects are crossbred to create new, hardier specimens. The new twist to the old technique is the use of genetic markers, which make it much easier to isolate plants with a positive trait and the gene that causes it. New plants can also be quickly tested for the presence of the isolated gene. The technology cuts traditional selective breeding time in half.

A number of companies are utilizing gene markers in their breeding programs. Arcadia Biosciences is hoping to develop seeds for wheat that can be eaten by people with the intestinal disorder called Celiac Disease. Monsanto is working on soybeans using the markers to make veggie burgers that taste more like beef. Syngenta is also using the technology to develop soybeans that are resistant to aphids.

Genetic markers are not new, but the ability to use them in a cost effective manner is relatively recent thanks to falling costs since the year 2000. Where it once took several dollars to conduct a plant scan, the same test can now be conducted for pennies, making testing on a large scale possible. Look for crop biotechnology companies including DuPont, Monsanto and Syngenta to invest millions of dollars in selective breeding assisted by gene markers over the near- to mid-term.

Company to Watch:
Sacramento, California-based Ventria, www.ventria.com, has received approval from the USDA every year since 1999 to produce crops for use in biopharmaceuticals. Ventria plants self-pollinating rice or barley specifically because they produce large quantities of proteins by nature and because they are not pollinated by wind or insect activity. Thus, they theoretically should have no effect on nearby traditional plantings.

19) Focus on Vaccines to Counter Potential Bioterror Attacks in the U.S.

In May 2004, the U.S. Congress passed a bill with great bipartisan support that provided for $5.6 billion in funding over ten years for stockpiling vaccines and other medicines in defense of possible bioterror attacks on U.S. population centers. The BioShield bill further enhances the possibility of fast track research in the event of a national emergency, and allows government officials to distribute certain treatments even if they have not yet been approved by the FDA. The intent is to create effective responses to attacks from chemical or biological weapons. Among the greatest concerns are vaccines against and treatments for anthrax and smallpox. Over 100 biotechnology firms are already working on products that could be candidates for BioShield contracts.

The biggest single bioterror threat to U.S. population centers is considered to be anthrax. A small parcel of anthrax sprayed over a major city could kill hundreds of thousands of people, and it is relatively easy to manufacture and hard to combat. If the release is detected or the first cases are rapidly diagnosed, quick action could save many lives. Providing the exposed population with antibiotics followed by vaccination could be lifesaving for exposed persons who would otherwise become ill with untreatable inhalation anthrax in the subsequent few weeks. Prophylactic antibiotics alone will prevent disease in persons exposed to antibiotic-susceptible organisms, but incorporating vaccination into the treatment regime can greatly reduce the length of treatment with antibiotics. Without vaccination, antibiotics must be continued for 60 days. However, if effective vaccination can be provided, antibiotic treatment can be reduced to 30 days.

Smallpox is also considered to be a major threat. The national stockpile (fewer than 7 million doses of vaccinia virus vaccine) is insufficient to meet national and international needs in the event of a major bioterror attack. The stockpile is also deteriorating and has a finite life span. The vaccine was made using the traditional method of scarifying and infecting the flanks and bellies of calves and harvesting the infected lymph. No manufacturer exists today with the capability to manufacture calf lymph vaccine by the traditional method. Replacing the stockpile will require the development and licensure of a new vaccine using modern cell-culture methods. This development program, which will include process development, validation of a new manufacturing process, and extensive clinical testing, will be expensive and may take several years.

Dozens of firms, particularly startups, are working on vaccines and antidotes for biological weapons with hopes of obtaining BioShield funds. Overall, the NIH grants $500 to $600 million per year (approximately one-third of the U.S. annual biodefense budget) into product development. The Department of Defense Joint Vaccine Acquisition Program has several experimental vaccines in development.

A total of seven major new, high-security, infectious disease research centers have been funded in the United States, accomplished largely with the assistance of government grants, including military funds. Much of the focus of these labs will be on biodefense, including vaccines. These centers include a new 15,000 square foot lab at the CDC in Atlanta, Georgia. Additional new labs include three facilities in Fort Detrick, Maryland; as well as one each in Galveston, Texas (at the University of Texas Medical Branch-UTMB); Hamilton, Montana (Rocky Mountain Laboratories); and Boston, Massachusetts (Boston University Medical Center).

20) Ethical Issues Abound

Significant ethical issues face the biotech industry as it moves forward. They include, for example, the ability to determine an individual's likelihood to develop a disease in the future, based on his or her genetic makeup today; the potential to harvest replacement organs and tissues from animals or from cloned human genetic material; and the ability to alter genetically the basic foods that we eat. These are only a handful of the powers of

biotechnology that must be dealt with by society. Watch for intense, impassioned discussion of such issues and a raft of governmental regulation as new technologies and therapies emerge.

The biggest single issue may be privacy. Who should have access to your personal genetic records? Where should they be stored? How should they be accessed? Can you be denied employment or insurance coverage due to your genetic makeup?

Internet Research Tip:
For the latest biotech developments check out www.biospace.com, a private sector portal for the biotech community, and www.bio.org, the web site of the highly regarded Biotechnology Industry Organization.

---

21) Technology Discussion—Genomics

The study of genes as a resource for the commercial development of new drugs received a significant boost from computer technology in the late 1970s. Frederick Sanger, a chemist, and Walter Gilbert, a biochemist, developed what is known as DNA sequencing technology, receiving a Nobel Prize in 1980 for their effort. In the same way that computer technology enabled the rapid growth of the Internet industry, computerization has been the catalyst for the booming biotech industry. For example, Sanger and Gilbert's computerized DNA sequencing technology enables scientists to collect massive amounts of data on human genes at high speed, analyzing how certain genes are connected with specific diseases. Using this technology, the gene responsible for Parkinson's disease was mapped in only nine days—a stark contrast to the nine years required to determine the gene connected with cystic fibrosis using traditional methods. Genomics, the mapping and analysis of genes and their uses, is the basic building block of the biopharmaceuticals business. Pharmacogenomics is the study of genomics for the purpose of creating new pharmaceuticals.

Genes are made up of DNA and reside on densely packed fibers called chromosomes. The genetic information encoded in a gene is copied into RNA (ribonucleic acid—closely related to DNA) and then used to assemble proteins. Think of DNA as a blueprint and RNA as the builder.

The human body contains about 75 trillion cells. Each cell contains 46 chromosomes, arranged in 23 pairs. Each chromosome is a strand of DNA. Each strand of DNA is composed of thousands of segments representing different genes. The sum total of the DNA contained in all 46 chromosomes is called the human genome. Until recently, it was estimated that the number of different genes in the human body would be in the 100,000 range, due to the complexity of systems within human beings. However, as genome mapping neared completion, the number appears to be a surprisingly small 20,000 to 25,000.

In October 2006, the X Prize Foundation announced a prize of $10 million for the first group to accurately produce complete genomes of 100 humans in 10 days or less and while spending less than $10,000 per genome. One of the genomics industry's goals is to able to market sub-$1,000 genomic studies to people who want a complete map of their DNA. This prize will be a big boost.

> *Internet Research Tip:*
> For a superior, easy-to-understand, illustrated resource on genetics and molecular biology, see: www.dnaftb.org/dnaftb/15/concept/.

22) Technology Discussion—Proteomics

Proteomics is the study of the proteins that a gene produces. A complete set of genetic information is contained in each cell, and this information provides a specific set of instructions to the body. The body carries out these instructions via proteins. Genes encode the genetic information for proteins.

All living organisms are composed largely of proteins. Each protein is a large, complex molecule composed of amino acids. Proteins have three main cellular functions: 1) they act as enzymes, hormones and antibodies, 2) they provide cell structure, and 3) they are involved in cell signaling and cell communication functions.

Proteins are important to researchers because they are the link between genes and pharmaceutical development. They indicate which genes are expressed or are being used. They are important for understanding gene function. They also have unique shapes or structures. Understanding these structures and how potential pharmaceuticals will bind to them is a key element in drug design. Proteomic researchers seek to determine the unique role of each of the hundreds of thousands of proteins in the human body, as well as the relationships that such proteins have with each other and with various diseases. Microarrays enable the high-speed analysis of these proteins and the discovery of SNPs (see below).

23) Technology Discussion—Microarrays

In the laboratory, scientists use microarrays to map the DNA of a patient's tissues. For example, a physician may take a small biopsy of a cancer patient's tumor. That biopsy would be placed into microarray equipment. There, the microarray would arrange hundreds or thousands of microscopic dots of tumor material onto glass laboratory slides. Lasers would scan the slides, and computer software would compare the contents of the slides to vast databases, in order to determine the exact genes in the tumor. Such information may be extremely useful in treating the patient. Genetic information scanned from microarrays has been used to create giant bioinformatics databases. Ideally, such databases could analyze the genetic blueprint of a patient's tumor in order to assist a physician in determining the drugs that would best treat the specific genes that have mutated into cancerous material, and thereby have the best opportunity to cure the patient.

24) Technology Discussion—DNA Chips

Several chip platforms have been developed to facilitate molecular biology research. For example, industry leader Affymetrix (www.affymetrix.com) has developed chips programmed to act like living cells. These chips can run thousands of tests in short order to help scientists determine a specific gene's expression.

Scientists know that a mutation—or alteration—in a particular gene's DNA often results in a certain disease. However, it can be very difficult to develop a test to detect these mutations, because most large genes have many regions where mutations can occur. For example, researchers believe that mutations in the genes BRCA1 and BRCA2 cause as many as 60 percent of all cases of hereditary breast and ovarian cancers. But there is not one specific mutation responsible for all of these cases. Researchers have already discovered over 800 different mutations in BRCA1 alone.

The DNA chip is a tool used to identify mutations in genes like BRCA1 and BRCA2. The chip, which consists of a small glass plate encased in plastic, is manufactured somewhat like a computer microchip. On the surface, each chip contains thousands of short, synthetic, single-stranded DNA sequences, which together, add up to the normal gene in question. To determine whether an individual possesses a mutation for BRCA1 or BRCA2, a scientist first obtains a sample of DNA from the patient's blood as well as a control sample—one that does not contain a mutation in either gene.

The researcher then denatures the DNA in the samples, a process that separates the two complementary strands of DNA into single-stranded molecules. The next step is to cut the long strands of DNA into smaller, more manageable fragments and then to label each fragment by attaching a fluorescent dye. The individual's DNA is labeled with green dye and the control, or normal, DNA is labeled with red dye. Both sets of labeled DNA are then inserted into the chip and allowed to hybridize (bind) to the synthetic BRCA1 or BRCA2 DNA on the chip. If the individual does not have a mutation for the gene, both the red and green samples will bind to the sequences on the chip.

If the individual does possess a mutation, the individual's DNA will not bind properly in the region where the mutation is located. The scientist can then examine this area more closely to confirm that a mutation is present. (Explanation provided courtesy of National Institutes of Health.)

25) Technology Discussion—SNPs ("Snips")

The holy grail of genomics is the search for single nucleotide polymorphisms (SNPs). These are DNA sequence variations that occur when a single nucleotide (A, T, C or G) in the genome sequence is altered. For example an SNP might change the DNA sequence AAGGCTAA to ATGGCTAA. SNPs occur in every 100 to 1,000 bases along the 3 billion-base human genome. SNPs can occur in both coding (gene) and noncoding regions of the genome. Many SNPs have no effect on cell function, but scientists believe others could predispose people to diseases or influence their response to a drug.

Variations in DNA sequence can have a major impact on how humans react to disease; environmental factors such as bacteria, viruses, toxins and chemicals; and drug therapies. This makes SNPs of great value for biomedical research and for developing pharmaceutical products or medical diagnostics. Scientists believe SNP maps will help them identify the multiple genes associated with such complex diseases as cancer, diabetes, vascular disease and some forms of mental illness.

26) Technology Discussion—Combinatorial Chemistry

After researchers determine which protein is involved in a specific disease, they use combinatorial chemistry to find the molecule that controls the protein. High-throughput screening techniques allow scientists to use automatic fluid handlers to mix chemicals under specific test conditions at extremely high speed. Combinatorial chemistry can generate thousands of chemical compound variations in a few hours. Previously, traditional chemistry methods could have required several weeks or even months to do the same work. High-throughput screening enables automated, computerized machines to screen as many as 100,000 chemical compounds daily, seeking potential molecules as drug candidates.

One company, CombinatoRx (www.combinatorx.com), uses a combinatorial array to mix together drugs already on the market to find combinations that could have new and unforeseen uses. The company found hundreds of potential uses for combinations of drugs whose patents had expired and has patented these precise mixtures for new uses. It has received rapid FDA approval in many cases because all the drugs being combined have proven records of safety. In one particularly bizarre instance, the firm found that by combining a particular sedative with an antibiotic, it had made an effective cancer-fighting agent.

27) Technology Discussion--Synthetic Biology

Scientists have followed up on the task of mapping genomes by attempting to directly alter them. This effort has gone past the point of injecting a single gene into a plant cell in order to provide a single trait, as in many agricultural biotech efforts. There are now several projects underway to create entirely new versions of life forms, such as bacteria, with genetic material inserted in the desired combination in the laboratory. These include the BioBricks Foundation, the Registry of Standard Biological Parts and the Synthetic Genomics Study, all of which are part of the scientific community at MIT.

Synthetic biology can be defined as the design and construction of new entities, including enzymes and cells, or the reformatting of existing biological systems. This science capitalizes on previous advances in molecular biology and systems biology, by applying a focus on the design and construction of unique core components that can be integrated into larger systems in order to solve specific problems.

In June 2004, the first international meeting on synthetic biology was held at MIT. Called Synthetic Biology 1.0, the conference brought together researchers who are working to design and build biological parts, devices and integrated biological systems; develop technologies that enable such work; and place this research within its current and future social context. Synthetic Biology 2.0 was held in May 2006 at UC Berkeley. In fact, the National

Science Foundation agreed, in mid 2006, to invest $16 million in a five-year grant to fund a new Synthetic Biology Engineering Research Center ("SynBERC) at UC Berkeley. An additional $4 million for the project has been raised from other sources, and the NSF is offering the possibility of a further five-year grant.

Leading proponents of synthetic biology include Dr. Craig Venter, well known for his efforts in sequencing the human genome. Elsewhere, at MIT, Dr. Drew Endy and colleague Tom Knight are working on a concept called BioBricks, which are strands of DNA with connectors at each end. They can be assembled into higher-level components.

28) Technology Discussion—Recombinant DNA

In today's recombinant DNA therapies, the implantation of selected strands of DNA into bacteria allows large-scale production of hormones, antibodies and other exciting new drugs. Additions to the medical arsenal developed from this technique include interferons, growth factor hormone, anticoagulants, human insulin (as opposed to the cow and pig insulin that diabetics have used for 60 years) and more effective immunosuppressive drugs. An anticoagulant proven highly effective for some heart attack patients, TPA, may dramatically reduce heart damage during an attack and cut hospital stays. These drugs are often expensive, but they may greatly reduce the duration of hospital confinement and, in some cases, the need for any hospitalization at all.

Long-term, potential applications of this technology include genetically-modified seeds, vaccines, production of insulin, production of blood clotting factors and the production of recombinant pharmaceuticals.

29) Technology Discussion—Polymerase Chain Reaction (PCR)

PCR, also called "molecular photocopying", is a fast, inexpensive method of replicating genetic material in the form of small segments of DNA. The technique was created by Kary B. Mullis of La Jolla, California. His efforts were awarded one-half of the Nobel Prize in Chemistry in 1993, which he shared with another leading DNA researcher, Michael Smith of the University of British Columbia, Canada.

PCR creates sufficient amounts of DNA to allow detailed analysis, which can be vital to researchers. This may lead to better diagnosis and treatment of diseases such as AIDS, Lyme disease, tuberculosis, viral meningitis, cystic fibrosis, Huntington's chorea,

sickle cell anemia and many others. PCR also shows promise in providing cancer susceptibility warnings.

Once amplified, the DNA produced by PCR can be used in many different laboratory procedures. For example, most mapping techniques in the Human Genome Project (HGP) rely on PCR.

PCR is also valuable in a number of newly emerging laboratory and clinical techniques, including DNA fingerprinting, detection of bacteria or viruses (particularly AIDS), and diagnosis of genetic disorders.

Chapter 2

BIOTECH & GENETICS INDUSTRY STATISTICS

Biotech Industry Overview

Global	Amount	Units	Year	Source
Public & Private Biotech Companies	4,275	Companies	2006	E&Y
Public Biotech Companies	710	Companies	2006	E&Y
Revenues	73.5	Bil. US$	2006	E&Y
Increase from 2005	14	%	2006	E&Y
R&D Expenses	27.8	Bil. US$	2006	E&Y
Net Income	-5.4	Bil. US$	2006	E&Y
Number of Employees	190.5	Thousand	2006	E&Y
Total Pharmaceutical Sales, Worldwide	643.0	Bil. US$	2006	IMS
Total R&D Expenses, PhRMA Member Companies	43.0	Bil. US$	2006	PhRMA
Total R&D Expenses, All Pharmaceutical Companies	55.2	Bil. US$	2006	Burrill
U.S.				
Public & Private Biotech Companies	1,452	Companies	2006	E&Y
Public Biotech Companies	336	Companies	2006	E&Y
Revenues	55.5	Bil. US$	2006	E&Y
R&D Expenses	22.9	Bil. US$	2006	E&Y
Net Income	-3.5	Bil. US$	2006	E&Y
Number of Employees	130.6	Thousand	2006	E&Y
Total Biotech Company Financing	20.3	Bil. US$	2006	E&Y
Pharmaceutical Sales[1], Domestic	174.7	Bil. US$	2006	PhRMA
% Generic (by volume)	58	%	2007	PhRMA
Pharmaceutical Sales[1], Foreign[2]	71.1	Bil. US$	2006	PhRMA
Branded Biotech Drug Sales, U.S.	40.3	Bil. US$	2006	IMS
Prescriptions Filled, U.S., 12 month period ending Sept.	3.6	Bil.	2005	PhRMA
Total Pharmaceutical Company Spending on R&D[1]	43.0	Bil. US$	2006	PhRMA
as a Percentage of All Sales[1]	16.9	%	2005	PhRMA
Share of R&D Spending by Function:				
Prehuman/Preclinical	25.7	%	2005	PhRMA
Phase I	5.8	%	2005	PhRMA
Phase II	11.7	%	2005	PhRMA
Phase III	25.5	%	2005	PhRMA
Approval	6.9	%	2005	PhRMA
Phase IV	13.3	%	2005	PhRMA
Uncategorized	11.0	%	2005	PhRMA
Average Time Required for Clinical Trials, from Discovery through Phase IV	16	Years	2007	PhRMA
Average Cost of Developing a Biologic Drug	1.2	Bil. US$	2005	Tufts
Number of FDA Approvals for New Drugs (NDAs)	101	Approvals	2006	FDA
Number of Approvals for New Molecular Entities (NMEs)	22	Approvals	2006	FDA
Patents Granted for "Multicellular Living Organisms and Unmodified Parts Thereof" Since 1970	5,644	Patents	2005	USPTO
Total Requested Proposed for Biological Science Research, U.S. National Science Foundation	633.0	Mil. US$	2008	NSF
Average Annual Salary for Biochemists & Biophysicists	80,900	US$	May-06	BLS

[1] PhRMA Member Companies. [2] Not including foreign divisions of foreign companies.

E&Y = Ernst & Young; PhRMA = Pharmaceutical Research and Manufacturers Association; Burrill = Burrill & Company; IMS = IMS Health; Tufts = Tufts Center for the Study of Drug Development; FDA = U.S. Food & Drug Administration; USPTO = U.S. Patent & Trademark Office; NSF = U.S. National Science Foundation; BLS = U.S. Bureau of Labor Statistics

Global Biotechnology at a Glance: 2006

(Revenues, Expenses & Losses in Millions of US$)

(Public Company Data)	Global	U.S.	Europe	Canada	Asia-Pacific
Revenues	73,478	55,458	11,489	3,242	3,289
R&D Expense	27,782	22,865	3,631,885	885	401
Net Loss	5,446	3,466	1,125	524	331
Number of Employees	190,500	130,600	39,740	7,190	12,970

(Number of Companies)					
Public companies	710	336	156	82	136
Public & private companies	4,275	1,452	1,621	465	737

Numbers may appear inconsistent because of rounding. Employment totals rounded to the nearest hundred in the U.S., and to the nearest 10 in other regions.

Source: Ernst & Young, Beyond Borders: The Global Perspective (2007)

Plunkett's Biotech & Genetics Industry Almanac 2008

Growth in Global Biotechnology:
2005-2006

(Revenues, Expenses & Losses in Millions of US$)

(Public Company Data)	2006	2005	Change
Revenues	73,478	64,213	14%
R&D Expense	27,782	20,934	33%
Net Loss	5,446	4,039	35%
Number of Employees	190,500	146,010	30%

(Number of Companies)			
Public Companies	710	673	5%
Public & Private Companies	4,275	7,263	1%

2006 financials largely represent data from 1 January 2006 through 31 December 2006. 2005 financials largely represent data from 1 January 2005 through 31 December 2005. Numbers may appear inconsistent because of rounding.

Source: Ernst & Young, Beyond Borders: The Global Perspective (2007)

Plunkett's Biotech & Genetics Industry Almanac 2008

Biotechnology Financing, U.S., Europe & Canada: 2005-2006

(In Millions of US$)

2006	U.S.	Europe	Canada
IPO	944	907	9
Follow-on & other offerings	16,067	3,069	1,589
Venture financing	3,302	1,907	205
Total	**20,313**	**5,883**	**1,803**

2005	U.S.	Europe	Canada
IPO	626	691	160
Follow-on & other offerings	10,740	1,577	608
Venture financing	3,328	1,738	242
Total	**14,694**	**4,006**	**1,010**

Change	U.S.	Europe	Canada
IPO	51%	31%	-94%
Follow-on & other offerings	50%	95%	161%
Venture financing	-1%	10%	-15%
Total	**38%**	**47%**	**79%**

Numbers may appear inconsistent because of rounding. Percentage changes for Europe and Canada based on conversion of currency to U.S. dollars.

Source: Ernst & Young, Beyond Borders: The Global Perspective (2007)

Plunkett's Biotech & Genetics Industry Almanac 2008

U.S. Biotechnology at a Glance: 2005-2006

(In Billions of US$)

	Public Companies			Industry Total		
	2006	2005	% Change	2006	2005	% Change
Product Sales	48.4	42.4	14.2	50.8	44.8	13.4
Revenues	55.5	48.5	14.3	58.8	51.8	13.4
R&D Expense	22.9	16.6	38.1	27.1	20.8	30.2
Net Loss	3.5	1.4	151.4	5.6	3.6	58.5
Market Capitalization	392.4	408.4	-3.9			
Total Financings	17.0	11.4	49.3	20.3	14.7	38.2
Number of IPOs	20	13	53.8	20	13	53.8
Number of Companies	336	331	1.5	1,452	1,475	-1.6
Employees	130,600	119,000	9.7	180,800	170,500	6.6

Data were generally derived from year-end information (31 December). 2006 data are estimates based on January-September quarterly filings and preliminary annual financial performance data for some companies. 2006 employee data are obtained from SEC filings at time of publishing and include a combination of 2005 and 2006 employee data. The 2005 estimates have been revised for compatibility with 2006 data. Numbers may appear inconsistent because of rounding.

Source: Ernst & Young, Beyond Borders: The Global Perspective (2007)

Plunkett's Biotech & Genetics Industry Almanac 2008

Quarterly Breakdown of Biotechnology Financings, U.S. & Canada: 2006

(In Millions of US$)

	1st Quarter		2nd Quarter		3rd Quarter		4th Quarter		Total	
	U.S.	Canada	U.S.	Canada	U.S.	Canada	U.S.	Canada	U.S.	Canada
IPO	293	3	248	0	49	0	355	6	944	9
# of Financings	6	1	6	0	2	0	6	1	20	2
Follow-on	1,229	217	991	245	439	81	2,455	382	5,114	925
# of Financings	20	19	8	18	3	4	16	20	47	61
Venture	596	56	1,001	79	776	15	929	55	3,302	205
# of Financings	42	11	41	12	32	8	38	11	153	42
Other	6,208	273	2,189	3	784	6	1,772	382	10,953	664
# of Financings	70	5	31	2	46	3	58	9	205	19
Total	8,326	549	4,429	327	2,047	825	5,510	825	20,313	1,803
# of Financings	138	36	86	32	83	41	118	41	425	124

Numbers may appear inconsistent because of rounding.

Source: Ernst & Young, BioCentury, BioWorld, and Venture One (Approaching Profitability: The Americas Perspective 2007)
Plunkett's Biotech & Genetics Industry Almanac 2008

Ernst & Young Survival Index, U.S. & Canadian Biotechnology Companies: 2005-2006

	U.S.			
	2005		2006	
	Number of Companies	Percent of Total	Number of Companies	Percent of Total
More than 5 years of cash	110	33%	100	30%
3-5 years of cash	25	7%	34	10%
2-3 years of cash	44	13%	56	17%
1-2 years of cash	77	23%	66	20%
Less than 1 year of cash	80	24%	76	23%
Total public companies	**336**		**332**	

	Canada			
	2005		2006	
	Number of Companies	Percent of Total	Number of Companies	Percent of Total
More than 5 years of cash	34	41%	31	38%
3-5 years of cash	2	3%	0	0%
2-3 years of cash	4	5%	6	7%
1-2 years of cash	21	25%	8	10%
Less than 1 year of cash	21	25%	36	45%
Total public companies	**82**		**81**	

Numbers may appear inconsistent because of rounding.

Sources: Ernst & Young, Approaching Profitability: The Americas Perspective (2007); company financial data

Plunkett's Biotech & Genetics Industry Almanac 2008

Global Area of Biotech Crops by Country: 2006

(In Millions of Hectares)

Rank	Country	Area	Biotech Crops
1*	USA	54.6	Soybean, maize, cotton, canola, squash, papaya, alfalfa
2*	Argentina	18.0	Soybean, maize, cotton
3*	Brazil	11.5	Soybean, cotton
4*	Canada	6.1	Canola, maize, soybean
5*	India	3.8	Cotton
6*	China	3.5	Cotton
7*	Paraguay	2.0	Soybean
8*	South Africa	1.4	Maize, soybean, cotton
9*	Uruguay	0.4	Soybean, maize
10*	Philippines	0.2	Maize
11*	Australia	0.2	Cotton
12*	Romania	0.1	Soybean
13*	Mexico	0.1	Cotton, soybean
14*	Spain	0.1	Maize
15	Colombia	<0.1	Cotton
16	France	<0.1	Maize
17	Iran	<0.1	Rice
18	Honduras	<0.1	Maize
19	Czech Republic	<0.1	Maize
20	Portugal	<0.1	Maize
21	Germany	<0.1	Maize
22	Slovakia	<0.1	Maize

* 14 biotech mega-countries growing 50,000 hectares, or more, of biotech crops.

Source: Clive James, ISAAA

Plunkett's Biotech & Genetics Industry Almanac 2008

Total U.S. Biotechnology Patents Granted per Year by Patent Class: 1977-2005

(Original & Cross-Reference Classifications;
Duplicates Eliminated; Latest Year Available)

Year	Patent Class*						
	47	71	119	426	435	800	930
1977-84	1,385	708	1,656	5,404	5,031	23	914
1985	137	107	235	797	735	7	170
1986	168	61	189	648	806	17	180
1987	163	62	191	733	1,095	19	309
1988	183	61	234	793	1,070	23	317
1989	221	80	265	1,157	1,433	22	255
1990	235	60	311	1,030	1,405	13	6
1991	244	68	391	1,003	1,562	29	12
1992	241	94	399	908	1,969	48	58
1993	226	73	341	899	2,259	38	76
1994	273	78	355	823	2,179	99	66
1995	271	93	313	884	2,252	91	38
1996	345	75	348	902	3,088	253	59
1997	296	100	362	814	4,142	282	85
1998	469	89	553	980	6,134	498	79
1999	325	87	501	1,261	6,219	670	79
2000	351	64	469	1,267	5,606	631	48
2001	305	97	447	1,226	6,281	666	46
2002	306	112	539	1,064	5,740	555	33
2003	283	69	532	1,015	5,304	525	31
2004	185	60	396	874	4,624	614	40
2005	159	46	406	519	4,131	521	29
Total	6,771	2,344	9,433	25,001	73,065	5,644	2,930

* The patent classes are as follows:
47: Plant Husbandry
71: Chemical Fertilizers
119: Animal Husbandry
426: Food or Edible Material: Processes, Compositions and Products
435: Chemistry: Molecular Biology & Microbiology
800: Multicellular Living Organisms and Unmodified Parts Thereof
930: Peptide or Protein Sequence
The agricultural classes 47, 71, 119 and 426 include only a small portion of patents that are related to biotechnology. All patents in classes 435, 800 and 930 are biotechnology-related.

Source: U.S. Patent & Trademark Office (USPTO)

Plunkett's Biotech & Genetics Industry Almanac 2008

Plant-Derived Pharmaceuticals for the Treatment of Human Diseases that are in the Pipeline for Commercialization: 2007

Product	Class	Indication	Crop
Various single-chain Fv antibody fragments	Antibody	Non-Hodgkin's lymphoma	Viral vectors in tobacco
CaroRx	Antibody	Dental caries	Transgenic tobacco
E. coli heat-labile toxin	Vaccine	Diarrhea	Transgenic maize
			Transgenic potato
Gastric lipase	Therapeutic enzyme	Cystic fibrosis, pancreatitis	Transgenic maize
Hepatitis B virus surface antigen	Vaccine	Hepatitis B	Transgenic potato
			Transgenic lettuce
Human intrinsic factor	Dietary	Vitamin B12 deficiency	Transgenic Arabidopsis
Lactoferrin	Dietary	Gastrointestinal infection	Transgenic maize
Norwalk virus capsid protein	Vaccine	Norwalk virus infection	Transgenic potato
Rabies glycoprotein	Vaccine	Rabies	Viral vectors in spinach
Cyanoverin-N	Microbicide	HIV	Transgenic tobacco
Insulin	Hormone	Diabetes	Transgenic safflower
Lysozyme, Lactoferrin, Human serum albumin	Dietary	Diarrhea	Transgenic rice

Source: ISAAA

Plunkett's Biotech & Genetics Industry Almanac 2008

The U.S. Drug Discovery & Approval Process

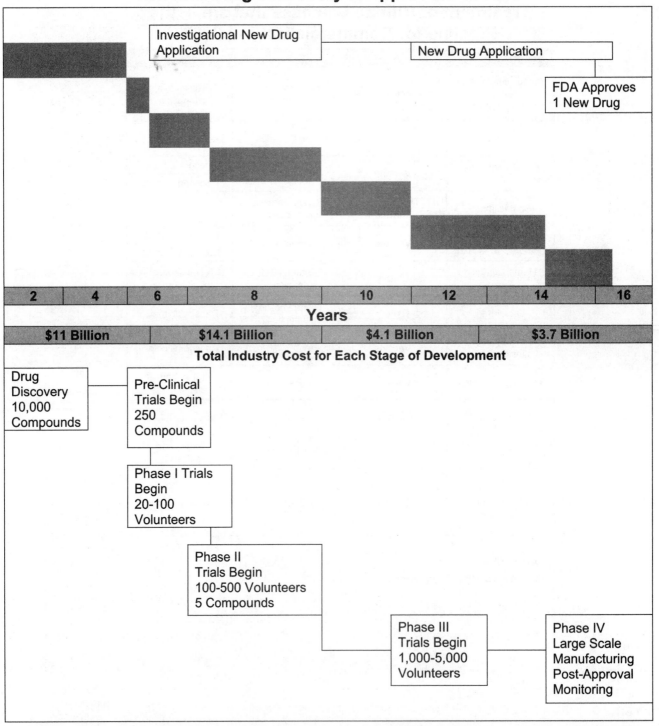

Source: Pharmaceutical Research and Manufacturers Association (PhRMA)

Plunkett's Biotech & Genetics Industry Almanac 2008

U.S. FDA Approval Times for New Drugs: 1993-2006

Calendar Year	Priority			Standard		
	Number Approved	Median FDA Review Time (months)	Median Total Approval Time (months)	Number Approved	Median FDA Review Time (months)	Median Total Approval Time (months)
1993	19	16.3	20.5	51	20.8	26.9
1994	17	15.0	15.0	45	16.8	22.1
1995	15	6.0	6.0	67	16.2	18.7
1996	29	7.8	7.8	102	15.1	17.8
1997	20	6.2	6.4	101	14.7	15.0
1998	25	6.2	6.4	65	12.0	12.0
1999	28	6.1	6.1	55	12.0	13.8
2000	20	6.0	6.0	78	12.0	12.0
2001	10	6.0	6.0	56	12.0	14.0
2002	11	13.8	19.1	67	12.7	15.3
2003	14	7.7	7.7	58	11.9	15.4
2004*	29	6.0	6.0	90	11.9	12.9
2005*	22	6.0	6.0	58	11.8	13.1
2006*	21	6.0	6.0	80	12.0	13.0

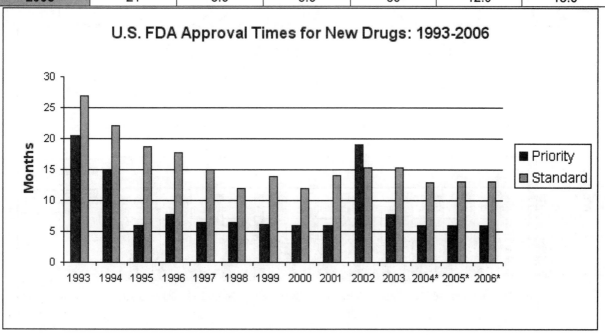

* Beginning in 2004, these figures include new Biologic License Applications (BLAs) for therapeutic biologic products transferred from the Center for Biologics Evaluation & Research (CBER) to the Center for Drug Evaluation & Research (CDER).

Source: U.S. Food & Drug Administration (FDA)

Plunkett's Biotech & Genetics Industry Almanac 2008

U.S. FDA Approval Times for New Molecular Entities (NMEs): 1993-2006

Calendar Year	Priority			Standard		
	Number Approved	Median FDA Review Time (months)	Median Total Approval Time (months)	Number Approved	Median FDA Review Time (months)	Median Total Approval Time (months)
1993	13	13.9	14.9	12	27.2	27.1
1994	12	13.9	14.0	9	22.2	23.7
1995	10	7.9	7.9	19	15.9	17.8
1996	18	7.7	9.6	35	14.6	15.1
1997	9	6.4	6.7	30	14.4	15.0
1998	16	6.2	6.2	14	12.3	13.4
1999	19	6.3	6.9	16	14.0	16.3
2000	9	6.0	6.0	18	15.4	19.9
2001	7	6.0	6.0	17	15.7	19.0
2002	7	13.8	16.3	10	12.5	15.9
2003	9	6.7	6.7	12	13.8	23.1
2004*	21	6.0	6.0	15	16.0	24.7
2005*	15	6.0	6.0	5	15.8	23.0
2006*	10	6.0	6.0	12	12.5	13.7

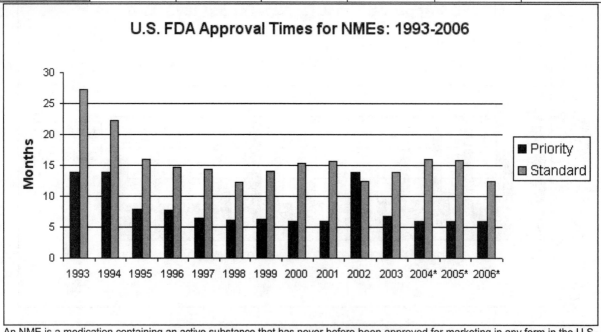

An NME is a medication containing an active substance that has never before been approved for marketing in any form in the U.S.

* Beginning in 2004, these figures include new Biologic License Applications (BLAs) for therapeutic biologic products transferred from the Center for Biologics Evaluation & Research (CBER) to the Center for Drug Evaluation & Research (CDER).

Source: U.S. Food & Drug Administration (FDA)

Plunkett's Biotech & Genetics Industry Almanac 2008

Domestic Biopharmaceutical R&D & R&D Abroad, PhRMA
Member Companies: 1970–2006

(In Millions of US$)

Year	Domestic R&D	Annual % Chg.	R&D Abroad**	Annual % Chg.	Total R&D	Annual % Chg.
2006*	33,967.9	9.7	9,005.6	1.3	42,973.5	7.8
2005	30,969.0	4.8	8,888.9	19.1	39,857.9	7.7
2004	29,555.5	9.2	7,462.6	1.0	37,018.1	7.4
2003	27,064.9	5.5	7,388.4	37.9	34,453.3	11.1
2002	25,655.1	9.2	5,357.2	-13.9	31,012.2	4.2
2001	23,502.0	10.0	6,220.6	33.3	29,772.7	14.4
2000	21,363.7	15.7	4,667.1	10.6	26,030.8	14.7
1999	18,471.1	7.4	4,219.6	9.9	22,690.7	8.2
1998	17,127.9	11.0	3,839.0	9.9	20,966.9	10.8
1997	15,466.0	13.9	3,492.1	6.5	18,958.1	12.4
1996	13,627.1	14.8	3,278.5	-1.6	16,905.6	11.2
1995	11,874.0	7.0	3,333.5	***	15,207.4	***
1994	11,101.6	6.0	2,347.8	3.8	13,449.4	5.6
1993	10,477.1	12.5	2,262.9	5.0	12,740.0	11.1
1992	9,312.1	17.4	2,155.8	21.3	11,467.9	18.2
1991	7,928.6	16.5	1,776.8	9.9	9,705.4	15.3
1990	6,802.9	13.0	1,617.4	23.6	8,420.3	14.9
1989	6,021.4	15.0	1,308.6	0.4	7,330.0	12.1
1988	5,233.9	16.2	1,303.6	30.6	6,537.5	18.8
1987	4,504.1	16.2	998.1	15.4	5,502.2	16.1
1986	3,875.0	14.7	865.1	23.8	4,740.1	16.2
1985	3,378.7	13.3	698.9	17.2	4,077.6	13.9
1984	2,982.4	11.6	596.4	9.2	3,578.8	11.2
1983	2,671.3	17.7	546.3	8.2	3,217.6	16.0
1982	2,268.7	21.3	505.0	7.7	2,773.7	18.6
1981	1,870.4	20.7	469.1	9.7	2,339.5	18.4
1980	1,549.2	16.7	427.5	42.8	1,976.7	21.5
1979	1,327.4	13.8	299.4	25.9	1,626.8	15.9
1978	1,166.1	9.7	237.9	11.6	1,404.0	10.0
1977	1,063.0	8.1	213.1	18.2	1,276.1	9.7
1976	983.4	8.8	180.3	14.1	1,163.7	9.6
1975	903.5	13.9	158.0	7.0	1,061.5	12.8
1974	793.1	12.0	147.7	26.3	940.8	14.0
1973	708.1	8.1	116.9	64.0	825.0	13.6
1972	654.8	4.5	71.3	24.9	726.1	6.2
1971	626.7	10.7	57.1	9.2	683.8	10.6
1970	566.2	-----	52.3	-----	618.5	-----
Average		12.2%		16.0%		12.7%

Note: All figures include company-financed R&D only. Total values may be affected by rounding.

* Estimated. ** R&D Abroad includes expenditures outside the United States by U.S.-owned PhRMA member companies and R&D conducted abroad by the U.S. divisions of foreign-owned PhRMA member companies. R&D performed abroad by the foreign divisions of foreign-owned PhRMA member companies is excluded. Domestic R&D, however, includes R&D expenditures within the United States by all PhRMA member companies. *** R&D Abroad affected by merger and acquisition activity.

Source: Pharmaceutical Research and Manufacturers of America (PhRMA)

Plunkett's Biotech & Genetics Industry Almanac 2008

Domestic Biopharmaceutical R&D Breakdown, PhRMA Member Companies: 2005

(In Millions of US$; Latest Year Available)

	Dollars	Share (%)
R&D Expenditures for Human-Use Pharmaceuticals		
Domestic	30,651.0	76.9
Abroad*	8,757.4	22
Total Human-Use R&D	39,408.4	98.9
R&D Expenditures for Veterinary-Use Pharmaceuticals		
Domestic	318.0	0.8
Abroad*	131.5	0.3
Total Vet-Use R&D	449.5	1.1
Total R&D	**39,857.9**	**100.0**

Domestic R&D by Source

	Dollars	Share (%)
Licensed-In	4,954.8	16.0
Self-Originated	22,349.3	72.2
Uncategorized	3,664.9	11.8
Total R&D	**30,969.0**	**100.0**

R&D By Function

	Dollars	Share (%)
Prehuman/Preclinical	10,258.1	25.7
Phase I	2,318.9	5.8
Phase II	4,670.9	11.7
Phase III	10,176.4	25.5
Approval	2,750.0	6.9
Phase IV	5,284.2	13.3
Uncategorized	4,399.4	11.0
Total R&D	**39,857.9**	**100.0**

Note: All figures include company-financed R&D only. Total values may be affected by rounding.

* R&D Abroad includes expenditures outside the United States by U.S.-owned PhRMA member companies and R&D conducted abroad by the U.S. divisions of foreign-owned PhRMA member companies. R&D performed abroad by the foreign divisions of foreign-owned PhRMA member companies is excluded. Domestic R&D, however, includes R&D expenditures within the United States by all PhRMA member companies.

Source: Pharmaceutical Research and Manufacturers Association (PhRMA)

Plunkett's Biotech & Genetics Industry Almanac 2008

R&D By Global Geographic Area, PhRMA Member Companies: 2005

(In Millions of US$; Latest Year Available)

Geographic Area*	Dollars	Share
Africa	28.0	0.1
Americas		
United States	30,969.0	77.7
Canada	479.3	1.2
Latin America[1]	174.9	0.4
Asia-Pacific		
Asia-Pacific (except Japan)	117.5	0.3
India & Pakistan	10.9	0.0
Japan	1,025.4	2.6
Australia & New Zealand	144.6	0.4
Europe		
France	498.8	1.3
Germany	548.2	1.4
Italy	342.1	0.9
Spain	208.8	0.5
United Kingdom	2,090.9	5.2
Other Western European nations	2,835.9	7.1
Central and Eastern European nations[2]	131.2	0.3
Other Eastern European nations[3]	113.4	0.3
Middle East[4]	37.7	0.1
Uncategorized	101.3	0.3
Total R&D	39,857.9	100.0

Note: All figures include company-financed R&D only. Total values may be affected by rounding.

* R&D Abroad includes expenditures outside the United States by U.S.-owned PhRMA member companies and R&D conducted abroad by the U.S. divisions of foreign-owned PhRMA member companies. R&D performed abroad by the foreign divisions of foreign-owned PhRMA member companies is excluded. Domestic R&D, however, includes R&D expenditures within the United States by all PhRMA member companies.

[1] South and Central America, Mexico and all Caribbean nations.
[2] Cyprus, Czech Republic, Estonia, Hungary, Poland, Slovenia, Bulgaria, Lithuania, Latvia, Romania, Slovakia and Malta.
[3] Including Russia and the Newly Independent States.
[4] Saudi Arabia, Yemen, United Arab Emirates, Iraq, Iran, Kuwait, Israel, Jordan, Syria, Afghanistan, Turkey, and Qatar.

Source: Pharmaceutical Research and Manufacturers of America (PhRMA)

Plunkett's Biotech & Genetics Industry Almanac 2008

R&D as a Percentage of U.S. Biopharmaceutical Sales, PhRMA Member Companies: 1970-2006

Year	Domestic R&D as a % of Domestic Sales	Total R&D as a % of Total Sales
2006*	19.4	17.5
2005	18.6	16.9
2004	18.4	16.1
2003	18.3	16.5
2002	18.4	16.1
2001	18.0	16.7
2000	18.4	16.2
1999	18.2	15.5
1998	21.1	16.8
1997	21.6	17.1
1996	21.0	16.6
1995	20.8	16.7
1994	21.9	17.3
1993	21.6	17.0
1992	19.4	15.5
1991	17.9	14.6
1990	17.7	14.4
1989	18.4	14.8
1988	18.3	14.1
1987	17.4	13.4
1986	16.4	12.9
1985	16.3	12.9
1984	15.7	12.1
1983	15.9	11.8
1982	15.4	10.9
1981	14.8	10.0
1980	13.1	8.9
1979	12.5	8.6
1978	12.2	8.5
1977	12.4	9.0
1976	12.4	8.9
1975	12.7	9.0
1974	11.8	9.1
1973	12.5	9.3
1972	12.6	9.2
1971	12.2	9.0
1970	12.4	9.3

*Estimated

Source: Pharmaceutical Research and Manufacturers of America (PhRMA)
Plunkett's Biotech & Genetics Industry Almanac 2008

U.S. Pharmaceutical R&D Spending Versus the Number of New Molecular Entity (NME) Approvals: 1993-2006

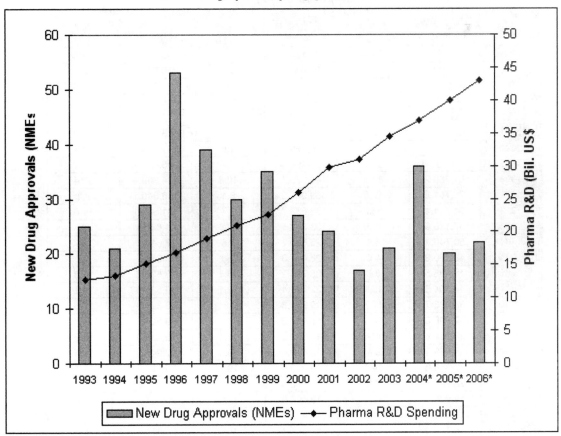

* Beginning in 2004, these figures include new BLAs for therapeutic biologic products transferred from CBER to CDER.

Notes: The FDA defines a New Molecular Entity (NME) as a medication containing an active substance that has never before been approved for marketing in any form in the U.S. Pharmaceutical R&D Spending includes expenditures inside and outside the U.S. by U.S.-owned PhRMA member companies and R&D conducted inside and outside the U.S. by the U.S. divisions of foreign-owned PhRMA member companies. R&D performed by the foreign divisions of foreign-owned PhRMA member companies is excluded.

Source: Pharmaceutical Research and Manufacturers Association (PhRMA); U.S. Food and Drug Administration

Plunkett's Biotech & Genetics Industry Almanac 2008

Domestic & Foreign Pharmaceutical Sales, PhRMA
Member Companies: 1970-2006
(In Millions of US$)

Year	Domestic Sales	APC	Sales Abroad[1]	APC	Total Sales	APC
2006[2]	174,667.4	5.1%	71,133.6	1.80%	245,801.0	4.1%
2005	166,155.5	3.4	69,881.0	0.1	236,036.5	2.4
2004[3]	160,751.0	8.6	69,806.9	14.6	230,557.9	10.3
2003[3]	148,038.6	6.4	60,914.4	13.4	208,953.0	8.4
2002	139,136.4	6.4	53,697.4	12.1	192,833.8	8.0
2001	130,715.9	12.8	47,886.9	5.9	178,602.8	10.9
2000	115,881.8	14.2	45,199.5	1.6	161,081.3	10.4
1999	101,461.8	24.8	44,496.6	2.7	145,958.4	17.1
1998	81,289.2	13.3	43,320.1	10.8	124,609.4	12.4
1997	71,761.9	10.8	39,086.2	6.1	110,848.1	9.1
1996	64,741.4	13.3	36,838.7	8.7	101,580.1	11.6
1995	57,145.5	12.6	33,893.5	(4)	91,039.0	(4)
1994	50,740.4	4.4	26,870.7	1.5	77,611.1	3.4
1993	48,590.9	1.0	26,467.3	2.8	75,058.2	1.7
1992	48,095.5	8.6	25,744.2	15.8	73,839.7	11.0
1991	44,304.5	15.1	22,231.1	12.1	66,535.6	14.1
1990	38,486.7	17.7	19,838.3	18	58,325.0	17.8
1989	32,706.6	14.4	16,817.9	-4.7	49,524.5	7.1
1988	28,582.6	10.4	17,649.3	17.1	46,231.9	12.9
1987	25,879.1	9.4	15,068.4	15.6	40,947.5	11.6
1986	23,658.8	14.1	13,030.5	19.9	36,689.3	16.1
1985	20,742.5	9.0	10,872.3	4	31,614.8	7.3
1984	19,026.1	13.2	10,450.9	0.4	29,477.0	8.3
1983	16,805.0	14.0	10,411.2	-2.4	27,216.2	7.1
1982	14,743.9	16.4	10,667.4	0.1	25,411.3	9.0
1981	12,665.0	7.4	10,658.3	1.4	23,323.3	4.6
1980	11,788.6	10.7	10,515.4	26.9	22,304.0	17.8
1979	10,651.3	11.2	8,287.8	21	18,939.1	15.3
1978	9,580.5	12.0	6,850.4	22.2	16,430.9	16.1
1977	8,550.4	7.5	5,605.0	10.2	14,155.4	8.6
1976	7,951.0	11.4	5,084.3	9.7	13,035.3	10.8
1975	7,135.7	5.9	4,633.3	19.1	11,769.0	13.6
1974	6,740.4	18.5	3,891.0	23.4	10,361.4	17.2
1973	5,686.5	9.1	3,152.5	15.9	8,839.0	11.5
1972	5,210.1	1.3	2,720.2	10.6	7,930.3	4.3
1971	5,144.9	13.0	2,459.7	18	7,604.6	14.6
1970	4,552.5	-----	2,084.0	-----	6,636.5	-----
Total		10.8%		10.2%		10.5%

Notes: Total values may be affected by rounding. APC = Annual Percent Change.
[1] Sales Abroad includes sales generated outside the United States by U.S.-owned PhRMA member companies and sales generated abroad by the U.S. divisions of foreign-owned PhRMA member companies. Sales generated abroad by the foreign divisions of foreign-owned PhRMA member companies are excluded. Domestic sales, however, includes sales generated within the United States by all PhRMA member companies.
[2] Estimated.
[3] Recalculated for updated data.
[4] Sales Abroad affected by merger and acquisition activity.

Source: Pharmaceutical Research and Manufacturers of America (PhRMA)
Plunkett's Biotech & Genetics Industry Almanac 2008

Domestic & Foreign Pharmaceutical Sales by End Use & Customer, PhRMA Member Companies: 2005

(In Millions of US$)

	Human Use	Vet Use	Total
To Private Sector	123,592.3	2,113.1	125,705.4
To Public Sector	30,984.9	0.5	30,985.4
Uncategorized	9,464.7		9,464.7
Total Domestic Sales	**164,091.9**	**2,113.6**	**166,155.5**

	Human Use	Vet Use	Total
Exports	1,181.5	38.6	1,220.1
Foreign Sales	65,486.6	2,311.9	67,818.5
Uncategorized	842.4		842.4
Total Sales Abroad*	**67,510.5**	**2,370.5**	**69,881.0**
TOTAL SALES	**231,552.4**	**4,481.1**	**236,036.5**

*Sales Abroad includes sales generated outside the United States by U.S.-owned PhRMA member companies and sales generated abroad by the U.S. divisions of foreign-owned PhRMA member companies. Sales generated abroad by the foreign divisions of foreign-owned PhRMA member companies are excluded. Domestic sales, however, includes sales generated within the United States by all PhRMA member companies.

Note: Total values may be affected by rounding

Source: Pharmaceutical Research and Manufacturers of America (PhRMA)

Plunkett's Biotech & Genetics Industry Almanac 2008

Generics' Share of the U.S. Drug Market by Volume: 1984-2007

Year	Percent	Year	Percent
1984	19	1996	43
1985	22	1997	44
1986	23	1998	45
1987	23	1999	47
1988	28	2000	47
1989	32	2001	48
1990	33	2002	51
1991	36	2003	52
1992	36	2004	55
1993	40	2005	57
1994	42	2006	54
1995	43	2007	58

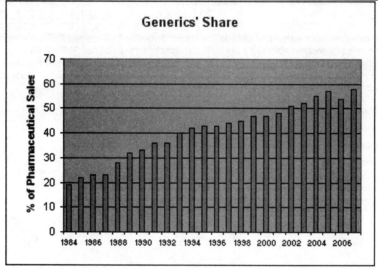

Source: Pharmaceutical Research and Manufacturers Association (PhRMA)

Plunkett's Biotech & Genetics Industry Almanac 2008

National Health Expenditure Amounts by Type of Expenditure, U.S.: Selected Calendar Years, 2001-2016[1]

(In Billions of US$)

Type of Expenditure	2001	2006	2007	2008	2009	2010	2015	2016
National Health Expenditures	1469.6	2122.5	2262.3	2420.0	2596.0	2776.4	3874.6	4136.9
Health Services & Supplies	1376.2	1987.7	2118.9	2267.3	2432.2	2600.8	3625.7	3869.9
Personal Health Care	1239.0	1769.2	1885.3	2016.6	2161.2	2312.9	3227.9	3449.4
Hospital Care	451.4	651.8	697.5	747.2	802.7	860.9	1206.7	1287.8
Professional Services	465.3	662.8	703.9	753.2	806.9	862.3	1179.3	1253.2
Physician & Clinical Services	313.2	447	474.2	506.2	541.4	577.1	774.9	819.9
Other Professional Services	42.8	60.9	64.9	69.1	73.4	78	104.6	111
Dental Services	67.5	92.8	98.6	104.9	111.6	118.4	155.4	163.4
Other Personal Health Care	41.9	62	66.2	73	80.5	88.8	144.4	159
Nursing Home & Home Health	133.7	179.4	190	201.5	213.7	226.1	302.6	322
Home Health Care	32.2	53.4	57.9	62.7	67.7	72.7	103.3	111.1
Nursing Home Care	101.5	126.1	132.1	138.8	146.1	153.4	199.2	210.9
Retail Outlet Sales of Medical Products	188.5	275.2	293.9	314.7	337.9	363.6	539.3	586.4
Prescription Drugs	138.6	213.7	229.5	247.6	268.3	291.5	453.6	497.5
Other Medical Products	49.9	61.5	64.3	67.1	69.5	72.2	85.7	88.9
Durable Medical Equipment	19.6	25.2	26.3	27.4	28.2	29.4	36.1	37.6
Other Non-Durable Medical Products	30.3	36.3	38	39.7	41.3	42.8	49.7	51.3
Program Administration & Net Cost of Private Health								
Insurance	90.4	156.8	167.4	179.8	194.9	206.2	281.5	295.7
Government Public Health Activities	46.8	61.7	66.2	70.9	76.1	81.7	116.2	124.8
Investment	93.4	134.8	143.4	152.8	163.8	175.6	248.9	267
Research[2]	28.8	41.7	43.9	46.3	49.1	52.1	70.9	75
Structures & Equipment	64.7	93.1	99.5	106.4	114.7	123.5	178	191.9

Note: Numbers may not add to totals because of rounding.

[1] The health spending projections were based on the 2005 version of the National Health Expenditures (NHE) released in January 2007.
[2] Research and development expenditures of drug companies and other manufacturers and providers of medical equipment and supplies are excluded from research expenditures. These research expenditures are implicitly included in the expenditure class in which the product falls, in that they are covered by the payment received for that product.

Source: Centers for Medicare & Medicaid Services (CMS), Office of the Actuary

Plunkett's Biotech & Genetics Industry Almanac 2008

Prescription Drug Expenditures, U.S.: 1965-2016

(In Millions of US$)

| Year | TOTAL | Private | | | Public | | | | | | | | |
| | | Total Private | Out-of-Pocket | Insurance | Total Public | Federal | | | | State & Local | | |
						Total Fed	Medicare	Medicaid	Other	Total S&L	Medicaid	Other
1965	3,715	3,571	3,441	130	144	58	0	0	58	85	0	85
1966	3,985	3,776	3,594	182	210	95	0	50	45	115	53	62
1967	4,227	3,950	3,712	238	277	127	0	99	29	150	106	44
1968	4,742	4,437	4,120	317	306	141	0	114	27	165	106	59
1969	5,149	4,761	4,362	399	389	191	0	165	26	197	135	62
1970	5,497	5,015	4,531	484	483	237	0	224	13	245	193	53
1971	5,877	5,309	4,752	558	568	297	0	281	16	271	214	57
1972	6,325	5,678	5,035	644	646	328	0	308	20	318	255	63
1973	6,817	6,083	5,341	742	735	355	0	336	19	379	308	71
1974	7,422	6,567	5,714	854	855	443	0	419	24	412	321	91
1975	8,052	7,032	6,068	963	1,021	508	0	478	30	512	392	120
1976	8,723	7,566	6,476	1,089	1,157	610	0	576	35	547	394	153
1977	9,196	7,946	6,754	1,192	1,250	624	0	587	37	626	449	177
1978	9,891	8,531	7,115	1,416	1,361	655	0	613	41	706	493	213
1979	10,744	9,177	7,470	1,707	1,567	743	0	701	42	823	536	288
1980	12,049	10,249	8,466	1,783	1,800	857	0	813	44	943	595	348
1981	13,398	11,338	8,844	2,494	2,060	979	0	932	47	1,081	679	402
1982	15,029	12,840	10,272	2,568	2,189	990	0	934	56	1,199	735	464
1983	17,323	14,808	11,254	3,554	2,515	1,133	0	1,067	66	1,382	841	541
1984	19,618	16,670	12,503	4,168	2,948	1,311	2	1,238	71	1,637	982	655
1985	21,795	18,566	13,609	4,957	3,229	1,421	21	1,323	78	1,808	1,009	799
1986	24,290	20,197	15,451	4,746	4,093	1,834	44	1,707	84	2,259	1,279	980
1987	26,889	22,261	16,406	5,855	4,628	2,077	75	1,910	92	2,550	1,486	1,065
1988	30,646	25,325	18,335	6,990	5,321	2,398	102	2,184	111	2,923	1,630	1,293
1989	34,758	28,831	20,153	8,678	5,927	2,617	139	2,353	124	3,309	1,729	1,580
1990	40,291	33,002	22,376	10,627	7,288	3,247	185	2,915	147	4,041	2,162	1,879
1991	44,381	35,951	23,047	12,904	8,431	3,832	226	3,437	169	4,599	2,607	1,992
1992	47,573	38,071	23,417	14,654	9,502	4,463	281	3,990	193	5,038	2,894	2,145
1993	50,991	40,476	24,097	16,379	10,515	5,122	352	4,540	230	5,393	3,181	2,212
1994	54,302	42,653	23,384	19,269	11,649	5,739	468	4,990	282	5,910	3,638	2,272
1995	60,876	47,790	23,349	24,441	13,086	6,614	680	5,557	377	6,472	4,143	2,330
1996	68,536	53,873	24,180	29,694	14,662	7,832	967	6,308	558	6,830	4,517	2,314
1997	77,666	61,244	25,670	35,574	16,422	9,027	1,206	7,025	796	7,395	5,077	2,317
1998	88,595	69,589	27,477	42,112	19,006	10,672	1,499	8,108	1,065	8,334	5,976	2,358
1999	104,684	81,597	30,410	51,187	23,087	13,163	1,866	9,805	1,492	9,924	7,257	2,668
2000	120,803	93,166	33,444	59,722	27,638	15,847	2,055	11,727	2,066	11,790	8,575	3,215
2001	138,559	105,543	36,206	69,337	33,016	19,191	2,415	13,919	2,857	13,825	10,081	3,744
2002	157,941	118,605	40,389	78,217	39,336	23,071	2,417	16,366	4,288	16,265	11,645	4,620
2003	174,639	128,577	44,437	84,140	46,062	27,665	2,366	19,569	5,731	18,397	13,197	5,200
2004	189,651	137,826	47,864	89,962	51,825	31,413	3,340	21,470	6,602	20,412	15,082	5,330
2005	200,716	146,110	50,906	95,204	54,606	32,872	3,999	21,522	7,351	21,733	16,075	5,658
2006	213,714	129,653	40,376	89,278	84,060	67,976	46,058	13,829	8,089	16,084	10,064	6,020

(Continued on next page)

Prescription Drug Expenditures, U.S.: 1965-2016 (cont.)

(In Millions of US$)

Year	TOTAL	Private			Public								
		Total Private	Out-of-Pocket	Insurance	Total Public	Federal				State & Local			
						Total Fed	Medicare	Medicaid	Other	Total S&L	Medicaid	Other	
2007	229,547	137,357	43,297	94,060	92,189	74,766	51,058	14,821	8,887	17,423	10,902	6,521
2008	247,612	144,278	46,119	98,158	103,334	84,726	59,222	15,600	9,904	18,609	11,466	7,143
2009	268,331	153,448	49,832	103,616	114,883	94,894	67,216	16,517	11,160	19,989	12,134	7,856
2010	291,492	163,299	53,967	109,332	128,193	106,510	76,199	17,770	12,541	21,682	13,050	8,633
2011	317,470	174,137	58,398	115,739	143,332	119,796	86,581	19,172	14,042	23,536	14,075	9,461
2012	346,496	185,564	63,253	122,311	160,931	135,357	98,873	20,708	15,775	25,575	15,198	10,376
2013	378,629	200,603	69,053	131,550	178,026	150,252	110,174	22,362	17,716	27,773	16,407	11,366
2014	414,162	217,271	75,548	141,723	196,890	166,738	122,687	24,147	19,904	30,153	17,712	12,441
2015	453,602	235,522	82,657	152,865	218,080	185,330	136,862	26,094	22,374	32,750	19,135	13,614
2016	497,526	255,547	90,454	165,093	241,978	206,420	153,054	28,174	25,191	35,559	20,655	14,903

Notes: Federal and State and Local Medicaid expenditures include Medicaid SCHIP Expansion. Federal and State and Local "Other" funds include SCHIP. The health spending projections were based on the 2005 version of the NHE released in January 2007. 2006-2016 are projections.

Source: Centers for Medicare & Medicaid Services (CMS), Office of the Actuary

Plunkett's Biotech & Genetics Industry Almanac 2008

U.S. Prescription Drug Expenditures, Aggregate & Per Capita Amounts, Percent Distribution: Selected Calendar Years, 2001-2016

(By Source of Funds)

Year	Total	Out-of-Pocket Payments	Third-Party Payments					Medicare[3]	Medicaid[4]
			Total	Private Health Insurance	Public				
					Total	Federal[2]	State and Local[2]		
Historical Estimates *(In Billions of US$)*									
2001	138.6	36.2	102.4	69.3	33.0	19.2	13.8	2.4	24.0
2002	157.9	40.4	117.6	78.2	39.3	23.1	16.3	2.4	28.0
2003	174.6	44.4	130.2	84.1	46.1	27.7	18.4	2.4	32.8
2004	189.7	47.9	141.8	90.0	51.8	31.4	20.4	3.3	36.6
2005	200.7	50.9	149.8	95.2	54.6	32.9	21.7	4.0	37.6
Projected									
2006	213.7	40.4	173.3	89.3	84.1	68.0	16.1	46.1	23.9
2007	229.5	43.3	186.2	94.1	92.2	74.8	17.4	51.1	25.7
2008	247.6	46.1	201.5	98.2	103.3	84.7	18.6	59.2	27.1
2009	268.3	49.8	218.5	103.6	114.9	94.9	20.0	67.2	28.7
2010	291.5	54.0	237.5	109.3	128.2	106.5	21.7	76.2	30.8
2011	317.5	58.4	259.1	115.7	143.3	119.8	23.5	86.6	33.2
2012	346.5	63.3	283.2	122.3	160.9	135.4	25.6	98.9	35.9
2013	378.6	69.1	309.6	131.6	178.0	150.3	27.8	110.2	38.8
2014	414.2	75.5	338.6	141.7	196.9	166.7	30.2	122.7	41.9
2015	453.6	82.7	370.9	152.9	218.1	185.3	32.7	136.9	45.2
2016	497.5	90.5	407.1	165.1	242.0	206.4	35.6	153.1	48.8
Historical Estimates *(Percent Distribution)*									
2001	485.0	127.0	359.0	243.0	116.0	67.0	48.0	(5)	(5)
2002	548.0	140.0	408.0	271.0	136.0	80.0	56.0	(5)	(5)
2003	600.0	153.0	447.0	289.0	158.0	95.0	63.0	(5)	(5)
2004	645.0	163.0	482.0	306.0	176.0	107.0	69.0	(5)	(5)
2005	676.0	172.0	505.0	321.0	184.0	111.0	73.0	(5)	(5)
Projected									
2006	714.0	135.0	579.0	298.0	281.0	227.0	54.0	(5)	(5)
2007	761.0	143.0	617.0	312.0	306.0	248.0	58.0	(5)	(5)
2008	814.0	152.0	663.0	323.0	340.0	279.0	61.0	(5)	(5)
2009	875.0	163.0	713.0	338.0	375.0	310.0	65.0	(5)	(5)
2010	943.0	175.0	769.0	354.0	415.0	345.0	70.0	(5)	(5)
2011	1019.0	188.0	832.0	372.0	460.0	385.0	76.0	(5)	(5)
2012	1104.0	202.0	902.0	390.0	513.0	431.0	81.0	(5)	(5)
2013	1197.0	218.0	979.0	416.0	563.0	475.0	88.0	(5)	(5)
2014	1299.0	237.0	1062.0	445.0	618.0	523.0	95.0	(5)	(5)
2015	1412.0	257.0	1155.0	476.0	679.0	577.0	102.0	(5)	(5)
2016	1537.0	279.0	1258.0	510.0	748.0	638.0	110.0	(5)	(5)

Notes: Per capita amounts based on July 1 Census resident based population estimates. Numbers and percents may not add to totals because of rounding. The health spending projections were based on the 2005 version of the National Health Expenditures (NHE) released in January 2007.

[1] Includes Medicaid SCHIP Expansion and SCHIP. 2 Subset of Federal funds. [3] Subset of Federal and State and local funds. Includes Medicaid SCHIP Expansion. [4] Calculation of per capita estimates is inappropriate.

Source: Centers for Medicare & Medicaid Services (CMS), Office of the Actuary

Plunkett's Biotech & Genetics Industry Almanac 2008

20 Largest Global Pharmaceutical Companies: 2005-2006

(In Thousands of US$)

Rank	Company	2006 Sales	2006 Profits	2005 Sales	2005 Profits
1	JOHNSON & JOHNSON	53,324,000	11,053,000	50,514,000	10,060,000
2	PFIZER INC	48,371,000	19,337,000	47,405,000	8,085,000
3	GLAXOSMITHKLINE PLC	45,595,800	10,793,000	37,783,631	8,400,952
4	SANOFI-AVENTIS	38,722,100	6,003,540	37,272,700	3,538,800
5	NOVARTIS AG	36,031,000	7,202,000	31,005,000	6,141,000
6	ROCHE HOLDING LTD	34,851,500	7,116,030	27,385,668	5,189,777
7	ASTRAZENECA PLC	26,475,000	4,392,000	23,950,000	3,881,000
8	MERCK & CO INC	22,636,000	4,433,800	22,011,900	4,631,300
9	ABBOTT LABORATORIES	22,476,322	1,716,755	22,337,808	3,372,065
10	WYETH	20,350,655	4,196,706	18,755,790	3,656,298
11	BRISTOL MYERS SQUIBB CO	17,914,000	1,585,000	19,207,000	3,000,000
12	ELI LILLY AND COMPANY	15,691,000	2,662,700	14,645,300	1,979,600
13	AMGEN INC	14,268,000	2,950,000	12,430,000	3,674,000
14	SCHERING-PLOUGH CORP	10,594,000	1,143,000	9,508,000	269,000
15	TAKEDA PHARMACEUTICAL COMPANY LTD	10,360,744	2,677,342	10,441,300	2,579,600
16	GENENTECH INC	9,284,000	2,113,000	6,633,372	1,278,991
17	MERCK KGAA	8,352,950	1,335,880	7,697,680	898,150
18	NOVO-NORDISK AS	6,913,700	1,126,020	5,446,472	946,073
19	ALTANA AG	5,234,410	1,008,440	4,429,010	915,040
20	ALCON INC	4,896,600	1,348,100	4,368,500	931,000

Note: All revenues are for the total company.

Source: Plunkett Research Ltd.
Plunkett's Biotech & Genetics Industry Almanac 2008
Copyright© 2007, All Rights Reserved

Research Funding for Biological Sciences, U.S. National Science Foundation: Fiscal Year 2006-2008

(In Millions of US$)

	FY 2006 Actual	FY 2007 Estimated	FY 2008 Requested	Change over 2007 Request	
				Amount	Percent
Molecular and Cellular Biosciences (MCB)	$108.46	$111.22	$116.37	$5.15	4.6%
Integrative Organismal Systems (IOS)	100.83	100.74	105.49	4.75	4.7%
Environmental Biology (EB)	107.21	109.61	114.66	5.05	4.6%
Biological Infrastructure (BI)	82.02	85.90	96.10	10.20	11.9%
Emerging Frontiers (EF)	81.87	99.16	99.16	-	-
Plant Genome (PG)	100.51	101.22	101.22	-	-
Total Biological Sciences Activity	**$580.90**	**$607.85**	**$633.00**	$25.15	4.1%

Totals may not add due to rounding.

Source: U.S. National Science Foundation
Plunkett's Biotech & Genetics Industry Almanac 2008

Federal R&D & R&D Plant Funding for General Science & Basic Research, U.S.: Fiscal Years 2005-2007

(In Millions of US$)

Funding Category and Agency	2005 Actual	2006 Prelim.	2007 Proposed	% Change (06-07)
Total	7,477	7,495	8,321	11.0
National Science Foundation (NSF)	4,102	4,175	4,523	8.3
Mathematical and physical sciences	1,069	1,085	1,150	6.0
Geosciences	697	703	745	6.0
Engineering	557	581	629	8.2
Biological sciences	577	577	608	5.4
Computer and information science and engineering	490	496	527	6.1
U.S. polar research programs	349	389	438	12.5
Major research equipment	174	191	240	26.0
Social, behavioral, and economic sciences	197	200	214	6.9
Office of cyberinfrastructure	123	127	182	43.5
Integrative activities	131	137	131	-4.2
Education and human resources	150	140	144	2.9
Budget authority adjustment[1]	-411	-452	-486	7.5
Department of Energy (DOE)	3,375	3,320	3,798	14.4
Basic energy sciences	1,084	1,134	1,421	25.3
High energy physics	723	716	775	8.2
Biological and environmental research	567	580	510	-12.1
Human genome	63	63	75	18.6
All other research	504	517	435	-15.8
Nuclear physics	394	367	454	23.7
Fusion energy sciences	267	288	319	10.8
Advanced scientific computing research	226	235	319	35.7
Small business innovation research[2]	114	0	0	N/A

Notes: Detail may not add to total because of rounding. Percent change derived from unrounded data. Not all federally sponsored basic research is categorized in subfunction 251. Data derived from agencies' submissions to Office of Management and Budget per MAX Schedule C, agencies' budget justification documents, and supplemental data obtained from agencies' budget offices.

[1] Budget authority adjustment subtracts costs for research facilities, major equipment support, and other non-R&D from total NSF budget authority.
[2] DOE treats this activity as a budget execution program (i.e., funds are collected from existing appropriations and are not allocated until three-quarters into the fiscal year).

N/A = Not applicable.

Source: U.S. National Science Foundation
Plunkett's Biotech & Genetics Industry Almanac 2008

Employment in Life & Physical Science Occupations
by Business Type, U.S.: May 2006

(Wage & Salary in US$)	Employ-ment[1]	Median Hourly Wage	Mean Hourly Wage	Mean Annual Salary[2]	Mean RSE[3] (%)
Animal Scientists	3,930	22.98	25.59	53,230	2.6
Food Scientists and Technologists	8,770	25.87	28.49	59,260	1.5
Soil and Plant Scientists	10,720	26.96	28.52	59,330	1.2
Biochemists and Biophysicists	18,680	36.69	38.90	80,900	1.7
Microbiologists	15,730	27.87	31.35	65,200	1.7
Zoologists and Wildlife Biologists	18,000	25.63	26.98	56,120	0.7
Biological Scientists, All Other	25,220	29.30	30.56	63,560	1.2
Conservation Scientists	16,000	26.43	26.64	55,410	0.7
Foresters	10,760	24.61	25.22	52,450	1.1
Epidemiologists	4,120	27.25	28.99	60,290	1.6
Medical Scientists, Except Epidemiologists	78,210	29.66	33.82	70,350	1.6
Life Scientists, All Other	12,830	27.39	31.00	64,480	1.9
Astronomers	1,430	46.03	45.67	95,000	2.1
Physicists	15,420	45.31	45.95	95,580	1.6
Atmospheric and Space Scientists	8,250	37.09	37.41	77,810	2.0
Chemists	80,500	28.78	31.75	66,040	1.1
Materials Scientists	9,390	35.87	37.02	77,010	1.3
Environmental Scientists and Specialists, Including Health	77,720	26.97	29.38	61,120	1.0
Geoscientists, Except Hydrologists and Geographers	28,980	34.93	38.41	79,890	1.4
Hydrologists	7,740	31.86	32.80	68,230	1.0
Physical Scientists, All Other	21,380	40.12	41.01	85,310	2.0
Agricultural and Food Science Technicians	19,220	15.26	16.20	33,700	0.9
Biological Technicians	71,590	17.17	18.38	38,240	0.6
Chemical Technicians	59,900	18.87	19.70	40,970	0.6
Geological and Petroleum Technicians	11,280	22.19	24.76	51,490	4.3
Nuclear Technicians	15,840	16.28	17.23	35,840	2.4
Environmental Science and Protection Technicians, Including Health	34,790	18.31	19.36	40,260	0.9
Forensic Science Technicians	12,310	21.79	23.14	48,130	1.5
Forest and Conservation Technicians	30,580	14.84	16.24	33,780	0.3

[1] Estimates for detailed occupations do not sum to the totals because the totals include occupations not shown separately. Estimates do not include self-employed workers.
[2] Annual wages have been calculated by multiplying the hourly mean wage by a "year-round, full-time" hours figure of 2,080 hours; for those occupations where there is not an hourly mean wage published, the annual wage has been directly calculated from the reported survey data.
[3] The relative standard error (RSE) is a measure of the reliability of a survey statistic. The smaller the relative standard error, the more precise the estimate.

Source: U.S. Bureau of Labor Statistics

Plunkett's Biotech & Genetics Industry Almanac 2008

Domestic Biopharmaceutical R&D Scientific, Professional, and Technical Personnel by Function, PhRMA Member Companies: 2005

(Latest Year Available)

Function	Personnel	Share (%)
Prehuman/Preclinical	25,940	31.2
Phase I	4,738	5.7
Phase II	8,491	10.2
Phase III	17,225	20.7
Approval	4,442	5.3
Phase IV	9,746	11.7
Uncategorized	4,075	4.9
Total R&D Staff	74,657	89.9
Supported R&D Nonstaff	8,420	10.1
TOTAL R&D PERSONNEL	83,077	100.0

Source: Pharmaceutical Research and Manufacturers of America (PhRMA)

Plunkett's Biotech & Genetics Industry Almanac 2008

Chapter 3

IMPORTANT BIOTECH & GENETICS INDUSTRY CONTACTS

Contents:

I.   Alternative Energy-Ethanol

Renewable Fuels Association (RFA)
1 Massachusetts Ave., Ste. 820
Washington, DC 20001 US
Phone: 202-289-3835
E-mail Address: *info@ethanolrfa.org*
Web Address: www.ethanolrfa.org
The Renewable Fuels Association (RFA) is a trade
organization representing the ethanol industry. It
publishes a wealth of useful information, including a
listing of biorefineries and monthly U.S. fuel ethanol
production and demand.

II.   Biotech Associations

Association of Biotechnology Led Enterprises (ABLE)
No. 13, 2nd Fl., 4th C Block, 10th Main Rd.
Koramangala, Bangalore 560034 India
Fax: 91-80-2553-3938
E-mail Address: *info@ableindia.org*
Web Address: www.ableindia.org
The Association of Biotechnology Led Enterprises (ABLE) is an organization focused on accelerating the pace of Biotechnology in India by enabling strategic alliances between researchers, the government and the global Biotech industry.

BIOCOM
4510 Executive Dr., Plaza 1
San Diego, CA 92121 US
Phone: 858-455-0300
Fax: 858-455-0022
Web Address: www.biocom.org
BIOCOM is a trade association for the life science industry in San Diego and Southern California.

BioIndustry Association
14/15 Belgrave Sq.
London, SW1X 8PS UK
Phone: 44-207-565-7190
Fax: 44-207-565-7191
E-mail Address: *admin@bioindustry.org*
Web Address: www.bioindustry.org
The BioIndustry Association promotes bioscience development in the U.K. The organization operates a public affairs program, a conference and seminar program, trade missions and publications for internal and external audiences.

Biotechnology Association of Alabama
500 Beacon Pkwy. W.
Birmingham, AL 35209 US
Phone: 205-290-9593
E-mail Address: *knugentphd@yahoo.com*
Web Address: www.bioalabama.com
The Biotechnology Association of Alabama is a nonprofit association that represents the biotech industry in Alabama.

Biotechnology Industry Organization (BIO)
1201 Maryland Ave., SW, Ste. 900
Washington, DC 20024 US
Phone: 202-962-9200
E-mail Address: *info@bio.org*
Web Address: www.bio.org
The Biotechnology Industry Organization (BIO) is involved in the research and development of health care, agricultural, industrial and environmental biotechnology products. BIO has both small and large member organizations.

California Healthcare Institute (CHI)
1020 Prospect St., Ste. 310
La Jolla, CA 92037 US
Phone: 858-551-6677
Fax: 858-551-6688
E-mail Address: *chi@chi.org*
Web Address: www.chi.org
California Healthcare Institute (CHI) is an independent organization that represents the biotech industry in California.

Connecticut BioScience Cluster (CURE)
300 George St., Ste. 561
New Haven, CT 06511 US
Phone: 203-777-8747
Fax: 203-777-8754
E-mail Address: *twallace@curenet.org*
Web Address: www.curenet.org
Connecticut BioScience Cluster (CURE) is a partnership with the Department of Economic and Community Development to promote Connecticut's biotech clusters.

European Society of Human Genetics (ESHG)
Vienna Medical Academy, Alser Strasse 4
Vienna, 1090 Austria
Phone: 43-1-405-13-83-20
Fax: 43-1-405-13-83-23
E-mail Address: *office@eshg.org*
Web Address: www.eshg.org
The European Society of Human Genetics (ESHG) is a organization that promotes the sharing of information among genetics societies in Europe.

Genetics Society of America (GSA)
9650 Rockville Pike
Bethesda, MD 20814-3998 US
Phone: 301-634-7300
Fax: 301-634-7079
Toll Free: 866-486-4343
E-mail Address: *GSA2008@ciwemb.edu*
Web Address: www.genetics-gsa.org
The Genetics Society of America (GSA) includes over 4,000 scientists and educators interested in the field of genetics. The society promotes the communication of advances in genetics through publication of the journal GENETICS, and by

sponsoring scientific meetings focused on key
organisms widely used in genetic research.

International Society for Stem Cell Research (ISSCR)
60 Revere Dr., Ste. 500
Northbrook, IL 60062 US
Phone: 847-509-1944
Fax: 847-480-9282
E-mail Address: *isscr@isscr.org*
Web Address: www.isscr.org
The International Society for Stem Cell Research is
an independent, nonprofit organization established to
promote the exchange and dissemination of
information and ideas relating to stem cells, to
encourage the general field of research involving
stem cells and to promote professional and public
education in all areas of stem cell research and
application.

Iowa Biotechnology Association
900 E. Des Moines St.
Des Moines, IA 50309 US
Phone: 515-327-9156
Fax: 515-327-1407
E-mail Address: *dgetter@netins.net*
Web Address: www.iowabiotech.com
The Iowa Biotech Association promotes research and
education in biotechnology in Iowa.

Korea Bio Venture Association (KOBIOVEN)
F8th 703- 5 Seo-il plaza
Yoeksam-dong, Kangnam-gu
Seoul, 135-513 Korea
Phone: 82-2-552-4771
Fax: 82-2-552-4840
E-mail Address: *kobioven@kobioven.or.kr*
Web Address: eng.kobioven.or.kr
KOBIOVEN was created to provide support to the
biotech industry in Korea. It is affiliated with the
Ministry of Commerce, Industry and Energy.

Massachusetts Biotechnology Council (MBC)
1 Cambridge Center, 9th Fl.
Cambridge, MA 02142 US
Phone: 617-674-5100
Fax: 617-674-5101
E-mail Address: *info_request@massbio.org*
Web Address: www.massbio.org
The Massachusetts Biotechnology Council (MBC) is
a nonprofit organization that promotes the
Massachusetts biotech industry.

Michigan Biotechnology Association
330 E. Liberty
Ann Arbor, MI 48107 US
Phone: 734-615-9670
Fax: 734-623-8289
E-mail Address: *info@michbio.org*
Web Address: www.michbio.org
The Michigan Biotechnology Association is a
nonprofit organization dedicated to promoting the
biotech industry in Michigan.

Missouri Biotechnology Association (MBA)
428 E. Capitiol
P.O. Box 148
Jefferson City, MO 65102-0148 US
Phone: 573-761-7600
Fax: 573-761-7601
E-mail Address: *gillespie@mobio.org*
Web Address: www.mobio.org
The Missouri Biotechnology Association (MBA) is
an industry organization that promotes the biotech
industry in Missouri.

New York Biotechnology Association (NYBA)
25 Health Sciences Dr., Ste. 203
Stony Brook, NY 11790 US
Phone: 631-444-8895
E-mail Address: *info@nyba.org*
Web Address: www.nyba.org
The New York Biotechnology Association (NYBA)
is a not-for-profit trade association dedicated to the
development and growth of biotechnology-related
industries and institutions in New York State.

Pennsylvania Bioscience Association (PBA)
7 Great Valley Pkwy., Ste. 290
Malvern, PA 19355 US
Phone: 610-578-9220
Fax: 610-578-9219
E-mail Address: *info@pennsylvaniabio.org*
Web Address: www.pennsylvaniabio.com
The Pennsylvania Bioscience Association (PBA) is a
nonprofit organization promoting bioscience in
Pennsylvania.

Society for Biomaterials
15000 Commerce Pkwy., Ste. C
Mt. Laurel, NJ 08054 US
Phone: 856-439-0826
Fax: 856-439-0525
E-mail Address: *info@biomaterials.org*
Web Address: www.biomaterials.org

The Society for Biomaterials is a professional society that promotes advances in all phases of materials research and development by encouraging cooperative educational programs, clinical applications and professional standards in the biomaterials field.

Tennessee Biotechnology Association (TBA)
Phone: 615-255-6270
Fax: 615-255-0094
E-mail Address: *jrolwing@tnbio.org*
Web Address: www.tnbio.org
The Tennessee Biotechnology Association (TBA) is a statewide clearinghouse that supports the biotech industry in Tennessee.

Virginia Biotechnology Association
800 E. Leigh St., Ste. 14
Richmond, VA 23219-1534 US
Phone: 804-643-6360
Fax: 804-643-6361
E-mail Address: *questions@vabio.org*
Web Address: www.vabio.org
The Virginia Biotechnology Association is a nonprofit organization that promotes biotechnology in Virginia.

Wisconsin Biotechnology and Medical Devices Association (WBMDA)
2 E. Mifflin St., Ste. 600
Madison, WI 53703 US
Phone: 608-252-9393
Fax: 608-283-5508
E-mail Address: *wisbiomed@dewittross.com*
Web Address: www.wisconsinbiotech.org
The Wisconsin Biotechnology and Medical Devices Association (WBMDA) is a professional organization devoted to the promotion of the biotech industry in Wisconsin.

III.   Biotech Investing

BioTech Stock Report
P.O. Box 7274
Beaverton, OR 97007-7274 US
Phone: 503-649-1355
Fax: 503-649-4490
E-mail Address: *info@biostockreport.com*
Web Address: www.biotechnav.com
The BioTech Stock Report is a monthly newsletter that provides analysis, commentary, news and company developments for biotechnology investors.

Biotechnology and Biological Sciences Research Council (BBSRC)
Polaris House, N. Star Ave.
Swindon, SN2 1UH UK
Phone: 44-0-1793-413200
Fax: 44-0-1793-413201
Web Address: www.bbsrc.ac.uk
The Biotechnology and Biological Sciences Research Council (BBSRC) provides funding for biotech research in the U.K.

Medical Technology Stock Letter
P.O. Box 40460
Berkeley, CA 94704 US
Phone: 510-843-1857
Fax: 510-843-0901
E-mail Address: *mtsl@bioinvest.com*
Web Address: www.bioinvest.com
The Medical Technology Stock Letter is a newsletter that provides financial advice about investing in biotechnology. It is distributed by mail and electronically.

IV.   Biotech Resources

About Biotech
Web Address: biotech.about.com
About Biotech provides news and information on the biotech industry, expertly compiled and edited by Theresa Phillips, PhD.

American College of Medical Genetics (ACMG)
9650 Rockville Pike
Bethesda, MD 20814-3998 US
Phone: 301-634-7127
Fax: 301-634-7275
E-mail Address: *acmg@acmg.net*
Web Address: www.acmg.net
The American College of Medical Genetics (ACMG) provides education, resources and a voice for the medical genetics profession. The ACMG promotes the development and implementation of methods to diagnose, treat and prevent genetic disease.

AusBiotech
322 Glenferrie Rd., Lv. 1
Malvern, VIC 3144 Australia
Phone: 03-9828-1400
Fax: 03-9824-5188
Web Address: www.ausbiotech.org
AusBiotech is a professional organization for the biotech industry in Australia, with members in the

human health, agricultural, medical device, bioinformatics, environmental and industrial sectors.

Bio Online
1900 Powell St., Ste. 230
Emeryville, CA 94608 US
Phone: 510-601-7194
Fax: 510-601-1862
E-mail Address: *corp@bio.com*
Web Address: www.bio.com
Bio Online is an online community of scientists, professionals, businesses and organizations supporting life science for the purpose of an exchange of information.

BioAbility
3200 Chapel Hill/Nelson Blvd., Ste. 201
Research Triangle Park, NC 27709-4569 US
Phone: 919-544-5111
Fax: 919-544-5401
E-mail Address: *info@bioability.com*
Web Address: www.bioability.com
BioAbility provides strategic business information to the biotechnology, pharmaceutical and life science industries.

BioAlberta
Phipps McKinnon Bldg., 10020-101A Ave. NW, Ste. 250
Edmonton, AB T5J 3G2 Canada
Phone: 780-425-3804
Fax: 780-409-9263
E-mail Address: *info@bioalberta.com*
Web Address: www.bioalberta.com
BioAlberta is a private, nonprofit industry association representing Alberta, Canada's biotech industry.

BioBasics
Canada
E-mail Address: *info@biotech.gc.ca*
Web Address: biobasics.gc.ca
BioBasics is a Canadian web site that offers information and links related to gene therapy, genetic testing and xenotransplantation. It also contains information on food, health, industrial biotechnology, natural resources and sustainable development.

Bioengineering Industry Links
Web Address:
www.seas.upenn.edu/be/misc/bmelink/cell.html
Bioengineering Industry Links is a web site provided by the University of Pennsylvania's Department of

Bioengineering. This site features links to companies involved in cell and tissue engineering.

Biofind
E-mail Address: *info@biofind.com*
Web Address: www.biofind.com
Biofind offers a biotech news directory, job search, chat room and event announcements, as well as a place to post announcements about biotech innovations.

BioMed Central
Middlesex House, 34-42 Cleveland St.
London, W1T 4LB UK
Phone: 44-20-7323-0323
Fax: 44-20-7631-9926
E-mail Address: *info@biomedcentral.com*
Web Address: www.biomedcentral.com
BioMed Central is an independent publishing house that prints approximately 160 peer-reviewed journals for the medical industry. Its web site provides free, open access to all of its research.

Biospace.com
90 New Montgomery St., Ste. 414
San Francisco, CA 94104 US
Phone: 732-528-3688
Fax: 732-528-3668
Toll Free: 888-246-7722
Web Address: www.biospace.com
Biospace.com offers information, news and profiles on biotech companies. It also provides an outlet for business and scientific leaders in bioscience to communicate with each other.

BIOTECanada
130 Albert St., Ste. 420
Ottawa, ON K1P 5G4 Canada
Phone: 613-230-5585
Fax: 613-563-8850
E-mail Address: *info@biotech.ca*
Web Address: www.biotech.ca
BIOTECanada is a trade organization that promotes the Canadian biotech industry.

BioTech
Austin, TX US
E-mail Address: *feedback@biotech.icmb.utexas.edu*
Web Address: biotech.icmb.utexas.edu
The BioTech web site offers a comprehensive dictionary of biotech terms, plus extensive research data regarding biotechnology.

Biotech Rumor Mill
E-mail Address: *info@biofind.com*
Web Address: www.biofind.com/rumor
The Biotech Rumor Mill is an online discussion
forum that attracts participants from many biotech
disciplines.

Biotechnology Australia
GPO Box 9839
Canberra, ACT 2601 Australia
Phone: 61-2-6213-6000
Fax: 61-2-6213-7615
Toll Free: 1-800-631-276
E-mail Address: *ba@biotechnology.gov.au*
Web Address: www.biotechnology.gov.au
Biotechnology Australia, a multi-departmental
government agency, is responsible for coordinating
non-regulatory biotechnology issues for the
Australian Government, and for developing the
National Biotechnology Strategy.

Biotechnology Information Directory Section
4364 S. Alston Ave.
Durham, NC 27713-2280 US
Phone: 919-361-2286
Fax: 919-361-2290
E-mail Address: *dhopp@cato.com*
Web Address: www.cato.com/biotech
The Biotechnology Information Directory Section
contains more than 3,000 links to companies,
research institutes, universities, sources of
information and other directories related to
biotechnology, pharmaceutical development and
similar fields.

Biotechnology Knowledge Center
800 N. Lindbergh Blvd.
St. Louis, MO 63167 US
Phone: 314-694-1000
Web Address: www.biotechknowledge.com
The Biotechnology Knowledge Center, supported by
Monsanto, is an online resource containing news,
technical documents, a discussion board and a
glossary.

Biotechterms.org
E-mail Address: *webmaster@biotechterms.org*
Web Address: biotechterms.org
Biotechterms.org is an online version of Technomic
Publishing's Glossary of Biotechnology Terms by
Kimball R. Nill.

BioWorld Online
3525 Piedmont Rd., Bldg. 6, Ste. 400
Atlanta, GA 30305 US
Toll Free: 800-688-2421
E-mail Address: *customerservice@bioworld.com*
Web Address: www.bioworld.com
BioWorld Online is a news and information site that
offers in-depth resources about the biotech industry
and leading companies.

Burrill & Company
1 Embarcadero Center, Ste. 2700
San Francisco, CA 94111 US
Phone: 415-591-5400
Fax: 415-591-5401
E-mail Address: *burrill@b-c.com*
Web Address: www.burrillandco.com
Burrill & Company is a leading private merchant
bank concentrated on companies in the life sciences
industries: biotechnology, pharmaceuticals, medical
technologies, agricultural technologies, animal health
and nutraceuticals.

Controlled Release Society (CRS)
3340 Pilot Knob Rd.
St. Paul, MN 55421 US
Phone: 651-454-7250
Fax: 651-454-0766
E-mail Address: *crs@scisoc.org*
Web Address: www.controlledrelease.org
The Controlled Release Society (CRS) is an
organization that promotes the science of the
controlled delivery of bioactive substances.

Council for Biotechnology Information (CBI)
1201 Maryland Ave., S.W., Ste. 900
Washington, DC 20024 US
Phone: 202-962-9200
Web Address: www.whybiotech.com
The Council for Biotechnology Information (CBI) is
a trade organization dedicated to promoting biotech.

DNAPatent.com
Web Address:
www.dnapatent.com/science/index.html
DNAPatent.com provides a web site offering
information on the basics of genetic engineering.

Electronic Journal of Biotechnology
Av. Brasil 2950
PO Box 4059
Valparaíso, Chile
E-mail Address: *edbiotec@ecv.cl*

Web Address: www.ejbiotechnology.info
The Electronic Journal of Biotechnology is an online
journal that publishes information about the biotech
industry.

Environmental Mutagen Society (EMS)
1821 Michael Faraday Dr., Ste. 300
Reston, VA 20190 US
Phone: 703-438-8220
Fax: 703-438-3113
E-mail Address: *emshq@ems-us.org*
Web Address: www.ems-us.org
Environmental Mutagen Society (EMS) provides
information on the process of biological mutagenesis,
and the application of this process in the field of
genetic toxicology.

Genetic Education Center: Human Genome Project Resources (GEC)
3901 Rainbow Blvd.
Kansas City, KS 66160 US
Phone: 913-588-5000
E-mail Address: *dcollins@kumc.edu*
Web Address: www.kumc.edu/gec/prof/hgc.html
The Genetic Education Center: Human Genome
Project Resources (GEC) provides a list of links
related to the human genome including human
genome centers, information sites and news articles.

Genetic Engineering News
140 Huguenot St., 3rd Fl.
New Rochelle, NY 10801-5215 US
Phone: 914-740-2100
Fax: 914-740-2201
Toll Free: 800-799-9436
Web Address: www.genengnews.com
Genetic Engineering News is a widely read magazine
that offers weekly news on topics in biotechnology,
bioregulation, bioprocess, bioresearch and
technology transfer.

GrantsNet
1200 New York Ave., NW
Washington, DC 20005 US
Phone: 202-326-6550
Web Address: www.grantsnet.org
GrantsNet is a free online service to locate funding
for training in the biomedical science industry and
undergraduate science education, provided through
ScienceCareers.org.

International Communication Forum in Human Genetics (ICFHG)
Denmark
E-mail Address: *admin@hum-molgen.de*
Web Address: www.hum-molgen.de
The International Communication Forum in Human
Genetics (ICFHG) contains news articles, a bulletin
board and a variety of other services related to human
molecular genetics.

LifeSciences World
725, rue Saint-Thomas
Longueuil, Quebec J4H 3A8 Canada
Phone: 450-616-9974
Fax: 450-616-9975
E-mail Address: *info@lifesciencesworld.com*
Web Address: www.biotechfind.com
LifeSciences World is a directory of life science
news, jobs, events, articles, reports and links to
information on biotechnology, pharmaceuticals and
medical devices.

MdBio
1003 W. 7th St., Ste. 202
Frederick, MD 21701 US
Phone: 301-228-2445
Fax: 800-863- Free:
E-mail Address: *jcoons@techcouncilmd.com*
Web Address: www.mdbio.org
MdBio is a nonprofit organization that promotes
biotech in Maryland. Areas of emphasis include
corporate and business development, networking and
community building, education and workforce
development and communications.

Medical Biochemistry Subject List
E-mail Address: *miking@iupui.edu*
Web Address:
web.indstate.edu/thcme/mwking/subjects.html
The Medical Biochemistry Subject List, produced by
Indiana State University, is a text-based introduction
to biochemistry.

MedWeb: Biomedical Internet Resources
Web Address: www.medweb.emory.edu/medweb
MedWeb: Biomedical Internet Resources is a web
site that lists resources by medical field, and allows
users to search for articles by topic or date.

Microbiology Network
150 Parkway Dr.
N. Chili, NY 14514 US
Phone: 585-594-8273

Fax: 585-594-3338
E-mail Address: *scott.sutton@microbiol.org*
Web Address: www.microbiol.org
The Microbiology Network is a communication
starting point for microbiologists, this site includes a
discussion forum, user's groups and extensive file
libraries.

MIT Synthetic Biology Working Group

Web Address: syntheticbiology.org
MIT Synthetic Biology Working Group is a
consortium of departments at MIT focused on
working together to advance the development of
synthetic biology. The site's FAQ includes
discussions of synthetic biology, its applications and
the difference between synthetic and systems
biology.

National Human Genome Research Institute
(NHGRI)

31 Center Dr., Bldg. 31, Rm. 4B09
Bethesda, MD 20892-2152 US
Phone: 301-402-0911
Fax: 301-402-2218
Web Address: www.nhgri.nih.gov
The National Human Genome Research Institute
(NHGRI) led the human genome project until its
completion in April 2003. The agency, a division of
the National Institutes of Health, now provides
research news and information about the field of
human genetics.

Recombinant Capital

2033 N. Main St., Ste. 1050
Walnut Creek, CA 94596-3722 US
Phone: 925-952-3870
Fax: 925-952-3871
E-mail Address: *info@recap.com*
Web Address: www.recap.com
Recombinant Capital provides databases of public
information, including press releases, clinical trial
references and a library of analyzed collaborations. It
hopes to facilitate alliances between biotechnology
companies.

Scotland Biotechnology

150 Broomielaw
Glasgow, G2 8LU UK
Phone: 0141 248 2700
E-mail Address: *lifesciences@scotent.co.uk*
Web Address: www.scottish-
enterprise.com/sedotcom_home/sig/life-sciences.htm

Scotland Lifesciences promotes the biotech industry
of Scotland, as well providing information on
medical devices and pharmaceuticals.

Signals

2033 N. Main St., Ste. 1050
Walnut Creek, CA 94596-3722 US
Phone: 925-952-3870
Fax: 925-952-3871
E-mail Address: *signals_edit@recap.com*
Web Address: www.signalsmag.com
Signals is an online magazine of analysis for the
biotechnology industry. The site provides trends
dealing with biotechnology, medical device and
pharmaceutical companies involved in the
development of medical products.

The Microbiology Network

150 Parkway Dr.
N. Chili, NY 14514 US
Phone: 585-594-8273
Fax: 585-594-3338
E-mail Address: *scott.sutton@microbiol.org*
Web Address: www.microbiol.org
The Microbiology Network is a virtual library
containing lists of organizations and associations in
the fields of microbiology, biology and general
science. Users can also submit new links.

Tufts Center for the Study of Drug Development

75 Kneeland St., 11th Fl.
Boston, MA 02111 US
Phone: 617-636-2170
Fax: 617-636-2425
E-mail Address: *csdd@tufts.edu*
Web Address: csdd.tufts.edu
The Tufts Center for the Study of Drug Development,
an affiliate of Tuft's University, provides analyses
and commentary on pharmaceutical issues. Its
mission is to improve the quality and efficiency of
pharmaceutical development, research and
utilization. It is famous, among other things, for its
analysis of the true total costs of developing and
commercializing a new drug.

University of California at Davis Biotechnology
Program

E-mail Address: *biotechprogram@ucdavis.edu*
Web Address: www.biotech.ucdavis.edu
The University of California at Davis Biotechnology
Program provides useful biotech information and
links, as well as the administrative home for UC
Davis' Biotechnology Program.

University of Pennsylvania's Center for Bioethics
3401 Market St., Ste. 320
Philadelphia, PA 19104 US
Phone: 215-898-7136
Web Address: www.bioethics.upenn.edu
The University of Pennsylvania's Center for Bioethics is a world-renowned resource. It incorporates the work of more than twenty people from the university's schools of law, medicine, business, philosophy, public policy and religious studies, as well as other departments. Resources include the PennBioethics newsletter.

Windhover Information
10 Hoyt St.
Norwalk, CT 06851 US
Phone: 203-838-4401
Fax: 203-838-3214
E-mail Address: *custserv@windhover.com*
Web Address: www.windhoverinfo.com
Windhover Information provides analysis and commentary on health care and biotech business strategy.

V. Biotech Resources-Agriculture

Ag BioTech Infonet
Web Address: www.biotech-info.net
Ag BioTech Infonet provides information on the application of biotechnology and genetic engineering in agricultural production, food processing and marketing.

Agricultural Biotechnology in Europe (ABE)
E-mail Address: *info@abeurope.info*
Web Address: www.abeurope.info
Agricultural Biotechnology in Europe (ABE) publishes science-based information about agricultural biotechnology, including country-by-country statistics and news. Members of ABE are companies that have pioneered the development of agriculture biotechnology and that strongly believe that biotechnology has the potential to be of great value to Europe's agricultural and food industries.

International Service for the Acquisition of Agri-Biotech Applications (ISAAA)
417 Bradfield Hall, Cornell University
Ithaca, NY 14853 US
Phone: 607-255-1724
Fax: 607-255-1215
E-mail Address: *americenter@isaaa.org*
Web Address: www.isaaa.org

The International Service for the Acquisition of Agri-Biotech Applications (ISAAA) is a not-for-profit organization that provides bioengineered seeds to poor and developing countries. In general, such seeds will enhance production per acre due to resistance to drought, insects and disease, and will offer additional crop enhancements.

UK Agricultural Biodiversity Coalition (UKABC)
UK
E-mail Address: *ukabc@ukabc.org*
Web Address: www.ukabc.org
The UK Agricultural Biodiversity Coalition (UKABC) provides links to life science and seed companies, databases and information resources, publicly funded research bodies and industry associations.

VI. Canadian Government Agencies

Biotechnology Research Institute of the National Research Council Canada (NRC-BRI)
6100 Royalmount avenue
Montréal, QC H4P 2R2 Canada
Phone: 514-496-6100
E-mail Address: *bri-info@cnrc-nrc.gc.ca*
Web Address: www.bri.nrc-cnrc.gc.ca
The NRC-BRI is one of the largest Canadian research facilities that focuses solely on biotechnology.

Institute for Biological Sciences (IBS)
Canada
E-mail Address: *Micha.Hage-Badr@nrc-cnrc.gc.ca*
Web Address: ibs-isb.nrc-cnrc.gc.ca
The IBS is a branch of Canada's National Research Council and focuses its research and development programs on life sciences operations designed to fight age-related and infectious diseases.

National Research Council (NRC)
1200 Montreal Road, Bldg. M-58
Ottawa, ON K1A 0R6 Canada
Phone: 613-993-9101
Fax: 613-952-9907
Toll Free: 877-672-2672
E-mail Address: *info@nrc-cnrc.gc.ca*
Web Address: www.nrc-cnrc.gc.ca
Canada's National Research Council (NRC) is an government organization of 20 research institutes that carry out multidisciplinary research with partners in industries and sectors key to Canada's economic development.

Plant Biotechnology Institute (PBI)
110 Gymnasium Place
Saskatoon, SK S7N 0W9 Canada
Phone: 306-975-5248
Fax: 306-975-4839
Web Address: pbi-ibp.nrc-cnrc.gc.ca
The PBI is a member of Canada's National Research
Council and is engaged in research regarding the
genomics, metabolic pathways, gene expression,
genetic transformation and structured biology of
plants and crops.

Steacie Institute for Molecular Sciences (SIMS)
100 Sussex Drive
Room 1141
Ottawa, ON K1A 0R6 Canada
Phone: 613-991-5419
Fax: 613-954-5242
Web Address: steacie.nrc-cnrc.gc.ca
The SIMS was created to facilitate the collaboration
of scientific communities researching molecular
sciences both within Canada and internationally.

VII.   Careers-Biotech

Bio Online Career Center
1900 Powell St., Ste. 230
Emeryville, CA 94608 US
Phone: 510-601-7194
Fax: 510-601-1862
E-mail Address: *careers@bio.com*
Web Address:
career.bio.com/careercenter/index.jhtml
The Bio Online Career Center enables the exchange
of information within the life sciences, biotechnology
and pharmaceutical industries. The center publishes
daily news, information and research tools for
professionals and students.

Biotechemployment.com
Phone: 561-630-5201
E-mail Address: *jobs@healthcarejobstore.com*
Web Address: www.biotechemployment.com
Biotechemployment.com is an online resource for job
seekers in biotechnology. The site's features includes
resume posting, job search agents and employer
profiles.

VIII. Careers-First Time Jobs/New Grads

Black Collegian Home Page
140 Carondelet St.

New Orleans, LA 70130 US
Phone: 504-523-0154
Web Address: www.black-collegian.com
Black Collegian Home Page features listings for job
and internship opportunities. The site includes a list
of the top 100 minority corporate employers and an
assessment of job opportunities.

Collegegrad.com
576 N. Washington Ave.
Cedarburg, WI 53012 US
Phone: 262-375-6700
Web Address: www.collegegrad.com
Collegegrad.com offers in-depth resources for
college students and recent grads seeking entry-level
jobs.

Job Web
62 Highland Ave.
Bethlehem, PA 18017-9085 US
Phone: 610-868-1421
Fax: 610-868-0208
Toll Free: 800-544-5272
Web Address: www.jobweb.com
Job Web, owned and sponsored by National
Association of Colleges and Employers (NACE),
displays job openings and employer descriptions. The
site also offers a database of career fairs, searchable
by state or keyword, with contact information.

MBAjobs.net
Fax: 413-556-8849
E-mail Address: *contact@mbajobs.net*
Web Address: www.mbajobs.net
MBAjobs.net is a unique international service for
MBA students and graduates, employers, recruiters
and business schools.

MonsterTrak
11845 W. Olympic Blvd., Ste. 500
Los Angeles, CA 90064 US
Toll Free: 800-999-8725
E-mail Address: *college.monstertrak@monster.com*
Web Address: www.monstertrak.monster.com
MonsterTrak features links to hundreds of university
and college career centers across the U.S. with entry-
level job listings categorized by industry. Major
companies can also utilize MonsterTrak.

**National Association of Colleges and Employers
(NACE)**
62 Highland Ave.
Bethlehem, PA 18017-9085 US

Phone: 610-868-1421
Fax: 610-868-0208
Toll Free: 800-544-5272
Web Address: www.naceweb.org
The National Association of Colleges and Employers
(NACE) is a premier U.S. organization representing
college placement offices and corporate recruiters
who focus on hiring new grads. The site offers in-
depth resources.

IX.   Careers-General Job Listings

America's Job Bank
Toll Free: 877-348-0502
E-mail Address: *info@careeronestop.org*
Web Address: www.jobsearch.org
America's Job Bank was developed by the U.S.
Department of Labor as part of an array of web-based
job tools. It offers an extensive list of searchable
employment vacancies as well as other job resources
for employers and job seekers.

Career Exposure, Inc.
805 SW Broadway, Ste. 2250
Portland, OR 97205 US
Phone: 503-221-7779
Fax: 503-221-7780
E-mail Address: *feedback@CareerExposure.com*
Web Address: www.careerexposure.com
Career Exposure is an online career center and job
placement service, with resources for employers,
recruiters and job seekers.

CareerBuilder
200 N. LaSalle St., Ste. 1100
Chicago, IL 60631 US
Phone: 773-527-3600
Fax: 773-399-6313
Toll Free: 800-638-4212
Web Address: www.careerbuilder.com
CareerBuilder focuses on the needs of companies and
also provides a database of job openings, called the
Mega Job Search. Hundreds of thousands of job
openings are posted. Resumes are sent directly to the
company, and applicants can set up a special e-mail
account for job-seeking purposes. CareerBuilder, Inc.
is a joint venture of three newspaper giants: Knight
Ridder, Gannett and Tribune Company.

**Careers.wsj.com from the Publishers of the Wall
Street Journal**
P.O. Box 300
Princeton, NJ 08543-0300 US

Web Address: www.careers.wsj.com
The Wall Street Journal's executive career site, called
CareerJournal.com, features a job database with
more than 100,000 available positions. It provides a
weekly career column and a range of articles about
topics including diversity issues and promotion.

HotJobs
45 W. 18th St., 6th Fl.
New York, NY 10011 US
Phone: 646-351-5300
Fax: 212-944-8962
Web Address: hotjobs.yahoo.com
HotJobs, designed for experienced professionals,
employers and job seekers, is a Yahoo-owned site
that provides company profiles, a resume posting
service and a resume workshop. The site allows
posters to block resumes from being viewed by
certain companies and provides a notification service
of new jobs.

HRS Federal Job Search
Web Address: www.hrsjobs.com
HRS Federal Job Search features a database of
federal jobs available across the U.S. Most jobs are
within the public sector. The job seeker creates a
profile with desired job type, salary and location to
receive applicable postings by e-mail.

JobCentral
DirectEmployers Association, Inc.
9002 N. Purdue Rd., Quad III, Ste. 100
Indianapolis, IN 46268 US
Phone: 317-874-9000
Fax: 317-874-9100
Toll Free: 866-268-6206
E-mail Address: *info@jobcentral.com*
Web Address: www.jobcentral.com
JobCentral, operated by the nonprofit
DirectEmployers Association, links users directly to
hundreds of thousands of job opportunities posted on
the sites of participating employers, thus bypassing
the usual job search sites. This saves employers
money and allows job seekers to access many more
job opportunities.

LaborMarketInfo
7000 Franklin Blvd., Ste. 1100
Sacramento, CA 95823 US
Phone: 916-262-2162
Fax: 916-262-2352
Web Address: www.labormarketinfo.edd.ca.gov

LaborMarketInfo, formerly the California Cooperative Occupational Information System, is sponsored by California's Economic Development Office. The web site is geared to providing job seekers and employers a wide range of resources, namely the ability to find, access and use labor market information and services. It provides demographical statistics for employment on both a local and regional level, as well as career searching tools for California residents.

Monster Worldwide, Inc.
622 Third Ave., 39th Fl.
New York, NY 10017 US
Phone: 212-351-7000
Toll Free: 800-666-7837
Web Address: www.monster.com
Monster Worldwide, Inc. is an electronic career center that displays hundreds of thousands of job opportunities in 23 countries around the world. Job seekers can build and store a resume online and find job listings that match their profiles. Monster e-mails the results once per week.

Recruiters Online Network
Web Address: www.recruitersonline.com
The Recruiters Online Network provides job postings from thousands of recruiters, Careers Online Magazine, a resume database, as well as other career resources.

TrueCareers, Inc.
Web Address: www.truecareers.com
TrueCareers, Inc. offers job listings and provides an array of career resources. The company also offers a search of over 2 million scholarships.

X.   Careers-Health Care

Health Care Source
Web Address: www.healthcaresource.com
Health Care Source offers career-related information and job finding tools for health care professionals.

HMonster
Web Address: healthcare.monster.com
HMonster, managed by monster.com, provides job listings, job searches and search agents for the medical field.

Medzilla.com
Web Address: www.medzilla.com

Medzilla.com offers job searches, salary surveys, a search agent and information on health care employment.

PracticeLink
415 2nd Ave.
P.O. Box 100
Hinton, WV 25951 US
Fax: 877-847-0120
Toll Free: 800-776-8383
E-mail Address: info@practicelink.com
Web Address: www.practicelink.com
PracticeLink is one of the largest physician employment web sites. It is a free service used by more than 18,000 practice-seeking physicians annually to quickly search and locate potential physician practice opportunities. PracticeLink is financially supported by more than 700 hospitals, medical groups, private practices and health care systems that advertise more than 5,000 opportunities.

XI.   Careers-Job Reference Tools

Newspaperlinks.com
E-mail Address: sally.clarke@naa.org
Web Address: www.newspaperlinks.com
Newspaperlinks.com, a service of the Newspaper Association of America, links individuals to local, national and international newspapers. Job seekers can search through thousands of classified sections.

Vault.com
150 W. 22nd St., 5th Fl.
New York, NY 10011 US
Phone: 212-366-4212
Web Address: www.vault.com
Vault.com is a comprehensive career web site for employers and employees, with job postings and valuable information on a wide variety of industries. Vault gears many of its features toward MBAs. The site has been recognized by Forbes and Fortune Magazines.

XII.   Careers-Science

Chem Jobs
ChemIndustry.com, 730 E. Cypress Ave.
Monrovia, CA 91016 US
Phone: 626-930-0808
Fax: 626-930-0102
E-mail Address: info@chemindustry.com
Web Address: www.chemjobs.net

Chem Jobs is a leading Internet site for job seekers in chemistry and related fields, aimed at chemists, biochemists, pharmaceutical scientists and chemical engineers looking for work. The web site is powered by chemindustry.com.

Employment Links for the Biomedical Scientist
E-mail Address: *graeme@his.com*
Web Address: www.his.com/~graeme/employ.html
Employment Links for the Biomedical Scientist offers a list of links for job seekers using the Internet to locate work in the biomedical field. It is compiled by a fellow scientist.

Science Jobs
E-mail Address: *webmaster@science-jobs.org*
Web Address: www.science-jobs.org/index.htm
Science Jobs is a web site that contains many useful categories of links, including employment newsgroups, scientific journals and placement agencies.

Sciencejobs.com
Web Address: www.sciencejobs.com
Sciencejobs.com is a web site produced by the publishers of New Scientist Magazine, which helps jobseekers and employers in the bioscience fields find each other. The site includes a job search engine and a free-of-charge e-mail job alert service.

XIII. Clinical Trials

Clinical Trials
8600 Rockville Pike
Bethesda, MD 20894 US
Phone: 301-594-5983
Fax: 301-402-1384
Toll Free: 888-346-3656
Web Address: www.clinicaltrials.gov
Clinical Trials offers up-to-date information for locating federally and privately supported clinical trials for a wide range of diseases and conditions. It is a service of the National Institutes of Health (NIH).

Office of Biotechnology Activities, NIH
6705 Rockledge Drive, Ste. 750, MSC 7985
Bethesda, MD 20892-7985 US
Phone: 301-496-9838
Fax: 301-496-9839
E-mail Address: *oba@od.nih.gov*
Web Address: http://www4.od.nih.gov/oba/
This unit of the U.S. National Institutes of Health operates a web site with links to clinical research in

recombinant DNA and gene transfer, along with information on the National Science Advisory Board for Biosecurity.

XIV. Corporate Information Resources

bizjournals.com
120 W. Morehead St., Ste. 400
Charlotte, NC 28202 US
Web Address: www.bizjournals.com
bizjournals.com is the online media division of American City Business Journals, the publisher of dozens of leading city business journals nationwide. It provides access to research into the latest news regarding companies small and large.

Business Wire
44 Montgomery St., 39th Fl.
San Francisco, CA 94104 US
Phone: 415-986-4422
Fax: 415-788-5335
Toll Free: 888-381-9473
Web Address: www.businesswire.com
Business Wire offers news releases, industry- and company-specific news, top headlines, conference calls, IPOs on the Internet, media services and access to tradeshownews.com and BW Connect On-line through its informative and continuously updated web site.

Edgar Online
50 Washington St., 11th Fl.
Norwalk, CT 06854 US
Phone: 203-852-5666
Fax: 203-852-5667
Toll Free: 800-416-6651
Web Address: www.edgar-online.com
Edgar Online is a gateway and search tool for viewing corporate documents, such as annual reports on Form 10-K, filed with the U.S. Securities and Exchange Commission.

PRNewswire
810 7th Ave., 32nd Fl.
New York, NY 10019 US
Phone: 212-596-1500
Toll Free: 800-832-5522
E-mail Address: *information@prnewswire.com*
Web Address: www.prnewswire.com
PRNewswire provides comprehensive communications services for public relations and investor relations professionals ranging from information distribution and market intelligence to

the creation of online multimedia content and investor relations web sites. Users can also view recent corporate press releases.

Silicon Investor
Web Address: www.siliconinvestor.com
Silicon Investor is focused on technology companies. The site serves as a financial discussion forum and offers quotes, profiles and charts.

XV. Economic Data & Research

STAT-USA
U.S. Department of Commerce, Room H-4885
Washington, DC 20230 US
Phone: 202-482-1986
Fax: 202-482-2164
Toll Free: 800-742-8872
E-mail Address: *statmail@doc.gov*
Web Address: www.stat-usa.gov
STAT-USA is an agency in the Economics and Statistics Administration of the U.S. Department of Commerce. The site offers daily economic news, statistical releases, and databases relating to export and trade, as well as the domestic economy.

XVI. Engineering, Research & Scientific Associations

American Association for the Advancement of Science (AAAS)
1200 New York Ave. NW
Washington, DC 20005 US
Phone: 202-326-6400
E-mail Address: *webmaster@aaas.org*
Web Address: www.aaas.org
The American Association for the Advancement of Science (AAAS) is the world's largest scientific society and the publisher of Science magazine. It is an international non-profit organization dedicating to advancing science.

American National Standards Institute (ANSI)
1819 L St. NW, 6th Fl.
Washington, DC 20036 US
Phone: 202-293-8020
Fax: 202-293-9287
E-mail Address: *info@ansi.org*
Web Address: www.ansi.org
The American National Standards Institute (ANSI), founded in 1918, is a private, nonprofit organization that administers and coordinates the U.S. voluntary

standardization and conformity assessment system. Its mission is to enhance both the global competitiveness of U.S. business and the quality of life by promoting and facilitating voluntary consensus standards and conformity assessment systems and safeguarding their integrity.

American Physical Society (APS)
1 Physics Ellipse
College Park, MD 20740-3844 US
Phone: 301-209-3200
Fax: 301-209-0865
E-mail Address: *exoffice@aps.org*
Web Address: www.aps.org
The American Physical Society (APS) develops and implements effective programs in physics education and outreach.

American Society for Healthcare Engineering (ASHE)
1 N. Franklin, 28th Fl.
Chicago, IL 60606 US
Phone: 312-422-3800
Fax: 312-422-4571
E-mail Address: *ashe@aha.org*
Web Address: www.ashe.org
The American Society for Healthcare Engineering (ASHE) is the advocate and resource for continuous improvement in the health care engineering and facilities management professions.

American Society of Agricultural and Biological Engineers (ASABE)
2950 Niles Rd.
St. Joseph, MI 49085 US
Phone: 269-429-0300
Fax: 269-429-3852
E-mail Address: *hq@asabe.org*
Web Address: www.asabe.org
The American Society of Agricultural and Biological Engineers (ASABE) is a nonprofit professional and technical organization interested in engineering knowledge and technology for food and agriculture and associated industries.

IEEE (Institute of Electrical and Electronics Engineers)
3 Park Ave., 17th Fl.
New York, NY 10016-5997 US
Phone: 212-419-7900
Fax: 212-752-4929
Toll Free: 800-701-4333
E-mail Address: *ieeeusa@ieee.org*

Web Address: www.ieee.org
The IEEE (Institute of Electrical and Electronics Engineers) is a nonprofit, technical professional association of more than 370,000 individual members in approximately 160 countries. The IEEE sets global technical standards and acts as an authority in technical areas ranging from computer engineering, biomedical technology and telecommunications, to electric power, aerospace and consumer electronics.

Industrial Research Institute (IRI)
2200 Clarendon Blvd., Ste. 1102
Arlington, VA 22201 US
Phone: 703-647-2580
Fax: 703-647-2581
Web Address: www.iriinc.org
The Industrial Research Institute (IRI) is a nonprofit organization of over 200 leading industrial companies, representing industries such as aerospace, automotive, chemical, computers and electronics, which carry out industrial research efforts in the U.S. manufacturing sector. IRI helps members improve research and development capabilities.

Institute of Biological Engineering (IBE)
P.O. Box 24267
Minneapolis, MN 55424-0267 US
Phone: 763-765-2388
Fax: 763-765-2329
E-mail Address: *director@ibeweb.org*
Web Address: www.ibeweb.org
The Institute of Biological Engineering (IBE) is a professional organization encouraging inquiry and interest in biological engineering and professional development for its members.

International Society of Pharmaceutical Engineers (ISPE)
3109 W. Dr. Martin Luther King, Jr. Blvd., Ste. 250
Tampa, FL 33607 US
Phone: 813-960-2105
Fax: 813-264-2816
E-mail Address: *customerservice@ispe.org*
Web Address: www.ispe.org
The International Society of Pharmaceutical Engineers (ISPE) is a worldwide nonprofit society dedicated to educating and advancing pharmaceutical manufacturing professionals and the pharmaceutical industry.

International Standards Organization (ISO)
1, ch de la Voie-Creuse, Case postale 56
Geneva, CH-1211 Switzerland

Phone: 41-22-749-01-11
Fax: 41-22-733-34-30
Web Address: www.iso.org
The International Standards Organization (ISO) is a global consortium national standards institutes from 157 countries. The established International Standards are designed to make products and services more efficient, safe and clean.

National Academy of Sciences (The National Academies)
Keck Center of the National Academies
500 Fifth St. N.W.
Washington, DC 20001 US
Phone: 202-334-2000
Web Address: www.nationalacademies.com
The National Academy of Sciences (NAS) is an honorific society of distinguished American scholars engaged in scientific and engineering research, dedicated to the furtherance of science and technology and to their use for the general welfare. The NAS was signed into being by President Abraham Lincoln on March 3, 1863, at the height of the Civil War. The four units of the National Academies include the National Academy of Sciences, the National Academy of Engineering, the Institute of Medicine and the National Research Council. Collectively, these units are know as the National Academies.

Netherlands Organization for Applied Scientific Research (TNO)
P.O. Box 6000
JA Delft, NL-2600 The Netherlands
Phone: 31-15-269-69-00
Fax: 31-15-261-24-03
E-mail Address: *infodesk@tno.nl*
Web Address: www.tno.nl
The Netherlands Organization for Applied Scientific Research (TNO) is a contract research organization that provides a link between fundamental research and practical application.

Royal Society (The)
6-9 Carlton House Terrace
London, SW1Y 5AG UK
Phone: 44 (0)20 7451 2500
Fax: 44 (0)20 7930 2170
Web Address: www.royalsoc.ac.uk
The Royal Society is the UK's leading scientific organization. It operates as a national academy of science, supporting scientists, engineers, technologists and research. On its website, you will

find a wealth of data about the research and
development initiatives of its Fellows and Foreign
Members.

Royal Society of Chemistry
Burlington House, Piccadilly
London, W1J 0BA UK
Phone: 44-0-20-7437-8656
Fax: 44-0-20-7437-8883
Web Address: www.rsc.org
The Royal Society of Chemistry is Europe's largest
organization for advancing the chemical sciences.

XVII.   European Government Agencies

Eurpoean Medicines Agency (EMEA)
7 Westferry Circus
Canary Wharf
London, E14 4HB UK
Phone: 44-20 74 18 84 00
Fax: (44 20 74 18 84 09
E-mail Address: *info@emea.europa.eu*
Web Address: www.emea.europa.eu
The EMEA is the European agency charged with
approving new drugs and monitoring the efficacy of
existing drugs.

XVIII. Gene Therapy

American Society of Gene Therapy (ASGT)
555 East Wells St., Ste. 1100
Milwaukee, WI 53202 US
Phone: 414-278-1341
Fax: 414-276-3349
E-mail Address: *mdean@asgt.org*
Web Address: www.asgt.org
The American Society of Gene Therapy (ASGT) is a
non-profit medical and professional organization that
represents researchers and scientists devoted to the
discovery of new gene therapies. ASGT was
established in 1996 by Dr. George
Stamatoyannopoulos, professor of medicine at the
University of Washington 's School of Medicine and
a group of the country's leading researchers in gene
therapy. With more than 2,000 members in the
United States and worldwide, ASGT is the largest
association of individuals involved in gene
therapeutics.

XIX. Health Care Business & Professional Associations

American Medical Technologists (AMT)
10700 W. Higgins Rd., Ste. 150
Rosemont, IL 60018 US
Phone: 847-823-5169
Fax: 847-823-0458
Toll Free: 800-275-1268
Web Address: www.amt1.com
American Medical Technologists (AMT) is a
nonprofit certification agency and professional
membership association representing individuals in
health care.

American Society of Clinical Oncology (ASCO)
1900 Duke Street, Ste. 200
Alexandria, VA 22314 US
Phone: 703-299-0150
Fax: 703-299-1044
E-mail Address: *asco@asco.org*
Web Address: www.asco.org
The American Society of Clinical Oncology (ASCO)
is a non-profit organization, founded in 1964, with
overarching goals of improving cancer care and
prevention and ensuring that all patients with cancer
receive care of the highest quality. Nearly 25,000
oncology practitioners belong to ASCO, representing
all oncology disciplines.

**Health and Science Communications Association
(H&SCA)**
39 Wedgewood Dr., Ste. A
Jewett City, CT 06351 US
Phone: 860-376-5915
E-mail Address: *hesca@hesca.org*
Web Address: www.hesca.org
The Health and Science Communications Association
(H&SCA) is an association of communications
professionals committed to sharing knowledge and
resources in the health sciences arena.

Health Industry Distributors Association (HIDA)
310 Montgomery St.
Alexandria, VA 22314-1516 US
Phone: 703-549-4432
Fax: 703-549-6495
Web Address: www.hida.org
The Health Industry Distributors Association (HIDA)
is the international trade association representing
medical products distributors.

Medical Device Manufacturers Association (MDMA)
1350 I St., NW, Ste. 540
Washington, DC 20005 US
Phone: 202-349-7171
Web Address: www.medicaldevices.org
The Medical Device Manufacturers Association (MDMA) is a national trade association that represents independent manufacturers of medical devices, diagnostic products and health care information systems.

Michigan Medical Device Association (MMDA)
P.O. Box 170
Howell, MI 48844 US
Fax: 517-546-3356
Toll Free: 800-930-5698
E-mail Address: *info@mmda.org*
Web Address: www.mmda.org
The Michigan Medical Device Association (MMDA) sponsors educational seminars and informational programs; is active in the areas of government relations, networking and business development; and acts as a source for the dissemination of matters of interest to its members.

XX. Health Care Resources

Access Excellence
1411 K St. NW, Ste. 1300
Washington, DC 20005 US
Phone: 650-712-1723
Web Address: www.accessexcellence.org
Access Excellence provides information for high school biology and life science teachers. It is produced by the National Health Museum.

XXI. Health Facts-Global

OECD Health Statistics
2, rue André Pascal
Paris, Cedex 16 France
Phone: 33-145-24-8200
Fax: 33-145-24-8500
E-mail Address: *health.contact@oecd.org*
Web Address: www.oecd.org
Health statistics on a country-by-country basis are offered by the OECD (Organisation for Economic Co-operation and Development). First, go to their main web site, then to the Statistics page, then to Health. Data ranges from health expenditures per capita to health expenditures as percent of GDP for the 30 nations with the world's largest economies.

XXII. Human Resources Industry Associations

Society of Human Resource Management (SHRM)
1800 Duke St.
Alexandria, VA 22314 US
Phone: 703-548-3440
Fax: 703-535-6490
Toll Free: 800-283-7476
Web Address: www.shrm.org
The Society of Human Resource Management (SHRM) addresses the interests and needs of HR professionals through its resource materials.

XXIII. Immunization

CDC National Immunization Information Hotline (NIIH)
Toll Free: 800-232-2522
Web Address: www.vaccines.ashastd.org
The CDC National Immunization Information Hotline (NIIH) offers up-to-date immunization information, including vaccine schedules, side effects, contraindications, recommendations and more.

XXIV. Industry Research/Market Research

Forrester Research
400 Technology Sq.
Cambridge, MA 02139 US
Phone: 617-613-6000
Fax: 617-613-5200
Web Address: www.forrester.com
Forrester Research identifies and analyzes emerging trends in technology and their impact on business. Among the firm's specialties are the financial services, retail, health care, entertainment, automotive and information technology industries.

Marketresearch.com
11200 Rockville Pike, Ste. 504
Rockville, MD 20852 US
Phone: 240-747-3000
Fax: 240-747-3004
Toll Free: 800-298-5699
E-mail Address:
customerservice@marketresearch.com
Web Address: www.marketresearch.com

Marketresearch.com is a leading broker for professional market research and industry analysis. Users are able to search the company's database of research publications including data on global industries, companies, products and trends.

Plunkett Research, Ltd.
P.O. Drawer 541737
Houston, TX 77254-1737 US
Phone: 713-932-0000
Fax: 713-932-7080
E-mail Address: *info@plunkettresearch.com*
Web Address: www.plunkettresearch.com
Plunkett Research, Ltd. is a leading provider of market research, industry trends analysis and business statistics. Since 1985, it has served clients worldwide, including corporations, universities, libraries, consultants and government agencies. At the firm's web site, visitors can view product information and pricing and access a great deal of basic market information on industries such as financial services, InfoTech, e-commerce, health care and biotech.

Reuters Investor
Web Address: www.investor.reuters.com
Reuters Investor is an excellent source for industry and company reports written by professional stock and business analysts. It also offers news and advice on stocks, funds and personal finance, and allows users to screen a database of major corporations and view pertinent financial and business data on selected firms.

XXV. Libraries

Library and Info Systems
Web Address: www.cellbio.wustl.edu/library.htm
Library and Info Systems provides information on libraries at various higher institutions of learning in the United States.

Library of the National Medical Society
Web Address: www.medical-library.org
The Library of the National Medical Society provides a free resource of medical information for both health care consumers and medical professionals.

Weill Cornell Medical Library
Weill Medical College, Cornell University, 1300 York Ave.
New York, NY 10021-4896 US
Phone: 212-746-6055

E-mail Address: *infodesk@med.cornell.edu*
Web Address: library.med.cornell.edu
The Weill Cornell Medical Library houses information on the biomedical sciences, as well as performing data retrieval, management and evaluation.

XXVI. MBA Resources

MBA Depot
1781 Spyglass Ln., Ste. 198
Austin, TX 78746 US
Phone: 512-499-8728
Fax: 847-556-0608
Toll Free: 888-858-8806
E-mail Address: *contact@mbadepot.com*
Web Address: www.mbadepot.com
MBA Depot is an online community for MBA professionals.

XXVII. Nanotechnology Associations

Nano Science and Technology Institute (NSTI)
1 Kendall Sq., PMB 308
Cambridge, MA 02139 US
Phone: 508-357-2925
Fax: 508-251-1665
E-mail Address: *mlaudon@nsti.org*
Web Address: www.nsti.org
The Nano Science and Technology Institute (NSTI) is engaged in the promotion and integration of nano and other advanced technologies through education, technology and business development. NSTI offers consulting services, continuing education programs, scientific and business publishing and community outreach.

NCI Alliance for Nanotechnology in Cancer
ATTN: NCI Alliance for Nanotechnology in Cancer
31 Center Dr., Bldg. 31, Rm 10A49
Bethesda, MD 20892-2580 US
E-mail Address: *cancer.nano@mail.nih.gov*
Web Address: nano.cancer.gov
The NCI Alliance for Nanotechnology in Cancer's major goal is to catalyze targeted discovery and development efforts that offer the greatest opportunity for advances in the near and medium terms and to lower the barriers for those advances to be handed off to the private sector for commercial development. The Alliance focuses on translational research and development work in six major

challenge areas, where nanotechnology can have the biggest and fastest impact on cancer treatment.

XXVIII.   Online Health Data

Medscape
76 Ninth Ave., Ste. 719
New York, NY 10011 US
Phone: 212-624-3700
Toll Free: 888-506-6098
E-mail Address: *editor2@medscape.net*
Web Address: www.medscape.com
Medscape, an online resource for better patient care, provides links to journal articles, health care-related sites and health care information.

PubMed
Web Address: www.ncbi.nlm.nih.gov/entrez/query
PubMed provides access to over 17 million MEDLINE citations dating back to the mid-1960s and additional life science journals. PubMed includes links to open access full text articles.

XXIX.   Patent Organizations-Global

World Intellectual Property Organization (WIPO)
P. O. Box 18
Geneva 20, CH-1211 Switzerland
Phone: 41-22-338-91-11
Fax: 41-22-733-54-28
Web Address: www.wipo.int
The WIPO has a United Nations mandate to assist organizations and companies in filing patents and other intellectual property data on a global basis. At its website, you can download free copies of its WIPO magazine, and you can search its international patent applications.

XXX.   Patent Resources

Patent Docs
E-mail Address: *patentdocs@gmail.com*
Web Address:
http://patentdocs.typepad.com/patent_docs/
An excellent blog about patent law and patent news in the fields of biotechnology and pharmaceuticals.

Patent Law for Non-Lawyers
Web Address: www.dnapatent.com/law/index.html
Patent Law for Non-Lawyers is an informative site detailing the patent process in the fields of

biotechnology and engineering. The site assumes a working knowledge of the industry.

XXXI.   Pharmaceutical Industry Associations

Academy of Pharmaceutical Physicians and Investigators (APPI)
500 Montgomery St., Ste. 800
Alexandria, VA 22314 US
Fax: 703-254-8101
Toll Free: 866-225-2779
E-mail Address: *andrea@acrpnet.org*
Web Address: www.aapp.org
The Academy of Pharmaceutical Physicians and Investigators (APPI) is an association that arose when the American Academy of Pharmaceutical Physicians and the Association of Clinical Research Professionals merged. It is a membership organization that provides scientific and educational activities on issues concerning pharmaceutical medicine.

Accreditation Council for Pharmacy Education (ACPE)
20 N. Clark St., Ste. 2500
Chicago, IL 60602-5109 US
Phone: 312-664-3575
Fax: 312-664-4652
E-mail Address: *info@acpe-accredit.org*
Web Address: www.acpe-accredit.org
The Accreditation Council for Pharmacy Education (ACPE) provides accreditation for pharmaceutical programs.

American Pharmaceutical Association (APhA)
1100 15th St. NW, Ste. 400
Washington, DC 20005-1707 US
Phone: 202-628-4410
Fax: 202-783-2351
Toll Free: 800-237-2742
Web Address: www.aphanet.org
American Pharmaceutical Association (APhA) is a national professional society that provides news and information to pharmacists.

Association of the British Pharmaceutical Industry (ABPI)
12 Whitehall
London, SW1A 2DY UK
Phone: 44-870-890-4333
Fax: 44-20-7747-1414
Web Address: www.abpi.org.uk

The Association of the British Pharmaceutical Industry (ABPI) is a trade association that provides research and information for the British pharmaceuticals industry.

Canadian Pharmacists Association (CPHA)
1785 Alta Vista Dr.
Ottawa, ON K1G 3Y6 Canada
Phone: 613-523-7877
Fax: 613-523-0445
Toll Free: 800-917-9489
E-mail Address: *info@pharmacists.ca*
Web Address: www.pharmacists.ca
The Canadian Pharmacists Association (CPHA) is a professional organization providing drug information, pharmacy practice support material, patient information and news about the pharmacy industry.

Canadian Research-Based Pharmaceutical Companies Association (CRBPCA)
55 Metcalfe St., Ste. 1220
Ottawa, ON K1P 6L5 Canada
Phone: 613-236-0455
Fax: 613-236-6756
E-mail Address: *info@canadapharma.org*
Web Address: www.canadapharma.org
The Canadian Research-Based Pharmaceutical Companies Association (CRBPCA) is a trade organization providing news and information to the Canadian biotech industry.

Institute of Clinical Research in the Pharmaceutical Industry (ICRPI)
Thames House, Mere Park, Dedmere Rd.
Marlow, Buckinghmashire SL7 1PB UK
Phone: 44-0-1628-899755
Fax: 44-0-1628-899766
E-mail Address: *info@instituteofclinicalresearch.org*
Web Address: www.instituteofclinicalresearch.org
The Institute of Clinical Research in the Pharmaceutical Industry (ICRPI) is a professional organization for clinical researchers in the pharmaceutical industry in the U.K.

International Federation of Pharmaceutical Manufacturers Associations (IFPMA)
15 Ch. Louis-Dunant
P.O. Box 195, 1211
Geneva, 20 Switzerland
Phone: 41-22-338-32-00
Fax: 41-22-338-32-99
E-mail Address: *admin@ifpma.org*
Web Address: www.ifpma.org

The International Federation of Pharmaceutical Manufacturers Associations (IFPMA) is a nonprofit organization that represents the world's research-based pharmaceutical and biotech companies.

International Pharmaceutical Federation (FIP)
Andries Bickerweg 5
P.O. Box 84200
The Hague, 2508 The Netherlands
Phone: 31-70-3021970
Fax: 31-70-3021999
E-mail Address: *fip@fip.org*
Web Address: www.fip.org
The International Pharmaceutical Federation (FIP) is a worldwide organization of pharmaceutical professional and scientific associations.

Korean Research-based Pharmaceutical Industry Association (KRPIA)
5th Fl., Dae Han Bldg.
201-6 Guui-Dong, Kwangjin-Gu
Seoul, 143-200 Korea
Phone: 82-2-456-8553
Fax: 82-2-456-8320
Web Address: www.krpia.or.kr
The KRPIA is an association of 24 research-based pharmaceutical companies operating in Korea.

Pharmaceutical Information and Pharmacovigilance Association (PIPA)
P.O. Box 254
Haslemere, Surrey GU27 9AF UK
Phone: 07726-221239
E-mail Address: *pipa@pipaonline.org*
Web Address: www.aiopi.org.uk
The Pharmaceutical Information and Pharmacovigilance Association (PIPA) is a professional organization that promotes the advancement of information in the pharmaceutical industry in the U.K.

Pharmaceutical Research and Manufacturers of America (PhRMA)
950 F St. NW, Ste. 300
Washington, DC 20004 US
Phone: 202-835-3400
Fax: 202-835-3414
Web Address: www.phrma.org
Pharmaceutical Research and Manufacturers of America (PhRMA) represents the nation's leading research-based pharmaceutical and biotechnology companies.

Royal Pharmaceutical Society of Great Britain (RPSGB)
1 Lambeth High St.
London, SE1 7JN UK
Phone: 44-20-7735-9141
Fax: 44-20-7572-2499
E-mail Address: *enquiries@rpsgb.org*
Web Address: www.rpsgb.org.uk
The Royal Pharmaceutical Society of Great Britain (RPSGB) is the regulatory agency for pharmacists in England, Wales and Scotland.

Pharmaportal.com
Web Address: www.pharmaportal.com
Pharmaportal.com is a pharmaceutical portal containing information about the Pharmaceutical Magazine, as well as links for executives in the industry to meet each other.

XXXII.   Pharmaceutical Resources

PharmaSUG
421 New Parkside Dr.
Chapel Hill, NC 27516 US
E-mail Address: *margaretlhung@yahoo.com*
Web Address: www.pharmasug.org
PharmaSUG promotes biotech information technology using SAS software.

XXXIII.   Research & Development, Laboratories

Battelle Memorial Institute
505 King Ave.
Columbus, OH 43201 US
Phone: 614-424-6424
Toll Free: 800-201-2011
Web Address: www.battelle.org
Battelle Memorial Institute serve commercial and governmental customers in developing new technologies and products. The institute adds technology to systems and processes for manufacturers; pharmaceutical and agrochemical industries; trade associations; and government agencies supporting energy, the environment, health, national security and transportation.

Commonwealth Scientific and Industrial Research Organization (CSRIO)
Bag 10
Clayton South, VIC 3169 Australia
Phone: 61-3-9545-2176
Fax: 61-3-9545-2175
E-mail Address: *enquiries@csiro.au*
Web Address: www.csiro.au
The Commonwealth Scientific and Industrial Research Organization (CSRIO) is Australia's national science agency and a leading international research agency. CSRIO performs research in Australia over a broad range of areas including agriculture, minerals and energy, manufacturing, communications, construction, health and the environment.

Computational Neurobiology Laboratory
The Salk Institute
10010 N. Torrey Pines Rd.
La Jolla, CA 92037 US
Phone: 858-453-4100
Web Address: www.cnl.salk.edu
The Computational Neurobiology Laboratory strives to understand the computational resources of the brain from the biophysical to the systems levels.

Council of Scientific & Industrial Research (CSIR)
Anusandhan Bhawan
2 Rafi Marg
New Delhi, 110001 India
Phone: 011-23710618
Fax: 011-23713011
Web Address: www.csir.res.in
The Council of Scientific & Industrial Research (CSIR) is a government-funded organization that promotes research and development initiatives in India. It operates in the fields of energy, biotechnology, space, science and technology.

SRI International
333 Ravenswood Ave.
Menlo Park, CA 94025-3493 US
Phone: 650-859-2000
E-mail Address: *webmaster@sri.com*
Web Address: www.sri.com
SRI International is a nonprofit organization offering a wide range of services, including engineering services, information technology, pure and applied physical sciences, product development, pharmaceutical discovery, biopharmaceutical discovery and policy issues. SRI conducts research for commercial and governmental customers.

XXXIV. Robotics Associations

Laboratory Robotics Interest Group (LRIG)
1730 W. Circle Dr.
Martinsville, NJ 08836-2147 US
Phone: 732-302-1038
Fax: 732-875-0270
E-mail Address: *andy.zaayenga@lab-robotics.org*
Web Address: www.lab-robotics.org
Laboratory Robotics Interest Group (LRIG) is a membership group focused on the application of robotics in the laboratory.

XXXV. Synthetic Biology

BioBricks Foundation (BBF)
1 Kendall Square
PMB 126
Cambridge, MA 02139 US
E-mail Address: *infobbf@gmail.com*
Web Address: www.biobricks.org
The BioBricks Foundation (BBF) is a not-for-profit organization founded by engineers and scientists from MIT, Harvard, and UCSF with significant experience in both non-profit and commercial biotechnology research. BBF encourages the development and responsible use of technologies based on BioBrick standard DNA parts that encode basic biological functions.

Syntheticbiology.org
E-mail Address: *synbioadmin@openwetware.org*
Web Address: www.syntheticbiology.org
An open source web site dedicated to disseminating news about synthetic biology. It was originally founded by a group of students, faculty and staff from MIT and Harvard.

XXXVI. Technology Transfer Associations

Association of University Technology Managers (AUTM)
60 Revere Dr., Ste. 500
Northbrook, IL 60062 US
Phone: 847-559-0846
Fax: 847-480-9282
E-mail Address: *info@autm.net*
Web Address: www.autm.net
The Association of University Technology Managers (AUTM) is a nonprofit professional association with membership of more than 3,000 intellectual property managers and business executives from 45 countries.

The association's mission is to advance the field of technology transfer, and enhance our ability to bring academic and nonprofit research to people around the world.

XXXVII. Trade Associations-Global

World Trade Organization (WTO)
Centre William Rappard
Rue de Lausanne 154
Geneva 21, CH-1211 Switzerland
Phone: 41-22-739-51-11
Fax: 41-22-731-42-06
E-mail Address: *enquiries@wto.og*
Web Address: www.wto.org
The World Trade Organization (WTO) is a global organization dealing with the rules of trade between nations. To become a member, nations must agree to abide by certain guidelines. Membership increases a nation's ability to import and export efficiently.

XXXVIII. U.S. Government Agencies

Bureau of Economic Analysis (BEA)
1441 L St. NW
Washington, DC 20230 US
Phone: 202-606-9900
E-mail Address: *customerservice@bea.gov*
Web Address: www.bea.gov/index.htm
The Bureau of Economic Analysis (BEA), an agency of the U.S. Department of Commerce, is the nation's economic accountant, preparing estimates that illuminate key national, international and regional aspects of the U.S. economy.

Bureau of Labor Statistics (BLS)
2 Massachusetts Ave. NE
Washington, DC 20212-0001 US
Phone: 202-691-5200
Fax: 202-691-6325
Web Address: stats.bls.gov
The Bureau of Labor Statistics (BLS) is the principal fact-finding agency for the Federal Government in the field of labor economics and statistics. It is an independent national statistical agency that collects, processes, analyzes and disseminates statistical data to the American public, U.S. Congress, other federal agencies, state and local governments, business and labor. The BLS also serves as a statistical resource to the Department of Labor.

Center for Biologics Evaluation and Research (CBER)

1401 Rockville Pike, Ste. 200N
Rockville, MD 20852-1448 US
Phone: 301-827-1800
Toll Free: 800-835-4709
Web Address: www.fda.gov/Cber
The Center for Biologics Evaluation and Research (CBER) regulates biologic products for use in humans. It is a source for a broad variety of data on drugs, including blood products, counterfeit drugs, exports, drug shortages, recalls and drug safety.

Center for Devices and Radiological Health (CDRH)

5600 Fishers Ln.
Rockville, MD 20857 US
Phone: 301-276-3103
Fax: 301-443-8818
Toll Free: 800-638-2041
E-mail Address: *dsmica@cdrh.fda.gov*
Web Address: www.fda.gov/cdrh
The Center for Devices and Radiological Health (CDRH) is a unit of the FDA that regulates medical devices and radiation-emitting products.

Center for Drug Evaluation and Research (CDER)

5600 Fishers Ln., HFD-240
Rockville, MD 20857 US
Phone: 301-827-4570
Toll Free: 888-463-6332
E-mail Address: *druginfo@cder.fea.gov*
Web Address: www.fda.gov/cder
The Center for Drug Evaluation and Research (CDER) is a division of the FDA that offers a wealth of information on new drug approval statistics and the approval process.

Center for Food Safety and Applied Nutrition-Biotechnology (CFSAN)

5100 Paint Branch Pkwy. HFS-555
College Park, MD 20740-3835 US
Toll Free: 888-723-3366
Web Address: www.cfsan.fda.gov/list.html
The Center for Food Safety and Applied Nutrition-Biotechnology (CFSAN) is an FDA site that provides information about genetically engineered food products.

Centers for Disease Control and Prevention (CDC)

1600 Clifton Rd.
Atlanta, GA 30333 US
Phone: 404-639-3311
Toll Free: 800-311-3435
E-mail Address: *cdcinfo@cdc.gov*
Web Address: www.cdc.gov
The CDC, headquartered in Atlanta and established as an operating health agency within the U.S. Public Health Service on July 1, 1973, is the federal agency charged with protecting the public health of the nation by providing leadership and direction in the prevention and control of diseases and other preventable conditions and responding to public health emergencies.

FedWorld

5285 Port Royal Rd.
Springfield, VA 22161 US
Phone: 703-605-6000
E-mail Address: *helpdesk@fedworld.gov*
Web Address:
www.fedworld.gov/jobs/jobsearch.html
FedWorld, a program of the U.S. Department of Commerce, provides an annotated index of links to job-, labor- and management-related U.S. government web sites. Employment opportunities, labor statistics and links to other government information sites are also offered. The site is managed by the National Technical Information Service (NTIS).

Government Printing Office (GPO)

732 N. Capitol St. NW
Washington, DC 20401 US
Phone: 202-512-0000
Fax: 202-512-2104
E-mail Address: *contactcenter@gpo.gov*
Web Address: www.gpo.gov
The U.S. Government Printing Office (GPO) is the primary information source concerning the activities of Federal agencies. GPO gathers, catalogues, produces, provides, authenticates and preserves published information.

National Cancer Institute (NCI)

6116 Executive Blvd., Ste. 3036A
Bethesda, MD 20892-8322 US
Toll Free: 800-422-6237
Web Address: www.cancer.gov
The National Cancer Institute (NCI) is the Federal Government's principal agency for cancer research and training.

National Center for Biotechnology Information (NCBI)
8600 Rockville Pike, Bldg. 38A
Bethesda, MD 20894 US
Phone: 301-496-2475
Fax: 301-480-9241
E-mail Address: *info@ncbi.nlm.nih.gov*
Web Address: www.ncbi.nlm.nih.gov
The National Center for Biotechnology Information (NCBI) creates public databases, conducts research in computational biology, develops software for analyzing genome data and disseminates biomedical information.

National Center for Infectious Diseases (NCID)
1600 Clifton Rd., MSC 14
Atlanta, GA 30333 US
Phone: 404-639-3311
Toll Free: 800-232-4696
E-mail Address:
Web Address: www.cdc.gov/ncidod
The National Center for Infectious Diseases (NCID) tracks infections disease out breaks world wide.

National Center for Research Resources (NCRR)
1 Democracy Plaza, 9th Fl.
6701 Democracy Blvd., MSC 4874
Bethesda, MD 20892-4874 US
Phone: 301-435-0888
Fax: 301-480-3558
E-mail Address: *info@ncrr.nih.gov*
Web Address: www.ncrr.nih.gov
The National Center for Research Resources (NCRR) supports primary health and life sciences research to create and develop critical resources, models and technologies.

National Center for Toxicological Research
3900 NCTR Rd.
Jefferson, AR 72079 US
Phone: 870-543-7130
E-mail Address: *rhuber@fda.hhs.gov*
Web Address: www.fda.gov/nctr
The mission of the National Center for Toxicological Research is to conduct peer-reviewed scientific research that supports and anticipates the FDA's current and future regulatory needs.

National Heart, Lung, and Blood Institute (NHLBI)
P.O. Box 30105
Bethesda, MD 20824-0105 US
Phone: 301-592-8573

Fax: 240-629-3246
E-mail Address: *nhlbiinfo@nhlbi.nih.gov*
Web Address: www.nhlbi.nih.gov
The National Heart, Lung, and Blood Institute (NHLBI) provides leadership for a national program in diseases of the heart, blood vessels, lung and blood; blood resources; and sleep disorders.

National Institute of Standards and Technology (NIST)
100 Bureau Dr., Stop 1070
Gaithersburg, MD 20899-1070 US
Phone: 301-975-6478
E-mail Address: *inquiries@nist.gov*
Web Address: www.nist.gov
The National Institute of Standards and Technology (NIST) is an agency of the U.S. Department of Commerce's Technology Administration. It works with various industries to develop and apply technology, measurements and standards.

National Institutes of Health (NIH)
9000 Rockville Pike
Bethesda, MD 20892 US
Phone: 301-496-4000
E-mail Address: *nihinfo@od.nih.gov*
Web Address: www.nih.gov
The National Institutes of Health (NIH) is the leader of medical and behavioral research for the nation, with over 15 institutes ranging from the National Cancer Institute to the National Institute of Mental Health.

National Science Foundation (NSF)
4201 Wilson Blvd.
Arlington, VA 22230 US
Phone: 703-292-5111
Toll Free: 800-877-8339
E-mail Address: *info@nsf.gov*
Web Address: www.nsf.gov
The National Science Foundation (NSF) is an independent U.S. government agency responsible for promoting science and engineering. The foundation provides grants and funding for research.

U.S. Business Advisor
Web Address: www.business.gov
U.S. Business Advisor offers a searchable directory of business-specific government information. Topics include taxes, regulations, international trade, financial assistance and business development. U.S. Business Advisor was created by the Small Business

Administration in a partnership with 21 other federal agencies.

U.S. Census Bureau
4700 Silver Hill Rd.
Washington, DC 20233 US
E-mail Address: *pio@census.gov*
Web Address: www.census.gov
The U.S. Census Bureau is the official collector of data about the people and economy of the U.S. It provides official social, demographic and economic information.

U.S. Department of Commerce (DOC)
1401 Constitution Ave. NW
Washington, DC 20230 US
Phone: 202-482-2000
E-mail Address: *cgutierrez@doc.gov*
Web Address: www.doc.gov
The U.S. Department of Commerce (DOC) regulates trade and provides valuable economic analysis of the economy.

U.S. Department of Labor (DOL)
Frances Perkins Building
200 Constitution Ave. NW
Washington, DC 20210 US
Toll Free: 866-487-2365
E-mail Address:
Web Address: www.dol.gov
The U.S. Department of Labor (DOL) is the government agency responsible for labor regulations. This site provides tools to help citizens find out whether companies are complying with family and medical-leave requirements.

U.S. Food and Drug Administration (FDA)
5600 Fishers Ln.
Rockville, MD 20857 US
Toll Free: 888-463-6332
Web Address: www.fda.gov
The U.S. Food and Drug Administration (FDA) promotes and protects the public health by helping safe and effective products reach the market in a timely way and by monitoring products for continued safety after they are in use. It regulates both prescription and over-the-counter drugs as well as medical devices and food products.

U.S. Patent and Trademark Office (PTO)
Public Search Facility
Dulany St., 1st Fl.
Alexandria, VA 22314 US

Phone: 571-272-1000
Fax: 571-273-8300
Toll Free: 800-786-9199
Web Address: www.uspto.gov
The U.S. Patent and Trademark Office (PTO) administers patent and trademark laws for the U.S. and enables registration of patents and trademarks.

U.S. Securities and Exchange Commission (SEC)
Office of Investor Education and Assistance
100 F St. NE
Washington, DC 20549 US
Phone: 202-551-6551
Toll Free: 800-732-0330
E-mail Address: *help@sec.gov*
Web Address: www.sec.gov
The U.S. Securities and Exchange Commission (SEC) is a nonpartisan, quasi-judicial regulatory agency responsible for administering federal securities laws. These laws are designed to protect investors in securities markets and ensure that they have access to disclosure of all material information concerning publicly traded securities. Visitors to the web site can access the EDGAR database of corporate financial and business information.

U.S. Technology Administration
U.S. Department of Commerce
1401 Constitution Ave. NW
Washington, DC 20230 US
Phone: 202-482-1575
Fax: 202-482-5687
E-mail Address: *Public_affairs@technology.gov*
Web Address: www.technology.gov
The U.S. Technology Administration seeks to maximize technology's contribution to economic growth, high-wage job creation and the social well-being of the United States. Its web site offers publications as well as information about events and services. Departments of this agency include the Office of Technology Policy, the National Institute of Standards & Technology and the National Technical Information Service.

White House (The)
1600 Pennsylvania Ave. NW
Washington, DC 20500 US
Phone: 202-456-1414
Fax: 202-456-2461
Web Address: www.whitehouse.gov
The White House site was designed as a communication channel between the Federal Government and the American people. It provides

access to all government information and services
that are available on the Internet.

Chapter 4

THE BIOTECH 400:
WHO THEY ARE AND HOW THEY WERE
CHOSEN

**Includes Indexes by Company Name, Industry & Location,
And a Complete Table of Sales, Profits and Ranks**

The companies chosen to be listed in PLUNKETT'S BIOTECH & GENETICS INDUSTRY ALMANAC comprise a unique list. THE BIOTECH 400 (the actual count is 412 companies) were chosen specifically for their dominance in the many facets of biotechnology and genetics in which they operate. Complete information about each firm can be found in the "Individual Profiles," beginning at the end of this chapter. These profiles are in alphabetical order by company name.

THE BIOTECH 400 includes leading companies from all parts of the United States as well as many other nations, and from all biotech and genetics related industry segments: pharmaceuticals; diagnostics; research and development; and support services.

Simply stated, the list contains 412 of the largest, most successful, fastest growing firms in biotech and related industries in the world. To be included in our list, the firms had to meet the following criteria:

1) Generally, these are corporations based in the U.S., however, the headquarters of 75 firms are located in other nations.
2) Prominence, or a significant presence, in biotech, genetics and supporting fields. (See the following Industry Codes section for a complete list of types of businesses that are covered).
3) The companies in THE BIOTECH 400 do not have to be exclusively in the biotech and genetics field.
4) Financial data and vital statistics must have been available to the editors of this book, either directly from the company being written about or from outside sources deemed reliable and accurate by the editors. A small number of companies that we would like to have included are not listed because of a lack of sufficient, objective data.

INDUSTRY LIST, WITH CODES

This book refers to the following list of unique industry codes, based on the 1997 NAIC code system (NAIC is used by many analysts as a replacement for older SIC codes because NAIC is more specific to today's industry sectors). Companies profiled in this book are given a primary NAIC code, reflecting the main line of business of each firm.

Financial Services

Stocks & Investments
523110   Investment Banking

Health Care

Health Products, Manufacturing
325411   Medicinals & Botanicals, Manufacturing
325412   Drugs (Pharmaceuticals), Discovery & Manufacturing
325412A   Drug Delivery Systems
325412B   Veterinary Products Manufacturing
325413   Diagnostic Services and Substances Manufacturing
325414   Biological Products, Manufacturing
325416   Drugs (Pharmaceuticals), Generic Manufacturing
339113   Medical/Dental/Surgical Equipment & Supplies, Manufacturing
Health Products, Wholesale Distribution
421450   Medical/Dental/Surgical Equipment & Supplies, Distribution
Health Care-Clinics, Labs and Organizations
621511   Laboratories & Diagnostic Services--Medical

InfoTech

Computers & Electronics Manufacturing
334500   Instrument Manufacturing, including Measurement, Control, Test & Navigational
Software
511212   Computer Software, Healthcare & Biotechnology

Manufacturing

Textiles Manufacturing
313000   Textiles, Fabrics, Sheets/Towels, Manufacturing
Chemicals
325000   Chemicals, Manufacturing

Nanotechnology

Nanotechnology
541710   Research and Development/Physical,
Engineering and Life Sciences
541710G   Nanotechnology-Optics

Services

Agriculture
115112   Agricultural Crop Production Support, Seeds,
Fertilizers
Consulting & Professional Services
541910   Market Research
Management
551110   Management of Companies & Enterprises

INDEX OF RANKINGS WITHIN INDUSTRY GROUPS

Company	Industry Code	2006 Sales (U.S. $ thousands)	Sales Rank	2006 Profits (U.S. $ thousands)	Profits Rank
Agricultural Crop Production Support, Seeds, Fertilizers					
ARCADIA BIOSCIENCES	115112				
DOW AGROSCIENCES LLC	115112				
DUPONT AGRICULTURE & NUTRITION	115112				
EXELIXIS PLANT SCIENCES INC	115112				
MONSANTO CO	115112	7,344,000	2	689,000	1
SYNGENTA AG	115112	8,046,000	1	634,000	2
Biological Products, Manufacturing					
ALPHARMA INC	325414	653,828	4	82,544	4
ANIKA THERAPEUTICS INC	325414	26,841	7	4,604	6
CSL LIMITED	325414	2,146,111	2	88,406	3
EISAI CO LTD	325414	5,113,100	1	539,200	1
GENENCOR INTERNATIONAL	325414				
GTC BIOTHERAPEUTICS INC	325414	6,128	8	-33,345	11
LIFECELL CORPORATION	325414	141,680	5	20,469	5
NORTHFIELD LABORATORIES	325414	0		-26,775	10
NOVOZYMES	325414	1,251,490	3	167,610	2
ORGANOGENESIS INC	325414				
POLYDEX PHARMACEUTICALS	325414	5,265	9	-1,489	7
SERACARE LIFE SCIENCES INC	325414				
STEMCELLS INC	325414	93	11	-18,948	8
VI TECHNOLOGIES INC	325414	300	10	-38,100	12
VIACELL INC	325414	54,426	6	-21,330	9
Chemicals, Manufacturing					
AKZO NOBEL NV	325000	19,659,800	4	1,566,190	4
BASF AG	325000	69,448,400	1	4,575,330	1
BAYER AG	325000	38,710,400	2	2,249,950	3
BAYER CORP	325000				
E I DU PONT DE NEMOURS & CO (DUPONT)	325000	27,421,000	3	3,148,000	2
INTERNATIONAL ISOTOPES	325000	4,470	7	-1,037	7
LONZA GROUP	325000	2,371,810	5	180,690	6
SIGMA ALDRICH CORP	325000	1,797,500	6	276,800	5
Computer Software, Healthcare & Biotechnology					
ACCELRYS INC	511212	82,001	3	-7,739	2
DECODE GENETICS INC	511212	40,510	4	-85,473	5
DENDRITE INTERNATIONAL INC	511212	423,958	1	-26,745	3
ELSEVIER MDL	511212				
ERESEARCH TECHNOLOGY INC	511212	86,368	2	8,310	1
TRIPOS INC	511212	27,384	5	-38,593	4
Diagnostic Services and Substances Manufacturing					
AFFYMETRIX INC	325413	355,317	4	-13,704	18
AVIVA BIOSCIENCES CORP	325413				
BIOSITE INC	325413	308,592	6	39,994	5
CALIPER LIFE SCIENCES	325413	107,871	12	-28,934	24

Company	Industry Code	2006 Sales (U.S. $ thousands)	Sales Rank	2006 Profits (U.S. $ thousands)	Profits Rank
CELSIS INTERNATIONAL PLC	325413	33,104	18	4,603	11
CEPHEID	325413	87,352	14	-25,985	22
CIPHERGEN BIOSYSTEMS INC	325413	18,215	22	-22,066	21
DIAGNOSTIC PRODUCTS CORP	325413				
DIGENE CORPORATION	325413	152,900	8	8,400	8
EPIX PHARMACEUTICALS INC	325413	6,041	27	-157,393	26
E-Z-EM INC	325413	138,369	9	9,766	7
GEN-PROBE INC	325413	354,764	5	59,498	4
HEMAGEN DIAGNOSTICS INC	325413	7,250	24	313	14
HYCOR BIOMEDICAL INC	325413				
IDEXX LABORATORIES INC	325413	739,117	2	93,678	1
IMMUNOMEDICS INC	325413	4,353	28	-28,764	23
INVITROGEN CORPORATION	325413	1,263,485	1	-191,049	27
LUMINEX CORPORATION	325413	52,989	17	1,507	13
MALLINCKRODT INC	325413				
MATRITECH INC	325413	12,195	23	-11,935	17
MERIDIAN BIOSCIENCE INC	325413	108,413	11	18,325	6
NANOGEN INC	325413	26,852	21	-49,070	25
NEOGEN CORPORATION	325413	72,433	15	7,941	9
NEOPROBE CORPORATION	325413	6,051	26	-4,741	16
ORASURE TECHNOLOGIES INC	325413	68,155	16	5,268	10
PERBIO SCIENCE AB	325413				
PROMEGA CORP	325413				
QIAGEN NV	325413	465,778	3	70,539	3
SEQUENOM INC	325413	28,496	19	-17,577	19
SPECTRAL DIAGNOSTICS INC	325413	6,200	25		
STRATAGENE CORP	325413	95,557	13	59	15
TECHNE CORP	325413	202,617	7	73,351	2
THIRD WAVE TECHNOLOGIES	325413	28,027	20	-18,887	20
TRINITY BIOTECH PLC	325413	118,674	10	3,276	12
VYSIS INC	325413				
Drug Delivery Systems					
ALKERMES INC	325412A	166,601	4	3,818	4
ALZA CORP	325412A				
ARADIGM CORPORATION	325412A	4,814	14	-13,027	10
BENTLEY PHARMACEUTICALS	325412A	109,471	5	974	6
BIOVAIL CORPORATION	325412A	1,070,500	1	203,900	1
DEPOMED INC	325412A	9,551	12	-39,659	19
DOR BIOPHARMA INC	325412A	2,313	17	-8,163	9
DURECT CORP	325412A	21,894	10	-33,327	17
EMISPHERE TECHNOLOGIES	325412A	7,259	13	-41,766	20
FLAMEL TECHNOLOGIES SA	325412A	23,020	8	-35,201	18
GENEREX BIOTECHNOLOGY	325412A	200	19	-68,000	21
IMPAX LABORATORIES INC	325412A				
INSITE VISION INC	325412A	2	21	-16,611	11
IOMED INC	325412A	10,843	11	600	7
KV PHARMACEUTICAL CO	325412A	367,618	2	15,787	3
MACROCHEM CORPORATION	325412A	0		1,187	5

Company	Industry Code	2006 Sales (U.S. $ thousands)	Sales Rank	2006 Profits (U.S. $ thousands)	Profits Rank
NANOBIO CORPORATION	325412A				
NASTECH PHARMACEUTICAL	325412A	28,490	7	-26,877	14
NEKTAR THERAPEUTICS	325412A	217,718	3	-154,761	22
NEOPHARM INC	325412A	11	20	-33,208	16
NEXMED INC	325412A	1,867	18	-8,043	8
NOVAVAX INC	325412A	4,683	15	-23,068	12
NOVEN PHARMACEUTICALS	325412A	60,689	6	15,988	2
PENWEST PHARMACEUTICALS	325412A	3,499	16	-31,312	15
SKYEPHARMA PLC	325412A				
SONUS PHARMACEUTICALS	325412A	22,392	9	-23,551	13
Drugs (Pharmaceuticals), Discovery & Manufacturing					
4SC AG	325412				
ABBOTT LABORATORIES	325412	22,476,322	9	1,716,755	14
ABRAXIS BIOSCIENCE INC	325412	765,488	36	-46,897	152
ACAMBIS PLC	325412	62,530	74	-34,800	137
ACCESS PHARMACEUTICALS	325412	0		-12,874	81
ACTELION LTD	325412	776,912	35	198,066	27
ADOLOR CORP	325412	15,087	116	-69,738	170
AEOLUS PHARMACEUTICALS	325412	92	186	-5,728	65
AETERNA ZENTARIS INC	325412	41,390	83	33,390	41
AKORN INC	325412	71,250	71	-5,963	66
ALCON INC	325412	4,896,600	21	1,348,100	16
ALEXION PHARMACEUTICALS	325412	1,558	153	-131,514	193
ALFACELL CORPORATION	325412	107	184	-7,810	68
ALIZYME PLC	325412	2,296	150	-39,820	144
ALLERGAN INC	325412	3,010,100	24	-127,400	189
ALLIANCE PHARMACEUTICAL	325412	129	181	-9,575	73
ALLOS THERAPEUTICS INC	325412	0		-21,837	109
ALSERES PHARMACEUTICALS	325412	0		-26,355	117
ALTANA AG	325412	5,234,410	20	1,008,440	21
ALTEON INC	325412	251	176	-17,680	100
AMARIN CORPORATION PLC	325412	500	170	-26,920	119
AMGEN INC	325412	14,268,000	13	2,950,000	10
AMYLIN PHARMACEUTICALS	325412	510,875	40	-218,856	198
ANGIOTECH PHARMACEUTICALS	325412	315,070	48	4,580	52
ANTIGENICS INC	325412	692	166	-51,881	157
ARENA PHARMACEUTICALS	325412	30,569	96	-86,279	178
ARIAD PHARMACEUTICALS INC	325412	896	161	-61,928	166
ARQULE INC	325412	6,626	135	-31,440	133
ASTELLAS PHARMA INC	325412	7,758,480	18	1,106,390	20
ASTRAZENECA PLC	325412	26,475,000	7	4,392,000	8
ATHEROGENICS INC	325412	22,917	108	-67,322	169
AUTOIMMUNE INC	325412	401	173	-481	58
AVANIR PHARMACEUTICALS	325412	15,186	115	-62,553	168
AVANT IMMUNOTHERAPEUTICS	325412	4,931	141	-20,374	106
AVAX TECHNOLOGIES INC	325412	735	163	-5,356	64
AVI BIOPHARMA INC	325412	115	183	-31,073	130
AVIGEN INC	325412	103	185	-24,256	114

Company	Industry Code	2006 Sales (U.S. $ thousands)	Sales Rank	2006 Profits (U.S. $ thousands)	Profits Rank
AXCAN PHARMA INC	325412	292,320	50	39,120	39
BAYER SCHERING PHARMA AG	325412				
BIOCRYST PHARMACEUTICALS	325412	6,212	136	-43,618	148
BIOGEN IDEC INC	325412	2,683,049	27	217,511	26
BIOMARIN PHARMACEUTICAL	325412	84,209	69	-28,533	122
BIOMIRA INC	325412	4,199	143	-16,591	95
BIOPURE CORPORATION	325412	1,715	152	-26,454	118
BIOTECH HOLDINGS LTD	325412	480	171	-1,940	60
BIOTIME INC	325412	1,162	156	-1,864	59
BRADLEY PHARMACEUTICALS	325412	144,807	60	9,668	48
BRISTOL MYERS SQUIBB CO	325412	17,914,000	11	1,585,000	15
CAMBREX CORP	325412	452,255	42	-30,100	127
CAMBRIDGE ANTIBODY TECHNOLOGY LIMITED	325412				
CANGENE CORP	325412	96,600	65	11,600	45
CARDIOME PHARMA CORP	325412	18,000	112	-30,700	129
CARRINGTON LABORATORIES	325412	27,406	100	-7,607	67
CELGENE CORP	325412	898,873	34	68,981	33
CELL GENESYS INC	325412	1,364	154	-82,929	176
CELL THERAPEUTICS INC	325412	80	187	-135,819	194
CELLEGY PHARMACEUTICALS	325412	2,660	148	9,672	47
CEL-SCI CORPORATION	325412	125	182	-7,939	69
CEPHALON INC	325412	1,764,069	29	144,816	30
CERUS CORPORATION	325412	35,580	91	-4,779	62
CHIRON CORP	325412				
COLLAGENEX PHARMACEUTICALS INC	325412	26,373	101	-33,434	135
COLLATERAL THERAPEUTICS	325412				
COLUMBIA LABORATORIES INC	325412	17,393	113	-12,612	80
COMPUGEN LTD	325412	215	178	-13,020	82
CORTEX PHARMACEUTICALS	325412	1,177	155	-16,055	91
CUBIST PHARMACEUTICALS	325412	194,748	55	-376	57
CURAGEN CORPORATION	325412	39,587	85	-59,839	165
CURIS INC	325412	14,936	117	-8,829	72
CV THERAPEUTICS INC	325412	36,785	90	-274,320	200
CYPRESS BIOSCIENCE INC	325412	4,322	142	-8,247	70
CYTOGEN CORPORATION	325412	17,307	114	-15,103	85
CYTRX CORPORATION	325412	2,066	151	-16,752	97
DENDREON CORPORATION	325412	273	175	-91,642	180
DISCOVERY LABORATORIES	325412	0		-46,333	151
DOV PHARMACEUTICAL INC	325412	25,951	102	-38,368	142
DR REDDY'S LABORATORIES	325412	541,304	39	36,620	40
DRAXIS HEALTH INC	325412	89,000	68	11,600	46
DSM PHARMACEUTICALS INC	325412				
DUSA PHARMACEUTICALS INC	325412	25,583	104	-31,350	131
DYAX CORP	325412	12,776	123	-50,323	154
ELAN CORP PLC	325412	560,400	38	-267,300	199
ELI LILLY AND COMPANY	325412	15,691,000	12	2,662,700	12
ENCYSIVE PHARMACEUTICALS	325412	18,995	110	-109,283	187

Company	Industry Code	2006 Sales (U.S. $ thousands)	Sales Rank	2006 Profits (U.S. $ thousands)	Profits Rank
ENDO PHARMACEUTICALS HOLDINGS INC	325412	909,659	32	137,839	31
ENTREMED INC	325412	6,894	134	-49,889	153
ENZO BIOCHEM INC	325412	39,826	84	-15,667	89
ENZON PHARMACEUTICALS INC	325412	185,653	56	21,309	44
EXELIXIS INC	325412	98,670	64	-101,492	182
FOREST LABORATORIES INC	325412	2,793,934	26	708,514	22
GENAERA CORPORATION	325412	892	162	-21,234	107
GENELABS TECHNOLOGIES	325412	11,209	127	-8,685	71
GENENTECH INC	325412	9,284,000	16	2,113,000	13
GENTA INC	325412	708	165	-56,781	162
GENVEC INC	325412	18,923	111	-19,272	103
GENZYME CORP	325412	3,187,013	22	-16,797	98
GENZYME ONCOLOGY	325412				
GERON CORPORATION	325412	3,277	146	-31,365	132
GILEAD SCIENCES INC	325412	3,026,139	23	-1,189,957	202
GLAXOSMITHKLINE PLC	325412	45,595,800	3	10,793,000	3
GLYCOGENESYS INC	325412				
HEMISPHERX BIOPHARMA INC	325412	933	159	-19,399	105
HOLLIS-EDEN PHARMACEUTICALS	325412	444	172	-30,231	128
HUMAN GENOME SCIENCES	325412	25,755	103	-210,327	197
ICOS CORPORATION	325412				
IDERA PHARMACEUTICALS INC	325412	2,421	149	-16,525	94
IDM PHARMA INC	325412	11,286	126	-23,455	113
IMCLONE SYSTEMS INC	325412	677,847	37	370,674	24
IMMTECH INTERNATIONAL	325412	3,575	145	-15,526	88
IMMUNE RESPONSE CORP	325412	932	160	195,290	28
IMMUNOGEN INC	325412	32,088	95	-17,834	101
INCYTE CORP	325412	27,643	99	-74,166	172
INDEVUS PHARMACEUTICALS	325412	50,452	77	-50,554	155
INSMED INCORPORATED	325412	991	158	-56,139	160
INSPIRE PHARMACEUTICALS	325412	37,059	88	-42,115	146
INTERMUNE PHARMACEUTICALS INC	325412	90,784	66	-107,206	186
INTROGEN THERAPEUTICS INC	325412	1,151	157	-28,801	123
ISIS PHARMACEUTICALS INC	325412	24,532	107	-45,903	150
JAZZ PHARMACEUTICALS	325412	44,856	82	-59,391	164
JOHNSON & JOHNSON	325412	53,324,000	1	11,053,000	2
KENDLE INTERNATIONAL INC	325412	373,936	45	8,530	49
KERYX BIOPHARMACEUTICALS	325412	534	169	-73,764	171
KING PHARMACEUTICALS INC	325412	1,988,500	28	288,949	25
KOS PHARMACEUTICALS INC	325412				
KOSAN BIOSCIENCES INC	325412	13,506	120	-29,469	126
LA JOLLA PHARMACEUTICAL	325412	0		-39,445	143
LARGE SCALE BIOLOGY CORP	325412				
LEXICON PHARMACEUTICALS	325412	72,798	70	-54,311	159
LIGAND PHARMACEUTICALS	325412	140,960	61	-31,743	134
LORUS THERAPEUTICS INC	325412	0		-14,900	84
MANHATTAN PHARMACEUTICALS INC	325412	0		-9,695	74
MAXYGEN INC	325412	25,021	106	-16,482	92

Company	Industry Code	2006 Sales (U.S. $ thousands)	Sales Rank	2006 Profits (U.S. $ thousands)	Profits Rank
MEDAREX INC	325412	48,646	78	-181,701	195
MEDICINES CO	325412	213,952	53	63,726	35
MEDICIS PHARMACEUTICAL	325412	349,242	46	-75,849	173
MEDIMMUNE INC	325412	1,276,800	30	48,700	37
MERCK & CO INC	325412	22,636,000	8	4,433,800	7
MERCK KGAA	325412	8,352,950	17	1,335,880	17
MERCK SERONO SA	325412				
MGI PHARMA INC	325412	342,788	47	-40,161	145
MIGENIX INC	325412	538	168	-10,720	77
MILLENNIUM PHARMACEUTICALS INC	325412	486,830	41	-43,953	149
MIRAVANT MEDICAL TECHNOLOGIES	325412				
MYRIAD GENETICS INC	325412	114,279	63	-38,189	141
NABI BIOPHARMACEUTICALS	325412	89,868	67	-58,703	163
NEOSE TECHNOLOGIES INC	325412	6,184	137	-27,107	120
NEUROBIOLOGICAL TECHNOLOGIES INC	325412	12,339	124	-27,839	121
NEUROCRINE BIOSCIENCES	325412	39,234	86	-107,205	185
NEUROGEN CORP	325412	9,813	131	-53,776	158
NOVARTIS AG	325412	36,031,000	5	7,202,000	4
NOVO-NORDISK AS	325412	6,913,700	19	1,126,020	19
NPS PHARMACEUTICALS INC	325412	48,502	79	-112,668	188
NUVELO INC	325412	3,888	144	-130,553	192
NYCOMED	325412				
ONYX PHARMACEUTICALS INC	325412	250	177	-92,681	181
OSCIENT PHARMACEUTICALS	325412	46,152	81	-78,477	174
OSI PHARMACEUTICALS INC	325412	375,696	44	-582,184	201
OXIGENE INC	325412	0		-15,457	87
OXIS INTERNATIONAL INC	325412	5,776	139	-4,940	63
PAIN THERAPEUTICS	325412	53,918	76	6,188	51
PALATIN TECHNOLOGIES INC	325412	19,749	109	-28,959	124
PDL BIOPHARMA	325412	414,770	43	-130,020	191
PEREGRINE PHARMACEUTICALS INC	325412	3,193	147	-17,061	99
PFIZER INC	325412	48,371,000	2	19,337,000	1
PHARMACYCLICS INC	325412	181	180	-42,158	147
PHARMION CORP	325412	238,646	51	-91,012	179
PHARMOS CORPORATION	325412	0		-35,137	138
PONIARD PHARMACEUTICALS	325412	0		-23,294	112
POZEN INC	325412	13,517	119	-19,310	104
PROGENICS PHARMACEUTICALS	325412	69,906	72	-21,618	108
QLT INC	325412	175,100	57	-101,600	183
QUESTCOR PHARMACEUTICALS	325412	12,788	122	-10,109	76
REGENERON PHARMACEUTICALS INC	325412	63,447	73	-102,337	184
REPLIGEN CORPORATION	325412	12,911	121	697	56
REPROS THERAPEUTICS INC	325412	596	167	-14,195	83
RIGEL PHARMACEUTICALS INC	325412	33,473	92	-37,637	140
ROCHE HOLDING LTD	325412	34,851,500	6	7,116,030	5
SALIX PHARMACEUTICALS	325412	208,533	54	31,510	42
SANKYO CO LTD	325412				
SANOFI-AVENTIS	325412	38,722,100	4	6,003,540	6

Company	Industry Code	2006 Sales (U.S. $ thousands)	Sales Rank	2006 Profits (U.S. $ thousands)	Profits Rank
SAVIENT PHARMACEUTICALS	325412	47,514	80	60,325	36
SCHERING-PLOUGH CORP	325412	10,594,000	14	1,143,000	18
SCICLONE PHARMACEUTICALS	325412	32,662	94	727	55
SCIELE PHARMA INC	325412	293,181	49	45,244	38
SCIOS INC	325412				
SEATTLE GENETICS	325412	10,005	128	-36,015	139
SENETEK PLC	325412	8,431	132	1,883	54
SEPRACOR INC	325412	1,196,534	31	184,562	29
SHIRE PHARMACEUTICALS PLC	325412				
SHIRE-BIOCHEM INC	325412				
SICOR INC	325412				
SIGA TECHNOLOGIES INC	325412	7,258	133	-9,899	75
SIMCERE PHARMACEUTICAL GROUP	325412	121,800	62	22,100	43
SPECTRUM PHARMACEUTICALS	325412	5,673	140	-23,284	111
STIEFEL LABORATORIES INC	325412				
SUPERGEN INC	325412	38,083	87	-16,487	93
TAKEDA PHARMACEUTICAL COMPANY LTD	325412	10,360,744	15	2,677,342	11
TANOX INC	325412	56,137	75	-2,568	61
TAPESTRY PHARMACEUTICALS	325412	0		-16,652	96
TARGETED GENETICS CORP	325412	9,864	129	-33,990	136
TELIK INC	325412	0		-79,624	175
TITAN PHARMACEUTICALS	325412	32	188	-15,737	90
TORREYPINES THERAPEUTICS	325412	9,850	130	-25,377	116
TRIMERIS INC	325412	36,980	89	7,384	50
UCB SA	325412	2,987,500	25	501,100	23
UNIGENE LABORATORIES	325412	6,059	138	-11,784	79
UNITED THERAPEUTICS CORP	325412	159,632	59	73,965	32
UNITED-GUARDIAN INC	325412	12,195	125	2,737	53
VALEANT PHARMACEUTICALS INTERNATIONAL	325412	907,238	33	-56,565	161
VALENTIS INC	325412	727	164	-15,337	86
VASOGEN INC	325412	0		-62,319	167
VAXGEN INC	325412				
VERNALIS PLC	325412	33,040	93	-85,880	177
VERTEX PHARMACEUTICALS	325412	216,356	52	-206,891	196
VICAL INC	325412	14,740	118	-23,148	110
VION PHARMACEUTICALS INC	325412	22	189	-25,347	115
VIRAGEN INC	325412	391	174	-18,215	102
VIROPHARMA INC	325412	167,181	58	66,666	34
WARNER CHILCOTT PLC	325412				
WYETH	325412	20,350,655	10	4,196,706	9
XECHEM INTERNATIONAL	325412	202	179	-11,130	78
XOMA LTD	325412	29,498	97	-51,841	156
ZILA INC	325412	28,188	98	-29,346	125
ZYMOGENETICS INC	325412	25,380	105	-130,002	190
Drugs (Pharmaceuticals), Generic Manufacturing					
BARR PHARMACEUTICALS INC	325416	1,314,465	5	336,477	2

Company	Industry Code	2006 Sales (U.S. $ thousands)	Sales Rank	2006 Profits (U.S. $ thousands)	Profits Rank
CARACO PHARMACEUTICAL LABORATORIES	325416	82,789	7	-10,423	8
HI-TECH PHARMACAL CO INC	325416	78,020	8	11,453	6
LANNETT COMPANY INC	325416	64,060	9	4,969	7
MYLAN LABORATORIES INC	325416	1,257,164	6	184,542	3
PAR PHARMACEUTICAL COMPANIES INC	325416				
PERRIGO CO	325416	1,366,821	4	71,400	5
RANBAXY LABORATORIES	325416	1,405,500	3	116,800	4
TARO PHARMACEUTICAL INDUSTRIES	325416				
TEVA PHARMACEUTICAL INDUSTRIES	325416	8,408,000	1	546,000	1
WATSON PHARMACEUTICALS	325416	1,979,244	2	-445,005	9
Instrument Manufacturing, including Measurement, Control, Test & Navigational					
AGILENT TECHNOLOGIES INC	334500	4,973,000	1	3,307,000	1
APPLIED BIOSYSTEMS GROUP	334500	1,911,226	2	275,117	2
HARVARD BIOSCIENCE INC	334500	76,181	5	-2,341	5
ILLUMINA INC	334500	184,586	4	39,968	4
MILLIPORE CORP	334500	1,255,371	3	96,984	3
WHATMAN PLC	334500				
Investment Banking					
BURRILL & COMPANY	523110				
Laboratories & Diagnostic Services--Medical					
BIORELIANCE CORP	621511				
MDS INC	621511	1,017,200	2	123,100	2
MEDTOX SCIENTIFIC INC	621511	69,804	3	4,548	3
ORCHID CELLMARK INC	621511	56,854	4	-11,271	4
QUEST DIAGNOSTICS INC	621511	6,268,659	1	586,421	1
SPECIALTY LABORATORIES INC	621511				
Management of Companies & Enterprises					
DAIICHI SANKYO CO LTD	551110				
Market Research					
IMS HEALTH INC	541910	1,958,588	1	315,511	1
Medical/Dental/Surgical Equipment & Supplies, Distribution					
THERMO FISHER SCIENTIFIC INC	421450	3,791,600	1	168,900	1
Medical/Dental/Surgical Equipment & Supplies, Manufacturing					
AASTROM BIOSCIENCES INC	339113	863	16	-16,475	14
ABAXIS INC	339113	68,928	8	7,475	8
ADVANCED BIONICS CORP	339113				
APPLERA CORPORATION	339113	1,949,400	5	212,500	4
BAUSCH & LOMB INC	339113	2,293,400	3		
BAXTER INTERNATIONAL INC	339113	10,378,000	1	1,397,000	1
BIO RAD LABORATORIES INC	339113	1,273,930	6	103,263	5
BIOCOMPATIBLES INTERNATIONAL PLC	339113	12,070	15	10,170	7
BIOMET INC	339113	2,025,739	4	406,144	2
BIOSPHERE MEDICAL INC	339113	22,891	13	-2,324	12
CARDIOTECH INTERNATIONAL	339113	22,381	14	-5,069	13
GE HEALTHCARE	339113				
GENZYME BIOSURGERY	339113				
HOSPIRA INC	339113	2,688,505	2	237,679	3

Company	Industry Code	2006 Sales (U.S. $ thousands)	Sales Rank	2006 Profits (U.S. $ thousands)	Profits Rank
INSTITUT STRAUMANN AG	339113				
INTEGRA LIFESCIENCES HOLDINGS CORP	339113	419,297	7	29,407	6
LIFECORE BIOMEDICAL INC	339113	63,097	9	7,040	9
NORTH AMERICAN SCIENTIFIC	339113	28,988	12	-17,130	15
SYNOVIS LIFE TECHNOLOGIES	339113	55,835	10	-1,481	11
THERAGENICS CORP	339113	54,096	11	6,865	10
TYCO HEALTHCARE GROUP	339113				
Medicinals & Botanicals, Manufacturing					
CHATTEM INC	325411	300,548	3	45,112	2
CYANOTECH CORPORATION	325411	11,131	6	-268	5
EXPERIENCE & APPLIED SCIENCES	325411				
FORBES MEDI-TECH INC	325411	6,200	8	-9,300	8
LEINER HEALTH PRODUCTS INC	325411	669,561	2	-3,768	6
MARTEK BIOSCIENCES CORP	325411	270,654	4	17,811	3
NBTY INC	325411	1,880,222	1	111,785	1
NUTRITION 21 INC	325411	10,664	7	-10,317	9
PROTEIN POLYMER TECHNOLOGIES	325411	605	9	-7,878	7
SCHIFF NUTRITION INTERNATIONAL INC	325411	178,372	5	15,839	4
VENTRIA BIOSCIENCE	325411				
Nanotechnology-Optics					
ARRYX INC	541710G				
Research & Development--Physical, Engineering & Life Sciences					
ADVANCED CELL TECHNOLOGY	541710	441	22	-18,720	14
ALBANY MOLECULAR RESEARCH	541710	179,807	9	2,183	7
APPLIED MOLECULAR EVOLUTION INC	541710				
ARRAY BIOPHARMA INC	541710	45,003	12	39,614	3
BIOANALYTICAL SYSTEMS INC	541710	43,048	14	-2,609	11
CELERA GENOMICS GROUP	541710	44,200	13	-62,700	22
CHARLES RIVER LABORATORIES INTERNATIONAL INC	541710	1,058,385	3	-55,783	21
CHESAPEAKE BIOLOGICAL LAB	541710				
COMMONWEALTH BIOTECHNOLOGIES INC	541710	6,532	21	-1,153	10
COVANCE INC	541710	1,406,058	1	144,998	2
DIVERSA CORPORATION	541710	49,198	11	-39,271	18
ENCORIUM GROUP INC	541710	15,326	18	-494	9
EVOTEC OAI AG	541710	111,700	10	-42,800	19
GENE LOGIC INC	541710	24,346	15	-54,710	20
ICON PLC	541710	455,597	5	38,304	4
INFINITY PHARMACEUTICALS	541710	18,495	16	-28,448	16
LIFE SCIENCES RESEARCH	541710	192,217	8	-14,872	12
PACIFIC BIOMETRICS INC	541710	10,750	19	179	8
PAREXEL INTERNATIONAL	541710	614,947	4	23,544	6
PHARMACEUTICAL PRODUCT DEVELOPMENT INC	541710	1,247,682	2	156,652	1
PHARMACOPEIA DRUG DISCOVERY	541710	16,936	17	-27,764	15
PRA INTERNATIONAL	541710	338,166	7	26,845	5
QUINTILES TRANSNATIONAL	541710				
ROSETTA INPHARMATICS LLC	541710				

Company	Industry Code	2006 Sales (U.S. $ thousands)	Sales Rank	2006 Profits (U.S. $ thousands)	Profits Rank
SANGAMO BIOSCIENCES INC	541710	7,885	20	-17,864	13
SFBC INTERNATIONAL INC	541710	406,955	6	-36,025	17
Textiles, Fabrics, Sheets/Towels, Manufacturing					
TOYOBO CO LTD	313000	3,418,200	1	107,100	1
Veterinary Products Manufacturing					
EMBREX INC	325412B				
HESKA CORP	325412B	75,060	1	1,828	1
IMMUCELL CORPORATION	325412B	4,801	2	647	2
SYNBIOTICS CORP	325412B				
VIRBAC CORP	325412B				

ALPHABETICAL INDEX

4SC AG
AASTROM BIOSCIENCES INC
ABAXIS INC
ABBOTT LABORATORIES
ABRAXIS BIOSCIENCE INC
ACAMBIS PLC
ACCELRYS INC
ACCESS PHARMACEUTICALS INC
ACTELION LTD
ADOLOR CORP
ADVANCED BIONICS CORPORATION
ADVANCED CELL TECHNOLOGY INC
AEOLUS PHARMACEUTICALS INC
AETERNA ZENTARIS INC
AFFYMETRIX INC
AGILENT TECHNOLOGIES INC
AKORN INC
AKZO NOBEL NV
ALBANY MOLECULAR RESEARCH
ALCON INC
ALEXION PHARMACEUTICALS INC
ALFACELL CORPORATION
ALIZYME PLC
ALKERMES INC
ALLERGAN INC
ALLIANCE PHARMACEUTICAL CORP
ALLOS THERAPEUTICS INC
ALPHARMA INC
ALSERES PHARMACEUTICALS INC
ALTANA AG
ALTEON INC
ALZA CORP
AMARIN CORPORATION PLC
AMGEN INC
AMYLIN PHARMACEUTICALS INC
ANGIOTECH PHARMACEUTICALS
ANIKA THERAPEUTICS INC
ANTIGENICS INC
APPLERA CORPORATION
APPLIED BIOSYSTEMS GROUP
APPLIED MOLECULAR EVOLUTION INC
ARADIGM CORPORATION
ARCADIA BIOSCIENCES
ARENA PHARMACEUTICALS INC
ARIAD PHARMACEUTICALS INC
ARQULE INC
ARRAY BIOPHARMA INC
ARRYX INC
ASTELLAS PHARMA INC
ASTRAZENECA PLC
ATHEROGENICS INC
AUTOIMMUNE INC
AVANIR PHARMACEUTICALS
AVANT IMMUNOTHERAPEUTICS
AVAX TECHNOLOGIES INC
AVI BIOPHARMA INC

AVIGEN INC
AVIVA BIOSCIENCES CORP
AXCAN PHARMA INC
BARR PHARMACEUTICALS INC
BASF AG
BAUSCH & LOMB INC
BAXTER INTERNATIONAL INC
BAYER AG
BAYER CORP
BAYER SCHERING PHARMA AG
BENTLEY PHARMACEUTICALS INC
BIO RAD LABORATORIES INC
BIOANALYTICAL SYSTEMS INC
BIOCOMPATIBLES INTERNATIONAL PLC
BIOCRYST PHARMACEUTICALS INC
BIOGEN IDEC INC
BIOMARIN PHARMACEUTICAL INC
BIOMET INC
BIOMIRA INC
BIOPURE CORPORATION
BIORELIANCE CORP
BIOSITE INC
BIOSPHERE MEDICAL INC
BIOTECH HOLDINGS LTD
BIOTIME INC
BIOVAIL CORPORATION
BRADLEY PHARMACEUTICALS INC
BRISTOL MYERS SQUIBB CO
BURRILL & COMPANY
CALIPER LIFE SCIENCES
CAMBREX CORP
CAMBRIDGE ANTIBODY TECHNOLOGY LIMITED
CANGENE CORP
CARACO PHARMACEUTICAL LABORATORIES
CARDIOME PHARMA CORP
CARDIOTECH INTERNATIONAL
CARDIUM THERAPEUTICS INC
CARRINGTON LABORATORIES INC
CELERA GENOMICS GROUP
CELGENE CORP
CELL GENESYS INC
CELL THERAPEUTICS INC
CELLEGY PHARMACEUTICALS
CEL-SCI CORPORATION
CELSIS INTERNATIONAL PLC
CEPHALON INC
CEPHEID
CERUS CORPORATION
CHARLES RIVER LABORATORIES INTERNATIONAL INC
CHATTEM INC
CHESAPEAKE BIOLOGICAL LAB
CHIRON CORP
CIPHERGEN BIOSYSTEMS INC
CML HEALTHCARE INCOME FUND
COLLAGENEX PHARMACEUTICALS INC
COLUMBIA LABORATORIES INC
COMMONWEALTH BIOTECHNOLOGIES INC

COMPUGEN LTD
CORTEX PHARMACEUTICALS INC
COVANCE INC
COVIDIEN LTD
CSL LIMITED
CUBIST PHARMACEUTICALS INC
CURAGEN CORPORATION
CURIS INC
CV THERAPEUTICS INC
CYANOTECH CORPORATION
CYPRESS BIOSCIENCE INC
CYTOGEN CORPORATION
CYTRX CORPORATION
DAIICHI SANKYO CO LTD
DECODE GENETICS INC
DENDREON CORPORATION
DENDRITE INTERNATIONAL INC
DEPOMED INC
DIAGNOSTIC PRODUCTS CORPORATION
DIGENE CORPORATION
DISCOVERY LABORATORIES INC
DIVERSA CORPORATION
DOR BIOPHARMA INC
DOV PHARMACEUTICAL INC
DOW AGROSCIENCES LLC
DR REDDY'S LABORATORIES LIMITED
DRAXIS HEALTH INC
DSM PHARMACEUTICALS INC
DUPONT AGRICULTURE & NUTRITION
DURECT CORP
DUSA PHARMACEUTICALS INC
DYAX CORP
E I DU PONT DE NEMOURS & CO (DUPONT)
EISAI CO LTD
ELAN CORP PLC
ELI LILLY AND COMPANY
ELSEVIER MDL
EMBREX INC
EMISPHERE TECHNOLOGIES INC
ENCORIUM GROUP INC
ENCYSIVE PHARMACEUTICALS INC
ENDO PHARMACEUTICALS HOLDINGS INC
ENTREMED INC
ENZO BIOCHEM INC
ENZON PHARMACEUTICALS INC
EPIX PHARMACEUTICALS INC
ERESEARCH TECHNOLOGY INC
EVOTEC OAI AG
EXELIXIS INC
EXELIXIS PLANT SCIENCES INC
EXPERIMENTAL & APPLIED SCIENCES INC
E-Z-EM INC
FLAMEL TECHNOLOGIES SA
FORBES MEDI-TECH INC
FOREST LABORATORIES INC
GE HEALTHCARE
GENAERA CORPORATION
GENE LOGIC INC

GENELABS TECHNOLOGIES INC
GENENCOR INTERNATIONAL INC
GENENTECH INC
GENEREX BIOTECHNOLOGY
GEN-PROBE INC
GENTA INC
GENVEC INC
GENZYME BIOSURGERY
GENZYME CORP
GENZYME ONCOLOGY
GERON CORPORATION
GILEAD SCIENCES INC
GLAXOSMITHKLINE PLC
GLYCOGENESYS INC
GTC BIOTHERAPEUTICS INC
HARVARD BIOSCIENCE INC
HEMAGEN DIAGNOSTICS INC
HEMISPHERX BIOPHARMA INC
HESKA CORP
HI-TECH PHARMACAL CO INC
HOLLIS-EDEN PHARMACEUTICALS
HOSPIRA INC
HUMAN GENOME SCIENCES INC
HYCOR BIOMEDICAL INC
ICON PLC
ICOS CORPORATION
IDERA PHARMACEUTICALS INC
IDEXX LABORATORIES INC
IDM PHARMA INC
ILLUMINA INC
IMCLONE SYSTEMS INC
IMMTECH INTERNATIONAL
IMMUCELL CORPORATION
IMMUNE RESPONSE CORP (THE)
IMMUNOGEN INC
IMMUNOMEDICS INC
IMPAX LABORATORIES INC
IMS HEALTH INC
INCYTE CORP
INDEVUS PHARMACEUTICALS INC
INFINITY PHARMACEUTICALS INC
INSITE VISION INC
INSMED INCORPORATED
INSPIRE PHARMACEUTICALS INC
INSTITUT STRAUMANN AG
INTEGRA LIFESCIENCES HOLDINGS CORP
INTERMUNE PHARMACEUTICALS INC
INTERNATIONAL ISOTOPES
INTROGEN THERAPEUTICS INC
INVITROGEN CORPORATION
IOMED INC
ISIS PHARMACEUTICALS INC
JAZZ PHARMACEUTICALS
JOHNSON & JOHNSON
KENDLE INTERNATIONAL INC
KERYX BIOPHARMACEUTICALS INC
KING PHARMACEUTICALS INC
KOS PHARMACEUTICALS INC

KOSAN BIOSCIENCES INC
KV PHARMACEUTICAL CO
LA JOLLA PHARMACEUTICAL
LANNETT COMPANY INC
LARGE SCALE BIOLOGY CORP
LEINER HEALTH PRODUCTS INC
LEXICON PHARMACEUTICALS INC
LIFE SCIENCES RESEARCH
LIFECELL CORPORATION
LIFECORE BIOMEDICAL INC
LIGAND PHARMACEUTICALS INC
LONZA GROUP
LORUS THERAPEUTICS INC
LUMINEX CORPORATION
MACROCHEM CORPORATION
MALLINCKRODT INC
MANHATTAN PHARMACEUTICALS INC
MARTEK BIOSCIENCES CORP
MATRITECH INC
MAXYGEN INC
MDS INC
MEDAREX INC
MEDICINES CO (THE)
MEDICIS PHARMACEUTICAL CORP
MEDIMMUNE INC
MEDTOX SCIENTIFIC INC
MERCK & CO INC
MERCK KGAA
MERCK SERONO SA
MERIDIAN BIOSCIENCE INC
MGI PHARMA INC
MIGENIX INC
MILLENNIUM PHARMACEUTICALS INC
MILLIPORE CORP
MIRAVANT MEDICAL TECHNOLOGIES
MONSANTO CO
MYLAN LABORATORIES INC
MYRIAD GENETICS INC
NABI BIOPHARMACEUTICALS
NANOBIO CORPORATION
NANOGEN INC
NASTECH PHARMACEUTICAL CO INC
NBTY INC
NEKTAR THERAPEUTICS
NEOGEN CORPORATION
NEOPHARM INC
NEOPROBE CORPORATION
NEOSE TECHNOLOGIES INC
NEUROBIOLOGICAL TECHNOLOGIES INC
NEUROCRINE BIOSCIENCES INC
NEUROGEN CORP
NEXMED INC
NORTH AMERICAN SCIENTIFIC
NOVARTIS AG
NOVAVAX INC
NOVEN PHARMACEUTICALS
NOVO-NORDISK AS
NOVOZYMES

NPS PHARMACEUTICALS INC
NUTRITION 21 INC
NUVELO INC
NYCOMED
ONYX PHARMACEUTICALS INC
ORASURE TECHNOLOGIES INC
ORCHID CELLMARK INC
ORGANOGENESIS INC
OSCIENT PHARMACEUTICALS INC
OSI PHARMACEUTICALS INC
OXIGENE INC
OXIS INTERNATIONAL INC
PACIFIC BIOMETRICS INC
PAIN THERAPEUTICS INC
PALATIN TECHNOLOGIES INC
PAR PHARMACEUTICAL COMPANIES INC
PAREXEL INTERNATIONAL
PDL BIOPHARMA
PENWEST PHARMACEUTICALS CO
PERBIO SCIENCE AB
PEREGRINE PHARMACEUTICALS INC
PERRIGO CO
PFIZER INC
PHARMACEUTICAL PRODUCT DEVELOPMENT INC
PHARMACOPEIA DRUG DISCOVERY
PHARMACYCLICS INC
PHARMANET DEVELOPMENT GROUP INC
PHARMION CORP
PHARMOS CORP
POLYDEX PHARMACEUTICALS
PONIARD PHARMACEUTICALS INC
POZEN INC
PRA INTERNATIONAL
PROGENICS PHARMACEUTICALS
PROMEGA CORP
PROTEIN POLYMER TECHNOLOGIES
QIAGEN NV
QLT INC
QUEST DIAGNOSTICS INC
QUESTCOR PHARMACEUTICALS
QUINTILES TRANSNATIONAL CORP
RANBAXY LABORATORIES LIMITED
REGENERON PHARMACEUTICALS INC
REPLIGEN CORPORATION
REPROS THERAPEUTICS INC
RIGEL PHARMACEUTICALS INC
ROCHE HOLDING LTD
ROSETTA INPHARMATICS LLC
SALIX PHARMACEUTICALS
SANGAMO BIOSCIENCES INC
SANKYO CO LTD
SANOFI-AVENTIS
SAVIENT PHARMACEUTICALS INC
SCHERING-PLOUGH CORP
SCHIFF NUTRITION INTERNATIONAL INC
SCICLONE PHARMACEUTICALS
SCIELE PHARMA INC
SCIOS INC

SEATTLE GENETICS
SENETEK PLC
SEPRACOR INC
SEQUENOM INC
SERACARE LIFE SCIENCES INC
SHIRE PLC
SHIRE-BIOCHEM INC
SICOR INC
SIGA TECHNOLOGIES INC
SIGMA ALDRICH CORP
SIMCERE PHARMACEUTICAL GROUP
SKYEPHARMA PLC
SONUS PHARMACEUTICALS
SPECIALTY LABORATORIES INC
SPECTRAL DIAGNOSTICS INC
SPECTRUM PHARMACEUTICALS INC
STEMCELLS INC
STIEFEL LABORATORIES INC
STRATAGENE CORP
SUPERGEN INC
SYNBIOTICS CORP
SYNGENTA AG
SYNOVIS LIFE TECHNOLOGIES INC
TAKEDA PHARMACEUTICAL COMPANY LTD
TANOX INC
TAPESTRY PHARMACEUTICALS INC
TARGETED GENETICS CORP
TARO PHARMACEUTICAL INDUSTRIES
TECHNE CORP
TELIK INC
TEVA PHARMACEUTICAL INDUSTRIES
THERAGENICS CORP
THERMO FISHER SCIENTIFIC INC
THIRD WAVE TECHNOLOGIES INC
TITAN PHARMACEUTICALS
TORREYPINES THERAPEUTICS INC
TOYOBO CO LTD
TRIMERIS INC
TRINITY BIOTECH PLC
TRIPOS INC
UCB SA
UNIGENE LABORATORIES
UNITED THERAPEUTICS CORP
UNITED-GUARDIAN INC
URIGEN PHARMACEUTICALS INC
VALEANT PHARMACEUTICALS INTERNATIONAL
VASOGEN INC
VAXGEN INC
VENTRIA BIOSCIENCE
VERNALIS PLC
VERTEX PHARMACEUTICALS INC
VI TECHNOLOGIES INC
VIACELL INC
VICAL INC
VION PHARMACEUTICALS INC
VIRAGEN INC
VIRBAC CORP
VIROPHARMA INC

VYSIS INC
WARNER CHILCOTT PLC
WATSON PHARMACEUTICALS INC
WHATMAN PLC
WYETH
XECHEM INTERNATIONAL
XOMA LTD
ZILA INC
ZYMOGENETICS INC

INDEX OF HEADQUARTERS LOCATION BY U.S. STATE

To help you locate members of THE BIOTECH 400 geographically, the city and state of the headquarters of each company are in the following index.

ALABAMA
BIOCRYST PHARMACEUTICALS INC; Birmingham

ARIZONA
MEDICIS PHARMACEUTICAL CORP; Scottsdale
ZILA INC; Phoenix

CALIFORNIA
ABAXIS INC; Union City
ABRAXIS BIOSCIENCE INC; Los Angeles
ACCELRYS INC; San Diego
ADVANCED BIONICS CORPORATION; Sylmar
ADVANCED CELL TECHNOLOGY INC; Alameda
AEOLUS PHARMACEUTICALS INC; Laguna Niguel
AFFYMETRIX INC; Santa Clara
AGILENT TECHNOLOGIES INC; Santa Clara
ALLERGAN INC; Irvine
ALLIANCE PHARMACEUTICAL CORP; San Diego
ALZA CORP; Mountain View
AMGEN INC; Thousand Oaks
AMYLIN PHARMACEUTICALS INC; San Diego
APPLIED BIOSYSTEMS GROUP; Foster City
APPLIED MOLECULAR EVOLUTION INC; San Diego
ARADIGM CORPORATION; Hayward
ARCADIA BIOSCIENCES; Davis
ARENA PHARMACEUTICALS INC; San Diego
AUTOIMMUNE INC; Pasadena
AVANIR PHARMACEUTICALS; Aliso Viejo
AVIGEN INC; Alameda
AVIVA BIOSCIENCES CORP; San Diego
BIO RAD LABORATORIES INC; Hercules
BIOMARIN PHARMACEUTICAL INC; Novato
BIOSITE INC; San Diego
BIOTIME INC; Emeryville
BURRILL & COMPANY; San Francisco
CARDIUM THERAPEUTICS INC; San Diego
CELL GENESYS INC; South San Francisco
CEPHEID; Sunnyvale
CERUS CORPORATION; Concord
CHIRON CORP; Emeryville
CIPHERGEN BIOSYSTEMS INC; Fremont
CORTEX PHARMACEUTICALS INC; Irvine
CV THERAPEUTICS INC; Palo Alto
CYPRESS BIOSCIENCE INC; San Diego
CYTRX CORPORATION; Los Angeles
DEPOMED INC; Menlo Park
DIAGNOSTIC PRODUCTS CORPORATION; Los Angeles
DIVERSA CORPORATION; San Diego

DURECT CORP; Cupertino
ELSEVIER MDL; San Ramon
EXELIXIS INC; So. San Francisco
GENELABS TECHNOLOGIES INC; Redwood City
GENENCOR INTERNATIONAL INC; Palo Alto
GENENTECH INC; South San Francisco
GEN-PROBE INC; San Diego
GERON CORPORATION; Menlo Park
GILEAD SCIENCES INC; Foster City
HOLLIS-EDEN PHARMACEUTICALS; San Diego
HYCOR BIOMEDICAL INC; Garden Grove
IDM PHARMA INC; Irvine
ILLUMINA INC; San Diego
IMMUNE RESPONSE CORP (THE); Carlsbad
IMPAX LABORATORIES INC; Hayward
INSITE VISION INC; Alameda
INTERMUNE PHARMACEUTICALS INC; Brisbane
INVITROGEN CORPORATION; Carlsbad
ISIS PHARMACEUTICALS INC; Carlsbad
JAZZ PHARMACEUTICALS; Palo Alto
KOSAN BIOSCIENCES INC; Hayward
LA JOLLA PHARMACEUTICAL; San Diego
LARGE SCALE BIOLOGY CORP; Vacaville
LEINER HEALTH PRODUCTS INC; Carson
LIGAND PHARMACEUTICALS INC; San Diego
MAXYGEN INC; Redwood City
MIRAVANT MEDICAL TECHNOLOGIES; Santa Barbara
NANOGEN INC; San Diego
NEKTAR THERAPEUTICS; San Carlos
NEUROBIOLOGICAL TECHNOLOGIES INC; Emeryville
NEUROCRINE BIOSCIENCES INC; San Diego
NORTH AMERICAN SCIENTIFIC; Chatsworth
NUVELO INC; San Carlos
ONYX PHARMACEUTICALS INC; Emeryville
OXIS INTERNATIONAL INC; Foster City
PAIN THERAPEUTICS INC; S. San Francisco
PDL BIOPHARMA; Fremont
PEREGRINE PHARMACEUTICALS INC; Tustin
PHARMACYCLICS INC; Sunnyvale
PONIARD PHARMACEUTICALS INC; S. San Francisco
PROTEIN POLYMER TECHNOLOGIES; San Diego
QUESTCOR PHARMACEUTICALS; Union City
RIGEL PHARMACEUTICALS INC; South San Francisco
SANGAMO BIOSCIENCES INC; Richmond
SCICLONE PHARMACEUTICALS; San Mateo
SCIOS INC; Mountain View
SENETEK PLC; Napa
SEQUENOM INC; San Diego
SICOR INC; Irvine
SPECIALTY LABORATORIES INC; Valencia
SPECTRUM PHARMACEUTICALS INC; Irvine
STEMCELLS INC; Palo Alto
STRATAGENE CORP; La Jolla
SUPERGEN INC; Dublin
TELIK INC; Palo Alto
TITAN PHARMACEUTICALS; S. San Francisco

TORREYPINES THERAPEUTICS INC; La Jolla
URIGEN PHARMACEUTICALS INC; Burlingame
VALEANT PHARMACEUTICALS INTERNATIONAL;
Aliso Viejo
VAXGEN INC; S. San Francisco
VENTRIA BIOSCIENCE; Sacramento
VICAL INC; San Diego
WATSON PHARMACEUTICALS INC; Corona
XOMA LTD; Berkeley

COLORADO
ALLOS THERAPEUTICS INC; Westminster
ARRAY BIOPHARMA INC; Boulder
EXPERIMENTAL & APPLIED SCIENCES INC; Golden
HESKA CORP; Loveland
PHARMION CORP; Boulder
TAPESTRY PHARMACEUTICALS INC; Boulder

CONNECTICUT
ALEXION PHARMACEUTICALS INC; Cheshire
APPLERA CORPORATION; Norwalk
CURAGEN CORPORATION; Branford
IMS HEALTH INC; Fairfield
NEUROGEN CORP; Branford
PENWEST PHARMACEUTICALS CO; Danbury
VION PHARMACEUTICALS INC; New Haven

DELAWARE
DUPONT AGRICULTURE & NUTRITION; Wilmington
E I DU PONT DE NEMOURS & CO (DUPONT);
Wilmington
INCYTE CORP; Wilmington

FLORIDA
DOR BIOPHARMA INC; Miami
NABI BIOPHARMACEUTICALS; Boca Raton
NOVEN PHARMACEUTICALS; Miami
STIEFEL LABORATORIES INC; Coral Gables
VIRAGEN INC; Plantation

GEORGIA
ATHEROGENICS INC; Alpharetta
SCIELE PHARMA INC; Atlanta
THERAGENICS CORP; Buford

HAWAII
CYANOTECH CORPORATION; Kailua-Kona

IDAHO
INTERNATIONAL ISOTOPES; Idaho Falls

ILLINOIS
ABBOTT LABORATORIES; Abbott Park
AKORN INC; Buffalo Grove
ARRYX INC; Chicago
BAXTER INTERNATIONAL INC; Deerfield
HOSPIRA INC; Lake Forest

NEOPHARM INC; Waukegan
VYSIS INC; Des Plaines

INDIANA
BIOANALYTICAL SYSTEMS INC; West Lafayette
BIOMET INC; Warsaw
DOW AGROSCIENCES LLC; Indianapolis
ELI LILLY AND COMPANY; Indianapolis

MAINE
IDEXX LABORATORIES INC; Westbrook
IMMUCELL CORPORATION; Portland

MARYLAND
BIORELIANCE CORP; Rockville
CELERA GENOMICS GROUP; Rockville
CHESAPEAKE BIOLOGICAL LAB; Baltimore
DIGENE CORPORATION; Gaithersburg
ENTREMED INC; Rockville
GENE LOGIC INC; Gaithersburg
GENVEC INC; Gaithersburg
HEMAGEN DIAGNOSTICS INC; Columbia
HUMAN GENOME SCIENCES INC; Rockville
MARTEK BIOSCIENCES CORP; Columbia
MEDIMMUNE INC; Gaithersburg
NOVAVAX INC; Rockville
UNITED THERAPEUTICS CORP; Silver Spring

MASSACHUSETTS
ALKERMES INC; Cambridge
ALSERES PHARMACEUTICALS INC; Hopkinton
ANIKA THERAPEUTICS INC; Woburn
ARIAD PHARMACEUTICALS INC; Cambridge
ARQULE INC; Woburn
AVANT IMMUNOTHERAPEUTICS; Needham
BIOGEN IDEC INC; Cambridge
BIOPURE CORPORATION; Cambridge
BIOSPHERE MEDICAL INC; Rockland
CALIPER LIFE SCIENCES; Hopkinton
CARDIOTECH INTERNATIONAL; Wilmington
CHARLES RIVER LABORATORIES
INTERNATIONAL INC; Wilmington
CUBIST PHARMACEUTICALS INC; Lexington
CURIS INC; Cambridge
DUSA PHARMACEUTICALS INC; Wilmington
DYAX CORP; Cambridge
EPIX PHARMACEUTICALS INC; Lexington
GENZYME BIOSURGERY; Cambridge
GENZYME CORP; Cambridge
GENZYME ONCOLOGY; Cambridge
GLYCOGENESYS INC; Boston
GTC BIOTHERAPEUTICS INC; Framingham
HARVARD BIOSCIENCE INC; Holliston
IDERA PHARMACEUTICALS INC; Cambridge
IMMUNOGEN INC; Cambridge
INDEVUS PHARMACEUTICALS INC; Lexington
INFINITY PHARMACEUTICALS INC; Cambridge
MACROCHEM CORPORATION; Lexington

MATRITECH INC; Newton
MILLENNIUM PHARMACEUTICALS INC; Cambridge
MILLIPORE CORP; Billerica
ORGANOGENESIS INC; Canton
OSCIENT PHARMACEUTICALS INC; Waltham
OXIGENE INC; Waltham
PAREXEL INTERNATIONAL; Waltham
REPLIGEN CORPORATION; Waltham
SEPRACOR INC; Marlborough
SERACARE LIFE SCIENCES INC; West Bridgewater
THERMO FISHER SCIENTIFIC INC; Waltham
VERTEX PHARMACEUTICALS INC; Cambridge
VI TECHNOLOGIES INC; Watertown
VIACELL INC; Cambridge

MICHIGAN
AASTROM BIOSCIENCES INC; Ann Arbor
CARACO PHARMACEUTICAL LABORATORIES;
Detroit
NANOBIO CORPORATION; Ann Arbor
NEOGEN CORPORATION; Lansing
PERRIGO CO; Allegan

MINNESOTA
LIFECORE BIOMEDICAL INC; Chaska
MEDTOX SCIENTIFIC INC; St. Paul
MGI PHARMA INC; Bloomington
SYNOVIS LIFE TECHNOLOGIES INC; St. Paul
TECHNE CORP; Minneapolis

MISSOURI
KV PHARMACEUTICAL CO; St. Louis
MALLINCKRODT INC; Hazelwood
MONSANTO CO; St. Louis
SIGMA ALDRICH CORP; St. Louis
SYNBIOTICS CORP; Kansas City
TRIPOS INC; St. Louis

NEW HAMPSHIRE
BENTLEY PHARMACEUTICALS INC; Exeter

NEW JERSEY
ALFACELL CORPORATION; Bloomfield
ALPHARMA INC; Bridgewater
ALTEON INC; Montvale
BRADLEY PHARMACEUTICALS INC; Fairfield
CAMBREX CORP; E. Rutherford
CELGENE CORP; Summit
COLUMBIA LABORATORIES INC; Livingston
COVANCE INC; Princeton
CYTOGEN CORPORATION; Princeton
DENDRITE INTERNATIONAL INC; Bedminster
DOV PHARMACEUTICAL INC; Somerset
ENZON PHARMACEUTICALS INC; Bridgewater
GENTA INC; Berkeley Heights
IMMUNOMEDICS INC; Morris Plains
INTEGRA LIFESCIENCES HOLDINGS CORP;
Plainsboro

JOHNSON & JOHNSON; New Brunswick
KOS PHARMACEUTICALS INC; Cranbury
LIFE SCIENCES RESEARCH; East Millstone
LIFECELL CORPORATION; Branchburg
MEDAREX INC; Princeton
MEDICINES CO (THE); Parsippany
MERCK & CO INC; Whitehouse Station
NEXMED INC; Robbinsville
NPS PHARMACEUTICALS INC; Parsippany
ORCHID CELLMARK INC; Princeton
PALATIN TECHNOLOGIES INC; Cranbury
PAR PHARMACEUTICAL COMPANIES INC;
Woodcliff Lake
PHARMACOPEIA DRUG DISCOVERY; Cranbury
PHARMANET DEVELOPMENT GROUP INC;
Princeton
PHARMOS CORP; Iselin
QUEST DIAGNOSTICS INC; Lyndhurst
SAVIENT PHARMACEUTICALS INC; E. Brunswick
SCHERING-PLOUGH CORP; Kenilworth
UNIGENE LABORATORIES; Fairfield
WARNER CHILCOTT PLC; Rockaway
WYETH; Madison
XECHEM INTERNATIONAL; New Brunswick

NEW YORK
ALBANY MOLECULAR RESEARCH; Albany
ANTIGENICS INC; New York
BARR PHARMACEUTICALS INC; Pomona
BAUSCH & LOMB INC; Rochester
BRISTOL MYERS SQUIBB CO; New York
EMISPHERE TECHNOLOGIES INC; Tarrytown
ENZO BIOCHEM INC; New York
E-Z-EM INC; Lake Success
FOREST LABORATORIES INC; New York
HI-TECH PHARMACAL CO INC; Amityville
IMCLONE SYSTEMS INC; New York
IMMTECH INTERNATIONAL; New York
KERYX BIOPHARMACEUTICALS INC; New York
MANHATTAN PHARMACEUTICALS INC; New York
NBTY INC; Bohemia
NUTRITION 21 INC; Purchase
OSI PHARMACEUTICALS INC; Melville
PFIZER INC; New York
PROGENICS PHARMACEUTICALS; Tarrytown
REGENERON PHARMACEUTICALS INC; Tarrytown
SIGA TECHNOLOGIES INC; New York
UNITED-GUARDIAN INC; Hauppauge

NORTH CAROLINA
DSM PHARMACEUTICALS INC; Greenville
EMBREX INC; Durham
INSPIRE PHARMACEUTICALS INC; Durham
PHARMACEUTICAL PRODUCT DEVELOPMENT
INC; Wilmington
POZEN INC; Chapel Hill
QUINTILES TRANSNATIONAL CORP; Durham
SALIX PHARMACEUTICALS; Morrisville

TRIMERIS INC; Morrisville

OHIO
KENDLE INTERNATIONAL INC; Cincinnati
MERIDIAN BIOSCIENCE INC; Cincinnati
NEOPROBE CORPORATION; Dublin

OREGON
AVI BIOPHARMA INC; Portland
EXELIXIS PLANT SCIENCES INC; Portland

PENNSYLVANIA
ADOLOR CORP; Exton
AVAX TECHNOLOGIES INC; Philadelphia
BAYER CORP; Pittsburgh
CELLEGY PHARMACEUTICALS; Quakertown
CEPHALON INC; Frazer
COLLAGENEX PHARMACEUTICALS INC; Newtown
DISCOVERY LABORATORIES INC; Warrington
ENCORIUM GROUP INC; Wayne
ENDO PHARMACEUTICALS HOLDINGS INC; Chadds Ford
ERESEARCH TECHNOLOGY INC; Philadelphia
GENAERA CORPORATION; Plymouth Meeting
HEMISPHERX BIOPHARMA INC; Philadelphia
LANNETT COMPANY INC; Philadelphia
MYLAN LABORATORIES INC; Canonsburg
NEOSE TECHNOLOGIES INC; Horsham
ORASURE TECHNOLOGIES INC; Bethlehem
VIROPHARMA INC; Exton

TENNESSEE
CHATTEM INC; Chattanooga
KING PHARMACEUTICALS INC; Bristol

TEXAS
ACCESS PHARMACEUTICALS INC; Dallas
CARRINGTON LABORATORIES INC; Irving
ENCYSIVE PHARMACEUTICALS INC; Houston
INTROGEN THERAPEUTICS INC; Austin
LEXICON PHARMACEUTICALS INC; The Woodlands
LUMINEX CORPORATION; Austin
REPROS THERAPEUTICS INC; The Woodlands
TANOX INC; Houston
VIRBAC CORP; Fort Worth

UTAH
IOMED INC; Salt Lake City
MYRIAD GENETICS INC; Salt Lake City
SCHIFF NUTRITION INTERNATIONAL INC; Salt Lake City

VIRGINIA
CEL-SCI CORPORATION; Vienna
COMMONWEALTH BIOTECHNOLOGIES INC; Richmond
INSMED INCORPORATED; Richmond
PRA INTERNATIONAL; Reston

WASHINGTON
CELL THERAPEUTICS INC; Seattle
DENDREON CORPORATION; Seattle
ICOS CORPORATION; Bothell
NASTECH PHARMACEUTICAL CO INC; Bothell
PACIFIC BIOMETRICS INC; Seattle
ROSETTA INPHARMATICS LLC; Seattle
SEATTLE GENETICS; Bothell
SONUS PHARMACEUTICALS; Bothell
TARGETED GENETICS CORP; Seattle
ZYMOGENETICS INC; Seattle

WISCONSIN
PROMEGA CORP; Madison
THIRD WAVE TECHNOLOGIES INC; Madison

INDEX OF NON-U.S. HEADQUARTERS
LOCATION BY COUNTRY

AUSTRALIA
CSL LIMITED; Parkville

BELGIUM
UCB SA; Brussels

BERMUDA
COVIDIEN LTD; Pembroke

CANADA
AETERNA ZENTARIS INC; Quebec, Quebec
ANGIOTECH PHARMACEUTICALS; Vancouver
AXCAN PHARMA INC; Mont-Saint-Hilaire
BIOMIRA INC; Edmonton
BIOTECH HOLDINGS LTD; Richmond
BIOVAIL CORPORATION; Mississauga
CANGENE CORP; Winnipeg
CARDIOME PHARMA CORP; Vancouver
CML HEALTHCARE INCOME FUND; Mississauga
DRAXIS HEALTH INC; Mississauga
FORBES MEDI-TECH INC; Vancouver
GENEREX BIOTECHNOLOGY; Toronto
LORUS THERAPEUTICS INC; Toronto
MDS INC; Mississauga
MIGENIX INC; Vancouver
POLYDEX PHARMACEUTICALS; Toronto
QLT INC; Vancouver
SHIRE-BIOCHEM INC; Ville Saint-Laurent
SPECTRAL DIAGNOSTICS INC; Toronto
VASOGEN INC; Mississauga

CHINA
SIMCERE PHARMACEUTICAL GROUP; Nanjing,
Jiangsu Province

DENMARK
NOVO-NORDISK AS; Basgvaerd
NOVOZYMES; Bagsvaerd
NYCOMED; Roskilde

FRANCE
FLAMEL TECHNOLOGIES SA; Venissieux
SANOFI-AVENTIS; Paris

GERMANY
4SC AG; Planegg-Martinsried
ALTANA AG; Bad Homburg
BASF AG; Ludwigshafen
BAYER AG; Leverkusen
BAYER SCHERING PHARMA AG; Berlin
EVOTEC OAI AG; Hamburg
MERCK KGAA; Darmstadt

ICELAND
DECODE GENETICS INC; Reykjavik

INDIA
DR REDDY'S LABORATORIES LIMITED; Hyderabad
RANBAXY LABORATORIES LIMITED; Gurgaon

IRELAND
ELAN CORP PLC; Dublin
TRINITY BIOTECH PLC; Bray

ISRAEL
COMPUGEN LTD; Tel Aviv
TARO PHARMACEUTICAL INDUSTRIES; Yakum
TEVA PHARMACEUTICAL INDUSTRIES; Petach
Tikva

JAPAN
ASTELLAS PHARMA INC; Tokyo
DAIICHI SANKYO CO LTD; Tokyo
EISAI CO LTD; Tokyo
SANKYO CO LTD; Tokyo
TAKEDA PHARMACEUTICAL COMPANY LTD;
Osaka
TOYOBO CO LTD; Osaka

SWEDEN
PERBIO SCIENCE AB; Helsingborg

SWITZERLAND
ACTELION LTD; Allschwil
ALCON INC; Hunenberg
INSTITUT STRAUMANN AG; Basel
LONZA GROUP; Basel
MERCK SERONO SA; Geneva
NOVARTIS AG; Basel
ROCHE HOLDING LTD; Basel
SYNGENTA AG; Basel

THE NETHERLANDS
AKZO NOBEL NV; Arnhem
QIAGEN NV; Venlo

UNITED KINGDOM
ACAMBIS PLC; Cambridge
ALIZYME PLC; Cambridge
AMARIN CORPORATION PLC; London
ASTRAZENECA PLC; London
BIOCOMPATIBLES INTERNATIONAL PLC; Farnham
CAMBRIDGE ANTIBODY TECHNOLOGY LIMITED;
Cambridge
CELSIS INTERNATIONAL PLC; Cambourne
GE HEALTHCARE; Chalfont St. Giles
GLAXOSMITHKLINE PLC; Middlesex
ICON PLC; Dublin
SHIRE PLC; Basingstoke
SKYEPHARMA PLC; London

VERNALIS PLC; Winnersh
WHATMAN PLC; Brentford

INDEX BY REGIONS OF THE U.S. WHERE THE FIRMS HAVE LOCATIONS

WEST

ABAXIS INC
ABBOTT LABORATORIES
ABRAXIS BIOSCIENCE INC
ACCELRYS INC
ADVANCED BIONICS CORPORATION
ADVANCED CELL TECHNOLOGY INC
AEOLUS PHARMACEUTICALS INC
AFFYMETRIX INC
AGILENT TECHNOLOGIES INC
ALBANY MOLECULAR RESEARCH
ALCON INC
ALLERGAN INC
ALLIANCE PHARMACEUTICAL CORP
ALLOS THERAPEUTICS INC
ALPHARMA INC
ALZA CORP
AMGEN INC
AMYLIN PHARMACEUTICALS INC
ANGIOTECH PHARMACEUTICALS
APPLERA CORPORATION
APPLIED BIOSYSTEMS GROUP
APPLIED MOLECULAR EVOLUTION INC
ARADIGM CORPORATION
ARCADIA BIOSCIENCES
ARENA PHARMACEUTICALS INC
ARRAY BIOPHARMA INC
ASTRAZENECA PLC
AUTOIMMUNE INC
AVANIR PHARMACEUTICALS
AVI BIOPHARMA INC
AVIGEN INC
AVIVA BIOSCIENCES CORP
BASF AG
BAUSCH & LOMB INC
BAXTER INTERNATIONAL INC
BAYER AG
BAYER CORP
BAYER SCHERING PHARMA AG
BIO RAD LABORATORIES INC
BIOANALYTICAL SYSTEMS INC
BIOGEN IDEC INC
BIOMARIN PHARMACEUTICAL INC
BIOMET INC
BIOSITE INC
BIOTIME INC
BRISTOL MYERS SQUIBB CO
BURRILL & COMPANY
CALIPER LIFE SCIENCES
CANGENE CORP
CARDIOTECH INTERNATIONAL
CARDIUM THERAPEUTICS INC
CELERA GENOMICS GROUP
CELL GENESYS INC
CELL THERAPEUTICS INC

CELLEGY PHARMACEUTICALS
CEPHALON INC
CEPHEID
CERUS CORPORATION
CHARLES RIVER LABORATORIES
INTERNATIONAL INC
CHIRON CORP
CIPHERGEN BIOSYSTEMS INC
COMPUGEN LTD
CORTEX PHARMACEUTICALS INC
COVANCE INC
CV THERAPEUTICS INC
CYANOTECH CORPORATION
CYPRESS BIOSCIENCE INC
CYTRX CORPORATION
DECODE GENETICS INC
DENDREON CORPORATION
DEPOMED INC
DIAGNOSTIC PRODUCTS CORPORATION
DISCOVERY LABORATORIES INC
DIVERSA CORPORATION
DUPONT AGRICULTURE & NUTRITION
DURECT CORP
E I DU PONT DE NEMOURS & CO (DUPONT)
ELAN CORP PLC
ELI LILLY AND COMPANY
ELSEVIER MDL
EXELIXIS INC
EXELIXIS PLANT SCIENCES INC
GENELABS TECHNOLOGIES INC
GENENCOR INTERNATIONAL INC
GENENTECH INC
GEN-PROBE INC
GENZYME CORP
GERON CORPORATION
GILEAD SCIENCES INC
GLAXOSMITHKLINE PLC
HARVARD BIOSCIENCE INC
HEMAGEN DIAGNOSTICS INC
HESKA CORP
HOLLIS-EDEN PHARMACEUTICALS
HOSPIRA INC
HYCOR BIOMEDICAL INC
ICON PLC
ICOS CORPORATION
IDEXX LABORATORIES INC
IDM PHARMA INC
ILLUMINA INC
IMMUNE RESPONSE CORP (THE)
IMPAX LABORATORIES INC
IMS HEALTH INC
INCYTE CORP
INSITE VISION INC
INSPIRE PHARMACEUTICALS INC
INTEGRA LIFESCIENCES HOLDINGS CORP
INTERMUNE PHARMACEUTICALS INC
INTERNATIONAL ISOTOPES
INVITROGEN CORPORATION

IOMED INC
ISIS PHARMACEUTICALS INC
JAZZ PHARMACEUTICALS
JOHNSON & JOHNSON
KOSAN BIOSCIENCES INC
LA JOLLA PHARMACEUTICAL
LARGE SCALE BIOLOGY CORP
LEINER HEALTH PRODUCTS INC
LIGAND PHARMACEUTICALS INC
MALLINCKRODT INC
MARTEK BIOSCIENCES CORP
MAXYGEN INC
MDS INC
MEDAREX INC
MEDIMMUNE INC
MERCK & CO INC
MERCK KGAA
MIGENIX INC
MILLENNIUM PHARMACEUTICALS INC
MILLIPORE CORP
MIRAVANT MEDICAL TECHNOLOGIES
MONSANTO CO
MYRIAD GENETICS INC
NANOGEN INC
NASTECH PHARMACEUTICAL CO INC
NBTY INC
NEKTAR THERAPEUTICS
NEUROBIOLOGICAL TECHNOLOGIES INC
NEUROCRINE BIOSCIENCES INC
NORTH AMERICAN SCIENTIFIC
NOVARTIS AG
NOVAVAX INC
NOVO-NORDISK AS
NOVOZYMES
NUVELO INC
ONYX PHARMACEUTICALS INC
OSI PHARMACEUTICALS INC
OXIS INTERNATIONAL INC
PACIFIC BIOMETRICS INC
PAIN THERAPEUTICS INC
PAREXEL INTERNATIONAL
PDL BIOPHARMA
PEREGRINE PHARMACEUTICALS INC
PFIZER INC
PHARMACEUTICAL PRODUCT DEVELOPMENT INC
PHARMACYCLICS INC
PHARMANET DEVELOPMENT GROUP INC
PHARMION CORP
PONIARD PHARMACEUTICALS INC
PRA INTERNATIONAL
PROMEGA CORP
PROTEIN POLYMER TECHNOLOGIES
QIAGEN NV
QLT INC
QUEST DIAGNOSTICS INC
QUESTCOR PHARMACEUTICALS
QUINTILES TRANSNATIONAL CORP
RIGEL PHARMACEUTICALS INC

ROCHE HOLDING LTD
ROSETTA INPHARMATICS LLC
SANGAMO BIOSCIENCES INC
SANKYO CO LTD
SANOFI-AVENTIS
SCHERING-PLOUGH CORP
SCHIFF NUTRITION INTERNATIONAL INC
SCICLONE PHARMACEUTICALS
SCIOS INC
SEATTLE GENETICS
SENETEK PLC
SEQUENOM INC
SICOR INC
SIGA TECHNOLOGIES INC
SONUS PHARMACEUTICALS
SPECIALTY LABORATORIES INC
SPECTRAL DIAGNOSTICS INC
SPECTRUM PHARMACEUTICALS INC
STEMCELLS INC
STIEFEL LABORATORIES INC
STRATAGENE CORP
SUPERGEN INC
SYNBIOTICS CORP
SYNGENTA AG
TAKEDA PHARMACEUTICAL COMPANY LTD
TANOX INC
TAPESTRY PHARMACEUTICALS INC
TARGETED GENETICS CORP
TELIK INC
TEVA PHARMACEUTICAL INDUSTRIES
THERAGENICS CORP
THERMO FISHER SCIENTIFIC INC
TITAN PHARMACEUTICALS
TORREYPINES THERAPEUTICS INC
TRINITY BIOTECH PLC
URIGEN PHARMACEUTICALS INC
VALEANT PHARMACEUTICALS INTERNATIONAL
VAXGEN INC
VENTRIA BIOSCIENCE
VERTEX PHARMACEUTICALS INC
VICAL INC
WATSON PHARMACEUTICALS INC
WYETH
XOMA LTD
ZYMOGENETICS INC

SOUTHWEST
ABBOTT LABORATORIES
ACCESS PHARMACEUTICALS INC
ALCON INC
ALLERGAN INC
ALPHARMA INC
APPLIED BIOSYSTEMS GROUP
ASTELLAS PHARMA INC
ASTRAZENECA PLC
BASF AG
BAXTER INTERNATIONAL INC
BAYER AG

BAYER CORP
BIO RAD LABORATORIES INC
BIOMET INC
BIOMIRA INC
BRISTOL MYERS SQUIBB CO
CARRINGTON LABORATORIES INC
CHARLES RIVER LABORATORIES
INTERNATIONAL INC
CIPHERGEN BIOSYSTEMS INC
COVANCE INC
DUPONT AGRICULTURE & NUTRITION
E I DU PONT DE NEMOURS & CO (DUPONT)
EISAI CO LTD
ELI LILLY AND COMPANY
ENCYSIVE PHARMACEUTICALS INC
GENZYME CORP
GLAXOSMITHKLINE PLC
HOSPIRA INC
IDEXX LABORATORIES INC
INTROGEN THERAPEUTICS INC
JOHNSON & JOHNSON
LEXICON PHARMACEUTICALS INC
LUMINEX CORPORATION
MDS INC
MEDICIS PHARMACEUTICAL CORP
MERCK & CO INC
MILLIPORE CORP
MONSANTO CO
MYLAN LABORATORIES INC
NBTY INC
NOVO-NORDISK AS
ORCHID CELLMARK INC
PFIZER INC
PONIARD PHARMACEUTICALS INC
PRA INTERNATIONAL
QUEST DIAGNOSTICS INC
QUINTILES TRANSNATIONAL CORP
REPROS THERAPEUTICS INC
ROCHE HOLDING LTD
SANOFI-AVENTIS
SCHERING-PLOUGH CORP
STRATAGENE CORP
SYNGENTA AG
TANOX INC
TEVA PHARMACEUTICAL INDUSTRIES
THERAGENICS CORP
THERMO FISHER SCIENTIFIC INC
VIRBAC CORP
WYETH
ZILA INC

MIDWEST
AASTROM BIOSCIENCES INC
ABBOTT LABORATORIES
ABRAXIS BIOSCIENCE INC
AKORN INC
AKZO NOBEL NV
ALCON INC

ALKERMES INC
ALPHARMA INC
ALTANA AG
ANGIOTECH PHARMACEUTICALS
ARRYX INC
ASTELLAS PHARMA INC
ASTRAZENECA PLC
AVANT IMMUNOTHERAPEUTICS
BARR PHARMACEUTICALS INC
BASF AG
BAUSCH & LOMB INC
BAXTER INTERNATIONAL INC
BAYER AG
BAYER CORP
BIOANALYTICAL SYSTEMS INC
BIOMET INC
BRISTOL MYERS SQUIBB CO
CAMBREX CORP
CARACO PHARMACEUTICAL LABORATORIES
CARDIOTECH INTERNATIONAL
CELSIS INTERNATIONAL PLC
CEPHALON INC
CHARLES RIVER LABORATORIES
INTERNATIONAL INC
CHIRON CORP
COVANCE INC
CSL LIMITED
DECODE GENETICS INC
DENDRITE INTERNATIONAL INC
DOW AGROSCIENCES LLC
DUPONT AGRICULTURE & NUTRITION
E I DU PONT DE NEMOURS & CO (DUPONT)
EISAI CO LTD
ELI LILLY AND COMPANY
ELSEVIER MDL
ENZON PHARMACEUTICALS INC
FOREST LABORATORIES INC
GE HEALTHCARE
GENENCOR INTERNATIONAL INC
GENZYME CORP
GLAXOSMITHKLINE PLC
HESKA CORP
HOSPIRA INC
ICON PLC
IDEXX LABORATORIES INC
IMMTECH INTERNATIONAL
INTEGRA LIFESCIENCES HOLDINGS CORP
INVITROGEN CORPORATION
JOHNSON & JOHNSON
KENDLE INTERNATIONAL INC
KING PHARMACEUTICALS INC
KV PHARMACEUTICAL CO
LARGE SCALE BIOLOGY CORP
LIFECORE BIOMEDICAL INC
LONZA GROUP
MALLINCKRODT INC
MARTEK BIOSCIENCES CORP
MDS INC

MEDTOX SCIENTIFIC INC
MERCK & CO INC
MERCK KGAA
MGI PHARMA INC
MILLIPORE CORP
MONSANTO CO
MYLAN LABORATORIES INC
NANOBIO CORPORATION
NBTY INC
NEOGEN CORPORATION
NEOPHARM INC
NEOPROBE CORPORATION
NOVARTIS AG
NOVO-NORDISK AS
ORCHID CELLMARK INC
PAREXEL INTERNATIONAL
PDL BIOPHARMA
PFIZER INC
PHARMACEUTICAL PRODUCT DEVELOPMENT INC
PHARMANET DEVELOPMENT GROUP INC
POLYDEX PHARMACEUTICALS
PRA INTERNATIONAL
PROMEGA CORP
QUEST DIAGNOSTICS INC
QUINTILES TRANSNATIONAL CORP
ROCHE HOLDING LTD
SANOFI-AVENTIS
SCHERING-PLOUGH CORP
SIGMA ALDRICH CORP
SPECTRAL DIAGNOSTICS INC
SYNGENTA AG
SYNOVIS LIFE TECHNOLOGIES INC
TAKEDA PHARMACEUTICAL COMPANY LTD
TECHNE CORP
TEVA PHARMACEUTICAL INDUSTRIES
THERMO FISHER SCIENTIFIC INC
THIRD WAVE TECHNOLOGIES INC
TRINITY BIOTECH PLC
TRIPOS INC
VIACELL INC
VIRBAC CORP
VYSIS INC
WATSON PHARMACEUTICALS INC
WYETH

SOUTHEAST
AKZO NOBEL NV
ALCON INC
ALPHARMA INC
AMGEN INC
ANGIOTECH PHARMACEUTICALS
ASTRAZENECA PLC
ATHEROGENICS INC
AXCAN PHARMA INC
BASF AG
BAUSCH & LOMB INC
BAXTER INTERNATIONAL INC
BAYER CORP

BIOCRYST PHARMACEUTICALS INC
BIOMET INC
BRISTOL MYERS SQUIBB CO
CANGENE CORP
CHARLES RIVER LABORATORIES INTERNATIONAL INC
CHATTEM INC
COVANCE INC
CSL LIMITED
DENDRITE INTERNATIONAL INC
DOR BIOPHARMA INC
DR REDDY'S LABORATORIES LIMITED
DUPONT AGRICULTURE & NUTRITION
E I DU PONT DE NEMOURS & CO (DUPONT)
EISAI CO LTD
ELAN CORP PLC
ELI LILLY AND COMPANY
ELSEVIER MDL
GENZYME CORP
GLAXOSMITHKLINE PLC
HOSPIRA INC
ICON PLC
IDEXX LABORATORIES INC
JOHNSON & JOHNSON
KING PHARMACEUTICALS INC
KOS PHARMACEUTICALS INC
MDS INC
MERCK & CO INC
MILLIPORE CORP
MONSANTO CO
NABI BIOPHARMACEUTICALS
NBTY INC
NEKTAR THERAPEUTICS
NEOGEN CORPORATION
NOVARTIS AG
NOVEN PHARMACEUTICALS
NOVO-NORDISK AS
ORCHID CELLMARK INC
PAREXEL INTERNATIONAL
PFIZER INC
PHARMACEUTICAL PRODUCT DEVELOPMENT INC
PHARMANET DEVELOPMENT GROUP INC
QUEST DIAGNOSTICS INC
QUINTILES TRANSNATIONAL CORP
RANBAXY LABORATORIES LIMITED
ROCHE HOLDING LTD
SANOFI-AVENTIS
SCHERING-PLOUGH CORP
SCIELE PHARMA INC
STIEFEL LABORATORIES INC
SYNGENTA AG
TEVA PHARMACEUTICAL INDUSTRIES
THERAGENICS CORP
THERMO FISHER SCIENTIFIC INC
UCB SA
UNITED THERAPEUTICS CORP
VIRAGEN INC
WATSON PHARMACEUTICALS INC

WYETH

NORTHEAST
ABBOTT LABORATORIES
ABRAXIS BIOSCIENCE INC
ACAMBIS PLC
ACCELRYS INC
ADOLOR CORP
ADVANCED CELL TECHNOLOGY INC
AFFYMETRIX INC
AGILENT TECHNOLOGIES INC
AKORN INC
AKZO NOBEL NV
ALBANY MOLECULAR RESEARCH
ALCON INC
ALEXION PHARMACEUTICALS INC
ALFACELL CORPORATION
ALKERMES INC
ALPHARMA INC
ALSERES PHARMACEUTICALS INC
ALTANA AG
ALTEON INC
AMGEN INC
ANGIOTECH PHARMACEUTICALS
ANIKA THERAPEUTICS INC
ANTIGENICS INC
APPLERA CORPORATION
APPLIED BIOSYSTEMS GROUP
ARIAD PHARMACEUTICALS INC
ARQULE INC
ASTELLAS PHARMA INC
ASTRAZENECA PLC
AVANT IMMUNOTHERAPEUTICS
AVAX TECHNOLOGIES INC
BARR PHARMACEUTICALS INC
BASF AG
BAUSCH & LOMB INC
BAXTER INTERNATIONAL INC
BAYER AG
BAYER CORP
BAYER SCHERING PHARMA AG
BENTLEY PHARMACEUTICALS INC
BIO RAD LABORATORIES INC
BIOCOMPATIBLES INTERNATIONAL PLC
BIOGEN IDEC INC
BIOMET INC
BIOMIRA INC
BIOPURE CORPORATION
BIORELIANCE CORP
BIOSPHERE MEDICAL INC
BIOVAIL CORPORATION
BRADLEY PHARMACEUTICALS INC
BRISTOL MYERS SQUIBB CO
CALIPER LIFE SCIENCES
CAMBREX CORP
CANGENE CORP
CARDIOTECH INTERNATIONAL
CELERA GENOMICS GROUP

CELGENE CORP
CELLEGY PHARMACEUTICALS
CEL-SCI CORPORATION
CELSIS INTERNATIONAL PLC
CEPHALON INC
CHARLES RIVER LABORATORIES
INTERNATIONAL INC
CHESAPEAKE BIOLOGICAL LAB
CHIRON CORP
CIPHERGEN BIOSYSTEMS INC
COLLAGENEX PHARMACEUTICALS INC
COLUMBIA LABORATORIES INC
COMMONWEALTH BIOTECHNOLOGIES INC
CORTEX PHARMACEUTICALS INC
COVANCE INC
CSL LIMITED
CUBIST PHARMACEUTICALS INC
CURAGEN CORPORATION
CURIS INC
CYTOGEN CORPORATION
CYTRX CORPORATION
DECODE GENETICS INC
DENDRITE INTERNATIONAL INC
DIAGNOSTIC PRODUCTS CORPORATION
DIGENE CORPORATION
DISCOVERY LABORATORIES INC
DOV PHARMACEUTICAL INC
DR REDDY'S LABORATORIES LIMITED
DSM PHARMACEUTICALS INC
DUPONT AGRICULTURE & NUTRITION
DUSA PHARMACEUTICALS INC
DYAX CORP
E I DU PONT DE NEMOURS & CO (DUPONT)
EISAI CO LTD
ELAN CORP PLC
ELI LILLY AND COMPANY
ELSEVIER MDL
EMBREX INC
EMISPHERE TECHNOLOGIES INC
ENCORIUM GROUP INC
ENDO PHARMACEUTICALS HOLDINGS INC
ENTREMED INC
ENZO BIOCHEM INC
ENZON PHARMACEUTICALS INC
EPIX PHARMACEUTICALS INC
ERESEARCH TECHNOLOGY INC
EVOTEC OAI AG
E-Z-EM INC
FOREST LABORATORIES INC
GENAERA CORPORATION
GENE LOGIC INC
GENENCOR INTERNATIONAL INC
GENEREX BIOTECHNOLOGY
GENTA INC
GENVEC INC
GENZYME BIOSURGERY
GENZYME CORP
GENZYME ONCOLOGY

GILEAD SCIENCES INC
GLAXOSMITHKLINE PLC
GLYCOGENESYS INC
GTC BIOTHERAPEUTICS INC
HARVARD BIOSCIENCE INC
HEMAGEN DIAGNOSTICS INC
HEMISPHERX BIOPHARMA INC
HI-TECH PHARMACAL CO INC
HOSPIRA INC
HUMAN GENOME SCIENCES INC
ICON PLC
IDERA PHARMACEUTICALS INC
IDEXX LABORATORIES INC
ILLUMINA INC
IMCLONE SYSTEMS INC
IMMTECH INTERNATIONAL
IMMUCELL CORPORATION
IMMUNE RESPONSE CORP (THE)
IMMUNOGEN INC
IMMUNOMEDICS INC
IMPAX LABORATORIES INC
IMS HEALTH INC
INCYTE CORP
INDEVUS PHARMACEUTICALS INC
INFINITY PHARMACEUTICALS INC
INSMED INCORPORATED
INSPIRE PHARMACEUTICALS INC
INSTITUT STRAUMANN AG
INTEGRA LIFESCIENCES HOLDINGS CORP
INVITROGEN CORPORATION
JOHNSON & JOHNSON
KENDLE INTERNATIONAL INC
KERYX BIOPHARMACEUTICALS INC
KING PHARMACEUTICALS INC
KOS PHARMACEUTICALS INC
LANNETT COMPANY INC
LEINER HEALTH PRODUCTS INC
LEXICON PHARMACEUTICALS INC
LIFE SCIENCES RESEARCH
LIFECELL CORPORATION
LONZA GROUP
MACROCHEM CORPORATION
MALLINCKRODT INC
MANHATTAN PHARMACEUTICALS INC
MARTEK BIOSCIENCES CORP
MATRITECH INC
MDS INC
MEDAREX INC
MEDICINES CO (THE)
MEDIMMUNE INC
MEDTOX SCIENTIFIC INC
MERCK & CO INC
MERCK KGAA
MERCK SERONO SA
MERIDIAN BIOSCIENCE INC
MGI PHARMA INC
MILLENNIUM PHARMACEUTICALS INC
MILLIPORE CORP

MONSANTO CO
MYLAN LABORATORIES INC
NABI BIOPHARMACEUTICALS
NASTECH PHARMACEUTICAL CO INC
NBTY INC
NEOSE TECHNOLOGIES INC
NEUROBIOLOGICAL TECHNOLOGIES INC
NEUROGEN CORP
NEXMED INC
NORTH AMERICAN SCIENTIFIC
NOVARTIS AG
NOVAVAX INC
NOVO-NORDISK AS
NOVOZYMES
NPS PHARMACEUTICALS INC
NUTRITION 21 INC
ORASURE TECHNOLOGIES INC
ORCHID CELLMARK INC
ORGANOGENESIS INC
OSCIENT PHARMACEUTICALS INC
OSI PHARMACEUTICALS INC
OXIGENE INC
PALATIN TECHNOLOGIES INC
PAR PHARMACEUTICAL COMPANIES INC
PAREXEL INTERNATIONAL
PDL BIOPHARMA
PENWEST PHARMACEUTICALS CO
PERRIGO CO
PFIZER INC
PHARMACEUTICAL PRODUCT DEVELOPMENT INC
PHARMACOPEIA DRUG DISCOVERY
PHARMANET DEVELOPMENT GROUP INC
PHARMOS CORP
POZEN INC
PRA INTERNATIONAL
PROGENICS PHARMACEUTICALS
PROMEGA CORP
QIAGEN NV
QUEST DIAGNOSTICS INC
QUINTILES TRANSNATIONAL CORP
RANBAXY LABORATORIES LIMITED
REGENERON PHARMACEUTICALS INC
REPLIGEN CORPORATION
ROCHE HOLDING LTD
SALIX PHARMACEUTICALS
SANOFI-AVENTIS
SAVIENT PHARMACEUTICALS INC
SCHERING-PLOUGH CORP
SCIELE PHARMA INC
SEPRACOR INC
SEQUENOM INC
SERACARE LIFE SCIENCES INC
SHIRE PLC
SIGA TECHNOLOGIES INC
SPECTRAL DIAGNOSTICS INC
STIEFEL LABORATORIES INC
SYNBIOTICS CORP
SYNGENTA AG

TAPESTRY PHARMACEUTICALS INC
TARO PHARMACEUTICAL INDUSTRIES
TEVA PHARMACEUTICAL INDUSTRIES
THERMO FISHER SCIENTIFIC INC
TOYOBO CO LTD
TRIMERIS INC
TRINITY BIOTECH PLC
UCB SA
UNIGENE LABORATORIES
UNITED THERAPEUTICS CORP
UNITED-GUARDIAN INC
VERNALIS PLC
VERTEX PHARMACEUTICALS INC
VI TECHNOLOGIES INC
VIACELL INC
VION PHARMACEUTICALS INC
VIROPHARMA INC
WARNER CHILCOTT PLC
WATSON PHARMACEUTICALS INC
WHATMAN PLC
WYETH
XECHEM INTERNATIONAL

INDEX OF FIRMS WITH OPERATIONS OUTSIDE THE U.S.

4SC AG
ABAXIS INC
ABBOTT LABORATORIES
ABRAXIS BIOSCIENCE INC
ACAMBIS PLC
ACCELRYS INC
ACCESS PHARMACEUTICALS INC
ACTELION LTD
ADVANCED BIONICS CORPORATION
AETERNA ZENTARIS INC
AFFYMETRIX INC
AGILENT TECHNOLOGIES INC
AKZO NOBEL NV
ALBANY MOLECULAR RESEARCH
ALCON INC
ALEXION PHARMACEUTICALS INC
ALIZYME PLC
ALKERMES INC
ALLERGAN INC
ALPHARMA INC
ALTANA AG
AMARIN CORPORATION PLC
AMGEN INC
AMYLIN PHARMACEUTICALS INC
ANGIOTECH PHARMACEUTICALS
APPLERA CORPORATION
APPLIED BIOSYSTEMS GROUP
ASTELLAS PHARMA INC
ASTRAZENECA PLC
AVAX TECHNOLOGIES INC
AXCAN PHARMA INC
BASF AG
BAUSCH & LOMB INC
BAXTER INTERNATIONAL INC
BAYER AG
BAYER SCHERING PHARMA AG
BENTLEY PHARMACEUTICALS INC
BIO RAD LABORATORIES INC
BIOANALYTICAL SYSTEMS INC
BIOCOMPATIBLES INTERNATIONAL PLC
BIOGEN IDEC INC
BIOMARIN PHARMACEUTICAL INC
BIOMET INC
BIOMIRA INC
BIORELIANCE CORP
BIOSITE INC
BIOSPHERE MEDICAL INC
BIOTECH HOLDINGS LTD
BIOVAIL CORPORATION
BRISTOL MYERS SQUIBB CO
CALIPER LIFE SCIENCES
CAMBREX CORP
CAMBRIDGE ANTIBODY TECHNOLOGY LIMITED
CANGENE CORP
CARDIOME PHARMA CORP

CARDIOTECH INTERNATIONAL
CARRINGTON LABORATORIES INC
CELGENE CORP
CELL THERAPEUTICS INC
CELSIS INTERNATIONAL PLC
CEPHALON INC
CEPHEID
CERUS CORPORATION
CHARLES RIVER LABORATORIES
INTERNATIONAL INC
CHATTEM INC
CHIRON CORP
CIPHERGEN BIOSYSTEMS INC
CML HEALTHCARE INCOME FUND
COLLAGENEX PHARMACEUTICALS INC
COLUMBIA LABORATORIES INC
COMPUGEN LTD
COVANCE INC
COVIDIEN LTD
CSL LIMITED
CV THERAPEUTICS INC
CYANOTECH CORPORATION
DAIICHI SANKYO CO LTD
DECODE GENETICS INC
DENDRITE INTERNATIONAL INC
DIAGNOSTIC PRODUCTS CORPORATION
DIGENE CORPORATION
DOW AGROSCIENCES LLC
DR REDDY'S LABORATORIES LIMITED
DRAXIS HEALTH INC
DSM PHARMACEUTICALS INC
DUPONT AGRICULTURE & NUTRITION
DUSA PHARMACEUTICALS INC
DYAX CORP
E I DU PONT DE NEMOURS & CO (DUPONT)
EISAI CO LTD
ELAN CORP PLC
ELI LILLY AND COMPANY
ELSEVIER MDL
EMBREX INC
ENCORIUM GROUP INC
EPIX PHARMACEUTICALS INC
ERESEARCH TECHNOLOGY INC
EVOTEC OAI AG
EXELIXIS INC
EXPERIMENTAL & APPLIED SCIENCES INC
E-Z-EM INC
FLAMEL TECHNOLOGIES SA
FORBES MEDI-TECH INC
FOREST LABORATORIES INC
GE HEALTHCARE
GENENCOR INTERNATIONAL INC
GENENTECH INC
GENEREX BIOTECHNOLOGY
GENZYME CORP
GILEAD SCIENCES INC
GLAXOSMITHKLINE PLC
HARVARD BIOSCIENCE INC

HEMAGEN DIAGNOSTICS INC
HEMISPHERX BIOPHARMA INC
HESKA CORP
HOSPIRA INC
HYCOR BIOMEDICAL INC
ICON PLC
IDEXX LABORATORIES INC
IDM PHARMA INC
ILLUMINA INC
IMMTECH INTERNATIONAL
IMMUNOMEDICS INC
IMS HEALTH INC
INSTITUT STRAUMANN AG
INTEGRA LIFESCIENCES HOLDINGS CORP
INVITROGEN CORPORATION
JOHNSON & JOHNSON
KENDLE INTERNATIONAL INC
LA JOLLA PHARMACEUTICAL
LEINER HEALTH PRODUCTS INC
LIFE SCIENCES RESEARCH
LIFECORE BIOMEDICAL INC
LONZA GROUP
LORUS THERAPEUTICS INC
LUMINEX CORPORATION
MALLINCKRODT INC
MATRITECH INC
MAXYGEN INC
MDS INC
MEDICINES CO (THE)
MEDIMMUNE INC
MERCK & CO INC
MERCK KGAA
MERCK SERONO SA
MERIDIAN BIOSCIENCE INC
MIGENIX INC
MILLENNIUM PHARMACEUTICALS INC
MILLIPORE CORP
MONSANTO CO
MYLAN LABORATORIES INC
NANOGEN INC
NBTY INC
NEKTAR THERAPEUTICS
NEOGEN CORPORATION
NEXMED INC
NORTH AMERICAN SCIENTIFIC
NOVARTIS AG
NOVO-NORDISK AS
NOVOZYMES
NPS PHARMACEUTICALS INC
NYCOMED
ORASURE TECHNOLOGIES INC
ORCHID CELLMARK INC
ORGANOGENESIS INC
OSI PHARMACEUTICALS INC
OXIGENE INC
PAREXEL INTERNATIONAL
PDL BIOPHARMA
PERBIO SCIENCE AB

PEREGRINE PHARMACEUTICALS INC   ZILA INC
PERRIGO CO
PFIZER INC
PHARMACEUTICAL PRODUCT DEVELOPMENT INC
PHARMANET DEVELOPMENT GROUP INC
PHARMION CORP
PHARMOS CORP
POLYDEX PHARMACEUTICALS
PRA INTERNATIONAL
PROMEGA CORP
QIAGEN NV
QLT INC
QUEST DIAGNOSTICS INC
QUINTILES TRANSNATIONAL CORP
RANBAXY LABORATORIES LIMITED
ROCHE HOLDING LTD
SANKYO CO LTD
SANOFI-AVENTIS
SAVIENT PHARMACEUTICALS INC
SCHERING-PLOUGH CORP
SCHIFF NUTRITION INTERNATIONAL INC
SCICLONE PHARMACEUTICALS
SENETEK PLC
SEQUENOM INC
SHIRE PLC
SHIRE-BIOCHEM INC
SICOR INC
SIGMA ALDRICH CORP
SIMCERE PHARMACEUTICAL GROUP
SKYEPHARMA PLC
SPECTRAL DIAGNOSTICS INC
STIEFEL LABORATORIES INC
STRATAGENE CORP
SYNBIOTICS CORP
SYNGENTA AG
TAKEDA PHARMACEUTICAL COMPANY LTD
TARO PHARMACEUTICAL INDUSTRIES
TECHNE CORP
TEVA PHARMACEUTICAL INDUSTRIES
THERMO FISHER SCIENTIFIC INC
TORREYPINES THERAPEUTICS INC
TOYOBO CO LTD
TRINITY BIOTECH PLC
UCB SA
UNITED THERAPEUTICS CORP
VALEANT PHARMACEUTICALS INTERNATIONAL
VASOGEN INC
VERNALIS PLC
VERTEX PHARMACEUTICALS INC
VIACELL INC
VIRAGEN INC
VIRBAC CORP
VYSIS INC
WARNER CHILCOTT PLC
WATSON PHARMACEUTICALS INC
WHATMAN PLC
WYETH
XECHEM INTERNATIONAL

Individual Profiles
On Each Of
THE BIOTECH 400

4SC AG

www.4sc.de

Industry Group Code: 325412 **Ranks within this company's industry group:** Sales: Profits:

Drugs:		Other:	Clinical:	Computers:	Other:
Discovery:	Y	AgriBio:	Trials/Services:	Hardware:	Specialty Services:
Licensing:	Y	Genomics/Proteomics:	Laboratories:	Software:	Consulting:
Manufacturing:		Tissue Replacement:	Equipment/Supplies:	Arrays:	Blood Collection:
Development:	Y		Research/Development Svcs.:	Database Management:	Drug Delivery:
Generics:			Diagnostics:		Drug Distribution:

TYPES OF BUSINESS:

Drug Discovery & Pre-Clinical Development
Drugs-Cancer
Drugs-Inflammation
Drugs-Autoimmune Diseases
High Throughput Screening

BRANDS/DIVISIONS/AFFILIATES:

4SCan Technology
SC12267

CONTACTS: *Note: Officers with more than one job title may be intentionally listed here more than once.*

Ulrich Dauer, CEO
Gerhard Keilhauer, COO
Enno Spillner, CFO
Britta-Andrea Jurecka, Mgr.-Human Resources
Daniel Vitt, Chief Science Officer
Gerhard Keilhauer, Chief Dev. Officer
Manfred Groppel, Dir.-Bus. Dev.
Bettina von Klitzing, Mgr.-Public Rel.
Bettina von Klitzing, Mgr.-Investor Rel.
Andrea Aschenbrenner, Mgr.-Customer Relations
Charlotte Herrlinger, VP-Medicine
Bernd Kramer, Head-Chem. & Bioinformatics
Robert Doblhofer, Head-Pharmacology & Preclinical Dev.
Jorg Neermann, Chmn.
Heike Fischer, Dir.-Purchasing

Phone: 49-89-700-763-0	Fax: 49-89-70-07-63-29
Toll-Free:	
Address: Am Klopferspitz 19a, Planegg-Martinsried, D-82152 Germany	

GROWTH PLANS/SPECIAL FEATURES:

4SC AG is a German-based drug discovery and development company which focuses on possible treatments for cancer and inflammatory diseases. The company develops its compounds to the clinical phase, after which they will be licensed to the biopharmaceutical industry for further testing, development and commercialization. 4SC's technology platform uses a virtual High Throughput Screening technology known as 4Scan, which allows the high volume analysis of molecules based on protein structures and/or biological activity on a computer, rather than in a laboratory. Using this technology the company has compiled a virtual library of more than 5 million small molecules drug candidates and additional combinatorial libraries of more than 10 million chemical scaffolds. It has collaborative efforts with Axxima, researching human cytomegalovirus treatments using small molecule kinase inhibitors; Beohringer Ingelheim, researching drugs based on ligands; Esteve, identifying drug candidates for CNS related diseases; Mutabilis S.A., researching infectious disease related drugs; ProQinase, researching anticancer drugs based on protein kinase inhibitors; Switch Biotech, developing active compounds to treat psoriasis; Wilex, researching second-generation urokinase inhibitors; and Schwarz Pharma, identifying pre-development drug candidates to prevent or treat urological disorders. It also has collaborations with Recordati, Sanofi-Aventis, Sanwa Kagaku Kenkyusho and Schering, working on undisclosed targets. Currently, the firm has six compounds in development: two for cancer, two for rheumatoid arthritis, one for multiple sclerosis and one for viral infection and oncology. 4SC's most advanced drug candidate, SC12267, currently targeting rheumatoid arthritis, has successfully completed Phase I studies, and preparations for Phase IIa trials are well under way. In June 2007, SC12267 received positive preclinical proof of concept results for the treatment of chronic inflammatory bowel diseases, suggesting it could be used in the treatment of other autoimmune diseases.

4SC offers internships and helps postgraduate students write diploma theses or dissertations on subjects relevant to ongoing research.

FINANCIALS: Sales and profits are in thousands of dollars—add 000 to get the full amount. 2006 Note: Financial information for 2006 was not available for all companies at press time.

2006 Sales: $	2006 Profits: $	U.S. Stock Ticker: Private
2005 Sales: $	2005 Profits: $	Int'l Ticker: Int'l Exchange:
2004 Sales: $	2004 Profits: $	Employees:
2003 Sales: $	2003 Profits: $	Fiscal Year Ends:
2002 Sales: $	2002 Profits: $	Parent Company:

SALARIES/BENEFITS:

Pension Plan:	ESOP Stock Plan:	Profit Sharing:	Top Exec. Salary: $184,952	Bonus: $35,886
Savings Plan:	Stock Purch. Plan:		Second Exec. Salary: $182,192	Bonus: $46,928

OTHER THOUGHTS:

Apparent Women Officers or Directors: 3
Hot Spot for Advancement for Women/Minorities: Y

LOCATIONS: ("Y" = Yes)

West:	Southwest:	Midwest:	Southeast:	Northeast:	International:
					Y

AASTROM BIOSCIENCES INC   www.aastrom.com

Industry Group Code: 339113 **Ranks within this company's industry group:** Sales: 16   Profits: 14

Drugs:		Other:		Clinical:		Computers:		Other:	
Discovery:		AgriBio:		Trials/Services:	Y	Hardware:		Specialty Services:	
Licensing:		Genomics/Proteomics:		Laboratories:		Software:		Consulting:	
Manufacturing:		Tissue Replacement:	Y	Equipment/Supplies:		Arrays:		Blood Collection:	
Development:				Research/Development Svcs.:		Database Management:		Drug Delivery:	
Generics:				Diagnostics:				Drug Distribution:	

TYPES OF BUSINESS:
Cell Products Development
Regenerative medicine products
Clinical & Pre-clinical Development

BRANDS/DIVISIONS/AFFILIATES:
Tissue Repair Cells

CONTACTS: *Note: Officers with more than one job title may be intentionally listed here more than once.*
George W. Dunbar, CEO
George W. Dunbar, Pres.
Gerald D. Brennan, Jr., CFO
Martin C. Peters, Sr. Dir.-Strategic Mktg.
Ronnda L. Bartel, VP-R&D
Gerald D. Brennan, Jr., VP-Admin.
Sheldon A. Schaffer, VP-Corp. Dev. & Intellectual Property
Gerald D. Brennan, Jr., VP-Finance
Robert J. Bard, VP-Regulatory/Clinical Affairs
Elmar R. Burchardt, VP-Medical Affairs
Stephen G. Sudovar, Chmn.

Phone: 734-930-5555	Fax: 734-665-0485
Toll-Free:	
Address: 24 Frank Lloyd Wright Dr., Ann Arbor, MI 48105 US	

GROWTH PLANS/SPECIAL FEATURES:
Aastrom Biosciences, Inc. is a development stage company focused on the development of autologous cell products for use in regenerative medicine. The company's pre-clinical and clinical products development programs utilize patient-derived bone marrow stem and progenitor cell populations, which are being investigated for their ability to aid in the regeneration of tissues such as vascular, bone, cardiac and neural. The firm's primary business is to develop Tissue Repair Cell (TRC) based products for use in multiple therapeutic areas. The TRC platform technology is based on its cell products (a unique cell mixture containing large numbers of stromal, stem and progenitor cells, produced outside the body from a small amount of bone marrow taken from a patient) and the means to produce these products in an automated process. TRC products were used in 225 patients. The pre-clinical data for the TRCs showed a substantial increase in the stem and progenitor cells that can develop into tissues such as hematopoietic (i.e., blood forming) or mesenchymal (i.e., developing into tissues characteristic of certain internal organs), as well as stromal progenitor cells that produce various growth factors. The company demonstrated in the laboratory that TRCs can progress into bone cell and blood vessel cell lineages. Based on these pre-clinical observations, the TRCs are currently in active clinical trials for bone regeneration and vascular regeneration applications. Aastrom reported positive interim clinical trial results for TRCs suggesting both the clinical safety and the ability of TRCs to induce tissue regeneration in long bone fractures and jaw bone reconstruction. The company currently intends to pursue TRC-based cell products for the following therapeutic areas: vascular tissue regeneration, bone tissue regeneration, cardiac tissue regeneration and neural tissue regeneration. The firm developed a patented manufacturing system to produce human cells for clinical use. Aastrom owns 25 U.S. patents.

FINANCIALS: Sales and profits are in thousands of dollars—add 000 to get the full amount. 2006 Note: Financial information for 2006 was not available for all companies at press time.
2006 Sales: $ 863	2006 Profits: $-16,475	U.S. Stock Ticker: ASTM
2005 Sales: $ 909	2005 Profits: $-11,811	Int'l Ticker: Int'l Exchange:
2004 Sales: $1,300	2004 Profits: $-10,500	Employees: 59
2003 Sales: $ 800	2003 Profits: $-9,600	Fiscal Year Ends: 6/30
2002 Sales: $ 900	2002 Profits: $-7,900	Parent Company:

SALARIES/BENEFITS:
Pension Plan:	ESOP Stock Plan:	Profit Sharing:	Top Exec. Salary: $345,000	Bonus: $60,000
Savings Plan:	Stock Purch. Plan:		Second Exec. Salary: $272,500	Bonus: $35,000

OTHER THOUGHTS:
Apparent Women Officers or Directors: 1
Hot Spot for Advancement for Women/Minorities:

LOCATIONS: ("Y" = Yes)
West:	Southwest:	Midwest:	Southeast:	Northeast:	International:
		Y			

Note: Financial information, benefits and other data can change quickly and may vary from those stated here.

ABAXIS INC   www.abaxis.com

Industry Group Code: 339113 Ranks within this company's industry group: Sales: 8   Profits: 8

Drugs:	Other:	Clinical:		Computers:	Other:
Discovery:	AgriBio:	Trials/Services:		Hardware:	Specialty Services:
Licensing:	Genomics/Proteomics:	Laboratories:		Software:	Consulting:
Manufacturing:	Tissue Replacement:	Equipment/Supplies:	Y	Arrays:	Blood Collection:
Development:		Research/Development Svcs.:		Database Management:	Drug Delivery:
Generics:		Diagnostics:	Y		Drug Distribution:

TYPES OF BUSINESS:
Point-of-Care Blood Analyzer Systems Equipment
Veterinary Blood Analyzer Systems
Reagents & Supplies

BRANDS/DIVISIONS/AFFILIATES:
VetScan VS2
VetScan Classic
Piccolo Classic
Piccolo Xpress
VetScan HM2
Orbos Discrete Lyophilization Process

CONTACTS: *Note: Officers with more than one job title may be intentionally listed here more than once.*
Clinton H. Severson, CEO
Robert B. Milder, COO
Clinton H.Severson, Pres.
Alberto R. Santa Ines, CFO
Christopher Bernard, VP-Mktg. & Sales, Medical Market
Kenneth P. Aron, VP-R&D
Alberto R. Santa Ines, VP-Finance
Martin Mulroy, VP-Mktg. & Sales, Veterinary Market
Clinton H. Severson, Chmn.
Vladimir E. Ostoich, VP-Government Affairs & Mktg., Pacific Rim

Phone: 510-675-6500	Fax: 510-441-6150
Toll-Free:	
Address: 3240 Whipple Rd., Union City, CA 94587 US	

GROWTH PLANS/SPECIAL FEATURES:
Abaxis, Inc. develops, manufactures and markets portable blood analysis systems for use in veterinary or human patient-care settings to provide clinicians with rapid blood constituent measurements. The systems are marketed as VetScan VS2 and VetScan Classic in the veterinary market and as the Piccolo Classic and Piccolo Xpress in the human medical market. Abaxis' blood analysis systems consist of a compact analyzer and a series of single-use plastic discs, called reagent discs, containing all the chemicals required to perform a panel of up to 27 diagnostic tests, 21 of which are also marketed for the veterinary market, with two unique to the veterinary market. In addition to blood analysis systems, Abaxis sells the VetScan HM2 (formerly VetScan HMII) hematology analyzer which provides an 18-parameter blood count analysis, including a three-part white blood cell differential. To produce the dry reagents used in the reagent disks, Abaxis uses its Orbos Discrete Lyophilization Process (Orbos process), a process which freeze dries reagents in small quantities, enabling efficient manufacturing of reagents in a convenient and stable format. Abaxis licenses its Orbos process to bioMerieux, Cepheid and GE Healthcare. In February 2007, Abaxis entered into a distribution agreement with Cardinal Health for its Piccolo Xpress and medical reagent disks and in July 2007, it signed a distribution agreement with McKesson Medical-Surgical for the same products.

Abaxis offers its employees flexible time off, a flexible spending account plan and memberships at financial institutions, retail outlets and clubs.

FINANCIALS: Sales and profits are in thousands of dollars—add 000 to get the full amount. 2006 Note: Financial information for 2006 was not available for all companies at press time.

2006 Sales: $68,928	2006 Profits: $7,475	**U.S. Stock Ticker:** ABAX
2005 Sales: $52,758	2005 Profits: $4,851	**Int'l Ticker:** Int'l Exchange:
2004 Sales: $46,874	2004 Profits: $24,033	Employees: 217
2003 Sales: $34,800	2003 Profits: $1,600	Fiscal Year Ends: 3/31
2002 Sales: $30,600	2002 Profits: $1,300	Parent Company:

SALARIES/BENEFITS:
Pension Plan:	ESOP Stock Plan:	Profit Sharing:	Top Exec. Salary: $311,000	Bonus: $556,000
Savings Plan: Y	Stock Purch. Plan:		Second Exec. Salary: $192,000	Bonus: $376,000

OTHER THOUGHTS:
Apparent Women Officers or Directors:
Hot Spot for Advancement for Women/Minorities:

LOCATIONS: ("Y" = Yes)
West:	Southwest:	Midwest:	Southeast:	Northeast:	International:
Y					Y

ABBOTT LABORATORIES
www.abbott.com

Industry Group Code: 325412 Ranks within this company's industry group: Sales: 9   Profits: 14

Drugs:		Other:		Clinical:		Computers:		Other:	
Discovery:	Y	AgriBio:		Trials/Services:		Hardware:		Specialty Services:	
Licensing:		Genomics/Proteomics:		Laboratories:		Software:		Consulting:	
Manufacturing:	Y	Tissue Replacement:		Equipment/Supplies:		Arrays:	Y	Blood Collection:	
Development:	Y			Research/Development Svcs.:		Database Management:		Drug Delivery:	
Generics:				Diagnostics:				Drug Distribution:	

TYPES OF BUSINESS:
Pharmaceuticals Manufacturing
Nutritional Products
Diagnostics
Consumer Health Products
Medical & Surgical Devices
Pharmaceutical Products
Animal Health

BRANDS/DIVISIONS/AFFILIATES:
GLYCO-FLEX
HUMIRA
FreeStyle Flash
AdvantEdge
PathVysion
SevoFlo
Similac
Ensure

CONTACTS: *Note: Officers with more than one job title may be intentionally listed here more than once.*
Miles D. White, CEO
Thomas C. Freyman, CFO
Stephen R. Fussell, Sr. VP-Human Resources
John C. Landgraf, Sr. VP Global Pharmaceutical Mgmt./Supply
Laura J. Schumacher, General Counsel/Exec. VP/Corp. Sec.
William G. Dempsey, Sr. VP-Pharm. Oper.
Richard Ashley, Exec. VP-Corp. Dev.
Catherine V. Babington, VP-Public Affairs
Thomas C. Freyman, Exec. VP-Finance
Jeffrey M. Leiden, Pres./COO-Pharmaceutical Prod./Dir.
William G. Dempsey, Exec. VP-Pharmaceutical Group
Holger Liepmann, Exec. VP-Global Nutrition
Olivier Bohuon, Sr. VP-Int'l Oper.
Miles D. White, Chmn.
Olivier Bohuon, Sr. VP-Int'l Oper.

Phone: 847-937-6100	Fax: 847-937-1511
Toll-Free:	
Address: 100 Abbott Park Rd., Abbott Park, IL 60064-3500 US	

GROWTH PLANS/SPECIAL FEATURES:

Abbott Laboratories' principal business is to discover, develop, manufacture and sell health care products and technologies ranging from pharmaceuticals, animal health products and medical devices. The pharmaceutical segment deals with adult and pediatric conditions such as rheumatoid arthritis, HIV, epilepsy and manic depression. The diagnostics segment deals with molecular diagnostics and diabetes care through glucose monitoring, while vascular devices target vessel closure using StarClose. Spinal implants for back problems include PathFinder. The company operates in and outside the U.S., in Europe, Asia, Africa, Latin and South America and the Middle East, marketing its products worldwide. Abbott's 50% shares in TAP Pharmaceutical Products, which makes the prostate cancer drug, Lupron, and Prevacid (Ogastro), a proton pump inhibitor for the short-term treatment of gastroesophageal reflux disease, gives it an edge. Abbott initiated the first complete blood glucose monitoring system designed for diabetic cats and dogs. In agreement with Boston Scientific, it acquired Guidant's vascular intervention and endovascular businesses. The company further acquired Kos Pharmaceuticals in 2006 through which it develops and commercializes hepatitis C virus protease inhibitors. It further received worldwide rights to market and distribute the Verax Platelet Test, used to detect the presence of a broad range of bacterial contaminants in platelets just prior to transfusion. In 2007, Abbott announced its intention to sell its core laboratory diagnostics business included in the Abbott Diagnostics Division and Abbott Point of Call to GE for $8.13 billion.

Abbott promotes employee diversity and supports working mothers. It offers internships and professional development programs for employees at all levels. Forms of compensation extend to savings and pension plans and profit sharing. Other benefits include child and elder care, wellness programs, health and dental insurance and tuition reimbursement.

FINANCIALS: Sales and profits are in thousands of dollars—add 000 to get the full amount. 2006 Note: Financial information for 2006 was not available for all companies at press time.

2006 Sales: $22,476,322	2006 Profits: $1,716,755	**U.S. Stock Ticker: ABT**
2005 Sales: $22,337,808	2005 Profits: $3,372,065	Int'l Ticker: Int'l Exchange:
2004 Sales: $19,680,016	2004 Profits: $3,235,851	Employees: 66,663
2003 Sales: $19,680,600	2003 Profits: $2,753,200	Fiscal Year Ends: 12/31
2002 Sales: $17,685,000	2002 Profits: $2,794,000	Parent Company:

SALARIES/BENEFITS:

Pension Plan: Y	ESOP Stock Plan:	Profit Sharing: Y	Top Exec. Salary: $1,605,990	Bonus: $2,650,000
Savings Plan: Y	Stock Purch. Plan:		Second Exec. Salary: $905,943	Bonus: $1,050,000

OTHER THOUGHTS:
Apparent Women Officers or Directors: 2
Hot Spot for Advancement for Women/Minorities: Y

LOCATIONS: ("Y" = Yes)

West:	Southwest:	Midwest:	Southeast:	Northeast:	International:
Y	Y	Y		Y	Y

Note: Financial information, benefits and other data can change quickly and may vary from those stated here.

ABRAXIS BIOSCIENCE INC

www.abraxisbio.com

Industry Group Code: 325412 **Ranks within this company's industry group:** Sales: 36   Profits: 152

Drugs:		Other:		Clinical:	Computers:		Other:	
Discovery:	Y	AgriBio:		Trials/Services:	Hardware:		Specialty Services:	
Licensing:	Y	Genomics/Proteomics:		Laboratories:	Software:		Consulting:	
Manufacturing:	Y	Tissue Replacement:		Equipment/Supplies:	Arrays:		Blood Collection:	
Development:	Y			Research/Development Svcs.:	Database Management:		Drug Delivery:	
Generics:				Diagnostics:			Drug Distribution:	

TYPES OF BUSINESS:

Pharmaceuticals Manufacturing
Injectable Oncology Drugs
Anti-Infectives Drugs
Critical Care Drugs

BRANDS/DIVISIONS/AFFILIATES:

American Pharmaceutical Partners, Inc.
American BioScience, Inc.
Abraxis BioScience
Abraxis Pharmaceutical Products
Abraxane
nab
Cruce Davila
AstraZeneca

CONTACTS: Note: Officers with more than one job title may be intentionally listed here more than once.

Patrick Soon-Shiong, CEO
Lisa Gopala, CFO/Exec. VP
Richard E. Maroun, Chief Admin. Officer
Richard E. Maroun, General Counsel/Sec.
Bruce Wendel, Exec. VP-Corp. Dev.
Thomas H. Silberg, Pres., Abraxis Pharmaceutical Products
Frank Harmon, COO/Exec. VP-Abraxis Pharmaceutical Products
Carlo Montagner, Pres., Abraxis Oncology
Patrick Soon-Shiong, Chmn.

Phone: 310-883-1300	Fax: 310-998-8553
Toll-Free:	
Address: 11755 Wilshire Blvd., 20th Fl., Los Angeles, CA 90025 US	

GROWTH PLANS/SPECIAL FEATURES:

Abraxis BioScience, Inc., formerly American Pharmaceutical Partners, Inc., is a global biopharmaceutical company focused on the injectable oncology, anti-infective and critical care markets. The company develops, manufactures and markets injectable products in each of the three basic forms: liquid, powder and lyophilized, or freeze-dried; and leverage revolutionary technology, such as the nab platform, to discover and deliver therapeutics used in the treatment of cancer and other life-threatening diseases. The firm operates in two business segments: Abraxis BioScience (ABI), representing the combined operations of Abraxis Oncology and Abraxis Research; and Abraxis Pharmaceutical Products (APP), representing the hospital-based operations. ABI focuses primarily on internally developed proprietary products, including Abraxane, which is used for the treatment of metastatic breast cancer, and the proprietary product pipeline. APP manufactures and markets a broad portfolio of injectable drugs, including oncology, critical care, anti-infectives and markets the proprietary products. Abraxis manufactures 16 injectable oncology products that include Carboplatin, Fluoroucacil, Ifosfamide and Pamidronate; 21 injectable anti-infective products that include ampicillin, azithromycin and cefoxiti; and 63 injectable critical care products that include Diprivan, heparin, Naropin and oxytacin. The company plans to maximize the commercial potential of Abraxane in various cancers including breast, lung, skin and stomach. The firm's products are used in hospitals, long-term care facilities, alternate care sites and clinics. In April 2006, the company merged with American BioScience, Inc., the firm's former parent, and changed its name to Abraxis BioScience, Inc. Later that month, Abraxis launched a partnership with AstraZeneca for both the co-promotion of Abraxane sales in the U.S. and the acquisition of certain anesthetic and analgesic product lines. In February 2007, Abraxis acquired Pfizer, Inc.'s Cruce Davila manufacturing facility in Barceloneta, Puerto Rico. In July 2007, the firm announced plans to split into two separate independent public companies, Abraxis BioScience and APP.

FINANCIALS: Sales and profits are in thousands of dollars—add 000 to get the full amount. 2006 Note: Financial information for 2006 was not available for all companies at press time.

		U.S. Stock Ticker: ABBI
2006 Sales: $765,488	2006 Profits: $-46,897	Int'l Ticker: Int'l Exchange:
2005 Sales: $520,757	2005 Profits: $17,657	Employees: 864
2004 Sales: $405,247	2004 Profits: $18,221	Fiscal Year Ends: 12/31
2003 Sales: $351,315	2003 Profits: $71,693	Parent Company:
2002 Sales: $	2002 Profits: $	

SALARIES/BENEFITS:

Pension Plan:	ESOP Stock Plan:	Profit Sharing:	Top Exec. Salary: $600,000	Bonus: $225,000
Savings Plan:	Stock Purch. Plan:		Second Exec. Salary: $553,846	Bonus: $

OTHER THOUGHTS:

Apparent Women Officers or Directors: 1
Hot Spot for Advancement for Women/Minorities:

LOCATIONS: ("Y" = Yes)

West:	Southwest:	Midwest:	Southeast:	Northeast:	International:
Y		Y		Y	Y

ACAMBIS PLC

www.acambis.com

Industry Group Code: 325412 Ranks within this company's industry group: Sales: 74 Profits: 137

Drugs:		Other:		Clinical:		Computers:		Other:	
Discovery:	Y	AgriBio:		Trials/Services:		Hardware:		Specialty Services:	
Licensing:		Genomics/Proteomics:	Y	Laboratories:		Software:		Consulting:	
Manufacturing:	Y	Tissue Replacement:		Equipment/Supplies:		Arrays:		Blood Collection:	
Development:	Y			Research/Development Svcs.:		Database Management:		Drug Delivery:	
Generics:				Diagnostics:				Drug Distribution:	Y

TYPES OF BUSINESS:

Vaccine Development & Manufacturing
Vaccine Sales & Distribution

BRANDS/DIVISIONS/AFFILIATES:

ACAM2000
MVA 3000
C-VIG
ACAM-FLU-A
Baxter Healthcare Corporation
ChimeriVax-JE
ChimeriVax-West Nile
ChimeriVax-Dengue

CONTACTS: *Note: Officers with more than one job title may be intentionally listed here more than once.*

an Garland, CEO
Elizabeth Brown, Acting CFO
Clement Lewin, VP-Mktg.
Michael Watson, Exec. VP-R&D
Joan Fusco, Sr. VP-Oper.
Paul Giannasca, VP-Dev.
Lyndsay Wright, VP-Corp. Comm.
Lyndsay Wright, VP-Investor Rel.
Elizabeth Brown, VP-Financial Mgmt.
Davin Wonnacott, Sr. VP-Regulatory Affairs & Quality Systems
Harry Kleanthous, VP-Research
Jayant Aphale, VP-Project Mgmt.
Clement Lewin, VP-Policy & Strategy
Peter Fellner, Chmn.

Phone: 44-1223-275-300	**Fax:** 44-1223-416-300

Toll-Free:

Address: Peterhouse Technology Park, 100 Fulbourn Rd., Cambridge, CB1 9PT UK

GROWTH PLANS/SPECIAL FEATURES:

Acambis plc is a major U.K.-based developer and producer of vaccines to prevent and treat infectious diseases. The company is internationally recognized as the leading producer of smallpox vaccines and a top supplier to governments wishing to build vaccine stockpiles, having supplied 200 million doses of the second-generation ACAM2000. Acambis also produces C-VIG, an antibody product developed with Cangene and used to prep recipients or to treat severe reactions to the smallpox vaccine. With Baxter Healthcare Corporation, the was developing a third-generation smallpox vaccine, MVA 3000, which is a weakened form of the current generation of smallpox vaccines, but Acambis is scaling down its MVA-related activities after the UD Department of Health and Human Services ceased to support the project in late 2006. Acambis is developing travel vaccines for Japanese encephalitis (ChimeriVax-JE, which is undergoing Phase III paediatric trials in India), West Nile (ChimeriVax-West Nile, which is undergoing Phase II clinical testing) and dengue fever (ChimeriVax-Dengue, a vaccine currently in Phase II clinical trials that is designed to protect against all for dengue virus serotypes). Moreover, Acambis has the only vaccine in development against the antibiotic-resistant bacteria Clostridium difficile, which causes diarrhea and is found in many hospitals. In October 2006, Acambis sold Berna Products Corporation (BPC) to Crucell NV for $16.5 million. BPC markets Vivotif, the world's only licensed oral typhoid vaccine, in North America, and has the U.S. sales and marketing rights to ARILVAX, a yellow fever vaccine developed and manufactured by Chiron Vaccines. In March 2007, the company announced plans to cut 15% of its work force, and 20% of its cost base by the end of the year. In July 2007, Acambis began Phase I trials of ACAM-FLU-A, a universal vaccine designed to target all strains of the influenza virus, both pandemic and seasonal.

FINANCIALS: Sales and profits are in thousands of dollars—add 000 to get the full amount. 2006 Note: Financial information for 2006 was not available for all companies at press time.

2006 Sales: $62,530	2006 Profits: $-34,800	**U.S. Stock Ticker: ACAM**	
2005 Sales: $70,400	2005 Profits: $-44,700	**Int'l Ticker: ACM** Int'l Exchange: London-LSE	
2004 Sales: $163,900	2004 Profits: $38,000	Employees: 320	
2003 Sales: $301,600	2003 Profits: $63,300	Fiscal Year Ends: 12/31	
2002 Sales: $128,500	2002 Profits: $15,400	Parent Company:	

SALARIES/BENEFITS:

Pension Plan:	ESOP Stock Plan:	Profit Sharing:	Top Exec. Salary: $	Bonus: $
Savings Plan:	Stock Purch. Plan:		Second Exec. Salary: $	Bonus: $

OTHER THOUGHTS:

Apparent Women Officers or Directors: 4
Hot Spot for Advancement for Women/Minorities: Y

LOCATIONS: ("Y" = Yes)

West:	Southwest:	Midwest:	Southeast:	Northeast:	International:
				Y	Y

ACCELRYS INC   www.accelrys.com

Industry Group Code: 511212 **Ranks within this company's industry group:** Sales: 3 Profits: 2

Drugs:	Other:	Clinical:	Computers:		Other:
Discovery:	AgriBio:	Trials/Services:	Hardware:		Specialty Services:
Licensing:	Genomics/Proteomics:	Laboratories:	Software:	Y	Consulting:
Manufacturing:	Tissue Replacement:	Equipment/Supplies:	Arrays:		Blood Collection:
Development:		Research/Development Svcs.:	Database Management:		Drug Delivery:
Generics:		Diagnostics:			Drug Distribution:

TYPES OF BUSINESS:

Software - Simulation & Informatics
Computational Nanotechnology Tools
Informatics Software
Modeling & Simulation Software

BRANDS/DIVISIONS/AFFILIATES:

Cheminformatics Programs
Material Studio 4.1
Discovery Studio 1.6
Accord
Catalyst
Insight II
QUANTA
SciTegic

CONTACTS: Note: Officers with more than one job title may be intentionally listed here more than once.

Mark J. Emkjer, CEO
Mark J. Emkjer, Pres.
Rick Russo, CFO/Sr. VP
Richard Murphy, VP-Worldwide Sales & Svcs.
Judith O. Hicks, VP-Human Resources
Nic Austin, VP-R&D
Matt Hahn, VP-Tech. & Platform Strategy
David Mersten, General Counsel/Corp. Sec./Sr. VP
R. William Taylor, VP-Corp. Dev. & Mktg.
Philomena Walsh, Dir.-Corp. Comm.
James Mihlik, Corp. Controller
Katie Hollister, VP-Worldwide Client Svcs.
Matt Hahn, General Mgr.-SciTegic
Frank Brown, Chief Science Officer
Kenneth L. Coleman, Chmn.

Phone: 858-799-5000	Fax: 858-799-5100
Toll-Free:	
Address: 10188 Telesis Ct., Ste. 100, San Diego, CA 92121 US	

GROWTH PLANS/SPECIAL FEATURES:

Accelrys, Inc. is a leading provider of modeling, simulation and informatics software and services to research and development organizations in the nanotechnology, biotechnical, pharmaceutical, chemical, petrochemical and materials sciences industries. The company's modeling and simulation software provides a spectrum of simulation technologies which include quantum mechanical simulation, molecular simulation and mesoscale simulation. These technologies predict properties of a molecule's shape, structure and reactivity. Accelrys' broad product suite consists of over 100 application molecules that employ advanced computer visualization, molecular modeling techniques and computational chemistry that can simulate, visualize and analyze chemical and biological systems. The firm's informatics software captures, stores, manages and mines scientific data and information. Accelrys' bioinformatics programs have the ability to search, edit, map and align sequence data for the analysis of DNA protein sequences and RNA secondary structures. Cheminformatics programs provide data visualization and analytical capabilities which make it possible to search, retrieve and harvest chemical data. In addition to software modules, the company offers customer support and training; contract research; and consulting services. Accelrys currently has locations in San Diego, California, the U.K., India as well as several global sales and support centers. The company also has key alliances with many research, hardware and information technology partners such as IBM, HP, Sun Microsystems, Oracle, Intel, Barnard Chemical Information, United Devices and Macrovision. In 2006, Accelrys introduced Materials Studio 4.1, which provides the most current version of its chemicals, materials and pharmaceutical development modeling and simulation tools. Accelrys also released Discovery Studio 1.6, which consists of life science modeling and simulation tools for lead discovery and optimization.

Accelrys offers U.S. employees tuition reimbursement; a computer loan program; publishing and patent bonuses; a referral bonus program; employee assistance programs; and flexible health and dependent care spending accounts.

FINANCIALS: Sales and profits are in thousands of dollars—add 000 to get the full amount. 2006 Note: Financial information for 2006 was not available for all companies at press time.

2006 Sales: $82,001	2006 Profits: $-7,739	U.S. Stock Ticker: ACCL	
2005 Sales: $79,030	2005 Profits: $-16,578	Int'l Ticker: Int'l Exchange:	
2004 Sales: $86,209	2004 Profits: $-4,596	Employees: 479	
2003 Sales: $85,561	2003 Profits: $-3,497	Fiscal Year Ends: 3/31	
2002 Sales: $95,100	2002 Profits: $	Parent Company:	

SALARIES/BENEFITS:

Pension Plan:	ESOP Stock Plan: Y	Profit Sharing:	Top Exec. Salary: $381,000	Bonus: $100,000
Savings Plan: Y	Stock Purch. Plan: Y		Second Exec. Salary: $238,918	Bonus: $48,020

OTHER THOUGHTS:

Apparent Women Officers or Directors: 3
Hot Spot for Advancement for Women/Minorities: Y

LOCATIONS: ("Y" = Yes)

West:	Southwest:	Midwest:	Southeast:	Northeast:	International:
Y				Y	Y

ACCESS PHARMACEUTICALS INC   www.accesspharma.com

Industry Group Code: 325412 Ranks within this company's industry group: Sales: Profits: 81

Drugs:		Other:		Clinical:	Computers:	Other:	
Discovery:	Y	AgriBio:		Trials/Services:	Hardware:	Specialty Services:	
Licensing:	Y	Genomics/Proteomics:		Laboratories:	Software:	Consulting:	
Manufacturing:	Y	Tissue Replacement:		Equipment/Supplies:	Arrays:	Blood Collection:	
Development:	Y			Research/Development Svcs.:	Database Management:	Drug Delivery:	Y
Generics:				Diagnostics:		Drug Distribution:	

TYPES OF BUSINESS:

Pharmaceutical Development
Drug Delivery Systems
Polymer Technology
Oncology Products

BRANDS/DIVISIONS/AFFILIATES:

MuGard
ProLindac

CONTACTS: Note: Officers with more than one job title may be intentionally listed here more than once.

Stephen R. Seiler, CEO
Stephen R. Seiler, Pres.
Stephen B. Thompson, CFO/VP
David P. Nowotnik, Sr. VP-R&D
Phillip Wise, VP-Bus. Dev. & Strategy
Donald C. Weinberger, Investor Rel.
Rosemary Mazanet, Vice Chmn.
Jeffrey Davis, Chmn.
Esteban Cvitkovic, Vice Chmn.-Europe

Phone: 214-905-5100	Fax: 214-905-5101
Toll-Free:	
Address: 2600 Stemmons Fwy., Ste. 176, Dallas, TX 752077 US	

GROWTH PLANS/SPECIAL FEATURES:

Access Pharmaceuticals, Inc. is an emerging biopharmaceutical company developing products for use in the treatment of cancer, the supportive care of cancer, and other disease states. The company has one technology approved by the FDA, MuGard, and three drug delivery technology platforms: synthetic polymer targeted delivery, which is designed to exploit enhanced permeability and retention at tumor sites to selectively accumulate drug and control drug release; Cobalamin-medicated oral delivery, which utilizes vitamin B12 to increase absorption of orally consumed medicines; and Cobalamin-medicated targeted delivery, which uses both active tumor targeting, such as attaching an additional fragment that will seek a complementary surface molecule to bind to, and passive tumor targeting, such as using a carrier molecule. MuGard is a viscous polymer solution that coats the oral cavity and is used in the treatment of mucositis, an issue that is faced by many chemo-therapy patients. Access is also in the process of developing its drug candidate ProLindac (AP5346), which links the DACH platform to a polymer and selectively releases the active drug to the tumor based on a difference between pH, as well as blood permeability, of a tumor and healthy tissue. The drug is in Phase 2 study. In December 2006, Access announced that the company has received marketing clearance for MuGard from the FDA for the indication of the management of oral wounds including mucositis, aphthous ulcers and traumatic ulcers. In April 2007, Access announced that it signed a definitive merger agreement with Somanta Pharmaceuticals, Inc. to acquire the company for 1.5 million shares of Access common stock for the entirety of the capital stock in Somanta. The acquisition will give Access four new anti-cancer drugs that are in development.

FINANCIALS: Sales and profits are in thousands of dollars—add 000 to get the full amount. 2006 Note: Financial information for 2006 was not available for all companies at press time.

2006 Sales: $	2006 Profits: $-12,874	U.S. Stock Ticker: ACCP.OB
2005 Sales: $	2005 Profits: $-1,700	Int'l Ticker: Int'l Exchange:
2004 Sales: $	2004 Profits: $-10,200	Employees: 9
2003 Sales: $1,295	2003 Profits: $-6,935	Fiscal Year Ends: 12/31
2002 Sales: $1,100	2002 Profits: $-9,400	Parent Company:

SALARIES/BENEFITS:

Pension Plan:	ESOP Stock Plan:	Profit Sharing:	Top Exec. Salary: $357,385	Bonus: $100,000
Savings Plan: Y	Stock Purch. Plan: Y		Second Exec. Salary: $253,620	Bonus: $20,000

OTHER THOUGHTS:

Apparent Women Officers or Directors:
Hot Spot for Advancement for Women/Minorities:

LOCATIONS: ("Y" = Yes)

West:	Southwest:	Midwest:	Southeast:	Northeast:	International:
	Y				Y

ACTELION LTD

www.actelion.com

Industry Group Code: 325412 Ranks within this company's industry group: Sales: 35   Profits: 27

Drugs:		Other:	Clinical:	Computers:	Other:
Discovery:	Y	AgriBio:	Trials/Services:	Hardware:	Specialty Services:
Licensing:		Genomics/Proteomics:	Laboratories:	Software:	Consulting:
Manufacturing:		Tissue Replacement:	Equipment/Supplies:	Arrays:	Blood Collection:
Development:	Y		Research/Development Svcs.:	Database Management:	Drug Delivery:
Generics:			Diagnostics:		Drug Distribution:

TYPES OF BUSINESS:

Drugs, Discovery & Development
Pharmaceutical Research
Cardiovascular Treatment
Genetic Disorder Treatment

BRANDS/DIVISIONS/AFFILIATES:

Curl Acquisition Subsidiary, Inc.
Zavesca
Tracleer
Clazosentan
Actelion Pharmaceuticals U.S., Inc.
Palosuran
Orexon RA
Actelion-1

CONTACTS: *Note: Officers with more than one job title may be intentionally listed here more than once.*

Jean-Paul Clozel, CEO
Andrew J. Oakley, CFO/VP
Frederic Bodin, Sr. VP-Head Global Medical Mktg.
Marian Borovsky, General Counsel/VP
Christian Chavy, Pres., Bus. Oper.
Simon Buckingham, Pres., Corp. & Bus. Dev.
Roland Haefeli, VP-Public Affairs
Roland Haefeli, VP-Investor Rel.
Louis de Lassence, VP-Corp. Svcs.
Isaac Kobrin, Sr. VP-Clinical Dev.
Martine Cozel, Sr. VP-Drug Discovery & Pharmacology
Walter Fischli, Sr. VP-Drug Discovery & Molecular Biology
Robert Cawthorn, Chmn.

Phone: 41-61-565-65-65	Fax: 41-61-565-65-00
Toll-Free:	
Address: Gewerbestrasse 16, Allschwil, Baselland 4123 Switzerland	

GROWTH PLANS/SPECIAL FEATURES:

Actelion, Ltd. is a biopharmaceutical company that focuses on the discovery, development and marketing of drugs for unaddressed medical needs. The firm typically focuses its treatments on diseases within limited patient populations and then expands and commercializes additional products to address a wider spectrum of ailments within the general practice market. The majority of Actelion's drug discovery products address medical needs in cardiovascular, central nervous system, oncology and immunology areas. The firm's research division uses a technology known as high-throughput screening, which is used to identify targeted compounds using molecular modeling and crystallography. Actelion's most recognized product is Tracleer, a dual entothelin receptor antagonist used for pulmonary arterial hypertension. Another popular Acetelion drug, Zavesca, is one of the first approved oral drug therapy treatments for a genetic lipid metabolic disorder called Gaucher disease. Additional drugs that are currently in the developmental stage include Clazosentan, Palosuran, Actelion-1 and Orexon RA. Clazosentan is intended for the prevention and treatment of vasospasms, a life threatening condition that leads to neurological deficits after a patient suffers an aneurysm. Palosuran is an oral form of Urotensin-II, one of the most potent vasoconstrictor substances, and is used in the treatment of cardiovascular and metabolic diseases. Actelion-1 is one of the first tissue targeting endothelin receptor antagonists. Orexon RA treats sleeping disorders. Actelion is based in Switzerland and has subsidiaries in 22 countries. Actelion and Roche recently entered into an autoimmune disorder collaboration in 2006 to jointly develop and commercialize Actelion's selective S1P1 receptor agonist. In 2007, a subsidiary of Actelion, Curl Acquisition Subsidiary, Inc., acquired CoTherix, Inc.

FINANCIALS: Sales and profits are in thousands of dollars—add 000 to get the full amount. 2006 Note: Financial information for 2006 was not available for all companies at press time.

2006 Sales: $776,912	2006 Profits: $198,066	**U.S. Stock Ticker: ALIOF**
2005 Sales: $545,168	2005 Profits: $103,135	**Int'l Ticker: ATLN** Int'l Exchange: Zurich-SWX
2004 Sales: $471,880	2004 Profits: $87,219	Employees: 400
2003 Sales: $247,600	2003 Profits: $-8,000	Fiscal Year Ends: 12/31
2002 Sales: $101,100	2002 Profits: $-29,400	Parent Company:

SALARIES/BENEFITS:

Pension Plan:	ESOP Stock Plan:	Profit Sharing:	Top Exec. Salary: $	Bonus: $
Savings Plan:	Stock Purch. Plan:		Second Exec. Salary: $	Bonus: $

OTHER THOUGHTS:

Apparent Women Officers or Directors: 1
Hot Spot for Advancement for Women/Minorities:

LOCATIONS: ("Y" = Yes)

West:	Southwest:	Midwest:	Southeast:	Northeast:	International:
					Y

ADOLOR CORP

www.adolor.com

Industry Group Code: 325412 Ranks within this company's industry group: Sales: 116 Profits: 170

Drugs:		Other:		Clinical:		Computers:		Other:	
Discovery:	Y	AgriBio:		Trials/Services:		Hardware:		Specialty Services:	
Licensing:	Y	Genomics/Proteomics:		Laboratories:		Software:		Consulting:	
Manufacturing:		Tissue Replacement:		Equipment/Supplies:		Arrays:		Blood Collection:	
Development:	Y			Research/Development Svcs.:		Database Management:		Drug Delivery:	
Generics:				Diagnostics:				Drug Distribution:	

TYPES OF BUSINESS:

Drugs, Discovery & Development
Pain Management Products
Gastrointestinal Products

BRANDS/DIVISIONS/AFFILIATES:

Entereg (alvimopan)
GlaxoSmithKline

CONTACTS: Note: Officers with more than one job title may be intentionally listed here more than once.

Michael R. Dougherty, CEO
Michael R. Dougherty, Pres.
Thomas P. Hess, CFO
Scott T. Megaffin, VP-Mktg.
Denise Kerton, VP-Human Resources
James E. Barrett, Chief Scientific Officer/Sr. VP/Pres., Research
George R. Maurer, VP-Commercial Mfg.
Martha Manning, General Counsel/Sr. VP/Corp. Sec.
Richard M. Mangano, VP-Clinical Oper.
Robert B. Jones, VP-Strategy & Bus. Analysis
Thomas P. Hess, VP-Finance
David Jackson, Chief Medical Officer
Randall J. Mack, VP-Project Mgmt.
Kevin G. Taylor, VP-Bus. Dev.
Linda Young, VP-Regulatory Affairs
David M. Madden, Chmn.

Phone: 484-595-1500	Fax: 484-595-1520
Toll-Free:	
Address: 700 Pennsylvania Dr., Exton, PA 19341 US	

GROWTH PLANS/SPECIAL FEATURES:

Adolor Corporation is a development stage biopharmaceutical corporation specializing in the discovery and development of prescription pain management products. The company's lead product candidate is Entereg, also known as alvimopan, which is being developed in collaboration with GlaxoSmithKline to selectively block the unwanted effects of opioid analgesics on the gastrointestinal (GI) tract, a chronic condition known as opioid-induced bowel dysfunction (OBD) characterized by painful GI conditions including constipation and resulting from the chronic use of opioid analgesics to treat persistent pain conditions. Entereg is additionally being developed by Adolor to treat the acute condition postoperative ileus (POI), a GI condition characterized by the slow return of gut function that can result from GI or other surgeries. For the treatment of POI, the company has completed four Phase III clinical studies of Entereg, and has submitted a New Drug Application (NDA) to the FDA. For the treatment of OBD, Adolor and GlaxoSmithKline have announced top-line results from two Phase III registration studies and completed last patient last visits for a phase III long-term safety study, the results of which are expected in second-quarter 2007. The company is also exploring the development of an analgesic product that would combine Entereg and an opioid. Through a proprietary research platform based on cloned, human opioid receptors, Adolor has also identified a series of novel, orally active delta agonists which selectively stimulate the delta opioid receptor, while all marketed opioid drugs currently interact with the mu receptors in the brain and spinal cord. Adolor is conducting Phase I clinical testing of its lead delta compound. In late 2006, Adolor decided to discontinue the clinical development of its sterile lidocaine patch for post-surgical incisional pain, which had been in Phase II clinical development.

FINANCIALS: Sales and profits are in thousands of dollars—add 000 to get the full amount. 2006 Note: Financial information for 2006 was not available for all companies at press time.

2006 Sales: $15,087	2006 Profits: $-69,738	U.S. Stock Ticker: ADLR	
2005 Sales: $15,719	2005 Profits: $-56,797	Int'l Ticker:	Int'l Exchange:
2004 Sales: $25,542	2004 Profits: $-43,586	Employees: 128	
2003 Sales: $20,727	2003 Profits: $-51,206	Fiscal Year Ends: 12/31	
2002 Sales: $28,409	2002 Profits: $-60,524	Parent Company:	

SALARIES/BENEFITS:

Pension Plan:	ESOP Stock Plan:	Profit Sharing:	Top Exec. Salary: $358,198	Bonus: $
Savings Plan: Y	Stock Purch. Plan: Y		Second Exec. Salary: $350,981	Bonus: $55,361

OTHER THOUGHTS:

Apparent Women Officers or Directors: 3
Hot Spot for Advancement for Women/Minorities: Y

LOCATIONS: ("Y" = Yes)

West:	Southwest:	Midwest:	Southeast:	Northeast:	International:
				Y	

Note: Financial information, benefits and other data can change quickly and may vary from those stated here.

ADVANCED BIONICS CORPORATION
www.advancedbionics.com

Industry Group Code: 339113 Ranks within this company's industry group: Sales: Profits:

Drugs:	Other:	Clinical:		Computers:		Other:	
Discovery:	AgriBio:	Trials/Services:		Hardware:		Specialty Services:	
Licensing:	Genomics/Proteomics:	Laboratories:		Software:	Y	Consulting:	
Manufacturing:	Tissue Replacement:	Equipment/Supplies:	Y	Arrays:		Blood Collection:	
Development:		Research/Development Svcs.:		Database Management:		Drug Delivery:	
Generics:		Diagnostics:				Drug Distribution:	

TYPES OF BUSINESS:

Medical Equipment-Manufacturing
Bionic Devices
Cochlear Implant Technology
Spinal Cord Stimulation Systems
Software
Chronic Pain Treatment

BRANDS/DIVISIONS/AFFILIATES:

HiResolution Bionic Ear System
Harmony
HiRes
Auria
Platinum
PrecisionPlus
Boston Scientific
Nihon Bionics Co., Ltd.

CONTACTS: *Note: Officers with more than one job title may be intentionally listed here more than once.*

Al Mann, Co-CEO
Jeff Greiner, Pres./Co-CEO
Jim Surek, VP-Sales
Tom Santogrossi, VP-Mfg.
Al Mann, Chmn.

Phone: 661-362-1400	Fax: 661-362-1500
Toll-Free: 800-678-2575	
Address: 12740 San Fernando Rd., Sylmar, CA 91342 US	

GROWTH PLANS/SPECIAL FEATURES:

Advanced Bionics Corporation, a subsidiary of Bostor Scientific, develops and markets bionic technologies tha employ implantable neutrostimulation devices to trea neurological conditions such as deafness and chronic pain Advanced Bionics is currently the only company in the U.S that develops cochlear implant technology, which restores hearing to deaf individuals.   Products in the company's HiResolution Bionic Ear System include Harmony, a behind the-ear processor that produces high-quality sound resolution; Auria, a line of sound processing products which include headpieces, earhooks, battery chargers, moisturizing kits and carrying cases; Platinum, a sound processor that is worn around a belt instead of behind the ears, and a 90k implant that uses integrated circuit computer technology witl an internal memory and Hifocus electrodes for neutra targeting. Chronic Pain sufferers are offered Precision Plus a spinal Cord stimulation therapy that masks pain signals by sending out doses of electricity which the brain interprets as pleasant sensations. PrecisionPlus is typically offered to patients with failed back surgeries, phantom limb pain sciatica, reflex sympathetic dystrophy (RSD) and/or comple regional pain syndrome (CRPS). Advanced Bionics currently has operations in California, France, Asia-Pacific and Latir America.   The firm additionally maintains a Japanese subsidiary company, Nihon Bionics Co., Ltd.

Advanced Bionics offers employees health, disability and life insurance; flexible and dependent care spending accounts and employee appreciation events such as picnics and volleyball tournaments.

FINANCIALS: Sales and profits are in thousands of dollars—add 000 to get the full amount. 2006 Note: Financial information for 2006 was not available for all companies at press time.

2006 Sales: $	2006 Profits: $	U.S. Stock Ticker: Subsidiary
2005 Sales: $	2005 Profits: $	Int'l Ticker: Int'l Exchange:
2004 Sales: $	2004 Profits: $	Employees: 500
2003 Sales: $56,400	2003 Profits: $	Fiscal Year Ends: 12/31
2002 Sales: $75,000	2002 Profits: $	Parent Company: BOSTON SCIENTIFIC CORP

SALARIES/BENEFITS:

Pension Plan:	ESOP Stock Plan: Y	Profit Sharing:	Top Exec. Salary: $	Bonus: $
Savings Plan: Y	Stock Purch. Plan:		Second Exec. Salary: $	Bonus: $

OTHER THOUGHTS:

Apparent Women Officers or Directors:
Hot Spot for Advancement for Women/Minorities:

LOCATIONS: ("Y" = Yes)

West:	Southwest:	Midwest:	Southeast:	Northeast:	International:
Y					Y

ADVANCED CELL TECHNOLOGY INC www.advancedcell.com

Industry Group Code: 541710   Ranks within this company's industry group: Sales: 22   Profits: 14

Drugs:		Other:		Clinical:	Computers:	Other:	
Discovery:	Y	AgriBio:		Trials/Services:	Hardware:	Specialty Services:	
Licensing:		Genomics/Proteomics:		Laboratories:	Software:	Consulting:	
Manufacturing:		Tissue Replacement:	Y	Equipment/Supplies:	Arrays:	Blood Collection:	
Development:	Y			Research/Development Svcs.:	Database Management:	Drug Delivery:	
Generics:				Diagnostics:		Drug Distribution:	

TYPES OF BUSINESS:
Human Stem Cell Research
Patent Licensing

BRANDS/DIVISIONS/AFFILIATES:
ACTCellerate
Infigen, Inc.

CONTACTS: Note: Officers with more than one job title may be intentionally listed here more than once.
William M. Caldwell, IV, CEO
Michael D. West, Pres.
James G. Stewart, CFO/Sr. VP
Michael D. West, Chief Scientific Officer
Pedro Huertas, Chief Dev. Officer
Jonathan F. Atzen, General Counsel/Sr. VP/Sec.
Pedro Huertas, Chief Dev. Officer
Jan Wolkind, VP-Finance/Chief Acct. Officer
Robert Lanza, VP-Medical & Scientific Dev.
Robert W. Peabody, VP-Grant Administration
William M. Caldwell, IV, Chmn.

Phone: 510-748-4900	Fax: 510-748-4950

Toll-Free: 800-218-4202
Address: 1201 Harbor Bay Pkwy., Ste. 120, Alameda, CA 95402 US

GROWTH PLANS/SPECIAL FEATURES:

Advanced Cell Technology, Inc. (ACT) is a biotechnology company that focuses on developing and commercializing human stem cell technology in the emerging field of regenerative medicine. Regenerative medicine treats a wide array of chronic degenerative diseases and facilitates regenerative repair of acute disease, such as trauma, infarction and burns. The company owns or licenses over 30 issued patents and over 280 patent applications related to the field of stem cell therapy; nuclear transfer, which allows the production of stem cells genetically matched to the patient; and a reduced complexity library of stem cells for acute clinical applications. ACT's technology platform is based on the use of embryonic stem cells derived from a novel technique that does not destroy embryos. It identifies and isolates these stem cell lines with its propriety ACTCellerate technology. The company's divides its research programs into three categories. Cellular reprogramming, which involves turning stem cells into one of over 200 different human cell types that may be therapeutically relevant in treating diseased or destroyed tissue, and which are tailored to each patient's needs. A reduced complexity program, which uses proprietary technology to generate cell therapy products for patients with acute medical needs that do not allow time for patient-specific reprogramming of cells. Finally, a stem cell differentiation segment, which controls the differentiation, culture and growth of the company's stem cells. The company is currently focused on developing eye treatments based on retinal pigment epithelium (RPE) cells; and is developing preclinical treatments for cardiovascular disease, stroke and cancer based on hemangioblast (HG) cells. It also has projects working on stem cell based dermal treatments that would provide scar free skin grafts. In February 2007, ACT acquired all of the intellectual property and assets owned by Infigen, Inc., including 26 patents, for a total consideration of $708,000.

FINANCIALS: Sales and profits are in thousands of dollars—add 000 to get the full amount. 2006 Note: Financial information for 2006 was not available for all companies at press time.

2006 Sales: $ 441	2006 Profits: $-18,720	**U.S. Stock Ticker: Private**
2005 Sales: $ 395	2005 Profits: $-9,394	**Int'l Ticker:** Int'l Exchange:
2004 Sales: $ 3	2004 Profits: $- 52	Employees: 40
2003 Sales: $ 3	2003 Profits: $ 30	Fiscal Year Ends: 12/31
2002 Sales: $	2002 Profits: $	Parent Company:

SALARIES/BENEFITS:

Pension Plan:	ESOP Stock Plan:	Profit Sharing:	Top Exec. Salary: $283,654	Bonus: $100,000
Savings Plan:	Stock Purch. Plan:		Second Exec. Salary: $264,422	Bonus: $65,000

OTHER THOUGHTS:
Apparent Women Officers or Directors:
Hot Spot for Advancement for Women/Minorities:

LOCATIONS: ("Y" = Yes)

West:	Southwest:	Midwest:	Southeast:	Northeast:	International:
Y				Y	

AEOLUS PHARMACEUTICALS INC   www.aeoluspharma.com

Industry Group Code: 325412 Ranks within this company's industry group: Sales: 186 Profits: 65

Drugs:		Other:	Clinical:	Computers:	Other:
Discovery:	Y	AgriBio:	Trials/Services:	Hardware:	Specialty Services:
Licensing:		Genomics/Proteomics:	Laboratories:	Software:	Consulting:
Manufacturing:		Tissue Replacement:	Equipment/Supplies:	Arrays:	Blood Collection:
Development:	Y		Research/Development Svcs.:	Database Management:	Drug Delivery:
Generics:			Diagnostics:		Drug Distribution:

TYPES OF BUSINESS:
Pharmaceutical Development
Catalytic Antioxidants

BRANDS/DIVISIONS/AFFILIATES:
AEOL 10150
AEOL 11207

CONTACTS: *Note: Officers with more than one job title may be intentionally listed here more than once.*
John L. McManus, CEO
John L. McManus, Pres.
Michael McManus, CFO
Brian J. Day, Chief Scientific Officer
Elaine Alexander, VP/Chief Medical Officer
David C. Cavalier, Chmn.

Phone: 949-481-9825	Fax: 949-481-9829
Toll-Free:	
Address: 23811 Inverness Place, Laguna Niguel, CA 92677 US	

GROWTH PLANS/SPECIAL FEATURES:
Aeolus Pharmaceuticals, Inc. is a Southern California-based biopharmaceutical company. It is developing a new class of catalytic antioxidant compounds for disease and disorders of the central nervous system, respiratory system, autoimmune system and oncology. The company's lead drug candidate is AEOL 10150, which is the first drug in its class of catalytic antioxidant compounds to enter human clinical evaluation. AEOL 10150 is a small molecule catalytic antioxidant that has shown the ability to scavenge a broad range of reactive oxygen species, or free radicals. As a catalytic antioxidant, AEOL 10150 mimics and thereby amplifies the body's natural enzymatic systems for eliminating these damaging compounds. AEOL 10150 is thought to be a treatment for among other diseases, amyotrophic lateral sclerosis, or ALS or also known as Lou Gehrig's disease. Aeolus boasts positive safety results from two completed Phase I single dose studies of AEOL 10150 in patients diagnosed with ALS. The drug has also shown promise in the field of radiation therapy. Tests on mice demonstrate that AEOL 10150 protects healthy lung tissue from radiation injury delivered either in a single dose or by fractionated radiation therapy doses, and that the drug does not negatively affect tumor radiotherapy. Aeolus has also selected AEOL 11207 as the company's second development candidate through the Aeolus Pipeline Initiative, an internal development initiative. Collected data suggests the compound may be useful as a potential once-every-other-day oral therapeutic treatment option for central nervous system disorders, most likely Parkinson's disease. In June 2006, Aeolus announced it had raised $5 million through the sale of newly issued shares of common stock and warrants to selected investors, led by Efficacy Capital. In March 2007, the company announced that it had completed the analysis of the results from the Phase 1 study of AEOL 10150, with no serious adverse effects reported.

FINANCIALS: Sales and profits are in thousands of dollars—add 000 to get the full amount. 2006 Note: Financial information for 2006 was not available for all companies at press time.
2006 Sales: $ 92	2006 Profits: $-5,728	U.S. Stock Ticker: AOLS.OB
2005 Sales: $ 252	2005 Profits: $-6,905	Int'l Ticker: Int'l Exchange:
2004 Sales: $ 305	2004 Profits: $-17,167	Employees: 4
2003 Sales: $	2003 Profits: $-2,976	Fiscal Year Ends: 9/30
2002 Sales: $ 100	2002 Profits: $-11,300	Parent Company:

SALARIES/BENEFITS:
Pension Plan:	ESOP Stock Plan:	Profit Sharing:	Top Exec. Salary: $281,132	Bonus: $164,413
Savings Plan:	Stock Purch. Plan: Y		Second Exec. Salary: $62,550	Bonus: $

OTHER THOUGHTS:
Apparent Women Officers or Directors: 1
Hot Spot for Advancement for Women/Minorities:

LOCATIONS: ("Y" = Yes)
West:	Southwest:	Midwest:	Southeast:	Northeast:	International:
Y					

AETERNA ZENTARIS INC

www.aeternazentaris.com

Industry Group Code: 325412 Ranks within this company's industry group: Sales: 83 Profits: 41

Drugs:		Other:	Clinical:	Computers:	Other:
Discovery:	Y	AgriBio:	Trials/Services:	Hardware:	Specialty Services:
Licensing:		Genomics/Proteomics:	Laboratories:	Software:	Consulting:
Manufacturing:	Y	Tissue Replacement:	Equipment/Supplies:	Arrays:	Blood Collection:
Development:	Y		Research/Development Svcs.:	Database Management:	Drug Delivery:
Generics:			Diagnostics:		Drug Distribution:

TYPES OF BUSINESS:

Drug Development
Oncology Products
Endocrine Therapy Products

BRANDS/DIVISIONS/AFFILIATES:

Cetrotide
Impavido
Merck Serono
Shionogi
Nippon Kayaku
Ozarelix
Perifosine
Cetrorelix

CONTACTS: Note: Officers with more than one job title may be intentionally listed here more than once.

David G. Mazzo, CEO
Jurgen Engel, COO
David G. Mazzo, Pres.
Dennis Turpin, CFO/Sr. VP
Jurgen Engel, Exec. VP-Global R&D
Mario Paradis, Sr. VP-Admin. Affairs
Mario Paradis, Sr. VP-Legal Affairs/Sec.
Ellen McDonald, Sr. VP-Bus. Oper./Chief Bus. Officer
Renene Thomas, Sr. Dir.-Corp. Comm.
Renene Thomas, Sr. Dir.-Investor Rel.
Nicholas J. Pelliccione, Sr. VP-Regulatory Affairs & Quality Assurance
Eric Dupont, Chmn.

Phone: 418-652-8525	Fax: 418-652-0881

Toll-Free:

Address: 1405 du Parc-Technologique Blvd., Quebec, Quebec, G1P 4P5 Canada

GROWTH PLANS/SPECIAL FEATURES:

AEterna Zentaris, Inc. is a Canadian biopharmaceutical company focused on endocrine therapy and oncology. The company is devoted to discovering and developing drugs for the treatment of certain forms of cancer, endocrine disorders and infectious diseases. The firm has two products on the market: Cetrotide and Impavido. Cetrotride was the first hormone antagonist treatment approved for in vitro fertilization. The drug is administered to women in order to prevent premature ovulation in order to increase the fertility success rate. Cetrotide is approved in over 80 countries. The drug is marketed worldwide by Merck Serono, except in Japan where it is marketed by Shionogi and Nippon Kayaku. Impavido is an oral drug used for the treatment of visceral and cutaneous leishmaniasis. Currently, the company has three drugs in Phase I of clinical trial, two in Phase II and one in Phase III. The drug in Phase III clinical trials is Cetrorelix, for benign prostatic hyperplasia and endometriosis. The firm's Phase II drugs are Ozarelix for prostate cancer and Perifosine for multiple cancers. In addition, the firm has two products in preclinical in vitro testing and three in preclinical development. In May 2006, AEterna began research with the University of Montreal to investigate the role of ghrelin, a hormone with effects on appetite and fat tissue, on trends in obesity. The company owns 100% of Zentaris GmbH, an integrated clinical research company. In January 2007, AEterna Zentaris became a pure play biopharmaceutical company, having completed the spin-off of Atrium Biotechnologies, Inc., the company's former subsidiary.

FINANCIALS: Sales and profits are in thousands of dollars—add 000 to get the full amount. 2006 Note: Financial information for 2006 was not available for all companies at press time.

2006 Sales: $41,390	2006 Profits: $33,390	U.S. Stock Ticker: AEZS
2005 Sales: $247,389	2005 Profits: $10,571	Int'l Ticker: AEZ Int'l Exchange: Toronto-TSX
2004 Sales: $179,212	2004 Profits: $-4,425	Employees: 500
2003 Sales: $128,587	2003 Profits: $-32,426	Fiscal Year Ends: 12/31
2002 Sales: $64,204	2002 Profits: $-16,748	Parent Company:

SALARIES/BENEFITS:

Pension Plan:	ESOP Stock Plan:	Profit Sharing:	Top Exec. Salary: $	Bonus: $
Savings Plan:	Stock Purch. Plan:		Second Exec. Salary: $	Bonus: $

OTHER THOUGHTS:

Apparent Women Officers or Directors: 1
Hot Spot for Advancement for Women/Minorities:

LOCATIONS: ("Y" = Yes)

West:	Southwest:	Midwest:	Southeast:	Northeast:	International:
					Y

Note: Financial information, benefits and other data can change quickly and may vary from those stated here.

AFFYMETRIX INC
www.affymetrix.com

Industry Group Code: 325413 Ranks within this company's industry group: Sales: 4   Profits: 18

Drugs:	Other:		Clinical:	Computers:		Other:
Discovery:	AgriBio:		Trials/Services:	Hardware:	Y	Specialty Services:
Licensing:	Genomics/Proteomics:	Y	Laboratories:	Software:		Consulting:
Manufacturing:	Tissue Replacement:		Equipment/Supplies:	Arrays:	Y	Blood Collection:
Development:			Research/Development Svcs.:	Database Management:		Drug Delivery:
Generics:			Diagnostics:			Drug Distribution:

TYPES OF BUSINESS:

Chips-Genetics
DNA Array Technology
Genomics

BRANDS/DIVISIONS/AFFILIATES:

GeneChip
GenFlex
CustomExpress
NimbleExpress
Perlegen Sciences, Inc.
Genetic MicroSystems, Inc.
Neomorphic, Inc.

CONTACTS: *Note: Officers with more than one job title may be intentionally listed here more than once.*

Stephen P. A. Fodor, CEO
Thane Kreiner, Sr. VP-Mktg. & Sales
Barbara A. Caulfield, Exec. VP/General Counsel
Kevin M. King, Pres., Life Sciences Business
Stephen P. A. Fodor, Chmn.

Phone: 408-731-5000	Fax: 508-731-5441
Toll-Free: 888-362-2447	
Address: 3420 Central Expwy., Santa Clara, CA 95051 US	

GROWTH PLANS/SPECIAL FEATURES:

Affymetrix, Inc. develops technology for examining an managing complex genetic information by applying the principles of semiconductor technology to molecular biolog research and the life sciences. The company's GeneChi technology is used for sequence analysis, genotyping and gene expression monitoring.   The company use photolithography and solid-phase chemistry to produc arrays containing hundreds of thousands of oligonucleotid (a short polymer of two to 20 nucleotides) probes packed a extremely high densities. Markets for its GeneChip, as we as related GenFlex tag array products and CustomExpres and NimbleExpress array programs, include all aspects c molecular biology research in the life sciences, such as basi human disease research, genetic analysis, pharmaceutica drug discovery and development, pharmacogenomics toxicogenomics and agricultural research.   Affymetri currently sells its products directly to pharmaceutica biotechnology, agrichemical, diagnostics and consume products companies as well as academic research centers government research laboratories, private foundatio laboratories and clinical reference laboratories in Nort America, Europe and Japan. More than 1,400 systems hav been installed around the world and nearly 4,000 peer reviewed papers have been published using the technology The company's Perlegen Sciences, Inc. subsidiary focuse on identifying the genetic variations among individuals an finding patterns in them. Subsidiary Genetic MicroSystems Inc. also specializes in DNA array technology. In additior the company owns Neomorphic, Inc., a computationa genomics company. The company has a few new projects i the works, including a 1-million SNP (single nucleotid polymorphism) product in 2007.

Affymetrix offers its employees an extensive benefit package that includes domestic partner benefits, legal an financial services, health fitness discounts, a tuitio assistance plan and a lunch program.

FINANCIALS: Sales and profits are in thousands of dollars—add 000 to get the full amount. 2006 Note: Financial information for 2006 was not available for all companies at press time.

2006 Sales: $355,317	2006 Profits: $-13,704	**U.S. Stock Ticker: AFFX**	
2005 Sales: $367,602	2005 Profits: $65,787	**Int'l Ticker:**	Int'l Exchange:
2004 Sales: $345,962	2004 Profits: $47,608	Employees: 1,128	
2003 Sales: $300,796	2003 Profits: $14,285	Fiscal Year Ends: 12/31	
2002 Sales: $289,947	2002 Profits: $-1,630	Parent Company:	

SALARIES/BENEFITS:

Pension Plan:	ESOP Stock Plan:	Profit Sharing:	Top Exec. Salary: $568,462	Bonus: $
Savings Plan: Y	Stock Purch. Plan:		Second Exec. Salary: $448,846	Bonus: $

OTHER THOUGHTS:

Apparent Women Officers or Directors: 3
Hot Spot for Advancement for Women/Minorities: Y

LOCATIONS: ("Y" = Yes)

West:	Southwest:	Midwest:	Southeast:	Northeast:	International:
Y				Y	Y

Note: Financial information, benefits and other data can change quickly and may vary from those stated here.

AGILENT TECHNOLOGIES INC
www.agilent.com

Industry Group Code: 334500 Ranks within this company's industry group: Sales: 1 Profits: 1

Drugs:	Other:	Clinical:		Computers:	Other:
Discovery:	AgriBio:	Trials/Services:		Hardware:	Specialty Services:
Licensing:	Genomics/Proteomics:	Laboratories:		Software:	Consulting:
Manufacturing:	Tissue Replacement:	Equipment/Supplies:	Y	Arrays:	Blood Collection:
Development:		Research/Development Svcs.:		Database Management:	Drug Delivery:
Generics:		Diagnostics:			Drug Distribution:

TYPES OF BUSINESS:
Test Equipment
Communications Equipment
Integrated Circuits
Optoelectronics
Image Sensors
Bioinstrumentation
Software Products
Informatics Products

BRANDS/DIVISIONS/AFFILIATES:
Silicon Genetics
Computational Biology Corp.
Eagleware-Elanix
Molecular Imaging Corp.
Yokogawa Analytical Systems
Verigy Pte. Ltd.

CONTACTS: Note: Officers with more than one job title may be intentionally listed here more than once.
William P. Sullivan, CEO
William P. Sullivan, Pres.
Adrian T. Dillon, CFO
Jean M. Halloran, Sr. VP-Human Resources
Darlene J. S. Solomon, CTO/VP
D. Craig Nordlund, General Counsel/Sr. VP
Shiela Robertson, VP-Corp. Dev.
Rodney Gonsalves, Dir.-Investor Rel.
Adrian T. Dillon, Exec. VP-Finance
Patrick Byrne, Pres., Electronics Measurement Group
Chris van Ingen, Pres., Life Sciences & Chemical Analysis
Keith Barnes, Pres./CEO-Semiconductor Test Verigy
Michael C. Gasparian, VP/Gen. Mgr.
VP./Gen Mgr-Customer/ Quality
James G. Cullen, Chmn.

Phone: 408-348-8886	Fax: 408-345-8474
Toll-Free: 877-424-4536	
Address: 5301 Stevens Creek Blvd., Santa Clara, CA 95051 US	

GROWTH PLANS/SPECIAL FEATURES:

Agilent Technologies, Inc. is an international diversified technology company that manufactures instruments, test equipment, pharmaceuticals and other products for the communications, electronics, life sciences, nanotechnology, homeland security and chemical analysis industries. Agilent, which separated from Hewlett-Packard as part of a corporate realignment, operates through three main businesses: test and measurement, semiconductor products, and life sciences and chemical analysis. The test and measurement business provides standard and customized products used in the design, development, manufacture and operation of electronic equipment and systems and communications networks and services. These products include test and measurement instruments, automated test equipment, communications network monitoring equipment, imaging and surveillance, and software design tools. The company's semiconductor business produces semiconductor components, modules and assemblies for the networking and personal systems markets. Agilent's life sciences and chemical analysis business produces a variety of product categories, including mass spectrometry, pharmaceuticals, gas chromatography, genomics and informatics products. The chemical business produces fuel cells, foods and flavors and drug testing products. The firm's primary research, development and manufacturing site is located in Santa Clara, California, with additional sites in Belgium, Scotland, China and Everett, Washington. In March 2007, Agilent announced a distribution agreement with BFi OPTiLAS, which will act as a stocking distributor in Europe for Agilent RF and microwave products. The company further announced plans to expand its investment in its Integrated Circuit Characterization and Analysis Program (IC-CAP) device modeling software in a new research and development modeling center in Beijing. Agilent plans to extend its India presence by laying a foundation for a 10-acre campus near Manesar, Haryana. In April 2007, the firm agreed to acquire Stratagene Corp. for $245.5 million.

Agilent's employees enjoy flexible work arrangements including part-time, telecommuting and variable schedules, as well as dependant care resources. Mothers returning to work have access to a mother's room.

FINANCIALS: Sales and profits are in thousands of dollars—add 000 to get the full amount. 2006 Note: Financial information for 2006 was not available for all companies at press time.

2006 Sales: $4,973,000	2006 Profits: $3,307,000	U.S. Stock Ticker: A
2005 Sales: $4,685,000	2005 Profits: $327,000	Int'l Ticker: Int'l Exchange:
2004 Sales: $4,556,000	2004 Profits: $369,000	Employees: 2,760
2003 Sales: $4,468,000	2003 Profits: $-2,058,000	Fiscal Year Ends: 10/31
2002 Sales: $6,010,000	2002 Profits: $-1,032,000	Parent Company:

SALARIES/BENEFITS:

Pension Plan: Y	ESOP Stock Plan:	Profit Sharing: Y	Top Exec. Salary: $906,247	Bonus: $1,139,953
Savings Plan: Y	Stock Purch. Plan: Y		Second Exec. Salary: $687,500	Bonus: $588,182

OTHER THOUGHTS:
Apparent Women Officers or Directors: 3
Hot Spot for Advancement for Women/Minorities: Y

LOCATIONS: ("Y" = Yes)

West:	Southwest:	Midwest:	Southeast:	Northeast:	International:
Y				Y	Y

Note: Financial information, benefits and other data can change quickly and may vary from those stated here.

AKORN INC

www.akorn.com

Industry Group Code: 325412 Ranks within this company's industry group: Sales: 71   Profits: 66

Drugs:		Other:		Clinical:		Computers:		Other:	
Discovery:	Y	AgriBio:		Trials/Services:		Hardware:		Specialty Services:	
Licensing:		Genomics/Proteomics:		Laboratories:		Software:		Consulting:	
Manufacturing:	Y	Tissue Replacement:		Equipment/Supplies:		Arrays:		Blood Collection:	
Development:	Y			Research/Development Svcs.:		Database Management:		Drug Delivery:	
Generics:	Y			Diagnostics:				Drug Distribution:	

TYPES OF BUSINESS:

Ophthalmic & Hospital Drugs & Injectables
Contract Services

BRANDS/DIVISIONS/AFFILIATES:

Akorn New Jersey, Inc.

CONTACTS: *Note: Officers with more than one job title may be intentionally listed here more than once.*

Arthur S. Przybyl, CEO
Arthur S. Przybyl, Pres.
Jeffrey A. Whitnell, CFO
Neil Stanahan, VP-Human Resources
Jay W. Stern, VP-Contract Mfg.
Michael P. Stehn, Sr. VP-Oper.
Abu Alam, Sr. VP-New Bus. Dev.
Jeffrey A. Whitnell, Sr. VP-Finance
John R. Sabat, Sr. VP-National Accounts & Trade Rel.
Mark M. Silverberg, Sr. VP-Global Quality Assurance & Decatur Oper.
Sam Boddapati, VP-Regulatory Affairs
John N. Kapoor, Chmn.

Phone: 847-279-6100	Fax: 800-943-3694
Toll-Free: 800-535-7155	
Address: 2500 Millbrook Dr., Buffalo Grove, IL 60089 US	

GROWTH PLANS/SPECIAL FEATURES:

Akorn, Inc. manufactures and markets diagnostic and therapeutic pharmaceuticals for specialty areas such as ophthalmology, rheumatology, anesthesia and antidota medicine. The company operates in three segments ophthalmic; hospital drugs and injectables; and contract services. Through the ophthalmic division, the firm markets a line of diagnostic and therapeutic ophthalmic pharmaceutical products. Diagnostic products, primarily used in the office setting, include mydriatics, cycloplegics anesthetics, topical stains, gonioscopic solutions and angiography dyes. Therapeutic products, sold primarily to wholesalers and other national account customers, include antibiotics, anti-infectives, steroids, steroid combinations glaucoma medications, decongestants/antihistamines and anti-edema medications. Non-pharmaceutical products include various artificial tear solutions, preservative-free lubricating ointments, eyelid cleansers, vitamin supplements and contact lens accessories. Through the hospital drugs and injectables segment, Akorn markets a line of specialty injectable pharmaceutical products, including antidotes anesthesia and products used in the treatment of rheumatoid arthritis and pain management. Through the contract services segment, the company manufactures products for third party pharmaceutical and biotechnology customers based on their specifications. Akorn New Jersey, Inc., a wholly-owned subsidiary, operating in New Jersey, is involved in manufacturing, product development and administrative activities related to the ophthalmic and hospital drugs and injectables segments. The firm has manufacturing facilities in Illinois and New Jersey. The Illinois facility manufactures products for all three segments The New Jersey facility manufactures ophthalmic solutions and ointment products. Customers include physicians optometrists, hospitals, wholesalers, group purchasing organizations and other pharmaceutical companies. Akorn owns seven U.S. patents.

The company offers its employees medical, dental, vision and life insurance; an employee assistance program; a 401(k) plan; and stock options.

FINANCIALS: Sales and profits are in thousands of dollars—add 000 to get the full amount. 2006 Note: Financial information for 2006 was not available for all companies at press time.

2006 Sales: $71,250	2006 Profits: $-5,963	**U.S. Stock Ticker: AKRX**
2005 Sales: $44,484	2005 Profits: $-8,609	Int'l Ticker: Int'l Exchange:
2004 Sales: $50,708	2004 Profits: $-3,026	Employees: 371
2003 Sales: $45,491	2003 Profits: $-12,325	Fiscal Year Ends: 12/31
2002 Sales: $51,419	2002 Profits: $-12,952	Parent Company:

SALARIES/BENEFITS:

Pension Plan:	ESOP Stock Plan:	Profit Sharing:	Top Exec. Salary: $400,000	Bonus: $255,000
Savings Plan: Y	Stock Purch. Plan:		Second Exec. Salary: $250,000	Bonus: $95,625

OTHER THOUGHTS:

Apparent Women Officers or Directors:
Hot Spot for Advancement for Women/Minorities:

LOCATIONS: ("Y" = Yes)

West:	Southwest:	Midwest:	Southeast:	Northeast:	International:
		Y		Y	

AKZO NOBEL NV

www.akzonobel.com

Industry Group Code: 325000 Ranks within this company's industry group: Sales: 4 Profits: 4

Drugs:		Other:	Clinical:	Computers:	Other:
Discovery:	Y	AgriBio:	Trials/Services:	Hardware:	Specialty Services:
Licensing:		Genomics/Proteomics:	Laboratories:	Software:	Consulting:
Manufacturing:		Tissue Replacement:	Equipment/Supplies:	Arrays:	Blood Collection:
Development:			Research/Development Svcs.:	Database Management:	Drug Delivery:
Generics:			Diagnostics:		Drug Distribution:

TYPES OF BUSINESS:

Specialty Chemicals-Coatings
Pharmaceuticals
Veterinary Pharmaceuticals
Over-the-Counter Drugs
Nanotechnology Research

BRANDS/DIVISIONS/AFFILIATES:

Organon International
Intervet, Inc.
Eka Chemicals
Sico, Inc.
The Flood Company

CONTACTS: Note: Officers with more than one job title may be intentionally listed here more than once.

G. J. Wijers, CEO
Rob Frohn, CFO
Leif Darner, Mgr.
Maarten van den Bergh, Chmn.

Phone: 31-26-366-4433	Fax: 31-26-366-3250
Toll-Free:	
Address: Velperweg 76, Arnhem, 6800 SB The Netherlands	

GROWTH PLANS/SPECIAL FEATURES:

Akzo Nobel N.V. produces health care products, coatings and chemicals, and operates in over 80 countries. The pharmaceuticals division provides products for the human health care market through Organon International, whose products include medicine for gynecology, including contraceptive NuvaRing, fertility, neuroscience, anesthesia and urology. Additionally, this division produces products for the animal health care market through Intervet, Inc., which is the third-largest animal health care company in the world. Its products include vaccines, antiparasitics, anti-infectives, endocrine products, feed additives and productivity enhancers. Akzo Nobel's coatings division makes a variety of chemical products including powder, wood, coil, and marine coatings; tile and wood adhesives; and a line of car refinishes. The company's chemical products division produces pulp and paper chemicals; polymer chemicals such as metal alkyls and suspending agents; surfactants used in hair and skincare products; base chemicals such as salt and chlor-alkali products used in the manufacture of glass and plastics; and functional chemicals used in toothpaste, ice cream and flame retardants. In early 2006, the company began a program to divest itself of its chemical production interests, including the sale of its oleochemicals joint ventures in Malaysia to the Lam Soon Group and its Electro Magnetic Compatibility business to ETL Semko KK. In 2006, the firm divested its Polymerization Catalysts and Components business to Basell Polyolefins (February), and sold the Technical Services Department of Akzo Nobel Base Chemichals to Stork NV (May). In June 2006, Akzo acquired Sico, Inc., and sold its Ink and Adhesives resins business to Hexion Specialty Chemicals. Also in 2006, Akzo Nobel acquired The Flood Company, a US woodcare business. In 2007, Schering-Plough Corp. agreed to acquire Akzo Nobel's Organon BioSciences unit $14.4 billion; the company sold its PVC additives business to GIL Investments; and the firm agreed to acquire Imperial Chemical Industries PLC for $16.18 billion.

FINANCIALS: Sales and profits are in thousands of dollars—add 000 to get the full amount. 2006 Note: Financial information for 2006 was not available for all companies at press time.

2006 Sales: $19,659,800	2006 Profits: $1,566,190	**U.S. Stock Ticker: AKZOY**
2005 Sales: $15,386,000	2005 Profits: $1,137,000	**Int'l Ticker: AKZA** Int'l Exchange: Amsterdam-Euronext
2004 Sales: $17,187,000	2004 Profits: $1,159,000	Employees: 61,340
2003 Sales: $16,408,000	2003 Profits: $757,000	Fiscal Year Ends: 12/31
2002 Sales: $14,706,000	2002 Profits: $859,000	Parent Company:

SALARIES/BENEFITS:

Pension Plan:	ESOP Stock Plan:	Profit Sharing:	Top Exec. Salary: $	Bonus: $
Savings Plan:	Stock Purch. Plan:		Second Exec. Salary: $	Bonus: $

OTHER THOUGHTS:

Apparent Women Officers or Directors: 1
Hot Spot for Advancement for Women/Minorities:

LOCATIONS: ("Y" = Yes)

West:	Southwest:	Midwest:	Southeast:	Northeast:	International:
		Y	Y	Y	Y

ALBANY MOLECULAR RESEARCH   www.albmolecular.com

Industry Group Code: 541710 Ranks within this company's industry group: Sales: 9 Profits: 7

Drugs:	Other:	Clinical:		Computers:		Other:
Discovery:	AgriBio:	Trials/Services:		Hardware:		Specialty Services:
Licensing:	Genomics/Proteomics:	Laboratories:		Software:		Consulting:
Manufacturing:	Tissue Replacement:	Equipment/Supplies:		Arrays:		Blood Collection:
Development:		Research/Development Svcs.:	Y	Database Management:		Drug Delivery:
Generics:		Diagnostics:				Drug Distribution:

TYPES OF BUSINESS:

Contract Drug Discovery & Development
Custom Biotech & Genomic Research
Chemistry Research
Manufacturing Services
Consulting Services
Analytical Chemistry Services

BRANDS/DIVISIONS/AFFILIATES:

Organichem Corporation
AMRI Hungary
ComGenex

CONTACTS: Note: Officers with more than one job title may be intentionally listed here more than once.

Thomas E. D'Ambra, CEO
Thomas E. D'Ambra, Pres.
Mark T. Frost, CFO
Brian Russell, Sr. Dir.-Human Resources
Bruce J. Sargent, VP-Discovery R&D
Harold Meckler, VP-Science & Tech.
James J. Grates, VP-Admin.
Eric W. Smart, VP-Bus. Dev.
Mark T. Frost, Treas./VP
Michael P. Trova, Sr. VP-Chemistry
Steven Hagen, VP-Quality & Analytical Chemistry
Patricia Ellis, VP-Quality Assurance & Regulatory Affairs
Michael D. Ironside, VP-Chemical Dev.
Thomas E. D'Ambra, Chmn.

Phone: 518-464-0279	Fax: 518-464-0289
Toll-Free:	
Address: 21 Corporate Cir., Albany, NY 12212-5098 US	

GROWTH PLANS/SPECIAL FEATURES:

Albany Molecular Research, Inc. (AMRI) is a chemistry-based drug discovery and development company, focusing on applications for new small-molecule and prescription drugs. In addition to developing its own drugs, AMRI has increasingly acted as a custom research and development source for the pharmaceutical, genomic and biotechnology industries. The company provides contract services across the entire product development cycle, from lead discovery to commercial manufacturing. AMRI recently expanded its commercial manufacturing operations through its wholly owned subsidiary, Organichem Corporation. The company's services allow a pharmaceutical company to outsource its chemistry department and simultaneously pursue a greater number of drug discovery and development opportunities. AMRI aims its proprietary research at discovering new compounds with commercial potential that it would then license for service fees, milestone and royalty payments. Some of the products of this research led to the development of the active ingredient for Allegra, a non-sedating antihistamine marketed by Sanofi-Aventis. The firm has research facilities in Albany, Syracuse and Rensselaer, New York and Bothell, Washington. AMRI recently began operations at new facilities in Singapore and Hyderabad, India as part of a strategic move to globalize its services. In 2006, the company completed its acquisition of ComGenex, which it later renamed AMRI Hungary. The acquisition gives Albany Molecular Research a laboratory in Europe, from which it can do business with European pharmaceutical companies.

The company provides its employees with bonuses and technology development incentive plans. Supplemental plans include an employee assistance plan, relocation support, tuition assistance and visa assistance.

FINANCIALS: Sales and profits are in thousands of dollars—add 000 to get the full amount. 2006 Note: Financial information for 2006 was not available for all companies at press time.

2006 Sales: $179,807	2006 Profits: $2,183	U.S. Stock Ticker: AMRI
2005 Sales: $183,906	2005 Profits: $16,321	Int'l Ticker: Int'l Exchange:
2004 Sales: $169,527	2004 Profits: $-11,691	Employees: 1,015
2003 Sales: $196,300	2003 Profits: $31,200	Fiscal Year Ends: 12/31
2002 Sales: $122,800	2002 Profits: $40,600	Parent Company:

SALARIES/BENEFITS:

Pension Plan: Y	ESOP Stock Plan:	Profit Sharing:	Top Exec. Salary: $330,769	Bonus: $100,000
Savings Plan: Y	Stock Purch. Plan:		Second Exec. Salary: $260,000	Bonus: $53,170

OTHER THOUGHTS:

Apparent Women Officers or Directors: 1
Hot Spot for Advancement for Women/Minorities:

LOCATIONS: ("Y" = Yes)

West:	Southwest:	Midwest:	Southeast:	Northeast:	International:
Y				Y	Y

ALCON INC   www.alconlabs.com

Industry Group Code: 325412   Ranks within this company's industry group: Sales: 21   Profits: 16

Drugs:		Other:		Clinical:		Computers:		Other:	
Discovery:	Y	AgriBio:		Trials/Services:		Hardware:		Specialty Services:	
Licensing:		Genomics/Proteomics:		Laboratories:		Software:		Consulting:	
Manufacturing:	Y	Tissue Replacement:		Equipment/Supplies:		Arrays:		Blood Collection:	
Development:	Y			Research/Development Svcs.:	Y	Database Management:		Drug Delivery:	
Generics:				Diagnostics:				Drug Distribution:	

TYPES OF BUSINESS:

Eye Care Products
Ophthalmic Products & Equipment
Contact Lens Care Products
Surgical Instruments

BRANDS/DIVISIONS/AFFILIATES:

Opti-Free
Patanol
AcrySof
Betoptic
Falcon Pharmaceuticals
Alcon Surgical
William C. Conner Research Center
Nestle Corporation

CONTACTS: *Note: Officers with more than one job title may be intentionally listed here more than once.*

Cary Rayment, CEO
Allen Baker, COO/Exec. VP
Cary Rayment, Pres.
Jacqualyn Fouse, CFO
Kevin J. Buehler, Chief Mktg. Officer
Gerald D. Cagle, Sr. VP-R&D
Andre Bens, Sr. VP-Global Mfg. & Tech. Support
Elaine E. Whitbeck, General Counsel/Chief Legal Officer
Cary Rayment, Sr. VP-U.S. Oper.
Doug MacHatton, Strategic Corp. Comm.
Doug MacHatton, VP-Investor Rel.
Jacqualyn Fouse, Sr. VP-Finance
Kevin Buehler, Sr. VP-Alcon U.S.
Cary Rayment, Chmn.
Fred Pettinato, Sr. VP-Intl. Oper.

Phone: 41-41-785-8888	Fax:
Toll-Free:	
Address: Bosch 69, Hunenberg, 6331 Switzerland	

GROWTH PLANS/SPECIAL FEATURES:

Alcon, Inc. is one of the world's largest eye care product companies. The company's three divisions (surgical, pharmaceutical and consumer vision care) develop, manufacture and market ophthalmic pharmaceuticals, surgical equipment and devices, contact lens care products and other consumer eye care products that treat diseases and conditions of the eye. Alcon maintains manufacturing plants, laboratories and offices in 50 countries and offers its products and services in over 75 countries. The company makes more than 10,000 unique products, including prescription and over-the-counter drugs, contact lens solutions, surgical instruments, intraocular lenses and office systems for ophthalmologists. Its brand names include Patanol solution for eye allergies, AcrySof intraocular lenses, Betoptic for glaucoma and the Opti-Free system for contact lens care. Alcon's research and development headquarters houses the 400,000-square-foot William C. Conner Research Center. Subsidiary Falcon Pharmaceuticals manufactures and markets generic ophthalmic and otic (ear-related) pharmaceuticals in the U.S. Falcon's main product is Timolol GFS, a patented gel forming solution used to treat glaucoma. Alcon Surgical creates implantable lenses, viscoelastics and medical tools specifically made for ocular surgeons, including phacoemulsification instruments for cataract removal and absorbable sutures. Alcon's global sales represent 22% of the ophthalmic pharmaceutical market, 51% of the ophthalmic surgical market and 21% of the ophthalmic consumer market. The FDA recently approved Alcon's AcrySof ReSTOR intraocular lens for us in the visual correction of aphakia following cataract surgery. In 2006, the firm signed a marketing agreement with Eli Lilly and Co. to co-promote ruboxistaurin mesylate (proposed brand name, Arxxant) in the U.S. and Puerto Rico. Nestle Corporation owns approximately 75% of the firm.

Alcon matches 240% of employee contributions to a 401(k), up to 5% of total compensation. Alcon has been named to FORTUNE magazine's list of the 100 Best Companies to Work for in the U.S. for eight consecutive years.

FINANCIALS: Sales and profits are in thousands of dollars—add 000 to get the full amount. 2006 Note: Financial information for 2006 was not available for all companies at press time.

2006 Sales: $4,896,600	2006 Profits: $1,348,100	U.S. Stock Ticker: ACL
2005 Sales: $4,368,500	2005 Profits: $931,000	Int'l Ticker: Int'l Exchange:
2004 Sales: $3,913,600	2004 Profits: $871,800	Employees: 13,500
2003 Sales: $3,406,900	2003 Profits: $595,400	Fiscal Year Ends: 12/31
2002 Sales: $3,009,100	2002 Profits: $466,900	Parent Company:

SALARIES/BENEFITS:

Pension Plan: Y	ESOP Stock Plan:	Profit Sharing:	Top Exec. Salary: $	Bonus: $
Savings Plan: Y	Stock Purch. Plan:		Second Exec. Salary: $	Bonus: $

OTHER THOUGHTS:

Apparent Women Officers or Directors: 2
Hot Spot for Advancement for Women/Minorities: Y

LOCATIONS: ("Y" = Yes)

West:	Southwest:	Midwest:	Southeast:	Northeast:	International:
Y	Y	Y	Y	Y	Y

ALEXION PHARMACEUTICALS INC
www.alexionpharmaceuticals.com
Industry Group Code: 325412 **Ranks within this company's industry group:** Sales: 153 Profits: 193

Drugs:		Other:	Clinical:	Computers:	Other:
Discovery:	Y	AgriBio:	Trials/Services:	Hardware:	Specialty Services:
Licensing:		Genomics/Proteomics:	Laboratories:	Software:	Consulting:
Manufacturing:	Y	Tissue Replacement:	Equipment/Supplies:	Arrays:	Blood Collection:
Development:	Y		Research/Development Svcs.:	Database Management:	Drug Delivery:
Generics:			Diagnostics:		Drug Distribution:

TYPES OF BUSINESS:
Therapeutic Products

BRANDS/DIVISIONS/AFFILIATES:
Soliris
Eculizumab

CONTACTS: *Note: Officers with more than one job title may be intentionally listed here more than once.*
Leonard Bell, CEO/Treas./Sec.
David W. Keiser, COO
David W. Keiser, Pres.
Vikas Sinha, CFO/Sr. VP
Paul W. Finnegan, VP-Global Strategic Mktg. & Dev.
Russell P. Rother, Sr. VP-Research
Scott A. Rollins, Sr. VP-Drug Dev. & Project Mgmt.
Daniel N. Caron, Exec. Dir.-Eng.
M. Stacy Hooks, VP-Mfg. & Tech. Svcs.
Thomas I.H. Dubin, General Counsel/Sr. VP
Daniel N. Caron, Exec. Dir.-Oper.
Barry P. Luke, VP-Finance
Nancy Motola, Sr. VP-Regulatory Affairs & Quality
David Hallal, VP-US Commercial Oper.
Max Link, Chmn.
Patrice Coissac, Sr. VP/General Manager/Pres., Alexion Europe SAS

Phone: 203-272-2596	Fax: 203-271-8198
Toll-Free:	
Address: 325 Knotter Dr., Cheshire, CT 06410 US	

GROWTH PLANS/SPECIAL FEATURES:
Alexion Pharmaceuticals, Inc. engages in the discovery and development of therapeutic products to treat patients with severe disease states, including hematologic diseases cancer and autoimmune disorders. The company devotes substantially all of its resources to drug discovery, research and product and clinical development. Wholly-owned subsidiary Alexion Antibody Technologies, Inc. endeavors to discover and develop a portfolio of additional antibody therapeutics targeting severe unmet medical needs. Alexion's lead product, eculizumab, marketed under the brand name Soliris, is a genetically altered antibody known as C5 complement inhibitor that is designed to selectively block the production of inflammation-causing protein in the complement cascade. The firm focuses its efforts on developing Soliris for the treatment of a rare blood disorder Paroxysmal Nocturnal Hemoglobinuria (PNH). The company, in collaboration with Procter & Gamble Pharmaceuticals, decided not to pursue further development on pexelizumab, a single-chain antibody that also inhibits complement immune response. In February 2006, Alexion announced that the Japanese Patent Office had issued a patent to the company for its Soliris brand of eculizumab. In July 2006, the firm acquired a manufacturing plant in Rhode Island for the future commercial production of its products. In March 2007, Alexion received approval from FDA to market Soliris for all patients with PNH in the U.S. In June 2007, the firm announced that Soliris was granted marketing approval in Europe for treatment of all patients with PNH.

FINANCIALS: Sales and profits are in thousands of dollars—add 000 to get the full amount. 2006 Note: Financial information for 2006 was not available for all companies at press time.

2006 Sales: $1,558	2006 Profits: $-131,514	**U.S. Stock Ticker:** ALXN	
2005 Sales: $1,064	2005 Profits: $-108,750	**Int'l Ticker:** Int'l Exchange:	
2004 Sales: $4,609	2004 Profits: $-74,095	Employees: 296	
2003 Sales: $ 900	2003 Profits: $-84,500	Fiscal Year Ends: 12/31	
2002 Sales: $6,500	2002 Profits: $-56,500	Parent Company:	

SALARIES/BENEFITS:
Pension Plan:	ESOP Stock Plan:	Profit Sharing:	Top Exec. Salary: $502,000	Bonus: $480,000
Savings Plan:	Stock Purch. Plan:		Second Exec. Salary: $339,082	Bonus: $198,000

OTHER THOUGHTS:
Apparent Women Officers or Directors: 1
Hot Spot for Advancement for Women/Minorities:

LOCATIONS: ("Y" = Yes)
West:	Southwest:	Midwest:	Southeast:	Northeast:	International:
				Y	Y

ALFACELL CORPORATION

www.alfacell.com

Industry Group Code: 325412 **Ranks within this company's industry group:** Sales: 184 Profits: 68

Drugs:		Other:		Clinical:		Computers:		Other:	
Discovery:	Y	AgriBio:		Trials/Services:		Hardware:		Specialty Services:	
Licensing:		Genomics/Proteomics:		Laboratories:		Software:		Consulting:	
Manufacturing:	Y	Tissue Replacement:		Equipment/Supplies:		Arrays:		Blood Collection:	
Development:	Y			Research/Development Svcs.:		Database Management:		Drug Delivery:	
Generics:				Diagnostics:				Drug Distribution:	

TYPES OF BUSINESS:

Cancer & Pathological Conditions Drugs
RNase-based Drugs

BRANDS/DIVISIONS/AFFILIATES:

Onconase

CONTACTS: *Note: Officers with more than one job title may be intentionally listed here more than once.*

Kuslima Shogen, CEO
Lawrence A. Kenyon, CFO/Exec. VP/Corp. Sec.
Andrew P. Aromando, Sr. VP-Oper.
Andrew P. Aromando, Sr. VP-Commercial Dev.
Diane Scudiery, Dir.-Clinical & Regulatory Oper.
Kuslima Shogen, Chmn.

Phone: 973-748-8082	Fax: 973-748-1355
Toll-Free:	
Address: 225 Belleville Ave., Bloomfield, NJ 07003 US	

GROWTH PLANS/SPECIAL FEATURES:

Alfacell Corporation is a biopharmaceutical company engaged in the discovery and development of new therapeutic drugs for the treatment of cancer and other pathological conditions. The company's drug discovery and development program consists of novel therapeutics that are being developed from amphibian ribonucleases (RNases). RNases are biologically active enzymes that split RNA molecules. The firm uses RNases for the development of therapeutics for cancer and other life-threatening diseases, including HIV and autoimmune diseases, that require anti-proliferative and apoptotic, or programmed cell death, properties. The company's proprietary product is Onconase, which targets solid tumors that have become resistant to other chemotherapeutic drugs. Onconase affects primarily exponentially growing malignant cells, with activity controlled through specific molecular mechanisms. The drug is currently being evaluated as a treatment for inoperable malignant mesotheliona, a rare cancer primarily affecting the pleura (lining of the lungs) usually caused by exposure to asbestos, in an international, centrally randomized, confirmatory Phase IIIb registration trial. The drug received orphan drug designation for malignant mesothelioma in Australia and the U.S. Onconase, in combination with other drugs, has also been seen to benefit patients with non small cell lung cancer, breast cancer and ovarian cancer. The firm was awarded a U.S. patent for a methodology for synthesizing gene sequences of ranpirnase, the active component of Onconase. In June 2007, the company announced that preclinical in vitro and in vivo data show Onconase to be active against naïve and chemoresistant neuroblastoma cells.

FINANCIALS: Sales and profits are in thousands of dollars—add 000 to get the full amount. 2006 Note: Financial information for 2006 was not available for all companies at press time.

2006 Sales: $ 107	2006 Profits: $-7,810	U.S. Stock Ticker: ACEL	
2005 Sales: $ 152	2005 Profits: $-6,462	Int'l Ticker:	Int'l Exchange:
2004 Sales: $ 42	2004 Profits: $-5,070	Employees: 14	
2003 Sales: $ 40	2003 Profits: $-2,412	Fiscal Year Ends: 7/31	
2002 Sales: $ 5	2002 Profits: $-2,591	Parent Company:	

SALARIES/BENEFITS:

Pension Plan:	ESOP Stock Plan:	Profit Sharing:	Top Exec. Salary: $215,000	Bonus: $
Savings Plan:	Stock Purch. Plan:		Second Exec. Salary: $180,000	Bonus: $

OTHER THOUGHTS:

Apparent Women Officers or Directors: 1
Hot Spot for Advancement for Women/Minorities:

LOCATIONS: ("Y" = Yes)

West:	Southwest:	Midwest:	Southeast:	Northeast:	International:
				Y	

ALIZYME PLC

www.alizyme.com

Industry Group Code: 325412 Ranks within this company's industry group: Sales: 150   Profits: 144

Drugs:		Other:		Clinical:	Computers:	Other:	
Discovery:	Y	AgriBio:		Trials/Services:	Hardware:	Specialty Services:	
Licensing:	Y	Genomics/Proteomics:		Laboratories:	Software:	Consulting:	
Manufacturing:		Tissue Replacement:		Equipment/Supplies:	Arrays:	Blood Collection:	
Development:	Y			Research/Development Svcs.:	Database Management:	Drug Delivery:	Y
Generics:				Diagnostics:		Drug Distribution:	

TYPES OF BUSINESS:

Drugs-Gastrointestinal
Colonic Drug Delivery System
Obesity Treatments

BRANDS/DIVISIONS/AFFILIATES:

COLAL
Renzapride
Cetilistat
ATL-104
COLAL-PRED

CONTACTS: *Note: Officers with more than one job title may be intentionally listed here more than once.*

Timothy P. McCarthy, CEO
David Campbell, Dir.-Finance/Sec.
Roger I. Hickling, Dir.-R&D
Brian Richards, Chmn.

Phone: 44-1223-896-000	Fax: 44-1223-896-001
Toll-Free:	
Address: Granta Park, Great Abington, Cambridge, CB21 6GX UK	

GROWTH PLANS/SPECIAL FEATURES:

Alizyme plc develops prescription drugs for the treatment c obesity and related diseases as well as gastrointestina disorders, including irritable bowel syndrome (IBS) inflammatory bowel diseases (IBD) and mucositis, gastrointestinal side effect of chemotherapy. Alizym generally takes products from an early stage into clinica development through the stage of proof of efficacy i humans, in Phase II clinical trials, with the intention o licensing out products for late-stage clinical trials and eventual marketing. The company owns the intellectua property rights to all four of the drug products it currently ha in clinical trials as well as for the recently acquired COLA drug delivery system, which delivers drugs directly to th colon. The firm has completed successful Phase IIb trials c renzapride for the treatment of IBS; and in April 200 received FDA approval to begin Phase III clinical trials fo Cetilistat (previously ATL-962) for the treatment of obesity Alizyme has a licensing agreement with Takeda Chemica Industries, Ltd. for Cetilistat in Japan. The company ha completed Phase IIa trials for ATL-104 in the U.K. for th treatment of mucositis, and is in preparation to begin Phas III trials. COLAL-PRED used for treatment of IBD is currentl in Phase III development. This product incorporates an anti inflammatory steroid with the colonic drug delivery systen COLAL and is under development for the management o ulcerative colitis. COLAL-PRED is also in Pre-Phase I development for the maintenance of remission.

FINANCIALS: Sales and profits are in thousands of dollars—add 000 to get the full amount. 2006 Note: Financial information for 2006 was not available for all companies at press time.

2006 Sales: $2,296	2006 Profits: $-39,820	**U.S. Stock Ticker: Private**	
2005 Sales: $	2005 Profits: $-38,260	**Int'l Ticker: AZM** Int'l Exchange: London-LSE	
2004 Sales: $3,333	2004 Profits: $-12,360	Employees: 15	
2003 Sales: $2,200	2003 Profits: $-17,800	Fiscal Year Ends: 12/31	
2002 Sales: $	2002 Profits: $-17,080	Parent Company:	

SALARIES/BENEFITS:

Pension Plan:	ESOP Stock Plan:	Profit Sharing:	Top Exec. Salary: $	Bonus: $
Savings Plan:	Stock Purch. Plan:		Second Exec. Salary: $	Bonus: $

OTHER THOUGHTS:

Apparent Women Officers or Directors:	**LOCATIONS:** ("Y" = Yes)					
Hot Spot for Advancement for Women/Minorities:	West:	Southwest:	Midwest:	Southeast:	Northeast:	International:
						Y

ALKERMES INC

www.alkermes.com

Industry Group Code: 325412A **Ranks within this company's industry group:** Sales: 4 Profits: 4

Drugs:		Other:		Clinical:	Computers:	Other:
Discovery:		AgriBio:		Trials/Services:	Hardware:	Specialty Services:
Licensing:		Genomics/Proteomics:		Laboratories:	Software:	Consulting:
Manufacturing:	Y	Tissue Replacement:		Equipment/Supplies:	Arrays:	Blood Collection:
Development:	Y			Research/Development Svcs.:	Database Management:	Drug Delivery:
Generics:				Diagnostics:		Drug Distribution:

TYPES OF BUSINESS:

Drug Delivery Systems
Pulmonary Drug Delivery Systems
Sustained Release Injection Delivery Systems

BRANDS/DIVISIONS/AFFILIATES:

ProLease
Medisorb
Advanced Inhalation Research (AIR)
Vivitrol
Risperdal Consta
AIR Insulin
Exenatide LAR
AIR Epinephrine

CONTACTS: *Note: Officers with more than one job title may be intentionally listed here more than once.*

David A. Broecker, CEO
Gordon G. Pugh, COO/Sr. VP
David A. Broecker, Pres.
James M. Frates, CFO/Sr. VP
F. Ken Andrews, VP-Mktg. & Sales/Chief Commercial Officer
Madeline D. Coffin, VP-Human Resources
Elliot W. Ehrich, Sr. VP-R&D/Chief Medical Officer
Kathryn L. Biberstein, General Counsel/Sr. VP/Corp. Sec.
Michael J. Landine, Sr. VP-Corp. Dev.
Rebecca J. Peterson, VP-Corp. Comm.
James M. Frates, Treas.
Richard Pops, Chmn.

Phone: 617-494-0171	Fax: 617-494-9263
Toll-Free:	
Address: 88 Sidney St., Cambridge, MA 02139-4234 US	

GROWTH PLANS/SPECIAL FEATURES:

Alkermes, Inc. is a biotechnology company that specializes in the development of sophisticated drug delivery technologies. The company currently markets two commercial products: RISPERDAL CONSTA, a long-acting atypical antipsychotic medication for schizophrenia, and VIVITROL, an injectable medication for the treatment of alcohol dependence. RISPERDAL CONSTA is administered through an intramuscular injection every two weeks and is currently approved for sale in approximately 80 countries. VIVITROL is aimed at prolonging abstinence in patients that abstain from alcohol a week prior to treatment and has also been shown to reduce the number of heavy drinking days in patients. In addition to its marketed products, Alkermes also develops extended-release injectable, pulmonary and oral products for the treatment of central nervous system disorders, addiction and diabetes. The company is also focused on the development of its Advanced Inhalation Research (AIR) technology, which maximizes drug delivery to the lungs through an inhaler-based delivery system. Through a collaboration with Eli Lilly and Company, Alkermes is currently developing AIR Insulin, which consists of an inhaled formulation of insulin that has the potential to improve the treatment of diabetes through the provision of a simpler dosing regimen. Additional developing products in the company's pipeline include Exenatide LAR for the treatment of type 2 diabetes; AIR parathyroid hormone for the treatment of osteoporosis; ALKS 27 for the treatment of COPD and ALKS 29 for treatment of addiction. In addition to its headquarters in Cambridge, Massaxhusetts, the company operates additional manufacturing facilities within Massachusetts and Ohio. In April 2007, the FDA approved a 12.5 mg dose of RESPERDAL CONSTA for the treatment of schizophrenia within specific patient populations, which will allow physicians more options in individualizing or adjusting treatment approaches for their patients.

Alkermes provides employees with tuition reimbursement, flexible spending accounts, an employee assistance program and transportation benefits.

FINANCIALS: **Sales and profits are in thousands of dollars—add 000 to get the full amount. 2006 Note: Financial information for 2006 was not available for all companies at press time.**

2006 Sales: $166,601	2006 Profits: $3,818	U.S. Stock Ticker: ALKS
2005 Sales: $76,126	2005 Profits: $-73,916	Int'l Ticker: Int'l Exchange:
2004 Sales: $39,054	2004 Profits: $-102,385	Employees: 760
2003 Sales: $47,300	2003 Profits: $-106,900	Fiscal Year Ends: 3/31
2002 Sales: $54,100	2002 Profits: $-61,400	Parent Company:

SALARIES/BENEFITS:

Pension Plan:	ESOP Stock Plan: Y	Profit Sharing:	Top Exec. Salary: $554,474	Bonus: $450,000
Savings Plan: Y	Stock Purch. Plan:		Second Exec. Salary: $379,377	Bonus: $240,000

OTHER THOUGHTS:

Apparent Women Officers or Directors: 3
Hot Spot for Advancement for Women/Minorities: Y

LOCATIONS: ("Y" = Yes)

West:	Southwest:	Midwest:	Southeast:	Northeast:	International:
		Y		Y	Y

ALLERGAN INC

www.allergan.com

Industry Group Code: 325412 Ranks within this company's industry group: Sales: 24   Profits: 189

Drugs:		Other:		Clinical:		Computers:		Other:	
Discovery:	Y	AgriBio:		Trials/Services:		Hardware:		Specialty Services:	
Licensing:		Genomics/Proteomics:		Laboratories:		Software:		Consulting:	
Manufacturing:	Y	Tissue Replacement:		Equipment/Supplies:		Arrays:		Blood Collection:	
Development:	Y			Research/Development Svcs.:		Database Management:		Drug Delivery:	
Generics:				Diagnostics:				Drug Distribution:	

TYPES OF BUSINESS:

Pharmaceutical Development
Eye Care Supplies
Dermatological Products
Neuromodulator Products

BRANDS/DIVISIONS/AFFILIATES:

Alphagan
Lumigan
Azelex
Ocuflux
Botox
Restasis
Inamed Corp.
EndoArt S.A.

CONTACTS: *Note: Officers with more than one job title may be intentionally listed here more than once.*

David E. I. Pyott, CEO
F. Michael Ball, Pres.
Jeffrey L. Edwards, CFO
Scott M. Whitcup, Exec. VP-R&D
Raymond H. Diradoorian, Exec. VP-Global Tech. Oper.
Douglas S. Ingram, General Counsel/Exec. VP/Corp. Sec.
Jeffrey L. Edwards, Exec. VP-Bus. Dev.
Jeffrey L. Edwards, Exec. VP-Finance
David E. I. Pyott, Chmn.

Phone: 714-246-4500	Fax: 714-246-4971
Toll-Free:	
Address: 2525 Dupont Dr., Irvine, CA 92612 US	

GROWTH PLANS/SPECIAL FEATURES:

Allergan, Inc. is a technology-driven global health care company that develops and commercializes specialty pharmaceutical products for the ophthalmic, neuromodulator, dermatological and other specialty markets. The company focuses on treatments for glaucoma and retinal disease, cataracts, dry eye, psoriasis, acne and neuromuscular disorders. Current research and development efforts are focused on gastroenterology, neuropathic pain and genitourinary diseases. Allergan's eye care pharmaceutical products include Alphagan, Alphagan P and Lumigan ophthalmic solutions, which are used for the treatment of open-angle glaucoma and ocular hypertension; Acular LS, which reduces ocular pain; and Ocuflux, Oflox and Exocin ophthalmic anti-infective solutions. The firm's neuromodulator products include Botox, which is used for the treatment of neuromuscular disorders, and Botox Cosmetic for the temporary improvement of wrinkles. Allergan's skin care product line is comprised of tazarotene products in cream and gel formulations for the treatment of acne, facial wrinkles and psoriasis, marketed under the names Tazorac and Avage; Azelex, an acne product; and M.D. Forte, a line of glycolic and alpha-hydroxy-acid-based products. The firm's fastest growing product is its Restasis dry-eye treatment. In March 2006, Allergan acquired Inamed Corp., maker of the Lap-Band gastric-banding system, for $3 billion dollars. In early 2007, Allergan acquired EndoArt S.A., a Swiss company with technology to remotely tighten or loosen a gastric band, for $97 million.

Allergan offers its employees benefits including a 401(k) plan, a defined benefit retirement contribution, adoption assistance, education assistance, before-tax flex dollars and flexible spending accounts, backup child care, company store, on-site gym and athletic fields, computer training facilities, an employee credit union, an employee assistance program, dependent scholarship awards and U.S. savings bond deductions.

FINANCIALS: Sales and profits are in thousands of dollars—add 000 to get the full amount. 2006 Note: Financial information for 2006 was not available for all companies at press time.

2006 Sales: $3,010,100	2006 Profits: $-127,400	U.S. Stock Ticker: AGN
2005 Sales: $2,319,200	2005 Profits: $403,900	Int'l Ticker: Int'l Exchange:
2004 Sales: $2,045,600	2004 Profits: $377,100	Employees: 6,772
2003 Sales: $1,171,400	2003 Profits: $-52,500	Fiscal Year Ends: 12/31
2002 Sales: $1,425,300	2002 Profits: $75,200	Parent Company:

SALARIES/BENEFITS:

Pension Plan: Y	ESOP Stock Plan:	Profit Sharing: Y	Top Exec. Salary: $1,125,769	Bonus: $1,243,000
Savings Plan: Y	Stock Purch. Plan: Y		Second Exec. Salary: $507,754	Bonus: $338,500

OTHER THOUGHTS:

Apparent Women Officers or Directors:
Hot Spot for Advancement for Women/Minorities:

LOCATIONS: ("Y" = Yes)

West:	Southwest:	Midwest:	Southeast:	Northeast:	International:
Y	Y				Y

ALLIANCE PHARMACEUTICAL CORP   www.allp.com

Industry Group Code: 325412   Ranks within this company's industry group: Sales: 181   Profits: 73

Drugs:		Other:		Clinical:		Computers:		Other:	
Discovery:	Y	AgriBio:		Trials/Services:		Hardware:		Specialty Services:	
Licensing:	Y	Genomics/Proteomics:		Laboratories:		Software:		Consulting:	
Manufacturing:		Tissue Replacement:		Equipment/Supplies:		Arrays:		Blood Collection:	
Development:	Y			Research/Development Svcs.:		Database Management:		Drug Delivery:	
Generics:				Diagnostics:				Drug Distribution:	

TYPES OF BUSINESS:
Drugs-Cardiovascular & Respiratory
Blood Substitutes
Immune Disorder Therapies
Intellectual Property Development

BRANDS/DIVISIONS/AFFILIATES:
Oxygent
Astral, Inc.
PFC Therapeutics

CONTACTS: *Note: Officers with more than one job title may be intentionally listed here more than once.*
Duane J. Roth, CEO
B. Jack Defranco, COO
B. Jack DeFranco, Pres.
Edward C. Hall, CFO
Duane J. Roth, Chmn.

Phone: 858-410-5200	Fax: 858-410-5201
Toll-Free:	
Address: 4660 La Jolla Village Dr., Ste. 825, San Diego, CA 92122 US	

GROWTH PLANS/SPECIAL FEATURES:

Alliance Pharmaceutical Corp. develops therapeutic and diagnostic products that utilize perfluorochemicals, which are chemical substances with high oxygen-carrying capacity. The firm's products are designed for use during acute care situations, such as heart surgery and during blood transfusions. The company's leading product candidate, Oxygent, is intended to be an emulsion-based red blood cell substitute. The blood substitute is sterile, universally compatible with all blood types and has a shelf-life of approximately two years. Oxygent has two indications, the first being to provide oxygen to tissues during elective surgeries where a blood transfusion is anticipated and thereby eliminating or reducing the need for blood transfusions. In addition, the drug is used as a therapeutic for organ perfusion during major elective surgeries to protect vital organs from hypoxic injury, meaning that the drug improves oxygenation to organs and tissues during surgery, and has the potential to reduce the time before organs are working properly after major surgeries, such as cardio pulmonary bypass surgery. Company subsidiary Astral, Inc., which is one-third owned by Alliance in conjunction with a collaborative relationship with MultiCell Technologies, Inc. and Astral Therapeutics, develops intellectual property for potential products to treat immune disorders, including autoimmune disease, cancer and infectious diseases. Alliance has formed many collaborative relationships with other companies to test, manufacture and eventually distribute Oxygent in other countries, including Beijing Double-Crane Pharmaceutical Co., Ltd. in China; Leo Pharma A/S in Europe, including the EU member and applicant countries, and Canada; and Il Yang Pharm. Co., Ltd. in South Korea. In December 2006, Alliance was approved to run a Phase 2 study of Oxygent in France and opened enrollment for the program in May 2007.

FINANCIALS: Sales and profits are in thousands of dollars—add 000 to get the full amount. 2006 Note: Financial information for 2006 was not available for all companies at press time.

2006 Sales: $ 129	2006 Profits: $-9,575	U.S. Stock Ticker: ALLP.OB
2005 Sales: $1,477	2005 Profits: $-5,743	Int'l Ticker: Int'l Exchange:
2004 Sales: $ 549	2004 Profits: $10,016	Employees: 7
2003 Sales: $ 59	2003 Profits: $-20,324	Fiscal Year Ends: 6/30
2002 Sales: $10,300	2002 Profits: $-34,200	Parent Company:

SALARIES/BENEFITS:

Pension Plan:	ESOP Stock Plan:	Profit Sharing:	Top Exec. Salary: $158,540	Bonus: $
Savings Plan:	Stock Purch. Plan:		Second Exec. Salary: $120,000	Bonus: $

OTHER THOUGHTS:
Apparent Women Officers or Directors:
Hot Spot for Advancement for Women/Minorities:

LOCATIONS: ("Y" = Yes)

West:	Southwest:	Midwest:	Southeast:	Northeast:	International:
Y					

ALLOS THERAPEUTICS INC

www.allos.com

Industry Group Code: 325412 Ranks within this company's industry group: Sales: Profits: 109

Drugs:		Other:	Clinical:	Computers:	Other:
Discovery:	Y	AgriBio:	Trials/Services:	Hardware:	Specialty Services:
Licensing:		Genomics/Proteomics:	Laboratories:	Software:	Consulting:
Manufacturing:	Y	Tissue Replacement:	Equipment/Supplies:	Arrays:	Blood Collection:
Development:	Y		Research/Development Svcs.:	Database Management:	Drug Delivery:
Generics:			Diagnostics:		Drug Distribution:

TYPES OF BUSINESS:

Cancer Treatment Drugs
Small-Molecule Therapies

BRANDS/DIVISIONS/AFFILIATES:

Efaproxyn
PDX
RH1

CONTACTS: *Note: Officers with more than one job title may be intentionally listed here more than once.*

Paul L. Berns, CEO
Paul L. Berns, Pres.
James V. Caruso, Exec. VP/Chief Commercial Officer
Pablo J. Cagnoni, Sr. VP/Chief Medical Officer
Douglas G. Johnson, VP-Mfg.
Marc H. Graboyes, General Counsel/VP
David C. Clark, VP-Finance
Stephen J. Hoffman, Chmn.

Phone: 303-426-6262	Fax: 303-426-4731
Toll-Free:	
Address: 11080 CirclePoint Rd., Ste. 200, Westminster, CO 80020 US	

GROWTH PLANS/SPECIAL FEATURES:

Allos Therapeutics, Inc. focuses on developing and commercializing small molecule drugs for cancer treatment. The company has three product candidates: Efaproxyn, PDX and RH1. Efaproxyn is a synthetic small molecule designed to sensitize oxygen-deprived areas of tumors during radiation therapy by facilitating the release of oxygen from hemoglobin, the oxygen-carrying protein contained within red blood cells. PDX is a small molecule chemotherapeutic agent that inhibits dihydrofolate reductase, a folic acid-dependent enzyme involved in the building of nucleic acid and other processes. PDX was designed for transport into tumor cells via the reduced folate carrier and effective intracellular drug retention. RH1 is a chemotherapeutic compound with the potential to target lung, colon, breast and liver tumors with limited toxicity to healthy tissue. RH1 is currently being evaluated in patients with advanced solid tumors refractory to other chemotherapy regimens in an open label, Phase I dose escalation. In August 2006, the firm initiated Propel, a Phase II clinical trial of PDX in patients with relapsed or refractory peripheral T-cell lymphoma. In September 2006, the company completed patient enrollment in Enrich, a Phase III clinical trial of Efaproxyn plus whole brain radiation therapy (WBRD) in women with brain metastases originating from breast cancer. In October 2006, FDA granted fast track designation to PDX for the treatment of patients with T-cell lymphoma. In June 2007, Allos announced that the Phase III Efaproxyn plus WBRT study failed to demonstrate a statistically significant improvement in overall survival of patients receiving the drug compared with patients receiving WBRT only. The company plans to discontinue the development of Efaproxyn and advance PDX to Phase II clinical studies.

The company offers its employees a 401(k) plan; health, dental and vision insurance; group life insurance; short- and long-term disability; educational reimbursement; an employee stock purchase plan; and flexible spending accounts.

FINANCIALS: Sales and profits are in thousands of dollars—add 000 to get the full amount. 2006 Note: Financial information for 2006 was not available for all companies at press time.

2006 Sales: $	2006 Profits: $-21,837	U.S. Stock Ticker: ALTH
2005 Sales: $	2005 Profits: $-20,137	Int'l Ticker: Int'l Exchange:
2004 Sales: $	2004 Profits: $-21,800	Employees: 57
2003 Sales: $	2003 Profits: $-23,127	Fiscal Year Ends: 12/31
2002 Sales: $	2002 Profits: $-25,800	Parent Company:

SALARIES/BENEFITS:

Pension Plan:	ESOP Stock Plan:	Profit Sharing:	Top Exec. Salary: $366,900	Bonus: $184,900
Savings Plan: Y	Stock Purch. Plan: Y		Second Exec. Salary: $263,000	Bonus: $62,500

OTHER THOUGHTS:

Apparent Women Officers or Directors:
Hot Spot for Advancement for Women/Minorities:

LOCATIONS: ("Y" = Yes)

West:	Southwest:	Midwest:	Southeast:	Northeast:	International:
Y					

ALPHARMA INC   www.alpharma.com

Industry Group Code: 325414 Ranks within this company's industry group: Sales: 4   Profits: 4

Drugs:		Other:		Clinical:	Computers:	Other:	
Discovery:	Y	AgriBio:	Y	Trials/Services:	Hardware:	Specialty Services:	
Licensing:		Genomics/Proteomics:		Laboratories:	Software:	Consulting:	
Manufacturing:	Y	Tissue Replacement:		Equipment/Supplies:	Arrays:	Blood Collection:	
Development:	Y			Research/Development Svcs.:	Database Management:	Drug Delivery:	
Generics:				Diagnostics:		Drug Distribution:	

TYPES OF BUSINESS:

Drugs-Animal Health
Human Pharmaceuticals
Animal Feed Additives
Active Pharmaceutical Ingredients

BRANDS/DIVISIONS/AFFILIATES:

Kadian
Bacitracin
BMD
Albac
3-Nitro
Histostat
Zoamix
Nystatin

CONTACTS: *Note: Officers with more than one job title may be intentionally listed here more than once.*

Dean J. Mitchell, CEO
Dean J. Mitchell, Pres.
Jeffrey S. Campbell, CFO
Peter Watts, Exec. VP-Human Resources
Thomas J. Spellman III, Chief Legal Officer/Exec. VP/Corp. Sec.
Stefan Aigner, Exec. VP-Corp. Dev.
Peter Watts, Exec. VP-Comm.
Carl-Ake Carlsson, Pres., API Active Pharmaceutical Ingredients
Carol A. Wrenn, Pres., Animal Health Div.
Ronald N. Warner, Pres., Pharmaceuticals
Peter G. Tombros, Chmn.

Phone: 908-566-3800	Fax: 908-566-4137

Toll-Free: 866-322-2525

Address: 440 U.S. Highway 22 East, 3rd Fl., Bridgewater, NJ 08807 US

GROWTH PLANS/SPECIAL FEATURES:

Alpharma, Inc. is a multinational pharmaceutical company that manufactures specialty and proprietary human pharmaceutical and animal health products. The company operates in three business segments: pharmaceuticals, active pharmaceutical ingredients (API) and animal health. The pharmaceuticals unit manufactures one branded product, Kadian, which is a morphine sulfate sustained release capsule. The medication accounted for 21% of the company's total revenues in 2006. Alpharma is also a leading producer of APIs, marketing and selling 14 APIs, specializing in fermented antibiotics in including Bacitracin, Polymyxin B, Vancomycin, Amphotericin B and Colistin. Alpharma's animal health segment is a leading provider of animal feed additives and water soluable therapeutics for poultry and livestock. It provides over 100 medicated feed-additive products in over 80 countries. Key products include BMD, a feed additive that promotes growth and feed efficiency, as well as prevents or treats diseases in poultry and swine; Albac, a feed additive for poultry, swine and calves; and 3-Nitro, Histostat, Zoamix and CTC feed grade antibiotics, all of which are used in combination or sequentially with BMD. In March 2006, Alpharma sold ParMed, its generics pharmaceutical telemarketing business, to Cardinal Health for $40 million. In October 2006, the FDA approved the Kaidian 80 mg dose, and in February 2007, it approved the 200mg dose strength. In June 2007, the company announced that it has acquired certain assets of Yantai JinHai Pharmaceutical Co. Ltd. in China and plans to use it to expand the animal health division.

Alpharma offers employees benefits including medical, dental, prescription and vision coverage; life and disability insurance; paid time off; and tuition reimbursement.

FINANCIALS: Sales and profits are in thousands of dollars—add 000 to get the full amount. 2006 Note: Financial information for 2006 was not available for all companies at press time.

2006 Sales: $653,828	2006 Profits: $82,544	**U.S. Stock Ticker: ALO**
2005 Sales: $553,617	2005 Profits: $133,769	**Int'l Ticker:** Int'l Exchange:
2004 Sales: $513,329	2004 Profits: $-314,737	Employees: 1,400
2003 Sales: $1,297,285	2003 Profits: $16,936	Fiscal Year Ends: 12/31
2002 Sales: $1,238,000	2002 Profits: $-98,800	Parent Company:

SALARIES/BENEFITS:

Pension Plan:	ESOP Stock Plan:	Profit Sharing:	Top Exec. Salary: $410,000	Bonus: $429,067
Savings Plan: Y	Stock Purch. Plan: Y		Second Exec. Salary: $400,000	Bonus: $619,067

OTHER THOUGHTS:

Apparent Women Officers or Directors: 1
Hot Spot for Advancement for Women/Minorities:

LOCATIONS: ("Y" = Yes)

West:	Southwest:	Midwest:	Southeast:	Northeast:	International:
Y	Y	Y	Y	Y	Y

ALSERES PHARMACEUTICALS INC www.alseres.com

Industry Group Code: 325412 **Ranks within this company's industry group:** Sales: Profits: 117

Drugs:		Other:	Clinical:	Computers:	Other:
Discovery:		AgriBio:	Trials/Services:	Hardware:	Specialty Services:
Licensing:	Y	Genomics/Proteomics:	Laboratories:	Software:	Consulting:
Manufacturing:		Tissue Replacement:	Equipment/Supplies:	Arrays:	Blood Collection:
Development:	Y		Research/Development Svcs.:	Database Management:	Drug Delivery:
Generics:			Diagnostics:		Drug Distribution:

TYPES OF BUSINESS:
Pharmaceutical Discovery & Development
Patent & Development Rights
Radio-Imaging Agents
CNS Disorder Treatments

BRANDS/DIVISIONS/AFFILIATES:
Boston Life Sciences, Inc.
Inosine
Cetherin
DAT Blocker
ALTROPANE

CONTACTS: *Note: Officers with more than one job title may be intentionally listed here more than once.*
Peter G. Savas, CEO
Mark J. Pykett, COO
Mark J. Pykett, Pres.
Kenneth L. Rice, Jr., CFO
Noel J. Cusack, Sr. VP-Preclinical Dev.
Kennith L. Rice, Jr., Exec. VP-Admin.
Kenneth L. Rice, Jr., In House Counsel
Richard M. Thorn, Sr. VP-Program Oper.
Frank Bobe, Chief Bus. Officer/Exec. VP
Kenneth L. Rice, Jr., Exec. VP-Finance
Irene Gonzalez, Sr. VP-Process Dev.
Mark Hurtt, Chief Medical Officer
Peter G. Savas, Chmn.

Phone: 508-497-2360	Fax: 508-497-9964
Toll-Free:	
Address: 85 Main St., Hopkinton, MA 01748 US	

GROWTH PLANS/SPECIAL FEATURES:
Alseres Pharmaceuticals, Inc., formerly Boston Life Sciences, Inc., is a development-stage biotechnology company engaged in the research and development of novel diagnostics and therapeutic products to treat and diagnose central nervous system (CNS) disorders. The company's research and development is based on three technology platforms: Nerve repair program focused on the functional recovery from CNS disorders resulting from traumas and utilizing axon regeneration technology; Molecular imaging program focused on the diagnosis of Parkinsonian Syndromes, including Parkinson's Disease (PD) and Attention Deficit Hyperactivity Disorder (ADHD); and Neurodegenerative program focused on treating the symptoms of PD and slowing or stopping the progression of PD. In the axon regeneration program, also known as the nerve repair program, Alseres has Cetherin for spinal cord injury entering Phase II of development and Inosine for stroke entering Phase I. Within the molecular imaging program, the company has two products in the pre-clinical stages of development, as well as ALTROPANE for PD in Phase III clinical trials and for ADHD in Phase II of clinical trials. In the nerodegenerative, or Dopamine Transporter (DAT) Blocker, program, the company is in the preclinical developmental stages for a novel DAT Blocker treatment for PD. Alseres is also working on an ocular therapeutics program for the treatment of optic nerve injury and glaucoma. In March 2006, Alseres announced that it had elected to terminate its current special protocol assessment and end its Phase III, Parkinson's or Essential Tremor (POET-1) trial for ALTROPANE for PD earlier than planned in order to analyze the full set of data for efficacy. In June 2007, the company changed its name from Boston Life Sciences, Inc. to Alseres Pharmaceuticals, Inc. as part of its change of focus from life sciences to biopharmaceuticals.

FINANCIALS: Sales and profits are in thousands of dollars—add 000 to get the full amount. 2006 Note: Financial information for 2006 was not available for all companies at press time.

2006 Sales: $	2006 Profits: $-26,355	U.S. Stock Ticker: ALSE
2005 Sales: $	2005 Profits: $-11,501	Int'l Ticker: Int'l Exchange:
2004 Sales: $	2004 Profits: $-11,251	Employees: 27
2003 Sales: $	2003 Profits: $-8,400	Fiscal Year Ends: 12/31
2002 Sales: $	2002 Profits: $-11,000	Parent Company:

SALARIES/BENEFITS:
| Pension Plan: | ESOP Stock Plan: | Profit Sharing: | Top Exec. Salary: $400,000 | Bonus: $100,000 |
| Savings Plan: Y | Stock Purch. Plan: | | Second Exec. Salary: $300,000 | Bonus: $75,000 |

OTHER THOUGHTS:
Apparent Women Officers or Directors: 1
Hot Spot for Advancement for Women/Minorities:

LOCATIONS: ("Y" = Yes)
West:	Southwest:	Midwest:	Southeast:	Northeast:	International:
				Y	

ALTANA AG

www.altana.de

Industry Group Code: 325412 Ranks within this company's industry group: Sales: 20 Profits: 21

Drugs:		Other:	Clinical:		Computers:		Other:
Discovery:		AgriBio:	Trials/Services:		Hardware:		Specialty Services:
Licensing:		Genomics/Proteomics:	Laboratories:		Software:		Consulting:
Manufacturing:	Y	Tissue Replacement:	Equipment/Supplies:	Y	Arrays:		Blood Collection:
Development:			Research/Development Svcs.:		Database Management:		Drug Delivery:
Generics:			Diagnostics:				Drug Distribution:

TYPES OF BUSINESS:

Specialty Chemical Manufacturing
Imaging Products
Electrical Insulation
Coatings

BRANDS/DIVISIONS/AFFILIATES:

Altana Chemie AG
BYK-Chemie
Altana Coatings and Sealants
ECKART Group
American Rad-Cure Corporation
ALTANA Pharma AG

CONTACTS: *Note: Officers with more than one job title may be intentionally listed here more than once.*

Nikolaus Schweickart, CEO
Nikolaus Schweickart, Pres.
Hermann Kullmer, CFO
Till Isensee, Dir.-Sales & Mktg.
Matthias L. Wolfgruber, CEO/Pres., Altana Chemie AG
Justus Mische, Chmn.

Phone: 49-0-6172-1712-0	Fax: 49-0-6172-1712-365
Toll-Free:	
Address: Herbert-Quandt-Haus, Am Pilgerrain 15, Bad Homburg, VD Hohe M2 D61352 Germany	

GROWTH PLANS/SPECIAL FEATURES:

Altana AG is an international pharmaceuticals and chemicals company that develops, manufactures and markets products for a range of targeted, highly specialized applications. Until December 2006, the company had two operating divisions: Pharmaceuticals and Chemicals. Due to the sale of its pharmaceutical division, which focused on therapeutics, imaging and over-the-counter self-medication products for the treatment of gastrointestinal and respiratory tract diseases, the firm now operates in a single division. Altana's chemical division is engaged in the production of specialty chemicals. This division offers products and services in four categories: Additives and Instruments, which is represented through subsidiary BYK-Chemie, manufactures additives and coatings for the paint, printing and plastics industries as well as a line of measuring instruments for quality analysis; Effect Pigments, which is represented by the ECKART group, manufactures metallic effect pigments for use in paints, plastics, cosmetics and printing inks; Electrical Insulation, which produces wire enamels and impregnating resins for electrical insulation; and Coatings and Sealants, which, in addition to being the world's market leader in bottle top sealants, develops paper and plastic sealants as well as flexible packaging. In January 2006, the firm acquired the captive production of resins for insulation of magnet wire activities of INVEX SpA. The deal included a chemical production site in Cerquilho, Brazil. Later in 2006, Altana acquired U.S.-based American Rad-Cure Corporation, a specialist in the field of overprint UV curable coatings and adhesives for the paper and board packaging sector. In December 2006, Altana sold its pharmaceuticals division, ALTANA Pharma AG, to Nycomed, a Danish pharmaceuticals company, for approximately $6 billion.

FINANCIALS: Sales and profits are in thousands of dollars—add 000 to get the full amount. 2006 Note: Financial information for 2006 was not available for all companies at press time.

2006 Sales: $5,234,410	2006 Profits: $1,008,440	U.S. Stock Ticker: AAA
2005 Sales: $4,429,010	2005 Profits: $915,040	Int'l Ticker: ALT Int'l Exchange: Fankfurt-Euronext
2004 Sales: $4,013,500	2004 Profits: $529,200	Employees: 13,404
2003 Sales: $3,438,200	2003 Profits: $434,000	Fiscal Year Ends: 12/31
2002 Sales: $2,739,900	2002 Profits: $340,700	Parent Company:

SALARIES/BENEFITS:

Pension Plan:	ESOP Stock Plan:	Profit Sharing:	Top Exec. Salary: $	Bonus: $
Savings Plan:	Stock Purch. Plan:		Second Exec. Salary: $	Bonus: $

OTHER THOUGHTS:

Apparent Women Officers or Directors:
Hot Spot for Advancement for Women/Minorities:

LOCATIONS: ("Y" = Yes)

West:	Southwest:	Midwest:	Southeast:	Northeast:	International:
		Y		Y	Y

ALTEON INC
www.alteon.com

Industry Group Code: 325412 Ranks within this company's industry group: Sales: 176   Profits: 100

Drugs:		Other:	Clinical:	Computers:	Other:
Discovery:	Y	AgriBio: Y	Trials/Services:	Hardware:	Specialty Services:
Licensing:	Y	Genomics/Proteomics:	Laboratories:	Software:	Consulting:
Manufacturing:		Tissue Replacement:	Equipment/Supplies:	Arrays:	Blood Collection:
Development:	Y		Research/Development Svcs.:	Database Management:	Drug Delivery:
Generics:			Diagnostics:		Drug Distribution:

TYPES OF BUSINESS:

Pharmaceutical Discovery & Development
Cardiovascular Drugs
Diabetes Drugs

BRANDS/DIVISIONS/AFFILIATES:

Haptoguard, Inc.

CONTACTS: Note: Officers with more than one job title may be intentionally listed here more than once.

Noah Berkowitz, CEO
Noah Berkowitz, Pres.
Jacob Victor, Exec. Dir.-Prod. Dev.
Nancy Regan, Mgr.-Exec. & Oper.
Jacob Victor, Exec. Dir.-Clinical Diagnostics
Malcolm W. MacNab, VP-Clinical Dev.

Phone: 201-934-5000	Fax: 201-934-8880
Toll-Free:	
Address: 221 W. Grand Ave., Montvale, NJ 07645 US	

GROWTH PLANS/SPECIAL FEATURES:

Alteon, Inc. is a product-based biopharmaceutical company primarily engaged in the discovery and development of oral drugs to reverse or inhibit cardiovascular aging and diabetic complications. These candidates were developed as a result of research on the advanced glycosylation end-product (AGE) pathway, a fundamental process and consequence of aging and diabetes that causes or contributes to many medical disorders. Alteon's current research and drug development activities targeting this pathway take three directions: the breaking of AGE crosslinks between proteins; the prevention or inhibition of AGE formation; and the reduction of the AGE burden through a novel class of anti-hyperglycemic agents, or glucose-lowering agents. Since its inception, Alteon has created an extensive library of compounds targeting the AGE pathway. The company's lead product candidate is alagebrium chloride, formerly ALT-711, an AGE crosslink breaker that offers the first therapeutic approach to breaking these crosslinks, which may reverse tissue damage caused by aging and diabetes, thereby restoring flexibility and function to blood vessels and organs of the body. Results suggest that alagebrium is a novel potential therapy for many conditions that are associated with diabetes and aging. Alteon is also working on its compound ALT-2074, which uses a certain form of haptoglobin which is present in 40% of the human population. With this technology, it will be able to offer directed therapy based on variability in the form of haptoglobin in the blood. The compound is currently in testing. Alteon maintains a number of licensing and collaboration agreements relating to the development and distribution of its AGE technology. In July 2006, Alteon acquired and merged HaptoGuard, Inc., for nearly $8.8 million in stocks. As a result of the merger, Alteon gained new additions to its product portfolio, including ALT-2074, which gives the company two lead candidate products in clinical development.

FINANCIALS: Sales and profits are in thousands of dollars—add 000 to get the full amount. 2006 Note: Financial information for 2006 was not available for all companies at press time.

2006 Sales: $ 251	2006 Profits: $-17,680	U.S. Stock Ticker: ALT
2005 Sales: $ 458	2005 Profits: $-12,614	Int'l Ticker: Int'l Exchange:
2004 Sales: $ 334	2004 Profits: $-13,959	Employees: 7
2003 Sales: $ 200	2003 Profits: $-14,500	Fiscal Year Ends: 12/31
2002 Sales: $	2002 Profits: $-16,900	Parent Company:

SALARIES/BENEFITS:

Pension Plan:	ESOP Stock Plan:	Profit Sharing:	Top Exec. Salary: $240,000	Bonus: $54,000
Savings Plan: Y	Stock Purch. Plan:		Second Exec. Salary: $240,000	Bonus: $36,000

OTHER THOUGHTS:

Apparent Women Officers or Directors: 1
Hot Spot for Advancement for Women/Minorities:

LOCATIONS: ("Y" = Yes)

West:	Southwest:	Midwest:	Southeast:	Northeast:	International:
				Y	

ALZA CORP

www.alza.com

Industry Group Code: 325412A Ranks within this company's industry group: Sales: Profits:

Drugs:		Other:		Clinical:		Computers:		Other:	
Discovery:		AgriBio:		Trials/Services:		Hardware:		Specialty Services:	
Licensing:		Genomics/Proteomics:		Laboratories:		Software:		Consulting:	
Manufacturing:	Y	Tissue Replacement:		Equipment/Supplies:		Arrays:		Blood Collection:	
Development:	Y			Research/Development Svcs.:		Database Management:		Drug Delivery:	Y
Generics:				Diagnostics:				Drug Distribution:	

TYPES OF BUSINESS:

Drug Delivery Systems
Pharmaceutical Development & Marketing

BRANDS/DIVISIONS/AFFILIATES:

Johnson & Johnson
OROS
L-OROS
D-TRANS
STEALTH
DUROS
E-TRANS
IONSYS

CONTACTS: Note: Officers with more than one job title may be intentionally listed here more than once.

Michael R. Jackson, Pres.

Phone: 650-564-5000	Fax: 650-564-7070
Toll-Free:	
Address: 1900 Charleston Rd., Mountain View, CA 94039-7120 US	

GROWTH PLANS/SPECIAL FEATURES:

ALZA Corp., a subsidiary of Johnson & Johnson, develops and markets pharmaceutical products involving advanced drug delivery technologies. The company focuses on several therapeutic areas, including oncology, AIDS, pain management, endocrinology and urology. ALZA partners with other companies to combine new pharmaceutical compounds with its oral, transdermal, implantable and liposomal delivery systems. It has five key brand name technologies. OROS oral delivery technology, currently incorporated into 13 different products worldwide including Sudafed 24 Hour, uses osmosis to provide controlled drug delivery for up to 24 hours, with L-OROS available for liquid formulations. D-TRANS transdermal technology is a patch drug delivery system currently used for seven different products, including Nicoderm CQ. Its STEALTH liposomal technology consists of microscopic lipid particles that incorporate a polyethylene glycol coating, which allows cancer therapeutics and gene therapy vectors to evade detection by the immune system, targeting specific areas of disease within the body. DUROS implants are miniature titanium cylinders which deliver therapeutics including small drugs, proteins, DNA and other bioactive macromolecules at a continuous rate for up to one year. E-TRANS is an electron transport technology which actively transports drugs through intact skin, both locally and systemically, using low-level electrical energy. Formerly, the company offered Macroflux transdermal technology, which utilized a thin screen of microprojections as an alternative to standard injections; however, in October 2006, Johnson & Johnson turned that division into Macroflux Corporation. In May 2006, ALZA received FDA approval for fentanyl-based analgesic IONSYS, its first product to make use of the E-TRANS technology.

Alza provides its employees with a wide range of benefits including pre-tax accounts for health and dependant care; medical, dental, life and accident insurance; adoption assistance; tuition reimbursement; short- and long-term disability care; daycare discounts; exercise programs; and a charitable matching gift program.

FINANCIALS: Sales and profits are in thousands of dollars—add 000 to get the full amount. 2006 Note: Financial information for 2006 was not available for all companies at press time.

2006 Sales: $	2006 Profits: $	U.S. Stock Ticker: Subsidiary
2005 Sales: $	2005 Profits: $	Int'l Ticker: Int'l Exchange:
2004 Sales: $	2004 Profits: $	Employees: 2,442
2003 Sales: $	2003 Profits: $	Fiscal Year Ends: 12/31
2002 Sales: $	2002 Profits: $	Parent Company: JOHNSON & JOHNSON

SALARIES/BENEFITS:

Pension Plan: Y	ESOP Stock Plan:	Profit Sharing:	Top Exec. Salary: $	Bonus: $
Savings Plan: Y	Stock Purch. Plan:		Second Exec. Salary: $	Bonus: $

OTHER THOUGHTS:

Apparent Women Officers or Directors:
Hot Spot for Advancement for Women/Minorities:

LOCATIONS: ("Y" = Yes)

West:	Southwest:	Midwest:	Southeast:	Northeast:	International:
Y					

AMARIN CORPORATION PLC   www.amarincorp.com

Industry Group Code: 325412 Ranks within this company's industry group: Sales: 170   Profits: 119

Drugs:		Other:	Clinical:	Computers:	Other:
Discovery:	Y	AgriBio:	Trials/Services:	Hardware:	Specialty Services:
Licensing:	Y	Genomics/Proteomics:	Laboratories:	Software:	Consulting:
Manufacturing:	Y	Tissue Replacement:	Equipment/Supplies:	Arrays:	Blood Collection:
Development:	Y		Research/Development Svcs.:	Database Management:	Drug Delivery:
Generics:			Diagnostics:		Drug Distribution:

TYPES OF BUSINESS:

Drugs, Neurology
Drugs, Huntington's Disease
Drugs, Depression
Drugs, Parkinson's Disease
Drugs, CNS Disorders
Drugs, Multiple Sclerosis Fatigue

BRANDS/DIVISIONS/AFFILIATES:

Miraxion

CONTACTS: Note: Officers with more than one job title may be intentionally listed here more than once.

Rick Stewart, CEO
Alan Cooke, Pres.
Alan Cooke, CFO
Declan Doogan, Pres., R&D
Tom Maher, General Counsel/Corp. Sec.
Darren Cunningham, Exec. VP-Strategic Dev.
Conor Dalton, VP-Finance
Anthony Clarke, VP-Clinical Dev.
David Boal, VP-Bus. Dev.
Paul F. Duffy, Pres., US Commercial Oper.
Mehar Manku, VP-R&D
Thomas G. Lynch, Chmn.

Phone: 44-20-7499-9009	Fax: 44-20-7499-9004
Toll-Free:	
Address: 7 Curzon St., London, W1J 5HG UK	

GROWTH PLANS/SPECIAL FEATURES:

Amarin Corporation plc is a specialty pharmaceutical company focused on neuroscience. The company's primary platforms utilize the central nervous system's innate quality of being affected by Polyunsaturated Fatty Acids. Amarin's lipophilic drugs are fat-soluble, which allows easy transportation across the blood-brain barrier. The company is also utilizing a combinational lipid platform, where bioactive lipids are attached either to either other drugs or other lipids. The company's leading product is Miraxion, currently in Phase III clinical studies as a fat soluble, or lipophilic, treatment for Huntington's disease. Miraxion has been granted fast track status by the FDA and orphan drug designation in the U.S. and Europe. Amarin is also conducting Phase II studies of Miraxion for both treatment of chronic depression that fails to respond to standard drugs and Parkinson's disease, respectively. The Parkinson's disease product, AMR-101, was acquired from a company researcher, who developed and applied for patents prior to joining the company, and involves an oral application of apomorphine providing a simpler alternative to the more usual injectable form. Amarin has a total of nine products in its pipeline, including Maraxion for Huntington's disease, depressive disorders and Parkinson's disease; two others for Parkinson's disease; and one for each Epilepsy, memory cognition, MS Fatigue and depression in women. The company's general strategy is to commercialize its neurology products directly in the U.S. and to out-license or partner its depression treatment pipeline and its European and Japanese product rights.

FINANCIALS: Sales and profits are in thousands of dollars—add 000 to get the full amount. 2006 Note: Financial information for 2006 was not available for all companies at press time.

2006 Sales: $ 500	2006 Profits: $-26,920	U.S. Stock Ticker: AMRN
2005 Sales: $ 500	2005 Profits: $-20,547	Int'l Ticker: Int'l Exchange:
2004 Sales: $1,000	2004 Profits: $4,700	Employees: 23
2003 Sales: $7,400	2003 Profits: $-19,200	Fiscal Year Ends: 8/31
2002 Sales: $65,400	2002 Profits: $-37,200	Parent Company:

SALARIES/BENEFITS:

Pension Plan: Y	ESOP Stock Plan:	Profit Sharing:	Top Exec. Salary: $515,000	Bonus: $291,000
Savings Plan:	Stock Purch. Plan:		Second Exec. Salary: $482,000	Bonus: $

OTHER THOUGHTS:

Apparent Women Officers or Directors:
Hot Spot for Advancement for Women/Minorities:

LOCATIONS: ("Y" = Yes)

West:	Southwest:	Midwest:	Southeast:	Northeast:	International:
					Y

AMGEN INC

www.amgen.com

Industry Group Code: 325412 Ranks within this company's industry group: Sales: 13 Profits: 10

Drugs:		Other:		Clinical:	Computers:		Other:	
Discovery:	Y	AgriBio:		Trials/Services:	Hardware:		Specialty Services:	
Licensing:		Genomics/Proteomics:		Laboratories:	Software:		Consulting:	
Manufacturing:	Y	Tissue Replacement:		Equipment/Supplies:	Arrays:		Blood Collection:	
Development:	Y			Research/Development Svcs.:	Database Management:		Drug Delivery:	
Generics:				Diagnostics:			Drug Distribution:	

TYPES OF BUSINESS:

Drugs-Diversified
Oncology Drugs
Nephrology Drugs
Inflammation Drugs
Neurology Drugs
Metabolic Drugs

BRANDS/DIVISIONS/AFFILIATES:

Neupogen
Epogen
Aranesp
Enbrel
Kineret
Sensipar
Neulasta
Avidia

CONTACTS: *Note: Officers with more than one job title may be intentionally listed here more than once.*

Kevin W. Sharer, CEO
Kevin W. Sharer, Pres.
Richard D. Nanula, CFO/Exec. VP
Craig Brooks, VP-Global Mktg.
Brian McNamee, Sr. VP-Human Resources
Roger Perlmutter, Exec. VP-R&D
Tom Flanagan, CIO/Sr. VP
Fabrizio Bonanni, Sr. VP-Mfg.
David J. Scott, General Counsel/Sr. VP/Corp. Sec.
Dennis M. Fenton, Exec. VP-Oper.
Sean Harper, Sr. VP-Global Dev./Chief Medical Officer
Phyllis Piano, VP-Corp. Comm./Philanthropy
Steve Schoch, VP-Finance
Nahed Ahmed, VP-Project Mgmt. R&D
Tim Daly, Sr. VP-Commercial Oper., N. America
David Beier, Sr. VP-Global Government Affairs
George Morrow, Exec. VP-Global Commercial Oper.
Kevin W. Sharer, Chmn.
Rollf Hoffmann, Sr. VP-Amgen Int'l Oper.
Laurel Junk, VP-Supply Chain

Phone: 805-447-1000	Fax: 805-447-1010
Toll-Free:	
Address: 1 Amgen Center Dr., Thousand Oaks, CA 91320-1799 US	

GROWTH PLANS/SPECIAL FEATURES:

Amgen, Inc. is a global biotechnology company that develops, manufactures and markets human therapeutics based on advanced cellular and molecular biology. Its products are used for treatment in the fields of nephrology, oncology, inflammation, neurology and metabolic disorders. The company manufactures and markets a line of human therapeutic products including Neupogen, Epogen, Aranesp, Enbrel, Kineret, Neulasta and Sensipar. Neupogen and Neulasta selectively stimulate the growth of infection-fighting white blood cells. Kineret and Enbrel reduce the signs and symptoms of moderately to severely active rheumatoid arthritis. Epogen and Aranesp stimulate the production of red blood cells. Amgen's research and development efforts are focused on human therapeutics delivered in the form of proteins, monoclonal antibodies and small molecules in the areas of hematology; oncology; inflammation; metabolic and bone disorders; and neuroscience. Sensipar is the firm's first small-molecule drug. It is used to treat forms of hyperparathyroidism and is licensed from NPS Pharmaceuticals, Inc. Amgen plans to complete an additional large-scale cell culture commercial manufacturing facility adjacent to the current Rhode Island facility. In February 2006, the firm announced that its plans to invest more than $1 billion over the next four years to expand its manufacturing capacity in Puerto Rico. In 2006, Amgen acquired Avidia, a privately held biopharmaceutical company, and Abgenix, a manufacturer of human therapeutic antibodies.

Amgen offers an employee benefits package that includes an education reimbursement plan, a company pension, a voluntary retirement savings plan, employee stock purchase plan, deferred compensation plan, and, in some locations, on-site health clubs, yoga and pilates classes, massages, hair salons, gift shops, discount ski lift tickets, day care centers, dry cleaners, cafeterias and shoe repair services.

FINANCIALS: Sales and profits are in thousands of dollars—add 000 to get the full amount. 2006 Note: Financial information for 2006 was not available for all companies at press time.

2006 Sales: $14,268,000	2006 Profits: $2,950,000	**U.S. Stock Ticker: AMGN**
2005 Sales: $12,430,000	2005 Profits: $3,674,000	**Int'l Ticker:** Int'l Exchange:
2004 Sales: $10,550,000	2004 Profits: $2,363,000	Employees: 14,000
2003 Sales: $8,356,000	2003 Profits: $2,259,500	Fiscal Year Ends: 12/31
2002 Sales: $5,523,000	2002 Profits: $-1,392,000	Parent Company:

SALARIES/BENEFITS:

Pension Plan: Y	ESOP Stock Plan:	Profit Sharing:	Top Exec. Salary: $1,390,385	Bonus: $4,500,000
Savings Plan: Y	Stock Purch. Plan: Y		Second Exec. Salary: $887,385	Bonus: $1,800,000

OTHER THOUGHTS:

Apparent Women Officers or Directors: 3
Hot Spot for Advancement for Women/Minorities: Y

LOCATIONS: ("Y" = Yes)

West:	Southwest:	Midwest:	Southeast:	Northeast:	International:
Y			Y	Y	Y

Note: Financial information, benefits and other data can change quickly and may vary from those stated here.

AMYLIN PHARMACEUTICALS INC

www.amylin.com

Industry Group Code: 325412 Ranks within this company's industry group: Sales: 40   Profits: 198

Drugs:		Other:	Clinical:	Computers:	Other:
Discovery:	Y	AgriBio:	Trials/Services:	Hardware:	Specialty Services:
Licensing:	Y	Genomics/Proteomics:	Laboratories:	Software:	Consulting:
Manufacturing:		Tissue Replacement:	Equipment/Supplies:	Arrays:	Blood Collection:
Development:	Y		Research/Development Svcs.:	Database Management:	Drug Delivery:
Generics:			Diagnostics:		Drug Distribution:

TYPES OF BUSINESS:

Pharmaceutical Discovery & Development
Drugs, Obesity
Drugs, Diabetes

BRANDS/DIVISIONS/AFFILIATES:

SYMLIN
BYETTA
Exenatide LAR
pramlintide
Integrated Neurohormonal Therapy for Obesity

CONTACTS: *Note: Officers with more than one job title may be intentionally listed here more than once.*

Daniel M. Bradbury, CEO
Daniel M. Bradbury, COO
Daniel M. Bradbury, Pres.
Mark G. Foletta, CFO
Craig Eberhard, VP-Sales
Roger Marchetti, Sr. VP-Human Resources & Corp. Svcs.
Alain D. Baron, Sr. VP-Research
Gregg Stetsko, VP-Strategy & Tech. Planning
Reed Vickerman, VP-Corp. Svcs.
Marcea B. Lloyd, General Counsel/Sr. VP-Legal & Corp. Affairs
Paul Marshall, VP-Oper.
Mark J. Gergen, VP-Corp. Dev.
Mark G. Foletta, Sr. VP-Finance
David Maggs, VP-Medical Affairs
Orville G. Kolterman, Sr. VP-Clinical & Regulatory Affairs
Anna E. Crivici, VP-Project Mgmt. & Bus. Process Dev.
Lisa Porter, VP-Clinical Dev.
Joseph C. Cook, Jr., Chmn.

Phone: 858-552-2200	Fax: 858-552-2212
Toll-Free:	
Address: 9360 Towne Centre Dr., Ste. 110, San Diego, CA 92121 US	

GROWTH PLANS/SPECIAL FEATURES:

Amylin Pharmaceuticals, Inc. is engaged in the discovery, development and commercialization of drug candidates for the treatment of diabetes, obesity and other diseases. The company has ten drugs and drug candidates in its pipeline, including two in commercialization, two in phase 3, three in phase 2 and three in phase 1. SYMLIN, amylinomimetics class, and BYETTA, incretin mimetics class, injections are both first-in-class diabetes medicines that are commercially marketed in the US. BYETTA is also approved in the European Union and the company expects it to be marketed by Eli Lilly. Amylin's phase 3 drugs include Exenatide LAR for diabetes and Pramlintide (AC137) for obesity. Pramlintide is the active ingredient in SYMLIN and Exenatide LAR is the active ingredient in BYETTA. Amylin launched an expansion to the company's clinical program, called Integrated Neurohormonal Therapy for Obesity (INTO), which will assess the safety and efficacy of multiple neurohormones used in combination with Pramlintide to treat obesity. The phase 2 and phase 1 drugs are primarily part of that expansion. The company's phase 2 drugs include PYY 3-36; Pramlintide + Leptin; and Pramlintide + Oral Obesity Agents. Phase 1 drugs include Second Generation Amylinomimetic and Pramlintide + PYY3-36 for obesity and Exenatide Nasal for diabetes. Amylin has strategic alliance partnerships with Alkermes, Inc. and Eli Lilly. In October 2006, Amylin announced the expansion of its obesity program to include multiple drugs used in conjunction to treat obesity. In November 2006, the European Commission announced that it has granted marketing authorization for BYETTA.

Amylin provides its employees with a medical, dental and voluntary vision plan; disability and life insurance; flexible spending accounts; employee assistance and education assistance programs; as well as discounted gym memberships and online concierge services.

FINANCIALS: Sales and profits are in thousands of dollars—add 000 to get the full amount. 2006 Note: Financial information for 2006 was not available for all companies at press time.

2006 Sales: $510,875	2006 Profits: $-218,856	U.S. Stock Ticker: AMLN
2005 Sales: $140,474	2005 Profits: $-206,832	Int'l Ticker: Int'l Exchange:
2004 Sales: $34,268	2004 Profits: $-157,157	Employees: 1,550
2003 Sales: $85,700	2003 Profits: $-122,800	Fiscal Year Ends: 12/31
2002 Sales: $13,400	2002 Profits: $-109,800	Parent Company:

SALARIES/BENEFITS:

Pension Plan:	ESOP Stock Plan:	Profit Sharing:	Top Exec. Salary: $559,167	Bonus: $754,880
Savings Plan: Y	Stock Purch. Plan: Y		Second Exec. Salary: $456,667	Bonus: $462,600

OTHER THOUGHTS:

Apparent Women Officers or Directors: 5
Hot Spot for Advancement for Women/Minorities: Y

LOCATIONS: ("Y" = Yes)

West:	Southwest:	Midwest:	Southeast:	Northeast:	International:
Y					Y

Note: Financial information, benefits and other data can change quickly and may vary from those stated here.

ANGIOTECH PHARMACEUTICALS   www.angiotech.com

industry Group Code: 325412   Ranks within this company's industry group: Sales: 48   Profits: 52

Drugs:		Other:		Clinical:		Computers:		Other:	
Discovery:		AgriBio:		Trials/Services:	Y	Hardware:		Specialty Services:	
Licensing:		Genomics/Proteomics:		Laboratories:		Software:		Consulting:	
Manufacturing:	Y	Tissue Replacement:		Equipment/Supplies:		Arrays:	Y	Blood Collection:	
Development:	Y			Research/Development Svcs.:		Database Management:		Drug Delivery:	
Generics:				Diagnostics:				Drug Distribution:	

TYPES OF BUSINESS:

Medical Device Coatings
Drugs, Inflammatory Disease
Surgical Equipment & Technology

BRANDS/DIVISIONS/AFFILIATES:

TAXUS
Adhibit
CoSeal
Vitagel Surgical Hemostat
Paxceed
Angiotech BioCoatings Corp.
American Medical Instruments Holdings, Inc.
Angiotech Vascular Graft

CONTACTS: *Note: Officers with more than one job title may be intentionally listed here more than once.*

William L. Hunter, CEO
William L. Hunter, Pres.
Tom Bailey, CFO
Chris Dennis, Sr. VP-Mktg. & Sales
Jeffery P. Walker, Sr. VP-R&D
David D. McMasters, Esq., General Counsel/Sr. VP-Legal
David M. Hall, Sr. VP-Gov't & Comm. Rel.
Jodi Regts, Mgr.-Investor Rel.
David M. Hall, Chief Compliance Officer
Gary Ingenito, Chief Clinical & Regulatory Affairs Officer
Rui Avelar, Chief Medical Officer
Santi Corsaro, VP-Sales & Mktg., OUS
David T. Howard, Chmn.

Phone: 604-221-7676	**Fax:** 604-221-2330

Toll-Free:

Address: 1618 Station St., Vancouver, BC V6A 1B6 Canada

GROWTH PLANS/SPECIAL FEATURES:

Angiotech Pharmaceuticals, Inc. is a Canadian pharmaceutical company that develops medical products for complications associated with medical device implants, surgical interventions and acute injury. The company consists of two segments: pharmaceutical technologies and medical products. The pharmaceutical technologies segment develops, licenses and sells technologies that improve the performances of medical devices and the outcomes of surgical procedures. The medical products segment manufactures and markets a wide range of single use, specialty medical devices. The firm's products address restenosis treatment, surgical adhesions, surgical sealants, surgical hemostats, systemic programs and clinical programs. The firm's flagship product is TAXUS, a polymeric formulation that delivers paclitaxel as a stent coating for restenosis. The firm uses paclitaxel, one of the most commercially successful and clinically effective anticancer drugs ever produced, to stop restenosis since it has been found to also block an important cellular pathway involved in inflammation. To treat surgical adhesions, which can lead to infertility and bowel obstructions, Angiotech has created Adhibit, a synthetic sprayable barrier that keeps tissue separate during healing. CoSeal is the firm's surgical sealant, which aids in the healing of suture lines and synthetic grafts during surgery. Hemostatic devices, like Vitagel Surgical Hemostat, control bleeding during surgery and help prevent complications including blood loss and ineffective closure. The firm has systematic programs for rheumatoid arthritis and severe psoriasis, and its drug Paxceed, an anti-inflammatory, that are undergoing trials. Angiotech owns several American subsidiaries to supplement its primary business, including Angiotech BioCoatings, Angiotech Clinical and Angiotech Vascular Graft. In March 2006, Angiotech acquired American Medical Instruments Holdings, Inc., a U.S. producer of specialty devices for surgery, implanting and the treatment of injury and trauma, for $750 million. In May 2006, the firm signed a contract to purchase Quill Medical, Inc., a specialist in minimally invasive, aesthetic surgical technology, for $40 million.

FINANCIALS: Sales and profits are in thousands of dollars—add 000 to get the full amount. 2006 Note: Financial information for 2006 was not available for all companies at press time.

2006 Sales: $315,070	2006 Profits: $4,580	**U.S. Stock Ticker: ANPI**
2005 Sales: $199,600	2005 Profits: $-1,200	**Int'l Ticker: ANP** Int'l Exchange: Toronto-TSX
2004 Sales: $130,800	2004 Profits: $52,500	Employees: 50
2003 Sales: $8,900	2003 Profits: $-34,400	Fiscal Year Ends: 12/31
2002 Sales: $4,600	2002 Profits: $-12,700	Parent Company:

SALARIES/BENEFITS:

Pension Plan:	ESOP Stock Plan:	Profit Sharing:	Top Exec. Salary: $	Bonus: $
Savings Plan:	Stock Purch. Plan:		Second Exec. Salary: $	Bonus: $

OTHER THOUGHTS:

Apparent Women Officers or Directors: 1
Hot Spot for Advancement for Women/Minorities:

LOCATIONS: ("Y" = Yes)

West:	Southwest:	Midwest:	Southeast:	Northeast:	International:
Y		Y	Y	Y	Y

ANIKA THERAPEUTICS INC   www.anikatherapeutics.com

Industry Group Code: 325414 Ranks within this company's industry group: Sales: 7   Profits: 6

Drugs:		Other:		Clinical:		Computers:		Other:	
Discovery:	Y	AgriBio:		Trials/Services:		Hardware:		Specialty Services:	
Licensing:		Genomics/Proteomics:		Laboratories:		Software:		Consulting:	
Manufacturing:	Y	Tissue Replacement:		Equipment/Supplies:		Arrays:		Blood Collection:	
Development:	Y			Research/Development Svcs.:		Database Management:		Drug Delivery:	
Generics:				Diagnostics:				Drug Distribution:	

TYPES OF BUSINESS:

Tissue Protection, Healing & Repair Drugs
Hyaluronic Acid Based Drugs
Tissue Augmentation Products

BRANDS/DIVISIONS/AFFILIATES:

Orthovisc
Elevess
Amvisc
Staarvisc-II
ShellGel
Incert
Hyvisc
DePuy Mitek

CONTACTS: *Note: Officers with more than one job title may be intentionally listed here more than once.*

Charles H. Sherwood, CEO
Charles H. Sherwood, Pres.
Kevin W. Quinlan, CFO
William J. Mrachek, VP-Human Resources
Frank Luppino, VP-Oper.

Phone: 781-932-6616	**Fax:** 781-935-7803
Toll-Free:	
Address: 160 New Boston St., Woburn, MA 01801 US	

GROWTH PLANS/SPECIAL FEATURES:

Anika Therapeutics, Inc. develops, manufactures and commercializes therapeutic products for tissue protection, healing and repair. These products are based on hyaluronic acid (HA), a naturally occurring, biocompatible polymer found throughout the body that plays an important role in a number of physiological functions such as the protection and lubrication of soft tissues and joints; the maintenance of the structural integrity of tissues; and the transport of molecule to and within cells. The firm's currently marketed products consist of Orthovisc, an HA product used in the treatment of some forms of osteoarthritis in humans; Amvisc, Amvisc Plus, Staarvisc-II and ShellGel, each an injectable ophthalmic viscoelastic HA product; Hyvisc, an HA product used in the treatment of equine osteoarthritis; and Incert, an HA based anti-adhesive for surgical applications. Products in development include Elevess, formerly known as Redefyne, an HA based dermal filler used for cosmetic tissue augmentation applications and osteoarthritis/joint health related products. Orthovisc is marketed in the U.S. by DePuy Mitek, a subsidiary of Johnson & Johnson and outside the U.S. by distributors in roughly 20 countries. Hyvisc is marketed in the U.S. through Boehringer Ingelheim Vetmedica, Inc. Currently, Incert is only marketed in three countries outside the U.S. Anika develops and manufactures Amvisc and Amvisc Plus for Bausch & Lomb, Inc. under a multiyear supply agreement. Orthovisc contributed roughly 47% of the company's product revenue in 2006 and Hyvisc 8%. In January 2007, Anika imitated a clinical study designed to evaluate the effectiveness of Orthovisc to treat pain caused by osteoarthritis in the shoulder joint. In April 2007, the company received Conformite Europeene (CE) mark approval for Elevess, which ill be marketed by Galderma.

The company offers its employees benefits that include health, dental life and disability insurance and a 401(k) plan.

FINANCIALS: Sales and profits are in thousands of dollars—add 000 to get the full amount. 2006 Note: Financial information for 2006 was not available for all companies at press time.

2006 Sales: $26,841	2006 Profits: $4,604	**U.S. Stock Ticker:** ANIK	
2005 Sales: $29,835	2005 Profits: $5,893	**Int'l Ticker:** Int'l Exchange:	
2004 Sales: $26,466	2004 Profits: $11,190	**Employees:** 64	
2003 Sales: $15,404	2003 Profits: $ 827	**Fiscal Year Ends:** 12/31	
2002 Sales: $13,200	2002 Profits: $-3,000	**Parent Company:**	

SALARIES/BENEFITS:

Pension Plan:	ESOP Stock Plan:	Profit Sharing:	Top Exec. Salary: $375,000	Bonus: $150,000
Savings Plan: Y	Stock Purch. Plan:		Second Exec. Salary: $224,000	Bonus: $42,013

OTHER THOUGHTS:

Apparent Women Officers or Directors:
Hot Spot for Advancement for Women/Minorities:

LOCATIONS: ("Y" = Yes)

West:	Southwest:	Midwest:	Southeast:	Northeast:	International:
				Y	

Note: Financial information, benefits and other data can change quickly and may vary from those stated here.

ANTIGENICS INC   www.antigenics.com

Industry Group Code: 325412   Ranks within this company's industry group: Sales: 166   Profits: 157

Drugs:		Other:		Clinical:	Computers:		Other:	
Discovery:	Y	AgriBio:		Trials/Services:	Hardware:		Specialty Services:	
Licensing:		Genomics/Proteomics:		Laboratories:	Software:		Consulting:	
Manufacturing:	Y	Tissue Replacement:		Equipment/Supplies:	Arrays:		Blood Collection:	
Development:	Y			Research/Development Svcs.:	Database Management:		Drug Delivery:	
Generics:				Diagnostics:			Drug Distribution:	

TYPES OF BUSINESS:
Cancers & Infectious Diseases Products & Technologies

BRANDS/DIVISIONS/AFFILIATES:
Oncophage
Aroplatin
AG-707
Aronex Pharmaceuticals, Inc.
Aquila Biopharmaceuticals, Inc.

CONTACTS: Note: Officers with more than one job title may be intentionally listed here more than once.
Garo H. Armen, CEO
Shalini Sharp, CFO/VP
John Cerio, VP-Human Resources
Pramod K. Srivastava, Founding Scientist
Terry Higgins Valentine, VP-Legal
Deanna M. Petersen, VP-Bus. Dev.
Gunny Uberoi, VP-Corp. Comm.
Pierre Champagne, Head-Clinical & Medical Affairs
Terry A. Wentworth, VP-Clinical Oper. & Regulatory Affairs
Garo H. Armen, Chmn.

Phone: 212-994-8200	Fax: 212-994-8299

Toll-Free:
Address: 162 5th Ave., Ste. 900, New York, NY 10010 US

GROWTH PLANS/SPECIAL FEATURES:

Antigenics, Inc. develops technologies and products to treat cancers and infectious diseases, primarily based on immunological approaches. The company's most advanced product candidate is Oncophage, a personalized therapeutic cancer vaccine candidate based on a heat shock protein called gp96 that is currently tested in several cancer indications, including in Phase III clinical trials for the treatment of renal cell carcinoma, the most common type of kidney cancer, and for metastatic melanoma. Oncophage is also being tested in Phase II and Phase I clinical trials in a range of indications and has received fast-track designation from the FDA. The product candidate portfolio also includes QS-21, an adjuvant used in numerous vaccines including hepatitis, lyme disease, HIV, influenza, cancer, Alzheimer's disease and malaria; AG-707, a therapeutic vaccine program in Phase I clinical trial for the treatment of genital herpes; and Aroplatin, a liposomal chemotherapeutic in Phase I clinical trial for the treatment of solid tumors and non-Hodgkin's lymphoma. Aronex Pharmaceuticals, Inc., the firm's subsidiary, has a portfolio of two advanced-stage cancer products; and subsidiary Aquila Biopharmaceuticals produces a line of products built on an immunology technology platform.

The company offers its employees medical, dental, life, AD&D and travel accident insurance; short- and long-term disability; flexible spending accounts; a 401(k) plan; an educational assistance program; an employee stock purchase program; and an employee assistance program.

FINANCIALS: Sales and profits are in thousands of dollars—add 000 to get the full amount. 2006 Note: Financial information for 2006 was not available for all companies at press time.

2006 Sales: $ 692	2006 Profits: $-51,881	U.S. Stock Ticker: AGEN
2005 Sales: $ 630	2005 Profits: $-74,104	Int'l Ticker: Int'l Exchange:
2004 Sales: $ 707	2004 Profits: $-56,162	Employees: 100
2003 Sales: $4,500	2003 Profits: $-65,900	Fiscal Year Ends: 12/31
2002 Sales: $3,400	2002 Profits: $-55,900	Parent Company:

SALARIES/BENEFITS:

Pension Plan:	ESOP Stock Plan:	Profit Sharing:	Top Exec. Salary: $440,000	Bonus: $
Savings Plan: Y	Stock Purch. Plan: Y		Second Exec. Salary: $291,447	Bonus: $70,000

OTHER THOUGHTS:
Apparent Women Officers or Directors: 5
Hot Spot for Advancement for Women/Minorities: Y

LOCATIONS: ("Y" = Yes)

West:	Southwest:	Midwest:	Southeast:	Northeast:	International:
				Y	

APPLERA CORPORATION

www.applera.com

Industry Group Code: 339113 Ranks within this company's industry group: Sales: 5 Profits: 4

Drugs:		Other:		Clinical:		Computers:		Other:	
Discovery:	Y	AgriBio:		Trials/Services:		Hardware:		Specialty Services:	
Licensing:		Genomics/Proteomics:	Y	Laboratories:		Software:	Y	Consulting:	
Manufacturing:		Tissue Replacement:		Equipment/Supplies:	Y	Arrays:		Blood Collection:	
Development:	Y			Research/Development Svcs.:		Database Management:	Y	Drug Delivery:	
Generics:				Diagnostics:	Y			Drug Distribution:	

TYPES OF BUSINESS:

Equipment-Life Sciences & Genomics
Genetic Database Management
Proteomics
Medical Software
DNA Sequencing
Drug Discovery & Development
Diagnostics

BRANDS/DIVISIONS/AFFILIATES:

Applied Biosystems
Celera Genomics
Celera Discovery System
Celera Diagnostics
Abbott Molecular

CONTACTS: Note: Officers with more than one job title may be intentionally listed here more than once.

Tony L. White, CEO
Tony L. White, Pres.
Dennis L. Winger, CFO/Sr. VP
Barbara J. Kerr, VP-Human Resources
Tama Olver, CIO/VP
William B. Sawch, General Counsel/Sr. VP
Peter Dworkin, VP-Corp. Comm.
Peter Dworkin, VP-Investor Rel.
Ugo D. DeBlasi, Controller/VP
Dennis Gilbert, Chief Scientific Officer-Applied Biosystem
Kathy P. Ordonez, Sr. VP/Pres., Celera Genomics & Celera Diagnostics
Robert F. G. Booth, Chief Scientific Officer, Celera Genomics
Tony L. White, Chmn.

Phone: 203-840-2000	Fax: 203-840-2312
Toll-Free: 800-761-5381	
Address: 301 Merritt 7, Norwalk, CT 06851-1070 US	

GROWTH PLANS/SPECIAL FEATURES:

Applera Corporation provides technology and information solutions for genomics research. The company operates as an administrative parent for two publicly traded subsidiaries Applied Biosystems and Celera Genomics. Applied Biosystems develops, manufactures, sells and services instrument systems, consumables and informatics products for life science research and related applications. The division has installed approximately 180,000 instrument systems in nearly 100 countries. Its products are used for synthesis, amplification, purification, isolation, analysis and sequencing of DNA, RNA, proteins and other biological molecules. Customers use these products for applications including research, pharmaceutical discovery and development, biosecurity, food and environmental testing, analysis of infectious diseases, human identification and forensic DNA analysis. Celera Genomics works internally and through collaborations to discover and develop new small molecule and antibody-based therapies and diagnostics for cancer, autoimmune and inflammatory diseases, with a focus on tumor surface cell proteins a potential antibody targets. The Celera Discovery System, an online service through which users can access the firm's database of genomic and medical information, is now marketed by Applied Biosciences. Celera's collaborator include Abbott Laboratories, General Electric Company, Merck and Bristol-Myers Squibb. Celera Diagnostics, a full subsidiary of Celera since early 2006, focuses on discovering, developing and commercializing diagnostic tests. Its products, mostly marketed by Abbott Molecular include tests relating to HIV, cystic fibrosis, tissue transplant rejection and hepatitis C. Its research activities are also focused on heart disease, breast cancer, Alzheimer's and rheumatoid arthritis. In March 2006, Applied Biosystems acquired the research products division of Ambion, Inc., a RNA technology company. In May 2007, Celera and Abbott announced the FDA has approved Abbott's RealTime HIV-viral load test, which uses Celera's m2000 automated instrument system.

Applera offers its employees health coverage, an educational assistance plan, employee assistance programs, gift matching programs and adoption assistance.

FINANCIALS: Sales and profits are in thousands of dollars—add 000 to get the full amount. 2006 Note: Financial information for 2006 was not available for all companies at press time.

2006 Sales: $1,949,400	2006 Profits: $212,500	**U.S. Stock Ticker: Private**
2005 Sales: $1,845,140	2005 Profits: $159,795	**Int'l Ticker:** Int'l Exchange:
2004 Sales: $1,825,200	2004 Profits: $114,953	Employees: 220
2003 Sales: $1,777,232	2003 Profits: $118,480	Fiscal Year Ends: 6/30
2002 Sales: $1,701,218	2002 Profits: $-40,581	Parent Company:

SALARIES/BENEFITS:

Pension Plan:	ESOP Stock Plan:	Profit Sharing:	Top Exec. Salary: $1,096,154	Bonus: $2,107,432
Savings Plan: Y	Stock Purch. Plan: Y		Second Exec. Salary: $571,154	Bonus: $717,978

OTHER THOUGHTS:

Apparent Women Officers or Directors: 4
Hot Spot for Advancement for Women/Minorities: Y

LOCATIONS: ("Y" = Yes)

West:	Southwest:	Midwest:	Southeast:	Northeast:	International:
Y				Y	Y

APPLIED BIOSYSTEMS GROUP www.appliedbiosystems.com

Industry Group Code: 334500 Ranks within this company's industry group: Sales: 2 Profits: 2

Drugs:	Other:		Clinical:		Computers:		Other:	
Discovery:	AgriBio:		Trials/Services:		Hardware:		Specialty Services:	Y
Licensing:	Genomics/Proteomics:	Y	Laboratories:		Software:	Y	Consulting:	
Manufacturing:	Tissue Replacement:		Equipment/Supplies:	Y	Arrays:		Blood Collection:	
Development:			Research/Development Svcs.:		Database Management:		Drug Delivery:	
Generics:			Diagnostics:	Y			Drug Distribution:	

TYPES OF BUSINESS:

Biosciences Testing Equipment
DNA & RNA Sequencing Equipment & Consumables
Food Testing Systems
Paternity Testing Systems
Consulting & Technical Support
Life Science Software

BRANDS/DIVISIONS/AFFILIATES:

Applera Corp.
Expedite 8909 DNA Synthesis Software
MessageAMP
Amino ALLYL MessageAMP
Geozipa Finch Suite Software
Agencourt Personal Genomics
BigDye XTerminator Purification Kit

CONTACTS: Note: Officers with more than one job title may be intentionally listed here more than once.

Tony L. White, CEO
Tony L. White, Pres.
Joan Cronin, VP-Human Resources
Dennis A. Gilbert, Chief Scientific Officer
Sandeep Nayyar, VP-Finance
John D'Angelo, VP-Quality & Regulatory Compliance
Leonard Klevan, Div. Pres., Applied Markets Div.
Laura C. Lauman, Div. Pres., Proteomics & Small Molecule
Mark P. Stevenson, Div. Pres., Molecular & Cellular Biology Div.
Tony L. White, Chmn.

Phone: 650-638-5800	Fax: 650-638-5998
Toll-Free: 800-327-3002	
Address: 850 Lincoln Centre Dr., Foster City, CA 94404 US	

GROWTH PLANS/SPECIAL FEATURES:

Applied Biosystems Group, (ABG) a subsidiary of Applera Corp., develops and markets instrument-based systems, software and testing services for research, pharmaceutical and biotechnological industries. The company currently has an installed base of approximately 180,000 instrument systems in 100 countries. ABG focuses primarily on three sectors in the life sciences marketplace: basic research, commercial research and standardized testing. Basic research includes work performed at universities, governments and non-profit institutions with the primary goal of uncovering basic laws of nature and human disease. Commercial research is conducted to aid biotechnology and pharmaceutical companies in drug discovery and development. Standardized testing is provided for forensic human identification, paternity testing and food testing. Products include a range of DNA analysis tools for use in sequencing, resequencing assays and SNP (single-nucleotide polymorphism) genotyping. The firm also offers a range of tools for DNA/RNA modification and labeling: the Expedite 8909 DNA Synthesis Software; custom primers and probes for DNA synthesis; and an array of enzymes, hybridization probe labeling and mRNA amplification kits. Products and services are offered for everything from genotyping and gene expression to peptide synthesis and transcription and translation systems. Through professional consulting and technical support and service, the company helps its customers develop and implement unique systems for genetic testing. ABG and Geozipa are currently working on a collaboration to produce Geozipa Finch Suite Software, a web-based system that manages genetic data, laboratory workflows and results analysis for genetic testing processes. In 2006, the company acquired Agencourt Personal Genomics, a developer of genetic analysis technologies, for $120 million. It also acquired the Research Products Division of Ambion, Inc. for $279 million. In 2007, ABG introduced the BigDye XTerminator Purification Kit, which removes byproducts of DNA sequencing.

ABG offers its employees educational assistance plans, an employee assistance program, gift matching programs and adoption assistance.

FINANCIALS: Sales and profits are in thousands of dollars—add 000 to get the full amount. 2006 Note: Financial information for 2006 was not available for all companies at press time.

2006 Sales: $1,911,226	2006 Profits: $275,117	U.S. Stock Ticker: Subsidiary
2005 Sales: $1,787,083	2005 Profits: $236,894	Int'l Ticker: Int'l Exchange:
2004 Sales: $1,741,098	2004 Profits: $172,253	Employees: 4,030
2003 Sales: $1,682,900	2003 Profits: $183,200	Fiscal Year Ends: 6/30
2002 Sales: $1,604,000	2002 Profits: $168,500	Parent Company: APPLERA CORPORATION

SALARIES/BENEFITS:

Pension Plan:	ESOP Stock Plan: Y	Profit Sharing:	Top Exec. Salary: $507,330	Bonus: $675,000
Savings Plan: Y	Stock Purch. Plan:		Second Exec. Salary: $529,307	Bonus: $462,977

OTHER THOUGHTS:

Apparent Women Officers or Directors: 2
Hot Spot for Advancement for Women/Minorities: Y

LOCATIONS: ("Y" = Yes)

West:	Southwest:	Midwest:	Southeast:	Northeast:	International:
Y	Y			Y	Y

Note: Financial information, benefits and other data can change quickly and may vary from those stated here.

APPLIED MOLECULAR EVOLUTION INCwww.amevolution.com

Industry Group Code: 541710 **Ranks within this company's industry group:** Sales: Profits:

Drugs:		Other:		Clinical:	Computers:		Other:	
Discovery:	Y	AgriBio:		Trials/Services:	Hardware:		Specialty Services:	
Licensing:		Genomics/Proteomics:	Y	Laboratories:	Software:		Consulting:	
Manufacturing:		Tissue Replacement:		Equipment/Supplies:	Arrays:		Blood Collection:	
Development:				Research/Development Svcs.:	Database Management:		Drug Delivery:	
Generics:				Diagnostics:			Drug Distribution:	

TYPES OF BUSINESS:

Research-Directed Molecular Evolution
Human Biotherapeutics
Small-Molecule Drugs

BRANDS/DIVISIONS/AFFILIATES:

AME System
DirectAME
ExpressAME
ScreenAME

CONTACTS: *Note: Officers with more than one job title may be intentionally listed here more than once.*

William D. Huse, Pres.
Jeffry D. Watkins, Chief Scientific Officer

Phone: 858-597-4990	Fax: 858-597-4950
Toll-Free:	
Address: 3520 Dunhill St., San Diego, CA 92121 US	

GROWTH PLANS/SPECIAL FEATURES:

Applied Molecular Evolution, Inc. (AME), a subsidiary of Eli Lilly, is a drug development company and a leader in the application of directed molecular evolution to the development of biotherapeutics. The company uses its proprietary AMEsystem technology to develop improved versions of currently marketed, FDA-approved biopharmaceuticals as well as novel human biotherapeutics. It does so by adjusting amino acids in individual proteins to create positive characteristics. There are three components to the AMEsystem technology. DirectAME is a gene synthesis process that enables the rapid production of a library of variant genes based on an initial gene. ExpressAME consists of gene expression systems that produce proteins from the genes generated using DirectAME. ScreenAME is a series of tests that facilitate the selection and identification of proteins with the desired commercial properties from the protein libraries produced using ExpressAME. While AME had collaborative projects in the past, all of them were completed before its acquisition by Eli Lilly, and its employees now devote themselves entirely to working on projects for Eli Lilly.

AME employees, as part of Eli Lilly, enjoy flexible schedules, a business casual dress code, educational assistance, prescription drug benefits, employee health services, an employee assistance program and a 401(k) savings plan.

FINANCIALS: Sales and profits are in thousands of dollars—add 000 to get the full amount. 2006 Note: Financial information for 2006 was not available for all companies at press time.

2006 Sales: $	2006 Profits: $	U.S. Stock Ticker: Subsidiary	
2005 Sales: $	2005 Profits: $	Int'l Ticker: Int'l Exchange:	
2004 Sales: $	2004 Profits: $	Employees: 110	
2003 Sales: $	2003 Profits: $	Fiscal Year Ends: 12/31	
2002 Sales: $7,900	2002 Profits: $-17,400	Parent Company: ELI LILLY & CO	

SALARIES/BENEFITS:

Pension Plan:	ESOP Stock Plan:	Profit Sharing:	Top Exec. Salary: $350,000	Bonus: $70,000
Savings Plan: Y	Stock Purch. Plan:		Second Exec. Salary: $300,000	Bonus: $81,000

OTHER THOUGHTS:

Apparent Women Officers or Directors:
Hot Spot for Advancement for Women/Minorities:

LOCATIONS: ("Y" = Yes)

West:	Southwest:	Midwest:	Southeast:	Northeast:	International:
Y					

ARADIGM CORPORATION

www.aradigm.com

Industry Group Code: 325412A Ranks within this company's industry group: Sales: 14 Profits: 10

Drugs:		Other:	Clinical:		Computers:		Other:	
Discovery:		AgriBio:	Trials/Services:		Hardware:		Specialty Services:	
Licensing:		Genomics/Proteomics:	Laboratories:		Software:		Consulting:	
Manufacturing:	Y	Tissue Replacement:	Equipment/Supplies:	Y	Arrays:		Blood Collection:	
Development:	Y		Research/Development Svcs.:		Database Management:		Drug Delivery:	Y
Generics:			Diagnostics:				Drug Distribution:	

TYPES OF BUSINESS:

Drug Delivery Systems
Pulmonary Drug Delivery Systems

GROWTH PLANS/SPECIAL FEATURES:

Aradigm Corporation develops advanced pulmonary drug delivery systems for the treatment of systemic conditions and lung diseases. The company is focused on improving the quality and cost-effectiveness of medical treatment by enabling patients to self-administer drugs without needles or lengthy nebulizer treatments. Aradigm's operations are centered on its AERx pulmonary drug delivery system, which creates aerosols from liquid drug formulations. This system is marketed as a replacement for medical devices, such as nebulizers, metered-dose inhalers and dry powder inhalers. AERx also has a wide array of possible applications, including cardiovascular health, oncology, respiratory issues, endocrinology, infection and neurology. AERx's current product pipeline consists of AERx for Asthma and other respiratory uses, which are undergoing clinical trials, and AERx applications of liposomal ciprofloxacin and liposomal treprostinil, as well as treatments for nicotine addiction. In August 2006, Aradigm sold its Intraject subcutaneous delivery technology to Zogenix for $4 million, plus royalties from future products based on the technology. The technology exploits the lungs' natural ability to absorb and rapidly transfer molecules into the bloodstream, for delivery of medication. Zogenix plans to complete development of the Intraject technology and commercialize the Intraject Sumatriptan product for migraine, and other uses. In January 2007, Aradigm received orphan drug designation from the FDA for an inhaled liposomal formulation of ciprofloxacin for the management of bronchiectasis (BE). Aradigm previously received orphan drug designation for the same formulation of ciprofloxacin for the management of cystic fibrosis in April 2006.

Aradigm offers employees medical and dental coverage; life and disability insurance; company-wide bonus programs; a company matching 401(K) savings plan; stock options; a 529 college saving plan; tuition reimbursement; and an educational rewards program.

BRANDS/DIVISIONS/AFFILIATES:

Intraject
AERx
AERx Insulin Diabetes Management System

CONTACTS: Note: Officers with more than one job title may be intentionally listed here more than once.

Igor Gonda, CEO
Igor Gonda, Pres.
Norman Halleen, Interim CFO
Babatunde A. Otulana, Sr. VP-Dev.
Jeffery Grimes, General Counsel
Babatunde A. Otulana, Chief Medical Officer
Virgil D. Thompson, Chmn.

Phone: 510-265-9000	Fax: 510-265-0277
Toll-Free:	
Address: 3929 Point Eden Way, Hayward, CA 94545 US	

FINANCIALS: Sales and profits are in thousands of dollars—add 000 to get the full amount. 2006 Note: Financial information for 2006 was not available for all companies at press time.

2006 Sales: $4,814	2006 Profits: $-13,027	U.S. Stock Ticker: ARDM
2005 Sales: $10,507	2005 Profits: $-29,215	Int'l Ticker: Int'l Exchange:
2004 Sales: $28,045	2004 Profits: $-30,189	Employees: 54
2003 Sales: $33,857	2003 Profits: $-25,970	Fiscal Year Ends: 12/31
2002 Sales: $29,000	2002 Profits: $-35,900	Parent Company:

SALARIES/BENEFITS:

Pension Plan:	ESOP Stock Plan:	Profit Sharing:	Top Exec. Salary: $287,038	Bonus: $45,352
Savings Plan: Y	Stock Purch. Plan: Y		Second Exec. Salary: $274,139	Bonus: $

OTHER THOUGHTS:

Apparent Women Officers or Directors:
Hot Spot for Advancement for Women/Minorities:

LOCATIONS: ("Y" = Yes)

West:	Southwest:	Midwest:	Southeast:	Northeast:	International:
Y					

ARCADIA BIOSCIENCES　　　www.arcadiabiosciences.com

Industry Group Code: 115112 Ranks within this company's industry group: Sales:　Profits:

Drugs:	Other:		Clinical:	Computers:	Other:
Discovery:	AgriBio:	Y	Trials/Services:	Hardware:	Specialty Services:
Licensing:	Genomics/Proteomics:		Laboratories:	Software:	Consulting:
Manufacturing:	Tissue Replacement:		Equipment/Supplies:	Arrays:	Blood Collection:
Development:			Research/Development Svcs.:	Database Management:	Drug Delivery:
Generics:			Diagnostics:		Drug Distribution:

TYPES OF BUSINESS:

Agricultural-Based Technologies
Environment Health Technologies
Human Health Technologies

BRANDS/DIVISIONS/AFFILIATES:

TILLING
GLA safflower oil

CONTACTS: *Note: Officers with more than one job title may be intentionally listed here more than once.*

Eric Rey, Pres.
Vic Knauf, Chief Science Officer
Steve Brandwein, VP-Admin.
Steve Brandwein, VP-Finance
Roger Salameh, Mgr.-Bus. Dev., Input Traits
Don Emlay, Dir.-Regulatory Affairs & Compliance
Frank Flider, VP-Bus. Dev., Nutrition

Phone: 530-756-7077	Fax: 530-756-7027
Toll-Free:	
Address: 202 Cousteau Place, Ste. 200, Davis, CA 95618 US	

GROWTH PLANS/SPECIAL FEATURES:

Arcadia Biosciences, Inc. specializes in developing and commercializing technologies based in agriculture which target the environment as well as human health. The company uses advance breeding techniques, genetic screening and genetic engineering to develop its product portfolio. Its current environmental products target three areas: Nitrogen use efficiency (NUE), salt tolerance and improved processing efficiency produce. Its NUE project seeks to minimize the amount of nitrogen fertilizer required to produce crops, in some cases utilizing up to 66% less nitrogen than conventional fertilizers to produce an equivalent yield. Its salt tolerance project aims to develop plants able to produce normal quality and yields in high saline conditions, and it is currently available for corn, rice, alfalfa, soybeans, turf, wheat and vegetables. These plants not only produce normal results in high saline areas, they also fix the salt, thus reducing the saline levels over time. Its improved processing efficiency project aims to produce tomato varieties, using is TILLING breeding technology, with low water, high solid composition, unlike standard tomato varieties which are 90% water, most of which is lost in processing. The company's human health projects also have three branches: GLA safflower oil, extended shelf-life produce and improved nutrition produce. The GLA safflower oil project aims to breed new varieties of safflower whose seeds will have as much as 40% GLA (gamma linolenic acid), an omega-6 fatty acid believed to have therapeutic benefits, which could be utilized to manufacture supplements, functional foods and nutraceuticals. Its extended shelf-life produce also uses TILLING technology to seek new genetic varieties of tomatoes, lettuce, melons and strawberries. Additionally, its improved nutrition produce project, using TILLING, seeks to develop tomato varieties with higher levels of lycopene and other natural antioxidants than other tomatoes. In the pipeline is a celiac-safe wheat variety for people with gluten intolerance.

FINANCIALS: Sales and profits are in thousands of dollars—add 000 to get the full amount. 2006 Note: Financial information for 2006 was not available for all companies at press time.

2006 Sales: $	2006 Profits: $	U.S. Stock Ticker: Private
2005 Sales: $	2005 Profits: $	Int'l Ticker:　Int'l Exchange:
2004 Sales: $	2004 Profits: $	Employees:
2003 Sales: $	2003 Profits: $	Fiscal Year Ends:
2002 Sales: $	2002 Profits: $	Parent Company:

SALARIES/BENEFITS:

Pension Plan:	ESOP Stock Plan:	Profit Sharing:	Top Exec. Salary: $	Bonus: $
Savings Plan: Y	Stock Purch. Plan:		Second Exec. Salary: $	Bonus: $

OTHER THOUGHTS:

Apparent Women Officers or Directors:
Hot Spot for Advancement for Women/Minorities:

LOCATIONS: ("Y" = Yes)

West:	Southwest:	Midwest:	Southeast:	Northeast:	International:
Y					

ARENA PHARMACEUTICALS INC   www.arenapharm.com

Industry Group Code: 325412 Ranks within this company's industry group: Sales: 96   Profits: 178

Drugs:		Other:	Clinical:	Computers:		Other:	
Discovery:	Y	AgriBio:	Trials/Services:	Hardware:		Specialty Services:	
Licensing:		Genomics/Proteomics:	Laboratories:	Software:		Consulting:	
Manufacturing:		Tissue Replacement:	Equipment/Supplies:	Arrays:		Blood Collection:	
Development:	Y		Research/Development Svcs.:	Y	Database Management:	Drug Delivery:	
Generics:			Diagnostics:			Drug Distribution:	

TYPES OF BUSINESS:
Oral Drugs Discovery, Development & Commercialization

BRANDS/DIVISIONS/AFFILIATES:
Lorcaserin Hydrochloride
APD 125
APD 791
APD668
Ortho-McNeil Pharmaceutical, Inc.
Merck & Co., Inc.

CONTACTS: Note: Officers with more than one job title may be intentionally listed here more than once.
Jack Lief, CEO
Jack Lief, Pres.
Robert E. Hoffman, CFO
Louis J. Scotti, VP-Mktg.
Dominic P. Behan, Chief Scientific Officer/Sr. VP
Steven W. Spector, General Counsel/Sr. VP/Sec.
Louis J. Scotti, VP-Bus. Dev.
Robert E. Hoffman, VP-Finance
C. A. Ajit-Simh, VP-Quality Systems
William R. Shanahan, VP/Chief Medical Officer

Phone: 858-453-7200	Fax: 858-453-7210
Toll-Free:	
Address: 6166 Nancy Ridge Dr., San Diego, CA 92121 US	

GROWTH PLANS/SPECIAL FEATURES:

Arena Pharmaceuticals, Inc. is a clinical-stage biopharmaceutical company that focuses on the discovery, development and commercialization of oral drugs in four major therapeutic areas: cardiovascular, central nervous system, inflammatory and metabolic diseases. The company's most advanced drug candidate, lorcaserin hydrochloride, is currently in a Phase III clinical trial program for the treatment of obesity. The firm has a broad pipeline of compounds that target known and orphan G protein-coupled receptors and includes compounds being evaluated independently and with the company's partners, Ortho-McNeil Pharmaceutical and Merck & Co. In 2006, Arena initiated the first of three planned Phase III trials evaluating the efficacy and safety of lorcaserin. The first trial is known as BLOOM (Behavioral modification and Lorcaserin for Overweight and Obesity Management). In addition to lorcaserin, the company's internal development programs include APD125 and APD791. The firm initiated dosing in a Phase II clinical trial of APD 125 in March 2007, which is an orally available drug candidate that has the potential to reduce insomnia symptoms and improve sleep maintenance. Arena expects to initiate a Phase I clinical trial of APD791, an orally available drug candidate that the company is investigating for the treatment and prevention of arterial thromboembolic diseases such as acute coronary syndrome. In addition to internal programs, the firm has partnerships with two pharmaceutical companies: Ortho-McNeil, with which it is focused on diabetes and its most advanced drug is APD668, an orally administered drug that is in clinical development for the treatment of type 2 diabetes; and Merck, with which it is focused on niacin receptor agonists as treatments for atherosclerosis and other disorders.

Arena Pharmaceuticals offers its employees medical, dental and vision insurance; life, AD&D and disability insurance; flexible spending accounts; a 401(k) plan; an employee assistance program; stock options; and a stock purchase plan.

FINANCIALS: Sales and profits are in thousands of dollars—add 000 to get the full amount. 2006 Note: Financial information for 2006 was not available for all companies at press time.

2006 Sales: $30,569	2006 Profits: $-86,279	**U.S. Stock Ticker: ARNA**
2005 Sales: $23,233	2005 Profits: $-67,901	Int'l Ticker: Int'l Exchange:
2004 Sales: $13,686	2004 Profits: $-57,992	Employees: 371
2003 Sales: $12,834	2003 Profits: $-47,059	Fiscal Year Ends: 12/31
2002 Sales: $19,400	2002 Profits: $-32,800	Parent Company:

SALARIES/BENEFITS:

Pension Plan:	ESOP Stock Plan:	Profit Sharing:	Top Exec. Salary: $619,431	Bonus: $260,000
Savings Plan: Y	Stock Purch. Plan: Y		Second Exec. Salary: $358,033	Bonus: $97,000

OTHER THOUGHTS:
Apparent Women Officers or Directors: 2
Hot Spot for Advancement for Women/Minorities: Y

LOCATIONS: ("Y" = Yes)

West:	Southwest:	Midwest:	Southeast:	Northeast:	International:
Y					

Note: Financial information, benefits and other data can change quickly and may vary from those stated here.

ARIAD PHARMACEUTICALS INC   www.ariad.com

Industry Group Code: 325412 Ranks within this company's industry group: Sales: 161   Profits: 166

Drugs:		Other:	Clinical:	Computers:	Other:
Discovery:	Y	AgriBio:	Trials/Services:	Hardware:	Specialty Services:
Licensing:		Genomics/Proteomics:	Laboratories:	Software:	Consulting:
Manufacturing:		Tissue Replacement:	Equipment/Supplies:	Arrays:	Blood Collection:
Development:	Y		Research/Development Svcs.: Y	Database Management:	Drug Delivery:
Generics:			Diagnostics:		Drug Distribution:

TYPES OF BUSINESS:

Signaling Inhibitors Drugs

BRANDS/DIVISIONS/AFFILIATES:

Ariad Gene Therapeutics, Inc.
AP24534
AP23573

CONTACTS: *Note: Officers with more than one job title may be intentionally listed here more than once.*

Harvey J. Berger, CEO
Edward M. Fitzgerald, CFO/Sr. VP-Finance/Treas.
Kathy Lawton, Manager-Human Resources
Timothy P. Clackson, Chief Scientific Officer/Sr. VP
David C. Dalgarno, VP-Research Technologies
Laurie A. Allen, Chief Legal Officer/Sr. VP-Legal Dev./Sec.
Edward M. Fitzgerald, Sr. VP-Corp. Oper.
John D. Luliucci, Sr. VP/Chief Dev. Officer
Joseph Bratica, VP-Finance/Controller
Camille Bedrosian, VP/Chief Medical Officer
John W. Loewy, VP-Biostatistics & Outcomes Research
Laurie A. Allen, Sr. VP-Bus. Dev.
Shirish Hirani, VP-Dev. Oper. & Planning
Harvey J. Berger, Chmn.

Phone: 617-494-0400	Fax: 617-494-8144
Toll-Free:	
Address: 26 Landsdowne St., Cambridge, MA 02139 US	

GROWTH PLANS/SPECIAL FEATURES:

Ariad Pharmaceuticals, Inc. is engaged in the discovery and development of medicines to treat cancers by regulating cell signaling with small molecules. The company's lead product candidate is AP23573, an inhibitor of the protein mTOR. mTOR serves as a 'master switch' and appears to have a central function in cancer cells. Blocking mTOR creates a starvation-like effect in cancer cells by interfering with cell growth, division, metabolism and angiogenesis. As a single agent, the firm completed enrollment in Phase II studies of patients with sarcomas, hormone refractory prostrate cancer, endometrial cancer and certain leukemias and lymphomas. Ariad also completed enrollment in a Phase Ib trial of patients with brain cancer. The company also studies the oral tablet formulation of the drug as a single agent in a Phase Ib trial of patients with various solid tumors. In addition to clinical development programs, the firm's preclinical programs include the development of orally active inhibitors of protein kinases that are validated targets in oncology. Ariad's second product candidate, AP24534 is an orally active kinase inhibitor, which is in development for the treatment of chronic myeloid leukemia. Ariad owns 80% of Ariad Gene Therapeutics, Inc., which owns or licenses from others the intellectual property related to the company's Argent technology as well as the product candidates developed from the application of this technology. In June 2007, Ariad announced that AP23573 demonstrated efficacy and was tolerated as a single agent in a Phase II trial in metastatic endometrial cancer.

The company offers its employees medical, dental, life and short- and long-term disability insurance; a 401(k) plan; stock options; an employee stock purchase plan; tuition reimbursement; and a parking or public transportation pass.

FINANCIALS: Sales and profits are in thousands of dollars—add 000 to get the full amount. 2006 Note: Financial information for 2006 was not available for all companies at press time.

2006 Sales: $ 896	2006 Profits: $-61,928	U.S. Stock Ticker: ARIA
2005 Sales: $1,217	2005 Profits: $-55,482	Int'l Ticker: Int'l Exchange:
2004 Sales: $ 742	2004 Profits: $-35,573	Employees: 103
2003 Sales: $ 660	2003 Profits: $-19,776	Fiscal Year Ends: 12/31
2002 Sales: $ 100	2002 Profits: $-27,900	Parent Company:

SALARIES/BENEFITS:

Pension Plan:	ESOP Stock Plan:	Profit Sharing:	Top Exec. Salary: $544,000	Bonus: $
Savings Plan: Y	Stock Purch. Plan: Y		Second Exec. Salary: $309,000	Bonus: $185,000

OTHER THOUGHTS:

Apparent Women Officers or Directors: 3
Hot Spot for Advancement for Women/Minorities: Y

LOCATIONS: ("Y" = Yes)

West:	Southwest:	Midwest:	Southeast:	Northeast:	International:
				Y	

Note: Financial information, benefits and other data can change quickly and may vary from those stated here.

ARQULE INC

www.arqule.com

Industry Group Code: 325412 Ranks within this company's industry group: Sales: 135 Profits: 133

Drugs:		Other:		Clinical:		Computers:		Other:	
Discovery:	Y	AgriBio:		Trials/Services:		Hardware:	Y	Specialty Services:	Y
Licensing:		Genomics/Proteomics:	Y	Laboratories:		Software:	Y	Consulting:	
Manufacturing:		Tissue Replacement:		Equipment/Supplies:		Arrays:		Blood Collection:	
Development:				Research/Development Svcs.:	Y	Database Management:	Y	Drug Delivery:	
Generics:				Diagnostics:				Drug Distribution:	

TYPES OF BUSINESS:

Research-Drug Discovery
Small-Molecule Compounds
Systems & Software
Predictive Modeling

BRANDS/DIVISIONS/AFFILIATES:

AMAP Chemistry Operating System
Optimal Chemical Entities
Activated Checkpoint Therapy
ARQ 501
ARQ 101
ARQ-650RP
ARQ 197
ARQ-550RP

CONTACTS: Note: Officers with more than one job title may be intentionally listed here more than once.

Stephen Hill, CEO
Stephen Hill, Pres.
Richard H. Woodrich, CFO
Peter S. Lawrence, General Counsel/Exec. VP
Peter S. Lawrence, Chief Business Officer
Nigel J. Rulewski, Chief Medical Officer
Patrick Zenner, Chmn.

Phone: 781-994-0300	Fax: 781-376-6019
Toll-Free:	
Address: 19 Presidential Way, Woburn, MA 01801-5140 US	

GROWTH PLANS/SPECIAL FEATURES:

ArQule, Inc. seeks to bring together genomics and clinical development by applying its proprietary technology platform and world-class chemistry capabilities to drug discovery. It is committed to developing cancer medicine that is less toxic than chemotherapy and effective on more cancer types. ArQule provides library design and compound production to pharmaceutical collaborators (such as Pfizer) and uses the gains from these endeavors to fund its internal cancer drug discovery. The firm designs small-molecule compounds called Optimal Chemical Entities (OCEs), which have a greater chance of success in clinical trials. ArQule uses a multi-disciplinary approach consisting of intelligent design of molecules, high-throughput automated chemistry and experimental/prognostic analysis of absorption, distribution, metabolism and elimination (ADME) properties. The company researches and develops small-molecule cancer therapeutics based on its Activated Checkpoint Therapy (ACT). ACT compounds selectively kill cancer cells by restoring and activating defective cellular checkpoints. The firm's chief compounds under investigation are ARQ 501 for solid tumors and ARQ 101 for rheumatoid arthritis. ArQule's Automated Molecular Assembly Plant (AMAP) Chemistry Operating System allows it to perform high-throughput, automated production of new compounds. AMAP consists of an integrated series of automated workstations that perform tasks including weighing and dissolution, chemical synthesis, thermally controlled agitation and reaction process development.

ArQule offers its employees counseling; legal services; tuition reimbursement; performance based stock option and cash bonus plans; college savings plans; aid in seeking permanent resident status for foreign nationals; dry cleaning services; mortgage services; discounted ski vouchers; yoga classes; a subsidized cafeteria; an onsite fitness center; ping pong and pool tables; discounted Six Flags Tickets; and movie passes.

FINANCIALS: Sales and profits are in thousands of dollars—add 000 to get the full amount. 2006 Note: Financial information for 2006 was not available for all companies at press time.

2006 Sales: $6,626	2006 Profits: $-31,440	U.S. Stock Ticker: ARQL	
2005 Sales: $6,628	2005 Profits: $-7,520	Int'l Ticker: Int'l Exchange:	
2004 Sales: $5,012	2004 Profits: $-4,921	Employees: 98	
2003 Sales: $65,500	2003 Profits: $-34,800	Fiscal Year Ends: 12/31	
2002 Sales: $62,800	2002 Profits: $-77,900	Parent Company:	

SALARIES/BENEFITS:

Pension Plan:	ESOP Stock Plan:	Profit Sharing:	Top Exec. Salary: $427,212	Bonus: $203,528
Savings Plan: Y	Stock Purch. Plan: Y		Second Exec. Salary: $304,640	Bonus: $107,015

OTHER THOUGHTS:

Apparent Women Officers or Directors:
Hot Spot for Advancement for Women/Minorities:

LOCATIONS: ("Y" = Yes)

West:	Southwest:	Midwest:	Southeast:	Northeast:	International:
				Y	

ARRAY BIOPHARMA INC

www.arraybiopharma.com

Industry Group Code: 541710 **Ranks within this company's industry group:** Sales: 12   Profits: 3

Drugs:		Other:		Clinical:		Computers:		Other:	
Discovery:	Y	AgriBio:		Trials/Services:		Hardware:		Specialty Services:	
Licensing:		Genomics/Proteomics:		Laboratories:		Software:		Consulting:	
Manufacturing:		Tissue Replacement:		Equipment/Supplies:		Arrays:	Y	Blood Collection:	
Development:	Y			Research/Development Svcs.:	Y	Database Management:		Drug Delivery:	
Generics:				Diagnostics:				Drug Distribution:	

TYPES OF BUSINESS:
Drug Development & Research Services
Small-Molecule Drugs
Arrays

BRANDS/DIVISIONS/AFFILIATES:
Array Discovery Platform
MEK
Optimer building blocks
Toll-like Receptor

CONTACTS: *Note: Officers with more than one job title may be intentionally listed here more than once.*
Robert E. Conway, CEO
David L. Snitman, COO
Kevin Koch, Pres.
R. Michael Carruthers, CFO
Kevin Koch, Chief Scientific Officer
John R. Moore, General Counsel/VP
David L. Snitman, VP-Bus. Dev.
John A. Josey, VP-Discovery Chem.
James D. Winkler, Sr. Dir.-Discovery Biology
Kyle A. Lefkoff, Chmn.

Phone: 303-381-6600	Fax: 303-386-1390
Toll-Free: 877-633-2436	
Address: 3200 Walnut St., Boulder, CO 80301 US	

GROWTH PLANS/SPECIAL FEATURES:
Array BioPharma, Inc. is a drug discovery company creating small-molecule drugs through the integration of chemistry, biology and informatics. The firm's scientists use its Array Discovery Platform, an integrated set of drug discovery technologies including predictive informatics and high throughput screening, to invent novel small-molecule drugs in collaboration with leading pharmaceutical and biotechnology companies, as well as for its own pipeline of proprietary drugs. The company holds collaborative partnerships with AstraZeneca; Genentech; InterMune; Ono Pharmaceutical Co., Ltd.; Amgen Inc.; Eli Lilly and Company; Japan Tobacco, Inc.; and Takeda Pharmaceutical Company, Ltd. Array's prime research focuses are cancer and inflammatory diseases. The company has seven cancer drugs and three inflammation drugs in its advanced pipeline, and holds the marketing rights to eight of those. The remaining marketing rights belong to AstraZeneca. The clinical candidate Arry-886, or MEK for Cancer, is an orally active drug that is currently in Phase 2 of study, with additional plans for other Phase 2 studies by Astra Zeneca. All other current candidates are either in Phase 1 or are addressing regulated safety. The company also sells its Optimer building blocks, which are the starting materials used to create more complex chemical compounds in the drug discovery process, on a per-compound basis without any restrictions on use. In March 2007, Array announced a license agreement with VentiRX Pharmaceuticals, Inc. to allow VentiRX exclusive worldwide rights to Array's Toll-like Receptor program, which VentiRX plans to use to develop candidates in allergy and oncology.

Array's employees enjoy many benefits including medical, dental, vision and prescription coverage; life insurance and disability coverage; an employee assistance program; flexible spending accounts; educational assistance; time off and a yearly summer picnic and winter holiday party.

FINANCIALS: Sales and profits are in thousands of dollars—add 000 to get the full amount. 2006 Note: Financial information for 2006 was not available for all companies at press time.

2006 Sales: $45,003	2006 Profits: $39,614	U.S. Stock Ticker: ARRY
2005 Sales: $45,505	2005 Profits: $-23,244	Int'l Ticker: Int'l Exchange:
2004 Sales: $34,831	2004 Profits: $-25,504	Employees: 276
2003 Sales: $35,125	2003 Profits: $-19,574	Fiscal Year Ends: 6/30
2002 Sales: $35,089	2002 Profits: $-4,481	Parent Company:

SALARIES/BENEFITS:
Pension Plan:	ESOP Stock Plan:	Profit Sharing:	Top Exec. Salary: $375,000	Bonus: $173,400
Savings Plan: Y	Stock Purch. Plan: Y		Second Exec. Salary: $320,000	Bonus: $135,000

OTHER THOUGHTS:
Apparent Women Officers or Directors:
Hot Spot for Advancement for Women/Minorities:

LOCATIONS: ("Y" = Yes)
West:	Southwest:	Midwest:	Southeast:	Northeast:	International:
Y					

ARRYX INC

www.arryx.com

Industry Group Code: 541710G **Ranks within this company's industry group:** Sales: Profits:

Drugs:	Other:	Clinical:	Computers:		Other:
Discovery:	AgriBio:	Trials/Services:	Hardware:	Y	Specialty Services:
Licensing:	Genomics/Proteomics:	Laboratories:	Software:		Consulting:
Manufacturing:	Tissue Replacement:	Equipment/Supplies:	Arrays:		Blood Collection:
Development:		Research/Development Svcs.:	Database Management:		Drug Delivery:
Generics:		Diagnostics:			Drug Distribution:

TYPES OF BUSINESS:

Holographic Laser Steering
Optics
Nanomaterial Manipulation Technology

BRANDS/DIVISIONS/AFFILIATES:

BioRyx 200

CONTACTS: *Note: Officers with more than one job title may be intentionally listed here more than once.*

Kenneth Bradley, CEO
Michael Reese, CFO
Joseph Plewa, Chief Science Officer
Daniel Mueth, CTO
Nicole S. Williams, Chmn.

Phone: 312-726-6675	Fax: 312-726-6652

Toll-Free:

Address: 316 N. Michigan Ave., Ste. CL20, Chicago, IL 60601 US

GROWTH PLANS/SPECIAL FEATURES:

Arryx, Inc. develops advanced systems and products for the optoelectronics and biotechnology industries. The company created the BioRyx 200 system, which employs holographic laser steering to improve productivity and profitability for manufacturing and processing in industries ranging from pharmaceuticals to integrated circuit manufacturing. The system integrates proprietary laser steering technology with easy-to-use software to give researchers the ability to manipulate 200 microscopic objects independently and simultaneously in three dimensions. The firm is the exclusive licensee of holographic optical trapping (HOT) technology, originally developed at the University of Chicago. HOT uses a holographic device, such as a spatial light modulator, to manipulate beams of light to capture and manipulate microscopic and nanoscopic objects such as carbon nanotubes in optical traps. The optical traps are also called Laser Tweezers. The system can grab, move, spin, assemble, stretch, join, separate and otherwise control materials ranging in size from 1/1000th the diameter of a human hair to the size of human cells. HOT uses non-standard beam profiles such as Bessel beams, which are non-diffracting, and optical vortices, which are optical fields with phase singularities. Applications include isolating valuable cells or molecules from other cells, tissues and contaminants; detecting and measuring the presence of materials to increase test sophistication and sensitivity; and manufacturing sensors to detect biohazards, chemical hazards and other contaminants on a universal basis for the rapidly growing homeland security industry. Clients of the firm have included the National Institute of Standards and Technology, a number of universities and a government lab in Japan. In July 2006, Arryx was acquired by Haemonetics Corporation. It remains a separate operating entity.

Arryx offers employees medical, dental and vision plans; a 401(k) with company match; flex time; a stock purchase plan; and bonuses.

FINANCIALS: Sales and profits are in thousands of dollars—add 000 to get the full amount. 2006 Note: Financial information for 2006 was not available for all companies at press time.

2006 Sales: $	2006 Profits: $	**U.S. Stock Ticker: Subsidiary**
2005 Sales: $	2005 Profits: $	**Int'l Ticker:** Int'l Exchange:
2004 Sales: $	2004 Profits: $	Employees:
2003 Sales: $	2003 Profits: $	Fiscal Year Ends: 12/31
2002 Sales: $	2002 Profits: $	Parent Company: HAEMONETICS CORPORATION

SALARIES/BENEFITS:

Pension Plan:	ESOP Stock Plan:	Profit Sharing:	Top Exec. Salary: $	Bonus: $
Savings Plan: Y	Stock Purch. Plan: Y		Second Exec. Salary: $	Bonus: $

OTHER THOUGHTS:

Apparent Women Officers or Directors: 1
Hot Spot for Advancement for Women/Minorities:

LOCATIONS: ("Y" = Yes)

West:	Southwest:	Midwest:	Southeast:	Northeast:	International:
		Y			

ASTELLAS PHARMA INC

www.astellas.com

Industry Group Code: 325412 Ranks within this company's industry group: Sales: 18 Profits: 20

Drugs:		Other:		Clinical:		Computers:		Other:	
Discovery:	Y	AgriBio:		Trials/Services:		Hardware:		Specialty Services:	
Licensing:		Genomics/Proteomics:	Y	Laboratories:		Software:		Consulting:	
Manufacturing:	Y	Tissue Replacement:		Equipment/Supplies:	Y	Arrays:		Blood Collection:	
Development:	Y			Research/Development Svcs.:	Y	Database Management:		Drug Delivery:	
Generics:				Diagnostics:				Drug Distribution:	

TYPES OF BUSINESS:

Drugs, Manufacturing
Immunological Pharmaceuticals
Over-the-Counter Products
Reagents
Genomic Research
Venture Capital
Drug Licensing

BRANDS/DIVISIONS/AFFILIATES:

Yamanouchi Pharmaceutical Co., Ltd.
Fujisawa Pharmaceutical Co., Ltd.
Prograf
Protopic
Lipitor
Flomax
Pepcid
Astellas Venture Capital, LLC

CONTACTS: Note: Officers with more than one job title may be intentionally listed here more than once.

Masafumi Nogimori, CEO
Masafumi Nogimori, Pres.
Kunihide Ichikawa, Sr. VP-Sales & Mktg.
Toshinari Tamura, Exec. VP/Chief Science Officer
Isao Kishi, Sr. VP-Info. Sys.
Hitoshi Ohta, Sr. VP-Tech.
Masao Shimizu, Sr. VP-Dev.
Toshio Ohsawa, Sr. VP-Corp. Admin.
Hirofumi Onosaka, Sr. VP-Corp. Strategy
Osamu Nagai, Sr. VP-Corp. Finance & Acct.
Toshinari Tamura, Exec. VP
Makoto Nishimura, CEO/Chmn., Astellas Pharma U.S. Inc.
Iwaki Miyazaki, Sr. VP-QA, RA & Pharmacovigilance
Toichi Takenaka, Co-Chmn.
Yasuo Ishii, CEO/Chmn., Astellas Pharma Europe Ltd.

Phone: 81-3-3244-3000	Fax: 80-3-3244-3272
Toll-Free:	
Address: 2-3-11 Nihonbashi-Honcho, Chuo-ku, Tokyo, 103-8411 Japan	

GROWTH PLANS/SPECIAL FEATURES:

Astellas Pharma, Inc., the result of a 2005 merger of Yamanouchi Pharmaceutical Co., Ltd. and Fujisawa Pharmaceutical Co., Ltd., is one of the top 20 pharmaceuticals manufacturers in the world and the second-largest in Japan (behind Takeda Chemical Industries, Ltd.). The company has operations in Europe and North America, as well as in Taiwan, Hong Kong, China, the Philippines, Thailand, Indonesia and South Korea. Roughly 90% of Astellas's revenue relates to sales of pharmaceuticals, led by Prograf, which is used as an immunosuppressant in conjunction with organ transplantation. Other products target needs in dermatology, urology, immunology and cardiology, including Protopic for the treatment of atopic dermatitis; Mycamine for the treatment of fungal infections; VESIcare for treating overactive bladders; Lipitor for high cholesterol; Flomax for symptoms caused by enlarged prostates; and Pepcid for heartburn. The firm's research and development budget is over $1.3 billion (exceeding 17% of sales). In addition to developing its own pharmaceuticals, Astellas pursues in-licensing and co-promotion agreements with biotechnology firms and a host of other pharmaceutical companies such as Bristol-Meyers Squibb and GlaxoSmithKline. Moreover, subsidiary Astellas Venture Capital, LLC is engaged in investing in biotechnology companies, starting with $30 million in initial capitalization. In December 2006, Astellas Pharma dissolved its insurance subsidiary, Astellas Insurance Service Co., Ltd, and transferred the business to Ginsen Co., Ltd. In February 2007, the company began construction of new research laboratory buildings at the Miyukigaoka Research Center in Ibaraki, Japan.

FINANCIALS: Sales and profits are in thousands of dollars—add 000 to get the full amount. 2006 Note: Financial information for 2006 was not available for all companies at press time.

2006 Sales: $7,758,480	2006 Profits: $1,106,390	U.S. Stock Ticker: ALPMF.PK
2005 Sales: $7,410,740	2005 Profits: $873,569	Int'l Ticker: 4503 Int'l Exchange: Tokyo-TSE
2004 Sales: $4,839,100	2004 Profits: $568,500	Employees: 9,500
2003 Sales: $	2003 Profits: $	Fiscal Year Ends: 3/31
2002 Sales: $	2002 Profits: $	Parent Company:

SALARIES/BENEFITS:

Pension Plan:	ESOP Stock Plan:	Profit Sharing:	Top Exec. Salary: $	Bonus: $
Savings Plan:	Stock Purch. Plan:		Second Exec. Salary: $	Bonus: $

OTHER THOUGHTS:

Apparent Women Officers or Directors:
Hot Spot for Advancement for Women/Minorities:

LOCATIONS: ("Y" = Yes)

West:	Southwest:	Midwest:	Southeast:	Northeast:	International:
	Y	Y		Y	Y

ASTRAZENECA PLC

www.astrazeneca.com

Industry Group Code: 325412 Ranks within this company's industry group: Sales: 7 Profits: 8

Drugs:		Other:		Clinical:		Computers:		Other:	
Discovery:	Y	AgriBio:		Trials/Services:		Hardware:		Specialty Services:	
Licensing:		Genomics/Proteomics:		Laboratories:		Software:		Consulting:	
Manufacturing:	Y	Tissue Replacement:		Equipment/Supplies:		Arrays:		Blood Collection:	
Development:	Y			Research/Development Svcs.:		Database Management:		Drug Delivery:	
Generics:				Diagnostics:				Drug Distribution:	

TYPES OF BUSINESS:

Drugs-Diversified
Pharmaceutical Research & Development

BRANDS/DIVISIONS/AFFILIATES:

Rhinocort
Zomig
Nolvadex
Prilosec
Nexium
Pulmicort
Cambridge Antibody Technology Group
Arrow Therapeutics Ltd.

CONTACTS: Note: Officers with more than one job title may be intentionally listed here more than once.

David Brennan, CEO
Martin Nicklasson, Exec. VP-Global Mktg.
Tony Bloxham, Exec. VP-Human Resources
Jan Lundberg, Exec. VP-Discovery Research
David Smith, Exec. VP-Oper.
John Patterson, Exec. Dir.-Dev.
Edel McCaffrey, Media
Mina Blair, Investor Rel.
John Patterson, VP-Prod. Strategy & Licensing
Tony Zook, Exec. VP-North America
Louis Schweitzer, Chmn.
Bruno Angelici, VP-Europe, Japan & Asia

Phone: 44-20-7304-5000	Fax: 44-20-7304-5151
Toll-Free:	
Address: 15 Stanhope Gate, London, W1K 1LN UK	

GROWTH PLANS/SPECIAL FEATURES:

AstraZeneca plc is a leading global pharmaceutical company that provides products to fight disease in areas of medical necessity. The company is the result of the merger of the Zeneca Group with Astra. The firm invests approximately $3.9 billion in annual research and development and enjoys sales in over 100 countries. It operates 27 manufacturing sites in 19 countries and 16 major research centers in 8 countries.   The company focuses on therapeutic interventions in six therapy areas: cancer, respitory and inflammation, cardiovascular, gastrointestinal, infection and neuroscience.   AstraZeneca's cardiovascular products include Seloken ZOK, Crestor, Atacand, Plendil, Zestril and Tenormin.   The firm's gastrointestinal products include Nexium, Entocort and Prilosec. Merrem, its primary infection product, is an antibiotic for serious hospital-acquired infections. The company's primary neuroscience offering is Zomig, an anti-migraine drug.   AstraZeneca's cancer treatments include Casodex for prostate cancer, Zoladex, Armidex for breast cancer, Iressa for lung cancer, Faslodex and Nolvadex.   The firm's respiratory and inflammation brands include Pulmicort, Symbicort, Rhinocort, Accolate and Oxis. Seloken, Seroquel, Nexium, Prilosec, Casodex and Pulmicort all have sales in excess of $1 billion. In 2006, AstraZeneca acquired Cambridge Antibody Technology Group. In 2007, the company acquired Arrow Therapeutics Ltd., a developer of anti-viral therapies. The May 2007, the firm agreed to acquire MedImmune for $15.6 billion. MedImmune is a biotech company, and the deal would push AstraZeneca into vaccines with MedImmune's FluMist.

FINANCIALS: Sales and profits are in thousands of dollars—add 000 to get the full amount. 2006 Note: Financial information for 2006 was not available for all companies at press time.

2006 Sales: $26,475,000	2006 Profits: $4,392,000	U.S. Stock Ticker: AZN
2005 Sales: $23,950,000	2005 Profits: $3,881,000	Int'l Ticker: AZN Int'l Exchange: London-LSE
2004 Sales: $21,426,000	2004 Profits: $3,813,000	Employees: 66,000
2003 Sales: $18,849,000	2003 Profits: $3,036,000	Fiscal Year Ends: 12/31
2002 Sales: $17,841,000	2002 Profits: $2,836,000	Parent Company:

SALARIES/BENEFITS:

Pension Plan:	ESOP Stock Plan:	Profit Sharing:	Top Exec. Salary: $1,191,000	Bonus: $588,000
Savings Plan:	Stock Purch. Plan:		Second Exec. Salary: $732,000	Bonus: $347,000

OTHER THOUGHTS:

Apparent Women Officers or Directors: 5
Hot Spot for Advancement for Women/Minorities: Y

LOCATIONS: ("Y" = Yes)

West:	Southwest:	Midwest:	Southeast:	Northeast:	International:
Y	Y	Y	Y	Y	Y

Note: Financial information, benefits and other data can change quickly and may vary from those stated here.

ATHEROGENICS INC

www.atherogenics.com

Industry Group Code: 325412 Ranks within this company's industry group: Sales: 108   Profits: 169

Drugs:		Other:	Clinical:	Computers:	Other:
Discovery:	Y	AgriBio:	Trials/Services:	Hardware:	Specialty Services:
Licensing:		Genomics/Proteomics:	Laboratories:	Software:	Consulting:
Manufacturing:		Tissue Replacement:	Equipment/Supplies:	Arrays:	Blood Collection:
Development:	Y		Research/Development Svcs.:	Database Management:	Drug Delivery:
Generics:			Diagnostics:		Drug Distribution:

TYPES OF BUSINESS:

Chronic Anti-inflammatory Drugs

BRANDS/DIVISIONS/AFFILIATES:

v-protectant
MEKK Technology
AGI-1067
AGI-1096
ARISE
Astellas Pharma, Inc.

CONTACTS: *Note: Officers with more than one job title may be intentionally listed here more than once.*

Russell M. Medford, CEO
Russell M. Medford, Pres.
Mark P. Colonnese, CFO
Robert A. D. Scott, Exec. VP-R&D/Chief Medical Officer
Joseph M. Gaynor, Jr., General Counsel/Sr. VP
Mark P. Colonnese, Exec. VP-Commercial Oper.
W. Charles Montgomery, VP-Bus. Dev. & Alliance Mgmt.
Michael A. Henos, Chmn.

Phone: 678-336-2500	Fax: 678-336-2501
Toll-Free:	
Address: 8995 Westside Pkwy., Alpharetta, GA 30004 US	

GROWTH PLANS/SPECIAL FEATURES:

AtheroGenics, Inc. is a research-based pharmaceutical company focused on the discovery, development and commercialization of drugs for the treatment of chronic inflammatory diseases, including coronary heart disease, organ transplant rejection, rheumatoid arthritis and asthma. The company developed a proprietary vascular protectant (v-protectant) technology platform to discover drugs to treat these types of diseases. V-protectant technology exploits the observation that the endothelial cells that line the interior wall of the blood vessel play an active role in recruiting white blood cells from the blood to the site of chronic inflammation. V-protectants are intended to block harmful effects of oxygen and other similar molecules. The firm has two drug development programs in clinical trials. AGI-1067 is designed to benefit patients with coronary heart diseases, which is atherosclerosis of the blood vessels of the heart. AtheroGenics completed the Phase III trial ARISE (Aggressive Reduction of Inflammation Stops Events) of the drug and in 2007 the company announced that the trial showed that AGI-1067 demonstrated considerable efficacy in several therapeutic areas such as heart disease and diabetes. AGI-1096 is an antioxidant and selective anti-inflammatory agent that is being developed to address the accelerated inflammation of grafted blood vessels, known as transplant arthritis, common in chronic organ transplant rejection. The firm is working with Astellas Pharma, Inc. to further develop AGI-1096 in preclinical and early-stage clinical trials. AtheroGenics has an exclusive license from National Jewish Medical and Research Center for its MEKK Technology, a group of patents and technical information for drug development. In May 2007, the company announced that it would conduct a Phase III clinical trial studying the effect of AGI-1067 in patients with diabetes.

The company offers its employees a 401(k) plan; stock options; health, dental and life insurance; short- and long-term disability insurance; and a cafeteria flexible spending account.

FINANCIALS: Sales and profits are in thousands of dollars—add 000 to get the full amount. 2006 Note: Financial information for 2006 was not available for all companies at press time.

2006 Sales: $22,917	2006 Profits: $-67,322	U.S. Stock Ticker: AGIX
2005 Sales: $	2005 Profits: $-82,554	Int'l Ticker: Int'l Exchange:
2004 Sales: $	2004 Profits: $-69,600	Employees: 127
2003 Sales: $	2003 Profits: $-53,288	Fiscal Year Ends: 12/31
2002 Sales: $	2002 Profits: $-27,966	Parent Company:

SALARIES/BENEFITS:

Pension Plan:	ESOP Stock Plan:	Profit Sharing:	Top Exec. Salary: $383,454	Bonus: $
Savings Plan: Y	Stock Purch. Plan:		Second Exec. Salary: $310,324	Bonus: $

OTHER THOUGHTS:

Apparent Women Officers or Directors: 1
Hot Spot for Advancement for Women/Minorities:

LOCATIONS: ("Y" = Yes)

West:	Southwest:	Midwest:	Southeast:	Northeast:	International:
			Y		

AUTOIMMUNE INC

www.autoimmuneinc.com

Industry Group Code: 325412 Ranks within this company's industry group: Sales: 173 Profits: 58

Drugs:		Other:		Clinical:	Computers:		Other:	
Discovery:	Y	AgriBio:		Trials/Services:	Hardware:		Specialty Services:	
Licensing:	Y	Genomics/Proteomics:		Laboratories:	Software:		Consulting:	
Manufacturing:		Tissue Replacement:		Equipment/Supplies:	Arrays:		Blood Collection:	
Development:	Y			Research/Development Svcs.:	Database Management:		Drug Delivery:	
Generics:				Diagnostics:			Drug Distribution:	

TYPES OF BUSINESS:

Drugs-Immune System & Inflammatory Disease
Drug Development & Licensing
Nutraceuticals

BRANDS/DIVISIONS/AFFILIATES:

Colloral
Colloral, LLC
Coral

CONTACTS: *Note: Officers with more than one job title may be intentionally listed here more than once.*

Robert C. Bishop, CEO
Robert C. Bishop, Pres.
Diane M. McClintock, Dir.-Finance
Diane M. McClintock, Treas.
Suzanne Glassburn, Corp. Sec.
Robert C. Bishop, Chmn.

Phone: 626-792-1235	Fax: 626-792-1236
Toll-Free:	
Address: 1199 Madia St., Pasadena, CA 91103 US	

GROWTH PLANS/SPECIAL FEATURES:

AutoImmune, Inc. owns and licenses rights to technology which it hopes to develop into a new class of therapeutics for the treatment of autoimmune and other cell-mediated inflammatory diseases and conditions. The company's products are based upon the principles of mucosal tolerance. When proteins are administered by a mucosal route (e.g., oral, nasal or by aerosol to the lungs), the body's natural immune system mechanisms suppress the response that would otherwise arise against a foreign substance. This immune suppression can be directed toward a specific tissue through appropriate selection and dosing of the protein in a mucosally-delivered product. AutoImmune is currently attempting to demonstrate to the FDA that Colloral, the company's main product initially developed as a rheumatoid arthritis medication, meets the statutory definition of a dietary supplement, as it failed to gain FDA approval as a pharmaceutical product. AutoImmune has license agreements with Teva Pharmaceutical Industries, Ltd. and BioMS Medical Corporation. Using AutoImmune's technology, an oral formulation of Copaxone, Teva's injectable multiple sclerosis drug, is being developed by Teva under the name of Coral. Although Teva announced ceasing development of Coral in March 2006, in December 2006 Teva announced its development of a new and potentially improved version of Coral, which may or may not involve intellectual property licensed by AutoImmune to Teva. Following its commencement of Phase II trials in November 2006, BioMS Medical Corporation received FDA approval in January 2007 to initiate a second trial in secondary progressive multiple sclerosis for AutoImmune's MBP8298 injectable therapy for multiple sclerosis treatment. The firm has also licensed development of AI 401, an oral diabetes drug, to Eli Lilly. In February 2007, the National Institute of Health initiated a multicenter Phase III clinical trial on whether treatment with AI 401 can delay or prevent Type 1 diabetes.

FINANCIALS: Sales and profits are in thousands of dollars—add 000 to get the full amount. 2006 Note: Financial information for 2006 was not available for all companies at press time.

2006 Sales: $ 401	2006 Profits: $- 481	U.S. Stock Ticker: AIMM	
2005 Sales: $ 179	2005 Profits: $- 666	Int'l Ticker:	Int'l Exchange:
2004 Sales: $ 130	2004 Profits: $- 761	Employees: 2	
2003 Sales: $1,445	2003 Profits: $ 628	Fiscal Year Ends: 12/31	
2002 Sales: $ 100	2002 Profits: $- 900	Parent Company:	

SALARIES/BENEFITS:

Pension Plan:	ESOP Stock Plan:	Profit Sharing:	Top Exec. Salary: $87,300	Bonus: $
Savings Plan:	Stock Purch. Plan:		Second Exec. Salary: $51,400	Bonus: $

OTHER THOUGHTS:

Apparent Women Officers or Directors: 2
Hot Spot for Advancement for Women/Minorities: Y

LOCATIONS: ("Y" = Yes)

West:	Southwest:	Midwest:	Southeast:	Northeast:	International:
Y					

Note: Financial information, benefits and other data can change quickly and may vary from those stated here.

AVANIR PHARMACEUTICALS   www.avanir.com

Industry Group Code: 325412 Ranks within this company's industry group: Sales: 115   Profits: 168

Drugs:		Other:	Clinical:	Computers:	Other:
Discovery:	Y	AgriBio:	Trials/Services:	Hardware:	Specialty Services:
Licensing:	Y	Genomics/Proteomics:	Laboratories:	Software:	Consulting:
Manufacturing:		Tissue Replacement:	Equipment/Supplies:	Arrays:	Blood Collection:
Development:	Y		Research/Development Svcs.:	Database Management:	Drug Delivery:
Generics:			Diagnostics:		Drug Distribution:

TYPES OF BUSINESS:

Pharmaceutical Discovery & Development
Human Antibody Technology Research
Central Nervous System Research
Allergy & Asthma Drugs
Antibody Generation
Drugs - HSV1 Treatment

BRANDS/DIVISIONS/AFFILIATES:

Xenerex
Abreva
Zeniva

CONTACTS: *Note: Officers with more than one job title may be intentionally listed here more than once.*

Keith A. Katkin, CEO
Keith A. Katkin, Pres.
Martin Sturgeon, Interim CFO
Theresa Hope-Reese, VP-Human Resources
Jagadish Sircar, VP-Drug Discovery
Gregory J. Flesher, Exec. Dir.-Bus. Dev. & Portfolio Strategy
Randall E. Kaye, VP-Medical Affairs

Phone: 949-389-6700	Fax:
Toll-Free:	
Address: 101 Enterprise, Ste. 300, Aliso Viejo, CA 92656 US	

GROWTH PLANS/SPECIAL FEATURES:

Avanir Pharmaceuticals is a biopharmaceutical company engaged in research, development, commercialization, licensing and sales of innovative drug products and antibody generation services. The firm developed Abreva (docosonal 10% cream), the only over-the-counter, FDA-approved treatment for Type-1 Herpes Simplex (HSV1, more commonly known as cold sores or fever blisters). GlaxoSmithKline is the company's marketing partner for Abreva in the U.S. and Canada. Avanir's lead drug candidate, Zeniva, is in Phase III clinical trials for the treatment of involuntary emotional expression disorder (IEED) and pseudobulbar affect. Pseudobulbar effect is a symptom that is characterized by unprovoked and uncontrollable episodes of crying or laughing, and afflicts patients with neurological disorders such as Lou Gehrig's disease, Alzheimer's disease, MS, stroke and traumatic brain injury. Zeniva is also in Phase III clinical trials for the treatment of neuropathic pain. In addition, the firm has a drug discovery program in Phase I clinical trials for the treatment of the underlying biological causes of allergy and asthma called AVP-13358. The company is also engaged in small-molecule research to develop treatments for central nervous system disorder and inflammatory diseases. Other areas of development include the company's patented Xenerex antibody technology for discovery of fully human monoclonal antibodies, clinical development work on new topical formulations of docosonal for the treatment of genital herpes and pre-clinical work on cholesterol and inflammation. In May 2006, Avanir acquired Alamo Pharmaceuticals, which produces FazaClo, the only orally-disintegrating formulation of clozapine, a treatment for severely ill schizophrenic patients who fail to respond to standard schizophrenic drug treatments. In June 2007, in order to fund the continued development of Zeniva, Avanir sold its FazaClo antipsychotic drug to Azur Pharma for $42 million.

Avanir offers employees major medical, dental, vision and disability insurance; life and AD&D insurance; and a 401(k) plan.

FINANCIALS: Sales and profits are in thousands of dollars—add 000 to get the full amount. 2006 Note: Financial information for 2006 was not available for all companies at press time.

2006 Sales: $15,186	2006 Profits: $-62,553	**U.S. Stock Ticker: AVNR**
2005 Sales: $16,691	2005 Profits: $-30,607	**Int'l Ticker:** Int'l Exchange:
2004 Sales: $3,589	2004 Profits: $-28,155	Employees: 150
2003 Sales: $2,439	2003 Profits: $-23,236	Fiscal Year Ends: 9/30
2002 Sales: $8,900	2002 Profits: $-10,200	Parent Company:

SALARIES/BENEFITS:

Pension Plan:	ESOP Stock Plan:	Profit Sharing:	Top Exec. Salary: $500,803	Bonus: $645,000
Savings Plan: Y	Stock Purch. Plan:		Second Exec. Salary: $272,648	Bonus: $109,762

OTHER THOUGHTS:

Apparent Women Officers or Directors: 1
Hot Spot for Advancement for Women/Minorities:

LOCATIONS: ("Y" = Yes)

West:	Southwest:	Midwest:	Southeast:	Northeast:	International:
Y					

AVANT IMMUNOTHERAPEUTICS   www.avantimmune.com

Industry Group Code: 325412 Ranks within this company's industry group: Sales: 141 Profits: 106

Drugs:		Other:		Clinical:	Computers:		Other:	
Discovery:	Y	AgriBio:		Trials/Services:	Hardware:		Specialty Services:	
Licensing:		Genomics/Proteomics:	Y	Laboratories:	Software:		Consulting:	
Manufacturing:	Y	Tissue Replacement:		Equipment/Supplies:	Arrays:		Blood Collection:	
Development:	Y			Research/Development Svcs.:	Database Management:		Drug Delivery:	
Generics:				Diagnostics:			Drug Distribution:	

TYPES OF BUSINESS:
Drugs - Vaccines & Immunotherapeutics
Drugs - Cardiovascular, Pulmonary & Autoimmune Disorder

BRANDS/DIVISIONS/AFFILIATES:
CETi
ETEC
Shigella
Megan Egg
Therapore
Megan Vac 1
Rotarix
Oral Anthrax

CONTACTS: Note: Officers with more than one job title may be intentionally listed here more than once.
Una S. Ryan, CEO
M. Timothy Cooke, COO/Sr. VP
Avery W. Catlin, CFO/Sr. VP
Ronald W. Ellis, Sr. VP-R&D
Taha Keilani, VP-Medical & Regulatory Affairs
Henry C. Marsh, Jr., VP-Research
Harry H. Penner, Jr., Chmn.

Phone: 781-433-0771	Fax: 781-433-0262
Toll-Free:	
Address: 119 4th Ave., Needham, MA 02494-2725 US	

GROWTH PLANS/SPECIAL FEATURES:
AVANT Immunotherapeutics, Inc. is a biopharmaceutical company that uses novel applications of immunology to develop products for the prevention and treatment of diseases. It is developing a broad portfolio of vaccines and immunotherapeutics addressing a range of applications including cardiovascular disease, bacterial and viral diseases, biodefense and food safety. Using its expertise in immunology, AVANT is building business franchises in three major disease areas: bacterial vaccines, viral vaccines and immunotherapies for cardiovascular diseases including cholesterol management. The immunotherapeutic area includes two drugs: TP10, for reduced tissue damage during cardiac bypass surgery; and CETi, a vaccine to stimulate an immune response against cholesteryl ester transfer protein (CETP), which mediates the balance between HDL and LDL cholesterol and may increase the risk of atherosclerotic lesions. Both drugs are in Phase II trials. The bacterial vaccine segment includes ETEC (e. coli), Shigella (dysentery), Campylobacter (campylobacter) and Oral Anthrax and Plague (anthrax and plague) in pre-clinical trials. TY800 (typhoid) is in Phase 1/2 and CholeraGarde (cholera) is in Phase II. Megan Vac 1 (salmonella in chicken) and Megan Egg (salmonella in laying hens and eggs) are marketed through Lohmann Animal Health International. The company also has other food and animal safety vaccines under development with Pfizer. The viral vaccine segment has Rotarix (rotavirus) marketed through GlaxoSmithKline in over 50 countries in addition to the European Union. In June 2007, Glaxo filed for marketing approval of Rotarix in the United States with the FDA. AVANT also has collaboration agreements with the National Institute of Health, the U.S. Army and several other pharmaceutical companies.

FINANCIALS: Sales and profits are in thousands of dollars—add 000 to get the full amount. 2006 Note: Financial information for 2006 was not available for all companies at press time.

		U.S. Stock Ticker: AVAN
2006 Sales: $4,931	2006 Profits: $-20,374	Int'l Ticker: Int'l Exchange:
2005 Sales: $3,088	2005 Profits: $-18,097	Employees: 99
2004 Sales: $6,859	2004 Profits: $-13,204	Fiscal Year Ends: 12/31
2003 Sales: $4,633	2003 Profits: $-12,669	Parent Company:
2002 Sales: $6,700	2002 Profits: $-13,800	

SALARIES/BENEFITS:
Pension Plan:	ESOP Stock Plan:	Profit Sharing:	Top Exec. Salary: $415,000	Bonus: $73,040
Savings Plan:	Stock Purch. Plan:		Second Exec. Salary: $262,500	Bonus: $28,875

OTHER THOUGHTS:
Apparent Women Officers or Directors: 1
Hot Spot for Advancement for Women/Minorities:

LOCATIONS: ("Y" = Yes)
West:	Southwest:	Midwest:	Southeast:	Northeast:	International:
		Y		Y	

AVAX TECHNOLOGIES INC

www.avax-tech.com

Industry Group Code: 325412 Ranks within this company's industry group: Sales: 163 Profits: 64

Drugs:		Other:		Clinical:	Computers:		Other:	
Discovery:	Y	AgriBio:		Trials/Services:	Hardware:		Specialty Services:	
Licensing:		Genomics/Proteomics:		Laboratories:	Software:		Consulting:	
Manufacturing:	Y	Tissue Replacement:		Equipment/Supplies:	Arrays:		Blood Collection:	
Development:	Y			Research/Development Svcs.:	Database Management:		Drug Delivery:	
Generics:				Diagnostics:			Drug Distribution:	

TYPES OF BUSINESS:

Drugs-Cancer
Melanoma Treatment
Non-Small Cell Lung Cancer Treatment
Ovarian Cancer Treatment
Vaccine Therapies

BRANDS/DIVISIONS/AFFILIATES:

AC Vaccine
M-Vax
O-Vax
Genopoeitic
L-Vax

CONTACTS: Note: Officers with more than one job title may be intentionally listed here more than once.

Richard P. Rainey, Pres.
Richard P. Rainey, VP-Admin.
Richard P. Rainey, VP-Finance
Richard P. Rainey, Corp. Sec.
David Berd, Chief Medical Officer
Andres Crespo, General Mgr.-Genopoietic
Henry E. Shea, III, Dir.-Quality Sys.
John K. Prendergast, Chmn.

Phone: 215-241-9760	Fax: 215-241-9684
Toll-Free:	
Address: 2000 Hamilton St., Ste. 204, Philadelphia, PA 19130 US	

GROWTH PLANS/SPECIAL FEATURES:

AVAX Technologies, Inc. is a development stage biotechnology company that specializes in the development and commercialization of individualized vaccine therapies and other technologies for the treatment of cancer. The company is primarily focusing its efforts on the development of immunotherapies for the treatment of cancer. The company's vaccine consists of autologous (the patient's own) cancer cells that have been treated with a chemical (haptenized) to make them more visible to the patient's immune system. AVAX refers to its cancer vaccine technology as autologous cell vaccine immunotherapy and to the vaccine as AC Vaccine. Previous clinical trials have focused on melanoma, ovarian carcinoma and non-small cell lung cancer. AVAX's AC Vaccine candidates are M-Vax, currently in Phase III trial for the treatment of melanoma; L-Vax, in Phase I/II trial for the treatment of non-small cell lung cancer; and O-Vax, in Phase I/II trial for the treatment of ovarian cancer. The company's leading AC Vaccine is M-Vax, which is designed as an immunotherapy for the post-surgical treatment of late stage (stages three and four) melanoma. AVAX believes that M-Vax is the first immunotherapy to show a substantial increase in the survival rate for patients with this type of melanoma. Of 214 stage three melanoma patients treated with M-Vax, mature studies (in which all surviving patients completed the five-year follow-up) evidenced a five-year overall survival rate of 44%, as opposed to the historical post-surgical survival rates of approximately 22-32%. In total studies of over 480 patients, no serious side effects have been reported.

FINANCIALS: Sales and profits are in thousands of dollars—add 000 to get the full amount. 2006 Note: Financial information for 2006 was not available for all companies at press time.

2006 Sales: $ 735	2006 Profits: $-5,356	U.S. Stock Ticker: AVXT	
2005 Sales: $1,624	2005 Profits: $-3,704	Int'l Ticker:	Int'l Exchange:
2004 Sales: $1,691	2004 Profits: $-3,457	Employees: 24	
2003 Sales: $ 931	2003 Profits: $-3,286	Fiscal Year Ends: 12/31	
2002 Sales: $ 800	2002 Profits: $-9,400	Parent Company:	

SALARIES/BENEFITS:

Pension Plan:	ESOP Stock Plan:	Profit Sharing:	Top Exec. Salary: $275,000	Bonus: $
Savings Plan: Y	Stock Purch. Plan:		Second Exec. Salary: $220,000	Bonus: $

OTHER THOUGHTS:

Apparent Women Officers or Directors:
Hot Spot for Advancement for Women/Minorities:

LOCATIONS: ("Y" = Yes)

West:	Southwest:	Midwest:	Southeast:	Northeast:	International:
				Y	Y

AVI BIOPHARMA INC

www.avibio.com

Industry Group Code: 325412 **Ranks within this company's industry group:** Sales: 183 Profits: 130

Drugs:		Other:	Clinical:	Computers:	Other:
Discovery:	Y	AgriBio:	Trials/Services:	Hardware:	Specialty Services:
Licensing:		Genomics/Proteomics:	Laboratories:	Software:	Consulting:
Manufacturing:		Tissue Replacement:	Equipment/Supplies:	Arrays:	Blood Collection:
Development:	Y		Research/Development Svcs.:	Database Management:	Drug Delivery:
Generics:			Diagnostics:		Drug Distribution:

TYPES OF BUSINESS:

Gene-Targeted Pharmaceuticals
Drugs - Cardiovascular Disease
Drugs - Cancer
Drugs - Infectious Disease

BRANDS/DIVISIONS/AFFILIATES:

AVICINE
Resten-NG
Cook Group, Inc.
NeuGene
Antivirals, Inc.
AVI-6001
AVI-6002
SuperGen, Inc.

CONTACTS: Note: Officers with more than one job title may be intentionally listed here more than once.

K. Michael Forrest, Interim CEO
Alan P. Timmins, COO
Alan P. Timmins, Pres.
Mark M. Webber, CFO
Patrick L. Iversen, Sr. VP-R&D
Mark M. Webber, CIO
Dwight D. Weller, Sr. VP-Mfg. & Chemistry
R. Ray Cummings, VP-Bus. Dev.
Peter D. O'Hanley, VP-Regulatory Affairs & Clinical Dev.
Janet Rose Christensen, VP-Regulatory Affairs & Quality
Jack L. Bowman, Chmn.

Phone: 503-227-0554	Fax: 503-227-0751
Toll-Free:	
Address: 1 SW Columbia, Ste. 1105, Portland, OR 97258 US	

GROWTH PLANS/SPECIAL FEATURES:

AVI BioPharma, Inc., formerly Antivirals, Inc., is a biopharmaceutical company that develops therapeutic products principally based on third-generation NEUGENE antisense technology. The company's principal products in development target life-threatening diseases, including cardiovascular and infectious diseases. The firm's NEUGENE antisense products include Resten-NG, the company's lead drug candidate for the treatment of cardiovascular restenosis (the re-narrowing of a coronary artery after angioplasty). In October 2006, Cook Group, Inc., the company's licensee and development partner, announced interim Phase II clinical trial data treating cardiovascular restenosis by delivering Resten-NG systemically using the proprietary microparticle delivery technology. Cook Group indicated plans for additional clinical studies with products based on the Resten-NG platform. Resten-CP, a drug for treating coronary artery bypass grafting, entered a human clinical trial in Eastern Europe. In addition, AVI BioPharma's infectious disease program is currently focusing on single-strand RNA viruses using its proprietary NEUGENE antisense agents targeting West Nile virus, hepatitis C, dengue virus, the SARS corona virus and Ebola virus. Other drugs include AVI-6001 for influenza/avian flu and AVI-6002 for Ebola virus, both of which are in preclinical studies; and AVI-4658 for muscular dystrophy, which completed preclinical trials and is in planned Phase I trials. The firm is involved in a strategic alliance with SuperGen, Inc. for the shared development and marketing of AVICINE. The company has 202 foreign and domestic issued or licensed patents and 198 foreign and domestic pending patent applications. In April 2007, AVI BioPharma acquired a facility in Oregon that will house additional capability for the large-scale GMP production of the company's proprietary phosphorodiamidate morpholino oligomers (PMOs) and for the recovery and purification of PMO precursors.

FINANCIALS: Sales and profits are in thousands of dollars—add 000 to get the full amount. 2006 Note: Financial information for 2006 was not available for all companies at press time.

2006 Sales: $ 115	2006 Profits: $-31,073	U.S. Stock Ticker: AVII
2005 Sales: $4,783	2005 Profits: $-16,676	Int'l Ticker: Int'l Exchange:
2004 Sales: $ 430	2004 Profits: $-24,778	Employees: 117
2003 Sales: $ 970	2003 Profits: $-14,617	Fiscal Year Ends: 12/31
2002 Sales: $ 800	2002 Profits: $-29,400	Parent Company:

SALARIES/BENEFITS:

Pension Plan:	ESOP Stock Plan:	Profit Sharing:	Top Exec. Salary: $375,000	Bonus: $
Savings Plan:	Stock Purch. Plan:		Second Exec. Salary: $300,000	Bonus: $110,000

OTHER THOUGHTS:

Apparent Women Officers or Directors: 1
Hot Spot for Advancement for Women/Minorities:

LOCATIONS: ("Y" = Yes)

West:	Southwest:	Midwest:	Southeast:	Northeast:	International:
Y					

Note: Financial information, benefits and other data can change quickly and may vary from those stated here.

AVIGEN INC

www.avigen.com

Industry Group Code: 325412 Ranks within this company's industry group: Sales: 185   Profits: 114

Drugs:	Other:		Clinical:	Computers:	Other:	
Discovery:	AgriBio:		Trials/Services:	Hardware:	Specialty Services:	
Licensing:	Genomics/Proteomics:	Y	Laboratories:	Software:	Consulting:	
Manufacturing:	Tissue Replacement:		Equipment/Supplies:	Arrays:	Blood Collection:	
Development: Y			Research/Development Svcs.:	Database Management:	Drug Delivery:	
Generics:			Diagnostics:		Drug Distribution:	

TYPES OF BUSINESS:

Neurological & Neuromuscular Therapeutics

BRANDS/DIVISIONS/AFFILIATES:

AV650
AV411
AV513

CONTACTS: *Note: Officers with more than one job title may be intentionally listed here more than once.*

Kenneth Chahine, CEO
Kenneth Chahine, Pres.
Kirk W. Johnson, VP-R&D
Christina Thomson, VP-Corp. Counsel
Michael D. Coffee, Chief Bus. Officer
Andrew A. Sauter, VP-Finance
Zola Horovitz, Chmn.

Phone: 510-748-7150	Fax: 510-748-7155
Toll-Free:	
Address: 1301 Harbor Bay Pkwy., Alameda, CA 94502 US	

GROWTH PLANS/SPECIAL FEATURES:

Avigen, Inc. develops and commercializes small molecule therapeutics and biologics to treat serious neurological and neuromuscular disorders. The company's current lead product candidates primarily address spasticity and neuromuscular spasm and neuropathic pain. Products in development include AV650 for the treatment of disabling neuromuscular spacticity and spasm; AV411, an oral therapy for the treatment of neuropathic pain based on the approved drug ibudilast; and AV513, an oral therapy for the treatment of bleeding disorders. The firm maintains a small ongoing preclinical research effort to identify additional opportunities to expand its product development pipeline. The efforts primarily focus on additional treatments for neuropathic pain and include, through external contract laboratories, a medicinal chemistry optimization effort focused on identification of new chemical entities with glia-attenuating characteristics similar to those of AV411, but with improved physiocochemical properties. The company is also pharmacologically testing additional therapeutic indications for AV411. In 2006, Avigen acquired exclusive license rights to develop and commercialize proprietary formulations of the compound tolperisone, which it has named AV650, for the North American markets. In March 2007, the company received approval to begin Phase II clinical development of AV650.

The company offers its employees medical, dental, vision, life and short- and long-term disability insurance; a 401(k) plan; an employee assistance program; stock options; and flexible spending accounts.

FINANCIALS: Sales and profits are in thousands of dollars—add 000 to get the full amount. 2006 Note: Financial information for 2006 was not available for all companies at press time.

2006 Sales: $ 103	2006 Profits: $-24,256	**U.S. Stock Ticker:** AVGN
2005 Sales: $12,026	2005 Profits: $-14,696	**Int'l Ticker:** Int'l Exchange:
2004 Sales: $2,195	2004 Profits: $-23,923	Employees: 33
2003 Sales: $ 463	2003 Profits: $-25,774	Fiscal Year Ends: 12/31
2002 Sales: $ 100	2002 Profits: $-27,700	Parent Company:

SALARIES/BENEFITS:

Pension Plan:	ESOP Stock Plan:	Profit Sharing:	Top Exec. Salary: $407,239	Bonus: $95,000
Savings Plan: Y	Stock Purch. Plan:		Second Exec. Salary: $288,017	Bonus: $28,000

OTHER THOUGHTS:

Apparent Women Officers or Directors: 2
Hot Spot for Advancement for Women/Minorities: Y

LOCATIONS: ("Y" = Yes)

West:	Southwest:	Midwest:	Southeast:	Northeast:	International:
Y					

AVIVA BIOSCIENCES CORP   www.avivabio.com

ndustry Group Code: 325413 Ranks within this company's industry group: Sales: Profits:

Drugs:	Other:	Clinical:	Computers:		Other:	
Discovery:	AgriBio:	Trials/Services:	Hardware:	Y	Specialty Services:	Y
Licensing:	Genomics/Proteomics:	Laboratories:	Software:		Consulting:	
Manufacturing:	Tissue Replacement:	Equipment/Supplies:	Arrays:		Blood Collection:	
Development:		Research/Development Svcs.:	Database Management:		Drug Delivery:	
Generics:		Diagnostics:			Drug Distribution:	

TYPES OF BUSINESS:

Cellular Biology Equipment
Cancer Cell Isolation Technology
Drug Candidate Screening Technology
Automated Patch Clamp Electrophysiology
Rare Cell Enrichment
Multiple Force Biochips

BRANDS/DIVISIONS/AFFILIATES:

Sealchip
Electrophysiology on Demand
Fetal Cell Enrichment Kit
hERGexpress
China Development Industrial Bank
CapityalBio Corporation
Pac-Link
WI Harper Group

CONTACTS: Note: Officers with more than one job title may be intentionally listed here more than once.

Julian Yuan, CEO
Jia Xu, VP-R&D
Lei Wu, VP-Mfg.
Lei Wu, VP-Oper.
Mingxian Huang, Sr. Dir.-Chemistry
Antonio Guia, Dir.-Ion Channel Tech.
Andrea Ghetti, Sr. Dir.-Pharma Svcs.
Ping Lin, Dir.-Cancer Cell Project
Vytas P. Ambutas, Chmn.

Phone: 858-522-0888	Fax: 858-522-9040
Toll-Free:	
Address: 11180 Roselle St., Ste. 200, San Diego, CA 92121 US	

GROWTH PLANS/SPECIAL FEATURES:

AVIVA Biosciences Corporation develops and integrates biochips for electrophysiology research, ion channel drug screening and rare-cell isolation. In addition, the company has developed proprietary surface chemistries, microbeads and reagent compositions as solutions to critical cell biology and bioassay applications. AVIVA's introduction of improved and automated cell manipulation platforms has enabled greater productivity of cell-based assays for use in the biotechnological development of medicinal products. The firm offers Electrophysiology on Demand (EPOD), a drug discovery service that utilizes ion channel drug screening in order to provide clients with a full line of automated electrophysiology instruments, experienced personnel and customer service. The main product line within the EPOD service is Sealchip, a single-use disposable cartridge designed for use in high fidelity ion channel measurements and higher throughput screening for drug discovery customers. Optimized and validated ion channel cell lines are also provided for patch clamp electrophysiology experiments. Additionally, AVIVA develops cancer cell isolation systems that reliably detect targeted tumor cells from blood samples. Through a proprietary depletion approach, the system removes non-relevant cells and exposes target cells that can then be analyzed and quantified. In addition to its product lines, the firm offers hERGexpress, a screening service for medicinal chemists and toxicologists that provides high quality data as guidance in assessing the cardiac safety of certain pharmaceutical compounds. AVIVA is owned by four investors: CapitalBio Corporation, China Development Industrial Bank, WI Harper Group and Pac-Link. In October 2007, AVIVA signed an agreement with Cosmo Bio Co., Ltd., which will allow the distribution of AVIVA's electrophysiology services to Cosmo operations in Tokyo, Japan.

FINANCIALS: Sales and profits are in thousands of dollars—add 000 to get the full amount. 2006 Note: Financial information for 2006 was not available for all companies at press time.

2006 Sales: $	2006 Profits: $	U.S. Stock Ticker: Private
2005 Sales: $	2005 Profits: $	Int'l Ticker: Int'l Exchange:
2004 Sales: $	2004 Profits: $	Employees:
2003 Sales: $	2003 Profits: $	Fiscal Year Ends:
2002 Sales: $	2002 Profits: $	Parent Company:

SALARIES/BENEFITS:

Pension Plan:	ESOP Stock Plan:	Profit Sharing:	Top Exec. Salary: $	Bonus: $
Savings Plan:	Stock Purch. Plan:		Second Exec. Salary: $	Bonus: $

OTHER THOUGHTS:

Apparent Women Officers or Directors:
Hot Spot for Advancement for Women/Minorities:

LOCATIONS: ("Y" = Yes)

West:	Southwest:	Midwest:	Southeast:	Northeast:	International:
Y					

AXCAN PHARMA INC

www.axcan.com

Industry Group Code: 325412 **Ranks within this company's industry group:** Sales: 50   Profits: 39

Drugs:	Other:		Clinical:	Computers:	Other:
Discovery:	AgriBio:		Trials/Services:	Hardware:	Specialty Services:
Licensing:	Genomics/Proteomics:		Laboratories:	Software:	Consulting:
Manufacturing: Y	Tissue Replacement:		Equipment/Supplies:	Arrays:	Blood Collection:
Development: Y			Research/Development Svcs.:	Database Management:	Drug Delivery:
Generics:			Diagnostics:		Drug Distribution:

TYPES OF BUSINESS:

Pharmaceutical Manufacturing
Gastroenterology Treatment Products

BRANDS/DIVISIONS/AFFILIATES:

Axcan Scandipharm, Inc.
Axcan Pharma S.A.
SALOFALK
CANASA
SUDCA
URSO
DELURSAN
ULTRASE

CONTACTS: *Note: Officers with more than one job title may be intentionally listed here more than once.*

Frank A. G. M. Verwiel, CEO
David W. Mims, Exec. VP/COO
Frank A. G. M. Verwiel, Pres.
Steve Gannon, CFO/Exec. VP
Alexandre P. LeBeaut, Sr. VP/Chief Scientific Officer
Jean-Francois Hebert, Sr. VP-Mfg. Oper.
Martha Donze, VP-Corp. Admin.
Richard Tarte, General Counsel
Richard Tarte, VP-Corp. Dev.
Jean Vezina, VP-Finance
Patrick Colin, VP-R&D
Darcy Toms, VP-Bus. Dev.
Michael E. Thiel, VP-North American Mktg. Oper.
Leon F. Gosselin, Chmn.
Jocelyn Pelchat, Sr. VP- Int'l Commercial Oper.
Jean-Francois Herbert, Sr. VP-Procurement

Phone: 450-467-5138	Fax: 450-464-9979
Toll-Free: 800-809-4950	
Address: 597 Laurier Blvd., Mont-Saint-Hilaire, QC J3H 6C4 Canada	

GROWTH PLANS/SPECIAL FEATURES:

Axcan Pharma, Inc. is a leading specialty pharmaceutical company concentrating in the field of gastroenterology, with operations in North America and Europe. Axcan markets and sells pharmaceutical products used in the treatment of a variety of gastrointestinal diseases and disorders, including inflammatory bowel disease, cholestatic liver diseases, irritable bowel syndrome and complications related to pancreatic insufficiency.   The firm seeks to expand its gastrointestinal franchise by in-licensing products and acquiring products or companies, as well as developing additional products and expanding indications for existing products. For the treatment of inflammatory bowel diseases, such as Ulcerative Colitis and Ulcerative Proctitis, Axcan markets mesalamine-based products SALOFALK and CANASA. Axcan is currently developing products for the prevention and treatment of colorectal cancer, including SUDCA, which is currently in Phase I trial for the prevention of the recurrence of colorectal polyps. For the treatment of the cholestatic liver disease Primary Biliary Cirrhosis (PBC), a condition which causes the slow destruction of bile ducts in the liver, Axcan markets URSO 250 and URSO Forte. For the treatment of both PBC and the cholestatic liver disease Primary Sclerosing Cholangitis (PSC), a condition which narrows the bile ducts inside and outside of the liver through inflammation and scarring, Axcan markets URSO, URSO DS and DELURSAN. Axcan is currently developing NCX-1000 for the treatment of portal hypertension, the most common symptom of chronic liver disease, and it is currently undergoing Phase II trials. For the treatment of pancreatic insufficiency, Axcan markets ULTRASE and VIOKASE, both of which were first made available in the United States before review and approval were required by the FDA. Both products are currently in Phase III trials. For the treatment of acid related disorders, Axcan markets CARAFATE and SULCRATE.   In May 2007, Axcan released PYLERA, a therapy for the eradication of Helicobacter pylori.

FINANCIALS: Sales and profits are in thousands of dollars—add 000 to get the full amount. 2006 Note: Financial information for 2006 was not available for all companies at press time.

2006 Sales: $292,320	2006 Profits: $39,120	U.S. Stock Ticker: AXCA
2005 Sales: $251,300	2005 Profits: $26,400	Int'l Ticker: AXP Int'l Exchange: Toronto-TSX
2004 Sales: $243,800	2004 Profits: $44,500	Employees: 425
2003 Sales: $179,084	2003 Profits: $19,925	Fiscal Year Ends: 9/30
2002 Sales: $133,175	2002 Profits: $21,188	Parent Company:

SALARIES/BENEFITS:

Pension Plan:	ESOP Stock Plan:	Profit Sharing:	Top Exec. Salary: $	Bonus: $
Savings Plan:	Stock Purch. Plan:		Second Exec. Salary: $	Bonus: $

OTHER THOUGHTS:

Apparent Women Officers or Directors: 2
Hot Spot for Advancement for Women/Minorities: Y

LOCATIONS: ("Y" = Yes)

West:	Southwest:	Midwest:	Southeast:	Northeast:	International:
			Y		Y

Note: Financial information, benefits and other data can change quickly and may vary from those stated here.

BARR PHARMACEUTICALS INC   www.barrlabs.com

Industry Group Code: 325416 Ranks within this company's industry group: Sales: 5   Profits: 2

Drugs:		Other:		Clinical:	Computers:		Other:	
Discovery:	Y	AgriBio:		Trials/Services:	Hardware:		Specialty Services:	Y
Licensing:		Genomics/Proteomics:		Laboratories:	Software:		Consulting:	
Manufacturing:	Y	Tissue Replacement:		Equipment/Supplies:	Arrays:		Blood Collection:	
Development:	Y			Research/Development Svcs.:	Database Management:		Drug Delivery:	
Generics:	Y			Diagnostics:			Drug Distribution:	

TYPES OF BUSINESS:

Drugs-Generic Pharmaceuticals
Contraceptives
Hormone Therapy Drugs
Female Healthcare Pharmaceuticals

BRANDS/DIVISIONS/AFFILIATES:

Barr Laboratories
Duramed Pharmaceuticals, Inc.
Seasonale
Sesonique
Mircette
Cenestin
Enjuvia
Plan B

CONTACTS: *Note: Officers with more than one job title may be intentionally listed here more than once.*

Bruce L. Downey, CEO
Paul M. Bisaro, COO
Paul M. Bisaro, Pres.
William T. McKee, CFO/Treas./Sr. VP
Timothy P. Catlett, Sr. VP-Generic Mktg. & Sales
Catherine F. Higgins, Sr. VP-Human Resources
Salah U. Ahmed, Sr. VP-R&D
Michael J. Bogda, Sr. VP-Eng.
Michael J. Bogda, Sr. VP-Mfg.
Fredrick J. Killion, General Counsel/Sr. VP
Christopher Mengler, Sr. VP-Corp. Dev.
Carol A. Cox, Sr. VP-Corp. Comm.
Carol A. Cox, Sr. VP-Global Investor Rel.
Sigurd C. Kirk, Controller/VP
Christine A. Mundkur, Sr. VP-Quality & Regulatory Counsel
Emad M. Alkhawan, VP-Analytical R&D
Charles E. Diliberti, VP-Scientific Affairs
G. Frederick Wilkinson, COO/Pres., Duramed Pharmaceuticals, Inc.
Bruce L. Downey, Chmn.
Timothy B. Sawyer, Sr. VP/Head-European Comm. Dev.

Phone: 845-362-1100	Fax: 845-362-2774
Toll-Free: 800-222-0190	
Address: 223 Quaker Rd., Pomona, NY 10970 US	

GROWTH PLANS/SPECIAL FEATURES:

Barr Pharmaceuticals, Inc. engages in the development, manufacture and marketing of generic and proprietary prescription pharmaceuticals. The company manufactures generic products under the Barr label through its Barr Laboratories subsidiary, and it produces proprietary products under the Duramed label through subsidiary Duramed Pharmaceuticals, Inc. Part of the firm's business strategy is to develop generic versions of other companies' drugs and then challenge the patents that protect them under the claim that such patents are invalid, unenforceable or not infringed by the company's product. The company's generic business segment manufactures and distributes over 100 dosage forms and strengths of over 75 different generic pharmaceutical products, including 22 oral contraceptive products, which represent the largest category of the generic portfolio, and an anticoagulant, Warfarin Sodium, for heart disease patients and those at high risk of stroke. The company currently has 19 proprietary products, which include the Seasonale, Seasonique and Mircette oral contraceptive product lines; the Cenestin and Enjuvia lines of hormone therapy products; and the Plan B emergency contraceptive product, approved by the FDA in August 2006 for non-prescription sale to people 18 years or older. In July 2007, the company received FDA approval to manufacture and market a generic version of Dostinex, currently supplied by Pharmacia and Upjohn Company, which it will supply through an agreement it has with Teva Pharmaceuticals Industries Ltd. The drug treats hyperprolactinemic disorders, a condition caused by an excess of the hormone prolactin in the blood often due to a tumor and sometimes by some other cause. In July 2007, the firm received FDA approval to market a generic form of the antifungal drug, Lamisil, currently supplied by Novartis Pharmaceutical Corp., which will be manufactured by Gedeon Richter Plc. Barr will only market the product in the U.S.

Barr Laboratories provides employees with tuition reimbursement programs.

FINANCIALS: Sales and profits are in thousands of dollars—add 000 to get the full amount. 2006 Note: Financial information for 2006 was not available for all companies at press time.

2006 Sales: $1,314,465	2006 Profits: $336,477	**U.S. Stock Ticker:** BRL	
2005 Sales: $1,047,399	2005 Profits: $214,988	**Int'l Ticker:** Int'l Exchange:	
2004 Sales: $1,309,088	2004 Profits: $123,103	Employees: 2,040	
2003 Sales: $902,900	2003 Profits: $167,600	Fiscal Year Ends: 6/30	
2002 Sales: $1,189,000	2002 Profits: $212,200	Parent Company:	

SALARIES/BENEFITS:

Pension Plan:	ESOP Stock Plan:	Profit Sharing:	Top Exec. Salary: $548,077	Bonus: $300,000
Savings Plan: Y	Stock Purch. Plan:		Second Exec. Salary: $422,115	Bonus: $250,000

OTHER THOUGHTS:

Apparent Women Officers or Directors: 3
Hot Spot for Advancement for Women/Minorities: Y

LOCATIONS: ("Y" = Yes)

West:	Southwest:	Midwest:	Southeast:	Northeast:	International:
		Y		Y	

Note: Financial information, benefits and other data can change quickly and may vary from those stated here.

BASF AG

www.basf.com

Industry Group Code: 325000 Ranks within this company's industry group: Sales: 1 Profits: 1

Drugs:	Other:	Clinical:	Computers:	Other:
Discovery:	AgriBio:	Trials/Services:	Hardware:	Specialty Services: Y
Licensing:	Genomics/Proteomics:	Laboratories:	Software:	Consulting:
Manufacturing:	Tissue Replacement:	Equipment/Supplies: Y	Arrays:	Blood Collection:
Development:		Research/Development Svcs.: Y	Database Management:	Drug Delivery:
Generics:		Diagnostics:		Drug Distribution:

TYPES OF BUSINESS:

Chemicals Manufacturing
Agricultural Products
Oil & Gas Production
Plastics
Coatings
Nanotechnology Research
Nutritional Products
Agricultural Biotechnology

BRANDS/DIVISIONS/AFFILIATES:

Wintershall AG
Orgamol SA
BASF Catalysts LLC
Johnson Polymer
CropDesign
Cehmische Fabrik WIBARCO GmBH
Hansa Chemie International

CONTACTS: *Note: Officers with more than one job title may be intentionally listed here more than once.*

Jurgen Hambrect, CEO
Kurt W. Bock, CFO
Eggert Voscherau, Dir.-Human Resources
Stefan Marcinowski, Dir.-Research Planning
Kurt W. Bock, Dir.-Info. Svcs.
Stefan Marcinowski, Dir.-Corp. Eng.
Magdalena Moll, Sr. VP-Investor Rel.
Kurt W. Bock, Dir.-Finance
Andreas Kreimeyer, Dir.-Performance Chemicals
John Feldmann, Dir.-Oil & Gas
Stefan Marcinowski, Dir.-Inorganics & Petrochemicals
Peter Oakley, Dir.-Agricultural Prod.
Juergen Hambrecht, Chmn.
Martin Brudermueller, Dir.-APAC
Kurt W. Bock, Dir.-Procurement & Logistics

Phone: 49-621-60-0	Fax: 49-621-60-42525
Toll-Free:	
Address: 38 Carl-Bosch St., Ludwigshafen, 67056 Germany	

GROWTH PLANS/SPECIAL FEATURES:

BASF is a chemical manufacturing company that operates production facilities in 38 countries, owns 159 subsidiaries and serves customers in more than 170 countries. Around 22% of BASF sales are made to North American industries. BASF operates in five business segments: chemicals; plastics; performance products; agricultural products and nutrition; and oil and gas. The chemicals segment manufactures over 1,500 inorganic, petrochemical and intermediate chemicals for the pharmaceutical, construction, textile and automotive industries. The plastics segment primarily manufactures polystyrene, styrenics and performance polymers for the manufacturing and packaging industries. The performance polymers segment produces pigments, inks, printing supplies, coatings and polymers for the automotive, coatings, oil, paper, packaging, textile, leather, detergent, sanitary care, construction and chemical industries. The firm's agricultural and nutritional products segment produces and markets genetically engineered plants, nutritional supplements, herbicides, fungicides and insecticides for use in agriculture, public health and pest control. Lastly, the oil and gas segment is operated through the BASF subsidiary, Wintershall AG, which focuses on petroleum and natural gas exploration and production in North America, Asia, Europe, the Middle East and Africa. BASF also employs chemical nanotechnology in pigments that are used to color coatings, paints and plastics; aqueous polymer dispersions; and sunscreen. BASF is currently investing over $244 million in the research and development of nanostructured materials, nanostructured surfaces and nanoparticles in hopes of developing nanoporous foams with greater insulating properties and nanocubes as storage mediums for large amounts of gases. BASF discontinued its lysine business in 2007 and shut down its production facility in Gunsan, South Korea. In April 2007, BASF signed an agreement to sell its subsidiary, Cehmische Fabrik WIBARCO GmBH to Hansa Chemie International in Zurich, Switzerland.

BASF offers U.S. employees reimbursement accounts, tuition reimbursement, employee savings plans, adoption assistance and a preferred supplier discount.

FINANCIALS: Sales and profits are in thousands of dollars—add 000 to get the full amount. 2006 Note: Financial information for 2006 was not available for all companies at press time.

2006 Sales: $69,448,400	2006 Profits: $4,575,330	**U.S. Stock Ticker:** BF
2005 Sales: $52,080,500	2005 Profits: $3,663,700	**Int'l Ticker:** BAS Int'l Exchange: Frankfurt-Euronext
2004 Sales: $51,572,600	2004 Profits: $2,550,700	Employees: 95,247
2003 Sales: $42,575,500	2003 Profits: $1,144,300	Fiscal Year Ends: 12/31
2002 Sales: $35,031,700	2002 Profits: $1,635,400	Parent Company:

SALARIES/BENEFITS:

Pension Plan: Y	ESOP Stock Plan:	Profit Sharing:	Top Exec. Salary: $	Bonus: $
Savings Plan: Y	Stock Purch. Plan: Y		Second Exec. Salary: $	Bonus: $

OTHER THOUGHTS:

Apparent Women Officers or Directors: 1
Hot Spot for Advancement for Women/Minorities:

LOCATIONS: ("Y" = Yes)

West:	Southwest:	Midwest:	Southeast:	Northeast:	International:
Y	Y	Y	Y	Y	Y

BAUSCH & LOMB INC

www.bausch.com

Industry Group Code: 339113 Ranks within this company's industry group: Sales: 3 Profits:

Drugs:		Other:		Clinical:		Computers:		Other:	
Discovery:		AgriBio:		Trials/Services:		Hardware:		Specialty Services:	
Licensing:		Genomics/Proteomics:		Laboratories:		Software:		Consulting:	
Manufacturing:	Y	Tissue Replacement:		Equipment/Supplies:	Y	Arrays:		Blood Collection:	
Development:	Y			Research/Development Svcs.:		Database Management:		Drug Delivery:	
Generics:	Y			Diagnostics:				Drug Distribution:	

TYPES OF BUSINESS:

Supplies-Eye Care
Contact Lens Products
Ophthalmic Pharmaceuticals
Surgical Products

BRANDS/DIVISIONS/AFFILIATES:

Bausch & Lomb
ReNu
Alrex
SofLens
Lotemax
Ocuvite
ReNu with MoistureLoc
ReNu MultiPlus

CONTACTS: Note: Officers with more than one job title may be intentionally listed here more than once.

Ronald L. Zarrella, CEO
Efrain Rivera, CFO/Sr. VP
David R. Nachbar, Sr. VP-Human Resources
Praveen Tyle, Chief Scientific Officer/Sr. VP-R&D
Evon L. Jones, CIO/VP
Gerhard Bauer, Sr. VP-Eng.
Robert B. Stiles, General Counsel/Sr. VP
Gehard Bauer, Sr. VP-Global Oper.
Stephen McCluski, Sr. VP-Corp. Strategy
Barbara M. Kelley, Corp. VP-Comm.
Barbara M. Kelley, Corp. VP-Investor Rel.
Efrain Riviera, Treas./VP
Brian Levy, Chief Medical Officer/VP
Jurij Z. Kushner, Controller/VP
Angela J. Panzarella, VP-Global Vision Care
Henry Tung, VP-Global Surgical
Ronald L. Zarrella, Chmn.
Alan H. Farnsworth, Sr. VP/Pres., EMEA

Phone: 585-338-6000	Fax: 585-338-6007

Toll-Free: 800-344-8815
Address: One Bausch & Lomb Pl., Rochester, NY 14604-2701 US

GROWTH PLANS/SPECIAL FEATURES:

Bausch & Lomb, Inc. (B&L) is a world leader in the development, marketing and manufacturing of eye care products. The firm's products are marketed in over 100 countries and in five categories: contact lenses; lens care products; ophthalmic pharmaceuticals; cataract and vitreoretinal surgery; and refractive surgery. In its contact lens category, which generates the largest percentage of revenues, B&L's product portfolio includes traditional, planned replacement disposable, daily disposable, multifocal, continuous wear, toric soft contact lenses and rigid gas-permeable materials. The firm's lens care products include multi-purpose solutions, enzyme cleaners and saline solutions. These products are marketed to licensed eye care professionals, health product retailers, independent pharmacies, drug stores, food stores and mass merchandisers. The firm's pharmaceuticals include generic and branded prescription ophthalmic pharmaceuticals, ocular vitamins, over-the-counter medications and vision accessories. Key pharmaceutical trademarks of the firm are Bausch & Lomb, Alrex, Liposic, Lotemax, Ocuvite, PreserVision and Zylet. B&L's cataract and vitreoretinal division offers a broad line of intraocular lenses as well as the Millennium line of phacoemulsification equipment used in the extraction of the patient's natural lens during cataract surgery. The company's refractive surgery products include lasers and diagnostic equipment used in the LASIK surgical procedure. B&L's global operations include research and development units across the world, all of which are dedicated to product research across all five segments of the firm. In 2006, the firm announced a number of new technologies in the lens, lens care and surgical products segments of the firm. The firm also announced an equity investment and an exclusive option to acquire AcuFocus, a privately held company. Lastly, the firm issued a worldwide recall of its ReNu with MoistureLoc contact lens solution in 2006, and also recalled over one million bottles of its ReNu MultiPlus contact lens solution in early 2007. In May 2007, the firm agreed to be acquired by private equity group Warburg Pincus for $4.5 billion.

FINANCIALS: Sales and profits are in thousands of dollars—add 000 to get the full amount. 2006 Note: Financial information for 2006 was not available for all companies at press time.

2006 Sales: $2,293,400	2006 Profits: $	U.S. Stock Ticker: BOL
2005 Sales: $2,353,800	2005 Profits: $19,200	Int'l Ticker: Int'l Exchange:
2004 Sales: $2,233,500	2004 Profits: $153,900	Employees: 13,700
2003 Sales: $2,019,500	2003 Profits: $125,500	Fiscal Year Ends: 12/26
2002 Sales: $1,816,700	2002 Profits: $72,500	Parent Company:

SALARIES/BENEFITS:

Pension Plan: Y	ESOP Stock Plan:	Profit Sharing:	Top Exec. Salary: $1,100,000	Bonus: $
Savings Plan: Y	Stock Purch. Plan:		Second Exec. Salary: $410,001	Bonus: $295,000

OTHER THOUGHTS:

Apparent Women Officers or Directors: 3
Hot Spot for Advancement for Women/Minorities: Y

LOCATIONS: ("Y" = Yes)

West:	Southwest:	Midwest:	Southeast:	Northeast:	International:
Y		Y	Y	Y	Y

Note: Financial information, benefits and other data can change quickly and may vary from those stated here.

BAXTER INTERNATIONAL INC
www.baxter.com

Industry Group Code: 339113 Ranks within this company's industry group: Sales: 1   Profits: 1

Drugs:	Other:	Clinical:	Computers:	Other:
Discovery:	AgriBio:	Trials/Services:	Hardware:	Specialty Services:
Licensing:	Genomics/Proteomics:	Laboratories:	Software:	Consulting:
Manufacturing:	Tissue Replacement:	Equipment/Supplies: Y	Arrays:	Blood Collection:
Development:		Research/Development Svcs.:	Database Management:	Drug Delivery:
Generics:		Diagnostics:		Drug Distribution:

TYPES OF BUSINESS:

Medical Equipment Manufacturing
Supplies-Intravenous & Renal Dialysis Systems
Medication Delivery Products & IV Fluids
Biopharmaceutical Products
Plasma Collection & Processing
Vaccines
Software
Contract Research

BRANDS/DIVISIONS/AFFILIATES:

Colleague CX
BioPharma Solutions
Advate
RenalSoft HD
Global Technical Services
BioLife Plasma Services
Fenwal, Inc.
Guangzhou Baiyunshan Pharmaceutical Co. Ltd.

CONTACTS: *Note: Officers with more than one job title may be intentionally listed here more than once.*

Robert L. Parkinson, CEO
Robert L. Parkinson, Pres.
Robert M. Davis, CFO/VP
Karen J. May, VP-Human Resources
Norbert G. Riedel, Chief Scientific Officer
Karenann Terrell, CIO/VP
J. Michael Gatling, VP-Mfg.
Susan R. Lichtenstein, General Counsel/VP
Joy A. Amundson, Pres., Bioscience
Bruce McGillivray, Pres., Renal
Peter J. Arduini, Pres., Medication Delivery
Carlos Alonso, Pres., Latin American Region
Robert L. Parkinson, Chmn.
John J. Greisch, Pres., Int'l

Phone: 847-948-2000	Fax: 847-948-3642
Toll-Free: 800-422-9837	
Address: 1 Baxter Pkwy., Deerfield, IL 60015-4625 US	

GROWTH PLANS/SPECIAL FEATURES:

Baxter International, Inc. is a global medical products, software and services company with expertise in medical devices, pharmaceuticals and biotechnology. Baxter markets its offerings to hospitals; clinical and medical research labs; blood and blood dialysis centers; rehab facilities; nursing homes; doctor's offices; and patients undergoing supervised home care. The firm has manufacturing facilities in 28 countries and offers products and services in 100 countries. Baxter operates in three segments: Medication Delivery, its largest sector, which provides a range of intravenous solutions and specialty products that are used in combination for fluid replenishment, nutrition therapy, pain management, antibiotic therapy and chemotherapy; Bioscience, which develops biopharmaceuticals, biosurgery products, vaccines, blood collection, processing and storage products and technologies; and Renal, which develops products and provides services to treat end-stage kidney disease. Products include the Colleague CX infusion pump; the Enlightened bar-coding system for flexible IV containers; Advate, a coagulant for hemophilia patients; and RenalSoft HD, a software module for the management of prescription, therapy and monitoring information relating to patients suffering from kidney failure. In addition, the company provides the following services: BioLife Plasma Services, a plasma collection and processing business; BioPharma Solutions, biotechnology; Global Technical Services, providing instrument service and support for devices manufactured and marketed by Baxter; Renal Clinical Helpline; Renal Services, an education and research operation; and Training and Education, a portfolio of interactive clinical web sites. In November 2006, Baxter entered a joint venture with Guangzhou Baiyunshan Pharmaceutical Co. Ltd. to produce and sell parenteral nutrition products in China. Also in 2006, Baxter expanded its agreement with Halozyme Therapeutics, Inc. to include the use of the HYLENEX recombinant with Baxter small molecule drugs. In March 2007, the firm sold its Transfusion Therapies business (now called Fenwal, Inc.) to Texas Pacific Group and Maverick Capital, Ltd. for $540 million.

FINANCIALS: Sales and profits are in thousands of dollars—add 000 to get the full amount. 2006 Note: Financial information for 2006 was not available for all companies at press time.

2006 Sales: $10,378,000	2006 Profits: $1,397,000	U.S. Stock Ticker: BAX
2005 Sales: $9,849,000	2005 Profits: $956,000	Int'l Ticker: Int'l Exchange:
2004 Sales: $9,509,000	2004 Profits: $388,000	Employees: 48,000
2003 Sales: $8,916,000	2003 Profits: $881,000	Fiscal Year Ends: 12/31
2002 Sales: $8,110,000	2002 Profits: $778,000	Parent Company:

SALARIES/BENEFITS:

Pension Plan: Y	ESOP Stock Plan:	Profit Sharing:	Top Exec. Salary: $1,133,651	Bonus: $1,941,420
Savings Plan: Y	Stock Purch. Plan: Y		Second Exec. Salary: $810,600	Bonus: $418,700

OTHER THOUGHTS:

Apparent Women Officers or Directors: 5
Hot Spot for Advancement for Women/Minorities: Y

LOCATIONS: ("Y" = Yes)

West:	Southwest:	Midwest:	Southeast:	Northeast:	International:
Y	Y	Y	Y	Y	Y

BAYER AG

Industry Group Code: 325000 Ranks within this company's industry group: Sales: 2   Profits: 3

Drugs:		Other:		Clinical:		Computers:		Other:	
Discovery:		AgriBio:	Y	Trials/Services:		Hardware:		Specialty Services:	Y
Licensing:		Genomics/Proteomics:		Laboratories:		Software:		Consulting:	
Manufacturing:	Y	Tissue Replacement:		Equipment/Supplies:	Y	Arrays:		Blood Collection:	
Development:	Y			Research/Development Svcs.:		Database Management:		Drug Delivery:	
Generics:				Diagnostics:				Drug Distribution:	

TYPES OF BUSINESS:

Chemicals Manufacturing
Pharmaceuticals
Animal Health Products
Synthetic Materials
Crop Science
Plant Biotechnology
Health Care Products

BRANDS/DIVISIONS/AFFILIATES:

Lanxess
Bayer CropScience
Bayer HealthCare
Bayer MaterialScience
Cipro
Levitra
Aleve
Schering AG

CONTACTS: *Note: Officers with more than one job title may be intentionally listed here more than once.*

Werner Wenning, Chmn.-Mgmt. Board
Klaus Kuhn, Dir.-Finance
Richard Pott, Dir.-Human Resources & Labor
Udo Oels, Dir.-Innovation
Udo Oels, Dir.-Tech.
Richard Pott, Dir.-Strategy
Udo Oels, Dir.-Environment & Asia-Pacific
Richard Pott, Dir.-North, Central & South America
Manfred Schneider, Chmn.-Supervisory Board
Klaus Kuhn, Dir.-EMEA

Phone: 49-214-30-1	Fax: 49-214-30-66328
Toll-Free: 800-269-2377	
Address: Bayerwerk Gebaeude W11, Leverkusen, D-51368 Germany	

GROWTH PLANS/SPECIAL FEATURES:

The Bayer Group is a German holding company encompassing some 280 consolidated subsidiaries on five continents. The company has three business segments: Bayer HealthCare, Bayer CropScience and Bayer MaterialScience. The health care segment (consisting of the pharmaceuticals and biological products divisions, consumer care and diagnostics, and animal health) develops, produces and markets products for the prevention, diagnosis and treatment of human and animal diseases. The firm's pharmaceutical and biological products division manufactures prescription drugs and treatments including: Cipro, a wide-spectrum antibiotic; Kogenate, a biological treatment for hemophilia; and Levitra, a treatment for impotence. The consumer care and diagnostic division provides professional and home diagnostic testing equipment and over-the-counter medications. Its products include Bayer Aspirin, Aleve, Midol, Alka-Seltzer and One-a-Day Vitamins. The animal health segment manufactures livestock and companion animal medicines, nutritional supplements and pesticides. The Bayer HealthCare segment accounts for approximately 47% of the company's total revenue. Bayer CropScience is active in the areas of chemical crop protection and seed treatment, non-agricultural pest and weed control and plant biotechnology. The segment accounts for approximately 33% of total revenue. Bayer MaterialScience develops, manufactures and markets polyurethane, polycarbonate, cellulose derivatives and special metals products. The segment accounts for approximately 18% of total revenue. The company has an annual research and development budget of approximately $2.1 billion. In 2006, Bayer purchased Schering AG for roughly $22.4 billion. In Febuary 2007, Bayer completed the sale of its chemical unit, H.C. Starck, to a consortium formed by the Carlyle Group and Advent International for $899 million and the assumption of $588.8 million of debt. In March 2007, Bayer announced that it planned to cut 6,100 jobs pursuant to its acquisition of Schering. Over half of the jobs will be cut from its Europe operations.

Bayer offers its employees individual training and development opportunities, deferred compensation, a defined benefit pension fund, sports amenities, flexible work schedules and a varied program of cultural events.

FINANCIALS: Sales and profits are in thousands of dollars—add 000 to get the full amount. 2006 Note: Financial information for 2006 was not available for all companies at press time.

2006 Sales: $38,710,400	2006 Profits: $2,249,950	**U.S. Stock Ticker: BAY**
2005 Sales: $32,662,374	2005 Profits: $1,902,517	**Int'l Ticker: BAY GR** Int'l Exchange: Frankfurt-Euronext
2004 Sales: $27,731,937	2004 Profits: $816,045	Employees: 93,300
2003 Sales: $35,914,000	2003 Profits: $-1,585,000	Fiscal Year Ends: 12/31
2002 Sales: $32,172,327	2002 Profits: $1,338,679	Parent Company:

SALARIES/BENEFITS:

Pension Plan: Y	ESOP Stock Plan:	Profit Sharing:	Top Exec. Salary: $862,878	Bonus: $1,640,607
Savings Plan: Y	Stock Purch. Plan: Y		Second Exec. Salary: $499,286	Bonus: $935,368

OTHER THOUGHTS:

Apparent Women Officers or Directors: 1
Hot Spot for Advancement for Women/Minorities:

LOCATIONS: ("Y" = Yes)

West:	Southwest:	Midwest:	Southeast:	Northeast:	International:
Y	Y	Y		Y	Y

Note: Financial information, benefits and other data can change quickly and may vary from those stated here.

BAYER CORP   www.bayerus.com

Industry Group Code: 325000 Ranks within this company's industry group: Sales: Profits:

Drugs:		Other:		Clinical:		Computers:		Other:	
Discovery:	Y	AgriBio:	Y	Trials/Services:		Hardware:		Specialty Services:	
Licensing:		Genomics/Proteomics:		Laboratories:		Software:		Consulting:	
Manufacturing:	Y	Tissue Replacement:		Equipment/Supplies:	Y	Arrays:		Blood Collection:	
Development:	Y			Research/Development Svcs.:		Database Management:		Drug Delivery:	
Generics:				Diagnostics:	Y			Drug Distribution:	

TYPES OF BUSINESS:

Chemicals Manufacturing
Animal Health Products
Over-the-Counter Drugs
Diagnostic Products
Coatings, Adhesives & Sealants
Polyurethanes & Plastics
Herbicides, Fungicides & Insecticides

BRANDS/DIVISIONS/AFFILIATES:

Bayer AG
Bayer HealthCare AG
Bayer MaterialSciences, LLC
Bayer CropScience, LP
Schering AG
California Planting Cotton Seed Distributors
Reliance Genetics LLC
Ure-Tech Group

CONTACTS: *Note: Officers with more than one job title may be intentionally listed here more than once.*

Attila Molnar, CEO
Attila Molnar, Pres.
Joyce Burgess, Dir.-Human Resources
Claudio Abreu, CIO
George J. Lykos, Chief Legal Officer
Mark Ryan, Chief Comm. Officer
Andreas Beier, Chief Acct. Officer
Willy Scherf, CEO/Pres., Bayer Corp. & Bus. Svcs.
Arthur Higgins, Chmn.-Bayer HealthCare AG
William Buckner, CEO/Pres., Bayer CropScience, LP
Gregory Babe, CEO/Pres., Bayer MaterialScience, LLC
Timothy Roseberry, Chief Procurement Officer/VP-Corp. Materials Mgmt.

Phone: 412-777-2000	Fax: 412-777-2034
Toll-Free:	
Address: 100 Bayer Rd., Pittsburgh, PA 15205-9741 US	

GROWTH PLANS/SPECIAL FEATURES:

Bayer Corporation is the U.S. subsidiary of chemical and pharmaceutical giant Bayer AG. The company operates through four subsidiaries: Bayer HealthCare; Bayer MaterialScience; Bayer Corporate and Business Services; and Bayer CropScience. Bayer HealthCare operates through five divisions: pharmaceuticals, consumer care, diagnostics, diabetes care and animal health. Its animal health products include vaccines and other preventative measures for farm and domestic animals. Its consumer care products include analgesics (Aleve and Bayer); cold and cough treatments (Alka-Seltzer Plus and Talcio); digestive relief products (Alka-Mints and Phillips' Milk of Magnesia); topical skin preparations (Domeboro and Bactine); and vitamins (One-A-Day and Flintstones). The diabetes care division is a leader in self-test blood glucose diagnostic systems, and has recently released the BREEZE product family that offers alternate site testing and automatic coding and requires smaller blood samples. Bayer HealthCare's diagnostics division, now called Siemens Medical Solutions Diagnostics, produces diagnostic systems for critical and intensive care, hematology, urinalysis, immunology, clinical chemistry and molecular testing. The Advia Centaur system is used for the diagnosis of diseases like cancer, cardiovascular diseases, allergies and infections; the Versant and Trugent brands of assays are used for the detection of HIV and hepatitis virus. Bayer's MaterialScience segment produces coatings, adhesives and sealant raw materials; polyurethanes; and plastics. Bayer CropScience makes products directed toward crop protection, environmental science and bioscience, which include herbicides, fungicides and insecticides. Bayer Corporate and Business Services provides business services to the aforementioned Bayer subsidiaries, such as administration, technology services, mergers/acquisitions and internal auditing. In 2006, the firm strengthened its pharmaceuticals segment with the acquisition of Schering AG, previously based in Berlin; the CropScience segment acquired California Planting Cotton Seed Distributors and Reliance Genetics LLC; and the firm announced plans to acquire Taiwan's Ure-Tech Group, the largest thermoplastic polyurethane producer in the Asia Pacific, in 2007.

FINANCIALS: Sales and profits are in thousands of dollars—add 000 to get the full amount. 2006 Note: Financial information for 2006 was not available for all companies at press time.

2006 Sales: $	2006 Profits: $	**U.S. Stock Ticker: Subsidiary**
2005 Sales: $8,747,200	2005 Profits: $	**Int'l Ticker:** Int'l Exchange:
2004 Sales: $11,504,000	2004 Profits: $	Employees: 16,200
2003 Sales: $10,999,300	2003 Profits: $	Fiscal Year Ends: 12/31
2002 Sales: $9,424,500	2002 Profits: $	Parent Company: BAYER AG

SALARIES/BENEFITS:

Pension Plan:	ESOP Stock Plan:	Profit Sharing:	Top Exec. Salary: $	Bonus: $
Savings Plan: Y	Stock Purch. Plan: Y		Second Exec. Salary: $	Bonus: $

OTHER THOUGHTS:

Apparent Women Officers or Directors: 1
Hot Spot for Advancement for Women/Minorities:

LOCATIONS: ("Y" = Yes)

West:	Southwest:	Midwest:	Southeast:	Northeast:	International:
Y	Y	Y	Y	Y	

BAYER SCHERING PHARMA AG   www.schering.de

Industry Group Code: 325412 **Ranks within this company's industry group:** Sales: Profits:

Drugs:		Other:		Clinical:		Computers:		Other:	
Discovery:		AgriBio:		Trials/Services:		Hardware:		Specialty Services:	
Licensing:		Genomics/Proteomics:	Y	Laboratories:		Software:		Consulting:	
Manufacturing:	Y	Tissue Replacement:		Equipment/Supplies:		Arrays:		Blood Collection:	
Development:	Y			Research/Development Svcs.:		Database Management:		Drug Delivery:	
Generics:				Diagnostics:			Y	Drug Distribution:	

TYPES OF BUSINESS:
Pharmaceuticals Discovery, Development & Manufacturing
Gynecology & Andrology Treatments
Contraceptives
Cancer Treatments
Multiple Sclerosis Treatments
Circulatory Disorder Treatments
Diagnostic & Radiopharmaceutical Agents
Proteomics

BRANDS/DIVISIONS/AFFILIATES:
Yasmin
Angeliq
Testogel
Androcur
Betaseron
Fludara
Illomedin
Bayer AG

CONTACTS: *Note: Officers with more than one job title may be intentionally listed here more than once.*
Arthur J. Higgins, Chmn.-Exec. Board
Werner Baumann, Member-Exec. Board, Human Resources
Werner Baumann, Member-Exec. Board, Production
Werner Baumann, Member-Exec. Board, Admin. & Organization
Andreas Busch, Member-Exec. Board
Ulrich Kostlin, Member-Exec. Board
Kemal Malik, Member-Exec. Board
Gunnar Riemann, Member-Exec. Board
Werner Wenning, Chmn.-Supervisory. Board

Phone: 49-30-468-1111	Fax: 49-30-468-15305

Toll-Free:
Address: Mullerstrasse 178, Berlin, 13353 Germany

GROWTH PLANS/SPECIAL FEATURES:
Bayer Schering Pharma AG, formerly Schering AG and a subsidiary of Bayer, is a major global research-based pharmaceutical company that operates through more than 140 subsidiaries. The firm concentrates its activities on four business areas: gynecology and andrology, oncology, specialized therapeutics, and diagnostics and radiopharmaceuticals. Schering's gynecology and andrology products include birth control pills (Yasmin), hormone therapy (Angeliq and Menostar) and other contraceptives for women (Mirena); products for the treatment of testosterone deficiency in men (Testoviron, Testogel and Nebido); and prostate cancer (Androcur). The firm's oncology unit has introduced the drug Fludara, to provide treatment for chronic lymphocytic leukemia, a variety of leukemia. Another product, Leukine, is a drug administered to treat the immune system weakened by chemotherapy. Zevalin is a readioimmunotherapy for follicular B-cell non-Hodgkin's lymphoma in E.U. countries, and MabCampath/Campath is a chemotherapy drug often used on those patients who do not respond to traditional chemotherapy. Schering's specialized therapeutic products focus on treating multiple sclerosis (MS). The firm has contributed to the body of research on MS through its Beyond and Benefit studies. Its Betaferon drug reduces the frequency of MS episodes significantly. In addition, the division produces Ilomedin to improve blood flow for those suffering with peripheral arterial occlusive disease; and Ventavis, to increase physical capability of patients suffering from primary pulmonary hypertension. Schering's diagnostics imaging products include a range of contrast media, such as Magnevist, a general MRI contrast agent; Resovist, a liver-specific MRI contrast agent; and Gadovist, a central nervous system MRI contrast agent. The company recently transferred its dermatology business to an independent company, Intendis GmbH. In June 2006, Bayer AG acquired sufficient stock in Schering to give Bayer effective control of the firm. In March 2007, Bayer announced that it would cut 6,100 jobs as a result of its purchase of Schering.

FINANCIALS: Sales and profits are in thousands of dollars—add 000 to get the full amount. 2006 Note: Financial information for 2006 was not available for all companies at press time.

2006 Sales: $	2006 Profits: $	**U.S. Stock Ticker:** Subsidiary
2005 Sales: $6,393,216	2005 Profits: $745,554	**Int'l Ticker: SCH** Int'l Exchange: Berlin-BBB
2004 Sales: $6,647,000	2004 Profits: $677,000	Employees: 24,658
2003 Sales: $6,070,000	2003 Profits: $557,000	Fiscal Year Ends: 12/31
2002 Sales: $5,267,000	2002 Profits: $910,000	Parent Company: BAYER AG

SALARIES/BENEFITS:

Pension Plan:	ESOP Stock Plan:	Profit Sharing:	Top Exec. Salary: $	Bonus: $
Savings Plan:	Stock Purch. Plan:		Second Exec. Salary: $	Bonus: $

OTHER THOUGHTS:
Apparent Women Officers or Directors:
Hot Spot for Advancement for Women/Minorities:

LOCATIONS: ("Y" = Yes)

West:	Southwest:	Midwest:	Southeast:	Northeast:	International:
Y				Y	Y

BENTLEY PHARMACEUTICALS INC   www.bentleypharm.com

Industry Group Code: 325412A Ranks within this company's industry group: Sales: 5   Profits: 6

Drugs:		Other:		Clinical:	Computers:		Other:	
Discovery:	Y	AgriBio:		Trials/Services:	Hardware:		Specialty Services:	
Licensing:	Y	Genomics/Proteomics:		Laboratories:	Software:		Consulting:	
Manufacturing:	Y	Tissue Replacement:		Equipment/Supplies:	Arrays:		Blood Collection:	
Development:	Y			Research/Development Svcs.:	Database Management:		Drug Delivery:	Y
Generics:	Y			Diagnostics:			Drug Distribution:	

TYPES OF BUSINESS:
Drug Delivery Systems Technologies
Drugs-Diversified
Generic Drugs
Drug Delivery Technology

BRANDS/DIVISIONS/AFFILIATES:
Laboratorios Rimafar
Laboratorios Belmac
Laboratorios Davur
Belmazol
Mio Relax
Enalapril Belmac
Senioral
Citalopram

CONTACTS: Note: Officers with more than one job title may be intentionally listed here more than once.
James R. Murphy, CEO
John A. Sedor, Pres.
Richard P. Lindsay, CFO/VP
Fred Feldman, VP-R&D
David C. Brush, VP-Bus. Dev. & Strategic Planning
Robert M. Stote, Sr. VP/Chief Medical Officer
James R. Murphy, Chmn.
Adolfo Herrera Malaga, Managing Dir-European Subsidiaries

Phone: 603-658-6100	Fax: 603-658-6101
Toll-Free:	
Address: Bentley Park, 2 Holland Way, Exeter, NH 03833 US	

GROWTH PLANS/SPECIAL FEATURES:

Bentley Pharmaceuticals, Inc. is an international specialty pharmaceutical company focused on advanced drug delivery technologies and generic pharmaceutical products. The company has U.S. and international patents and other proprietary rights to technologies that enhance or facilitate the absorption of drugs. Bentley is developing products incorporating these technologies and has licensed applications of its CPE-215 drug delivery technology to Auxilium Pharmaceuticals, Inc. for Testim, a gel indicated for testosterone replacement therapy. Bentley continues to seek other pharmaceutical and biotechnology companies to form additional strategic alliances to facilitate the development and commercialization of other products using its drug delivery technologies, including product candidates that deliver insulin to diabetic patients intranasally, deliver macromolecule therapeutics using a biodegradable Nanocaplet technology and treat nail fungus infections topically. The firm has a significant commercial presence in Spain, where it manufactures and markets more than 118 pharmaceutical products through subsidiaries Laboratorios Belmac, Laboratorios Davur and Laboratorios Rimafar. The bulk of the company's product line consists of generic and proprietary products for the treatment of cardiovascular, gastrointestinal, infectious and neurological diseases. The company's drug technologies include its CPE-215 technology, which enhances the absorption of drugs across skin membranes; solubility enhancement technology; oral formulation technologies; and nanocaplet technology for the delivery of macromolecule therapeutics. The company's branded products include Belmazol, sold by AstraZeneca as Prilosec; Senioral, sold by Aventis as Denoral; and Enalapril Belmac, sold by Merck as Vasotec. In February 2006, Bentley received 12 marketing approvals in Spain for various dosage forms of Citalopram (a generic version of Celexa), Finasteride (a generic version of Proscar) and Pravastatin (a generic version of Pravachol). In July 2007, Bentley received U.S. patent protection which extends coverage for its current intranasal drug delivery technology utilizing CPE-215 beyond insulin to include delivery of other therapeutically effective, pharmaceutically active peptides, peptidomimetics and proteins.

FINANCIALS: Sales and profits are in thousands of dollars—add 000 to get the full amount. 2006 Note: Financial information for 2006 was not available for all companies at press time.

2006 Sales: $109,471	2006 Profits: $ 974	U.S. Stock Ticker: BNT
2005 Sales: $97,730	2005 Profits: $10,919	Int'l Ticker: Int'l Exchange:
2004 Sales: $73,393	2004 Profits: $5,690	Employees: 442
2003 Sales: $64,676	2003 Profits: $6,097	Fiscal Year Ends: 12/31
2002 Sales: $39,100	2002 Profits: $1,600	Parent Company:

SALARIES/BENEFITS:

Pension Plan:	ESOP Stock Plan:	Profit Sharing:	Top Exec. Salary: $653,125	Bonus: $
Savings Plan: Y	Stock Purch. Plan:		Second Exec. Salary: $470,250	Bonus: $

OTHER THOUGHTS:

Apparent Women Officers or Directors:
Hot Spot for Advancement for Women/Minorities:

LOCATIONS: ("Y" = Yes)

West:	Southwest:	Midwest:	Southeast:	Northeast:	International:
				Y	Y

Note: Financial information, benefits and other data can change quickly and may vary from those stated here.

BIO RAD LABORATORIES INC

www.bio-rad.com

Industry Group Code: 339113   Ranks within this company's industry group: Sales: 6   Profits: 5

Drugs:		Other:		Clinical:		Computers:		Other:	
Discovery:		AgriBio:	Y	Trials/Services:		Hardware:		Specialty Services:	
Licensing:		Genomics/Proteomics:	Y	Laboratories:	Y	Software:	Y	Consulting:	
Manufacturing:		Tissue Replacement:		Equipment/Supplies:	Y	Arrays:		Blood Collection:	
Development:				Research/Development Svcs.:		Database Management:		Drug Delivery:	
Generics:				Diagnostics:	Y			Drug Distribution:	

TYPES OF BUSINESS:

Equipment-Life Sciences Research
Clinical Diagnostics Products
Analytical Instruments
Laboratory Devices
Biomaterials
Imaging Products
Assays
Software

BRANDS/DIVISIONS/AFFILIATES:

KnowItAll
HaveItAll
PhD Workstation
VersArray
SmartSpec Plus
Variant II Turbo Hemoglobin Testing System
MiniOpticon
Blackhawk BioSystems

CONTACTS: Note: Officers with more than one job title may be intentionally listed here more than once.

Norman Schwartz, CEO
Norman Schwartz, Pres.
Christine Tsingos, CFO/VP
Sanford S. Wadler, General Counsel/VP
Tina Cuccia, Mgr.-Corp. Comm.
John Goetz, VP
David Schwartz, Chmn.

Phone: 510-724-7000	Fax: 510-741-5817
Toll-Free: 800-424-6723	
Address: 1000 Alfred Nobel Dr., Hercules, CA 94547 US	

GROWTH PLANS/SPECIAL FEATURES:

Bio-Rad Laboratories supplies the life science research, health care and analytical chemistry markets with a broad range of products and systems used to separate complex chemical and biological materials and to identify, analyze and purify components. The company operates through two industry segments: Life Sciences and Clinical Diagnostics. The firm's Life Sciences division develops laboratory devices, biomaterials, imaging products and microscopy systems. The division uses electrophoresis, image analysis, microplate readers, chromatography, gene transfer and sample preparation and amplification as its primary technological applications. Bio-Rad Life Sciences provides its services to universities and medical schools, industrial research organizations, government agencies and biotechnology researchers. The company's Clinical Diagnostics division encompasses a broad array of technologies incorporated into a variety of tests used to detect, identify and quantify substances in blood or other body fluids and tissues. The test results are used as aids for medical diagnosis, detection, evaluation, monitoring and treatment of diseases and other medical conditions. In addition, Bio-Rad is a leading provider of bovine spongiform encephalopathy (mad cow disease) tests throughout the world. Some of Bio-Rad's numerous brand name systems include: KnowItAll and HaveItAll, which are informatics systems integrating software and database management for a variety of biological information; the PhD System, for autoimmune detection; the VersArray hybridization chamber; the SmartSpec Plus spectrophotometer; the MiniOpticon PCR (polymerase chain reaction) detection system; and the Variant II Turbo hemoglobin testing system. In October 2006, Bio-Rad acquired Blackhawk BioSystems, a producer of quality control products for laboratories that work with infectious diseases. In November 2006, the company purchased a life sciences research business from Ciphergen Biosystems, Inc. for $20 million. In May 2007, Bio-Rad announced that it agreed to acquire DiaMed Holding AG, which develops, manufactures and markets products used in blood typing and screening, for approximately $397.55 million.

FINANCIALS: Sales and profits are in thousands of dollars—add 000 to get the full amount. 2006 Note: Financial information for 2006 was not available for all companies at press time.

2006 Sales: $1,273,930	2006 Profits: $103,263	U.S. Stock Ticker: BIO
2005 Sales: $1,180,985	2005 Profits: $81,553	Int'l Ticker: Int'l Exchange:
2004 Sales: $1,090,012	2004 Profits: $68,242	Employees: 5,400
2003 Sales: $1,003,382	2003 Profits: $76,171	Fiscal Year Ends: 12/31
2002 Sales: $892,700	2002 Profits: $67,900	Parent Company:

SALARIES/BENEFITS:

Pension Plan:	ESOP Stock Plan:	Profit Sharing:	Top Exec. Salary: $526,065	Bonus: $266,938
Savings Plan:	Stock Purch. Plan:		Second Exec. Salary: $515,760	Bonus: $261,709

OTHER THOUGHTS:

Apparent Women Officers or Directors: 2
Hot Spot for Advancement for Women/Minorities: Y

LOCATIONS: ("Y" = Yes)

West:	Southwest:	Midwest:	Southeast:	Northeast:	International:
Y	Y			Y	Y

Note: Financial information, benefits and other data can change quickly and may vary from those stated here.

BIOANALYTICAL SYSTEMS INC   www.bioanalytical.com

Industry Group Code: 541710 **Ranks within this company's industry group:** Sales: 14   Profits: 11

Drugs:	Other:	Clinical:		Computers:		Other:	
Discovery:	AgriBio:	Trials/Services:	Y	Hardware:		Specialty Services:	Y
Licensing:	Genomics/Proteomics:	Laboratories:	Y	Software:		Consulting:	Y
Manufacturing:	Tissue Replacement:	Equipment/Supplies:	Y	Arrays:		Blood Collection:	
Development:		Research/Development Svcs.:	Y	Database Management:		Drug Delivery:	
Generics:		Diagnostics:				Drug Distribution:	

TYPES OF BUSINESS:

Contract Research Services
Screening & Testing Services
Bioanalytical Instruments
Regulatory & Compliance Consulting
Toxicology Testing
Formulation Development
Immunochemistry

BRANDS/DIVISIONS/AFFILIATES:

Culex APS
epsilon

CONTACTS: Note: Officers with more than one job title may be intentionally listed here more than once.

Richard M. Shepperd, CEO
Ronald E. Shoup, COO
Richard M. Shepperd, Pres.
Michael R. Cox, CFO
Candice B. Kissinger, Sr. VP-Mktg.
Lina L. Reeves-Kerner, VP-Human Resources
Candice B. Kissinger, Sr. VP-R&D/Corp. Sec.
Peter T. Kissinger, Chief Scientific Officer
Michael R. Cox, VP-Finance
Craig S. Bruntlett, Sr. VP-Sales Dev.
Peter T. Kissinger, Chmn.

Phone: 765-463-4527	Fax: 765-497-1102
Toll-Free: 800-845-4246	
Address: 2701 Kent Ave., Purdue Research Pk., West Lafayette, IN 47906 US	

GROWTH PLANS/SPECIAL FEATURES:

Bioanalytical Systems, Inc. (BASi) is a drug development firm that provides contract services to global pharmaceutical, medical device and biotechnology companies. Principal clients of BASi include scientists engaged in drug metabolism studies, pharmacokinetics and neuroscience research at pharmaceutical organizations. The company operates under two business segments: contract research services and research products to address bioanalytical, preclinical and clinical research needs of drug developers. The contract research services segment of BASi provides screening and pharmacological testing, preclinical safety testing, formulation development, clinical trials, regulatory compliance and quality control testing. Analytical methods such as bioanalytical and stability testing are employed to determine the potency, purity and composition of compounds. In addition, BASi manufactures specialized instruments and accessories for liquid chromatography and phase I and II clinical trials. The company's research products include robotic sampling systems, in vivo microdialysis collection systems, physiology monitoring tools, liquid chromatography and electrochemical instruments. Patented products include the Culex APS robotic pharmacology system, an automated program used by pharmaceutical researchers to monitor drug concentrations, and epsilon, a liquid chromatography instrument used for separation systems and chemical analysis. BASI research has significantly aided in the understanding of central nervous system disorders, diabetes, osteoporosis and other diseases. The firm operates labs in the U.S., U.K. and a lab in France in partnership with Biotec Centre. BASi has formed contract research partnerships with MicaGenix, Inc. and INCAPS, a BioCrossroads Company, in order to provide complementary preclinical research services for pharmaceutical, medical device and biotechnology companies.

FINANCIALS: Sales and profits are in thousands of dollars—add 000 to get the full amount. 2006 Note: Financial information for 2006 was not available for all companies at press time.

2006 Sales: $43,048	2006 Profits: $-2,609	**U.S. Stock Ticker: BASI**
2005 Sales: $42,395	2005 Profits: $- 101	Int'l Ticker: Int'l Exchange:
2004 Sales: $37,152	2004 Profits: $- 203	Employees: 330
2003 Sales: $29,838	2003 Profits: $ 87	Fiscal Year Ends: 9/30
2002 Sales: $26,500	2002 Profits: $1,100	Parent Company:

SALARIES/BENEFITS:

Pension Plan:	ESOP Stock Plan:	Profit Sharing: Y	Top Exec. Salary: $153,000	Bonus: $
Savings Plan: Y	Stock Purch. Plan:		Second Exec. Salary: $150,461	Bonus: $

OTHER THOUGHTS:

Apparent Women Officers or Directors: 2
Hot Spot for Advancement for Women/Minorities: Y

LOCATIONS: ("Y" = Yes)

West:	Southwest:	Midwest:	Southeast:	Northeast:	International:
Y		Y			Y

BIOCOMPATIBLES INTERNATIONAL PLC
www.biocompatibles.com

Industry Group Code: 339113 Ranks within this company's industry group: Sales: 15 Profits: 7

Drugs:		Other:		Clinical:		Computers:		Other:	
Discovery:		AgriBio:		Trials/Services:		Hardware:		Specialty Services:	
Licensing:		Genomics/Proteomics:		Laboratories:		Software:		Consulting:	
Manufacturing:	Y	Tissue Replacement:		Equipment/Supplies:	Y	Arrays:		Blood Collection:	
Development:	Y			Research/Development Svcs.:		Database Management:		Drug Delivery:	Y
Generics:				Diagnostics:				Drug Distribution:	

TYPES OF BUSINESS:
Medical Implant Technology
Supplies-Phosphorycholine Coatings
Embolisation Microspheres
Biomaterials
Cancer Treatment
Drug Delivery Platforms

BRANDS/DIVISIONS/AFFILIATES:
CellMed
Bead Block
Alginate Bead
NFil Technology
CellBeads
Phosphorylcholine (PC) Technology
Medtronic
PRECISION Drug Eluting Bead

CONTACTS: *Note: Officers with more than one job title may be intentionally listed here more than once.*
Crispin Simon, CEO
Ian Ardill, Dir.-Finance
Tim Maloney, Dir.-Sales
Geoff Tompsett, Dir.-Human Resources
Andy Lewis, Dir.-Research & Tech.
Geoff Tompsett, Dir.-IT
Ian Ardill, Corp. Sec.
Mike Motion, Dir.-Sales
Alistair Taylor, Dir.-Clinical & Regulatory Affairs
Paul Baxter, Dir.-Intellectual Property
Gerry Brown, Chmn.

Phone: 44-1252-732-732	Fax: 44-1252-732-777

Toll-Free:

Address: Chapman House, Farnham Business Park, Weydon Ln., Farnham, Surrey GU9 8QL UK

GROWTH PLANS/SPECIAL FEATURES:
Biocompatibles International plc is an international provider of medical products that combine medical devices with ancillary therapeutic drugs. The company has two businesses: Biocompatibles in Farnham, UK and CellMed in Alzenau, Germany. Biocompatibles uses bead technology to deliver therapeutic agents in its products, as well as providing Bead Block and Alginate Bead, two bead products which contain no drug. The company's three biomedical polymer systems are its NFil technology; CellBeads, which are used in the delivery of biological agents in its CellMed programs; and its Phosphorylcholine (PC) Technology, used for such applications as Medtronic's Endeavor Drug Eluting Stint. Four products comprise the company's product portfolio: Bead Block for use in embolisation therapy, a non-surgical procedure in which the blood supply is blocked to a target area; DC Bead drug delivery embolisation system; LC Bead controlled embolisation system; and PRECISION Drug Eluting Bead, an embolisation system which releases a local, controlled and sustained dose of doxorubicin for the treatment of Hepatocellular Carcinoma (HCC). The company's N-fil technology uses calibrated microspheres to treat cancer by physically blocking the blood vessels that feed tumors. Microspheres are the preferred method for uterine fibroid embolisation, an increasingly employed alternative to hysterectomy for treatment of uterine fibroids. In May 2007, Biocompatibles announced positive survival data from a trial of its PRECISION Drug Eluting Bead, in which 62 patients were treated in an open label study of cirrhosis-related HCC.

FINANCIALS: Sales and profits are in thousands of dollars—add 000 to get the full amount. 2006 Note: Financial information for 2006 was not available for all companies at press time.

2006 Sales: $12,070	2006 Profits: $10,170	**U.S. Stock Ticker:**	
2005 Sales: $6,880	2005 Profits: $-9,490	**Int'l Ticker: BII** Int'l Exchange: London-LSE	
2004 Sales: $5,000	2004 Profits: $- 300	Employees: 72	
2003 Sales: $3,700	2003 Profits: $28,500	Fiscal Year Ends: 12/31	
2002 Sales: $34,554	2002 Profits: $63,154	Parent Company:	

SALARIES/BENEFITS:

Pension Plan:	ESOP Stock Plan:	Profit Sharing:	Top Exec. Salary: $	Bonus: $
Savings Plan:	Stock Purch. Plan:		Second Exec. Salary: $	Bonus: $

OTHER THOUGHTS:
Apparent Women Officers or Directors:
Hot Spot for Advancement for Women/Minorities:

LOCATIONS: ("Y" = Yes)

West:	Southwest:	Midwest:	Southeast:	Northeast:	International:
				Y	Y

Note: Financial information, benefits and other data can change quickly and may vary from those stated here.

BIOCRYST PHARMACEUTICALS INC   www.biocryst.com

Industry Group Code: 325412 **Ranks within this company's industry group:** Sales: 136   Profits: 148

Drugs:		Other:	Clinical:	Computers:	Other:
Discovery:	Y	AgriBio:	Trials/Services:	Hardware:	Specialty Services:
Licensing:	Y	Genomics/Proteomics:	Laboratories:	Software:	Consulting:
Manufacturing:		Tissue Replacement:	Equipment/Supplies:	Arrays:	Blood Collection:
Development:	Y		Research/Development Svcs.:	Database Management:	Drug Delivery:
Generics:			Diagnostics:		Drug Distribution:

TYPES OF BUSINESS:
Small-Molecule Pharmaceutical Products
Drugs-Immunological, Infectious & Inflammatory Disease

BRANDS/DIVISIONS/AFFILIATES:
Fodosine
Tissue Factor VIIa
peramir

CONTACTS: *Note: Officers with more than one job title may be intentionally listed here more than once.*
Jon P. Stonehouse, CEO
J. Claude Bennett, COO
J. Claude Bennett, Pres.
Michael A. Darwin, CFO
W. James Alexander, Chief Medical Officer
Randall B. Riggs, Sr. VP-Corp. Dev.
Jonathan M. Nugent, VP-Corp. Comm.
Michael A. Darwin, Treas./Corp. Sec.
Yarlagadda S. Babu, VP-Drug Discovery
W. James Alexander, Sr. VP-Clinical & Regulatory Oper.
David S. McCullough, VP-Commercialization & Strategic Planning
Charles E. Bugg, Chmn.

Phone: 205-444-4600	Fax: 205-444-4640
Toll-Free:	
Address: 2190 Parkway Lake Dr., Birmingham, AL 35244 US	

GROWTH PLANS/SPECIAL FEATURES:
BioCryst Pharmaceuticals, Inc. is a biotechnology company that designs, optimizes and develops novel small-molecule drugs that treat cancer, cardiovascular and autoimmune diseases, as well as viral infections. The company's technology is designed to stifle the generation of malignant cells at their origin, which is the site of chemical reactions in enzymes. BioCryst integrates the disciplines of biology, crystallography, medicinal chemistry and computer modeling to design structure-based drugs and develop pharmaceuticals. BioCryst has three main drugs in development, with several smaller ones in trials. The drugs are Fodosine (BCX-1777), peramivir and BCX-4208. Fodosine is in Phase 2 trials for the treatment of patients with T-cell leukemia. The FDA has granted orphan drug status for Fodosine in three indications, including T-cell non-Hodgkins lymphoma, chronic lymphocytic leukemia and B-cell acute lymphoblastic leukemia. Peramivir, a neuraminidase inhibitor, has just entered into Phase 2 trials for seasonal and life-threatening influenza. BCX-4208 is a second-generation PNP inhibitor in development for the treatment of T-cell mediated autoimmune disease, including psoriasis, rheumatoid arthritis, multiple sclerosis and Crohn's disease. The drug is licensed to Roche for the prevention of acute rejection in transplantation and for the treatment of autoimmune diseases. The company is also developing Tissue Factor VIIa inhibitors that minimize blood clotting and cardiovascular inhibitors for the treatment of cardiovascular and acute coronary events. In addition, BioCryst has been working on an RHA polymerase inhibitor for the treatment of hepatitis C. The firm has corporate alliances with Roche, Mundipharma, Green Cross and Shionogi & Co., Ltd. In June 2006, BioCryst agreed to allow the Korean company Green Cross Corporation to develop and commercialize peramir, a potent influenza neutaminidase inhibitor. In March 2007, the company announced that it has licensed Shionogi & Co., Ltd. to develop and commercialize peramir in Japan.

FINANCIALS: Sales and profits are in thousands of dollars—add 000 to get the full amount. 2006 Note: Financial information for 2006 was not available for all companies at press time.

2006 Sales: $6,212	2006 Profits: $-43,618	U.S. Stock Ticker: BCRX
2005 Sales: $ 152	2005 Profits: $-26,099	Int'l Ticker: Int'l Exchange:
2004 Sales: $ 337	2004 Profits: $-21,104	Employees: 85
2003 Sales: $ 635	2003 Profits: $-12,700	Fiscal Year Ends: 12/31
2002 Sales: $	2002 Profits: $-16,900	Parent Company:

SALARIES/BENEFITS:

Pension Plan:	ESOP Stock Plan:	Profit Sharing:	Top Exec. Salary: $453,384	Bonus: $
Savings Plan: Y	Stock Purch. Plan: Y		Second Exec. Salary: $347,016	Bonus: $

OTHER THOUGHTS:
Apparent Women Officers or Directors:
Hot Spot for Advancement for Women/Minorities:

LOCATIONS: ("Y" = Yes)

West:	Southwest:	Midwest:	Southeast:	Northeast:	International:
			Y		

BIOGEN IDEC INC   www.biogenidec.com

Industry Group Code: 325412 Ranks within this company's industry group: Sales: 27 Profits: 26

Drugs:		Other:		Clinical:		Computers:		Other:	
Discovery:	Y	AgriBio:		Trials/Services:		Hardware:		Specialty Services:	
Licensing:		Genomics/Proteomics:	Y	Laboratories:	Y	Software:		Consulting:	
Manufacturing:	Y	Tissue Replacement:		Equipment/Supplies:		Arrays:		Blood Collection:	
Development:	Y			Research/Development Svcs.:		Database Management:		Drug Delivery:	
Generics:				Diagnostics:				Drug Distribution:	

TYPES OF BUSINESS:

Drugs-Immunology, Neurology & Oncology
Autoimmune & Inflammatory Disease Treatments
Drugs-Multiple Sclerosis
Drugs-Cancer
Genetic Engineering
Vaccines
Cell Cultures
Bulk Manufacturing

BRANDS/DIVISIONS/AFFILIATES:

AVONEX
AMEVIVE
TYSABRI
ANTEGREN
RITUXAN
Fumapharm AG
Fumarderm

CONTACTS: Note: Officers with more than one job title may be intentionally listed here more than once.

James C. Mullen, CEO
William R. Rohn, COO
James C. Mullen, Pres.
Craig E. Schneier, Exec. VP-Human Resources & Public Affairs
Cecil B. Pickett, Pres., R&D
Michael D. Kowolenko, Sr. VP-Pharmaceutical Oper. & Tech.
Susan H. Alexander, General Counsel/Exec. VP/Corp. Sec.
Mark Wiggins, Exec. VP-Corp. & Bus. Dev.
Craig E. Schneier, Exec. VP-Corp. Comm.
Michael F. MacLean, Chief Acct. Officer/Sr. VP/Controller
John M. Dunn, Exec. VP-New Ventures
Burt A. Adelman, Exec. VP-Portfolio Strategy
Robert A. Hamm, Sr. VP-Neurology Strategic Bus.
Faheem Hasnain, Sr. VP-Oncology Rheumatology Strategic Bus.
Bruce R. Ross, Chmn.
Hans Peter Hasler, Sr. VP-Int'l Strategic Business

Phone: 617-679-2000   Fax: 617-679-2617
Toll-Free:
Address: 14 Cambridge Ctr., Cambridge, MA 02142-1481 US

GROWTH PLANS/SPECIAL FEATURES:

Biogen IDEC, Inc. (Biogen), a leading biotech firm formed through the merger of IDEC Pharmaceuticals and Biogen, develops, manufactures and markets therapeutic pharmaceuticals for immunology, neurology and oncology. Biogen currently has five products: AVONEX is used to decrease the frequency of neurological attacks in patients with relapsing forms of multiple sclerosis (MS) and is used by 135,000 patients globally; TYSABRI is approved for the treatment of relapsing forms of MS; RITUXAN is globally approved for the treatment of relapsed or refractory low-grade or follicular, CD20-positive, B-cell non-Hodgkin's lymphomas (NHLs), or B-cell NHLs; ZEVALIN is a radioimmuno therapy approved for the treatment of patients with relapsed or refractory low-grade, follicular, or transformed B-cell NHL; and FUMADERM, acquired with the purchase of Fumapharm AG in June 2006, acts as an immunomodulator and is approved in Germany for the treatment of severe psoriasis. In February 2006, the FDA approved the RITUXAN supplemental Biologics License Application for use of RITUXAN in combination with methotrexate, for reducing signs and symptoms in adult patients with moderately-to-severely active rheumatoid arthritis who have had an inadequate response to one or more tumor necrosis factor antagonist therapies. Biogen is working with Genentech and Roche on the development of RITUXAN in additional oncology and other indications. Biogen also generates revenue by licensing drugs it has developed to other companies, including Schering-Plough, Merck and GlaxoSmithKline. In January 2007, Biogen agreed to acquire Syntonix Pharmaceuticals, a privately held biopharmaceutical company focused on discovering and developing long-acting therapeutic products to improve treatment regimens for chronic diseases.

Biogen offers its employees dental insurance, tuition reimbursement, commuter benefits, fitness benefits, an employee assistance program and concierge services.

FINANCIALS: Sales and profits are in thousands of dollars—add 000 to get the full amount. 2006 Note: Financial information for 2006 was not available for all companies at press time.

2006 Sales: $2,683,049	2006 Profits: $217,511	U.S. Stock Ticker: BIIB
2005 Sales: $2,422,500	2005 Profits: $160,711	Int'l Ticker: Int'l Exchange:
2004 Sales: $2,211,562	2004 Profits: $25,086	Employees: 3,750
2003 Sales: $679,183	2003 Profits: $-875,097	Fiscal Year Ends: 12/31
2002 Sales: $404,222	2002 Profits: $148,090	Parent Company:

SALARIES/BENEFITS:

Pension Plan:	ESOP Stock Plan:	Profit Sharing:	Top Exec. Salary: $1,084,616	Bonus: $2,000,000
Savings Plan: Y	Stock Purch. Plan: Y		Second Exec. Salary: $568,387	Bonus: $271,256

OTHER THOUGHTS:

Apparent Women Officers or Directors: 1
Hot Spot for Advancement for Women/Minorities:

LOCATIONS: ("Y" = Yes)

West:	Southwest:	Midwest:	Southeast:	Northeast:	International:
Y				Y	Y

Note: Financial information, benefits and other data can change quickly and may vary from those stated here.

BIOMARIN PHARMACEUTICAL INC www.biomarinpharm.com

Industry Group Code: 325412 Ranks within this company's industry group: Sales: 69 Profits: 122

Drugs:		Other:		Clinical:		Computers:		Other:	
Discovery:	Y	AgriBio:		Trials/Services:		Hardware:		Specialty Services:	
Licensing:	Y	Genomics/Proteomics:		Laboratories:		Software:		Consulting:	
Manufacturing:	Y	Tissue Replacement:		Equipment/Supplies:		Arrays:		Blood Collection:	
Development:	Y			Research/Development Svcs.:		Database Management:		Drug Delivery:	Y
Generics:				Diagnostics:				Drug Distribution:	

TYPES OF BUSINESS:

Biopharmaceutical Product Development
Drugs-Severe Conditions
Pediatric Disease Treatments
Asthma Treatments
Drug Delivery Technologies

BRANDS/DIVISIONS/AFFILIATES:

Orapred
Aldurazyme
Naglazyme
Phenoptin
Phenylase
Kuvan
Immune Tolerance
Ascent Pediatrics

CONTACTS: Note: Officers with more than one job title may be intentionally listed here more than once.

Jean-Jacques Bienaime, CEO
Jeffrey Cooper, CFO/VP
Jeff Ajer, VP-Mktg. & Sales
Mark Wood, VP-Human Resources
Emil D. Kakkis, Chief Medical Officer
Robert A. Baffi, Sr. VP-Tech. Oper.
Daniel P. Maher, VP-Prod. Dev.
R. Andrew Ramelmeier, VP-Mfg. & Process Dev.
G. Eric Davis, General Counsel/VP/Sec.
Stephen Aselage, Sr. VP-Global Commercial Oper.
Amy Waterhouse, VP-Regulatory & Gov't Affairs
Stuart J. Swiedler, Sr. VP-Clinical Affairs
Victoria Sluzky, VP-Quality & Analytical Chemistry
Elaine Heron, Chmn.
William E. Aliski, General Mgr-European Oper./VP
Steven Jungles, VP-Supply Chain

Phone: 415-506-6700	Fax: 415-382-7889
Toll-Free:	
Address: 105 Digital Dr., Novato, CA 94949 US	

GROWTH PLANS/SPECIAL FEATURES:

BioMarin Pharmaceutical, Inc. develops and commercializes biopharmaceutical products for serious diseases and medical conditions. BioMarin has three approved products: Orapred, for the treatment of severe asthma; Aldurazyme, for the treatment of mucopolysaccharidosis I (MPS-I); and Naglazyme, for the treatment of mucopolysaccharidosis VI (MPS-VI). MPS is a life-threatening genetic disorder caused by the lack of a sufficient quantity of the enzyme (alpha)-L-iduronidase. Patients with MPS have multiple debilitating symptoms, including delayed physical and mental growth, enlarged liver and spleen, skeletal and joint deformities, airway obstruction, heart disease, reduced endurance and pulmonary function and impaired hearing and vision. BioMarin has a joint venture with Genzyme for the worldwide development and commercialization of Aldurazyme. Kuvan (formerly Phenoptin) and Phenylase, two of BioMarin's clinical stage products being developed in collaboration with Merck Serono, treat phenyketonuria (PKU), an inherited metabolic disease for which there are currently no approved drug therapies. The company is also evaluating preclinical development of several other enzyme product candidates for genetic and other diseases as well as immune tolerance platform technology to overcome limitations associated with the delivery of existing pharmaceuticals. BioMarin also controls Ascent Pediatrics, a pediatrics business. In June 2007, BioMarin announced initiating the expanded access program, a program under which the FDA allows early access to investigational drugs being developed to treat serious diseases for which there is no satisfactory alternative therapy, for Kuvan.

BioMarin offers its employees dental insurance; a flexible spending plan; an employee assistance program; an education assistance program; on-site Weight Watchers meetings; Toastmasters; bagels, donuts and fruit on Wednesdays; and bi-weekly on-site chair massage.

FINANCIALS: Sales and profits are in thousands of dollars—add 000 to get the full amount. 2006 Note: Financial information for 2006 was not available for all companies at press time.

2006 Sales: $84,209	2006 Profits: $-28,533	U.S. Stock Ticker: BMRN
2005 Sales: $25,669	2005 Profits: $-74,270	Int'l Ticker: Int'l Exchange:
2004 Sales: $18,641	2004 Profits: $-187,443	Employees: 410
2003 Sales: $12,100	2003 Profits: $-75,798	Fiscal Year Ends: 12/31
2002 Sales: $13,900	2002 Profits: $-77,500	Parent Company:

SALARIES/BENEFITS:

Pension Plan:	ESOP Stock Plan:	Profit Sharing:	Top Exec. Salary: $552,393	Bonus: $662,458
Savings Plan: Y	Stock Purch. Plan: Y		Second Exec. Salary: $290,197	Bonus: $81,750

OTHER THOUGHTS:

Apparent Women Officers or Directors: 3
Hot Spot for Advancement for Women/Minorities: Y

LOCATIONS: ("Y" = Yes)

West:	Southwest:	Midwest:	Southeast:	Northeast:	International:
Y					Y

BIOMET INC

www.biomet.com

Industry Group Code: 339113 Ranks within this company's industry group: Sales: 4 Profits: 2

Drugs:		Other:		Clinical:		Computers:		Other:	
Discovery:		AgriBio:		Trials/Services:		Hardware:		Specialty Services:	
Licensing:		Genomics/Proteomics:		Laboratories:		Software:		Consulting:	
Manufacturing:		Tissue Replacement:	Y	Equipment/Supplies:	Y	Arrays:		Blood Collection:	
Development:				Research/Development Svcs.:		Database Management:		Drug Delivery:	
Generics:				Diagnostics:				Drug Distribution:	

TYPES OF BUSINESS:

Orthopedic Supplies
Electrical Bone Growth Stimulators
Orthopedic Support Devices
Operating Room Supplies
Powered Surgical Instruments
Arthroscopy Products
Imaging Equipment
Human Bone Joint Replacement Systems

BRANDS/DIVISIONS/AFFILIATES:

Anthrotek, Inc.
Walter Lorenz Surgical, Inc.
EBI, LP
Implant Innovations, Inc.
Regenerex
Biomet Orthopedics, Inc.
Maxim Total Knee System
Vanguard System

CONTACTS: Note: Officers with more than one job title may be intentionally listed here more than once.

Jeffrey R. Binder, CEO
Jeffrey R. Binder, Pres.
Daniel P. Florin, CFO/Sr. VP
Wilber C. Boren, Corp. VP-Contract Sales Admin.
Darlene Whaley, Sr. VP-Human Resources
Richard J. Borror, CIO
Lance Perry, Corp. VP-Global Prod. Dev. Reconstructive Devices
Richard J. Borror, Sr. VP-Mfg. Oper.
Daniel P. Hann, Exec. VP-Admin.
Bradley J. Tandy, General Counsel/Sr. VP/Sec.
Greg W. Sasso, Sr. VP-Corp. Dev.
Greg W. Sasso, Sr. VP-Corp. Comm.
J. Pat Richardson, VP-Finance/Treas.
Glen A. Kashuba, Sr. VP/Pres., Biomet Trauma & Biomet Spine
Steven F. Schiess, VP/Pres., Biomet 3i
Thomas R. Allen, Pres., Int'l Oper.-Americas & Asia Pacific
William C. Kolter, Pres., Biomet Orthopedics
Niles L. Noblitt, Chmn.
Roger P. van Broek, VP/Pres., Int'l Oper.

Phone: 574-267-6639	Fax: 574-267-8137

Toll-Free:

Address: 56 E. Bell Dr., Warsaw, IN 46582 US

GROWTH PLANS/SPECIAL FEATURES:

Biomet, Inc., founded in 1977, designs, manufactures and markets products that are used primarily by musculoskeletal medical specialists in both surgical and non-surgical therapy. The company's product portfolio encompasses reconstructive products, fixation devices, spinal products and other products. Biomet has four major market segments: reconstructive products, which accounted for 68% of the company's net revenues in 2006; fixation devices, which accounted for 12%; spinal products, which accounted for 11%, and other products, which account for 9%. Reconstructive products include knee, hip and extremity joint replacement systems, as well as dental reconstructive implants, bone cements and accessories. Fixation devices include electrical stimulation systems, internal and external fixation devices, craniomaxillofacial fixation systems and bone substitution materials. Spinal products include spinal fusion stimulation systems, spinal fixation systems, spinal bone substitution materials, precision machine allograft and motion preservation products. The other product market segment includes arthroscopy products, orthopedic support products, operating room supplies, casting materials, general surgical instruments and wound care products. Biomet manufactures numerous knee systems, including the Vanguard System, the Oxford Unicompartmental Knee, the Alpina Unicompartmental Knee, the Vanguard M System, the Repicci II Unicondylar Knee System and the Biomet OSS Orthopaedic Salvage System. Biomet's Arthrotek, Inc. subsidiary manufactures arthroscopy products in five product categories: power instruments, manual instruments, visualization products, soft tissue anchors and procedure-specific instruments and implants. In December 2006, Biomet agreed to be acquired for $10.9 billion by a private-equity group that includes affiliates of Blackstone Group, Goldman Sachs Capital Partners, Kohlberg Kravis Roberts & Co. and Texas Pacific Group. The private equity consortium increased its offer to $11.4 billion in June 2007. In May 2007, Biomet introduced the first vitamin E stabilized acetabular hip liners, which are expected to improve the longevity of the implant bearings used in total joint replacements.

FINANCIALS: Sales and profits are in thousands of dollars—add 000 to get the full amount. 2006 Note: Financial information for 2006 was not available for all companies at press time.

2006 Sales: $2,025,739	2006 Profits: $406,144	**U.S. Stock Ticker: BMET**
2005 Sales: $1,879,950	2005 Profits: $351,616	**Int'l Ticker:** Int'l Exchange:
2004 Sales: $1,615,751	2004 Profits: $325,627	Employees: 4,075
2003 Sales: $1,390,300	2003 Profits: $286,700	Fiscal Year Ends: 5/31
2002 Sales: $1,191,900	2002 Profits: $239,700	Parent Company:

SALARIES/BENEFITS:

Pension Plan:	ESOP Stock Plan:	Profit Sharing:	Top Exec. Salary: $358,800	Bonus: $250,000
Savings Plan:	Stock Purch. Plan:		Second Exec. Salary: $341,300	Bonus: $289,200

OTHER THOUGHTS:

Apparent Women Officers or Directors: 1

Hot Spot for Advancement for Women/Minorities:

LOCATIONS: ("Y" = Yes)

West:	Southwest:	Midwest:	Southeast:	Northeast:	International:
Y	Y	Y	Y	Y	Y

BIOMIRA INC

www.biomira.com

Industry Group Code: 325412 Ranks within this company's industry group: Sales: 143 Profits: 95

Drugs:		Other:	Clinical:	Computers:	Other:
Discovery:	Y	AgriBio:	Trials/Services:	Hardware:	Specialty Services:
Licensing:	Y	Genomics/Proteomics:	Laboratories:	Software:	Consulting:
Manufacturing:		Tissue Replacement:	Equipment/Supplies:	Arrays:	Blood Collection:
Development:	Y		Research/Development Svcs.:	Database Management:	Drug Delivery:
Generics:			Diagnostics:		Drug Distribution:

TYPES OF BUSINESS:

Drugs-Cancer
Cancer Vaccines
Drugs-Immunological

BRANDS/DIVISIONS/AFFILIATES:

STIMUVAX
ProlX Pharmaceutical Company

CONTACTS: *Note: Officers with more than one job title may be intentionally listed here more than once.*

Robert L. Kirkman, CEO
Gary Christianson, COO
Robert L. Kirkman, Pres.
Edward A. Taylor, CFO
D. Lynn Kirkpatrick, Chief Scientific Officer
Edward A. Taylor, VP-Admin.
Stephanie H. Seiler, Media Rel. Contact
Stephanie H. Seiler, Investor Rel. Contact
Edward A. Taylor, VP-Finance
Marilyn Olson, VP-Quality & Regulatory Affairs
R. Rao Koganty, VP/General Mgr.-Synthetic Biologics
Christopher S. Henney, Chmn.

Phone: 780-450-3761	Fax: 780-450-4772
Toll-Free: 877-234-0444	
Address: 2011 94th St., Edmonton, AB T6N 1H1 Canada	

GROWTH PLANS/SPECIAL FEATURES:

Biomira, Inc. focuses its biotechnology expertise on the development of cancer therapeutics. The company is seeking to develop vaccine-like treatments for cancer that will stimulate a patient's immune system to recognize and fight malignant cells. Stimuvax, a liposome vaccine and the firm's leading product candidate, is being developed in collaboration with Merck to generate a cellular immune response to the tumor-associated antigen mucin MUC1. It has completed its Phase 3 trials for metastatic non-small cell lung cancer and potential prostate cancer treatment, and has received fast track status from the FDA. In pre-clinical studies Stimuvax was able to prevent metastasis in lung cancer in mice. Biomira and EMD Pharmaceuticals, Inc., Merck's U.S. affiliate, intend to market the drug jointly in the U.S. Biomira's BGLP40 liposome vaccine has the potential for use in multiple forms of cancer and is in preclinical development. In conjunction with the merger of ProlX Pharmaceutical Company, Biomira has begun to develop small molecule drugs for the treatment of cancer. The company currently has three in its pipeline, with two in pre-clinical trials and one in Phase 2 study. In October 2006, Biomira acquired ProlX Pharmaceuticals for $3 million and a portion of common stock. With this acquisition, the company expanded its business to small molecule drugs used for the treatment of cancer. In February 2007, the first patient was enrolled into the Stimuvax Phase 3 trial that is being run by Merck KGaA, a German subsidiary of EMD Pharmaceuticals Inc.

FINANCIALS: Sales and profits are in thousands of dollars—add 000 to get the full amount. 2006 Note: Financial information for 2006 was not available for all companies at press time.

2006 Sales: $4,199	2006 Profits: $-16,591	**U.S. Stock Ticker: BIOM**
2005 Sales: $3,800	2005 Profits: $-16,400	**Int'l Ticker: BRA** Int'l Exchange: Toronto-TSX
2004 Sales: $7,500	2004 Profits: $-10,200	Employees: 235
2003 Sales: $2,700	2003 Profits: $-14,700	Fiscal Year Ends: 12/31
2002 Sales: $3,400	2002 Profits: $-19,900	Parent Company:

SALARIES/BENEFITS:

Pension Plan:	ESOP Stock Plan:	Profit Sharing:	Top Exec. Salary: $249,757	Bonus: $80,422
Savings Plan:	Stock Purch. Plan:		Second Exec. Salary: $209,818	Bonus: $51,143

OTHER THOUGHTS:

Apparent Women Officers or Directors: 3
Hot Spot for Advancement for Women/Minorities: Y

LOCATIONS: ("Y" = Yes)

West:	Southwest:	Midwest:	Southeast:	Northeast:	International:
	Y			Y	Y

BIOPURE CORPORATION   www.biopure.com

ndustry Group Code: 325412   Ranks within this company's industry group: Sales: 152   Profits: 118

Drugs:		Other:	Clinical:	Computers:	Other:
iscovery:	Y	AgriBio:	Trials/Services:	Hardware:	Specialty Services:
icensing:		Genomics/Proteomics:	Laboratories:	Software:	Consulting:
Manufacturing:	Y	Tissue Replacement:	Equipment/Supplies:	Arrays:	Blood Collection:
evelopment:	Y		Research/Development Svcs.:	Database Management:	Drug Delivery:
enerics:			Diagnostics:		Drug Distribution:

TYPES OF BUSINESS:

lood Transfusion Products
Oxygen Therapeutics
eterinary Drugs

BRANDS/DIVISIONS/AFFILIATES:

Hemopure
Oxyglobin
Restore Effective Survival in Shock (RESUS)

CONTACTS: Note: Officers with more than one job title may be intentionally listed here more than once.

afiris G. Zafirelis, CEO
afiris G. Zafirelis, Pres.
rancis H. Murphy, CFO
V. Richard Light, VP-Tech. Dev.
Geoffrey J. Filbey, VP-Eng.
ane Kober, General Counsel/Sr. VP/Sec.
arry L. Scott, VP-Bus. Dev.
irginia T. Rentko, VP-Preclinical Dev.
. Gerson Greenburg, VP-Medical Affairs
afiris G. Zafirelis, Chmn.

Phone: 617-234-6500	Fax: 617-234-6505

oll-Free:
ddress: 11 Hurley St., Cambridge, MA 02141 US

GROWTH PLANS/SPECIAL FEATURES:

Biopure Corporation develops, manufactures and markets oxygen therapeutics, a class of intravenous pharmaceuticals that increase oxygen transport to the body's tissues. Products include Hemopure for human use and Oxyglobin for veterinary use. Hemopure is approved in South Africa for treating adult surgical patients who are anemic and for eliminating, reducing or delaying the need for red blood cell transfusion in these patients. Biopure's current clinical development efforts for Hemopure are focused on a potential indication in cardiovascular ischemia, on supporting the U.S. Navy's government-funded efforts to develop a potential out-of-hospital trauma indication and include applying in the United Kingdom for regulatory approval of a proposed orthopedic surgical anemia indication. Once infused into a patient, Hemopure molecules disperse throughout the plasma space and are in continuous contact with the blood vessel wall, turning plasma into an oxygen-delivering substance which can bypass partial blockages or pass through constricted vessels that impede the normal passage of red blood cells. As an alternative to red blood cell transfusion, Hemopure's advantages include immediate onset of action, more efficient oxygen release, long shelf life, universal compatibility, low viscosity and room temperature storage. In December 2006, the Navy began discussions with the FDA on adapting its previous plans for RESUS (Restore Effective Survival in Shock), in which Hemopure would be developed for use in trauma patients at accident scenes or on the battlefield, as a Phase IIb/III trial to a smaller Phase II trial. Oxyglobin, the firm's other product, is used for the treatment of anemia in dogs and is marketed and sold to veterinary hospitals and small animal veterinary practices in the U.S. and Europe.

Biopure employee benefits include tuition reimbursement, dependent care reimbursement, paid personal days, transportation assistance and access to a credit union.

FINANCIALS: Sales and profits are in thousands of dollars—add 000 to get the full amount. 2006 Note: Financial information for 2006 was not available for all companies at press time.

006 Sales: $1,715	2006 Profits: $-26,454	U.S. Stock Ticker: BPUR
005 Sales: $2,110	2005 Profits: $-28,671	Int'l Ticker: Int'l Exchange:
004 Sales: $3,750	2004 Profits: $-41,665	Employees: 76
003 Sales: $4,019	2003 Profits: $47,042	Fiscal Year Ends: 10/31
002 Sales: $2,000	2002 Profits: $-45,800	Parent Company:

SALARIES/BENEFITS:

ension Plan:	ESOP Stock Plan:	Profit Sharing:	Top Exec. Salary: $253,016	Bonus: $
avings Plan: Y	Stock Purch. Plan:		Second Exec. Salary: $231,036	Bonus: $

OTHER THOUGHTS:

pparent Women Officers or Directors: 2
ot Spot for Advancement for Women/Minorities: Y

LOCATIONS: ("Y" = Yes)

West:	Southwest:	Midwest:	Southeast:	Northeast:	International:
				Y	

Note: Financial information, benefits and other data can change quickly and may vary from those stated here.

BIORELIANCE CORP   www.bioreliance.com

Industry Group Code: 621511 **Ranks within this company's industry group:** Sales: Profits:

Drugs:		Other:	Clinical:		Computers:		Other:
Discovery:		AgriBio:	Trials/Services:		Hardware:		Specialty Services:
Licensing:	Y	Genomics/Proteomics:	Laboratories:	Y	Software:		Consulting:
Manufacturing:		Tissue Replacement:	Equipment/Supplies:		Arrays:		Blood Collection:
Development:			Research/Development Svcs.:	Y	Database Management:		Drug Delivery:
Generics:			Diagnostics:				Drug Distribution:

TYPES OF BUSINESS:

Research-Nonclinical Product Testing
Contract Biologics Manufacturing
Biologics Pharmaceutical Services

BRANDS/DIVISIONS/AFFILIATES:

Invitrogen Corporation
BioReliance Laboratory Animal Diagnostic Services

CONTACTS: Note: Officers with more than one job title may be intentionally listed here more than once.

Ted Maker, General Mgr.
John Green, CFO
Diana Morgan, VP-Mktg. & Sales

Phone: 301-738-1000	Fax: 301-610-2590
Toll-Free: 800-553-5372	
Address: 14920 Broschart Rd., Rockville, MD 20850 US	

GROWTH PLANS/SPECIAL FEATURES:

BioReliance Corp., a subsidiary of Invitrogen, provides contract testing, development and manufacturing services for biotechnology and pharmaceutical companies worldwide. It also supports early-stage companies, which often lack the staff, expertise and financial resources to conduct many aspects of product development internally. BioReliance is the largest provider of outsourcing services focused on biologics, one of the pharmaceutical industry's fastest growing sectors. The company provides its services throughout the product cycle, starting from early preclinical development through licensed production. Through its testing and development division, the firm evaluates products to ensure that they are free of disease-causing agents and do not cause adverse effects; characterizes products chemical structures; develops formulations for long-term stability; and validates purification processes under regulatory guidelines. Testing services include assessments of the cell banks used to manufacture biologics; validation of purification processes for clearance of adventitious agents such as viruses; and testing of in-process and final products. Its Laboratory Animal Diagnostics Services department offers routine and emergency testing programs for both rodents and simians. In its manufacturing division, BioReliance develops unique production processes and manufactures biologics on behalf of clients both for use in clinical trials and for the worldwide commercial market.

As employees of Invitrogen, BioReliance's workers receive a full health package, education reimbursement, credit union membership and an employee assistance plan.

FINANCIALS: Sales and profits are in thousands of dollars—add 000 to get the full amount. 2006 Note: Financial information for 2006 was not available for all companies at press time.

2006 Sales: $	2006 Profits: $	**U.S. Stock Ticker:** Subsidiary		
2005 Sales: $	2005 Profits: $	**Int'l Ticker:** Int'l Exchange:		
2004 Sales: $	2004 Profits: $	Employees: 638		
2003 Sales: $	2003 Profits: $	Fiscal Year Ends: 12/31		
2002 Sales: $82,400	2002 Profits: $10,800	Parent Company: INVITROGEN CORPORATION		

SALARIES/BENEFITS:

Pension Plan:	ESOP Stock Plan:	Profit Sharing:	Top Exec. Salary: $401,923	Bonus: $181,149
Savings Plan: Y	Stock Purch. Plan: Y		Second Exec. Salary: $229,391	Bonus: $76,991

OTHER THOUGHTS:

Apparent Women Officers or Directors: 1
Hot Spot for Advancement for Women/Minorities:

LOCATIONS: ("Y" = Yes)

West:	Southwest:	Midwest:	Southeast:	Northeast:	International:
				Y	Y

Note: Financial information, benefits and other data can change quickly and may vary from those stated here.

BIOSITE INC

www.biosite.com

Industry Group Code: 325413   Ranks within this company's industry group: Sales: 6   Profits: 5

Drugs:	Other:	Clinical:		Computers:		Other:	
Discovery:	AgriBio:	Trials/Services:		Hardware:		Specialty Services:	Y
Licensing:	Genomics/Proteomics:	Laboratories:		Software:		Consulting:	Y
Manufacturing:	Tissue Replacement:	Equipment/Supplies:	Y	Arrays:		Blood Collection:	
Development:		Research/Development Svcs.:	Y	Database Management:		Drug Delivery:	
Generics:		Diagnostics:	Y			Drug Distribution:	

TYPES OF BUSINESS:

Medical Diagnostics Products
Rapid Immunoassays
Antibody Development Services

BRANDS/DIVISIONS/AFFILIATES:

Biosite Discovery
Triage Drugs of Abuse Panel
Triage Cardiac Panel
Triage TOX Drug Screen
Triage BNP Test
Triage Profiler Panels
Triage Parasite Panel
Triage D-Dimer Test

CONTACTS: Note: Officers with more than one job title may be intentionally listed here more than once.

Kim D. Blickenstaff, CEO
Kenneth F. Buechler, Pres./Chief Scientific Officer
Christopher J. Twomey, CFO
Robert Anacone, Sr. VP-Worldwide Mktg. & Sales
Paul H. McPherson, VP-R&D
Stephen Lesefko, VP-Eng.
David Berger, VP-Legal Affairs
Christopher R. Hibberd, Sr. VP-Corp. Dev.
Nadine E. Padilla, VP-Corp. Rel.
Nadine E. Padilla, VP-Corp. & Investor Rel.
Christopher J. Twomey, Sr. VP-Finance/Sec.
Robin G. Weiner, VP-Quality Assurance & Govt. Affairs
Gunars E. Valkirs, Sr. VP-Biosite Discovery
Thomas G. Blassey, VP-U.S. Sales
S. Elaine Walton, VP-Quality Assurance & Program Mgt.
Kim D. Blickenstaff, Chmn.
Gary A. King, VP-Int'l Oper.

Phone: 858-805-4808	Fax:

Toll-Free: 888-246-7483
Address: 9975 Summers Ridge Rd., San Diego, CA 92121 US

GROWTH PLANS/SPECIAL FEATURES:

Biosite, Inc. is a global diagnostics company dedicated to utilizing biotechnology in the development of diagnostic products. The firm validates and patents novel protein biomarkers and panels of biomarkers, develops and markets products, conducts strategic research on its products and educates healthcare providers about its products. Biosite markets immunoassay diagnostics in the areas of cardiovascular disease, drug overdose and infectious disease. Cardiovascular products account for 83% of product sales and include the Triage BNP Test, Triage Cardiac Panel, Triage Profiler Panels, Triage D-Dimer Test and Triage Stroke Panel. The Triage Drugs of Abuse Panel and Triage TOX Drug Screen are rapid, qualitative urine screens that test for up to nine different illicit and prescription drugs or drug classes and provide results in less than 15 minutes. The Triage C. Difficile Panel and Triage Parasite Panel aid in the diagnosis of infectious diseases. The firm's Biosite Discovery research business seeks to identify new protein markers of diseases that lack effective diagnostic tests. Additionally, with Biosite Discovery, the company has the capacity to offer antibody development services to companies seeking high-affinity antibodies for use in drug research. In return, Biosite seeks diagnostic licenses. In May 2006, the company withdrew its pre-market approval submission to the FDA for the Triage Stroke Panel. The withdrawal will allow the company time for an additional clinical study of the product, which the firm hopes will support a new U.S. regulatory submission. In May 2007, Biosite agreed to be acquired by Inverness Medical Innovations, Inc., a diagnostics developing company, for $92.50 per share.

Biosite offers employees a benefits package including flexible spending accounts, an employee assistance program, bereavement leave, education reimbursement, a 401(k) plan and an employee stock purchase plan.

FINANCIALS: Sales and profits are in thousands of dollars—add 000 to get the full amount. 2006 Note: Financial information for 2006 was not available for all companies at press time.

2006 Sales: $308,592	2006 Profits: $39,994	**U.S. Stock Ticker: BSTE**
2005 Sales: $287,699	2005 Profits: $54,029	**Int'l Ticker:** Int'l Exchange:
2004 Sales: $244,900	2004 Profits: $41,400	Employees: 1,036
2003 Sales: $173,364	2003 Profits: $24,763	Fiscal Year Ends: 12/31
2002 Sales: $105,200	2002 Profits: $13,400	Parent Company:

SALARIES/BENEFITS:

Pension Plan:	ESOP Stock Plan:	Profit Sharing:	Top Exec. Salary: $553,500	Bonus: $264,424
Savings Plan: Y	Stock Purch. Plan: Y		Second Exec. Salary: $430,961	Bonus: $203,944

OTHER THOUGHTS:

Apparent Women Officers or Directors: 3
Hot Spot for Advancement for Women/Minorities: Y

LOCATIONS: ("Y" = Yes)

West:	Southwest:	Midwest:	Southeast:	Northeast:	International:
Y					Y

Note: Financial information, benefits and other data can change quickly and may vary from those stated here.

BIOSPHERE MEDICAL INC   www.biospheremed.com

Industry Group Code: 339113 Ranks within this company's industry group: Sales: 13   Profits: 12

Drugs:		Other:		Clinical:		Computers:		Other:	
Discovery:	Y	AgriBio:		Trials/Services:		Hardware:		Specialty Services:	
Licensing:		Genomics/Proteomics:		Laboratories:		Software:		Consulting:	
Manufacturing:	Y	Tissue Replacement:		Equipment/Supplies:	Y	Arrays:		Blood Collection:	
Development:	Y			Research/Development Svcs.:		Database Management:		Drug Delivery:	
Generics:				Diagnostics:				Drug Distribution:	

TYPES OF BUSINESS:

Drugs-Bioengineered Microspheres
Cancer Treatments
Microsphere Delivery Systems

BRANDS/DIVISIONS/AFFILIATES:

Embosphere
EmboGold
HepaSphere SAP
EmboCath
Segway
Radiospheres
TempRx

CONTACTS: *Note: Officers with more than one job title may be intentionally listed here more than once.*

Richard J. Faleschini, CEO
Richard J. Faleschini, COO
Richard J. Faleschini, Pres.
Martin J. Joyce, CFO/Exec. VP
Peter Sutcliffe, VP-Mfg.
Martin J. Joyce, VP-Admin.
David P. Southwell, Chmn.

Phone: 781-681-7900	**Fax:** 781-792-2745
Toll-Free: 800-394-0295	
Address: 1050 Hingham St., Rockland, MA 02370 US	

GROWTH PLANS/SPECIAL FEATURES:

Biosphere Medical, Inc. develops, manufactures and markets products for medical procedures using embolotherapy techniques. Biosphere's core technology consists of patented bioengineered polymers, which are used to produce miniature spherical embolic particles called microspheres. Embosphere Microspheres and EmboGold Microspheres, the company's pioneering embolic products, are made of an acrylic co-polymer that is cross-linked with gelatin. Due to their uniform, spherical shape and soft slippery surface, microspheres are easy to inject through microcatheters, resulting in an even distribution within the vessel network. The microspheres selectively block the target tissue's blood supply, which destroys or devitalizes its target. Biosphere provides its products in calibrated size ranges, so they can be selected to target occlusion of specific sized vessels. The company's principle focus is the treatment of symptomatic uterine fibroids, which are noncancerous tumors growing within or on the wall of the uterus, using a procedure called uterine fibroid embolization. Uterine fibroid embolization is an alternative to hysterectomy and myomectomy that has been shown to allow faster recovery time and provide equivalent quality of life benefit. In addition to its microspheres, Biosphere produces two microcatheters: the Embocath Plus for general application and the Sequitor Guidewire for increased access to pelvic and visceral anatomy. Biosphere also produces the QuatraSphere product line for a number of other medical treatments, including the use of microspheres in the treatment of peripheral arteriovenous malformations and hypervascularized tumors like primary cancer of the liver. This product is known as Hepasphere in Japan and Europe. QuatraSphere microspheres work by absorbing contrast media and by expanding up to four times its dry diameter in order to increase surface area contact. In June 2006 BioSphere began participation in a Phase II clinical study for the use of its QuatraSpheres to treat patients with primary liver cancer in the US.

FINANCIALS: Sales and profits are in thousands of dollars—add 000 to get the full amount. 2006 Note: Financial information for 2006 was not available for all companies at press time.

2006 Sales: $22,891	2006 Profits: $-2,324	**U.S. Stock Ticker: BSMD**
2005 Sales: $18,484	2005 Profits: $-2,801	**Int'l Ticker:** Int'l Exchange:
2004 Sales: $14,158	2004 Profits: $-6,841	Employees: 83
2003 Sales: $12,803	2003 Profits: $-7,352	Fiscal Year Ends: 12/31
2002 Sales: $12,200	2002 Profits: $-6,400	Parent Company:

SALARIES/BENEFITS:

Pension Plan:	ESOP Stock Plan:	Profit Sharing:	Top Exec. Salary: $375,000	Bonus: $184,201
Savings Plan: Y	Stock Purch. Plan:		Second Exec. Salary: $260,018	Bonus: $99,125

OTHER THOUGHTS:

Apparent Women Officers or Directors:
Hot Spot for Advancement for Women/Minorities:

LOCATIONS: ("Y" = Yes)

West:	Southwest:	Midwest:	Southeast:	Northeast:	International:
				Y	Y

BIOTECH HOLDINGS LTD   www.biotechltd.com

Industry Group Code: 325412 Ranks within this company's industry group: Sales: 171   Profits: 60

Drugs:		Other:		Clinical:	Computers:		Other:
Discovery:	Y	AgriBio:		Trials/Services:	Hardware:		Specialty Services:
Licensing:		Genomics/Proteomics:		Laboratories:	Software:		Consulting:
Manufacturing:	Y	Tissue Replacement:		Equipment/Supplies:	Arrays:		Blood Collection:
Development:	Y			Research/Development Svcs.:	Database Management:		Drug Delivery:
Generics:				Diagnostics:			Drug Distribution:

TYPES OF BUSINESS:
Pharmaceuticals Development & Manufacturing
Drugs-Diabetes

BRANDS/DIVISIONS/AFFILIATES:
Sucanon
Glucanin
Diab II

CONTACTS: *Note: Officers with more than one job title may be intentionally listed here more than once.*
Robert B. Rieveley, CEO
Robert B. Rieveley, Pres.
Lorne D. Brown, CFO
Gale Belding, Exec. VP-Admin. & Regulatory Affairs
Robert B. Rieveley, Chmn.
Luis M. Ornelas, VP-Latin American Oper.

Phone: 604-295-1119	**Fax:** 604-295-1110

Toll-Free: 888-216-1111
Address: 3751 Shell Rd., Ste. 160, Richmond, BC V6X 2W2
Canada

GROWTH PLANS/SPECIAL FEATURES:
Biotech Holdings, Ltd., based in Canada, researches, develops, manufactures and markets pharmaceutical products. The firm's only marketed drug, trademarked as Sucanon and Diab II, treats Type II diabetes mellitus, known as non-insulin-dependent, the most common form of diabetes caused by an inborn lack or complete absence of natural insulin. Sucanon is an insulin sensitizer, so named because it increases the body's ability to absorb and use insulin, which results in decreased blood sugar levels, causing improved metabolism and an enhanced sense of well-being in patients. Sucanon is primarily marketed in China, Mexico and Brazil, but is licensed for sale in several other countries as well. The drug is one of only three of its kind approved anywhere in the world. The firm has product facilities in Canada and Mexico and is seeking drug acceptance in the U.S. Sucanon is marketed and distributed through retail drug stores throughout Latin America and Mexico, including Mexico's largest chain drug store. Biotech Holdings entered into a major sales agreement for Sucanon with Wal-Mart's Mexican division in April 2005, and the drug is now the subject of a national advertising campaign. In June 2007, the company launched an Internet website to sell its drug in the U.S. and Canada, making the drug available for personal use.

FINANCIALS: Sales and profits are in thousands of dollars—add 000 to get the full amount. 2006 Note: Financial information for 2006 was not available for all companies at press time.

2006 Sales: $ 480	2006 Profits: $-1,940	**U.S. Stock Ticker: BIOHF.OB**
2005 Sales: $	2005 Profits: $-1,520	**Int'l Ticker: BIO** Int'l Exchange: Toronto-TSX
2004 Sales: $	2004 Profits: $-1,800	Employees: 4
2003 Sales: $	2003 Profits: $-2,000	Fiscal Year Ends: 3/31
2002 Sales: $1,006	2002 Profits: $-2,140	Parent Company:

SALARIES/BENEFITS:

Pension Plan:	ESOP Stock Plan:	Profit Sharing:	Top Exec. Salary: $	Bonus: $
Savings Plan:	Stock Purch. Plan:		Second Exec. Salary: $	Bonus: $

OTHER THOUGHTS:
Apparent Women Officers or Directors: 2
Hot Spot for Advancement for Women/Minorities: Y

LOCATIONS: ("Y" = Yes)

West:	Southwest:	Midwest:	Southeast:	Northeast:	International:
					Y

BIOTIME INC

www.biotimeinc.com

Industry Group Code: 325412 Ranks within this company's industry group: Sales: 156   Profits: 59

Drugs:		Other:		Clinical:		Computers:		Other:	
Discovery:	Y	AgriBio:		Trials/Services:		Hardware:		Specialty Services:	
Licensing:	Y	Genomics/Proteomics:		Laboratories:		Software:		Consulting:	
Manufacturing:		Tissue Replacement:		Equipment/Supplies:	Y	Arrays:		Blood Collection:	
Development:	Y			Research/Development Svcs.:		Database Management:		Drug Delivery:	
Generics:				Diagnostics:				Drug Distribution:	

TYPES OF BUSINESS:

Drugs-Surgical
Blood Plasma Expanders
Blood Replacement Solutions

BRANDS/DIVISIONS/AFFILIATES:

Hextend
PentaLyte
HetaCool

CONTACTS: *Note: Officers with more than one job title may be intentionally listed here more than once.*

Steven A. Seinberg, CFO/Treas.
Jeffrey Nickel, VP-Mktg.
Hal Sternberg, VP-Research
Harold Waitz, VP-Eng. & Regulatory Affairs
Judith Segall, VP-Oper./Corp. Sec.
Jeffrey Nickel, VP-Bus. Dev.
Judith Segall, Press Contact
Mal Sternberg, Member-Office of the Pres.
Harold Waitz, Member-Office of the Pres.
Judith Segall, Member-Office of the Pres.
Steven A. Seinberg, Member-Office of the Pres.

Phone: 510-350-2940	Fax: 510-350-2948
Toll-Free:	
Address: 6121 Hollis St., Emeryville, CA 94608 US	

GROWTH PLANS/SPECIAL FEATURES:

BioTime, Inc. is a development-stage company engaged in doing research and development on aqueous-based synthetic surgery solutions. These products can be used as blood plasma expanders and blood replacement solutions during hypothermic surgery and organ preservation. The firm's specially formulated hypothermic blood substitute solution would be used for the replacement of very large volumes of patients' blood during cardiac surgery, neurosurgery and other surgeries that involve lowering body temperature to hypothermic levels. Hextend, the company's first product, is a physiologically-balanced blood plasma volume expander for the treatment of hypovolemia, a condition associated with blood loss during surgery. Hextend maintains circulatory system fluid volume and keeps vital organs perfused during surgery. The product is sold in the U.S. and Canada through Abbott Laboratories' spin-off business called Hospira, Inc. and in South Korea by C. Corp. under exclusive licenses from the company. BioTime is developing two other blood volume replacement products, PentaLyte and HetaCool. The company in the midst of Phase 2 clinical trials of PentaLyte and intends to test the product for the treatment of hypovolemia. HetaCool, a modified formulation of Hextend specifically designed for use at low temperatures, is currently in the discovery phase, as is HetaFreeze, which may extend the time during which organs and tissues can be stored for future transplant or surgical grafting. BioTime has an agreement with Summit Pharmaceuticals International Corporation to develop Hextend and PentaLyte in Japan, the People's Republic of China and Taiwan. Other potential uses of BioTime's products include use during anti-cancer therapies. In December 2006, the company announced that a Phase 2 clinical trial for Hextend is being conducted in Japan by Summit.

FINANCIALS: Sales and profits are in thousands of dollars—add 000 to get the full amount. 2006 Note: Financial information for 2006 was not available for all companies at press time.

2006 Sales: $1,162	2006 Profits: $-1,864	U.S. Stock Ticker: BTIM.OB
2005 Sales: $ 903	2005 Profits: $-2,074	Int'l Ticker: Int'l Exchange:
2004 Sales: $ 688	2004 Profits: $-3,085	Employees: 9
2003 Sales: $ 556	2003 Profits: $-1,742	Fiscal Year Ends: 12/31
2002 Sales: $ 400	2002 Profits: $-2,800	Parent Company:

SALARIES/BENEFITS:

Pension Plan:	ESOP Stock Plan:	Profit Sharing:	Top Exec. Salary: $108,000	Bonus: $
Savings Plan:	Stock Purch. Plan:		Second Exec. Salary: $100,000	Bonus: $

OTHER THOUGHTS:

Apparent Women Officers or Directors: 1
Hot Spot for Advancement for Women/Minorities:

LOCATIONS: ("Y" = Yes)

West:	Southwest:	Midwest:	Southeast:	Northeast:	International:
Y					

BIOVAIL CORPORATION

www.biovail.com

Industry Group Code: 325412A Ranks within this company's industry group: Sales: 1   Profits: 1

Drugs:		Other:		Clinical:		Computers:		Other:	
Discovery:		AgriBio:		Trials/Services:	Y	Hardware:		Specialty Services:	
Licensing:		Genomics/Proteomics:		Laboratories:	Y	Software:		Consulting:	
Manufacturing:	Y	Tissue Replacement:		Equipment/Supplies:		Arrays:		Blood Collection:	
Development:	Y			Research/Development Svcs.:	Y	Database Management:		Drug Delivery:	Y
Generics:	Y			Diagnostics:				Drug Distribution:	

TYPES OF BUSINESS:

Drug Delivery Systems Technologies
Generic Drugs
Drugs-Hypertension
Drugs-Antidepressants
Nutraceuticals
Contract Research Services

BRANDS/DIVISIONS/AFFILIATES:

Biovail Pharmaceuticals, Inc.
Biovail Pharmaceuticals Canada
Shearform
Zero Order Release System
Wellbutrin XL
FlashDose
Cardizem LA
Zovirax

CONTACTS: Note: Officers with more than one job title may be intentionally listed here more than once.

Douglas J. P. Squires, CEO
Gilbert Godin, COO/Exec. VP
Kenneth G. Howling, CFO/Sr. VP
Mark Durham, Sr. VP-Human Resources
Peter Silverstone, Sr. VP-Scientific & Medical Affairs
Mark Durham, Sr. VP-IT
John Sebben, VP-Mfg.
Wendy Kelley, General Counsel/Sr. VP/Corp. Sec.
Gregory Gubitz, Sr. VP-Corp. Dev.
Nelson F. Isabel, VP-Corp. Comm.
Nelson F. Isabel, VP-Investor Rel.
Christopher Bovaird, VP-Corp. Finance
John Miszuk, VP/Controller
Adrian de Saldanha, VP-Finance/Treas.
David Keefer, Sr. VP-Commercial Oper.
Christine Mayer, Sr. VP-Bus. Dev., Biovail Pharmaceuticals, Inc.
Douglas J.P. Squires, Interim Chmn.
Michel Chouinard, COO-Biovail Laboratories Int'l SRL

Phone: 905-286-3000	Fax: 905-286-3050

Toll-Free:

Address: 7150 Mississauga Rd., Mississauga, ON L5N 8M5 Canada

GROWTH PLANS/SPECIAL FEATURES:

Biovail Corporation is a full-service pharmaceutical company that develops, tests and commercializes its proprietary drug delivery technologies to improve the clinical effectiveness of medications. The firm's primary areas of focus include cardiovascular disease, Type II diabetes, central nervous system disorders and pain management. Biovail markets its products through its marketing divisions, Biovail Pharmaceuticals, Inc. and Biovail Pharmaceuticals Canada, and through other strategic partners to health care professionals. The company primarily employs its drug delivery technologies to develop enhanced formulations and controlled-release generic versions of existing and pre-market oral medications. These delivery technologies include controlled release, graded release, enhanced absorption, rapid absorption, taste masking and oral disintegration processes. Trademarks for these technologies include FlashDose, CEFORM, Shearform, Consurf and Zero Order Release System (ZORS). Biovail's products include Wellbutrin XL, which is marketed worldwide through GlaxoSmithKline to treat depression through a once-daily formulation. Other noteworthy products produced by the firm include Cardizem LA, a hypertension medication, and Zovirax, an antiviral ointment. Biovail currently maintains fully integrated pharmaceutical manufacturing facilities in Canada, Ireland, North Carolina, Barbados and Puerto Rico. In addition, the company operates a contract research division that provides Biovail and other pharmaceutical companies with a broad range of Phase I/II clinical research services in pharmacokinetic studies and bioanalytical laboratory testing. Biovail also owns Nutravail Technologies, Inc., which specializes in the development of over 80 patented nutraceuticals, nutritional products and functional foods. In May 2006, Biovail Pharmaceuticals Canada entered into an agreement with Novartis Pharmaceuticals Canada, Inc. to market and promote Lescol fluvastatin sodium capsules and once-daily Lescol XL to Canadian physicians. Biovail's subsidiary, Biovail Pharmaceuticals, Inc. entered into an exclusive promotional services agreement in late 2006 with Sciele Pharma, Inc. to promote the company's Zovirax Ointment and Zovirax Cream to the U.S.

FINANCIALS: Sales and profits are in thousands of dollars—add 000 to get the full amount. 2006 Note: Financial information for 2006 was not available for all companies at press time.

2006 Sales: $1,070,500	2006 Profits: $203,900	U.S. Stock Ticker: BVF
2005 Sales: $935,500	2005 Profits: $89,000	Int'l Ticker: BVF Int'l Exchange: Toronto-TSX
2004 Sales: $886,500	2004 Profits: $161,000	Employees: 1,734
2003 Sales: $823,700	2003 Profits: $-27,300	Fiscal Year Ends: 10/31
2002 Sales: $788,000	2002 Profits: $87,800	Parent Company:

SALARIES/BENEFITS:

Pension Plan:	ESOP Stock Plan:	Profit Sharing:	Top Exec. Salary: $750,607	Bonus: $
Savings Plan:	Stock Purch. Plan:		Second Exec. Salary: $700,000	Bonus: $525,000

OTHER THOUGHTS:

Apparent Women Officers or Directors: 3
Hot Spot for Advancement for Women/Minorities: Y

LOCATIONS: ("Y" = Yes)

West:	Southwest:	Midwest:	Southeast:	Northeast:	International:
				Y	Y

Note: Financial information, benefits and other data can change quickly and may vary from those stated here.

BRADLEY PHARMACEUTICALS INC   www.bradpharm.com

Industry Group Code: 325412 Ranks within this company's industry group: Sales: 60   Profits: 48

Drugs:	Other:	Clinical:	Computers:	Other:	
Discovery:	AgriBio:	Trials/Services:	Hardware:	Specialty Services:	Y
Licensing:	Genomics/Proteomics:	Laboratories:	Software:	Consulting:	
Manufacturing:	Tissue Replacement:	Equipment/Supplies:	Arrays:	Blood Collection:	
Development: Y		Research/Development Svcs.:	Database Management:	Drug Delivery:	
Generics:		Diagnostics:		Drug Distribution:	

TYPES OF BUSINESS:

Pharmaceutical Acquisitions & Marketing

BRANDS/DIVISIONS/AFFILIATES:

Kenwood Therapeutics
Doak Dermatologics
A. Aarons, Inc.
MediGene AG
BioSante Pharmaceuticals
Veregen
Elestrin
Solaraze

CONTACTS: *Note: Officers with more than one job title may be intentionally listed here more than once.*

Daniel Glassman, CEO
Daniel Glassman, Pres.
R. Brent Lenczycki, CFO/VP
Bradley Glassman, Sr. VP-Sales & Mktg.
Ralph Landau, Chief Scientific Officer/VP
Alan Goldstein, VP-Corp. Dev.
Alton Delane, VP-Bus. Dev.
Seth W. Hamot, Interim Chmn.

Phone: 973-882-1505	Fax: 973-575-5366
Toll-Free:	
Address: 383 Route 46 W., Fairfield, NJ 07004 US	

GROWTH PLANS/SPECIAL FEATURES:

Bradley Pharmaceuticals, Inc. is a specialty pharmaceutica company that acquires, develops and markets prescription and over-the-counter products in niche therapeutic markets including dermatology, podiatry, gastroenterology and women's health. The company operates in two business segments: the Doak Dermatologics subsidiary, which specializes in therapies for dermatology and podiatry; and the Kentwood Therapeutics division, which provides gastroenterology, women's health, respiratory and other internal medicine brands. Kentwood Therapeutics, to a lesser extent, also markets nutritional supplements and respiratory products. Bradley's core branded products include Adoxa, used for the treatment of severe acne; Kero and Keralac, topical moisturizing therapies; Solaraze Gel, a dermatological product used to treat actinic keratosis Zoderm cleanser, cream, gel and redi-pads, benzoy peroxide topical prescription products that help treat acne Lidamantle and Lidamantle HC, prescription products with a topical anesthetic that help relive pain, soreness, itching and irritation caused by insect bites, eczema and other skin conditions; and Pamine and Pamine Forte, prescription lactose-free, antispasmodic oral therapies that are indicated for the adjunctive therapy of peptic ulcer. In 2006, Bradley launched the A. Aarons, Inc. subsidiary, which markets authorized genetic versions of Doak's and Kentwood's products. In addition, the company entered into agreements with MediGene AG and BioSante Pharmaceuticals respectively, to commercialize Veregen, a topical ointmen indicated for the treatment of external genital and perianal warts, and Elestrin, an estradiol transdermal gel indicated for the treatment of moderate-to-severe vasomotor symptoms (hot flashes) in menopausal women. In June 2007, Doak Dermatologics launched two new product line extensions from its Kerol and Rosular brand portfolios.

The company offers its employees health, dental, vision and life insurance; short- and long-term disability; and a 401(k plan.

FINANCIALS: Sales and profits are in thousands of dollars—add 000 to get the full amount. 2006 Note: Financial information for 2006 was not available for all companies at press time.

2006 Sales: $144,807	2006 Profits: $9,668	**U.S. Stock Ticker: BDY**
2005 Sales: $133,382	2005 Profits: $7,962	**Int'l Ticker:** Int'l Exchange:
2004 Sales: $96,694	2004 Profits: $7,954	Employees: 300
2003 Sales: $74,679	2003 Profits: $16,825	Fiscal Year Ends: 12/31
2002 Sales: $39,700	2002 Profits: $7,600	Parent Company:

SALARIES/BENEFITS:

Pension Plan:	ESOP Stock Plan:	Profit Sharing:	Top Exec. Salary: $691,996	Bonus: $140,941
Savings Plan: Y	Stock Purch. Plan:		Second Exec. Salary: $278,114	Bonus: $65,514

OTHER THOUGHTS:

Apparent Women Officers or Directors:
Hot Spot for Advancement for Women/Minorities:

LOCATIONS: ("Y" = Yes)

West:	Southwest:	Midwest:	Southeast:	Northeast:	International:
				Y	

BRISTOL MYERS SQUIBB CO   www.bms.com

Industry Group Code: 325412 Ranks within this company's industry group: Sales: 11   Profits: 15

Drugs:		Other:		Clinical:		Computers:		Other:	
Discovery:	Y	AgriBio:		Trials/Services:		Hardware:		Specialty Services:	
Licensing:		Genomics/Proteomics:		Laboratories:		Software:		Consulting:	
Manufacturing:	Y	Tissue Replacement:		Equipment/Supplies:		Arrays:	Y	Blood Collection:	
Development:	Y			Research/Development Svcs.:		Database Management:		Drug Delivery:	
Generics:				Diagnostics:				Drug Distribution:	

TYPES OF BUSINESS:
Drugs-Diversified
Medical Imaging Products
Nutritional Products
Wound Care Products
Over-the-Counter Medicines

BRANDS/DIVISIONS/AFFILIATES:
Mead Johnson Nutritionals
ConvaTec
Enfamil
Cardiolite
Excedrin
TAXOL
Pravachol
Glucophage

CONTACTS: *Note: Officers with more than one job title may be intentionally listed here more than once.*
James M. Cornelius, CEO
Andrew R. J. Bonfield, CFO/Exec. VP
Wendy L. Dixon, Chief Mktg. Officer/Pres., Global Mktg.
Stephen E. Bear, Sr. VP-Human Resources
Elliott Sigal, Chief Scientific Officer
Susan O'Day, CIO
Carlo de Notaristefani, Pres., Tech. Oper.
Sandra Leung, Sr. VP/General Counsel
Robert T. Zito, Sr. VP-Corp. & Bus. Comm.
Anthony C. Hooper, Pres., U.S. Pharmaceuticals
Elliott Sigal, Pres., Pharmaceutical Research Institute
Jonathan K. Sprole, VP/Chief Compliance Officer
Pete Paradossi, Health Care Group
James D. Robinson, III, Chmn.
Lamberto Andreotti, Exec. VP/Pres., Worldwide Pharmaceuticals

Phone: 212-546-4000	Fax: 212-546-4020
Toll-Free:	
Address: 345 Park Ave., New York, NY 10154-0037 US	

GROWTH PLANS/SPECIAL FEATURES:
Bristol-Myers Squibb is one of the largest pharmaceutical and health care products companies in the world. The company operates through three segments: pharmaceuticals, nutritionals and other health care. The pharmaceutical segment, which accounts for 77% of total sales, discovers, develops, licenses, manufactures, markets, distributes and sells branded pharmaceuticals worldwide. It is a leading provider of drugs for anti-cancer therapies (TAXOL, Paraplatin and Erbitux) and treatments for high blood pressure (Avapro and Monopril), high cholesterol (Pravachol), stroke (Plavix), deep venous thrombosis/pulmonary embolism (Coumadin), type-2 diabetes (Glucophage), HIV/AIDS (Sustiva, Reyataz and Zerit), infectious diseases (Cefzil and Tequin) and schizophrenia (Abilify). The nutritionals segment, through Mead Johnson Nutritionals, manufactures and markets infant formulas, including the Enfamil line of products and children's nutritionals. The other health care segment consists of the ConvaTec, Medical Imaging and Consumer Medicines businesses. ConvaTec provides ostomy care and modern wound care products under the Natura, Sur-Fit, Esteem, Aquacel and DuoDerm brands. Medical Imaging manufactures, distributes and sells imaging products under the Cardiolite and Definity brands. Consumer Medicines manufactures, distributes and sells over-the-counter products including Excedrin, Bufferin, the Keri line of moisturizers and Comtrex for cold, cough and flu. Recently, the company partnered with AstraZeneca to develop Type 2 diabetes compounds; with Gilead Sciences to commercialize Atripla (the first once-daily single tablet HIV-1 treatment) in Canada; and with Adnexus to develop adnectin-based therapeutics for oncology-related patients. In addition, the company partnered with Onmark in 2007 to develop anti-cancer products. In recent news, the firm sold its inventory, trademark, patent and intellectual property rights related to DOVONEX, a treatment for psoriasis, to Warner Chilcott Company for $200 million in cash. In 2007, the company bought 89 acres of land for a new biologics facility that will be complete in 2010.

Bristol-Myers Squibb offers its employees medical plans and dependent life insurance.

FINANCIALS: Sales and profits are in thousands of dollars—add 000 to get the full amount. 2006 Note: Financial information for 2006 was not available for all companies at press time.

2006 Sales: $17,914,000	2006 Profits: $1,585,000	**U.S. Stock Ticker: BMY**
2005 Sales: $19,207,000	2005 Profits: $3,000,000	**Int'l Ticker:** Int'l Exchange:
2004 Sales: $19,380,000	2004 Profits: $2,388,000	Employees: 43,000
2003 Sales: $20,671,000	2003 Profits: $2,952,000	Fiscal Year Ends: 12/31
2002 Sales: $18,119,000	2002 Profits: $2,066,000	Parent Company:

SALARIES/BENEFITS:
Pension Plan: Y	ESOP Stock Plan:	Profit Sharing:	Top Exec. Salary: $1,250,000	Bonus: $2,224,875
Savings Plan: Y	Stock Purch. Plan:		Second Exec. Salary: $829,806	Bonus: $723,701

OTHER THOUGHTS:
Apparent Women Officers or Directors: 3
Hot Spot for Advancement for Women/Minorities: Y

LOCATIONS: ("Y" = Yes)
West:	Southwest:	Midwest:	Southeast:	Northeast:	International:
Y	Y	Y	Y	Y	Y

Note: Financial information, benefits and other data can change quickly and may vary from those stated here.

BURRILL & COMPANY   www.burrillandco.com

Industry Group Code: 523110 Ranks within this company's industry group: Sales: Profits:

Drugs:	Other:	Clinical:	Computers:	Other:	
Discovery:	AgriBio:	Trials/Services:	Hardware:	Specialty Services:	Y
Licensing:	Genomics/Proteomics:	Laboratories:	Software:	Consulting:	Y
Manufacturing:	Tissue Replacement:	Equipment/Supplies:	Arrays:	Blood Collection:	
Development:		Research/Development Svcs.:	Database Management:	Drug Delivery:	
Generics:		Diagnostics:		Drug Distribution:	

TYPES OF BUSINESS:

Investment Banking-Life Sciences
Strategic Partnership & Spin-Off/Outlicensing Consulting
Life Sciences Publications
Industry Conferences
Venture Capital Funds

BRANDS/DIVISIONS/AFFILIATES:

Burrill Life Sciences Capital Fund
Burrill Biotechnology Capital Fund
Burrill Agbio Capital Fund
Biotech Meeting at Laguna Niguel
Burrill Nutraceuticals Capital Fund
China Life Sciences Partnership Meeting
India Life Sciences Partnering Meeting
Stem Cell Summit

CONTACTS: Note: Officers with more than one job title may be intentionally listed here more than once.

G. Steven Burrill, CEO
Giovanni A. Ferrara, Managing Dir.-Venture Group
James D. Watson, Head-Merchant Banking
Jeff Miller, Dir.-Media Group
Ann F. Hanham, Managing Dir.-Venture Group

Phone: 415-591-5400	Fax: 415-591-5401
Toll-Free:	
Address: 1 Embarcadero Center, Ste. 2700, San Francisco, CA 94111 US	

GROWTH PLANS/SPECIAL FEATURES:

Burrill & Company is a life sciences merchant bank focused exclusively on companies involved in biotechnology, pharmaceuticals, drug delivery devices, diagnostics, medical devices, human health care and related medical technologies, nutraceuticals, agricultural biotechnologies and industrial biomaterials and bioprocesses. The company operates through several business units: venture capital, merchant banking, publications and conferences. The venture capital unit manages and offers various capital funds, including the Burrill Life Sciences Capital Fund, Burrill Biotechnology Capital Fund, Burrill Agbio Capital Fund I and II and Burrill Nutraceuticals Capital Fund. The merchant banking unit assists life science companies in identifying, negotiation and forming strategic partnerships with other companies for access to resources, technologies or collaborations. The unit also works with major life sciences companies to spin off divisions or outlicense technologies. The publications unit publishes monthly indices on biotech industry stock market performance; quarterly reports that highlight important industry developments such as advancements in science, technology breakthroughs and important business transactions and deals; articles and commentary on the biotechnology industry for various publications; and annual biotechnology industry reports. The conference unit annually hosts and sponsors various industry conferences including the Biotech Meeting at Laguna Niguel, the India Life Sciences Partnering Meeting, the Stem Cell Summit, the Japan Biotech Meeting, The China Life Sciences Partnership Meeting and the Indiana Life Sciences Forum.

FINANCIALS: Sales and profits are in thousands of dollars—add 000 to get the full amount. 2006 Note: Financial information for 2006 was not available for all companies at press time.

2006 Sales: $	2006 Profits: $	U.S. Stock Ticker: Private
2005 Sales: $	2005 Profits: $	Int'l Ticker: Int'l Exchange:
2004 Sales: $	2004 Profits: $	Employees: 40
2003 Sales: $13,300	2003 Profits: $	Fiscal Year Ends: 12/31
2002 Sales: $	2002 Profits: $	Parent Company:

SALARIES/BENEFITS:

Pension Plan:	ESOP Stock Plan:	Profit Sharing:	Top Exec. Salary: $	Bonus: $
Savings Plan:	Stock Purch. Plan:		Second Exec. Salary: $	Bonus: $

OTHER THOUGHTS:

Apparent Women Officers or Directors: 1
Hot Spot for Advancement for Women/Minorities:

LOCATIONS: ("Y" = Yes)

West:	Southwest:	Midwest:	Southeast:	Northeast:	International:
Y					

CALIPER LIFE SCIENCES

www.calipertech.com

Industry Group Code: 325413 **Ranks within this company's industry group:** Sales: 12 Profits: 24

Drugs:	Other:	Clinical:	Computers:		Other:	
Discovery:	AgriBio:	Trials/Services:	Hardware:	Y	Specialty Services:	
Licensing:	Genomics/Proteomics:	Laboratories:	Software:	Y	Consulting:	
Manufacturing:	Tissue Replacement:	Equipment/Supplies:	Y	Arrays:	Y	Blood Collection:	
Development:		Research/Development Svcs.:	Database Management:		Drug Delivery:	
Generics:		Diagnostics:			Drug Distribution:	

TYPES OF BUSINESS:

Bioanalysis Equipment
Microfluidic Systems
High-Throughput Screening Machines
Liquid Handling Systems
Drug Discovery Platforms
Laboratory Automation Solutions
Software

BRANDS/DIVISIONS/AFFILIATES:

Zymark Corp.
Caliper Technologies
LabChip 90
Lab Chip 3000
NovaScreen Biosciences Corp.
Xenogen Corp.

CONTACTS: *Note: Officers with more than one job title may be intentionally listed here more than once.*

E. Kevin Hrusovsky, CEO
E. Kevin Hrusovsky, Pres.
Thomas T. Higgins, CFO/Exec. VP
Mark Roskey, VP-Worldwide Mktg.
Paula Cassidy, VP-Human Resources
Enrique Bernal, VP-Instrument R&D
Steve Creager, General Counsel/Corp. Sec./Sr. VP
Bruce Bal, Sr. VP-Oper. & Aftermarket Bus.
William C. Kruka, Sr. VP-Bus. Dev.
David Manyak, Exec. VP-Drug Discovery Svcs.
Andrea Chow, VP-Microfluidics R&D
Auro Nair, VP-North American Sales
Mark T. Roskey, VP-Reagents and Applied Biology
Daniel L. Kisner, Chmn.
Jean-Louis Rufener, VP-Int'l Oper.

Phone: 508-435-9500	Fax: 508-435-3439
Toll-Free:	
Address: 68 Elm St., Hopkinton, MA 01748 US	

GROWTH PLANS/SPECIAL FEATURES:

Caliper Life Sciences uses its core technologies of liquid handling, automation and LabChip microfluidics to foster developments in the life sciences industry. The company manufactures high-throughput screening machines, automated liquid handling machines, micro-plate management, pharmaceutical development and quality control systems. Caliper is best known for its LabChip systems, which are designed to accelerate laboratory experimentation with applicability in the pharmaceutical and diagnostics industries. The firm makes the following two types of LabChip systems: LabChip 90 and LabChip 3000. LabChip 90, which is designed to meet the high through put needs of laboratories, uses microfluidic technology to automate the analysis of proteins and DNA fragments. The LabChip 3000 drug discovery system miniaturizes, integrates and automates enzymatic and cell-based assays even when unattended. LabChip assays are separations-based, so the quality of results exceeds what is achievable in homogeneous, well-based assays. Other products include various plate management; pharmaceutical development; evaporation and solid phase extraction devices; software; and workstations. In early 2006, Caliper acquired Xenogen Corp., a maker of advanced imaging systems including instruments, biological solutions and software designed to for drug development and discovery.

Caliper offers employees tuition reimbursement for job-related courses.

FINANCIALS: Sales and profits are in thousands of dollars—add 000 to get the full amount. 2006 Note: Financial information for 2006 was not available for all companies at press time.

2006 Sales: $107,871	2006 Profits: $-28,934	U.S. Stock Ticker: CALP	
2005 Sales: $87,009	2005 Profits: $-14,457	Int'l Ticker:	Int'l Exchange:
2004 Sales: $80,127	2004 Profits: $-31,600	Employees: 550	
2003 Sales: $49,411	2003 Profits: $-49,337	Fiscal Year Ends: 12/31	
2002 Sales: $25,800	2002 Profits: $-41,000	Parent Company:	

SALARIES/BENEFITS:

Pension Plan:	ESOP Stock Plan:	Profit Sharing:	Top Exec. Salary: $413,333	Bonus: $312,000
Savings Plan: Y	Stock Purch. Plan: Y		Second Exec. Salary: $265,833	Bonus: $85,233

OTHER THOUGHTS:

Apparent Women Officers or Directors: 2
Hot Spot for Advancement for Women/Minorities: Y

LOCATIONS: ("Y" = Yes)

West:	Southwest:	Midwest:	Southeast:	Northeast:	International:
Y				Y	Y

Note: Financial information, benefits and other data can change quickly and may vary from those stated here.

CAMBREX CORP

www.cambrex.com

Industry Group Code: 325412 Ranks within this company's industry group: Sales: 42   Profits: 127

Drugs:		Other:	Clinical:		Computers:		Other:
Discovery:		AgriBio:	Trials/Services:	Y	Hardware:		Specialty Services:
Licensing:		Genomics/Proteomics:	Laboratories:		Software:		Consulting:
Manufacturing:	Y	Tissue Replacement:	Equipment/Supplies:	Y	Arrays:		Blood Collection:
Development:	Y		Research/Development Svcs.:	Y	Database Management:		Drug Delivery:
Generics:			Diagnostics:				Drug Distribution:

TYPES OF BUSINESS:

Contract Pharmaceutical Manufacturing
Contract Research
Pharmaceutical Ingredients
Testing Products & Services
Technical Support

BRANDS/DIVISIONS/AFFILIATES:

Cambrex Bio Science Walkersville, Inc.
FlashGel
Platimun UltraPAK

CONTACTS: Note: Officers with more than one job title may be intentionally listed here more than once.

James A. Mack, CEO
Gary L. Mossman, COO/Exec. VP
James A. Mack, Pres.
Greg Sargen, CFO/Exec. VP
Melissa M. Lesko, VP-Human Resources
Robert J. Conguisti, VP-IT
Peter E. Thauer, General Counsel/Sr. VP/Corp. Sec.
Luke M. Beshar, VP-Corp. Dev.
Anup Gupta, VP-Finance
Steven M. Klosk, Exec. VP/COO-Biopharma Business Unit
Paolo Russolo, Pres., Cambrex Profarmaco Milano
Charles W. Silvey, VP-Internal Audit
James A. Mack, Chmn.

Phone: 201-804-3000	Fax: 201-804-9852
Toll-Free:	
Address: 1 Meadowlands Plaza, E. Rutherford, NJ 07073 US	

GROWTH PLANS/SPECIAL FEATURES:

Cambrex Corporation provides products and services to aid and enhance the discovery and commercialization of therapeutics. The company offers a variety of outsourcing products and services for drug discovery research and therapeutic testing. The firm manufactures more than 1,800 products, which are sold to more than 14,000 customers worldwide, including research organizations, pharmaceutical, biopharmaceutical and generic drug companies. Outsourcing options include bulk biologics manufacturing; development, manufacturing and commercialization services for cell-based therapeutics and pharmaceutical products; and testing services including assays for microbiology, sterility and veterinary services. Products offered for drug discovery research include bioassays; cell model systems; cell analysis stains; electrophoresis products, including the FlashGel Rapid Electrophoresis System; and protein analysis products. Cambrex offers technical support for all its research products. In addition, Cambrex offers Platinum UltraPAK, a line of flexible packaging systems that can be modified to fit specific customer needs. Therapeutic testing products include a range of endotoxin services and products, including endotoxin detection assays, removal products, testing services accessory products, instrumentation and software. The company also offers testing products using a wide range of assays. Recently, Cambrex Bio Science Walkersville, Inc., a subsidiary of the company, announced its intent to purchase Cutanogen Corporation, a company specializing in the treatment of severe burns. In February 2007, Camrex announced that the sale of Bioproducts and Biopharma subsidiaries to Lonza Group AG was completed.

Cambrex provides its employees with complete health coverage, flexible spending plans, tuition reimbursement, a 401(k) and an employee assistance program.

FINANCIALS: Sales and profits are in thousands of dollars—add 000 to get the full amount. 2006 Note: Financial information for 2006 was not available for all companies at press time.

2006 Sales: $452,255	2006 Profits: $-30,100	**U.S. Stock Ticker: CBM**
2005 Sales: $414,761	2005 Profits: $-110,458	**Int'l Ticker:** Int'l Exchange:
2004 Sales: $395,906	2004 Profits: $-26,870	Employees: 1,916
2003 Sales: $405,600	2003 Profits: $-54,100	Fiscal Year Ends: 12/31
2002 Sales: $526,900	2002 Profits: $36,200	Parent Company:

SALARIES/BENEFITS:

Pension Plan:	ESOP Stock Plan:	Profit Sharing:	Top Exec. Salary: $458,333	Bonus: $
Savings Plan: Y	Stock Purch. Plan:		Second Exec. Salary: $382,000	Bonus: $100,000

OTHER THOUGHTS:

Apparent Women Officers or Directors: 2
Hot Spot for Advancement for Women/Minorities: Y

LOCATIONS: ("Y" = Yes)

West:	Southwest:	Midwest:	Southeast:	Northeast:	International:
		Y		Y	Y

Plunkett Research, Ltd. 213

CAMBRIDGE ANTIBODY TECHNOLOGY LIMITED
www.cambridgeantibody.com

Industry Group Code: 325412 Ranks within this company's industry group: Sales: Profits:

Drugs:		Other:		Clinical:	Computers:		Other:	
Discovery:	Y	AgriBio:		Trials/Services:	Hardware:		Specialty Services:	
Licensing:	Y	Genomics/Proteomics:	Y	Laboratories:	Software:		Consulting:	
Manufacturing:		Tissue Replacement:	Y	Equipment/Supplies:	Arrays:		Blood Collection:	
Development:	Y			Research/Development Svcs.:	Database Management:		Drug Delivery:	
Generics:				Diagnostics:			Drug Distribution:	

TYPES OF BUSINESS:
Drug Discovery & Research Services
Drugs-Human Antibodies
Biotech Licensing

BRANDS/DIVISIONS/AFFILIATES:
HUMIRA
Abthrax
LymphoStat-B
CAT-354
Cambridge Antibody Technology Group plc

CONTACTS: Note: Officers with more than one job title may be intentionally listed here more than once.
Hamish Cameron, CEO
Lynn Lester, Sr. VP-Human Resources
Alex Duncan, Sr. VP-Tech.
Adrian Kemp, General Counsel
Lynn Lester, Sr. VP-Bus. Oper.
Patrick Round, Sr. VP-Dev.
Lizzie Dant, VP-Finance
Alex Duncan, Sr. VP-Alliances

Phone: 44-1223-471-471	Fax: 44-1223-471-472
Toll-Free:	
Address: Milstein Bldg., Granta Park, Cambridge, CB1 6GH UK	

GROWTH PLANS/SPECIAL FEATURES:
Cambridge Antibody Technology Limited (CAT), formerly Cambridge Antibody Technology Group plc, uses its proprietary technologies in human monoclonal antibodies to conduct drug discovery and development research. The firm's technology aids in the validation of drug targets and is intended to generate short-term revenue through partnerships with pharmaceutical and biotechnology companies. CAT uses the phage display method, which employs bacterial viruses to create libraries of antibodies. These antibodies are then made to target chosen molecules. Currently, CAT's libraries incorporate more than 100 billion distinct antibodies. The firm's business has two strands: developing proprietary products; and licensing technologies and capabilities to enable others to develop products. The company has formed alliances with a number of pharmaceutical and biotechnology companies to discover, develop and commercialize human monoclonal antibody-based products. In particular, CAT has a continuing collaboration with Genzyme for the development and commercialization of antibodies directed against TGFB, a family of proteins associated with pulmonary fibrosis and scarring. The company has also licensed its proprietary technologies to other companies, including Abbott, Amgen, AstraZeneca, Chugai, Human Genome Sciences, Merck & Co., Pfizer and Wyeth Research. CAT has a strong patent portfolio in both the U.S. and Europe. HUMIRA, an antibody for the treatment of rheumatoid arthritis, isolated and optimized in collaboration with Abbott, is now marketed in the U.S. and 40 other countries and has exceeded $1 billion in sales. Some products under development include Abthrax, a treatment for anthrax infection; LymphoStat-B, for both lupus erythematosus and rheumatoid arthritis; and CAT-354, which could potentially provide a treatment for severe cases of asthma. In September 2006, AstraZeneca plc acquired CAT for $1.3 billion.

CAT ranked number 45 in the Financial Times list of the 2006 Best Workplaces in the UK. CAT was the only UK biotechnology company listed in the top 50.

FINANCIALS: Sales and profits are in thousands of dollars—add 000 to get the full amount. 2006 Note: Financial information for 2006 was not available for all companies at press time.

2006 Sales: $	2006 Profits: $	U.S. Stock Ticker: Subsidiary
2005 Sales: $49,664	2005 Profits: $-9,226	Int'l Ticker: Int'l Exchange:
2004 Sales: $16,347	2004 Profits: $-40,751	Employees: 283
2003 Sales: $-9,165	2003 Profits: $-46,552	Fiscal Year Ends: 9/30
2002 Sales: $14,800	2002 Profits: $-44,200	Parent Company: ASTRAZENECA PLC

SALARIES/BENEFITS:
Pension Plan: Y	ESOP Stock Plan:	Profit Sharing:	Top Exec. Salary: $686,228	Bonus: $250,169
Savings Plan:	Stock Purch. Plan:		Second Exec. Salary: $305,762	Bonus: $111,012

OTHER THOUGHTS:
Apparent Women Officers or Directors: 2
Hot Spot for Advancement for Women/Minorities: Y

LOCATIONS: ("Y" = Yes)
West:	Southwest:	Midwest:	Southeast:	Northeast:	International: Y

Note: Financial information, benefits and other data can change quickly and may vary from those stated here.

CANGENE CORP

www.cangene.com

Industry Group Code: 325412 Ranks within this company's industry group: Sales: 65   Profits: 45

Drugs:		Other:		Clinical:		Computers:		Other:	
Discovery:	Y	AgriBio:		Trials/Services:		Hardware:		Specialty Services:	
Licensing:		Genomics/Proteomics:	Y	Laboratories:		Software:		Consulting:	
Manufacturing:	Y	Tissue Replacement:		Equipment/Supplies:		Arrays:		Blood Collection:	
Development:	Y			Research/Development Svcs.:	Y	Database Management:		Drug Delivery:	
Generics:	Y			Diagnostics:				Drug Distribution:	

TYPES OF BUSINESS:

Drugs-Hyperimmunes
Generic Drugs
Contract Manufacturing
Vaccines

BRANDS/DIVISIONS/AFFILIATES:

WinRho SDF
VariZIG
CANGENUS
VIG
HepaGam B
Leucotropin
Accretropin
CANGENUS

CONTACTS: Note: Officers with more than one job title may be intentionally listed here more than once.

John Langstaff, CEO
John Langstaff, Pres.
Michael Graham, CFO
Grant McClarty, VP-R&D
Bill L. Bees, Sr. VP-Oper.
John W. McMillan, VP-Commercial Dev.
Andrew Storey, VP-Quality Assurance
Andrew Storey, VP-Clinical & Regulatory Affairs
Jack Kay, Chmn.

Phone: 204-275-4200	Fax: 204-269-7003
Toll-Free:	
Address: 155 Innovation Dr., Winnipeg, MB R3T 5Y3 Canada	

GROWTH PLANS/SPECIAL FEATURES:

Cangene Corporation is a world leader in hyperimmune plasma and recombinant biopharmaceuticals. It specializes in manufacturing injectable products, and offers contract research and process development services, bulk products manufacturing and finished products manufacturing services to biopharmaceutical companies. Cangene's leading product offering is WinRho SDF, a hyperimmune used to prevent hemolytic disease in newborns and to treat immune thrombocytopenic purpura (ITP), a clotting disorder. The company's other approved hyperimmune products include VariZIG, a chicken pox vaccine, developed to prevent chicken pox in pregnant women; VIG, an antibody product used to prep recipients for or to treat severe reactions to the smallpox vaccine that was recently approved by the FDA; and HepaGam B, a specialized antibody for treatment following acute exposure to the hepatitis B virus. Its products currently under development include botulinum immune globulin, a specialized antibody that counteracts botulinum toxin; anthrax immune globulin, an adjunct to antibiotic therapy in critically ill patients with anthrax; Leucotropin, a recombinant version of a protein that stimulates the production of certain white blood cells, for the treatment of acute radiation syndrome; Accretropin, a treatment for children with growth hormone deficiency and girls with Turner Syndrome which stimulates long bone growth prior to puberty; and Burkholderia antibodies to treat infection by Burkholderia bacteria. The firm's proprietary CANGENUS expression system produces recombinant proteins and simplifies protein recovery, keeping costs low. In March 2007, Cangene received an approval letter from the FDA for Accretropin, and in April the FDA sent an approval letter for the use of HepaGam B in hepatitis B-positive liver transplant recipients. In May 2007, Health Canada approved Cangene's VIG, which confers marketing approval of the drug in Canada.

FINANCIALS: Sales and profits are in thousands of dollars—add 000 to get the full amount. 2006 Note: Financial information for 2006 was not available for all companies at press time.

2006 Sales: $96,600	2006 Profits: $11,600	**U.S. Stock Ticker:**
2005 Sales: $9,062	2005 Profits: $-13,638	**Int'l Ticker: CNJ** Int'l Exchange: Toronto-TSX
2004 Sales: $118,000	2004 Profits: $24,500	Employees: 617
2003 Sales: $130,300	2003 Profits: $28,600	Fiscal Year Ends: 7/31
2002 Sales: $54,100	2002 Profits: $6,600	Parent Company:

SALARIES/BENEFITS:

Pension Plan:	ESOP Stock Plan:	Profit Sharing:	Top Exec. Salary: $	Bonus: $
Savings Plan:	Stock Purch. Plan:		Second Exec. Salary: $	Bonus: $

OTHER THOUGHTS:

Apparent Women Officers or Directors:
Hot Spot for Advancement for Women/Minorities:

LOCATIONS: ("Y" = Yes)

West:	Southwest:	Midwest:	Southeast:	Northeast:	International:
Y			Y	Y	Y

CARACO PHARMACEUTICAL LABORATORIES

www.caraco.com

Industry Group Code: 325416 Ranks within this company's industry group: Sales: 7   Profits: 8

Drugs:		Other:		Clinical:		Computers:		Other:	
Discovery:	Y	AgriBio:		Trials/Services:		Hardware:		Specialty Services:	
Licensing:		Genomics/Proteomics:		Laboratories:		Software:		Consulting:	
Manufacturing:	Y	Tissue Replacement:		Equipment/Supplies:		Arrays:		Blood Collection:	
Development:	Y			Research/Development Svcs.:		Database Management:		Drug Delivery:	
Generics:	Y			Diagnostics:				Drug Distribution:	Y

TYPES OF BUSINESS:

Drugs-Generic

BRANDS/DIVISIONS/AFFILIATES:

Sun Pharmaceutical Industries
Glucophage
Tegretol
Entex PSE
Mirtazapine
Lopressor
Ambien
Nimotop

CONTACTS: *Note: Officers with more than one job title may be intentionally listed here more than once.*

Daniel H. Movens, CEO
Mukul Rathi, Interim CFO
Thomas Larkin, Dir.-Mktg.
Tammy Bitterman, Dir.-Human Resources
Daniel Barone, Dir.-Tech.
Kaushikkumar Gandhi, VP-Mfg.
Steven Walker, General Counsel/Sec.
Gurpartap Singh, VP-Bus. Strategies
Ann Zajac, Controller
Robert Kurkiewicz, Sr. VP- Regulatory Affairs
Jayesh Shah, Dir.-Commercial
David Risk, Dir.-Bus. Dev.
Derrick Mann, Dir.-Regulatory
Dilip S. Shanghvi, Chmn.

Phone: 313-871-8400	Fax: 313-871-8314

Toll-Free: 800-818-4555
Address: 1150 Elijah McCoy Dr., Detroit, MI 48202 US

GROWTH PLANS/SPECIAL FEATURES:

Caraco Pharmaceutical Laboratories, Ltd. develops, manufactures and markets generic drugs for prescription and over-the-counter markets. The company's product portfolio includes 35 products in 73 strengths and various package sizes. These drugs relate to a variety of therapeutic segments, including arthritis, pain control, seizures, diabetes, cardiac and neurological disorders. Pharmaceutical products that the company produces, with each brand name, include metoprolol tartrate, Lopressor; metformin hydrochloride, Glucophage; carbamazephine, Tegretol; and Citalopram HBr, Celexa. The company also has several drugs awaiting FDA approval. The company has collaborative agreements with several companies, with the most prominent being Sun Pharmaceutical Industries (Sun Pharma), which is the majority stock holder of Caraco. Under these agreements, Caraco develops generic drugs for each company to market as its own brand. Caraco has entered an agreement with Sun Pharma Global, Inc., a wholly-owned subsidiary of Sun Pharma, to develop 25 products over the next five years, and has approval to manufacture and market Mirtazapine, a generic form of Remeron. It distributes its products through wholesalers, chain drug stores, retail pharmacies, mail-order companies, managed care organizations, hospital groups and nursing homes. Some of the wholesalers that distribute Caraco's products include Amerisource-Bergen Corporation, McKesson Corporation and Cardinal Health. In the seven months of 2007, Caraco received FDA approval for six generic drugs including the generics for Ambien, Nimotop, Zyrtec, Paxil, Tenormin and Provigil.

Caraco offers its employees many benefits including, but not limited to, medical, dental, and vision care; paid time off and holiday pay; life insurance provided by the company; and health and dependent care accounts.

FINANCIALS: **Sales and profits are in thousands of dollars—add 000 to get the full amount. 2006 Note: Financial information for 2006 was not available for all companies at press time.**

2006 Sales: $82,789	2006 Profits: $-10,423	**U.S. Stock Ticker: CPD**	
2005 Sales: $64,116	2005 Profits: $-2,278	**Int'l Ticker:** Int'l Exchange:	
2004 Sales: $60,340	2004 Profits: $- 199	Employees: 272	
2003 Sales: $45,498	2003 Profits: $11,222	Fiscal Year Ends: 3/31	
2002 Sales: $22,400	2002 Profits: $-2,300	Parent Company: SUN PHARMACEUTICAL INDUSTRIES LTD	

SALARIES/BENEFITS:

Pension Plan:	ESOP Stock Plan:	Profit Sharing: Y	Top Exec. Salary: $352,900	Bonus: $146,250
Savings Plan: Y	Stock Purch. Plan:		Second Exec. Salary: $201,200	Bonus: $45,000

OTHER THOUGHTS:

Apparent Women Officers or Directors: 2
Hot Spot for Advancement for Women/Minorities: Y

LOCATIONS: ("Y" = Yes)

West:	Southwest:	Midwest:	Southeast:	Northeast:	International:
		Y			

CARDIOME PHARMA CORP

www.cardiome.com

Industry Group Code: 325412 Ranks within this company's industry group: Sales: 112   Profits: 129

Drugs:		Other:	Clinical:	Computers:	Other:
Discovery:	Y	AgriBio:	Trials/Services:	Hardware:	Specialty Services:
Licensing:		Genomics/Proteomics:	Laboratories:	Software:	Consulting:
Manufacturing:		Tissue Replacement:	Equipment/Supplies:	Arrays:	Blood Collection:
Development:	Y		Research/Development Svcs.:	Database Management:	Drug Delivery:
Generics:			Diagnostics:		Drug Distribution:

TYPES OF BUSINESS:

Drugs-Cardiac

BRANDS/DIVISIONS/AFFILIATES:

Astellas Pharma US, Inc.
Artesian Therapeutics, Inc.
vernakalant (iv)
vernakalant (oral)

CONTACTS: *Note: Officers with more than one job title may be intentionally listed here more than once.*

Bob Rieder, CEO
Doug Janzen, Pres.
Curtis Sikorsky, CFO
Donald A. McAfee, Chief Scientific Officer
Sheila M. Grant, VP-Product Dev.
Taryn Boivin, VP-Mfg. & Pharmaceutical Sciences
Doug Janzen, Chief Bus. Officer
Karim Lalji, Sr. VP-Comm. Affairs
Charles Fisher, Chief Medical Officer/Exec. VP-Clinical Affairs
Guy F. Cipriani, VP-Bus. Dev.
Gregory N. Beatch, VP-Scientific Affairs
Bob Rieder, Chmn.

Phone: 604-677-6905	Fax: 604-677-6915
Toll-Free: 800-330-9928	
Address: 6190 Agronomy Rd., 6th Fl., Vancouver, BC V6T 1Z3 Canada	

GROWTH PLANS/SPECIAL FEATURES:

Cardiome Pharma Corp. is a drug discovery and development company that is focused on the development of proprietary drugs for the treatment and prevention of cardiovascular diseases. The company currently has two late-stage clinical drug programs for atrial arrhythmia: a Phase I program for GED-aPC, an engineered analog of recombinant human activated Protein C, and a pre-clinical program that specifically focuses on the improvement of cardiovascular function. Cardiome's main product focus is on Vernakalant Hydrochloride, formerly known as RSD1235, which consists of an intravenous, vernakalant (iv), or oral drug formulation, vernakalant (oral), for the acute conversion of atrial fibrillation to a normal heart rhythm. The drug works by selectively blocking ion channels in the heart that become active during episodes of atrial fibrillation. Cardiome is currently involved in academic collaborations with Johns Hopkins University School of Medicine, the Montreal Heart Institute, the University of British Columbia and several other major medical institutions. In June 2007, the firm and its co-development partner, Astellas Pharma US, Inc., completed the Phase III clinical study of the efficacy and safety of vernakalant (iv), which was conducted in 42 centers in the U.S., Argentina, India and Europe. Astellas currently has the exclusive rights to develop and commercialize vernakalant (iv) within North America.

Cardiome offers employees medical and dental benefits; an employee incentive stock option program; professional development support; and activities and social events that are organized by an interdepartmental social committee.

FINANCIALS: Sales and profits are in thousands of dollars—add 000 to get the full amount. 2006 Note: Financial information for 2006 was not available for all companies at press time.

2006 Sales: $18,000	2006 Profits: $-30,700	**U.S. Stock Ticker: CRME**
2005 Sales: $13,900	2005 Profits: $-45,900	**Int'l Ticker: COM** Int'l Exchange: Toronto-TSX
2004 Sales: $22,000	2004 Profits: $-23,100	Employees: 63
2003 Sales: $4,673	2003 Profits: $-15,436	Fiscal Year Ends: 12/31
2002 Sales: $1,308	2002 Profits: $-10,700	Parent Company:

SALARIES/BENEFITS:

Pension Plan:	ESOP Stock Plan: Y	Profit Sharing:	Top Exec. Salary: $258,836	Bonus: $
Savings Plan:	Stock Purch. Plan:		Second Exec. Salary: $241,250	Bonus: $

OTHER THOUGHTS:

Apparent Women Officers or Directors: 2
Hot Spot for Advancement for Women/Minorities: Y

LOCATIONS: ("Y" = Yes)

West:	Southwest:	Midwest:	Southeast:	Northeast:	International:
					Y

CARDIOTECH INTERNATIONAL
www.cardiotech-inc.com

Industry Group Code: 339113 Ranks within this company's industry group: Sales: 14 Profits: 13

Drugs:	Other:	Clinical:		Computers:	Other:
Discovery:	AgriBio:	Trials/Services:		Hardware:	Specialty Services:
Licensing:	Genomics/Proteomics:	Laboratories:		Software:	Consulting:
Manufacturing:	Tissue Replacement: Y	Equipment/Supplies:	Y	Arrays:	Blood Collection:
Development:		Research/Development Svcs.:		Database Management:	Drug Delivery:
Generics:		Diagnostics:			Drug Distribution:

TYPES OF BUSINESS:

Medical Device Manufacturing
Disposable Medical Devices
Polyurethane-Based Biomaterials
Moist Wound Care Products
Catheters

BRANDS/DIVISIONS/AFFILIATES:

CardioTech International, Ltd.
CT Biomaterials
Catheter & Disposables Technology, Inc.
Gish Biomedical, Inc.
Dermaphylyx, Inc.
HydroThane
CardioPass
AlgiMed

CONTACTS: Note: Officers with more than one job title may be intentionally listed here more than once.

Michael Szycher, CEO
Michael Szycher, Pres.
Leslie M. Taeger, CFO
Thomas Lovett, VP-Finance
Jill Knudson, General Mgr.-Catheters & Disposable Technologies
Michael Szycher, Chmn./Treas.

Phone: 978-657-0075	Fax: 978-657-0074

Toll-Free:
Address: 229 Andover St., Wilmington, MA 01887 US

GROWTH PLANS/SPECIAL FEATURES:

CardioTech International, Inc. develops and markets medical devices used in the treatment of late-stage cardiovascular disease. Its products are designed to reduce risk, trauma, cost and surgical procedure time for cardiac patients. The company operates through a single operating segment that is divided into four divisions: Gish Biomedical, Inc.; Catheter and Disposables Technology, Inc.; Dermaphylyx, Inc. and CT Biomaterials. Gish designs, manufactures and markets disposable medical devices for various surgical specialties, including cardiovascular surgery, orthopedics and oncology. All of its products are single-use disposable products or have a disposable component. The division's principal products include custom cardiovascular tubing systems, blood reservoirs, oxygenators, central venous access catheters and ports, and blood recovery devices for post-operative use in orthopedic surgery. Catheter and Disposables Technology, Inc. is a full-service medical device design, development and manufacturing company that focuses on catheters and medical products based on molded and extruded components for the cardiovascular, peripheral, urological, gastroenterology and minimally invasive device markets. Dermaphylyx has three advanced moist wound care products on the market: HydroMed, an absorbent wound dressing; AlgiMed, a dressing for oozing and bleeding wounds; and BlistRX, a blister-prevention dressing. In recent news, CardioTech's CardioPass synthetic coronary artery bypass graft is undergoing a clinical feasibility study in Brazil on patients suffering from coronary artery disease. CT Biomaterials is a division of the company that markets medical polyurethanes under the ChronoFlex, ChronoThane and several other brand names.

FINANCIALS: Sales and profits are in thousands of dollars—add 000 to get the full amount. 2006 Note: Financial information for 2006 was not available for all companies at press time.

2006 Sales: $22,381	2006 Profits: $-5,069	**U.S. Stock Ticker: CTE**
2005 Sales: $21,841	2005 Profits: $-1,595	**Int'l Ticker:** Int'l Exchange:
2004 Sales: $21,799	2004 Profits: $-1,515	Employees: 16
2003 Sales: $3,000	2003 Profits: $-1,000	Fiscal Year Ends: 3/31
2002 Sales: $3,200	2002 Profits: $-2,000	Parent Company:

SALARIES/BENEFITS:

Pension Plan:	ESOP Stock Plan:	Profit Sharing:	Top Exec. Salary: $325,000	Bonus: $
Savings Plan: Y	Stock Purch. Plan:		Second Exec. Salary: $94,894	Bonus: $

OTHER THOUGHTS:

Apparent Women Officers or Directors: 1
Hot Spot for Advancement for Women/Minorities:

LOCATIONS: ("Y" = Yes)

West:	Southwest:	Midwest:	Southeast:	Northeast:	International:
Y		Y		Y	Y

Note: Financial information, benefits and other data can change quickly and may vary from those stated here.

CARDIUM THERAPEUTICS INC

www.cardiumthx.com

Industry Group Code: 325412 **Ranks within this company's industry group:** Sales: 100   Profits: 67

Drugs:		Other:		Clinical:		Computers:		Other:	
Discovery:		AgriBio:		Trials/Services:		Hardware:		Specialty Services:	
Licensing:		Genomics/Proteomics:	Y	Laboratories:		Software:		Consulting:	
Manufacturing:		Tissue Replacement:		Equipment/Supplies:	Y	Arrays:		Blood Collection:	
Development:	Y			Research/Development Svcs.:	Y	Database Management:		Drug Delivery:	
Generics:				Diagnostics:				Drug Distribution:	

TYPES OF BUSINESS:

Biologic Therapeutics & Medical Devices

BRANDS/DIVISIONS/AFFILIATES:

Generx
Corgentin
Genvascor
Excellerate
Cardium Biologics
InnerCool Therapeutics
Tissue Repair Co.

CONTACTS: *Note: Officers with more than one job title may be intentionally listed here more than once.*

Christopher J. Reinhard, CEO
Christopher J. Reinhard, Pres.
Dennis M. Mulroy, CFO
Gabor M. Rubanyi, Chief Scientific Officer
Ted Williams, VP-Tech. Oper.
Ted Williams, VP-Mfg.
Tyler M. Dylan, General Counsel/Exec. VP/Sec./Chief Bus. Officer
Mark McCutchen, VP-Bus. Dev.
Christopher J. Reinhard, Treas.
Randall W. Moreadith, Exec. VP/Chief Medical Officer
Anthony Andrasfay, VP-Clinical Oper.
Patricia L. Novak, VP-Program Dev.
Jennifer A. Spinella, VP-Regulatory Affairs & Quality Assurance
Christopher J. Reinhard, Chmn.

Phone: 858-436-1000	Fax: 858-436-1001
Toll-Free:	
Address: 3611 Valley Ctr., Ste. 525, San Diego, CA 92130 US	

GROWTH PLANS/SPECIAL FEATURES:

Cardium Therapeutics, Inc. is a medical technology company primarily focused on the development and commercialization of biologic therapeutics and medical devices for cardiovascular and ischemic diseases. The company's pipeline of products is divided between three companies: Cardium Biologics, InnerCool Therapeutics and Tissue Repair Co. Product candidates for Cardium Biologics include Generx, a late-stage DNA-based growth factor therapeutic that is being developed as a one-time treatment to promote and stimulate the growth of collateral circulation in the hearts of patients with ischemic conditions such as recurrent angina; Corgentin, a DNA-based therapeutic based on myocardial produced insulin-like growth factor-I which could be developed for administration in an acute care setting by interventional cardiologists as a treatment for heart attack patients immediately following percutaneous coronary intervention; and Genvascor, a DNA-based therapeutic intended to induce the localized the sustained production of nitric oxide directed at mediating the effects of multiple growth factors to enhance neovascularization and increased blood flow for the treatment of patients with critical limb ischemia due to advanced peripheral arterial occlusive diseases. In March 2006, the company acquired the technologies and products of InnerCool Therapies, Inc., including the Celsius Control System, designed to rapidly and controllably cool the body in order to reduce cell death and damage following acute ischemic events such as cardiac arrest or stroke. The system received regulatory clearance in the U.S., Europe and Australia. In August 2006, Cardium Therapeutics obtained rights to develop various technologies and products that are part of the Tissue Repair Co. The lead product candidate, Excellarate, is a DNA-activated collagen gel for topical treatment formulated with an adenovector delivery carrier for the treatment of non-healing diabetic foot ulcers. In July 2007, the FDA granted fast track designation to Generx for the potential treatment of myocardial ischemia.

FINANCIALS: Sales and profits are in thousands of dollars—add 000 to get the full amount. 2006 Note: Financial information for 2006 was not available for all companies at press time.

2006 Sales: $	2006 Profits: $	**U.S. Stock Ticker:** CXM
2005 Sales: $	2005 Profits: $	**Int'l Ticker:** Int'l Exchange:
2004 Sales: $	2004 Profits: $	Employees:
2003 Sales: $	2003 Profits: $	Fiscal Year Ends: 12/31
2002 Sales: $	2002 Profits: $	Parent Company:

SALARIES/BENEFITS:

Pension Plan:	ESOP Stock Plan:	Profit Sharing:	Top Exec. Salary: $350,000	Bonus: $
Savings Plan:	Stock Purch. Plan:		Second Exec. Salary: $325,000	Bonus: $

OTHER THOUGHTS:

Apparent Women Officers or Directors: 3
Hot Spot for Advancement for Women/Minorities: Y

LOCATIONS: ("Y" = Yes)

West:	Southwest:	Midwest:	Southeast:	Northeast:	International:
Y					

Note: Financial information, benefits and other data can change quickly and may vary from those stated here.

CARRINGTON LABORATORIES INC   www.carringtonlabs.com

Industry Group Code: 325412 Ranks within this company's industry group: Sales: 34   Profits: 33

Drugs:	Other:		Clinical:		Computers:		Other:	
Discovery:	AgriBio:		Trials/Services:		Hardware:		Specialty Services:	
Licensing:	Genomics/Proteomics:		Laboratories:		Software:		Consulting:	
Manufacturing: Y	Tissue Replacement:		Equipment/Supplies:	Y	Arrays:		Blood Collection:	
Development: Y			Research/Development Svcs.:	Y	Database Management:		Drug Delivery:	
Generics:			Diagnostics:				Drug Distribution:	

TYPES OF BUSINESS:

Drugs-Carbohydrate-Based
Skin Care Products
Wound Care Products
Dietary Supplements
Veterinary Products
Bulk Raw Materials
Topical Dressings
Manufacturing & Product Development Services

BRANDS/DIVISIONS/AFFILIATES:

Sabila Industrial SA
Caraloe, Inc.
DelSite Biotechnologies, Inc.
Manapol
SaliCept
CarraSmart
Acemannan Hydrogel
GelSite

CONTACTS: *Note: Officers with more than one job title may be intentionally listed here more than once.*

Carlton E. Turner, CEO
Carlton E. Turner, Pres.
Robert W. Schnitzius, CFO
Doug Golwas, VP-Sales & Mktg.
Carol Kitchell, Dir-Human Resources
Jose Zunica,
Robert W. Schnitzius, Treas.
Doug Talley, Dir.-Quality Control
George DeMott, Chmn.
Jose Zuniga, Mgr.-South American Oper.

Phone: 972-518-1300	Fax: 972-518-1020
Toll-Free: 800-527-5216	
Address: 2001 Walnut Hill Ln., Irving, TX 75038 US	

GROWTH PLANS/SPECIAL FEATURES:

Carrington Laboratories, Inc. develops, manufactures and markets naturally derived complex carbohydrates and other natural product therapeutics for the treatment of major illnesses, the dressing and management of wounds and nutritional supplements. The company formulates Aloe Vera L. into products including oral, topical and injectable forms. Carrington grows its Aloe Vera L. in Costa Rica, and manufactures its products through Sabila Industrial. Subsidiary, Caraloe, Inc., markets consumer and bulk raw material products, including Manapol powder, its premier product, and provides product development and manufacturing services to customers in the cosmetic, nutraceutical and medical markets. Carrington products include its CarraSmart water-resistant wound care line; its Acemannan Hydrogel wound dressing; its Acemannan Immunostimulant adjuvant therapy for certain cancers in dogs and cats; its SaliCept oral care line; its DiaB, diabetic skin care line; and its Radia radiation skin care line. Subsidiary DelSite Biotechnologies, Inc. develops and markets the company's proprietary GelSite technology for controlled release and delivery of bioactive pharmaceutical ingredients. Since January 2007, DelSite has completed agreements with various companies to develop GelSite for uses such as a vaccine against HIV infection, a typhoid vaccine, sublingual vaccines for needle-free immunization programs, a matrix for injectable applications, and as a way of enhancing the intranasal delivery of peptide and protein therapeutics. In April 2007, DelSite secured a source of influenza antigen for a planned Phase I clinical trial of a nasal powder influenza vaccine based on its GelVac dry powder technology. In July 2007, DelSite entered into a consulting agreement with Biologics Consulting Group, Inc., for regulatory and product development advice with the H5N1 bird influenza vaccine.

Carrington offers employees health, dental, life, and AD&D coverage; short and long term disability; an employer matching 401(k) plan; an Employee Stock Purchase Plan; a Section 125 Medical Reimbursement Plan; and tuition assistance.

FINANCIALS: Sales and profits are in thousands of dollars—add 000 to get the full amount. 2006 Note: Financial information for 2006 was not available for all companies at press time.

2006 Sales: $27,406	2006 Profits: $-7,607	U.S. Stock Ticker: CARN
2005 Sales: $27,961	2005 Profits: $-5,336	Int'l Ticker: Int'l Exchange:
2004 Sales: $30,821	2004 Profits: $ 36	Employees: 260
2003 Sales: $29,100	2003 Profits: $-1,500	Fiscal Year Ends: 12/31
2002 Sales: $18,041	2002 Profits: $-3,378	Parent Company:

SALARIES/BENEFITS:

Pension Plan:	ESOP Stock Plan:	Profit Sharing:	Top Exec. Salary: $384,780	Bonus: $
Savings Plan: Y	Stock Purch. Plan:		Second Exec. Salary: $184,750	Bonus: $20,000

OTHER THOUGHTS:

Apparent Women Officers or Directors: 1
Hot Spot for Advancement for Women/Minorities:

LOCATIONS: ("Y" = Yes)

West:	Southwest:	Midwest:	Southeast:	Northeast:	International:
	Y				Y

Note: Financial information, benefits and other data can change quickly and may vary from those stated here.

CELERA GENOMICS GROUP   www.celera.com

Industry Group Code: 541710 **Ranks within this company's industry group:** Sales: 13   Profits: 22

Drugs:	Other:		Clinical:		Computers:		Other:	
Discovery:	AgriBio:		Trials/Services:		Hardware:		Specialty Services:	
Licensing:	Genomics/Proteomics:	Y	Laboratories:		Software:	Y	Consulting:	
Manufacturing:	Tissue Replacement:	Y	Equipment/Supplies:		Arrays:		Blood Collection:	
Development:			Research/Development Svcs.:	Y	Database Management:	Y	Drug Delivery:	
Generics:			Diagnostics:				Drug Distribution:	

TYPES OF BUSINESS:

Research-Human Genome Mapping
Information Management & Analysis Software
Consulting, Research & Development Services

BRANDS/DIVISIONS/AFFILIATES:

Celera Discovery System
Applera Corp.
Celera Diagnostics
Applied Biosystems
Paracel

CONTACTS: *Note: Officers with more than one job title may be intentionally listed here more than once.*

Kathy Ordonez, Pres./Pres., Celera Diagnostics
Tom White, Chief Scientific Officer/VP-R&D
Victor K. Lee, Chief Group Counsel/VP
James Yee, VP-Dev.
David P. Speechly, Sr. Dir.-Corp. Comm.
David P. Speechly, Sr. Dir.-Investor Rel.
Joel Jung, VP-Finance
Samuel Broder, Chief Medical Officer
Stacey Sias, Chief Bus. Officer
Victoria Mackinnon, VP-Regulatory Affairs
Steven M. Ruben, VP-Protein Therapeutics

Phone: 240-453-3000	Fax: 240-453-4000
Toll-Free: 877-235-3721	
Address: 45 W. Gude Dr., Rockville, MD 20850 US	

GROWTH PLANS/SPECIAL FEATURES:

Celera Genomics Group, a division of Applera Corporation, is engaged in the discovery and development of targeted therapeutics for cancer, autoimmune and inflammatory diseases. The company utilizes genomics, proteomics and bioinformatics platforms to identify gene markers linked to disease. Celera divides its research into two sectors: Genomics R&D and Proteomics R&D. In the Genomics R&D sector, Celera houses an industrial-scale facility that can perform high volume genotyping and gene expression analyses through it ABI PRISM 7900 HT Sequence Detection System. Discovery efforts are focused on the identification of genetic variations associated with diseases, which are useful in predicting the predisposition or severity of a disease, the progression of a disease, and any predictable monitoring responses that can be achieved through drug therapy. In Proteomics R&D, Celera has developed proprietary techniques for capturing cell surface proteins in order to identify and quantify the proteins using mass spectrometry and informatics platforms. Areas that this sector is currently investigating include cancer stem cells and tumor angiogenesis (the growth of new blood vessels). In recent news, the company identified novel genes that were associated with late-onset Alzheimer's disease and two genetic variations that put individuals at a higher risk for psoriasis. In 2006, Celera received a NIH Grant to develop and commercialize Avian Flu diagnostic tests. The company sold its Cathepsin S Inhibitor Program to Schering AG and also sold its therapeutic program to Pharmacyclics.

Celera provides its employees a comprehensive benefits package that includes educational assistance, a survivor support program, adoption and employee assistance programs and a corporate matching gift program.

FINANCIALS: Sales and profits are in thousands of dollars—add 000 to get the full amount. 2006 Note: Financial information for 2006 was not available for all companies at press time.

2006 Sales: $44,200	2006 Profits: $-62,700	**U.S. Stock Ticker:** Subsidiary
2005 Sales: $31,000	2005 Profits: $-77,100	**Int'l Ticker:** Int'l Exchange:
2004 Sales: $60,100	2004 Profits: $-57,500	Employees: 300
2003 Sales: $88,300	2003 Profits: $-81,900	Fiscal Year Ends: 6/30
2002 Sales: $120,900	2002 Profits: $-211,800	Parent Company: APPLERA CORPORATION

SALARIES/BENEFITS:

Pension Plan:	ESOP Stock Plan: Y	Profit Sharing:	Top Exec. Salary: $468,077	Bonus: $451,250
Savings Plan: Y	Stock Purch. Plan:		Second Exec. Salary: $484,069	Bonus: $258,851

OTHER THOUGHTS:

Apparent Women Officers or Directors: 5
Hot Spot for Advancement for Women/Minorities: Y

LOCATIONS: ("Y" = Yes)

West:	Southwest:	Midwest:	Southeast:	Northeast:	International:
Y				Y	

Note: Financial information, benefits and other data can change quickly and may vary from those stated here.

CELGENE CORP
www.celgene.com

Industry Group Code: 325412 **Ranks within this company's industry group:** Sales: 154 Profits: 176

Drugs:		Other:	Clinical:	Computers:	Other:
Discovery:	Y	AgriBio:	Trials/Services:	Hardware:	Specialty Services:
Licensing:	Y	Genomics/Proteomics:	Laboratories:	Software:	Consulting:
Manufacturing:		Tissue Replacement:	Equipment/Supplies:	Arrays:	Blood Collection:
Development:	Y		Research/Development Svcs.: Y	Database Management:	Drug Delivery:
Generics:			Diagnostics:		Drug Distribution:

TYPES OF BUSINESS:
Cancer & Immune-Inflammatory Related Diseases Drugs

BRANDS/DIVISIONS/AFFILIATES:
Revlimid
Thalomid
Alkeran
Focalin
Focalin XR
Pharmion Corp.
Novartis Pharma AG
Siegfried, Ltd.

CONTACTS: *Note: Officers with more than one job title may be intentionally listed here more than once.*
Sol J. Barer, CEO
Robert J. Hugin, COO
Robert J. Hugin, Pres.
David Gryska, CFO/Sr. VP
Brian P. Gill, VP-Corp. Comm.
Sol J. Barer, Chmn.

Phone: 908-673-9000	Fax: 908-673-9001
Toll-Free:	
Address: 86 Morris Ave., Summit, NJ 07901 US	

GROWTH PLANS/SPECIAL FEATURES:
Celgene Corp. is primarily engaged in the discovery, development and commercialization of therapies designed to treat cancer and immune-inflammatory related diseases. The company's lead products are Revlimid, which FDA approved in 2006 for treatment in combination with dexamethasone for multiple myeloma (a blood cancer) and in December 2005 for treatment of patients with transfusion-dependent anemia due to low- or intermediate-1-risk myelodysplastic syndromes; and Thalomid, which FDA approved in May 2006 for treatment in combination with dexamethasone of newly diagnosed multiple myeloma patients and which is also approved for the treatment and suppression of cutaneous manifestations of erythema nodosum leprosum, an inflammatory complication of leprosy. The firm's portfolio of drug candidates includes IMiDs compounds, which have demonstrated certain immunomodulatory and other biologically important properties. In addition, Celgene is involved in research in several scientific areas that may deliver next-generation therapies, such as intracellular signaling, immunomodulation and placental stem cell research. Celgene's commercial programs include pharmaceutical product sales of Revlimid, Thalomid, Alkeran and Focalin to Novartis Pharma AG; a licensing agreement with Novartis that entitles the firm to royalties of Focalin XR and the entire Ritalin family of drugs; a licensing and product supply agreement with Pharmion Corp. for sales of Pharmion's thalidomide; and sales of bio-therapeutic products and services through subsidiary Cellular Therapeutics. In December 2006, Celgene purchased an active pharmaceutical ingredient (API) manufacturing facility from Siegfried, Ltd. and Siegfried Dienste AG located in Switzerland. The facility has the capability to produce multiple drug substances and will initially be used to produce Revlimid API. In June 2007, Revlimid was granted full marketing authorization in the EU.

The company offers employees medical, dental and vision insurance; healthcare flexible spending accounts; life and AD&D insurance; disability plans; a 401(k) plan; an employee assistance program; educational assistance; and access to two credit unions.

FINANCIALS: Sales and profits are in thousands of dollars—add 000 to get the full amount. 2006 Note: Financial information for 2006 was not available for all companies at press time.

2006 Sales: $898,873	2006 Profits: $68,981	U.S. Stock Ticker: CELG
2005 Sales: $536,941	2005 Profits: $63,656	Int'l Ticker: Int'l Exchange:
2004 Sales: $377,502	2004 Profits: $52,756	Employees: 1,287
2003 Sales: $271,475	2003 Profits: $25,693	Fiscal Year Ends: 12/31
2002 Sales: $135,700	2002 Profits: $-100,000	Parent Company:

SALARIES/BENEFITS:
Pension Plan:	ESOP Stock Plan:	Profit Sharing:	Top Exec. Salary: $811,000	Bonus: $2,422,600
Savings Plan: Y	Stock Purch. Plan:		Second Exec. Salary: $716,333	Bonus: $1,959,600

OTHER THOUGHTS:
Apparent Women Officers or Directors: 1
Hot Spot for Advancement for Women/Minorities:

LOCATIONS: ("Y" = Yes)
West:	Southwest:	Midwest:	Southeast:	Northeast:	International:
				Y	Y

CELL GENESYS INC

www.cellgenesys.com

Industry Group Code: 325412 Ranks within this company's industry group: Sales: 187   Profits: 194

Drugs:		Other:	Clinical:	Computers:	Other:
Discovery:	Y	AgriBio:	Trials/Services:	Hardware:	Specialty Services:
Licensing:	Y	Genomics/Proteomics:	Laboratories:	Software:	Consulting:
Manufacturing:	Y	Tissue Replacement:	Equipment/Supplies:	Arrays:	Blood Collection:
Development:	Y		Research/Development Svcs.:	Database Management:	Drug Delivery:
Generics:			Diagnostics:		Drug Distribution:

TYPES OF BUSINESS:

Cancer Immunotherapies
Oncolytic Virus Therapy Drugs

BRANDS/DIVISIONS/AFFILIATES:

GVAX
CG0070
CG5757

CONTACTS: Note: Officers with more than one job title may be intentionally listed here more than once.

Stephen A. Sherwin, CEO
Sharon E. Tetlow, CFO/Sr. VP
Christine McKinley, Sr. VP-Human Resources
Peter K. Working, Sr. VP-R&D
Michael W. Ramsay, Sr. VP-Oper.
Robert H. Tidwell, Sr. VP-Corp. Dev.
Robert J. Dow, Sr. VP-Medical Affairs
Carol C. Grundfest, Sr. VP-Regulatory Affairs & Portfolio Mgmt.
Kristen M. Hege, VP-Clinical Research
Stephen A. Sherwin, Chmn.

Phone: 650-266-3000	Fax: 650-266-3010
Toll-Free:	
Address: 500 Forbes Blvd., South San Francisco, CA 94080 US	

GROWTH PLANS/SPECIAL FEATURES:

Cell Genesys, Inc. is a biotechnology company that focuses on the development and commercialization of biological therapies for patients with cancer. The company currently develops cell-based immunoherapies and oncolytic virus therapies to threat different types of cancer. The firm's clinical stage cancer programs involve cell- or viral-based products that have been genetically modified to impart disease-fighting characteristics. Cell Genesys' lead program is the GVAX cell-based immunotherapy for cancer. The company is conducting Phase III clinical trials in prostrate cancer and Phase II trials in each of pancreatic cancer and leukemia. Ongoing clinical programs evaluating the company's oncolytic virus therapies focus on CG0070, a therapy for bladder cancer, which could be evaluated in multiple types of cancer in the future. In addition, Cell Genesys has preclinical oncolytic virus therapy programs, including CG5757, which the company is evaluating as potential therapies for multiple types of cancer. The company owns roughly 4122 U.S. and foreign patents issued or granted to it or available based on licensing arrangements and roughly 305 U.S. and foreign applications pending in its name or available based on licensing agreements. In May 2006, the company was granted Fast Track designation for GVAX immunotherapy for prostrate cancer by the FDA.

The company offers its employees medical, dental and vision insurance; short- and long-term disability; a 401(k) plan; an employee stock purchase plan; life, AD&D and travel accident insurance; and an employee assistance program. Additional perks include an alternative commute program, annual company events, same sex partner health insurance benefits, Friday raves, a fitness center and lunchtime seminars.

FINANCIALS: Sales and profits are in thousands of dollars—add 000 to get the full amount. 2006 Note: Financial information for 2006 was not available for all companies at press time.

2006 Sales: $1,364	2006 Profits: $-82,929	U.S. Stock Ticker: CEGE
2005 Sales: $4,584	2005 Profits: $-64,939	Int'l Ticker: Int'l Exchange:
2004 Sales: $11,458	2004 Profits: $-97,411	Employees: 296
2003 Sales: $18,128	2003 Profits: $-56,406	Fiscal Year Ends: 12/31
2002 Sales: $39,100	2002 Profits: $-26,600	Parent Company:

SALARIES/BENEFITS:

Pension Plan:	ESOP Stock Plan:	Profit Sharing:	Top Exec. Salary: $555,000	Bonus: $237,500
Savings Plan: Y	Stock Purch. Plan: Y		Second Exec. Salary: $330,000	Bonus: $155,757

OTHER THOUGHTS:

Apparent Women Officers or Directors: 5
Hot Spot for Advancement for Women/Minorities: Y

LOCATIONS: ("Y" = Yes)

West:	Southwest:	Midwest:	Southeast:	Northeast:	International:
Y					

CELL THERAPEUTICS INC

www.cticseattle.com

Industry Group Code: 325412 **Ranks within this company's industry group:** Sales: 148   Profits: 47

Drugs:		Other:		Clinical:	Computers:		Other:	
Discovery:	Y	AgriBio:		Trials/Services:	Hardware:		Specialty Services:	
Licensing:	Y	Genomics/Proteomics:		Laboratories:	Software:		Consulting:	
Manufacturing:		Tissue Replacement:		Equipment/Supplies:	Arrays:		Blood Collection:	
Development:	Y			Research/Development Svcs.:	Database Management:		Drug Delivery:	
Generics:				Diagnostics:			Drug Distribution:	

TYPES OF BUSINESS:

Cancer Treatment Drugs

BRANDS/DIVISIONS/AFFILIATES:

Cell Therapeutics Europe S.r.l.
XYOTAX
Pixantrone
CT-2106
Novartis International Pharmaceutical, Ltd.

CONTACTS: *Note: Officers with more than one job title may be intentionally listed here more than once.*

James A. Bianco, CEO
James A. Bianco, Pres.
Scott C. Stromatt, Exec. VP-Clinical Dev. & Regulatory Affairs
Louis A. Bianco, Exec. VP-Admin.
Dan Eramian, Exec. VP-Corp. Comm.
Louis A. Bianco, Exec. VP-Finance
Gabrielle Pezzoni, Scientific Dir.-Cell Therapeutics Europe
Jack W. Singer, Chief Medical Officer
Giovanni Ravaioli, Dir.-Human Resources-Cell Therapeutics Europe
Phillip Nudelman, Chmn.
Mauro G. Premi, Acting Managing Dir.-Cell Therapeutics Europe

Phone: 206-282-7100	Fax: 206-284-6206

Toll-Free: 800-215-2355
Address: 501 Elliott Ave. W., Ste. 400, Seattle, WA 98119 US

GROWTH PLANS/SPECIAL FEATURES:

Cell Therapeutics, Inc. (CTI) develops, acquires and commercializes treatments for cancer. The company's research and in-licensing activities are concentrated on identifying new, less toxic and more effective ways to treat cancer. The firm is developing Xyotax for the treatment of non-small cell lung cancer and ovarian cancer; pixantrone, an anthracycline derivative for the treatment of non-Hodgkin's lymphoa; and CT-2106, which is in Phase II component of a phase I/II trial, for the treatment of colorectal cancer relapsing following Folfox therapy. CTI is also working on a number of drug targets in discovery research. Among these programs are bisplatinum agents and proteasome inhibitors with indirect inhibition properties. In addition to discovery research, preclinical activities are focused on product lifecycle management, including the development of alternative dosage forms and routes of administration for existing products in the development pipeline. The company has exclusive rights to six issued U.S. patents and 126 U.S. and foreign pending or issued patent applications relating to its polymer drug delivery technology. The firm has a research and development facility in Europe, Cell Therapeutics Europe S.r.l. in Bresso, Italy. In September 2006, CTI entered into an exclusive worldwide licensing agreement with Novartis International Pharmaceutical, Ltd. for the development and commercialization of Xyotax.

The company offers its employees medical, dental, vision and prescription insurance; life, AD&D, short- and long-term disability insurance; an employee assistance program; travel accident insurance and travel assistance; a 401(k) plan; an employee stock purchase plan; and an educational assistance program. Additional perks include wellness seminars, on-site exercise program, discounted health club memberships, employee breakfasts and ice cream socials.

FINANCIALS: Sales and profits are in thousands of dollars—add 000 to get the full amount. 2006 Note: Financial information for 2006 was not available for all companies at press time.

2006 Sales: $ 80	2006 Profits: $-135,819	**U.S. Stock Ticker:** CTIC
2005 Sales: $16,092	2005 Profits: $-102,505	**Int'l Ticker:** Int'l Exchange:
2004 Sales: $29,594	2004 Profits: $-252,298	Employees: 197
2003 Sales: $24,765	2003 Profits: $-130,031	Fiscal Year Ends: 12/31
2002 Sales: $16,900	2002 Profits: $-49,900	Parent Company:

SALARIES/BENEFITS:

Pension Plan:	ESOP Stock Plan:	Profit Sharing:	Top Exec. Salary: $650,000	Bonus: $240,000
Savings Plan: Y	Stock Purch. Plan: Y		Second Exec. Salary: $340,000	Bonus: $102,000

OTHER THOUGHTS:

Apparent Women Officers or Directors: 2
Hot Spot for Advancement for Women/Minorities: Y

LOCATIONS: ("Y" = Yes)

West:	Southwest:	Midwest:	Southeast:	Northeast:	International:
Y					Y

CELLEGY PHARMACEUTICALS   www.cellegy.com

Industry Group Code: 325412 Ranks within this company's industry group: Sales: 182   Profits: 69

Drugs:	Other:	Clinical:	Computers:	Other:
Discovery:	AgriBio:	Trials/Services:	Hardware:	Specialty Services:
Licensing:	Genomics/Proteomics:	Laboratories:	Software:	Consulting:
Manufacturing:	Tissue Replacement:	Equipment/Supplies:	Arrays:	Blood Collection:
Development: Y		Research/Development Svcs.:	Database Management:	Drug Delivery:
Generics:		Diagnostics:		Drug Distribution:

TYPES OF BUSINESS:
Drugs
HIV Prevention

BRANDS/DIVISIONS/AFFILIATES:
Biosyn, Inc.
Cellegy Canada, Inc.
Savvy

CONTACTS: *Note: Officers with more than one job title may be intentionally listed here more than once.*
Richard C. Williams, Interim CEO
Robert J. Caso, CFO
Frank Malinoski, VP-R&D
John J. Chandler, VP-Corp. Dev.
Robert J. Caso, VP-Finance
Anne-Marie Corner, Sr. VP-Women's Preventative Health
Jung-Chung Lee, VP-Dev.
Richard C. Williams, Chmn.

Phone: 215-529-6084	Fax: 215-529-6086
Toll-Free:	
Address: 2085B Quaker Point Rd., Quakertown, PA 18951 US	

GROWTH PLANS/SPECIAL FEATURES:
Cellegy Pharmaceuticals, Inc. develops and commercializes prescription drugs for the treatment of sexually transmitted diseases and HIV prevention. After the elimination of its research activities in late 2006 and the sale of the majority of its patented product lines, Cellegy's operations are focused primarily on the ownership of its intellectual property rights of its Biosyn product candidates. The firm's subsidiary, Biosyn, Inc., produces a line of proprietary contraceptive and non-contraceptive microbides, which are used to reduce the transmission of sexually transmitted diseases such as AIDS. Biosyn's primary product is Savvy, which was originally undergoing Phase III clinical trials in Africa for its anti-HIV microbicidal efficacy before its recent termination by the Data Monitoring Committee, which declared that the trials were unlikely to provide any relevant statistically significant results. The firm is also conducting Phase I human safety studies on UC-781, a non-nucleoside RT inhibitor that is used to treat HIV-1 isolates in laboratory adapted strains, T cell and macrophage tropic isolates and primary isolates that are resistant to other RT inhibitors. In recent news, Epsilon Pharmaceuticals acquired all the shares of Cellegy Australia Pty Ltd for $1.33 million in April 2006. In July 2006, the FDA issued a conditional approval letter of Cellegy's Cellegesic product if the company would conduct additional clinical trial studies on the efficacy of the product. However, due to insufficient funds, the company sold its right to Cellagesic and Rectogesic, a nitroglycerin ointment for anal fissures; Fortigel and Tostrex, a topical testerone gel for the treatment of maly hypogonadism; and Tostrelle, a testosterone gel for the treatment of female sexual dysfunction in postmenopausal women, to Straken International Limited for $9 million in November 2006.

FINANCIALS: Sales and profits are in thousands of dollars—add 000 to get the full amount. 2006 Note: Financial information for 2006 was not available for all companies at press time.

2006 Sales: $2,660	2006 Profits: $9,672	U.S. Stock Ticker: CLGY
2005 Sales: $12,199	2005 Profits: $-5,008	Int'l Ticker: Int'l Exchange:
2004 Sales: $2,033	2004 Profits: $-28,154	Employees: 5
2003 Sales: $1,600	2003 Profits: $-13,500	Fiscal Year Ends: 12/31
2002 Sales: $1,400	2002 Profits: $-15,900	Parent Company:

SALARIES/BENEFITS:
Pension Plan:	ESOP Stock Plan:	Profit Sharing:	Top Exec. Salary: $398,000	Bonus: $179,100
Savings Plan: Y	Stock Purch. Plan:		Second Exec. Salary: $237,000	Bonus: $17,500

OTHER THOUGHTS:
Apparent Women Officers or Directors: 1
Hot Spot for Advancement for Women/Minorities:

LOCATIONS: ("Y" = Yes)
West:	Southwest:	Midwest:	Southeast:	Northeast:	International:
Y				Y	

CEL-SCI CORPORATION
www.cel-sci.com

ndustry Group Code: 325412 Ranks within this company's industry group: Sales: 29   Profits: 30

Drugs:		Other:		Clinical:		Computers:		Other:	
Discovery:		AgriBio:		Trials/Services:		Hardware:		Specialty Services:	
Licensing:		Genomics/Proteomics:		Laboratories:		Software:		Consulting:	
Manufacturing:	Y	Tissue Replacement:		Equipment/Supplies:		Arrays:		Blood Collection:	
Development:	Y			Research/Development Svcs.:		Database Management:		Drug Delivery:	
Generics:				Diagnostics:				Drug Distribution:	

TYPES OF BUSINESS:
Drugs-Cancer & Infectious Disease
Vaccines

BRANDS/DIVISIONS/AFFILIATES:
MultiKine
LEAPS
Cel-1000
AdapT

CONTACTS: *Note: Officers with more than one job title may be intentionally listed here more than once.*
Geert R. Kersten, CEO
Maximilian de Clara, Pres.
Eyal Talor, Sr. VP-Research
Eyal Talor, Sr. VP-Mfg.
Patricia B. Prichep, Sr. VP-Oper.
Geert R. Kersten, Treas.
Daniel H. Zimmerman, Sr. VP-Cellular Immunology Research
John Cipriano, Sr. VP-Regulatory Affairs
Maximilian de Clara, Chmn.

Phone: 703-506-9460	Fax: 703-506-9471
Toll-Free:	
Address: 8229 Boone Blvd., Ste. 802, Vienna, VA 22182 US	

GROWTH PLANS/SPECIAL FEATURES:
CEL-SCI Corp. researches and develops vaccines and drugs for the treatment of cancer and infectious diseases. Its flagship product, Multikine is an immunotherapeutic drug designed to target tumor micro-metastases primarily responsible for cancer treatment failure. When used to supplement standard form of treatment, Multikine has been proven to greatly increase treatment effectiveness. The company is currently planning to conduct an international Phase III clinical study, in which 800 patients will participate. Additionally, the company owns a pre-clinical T-cell modulation process technology called Ligand Epitope Antigen Presentations System (LEAPS), through which it has developed the Cel-1000 peptide. Cel-1000, designed to be a supplement to a vaccine, has proven to be effective in protecting against herpes, malaria, viral encephalitis and cancer. In partnership with NIAID, the U.S. Navy and the U.S. Army, the company is presently testing its effectiveness of the Cel-100 against avian flu, West Nile virus, SARS, Vaccinia and smallpox. In January 2007, CEL-SCI began plans to build a $15 million manufacturing facility for the production of its cancer product Multikine. In April 2007, CEL-SCI was issued a patent for a T-cell modulation platform technology called AdapT, (Antigen Directed Apoptosis) which can be used for the treatment of a number of major diseases including autoimmune disease, asthma, allergy and transplant rejection. In May 2007, CEL-SCI announced that Cel-1000 was shown to significantly enhance the immune response against avian flu and hepatitis B in animals. In June 2007, Cel-Sci announced that the FDA granted its cancer drug Multikine orphan-drug designation as neoadjuvant therapy in patients with squamous cell carcinoma of the head and neck.

Cel-Sci offers its employees health, dental, vision, and long term disability insurance; a 401(k) plan; flexible spending accounts that allow for pre-tax dependent care and non-reimbursed medical expenses; tuition reimbursement; and an employee referral program.

FINANCIALS: Sales and profits are in thousands of dollars—add 000 to get the full amount. 2006 Note: Financial information for 2006 was not available for all companies at press time.

2006 Sales: $ 125	2006 Profits: $-7,939	U.S. Stock Ticker: CVM
2005 Sales: $ 270	2005 Profits: $-3,040	Int'l Ticker: Int'l Exchange:
2004 Sales: $ 325	2004 Profits: $-2,952	Employees: 19
2003 Sales: $ 318	2003 Profits: $-7,986	Fiscal Year Ends: 9/30
2002 Sales: $ 400	2002 Profits: $-8,300	Parent Company:

SALARIES/BENEFITS:

Pension Plan:	ESOP Stock Plan:	Profit Sharing:	Top Exec. Salary: $386,693	Bonus: $
Savings Plan: Y	Stock Purch. Plan:		Second Exec. Salary: $363,000	Bonus: $200,000

OTHER THOUGHTS:
Apparent Women Officers or Directors: 1
Hot Spot for Advancement for Women/Minorities:

LOCATIONS: ("Y" = Yes)

West:	Southwest:	Midwest:	Southeast:	Northeast:	International:
				Y	

CELSIS INTERNATIONAL PLC

www.celsis.com

Industry Group Code: 325413 **Ranks within this company's industry group:** Sales: 18   Profits: 11

Drugs:	Other:	Clinical:		Computers:		Other:	
Discovery:	AgriBio:	Trials/Services:		Hardware:		Specialty Services:	Y
Licensing:	Genomics/Proteomics:	Laboratories:	Y	Software:		Consulting:	
Manufacturing:	Tissue Replacement:	Equipment/Supplies:	Y	Arrays:	Y	Blood Collection:	
Development:		Research/Development Svcs.:		Database Management:		Drug Delivery:	
Generics:		Diagnostics:	Y			Drug Distribution:	

TYPES OF BUSINESS:

Microbial Contamination Detection Instruments
Laboratory Services Outsourcing

BRANDS/DIVISIONS/AFFILIATES:

RapiScreen
AKuScreen
In Vitro Technologies, Inc.
Celsis Development Services
Celsis Laboratory Group

CONTACTS: *Note: Officers with more than one job title may be intentionally listed here more than once.*

Jay LeCoque, CEO
Christian Madrolle, Dir.-Finance
Christian Madrolle, Corp. Sec.
Jenny Parsons, Dir.-Corp. Comm.
Jack Rowell, Chmn.

Phone: 44-1223-597-851	Fax: 44-1223-597-985
Toll-Free:	
Address: 1010 Cambourne Business Park, Cambourne, Cambridge CB23 6DP UK	

GROWTH PLANS/SPECIAL FEATURES:

Celsis International plc is a leading global provider of rapid diagnostic and monitoring systems to detect and measure microbial contamination in finished products. Its core technology, allowing its products to work successfully, involves bioluminescence-based technology in which ATP (adenosine triphosphate), a nucleotide present in all biological matter, is catalyzed to produce light (as in the abdomen of a firefly). It offers its products and services for many industries, including dairy, beverage, food, pharmaceutical, personal care and cosmetic products. The firm has offices in the U.K., U.S., the Netherlands, Germany and Belgium, as well as a global network of distributors. Celsis's detection product lines include RapiScreen and AKuScreen assays. These systems are capable of reducing microbiological testing times from 5-7 days to 24-48 hours, offering manufacturers a means to ship stock sooner and reduce warehousing expenses. The company also offers validation support, training, instrument support and assistance to help its customers implement Celsis systems. Celsis Development Services, a subsidiary of Celsis, provides a broad range of contract research services that span the drug development process, including evaluations of the in vitro ADME-Tox properties of drug candidates, evaluations of the absorption, metabolism, potential drug-drug interactions and toxicity of compounds and evaluations of drug molecule physiochemical properties. The Celsis Laboratory Group (CLG) is a leading analytical laboratory services provider and has become an outsourcing partner with manufacturers, formulators, re-packagers and their suppliers in the pharmaceutical, biotech, cosmetics, medical devices, household products, specialty chemical and agriculture industries. CLG's services include method development, analytical chemistry, microbiology, stability storage and testing, sterility testing and quality assurance. In July 2006, the company acquired In Vitro Technologies, Inc. which produces in vitro testing products for pharmaceutical and biotechnology interests, for about $30 million.

FINANCIALS: Sales and profits are in thousands of dollars—add 000 to get the full amount. 2006 Note: Financial information for 2006 was not available for all companies at press time.

2006 Sales: $33,104	2006 Profits: $4,603	U.S. Stock Ticker:
2005 Sales: $30,397	2005 Profits: $7,999	Int'l Ticker: CEL Int'l Exchange: London-LSE
2004 Sales: $50,400	2004 Profits: $12,200	Employees: 179
2003 Sales: $42,872	2003 Profits: $26,352	Fiscal Year Ends: 3/31
2002 Sales: $27,999	2002 Profits: $17,225	Parent Company:

SALARIES/BENEFITS:

Pension Plan:	ESOP Stock Plan:	Profit Sharing:	Top Exec. Salary: $159,394	Bonus: $
Savings Plan: Y	Stock Purch. Plan:		Second Exec. Salary: $156,496	Bonus: $

OTHER THOUGHTS:

Apparent Women Officers or Directors: 1
Hot Spot for Advancement for Women/Minorities:

LOCATIONS: ("Y" = Yes)

West:	Southwest:	Midwest:	Southeast:	Northeast:	International:
		Y		Y	Y

CEPHALON INC

www.cephalon.com

Industry Group Code: 325412   Ranks within this company's industry group: Sales: 91   Profits: 62

Drugs:		Other:		Clinical:		Computers:		Other:	
Discovery:	Y	AgriBio:		Trials/Services:		Hardware:		Specialty Services:	
Licensing:	Y	Genomics/Proteomics:		Laboratories:		Software:		Consulting:	
Manufacturing:		Tissue Replacement:		Equipment/Supplies:		Arrays:		Blood Collection:	
Development:	Y			Research/Development Svcs.:	Y	Database Management:	Y	Drug Delivery:	
Generics:				Diagnostics:				Drug Distribution:	

TYPES OF BUSINESS:

Pharmaceutical Discovery & Development
Neurological Disorder Treatments
Cancer Treatments
Pain Medications
Addiction Treatment

BRANDS/DIVISIONS/AFFILIATES:

Provigil
Actiq
Spasfon
Myocet
Abelcet
Vivitrol
Fentora
Nuvigil

CONTACTS: Note: Officers with more than one job title may be intentionally listed here more than once.

Frank Baldino, Jr., CEO
J. Kevin Buchi, CFO/Exec. VP
Jeffry L. Vaught, Exec. VP-R&D
Peter E. Grebow, Sr. VP-Worldwide Tech. Oper.
Carl Savini, Chief Admin. Officer/Exec. VP
John E. Osborn, General Counsel/Exec. VP/Sec.
Lesley Russell, Exec. VP-Worldwide Medical & Regulatory Affairs
Frank Baldino, Jr., Chmn.
Robert P. Roche, Jr., Sr. VP-Worldwide Pharmaceutical Oper.

Phone: 610-344-0200	Fax: 610-738-6590
Toll-Free:	
Address: 41 Moores Rd., Frazer, PA 19355 US	

GROWTH PLANS/SPECIAL FEATURES:

Cephalon, Inc. is a biopharmaceutical company focused on the discovery, development and marketing of products in four core areas: central nervous system (CNS) disorders, pain, cancer and addiction. In addition to actively conducting research and development, the company markets six products in the U.S., as well as numerous other products throughout Europe. Cephalon's technology principally focuses on an understanding of kinases and the role they play in cellular survival and proliferation. The company's most significant products include Provigil tablets, which generated 43% of total revenue in 2006, and Actiq (oral transmucosal fentanyl citrate), which comprised roughly 34% of net sales in 2006. The firm discontinued studying Gabitril for the treatment of generalized anxiety disorders due to unsatisfactory Phase III clinical studies results. Cephalon markets and sells over 25 products in nearly 25 European countries. The five largest products in terms of sales in Europe are Spasfon, Provigil, Myocet, Actiq and Abelcet. Together, these products accounted for 64% of European revenue. During 2006, two products were approved by the FDA: Fentora, a pain product indicated for the management of pain in patients with cancer who are already receiving and are tolerant to opioid therapy for underlying persistent cancer pain; and Vivitrol, indicated for the treatment of alcohol dependant patients who are able to abstain from alcohol in an outpatient setting and are not actively drinking when initiating treatment. In March 2006, FDA did not approve Sparlon for the treatment of ADHD in children and adolescents. In June 2007, Cephalon received FDA approval to market Nuvigil tablets for the treatment of excessive sleepiness.

The company offers its employees a 401(k) plan; health, life and disability coverage; educational reimbursement; and an employee assistance program.

FINANCIALS: Sales and profits are in thousands of dollars—add 000 to get the full amount. 2006 Note: Financial information for 2006 was not available for all companies at press time.

2006 Sales: $1,764,069	2006 Profits: $144,816	U.S. Stock Ticker: CEPH
2005 Sales: $1,211,892	2005 Profits: $-174,954	Int'l Ticker: Int'l Exchange:
2004 Sales: $1,015,400	2004 Profits: $-73,800	Employees: 2,895
2003 Sales: $714,800	2003 Profits: $83,900	Fiscal Year Ends: 12/31
2002 Sales: $506,897	2002 Profits: $171,528	Parent Company:

SALARIES/BENEFITS:

Pension Plan:	ESOP Stock Plan:	Profit Sharing:	Top Exec. Salary: $1,129,000	Bonus: $1,772,500
Savings Plan: Y	Stock Purch. Plan:		Second Exec. Salary: $525,000	Bonus: $404,300

OTHER THOUGHTS:

Apparent Women Officers or Directors: 1
Hot Spot for Advancement for Women/Minorities:

LOCATIONS: ("Y" = Yes)

West:	Southwest:	Midwest:	Southeast:	Northeast:	International:
Y		Y		Y	Y

Note: Financial information, benefits and other data can change quickly and may vary from those stated here.

CEPHEID

www.cepheid.com

Industry Group Code: 325413 Ranks within this company's industry group: Sales: 14   Profits: 22

Drugs:	Other:	Clinical:		Computers:	Other:
Discovery:	AgriBio:	Trials/Services:		Hardware:	Specialty Services:
Licensing:	Genomics/Proteomics:	Laboratories:		Software:	Consulting:
Manufacturing:	Tissue Replacement:	Equipment/Supplies:	Y	Arrays:	Blood Collection:
Development:		Research/Development Svcs.:		Database Management:	Drug Delivery:
Generics:		Diagnostics:	Y		Drug Distribution:

TYPES OF BUSINESS:

Equipment-Biological Testing
Genetic Profiling
DNA Analysis Systems

BRANDS/DIVISIONS/AFFILIATES:

Smart Cycler
Smart Cycler II
GeneXpert
IVD
ASR
RUO
Sangtec Molecular Diagnostics AB
Actigenics

CONTACTS: *Note: Officers with more than one job title may be intentionally listed here more than once.*

John L. Bishop, CEO
John Sluis, CFO
Robert J. Koska, Sr. VP-Sales & Mktg.
Laurie King, VP-Human Resources
Peter J. Dailey, VP-R&D
David H. Persing, CTO/Exec. VP
Nicki Bowen, VP-Eng.
Joseph H. Smith, General Counsel/Sr. VP
Humberto Reyes, Exec. VP-Oper.
Emily S. Winn-Deen, VP-Strategic Planning & Bus. Dev.
John Sluis, Sr. VP-Finance
William McMillan, Sr. VP-Dev.
Russel K. Enns, Sr. VP-Regulatory, Quality & Reimbursement
Sandra Finley, VP-Mktg.
Kerry Flom, VP-Clinical Affairs
Thomas L. Gutshall, Chmn.

Phone: 408-541-4191	Fax: 408-541-4192
Toll-Free: 888-838-3222	
Address: 904 Caribbean Dr., Sunnyvale, CA 94089 US	

GROWTH PLANS/SPECIAL FEATURES:

Cepheid is a molecular diagnostics company that develops, manufactures and markets fully integrated systems for the clinical genetic assessment, life sciences, industrial and biothreat markets. Cephid systems enable rapid molecular testing for organisms and genetic-based diseases by implementing automated technology to reduce the complicated and time-intensive steps that are usually involved in molecular testing. Complex biological systems are analyzed in the company's proprietary biocartridges which eliminates any lengthy preparation, amplification and detection of targeted genes. The company's two principal system platforms are the SmartCycler and GeneXper systems. SmartCycles integrates DNA amplification and detection in order to rapidly analyze samples while the GeneXpert system incorporates sample preparation with DNA amplification and detection. Both systems are utilized in areas of critical infectious diseases, immunocompromised transplantations, women's health and oncology. Other products include the IVD line, ASR line, RUO products line biothreat products and life science products. The company has sold more than 4,300 units to a wide range of customers including the Center for Disease Control and Prevention, the U.S. Food and Drug Administration, Johns Hopkins University, Memorial Sloan-Kettering Cancer Center, the National Institute of Health, Stanford University and the U.S Army Medical Research Institute for Infectious Disease (USAMRIID). GeneXpert has also been incorporated into the United States Postal Service's Biohazard Detection System to identify the presence of anthrax from air samples. Cepheid recently required Actigenics, a microRNA company that has discovered and validated 88 novel microRNAs within the human genome. The company also acquired Sangtec Molecular Diagnostics AB in February 2007, which will add a line of products that manages infections in immunocompromised patients to Cepheid's technologies.

Cepheid offers employees flexible spending accounts income protections plans and an employee assistance program.

FINANCIALS: Sales and profits are in thousands of dollars—add 000 to get the full amount. 2006 Note: Financial information for 2006 was not available for all companies at press time.

2006 Sales: $87,352	2006 Profits: $-25,985	**U.S. Stock Ticker: CPHD**
2005 Sales: $85,010	2005 Profits: $-13,594	**Int'l Ticker:** Int'l Exchange:
2004 Sales: $52,968	2004 Profits: $-13,800	Employees: 307
2003 Sales: $18,534	2003 Profits: $-17,531	Fiscal Year Ends: 12/31
2002 Sales: $14,700	2002 Profits: $-19,700	Parent Company:

SALARIES/BENEFITS:

Pension Plan:	ESOP Stock Plan:	Profit Sharing:	Top Exec. Salary: $400,000	Bonus: $
Savings Plan: Y	Stock Purch. Plan: Y		Second Exec. Salary: $350,000	Bonus: $122,500

OTHER THOUGHTS:

Apparent Women Officers or Directors: 4
Hot Spot for Advancement for Women/Minorities: Y

LOCATIONS: ("Y" = Yes)

West:	Southwest:	Midwest:	Southeast:	Northeast:	International:
Y					Y

CERUS CORPORATION

www.cerus.com

Industry Group Code: 325412 Ranks within this company's industry group: Sales: Profits:

Drugs:		Other:		Clinical:		Computers:		Other:	
Discovery:	Y	AgriBio:		Trials/Services:		Hardware:		Specialty Services:	
Licensing:		Genomics/Proteomics:		Laboratories:		Software:		Consulting:	
Manufacturing:		Tissue Replacement:		Equipment/Supplies:		Arrays:		Blood Collection:	
Development:	Y			Research/Development Svcs.:		Database Management:		Drug Delivery:	
Generics:				Diagnostics:				Drug Distribution:	

TYPES OF BUSINESS:

Pharmaceutical Development
Cancer Vaccines
Blood Treatment Systems

BRANDS/DIVISIONS/AFFILIATES:

Helinx
INTERCEPT
Listeria
Baxter International, Inc.
Killed But Metabolically Active (KBMA)
MedImmune, Inc.

CONTACTS: Note: Officers with more than one job title may be intentionally listed here more than once.

Claes Glassell, CEO
Claes Glassell, Pres.
William J. Dawson, CFO
Laurence M. Corash, VP/Chief Medical Officer
Lori L. Roll, VP-Admin./Corp. Sec.
Howard G. Ervin, VP-Legal Affairs
Myesha Edwards, Dir.-Corp. Comm.
Myesha Edwards, Dir.-Investor Rel.
William J. Dawson, VP-Finance
Thomas W. Dubensky, VP-Vaccine Research
David N. Cook, Corp. Sr. VP
B. J. Cassin, Chmn.
William M. Greenman, Pres., Cerus Europe

Phone: 925-288-6000	Fax: 925-288-6001
Toll-Free:	
Address: 2411 Stanwell Dr., Concord, CA 94520 US	

GROWTH PLANS/SPECIAL FEATURES:

Cerus Corp. develops and commercializes novel, proprietary products and technologies within the fields of immunotherapy and blood safety. In the immunotherapy field, the company is employing its proprietary attenuated Listeria vaccine platform to develop a series of therapies to treat cancer. It currently has three immunotherapeutic cancer vaccine product candidates, one of which entered Phase I clinical trials in 2006. Two others are in preclinical development. These products are designed to stimulate both innate and adaptive immune pathways, generating highly specific and highly potent anti-tumor responses. Cerus is collaborating with Johns Hopkins University and MedImmune, Inc. in the development of these product candidates. Also in immunotherapy, the company is applying its proprietary Killed But Metabolically Active, or KBMA, technology platform in research and development of prophylactic and therapeutic vaccines for infectious diseases, including hepatitis C. In the field of blood safety, Cerus is developing and commercializing the INTERCEPT Blood System for platelets, plasma and red blood cells. INTERCEPT system, which is based on the company's proprietary Helinx technology for controlling biological replication, is designed to enhance the safety of donated blood components by inactivating viruses, bacteria, parasites and other pathogens, as well as potentially harmful white blood cells. Cerus' Helinx technology prevents the replication of the DNA or RNA of susceptible pathogens, such as hepatitis B, hepatitis C and HIV. INTERCEPT for platelets and plasma is marketed both in the US and the European Union, while INTERCEPT for red blood cells is currently in Phase I clinical trials. In February 2006, Cerus obtained the exclusive rights to INTERCEPT from its former partner, Baxter International, Inc.

Cerus offers its employees flexible spending accounts, domestic partner coverage, stock options, a health club membership discount, a tuition reimbursement program and paid time off.

FINANCIALS: Sales and profits are in thousands of dollars—add 000 to get the full amount. 2006 Note: Financial information for 2006 was not available for all companies at press time.

			U.S. Stock Ticker: CERS
2006 Sales: $35,580	2006 Profits: $-4,779		Int'l Ticker: Int'l Exchange:
2005 Sales: $24,371	2005 Profits: $13,064		Employees: 124
2004 Sales: $13,911	2004 Profits: $-31,153		Fiscal Year Ends: 12/31
2003 Sales: $9,665	2003 Profits: $-58,267		Parent Company:
2002 Sales: $8,500	2002 Profits: $-57,200		

SALARIES/BENEFITS:

Pension Plan:	ESOP Stock Plan:	Profit Sharing:	Top Exec. Salary: $447,917	Bonus: $221,238
Savings Plan: Y	Stock Purch. Plan: Y		Second Exec. Salary: $355,834	Bonus: $87,466

OTHER THOUGHTS:

Apparent Women Officers or Directors: 1
Hot Spot for Advancement for Women/Minorities:

LOCATIONS: ("Y" = Yes)

West:	Southwest:	Midwest:	Southeast:	Northeast:	International:
Y					Y

CHARLES RIVER LABORATORIES INTERNATIONAL INC
www.criver.com

Industry Group Code: 541710 **Ranks within this company's industry group:** Sales: 3   Profits: 21

Drugs:	Other:	Clinical:		Computers:		Other:	
Discovery:	AgriBio:	Trials/Services:		Hardware:		Specialty Services:	Y
Licensing:	Genomics/Proteomics:	Laboratories:	Y	Software:	Y	Consulting:	
Manufacturing:	Tissue Replacement:	Equipment/Supplies:	Y	Arrays:		Blood Collection:	
Development:		Research/Development Svcs.:		Database Management:		Drug Delivery:	
Generics:		Diagnostics:	Y			Drug Distribution:	

TYPES OF BUSINESS:
Animal Research Models
Consulting
Bioactivity Software
Biosafety Testing
Contract Staffing
Laboratory Diagnostics
Intellectual Property Consulting
Analytical Testing

BRANDS/DIVISIONS/AFFILIATES:
Endo-Scan-V
BioTrend
Inveresk Research Group, Inc.

CONTACTS: *Note: Officers with more than one job title may be intentionally listed here more than once.*
James C. Foster, CEO
James C. Foster, Pres.
Thomas F. Ackerman, CFO/Exec. VP
Stephanie Wells, Sr. VP-Mktg.
David P. Johst, Exec. VP-Human Resources
Real Renaud, Exec. VP-Research Model Products
Nicholas Ventresca, CIO/Sr. VP-IT
David P. Johst, Exec. VP-Admin.
Joanne Acford, General Counsel/Sr. VP/Corp. Sec.
John C. Ho, Sr. VP-Corp. Strategy
Susan Hardy, VP-Investor Rel.
David J. Elliott, Corp. Controller
Nancy Gillett, Sr. VP/Pres., Global Preclinical Svcs.
Christopher Perkin, VP/Pres., Canadian Preclinical Svcs.
James C. Foster, Chmn.

Phone: 978-658-6000	Fax: 978-658-7132
Toll-Free:	
Address: 261 Ballardvale St., Wilmington, MA 01867 US	

GROWTH PLANS/SPECIAL FEATURES:
Charles River Laboratories International, Inc. (CRL) is a provider of critical research tools and integrated support services that enable innovative and efficient drug discovery and development. The company is a global leader in providing animal research models required for the research and development of new drugs, devices and therapies. Animal disease model programs are focused on research into cardiovascular, metabolic, renal and oncology issues. The firm operates 80 production facilities in 15 countries. CRL is divided into three segments: research models and services; preclinical services; and clinical services. The research models and services division provides transgenic laboratory; preconditioning and surgical; consulting and staffing; vaccine support; and in-vitro technology services. The preclinical services division provides services in the discovery and development of new drugs, devices and therapies. The development services portion of the division enables customers to outsource critical regulatory-required toxicology and drug disposition activities to Charles River. The clinical services business is a new market originating through the acquisition of Inveresk Research Group, Inc. and includes a Phase I clinic in Europe and the international capability to manage Phase II-IV studies. Clinical trials management services include strategy development; investigator recruitment; study monitoring; data management; research and consulting; study design; patient recruitment; pharmaco-vigilance; biostatistical analysis; and post-marketing/Phase IV studies. In 2007, the company announced its global expansion into Asia, constructing a 50,000 square foot preclinical services facility in Shanghai. Charles River has also recently received FDA approval for the sale and marketing of the first portable endotoxin system.

FINANCIALS: Sales and profits are in thousands of dollars—add 000 to get the full amount. 2006 Note: Financial information for 2006 was not available for all companies at press time.
2006 Sales: $1,058,385	2006 Profits: $-55,783	U.S. Stock Ticker: CRL
2005 Sales: $993,328	2005 Profits: $141,999	Int'l Ticker: Int'l Exchange:
2004 Sales: $724,221	2004 Profits: $89,792	Employees: 8,000
2003 Sales: $613,723	2003 Profits: $80,151	Fiscal Year Ends: 12/31
2002 Sales: $554,600	2002 Profits: $50,200	Parent Company:

SALARIES/BENEFITS:
Pension Plan: Y	ESOP Stock Plan:	Profit Sharing:	Top Exec. Salary: $450,000	Bonus: $1,200,000
Savings Plan:	Stock Purch. Plan:		Second Exec. Salary: $340,385	Bonus: $2,500,000

OTHER THOUGHTS:
Apparent Women Officers or Directors: 4
Hot Spot for Advancement for Women/Minorities: Y

LOCATIONS: ("Y" = Yes)
West:	Southwest:	Midwest:	Southeast:	Northeast:	International:
Y	Y	Y	Y	Y	Y

CHATTEM INC

www.chattem.com

Industry Group Code: 325411 Ranks within this company's industry group: Sales: 3   Profits: 2

Drugs:		Other:	Clinical:	Computers:	Other:
Discovery:		AgriBio:	Trials/Services:	Hardware:	Specialty Services:
Licensing:	Y	Genomics/Proteomics:	Laboratories:	Software:	Consulting:
Manufacturing:	Y	Tissue Replacement:	Equipment/Supplies:	Arrays:	Blood Collection:
Development:			Research/Development Svcs.:	Database Management:	Drug Delivery:
Generics:			Diagnostics:		Drug Distribution:

TYPES OF BUSINESS:

Branded Consumer Products - Medicinal
Over-the-Counter Drugs
Toiletries
Skin Care Products

BRANDS/DIVISIONS/AFFILIATES:

Chattem (U.K.), Ltd.
Sundex, LLC
Pamprin
Icy Hot
Gold Bond
Bull Frog
Selsun Blue
Dexatrim

CONTACTS: Note: Officers with more than one job title may be intentionally listed here more than once.

Ian Guerry, CEO
Robert E. Bosworth, COO
Robert E. Bosworth, Pres.
Charles M. Stafford, VP-Sales
Theodore K. Whitfield, Jr., General Counsel/VP/Sec.
. Derrill Pitts, VP-Oper.
Ron Galante, VP-New Bus. Dev.
Robert B. Long, Chief Acct. Officer
Andrea M. Crouch, VP-Brand Mgmt.
Richard W. Kornhauser, VP-Mktg.
. Blair Ramey, VP-Brand Mgmt. & Media
Ian Guerry, Chmn.

Phone: 423-821-4571	Fax:

Toll-Free:

Address: 1715 W. 38th St., Chattanooga, TN 37409 US

GROWTH PLANS/SPECIAL FEATURES:

Chattem, Inc., founded in 1879 as Chatanooga Medicine Corp., is a leading marketer and manufacturer of branded consumer products, including over-the-counter drugs, toiletries, topical pain care products, medicated skin care products, medicated dandruff shampoos and dietary supplements. The company's products target niche market segments which are often overlooked by larger companies. Principally utilizing its own sales force, Chattem sells its products nationally through mass merchandisers, drug and food channels. Chattem's topical pain care brands are Icy Hot, Icy Hot Pro-Therapy, Aspercreme, Flexall, Capzasin, Sportscreme and Arthritis Hot. The company's medicated skin care brands are Gold Bond, Cortizone and Balmex, while Selsun Blue and Selsun Salon make up its medication dandruff shampoo line. Chattem offers dietary supplement brands Dexatrim, Garlique, Melatonex, New Phase and Omnigest EZ; Pamprin, a menstrual pain reliever; Bullfrog, a waterproof sunscreen; Unisom, a sleep aid; and Mudd, a line of specialty masque products. The company also operates an international division which represented approximately 8% of its 2006 sales and is concentrated in Canada, where brands such as Icy Hot, Selsun, Gold Bond and Pamprin are marketed through Chattem's subsidiary Chattem Canada. European business is conducted through Chattem Global Consumer Products Ltd., an Irish subsidiary of Chattem's, as well as Chattem (U.K.) Limited, a subsidiary located in England. Additionally, Chattem controls three other subsidiaries: Signal Investment & Management Co.; Sundex, LLC; and HBA Indemnity Company, Ltd. In January 2007, the company acquired brands ACT, a mouthwash; Unisom; Cortizone; Kaopectate, an anti-diarrhea product; and Balmex, a diaper rash product.

FINANCIALS: Sales and profits are in thousands of dollars—add 000 to get the full amount. 2006 Note: Financial information for 2006 was not available for all companies at press time.

2006 Sales: $300,548	2006 Profits: $45,112	U.S. Stock Ticker: CHTT
2005 Sales: $279,318	2005 Profits: $36,047	Int'l Ticker: Int'l Exchange:
2004 Sales: $258,155	2004 Profits: $1,451	Employees: 447
2003 Sales: $233,749	2003 Profits: $23,347	Fiscal Year Ends: 11/30
2002 Sales: $220,800	2002 Profits: $10,000	Parent Company:

SALARIES/BENEFITS:

Pension Plan:	ESOP Stock Plan:	Profit Sharing:	Top Exec. Salary: $539,785	Bonus: $150,000
Savings Plan:	Stock Purch. Plan:		Second Exec. Salary: $381,923	Bonus: $100,000

OTHER THOUGHTS:

Apparent Women Officers or Directors: 1
Hot Spot for Advancement for Women/Minorities:

LOCATIONS: ("Y" = Yes)

West:	Southwest:	Midwest:	Southeast:	Northeast:	International:
			Y		Y

CHESAPEAKE BIOLOGICAL LAB   www.cblinc.com

Industry Group Code: 541710 Ranks within this company's industry group: Sales: Profits:

Drugs:		Other:		Clinical:		Computers:		Other:	
Discovery:		AgriBio:		Trials/Services:	Y	Hardware:		Specialty Services:	Y
Licensing:		Genomics/Proteomics:		Laboratories:	Y	Software:		Consulting:	Y
Manufacturing:	Y	Tissue Replacement:		Equipment/Supplies:		Arrays:		Blood Collection:	
Development:				Research/Development Svcs.:	Y	Database Management:		Drug Delivery:	
Generics:				Diagnostics:				Drug Distribution:	

TYPES OF BUSINESS:

Research & Development
Contract Pharmaceutical Services
Regulatory & Compliance Consulting

BRANDS/DIVISIONS/AFFILIATES:

Cangene Corporation

CONTACTS: *Note: Officers with more than one job title may be intentionally listed here more than once.*

William Labossiere Bees, VP-Oper., Cangene
Steven E. Rowan, VP-Mktg.
Vicki Wolff-Long, VP-Oper.
Steven E. Rowan, VP-Bus. Dev.

Phone: 410-843-5000	Fax: 410-843-4414
Toll-Free: 800-441-4225	
Address: 1111 S. Paca St., Baltimore, MD 21230-2591 US	

GROWTH PLANS/SPECIAL FEATURES:

Chesapeake Biological Laboratories, Inc. (CBL), a subsidiary of Cangene Corporation, provides contract pharmaceutical and biopharmaceutical product development and production services to over 150 companies. Customers typically hire the company to produce developmental stage products for FDA clinical trials or to produce and manufacture FDA approved products for commercial sale. Since its inception CBL has developed over 175 therapeutic products and provides a variety of services in material processing; aseptic vial and syringe filling; clinical supply and commercial production; formulation and quality control testing. The firm recently introduced an accelerated clinical trial program which produces desired clinical products in only 120 days for R&D companies. The clinical trial drug manufacturing process involves a process of assay validation, lyophilization cycle development and stability testing. A variety of formulation processes are conducted at CBL's multiple suites in order to form aseptic suspensions and formulas intended for freeze-drying (lyophilization). Aseptic process capabilities of the company include dilutions of sterilized solutions formulation of sterile suspensions, hydrogel and micro sphere inclusion compounding. Products that require freeze drying are filled, partially stoppered and loaded into a pre chilled lyophilizer throughout the filling period. The CBL validation department then designs, calibrates, certifies and validates the systems used in aseptic filling and freeze drying so that products can proceed to stability testing and packaging. Analytical methods are also employed to ensure the identity, purity, quality and strength of each finished drug product. In the past, CBL has acted as an FDA-approved third part contractor in filling, freeze-drying and packaging CroFab for Protherics plc. In response to the recognition of demands for biodefense products, Chesapeake has also constructed the BL-2 vaccine plant, which has been used to fill substantial portions of the U.S. smallpox vaccine stockpile.

CBL offers its employees health, life and disability insurance and a 401(k) plan.

FINANCIALS: Sales and profits are in thousands of dollars—add 000 to get the full amount. 2006 Note: Financial information for 2006 was not available for all companies at press time.

2006 Sales: $	2006 Profits: $	**U.S. Stock Ticker: Subsidiary**
2005 Sales: $	2005 Profits: $	**Int'l Ticker:** Int'l Exchange:
2004 Sales: $	2004 Profits: $	Employees: 125
2003 Sales: $17,000	2003 Profits: $	Fiscal Year Ends: 7/31
2002 Sales: $	2002 Profits: $	Parent Company: CANGENE CORP

SALARIES/BENEFITS:

Pension Plan:	ESOP Stock Plan:	Profit Sharing:	Top Exec. Salary: $	Bonus: $
Savings Plan: Y	Stock Purch. Plan:		Second Exec. Salary: $	Bonus: $

OTHER THOUGHTS:

Apparent Women Officers or Directors: 1
Hot Spot for Advancement for Women/Minorities:

LOCATIONS: ("Y" = Yes)

West:	Southwest:	Midwest:	Southeast:	Northeast:	International:
				Y	

CHIRON CORP   www.chiron.com

ndustry Group Code: 325412   Ranks within this company's industry group: Sales: 101   Profits: 135

Drugs:		Other:		Clinical:		Computers:		Other:	
Discovery:	Y	AgriBio:		Trials/Services:		Hardware:		Specialty Services:	
Licensing:		Genomics/Proteomics:	Y	Laboratories:		Software:		Consulting:	
Manufacturing:	Y	Tissue Replacement:		Equipment/Supplies:		Arrays:		Blood Collection:	
Development:	Y			Research/Development Svcs.:		Database Management:		Drug Delivery:	
Generics:				Diagnostics:	Y			Drug Distribution:	

TYPES OF BUSINESS:

Pharmaceuticals Discovery & Development
Biopharmaceuticals
Vaccines
Blood Screening Assays

BRANDS/DIVISIONS/AFFILIATES:

Novartis AG
TOBI
Betaseron
Fluad
RabAvert
Chiron Blood Testing
Chiron Biopharmaceuticals
Chiron Vaccines

CONTACTS: *Note: Officers with more than one job title may be intentionally listed here more than once.*

Gene Walther, Pres.

Phone: 510-655-8730	Fax: 510-655-9910
Toll-Free:	
Address: 4560 Horton St., Emeryville, CA 94608-2916 US	

GROWTH PLANS/SPECIAL FEATURES:

Chiron, a subsidiary of Novatis AG, is a biotechnology company that participates in three global health care businesses: biopharmaceuticals, vaccines and blood testing. The company develops innovative products for preventing and treating cancer, infectious diseases and cardiovascular disease. Chiron BioPharmaceuticals discovers, develops, manufactures and markets a range of therapeutic products, including TOBI for lung infections in cystic fibrosis patients, Betaseron for multiple sclerosis and Proleukin for cancer. Chiron Vaccines, the fifth-largest vaccines business in the world, currently offers more than 30 vaccines. They include Menjugate for meningococcal meningitis, Fluad for influenza, Encepur for tick-borne encephalitis and RabAvert for exposure to rabies. Menjugate has not been approved for use in the U.S. Chiron Blood Testing provides screening products used by the blood bank industry. Its nucleic acid testing blood screening assays include the Procleix HIV-1/HCV and Procleix West Nile Virus assays. In addition, the firm produces the Ultrio assay, designed to detect HIV and hepatitis C and B viruses. Chiron maintains partnerships with several major pharmaceutical companies, including Schering AG/Berlex Laboratories, Gen-Probe, Nektar Therapeutics and Ortho-Clinical Diagnostics to expand its portfolio of intellectual properties, develop new products and license those products in key markets. In April 2006, Novartis AG, a major pharmaceutical company, purchased Chiron.

Chiron offers its employees educational assistance, credit union membership and access to prepaid legal services, as well as discounted auto and home insurance. The company also has a LifeCare service that provides information and support on topics such as adoption, prenatal planning and child care.

FINANCIALS: Sales and profits are in thousands of dollars—add 000 to get the full amount. 2006 Note: Financial information for 2006 was not available for all companies at press time.

2006 Sales: $	2006 Profits: $	U.S. Stock Ticker: Subsidiary	
2005 Sales: $1,921,000	2005 Profits: $187,000	Int'l Ticker: Int'l Exchange:	
2004 Sales: $1,723,355	2004 Profits: $78,917	Employees: 5,500	
2003 Sales: $1,776,361	2003 Profits: $227,313	Fiscal Year Ends: 12/31	
2002 Sales: $972,900	2002 Profits: $180,800	Parent Company: NOVARTIS AG	

SALARIES/BENEFITS:

Pension Plan:	ESOP Stock Plan: Y	Profit Sharing:	Top Exec. Salary: $800,000	Bonus: $897,293
Savings Plan: Y	Stock Purch. Plan: Y		Second Exec. Salary: $552,461	Bonus: $1,500,000

OTHER THOUGHTS:

Apparent Women Officers or Directors:
Hot Spot for Advancement for Women/Minorities:

LOCATIONS: ("Y" = Yes)

West:	Southwest:	Midwest:	Southeast:	Northeast:	International:
Y		Y		Y	Y

CIPHERGEN BIOSYSTEMS INC   www.ciphergen.com

Industry Group Code: 325413 Ranks within this company's industry group: Sales: 22   Profits: 21

Drugs:	Other:		Clinical:		Computers:		Other:	
Discovery:	AgriBio:		Trials/Services:		Hardware:		Specialty Services:	
Licensing:	Genomics/Proteomics:	Y	Laboratories:	Y	Software:		Consulting:	
Manufacturing:	Tissue Replacement:		Equipment/Supplies:	Y	Arrays:		Blood Collection:	
Development:			Research/Development Svcs.:	Y	Database Management:		Drug Delivery:	
Generics:			Diagnostics:	Y			Drug Distribution:	

TYPES OF BUSINESS:

Biomarkers
Translation Proteomics
Research Services
Specialty Diagnostics

BRANDS/DIVISIONS/AFFILIATES:

Biomarker Discovery Centers
Ciphergen Biosystems K.K.

CONTACTS: *Note: Officers with more than one job title may be intentionally listed here more than once.*

Gail S. Page, CEO
Gail S. Page, Pres.
Debra A. Young, CFO/VP
Steve Lundy, Sr. VP-Sales & Mktg.
Eric Fung, Chief Scientific Officer/VP
William C. Sullivan, VP-Corp. Oper.
Simon C. Shorter, VP-Bus. Dev., Diagnostics Div.
Daniel M. Caserza, Corp. Controller
James L. Rathmann, Chmn.

Phone: 510-505-2100	Fax: 510-505-2101
Toll-Free: 888-864-3770	
Address: 6611 Dumbarton Cir., Fremont, CA 94555 US	

GROWTH PLANS/SPECIAL FEATURES:

Ciphergen Biosystems, Inc. discovers, develops and commercializes specialty diagnostic tests with translation proteomics technologies, which utilizes advanced protein separation tools to identify and resolve variants of specific biomarkers, developing assays and tests. The company's most advanced products are focused on the accurate diagnosis of ovarian cancer, which is achieved through a panel of Ciphergen biomarkers that provides risk stratification information on ovarian cancer. In addition Ciphergen also provides diagnostic products in the areas of hematology/oncology, cardiovascular disease and women's health. Ciphergen is currently working in collaboration with Johns Hopkins to produce blood markers and molecular imaging targets for breast cancer diagnosis and has also discovered two biomarkers that will be able to predict the likelihood of the recurrence of prostate cancer. In addition Ciphergen has also established Biomarker Discovery Center facilities in California, Pennsylvania, Denmark and Japan which offers a variety of research and development services to solve complex biological research problems. Current company research is focused on methodologies for bead technologies, orthogonal chromatographic separation of proteomes and clinical assay development that utilizes novel proteomics technologies. In recent news, Ciphergen announced the discovery and validation of biomarkers in Alzheimer's disease and additionally discovered protein biomarkers that are essential for the diagnosis of early-stage ovarian cancer. In late 2006, the firm sold its proprietary proteomics instrument business, which includes Ciphergen's Surface Enhanced Laser Desorption/Ionization (SELDI) technology, ProteinChip Array and software for $20 million to Bio-Rad Laboratories, Inc. Ciphergen will continue to have a supply agreement with Bio-Rad in order to sell its SELDI instruments and consumables for the firm's own diagnostics business. In early 2007, Ciphergen established a development program with Quest Diagnostics to further develop a blood-based assay for the detection of peripheral artery disease (PAD).

FINANCIALS: Sales and profits are in thousands of dollars—add 000 to get the full amount. 2006 Note: Financial information for 2006 was not available for all companies at press time.

2006 Sales: $18,215	2006 Profits: $-22,066	U.S. Stock Ticker: CIPH
2005 Sales: $27,246	2005 Profits: $-35,433	Int'l Ticker: Int'l Exchange:
2004 Sales: $20,181	2004 Profits: $-19,841	Employees: 36
2003 Sales: $43,638	2003 Profits: $-36,747	Fiscal Year Ends: 12/31
2002 Sales: $39,300	2002 Profits: $-29,100	Parent Company:

SALARIES/BENEFITS:

Pension Plan:	ESOP Stock Plan:	Profit Sharing:	Top Exec. Salary: $275,000	Bonus: $
Savings Plan: Y	Stock Purch. Plan: Y		Second Exec. Salary: $262,538	Bonus: $174,750

OTHER THOUGHTS:

Apparent Women Officers or Directors: 2
Hot Spot for Advancement for Women/Minorities: Y

LOCATIONS: ("Y" = Yes)

West:	Southwest:	Midwest:	Southeast:	Northeast:	International:
Y	Y			Y	Y

Note: Financial information, benefits and other data can change quickly and may vary from those stated here.

CML HEALTHCARE INCOME FUND   www.cmlhealthcare.com

Industry Group Code: 621511   Ranks within this company's industry group: Sales: Profits:

Drugs:	Other:	Clinical:		Computers:	Other:
Discovery:	AgriBio:	Trials/Services:		Hardware:	Specialty Services:
Licensing:	Genomics/Proteomics:	Laboratories:	Y	Software:	Consulting:
Manufacturing:	Tissue Replacement:	Equipment/Supplies:		Arrays:	Blood Collection:
Development:		Research/Development Svcs.:		Database Management:	Drug Delivery:
Generics:		Diagnostics:	Y		Drug Distribution:

TYPES OF BUSINESS:

Medical Diagnostic Services
Laboratory Testing Services
Medical Imaging Services
Investments

BRANDS/DIVISIONS/AFFILIATES:

CML Healthcare Inc.

CONTACTS: Note: Officers with more than one job title may be intentionally listed here more than once.

Paul J. Bristow, CEO
Paul J. Bristow, Pres.
Tom Weber, CFO
Cameron Duff, VP-Corp. Dev.
Donald W. Kerr, Gen. Mgr.-Laboratory Svcs.
Kent Wentzell, Gen. Mgr.-Imaging Svcs.

Phone: 905-565-0043	Fax: 905-565-1776

Toll-Free: 800-263-0801
Address: 6560 Kennedy Rd., Mississauga, ON L5T 2X4 Canada

GROWTH PLANS/SPECIAL FEATURES:

CML HealthCare Income Fund is an open-ended trust that owns CML Healthcare, Inc., a Canadian diagnostic services business providing laboratory testing services in Ontario and medical imaging services across Canada. CML HealthCare operates in two divisions: laboratory services and imaging services. The laboratory services division conducts a wide range of medical tests, including tests for blood chemistry analysis; urinalysis; hematology; PAP smears; microbiology; and drug and HIV screening. All of the tests are used by physicians to assist in diagnosis and treatment. The laboratory services division operates through the company's network of two licensed medical diagnostic laboratories and 119 operating licensed specimen collection centers. The imaging services division provides medical imaging services, including x-ray and fluoroscopy, ultrasound, mammography, bone densitometry, nuclear medicine, magnetic resonance imaging and computed tomography through a network of 101 non-hospital based medical imaging clinics located in Ontario, British Columbia, Alberta, Manitoba and Quebec.

FINANCIALS: Sales and profits are in thousands of dollars—add 000 to get the full amount. 2006 Note: Financial information for 2006 was not available for all companies at press time.

		U.S. Stock Ticker:
2006 Sales: $272,105	2006 Profits: $87,056	Int'l Ticker: CLC.UN Int'l Exchange: Toronto-TSX
2005 Sales: $256,753	2005 Profits: $74,979	Employees: 2,037
2004 Sales: $	2004 Profits: $	Fiscal Year Ends: 12/31
2003 Sales: $	2003 Profits: $	Parent Company:
2002 Sales: $	2002 Profits: $	

SALARIES/BENEFITS:

Pension Plan:	ESOP Stock Plan:	Profit Sharing:	Top Exec. Salary: $	Bonus: $
Savings Plan:	Stock Purch. Plan:		Second Exec. Salary: $	Bonus: $

OTHER THOUGHTS:

Apparent Women Officers or Directors:
Hot Spot for Advancement for Women/Minorities:

LOCATIONS: ("Y" = Yes)

West:	Southwest:	Midwest:	Southeast:	Northeast:	International:
					Y

Note: Financial information, benefits and other data can change quickly and may vary from those stated here.

COLLAGENEX PHARMACEUTICALS INCwww.collagenex.com

Industry Group Code: 325412 **Ranks within this company's industry group:** Sales: Profits:

Drugs:		Other:	Clinical:	Computers:	Other:	
Discovery:	Y	AgriBio:	Trials/Services:	Hardware:	Specialty Services:	
Licensing:	Y	Genomics/Proteomics:	Laboratories:	Software:	Consulting:	
Manufacturing:		Tissue Replacement:	Equipment/Supplies:	Arrays:	Blood Collection:	
Development:	Y		Research/Development Svcs.:	Database Management:	Drug Delivery:	Y
Generics:			Diagnostics:		Drug Distribution:	Y

TYPES OF BUSINESS:

Pharmaceuticals Development
Drugs-Periodontal Disease
Drugs-Dermatology

BRANDS/DIVISIONS/AFFILIATES:

Pandel
Alcortin
Novacort
IMPACS
Oracea
Restoraderm
Periostat
SansRosa

CONTACTS: *Note: Officers with more than one job title may be intentionally listed here more than once.*

Colin W. Stewart, CEO
Colin W. Stewart, Pres.
Nancy C. Broadbent, CFO
David F. Pfeiffer, Sr. VP-Sales & Mktg.
Klaus P. J. Theobald, Sr. VP/Chief Medical Officer
Andrew K. W. Powell, General Counsel/VP/Corp. Sec.
J. Gregory Ford, VP-Bus. Dev. & Strategic Planning
Nancy C. Broadbent, Treas.
James E. Daverman, Chmn.

Phone: 215-579-7388	Fax: 215-579-8577
Toll-Free:	
Address: 41 University Dr., Newtown, PA 18940 US	

GROWTH PLANS/SPECIAL FEATURES:

CollaGenex Pharmaceuticals, Inc. is a specialty pharmaceutical company focused on providing innovative medical therapies to the dermatology market. Its chief products are Oracea for the treatment of inflammatory lesions of rosacea in adults; Pandel, used for the treatment of dermatitis and psoriasis; Alcortin, a mild dermatosis gel with combined anti-fungal, anti-bacterial and anti-inflammatory effects; and Novacort, a topical corticosteroid with anti-inflammatory and anesthetic treatment for certain dermatoses. CollaGenex's development products are based on its core technologies: IMPACS, Restoraderm and SansRosa. IMPACS compounds have the potential to treat diseases that cause inflammation and destruction of the connective tissues. Restoraderm delivers key ingredients to the dermal layers underlying the skin. The SansRosa technology, which was obtained in the acquisition of SansRosa Pharmaceutical Development, has shown promise in reducing redness associated with rosacea. CollaCenex has discontinued the development of periodontal products, but still continues to market the products Atridox, Atrisorb-Freeflow and Atrisorb-D, which are licensed from Tolmar, Inc., a subsidiary of Tecnofarma, S.A.; and Periostat, its own development. In May 2006, CollaGenex received FDA approval for Oracea, an IMPACS-based oral therapy for inflammatory lesions and began to market the product in July 2006. In December 2006, the company signed a license and supply agreement with Medigene AG, a German biopharmaceutical company, to market Oracea in the European Union and certain contiguous countries and Russia. In May 2007, the company announced that it had signed an agreement with QuatRx Pharmaceuticals Company to develop and commercialize QuatRx's product bencocalcidiol for psoriasis.

FINANCIALS: Sales and profits are in thousands of dollars—add 000 to get the full amount. 2006 Note: Financial information for 2006 was not available for all companies at press time.

2006 Sales: $26,373	2006 Profits: $-33,434	U.S. Stock Ticker: CGPI
2005 Sales: $26,405	2005 Profits: $-18,805	Int'l Ticker: Int'l Exchange:
2004 Sales: $52,146	2004 Profits: $6,528	Employees: 130
2003 Sales: $52,900	2003 Profits: $6,400	Fiscal Year Ends: 12/31
2002 Sales: $44,600	2002 Profits: $ 900	Parent Company:

SALARIES/BENEFITS:

Pension Plan:	ESOP Stock Plan:	Profit Sharing:	Top Exec. Salary: $377,000	Bonus: $258,246
Savings Plan: Y	Stock Purch. Plan:		Second Exec. Salary: $269,100	Bonus: $139,932

OTHER THOUGHTS:

Apparent Women Officers or Directors: 1
Hot Spot for Advancement for Women/Minorities:

LOCATIONS: ("Y" = Yes)

West:	Southwest:	Midwest:	Southeast:	Northeast:	International:
				Y	Y

COLUMBIA LABORATORIES INC   www.columbialabs.com

Industry Group Code: 325412 Ranks within this company's industry group: Sales: 113   Profits: 80

Drugs:		Other:		Clinical:	Computers:		Other:	
Discovery:	Y	AgriBio:		Trials/Services:	Hardware:		Specialty Services:	
Licensing:		Genomics/Proteomics:		Laboratories:	Software:		Consulting:	
Manufacturing:	Y	Tissue Replacement:		Equipment/Supplies:	Arrays:		Blood Collection:	
Development:	Y			Research/Development Svcs.:	Database Management:		Drug Delivery:	Y
Generics:				Diagnostics:			Drug Distribution:	

TYPES OF BUSINESS:
Women's Healthcare Drugs
Endocrine Drugs

BRANDS/DIVISIONS/AFFILIATES:
Prochieve
Crinone
Striant
RepHresh
Lidocaine
Bioadhesive Delivery System

CONTACTS: *Note: Officers with more than one job title may be intentionally listed here more than once.*
Robert S. Mills, CEO
RobertS. Mills, Pres.
James A. Meer, CFO
Carl Worrell, Head-Sales & Mktg.
George W. Creasy, VP-Clinical Research & Dev.
Michael McGrane, General Counsel/Sr. VP/Sec.
James A. Meer, Treas./Sr. VP
Stephen G. Kasnet, Chmn.

Phone: 973-994-3999	Fax: 973-994-3001

Toll-Free: 866-566-5636
Address: 354 Eisenhower Pkwy., Plaza 1, 2nd Fl., Livingston, NJ 07039 US

GROWTH PLANS/SPECIAL FEATURES:
Columbia Laboratories, Inc. develops, manufactures and markets drugs that target women's healthcare and endocrine-related disorders. The company's products are used for vaginal delivery of hormones, analgesics and other drugs; and for buccal delivery of hormones and peptides. The vaginal products adhere to the vaginal epithelium and the buccal products adhere to the mucosal membrane of the gum and cheek. Both forms provide sustained and controlled delivery of active drug ingredients. This delivery system is particularly useful for active drug ingredients that cannot be ingested. All of the firm's products utilize its patented Bioadhesive Delivery System (BDS) technology, which consists principally of a polymer and an active ingredient. Products include Crinone and Prochieve progesterone gels, designed to deliver progesterone directly to the uterus; Striant used to treat hypogonadism, a deficiency or absence of endogenous testosterone production; Replens vaginal moisturizer, which replenishes vaginal moisture on a sustained basis and relieves the discomfort associated with vaginal dryness; and RepHresh vaginal gel, a feminine hygiene product that can eliminate vaginal odor by maintaining pH in the normal physiological range of 4.5 or below. Outside the U.S., Crinone is approved for marketing in 56 countries for medical indications such as menstrual irregularities, dysmenorrheal and dysfunctional uterine bleeding. Columbia contracts its manufacturing activities to third parties in the U.K., Switzerland and Italy. Merck Serono S.A. sells Crinone outside of the U.S. The company's next product candidate is vaginally-administered Lidocaine for prevention and the treatment of dysmenorrhea, a condition that effects over 50% of all menstruating women.

The company offers employees benefits that include a 401(k) plan; medical, dental, life and disability insurance; and paid vacation and holidays.

FINANCIALS: Sales and profits are in thousands of dollars—add 000 to get the full amount. 2006 Note: Financial information for 2006 was not available for all companies at press time.

2006 Sales: $17,393	2006 Profits: $-12,612	U.S. Stock Ticker: CBRX	
2005 Sales: $22,041	2005 Profits: $-9,307	Int'l Ticker:	Int'l Exchange:
2004 Sales: $17,860	2004 Profits: $-25,130	Employees: 47	
2003 Sales: $22,415	2003 Profits: $-21,151	Fiscal Year Ends: 12/31	
2002 Sales: $9,400	2002 Profits: $-16,800	Parent Company:	

SALARIES/BENEFITS:

Pension Plan:	ESOP Stock Plan:	Profit Sharing:	Top Exec. Salary: $339,570	Bonus: $136,000
Savings Plan: Y	Stock Purch. Plan:		Second Exec. Salary: $279,142	Bonus: $89,440

OTHER THOUGHTS:
Apparent Women Officers or Directors: 1
Hot Spot for Advancement for Women/Minorities:

LOCATIONS: ("Y" = Yes)

West:	Southwest:	Midwest:	Southeast:	Northeast:	International:
				Y	Y

COMMONWEALTH BIOTECHNOLOGIES INC   www.cbi-biotech.com

Industry Group Code: 541710 Ranks within this company's industry group: Sales: 21   Profits: 10

Drugs:	Other:		Clinical:		Computers:		Other:	
Discovery:	AgriBio:		Trials/Services:		Hardware:		Specialty Services:	
Licensing:	Genomics/Proteomics:	Y	Laboratories:	Y	Software:		Consulting:	
Manufacturing:	Tissue Replacement:	Y	Equipment/Supplies:		Arrays:		Blood Collection:	
Development:			Research/Development Svcs.:	Y	Database Management:	Y	Drug Delivery:	
Generics:			Diagnostics:	Y			Drug Distribution:	

TYPES OF BUSINESS:

Contract Research-Biotech & Genetics
DNA & Genome Sequencing
Genetic Testing Services
Molecular Biology
Biophysical Analysis
Nucleic Acid Synthesis
Peptide Sequencing & Synthesis
Biodefense Services

BRANDS/DIVISIONS/AFFILIATES:

AccuTrac
Fairfax Identity Labs
HepArrest
Mimotopes Pty Ltd

CONTACTS: *Note: Officers with more than one job title may be intentionally listed here more than once.*

Paul D'Sylva, CEO
Richard J. Freer, COO
Robert B. Harris, Pres.
Thomas R. Reynolds, Exec. VP-Science/Corp. Sec.
Thomas R. Reynolds, Exec. VP-Tech.
James H. Brennan, VP-Financial Oper.
Russel L. Wolz, Sr. Research Scientist/Group Dir.-Biochemistry
Charles M. Kelly, Dir.-Fairfax Identity Laboratories
Joseph A. Buettner, Sr. Scientist
Joseph Chimera, Sr. VP-Bus. Dev., Sales & Mktg.
Richard J. Freer, Chmn.

Phone: 804-648-3820	Fax: 804-648-2641
Toll-Free: 800-735-9224	
Address: 601 Biotech Dr., Richmond, VA 23235 US	

GROWTH PLANS/SPECIAL FEATURES:

Commonwealth Biotechnologies, Inc. (CBI) provides early developmental contract research and services for global biotechnology industries, academic institutions, government agencies and pharmaceutical companies. The company offers a broad array of analytical and synthetic chemistries and biophysical analysis technologies.   CBI primarily concentrates in the fields of protein chemistry in the isolation analysis, characterization and production of proteins and peptides; nucleic acid chemistry for gene target identification using molecular biology-based assay development immunology for antibody production and ELISA development; and recombinant protein purification and bacteriology, virology and bio-safety testing in the field of microbiology.   CBI operates in four principal areas: bio-defense; laboratory support services for on-going clinical trials; comprehensive contract projects in private sectors and DNA reference lab activities. In the bio-defense area the firm offers environmental and laboratory testing for pathogens and holds a licensed bio-safety hazard laboratory staffed by scientists with government security clearances. DNA laboratories provide services such as DNA sequencing, nucleic acid synthesis, peptide and protein technologies, molecular biology, genetic testing services and biophysical analysis. Through its acquisition of Fairfax Identity Labs, CBI offers comprehensive genetic identity testing, which includes paternity, forensic, and Convicted Offender DNA Index System (CODIS) analyses. The firm has also introduced a fluorescent DNA sequencing loading buffer called AccuTrac for sample tracking in automated DNA sequencing applications. CBI has entered into an agreement with Prism Pharmaceuticals to develop, manufacture and commercialize CBI's helix-based peptide technologies for acute-care cardiovascular treatment. In February 2007, CBI completed its acquisition of Mimotopes Pty Ltd.

FINANCIALS: Sales and profits are in thousands of dollars—add 000 to get the full amount. 2006 Note: Financial information for 2006 was not available for all companies at press time.

2006 Sales: $6,532	2006 Profits: $-1,153	U.S. Stock Ticker: CBTE
2005 Sales: $7,803	2005 Profits: $ 79	Int'l Ticker: Int'l Exchange:
2004 Sales: $5,749	2004 Profits: $- 368	Employees: 34
2003 Sales: $5,104	2003 Profits: $- 81	Fiscal Year Ends: 12/31
2002 Sales: $4,400	2002 Profits: $- 600	Parent Company:

SALARIES/BENEFITS:

Pension Plan:	ESOP Stock Plan:	Profit Sharing:	Top Exec. Salary: $205,000	Bonus: $18,869
Savings Plan:	Stock Purch. Plan:		Second Exec. Salary: $190,000	Bonus: $8,579

OTHER THOUGHTS:

Apparent Women Officers or Directors: 3
Hot Spot for Advancement for Women/Minorities: Y

LOCATIONS: ("Y" = Yes)

West:	Southwest:	Midwest:	Southeast:	Northeast:	International:
				Y	

COMPUGEN LTD

www.cgen.com

Industry Group Code: 325412 Ranks within this company's industry group: Sales: 178 Profits: 82

Drugs:		Other:		Clinical:		Computers:		Other:	
Discovery:	Y	AgriBio:	Y	Trials/Services:		Hardware:	Y	Specialty Services:	
Licensing:		Genomics/Proteomics:	Y	Laboratories:		Software:	Y	Consulting:	
Manufacturing:		Tissue Replacement:		Equipment/Supplies:		Arrays:		Blood Collection:	
Development:				Research/Development Svcs.:		Database Management:	Y	Drug Delivery:	
Generics:				Diagnostics:	Y			Drug Distribution:	

TYPES OF BUSINESS:

Drug Discovery
Biotechnology Databases & Diagnostics
Genomics
Proteomics
Small Molecule Drug Discovery Technology
Agricultural Biotechnology

BRANDS/DIVISIONS/AFFILIATES:

LEADS
Keddem Bioscience Ltd.
Evogene Ltd.

CONTACTS: *Note: Officers with more than one job title may be intentionally listed here more than once.*

Alex Kotzer, CEO
Alex Kotzer, Pres.
Ronit Lerner, CFO
Ronit Weinstein, VP-Human Resources
Yossi Cohen, VP-R&D
Rachel Bart, General Counsel
Eli Zangvil, VP-Bus. Dev.
Anat Cohen-Dayag, VP-Diagnostic Biomarkers & Therapeutics
Martin S. Gerstel, Chmn.

Phone: 972-3-765-8585	Fax: 972-3-765-8555
Toll-Free:	
Address: 72 Pinchas Rosen St., Tel Aviv, 69512 Israel	

GROWTH PLANS/SPECIAL FEATURES:

Compugen, Ltd. is a pioneer in the fields of computational genomics and proteomics. Headquartered in Israel, it also works through a wholly-owned subsidiary in California. A synthetic, multi-disciplinary approach allows the company to create models for more accurate predictions, including what it believes is the most accurate representation of the transcriptome, the fundamental bridge between genome and proteome. The company constantly maintains a pipeline of potentially effective therapeutics, although the majority of these are licensed out to other companies. Compugen's core product is LEADS, a bioinformatics platform for analyzing genomic and protein data used by companies including Pfizer, Inc., Novartis Pharma A.G. and Abbott Laboratories. After developing a wide variety of hardware and software across the diagnostics and genomics board, the company eventually decided to focus on the discovery and development of novel therapeutic proteins and diagnostic markers. It currently has agreements with Diagnostic Products Corporation, Ortho-Clinical Diagnostics and Biosite to develop and commercialize in-vitro diagnostic assays for cancer and cell therapy products for the treatment of cardiovascular diseases. Keddem Bioscience Ltd., a full subsidiary, is involved in the discovery of small molecule-based pharmaceuticals; and Evogene Ltd., another subsidiary, focuses on agricultural biotechnology, especially plant genomics. In 2007, the company announced a collaboration with Mayo Clinic aimed at discovering biomarkers for diagnosing the presence of unstable atherosclerotic plaques in coronary artery disease and cerebrovascular disease.

Compugen offers employees a competitive benefits package and a casual work environment.

FINANCIALS: Sales and profits are in thousands of dollars—add 000 to get the full amount. 2006 Note: Financial information for 2006 was not available for all companies at press time.

2006 Sales: $ 215	2006 Profits: $-13,020	U.S. Stock Ticker: CGEN
2005 Sales: $ 646	2005 Profits: $-13,978	Int'l Ticker: CGEN Int'l Exchange: Tel Aviv-TASE
2004 Sales: $2,630	2004 Profits: $-13,722	Employees: 80
2003 Sales: $8,800	2003 Profits: $-11,400	Fiscal Year Ends: 12/31
2002 Sales: $11,100	2002 Profits: $-12,200	Parent Company:

SALARIES/BENEFITS:

Pension Plan:	ESOP Stock Plan:	Profit Sharing:	Top Exec. Salary: $	Bonus: $
Savings Plan:	Stock Purch. Plan:		Second Exec. Salary: $	Bonus: $

OTHER THOUGHTS:

Apparent Women Officers or Directors: 2
Hot Spot for Advancement for Women/Minorities: Y

LOCATIONS: ("Y" = Yes)

West:	Southwest:	Midwest:	Southeast:	Northeast:	International:
Y					Y

CORTEX PHARMACEUTICALS INC   www.cortexpharm.com

Industry Group Code: 325412 Ranks within this company's industry group: Sales: 155   Profits: 91

Drugs:		Other:	Clinical:	Computers:	Other:
Discovery:	Y	AgriBio:	Trials/Services:	Hardware:	Specialty Services:
Licensing:	Y	Genomics/Proteomics:	Laboratories:	Software:	Consulting:
Manufacturing:		Tissue Replacement:	Equipment/Supplies:	Arrays:	Blood Collection:
Development:	Y		Research/Development Svcs.:	Database Management:	Drug Delivery:
Generics:			Diagnostics:		Drug Distribution:

TYPES OF BUSINESS:
Drugs-Neurological
Psychiatric Drugs

BRANDS/DIVISIONS/AFFILIATES:
Ampakine
CX717
NV Organon
Azko Nobel

CONTACTS: *Note: Officers with more than one job title may be intentionally listed here more than once.*
Roger G. Stoll, CEO
Mark Varney, COO
Roger G. Stoll, Pres.
Maria S. Messinger, CFO/VP
Mark Varney, Chief Scientific Officer
James H. Coleman, Sr. VP-Bus. Dev.
Gary A. Rogers, Sr. VP-Pharmaceutical Research
Steven A. Johnson, VP-Preclinical Dev.
Leslie J. Street, Sr. Dir.-Medicinal Chemistry
Roger G. Stoll, Chmn.

Phone: 949-727-3157	Fax: 949-727-3657
Toll-Free:	
Address: 15231 Barranca Pkwy., Irvine, CA 92618 US	

GROWTH PLANS/SPECIAL FEATURES:
Cortex Pharmaceuticals, Inc. focuses on developing drug-like compounds that positively modulate AMPA-type glutamate receptors, a complex of proteins involved in the communication between nerve cells in the mammalian brain. These compounds, known as Ampakine compounds, enhance the activity of the AMPA receptor. The compounds act on chemical pathways that impact about 85% of all the neurotransmission occurring in the brain, with the benefit of enhancing memory and cognition. The drugs thus have the potential to treat mental illnesses such as Alzheimer's disease, schizophrenia and depression, and to impact disorders such as autism, fragile X syndrome and ADHD. The company is involved in a number of strategic partnerships with pharmaceutical leaders including NV Organon, a pharmaceutical business unit of Azko Nobel, which is licensed to market Cortex products. The firm's research areas span neurological diseases, central nervous system traumatic injuries, psychiatric disorders and cerebral ischemic disorders. Cortex owns or has exclusive rights to more than 50 issued or allowed U.S. and foreign patents. Over 40 of these are composition of matter patents that cover hundreds of the company's compounds. In July 2007, the company received FDA approval for resuming enrollment in the drug CX717 Alzheimer's positron emission tomography scan study at all requested dose levels.

The company offers its employees medical, dental and vision insurance; life insurance; a 401(k) plan; a cafeteria plan; and company stock options.

FINANCIALS: Sales and profits are in thousands of dollars—add 000 to get the full amount. 2006 Note: Financial information for 2006 was not available for all companies at press time.

2006 Sales: $1,177	2006 Profits: $-16,055	**U.S. Stock Ticker: COR**
2005 Sales: $2,577	2005 Profits: $-11,605	**Int'l Ticker:** Int'l Exchange:
2004 Sales: $6,973	2004 Profits: $-5,994	Employees: 25
2003 Sales: $5,231	2003 Profits: $-1,174	Fiscal Year Ends: 12/31
2002 Sales: $6,400	2002 Profits: $-1,000	Parent Company:

SALARIES/BENEFITS:

Pension Plan:	ESOP Stock Plan:	Profit Sharing:	Top Exec. Salary: $336,500	Bonus: $
Savings Plan: Y	Stock Purch. Plan:		Second Exec. Salary: $286,295	Bonus: $

OTHER THOUGHTS:
Apparent Women Officers or Directors: 1
Hot Spot for Advancement for Women/Minorities:

LOCATIONS: ("Y" = Yes)

West:	Southwest:	Midwest:	Southeast:	Northeast:	International:
Y				Y	

COVANCE INC

www.covance.com

Industry Group Code: 541710   Ranks within this company's industry group: Sales: 1   Profits: 2

Drugs:		Other:		Clinical:		Computers:		Other:	
Discovery:		AgriBio:		Trials/Services:	Y	Hardware:		Specialty Services:	Y
Licensing:		Genomics/Proteomics:		Laboratories:		Software:	Y	Consulting:	
Manufacturing:		Tissue Replacement:		Equipment/Supplies:		Arrays:		Blood Collection:	
Development:				Research/Development Svcs.:	Y	Database Management:		Drug Delivery:	
Generics:				Diagnostics:				Drug Distribution:	

TYPES OF BUSINESS:

Pharmaceutical Research & Development
Drug Preclinical/Clinical Trials
Laboratory Testing & Analysis
Approval Assistance
Health Economics & Outcomes Services
Online Tools

BRANDS/DIVISIONS/AFFILIATES:

LabLink
Study Tracker
Trial Tracker
Digitography
Signet Laboratories, Inc.

CONTACTS: Note: Officers with more than one job title may be intentionally listed here more than once.

Joseph L. Herring, CEO
William Klitgaard, CFO/Sr. VP
Donald Kraft, Sr. VP-Human Resources
James W. Lovett, General Counsel/Sr. VP
Joseph L. Herring, Chmn.

Phone: 609-452-4440	Fax: 609-452-9375
Toll-Free: 888-268-2623	
Address: 210 Carnegie Ctr., Princeton, NJ 08540 US	

GROWTH PLANS/SPECIAL FEATURES:

Covance, Inc. is a leading drug development services company and contract research organization. It provides a wide range of product development services to pharmaceutical, biotechnology and medical device industries across the globe. The company also provides laboratory testing services for clients in the chemical, agrochemical and food businesses. Covance's early development services include preclinical services (such as toxicology, pharmaceutical development, research products and BioLink, a bioanalytical testing service) and Phase I clinical services. Its late-stage development services cover clinical development and support; clinical trials; periapproval and market access; central laboratory operations; and central ECG diagnostics. Covance has also introduced several Internet-based products: Study Tracker is an Internet-based client access product, which permits customers of toxicology services to review study data and schedules on a near real-time basis; LabLink is a client access program that allows customers of central laboratory services to review and query lab data on a near real-time basis; and Trial Tracker is a web-enabled clinical trial project management and tracking tool intended to allow both employees and customers of its late-stage clinical business to review and manage all aspects of clinical-trial projects. Digitography, another electronic system, allows on-screen digital ECG waveform measurement. In 2006 Covance acquired two companies: Signet Laboratories, Inc., which is a leading provider of monoclonal antibodies used in the research of cancer, infectious disease and neurodegenerative disease; and the Early Phase Clinical Research Business Unit (Phase I/IIA) of Radiant Research Inc., which includes eight sites located in Texas, Idaho, Florida, Hawaii, Oregon and California. To complete its plans for the acquired sites, the company relocated and expanded the Austin- and Daytona Beach-based clinical research units.

Covance offers its employees benefits such as medical, dental and vision plans; a range of insurance benefits; employee assistance; financial planning services; and tuition reimbursement.

FINANCIALS: Sales and profits are in thousands of dollars—add 000 to get the full amount. 2006 Note: Financial information for 2006 was not available for all companies at press time.

2006 Sales: $1,406,058	2006 Profits: $144,998	U.S. Stock Ticker: CVD
2005 Sales: $1,250,400	2005 Profits: $119,600	Int'l Ticker: Int'l Exchange:
2004 Sales: $1,056,397	2004 Profits: $97,947	Employees: 8,100
2003 Sales: $974,210	2003 Profits: $76,136	Fiscal Year Ends: 12/31
2002 Sales: $924,700	2002 Profits: $63,800	Parent Company:

SALARIES/BENEFITS:

Pension Plan:	ESOP Stock Plan:	Profit Sharing:	Top Exec. Salary: $600,000	Bonus: $
Savings Plan: Y	Stock Purch. Plan: Y		Second Exec. Salary: $525,000	Bonus: $900,000

OTHER THOUGHTS:

Apparent Women Officers or Directors:
Hot Spot for Advancement for Women/Minorities:

LOCATIONS: ("Y" = Yes)

West:	Southwest:	Midwest:	Southeast:	Northeast:	International:
Y	Y	Y	Y	Y	Y

Note: Financial information, benefits and other data can change quickly and may vary from those stated here.

COVIDIEN LTD

www.covidien.com

Industry Group Code: 339113 Ranks within this company's industry group: Sales: Profits:

Drugs:		Other:		Clinical:		Computers:		Other:	
Discovery:	Y	AgriBio:		Trials/Services:		Hardware:		Specialty Services:	Y
Licensing:		Genomics/Proteomics:		Laboratories:		Software:		Consulting:	
Manufacturing:	Y	Tissue Replacement:		Equipment/Supplies:	Y	Arrays:		Blood Collection:	
Development:	Y			Research/Development Svcs.:	Y	Database Management:		Drug Delivery:	Y
Generics:				Diagnostics:				Drug Distribution:	Y

TYPES OF BUSINESS:

Medical Equipment & Supplies, Manufacturing
Imaging Agents
Pharmaceutical Products
Retail Products

BRANDS/DIVISIONS/AFFILIATES:

Kendall
Autosuture
Syneture
Valleylab
Mallinckrodt Pharmaceuticals
Nellcor
Puritan Bennett
Tyco Healthcare

CONTACTS: Note: Officers with more than one job title may be intentionally listed here more than once.

Richard J. Meelia, CEO
Richard J. Meelia, Pres.
Charles J. Dockendorff, CFO/Exec. VP
Karen A. Quinn-Quintin, Sr. VP-Human Resources
Steven M. McManama, CIO
Brian D. King, Sr. VP-Oper.
Amy A. Wendell, Sr. VP-Bus. Dev. & Strategy
Eric Kraus, Sr. VP-Corp. Comm.
Coleman Lannum, VP-Investor Rel.
Richard G. Brown, Jr., Chief Accounting Officer/Controller
Jose E. Almeida, Pres., Medical Devices
Timothy R. Wright, Pres., Pharmaceutical Products & Imaging Solutions
James C. Clemmer, Pres., Medical Supplies
Richard G. Brown, Jr., Pres., Retail Products

Phone: 441-292-8674	Fax:
Toll-Free:	
Address: 90 Pitts Bay Rd., 2nd Fl., Pembroke, HM 08 Bermuda	

GROWTH PLANS/SPECIAL FEATURES:

Covidien Ltd., formerly Tyco Healthcare, is a global healthcare products company that provides medical solutions. Covidien manufactures, distributes and services a diverse range of industry-leading brands such as Kendall, Autosuture, Syneture, Valleylab, Mallinckrodt Pharmaceuticals, Nellcor and Puritan Bennett. The company's business consists of five segments: medical devices, imaging solutions, pharmaceutical products, medical supplies and retail products. The medical devices segment includes surgical devices, energy-based devices, respiratory and monitoring solutions, patient care and safety products, in addition to laparoscopic instruments, surgical staplers, sutures, energy-based instruments, pulse oximeters, ventilators, vascular compression devices, needles and syringes and sharps collection systems. The imaging solutions segment includes contrast agents, contrast delivery systems and radiopharmaceuticals. The pharmaceutical products segment produces active pharmaceutical ingredients, dosage pharmaceuticals and specialty chemicals. The medical supplies segment provides traditional wound care products, absorbent hygiene products, operating room kits and OEM products. The retail products segment includes private label adult incontinence, feminine hygiene and infant care products. Covidien operates in approximately 57 countries, and its products are sold in over 130 countries. The company was formed from Tyco International's healthcare division, Tyco Healthcare, in July 2007.

FINANCIALS: Sales and profits are in thousands of dollars—add 000 to get the full amount. 2006 Note: Financial information for 2006 was not available for all companies at press time.

2006 Sales: $	2006 Profits: $	**U.S. Stock Ticker: COV**
2005 Sales: $	2005 Profits: $	**Int'l Ticker:** Int'l Exchange:
2004 Sales: $	2004 Profits: $	Employees:
2003 Sales: $	2003 Profits: $	Fiscal Year Ends: 9/30
2002 Sales: $	2002 Profits: $	Parent Company:

SALARIES/BENEFITS:

Pension Plan:	ESOP Stock Plan:	Profit Sharing:	Top Exec. Salary: $	Bonus: $
Savings Plan:	Stock Purch. Plan:		Second Exec. Salary: $	Bonus: $

OTHER THOUGHTS:

Apparent Women Officers or Directors: 2
Hot Spot for Advancement for Women/Minorities: Y

LOCATIONS: ("Y" = Yes)

West:	Southwest:	Midwest:	Southeast:	Northeast:	International:
					Y

CSL LIMITED

www.csl.com.au

Industry Group Code: 325414 Ranks within this company's industry group: Sales: 2   Profits: 3

Drugs:		Other:		Clinical:		Computers:		Other:	
Discovery:		AgriBio:		Trials/Services:		Hardware:		Specialty Services:	Y
Licensing:		Genomics/Proteomics:		Laboratories:		Software:		Consulting:	
Manufacturing:	Y	Tissue Replacement:		Equipment/Supplies:	Y	Arrays:		Blood Collection:	Y
Development:	Y			Research/Development Svcs.:		Database Management:		Drug Delivery:	
Generics:				Diagnostics:	Y			Drug Distribution:	

TYPES OF BUSINESS:

Human Blood-Plasma Collection
Plasma Products
Immunohematology Products
Vaccines
Pharmaceutical Marketing
Antivenom
Drugs-Cancer

BRANDS/DIVISIONS/AFFILIATES:

ZLB Behring
CSL Bioplasma
CSL Behring
CSL Pharmaceutical
CSL Biotherapies
Zenyth Therapeutics
CytoGam
ZLB Plasma Services

CONTACTS: Note: Officers with more than one job title may be intentionally listed here more than once.

Brian McNamee, CEO
Kevin Milroy, Gen. Mgr.-Human Resources
Andrew Cuthbertson, Chief Scientific Officer
Peter R. Turvey, General Counsel/Corp. Sec.
Peter Turner, Pres., ZLB Behring
Colin Armit, Pres., CSL Pharmaceutical
Paul Bordonaro, Pres., CSL Bioplasma
Peter H. Wade, Chmn.
T. Giarla, Pres., Bioplasma Asia Pacific

Phone: 61-3-9389-1911	Fax: 61-3-9389-1434
Toll-Free:	
Address: 45 Poplar Rd., Parkville, VIC 3052 Australia	

GROWTH PLANS/SPECIAL FEATURES:

CSL Limited develops, manufactures and markets pharmaceutical products of biological origin in 27 countries worldwide. The company operates through several subsidiaries that manufacture and distribute pharmaceuticals, vaccines and diagnostics derived from human plasma. Subsidiary CSL Behring is a world leader in the manufacture of plasma products such as hemophilia treatments, immunoglobulins and wound healing agents. CSL Limites operates one of the largest plasma collection networks in the world, named ZLB Plasma Services, which includes 65 collection centers in the U.S. and eight in Germany. CSL Bioplasma is one of the largest manufacturers of plasma products in the southern hemisphere and works with the Red Cross and government entities to supply such products in Australia, New Zealand, Singapore, Malaysia and Hong Kong. It also provides contract plasma fractionation services. CSL Biotherapies, formerly CSL Pharmaceutical, manufactures and markets vaccines for human use, including children's vaccines, travel vaccines, respiratory vaccines and antivenom. Currently, its primary focus is the manufacturing of flu vaccines. The company's research and development portfolio includes treatments for stroke, acute coronary syndromes, cervical cancer, melanoma, genital warts, papilloma viruses and hepatitis C, in addition to a method of topical delivery of drugs to the eye. In November 2006, Zenyth Therapeutics became a wholly-owned subsidiary of CSL when it was acquired for nearly $96 million. Also in November 2006, subsidiary ZLB Behring acquired CytoGam, an intravenous drug for the prevention of antibodies against cytomegalovirus in transplant patients, from MedImmune for $70 million.

FINANCIALS: Sales and profits are in thousands of dollars—add 000 to get the full amount. 2006 Note: Financial information for 2006 was not available for all companies at press time.

2006 Sales: $2,146,111	2006 Profits: $88,406	U.S. Stock Ticker: CSLX.PK
2005 Sales: $1,965,359	2005 Profits: $176,824	Int'l Ticker: CSL Int'l Exchange: Sydney-ASX
2004 Sales: $1,650,196	2004 Profits: $219,625	Employees: 7,000
2003 Sales: $1,313,000	2003 Profits: $47,000	Fiscal Year Ends: 6/30
2002 Sales: $753,100	2002 Profits: $	Parent Company:

SALARIES/BENEFITS:

Pension Plan:	ESOP Stock Plan:	Profit Sharing:	Top Exec. Salary: $782,154	Bonus: $782,735
Savings Plan:	Stock Purch. Plan: Y		Second Exec. Salary: $409,753	Bonus: $273,292

OTHER THOUGHTS:

Apparent Women Officers or Directors: 1
Hot Spot for Advancement for Women/Minorities:

LOCATIONS: ("Y" = Yes)

West:	Southwest:	Midwest:	Southeast:	Northeast:	International:
		Y	Y	Y	Y

Note: Financial information, benefits and other data can change quickly and may vary from those stated here.

CUBIST PHARMACEUTICALS INC

www.cubist.com

Industry Group Code: 325412 Ranks within this company's industry group: Sales: 55   Profits: 57

Drugs:		Other:	Clinical:	Computers:	Other:
Discovery:	Y	AgriBio:	Trials/Services:	Hardware:	Specialty Services:
Licensing:		Genomics/Proteomics:	Laboratories:	Software:	Consulting:
Manufacturing:		Tissue Replacement:	Equipment/Supplies:	Arrays:	Blood Collection:
Development:	Y		Research/Development Svcs.:	Database Management:	Drug Delivery:
Generics:			Diagnostics:		Drug Distribution:

TYPES OF BUSINESS:

Drugs-Infectious Disease
Antimicrobial Drugs
Antiviral Drugs

BRANDS/DIVISIONS/AFFILIATES:

CUBICIN
Novartis AG
Chiron Healthcare Ireland, Ltd.
Medion Pharma Co., Ltd.
Oryx Pharmaceuticals, Inc.
TTY BioPharma
Kuhnil Pharma Co., Ltd.

CONTACTS: Note: Officers with more than one job title may be intentionally listed here more than once.

Michael W. Bonney, CEO
Michael W. Bonney, Pres.
David W.J. McGirr, CFO/Sr. VP
Gregory Stea, VP-Sales
Oliver S. Fetzer, Sr. VP-R&D
Lindon Fellows, Sr. VP-Tech. Oper.
Christopher D.T. Guiffre, General Counsel/Sr. VP/Sec.
Oliver S. Fetzer, Sr. VP-Corp. Dev.
William Pullman, Sr. VP/Chief Medical Officer
Robert J. Perez, Sr. VP-Commercial Oper.
Thomas J. Slater, VP-Commercial Dev.
Barry I. Eisenstein, Sr. VP-Scientific Affairs
David W. Martin Jr., Lead Dir.

Phone: 781-860-8660	Fax: 781-861-0566
Toll-Free:	
Address: 65 Hayden Ave., Lexington, MA 02421 US	

GROWTH PLANS/SPECIAL FEATURES:

Cubist Pharmaceuticals, Inc. is a biopharmaceutical company focused on the research, development and commercialization of pharmaceutical products that address unmet medical needs. The company focuses on developing products for the anti-infective marketplace. The firm's one marketed product is CUBICIN, the first in a new class of antimicrobial drugs called lipopeptides, which is approved for treatment of skin and skin structure infections. The drug is also approved in the U.S., E.U., Israel, Taiwan and Argentina. Because CUBICIN attacks bacteria through a novel mechanism and has demonstrated the unique ability in vitro to rapidly kill virtually all clinically significant gram-positive bacteria, it may be particularly effective in treating infections caused by drug-resistant bacteria that cannot be eradicated by existing antibiotics. The firm has completed a successful Phase III trial of the drug for endocarditis (infection of the heart valves) and bacteremia (bacteria in the blood). The firm has alliances with other companies to market Cubicin outside of the U.S. Novartis AG, through its subsidiary Chiron Healthcare Ireland, Ltd., is responsible for regulatory filings, sales, marketing and distribution costs in Europe, Australia, New Zealand, India and certain Central American, South American and Middle Eastern countries. Other international partners include Medion Pharma, Ltd., for Israel; Oryx Pharmaceuticals, Inc. for Canada; TTY BioPharm for Taiwan; and Kuhnil Pharma Co., Ltd. for Korea. Cubist decided not to make further investment in the development of HepeX-B. In May 2006, Cubist announced that the FDA had approved the supplemental new drug application for Cubicin as a once-per-day therapy for the treatment of Staphylococcus aureaus bloodstream infections.

Cubist Pharmaceuticals offers its employees a 401(k) plan; a medical plan; an employee stock purchase program; dental, life and disability insurance; tuition reimbursement; and flexible spending accounts. Additional perks include on-site massage therapy, a fitness subsidy, celebratory parties, free parking and public transportation.

FINANCIALS: Sales and profits are in thousands of dollars—add 000 to get the full amount. 2006 Note: Financial information for 2006 was not available for all companies at press time.

2006 Sales: $194,748	2006 Profits: $- 376	U.S. Stock Ticker: CBST
2005 Sales: $120,645	2005 Profits: $-31,852	Int'l Ticker: Int'l Exchange:
2004 Sales: $68,071	2004 Profits: $-76,512	Employees: 410
2003 Sales: $3,716	2003 Profits: $-115,003	Fiscal Year Ends: 12/31
2002 Sales: $11,500	2002 Profits: $-82,400	Parent Company:

SALARIES/BENEFITS:

Pension Plan:	ESOP Stock Plan:	Profit Sharing:	Top Exec. Salary: $415,000	Bonus: $282,200
Savings Plan: Y	Stock Purch. Plan: Y		Second Exec. Salary: $375,000	Bonus: $126,300

OTHER THOUGHTS:

Apparent Women Officers or Directors: 1
Hot Spot for Advancement for Women/Minorities:

LOCATIONS: ("Y" = Yes)

West:	Southwest:	Midwest:	Southeast:	Northeast:	International:
				Y	

CURAGEN CORPORATION

www.curagen.com

Industry Group Code: 325412 Ranks within this company's industry group: Sales: 85 Profits: 165

Drugs:		Other:		Clinical:	Computers:	Other:
Discovery:	Y	AgriBio:		Trials/Services:	Hardware:	Specialty Services:
Licensing:		Genomics/Proteomics:	Y	Laboratories:	Software:	Consulting:
Manufacturing:		Tissue Replacement:		Equipment/Supplies:	Arrays:	Blood Collection:
Development:	Y			Research/Development Svcs.:	Database Management:	Drug Delivery:
Generics:				Diagnostics:		Drug Distribution:

TYPES OF BUSINESS:

Drug Discovery
Genomics Technology & Databases

BRANDS/DIVISIONS/AFFILIATES:

Pharmaceutically Tractable Genome
Predictive Toxicogenomic Screen
CuraChip
PXD101

CONTACTS: Note: Officers with more than one job title may be intentionally listed here more than once.

Frank M. Armstrong, CEO
Frank M. Armstrong, Pres.
David M. Wurzer, CFO/Exec. VP
Timothy Shannon, Exec. VP-R&D
Henri S. Lichenstein, VP-Prod. Dev.
Steven A. Henck, VP-Oper.
Elizabeth A. Whayland, VP-Finance
David M. Wurzer, Treas.
Mary Taylor, Sr. VP-Regulatory Affairs
Timothy Shannon, Chief Medical Officer
Johnathan M. Rothberg, Chmn.

Phone: 203-481-1104	Fax: 203-483-2552

Toll-Free:
Address: 322 E. Main St., Branford, CT 06405 US

GROWTH PLANS/SPECIAL FEATURES:

CuraGen is a genomics-based drug discovery and development company that researches, develops and uses technologies based on an understanding of gene function and relationships. The company develops products to treat metabolic diseases, cancer, inflammatory diseases and central nervous system (CNS) disorders. The firm has an array of technologies that facilitate every aspect of genetic research, including expression analysis, functional genomics, pathway analysis, functional proteomics, immunoassays, protein manufacturing and data analysis. CuraGen's genomic assets include the Pharmaceutically Tractable Genome (PTG), a database of 8,000 human genes identified as being suitable drug targets. The company and its collaborators utilize CuraChip, a microarray technology based on the PTG database, to identify new drug candidates. The firm collaborates with Abgenix to produce antibody drugs; TopoTarget, for HDAC inhibitors; Seattle Genetics for antibody-drug conjugation technology; and Bayer for diabetes development pharmacogenomics and toxicogenomics. CuraGen's subsidiary, 454 Life Sciences Corp., commercializes novel nanoscale instrumentation and technologies for determining the nucleotide sequence of entire genomes. Its proprietary technology is expected to have widespread applications in industrial processes, agriculture, animal health, biodefense and human health care, including drug discovery and development and disease diagnosis. The firm has also developed the Predictive Toxicogenomic Screen (PTS), a genomic tool that assesses the safety of candidate drugs. CuraGen continues to make progress with innovative drugs like PXD101 and CG53135 (an internally developed protein therapeutic that could prevent oral mucositis and has also moved to a Phase II clinical trial). In March 2007, CuraGen agreed to sell its 454 Life Sciences subsidiary to Roche for approximately $154.9 million.

FINANCIALS: Sales and profits are in thousands of dollars—add 000 to get the full amount. 2006 Note: Financial information for 2006 was not available for all companies at press time.

2006 Sales: $39,587	2006 Profits: $-59,839	U.S. Stock Ticker: CRGN
2005 Sales: $23,531	2005 Profits: $-73,244	Int'l Ticker: Int'l Exchange:
2004 Sales: $6,339	2004 Profits: $-90,397	Employees: 454
2003 Sales: $6,918	2003 Profits: $-74,497	Fiscal Year Ends: 12/31
2002 Sales: $18,200	2002 Profits: $-90,400	Parent Company:

SALARIES/BENEFITS:

Pension Plan:	ESOP Stock Plan:	Profit Sharing:	Top Exec. Salary: $389,635	Bonus: $175,000
Savings Plan: Y	Stock Purch. Plan:		Second Exec. Salary: $323,846	Bonus: $

OTHER THOUGHTS:

Apparent Women Officers or Directors: 2
Hot Spot for Advancement for Women/Minorities: Y

LOCATIONS: ("Y" = Yes)

West:	Southwest:	Midwest:	Southeast:	Northeast:	International:
				Y	

Note: Financial information, benefits and other data can change quickly and may vary from those stated here.

CURIS INC
www.curis.com

Industry Group Code: 325412 **Ranks within this company's industry group:** Sales: 117   Profits: 72

Drugs:		Other:	Clinical:	Computers:	Other:
Discovery:	Y	AgriBio:	Trials/Services:	Hardware:	Specialty Services:
Licensing:		Genomics/Proteomics:	Laboratories:	Software:	Consulting:
Manufacturing:		Tissue Replacement:	Equipment/Supplies:	Arrays:	Blood Collection:
Development:	Y		Research/Development Svcs.:	Database Management:	Drug Delivery:
Generics:			Diagnostics:		Drug Distribution:

TYPES OF BUSINESS:
Drugs-Regenerative Medicine
Signaling Pathway Therapeutics
Stem Cell Technologies

BRANDS/DIVISIONS/AFFILIATES:
Bone Morphogenic Protein
Hedgehog
OP-1
Curis Shanghai

CONTACTS: Note: Officers with more than one job title may be intentionally listed here more than once.
Daniel R. Passeri, CEO
Daniel R. Passeri, Pres.
Michael P. Gray, CFO
Changgeng Qian, VP-Discovery & Preclinical Dev.
Mark Noel, VP-Tech. Mgmt.
Mark Noel, VP-Bus. Dev.
James R. McNab, Jr., Chmn.

Phone: 617-503-6500	Fax: 617-503-6501
Toll-Free:	
Address: 45 Moulton St., Cambridge, MA 02138 US	

GROWTH PLANS/SPECIAL FEATURES:

Curis, Inc. focuses on the discovery, development and future commercialization of products that modulate key regulatory signaling pathways, which are systems used by the body to control tissue formation, maintenance and repair. Its product development approach involves using small molecules, proteins or antibodies to modulate these regulatory signaling pathways. Independently and through strategic alliances, the company focuses its research efforts on tissue repair regulators with applications including kidney disease, cancer and neurological disorders. Curis' proprietary signaling pathway and stem cell technologies include the Bone Morphogenic Protein (BMP) and the Hedgehog family of product candidates. The BMP family of regulators orchestrates the development and repair of bone and kidneys. OP-1, marketed by Stryker Corp., is approved in four major markets for the repair of long bone fractures. Curis is currently developing BMP-7 for the repair of damaged kidney tissue. The Hedgehog family of proteins contributes to the formation, maintenance and repair of tissues in the nervous system and cartilage, as well as in the control of blood vessel formation. The company has alliances with Genentech and Wyeth Pharmaceuticals to develop therapeutics that utilize these signaling pathways to repair and regenerate human tissues and organs. In January 2007, Curis provided an update on its Phase I clinical trial of its Hedgehog antagonist collaboration with Genentech, indicating that it had treated its first patient with the drug.

Curis offers its employees prescription drug benefits, an employee assistance program, education assistance, stock option grants, company-paid parking, public transportation and membership in company sports teams.

FINANCIALS: Sales and profits are in thousands of dollars—add 000 to get the full amount. 2006 Note: Financial information for 2006 was not available for all companies at press time.

2006 Sales: $14,936	2006 Profits: $-8,829	U.S. Stock Ticker: CRIS
2005 Sales: $6,002	2005 Profits: $-14,855	Int'l Ticker: Int'l Exchange:
2004 Sales: $3,699	2004 Profits: $-15,075	Employees: 51
2003 Sales: $10,378	2003 Profits: $-12,679	Fiscal Year Ends: 12/31
2002 Sales: $ 200	2002 Profits: $-82,300	Parent Company:

SALARIES/BENEFITS:

Pension Plan:	ESOP Stock Plan:	Profit Sharing:	Top Exec. Salary: $332,596	Bonus: $
Savings Plan: Y	Stock Purch. Plan: Y		Second Exec. Salary: $272,304	Bonus: $

OTHER THOUGHTS:

Apparent Women Officers or Directors:
Hot Spot for Advancement for Women/Minorities:

LOCATIONS: ("Y" = Yes)

West:	Southwest:	Midwest:	Southeast:	Northeast:	International:
				Y	

CV THERAPEUTICS INC   www.cvt.com

Industry Group Code: 325412   Ranks within this company's industry group: Sales: 90   Profits: 200

Drugs:		Other:		Clinical:		Computers:		Other:	
Discovery:	Y	AgriBio:		Trials/Services:		Hardware:		Specialty Services:	
Licensing:		Genomics/Proteomics:		Laboratories:		Software:		Consulting:	
Manufacturing:	Y	Tissue Replacement:		Equipment/Supplies:		Arrays:		Blood Collection:	
Development:	Y			Research/Development Svcs.:		Database Management:		Drug Delivery:	
Generics:				Diagnostics:				Drug Distribution:	

TYPES OF BUSINESS:
Drugs-Chronic Cardiovascular Diseases

BRANDS/DIVISIONS/AFFILIATES:
CV Therapeutics Europe, Ltd.
Ranexa
Regadenoson
Tecadenoson
CVT-6883

CONTACTS: *Note: Officers with more than one job title may be intentionally listed here more than once.*
Louis G. Lange, CEO
Daniel K. Spiegelman, CFO/Sr. VP
Lorenz Muller, VP-Mktg.
Diane L. Liguori, Sr. VP-Human Resources
Tricia Borga Suvari, General Counsel/Sr. VP
David Banks, Sr. VP-Oper.
Anders Waas, VP-Bus. Dev.
Michael M. Sweeney, VP-Medical Affairs
David C. McCaleb, Sr. VP-Commercial Oper.
Lewis J. Stuart, VP-Sales
Louis G. Lange, Chmn.
Diane L. Liguori, General Manager-EU

Phone: 650-384-8500	Fax: 650-858-0390

Toll-Free:

Address: 3172 Porter Dr., Palo Alto, CA 94304 US

GROWTH PLANS/SPECIAL FEATURES:

CV Therapeutics, Inc. is a biopharmaceutical company focused on the discovery, development and commercialization of new small molecule drugs for the treatment of cardiovascular diseases. The company currently promotes Ranexa for the treatment of chronic angina. Ranexa should be used in combination with amlodipine, beta-blockers or nitrates. The firm is no longer promoting Aceon, which was used for stable coronary artery disease. CV Therapeutics is developing other drug candidates, including regadenoson, a selective A2A-adenosine receptor agonist, for potential use as a pharmacologic agent in myocardial perfusion imaging studies; tecadenoson, a selective A1-adenosine receptor agonist for the potential reduction of rapid heart rate during acute atrial arrhythmias; and CVT-6883, a selective A2B-adenosine antagonist for treating conditions caused by inflammation and fibrosis. Drugs in preclinical development programs include ALDH2 antagonist for alcohol addiction; SC desaturase, for obesity/metabolic syndrome; ABCA1/ApoA-I for atherosclerosis; and late INA for coronary artery disease. The company has a wholly-owned subsidiary, CV Therapeutics Europe Limited in Hertfordshire in the U.K., that promotes the firm's activities in Europe. In May 2007, the company announced that it submitted a new drug application to the FDA seeking approval for regadenoson for use in myocardial perfusion imaging studies.

CV Therapeutics offers its employees a 401(k) plan; an employee stock purchase plan; flexible spending accounts; a college savings plan; medical and dental insurance; an employee assistance program; and an educational reimbursement program.

FINANCIALS: Sales and profits are in thousands of dollars—add 000 to get the full amount. 2006 Note: Financial information for 2006 was not available for all companies at press time.

2006 Sales: $36,785	2006 Profits: $-274,320	**U.S. Stock Ticker:** CVTX	
2005 Sales: $18,951	2005 Profits: $-227,995	**Int'l Ticker:** Int'l Exchange:	
2004 Sales: $20,428	2004 Profits: $-115,083	Employees: 746	
2003 Sales: $11,305	2003 Profits: $-110,951	Fiscal Year Ends: 12/31	
2002 Sales: $5,300	2002 Profits: $-107,800	Parent Company:	

SALARIES/BENEFITS:

Pension Plan:	ESOP Stock Plan:	Profit Sharing:	Top Exec. Salary: $624,000	Bonus: $762,500
Savings Plan: Y	Stock Purch. Plan: Y		Second Exec. Salary: $312,000	Bonus: $212,500

OTHER THOUGHTS:
Apparent Women Officers or Directors: 3
Hot Spot for Advancement for Women/Minorities: Y

LOCATIONS: ("Y" = Yes)

West:	Southwest:	Midwest:	Southeast:	Northeast:	International:
Y					Y

CYANOTECH CORPORATION

www.cyanotech.com

Industry Group Code: 325411 Ranks within this company's industry group: Sales: 6   Profits: 5

Drugs:	Other:		Clinical:		Computers:	Other:	
Discovery:	AgriBio:	Y	Trials/Services:		Hardware:	Specialty Services:	
Licensing:	Genomics/Proteomics:		Laboratories:		Software:	Consulting:	
Manufacturing:	Tissue Replacement:		Equipment/Supplies:		Arrays:	Blood Collection:	
Development:			Research/Development Svcs.:		Database Management:	Drug Delivery:	
Generics:			Diagnostics:	Y		Drug Distribution:	

TYPES OF BUSINESS:

Microalgae Products
Nutritional Supplements
Aquaculture Feed
Animal Nutrition Products
Immunological Diagnostics Products

BRANDS/DIVISIONS/AFFILIATES:

NatuRose
BioAstin
Spirulina Pacifica
Nutrex Hawaii, Inc.
Cyanotech Japan YK
PhytoDome CCS
Ocean-Chill Drying
Integrated Culture Biology Management

CONTACTS: Note: Officers with more than one job title may be intentionally listed here more than once.

Gerald R. Cysewski, CEO
Gerald R. Cysewski, Pres.
William R. Maris, CFO
Robert J. Capelli, VP-Sales
William R. Maris, VP-Admin.
Glenn D. Jensen, VP-Oper.
William R. Maris, VP-Finance
Gerald R. Cysewski, Chmn.

Phone: 808-326-1353	Fax: 808-329-4533
Toll-Free: 800-395-1353	
Address: 73-4460 Queen Kaahamanu Hwy., Ste. 102, Kailua-Kona, HI 96740 US	

GROWTH PLANS/SPECIAL FEATURES:

Cyanotech Corporation develops and commercializes natural products derived from microalgae. Microalgae are a diverse group of over 30,000 species of microscopic plants with a wide range of physiological and biochemical characteristics and naturally high levels of proteins, amino acids, vitamins, pigments and enzymes. Microalgae crops grow up to 100 times faster than land grown plants. The firm has developed proprietary production and harvesting processes to facilitate its production of microalgae, including Integrated Culture Biology Management, Ocean-Chill Drying and the PhytoDome Closed Culture System. Its 90-acre facility is the world's first microalgae production site to obtain ISO 9001:2000 compliancy. Cyanotech's natural products consist of human dietary supplements, aquaculture, animal feed, pigments, health and natural foods and immunological diagnostics. In the human dietary supplement market, Cyanotech produces BioAstin natural astaxanthin as a dietary antioxidant. BioAstin claims antioxidant properties surpassing those of leading vitamins and beta-carotene, and indications of benefits for carpal tunnel syndrome, arthritis and sunburn protection. For the animal nutrition market Cyanotech produces NatuRose natural astaxanthin, used in aquaculture to nourish animals and impart color to pen-raised fish and shrimp flesh. Spirulina Pacifica, Cyanotech's principal revenue source, provides a vegetable-based, highly absorbable source of protein, beta-carotene, amino acids and other phytonutrients. Cyanotech produces both all-natural grade and organic grade Spirulina Pacifica. The company sells Spirulina products in bulk under the Nutrex Hawaii label. Cyanotech also produces natural phycobiliproteins, highly fluorescent pigments for use as tags or markers in the immunological diagnostics market. Cyanotech's subsidiaries include Nutrex Hawaii and Cyanotech Japan YK.

FINANCIALS: Sales and profits are in thousands of dollars—add 000 to get the full amount. 2006 Note: Financial information for 2006 was not available for all companies at press time.

2006 Sales: $11,131	2006 Profits: $- 268	**U.S. Stock Ticker:** CYAN
2005 Sales: $11,445	2005 Profits: $ 486	**Int'l Ticker:** Int'l Exchange:
2004 Sales: $11,582	2004 Profits: $ 399	Employees: 62
2003 Sales: $9,000	2003 Profits: $-1,800	Fiscal Year Ends: 3/31
2002 Sales: $8,200	2002 Profits: $-2,600	Parent Company:

SALARIES/BENEFITS:

Pension Plan:	ESOP Stock Plan:	Profit Sharing:	Top Exec. Salary: $130,048	Bonus: $
Savings Plan:	Stock Purch. Plan:		Second Exec. Salary: $107,462	Bonus: $

OTHER THOUGHTS:

	LOCATIONS: ("Y" = Yes)					
Apparent Women Officers or Directors:	West:	Southwest:	Midwest:	Southeast:	Northeast:	International:
Hot Spot for Advancement for Women/Minorities:	Y					Y

Note: Financial information, benefits and other data can change quickly and may vary from those stated here.

CYPRESS BIOSCIENCE INC
www.cypressbio.com

Industry Group Code: 325412 Ranks within this company's industry group: Sales: 142 Profits: 70

Drugs:		Other:	Clinical:	Computers:	Other:
Discovery:	Y	AgriBio:	Trials/Services:	Hardware:	Specialty Services:
Licensing:	Y	Genomics/Proteomics:	Laboratories:	Software:	Consulting:
Manufacturing:	Y	Tissue Replacement:	Equipment/Supplies:	Arrays:	Blood Collection:
Development:	Y		Research/Development Svcs.:	Database Management:	Drug Delivery:
Generics:			Diagnostics:		Drug Distribution:

TYPES OF BUSINESS:
Drugs-Manufacturing
Drugs-Sleep Apnea
Drugs-Fibromyalgia Syndrome
Drugs-Functional Somatic Syndromes

BRANDS/DIVISIONS/AFFILIATES:
FMS Patient Registry
FMS Genomics Program
Milnacipran
Mirtazapine
Forest Laboratories
Organon

CONTACTS: Note: Officers with more than one job title may be intentionally listed here more than once.
Jay D. Kranzler, CEO
Sabrina M. Johnson, CFO/Exec. VP
Srinivas G. Rao, Chief Scientific Officer
R. Michael Gendreau, VP-Clinical Dev./Chief Medical Officer
Denise L. Wheeler, VP-Legal Affairs
Sabrina M. Johnson, Chief Bus. Officer
Jay D. Kranzler, Chmn.

Phone: 858-452-2323	Fax: 858-452-1222
Toll-Free:	
Address: 4350 Executive Dr., Ste. 325, San Diego, CA 92121 US	

GROWTH PLANS/SPECIAL FEATURES:
Cypress Bioscience, Inc. develops, manufactures and markets products that improve the diagnosis and treatment of patients with Functional Somatic Syndromes (FSSs), such as Fibromyalgia Syndrome (FMS) and other pain and central nervous system disorders. Cypress and its partner Forest Laboratories have signed a license agreement with Pierre Fabre Medicament to develop and market products with the compound milnacipran for the treatment of FMS in the U.S. and Canada. Milnacipran, a norepinephrine serotonin reuptake inhibitor which increases the level of norepinephrine more than it does serotonin, has been marketed outside the U.S. as an antidepressant since 1997; after a complete Phase III trial failed to show significant results, Cypress has engaged in two more such studies. Analogs of milnacipran have also been developed through collaboration with Collegium Pharmaceutical, Inc., and Cypress hopes to explore new indications for the drug. Cypress is currently considering milnacipran as a possible treatment for irritable bowel syndrome (IBS), a FSS. The company's second development program, in collaboration with the Dutch pharmaceutical company Organon, is evaluating potential pharmaceutical treatments for Obstructive Sleep Apnea (OSA). Although the Phase IIa trials conducted by Cypress and Organon, in which mirtazapine was evaluated in combination with another approved drug for the treatment of OSA, the companies are continuing to explore new potential opportunities to continue the collaboration. In May 2007, Cypress announced with Forest Laboratories that preliminary results from a Phase III study of milnacipran for the treatment of FMS are positive.

FINANCIALS: Sales and profits are in thousands of dollars—add 000 to get the full amount. 2006 Note: Financial information for 2006 was not available for all companies at press time.

2006 Sales: $4,322	2006 Profits: $-8,247	U.S. Stock Ticker: CYPB
2005 Sales: $8,384	2005 Profits: $-8,627	Int'l Ticker: Int'l Exchange:
2004 Sales: $14,415	2004 Profits: $-11,215	Employees: 15
2003 Sales: $	2003 Profits: $-21,742	Fiscal Year Ends: 12/31
2002 Sales: $6,400	2002 Profits: $-1,000	Parent Company:

SALARIES/BENEFITS:

Pension Plan:	ESOP Stock Plan:	Profit Sharing:	Top Exec. Salary: $529,406	Bonus: $
Savings Plan: Y	Stock Purch. Plan: Y		Second Exec. Salary: $286,275	Bonus: $

OTHER THOUGHTS:
Apparent Women Officers or Directors: 2
Hot Spot for Advancement for Women/Minorities: Y

LOCATIONS: ("Y" = Yes)

West:	Southwest:	Midwest:	Southeast:	Northeast:	International:
Y					

Note: Financial information, benefits and other data can change quickly and may vary from those stated here.

CYTOGEN CORPORATION

www.cytogen.com

Industry Group Code: 325412 Ranks within this company's industry group: Sales: 114   Profits: 85

Drugs:		Other:		Clinical:		Computers:		Other:	
Discovery:	Y	AgriBio:		Trials/Services:		Hardware:		Specialty Services:	
Licensing:	Y	Genomics/Proteomics:		Laboratories:		Software:		Consulting:	
Manufacturing:		Tissue Replacement:		Equipment/Supplies:		Arrays:		Blood Collection:	
Development:	Y			Research/Development Svcs.:		Database Management:		Drug Delivery:	
Generics:				Diagnostics:	Y			Drug Distribution:	

TYPES OF BUSINESS:

Drugs-Oncology
Diagnostic Imaging Agents

BRANDS/DIVISIONS/AFFILIATES:

ProstaScint
Caphosol
Quadramet
CYT-500
Soltamox
PSMA Development Company, LLC

CONTACTS: Note: Officers with more than one job title may be intentionally listed here more than once.

Michael D. Becker, CEO
Michael D. Becker, Pres.
Kevin J. Bratton, Sr. VP/CFO
Stephen A. Ross, Sr. VP-Sales & -Mktg.
William J. Thomas, General Counsel/Sr. VP
William F. Goeckeler, Sr. VP-Oper.
Susan M. Mesco, Dir.-Investor Rel.
Kevin J. Bratton, Sr. VP-Finance
James A. Grigsby, Chmn.

Phone: 609-750-8200	Fax: 609-452-2476
Toll-Free: 800-833-3533	
Address: 650 College Rd. E., Ste. 3100, Princeton, NJ 08540 US	

GROWTH PLANS/SPECIAL FEATURES:

Cytogen Corp. develops and commercializes oncology products. Its FDA-approved products include ProstaScint, a monoclonal antibody-based agent used to image the extent and spread of prostate cancer; Quadramet, a therapeutic agent marketed for the relief of pain in prostate, breast and other cancers that have spread to the bone; Soltamox, a liquid hormonal breast cancer therapy; and Caphosol, an advanced electrolyte solution for the treatment of oral mucositis and dry mouth. The company is also developing CYT-500, a third-generation radiolabeled antibody to treat prostate cancer. Cytogen's therapeutic 7E11-C5 monoclonal antibody is under new product development, and uses technology licensed from the Dow Chemical Company. The antibody product is designed to direct and deliver radiation to tumor cells. In February 2006, the company signed an agreement giving Savient Pharmaceuticals rights to market the drug Soltamox. In September 2006, Cytogen entered into a non-exclusive manufacturing agreement with Laureate Pharma, L.P. under which Laureate manufactures ProstaScint and its primary raw materials. In October 2006, the company entered into a license agreement with InPharma granting it exclusive rights to Caphosol in North America. In February 2007, Cytogen entered into a non-exclusive manufacturing agreement with Holopack Verpackungstechnik GmbH for the manufacture of Caphosol. In February 2007, Cytogen announced the initiation of the first human clinical study of CYT-500. The Phase I trial will investigate the safety and tolerability of CYT-500 and determine the optimal antibody mass and therapeutic dose for further studies.

Cytogen offers its employees medical, dental and life insurance; short and long-term disability; long-term care coverage; flexible spending accounts; a 401(k) plan incentive stock options; and an employee stock purchase plan.

FINANCIALS: Sales and profits are in thousands of dollars—add 000 to get the full amount. 2006 Note: Financial information for 2006 was not available for all companies at press time.

2006 Sales: $17,307	2006 Profits: $-15,103	**U.S. Stock Ticker:** CYTO	
2005 Sales: $15,946	2005 Profits: $-26,289	**Int'l Ticker:**	Int'l Exchange:
2004 Sales: $14,619	2004 Profits: $-20,540	Employees: 95	
2003 Sales: $13,842	2003 Profits: $-9,358	Fiscal Year Ends: 12/31	
2002 Sales: $12,900	2002 Profits: $-15,700	Parent Company:	

SALARIES/BENEFITS:

Pension Plan:	ESOP Stock Plan:	Profit Sharing:	Top Exec. Salary: $321,423	Bonus: $181,000
Savings Plan: Y	Stock Purch. Plan: Y		Second Exec. Salary: $229,007	Bonus: $91,664

OTHER THOUGHTS:

Apparent Women Officers or Directors: 1
Hot Spot for Advancement for Women/Minorities:

LOCATIONS: ("Y" = Yes)

West:	Southwest:	Midwest:	Southeast:	Northeast:	International:
				Y	

Note: Financial information, benefits and other data can change quickly and may vary from those stated here.

CYTRX CORPORATION

www.cytrx.com

Industry Group Code: 325412 **Ranks within this company's industry group:** Sales: 151 Profits: 97

Drugs:		Other:		Clinical:	Computers:	Other:	
Discovery:	Y	AgriBio:		Trials/Services:	Hardware:	Specialty Services:	
Licensing:	Y	Genomics/Proteomics:	Y	Laboratories:	Software:	Consulting:	
Manufacturing:	Y	Tissue Replacement:		Equipment/Supplies:	Arrays:	Blood Collection:	
Development:	Y			Research/Development Svcs.:	Database Management:	Drug Delivery:	Y
Generics:				Diagnostics:		Drug Distribution:	

TYPES OF BUSINESS:

Pharmaceutical Research & Development
Small-Molecule Drugs
RNAi-Based Products
Drug Delivery Technologies
Drugs-Sickle Cell Anemia

BRANDS/DIVISIONS/AFFILIATES:

CytRx Laboratories, Inc.
Araios, Inc.
Iroxanadine
TranzFect
Arimoclomol
FLOCOR

CONTACTS: Note: Officers with more than one job title may be intentionally listed here more than once.

Steven A. Kriegsman, CEO
Steven A. Kriegsman, Pres.
Matthew Natalizio, CFO
Jack R. Barber, Chief Scientific Officer
Benjamin S. Levin, General Counsel/VP-Legal Affairs/Corp. Sec.
Edward Umali, Dir.-Oper.
David Haen, Dir.-Bus. Dev.
Matthew Natalizio, Treas.
Mark A. Tepper, Sr. VP-Drug Discovery
Max Link, Chmn.

Phone: 310-826-5648	Fax: 310-826-6139

Toll-Free:

Address: 11726 San Vicente Blvd., Ste. 650, Los Angeles, CA 90049 US

GROWTH PLANS/SPECIAL FEATURES:

CytRx Corporation, located in Los Angeles, California, is a biopharmaceutical company engaged in the development of high-value therapeutics utilizing its core technologies in small molecule drugs, RNA interference (RNAi) drug discovery and DNA vaccines. CytRx's small molecule drug candidates provide cellular protection by activating molecular chaperone proteins that can repair or degrade proteins that are believed to cause many diseases. Molecular chaperones represent a therapeutic development platform from which CytRx may derive several drugs that address large markets and unmet medical needs. CytRx initiated a Phase II trial with arimociomol in ALS in 2005. The FDAA has granted fast track and orphan drug designation for arimociomol in ALS. The company has an obesity and type 2 diabetes research laboratory in Worcester, Massachusetts. The company's strategy is to seek partners and to acquire new technologies and products, including those that are already being marketed. The firm has an agreement with the University of Massachusetts Medical School (UMMS), whereby CytRx funds research for development of various potential products and in turn acquires the rights to them. Such products include a gene silencing technology treating cytomegalovirus and a DNA-based HIV vaccine, currently in clinical trials. Its other products are FLOCOR, an intravenous agent treating sickle cell anemia and other acute vaso-occlusive disorders; and TranzFect, a delivery technology for DNA and conventional-based vaccines. CytRx has sold the rights to FLOCOR to SynthRx, and TranzFect technology has been licensed to Merck & Co., Inc. and Vical, Inc. CytRx is also involved in the development of iroxanadine, for cardiovascular disease, and recently received FDA approval for the use of arimoclomol in the treatment of amyotrophic lateral sclerosis (Lou Gehrig's disease).

FINANCIALS: Sales and profits are in thousands of dollars—add 000 to get the full amount. 2006 Note: Financial information for 2006 was not available for all companies at press time.

2006 Sales: $2,066	2006 Profits: $-16,752	**U.S. Stock Ticker: CYTR**	
2005 Sales: $ 184	2005 Profits: $-15,093	**Int'l Ticker:** Int'l Exchange:	
2004 Sales: $ 428	2004 Profits: $-16,392	Employees: 25	
2003 Sales: $ 94	2003 Profits: $-17,845	Fiscal Year Ends: 12/31	
2002 Sales: $1,300	2002 Profits: $-6,200	Parent Company:	

SALARIES/BENEFITS:

Pension Plan:	ESOP Stock Plan:	Profit Sharing:	Top Exec. Salary: $417,175	Bonus: $400,000
Savings Plan:	Stock Purch. Plan:		Second Exec. Salary: $261,750	Bonus: $68,750

OTHER THOUGHTS:

Apparent Women Officers or Directors:
Hot Spot for Advancement for Women/Minorities:

LOCATIONS: ("Y" = Yes)

West:	Southwest:	Midwest:	Southeast:	Northeast:	International:
Y				Y	

Note: Financial information, benefits and other data can change quickly and may vary from those stated here.

DAIICHI SANKYO CO LTD

www.daiichisankyo.co.jp

Industry Group Code: 551110 Ranks within this company's industry group: Sales: Profits:

Drugs:		Other:	Clinical:	Computers:	Other:
Discovery:	Y	AgriBio:	Trials/Services:	Hardware:	Specialty Services:
Licensing:	Y	Genomics/Proteomics:	Laboratories:	Software:	Consulting:
Manufacturing:	Y	Tissue Replacement:	Equipment/Supplies:	Arrays:	Blood Collection:
Development:	Y		Research/Development Svcs.:	Database Management:	Drug Delivery:
Generics:			Diagnostics:		Drug Distribution:

TYPES OF BUSINESS:

Pharmaceuticals

BRANDS/DIVISIONS/AFFILIATES:

Daiichi Sankyo, Inc.
Daiichi Pharmaceutical Co., Ltd.
Pravastatin
Levofloxacin
Olmesartan
Omnipaque
Loxonin
Sonazoid for Injection

CONTACTS: *Note: Officers with more than one job title may be intentionally listed here more than once.*

Takashi Shoda, CEO
Takashi Shoda, Pres.
Ryuzo Takada, Chief Mktg. & Sales Officer
Akio Ozaki, Chief Human Resources Officer
Kazunori Hirokawa, General Mgr.-R&D
Takeshi Ogita, General Mgr.-Pharmaceutical Tech.
Tsutomu Une, Chief Corp. Strategy Officer
Hitoshi Matsuda, Chief Corp. Performance Mgmt. Officer
Yoshihiko Suzuki, General Mgr.-Domestic Sales & Mktg.
Akira Nagano, General Mgr.-Quality & Safety Mgmt.
Kazuhiko Tanzawa, Pres., Daiichi Sankyo Research Institute
Kiyoshi Morita, Chmn.
Toru Kuroda, General Mgr.-Supply Chain

Phone: 81-3-6225-1111	Fax:
Toll-Free:	
Address: 3-5-1, Nihonbashi-honcho, Chuo-ku, Tokyo, 103-8426 Japan	

GROWTH PLANS/SPECIAL FEATURES:

Daiichi Sankyo Co., Ltd., formed from the 2005 merger of the 106 year old Sankyo Co., Ltd. and the 90 year old Daiichi Pharmaceutical Co., Ltd, is primarily a pharmaceutical manufacturing firm. Its products, although utilized most extensively in Japan, are distributed in Asia, the US, Europe and elsewhere through 5,000 representatives, many in the US and Europe and half of them in Japan. Its research and development is focused on cardiovascular, bone/joint and infectious diseases; glucose metabolic disorders; cancer; immunity; and allergies. Available only in Japan, Daiichi Sankyo Healthcare offers self-medication options, such as over-the-counter medicine, skin care products and functional foods.   Some of its major drugs are Pravastatin, an antihyperlipidemic agent which is also sold under the name Mevalotin; Levofloxacin, an oral antibacterial agent, also called Cravit; and Olmesartan, an antihypertension agent sold in the US as Benicar, and sold in Japan and Europe as Olmetec. Some of its other drugs include Omnipaque, a non-ionicity contrast agent; Loxonin, a non-steroidal analgesic and anti-inflammatory agent; Panaldine, an anti-platelet agent; Artist, a long-acting beta-blocker; Venofer, an anemia treatment; WelChol, an antihyperlipidemic agent; and Kremezin, a chronic renal failure treatment. It non-drug products include Karoyan Gush, a hair-growth accelerator; Cystina C, a vitamin C supplement; and LamisilAT, an athlete's foot and ringworm treatment. Daiichi Sankyo, Inc., the company's US subsidiary, has 900 sales representatives, marketing Benicar, Evoxac, WelChol and Floxin Otic. In April 2007, it agreed to transfer all commercial rights of Panaldine to Sanofi-aventis, effective October 2007. In January 2007, the firm's wholly owned subsidiary, Daiichi Pharmaceutical Co., Ltd., released Sonazoid for Injection, an ultrasound contrast agent.

FINANCIALS: Sales and profits are in thousands of dollars—add 000 to get the full amount. 2006 Note: Financial information for 2006 was not available for all companies at press time.

2006 Sales: $	2006 Profits: $	**U.S. Stock Ticker:**
2005 Sales: $	2005 Profits: $	**Int'l Ticker: 4568** Int'l Exchange: Tokyo-TSE
2004 Sales: $	2004 Profits: $	Employees:
2003 Sales: $	2003 Profits: $	Fiscal Year Ends: 3/31
2002 Sales: $	2002 Profits: $	Parent Company:

SALARIES/BENEFITS:

Pension Plan: Y	ESOP Stock Plan:	Profit Sharing:	Top Exec. Salary: $	Bonus: $
Savings Plan:	Stock Purch. Plan:		Second Exec. Salary: $	Bonus: $

OTHER THOUGHTS:

Apparent Women Officers or Directors:
Hot Spot for Advancement for Women/Minorities:

LOCATIONS: ("Y" = Yes)

West:	Southwest:	Midwest:	Southeast:	Northeast:	International:
					Y

DECODE GENETICS INC

www.decode.com

Industry Group Code: 511212 **Ranks within this company's industry group:** Sales: 4 Profits: 5

Drugs:		Other:		Clinical:		Computers:		Other:	
Discovery:	Y	AgriBio:		Trials/Services:		Hardware:		Specialty Services:	Y
Licensing:		Genomics/Proteomics:	Y	Laboratories:		Software:	Y	Consulting:	
Manufacturing:		Tissue Replacement:		Equipment/Supplies:		Arrays:		Blood Collection:	
Development:	Y			Research/Development Svcs.:	Y	Database Management:		Drug Delivery:	
Generics:				Diagnostics:				Drug Distribution:	

TYPES OF BUSINESS:

Bioinformatics & Medical Records Databases
Bioinformatics Software Products
Genetic Disease Research
Clinical Testing

BRANDS/DIVISIONS/AFFILIATES:

Clinical Genome Miner
Encode
BioStructures Group
Secure Robotized Sample Vault

CONTACTS: *Note: Officers with more than one job title may be intentionally listed here more than once.*

Kari Stefansson, CEO
Axel Nielsen, COO
Kari Stefansson, Pres.
Lance Thibault, CFO
Jeffrey Gulcher, Chief Scientific Officer
Hakon Gudbjartsson, VP-Informatics
Daniel L. Hartman, Sr. VP-Product Dev.
Jakob Sigurdsson, Sr. VP-Corp. Dev.
Lance Thibault, Treas.
Mark Gurney, Sr. VP-Drug Discovery & Dev.
C. Augustine Kong, VP-Statistics
Kari Stefansson, Chmn.

Phone: 354-570-1900	Fax: 354-570-1981
Toll-Free:	
Address: Sturlugata 8, Reykjavik, IS-101 Iceland	

GROWTH PLANS/SPECIAL FEATURES:

deCODE Genetics is a biopharmaceutical company which performs population-based genetic and medical research in order to identify diseased genes, drug targets and diagnostic targets. By analyzing the genotypic and medical data from over 50% of Iceland's genetically homogenous adult population, combined with genealogical data which covers the past 1,100 years and links the entire population, the company is able to gain insights into the pathogenesis of certain illnesses, as well as the reasons why patients respond differently to the same drugs. The comparison of DNA samples from closely and distantly related individuals with the same disease helps expedite the process of identifying specific genes and specific markers with those genes that cause the illness. deCODE has actively studied the genetics and pathology of over 50 different common diseases. Several of deCODE's product candidates resulting from its research are currently in pre-clinical or clinical trials, including products for the treatment of heart attack, stroke and asthma. The company also provides services in genotyping, structural biology, chemistry and drug discovery and development. Subsidiaries deCODE chemistry and deCODE biostructures provide structural biology and drug discovery services, and subsidiary Encode offers services including Information-rich Clinical Trials. deCODE also maintains the first Secure Robotized Sample Vault (SRSV) which archives its 530,000 clinical specimens, and the company offers several product versions of the SRSV to the research and clinical communities. deCODE is involved in collaborations with F. Hoffmann-La Roche, Merck, the National Institute of Allergy and Infectious Diseases and Bayer HealthCare. In July 2007, deCODE found the first known gene variant association with Restless Legs Syndrome and Periodic Limb Movements, two major sleep disorders.

deCODE Genetics offer relocation assistance to new employees from overseas. Employees at deCODE Genetics enjoy company sports clubs and an all-staff social club that organizes festivities such as costume parties, white-river rafting and Friday happy-hours.

FINANCIALS: Sales and profits are in thousands of dollars—add 000 to get the full amount. 2006 Note: Financial information for 2006 was not available for all companies at press time.

2006 Sales: $40,510	2006 Profits: $-85,473	**U.S. Stock Ticker: DCGN**
2005 Sales: $43,955	2005 Profits: $-62,750	**Int'l Ticker:** Int'l Exchange:
2004 Sales: $42,127	2004 Profits: $-57,255	Employees: 429
2003 Sales: $46,800	2003 Profits: $-35,100	Fiscal Year Ends: 12/31
2002 Sales: $41,100	2002 Profits: $-131,900	Parent Company:

SALARIES/BENEFITS:

Pension Plan:	ESOP Stock Plan:	Profit Sharing:	Top Exec. Salary: $655,559	Bonus: $400,000
Savings Plan: Y	Stock Purch. Plan: Y		Second Exec. Salary: $360,000	Bonus: $190,000

OTHER THOUGHTS:

Apparent Women Officers or Directors:
Hot Spot for Advancement for Women/Minorities:

LOCATIONS: ("Y" = Yes)

West:	Southwest:	Midwest:	Southeast:	Northeast:	International:
Y		Y		Y	Y

Note: Financial information, benefits and other data can change quickly and may vary from those stated here.

DENDREON CORPORATION   www.dendreon.com

Industry Group Code: 325412 Ranks within this company's industry group: Sales: 175   Profits: 180

Drugs:		Other:	Clinical:	Computers:	Other:
Discovery:	Y	AgriBio:	Trials/Services:	Hardware:	Specialty Services:
Licensing:		Genomics/Proteomics:	Laboratories:	Software:	Consulting:
Manufacturing:		Tissue Replacement:	Equipment/Supplies:	Arrays:	Blood Collection:
Development:	Y		Research/Development Svcs.:	Database Management:	Drug Delivery:
Generics:			Diagnostics:		Drug Distribution:

TYPES OF BUSINESS:

Cancer Drugs

BRANDS/DIVISIONS/AFFILIATES:

Provenge
Neuvenge
Trp-p8
Genetech, Inc.
Amgen Fremont, Inc.
CA-p (MN)
CEA

CONTACTS: *Note: Officers with more than one job title may be intentionally listed here more than once.*

Mitchell H. Gold, CEO
Mitchell H. Gold, Pres.
Gregory T. Schiffman, CFO/Sr. VP
James V. Caggiano, VP-Sales & Mktg.
David L. Urdal, Chief Scientific Officer/Sr. VP
Rick Hamm, General Counsel/Sec.
Rick Hamm, Sr. VP-Corp. Dev.
Richard B. Brewer, Chmn.

Phone: 206-256-4545	Fax: 206-256-0571
Toll-Free:	
Address: 3005 First Ave., Seattle, WA 98121 US	

GROWTH PLANS/SPECIAL FEATURES:

Dendreon Corporation develops and commercializes cancer treatments. The company's portfolio includes active cellular immunotherapy, monoclonal antibody and small molecule product candidates to treat a wide range of cancers. The most advanced product candidate is Provenge, an active cellular immunotherapy that has completed two Phase III trials for the treatment of asymptomatic, metastatic, androgen-independent prostrate cancer. The company's second candidate, Neuvenge, is the investigational active immunotherapy for the treatment of patients with breast, ovarian and other solid tumor expressing HER2/neu. The firm is currently evaluating future development plans for Neuvenge. Dendreon also has a number of products in the pre-clinical stage. Dendreon has a few product candidates in the research and development program such as Trp-p8 for lung, breast, prostrate and colon cancer; CA-9 (MN) for kidney, colon and cervical cancer; and CEA for breast, lung and colon cancer. The company is in collaboration with Genentech, Inc. for the preclinical research, clinical development and commercialization of potential products derived from trp-p8, an ion channel found in prostrate cancer cells. The firm and Genetech are currently engaged in discovering, evaluating and developing small molecule and monoclonal antibody therapeutics that modify trp-p8 function. Dendreon also has preclinical collaborations with Amgen Fremont, Inc., the survivor of the acquisition of Abgenix., Inc. by Amgen, Inc., focused on the discovery, development and commercialization of fully-human monoclonal antibodies against a membrane-bound serine protease. Dendreon is currently awaiting FDA approval for the commercialization of Provenge.

The company offers its employees medical, dental and vision insurance; life, AD&D and disability insurance; a 401(k) plan; an employee assistance program; a public transportation subsidy; and a tuition subsidy program.

FINANCIALS: Sales and profits are in thousands of dollars—add 000 to get the full amount. 2006 Note: Financial information for 2006 was not available for all companies at press time.

2006 Sales: $ 273	2006 Profits: $-91,642	**U.S. Stock Ticker: DNDN**
2005 Sales: $ 210	2005 Profits: $-81,547	**Int'l Ticker:** Int'l Exchange:
2004 Sales: $5,035	2004 Profits: $-75,240	Employees: 232
2003 Sales: $27,041	2003 Profits: $-28,493	Fiscal Year Ends: 12/31
2002 Sales: $15,269	2002 Profits: $-24,669	Parent Company:

SALARIES/BENEFITS:

Pension Plan:	ESOP Stock Plan:	Profit Sharing:	Top Exec. Salary: $450,000	Bonus: $222,976
Savings Plan: Y	Stock Purch. Plan:		Second Exec. Salary: $375,000	Bonus: $148,650

OTHER THOUGHTS:

Apparent Women Officers or Directors: 2
Hot Spot for Advancement for Women/Minorities: Y

LOCATIONS: ("Y" = Yes)

West:	Southwest:	Midwest:	Southeast:	Northeast:	International:
Y					

DENDRITE INTERNATIONAL INC
www.drte.com

Industry Group Code: 511212 Ranks within this company's industry group: Sales: 1 Profits: 3

Drugs:	Other:	Clinical:		Computers:		Other:	
Discovery:	AgriBio:	Trials/Services:	Y	Hardware:		Specialty Services:	
Licensing:	Genomics/Proteomics:	Laboratories:		Software:	Y	Consulting:	Y
Manufacturing:	Tissue Replacement:	Equipment/Supplies:		Arrays:		Blood Collection:	
Development:		Research/Development Svcs.:		Database Management:	Y	Drug Delivery:	
Generics:		Diagnostics:				Drug Distribution:	

TYPES OF BUSINESS:
Software - Development & Marketing
Marketing & Sales Support & Outsourcing
Clinical Trials Support
Pharmaceuticals Marketing & Tracking Software
Compliance Services
Databases

BRANDS/DIVISIONS/AFFILIATES:
Optas, Inc.
Mobile Intelligence
Webforce
Analyzer
Organization Manager

CONTACTS: *Note: Officers with more than one job title may be intentionally listed here more than once.*
John E. Bailye, CEO
Joseph A. Ripp, COO
Joseph A. Ripp, Pres.
Jeffrey J. Bairstow, CFO/Exec. VP
Garry D. Johnson, CTO/Sr. VP
Christine A. Pellizzari, Sr. VP/General Counsel/Corp. Sec.
Mark H. Cieplik, Sr. VP-Clinical Svcs.
Jean-Paul Modde, Sr. VP-Sales & Mktg., Europe & Asia/Pacific
Mark Theilken, Sr. VP-Sales & Mktg., N. America
John E. Bailye, Chmn.
Natasha Giordano, Sr. VP-Global Accounts

Phone: 908-443-2000	Fax: 908-470-9900
Toll-Free:	
Address: 1405/1425 Rte. 206 S., Bedminster, NJ 07921 US	

GROWTH PLANS/SPECIAL FEATURES:
Dendrite International, Inc. provides software and service applications to enhance pharmaceutical and life science companies' sales, marketing and clinical processes. These offerings are organized into four categories: sales solutions, marketing solutions, compliance solutions and clinical solutions. Dendrite's sales services include software systems Mobile Intelligence, Analyzer, Organization Manager and Webforce; hardware and asset management; managed hosting services; regulatory consulting; and training services. The company's marketing services support companies in direct marketing initiatives. Marketing services include data systems for tracking prescriber information to enable targeted delivery of promotional products to health care professionals. The compliance offerings unite software tools with consulting and professional services to help pharmaceutical and life sciences companies meet federal and state regulatory compliance requirements. Offerings include sampling solutions such as FDA/PDMA consulting and reconciliation services and federal and state regulatory compliance services including training, data capture, computer systems auditing and validation security. Through its clinical solutions offerings, Dendrite helps customers streamline clinical trial processes for drug development. Dendrite serves approximately 150 companies in over 50 countries, including the world's top 20 pharmaceutical companies. Noteworthy customers include Bayer, Bristol-Meyers Squibb, Eli Lilly, Merck, Mitsubishi Tokyo, Pfizer and Proctor & Gamble. In recent years the firm has acquired Optas, Inc., a leading provider of relationship marketing solutions for patients and physicians, designed for patient acquisition. In March 2007, the company agreed to be acquired in a merger transaction by CEGEDIM S.A. for $751 million.

Dendrite's employees enjoy flexible and dependent care spending accounts; a prescription program; gym and nutrition center reimbursements; alternative work schedules; business casual dress; discounted movie and Broadway tickets; an employee stock purchase plan; a 401(k) plan; and on-site services including dry cleaning, automotive oil changes, vehicle repair and chair massages.

FINANCIALS: Sales and profits are in thousands of dollars—add 000 to get the full amount. 2006 Note: Financial information for 2006 was not available for all companies at press time.

2006 Sales: $423,958	2006 Profits: $-26,745	U.S. Stock Ticker: DRTE
2005 Sales: $437,240	2005 Profits: $21,447	Int'l Ticker: Int'l Exchange:
2004 Sales: $399,197	2004 Profits: $29,565	Employees: 2,534
2003 Sales: $321,107	2003 Profits: $21,060	Fiscal Year Ends: 12/31
2002 Sales: $225,756	2002 Profits: $15,398	Parent Company:

SALARIES/BENEFITS:
Pension Plan:	ESOP Stock Plan:	Profit Sharing:	Top Exec. Salary: $522,917	Bonus: $
Savings Plan: Y	Stock Purch. Plan: Y		Second Exec. Salary: $509,760	Bonus: $

OTHER THOUGHTS:
Apparent Women Officers or Directors: 2
Hot Spot for Advancement for Women/Minorities: Y

LOCATIONS: ("Y" = Yes)
West:	Southwest:	Midwest:	Southeast:	Northeast:	International:
		Y	Y	Y	Y

DEPOMED INC

www.depomedinc.com

Industry Group Code: 325412A Ranks within this company's industry group: Sales: 12   Profits: 19

Drugs:	Other:	Clinical:	Computers:	Other:
Discovery:	AgriBio:	Trials/Services:	Hardware:	Specialty Services:
Licensing:	Genomics/Proteomics:	Laboratories:	Software:	Consulting:
Manufacturing:	Tissue Replacement:	Equipment/Supplies:	Arrays:	Blood Collection:
Development: Y		Research/Development Svcs.:	Database Management:	Drug Delivery: Y
Generics:		Diagnostics:		Drug Distribution:

TYPES OF BUSINESS:

Drug Delivery System-Based Drugs

BRANDS/DIVISIONS/AFFILIATES:

AcuForm
Glumetza
ProQuin XR
Gabapentin GR
King Pharmaceutical, Inc.
Watson Pharmaceuticals, Inc.

CONTACTS: Note: Officers with more than one job title may be intentionally listed here more than once.

John W. Fara, CEO
Carl Pelzel, COO/Exec. VP
John W. Fara, Pres.
John F. Hamilton, CFO/VP
Matthew Gosling, General Counsel/VP-Legal
John N. Shell, VP-Oper.
Thadd Vargas, VP-Bus. Dev.
Jeffrey Miller, VP-Regulatory Affairs & Quality Assurance
John W. Fara, Chmn.

Phone: 650-462-5900	Fax: 650-462-9993
Toll-Free:	
Address: 1360 O'Brien Dr., Menlo Park, CA 94025 US	

GROWTH PLANS/SPECIAL FEATURES:

Depomed, Inc. focuses on the development and commercialization of differentiated products that are based on proprietary drug delivery technologies. The company developed two commercial products: Glumetza is a once-daily treatment for adults with type 2 diabetes that the firm jointly commercializes in the U.S. with King Pharmaceutical, Inc.; ProQuin XR is a once-daily treatment of uncomplicated urinary tract infections that Esprit Pharma, Inc. markets in the U.S. Depomed's most advanced product candidate in development is Gabapentin GR, an extended release form of gabapentin. The company submitted an investigational new drug application for Gabapentin GR for a Phase II clinical trial for the treatment of menopausal hot flashes. In addition, the firm has other product candidates in earlier stages of development, including a treatment for gastroesophageal reflux disease. Depomed's drugs are based on its proprietary drug delivery system, AcuForm. The AcuForm technology is a polymer-based drug delivery platform that provides targeted drug delivery solutions for a wide range of compounds. The technology embraces diffusional, erosional, bilayer and multi-drug systems that can optimize oral drug delivery for both soluble and insoluble drugs. One application of the technology allows standard-sized tablets to be retained in the stomach for six to eight hours after administration, thereby extending the time of drug delivery to the small intestine. In March 2006, Depomed announced the approval in Sweden of ProQuin XR for the treatment of uncomplicated urinary tract infections. In July 2007, the company entered into a co-promotion agreement with Watson Pharmaceuticals, Inc. for ProQuin XR. and announced negative results for Gabapentin GR Phase III clinical trials in postherpetic neuralgia. This will not affect the clinical development program in menopausal hot flashes.

The company offers its employees medical, dental, life and long-term disability insurance; flexible spending accounts; a 401(k) plan; and an employee stock purchase plan.

FINANCIALS: Sales and profits are in thousands of dollars—add 000 to get the full amount. 2006 Note: Financial information for 2006 was not available for all companies at press time.

2006 Sales: $9,551	2006 Profits: $-39,659	U.S. Stock Ticker: DEPO
2005 Sales: $4,405	2005 Profits: $-24,467	Int'l Ticker: Int'l Exchange:
2004 Sales: $ 203	2004 Profits: $-26,775	Employees: 105
2003 Sales: $ 982	2003 Profits: $-30,015	Fiscal Year Ends: 12/31
2002 Sales: $1,700	2002 Profits: $-13,500	Parent Company:

SALARIES/BENEFITS:

Pension Plan:	ESOP Stock Plan:	Profit Sharing:	Top Exec. Salary: $500,000	Bonus: $135,000
Savings Plan: Y	Stock Purch. Plan: Y		Second Exec. Salary: $325,000	Bonus: $110,000

OTHER THOUGHTS:

Apparent Women Officers or Directors:
Hot Spot for Advancement for Women/Minorities:

LOCATIONS: ("Y" = Yes)

West:	Southwest:	Midwest:	Southeast:	Northeast:	International:
Y					

Note: Financial information, benefits and other data can change quickly and may vary from those stated here.

DIAGNOSTIC PRODUCTS CORPORATION   www.dpcweb.com

Industry Group Code: 325413   Ranks within this company's industry group: Sales: Profits:

Drugs:	Other:	Clinical:		Computers:		Other:
Discovery:	AgriBio:	Trials/Services:		Hardware:		Specialty Services:
Licensing:	Genomics/Proteomics:	Laboratories:		Software:		Consulting:
Manufacturing:	Tissue Replacement:	Equipment/Supplies:	Y	Arrays:		Blood Collection:
Development:		Research/Development Svcs.:		Database Management:		Drug Delivery:
Generics:		Diagnostics:	Y			Drug Distribution:

TYPES OF BUSINESS:
Supplies-Immunodiagnostic Kits
Nonisotopic Diagnostic Tests
Immunoassay Analyzers
Allergy Testing

BRANDS/DIVISIONS/AFFILIATES:
IMMULITE
IMMULITE 1000
Bayer Diagnostics
IMMULITE Turbo
Sample Management System
IMMULITE 2500
3gAllergy
Diagnostic Products Corp.

CONTACTS: *Note: Officers with more than one job title may be intentionally listed here more than once.*
Anthony P. Bihl, CEO
Sidney A. Aroesty, COO
Jochen Schmitz, CFO

Phone: 310-645-8200   **Fax:** 310-645-9999
Toll-Free: 800-768-6699
Address: 5210 Pacific Concourse Dr., Los Angeles, CA 90045-6900 US

GROWTH PLANS/SPECIAL FEATURES:
Siemens Medical Solutions Diagnostics (SMSD), formed from the 2006 acquisition and subsumption of Diagnostic Products Corp. (DPC) and Bayer Diagnostics, is a wholly-owned subsidiary of Siemens Medical Solutions USA, Inc. SMSD offers a broad portfolio of performance-driven diagnostics solutions that provide more effective ways to assist in the diagnosis, monitoring and management of disease. Its products and services offer the right balance of science, technology and practicality across the healthcare continuum to provide healthcare professionals with the vital information they need to deliver better, more personalized healthcare to patients around the world. With the acquisition of DPC and Bayer, SMSD has bridged the gap between in-vivo and in-vitro diagnostics to become the first full service diagnostics company. By bringing together medical imaging, laboratory diagnostics and healthcare and healthcare information technology, SMSD offers a unique set of solutions that provide clinical, operational and financial outcomes. The portfolio of SMSD includes products and services designed to optimize efficiency, improve workflow, help ensure patient safety and provide the highest levels of productivity and flexibility. Its broad spectrum of immunoassay, chemistry, hematology, molecular, urinalysis, diabetes and blood gas testing systems, in conjunction with automation, informatics and consulting solutions, serve the needs of laboratories of any size. SMSD's proprietary products include: the IMMULITE, ADVIA, RapidLab and Clinitek families of products, along with the DCA 2000+, the MicroMix 5, 3gAllergy and SMS products. In 2007, Siemens Medical Solutions USA agreed to acquire Dade Behring, Inc., a leading clinical laboratory diagnostics company. In April 2006, Diagnostic Products Corp. (DPC) was acquired by Siemens Medical Solutions USA, Inc. for approximately $1.86 billion.

FINANCIALS: Sales and profits are in thousands of dollars—add 000 to get the full amount. 2006 Note: Financial information for 2006 was not available for all companies at press time.

2006 Sales: $	2006 Profits: $	**U.S. Stock Ticker: Subsidiary**
2005 Sales: $481,100	2005 Profits: $67,200	**Int'l Ticker:** Int'l Exchange:
2004 Sales: $446,800	2004 Profits: $61,700	Employees: 2,495
2003 Sales: $381,386	2003 Profits: $61,795	Fiscal Year Ends: 12/31
2002 Sales: $324,100	2002 Profits: $47,300	Parent Company: SIEMENS MEDICAL SOLUTIONS

SALARIES/BENEFITS:
Pension Plan:	ESOP Stock Plan:	Profit Sharing:	Top Exec. Salary: $600,000	Bonus: $
Savings Plan:	Stock Purch. Plan:		Second Exec. Salary: $450,000	Bonus: $30,000

OTHER THOUGHTS:
Apparent Women Officers or Directors: 1
Hot Spot for Advancement for Women/Minorities:

LOCATIONS: ("Y" = Yes)
West:	Southwest:	Midwest:	Southeast:	Northeast:	International:
Y				Y	Y

Note: Financial information, benefits and other data can change quickly and may vary from those stated here.

DIGENE CORPORATION   www.digene.com

Industry Group Code: 325413 Ranks within this company's industry group: Sales: 8   Profits: 8

Drugs:	Other:		Clinical:		Computers:		Other:
Discovery:	AgriBio:		Trials/Services:		Hardware:		Specialty Services:
Licensing:	Genomics/Proteomics:	Y	Laboratories:		Software:		Consulting:
Manufacturing:	Tissue Replacement:		Equipment/Supplies:	Y	Arrays:		Blood Collection:
Development:			Research/Development Svcs.:	Y	Database Management:		Drug Delivery:
Generics:			Diagnostics:	Y			Drug Distribution:

TYPES OF BUSINESS:
Gene-Based Diagnostic Tests
Diagnostics Test Kits

BRANDS/DIVISIONS/AFFILIATES:
Hybrid Capture

CONTACTS: *Note: Officers with more than one job title may be intentionally listed here more than once.*
Daryl J. Faulkner, CEO
Daryl J. Faulkner, Pres.
Joseph P. Slattery, CFO/Sr. VP
Robert McG. Lilley, Sr. VP-Global Sales & Mktg.
James H. Godsey, Sr. VP-R&D
Belinda O. Patrick, Sr. VP-Mfg. Oper.
Donna Marie Seyfried, VP-Bus. Dev.
Attila T. Lorincz, Chief Scientific Officer
C. Douglas White, Sr. VP-Sales & Mktg.-Americas & Asia Pacific

Phone: 301-944-7000	Fax: 240-632-7121
Toll-Free: 800-344-3631	
Address: 1201 Clopper Rd., Gaithersburg, MD 20878 US	

GROWTH PLANS/SPECIAL FEATURES:

Digene Corp. develops, manufactures and markets its proprietary gene-based diagnostic tests for the screening, monitoring and diagnosis of human diseases. The company's primary focus is women's cancers and infectious diseases. The firm applies its proprietary Hybrid Capture technology to develop successful diagnostic test for human papillomavirus (HPV), which is the primary cause of cervical cancer. The Hybrid Capture platform is a signal amplification technology that combines the convenience of a direct probe test with the sensitivity of an amplification test, requires minimal sample preparation and provides objective test results. Digene also developed its Hybrid Capture technology in the form of diagnostic test kits to include tests for the detection of chlamydia, gonorrhea and other sexually transmitted infections. Products in development include hc4, a fully automated screening and genotyping platform; HPV genotyping products, whose potential uses include HPV genotyping in connection with HPV vaccines and revised patient care management; and Signature cystic fibrosis tests, which are molecular diagnosis tests for cystic fibrosis. The firm's core research efforts include research program for improved molecular diagnosis assay systems for detection of HPV and other targets in the area of women's cancers and infectious diseases; and research on nucleic acid detection technology.

The company offers its employees medical, dental and vision insurance; a 401(k) plan; an employee assistance program; life, AD&D, short- and long-term disability coverage; tuition reimbursement; a college savings plan; flexible spending accounts; and access to credit unions.

FINANCIALS: Sales and profits are in thousands of dollars—add 000 to get the full amount. 2006 Note: Financial information for 2006 was not available for all companies at press time.

2006 Sales: $152,900	2006 Profits: $8,400	U.S. Stock Ticker: DIGE
2005 Sales: $115,142	2005 Profits: $-8,167	Int'l Ticker: Int'l Exchange:
2004 Sales: $90,160	2004 Profits: $21,542	Employees: 401
2003 Sales: $63,100	2003 Profits: $-4,300	Fiscal Year Ends: 6/30
2002 Sales: $48,800	2002 Profits: $-9,400	Parent Company:

SALARIES/BENEFITS:

Pension Plan:	ESOP Stock Plan:	Profit Sharing:	Top Exec. Salary: $449,533	Bonus: $505,300
Savings Plan: Y	Stock Purch. Plan:		Second Exec. Salary: $392,463	Bonus: $357,300

OTHER THOUGHTS:
Apparent Women Officers or Directors: 3
Hot Spot for Advancement for Women/Minorities: Y

LOCATIONS: ("Y" = Yes)

West:	Southwest:	Midwest:	Southeast:	Northeast:	International:
				Y	Y

Note: Financial information, benefits and other data can change quickly and may vary from those stated here.

DISCOVERY LABORATORIES INC   www.discoverylabs.com

Industry Group Code: 325412   Ranks within this company's industry group: Sales: Profits: 151

Drugs:		Other:		Clinical:		Computers:		Other:	
Discovery:	Y	AgriBio:		Trials/Services:		Hardware:		Specialty Services:	
Licensing:		Genomics/Proteomics:		Laboratories:		Software:		Consulting:	
Manufacturing:	Y	Tissue Replacement:		Equipment/Supplies:		Arrays:		Blood Collection:	
Development:	Y			Research/Development Svcs.:		Database Management:		Drug Delivery:	Y
Generics:				Diagnostics:				Drug Distribution:	

TYPES OF BUSINESS:

Respiratory Disease Treatments
Pulmonary Drug Delivery Products

BRANDS/DIVISIONS/AFFILIATES:

Surfaxin
Aerosurf
Chrysalis Technologies
Laboratorios del Dr. Esteve, S.A.

CONTACTS: *Note: Officers with more than one job title may be intentionally listed here more than once.*

Robert J. Capetola, CEO
Robert J. Capetola, Pres.
John G. Cooper, CFO/Exec. VP
Kathryn Cole, Sr. VP-Human Resources
Robert Segal, Sr. VP-Medical & Scientific Affairs
Charles F. Katzer, Sr. VP-Mfg. Oper.
David L. Lopez, General Counsel/Exec. VP
Thomas F. Miller, Sr. VP-Commercialization & Corp. Dev.
Robert Segal, Chief Medical Officer
Mary B. Templeton, Sr. VP/Deputy General Counsel
W. Thomas Amick, Chmn.

Phone: 215-488-9300	Fax: 215-488-9301
Toll-Free:	
Address: 2600 Kelly Road, Ste. 100, Warrington, PA 18976 US	

GROWTH PLANS/SPECIAL FEATURES:

Discovery Laboratories, Inc. (DLI) is a biotechnology company that develops proprietary surfactant technology as Surfactant Replacement Therapies (SRT) for respiratory disorders and diseases. Surfactants are produced naturally by the lungs and are essential for breathing. Discovery's technology produces a precision-engineered surfactant designed to closely mimic the essential properties of natural human lung surfactant. The SRT pipeline is focused initially on the most significant respiratory conditions prevalent in the neonatal intensive care unit. The firm's lead product, Surfaxin, is used in the prevention of respiratory distress syndrome in premature infants. DLI also develops Surfaxin for the prevention and treatment of bronchopulmonary dysplasia in premature infants. Aerosurf, the company's proprietary SRT in aerosolized form administered through nasal continuous positive airway pressure, is being developed for the prevention and treatment of infants at risk for respiratory failure. The firm has an alliance with Chrysalis Technologies, a division of Philip Morris USA, Inc., to develop and commercialize aerosol SRT to address a broad range of serious respiratory conditions such as neonatal respiratory failure, cystic fibrosis, chronic obstructive respiratory disorder and asthma; and Laboratorios del Dr. Esteve, S.A. for the development, marketing and sales of DLI's products. In June 2006, the company voluntarily withdrew its application for European Medicines Evaluation Agency approval of Surfaxin, due to irremediable manufacturing issues. The manufacturing issue surrounds inconstant stability and the failure to achieve certain stability parameters.

The company offers its employees benefits that include medical, dental and life insurance; short- and long-term disability coverage; a 401(k) plan; stock options; an employee assistance program; tuition reimbursement; a college savings plan; and access to a fitness center.

FINANCIALS: Sales and profits are in thousands of dollars—add 000 to get the full amount. 2006 Note: Financial information for 2006 was not available for all companies at press time.

2006 Sales: $	2006 Profits: $-46,333	**U.S. Stock Ticker: DSCO**	
2005 Sales: $ 134	2005 Profits: $-58,904	**Int'l Ticker:**	**Int'l Exchange:**
2004 Sales: $1,200	2004 Profits: $-46,200	Employees: 160	
2003 Sales: $1,037	2003 Profits: $-24,280	Fiscal Year Ends: 12/31	
2002 Sales: $1,800	2002 Profits: $-17,400	Parent Company:	

SALARIES/BENEFITS:

Pension Plan:	ESOP Stock Plan:	Profit Sharing:	Top Exec. Salary: $470,000	Bonus: $150,000
Savings Plan: Y	Stock Purch. Plan: Y		Second Exec. Salary: $292,000	Bonus: $120,000

OTHER THOUGHTS:

Apparent Women Officers or Directors: 2
Hot Spot for Advancement for Women/Minorities: Y

LOCATIONS: ("Y" = Yes)

West:	Southwest:	Midwest:	Southeast:	Northeast:	International:
Y				Y	

DIVERSA CORPORATION   www.diversa.com

Industry Group Code: 541710 Ranks within this company's industry group: Sales: 11   Profits: 18

Drugs:	Other:		Clinical:		Computers:		Other:
Discovery:	AgriBio:	Y	Trials/Services:		Hardware:		Specialty Services:
Licensing:	Genomics/Proteomics:	Y	Laboratories:		Software:		Consulting:
Manufacturing:	Tissue Replacement:		Equipment/Supplies:		Arrays:		Blood Collection:
Development:			Research/Development Svcs.:	Y	Database Management:	Y	Drug Delivery:
Generics:			Diagnostics:				Drug Distribution:

TYPES OF BUSINESS:

Genomics Research
Enzyme & Biological Compound Discovery
Gene Expression Libraries

BRANDS/DIVISIONS/AFFILIATES:

DiverseLibraries
PathwayLibraries
DirectEvolution
SciLect
SingleCell
Luminase
Cottonase
Tunable GeneReassembly

CONTACTS: *Note: Officers with more than one job title may be intentionally listed here more than once.*

Edward T. Shonsey, CEO
Anthony E. Altig, CFO
William H. Baum, Exec. VP-Bioscience Prod.
Patrick Simms, Sr. VP-Oper.
Wendy Kelley, Dir.-Investor Rel.
Anthony E. Altig, Sr. VP-Finance
James H. Cavanaugh, Chmn.

Phone: 858-526-5000	Fax: 858-526-5551
Toll-Free: 800-523-2990	
Address: 4955 Directors Pl., San Diego, CA 92121-1609 US	

GROWTH PLANS/SPECIAL FEATURES:

Diversa Corp. is a leader in utilizing proprietary genomic technologies for the rapid discovery and optimization of novel products from genes and gene pathways. The company directs its integrated portfolio of technologies towards the discovery, evolution and production of commercially valuable molecules with pharmaceutical applications such as monoclonal antibodies and orally active drugs, as well as enzymes and small molecules with agricultural, chemical and industrial applications. Diversa uses the large collection of genetic material captured in its DiverseLibraries and PathwayLibraries, estimated to contain the complete genomes of over 1 million microorganisms, to accelerate the discovery of small-molecule pharmaceuticals. The firm's proprietary DirectEvolution technologies, including Gene Site Saturation Mutagenesis (GSSM) and Tunable GeneReassembly, then allow scientists to create a family of variant genes and reassemble genes into new combinations, thus creating potentially usable new compounds. Finally, Diversa's SciLect and SingleCell screening programs analyze novel genes. The company's products include Luminase and Cottonase, enzymes improving pulp bleaching and textile processing, respectively; Phyzyme XP, which helps pigs and chickens to digest phosphorus; Pyrolase 160 and 200 enzymes for oil and gas well fracturing operations; and ThermalAce, a thermostable DNA polymerase. Diversa, in conjunction with Valley Research, has launched the Ultra-Thin enzyme, which improves the efficiency of ethanol production. In January 2007, the company announced a new research program with New Zealand Crown Research Institutes Scion and AgResearch, which aims to create New Zealand grown and manufactured biofuels. Also in 2007, Diversa signed a definitive merger agreement with Celuno Corp. Diversa stockholders will retain approximately 74% of the resulting cellulosic ethanol firm.

Diversa offers its employees health club reimbursement and flexible spending accounts. The company has been ranked one of the best places to work in the life science industry by Scientist magazine for two years in a row.

FINANCIALS: Sales and profits are in thousands of dollars—add 000 to get the full amount. 2006 Note: Financial information for 2006 was not available for all companies at press time.

2006 Sales: $49,198	2006 Profits: $-39,271	**U.S. Stock Ticker: DVSA**
2005 Sales: $54,303	2005 Profits: $-89,718	**Int'l Ticker:** Int'l Exchange:
2004 Sales: $57,600	2004 Profits: $-33,400	Employees: 287
2003 Sales: $48,959	2003 Profits: $-57,696	Fiscal Year Ends: 12/31
2002 Sales: $31,700	2002 Profits: $-28,000	Parent Company:

SALARIES/BENEFITS:

Pension Plan:	ESOP Stock Plan:	Profit Sharing:	Top Exec. Salary: $337,816	Bonus: $7,965
Savings Plan: Y	Stock Purch. Plan: Y		Second Exec. Salary: $335,500	Bonus: $

OTHER THOUGHTS:

Apparent Women Officers or Directors: 1
Hot Spot for Advancement for Women/Minorities:

LOCATIONS: ("Y" = Yes)

West:	Southwest:	Midwest:	Southeast:	Northeast:	International:
Y					

Note: Financial information, benefits and other data can change quickly and may vary from those stated here.

DOR BIOPHARMA INC

www.dorbiopharma.com

Industry Group Code: 325412A **Ranks within this company's industry group:** Sales: 17   Profits: 9

Drugs:		Other:		Clinical:	Computers:	Other:	
Discovery:	Y	AgriBio:		Trials/Services:	Hardware:	Specialty Services:	
Licensing:		Genomics/Proteomics:	Y	Laboratories:	Software:	Consulting:	
Manufacturing:		Tissue Replacement:		Equipment/Supplies:	Arrays:	Blood Collection:	
Development:	Y			Research/Development Svcs.:	Database Management:	Drug Delivery:	Y
Generics:				Diagnostics:		Drug Distribution:	

TYPES OF BUSINESS:

Drugs-Drug Delivery
Oral Formulations
Biodefense Vaccines

BRANDS/DIVISIONS/AFFILIATES:

orBec
RiVax
Endorex Corporation
Leuprolide
Oraprine
BT-VACC

CONTACTS: *Note: Officers with more than one job title may be intentionally listed here more than once.*

Christopher J. Schaber, CEO
Christopher J. Schaber, Pres.
Evan Myrianthopoulos, CFO
Robert N. Brey, Chief Scientific Officer
James Clavijo, Treas./Controller/Corp. Sec.
James S. Kuo, Chmn.

Phone: 786-425-3848	Fax: 786-425-3853
Toll-Free:	
Address: 1101 Brickell Ave., Ste. 701 South, Miami, FL 33131 US	

GROWTH PLANS/SPECIAL FEATURES:

DOR BioPharma, Inc. develops biodefense vaccines and oral formulations of small-molecule drugs that are traditionally delivered in non-oral formats. The company's leading oral product, orBec, which has completed Phase III trials and has filed a New Drug Application, is used to treat inflammation of tissues in the intestine and stomach that accompany certain types of bone marrow and stem cell transplants. It has also had positive results in other intestinal diseases, such as inflammatory bowel disease and idiopathic lymphocytic enterocolitis. DOR is developing an oral formulation of the peptide drug Leuprolide, using a Lipid Polymer Micelle (LPM) system for enhancing the intestinal absorption of water-soluble drugs/peptides that are not ordinarily absorbed or are degraded in the gastrointestinal tract. DOR is also working to develop Oraprine, (azathioprine) used as an immunosuppressant to inhibit rejection of transplanted organs, and as a second-line treatment for severe, active rheumatoid arthritis. Additionally, the company is developing vaccines to combat bioterrorism. DOR has licensed the worldwide rights to the University of Texas Southwestern Medical Center's ricin toxin vaccine, RiVax, and has entered into an agreement with Cambrex for the development and production of this vaccine. RiVax is in Phase I clinical trials and has received support from the NIH and the National Institute of Allergy and Infectious Diseases. The company has also entered into a cooperative research and development agreement with Thomas Jefferson University and the U.S. Army's Medical Research Institute of Infectious Diseases (USAMRIID) to create a mucosally delivered vaccine for botulinum toxin BT-VACC. DOR hopes its oral and nasal delivery technology will offer an advantage over delivery by injection in situations that require mass immunizations. The vaccine is currently in the animal testing stage of development. In February 2007, DOR announced it had raised approximately $5.5 million in private equity financing from institutional investors.

FINANCIALS: Sales and profits are in thousands of dollars—add 000 to get the full amount. 2006 Note: Financial information for 2006 was not available for all companies at press time.

2006 Sales: $2,313	2006 Profits: $-8,163	**U.S. Stock Ticker: DORB**	
2005 Sales: $3,076	2005 Profits: $-4,720	**Int'l Ticker:** Int'l Exchange:	
2004 Sales: $ 997	2004 Profits: $-5,872	Employees: 8	
2003 Sales: $ 84	2003 Profits: $-5,289	Fiscal Year Ends: 12/31	
2002 Sales: $	2002 Profits: $-5,000	Parent Company:	

SALARIES/BENEFITS:

Pension Plan:	ESOP Stock Plan:	Profit Sharing:	Top Exec. Salary: $300,000	Bonus: $100,000
Savings Plan:	Stock Purch. Plan:		Second Exec. Salary: $185,000	Bonus: $50,000

OTHER THOUGHTS:

Apparent Women Officers or Directors:
Hot Spot for Advancement for Women/Minorities:

LOCATIONS: ("Y" = Yes)

West:	Southwest:	Midwest:	Southeast:	Northeast:	International:
			Y		

DOV PHARMACEUTICAL INC   www.dovpharm.com

Industry Group Code: 325412 Ranks within this company's industry group: Sales: 102   Profits: 142

Drugs:		Other:	Clinical:	Computers:	Other:
Discovery:	Y	AgriBio:	Trials/Services:	Hardware:	Specialty Services:
Licensing:	Y	Genomics/Proteomics:	Laboratories:	Software:	Consulting:
Manufacturing:		Tissue Replacement:	Equipment/Supplies:	Arrays:	Blood Collection:
Development:	Y		Research/Development Svcs.:	Database Management:	Drug Delivery:
Generics:			Diagnostics:		Drug Distribution:

TYPES OF BUSINESS:

Drugs-Neurological
Nervous System Treatments
Analgesics

BRANDS/DIVISIONS/AFFILIATES:

Indiplon
Bicifadine
Ocinaplon

CONTACTS: *Note: Officers with more than one job title may be intentionally listed here more than once.*

Barbara G. Duncan, CEO
Phil Skolnick, Pres.
Phil Skolnick, Chief Scientific Officer/Exec. VP
Warren Stern, Sr. VP-Drug Dev.
Arnold Lippa, Chmn.

Phone: 732-907-3600	Fax: 732-907-3799
Toll-Free:	
Address: 150 Pierce St., Somerset, NJ 08873 US	

GROWTH PLANS/SPECIAL FEATURES:

DOV Pharmaceutical, Inc. is focused on the discovery, acquisition, development and commercialization of novel drug candidates for the central nervous system (CNS) and other abnormal neuronal processing disorders.   The firm currently has two product candidates named DOV 21,947 and DOV 102,677 that are in various stages of clinical development and an active preclinical discovery program in reuptake inhibitors and GABA modulators.   DOV has also entered into a sublicensing agreement with Neurocrine Biosciences, Inc. for FDA approval of indiplon, a treatment for insomnia, and with XTL Biopharmaceuticals Ltd. for the development and commercialization of bicifadine for the treatment of pain. Indiplon has been shown to decrease the amount of time before the onset of sleep whilst improving the overall quality in sleep duration without next day impairment in over 70 clinical trials that involve almost 8,000 patients. Bicifadine has been proven to effectively treat pain in clinical trials, DOV and XTL intend to pursue the utilization of bicifadine as a possible treatment for chronic neuropathic pain in the future.   DOV 21,947 and DOC 102, 677 are currently the firm's leading product candidates for the treatment of depression by acting as a triple reuptake inhibitor (TRI) of serotonin, norepinephrine and dopamine. In addition to its TRI platform, the company also develops reuptake inhibitor platforms that include NEDs (norepinephrine and dopamine reuptake inhibitors), SADs (serotonin and dopamine reuptake inhibitors) and SNRIs (serotonin and norepinephrine reuptake inhibitors).   These reuptake inhibitors can be modified to create compounds that treat a wide variety of neuropsychiatric disorders such as depression, attention deficit hyperactivity disorder, pain and obesity. DOV's GABA modulator program is focused on the development of a molecule that can treat anti-anxiety without any of the side effects that are typically attributed to benzodiazepines. In addition, GABA modulators also act as a sedative-hypnotic, anticonvulsant and muscle relaxant.

FINANCIALS: Sales and profits are in thousands of dollars—add 000 to get the full amount. 2006 Note: Financial information for 2006 was not available for all companies at press time.

2006 Sales: $25,951	2006 Profits: $-38,368	U.S. Stock Ticker: DOVP
2005 Sales: $8,647	2005 Profits: $-52,968	Int'l Ticker: Int'l Exchange:
2004 Sales: $2,542	2004 Profits: $-32,921	Employees: 41
2003 Sales: $2,969	2003 Profits: $-26,731	Fiscal Year Ends: 12/31
2002 Sales: $2,390	2002 Profits: $-16,820	Parent Company:

SALARIES/BENEFITS:

Pension Plan:	ESOP Stock Plan:	Profit Sharing:	Top Exec. Salary: $343,731	Bonus: $
Savings Plan: Y	Stock Purch. Plan:		Second Exec. Salary: $320,827	Bonus: $

OTHER THOUGHTS:

Apparent Women Officers or Directors: 1
Hot Spot for Advancement for Women/Minorities:

LOCATIONS: ("Y" = Yes)

West:	Southwest:	Midwest:	Southeast:	Northeast: Y	International:

Note: Financial information, benefits and other data can change quickly and may vary from those stated here.

DOW AGROSCIENCES LLC   www.dowagro.com

Industry Group Code: 115112 Ranks within this company's industry group: Sales: Profits:

Drugs:	Other:		Clinical:	Computers:	Other:
Discovery:	AgriBio:	Y	Trials/Services:	Hardware:	Specialty Services:
Licensing:	Genomics/Proteomics:	Y	Laboratories:	Software:	Consulting:
Manufacturing:	Tissue Replacement:		Equipment/Supplies:	Arrays:	Blood Collection:
Development:			Research/Development Svcs.:	Database Management:	Drug Delivery:
Generics:			Diagnostics:		Drug Distribution:

TYPES OF BUSINESS:

Agricultural Chemicals
Agricultural Biotechnology Products
Herbicides, Pesticides & Fungicides
Plant Genetics

BRANDS/DIVISIONS/AFFILIATES:

Dow Chemical Co.
DowElanco
Herculex
WideStrike
Mycogen
PhytoGen
Brazil Seeds
Cargill Hybrid Seeds

CONTACTS: Note: Officers with more than one job title may be intentionally listed here more than once.

Jerome A. Peribere, CEO
Jerome A. Peribere, Pres.
John Madia, VP-Human Resources
John Madia, VP-Productivity & Site Oper.
Kenda Resler-Friend, Dir.-Media Rel.
Rogelio Lara, VP-Finance

Phone: 317-337-3000	Fax: 317-337-4096
Toll-Free:	
Address: 9330 Zionsville Rd., Indianapolis, IN 46268 US	

GROWTH PLANS/SPECIAL FEATURES:

Dow AgroSciences, LLC, a wholly-owned subsidiary of Dow Chemical Co., is a leading global provider of pest management and biotechnology products for agricultural and specialty markets. The company was formed by a joint venture between Dow Chemical and Eli Lilly and was named DowElanco until Dow Chemical purchased Eli Lilly's share in 1997. The firm's products are broken into two categories: Agricultural chemicals, and plant genetics and biotechnology. The agricultural chemicals unit produces products in several categories: Herbicides; insecticides; fungicides; pest management solutions such as gas fumigants and termite detection tools; and Other, which encompasses N-serve nitrogen stabilizer and Telone soil fumigant. The plant genetics and biotechnology business develops agricultural products that protect crops against insects, boost nutritional value and increase crop yields. Products are broken into three segments: Traits, which encompasses the brands Herculex and WideStrike; seeds, which includes Mycogen brand seeds, Nexera canola and sunflower seeds and PhytoGen cottonseed; and oils. The company has continued to grow through mergers, acquisitions and alliances. Dow AgroSciences recently purchased Brazil Seeds, Cargill Hybrid Seeds and the Rohm and Haas agricultural chemicals business. Overall, the subsidiary has operations in 50 countries worldwide. Dow Agrosciences is currently involved in research and development collaborations with NRC, the Biodesign Institute at Arizona State University, Chlorogen, Sangamo BioSciences and the University of Melbourne. In November 2006, the firm sold its global propanil herbicide business to United Phosphorus, Ltd. for roughly $25 million.

Dow AgroSciences offers its employees benefits including an employee assistance program, stock options, a 401(k) plan, a defined benefit pension plan, continuing medical and life insurance coverage for retirees, financial planning services, dependent care reimbursement accounts, lactation support stations, adoption assistance, child daycare facilities, an on-site fitness center, nutrition consultants, credit union membership, discount event tickets and an on-site forest preserve.

FINANCIALS: Sales and profits are in thousands of dollars—add 000 to get the full amount. 2006 Note: Financial information for 2006 was not available for all companies at press time.

2006 Sales: $	2006 Profits: $	U.S. Stock Ticker: Subsidiary
2005 Sales: $3,364,000	2005 Profits: $	Int'l Ticker: Int'l Exchange:
2004 Sales: $3,368,000	2004 Profits: $	Employees: 5,500
2003 Sales: $3,008,000	2003 Profits: $	Fiscal Year Ends: 12/31
2002 Sales: $2,700,000	2002 Profits: $	Parent Company: DOW CHEMICAL COMPANY (THE)

SALARIES/BENEFITS:

Pension Plan: Y	ESOP Stock Plan:	Profit Sharing:	Top Exec. Salary: $	Bonus: $
Savings Plan: Y	Stock Purch. Plan: Y		Second Exec. Salary: $	Bonus: $

OTHER THOUGHTS:

Apparent Women Officers or Directors: 1
Hot Spot for Advancement for Women/Minorities:

LOCATIONS: ("Y" = Yes)

West:	Southwest:	Midwest:	Southeast:	Northeast:	International:
		Y			Y

Note: Financial information, benefits and other data can change quickly and may vary from those stated here.

DR REDDY'S LABORATORIES LIMITED   www.drreddys.com

Industry Group Code: 325412   Ranks within this company's industry group: Sales: 39   Profits: 40

Drugs:		Other:		Clinical:		Computers:		Other:	
Discovery:	Y	AgriBio:		Trials/Services:		Hardware:		Specialty Services:	
Licensing:		Genomics/Proteomics:		Laboratories:		Software:		Consulting:	
Manufacturing:	Y	Tissue Replacement:		Equipment/Supplies:		Arrays:		Blood Collection:	
Development:	Y			Research/Development Svcs.:		Database Management:		Drug Delivery:	
Generics:	Y			Diagnostics:				Drug Distribution:	Y

TYPES OF BUSINESS:

Pharmaceuticals
Generic Drugs
Active Ingredients
Drug Discovery & Development

BRANDS/DIVISIONS/AFFILIATES:

Reddy US Therapeutics, Inc.
New Life Colostrum
Plermin
Doxoboid

CONTACTS: Note: Officers with more than one job title may be intentionally listed here more than once.

G. V. Prasad, CEO
Satish Reddy, COO
Saumen Chakraborty, Pres.
Saumen Chakraborty, CFO
Prabir Jha, Global Chief-Human Resources/Sr. VP
Rajinder Kumar, Pres., R&D
K. B. Sankara Rao, Exec. VP-Integrated Prod. Dev.
Raghu Cidambi, Advisor-Legal Affairs
Raghu Cidambi, Advisor-Strategic Planning
M. Mythili, Comm. Affairs
Nikhil Shah, Investor Rel.
Jeffrey Wasserstein, Exec. VP-North America Specialty
Arun Sawhney, Pres., Active Pharmaceutical Ingredients
Jaspa S. Bajwa, Pres., Branded Formulations
Anji Reddy, Chmn.
V. S. Vasudevan, Pres./Head-European Oper.

Phone: 91-40-2373-1946	Fax: 91-40-2373-1955
Toll-Free:	
Address: 7-1-27 Ameerpet, Hyderabad, 500 016 India	

GROWTH PLANS/SPECIAL FEATURES:

Dr. Reddy's Laboratories Limited is a global pharmaceutical company that markets its products worldwide, with a main focus on India, the U.S., Europe, the former Soviet Union and portions of Africa. The company's income is derived from five of its six segments: formulations, which contributes 40.9%; active pharmaceutical ingredients and intermediates, 34%; generics, 16.7%; critical care and biotechnology, 2.8%; custom pharmaceutical services 5.5%; and drug discovery, which generates no revenue. Dr. Reddy's manufactures and markets its more than 100 bulk actives, which are the principal ingredients in the finished dosages of drugs, in over 80 countries worldwide. In the generics segment, the company markets 17 products in the U.S. and 36 products in the European Union. The critical care and biotechnology field is working to develop oncology generics for the U.S. and E.U. The drug discovery segment focuses on the areas of metabolic disorders, cardiovascular disorders, bacteria infections, inflammation and cancer. There are currently seven products in Dr. Reddy's pipeline, all in various stages of development. Dr. Reddy's has a dedicated research subsidiary, Reddy US Therapeutics, Inc., based outside of Atlanta, Georgia, that is investigating novel therapeutics in the areas of diabetes, inflammation, and cardiovascular disease. In February 2006, Dr. Reddy's and 3i, Europe's leading private equity house, jointly announced today that they have entered into a definitive agreement providing for the strategic investment by Dr. Reddy's to acquire 100% of Betapharm Group, the fourth-largest generic pharmaceuticals company in Germany, for a total enterprise value of $606 million in cash. In April 2006, the FDA granted approval for the company's fexofenadine hydrochloride tables, a generic equivalent of Allegra. In May 2006, the company introduced in India a novel, second-generation xanthine bronchodilator called Doxoboid. In June 2007, Dr. Reddy's opened a new office in Nigeria, therefore expanding its presence in Africa.

FINANCIALS: Sales and profits are in thousands of dollars—add 000 to get the full amount. 2006 Note: Financial information for 2006 was not available for all companies at press time.

2006 Sales: $541,304	2006 Profits: $36,620	U.S. Stock Ticker: RDY
2005 Sales: $438,000	2005 Profits: $4,800	Int'l Ticker: 500124 Int'l Exchange: Bombay-BSE
2004 Sales: $463,900	2004 Profits: $57,200	Employees: 7,525
2003 Sales: $380,200	2003 Profits: $74,300	Fiscal Year Ends: 3/31
2002 Sales: $340,800	2002 Profits: $100,900	Parent Company:

SALARIES/BENEFITS:

Pension Plan: Y	ESOP Stock Plan:	Profit Sharing:	Top Exec. Salary: $794,352	Bonus: $
Savings Plan: Y	Stock Purch. Plan:		Second Exec. Salary: $529,496	Bonus: $

OTHER THOUGHTS:

Apparent Women Officers or Directors:
Hot Spot for Advancement for Women/Minorities:

LOCATIONS: ("Y" = Yes)

West:	Southwest:	Midwest:	Southeast:	Northeast:	International:
			Y	Y	Y

DRAXIS HEALTH INC

www.draxis.com

ndustry Group Code: 325412 Ranks within this company's industry group: Sales: 68   Profits: 46

rugs:		Other:		Clinical:		Computers:		Other:	
iscovery:		AgriBio:		Trials/Services:		Hardware:		Specialty Services:	Y
icensing:	Y	Genomics/Proteomics:		Laboratories:		Software:		Consulting:	
lanufacturing:	Y	Tissue Replacement:		Equipment/Supplies:	Y	Arrays:		Blood Collection:	
evelopment:	Y			Research/Development Svcs.:		Database Management:		Drug Delivery:	
enerics:	Y			Diagnostics:	Y			Drug Distribution:	

TYPES OF BUSINESS:

Radiopharmaceuticals
Diagnostic Imaging Agents
Specialty Pharmaceuticals
Contract Manufacturing
Animal Health Products

BRANDS/DIVISIONS/AFFILIATES:

DRAXIS Specialty Pharmaceuticals, Inc.
DRAXIMAGE
DRAXIS Pharma
DRAXIMAGE Sestamibi

CONTACTS: *Note: Officers with more than one job title may be intentionally listed here more than once.*

Martin Barkin, CEO
an Brazier, COO
Martin Barkin, Pres.
Mark Oleksiw, CFO
ack A. Carter, VP-Admin. & Shared Services
lida Gualtieri, General Counsel/Corp. Sec.
erry Ormiston, Exec. Dir.-Investor Rel.
hien Huang, VP-Finance
ean-Pierre Robert, Pres., DRAXIS Specialty Pharmaceuticals
rian M. King, Chmn.

Phone: 905-677-5500	Fax: 905-677-5494
Toll-Free: 877-441-1984	
Address: 6870 Goreway Dr., Ste. 200, Mississauga, ON L4V 1P1 Canada	

GROWTH PLANS/SPECIAL FEATURES:

DRAXIS Health, Inc. is a pharmaceutical company focused on sterile products, non-sterile products and radiopharmaceuticals. DRAXIS's sterile products include liquid and freeze-dried injectables, sterile ointments and sterile creams, while the company's specialty non-sterile products consist of tablets, ointments, liquids and creams. Developed, produced and sold through its DRAXIMAGE division, DRAXIS's radiopharmaceuticals are used for both therapeutic and diagnostic molecular imaging applications. DRAXIMAGE is a leading supplier of radioactive Iodine-131 labeled radiopharmaceuticals for both diagnosis and therapy and also manufactures products for diagnosis based on Indium-111, Cobalt-57, Chromium-51 and Xenon-133 as well as radiotherapy products based on Iodine-125 and Phosphorous P-32. The company provides pharmaceutical contract manufacturing services through its DRAXIS Pharma division. Packaging, warehousing and distribution are coordinated through DRAXIS Pharma's logistics management division. Other divisions of DRAXIS Pharma include the solid dose products division, which manufactures tablets, caplets and capsules; the sterile injectable and drop products division; the non-sterile products division, which manufactures ointments, creams and liquids; the sterile lyophilized products division, which manufactures products in freeze-dried form; and the sterile ointment products division. In February 2007, DRAXIMAGE submitted a new drug application to the FDA for its generic kit for the preparation for injection of a nuclear medicine imaging agent, Tc-99m, used in evaluations of blood flow to the heart in patients undergoing cardiac tests, which will be marketed under the name DRAXIMAGE Sestamibi. In July 2007, DRAXIMAGE Sestamibi was filed by DRAXIMAGE with the European regulatory authorities.

FINANCIALS: Sales and profits are in thousands of dollars—add 000 to get the full amount. 2006 Note: Financial information for 2006 was not available for all companies at press time.

006 Sales: $89,000	2006 Profits: $11,600	U.S. Stock Ticker: DRAX
005 Sales: $79,400	2005 Profits: $7,800	Int'l Ticker: DAX Int'l Exchange: Toronto-TSX
004 Sales: $57,800	2004 Profits: $6,500	Employees: 404
003 Sales: $40,500	2003 Profits: $8,200	Fiscal Year Ends: 12/31
002 Sales: $30,300	2002 Profits: $2,200	Parent Company:

SALARIES/BENEFITS:

ension Plan:	ESOP Stock Plan:	Profit Sharing:	Top Exec. Salary: $401,784	Bonus: $81,635
avings Plan:	Stock Purch. Plan: Y		Second Exec. Salary: $267,992	Bonus: $13,980

OTHER THOUGHTS:

pparent Women Officers or Directors: 1
ot Spot for Advancement for Women/Minorities:

LOCATIONS: ("Y" = Yes)

West:	Southwest:	Midwest:	Southeast:	Northeast:	International:
					Y

Note: Financial information, benefits and other data can change quickly and may vary from those stated here.

DSM PHARMACEUTICALS INCwww.dsmpharmaceuticals.com

Industry Group Code: 325412 Ranks within this company's industry group: Sales: Profits:

Drugs:	Other:	Clinical:		Computers:		Other:	
Discovery:	AgriBio:	Trials/Services:	Y	Hardware:		Specialty Services:	Y
Licensing:	Genomics/Proteomics:	Laboratories:		Software:		Consulting:	
Manufacturing: Y	Tissue Replacement:	Equipment/Supplies:	Y	Arrays:		Blood Collection:	
Development: Y		Research/Development Svcs.:	Y	Database Management:		Drug Delivery:	
Generics:		Diagnostics:				Drug Distribution:	

TYPES OF BUSINESS:

Drugs-Custom Manufacturing
Oral & Topical Formulations
Sterile Liquids
Lyophilization Services
Pharmaceutical Packaging
Product Development & Clinical Trial Services
Product Packaging

BRANDS/DIVISIONS/AFFILIATES:

DSM
Lyo-Advantage
Liquid-Advantage
DSM Pharmaceuticals Packaging Group

CONTACTS: Note: Officers with more than one job title may be intentionally listed here more than once.

Hans Engels, COO
Leendert Staal, Dir.-Bus. Group
Rolf-Dieter Schwalb, CFO
Terry Novak, Chief Mktg. Officer
Cor Visker, Sr. VP-Human Resources
Feike Sijbesma, Chmn.

Phone: 252-707-2307	Fax: 973-257-8481
Toll-Free:	
Address: 5900 NW Greenville Blvd., Greenville, NC 27834 US	

GROWTH PLANS/SPECIAL FEATURES:

DSM Pharmaceuticals, Inc., a subsidiary of DSM NV, is an outsourcing partner for both large and small pharmaceutical companies that specializes in custom chemical manufacturing of synthetic drugs and biopharmaceuticals. The firm also provides products and services from early and preclinical development to initial synthesis of promising new drugs and dosage formulation and packaging. The company offers its products and services under five main headings: manufacturing; orals and topicals; sterile liquids; lyophilization (freeze drying); and packaging. DSM's manufacturing services convert active pharmaceutical ingredients (APIs) and excipients processed by DSM Pharma Chemicals and other suppliers into finished dosage forms such as solid-dose creams, ointments, steriles and liquids. Its orals and topicals products include creams and ointments, while services in this area include blending and granulation, coating, tablet compression, capsule filling, drying and all related packaging. Sterile liquid operations include the manufacturing, aseptic filling, terminal sterilization, cold storage and packaging of sterile liquids. The company's steriles processing facilities use state of the art Liquid-Advantage and Lyo-Advantage systems, both of which were developed in-house. These systems provide a precise level of manufacturing and environment control. The firm offers cutting edge lyophilization services, which involve freeze-drying of biological substances. The DSM Pharmaceuticals Packaging Group is capable of producing high-quality specified package engineering, graphics design and testing services. Its offerings include bottles, liners, stoppers, dosage cups and droppers, thermo films, pouches, labels, cartons, inserts, shrink/stretch/wrap films, shippers and IBC containers. Other services include clinical trial management and the development and manufacture of finished dosage forms for pharmaceutical companies.

DSM Pharmaceuticals offers its employees a benefits plan which includes prepaid legal assistance and education reimbursement.

FINANCIALS: Sales and profits are in thousands of dollars—add 000 to get the full amount. 2006 Note: Financial information for 2006 was not available for all companies at press time.

2006 Sales: $	2006 Profits: $	**U.S. Stock Ticker: Subsidiary**
2005 Sales: $	2005 Profits: $	**Int'l Ticker:** Int'l Exchange:
2004 Sales: $	2004 Profits: $	Employees:
2003 Sales: $	2003 Profits: $	Fiscal Year Ends: 12/31
2002 Sales: $	2002 Profits: $	Parent Company: DSM NV

SALARIES/BENEFITS:

Pension Plan:	ESOP Stock Plan:	Profit Sharing:	Top Exec. Salary: $265,000	Bonus: $170,000
Savings Plan: Y	Stock Purch. Plan: Y		Second Exec. Salary: $208,000	Bonus: $21,000

OTHER THOUGHTS:

Apparent Women Officers or Directors:
Hot Spot for Advancement for Women/Minorities:

LOCATIONS: ("Y" = Yes)

West:	Southwest:	Midwest:	Southeast:	Northeast:	International:
				Y	Y

DUPONT AGRICULTURE & NUTRITION   www.dupont.com/ag

Industry Group Code: 115112 Ranks within this company's industry group: Sales: Profits:

Drugs:	Other:		Clinical:	Computers:	Other:	
Discovery:	AgriBio:	Y	Trials/Services:	Hardware:	Specialty Services:	
Licensing:	Genomics/Proteomics:		Laboratories:	Software:	Consulting:	
Manufacturing:	Tissue Replacement:		Equipment/Supplies:	Arrays:	Blood Collection:	
Development:			Research/Development Svcs.:	Database Management:	Drug Delivery:	
Generics:			Diagnostics:		Drug Distribution:	

TYPES OF BUSINESS:
Agricultural Biotechnology Products & Chemicals Manufacturing
Insecticides
Herbicides
Fungicides
Genetically Modified Plants
Soy Products
Forage & Grain Additives

BRANDS/DIVISIONS/AFFILIATES:
Pioneer Hi-Bred International
DuPont Crop Protection
DuPont Nutrition and Health
Solae Company (The)
Supro
Agroproducts Corey, S.A. de C.V.
Nurish
Qualicon

CONTACTS: *Note: Officers with more than one job title may be intentionally listed here more than once.*
J. Erik Fyrwald, Group VP
John Bedbrook, VP-R&D
Don Wirth, VP-Oper.
Kathleen H. Forte, VP-Public Affairs
Tony L. Arnold, Pres./CEO-Solae Company (The)
James C. Collins, VP/General Mgr.-Crop Protection
William S. Niebur, VP-Crop Genetics, R&D
Dean Oestreich, Pres., Pioneer Hi-Bred/General Mgr.-DuPont

Phone: 302-774-1000	Fax:

Toll-Free: 888-638-7668
Address: DuPont Bldg., 1007 Market St., Wilmington, DE 19898
US

GROWTH PLANS/SPECIAL FEATURES:

DuPont Agriculture & Nutrition (DPAN) is a business unit of global chemical giant DuPont. The company oversees a number of business units covering many aspects of crop protection, optimization and additives. Pioneer Hi-Bred International develops advanced plant genetics including seeds and forage and grain additives. DuPont Crop Protection produces herbicide, fungicide and insecticide products and services. DuPont Nutrition and Health provides soy protein and soy fiber ingredients under brand names including: The Solae Company, a joint venture with Bunge; Supro; SuproSoy; and Nurish. DPAN has joint ventures in the U.S. and around the world, with projects such as: an Agricultural products venture with Agroproducts Corey, S.A. de C.V. in Mexico; a crop protection venture with AO Khimprom in Russia; a biofuel production partnership with British Petroleum (BP); and soy-based ventures with General Mills/PTI in Minnesota, So Good in the U.K and Syngenta in Illinois. DPAN also has a microbial testing branch, Qualicon. Recently a number of new products have been approved for registration by the Environmental Protection Agency (EPA), including the herbicides DuPont Canopy, DuPont Cimarron Plus, DuPont Cimarron X-tra and the fungicide DuPont Kocide 3000. In September 2006, the company sold its bensulfuron methyl rice and aquatic business assets outside of Asia Pacific to United Phosphorous Limited (UPL).

The company offers employees a benefits package that includes the LifeWorks employee resource program, dependent care spending accounts, flexible work practices, adoption assistance and emergency care options.

FINANCIALS: Sales and profits are in thousands of dollars—add 000 to get the full amount. 2006 Note: Financial information for 2006 was not available for all companies at press time.

2006 Sales: $	2006 Profits: $	**U.S. Stock Ticker: Subsidiary**	
2005 Sales: $	2005 Profits: $	**Int'l Ticker:** Int'l Exchange:	
2004 Sales: $6,247,000	2004 Profits: $	Employees:	
2003 Sales: $5,470,000	2003 Profits: $	Fiscal Year Ends: 12/31	
2002 Sales: $4,510,000	2002 Profits: $	Parent Company: DUPONT (E I DU PONT DE NEMOURS & CO)	

SALARIES/BENEFITS:

Pension Plan:	ESOP Stock Plan:	Profit Sharing:	Top Exec. Salary: $	Bonus: $
Savings Plan:	Stock Purch. Plan:		Second Exec. Salary: $	Bonus: $

OTHER THOUGHTS:
Apparent Women Officers or Directors: 3
Hot Spot for Advancement for Women/Minorities: Y

LOCATIONS: ("Y" = Yes)

West:	Southwest:	Midwest:	Southeast:	Northeast:	International:
Y	Y	Y	Y	Y	Y

Note: Financial information, benefits and other data can change quickly and may vary from those stated here.

DURECT CORP

www.durect.com

Industry Group Code: 325412A Ranks within this company's industry group: Sales: 10 Profits: 17

Drugs:		Other:	Clinical:	Computers:	Other:	
Discovery:	Y	AgriBio:	Trials/Services:	Hardware:	Specialty Services:	
Licensing:		Genomics/Proteomics:	Laboratories:	Software:	Consulting:	
Manufacturing:	Y	Tissue Replacement:	Equipment/Supplies:	Arrays:	Blood Collection:	
Development:	Y		Research/Development Svcs.:	Database Management:	Drug Delivery:	Y
Generics:			Diagnostics:		Drug Distribution:	

TYPES OF BUSINESS:

Drug Delivery Systems
Biodegradable Polymer Manufacturing
Pharmaceutical Products

BRANDS/DIVISIONS/AFFILIATES:

Saber
Oradur
Transdur
Durin
Microdur
Absorbable Polymer Technologies, Inc.
Corium International, Inc.

CONTACTS: *Note: Officers with more than one job title may be intentionally listed here more than once.*

James E. Brown, CEO
James E. Brown, Pres.
Matthew J. Hogan, CFO
Felix Theeuwes, Chief Scientific Officer
Harry Guy, VP-Eng. & Safety
Paula Mendenhall, Exec. VP-Admin.
Jean I. Liu, General Counsel/Sr. VP
Paula Mendenhall, Exec. VP-Oper.
Michael Arenberg, Exec. Dir.-Bus. Dev.
Jian Li, VP-Finance/Corp. Controller
Peter J. Langecker, Chief Medical Officer
Su IL Yum, Exec. VP-Pharmaceutical Systems R&D/Principal Eng.
Nacer E. Dean Abrouk, VP-Biostatistics
Andrew R. Miksztal, VP-Pharmaceutical R&D
Felix Theeuwes, Chmn.

Phone: 408-777-1417	**Fax:** 408-777-3577
Toll-Free:	
Address: 2 Results Way, Cupertino, CA 95014 US	

GROWTH PLANS/SPECIAL FEATURES:

Durect Corp. is an emerging specialty pharmaceutica company focused on the development of pharmaceutica products based on its proprietary drug delivery technology platforms. The company's pharmaceutical systems enable optimized therapy for a given disease or patient population by controlling the rate and duration of drug administration as well as, for certain applications, placement of the drug at the intended site of action. The firm's proprietary drug delivery technology platforms include: Saber Delivery System, a depot injectable useful for small molecule and protein delivery that can be formulated for systemic or local administration; Oradur, an oral sustained release gel-cap technology; Transdur Delivery System, a transdermal patch technology; Durin Biodegradable Implant, a proprietary biodegradable drug-loaded implant that is absorbed into the body; Duros System, an osmotic implant technology licensed to the company for specified fields from Alza Corp., a Johnson & Johnson company; and Microdur Biodegradable Microparticulates, a biodegradable microparticulate depot injectable. Durect's pharmaceutical systems combine engineering with proprietary small molecule pharmaceutica and biotechnology drug formulation to yield proprietary delivery technologies and products. The company's development efforts are focused on the application of its pharmaceutical systems technologies to potential products in a variety of chronic and episodic disease areas including pain, central nervous system disorders, cardiovascular disease and other chronic diseases. The firm produces biodegradable polymers through subsidiary Absorbable Polymer Technologies, Inc. In February 2007, Durec entered into a manufacturing and supply agreement with Corium International, Inc. for Transdur-Bupivacaine, a transdermal pain patch.

The company offers its employees medical, dental and vision insurance; life, AD&D, short- and long-term disability insurance; a 401(k) plan; an employee stock purchase plan stock options; an employee assistance program; flexible spending accounts; a wellness program; and annua company events.

FINANCIALS: Sales and profits are in thousands of dollars—add 000 to get the full amount. 2006 Note: Financial information for 2006 was not available for all companies at press time.

2006 Sales: $21,894	2006 Profits: $-33,327	**U.S. Stock Ticker: DRRX**
2005 Sales: $28,571	2005 Profits: $-18,128	**Int'l Ticker:** Int'l Exchange:
2004 Sales: $13,853	2004 Profits: $-27,637	Employees: 170
2003 Sales: $11,835	2003 Profits: $-22,698	Fiscal Year Ends: 12/31
2002 Sales: $7,200	2002 Profits: $-37,200	Parent Company:

SALARIES/BENEFITS:

Pension Plan:	ESOP Stock Plan:	Profit Sharing:	Top Exec. Salary: $425,250	Bonus: $150,698
Savings Plan: Y	Stock Purch. Plan: Y		Second Exec. Salary: $414,750	Bonus: $146,977

OTHER THOUGHTS:

Apparent Women Officers or Directors: 3
Hot Spot for Advancement for Women/Minorities: Y

LOCATIONS: ("Y" = Yes)

West:	Southwest:	Midwest:	Southeast:	Northeast:	International:
Y					

DUSA PHARMACEUTICALS INC
www.dusapharma.com

Industry Group Code: 325412 Ranks within this company's industry group: Sales: 104   Profits: 131

Drugs:		Other:		Clinical:		Computers:		Other:	
Discovery:	Y	AgriBio:		Trials/Services:		Hardware:		Specialty Services:	
Licensing:		Genomics/Proteomics:		Laboratories:		Software:		Consulting:	
Manufacturing:	Y	Tissue Replacement:		Equipment/Supplies:		Arrays:		Blood Collection:	
Development:	Y			Research/Development Svcs.:		Database Management:		Drug Delivery:	
Generics:				Diagnostics:				Drug Distribution:	

TYPES OF BUSINESS:
Photodynamic Therapy Products

BRANDS/DIVISIONS/AFFILIATES:
Levulan
BLU-U
Kerastick
Nicomide
Avar
Sirius Laboratories, Inc.
Daewoong Pharmaceutical Co., Ltd.
Coherent-AMT

CONTACTS: Note: Officers with more than one job title may be intentionally listed here more than once.
Robert F. Doman, CEO
Robert F. Doman, Pres.
Richard Christopher, CFO
William F. O'Dell, Exec. VP-Sales & Mktg.
Stuart Marcus, VP-Scientific Affairs/Chief Medical Officer
Mark Carota, VP-Oper.
D. Geoffrey Shulman, Chief Strategy Officer
Richard Christopher, VP-Finance
Michael J. Todisco, VP/Controller
Scott Lundahl, VP-Regulatory Affairs & Intellectual Property
Nanette W. Mantell, Sec.
D. Geoffrey Shulman, Chmn.

Phone: 978-657-7500	Fax: 978-657-9193
Toll-Free:	
Address: 25 Upton Dr., Wilmington, MA 01887 US	

GROWTH PLANS/SPECIAL FEATURES:

DUSA Pharmaceuticals, Inc. is a dermatology company that develops and markets Levulan photodynamic therapy (PDT) and other products for common skin conditions. The company's market products include among others Levulan Kerastick 20% topical solution with PDT, the BLU-U brand light source, Nicomide, Nicomide-T and the Avar line of products. The Levulan Kerastick 20% topical solution with PDT and the BLU-U brand light source are used for the treatment of actinic keratoses, precancerous skin lesions caused by chronic sun exposure, of the face or scalp. BLU-U is also used for the treatment of moderate inflammatory acne vulgaris and general dermatological conditions. In March 2006, the firm acquired, among others, Nicomide, an oral prescription vitamin supplement; Nicomide-T, a topical cosmetic product; and the Avar line of products, which target the treatment of acne vulgaris and acne rosacea, as well as psoriasis. DUSA has an agreement with Stiefel Laboratories to market the Levulan Kerastick in Latin America and with Coherent-AMT to market its products in Canada. In March 2006, DUSA acquired Sirius Laboratories, Inc., formerly a privately-held dermatology specialty pharmaceutical company. DUSA paid $8 million of cash and $14 million, represented by approximately 2 million shares of its common stock. In January 2007, the company entered into an exclusive marketing, distribution and supply agreement for certain Asian territories with Daewoong Pharmaceutical Co., Ltd.

The company offers its employees a 401(k) plan; health, dental and life insurance; flexible spending accounts; tuition reimbursement; short- and long-term disability; and business travel insurance.

FINANCIALS: Sales and profits are in thousands of dollars—add 000 to get the full amount. 2006 Note: Financial information for 2006 was not available for all companies at press time.

2006 Sales: $25,583	2006 Profits: $-31,350	U.S. Stock Ticker: DUSA	
2005 Sales: $11,337	2005 Profits: $-14,999	Int'l Ticker: Int'l Exchange:	
2004 Sales: $7,988	2004 Profits: $-15,629	Employees: 85	
2003 Sales: $ 970	2003 Profits: $-14,826	Fiscal Year Ends: 12/31	
2002 Sales: $25,500	2002 Profits: $5,800	Parent Company:	

SALARIES/BENEFITS:

Pension Plan:	ESOP Stock Plan:	Profit Sharing:	Top Exec. Salary: $416,000	Bonus: $221,104
Savings Plan: Y	Stock Purch. Plan:		Second Exec. Salary: $312,000	Bonus: $134,534

OTHER THOUGHTS:
Apparent Women Officers or Directors: 1
Hot Spot for Advancement for Women/Minorities:

LOCATIONS: ("Y" = Yes)

West:	Southwest:	Midwest:	Southeast:	Northeast:	International:
				Y	Y

DYAX CORP

www.dyax.com

Industry Group Code: 325412 Ranks within this company's industry group: Sales: 123 Profits: 154

Drugs:		Other:		Clinical:		Computers:		Other:	
Discovery:	Y	AgriBio:		Trials/Services:		Hardware:		Specialty Services:	
Licensing:	Y	Genomics/Proteomics:		Laboratories:		Software:	Y	Consulting:	
Manufacturing:		Tissue Replacement:		Equipment/Supplies:		Arrays:		Blood Collection:	
Development:	Y			Research/Development Svcs.:	Y	Database Management:		Drug Delivery:	
Generics:				Diagnostics:				Drug Distribution:	

TYPES OF BUSINESS:
Pharmaceuticals Discovery & Development
Proteins, Peptides & Antibodies
Drugs-Cancer
Drugs-Anti-Inflammatory

BRANDS/DIVISIONS/AFFILIATES:
DX-88
DX-2240
WebPhage

CONTACTS: Note: Officers with more than one job title may be intentionally listed here more than once.
Henry E. Blair, CEO
Henry E. Blair, Pres.
Stephen S. Galliker, CFO
Clive R. Wood, Exec. VP-Research/Chief Scientific Officer
Ivana Magovcevic-Liebisch, Exec. VP-Admin.
Ivana Magovcevic-Liebisch, General Counsel
Gustav A. Christensen, Chief Bus. Officer/Exec. VP
Nicole Jones, Associate Dir.-Corp. Comm. & Investor Rel.
Ivana Magovcevic - Liebisch, Primary Investor Rel. Contact
Stephen S. Galliker, Exec. VP-Finance
Ellen Flipse, Associate Dir.-Corp. Comm. & Investor Rel.
Peggy Berry, Sr. VP-Quality & Regulatory Affairs
Henry E. Blair, Chmn.

Phone: 617-225-2500	Fax: 617-225-2501
Toll-Free:	
Address: 300 Technology Sq., Cambridge, MA 02139 US	

GROWTH PLANS/SPECIAL FEATURES:
Dyax Corp. is a biopharmaceutical company principally focused on the discovery, development and commercialization of antibodies, proteins and peptides as therapeutic products, with an emphasis on cancer and inflammatory conditions. Its lead product candidate is DX-88, a recombinant form of a small protein that is being tested for applications in hereditary angioedema, a genetic disease that causes swelling of the larynx, gastrointestinal tract and extremities, and is in the midst of Phase III trials. DX-88 is also being tested for complications, such as blood loss and systemic inflammation, during on-pump coronary artery bypass grafts, and is Phase I and II of study. Other product candidates in the company's discovery and development pipeline include DX-2240, a fully human monoclonal antibody that targets the Tie-1 receptor, a protein receptor that scientists believe is important in the process of tumor blood vessel formation known as angiogenesis. With its proprietary phage display methodology, called WebPhage, the firm identifies a broad range of compounds with potential for the treatment of diseases. The process involves generating phage display libraries of protein variations, screening libraries to select binding compounds with high affinity and high specificity to a target and producing and evaluating the selected binding compounds. Through the company's Licensing and Funded Research Program, it allows other companies to use the phage display technology to discover new compounds. Over 70 companies have licenses to use this technology and its derived compounds. In 2006, the company announced the awarding of the fifth U.S. patent covering its phage display. In April 2007, DX-88 began a confirmatory Phase III trial for hereditary angioedema.

FINANCIALS: Sales and profits are in thousands of dollars—add 000 to get the full amount. 2006 Note: Financial information for 2006 was not available for all companies at press time.
2006 Sales: $12,776
2005 Sales: $19,859
2004 Sales: $16,590
2003 Sales: $16,853
2002 Sales: $23,158

2006 Profits: $-50,323
2005 Profits: $-30,944
2004 Profits: $-33,114
2003 Profits: $-7,415
2002 Profits: $-26,800

U.S. Stock Ticker: DYAX
Int'l Ticker: Int'l Exchange:
Employees: 161
Fiscal Year Ends: 12/31
Parent Company:

SALARIES/BENEFITS:
Pension Plan:	ESOP Stock Plan:	Profit Sharing: Y	Top Exec. Salary: $500,000	Bonus: $180,000
Savings Plan: Y	Stock Purch. Plan: Y		Second Exec. Salary: $340,000	Bonus: $115,600

OTHER THOUGHTS:
Apparent Women Officers or Directors: 3
Hot Spot for Advancement for Women/Minorities: Y

LOCATIONS: ("Y" = Yes)
West:	Southwest:	Midwest:	Southeast:	Northeast:	International:
				Y	Y

Note: Financial information, benefits and other data can change quickly and may vary from those stated here.

E I DU PONT DE NEMOURS & CO (DUPONT) www.dupont.com

Industry Group Code: 325000 **Ranks within this company's industry group:** Sales: 3 Profits: 2

Drugs:	Other:		Clinical:	Computers:	Other:
Discovery:	AgriBio:	Y	Trials/Services:	Hardware:	Specialty Services:
Licensing:	Genomics/Proteomics:		Laboratories:	Software:	Consulting:
Manufacturing:	Tissue Replacement:		Equipment/Supplies:	Arrays:	Blood Collection:
Development: Y			Research/Development Svcs.:	Database Management:	Drug Delivery:
Generics:			Diagnostics:		Drug Distribution:

TYPES OF BUSINESS:
Chemicals Manufacturing
Polymers
Performance Coatings
Nutrition & Health Products
Electronics Materials
Agricultural Seeds
Fuel-Cell, Biofuels & Solar Panel Technology
Contract Research & Development

BRANDS/DIVISIONS/AFFILIATES:
Teflon
Mylar
Corian
Sorona
Dacron
Kevlar
Cellophane
Fabrikoid

CONTACTS: *Note: Officers with more than one job title may be intentionally listed here more than once.*
Charles O. Holliday, Jr., CEO
Richard R. Goodmanson, COO/Exec. VP
Jeffrey L. Keefer, CFO/Exec. VP
David G. Bills, Chief Mktg. & Sales Officer
James C. Borel, Sr. VP-Global Human Resources
Uma Chowdhry, Chief Science Officer/Sr. VP
Robert R. Ridout, CIO/VP-IT
Uma Chowdhry, CTO/Sr. VP
James B. Porter Gr., VP-Eng., Safety, Health & Environment
Stacey J. Mobley, Chief Admin. Officer
Stacey J. Mobley, General Counsel/Sr. VP
Mathieu Vrijsen, Sr. VP-Oper.
Kathleen H. Forte, VP-Public Affairs
Carl J. Lukach, VP-Investor Rel.
Susan M. Stalnecker, VP/Treas.
Terry Caloghiris, Group VP-DuPont Coatings & Color Tech.
David G. Bills, Group VP-DuPont Global Biotechnology
Ellen J. Kullman, Group VP-DuPont Safety & Protection
W. Erik Fyrwald, Group VP-DuPont Agriculture & Nutrition
Charles O. Holliday, Jr., Chmn.
Jeffrey A. Coe, Chief Procurement Officer/VP-Sourcing & Logistics

Phone: 302-774-1000	Fax: 302-999-4399
Toll-Free:	
Address: 1007 Market St., Wilmington, DE 19898 US	

GROWTH PLANS/SPECIAL FEATURES:

E. I. du Pont de Nemours and Company (DuPont) develops and manufactures products in the biotechnology, electronics, materials science, safety and security and synthetic fibers sectors. Products and services are offered to markets including transportation; safety and protection; construction; agriculture; home furnishings; medical; communications; and nutrition. The company operates in five business segments: agriculture and nutrition (A&N); coatings and color technologies (C&CT); electronic and communication technologies (E&C); performance materials (PM); and safety and protection (S&P). A&N delivers Pioneer-brand seed products, insecticides, fungicides, herbicides, soy-based food ingredients, food quality diagnostic testing equipment and liquid food packaging systems. The C&CT segment supplies automotive coatings and titanium dioxide white pigments. The segment also offers specialty products such as pigment and dye-based inks for ink-jet digital printing. E&C provides advanced materials for the electronics industry. Products include flexographic printing; color communication systems; and fluoro-polymer and fluoro-chemical products. PM manufactures polymer-based materials, which includes engineered polymers and specialized resins and films for use in food packaging, sealants and adhesives, sporting goods and laminated safety glass. The S&P segment provides protective materials and safety consulting services. In the renewable energy field, the firm is currently developing biofuels such as butanol and solar power technologies. DuPont has also made several advances in the nanotechnology industry by utilizing DNA to sort carbon nanotubes by conductivity. In 2006, DuPont announced the possible commercialization of nano-coatings, which will greatly protect automotive components and reduce the amount of energy and materials utilized during production. Nano particles can be applied using spraying equipment and can be cured by UV light in less than ten seconds.

DuPont offers its employees personal family leave; adoption assistance; LifeWorks Employee Resource Program, which offers personalized and confidential consultations; and 'Just in Time' Care, an emergency backup service that links employees to a variety of dependent care options.

FINANCIALS: Sales and profits are in thousands of dollars—add 000 to get the full amount. 2006 Note: Financial information for 2006 was not available for all companies at press time.

2006 Sales: $27,421,000	2006 Profits: $3,148,000	**U.S. Stock Ticker: DD**
2005 Sales: $26,639,000	2005 Profits: $2,053,000	**Int'l Ticker:** Int'l Exchange:
2004 Sales: $27,340,000	2004 Profits: $1,780,000	Employees: 59,000
2003 Sales: $26,996,000	2003 Profits: $973,000	Fiscal Year Ends: 12/31
2002 Sales: $24,006,000	2002 Profits: $-1,103,000	Parent Company:

SALARIES/BENEFITS:

Pension Plan: Y	ESOP Stock Plan:	Profit Sharing:	Top Exec. Salary: $1,293,000	Bonus: $
Savings Plan: Y	Stock Purch. Plan:		Second Exec. Salary: $606,892	Bonus: $2,000,000

OTHER THOUGHTS:
Apparent Women Officers or Directors: 13
Hot Spot for Advancement for Women/Minorities: Y

LOCATIONS: ("Y" = Yes)

West:	Southwest:	Midwest:	Southeast:	Northeast:	International:
Y	Y	Y	Y	Y	Y

Note: Financial information, benefits and other data can change quickly and may vary from those stated here.

EISAI CO LTD

www.eisai.co.jp

Industry Group Code: 325414 **Ranks within this company's industry group:** Sales: 1   Profits: 1

Drugs:		Other:		Clinical:		Computers:		Other:	
Discovery:		AgriBio:		Trials/Services:		Hardware:		Specialty Services:	
Licensing:		Genomics/Proteomics:		Laboratories:		Software:		Consulting:	
Manufacturing:	Y	Tissue Replacement:		Equipment/Supplies:	Y	Arrays:		Blood Collection:	
Development:	Y			Research/Development Svcs.:		Database Management:		Drug Delivery:	
Generics:				Diagnostics:	Y			Drug Distribution:	

TYPES OF BUSINESS:

Pharmaceuticals Manufacturing
Over-the-Counter Pharmaceuticals
Pharmaceutical Production Equipment
Diagnostic Products
Food Additives
Personal Health Care Products
Chemicals
Vitamins & Nutritional Supplements

BRANDS/DIVISIONS/AFFILIATES:

Eisai Machinery Co., Ltd.
Aciphex/Pariet
Aricept
Chocola
Zonegran
Prialt
Eisai Manufacturing Ltd.

CONTACTS: *Note: Officers with more than one job title may be intentionally listed here more than once.*

Haruo Naito, CEO
Haruo Naito, Pres.
Hideaki Matsui, Exec. VP-Human Resources & Mgmt. Affairs
Jiro Hasegawa, Sr. VP-Global Clinical Research
Yukio Akada, VP-Info. Sys. & Corp. Planning
Toshio Arai, Sr. VP-Prod.
Nobuo Deguchi, Sr. VP-Legal Affairs & Internal Control
Makoto Shiina, Exec. VP-Strategy
Hiroyuki Mitsui, VP-Corp. Comm. & General Affairs
Hiroyuki Mitsui, VP-Investor Rel.
Soichi Matsuno, Deputy Pres., Global Pharmaceuticals Bus.
Yoji Takaoka, Sr. VP-Corp. Regulatory Compliance
Matsuo Ohara, Sr. VP-Japanese Pharmaceuticals Bus.
Kenji Toda, VP-Corp. Regulatory Compliance & Quality Assurance
Hiromasa Nakai, Chmn.
Yutaka Tsuchiya, VP/Pres., Eisai Europe Ltd.
Toshio Arai, Sr. VP-Logistics

Phone: 81-3-3817-3700	Fax: 81-3-3811-3077
Toll-Free:	
Address: 4-6-10 Koishikawa, Bunkyo-ku, Tokyo, 112-8088 Japan	

GROWTH PLANS/SPECIAL FEATURES:

Eisai Co., Ltd. develops, manufactures and distributes prescription and over-the-counter pharmaceuticals, consumer health care products, diagnostic products, pharmaceutical production equipment, and chemicals and food additives worldwide. It has a long-standing leading position in the Japanese vitamin and nutritional supplement market, focusing on synthetic and natural vitamin E products and derivatives. The company markets a full line of vitamin-enriched dietary supplements under the brand name Chocola. Prescription pharmaceuticals represent Eisai's largest source of revenue, with roughly half of current earnings relating to the company's Alzheimer's drug Aricept and proton pump inhibitor (PPI) Aciphex/Pariet, which is a gastroesophageal reflux disease and ulcer therapeutic. The firm's research teams are currently investigating new indications for Aricept including dementia associated with Parkinson's disease. Eisai also markets a full range of diagnostic products and manufactures pharmaceutical production systems including continuous sterilization devices, inspection systems and ampoule packing machines. The company is transferring its systems manufacturing operations to a subsidiary business, which will be renamed as Eisai Machinery Co., Ltd. when the restructuring is done. In February 2006, Eisai acquired the European production and distribution rights for severe chronic pain agent Prialt from Elan. In October 2006 Eisai acquired four oncology-related products from Ligand Pharmaceuticals, and in March 2007 the company established its pharmaceutical manufacturing subsidiary Eisai Manufacturing Ltd. in Hatfield, U.K.

FINANCIALS: Sales and profits are in thousands of dollars—add 000 to get the full amount. 2006 Note: Financial information for 2006 was not available for all companies at press time.

2006 Sales: $5,113,100	2006 Profits: $539,200	**U.S. Stock Ticker: ESALY**
2005 Sales: $4,530,600	2005 Profits: $471,760	**Int'l Ticker: 4523** Int'l Exchange: Tokyo-TSE
2004 Sales: $4,734,600	2004 Profits: $474,700	Employees: 3,945
2003 Sales: $3,893,400	2003 Profits: $342,300	Fiscal Year Ends: 3/31
2002 Sales: $3,254,400	2002 Profits: $275,300	Parent Company:

SALARIES/BENEFITS:

Pension Plan:	ESOP Stock Plan:	Profit Sharing:	Top Exec. Salary: $	Bonus: $
Savings Plan:	Stock Purch. Plan:		Second Exec. Salary: $	Bonus: $

OTHER THOUGHTS:

Apparent Women Officers or Directors:
Hot Spot for Advancement for Women/Minorities:

LOCATIONS: ("Y" = Yes)

West:	Southwest:	Midwest:	Southeast:	Northeast:	International:
	Y	Y	Y	Y	Y

ELAN CORP PLC
www.elan.com

Industry Group Code: 325412 Ranks within this company's industry group: Sales: 38 Profits: 199

Drugs:		Other:		Clinical:	Computers:		Other:	
Discovery:	Y	AgriBio:		Trials/Services:	Hardware:		Specialty Services:	
Licensing:		Genomics/Proteomics:		Laboratories:	Software:		Consulting:	
Manufacturing:	Y	Tissue Replacement:		Equipment/Supplies:	Arrays:		Blood Collection:	
Development:	Y			Research/Development Svcs.:	Database Management:		Drug Delivery:	
Generics:				Diagnostics:			Drug Distribution:	

TYPES OF BUSINESS:

Drugs-Neurology
Acute Care Drugs
Pain Management Drugs
Autoimmune Disease Drugs
Drug Delivery Technologies

BRANDS/DIVISIONS/AFFILIATES:

NanoCrystal
Maxipime
Prialt
Tysabri
Azactam
Elan Drug Technologies (EDT)

CONTACTS: Note: Officers with more than one job title may be intentionally listed here more than once.

Kelly Martin, CEO
Kelly Martin, Pres.
Shane Cooke, CFO/Exec. VP
Lars Ekman, Exec. VP/Pres., Global R&D/Head-Neurodegeneration
Richard T. Collier, General Counsel/Exec. VP
Karen Kim, Exec. VP-Strategy, Bus. Dev. & Brand Mgmt.
Chris Burns, Sr. VP-Investor Rel.
Nigel Clerkin, Sr. VP-Finance/Controller
Paul Breen, Exec. VP-Elan Drug Tech.
Dale Schenk, Sr. VP/Chief Scientific Officer
Lars Ekman, Head-Neurodegeneration Franchise
Allison Hulme, Exec. VP-Autoimmune & Tysabri Franchise
Kyran McLaughlin, Chmn.

Phone: 353-1-709-4444	Fax: 353-1-709-4108
Toll-Free:	
Address: Treasury Bldg., Lower Grand Canal St., Dublin, 2 Ireland	

GROWTH PLANS/SPECIAL FEATURES:

Elan Corporation, a leading global specialty pharmaceutical company, focuses on the discovery, development and marketing of therapeutic products and services in neurology, acute care and pain management. Elan is divided into two segments: Biopharmaceuticals and Elan Drug Technologies (EDT). The Irish company conducts its worldwide business through subsidiaries in Ireland, the U.S., the U.K. and other countries. Nearly half of its revenue is generated from EDT, which controls its four marketed products. The biopharmaceuticals division focuses its research and development on Alzheimer's disease, Parkinson's disease, multiple sclerosis, pain management and Crohn's. The company uses proprietary technologies to develop, market and license drug delivery products to its pharmaceutical clients. It does not manufacture any of its products. It has developed three novel therapeutic approaches in its breakthrough research program in Alzheimer's disease: immunotherapy, beta secretase inhibitors and gamma secretase inhibitors. The company's four products are: Tysabri, an alpha four integrin antagonist for the treatment of recurring multiple sclerosis; Prialt, used for the management of severe chronic pain; Azactam, which is used to treat gram-negative organism induced disorders such as urinary and lower respiratory tract infections and intra-abdominal infections (this product lost patent protection in 2005, and Elan expects a generic to appear on the market sometime in 2007); and Maxipime, which treats a number of disorders including urinary tract infections and pneumonia. Elan also owns proprietary rights to NanoCrystal technology, which it licenses to Roche and Johnson & Johnson. In the European Union, the firm's primary product is Prialt. In 2006, Elan and its partner Biogen Idec, Inc. received approval to reintroduce Tysabri as a treatment for MS to the U.S. and European markets. The firm is currently trying to have Tysabri approved for treatment of Crohn's Disease.

Employee benefits include an educational assistance program.

FINANCIALS: Sales and profits are in thousands of dollars—add 000 to get the full amount. 2006 Note: Financial information for 2006 was not available for all companies at press time.

2006 Sales: $560,400	2006 Profits: $-267,300	U.S. Stock Ticker: ELN
2005 Sales: $490,300	2005 Profits: $-383,600	Int'l Ticker: DRX Int'l Exchange: Dublin-ISE
2004 Sales: $481,700	2004 Profits: $-394,700	Employees: 1,734
2003 Sales: $746,000	2003 Profits: $-529,400	Fiscal Year Ends: 12/31
2002 Sales: $1,470,100	2002 Profits: $-2,394,800	Parent Company:

SALARIES/BENEFITS:

Pension Plan: Y	ESOP Stock Plan:	Profit Sharing:	Top Exec. Salary: $702,854	Bonus: $800,000
Savings Plan: Y	Stock Purch. Plan: Y		Second Exec. Salary: $266,666	Bonus: $150,000

OTHER THOUGHTS:

Apparent Women Officers or Directors: 2
Hot Spot for Advancement for Women/Minorities: Y

LOCATIONS: ("Y" = Yes)

West:	Southwest:	Midwest:	Southeast:	Northeast:	International:
Y			Y	Y	Y

Note: Financial information, benefits and other data can change quickly and may vary from those stated here.

ELI LILLY AND COMPANY   www.lilly.com

Industry Group Code: 325412 Ranks within this company's industry group: Sales: 12   Profits: 12

Drugs:		Other:		Clinical:		Computers:		Other:	
Discovery:	Y	AgriBio:		Trials/Services:		Hardware:		Specialty Services:	Y
Licensing:		Genomics/Proteomics:	·	Laboratories:		Software:		Consulting:	
Manufacturing:	Y	Tissue Replacement:		Equipment/Supplies:		Arrays:		Blood Collection:	
Development:	Y			Research/Development Svcs.:		Database Management:		Drug Delivery:	
Generics:				Diagnostics:				Drug Distribution:	

TYPES OF BUSINESS:

Pharmaceuticals Discovery & Development
Veterinary Products

BRANDS/DIVISIONS/AFFILIATES:

Zyprexa
Prozac
Humalog
Gemzar
Coban
Cialis
Icos Corp.
Xigris

CONTACTS: *Note: Officers with more than one job title may be intentionally listed here more than once.*

Sidney Taurel, CEO
John Lechleiter, COO
John Lechleiter, Pres.
Derica Rice, CFO/Sr. VP
Richard Pilnik, Chief Mktg. Officer/VP
Anthony Murphy, Sr. VP-Human Resources
Thomas Verhoeven, VP-R&D
Michael Heim, CIO/VP-IT
Steven M. Paul, Exec. VP-Science & Tech.
Thomas Verhoeven, VP-Prod. R&D
Robert A. Cole, VP-Eng./Environmental Health & Safety
Scott Canute, Pres., Mfg. Oper.
Robert A. Armitage, General Counsel/Sr. VP
Peter Johnson, Exec. Dir.-Corp. Strategic Dev.
Alex M. Azar II, Sr. VP-Corp. Affairs & Comm.
Thomas W. Grein, Treas./VP
Alan Brier, Chief Medical Officer/VP-Medical
Bryce Carmine, Pres., Global Brand Dev.
Jacques Tapiero, Pres., Intercontinental Oper.
Abbas Hussain, Pres., European Oper.
Sidney Taurel, Chmn.
Lorenzo Tallarigo, Pres., Int'l Oper.

Phone: 317-276-2000	Fax: 317-277-6579
Toll-Free:	
Address: Lilly Corporate Center, Drop Code 1112, Indianapolis, IN 46285-0001 US	

GROWTH PLANS/SPECIAL FEATURES:

Eli Lilly and Company researches, develops, manufactures and sells pharmaceuticals designed to treat a variety of conditions. The company manufactures and distributes its products through facilities in the U.S., Puerto Rico and 25 other countries, which are then sold to markets in 140 countries throughout the world. Most of Eli Lilly's products are developed by its in-house research staff, which primarily directs its research efforts towards the search for products to prevent and treat cancer and diseases of the central nervous, endocrine and cardiovascular systems. The firm's other research lies in anti-infectives and products to treat animal diseases. In 2006, research and development expenditures exceeded $3.1 billion. Major brands include neuroscience products Zyprexa, Strattera, Prozac, Cymbalta and Permax; endocrine products Humalog, Humulin and Actos; oncology products Gemzar and Alimta; animal health products Tylan, Rumensin and Coban; cardiovascular products ReoPro and Xigris; anti-infectives Ceclor and Vancocin; and Cialis, for erectile dysfunction. In the U.S., the company distributes pharmaceuticals primarily through independent wholesale distributors. Marketing is through an in-house sales force who call upon physicians, hospitals and managed care providers directly. Outside of the U.S., Eli Lilly products are primarily marketed either through this direct sales force or by way of standing partnerships with several companies, including Takeda Chemical Industries, ICOS Corporation, Quintiles Transnational and Boehringer Ingelheim. In January 2007 the company acquired Icos Corp., maker of the drug Cialis, for $2.3 billion. In March 2007, Eli Lily agreed to acquire Hypnion, Inc., a neuroscience drug company focused on sleep disorders.

Eli Lilly offers its employees domestic partner benefits and an employee assistance program, as well as up to 10 weeks of paid maternity leave. The firm also offers an on-site fitness center, flexible hours or telecommuting, parenting and dependant care leaves, adoption assistance and tuition reimbursement, among many other things.

FINANCIALS: Sales and profits are in thousands of dollars—add 000 to get the full amount. 2006 Note: Financial information for 2006 was not available for all companies at press time.

2006 Sales: $15,691,000	2006 Profits: $2,662,700	U.S. Stock Ticker: LLY
2005 Sales: $14,645,300	2005 Profits: $1,979,600	Int'l Ticker: Int'l Exchange:
2004 Sales: $13,857,900	2004 Profits: $1,810,100	Employees: 41,500
2003 Sales: $12,582,500	2003 Profits: $2,560,800	Fiscal Year Ends: 12/31
2002 Sales: $11,078,000	2002 Profits: $2,708,000	Parent Company:

SALARIES/BENEFITS:

Pension Plan: Y	ESOP Stock Plan:	Profit Sharing:	Top Exec. Salary: $1,650,333	Bonus: $2,764,308
Savings Plan: Y	Stock Purch. Plan:		Second Exec. Salary: $1,112,000	Bonus: $1,490,080

OTHER THOUGHTS:

Apparent Women Officers or Directors: 10
Hot Spot for Advancement for Women/Minorities: Y

LOCATIONS: ("Y" = Yes)

West:	Southwest:	Midwest:	Southeast:	Northeast:	International:
Y	Y	Y	Y	Y	Y

Note: Financial information, benefits and other data can change quickly and may vary from those stated here.

ELSEVIER MDL

www.mdli.com

ndustry Group Code: 511212 Ranks within this company's industry group: Sales: Profits:

rugs:	Other:	Clinical:	Computers:		Other:	
iscovery:	AgriBio:	Trials/Services:	Hardware:		Specialty Services:	Y
icensing:	Genomics/Proteomics:	Laboratories:	Software:	Y	Consulting:	Y
lanufacturing:	Tissue Replacement:	Equipment/Supplies:	Arrays:		Blood Collection:	
evelopment:		Research/Development Svcs.:	Database Management:		Drug Delivery:	
eneriCs:		Diagnostics:			Drug Distribution:	

TYPES OF BUSINESS:

Software-Biotech Research
Biotech Research Consulting
Software Training
Custom Software

BRANDS/DIVISIONS/AFFILIATES:

Discovery Framework
Discovery Experiment Management
Discovery Knowledge
Discovery Predictive Science
PharmaPendium
DiscoveryGate
MDL Isentris
MDL Available Chemicals Directory

CONTACTS: Note: Officers with more than one job title may be intentionally listed here more than once.

Lars Barfod, CEO
Carmel Andrews, CFO
Jean Colombel, Sr. VP-Sales & Services
Trevor Heritage, Chief Scientific Officer/Sr. VP-Workflow Bus Group
Howard Abels, CIO
Howard Abels, VP-Tech. & Info. Systems
David Hughes, VP-Core Tech. Prod. Mgmt.
Greg Bartolo, VP-Contract Mgmt. & Licensing
Carmel Andrews, Sr. VP-Oper.
Carol Dockery, VP-Bus. Oper.
W. Douglas Hounshell, VP-Content Dev.
John McCarthy, VP- Consulting
Phil McHale, VP- Solutions Prod. Mgmt.
Julie Page, VP-Global Customer Care

Phone: 925-543-5400	Fax: 925-543-5401
Toll-Free:	
Address: 2440 Camino Ramon, Ste. 300, San Ramon, CA 94583 US	

GROWTH PLANS/SPECIAL FEATURES:

Elsevier MDL Information Services, a subsidiary of the Reed Elsevier Group, is a global publisher of scientific, technical and medical information products and services. Founded in 1978, the company publishes over 2,000 journals and 1,900 new books every year. Covering such materials as bioactivity, chemical sourcing, synthetic methodology, pharmacology, metabolism and toxicology, Elsevier's factual databases and reference works can be locally installed and many are available through its hosted DiscoveryGate platform. The DiscoveryGate platform provides access to over 20,000 scientific, technical and medical journals as well as information from partner databases. The company's primary products, Discovery Framework, Discovery Experiment Management, Discovery Knowledge, and Discovery Predictive Science, are used by over 1,000 biotech research labs worldwide. Elsevier's MDL Isentris system features MDL Draw chemical drawing and rendering software, MDL Core Interface middleware and MDL Direct, a support system for storing, searching and retrieving molecules and reactions. The company also provides applications which assist in designing chemical syntheses, managing plates and analyzing assay results. In May 2006, the company launched PharmaPendium, an Internet-based application that allows pharmaceutical developers to access information on the FDA's drug approval process. In addition to these pre-designed packages, Elsevier builds custom software tools for research scientists, managers and information scientists, and provides consulting and education services for its clients. Elsevier has strategic alliances with such hardware, operating system and application vendors as Sun Microsystems, Microsoft Corporation, Oracle Corporation and Citrix Systems. In recent news, Elsevier announced its partnership with MathSpec, Inc., a provider of software tools for mass spectrometrists, combining MathSpec's Rational Numbers FragSearch program with Elsevier's MDL Available Chemicals Directory.

Elsevier MDL offers its employees internal education programs and tuition payment plans; has a casual work environment; and a parallel promotion structure with equivalent-level positions in both management and software development.

FINANCIALS: Sales and profits are in thousands of dollars—add 000 to get the full amount. 2006 Note: Financial information for 2006 was not available for all companies at press time.

2006 Sales: $	2006 Profits: $	U.S. Stock Ticker: Subsidiary
2005 Sales: $	2005 Profits: $	Int'l Ticker: Int'l Exchange:
2004 Sales: $	2004 Profits: $	Employees:
2003 Sales: $	2003 Profits: $	Fiscal Year Ends: 12/31
2002 Sales: $	2002 Profits: $	Parent Company: REED ELSEVIER GROUP PLC

SALARIES/BENEFITS:

Pension Plan:	ESOP Stock Plan:	Profit Sharing:	Top Exec. Salary: $	Bonus: $
Savings Plan: Y	Stock Purch. Plan:		Second Exec. Salary: $	Bonus: $

OTHER THOUGHTS:

Apparent Women Officers or Directors: 3
Hot Spot for Advancement for Women/Minorities: Y

LOCATIONS: ("Y" = Yes)

West:	Southwest:	Midwest:	Southeast:	Northeast:	International:
Y		Y	Y	Y	Y

EMBREX INC

www.embrex.com

Industry Group Code: 325412B Ranks within this company's industry group: Sales: Profits:

Drugs:	Other:		Clinical:	Computers:	Other:
Discovery:	AgriBio:	Y	Trials/Services:	Hardware:	Specialty Services:
Licensing:	Genomics/Proteomics:		Laboratories:	Software:	Consulting:
Manufacturing: Y	Tissue Replacement:		Equipment/Supplies: Y	Arrays:	Blood Collection:
Development: Y			Research/Development Svcs.:	Database Management:	Drug Delivery:
Generics:			Diagnostics:		Drug Distribution:

TYPES OF BUSINESS:

Equipment-Automated Poultry Vaccination Systems
Animal Health Products
Avian Vaccines

BRANDS/DIVISIONS/AFFILIATES:

Inovoject
Vaccine Saver
Egg Remover
Bursaplex
Newplex
Inovocox
Embrex Poultry Health LLC
Pfizer Animal Health

CONTACTS: *Note: Officers with more than one job title may be intentionally listed here more than once.*

Phone: 919-941-5185	Fax: 919-941-5186
Toll-Free: 800-849-3629	
Address: 1040 Swabia Ct., Durham, NC 27703 US	

GROWTH PLANS/SPECIAL FEATURES:

Embrex, Inc., a wholly-owned subsidiary of Pfizer, Inc., is an international agricultural biotechnology company specializing in the development of innovative in ovo (in the egg) solutions that meet the needs of the worldwide poultry industry. The firm focuses on developing patented biological and mechanical products that improve bird health, help reduce inoculation costs and provide other economic benefits to the industry. Embrex's proprietary Inovoject system is an automated, in-the-egg injection system that can inoculate 20,000 to 70,000 eggs per hour, eliminating the need for manual, post-hatch injection of certain vaccines. The system injects vaccines and other compounds in precisely calibrated volumes into targeted compartments within the egg. It is currently used to inoculate approximately 80% of broiler chickens in the U.S. against Marek's disease, and is deployed in 34 other countries. Embrex also offers the Vaccine Saver and Egg Remover modules to remove unviable eggs or selectively vaccinate only fertilized eggs. The firm uses its antigen-antibody complex technology (AAC- formerly Viral Neutralizing Factor) in the development of avian vaccines that immunize a bird for life with a single dose. Bursaplex is an AAC-based vaccine for protection against infectious bursal disease, and Newplex is a vaccine for Newcastle disease. Embrex is developing various other proprietary mechanical and biological products for the poultry industry, including an avian gender-sorting device and Inovocox, an in-egg vaccine for the control of coccidiosis. Inovocox is in the early stages of USDA review. Embrex was acquired by Pfizer Animal Health in early 2007.

The firm offers its employees reimbursement accounts for day care and healthcare, as well as an employee referral program.

FINANCIALS: Sales and profits are in thousands of dollars—add 000 to get the full amount. 2006 Note: Financial information for 2006 was not available for all companies at press time.

2006 Sales: $	2006 Profits: $	U.S. Stock Ticker: Subsidiary
2005 Sales: $52,592	2005 Profits: $2,947	Int'l Ticker: Int'l Exchange:
2004 Sales: $48,717	2004 Profits: $3,313	Employees: 309
2003 Sales: $46,025	2003 Profits: $7,611	Fiscal Year Ends: 12/31
2002 Sales: $45,300	2002 Profits: $7,200	Parent Company: PFIZER INC

SALARIES/BENEFITS:

Pension Plan:	ESOP Stock Plan:	Profit Sharing:	Top Exec. Salary: $335,000	Bonus: $64,000
Savings Plan: Y	Stock Purch. Plan: Y		Second Exec. Salary: $240,097	Bonus: $40,674

OTHER THOUGHTS:

Apparent Women Officers or Directors: 1
Hot Spot for Advancement for Women/Minorities:

LOCATIONS: ("Y" = Yes)

West:	Southwest:	Midwest:	Southeast:	Northeast:	International:
				Y	Y

EMISPHERE TECHNOLOGIES INC   www.emisphere.com

Industry Group Code: 325412A Ranks within this company's industry group: Sales: 13   Profits: 20

Drugs:		Other:		Clinical:		Computers:		Other:	
Discovery:	Y	AgriBio:		Trials/Services:		Hardware:		Specialty Services:	
Licensing:		Genomics/Proteomics:		Laboratories:		Software:		Consulting:	
Manufacturing:	Y	Tissue Replacement:		Equipment/Supplies:		Arrays:		Blood Collection:	
Development:	Y			Research/Development Svcs.:	Y	Database Management:		Drug Delivery:	Y
Generics:				Diagnostics:				Drug Distribution:	

TYPES OF BUSINESS:

Drug Delivery Systems
Cardiovascular Diseases Oral Drugs
Osteoarthritis & Osteoporosis Oral Drugs
Growth Disorders Oral Drugs
Diabetes Oral Drugs
Asthma & Allergies Oral Drugs
Obesity Oral Drugs
Infectious Diseases & Oncology Oral Drugs

BRANDS/DIVISIONS/AFFILIATES:

Eligen
Heparin
Acyclovir
Calcitonin
Genta, Inc.
Hoffman-La Roche, Inc.
F. Hoffman-La Roche, Ltd.
Novartis Pharma AG

CONTACTS: *Note: Officers with more than one job title may be intentionally listed here more than once.*

Michael V. Novinski, CEO
Michael V. Novinski, Pres.
Steven M. Dinh, VP-Research & Tech. Dev.
Lewis H. Bender, CTO
Paul R. Nemeth, VP-Regulatory Affairs
Richard E. Connor, Asst. VP-Quality & Analytical Resources
Laura Kragie, VP-Clinical Dev./Chief Medical Officer

Phone: 914-347-2220	**Fax:** 914-347-2498
Toll-Free:	
Address: 765 Old Saw Mill River Rd., Tarrytown, NY 10591 US	

GROWTH PLANS/SPECIAL FEATURES:

Emisphere Technologies, Inc. is a biopharmaceutical company specializing in the oral delivery of therapeutic macromolecules and other compounds. The company's product pipeline includes product candidates for the treatment of cardiovascular diseases, osteoarthritis, osteoporosis, growth disorders, diabetes, asthma/allergies, obesity, infectious diseases and oncology. The eligen technology, the firm's oral drug delivery technology, is based on the use of proprietary, synthetic chemical compounds known as Emisphere delivery agents. These delivery agents facilitate and/or enable the transport of therapeutic macromolecules (such as proteins, peptides and polysaccharides) and poorly absorbed small molecules across biological membranes such as the small intestine. Emisphere has numerous drug candidates in its pipeline: oral Heparin for cardiovascular disease is in Phase III; oral Salmon Calcitonin for osteoporosis and osteoarthritis is in Phase III; oral recombinant growth hormone for growth disorders is in Phase I; oral insulin and oral glucagon-like peptides for diabetes are in Phase II and Phase I, respectively; oral Acyclovir, an antiviral candidate, is in Phase I; oral Galium, an oncology product, is in preclinical studies; and oral Cromolyn sodium for asthma/allergies is in Phase I. Emisphere works with Novartis Pharma, Genta, Hoffman-La Roche and F. Hoffman-La Roche to discover and design some of its candidates. In May 2007, the company announced initiation of Phase III clinical trials for oral Calcitonin for the treatment of osteoarthritis.

The company offers its employees medical, dental and life insurance; long-term disability; a 401(k) plan; a college savings plan; tuition reimbursement; an employee stock purchase plan; and stock option programs.

FINANCIALS: Sales and profits are in thousands of dollars—add 000 to get the full amount. 2006 Note: Financial information for 2006 was not available for all companies at press time.

2006 Sales: $7,259	2006 Profits: $-41,766	**U.S. Stock Ticker:** EMIS	
2005 Sales: $3,540	2005 Profits: $-18,051	**Int'l Ticker:** Int'l Exchange:	
2004 Sales: $1,953	2004 Profits: $-37,522	Employees: 111	
2003 Sales: $ 400	2003 Profits: $-44,869	Fiscal Year Ends: 12/31	
2002 Sales: $3,400	2002 Profits: $-71,300	Parent Company:	

SALARIES/BENEFITS:

Pension Plan:	ESOP Stock Plan:	Profit Sharing:	Top Exec. Salary: $544,079	Bonus: $
Savings Plan: Y	Stock Purch. Plan: Y		Second Exec. Salary: $317,366	Bonus: $

OTHER THOUGHTS:

Apparent Women Officers or Directors: 1
Hot Spot for Advancement for Women/Minorities:

LOCATIONS: ("Y" = Yes)

West:	Southwest:	Midwest:	Southeast:	Northeast:	International:
				Y	

ENCORIUM GROUP INC

www.encorium.com

Industry Group Code: 541710 Ranks within this company's industry group: Sales: 18   Profits: 9

Drugs:	Other:	Clinical:		Computers:		Other:	
Discovery:	AgriBio:	Trials/Services:	Y	Hardware:		Specialty Services:	
Licensing:	Genomics/Proteomics:	Laboratories:		Software:	Y	Consulting:	
Manufacturing:	Tissue Replacement:	Equipment/Supplies:		Arrays:		Blood Collection:	Y
Development:		Research/Development Svcs.:		Database Management:		Drug Delivery:	
Generics:		Diagnostics:				Drug Distribution:	

TYPES OF BUSINESS:

Contract Research
Clinical Trial & Data Management
Disease Assessment Software
Biostatistical Analysis
Regulatory Affairs Services

BRANDS/DIVISIONS/AFFILIATES:

Covalent TeleTrial
Oracle Clinical
DataFax
Remedium Oy

CONTACTS: *Note: Officers with more than one job title may be intentionally listed here more than once.*

Kenneth M. Borow, CEO
Kenneth M. Borow, Pres.
Lawrence R. Hoffman, CFO/Exec. VP
Helen Springford, VP-Global Bus. Dev.
Alison O'Neill, Sr. VP-Clinical Oper., Americas
Anna Minkkinen, Sr. Dir.-Project Mgmt.
Kai E. Lindevall, Pres., European & Asian Oper.
Katarina Lonnquist, VP-Int'l Oper.

Phone: 610-975-9533	Fax: 610-975-9556
Toll-Free:	
Address: 1 Glenhardie Corp Ctr, 1275 Drummers Lane, Ste 100, Wayne, PA 19087 US	

GROWTH PLANS/SPECIAL FEATURES:

Encorium Group, Inc., formed in 2006 from a merger between Remedium Oy and Covalent Group, Inc., is a contract research organization (CRO) that designs and manages clinical trials in the pharmaceutical, biotechnology and medical device development process. It also designs and manages associated cost-containment and quality-of-care components. The company specializes in Phase I through IV clinical trials, cost-effectiveness studies and outcomes studies for pharmaceutical companies, managed-care organizations, insurers and employers. It offers a full array of integrated services, including study design, clinical trial monitoring and management, data management, biostatistical analysis and regulatory affairs services. Encorium's international clients include four of the world's top five pharmaceutical companies. To aid its customers in clinical trials and outcomes research projects, Encorium has developed several proprietary products incorporating interactive voice recognition, fax and Internet technology. Encorium's expertise covers a wide range of therapeutic areas, including oncology, neurology, gastroenterology, cardiovascular diseases, vaccines and infectious diseases. The company partners with pharmaceutical companies early in the development process to optimize the design and conduct of clinical trials.

Encorium offers its employees benefits including an employee assistance program; bereavement leave; election day and school visit time off; flexible spending accounts; flexible work schedules; a business casual dress code; referral programs; a dry cleaning service; stock options; and a 401(k) savings plan.

FINANCIALS: Sales and profits are in thousands of dollars—add 000 to get the full amount. 2006 Note: Financial information for 2006 was not available for all companies at press time.

2006 Sales: $15,326	2006 Profits: $- 494	U.S. Stock Ticker: ENCO
2005 Sales: $10,403	2005 Profits: $-1,484	Int'l Ticker: Int'l Exchange:
2004 Sales: $13,590	2004 Profits: $-4,223	Employees: 248
2003 Sales: $20,836	2003 Profits: $- 562	Fiscal Year Ends: 12/31
2002 Sales: $29,187	2002 Profits: $2,454	Parent Company:

SALARIES/BENEFITS:

Pension Plan:	ESOP Stock Plan:	Profit Sharing:	Top Exec. Salary: $344,970	Bonus: $
Savings Plan: Y	Stock Purch. Plan: Y		Second Exec. Salary: $216,666	Bonus: $

OTHER THOUGHTS:

Apparent Women Officers or Directors: 5
Hot Spot for Advancement for Women/Minorities: Y

LOCATIONS: ("Y" = Yes)

West:	Southwest:	Midwest:	Southeast:	Northeast:	International:
				Y	Y

ENCYSIVE PHARMACEUTICALS INC

www.encysive.com

ndustry Group Code: 325412 Ranks within this company's industry group: Sales: 110 Profits: 187

Drugs:		Other:		Clinical:		Computers:		Other:	
Discovery:	Y	AgriBio:		Trials/Services:		Hardware:		Specialty Services:	
Licensing:	Y	Genomics/Proteomics:		Laboratories:		Software:		Consulting:	
Manufacturing:		Tissue Replacement:		Equipment/Supplies:		Arrays:		Blood Collection:	
Development:	Y			Research/Development Svcs.:	Y	Database Management:		Drug Delivery:	
Generics:				Diagnostics:				Drug Distribution:	

TYPES OF BUSINESS:

Cardiovascular & Inflammatory Diseases Drugs

BRANDS/DIVISIONS/AFFILIATES:

Argatroban
Thelin
TBC3711
TBC4746
Bimosiamose
GlaxoSmithKline
Schering-Plough
Mitsubishi Pharma Corp.

CONTACTS: *Note: Officers with more than one job title may be intentionally listed here more than once.*

George W. Cole, CEO
George W. Cole, Pres.
Derek Maetzold, VP-Mktg. & Sales
Pamela Mabry, Dir.-Human Resources
Richard Dixon, Sr. VP-Research/Chief Scientific Officer
Paul S. Manierre, General Counsel/VP
Heather Giles, VP-Strategic Planning
Ann Tanabe, VP-Corp. Comm.
Ann Tanabe, VP-Investor Rel.
Tom Brock, VP-Pharmacology & Preclinical Dev.
D. Jeffrey Keyser, VP-Regulatory Affairs
John M. Pietruski, Chmn.

Phone: 713-796-8822	Fax: 713-796-8232
Toll-Free:	
Address: 4848 Loop Central Dr., 7th Fl., Houston, TX 77081 US	

GROWTH PLANS/SPECIAL FEATURES:

Encysive Pharmaceuticals, Inc. is a biopharmaceutical company that engages in the discovery, development and commercialization of synthetic, small molecule compounds for the treatment of vascular diseases. The company has developed one FDA approved drug, Argatroban, for the treatment of heparin-induced thrombocytopenia, an immune reaction to the widely used anticoagulant heparin that causes abnormal clotting and often leads to amputation and death. The drug is marketed in the U.S. and Canada through GlaxoSmithKline. Encysive's leading product candidate is Thelin, an endothelin receptor antagonist developed for the treatment of pulmonary arterial hypertension (PAH). Thelin received marketing authorization in the EU and Australia and is under review in the U.S. The firm has three additional compounds in clinical development: TBC3711, an endothelin receptor antagonist; TBC4746, a very late antigen-4 antagonist licensed to Schering Corp. and Schering-Plough, Ltd.; and bimosiamose, a selectin antagonist licensed to Revoltar Biopharmaceutical AG, the company's former majority-owned subsidiary. The company's clinical development program includes plans to develop oral Thelin in PAH and to explore other indications for both intravenous and oral Thelin. Encysive has collaborations with several corporations including Mitsubishi Pharma Corp., under which the company uses and sells Argatroban in the U.S. and Canada for all cardiovascular, renal, neurological and immunological purposes; GlaxoSmithKline, under which GlaxoSmithKline has an exclusive sublicense in the U.S. and Canada for the indication of Argatroban; and Schering-Plough, with which the firm discovers, develops and commercializes antigen-4 antagonists. In May 2007, Encysive announced that Canada approved Thelin for the treatment of primary PAH and pulmonary hypertension secondary to connective tissue disease.

FINANCIALS: Sales and profits are in thousands of dollars—add 000 to get the full amount. 2006 Note: Financial information for 2006 was not available for all companies at press time.

2006 Sales: $18,995	2006 Profits: $-109,283	**U.S. Stock Ticker: ENCY**
2005 Sales: $14,006	2005 Profits: $-74,877	**Int'l Ticker:** Int'l Exchange:
2004 Sales: $12,830	2004 Profits: $-54,660	Employees: 282
2003 Sales: $10,951	2003 Profits: $-35,293	Fiscal Year Ends: 12/31
2002 Sales: $6,900	2002 Profits: $-23,500	Parent Company:

SALARIES/BENEFITS:

Pension Plan:	ESOP Stock Plan:	Profit Sharing:	Top Exec. Salary: $466,667	Bonus: $52,800
Savings Plan:	Stock Purch. Plan:		Second Exec. Salary: $425,000	Bonus: $73,000

OTHER THOUGHTS:

Apparent Women Officers or Directors: 4
Hot Spot for Advancement for Women/Minorities: Y

LOCATIONS: ("Y" = Yes)

West:	Southwest:	Midwest:	Southeast:	Northeast:	International:
	Y				

ENDO PHARMACEUTICALS HOLDINGS INC   www.endo.com

Industry Group Code: 325412 Ranks within this company's industry group: Sales: 32   Profits: 31

Drugs:		Other:		Clinical:		Computers:		Other:	
Discovery:	Y	AgriBio:		Trials/Services:	Y	Hardware:		Specialty Services:	Y
Licensing:	Y	Genomics/Proteomics:		Laboratories:		Software:		Consulting:	
Manufacturing:	Y	Tissue Replacement:		Equipment/Supplies:		Arrays:		Blood Collection:	
Development:	Y			Research/Development Svcs.:		Database Management:		Drug Delivery:	
Generics:	Y			Diagnostics:				Drug Distribution:	

TYPES OF BUSINESS:

Drugs-Pain Management
Pharmaceutical Preparations

BRANDS/DIVISIONS/AFFILIATES:

Endo Pharmaceuticals, Inc.
Lidoderm
Percocet
Zydone
Percodan
Ketoprofen
Frova
Synera

CONTACTS: Note: Officers with more than one job title may be intentionally listed here more than once.

Peter A. Lankau, CEO
Peter A. Lankau, Pres.
Charles A. Rowland Jr., CFO/Exec. VP
David A. H. Lee, Exec. VP-R&D/Chief Scientific Officer
Caroline B. Manogue, Chief Legal Officer/Corp. Sec./Exec. VP
Charles A. Rowland Jr., Treas.
Roger H. Kimmel, Chmn.

Phone: 610-558-9800	Fax: 610-558-8979
Toll-Free:	
Address: 100 Endo Blvd., Chadds Ford, PA 19317 US	

GROWTH PLANS/SPECIAL FEATURES:

Endo Pharmaceuticals Holdings, Inc., through subsidiary Endo Pharmaceuticals, Inc., is a specialty pharmaceutical company with market leadership in pain management. Under both generic and brand names, the company discovers, produces and markets pharmaceutical products, principally for the treatment of pain. The branded products, Percodan, Zydone, Lidoderm, Frova and Percocet, are sold to health care professionals by a dedicated force of approximately 590 sales representatives in the U.S. Percocet and Zydone are used to treat moderate to severe pain; Lidoderm is used to treat postherpetic neuralgia; and Percodan is used to treat severe pain. Endo focuses on generic products that are challenging to bring to market due to complex formulation, regulatory or legal problems or barriers in raw material sourcing. Endo has a licensing agreement with Vernalis to market Frova, an acute migraine treatment for adults. In addition, the company has come to an agreement with the FDA over its design of clinical trials for oxymorphone extended-release tablets. The company has exclusive rights to develop and market Orexo AB's patented sublingual muco-adhesive fentanyl product, Rapinyl, in North America. Endo has licensed a topical Ketoprofen patch with ProEthic Pharmaceuticals for treating inflammation and pain, currently in Phase III trials. Endo launched three products in 2006: Opana ER, Opana and Synera. In late 2006, the company purchased RxKinetix, Inc., a privately held company headquartered in Boulder, Colorado, that develops new formulations of approved products for oral mucositis and other supportive care oncology conditions.

The firm offers employees dependent care and medical spending accounts, an enhanced vacation package and an educational assistance program.

FINANCIALS: Sales and profits are in thousands of dollars—add 000 to get the full amount. 2006 Note: Financial information for 2006 was not available for all companies at press time.

2006 Sales: $909,659	2006 Profits: $137,839	**U.S. Stock Ticker: ENDP**
2005 Sales: $820,164	2005 Profits: $202,295	**Int'l Ticker:** Int'l Exchange:
2004 Sales: $615,100	2004 Profits: $143,300	Employees: 1,024
2003 Sales: $595,608	2003 Profits: $69,790	Fiscal Year Ends: 12/31
2002 Sales: $399,000	2002 Profits: $30,800	Parent Company:

SALARIES/BENEFITS:

Pension Plan:	ESOP Stock Plan:	Profit Sharing:	Top Exec. Salary: $519,500	Bonus: $
Savings Plan: Y	Stock Purch. Plan: Y		Second Exec. Salary: $404,431	Bonus: $

OTHER THOUGHTS:

Apparent Women Officers or Directors: 1
Hot Spot for Advancement for Women/Minorities:

LOCATIONS: ("Y" = Yes)

West:	Southwest:	Midwest:	Southeast:	Northeast:	International:
				Y	

ENTREMED INC

www.entremed.com

Industry Group Code: 325412 Ranks within this company's industry group: Sales: 134 Profits: 153

Drugs:		Other:		Clinical:		Computers:		Other:	
Discovery:		AgriBio:		Trials/Services:		Hardware:		Specialty Services:	
Licensing:	Y	Genomics/Proteomics:		Laboratories:		Software:		Consulting:	
Manufacturing:		Tissue Replacement:		Equipment/Supplies:		Arrays:		Blood Collection:	
Development:	Y			Research/Development Svcs.:		Database Management:		Drug Delivery:	
Generics:				Diagnostics:				Drug Distribution:	

TYPES OF BUSINESS:

Drugs-Angiogenesis
Drugs-Cancer & Rheumatoid Arthritis

BRANDS/DIVISIONS/AFFILIATES:

Panzem
Panzem NCD
Miikana Therapeutics, Inc.

CONTACTS: Note: Officers with more than one job title may be intentionally listed here more than once.

James S. Burns, CEO
James S. Burns, Pres.
Dane R. Saglio, CFO
Mark R. Bray, VP-R&D
Anthony M. Treston, VP-Prod. Dev.
Anthony M. Treston, VP-Mfg.
Cynthia Wong, General Counsel/VP/Corp. Sec.
Marc Corrado, VP-Corp. Dev.
Carolyn F. Sidor, VP/Chief Medical Officer
Michael Tarnow, Chmn.

Phone: 240-864-2600	Fax: 240-864-2601
Toll-Free:	
Address: 9640 Medical Center Dr., Rockville, MD 20850 US	

GROWTH PLANS/SPECIAL FEATURES:

EntreMed, Inc. is a clinical-stage pharmaceutical development company that focuses on the development of multi-mechanism oncology drugs that either target diseases cells directly and/or the blood vessels that nourish diseased cells. The company's three clinical product candidates, Panzem NCD, MKC-1, ENMD-1198, are aimed at controlling angiogenesis, cell cycle regulation and inflammation to prevent the progression of cancer. Panzem NCD is the company's current lead product candidate, and works by inhibiting angiogenesis, disrupting microtubule formation, down regulating hypoxia inducible factor-one alpha and inducing apoptosis. In recent years, the FDA has approved the use of Panzem NMD, which utilizes NanoCrystal technology through its licensing agreement with Elan Corporation for the treatment of glioblastoma multiforme, multiple myeloma and ovarian cancer. MKC-1 is a drug candidate that arrests cellular mitosis by inhibiting a novel intracellular target involved in cell division during cellular trafficking in metastatic breast cancer patients, hematological cancers and non-small cell lung cancer. ENMD-1198 inhibits tumor growth through the modification of the chemical structure of 20methoxyestradiol, which increases the molecule's anti-tumor properties, anti-angiogenic properties and metabolic rate. In addition to its three products, the corporation is also investigating the possible use of Panzem, or 2ME2, for the treatment of rheumatoid arthritic and is also developing multi-kinase inhibitors, tubulin inhibitors and HDAS inhibitors for the treatment of cancer. In January 2006, the firm acquired Miikana Therapeutics, Inc., a clinical-stage biopharmaceutical company with research laboratories in Toronto, Canada. As a result of this transaction, EntreMed enhanced its pipeline with the addition of MKC-1 and two preclinical programs for the development of aurora kinase and HDAC inhibitors.

EntreMed offers employees tuition reimbursement; an employee assistance plan that includes dependent care and elder care referral programs; company paid life insurance and supplemental coverage of home and auto insurance; and a 529 savings plan.

FINANCIALS: Sales and profits are in thousands of dollars—add 000 to get the full amount. 2006 Note: Financial information for 2006 was not available for all companies at press time.

2006 Sales: $6,894	2006 Profits: $-49,889	U.S. Stock Ticker: ENMD
2005 Sales: $5,918	2005 Profits: $-16,313	Int'l Ticker: Int'l Exchange:
2004 Sales: $ 500	2004 Profits: $-12,600	Employees: 57
2003 Sales: $1,575	2003 Profits: $-19,494	Fiscal Year Ends: 12/31
2002 Sales: $1,200	2002 Profits: $-39,000	Parent Company:

SALARIES/BENEFITS:

Pension Plan:	ESOP Stock Plan:	Profit Sharing:	Top Exec. Salary: $400,000	Bonus: $
Savings Plan: Y	Stock Purch. Plan:		Second Exec. Salary: $260,000	Bonus: $

OTHER THOUGHTS:

Apparent Women Officers or Directors: 2
Hot Spot for Advancement for Women/Minorities: Y

LOCATIONS: ("Y" = Yes)

West:	Southwest:	Midwest:	Southeast:	Northeast:	International:
				Y	

ENZO BIOCHEM INC
www.enzo.com

Industry Group Code: 325412 Ranks within this company's industry group: Sales: 84   Profits: 89

Drugs:		Other:		Clinical:		Computers:		Other:	
Discovery:		AgriBio:		Trials/Services:		Hardware:		Specialty Services:	Y
Licensing:		Genomics/Proteomics:	Y	Laboratories:	Y	Software:		Consulting:	
Manufacturing:	Y	Tissue Replacement:		Equipment/Supplies:	Y	Arrays:		Blood Collection:	
Development:	Y			Research/Development Svcs.:		Database Management:		Drug Delivery:	
Generics:				Diagnostics:	Y			Drug Distribution:	

TYPES OF BUSINESS:

Pharmaceutical Discovery & Development
Genetic Analysis Products
Therapeutic Products
Diagnostic Products
Drug Development
Genetics Research
Clinical Laboratories
Biomedical Research Products & Tools

BRANDS/DIVISIONS/AFFILIATES:

Enzo Life Sciences, Inc.
Enzo Therapeutics, Inc.
Enzo Clinical Laboratores, Inc.

CONTACTS: Note: Officers with more than one job title may be intentionally listed here more than once.

Elazar Rabbani, CEO
Shahram K. Rabbani, COO/Treas./Corp. Sec.
Barry W. Weiner, Pres.
Barry W. Weiner, CFO
Debbie Sohmer, Dir.-Human Resources
Natalie Bogdanos, Corp. & Patent Counsel
Barbara E. Thalenfeld, VP-Corp. Dev.
Herbert B. Bass, VP-Finance
Dean Engelhardt, Exec. VP
Norman E. Kelker, Sr. VP
David C. Goldberg, VP-Bus. Dev.
Ronald Fedus, Corp. & Patent Counsel
Elazar Rabbani, Chmn.

Phone: 212-583-0100	Fax:
Toll-Free:	
Address: 527 Madison Ave., New York, NY 10022 US	

GROWTH PLANS/SPECIAL FEATURES:

Enzo Biochem, Inc. is a life sciences and biotechnology company that employs a variety of genetic processes to develop research tools, diagnostics and therapeutics. The firm has approximately 211 patents and more than 185 pending patent applications. The main focus of Enzo is to research the use of nucleic acids as informational molecules and to investigate the use of compounds for immune modulation. Enzo is particularly interested in the potency of gene-based diagnostics, which offers a greater advantage than immunoassay technology by allowing detection of diseases that are in the earlier stages of development. Operations are conducted under three subsidiary companies: Enzo Life Sciences, Inc.; Enzo Therapeutics, Inc.; and Enzo Clinical Labs, Inc. Enzo Life Sciences manufactures, develops and markets biomedical research products and tools to global research and pharmaceutical customers. Enzo Therapeutics works in collaboration with Enzo Life Sciences to develop multiple approaches in the areas of gastrointestinal, infectious, ophthalmic and metabolic diseases. This sector has also generated several clinical and preclinical pipelines in its effort to develop treatments for diseases and conditions where current treatment options are either ineffective, costly or cause undesired side effects. Additional technologies developed in therapeutics deal with gene regulation, 'smart' vectors with affinities for only certain cell surfaces and the efficient transduction of operating genes to target cells without the elicitation of an unfavorable immune response. Enzo Clinical Labs are located in New York and New Jersey and offer a variety of routine clinical laboratory tests and procedures for general patient care. In Farmingdale, New York, Enzo Clinical Labs offers a full-service clinical laboratory with 19 patient service centers, a rapid response laboratory and a phlebotomy department.

Enzo offers its employees health care coverage, disability plans, life insurance and paid entitlements such as vacation, holiday, sick personal, jury duty and bereavement time.

FINANCIALS: Sales and profits are in thousands of dollars—add 000 to get the full amount. 2006 Note: Financial information for 2006 was not available for all companies at press time.

2006 Sales: $39,826	2006 Profits: $-15,667	**U.S. Stock Ticker:** ENZ
2005 Sales: $43,403	2005 Profits: $3,004	**Int'l Ticker:** Int'l Exchange:
2004 Sales: $41,644	2004 Profits: $-6,232	Employees: 340
2003 Sales: $52,800	2003 Profits: $3,800	Fiscal Year Ends: 7/31
2002 Sales: $54,000	2002 Profits: $6,900	Parent Company:

SALARIES/BENEFITS:

Pension Plan:	ESOP Stock Plan:	Profit Sharing:	Top Exec. Salary: $442,200	Bonus: $283,250
Savings Plan: Y	Stock Purch. Plan:		Second Exec. Salary: $405,423	Bonus: $267,800

OTHER THOUGHTS:

Apparent Women Officers or Directors: 2
Hot Spot for Advancement for Women/Minorities: Y

LOCATIONS: ("Y" = Yes)

West:	Southwest:	Midwest:	Southeast:	Northeast:	International:
				Y	

Note: Financial information, benefits and other data can change quickly and may vary from those stated here.

ENZON PHARMACEUTICALS INC   www.enzon.com

Industry Group Code: 325412 Ranks within this company's industry group: Sales: 56   Profits: 44

Drugs:		Other:		Clinical:	Computers:	Other:
Discovery:	Y	AgriBio:		Trials/Services:	Hardware:	Specialty Services:
Licensing:	Y	Genomics/Proteomics:		Laboratories:	Software:	Consulting:
Manufacturing:		Tissue Replacement:		Equipment/Supplies:	Arrays:	Blood Collection:
Development:	Y			Research/Development Svcs.:	Database Management:	Drug Delivery:
Generics:				Diagnostics:		Drug Distribution:

TYPES OF BUSINESS:

Pharmaceutical Development
Drugs-Oncology & Hematology
Drugs-Transplantation
Drugs-Infectious Diseases
Single-Chain Antibody Technology
Polyethylene Glycol Technology

BRANDS/DIVISIONS/AFFILIATES:

ONCASPAR
ADAGEN
PEG-INTRON
ABELCET
DEPOCYT
ATG FRESENIUS S
PEG-SN38
Camptosar

CONTACTS: Note: Officers with more than one job title may be intentionally listed here more than once.

Jeffrey H. Buchalter, CEO
Jeffrey H. Buchalter, Pres.
Craig Tooman, CFO
Paul Davit, Exec. VP-Human Resources
Ivan Horak, Chief Scientific Officer/R&D
Ralph del Campo, Exec. VP-Technical Oper.
Craig Tooman, Exec. VP-Finance
Jeffrey H. Buchalter, Chmn.

Phone: 908-541-8600	Fax: 908-575-9457
Toll-Free:	
Address: 685 Rte. 202/206, Bridgewater, NJ 08807 US	

GROWTH PLANS/SPECIAL FEATURES:

Enzon, Inc. is a biopharmaceutical company that develops and commercializes pharmaceutical products in oncology and hematology, transplantation and infectious diseases. The firm advances its product pipeline through continued research and development of its proprietary PEG and SCA technologies. Enzon's PEG (polyethylene glycol) technology is used to improve the delivery, safety and efficacy of proteins and small molecules with known therapeutic value. The company has developed five products that use PEG technology, with several more in development. ADAGEN is a PEG treatment for severe combined immunodeficiency disease (SCID); ONCASPAR is used for the treatment of acute lymphoblastic leukemia; PEG-INTRON, PEGASYS and MACUGEN have been licensed out to other companies. Enzon has licensing agreements with Elan Corp. and SkyePharma to sell ABELCET, an intravenous antifungal, and DEPOCYT, a treatment for lymphomatous meningitis. SCA (single-chain antibody) technology is used to discover and produce antibody-like molecules that can offer the therapeutic benefits of monoclonal antibodies while overcoming some of their limitations. The firm has licensed its SCA technology to more than 15 companies. In January 2007, Enzon entered into an agreement with Ovation Pharmaceuticals, Inc. for the supply of the active ingredient used in the production of Enzon's Oncaspar. Oncaspar is a form of L-asparaginase enhanced with Enzon's PEGylation technology for the treatment of Acute Lymphoblastic Leukemia. In April 2007, U.S. Food and Drug Administration (FDA) approved Enzon's Investigational New Drug (IND) application for PEG-SN38, Enzon's PEGylated form of SN38, the active metabolite of the cancer drug Camptosar. The FDA also granted full approval for DepoCyt for the treatment of patients with lymphomatous meningitis, a life-threatening complication of lymphoma.

Enzon offers its employees medical, dental and life insurance; prescription and vision plans; dependent and healthcare flexible spending accounts; short- and long-term disability; a 401(k) plan; tuition reimbursement; and an employee assistance plan.

FINANCIALS: Sales and profits are in thousands of dollars—add 000 to get the full amount. 2006 Note: Financial information for 2006 was not available for all companies at press time.

2006 Sales: $185,653	2006 Profits: $21,309	U.S. Stock Ticker: ENZN
2005 Sales: $166,250	2005 Profits: $-89,606	Int'l Ticker: Int'l Exchange:
2004 Sales: $169,571	2004 Profits: $4,208	Employees: 359
2003 Sales: $146,406	2003 Profits: $45,726	Fiscal Year Ends: 12/31
2002 Sales: $75,800	2002 Profits: $45,800	Parent Company:

SALARIES/BENEFITS:

Pension Plan:	ESOP Stock Plan:	Profit Sharing:	Top Exec. Salary: $672,115	Bonus: $1,050,000
Savings Plan: Y	Stock Purch. Plan:		Second Exec. Salary: $461,538	Bonus: $300,000

OTHER THOUGHTS:

Apparent Women Officers or Directors:
Hot Spot for Advancement for Women/Minorities:

LOCATIONS: ("Y" = Yes)

West:	Southwest:	Midwest:	Southeast:	Northeast:	International:
		Y		Y	

Note: Financial information, benefits and other data can change quickly and may vary from those stated here.

EPIX PHARMACEUTICALS INC

www.epixmed.com

Industry Group Code: 325413 Ranks within this company's industry group: Sales: 27   Profits: 26

Drugs:		Other:		Clinical:		Computers:		Other:	
Discovery:		AgriBio:		Trials/Services:		Hardware:		Specialty Services:	
Licensing:		Genomics/Proteomics:		Laboratories:		Software:		Consulting:	
Manufacturing:		Tissue Replacement:		Equipment/Supplies:		Arrays:	Y	Blood Collection:	
Development:	Y			Research/Development Svcs.:		Database Management:		Drug Delivery:	
Generics:				Diagnostics:	Y			Drug Distribution:	

TYPES OF BUSINESS:

Medical Diagnostics Products
MRI Contrast Agents

BRANDS/DIVISIONS/AFFILIATES:

Vasovist
EPIX Medical, Inc.
Predix Pharmaceuticals Holdings, Inc.

CONTACTS: Note: Officers with more than one job title may be intentionally listed here more than once.

Michael G. Kauffman, CEO
Andrew C. G. Uprichard, Pres.
Kim C. Drapkin, CFO
Oren M. Becker, Chief Scientific Officer
Sharon Shacham, VP-Drug Dev.
Chen Schor, Chief Bus. Officer
Sheila DeWitt, VP-Discovery
Simon S. Jones, VP-Biology & ADMET
Frederick Frank, Chmn.

Phone: 781-761-7600	Fax: 781-761-7641
Toll-Free:	
Address: 4 Maguire Rd., Lexington, MA 02421 US	

GROWTH PLANS/SPECIAL FEATURES:

EPIX Pharmaceuticals, Inc. discovers, develops and commercializes novel pharmaceutical products and imaging agents to improve the quality of magnetic resonance imaging (MRI) as a tool for the diagnosis of human diseases. The company currently has four therapeutic product candidates in clinical trials that target the treatment of conditions such as depression, Alzheimer's disease, cardiovascular disease and obesity. EPIX's drug discovery and development is focused on two classes of drug targets: G-protein Coupled Receptors (GPCRs) and ion channels, which both involve classes of protein that mediate cellular biological signaling. In addition, EPIX develops imaging agents such as Vasovist, which consists of an injectable intravascular contrast agent that is designed for multiple cardiovascular imaging applications in peripheral vascular disease and coronary artery disease. EPIX believes that Vasovist will significantly enhance the quality of MRI images whilst providing physicians with a minimally-invasive and cost-effective diagnostic method without exposing the patient to ionizing radiation. EPIX currently collaborates with Schering AG for the development and commercialization of Vasovist, and also works with the three leading MRI equipment manufacturers, General Electric Medical Systems, Siemens Medical Systems and Philips Medical Systems. EPIX is also developing a contrast agent, EP-2104R, with shows detailed images of blood clots with MRI technologies. The firm continues to maintain its Israeli headquarters in Ramat-Gan. In August 2006, EPIX closed its merger with Predix Pharmaceuticals, Inc. to form a combined company that will continue to market under the EPIX name. Additionally, the company also entered into a discovery and development collaboration with GlaxoSmithKline to further develop novel therapeutics that target four GPCRs to treat Alzheimer's disease. Vasovist was also approved for marketing in Canada by Health Canada's Health Products and Food Branch in November 2006.

FINANCIALS: Sales and profits are in thousands of dollars—add 000 to get the full amount. 2006 Note: Financial information for 2006 was not available for all companies at press time.

2006 Sales: $6,041	2006 Profits: $-157,393	**U.S. Stock Ticker:** EPIX
2005 Sales: $7,190	2005 Profits: $-21,269	**Int'l Ticker:** Int'l Exchange:
2004 Sales: $12,259	2004 Profits: $-22,621	Employees: 89
2003 Sales: $13,525	2003 Profits: $-20,795	Fiscal Year Ends: 12/31
2002 Sales: $12,300	2002 Profits: $22,200	Parent Company:

SALARIES/BENEFITS:

Pension Plan:	ESOP Stock Plan:	Profit Sharing:	Top Exec. Salary: $326,552	Bonus: $
Savings Plan: Y	Stock Purch. Plan: Y		Second Exec. Salary: $154,111	Bonus: $

OTHER THOUGHTS:

Apparent Women Officers or Directors: 3
Hot Spot for Advancement for Women/Minorities: Y

LOCATIONS: ("Y" = Yes)

West:	Southwest:	Midwest:	Southeast:	Northeast:	International:
				Y	Y

ERESEARCH TECHNOLOGY INC

www.ert.com

Industry Group Code: 511212 Ranks within this company's industry group: Sales: 2 Profits: 1

Drugs:	Other:	Clinical:		Computers:		Other:	
Discovery:	AgriBio:	Trials/Services:		Hardware:		Specialty Services:	Y
Licensing:	Genomics/Proteomics:	Laboratories:	Y	Software:	Y	Consulting:	Y
Manufacturing:	Tissue Replacement:	Equipment/Supplies:		Arrays:		Blood Collection:	
Development:		Research/Development Svcs.:		Database Management:		Drug Delivery:	
Generics:		Diagnostics:				Drug Distribution:	

TYPES OF BUSINESS:

Software-Clinical Research Technology
Technology Consulting Services
Cardiac Safety Services
Service-Business Services, NEC
Service-Testing Laboratories

BRANDS/DIVISIONS/AFFILIATES:

EXPeRT
eResearch Network
eData Entry
eSafety Net
eResearch Community
eStudy Conduct
eData Management

CONTACTS: Note: Officers with more than one job title may be intentionally listed here more than once.

Michael J. McKelvey, CEO
Michael J. McKelvey, Pres.
Richard Baron, CFO/Exec. VP
Robert S. Brown, Sr. VP-Strategic Mktg.
Joel Morganroth, Chief Scientist
Thomas P. Devine, Sr. VP/Chief Dev. Officer
Robert S. Brown, Sr. VP-Planning & Partnerships
Amy Furlong, Exec. VP-Cardiac Safety Oper.
Jeffrey S. Litwin, Sr. VP/Chief Medical Officer
George Tiger, Sr. VP-Americas Sales
Joel Morganroth, Chmn.
John M. Blakeley, Sr. VP-Int'l Oper. & Sales

Phone: 215-972-0420	Fax: 215-972-0414

Toll-Free:

Address: 30 S. 17th St., Philadelphia, PA 19103-4001 US

GROWTH PLANS/SPECIAL FEATURES:

eResearch Technology, Inc. (ERT) is a provider of technology-based products and services that enable the pharmaceutical, biotechnology and medical device and contract resource companies to collect, interpret and distribute cardiac safety and clinical data more efficiently. The company also provides centralized electrocardiograph (ECG) services (through its EXPeRT Cardiac Safety Intelligent Data Management system) and clinical research technology and services, which include the development, marketing and support of clinical research technology. Clinical trial sponsors and clinical research organizations use the cardiac safety services during clinical trials. ERT's clinical research technology and services include the licensing of its proprietary software products and the provision of maintenance and consulting services supporting its proprietary software products. The eResearch Network (eResNet) technology provides an integrated end-to-end clinical research platform that includes trials, data and safety management modules. eResNet includes eData Management, eSafety Net and eStudy Conduct. The eResearch Community is a central command and control web portal that provides real-time information related to monitoring clinical trial activities, data quality and safety. eData Entry is a comprehensive electronic data capture system comprised of technology and consulting services formulated to deliver rapid time to benefit for electronic trial initiatives. In 2007, the company announced its entry into the electronic patient reported outcomes (ePRO) business via the launch of a new line of business which will initially concentrate on the central nervous system (CNS) therapeutic area.

FINANCIALS: Sales and profits are in thousands of dollars—add 000 to get the full amount. 2006 Note: Financial information for 2006 was not available for all companies at press time.

2006 Sales: $86,368	2006 Profits: $8,310	U.S. Stock Ticker: ERES
2005 Sales: $86,847	2005 Profits: $15,365	Int'l Ticker: Int'l Exchange:
2004 Sales: $109,393	2004 Profits: $29,724	Employees: 341
2003 Sales: $66,842	2003 Profits: $14,463	Fiscal Year Ends: 12/31
2002 Sales: $41,500	2002 Profits: $6,200	Parent Company:

SALARIES/BENEFITS:

Pension Plan:	ESOP Stock Plan:	Profit Sharing:	Top Exec. Salary: $300,041	Bonus: $
Savings Plan: Y	Stock Purch. Plan:		Second Exec. Salary: $249,999	Bonus: $

OTHER THOUGHTS:

Apparent Women Officers or Directors: 1
Hot Spot for Advancement for Women/Minorities: Y

LOCATIONS: ("Y" = Yes)

West:	Southwest:	Midwest:	Southeast:	Northeast:	International:
				Y	Y

Note: Financial information, benefits and other data can change quickly and may vary from those stated here.

EVOTEC OAI AG
www.evotecoai.com

Industry Group Code: 541710 Ranks within this company's industry group: Sales: 10   Profits: 19

Drugs:		Other:		Clinical:		Computers:		Other:	
Discovery:	Y	AgriBio:		Trials/Services:		Hardware:	Y	Specialty Services:	Y
Licensing:	Y	Genomics/Proteomics:		Laboratories:	Y	Software:	Y	Consulting:	
Manufacturing:	Y	Tissue Replacement:		Equipment/Supplies:		Arrays:		Blood Collection:	
Development:	Y			Research/Development Svcs.:	Y	Database Management:	Y	Drug Delivery:	
Generics:				Diagnostics:				Drug Distribution:	

TYPES OF BUSINESS:

Drug Discovery & Development Services
Assay Development
High-Throughput Screening
Chemical Compound Libraries
Medicinal Chemistry
Manufacturing Services
Drug Discovery Hardware & Software

BRANDS/DIVISIONS/AFFILIATES:

EVT 201
EVT 101
EVT 103
EVT 302
EVT 102
Evotec-RSIL
Research Support International Limited
Evotec Technologies GmbH

CONTACTS: *Note: Officers with more than one job title may be intentionally listed here more than once.*

Jorn Aldag, CEO
Mario Polywka, COO
Jorn Aldag, Pres.
Dirk H. Ehlers, CFO
John Kemp, Chief R&D Officer
Tim Tasker, Exec. VP-Clinical Dev.
David Brister, Chief Bus. Officer
Anne Hennecke, Sr. VP-Corp. Comm.
Anne Hennecke, Sr. VP-Investor Rel.
Klaus Maleck, Exec. VP-Finance
Mark Ashton, Exec. VP-Bus. Dev., Svcs. Div.
Erich Greiner, Chief Innovation Officer
Carsten Claussen, CEO-Evotec Technologies GmbH
Heinz Riesenhuber, Chmn.

Phone: 49-40-5-60-81-0	Fax: 49-40-5-60-81-222
Toll-Free:	
Address: Schnackenburgallee 114, Hamburg, D-22525 Germany	

GROWTH PLANS/SPECIAL FEATURES:

Evotec OAI AG provides investigational new drug programs for its biotechnology and pharmaceutical company clients. The company has specific expertise in the area of Central Nervous System (CNS) related diseases. Evotec's collaborative research division provides contract research and development services through its Innovation Centers, with capabilities including assay development; high-throughput screening; compound libraries; medicinal chemistry; chemical and pharmaceutical development; and formulation and manufacture. Assay development identifies chemical compounds that interact with a selected biological target. High-throughput screening takes the assay format and uses it to test up to 100,000 chemical compounds per day in Evotec's, and its clients', chemical libraries. Primary screens determine whether a compound shows biological activity worth exploring. The company fine-tunes its chemical library findings into pre-clinical candidates, and then develops processes for manufacturing enough of the test drug for Phase I-III clinical trials. Evotec's Proprietary CNS Pipeline division conducts internal research to discover compounds for the treatment of major CNS related disorders including sleep disorders, Alzheimer's disease and pain. For the treatment of insomnia, Evotec has developed EVT 201, which has completed phase II trials with positive results. EVT 101, in Phase II program planning, and EVT 103, in preclinical development, are the company's Alzheimer's disease and neuropathic pain products. Evotec has an MAO-B inhibitor, EVT 302, which will undergo Phase I clinical trials in 2007. For the treatment of post-operative pain, the company is developing an intravenous selective NMDA Antagonist, EVT 102, which is currently in late preclinical development. Evotec has completed more than 1,200 projects for over 150 clients, including Celgene, DuPont, GlaxoSmithKline, Merck, Pfizer, Proctor & Gamble and Roche. In July 2007, Evotec and Research Support International Limited announced the formation of a joint venture for the design, synthesis, management and commercialization of compound libraries, Evotec-RSIL.

FINANCIALS: Sales and profits are in thousands of dollars—add 000 to get the full amount. 2006 Note: Financial information for 2006 was not available for all companies at press time.

2006 Sales: $111,700	2006 Profits: $-42,800	**U.S. Stock Ticker:**	
2005 Sales: $101,630	2005 Profits: $-42,780	**Int'l Ticker: EVT** Int'l Exchange: Frankfurt-Euronext	
2004 Sales: $99,200	2004 Profits: $-114,900	Employees: 599	
2003 Sales: $96,900	2003 Profits: $-17,900	Fiscal Year Ends: 12/31	
2002 Sales: $73,400	2002 Profits: $-138,000	Parent Company:	

SALARIES/BENEFITS:

Pension Plan:	ESOP Stock Plan:	Profit Sharing:	Top Exec. Salary: $	Bonus: $
Savings Plan:	Stock Purch. Plan:		Second Exec. Salary: $	Bonus: $

OTHER THOUGHTS:

Apparent Women Officers or Directors: 1
Hot Spot for Advancement for Women/Minorities:

LOCATIONS: ("Y" = Yes)

West:	Southwest:	Midwest:	Southeast:	Northeast:	International:
				Y	Y

EXELIXIS INC

www.exelixis.com

Industry Group Code: 325412 **Ranks within this company's industry group:** Sales: 64   Profits: 182

Drugs:		Other:		Clinical:		Computers:		Other:	
Discovery:	Y	AgriBio:	Y	Trials/Services:		Hardware:		Specialty Services:	
Licensing:	Y	Genomics/Proteomics:	Y	Laboratories:		Software:		Consulting:	
Manufacturing:		Tissue Replacement:	Y	Equipment/Supplies:		Arrays:		Blood Collection:	
Development:	Y			Research/Development Svcs.:		Database Management:		Drug Delivery:	
Generics:				Diagnostics:				Drug Distribution:	

TYPES OF BUSINESS:

Genetic Research & Drug Development
Crop Protection Products
Genomics
Anti-Cancer Compounds

BRANDS/DIVISIONS/AFFILIATES:

Exelixis Plant Sciences, Inc.
Rebeccamycin
Genoptera, LLC
Agrinomica and Renessen, LLC

CONTACTS: *Note: Officers with more than one job title may be intentionally listed here more than once.*

George A. Scangos, CEO
George A. Scangos, Pres.
Frank Karbe, CFO/Sr. VP
Ian D. Malcolm, VP-Strategic Mktg.
Lupe M. Rivera, Sr. VP-Human Resources
Michael Morrissey, Pres., R&D
Peter Lamb, Chief Scientific Officer/Sr. VP-Discovery
Lupe M. Rivera, Sr. VP-Corp. Comm.
Ry Wagner, VP-Research, Elexis Plant Sciences
Pamela A. Simonton, Sr. VP-Patents & Licensing
Gisela M. Schwab, Chief Medical Officer/Exec. VP
Stelios Papadopoulos, Chmn.

Phone: 650-837-7000	Fax: 650-837-8300

Toll-Free:
Address: 210 E. Grand Ave., So. San Francisco, CA 94083-0511 US

GROWTH PLANS/SPECIAL FEATURES:

Exelixis, Inc. is a biotechnology company that focuses on the discovery and development of potential new drug therapies for cancer and other life-threatening diseases. The company has developed an integrated research and discovery platform utilizing proprietary technologies such as medicinal chemistry, bioinformatics, structural biology and early in vivo testing to provide an efficient and cost-effective process in gene analysis and drug development. Exelixis' translational research group employs knowledge generated in the discovery process to identify targeted patient populations for possible gene mutations or gene variants that impact response to therapy. The company's clinical program then conducts clinical trials to move candidate compounds through clinical registration phases in order to market newly discovered treatments. Exelixis Plant Sciences, a subsidiary of Exelixis, is working in collaboration with agricultural companies in plant biotechnology and crop protection. Research areas in crop protection focus on chemical products such as herbicides, insecticides and nematides designed specifically to target implicated crops. The plant biotechnology sector aims to develop crops with higher yields and improved nutritional profiles in oil content and protein composition. Exelixis has entered into research collaborations with Bristol-Myers Squibb to identify novel targets for new drugs in the fields of oncology and cardiovascular disease. Exelixis collaborates its research with GlaxoSmithKline in therapeutic areas such as vascular biology, inflammatory disease and oncology. In 2006, the company entered into an alliance with Sankyo Company to discover, develop and market new therapies for various cardiovascular diseases. In 2007, Exelixis signed a co-development agreement with Genentech to develop a small-molecular inhibitor of mitogen activated protein kinase, a key component in the growth of human tumors.

Exelixis offers its employees tuition reimbursement; a corporate fitness program; shuttle services; relocation assistance; adoption and infertility assistance programs; and entertainment events such as concerts, sporting events and picnic celebrations.

FINANCIALS: Sales and profits are in thousands of dollars—add 000 to get the full amount. 2006 Note: Financial information for 2006 was not available for all companies at press time.

			U.S. Stock Ticker: EXEL		
2006 Sales: $98,670	2006 Profits: $-101,492				
2005 Sales: $75,961	2005 Profits: $-84,404		Int'l Ticker: Int'l Exchange:		
2004 Sales: $52,857	2004 Profits: $-137,245		Employees: 651		
2003 Sales: $51,540	2003 Profits: $-94,774		Fiscal Year Ends: 12/31		
2002 Sales: $44,300	2002 Profits: $-86,200		Parent Company:		

SALARIES/BENEFITS:

Pension Plan:	ESOP Stock Plan: Y	Profit Sharing:	Top Exec. Salary: $750,000	Bonus: $400,000
Savings Plan: Y	Stock Purch. Plan:		Second Exec. Salary: $400,520	Bonus: $200,260

OTHER THOUGHTS:

Apparent Women Officers or Directors: 3
Hot Spot for Advancement for Women/Minorities: Y

LOCATIONS: ("Y" = Yes)

West:	Southwest:	Midwest:	Southeast:	Northeast:	International:
Y					Y

EXELIXIS PLANT SCIENCES INC   www.exelixis.com

Industry Group Code: 115112 Ranks within this company's industry group: Sales: Profits:

Drugs:	Other:		Clinical:		Computers:	Other:
Discovery:	AgriBio:	Y	Trials/Services:		Hardware:	Specialty Services:
Licensing:	Genomics/Proteomics:	Y	Laboratories:		Software:	Consulting:
Manufacturing:	Tissue Replacement:		Equipment/Supplies:		Arrays:	Blood Collection:
Development:			Research/Development Svcs.:		Database Management:	Drug Delivery:
Generics:			Diagnostics:			Drug Distribution:

TYPES OF BUSINESS:

Agricultural Biotechnology Products
Biochemical Compounds
Natural Flavors & Colorants
Pesticides
Pharmaceutical Drug Development

BRANDS/DIVISIONS/AFFILIATES:

Exelixis, Inc.
Agrinomics, LLC

CONTACTS: Note: Officers with more than one job title may be intentionally listed here more than once.

D. Ry Wagner, VP-Research
T. J. Paulsen, Controller

Phone: 503-670-7702	Fax: 503-403-5790
Toll-Free:	
Address: 16160 SW Upper Boones Ferry Rd., Portland, OR 97224-7744 US	

GROWTH PLANS/SPECIAL FEATURES:

Exelixis Plant Sciences, Inc. (EPS), a subsidiary of Exelixis, Inc., is an agricultural biotechnology company that develops improved plant products and technologies for the agricultural industry. The company uses several model plant species such as Arabidopsis thaliana, a plant related to canola, and tomato and rice plants for high-throughput gene discovery analyses. Each of these model plants contains attributes that make it uniquely suited for specific research purposes in the areas of improved oil content, protein composition and human nutrition. In its Portland, Oregon facilities, the company utilizes plants as factories to produce high-value compounds that are naturally produced in plants. Produced plant compounds include natural flavors and colorants for the packaged foods and cosmetics industries. The company is also investigating the use of a plant's biological machinery to produce economical pharmaceuticals, which would replace the standard production methods that utilize synthetic and biologically produced drugs. EPS also employs its expertise in target identification, high-throughput screening and chemistry to manufacture chemical products such as herbicides, insecticides and nematicides. Through joint ventures, EPS is additionally developing crops with a better yield and improved nutritional profiles in oil content and protein composition and plants with higher levels of valuable biochemical compounds. In July 1999, EPS and Bayer CropScience formed a joint venture company, Agrinomics, LLC, which conducts research, development and commercialization programs in the field of agricultural functional genomics. Since its inception, Agrinomics has created a fully indexed and archived collection of over 250,000 ACTTAG Arabidopsis lines, which covers the majority of the entire Arabidopsis genome. Agrinomics also established a joint venture with Renessen, LLC to enhance the oil content in commercially available seed oil crops.

Exelixis offers its employees an infertility assistance program, adoption assistance, tuition reimbursement, pet insurance, a college savings plan and shuttle service.

FINANCIALS: Sales and profits are in thousands of dollars—add 000 to get the full amount. 2006 Note: Financial information for 2006 was not available for all companies at press time.

2006 Sales: $	2006 Profits: $	**U.S. Stock Ticker: Subsidiary**
2005 Sales: $	2005 Profits: $	**Int'l Ticker:** Int'l Exchange:
2004 Sales: $32,800	2004 Profits: $	Employees: 437
2003 Sales: $	2003 Profits: $	Fiscal Year Ends: 12/31
2002 Sales: $	2002 Profits: $	Parent Company: EXELIXIS INC

SALARIES/BENEFITS:

Pension Plan:	ESOP Stock Plan:	Profit Sharing:	Top Exec. Salary: $	Bonus: $
Savings Plan: Y	Stock Purch. Plan: Y		Second Exec. Salary: $	Bonus: $

OTHER THOUGHTS:

Apparent Women Officers or Directors:
Hot Spot for Advancement for Women/Minorities:

LOCATIONS: ("Y" = Yes)

West:	Southwest:	Midwest:	Southeast:	Northeast:	International:
Y					

EXPERIMENTAL & APPLIED SCIENCES INC   www.eas.com

industry Group Code: 325411 Ranks within this company's industry group: Sales: Profits:

Drugs:		Other:		Clinical:	Computers:		Other:	
Discovery:		AgriBio:		Trials/Services:	Hardware:		Specialty Services:	Y
Licensing:		Genomics/Proteomics:		Laboratories:	Software:		Consulting:	
Manufacturing:	Y	Tissue Replacement:		Equipment/Supplies:	Arrays:		Blood Collection:	
Development:				Research/Development Svcs.:	Database Management:		Drug Delivery:	
Generics:				Diagnostics:			Drug Distribution:	Y

TYPES OF BUSINESS:
Nutritional Products
Supplements-Athletic Training
Supplements-Weight Loss
Books & Videos-Athletic Training
Research-Dietary & Sports Medicine

BRANDS/DIVISIONS/AFFILIATES:
AdvantEdge
DynamX
PhosphaGen
Myoplex
Piranha
Body-for-LIFE
Abbot Laboratories
Ross Products

CONTACTS: Note: Officers with more than one job title may be intentionally listed here more than once.
Chris Scoggins, Gen. Mgr.
Chris Hickey, Dir.-Mktg.

Phone: 303-384-0080	Fax: 303-279-6465
Toll-Free: 800-297-9776	
Address: 555 Corporate Cir., Golden, CO 80401 US	

GROWTH PLANS/SPECIAL FEATURES:
Experimental and Applied Sciences, Inc., (EAS) is a wholly-owned subsidiary of Abbot Laboratories' Ross Products Division that specializes in making and distributing weight management products and dietary supplements. The company's products, which are marketed worldwide, include Myoplex meal-replacement supplements, DynamX weight loss formula, Piranha energy drinks, AdvantEdge low-carbohydrate snacks, PhosphaGen muscle growth formula, and Body-for-LIFE food products. EAS, founded by Bill Phillips, author of the popular fitness manual Body for Life, sells training and exercise videos in addition to its primary line of foods and supplements. Its products have received a seal of approval from the National Football League, whose approval indicates that EAS products contain no illegal steroids or other illicit substances. EAS has research affiliations with Purdue University, Cornell University and the University of Colorado Health Sciences Center. In order to provide funding for further research and development in the sport nutrition field, the company spends a percentage of each dollar it receives on research and development. To further the field, EAS provides students at its affiliated universities with research grants via the American College of Sports Medicine grant program. Additionally, every year the company contributes large sums to private research programs all over the world seeking to further the sports medicine industry.

FINANCIALS: Sales and profits are in thousands of dollars—add 000 to get the full amount. 2006 Note: Financial information for 2006 was not available for all companies at press time.

			U.S. Stock Ticker: Subsidiary
2006 Sales: $	2006 Profits: $		Int'l Ticker: Int'l Exchange:
2005 Sales: $	2005 Profits: $		Employees:
2004 Sales: $	2004 Profits: $		Fiscal Year Ends: 12/31
2003 Sales: $	2003 Profits: $		Parent Company: ABBOTT LABORATORIES
2002 Sales: $	2002 Profits: $		

SALARIES/BENEFITS:

Pension Plan:	ESOP Stock Plan:	Profit Sharing:	Top Exec. Salary: $	Bonus: $
Savings Plan:	Stock Purch. Plan:		Second Exec. Salary: $	Bonus: $

OTHER THOUGHTS:
Apparent Women Officers or Directors:
Hot Spot for Advancement for Women/Minorities:

LOCATIONS: ("Y" = Yes)

West:	Southwest:	Midwest:	Southeast:	Northeast:	International:
					Y

Note: Financial information, benefits and other data can change quickly and may vary from those stated here.

E-Z-EM INC

www.ezem.com

Industry Group Code: 325413 Ranks within this company's industry group: Sales: 9 Profits: 7

Drugs:		Other:		Clinical:		Computers:		Other:	
Discovery:		AgriBio:		Trials/Services:		Hardware:	Y	Specialty Services:	
Licensing:		Genomics/Proteomics:		Laboratories:		Software:	Y	Consulting:	
Manufacturing:	Y	Tissue Replacement:		Equipment/Supplies:	Y	Arrays:		Blood Collection:	
Development:	Y			Research/Development Svcs.:		Database Management:		Drug Delivery:	
Generics:				Diagnostics:	Y			Drug Distribution:	

TYPES OF BUSINESS:

Medical Diagnostics Products
Virtual Colonoscopy Products
Diagnostic Contrast Media
Diagnostic Radiology Devices
Custom Pharmaceuticals
Gastrointestinal Diagnostic Products
Decontaminant Lotions

BRANDS/DIVISIONS/AFFILIATES:

NutraPrep
CT Smoothies
Readi-CAT
Varibar
innerviewGI
Reactive Skin Decontaminant Lotion (RSDL)
EmpowerSync

CONTACTS: Note: Officers with more than one job title may be intentionally listed here more than once.

Anthony A. Lombardo, CEO
Anthony A. Lombardo, Pres.
Joseph Cacchioli, Acting CFO
Peter J. Graham, Head-Human Resources
Jeffrey S. Peacock, Sr. VP-Global Scientific
Jeffrey S. Peacock, Sr. VP-Tech. Oper.
Peter J. Graham, General Counsel/VP/Sec.
Joseph J. Palma, Sr. VP-Corp. Rel.
Brad S. Schreck, Sr. VP-Global Mktg. Eng.
Paul S. Echenberg, Chmn.

Phone: 516-333-8230	Fax: 516-333-8278
Toll-Free: 800-544-4624	
Address: 1111 Marcus Ave., Ste. LL26, Lake Success, NY 11042 US	

GROWTH PLANS/SPECIAL FEATURES:

E-Z-EM, Inc. develops, manufactures and markets diagnostic products used by physicians during image-assisted procedures to detect anatomic abnormalities and diseases. Its products and services are designed for use in the radiology, gastroenterology, speech pathology and virtual colonoscopy industries for colorectal cancer screening and testing for other gastrointestinal disorders. E-Z-EM's products coat portions of the throat or digestive tract to help targeted systems appear under imaging systems such as X-ray fluoroscopy and CT imaging. The company's lead products include CT Smoothies, Readi-CAT contrast products, NutraPrep low residue pre-procedural foods, the Varibar line of swallowing evaluation agents, CT injector systems and the innerviewGI hardware and software system for conducting virtual colonoscopy. E-Z-EM provides contract manufacturing in the areas of diagnostic contrast media, pharmaceuticals and cosmetics. In November 2006, the Food and Drug Administration approved E-Z-EM's EmpowerSync system—a product based on the CAN-CiA (Controller Area Network-CAN in Automation) DSP 425 protocol. EmpowerSync permits synchronized operation of EmpowerCT and EmpowerCTA injector systems and CT scanners from all of the manufacturers who also adopt the DSP 425 standard. The company is currently working with O'Dell Engineering Ltd. to commercialize a product line of Reactive Skin Decontaminant Lotions (RSDL) for neutralizing chemical warfare agents. These RSDL products are in use in the armed forces of Canada, Australia, Ireland and the Netherlands. In April 2007, the U.S. Army Space & Missile Defense Command (USASMD) placed a $5 million procurement order for E-Z-EM's RSDL personal skin decontamination product. Recently, RSDL received Milestone C approval from the Joint Program Executive Office for Chemical Biological Defense (JPEO-CBD) clearing the way for procurement by the individual service branches of the DoD.

E-Z-EM offers its employees medical, dental, life and travel insurance; long term disability; a 401(k); a stock purchase plan; employee training and development programs; tuition reimbursement; and flexible spending accounts.

FINANCIALS: Sales and profits are in thousands of dollars—add 000 to get the full amount. 2006 Note: Financial information for 2006 was not available for all companies at press time.

2006 Sales: $138,369	2006 Profits: $9,766	U.S. Stock Ticker: EZEM
2005 Sales: $113,075	2005 Profits: $6,936	Int'l Ticker: Int'l Exchange:
2004 Sales: $148,771	2004 Profits: $6,726	Employees: 611
2003 Sales: $133,200	2003 Profits: $2,700	Fiscal Year Ends: 5/31
2002 Sales: $122,100	2002 Profits: $ 600	Parent Company:

SALARIES/BENEFITS:

Pension Plan:	ESOP Stock Plan:	Profit Sharing: Y	Top Exec. Salary: $350,784	Bonus: $249,769
Savings Plan: Y	Stock Purch. Plan: Y		Second Exec. Salary: $224,591	Bonus: $101,610

OTHER THOUGHTS:

Apparent Women Officers or Directors:
Hot Spot for Advancement for Women/Minorities:

LOCATIONS: ("Y" = Yes)

West:	Southwest:	Midwest:	Southeast:	Northeast:	International:
				Y	Y

Note: Financial information, benefits and other data can change quickly and may vary from those stated here.

FLAMEL TECHNOLOGIES SA   www.flamel-technologies.fr

Industry Group Code: 325412A Ranks within this company's industry group: Sales: 8 Profits: 18

Drugs:		Other:		Clinical:		Computers:		Other:	
Discovery:		AgriBio:		Trials/Services:		Hardware:		Specialty Services:	
Licensing:	Y	Genomics/Proteomics:		Laboratories:		Software:		Consulting:	
Manufacturing:	Y	Tissue Replacement:		Equipment/Supplies:	Y	Arrays:		Blood Collection:	
Development:	Y			Research/Development Svcs.:		Database Management:		Drug Delivery:	Y
Generics:				Diagnostics:				Drug Distribution:	

TYPES OF BUSINESS:
Drug Delivery Systems
Extended-Release Formulations
Collagen-Based Biomaterials
Long-Acting Insulin

BRANDS/DIVISIONS/AFFILIATES:
Medusa
Micropump
Asacard
Genvir
Colcys
Basulin
Trigger-Lock
Metformin XL

CONTACTS: Note: Officers with more than one job title may be intentionally listed here more than once.
Stephen H. Willard, CEO
Raphael Jorda, COO
Michel Finance, CFO
Michel Domenget, Dir.-Mktg.
Remi Meyrueix, Dir.-Scientific
Olivier Soula, Dir.-Nanotechnologies/Head-Medusa Team
Rafael Jorda, Dir.-Mfg. & Dev.
Sian Crouzet, Dir.-Admin.
Andy Francis, VP-Bus. Dev.
Charles Marlio, Dir.-Investor Rel./Strategic Planning
Sian Crouzet, Controller/Dir.-Finance
You-Ping Chan, Dir.-Chemistry Dept.
Catherine Castan, Dir.-R&D Micropump/Head-Micropump Team
Roger Kravtzoff, Dir.-Pre-clinical & Early Clinical Dev.
Katherine Hanras, Dir.-Analytical Department
Elie Vannier, Chmn.
David Weber, Dir.-Purchasing Oper.

Phone: 33-0-472-783-434	Fax: 33-0-472-783-435

Toll-Free:

Address: 33 Ave. du Docteur Georges Levy, Venissieux, 69693 France

GROWTH PLANS/SPECIAL FEATURES:
Flamel Technologies, S.A., is a drug delivery company that is engaged in the development of small molecule and protein therapeutic products for the biotechnology and pharmaceutical industries. The company operates in a 50,000 square foot cGMP (current Good Manufacturing Practice) pharmaceutical production facility in Pessac, France and produces over 2 billion tablets and capsules every year. Flamel currently holds two patented drug delivery platforms Medusa and Micropump. Medusa consists of a self-assembled poly-aminoacid nanoparticle system that enables the controlled delivery of fully human and non-denatured proteins. Medusa technology is currently being used in Basulin, a long-acting insulin for the treatment of type I diabetes; and in IFN alpha-2b XL for the treatment of hepatitis B, hepatitis C and some forms of cancers. Micropump allows controlled delivery of small molecule drugs and is geared specifically towards pediatric and geriatric customers. Micropump technology utilizes thousands of 200-500 micrometer-sized particles to release drugs through osmotic pressure at an adjustable rate. Since the active ingredients are enclosed within each micro-particle, drugs can easily be delivered in suspensions or syrups that completely mask any taste of the active pharmaceutical compound. The company has four products in the Micropump system: Genvir, a treatment for acute genital herpes; Metformin XL, a treatment for type II diabetes; Augmentin SR, an antibiotic; and ACE and protein pump inhibitors. In addition, the company owns ColCys, which consists of a family of proprietary collagen-based implantable biomaterials for post-surgical adhesion prevention. Flamel operates primarily through licensing agreements and collaborations with other pharmaceutical companies such as GlaxoSmithKline, Merck and Servier Monde. In May 2006, Flamel received a $1 million payment from GlaxoSmithKline (GSK) for the use of Flamel's microparticle technology in the development of a FDA-approved GSK drug, Coreg CR.

FINANCIALS: Sales and profits are in thousands of dollars—add 000 to get the full amount. 2006 Note: Financial information for 2006 was not available for all companies at press time.

2006 Sales: $23,020	2006 Profits: $-35,201	U.S. Stock Ticker: FLML
2005 Sales: $23,598	2005 Profits: $-27,377	Int'l Ticker: FL3 Int'l Exchange: Frankfurt-Euronext
2004 Sales: $55,410	2004 Profits: $12,499	Employees: 306
2003 Sales: $25,300	2003 Profits: $-3,300	Fiscal Year Ends: 12/31
2002 Sales: $18,400	2002 Profits: $3,000	Parent Company:

SALARIES/BENEFITS:

Pension Plan: Y	ESOP Stock Plan:	Profit Sharing:	Top Exec. Salary: $		Bonus: $
Savings Plan:	Stock Purch. Plan:		Second Exec. Salary: $		Bonus: $

OTHER THOUGHTS:
Apparent Women Officers or Directors: 2
Hot Spot for Advancement for Women/Minorities: Y

LOCATIONS: ("Y" = Yes)

West:	Southwest:	Midwest:	Southeast:	Northeast:	International:
					Y

Note: Financial information, benefits and other data can change quickly and may vary from those stated here.

FORBES MEDI-TECH INC
www.forbesmedi.com

Industry Group Code: 325411 Ranks within this company's industry group: Sales: 8   Profits: 8

Drugs:	Other:		Clinical:	Computers:	Other:
Discovery:	AgriBio:	Y	Trials/Services:	Hardware:	Specialty Services:
Licensing:	Genomics/Proteomics:		Laboratories:	Software:	Consulting:
Manufacturing: Y	Tissue Replacement:		Equipment/Supplies:	Arrays:	Blood Collection:
Development: Y			Research/Development Svcs.:	Database Management:	Drug Delivery:
Generics:			Diagnostics:		Drug Distribution:

TYPES OF BUSINESS:

Medicinals & Botanicals, Manufacturing
Natural Pharmaceuticals & Nutraceuticals
Food Additives

BRANDS/DIVISIONS/AFFILIATES:

Reducol
Vivola
Phyto-Source, LP
Forbes Fayrefield Ltd.
TheraPei Pharmaceuticals, Inc.

CONTACTS: *Note: Officers with more than one job title may be intentionally listed here more than once.*

Charles A. Butt, CEO
Charles A. Butt, Pres.
David Goold, CFO
Jeffrey J. E. Motley, VP-Mktg. & Sales
Don Nestor, Chief Scientific Officer
Laura Wessman, Sr. VP-Oper.
Jerzy Zawistowski, VP-Functional Foods & Nutraceuticals
David Stewart, VP-Regulatory Affairs & Scientific Svcs.
Don Buxton, Chmn.

Phone: 604-689-5899	Fax: 604-689-7641
Toll-Free:	
Address: 750 W. Pender St., Ste. 200, Vancouver, BC V6C 2T8 Canada	

GROWTH PLANS/SPECIAL FEATURES:

Forbes Medi-Tech, Inc. researches, develops and commercializes pharmaceutical and nutraceutical products for the prevention and treatment of cardiovascular disease. The company's core technology employs plant sterols, or phytosterols, which are derived from byproducts of the forest and other natural sources. These lipid-like compounds act as cholesterol-lowering agents and are utilized for pharmaceutical therapeutics and as functional food ingredients. Forbes operates through two sectors that consist of the nutraceuticals division and the prescription pharmaceuticals division. In nutraceuticals, the company has applied licensed technology from Novartis to develop its patented Reducol, which consists of a blend of plant sterols and plant stanols that reduce LDL-cholesterol. Reducol is currently approved for use in certain food groups and dietary supplements within the USA, Europe and Canada in vegetable spreads and oils, low-fat foods, cereals, confectionaries, juices, vegetable-fat spreads and milk drinks. In addition, Forbes has developed a phytosterol-based cooking oil called Vivola to help lower weight and cholesterol levels. Forbes's prescription pharmaceuticals operations are presently focused on the development of FM-VP4, a cholesterol-lowering amphipathic analog of phytostanol that has been proven to inhibit cholesterol absorption while lowering plasma LDL-cholesterol and total cholesterol levels. Forbes is further developing a proprietary FM-VPx library of compounds, which identifies synthetic compounds that have the potential to lower triglycerides and increase HDL. In addition, the firm also produces a FM-TP series of compounds, which targets metabolic syndromes such as diabetes and inflammatory lung diseases through various agonists and inhibitors. In June 2006, Forbes partnered with Fayrefield Foods Ltd., a U.K. producer and marketer of dairy products, to form Forbes Fayrefield Ltd., which will be devoted to distributing finished Reducol-containing food products throughout Europe. In October 2006, the firm closed its acquisition of TheraPei Pharmaceuticals, Inc., a company that focuses on the underlying causes of type II diabetes and metabolic diseases.

FINANCIALS: Sales and profits are in thousands of dollars—add 000 to get the full amount. 2006 Note: Financial information for 2006 was not available for all companies at press time.

2006 Sales: $6,200	2006 Profits: $-9,300	**U.S. Stock Ticker:** FMTI
2005 Sales: $17,700	2005 Profits: $-11,000	**Int'l Ticker:** FMI Int'l Exchange: Toronto-TSX
2004 Sales: $14,589	2004 Profits: $-5,032	Employees: 38
2003 Sales: $11,024	2003 Profits: $- 819	Fiscal Year Ends: 12/31
2002 Sales: $5,063	2002 Profits: $-2,537	Parent Company:

SALARIES/BENEFITS:

Pension Plan:	ESOP Stock Plan:	Profit Sharing:	Top Exec. Salary: $	Bonus: $
Savings Plan:	Stock Purch. Plan:		Second Exec. Salary: $	Bonus: $

OTHER THOUGHTS:

Apparent Women Officers or Directors: 2
Hot Spot for Advancement for Women/Minorities: Y

LOCATIONS: ("Y" = Yes)

West:	Southwest:	Midwest:	Southeast:	Northeast:	International: Y

FOREST LABORATORIES INC   www.frx.com

Industry Group Code: 325412   Ranks within this company's industry group: Sales: 26   Profits: 22

Drugs:		Other:		Clinical:	Computers:		Other:	
Discovery:		AgriBio:		Trials/Services:	Hardware:		Specialty Services:	
Licensing:		Genomics/Proteomics:		Laboratories:	Software:		Consulting:	
Manufacturing:	Y	Tissue Replacement:		Equipment/Supplies:	Arrays:		Blood Collection:	
Development:	Y			Research/Development Svcs.:	Database Management:		Drug Delivery:	
Generics:				Diagnostics:			Drug Distribution:	

TYPES OF BUSINESS:

Drugs, Manufacturing
Over-the-Counter Pharmaceuticals
Generic Pharmaceuticals
Antidepressants
Asthma Medications
Cardiovascular Products
OB/Gyn-Pediatrics
Endocrinology

BRANDS/DIVISIONS/AFFILIATES:

Lexapro
Namenda
Celexa
Benicar
Tiazac
AeroChamber Plus
Armour Thyroid
Aerobid

CONTACTS: Note: Officers with more than one job title may be intentionally listed here more than once.

Howard Solomon, CEO
Kenneth E. Goodman, COO
Kenneth E. Goodman, Pres.
Francis I. Perier, Jr., CFO
Elaine Hochberg, Sr. VP-Mktg.
Ivan Gergel, VP-Scientific Affairs
Charles E. Triano, VP-Investor Rel.
Francis I. Perier, Jr., Sr. VP-Finance
Ivan Gergel, Sr. VP/Pres., Forest Research Institute
Howard Solomon, Chmn.

Phone: 212-421-7850	Fax: 212-750-9152
Toll-Free: 800-947-5227	
Address: 909 3rd Ave., New York, NY 10022 US	

GROWTH PLANS/SPECIAL FEATURES:

Forest Laboratories, Inc., based in New York, identifies, develops and delivers pharmaceutical products that are used to treat a wide range of illnesses. Through its therapeutic franchises such as respiratory, pain management, ob/gyn-pediatrics, endocrinology, central nervous system and cardiovascular, the firm has come up with brands such as Lexapro, an antidepressant, Benicar for the treatment of hypertension, Namenda, a therapy for moderate or severe Alzheimer's disease and Tiazac for cardiovascular disease. The company has developed medications to treat asthma and an inhalant delivery system called AeroChamber Plus. It markets directly to physicians who have the most potential for growth and are agreeable to the introduction of new products. Forest has operations on Long Island, in New Jersey, Missouri, Ohio, Ireland and the United Kingdom. Forest Laboratories Europe is the company's European subsidiary, with offices in Dublin, Ireland and Bexley Kent, England. In the U.S., Forest Laboratories maintains eight separate labs in New York and one in New Jersey. Forest Pharmaceuticals, Inc., a wholly-owned subsidiary headquartered in St. Louis, Missouri, also manufactures and distributes branded and generic prescription and over-the counter drugs. The firm acquired Cerexa, Inc., a privately-held biopharmaceutical company based in Alameda, California, in a cash-for-stock transaction in January 2007. In February 2007, Forest began collaborating with Aurigene, a Bangalore-based discovery services company to discover small molecule drug candidates for a novel obesity and metabolic disorders target. That same month, Forest and Replidyne, Inc. terminated their collaboration agreement for the development of faropenem medoxomil, a novel oral, community antibiotic.

Employees at Forest receive financial assistance for adoption and fertility treatments, in addition to flexible spending accounts, child-care resource and referral and a commuter benefit program. The company also offers medical, dental and life insurance.

FINANCIALS: Sales and profits are in thousands of dollars—add 000 to get the full amount. 2006 Note: Financial information for 2006 was not available for all companies at press time.

2006 Sales: $2,793,934	2006 Profits: $708,514	U.S. Stock Ticker: FRX
2005 Sales: $3,052,408	2005 Profits: $838,805	Int'l Ticker: Int'l Exchange:
2004 Sales: $2,650,432	2004 Profits: $735,874	Employees: 5,050
2003 Sales: $2,206,700	2003 Profits: $622,000	Fiscal Year Ends: 3/31
2002 Sales: $1,566,600	2002 Profits: $338,000	Parent Company:

SALARIES/BENEFITS:

Pension Plan:	ESOP Stock Plan: Y	Profit Sharing: Y	Top Exec. Salary: $1,067,500	Bonus: $525,000
Savings Plan: Y	Stock Purch. Plan: Y		Second Exec. Salary: $725,500	Bonus: $330,000

OTHER THOUGHTS:

Apparent Women Officers or Directors: 3
Hot Spot for Advancement for Women/Minorities: Y

LOCATIONS: ("Y" = Yes)

West:	Southwest:	Midwest:	Southeast:	Northeast:	International:
		Y		Y	Y

Note: Financial information, benefits and other data can change quickly and may vary from those stated here.

GE HEALTHCARE   www.gehealthcare.com

Industry Group Code: 339113 Ranks within this company's industry group: Sales: Profits:

Drugs:		Other:		Clinical:		Computers:		Other:	
Discovery:	Y	AgriBio:		Trials/Services:		Hardware:		Specialty Services:	Y
Licensing:		Genomics/Proteomics:		Laboratories:		Software:	Y	Consulting:	
Manufacturing:		Tissue Replacement:		Equipment/Supplies:	Y	Arrays:	Y	Blood Collection:	
Development:				Research/Development Svcs.:		Database Management:		Drug Delivery:	
Generics:				Diagnostics:	Y			Drug Distribution:	

TYPES OF BUSINESS:

Medical Imaging & Information Technology
Magnetic Resonance Imaging Systems
Patient Monitoring Systems
Clinical Information Systems
Nuclear Medicine
Surgery & Vascular Imaging
X-Ray & Ultrasound Bone Densitometers
Clinical & Business Services

BRANDS/DIVISIONS/AFFILIATES:

GE Electric Co.
GE Healthcare Bio-Sciences
GE Healthcare Technologies
GE Healthcare Information Technologies
Centricity
Giraffe OmniBed
Innova
InstaTrak

CONTACTS: *Note: Officers with more than one job title may be intentionally listed here more than once.*

Joseph M. Hogan, CEO/Sr. VP
William Castell, Pres.
Kathryn McCarthy, CFO/Exec. VP
Jean-Michel Cossery, Chief Mktg. Officer
Mike Hanley, VP-Human Resources
James Rothman, Chief Scientific Advisor
Russel P. Meyer, CIO
William R. Clarke, CTO/Chief Medical Officer/Exec. VP
Peter Y. Solmssen, General Counsel/Exec. VP
Ralph Strosin, General Mgr.-Oper.
Michael A. Jones, Exec. VP-Bus. Dev.
Lynne Gailey, Dir.-Global Comm.
Dan Peters, CEO/Pres., Medical Diagnostics
Omar Ishrak, CEO/Pres., Clinical Systems
Peter Ehrenheim, CEO/Pres., Life Sciences
Mark Vachon, CEO/Pres., Global Diagnostic Imaging

Phone: 44-1494-544-000	Fax:
Toll-Free:	
Address: Pollards Wood, Nightingales Ln., Chalfont St. Giles, Buckinghamshire HP8 4SP UK	

GROWTH PLANS/SPECIAL FEATURES:

GE Healthcare, a subsidiary of GE, is a global leader in medical imaging and information technologies, patient monitoring systems and health care services. The company operates through seven divisions, including diagnostic imaging; global services; clinical systems; life systems; medical diagnostics; integrated information technology solutions; and interventional, cardiology and surgery. The diagnostic imaging business provides X-ray, digital mammography, computed tomography, magnetic resonance and molecular imaging technologies. GE Healthcare's global services business provides maintenance of a wide range of medical systems and devices. The clinical systems business provides technologies and services for clinicians and healthcare administrators, including ultrasound, ECG, bone densitometry, patient monitoring, incubators, infant warmers, respiratory care and anesthesia management. GE Healthcare's life sciences segment offers drug discovery, biopharmaceutical manufacturing and cellular technologies, enabling scientists and specialists around the world to discover new ways to predict, diagnose and treat disease earlier. The segment also makes systems and equipment for the purification of biopharmaceuticals. The firm's medical diagnostics business researches, manufactures and markets agents used during medical scanning procedures to highlight organs, tissue and functions inside the human body. The integrated information technology (IT) solutions business provides clinical and financial information technology solutions including enterprise and departmental IT products, revenue cycle management and practice applications, to help customers streamline healthcare costs and improve the quality of care. GE Healthcare's interventional, cardiology and surgery (ICS) business provides tools and technologies for fully integrated cardiac, surgical and interventional care. The company's major products include Centricity, a suite of applications designed to provide real-time patient information at the point of care; Giraffe OmniBed and Incubators for critically ill and premature infants; Innova 3100, a digital imaging system for visualizing the heart and fine vessels; and InstaTrak, which creates an anatomical roadmap and real-time visualization of surgical instrumentation to guide surgeons during life-critical procedures.

FINANCIALS: Sales and profits are in thousands of dollars—add 000 to get the full amount. 2006 Note: Financial information for 2006 was not available for all companies at press time.

2006 Sales: $	2006 Profits: $	U.S. Stock Ticker: Subsidiary
2005 Sales: $15,153,000	2005 Profits: $2,665,000	Int'l Ticker: Int'l Exchange:
2004 Sales: $13,456,000	2004 Profits: $2,286,000	Employees: 43,000
2003 Sales: $10,200,000	2003 Profits: $1,701,000	Fiscal Year Ends: 12/31
2002 Sales: $8,955,000	2002 Profits: $1,546,000	Parent Company: GENERAL ELECTRIC CO (GE)

SALARIES/BENEFITS:

Pension Plan: Y	ESOP Stock Plan:	Profit Sharing:	Top Exec. Salary: $	Bonus: $
Savings Plan: Y	Stock Purch. Plan:		Second Exec. Salary: $	Bonus: $

OTHER THOUGHTS:

Apparent Women Officers or Directors: 1
Hot Spot for Advancement for Women/Minorities:

LOCATIONS: ("Y" = Yes)

West:	Southwest:	Midwest:	Southeast:	Northeast:	International:
		Y			Y

Note: Financial information, benefits and other data can change quickly and may vary from those stated here.

GENAERA CORPORATION   www.genaera.com

ndustry Group Code: 325412 Ranks within this company's industry group: Sales: 162   Profits: 107

Drugs:		Other:	Clinical:	Computers:	Other:
Discovery:	Y	AgriBio:	Trials/Services:	Hardware:	Specialty Services:
Licensing:		Genomics/Proteomics:	Laboratories:	Software:	Consulting:
Manufacturing:	Y	Tissue Replacement:	Equipment/Supplies:	Arrays:	Blood Collection:
Development:	Y		Research/Development Svcs.:	Database Management:	Drug Delivery:
Generics:			Diagnostics:		Drug Distribution:

TYPES OF BUSINESS:

Drugs-Anti-Angiogenesis
Drugs-Respiratory Genomics
Drugs-Infectious Diseases
Drugs-Obesity

BRANDS/DIVISIONS/AFFILIATES:

Squalamine lactate
Trodusquemine
Lomucin
Evizon

CONTACTS: *Note: Officers with more than one job title may be intentionally listed here more than once.*

John L. Armstrong, Jr., CEO
John L. Armstrong, Jr., Pres.
Leanne M. Kelly, CFO/Sr. VP
Michael Gast, Sr. VP-Clinical Research & Dev.
Leanne M. Kelly, Sec..

Phone: 610-941-4020	Fax: 610-941-5399

Toll-Free:

Address: 5110 Campus Dr., Plymouth Meeting, PA 19462 US

GROWTH PLANS/SPECIAL FEATURES:

Genaera Corporation focuses its research and development efforts in anti-angiogenesis (prevention of new blood vessel growth), respiratory genomics, obesity and infectious diseases. Genaera has an active respiratory research and development program involving the development of a blocking antibody to interleukin-9 (IL9) to treat the root cause of asthma and the development of mucoregulators, which reduce mucin gene production in patients with chronic obstructive lung diseases. The company has ongoing clinical studies in non-small-cell lung cancer, ovarian cancer and certain other solid tumors as well. In January 2007, Genaera decided to terminate its Evizon (squalamine lactate) clinical development program in wet age-related macular degeneration. In July 2007, Genaera also decided to discontinue the Phase II study of Lomucin for the treatment of cystic fibrosis and the Phase II study of squalamine for prostate cancer and focus Company resources on the development of trodusquemine (MSI-1436) for the treatment of obesity. Trodusquemine, which is in Phase I trials, is a drug for use in appetite suppression and weight loss. The merits of this obesity drug have been demonstrated on over-sized mice.

Genaera offers its employees an educational assistance program and flexible spending accounts.

FINANCIALS: Sales and profits are in thousands of dollars—add 000 to get the full amount. 2006 Note: Financial information for 2006 was not available for all companies at press time.

2006 Sales: $ 892	2006 Profits: $-21,234	**U.S. Stock Ticker: GENR**
2005 Sales: $ 446	2005 Profits: $-26,361	**Int'l Ticker:** Int'l Exchange:
2004 Sales: $ 873	2004 Profits: $-17,873	Employees: 43
2003 Sales: $1,113	2003 Profits: $-9,393	Fiscal Year Ends: 12/31
2002 Sales: $1,500	2002 Profits: $-12,200	Parent Company:

SALARIES/BENEFITS:

Pension Plan:	ESOP Stock Plan:	Profit Sharing:	Top Exec. Salary: $400,000	Bonus: $100,000
Savings Plan: Y	Stock Purch. Plan: Y		Second Exec. Salary: $272,750	Bonus: $

OTHER THOUGHTS:

Apparent Women Officers or Directors: 1
Hot Spot for Advancement for Women/Minorities:

LOCATIONS: ("Y" = Yes)

West:	Southwest:	Midwest:	Southeast:	Northeast:	International:
				Y	

Note: Financial information, benefits and other data can change quickly and may vary from those stated here.

GENE LOGIC INC

www.genelogic.com

Industry Group Code: 541710 Ranks within this company's industry group: Sales: 15   Profits: 20

Drugs:	Other:		Clinical:		Computers:		Other:
Discovery:	AgriBio:		Trials/Services:		Hardware:		Specialty Services:
Licensing:	Genomics/Proteomics:	Y	Laboratories:		Software:		Consulting:
Manufacturing:	Tissue Replacement:		Equipment/Supplies:		Arrays:		Blood Collection:
Development:			Research/Development Svcs.:	Y	Database Management:	Y	Drug Delivery:
Generics:			Diagnostics:				Drug Distribution:

TYPES OF BUSINESS:

Gene Expression Data
Drug Discovery
Bioinformatics
Genomic Databases
Custom Discovery Databases
Contract Study Services

BRANDS/DIVISIONS/AFFILIATES:

Drug Repositioning and Selection (DRS) Program
Genomics Program
Pfizer, Inc.
Eli Lilly and Co.
Hoffman-La Roche

CONTACTS: *Note: Officers with more than one job title may be intentionally listed here more than once.*

Charles L. Dimmler, III, CEO
Charles L. Dimmler, III, Pres.
Philip L. Rohrer, Jr., CFO
Mark T. Crane, VP-Bus. Dev. & Mktg.
V. W. Brinkerhoff, III, Sr. VP/General Mgr.-Laboratories
F. Dudley Staples, Jr., General Counsel/Sr. VP/Sec.
Joanne M. Smith-Farrell, VP-Corp. Dev. & Strategy
Douglas Dolginow, Exec. VP-Pharmacogenomics
Louis A. Tartaglia, Sr. VP-Drug Repositioning & Selection
Dennis A. Rossi, Sr. VP-Genomics
Mark D. Gessler, Chmn.

Phone: 301-987-1700	**Fax:** 301-987-1701
Toll-Free: 800-436-3564	
Address: 50 W. Watkins Mill Rd., Gaithersburg, MD 20878 US	

GROWTH PLANS/SPECIAL FEATURES:

Gene Logic is a provider of drug discovery and drug development services for the pharmaceutical and biotechnology industries. The firm organizes its services into two segments: Drug Repositioning/Selection and Genomics Services. The Drug Repositioning and Selection (DRS) Program serves the pharmaceutical industry by finding new uses for discontinued compounds and finding the widest range of therapeutic applications for currently marketed compounds. Firms who partner with Gene Logic are able to share the risks of launching new compounds, and also have access to the resources of Gene Logic's drug developers. Pharmaceutical companies and biotechnology companies both make use of Gene Logic's Genomic Services, which help accelerate the drug development process through the firm's genomic informatics capabilities. Gene Logic's genomic services consist of gene reference and clinical database systems; software; toxicity consulting services; gene expression; microarray data generation; and outsourcing services. Gene Logic allows companies to purchase subscriptions to these gene expression and toxicogenomics databases. The firm is headquartered in Maryland and maintains partnerships with drug makers Pfizer, Inc.; Eli Lilly and Co.; and Hoffman-La Roche. In December 2006, the firm sold its Preclinical Division to Bridge Pharmaceuticals, a global contract research organization, for $15 million. Additionally, the firm entered into an agreement with the U.S. Food and Drug Administration (FDA), whereby the firm will provide the FDA with access to genomics and software data, including raw classical toxicology data. In early 2007, the firm announced plans to restructure its genomics segment, focusing chiefly on finding new therapeutic uses for drug candidates that are safe but have discontinued for various reasons.

Gene Logic offers its employees a health plan, tuition assistance/reimbursement and an annual bonus plan.

FINANCIALS: Sales and profits are in thousands of dollars—add 000 to get the full amount. 2006 Note: Financial information for 2006 was not available for all companies at press time.

2006 Sales: $24,346	2006 Profits: $-54,710	**U.S. Stock Ticker: GLGC**
2005 Sales: $57,190	2005 Profits: $-48,304	**Int'l Ticker:** Int'l Exchange:
2004 Sales: $52,171	2004 Profits: $-28,520	Employees: 151
2003 Sales: $69,519	2003 Profits: $-24,771	Fiscal Year Ends: 12/31
2002 Sales: $54,800	2002 Profits: $-24,100	Parent Company:

SALARIES/BENEFITS:

Pension Plan:	ESOP Stock Plan:	Profit Sharing:	Top Exec. Salary: $400,000	Bonus: $142,074
Savings Plan: Y	Stock Purch. Plan: Y		Second Exec. Salary: $280,000	Bonus: $75,000

OTHER THOUGHTS:

Apparent Women Officers or Directors: 1
Hot Spot for Advancement for Women/Minorities:

LOCATIONS: ("Y" = Yes)

West:	Southwest:	Midwest:	Southeast:	Northeast:	International:
				Y	

GENELABS TECHNOLOGIES INC

www.genelabs.com

ndustry Group Code: 325412 Ranks within this company's industry group: Sales: 127   Profits: 71

rugs:		Other:	Clinical:	Computers:	Other:
iscovery:	Y	AgriBio:	Trials/Services:	Hardware:	Specialty Services:
icensing:		Genomics/Proteomics:	Laboratories:	Software:	Consulting:
anufacturing:		Tissue Replacement:	Equipment/Supplies:	Arrays:	Blood Collection:
evelopment:	Y		Research/Development Svcs.:	Database Management:	Drug Delivery:
eneriics:			Diagnostics:		Drug Distribution:

YPES OF BUSINESS:

rugs-DNA-binding
rugs-Small Molecule
harmaceutical Preparations

BRANDS/DIVISIONS/AFFILIATES:

Prestara

CONTACTS: *Note: Officers with more than one job title may be ntentionally listed here more than once.*

ames A. D. Smith, CEO
ames A. D. Smith, Pres.
onald C. Griffith, Chief Scientific Officer
eather C. Keller, Corp. Sec.
eather C. Keller, Sr. Bus. Strategy Advisor
oy J. Wu, VP-Bus. Dev.
enneth E. Schwartz, VP-Medical Affairs
ene A. Chow, Chmn.

Phone: 650-369-9500	Fax: 650-368-0709

oll-Free:

ddress: 505 Penobscot Dr., Redwood City, CA 94063-4738 US

GROWTH PLANS/SPECIAL FEATURES:

Genelabs Technologies, Inc. focuses its research on the discovery of novel antibacterial, antifungal, and antiviral agents that function by regulating gene expression and uses its proprietary DNA-binding technology for the discovery and development of novel small-molecule compounds. It focuses on discovering compounds that target DNA and determining compounds that restrain the hepatitis C virus, as well as on therapeutic intervention through drugs that prevent protein binding in disease-invoking genes. The company's chief drug, Prestara, is a proprietary hormone treatment for systemic lupus in women, which is currently in Phase III FDA testing. Genelabs' core research capabilities include medicinal chemistry, combinatorial chemistry, computational modeling, molecular biology, assay development and high-throughput screening drug metabolism and pharmacokinetics. Genelabs collaborates in research and development with Watson Pharmaceuticals, Inc.; Genovate Biotechnology Co., Ltd.; Abbott Laboratories, Inc.; and recently with Novartis for a Hepatitis C Virus.

Genelabs Technologies offers employees a benefits package that includes medical, dental, vision, and life insurance; a flexible benefit plan; tuition reimbursement; bonus incentives and cash awards for employees who accomplish special achievements; an employee stock purchase plan; stock options; an employee referral program; a confidential employee assistance program, access to credit unions, personal and work-related counseling; and a fitness and recreation center.

FINANCIALS: Sales and profits are in thousands of dollars—add 000 to get the full amount. 2006 Note: Financial information for 2006 was not available for all companies at press time.

006 Sales: $11,209	2006 Profits: $-8,685	U.S. Stock Ticker: GNLB
005 Sales: $6,849	2005 Profits: $-10,842	Int'l Ticker: Int'l Exchange:
004 Sales: $5,556	2004 Profits: $-13,511	Employees: 67
003 Sales: $2,916	2003 Profits: $-19,807	Fiscal Year Ends: 12/31
002 Sales: $3,600	2002 Profits: $-16,000	Parent Company:

SALARIES/BENEFITS:

ension Plan:	ESOP Stock Plan:	Profit Sharing:	Top Exec. Salary: $334,500	Bonus: $200,233
avings Plan: Y	Stock Purch. Plan: Y		Second Exec. Salary: $284,750	Bonus: $111,523

OTHER THOUGHTS:

pparent Women Officers or Directors: 2
ot Spot for Advancement for Women/Minorities: Y

LOCATIONS: ("Y" = Yes)

West:	Southwest:	Midwest:	Southeast:	Northeast:	International:
Y					

Note: Financial information, benefits and other data can change quickly and may vary from those stated here.

GENENCOR INTERNATIONAL INC

www.genencor.com

Industry Group Code: 325414 Ranks within this company's industry group: Sales: Profits:

Drugs:		Other:		Clinical:	Computers:		Other:	
Discovery:	Y	AgriBio:		Trials/Services:	Hardware:		Specialty Services:	Y
Licensing:		Genomics/Proteomics:	Y	Laboratories:	Software:		Consulting:	
Manufacturing:	Y	Tissue Replacement:	Y	Equipment/Supplies:	Arrays:		Blood Collection:	
Development:	Y			Research/Development Svcs.:	Database Management:		Drug Delivery:	
Generics:				Diagnostics:			Drug Distribution:	

TYPES OF BUSINESS:

Biological Manufacturing
Biologicals Research & Discovery
Protein-Based Products
Proteomics
Cancer Treatments

BRANDS/DIVISIONS/AFFILIATES:

Danisco A/S
STARGEN
Clarase G Plus
Protax 6L
Spezyme Fred,
Optimax 4060 VHP
Gensweet IGI

CONTACTS: *Note: Officers with more than one job title may be intentionally listed here more than once.*

Tjerk De Ruiter, CEO
Thomas J. Pekich, Pres.
Jim Sjoerdsma, VP-Human Resources
Michael V. Arbige, Sr. VP-Tech.
Carole B. Cobb, VP-Global Supply

Phone: 650-846-7500	Fax: 650-845-6500
Toll-Free:	
Address: 925 Page Mill Rd., Palo Alto, CA 94304 US	

GROWTH PLANS/SPECIAL FEATURES:

Genencor International, Inc. discovers, develops and sells biocatalysts and other biochemicals for the health care, agricultural, industrial chemical and consumer markets. From its nine manufacturing centers in the U.S., Argentina, Belgium, China and Finland, the company is able to deliver 250 products to customers around the world. Genencor works in the Agri-processing market to serve customers who process agricultural raw materials such as barley, corn, wheat and soybeans to produce animal feeds, food, food ingredients and renewable fuels. Examples of products that Genecor has developed to enhance its client's agricultural products include Clarase G Plus, Protax 6L, Spezyme Fred, Optimax 4060 VHP and Gensweet IGI. For the consumer market, the company develops a variety of enzymes that are used in a variety of applications within fabric and household care, which include protein-degrading enzymes, such as proteases, starch degrading enzymes, such as amylases and cellulose degrading enzymes, such as cellulases to remove recalcitrant stains more efficiently than soaps and detergents alone. Within the consumer market, Genecor also serves textiles industries by producing cellulases, amylases and proteases for applications such as denim finishing, biofinishing of cotton and cellulosics and desizing and treatment of wool and silk. The company works to serve the industrial market by developing novel biochemicals, biomaterials and alternate sources of energy, such as ethanol, from natural resources. In April 2007, Genecor opened a new production facility in Wuxi, China.

FINANCIALS: Sales and profits are in thousands of dollars—add 000 to get the full amount. 2006 Note: Financial information for 2006 was not available for all companies at press time.

2006 Sales: $	2006 Profits: $	**U.S. Stock Ticker: Subsidiary**
2005 Sales: $	2005 Profits: $	**Int'l Ticker:** Int'l Exchange:
2004 Sales: $410,417	2004 Profits: $26,178	Employees: 1,271
2003 Sales: $383,162	2003 Profits: $22,808	Fiscal Year Ends: 12/31
2002 Sales: $350,100	2002 Profits: $5,800	Parent Company: DANISCO AS

SALARIES/BENEFITS:

Pension Plan:	ESOP Stock Plan:	Profit Sharing:	Top Exec. Salary: $540,346	Bonus: $370,900
Savings Plan: Y	Stock Purch. Plan:		Second Exec. Salary: $337,888	Bonus: $172,800

OTHER THOUGHTS:

Apparent Women Officers or Directors: 1
Hot Spot for Advancement for Women/Minorities:

LOCATIONS: ("Y" = Yes)

West:	Southwest:	Midwest:	Southeast:	Northeast:	International:
Y		Y		Y	Y

GENENTECH INC   www.gene.com

Industry Group Code: 325412   Ranks within this company's industry group: Sales: 16   Profits: 13

Drugs:		Other:		Clinical:		Computers:		Other:	
Discovery:	Y	AgriBio:		Trials/Services:		Hardware:		Specialty Services:	
Licensing:		Genomics/Proteomics:		Laboratories:		Software:		Consulting:	
Manufacturing:	Y	Tissue Replacement:		Equipment/Supplies:		Arrays:		Blood Collection:	
Development:	Y			Research/Development Svcs.:		Database Management:		Drug Delivery:	
Generics:				Diagnostics:				Drug Distribution:	

TYPES OF BUSINESS:
Drug Development & Manufacturing
Genetically Engineered Drugs

BRANDS/DIVISIONS/AFFILIATES:
Avastin
TNKase
Herceptin
Rituxan
Activase
Pulmozyme
Nutropin

CONTACTS: *Note: Officers with more than one job title may be intentionally listed here more than once.*
Arthur D. Levinson, CEO
Myrtle S. Potter, COO
Arthur D. Levinson, Pres.
David A. Ebersman, CFO/Exec. VP
Richard H. Scheller, Exec. VP-Research
Susan Desmond-Hellmann, Pres., Prod. Dev.
Stephen G. Juelsgaard, Exec. VP/General Counsel
Ian T. Clark, Exec. VP-Comm. Oper.
Stephen G. Juelsgaard, Corp. Sec.
Patrick Y. Yang, Exec. VP-Product Oper.
Arthur D. Levinson, Chmn.

Phone: 650-225-1000	Fax: 650-225-6000
Toll-Free:	
Address: 1 DNA Way, South San Francisco, CA 94080 US	

GROWTH PLANS/SPECIAL FEATURES:

Genentech, Inc. makes medicines by splicing genes into fast-growing bacteria that then produce therapeutic proteins and combat diseases on a molecular level. Genentech uses cutting-edge technologies such as computer visualization of molecules, micro arrays and sensitive assaying techniques to develop, manufacture and market pharmaceuticals for unmet medical needs. Genentech's research is directed toward the oncology, immunology and vascular biology fields. The company's products consist of a variety of cardio-centric medications, as well as cancer, growth hormone deficiency (GHD) and cystic fibrosis treatments. Biotechnology products offered by Genentech include Herceptin, used to treat metastatic breast cancers; Avastin, used to inhibit angiogenesis of solid-tumor cancers Nutropin, a growth hormone for the treatment of GHD in children and adults; TNKase, for the treatment of acute myocardial infarction; and Pulmozyme, for the treatment of cystic fibrosis. The company also produces the Rituxan antibody, used for the treatment of patients with non-Hodgkin's lymphoma. Through its long-standing Genentech Access to Care Foundation, Genentech assists those without sufficient health insurance to receive its medicines. The company's scientists publish around 250 scientific papers per year and are regarded as among the most prolific in the industry. In 2006, 18% of operating revenues (roughly $1.6 billion) was dedicated to research and development. In November 2006, the FDA approved Herceptin for treatment of a specific kind of breast cancer. The firm recently agreed to acquire Tanox, a firm that focuses on monoclonal antibody technology. Roche Holdings, Ltd. owns 55.8% of Genentech.

For the last nine years, the company has been named to Fortune Magazine's 100 Best Companies to Work For. Every Friday evening, Genentech hosts socials called Ho-Hos, providing free food, beverages and a chance to socialize with co-workers.

FINANCIALS: Sales and profits are in thousands of dollars—add 000 to get the full amount. 2006 Note: Financial information for 2006 was not available for all companies at press time.

2006 Sales: $9,284,000	2006 Profits: $2,113,000	U.S. Stock Ticker: DNA
2005 Sales: $6,633,372	2005 Profits: $1,278,991	Int'l Ticker: Int'l Exchange:
2004 Sales: $4,621,157	2004 Profits: $784,816	Employees: 10,533
2003 Sales: $2,799,400	2003 Profits: $562,527	Fiscal Year Ends: 12/31
2002 Sales: $2,252,300	2002 Profits: $63,800	Parent Company:

SALARIES/BENEFITS:

Pension Plan:	ESOP Stock Plan:	Profit Sharing:	Top Exec. Salary: $995,000	Bonus: $2,725,000
Savings Plan: Y	Stock Purch. Plan: Y		Second Exec. Salary: $625,000	Bonus: $870,000

OTHER THOUGHTS:
Apparent Women Officers or Directors: 2
Hot Spot for Advancement for Women/Minorities: Y

LOCATIONS: ("Y" = Yes)

West:	Southwest:	Midwest:	Southeast:	Northeast:	International:
Y					Y

Note: Financial information, benefits and other data can change quickly and may vary from those stated here.

GENEREX BIOTECHNOLOGY   www.generex.com

Industry Group Code: 325412A Ranks within this company's industry group: Sales: 19   Profits: 21

Drugs:		Other:	Clinical:	Computers:	Other:	
Discovery:	Y	AgriBio:	Trials/Services:	Hardware:	Specialty Services:	
Licensing:		Genomics/Proteomics:	Laboratories:	Software:	Consulting:	
Manufacturing:		Tissue Replacement:	Equipment/Supplies:	Arrays:	Blood Collection:	
Development:	Y		Research/Development Svcs.:	Database Management:	Drug Delivery:	Y
Generics:			Diagnostics:		Drug Distribution:	

TYPES OF BUSINESS:

Drug Delivery Systems
Buccal Drug Delivery Systems
Diabetes Treatment-Insulin
Infectious & Autoimmune Disease Treatments
Large-Molecule Drug Delivery Systems

BRANDS/DIVISIONS/AFFILIATES:

RapidMist
Oralin
Antigen Express, Inc.
BaBOOM!
GlucoBreak

CONTACTS: *Note: Officers with more than one job title may be intentionally listed here more than once.*

Anna E. Gluskin, CEO
Rose C. Perri, COO
Anna E. Gluskin, Pres.
Rose C. Perri, CFO/Treas./Corp. Sec.
Gerald Bernstein, VP-Medical Affairs
Mark Fletcher, General Counsel/Exec. VP
Slava Jarnitskii, Controller
Eric von Hofe, Pres., Antigen Express
Roberto Cid, VP-Mktg. & Sales Latin/South America

Phone: 416-364-2551	**Fax:** 416-364-9363
Toll-Free:	
Address: 33 Harbour Sq., Ste. 202, Toronto, ON M5J 2G2 Canada	

GROWTH PLANS/SPECIAL FEATURES:

Generex Biotechnology Corporation is engaged in the research and development of drug delivery technologies. The company is focused on formulations to administer large molecule drugs orally (through buccal mucosa, particularly the inner cheek walls) using its RapidMist hand-held aerosol applicator. Oral-Lyn is a formulation of human insulin designed for buccal delivery. It is commercially approved in Ecuador and is in Phase III studies in the U.S. and Canada. The company is also developing buccal formulations of morphine and of the synthetic opioid analgesic fentanyl. These formulations, like the others, are designed to ease patients onto medication without invasive procedures (injections) and with the benefit of being self-administered. Generex also owns Antigen Express, Inc., a Massachusetts-based company that develops treatments for malignant, infectious, autoimmune and allergic diseases, including melanoma, breast and prostate cancer, HIV, influenza, smallpox, SARS and Type I diabetes. Antigen's product candidates are in pre-clinical stages of development. In recent news, Generex entered into a licensing and distribution agreement with the Armenian Development Agency and Canada Armenia Trading House Ltd. for the commercialization of Oral-lyn in the Republic of Armenia, Georgia, and the Republic of Kazakhstan. The company launched two new products in May 2007: BaBOOM!, an energy spray, and GlucoBreak, a glucose-spray diet aid.

FINANCIALS: Sales and profits are in thousands of dollars—add 000 to get the full amount. 2006 Note: Financial information for 2006 was not available for all companies at press time.

2006 Sales: $ 200	2006 Profits: $-68,000	**U.S. Stock Ticker: GNBT**	
2005 Sales: $ 392	2005 Profits: $-24,002	**Int'l Ticker:** Int'l Exchange:	
2004 Sales: $ 627	2004 Profits: $-18,363	Employees: 23	
2003 Sales: $	2003 Profits: $-13,262	Fiscal Year Ends: 7/31	
2002 Sales: $	2002 Profits: $-13,700	Parent Company:	

SALARIES/BENEFITS:

Pension Plan:	ESOP Stock Plan:	Profit Sharing:	Top Exec. Salary: $425,000	Bonus: $206,125
Savings Plan:	Stock Purch. Plan:		Second Exec. Salary: $325,000	Bonus: $157,625

OTHER THOUGHTS:

Apparent Women Officers or Directors: 2
Hot Spot for Advancement for Women/Minorities: Y

LOCATIONS: ("Y" = Yes)

West:	Southwest:	Midwest:	Southeast:	Northeast:	International:
				Y	Y

Note: Financial information, benefits and other data can change quickly and may vary from those stated here.

GEN-PROBE INC   www.gen-probe.com

Industry Group Code: 325413 Ranks within this company's industry group: Sales: 5   Profits: 4

Drugs:	Other:	Clinical:		Computers:	Other:
Discovery:	AgriBio:	Trials/Services:		Hardware:	Specialty Services:
Licensing:	Genomics/Proteomics:	Laboratories:		Software:	Consulting:
Manufacturing:	Tissue Replacement:	Equipment/Supplies:		Arrays:	Blood Collection:
Development:		Research/Development Svcs.:		Database Management:	Drug Delivery:
Generics:		Diagnostics:	Y		Drug Distribution:

TYPES OF BUSINESS:

Medical Diagnostics Products
Diagnostic Tests
Blood Screening Assays
Services-Commercial Physical Research
Services-Commercial Biological Research

BRANDS/DIVISIONS/AFFILIATES:

TIGRIS
Novartis Corp.
Procleix
PROCELIEX ULTRIO

CONTACTS: Note: Officers with more than one job title may be intentionally listed here more than once.

Henry L. Nordhoff, CEO
Carl W. Hull, COO/Exec. VP
Henry L. Nordhoff, Pres.
Herm Rosenman, CFO/VP-Finance
Stephen J. Kondor, Sr. VP-Mktg. & Sales
Diana De Walt, VP-Human Resources
Daniel L. Kacian, Exec. VP/Chief Scientist
Valerie M. Day, VP-Product Dev.
R. William Bowen, General Counsel/VP/Secretary
Niall M. Conway, Exec. VP-Oper.
Martin B. Edelshain, VP-Strategic Planning & Bus. Dev.
Don Tartre, VP-Finance/Controller
Robert B. Blake, VP-Instrument Systems
Paul Gargan, VP-Bus. Dev.
Christina Yang, Sr. VP-Clinical/Regulatory/Quality
Pete Shearer, VP-Intellectual Property
Henry L. Nordhoff, Chmn.
Gurney Lashley, VP-Supply Chain Mgmt.

Phone: 858-410-8000	Fax: 800-288-3141

Toll-Free: 800-523-5001
Address: 10210 Genetic Center Dr., San Diego, CA 92121-4362
US

GROWTH PLANS/SPECIAL FEATURES:

Gen-Probe, Inc. develops, manufactures and markets nucleic acid testing (NAT) products for clinical diagnosis of human diseases and screening of human blood donations. Gen-Probe has over 200 U.S. patents including one for its automated nucleic acid sequence isolation system, TIGRIS; and has received FDA approval for over 60 products, including the company's WNV (West Nile Virus) assay on aSAS to screen donated human blood, and tests to detect microorganisms indicative of sexually transmitted diseases, tuberculosis, strep throat, pneumonia and fungal infections. The company also developed, and now manufactures, the only FDA-approved blood screening assay for the simultaneous detection of HIV-1 and HCV (hepatitis C virus), presently marketed by Novartis Corporation and used by blood collection agencies like the American Red Cross and America's Blood Centers. Products are also marketed to clinical laboratories, public health institutions and hospitals in the U.S. and Canada. Gen-Probe plans to continue developing its current market positions, while also seeking out applications for its proprietary NAT technologies, potentially to include viral load testing, cancer monitoring, pharmacogenomics, industrial uses and bioterrorism-related testing. In partnership with Chiron, Gen-Probe developed the Procleix WNV blood screening assay, approved by the FDA in March 2007. In May 2007, the FDA approved the PROCLEIX TIGRIS system for use with the PROCELIEX ULTRIO assay. The system can process 1,000 blood samples in 14 hours.

Gen-Probe offers a number of employee benefits, including income protection, paid time off, an employee assistance plan, flexible spending accounts and an on-site cafeteria. In addition, the company has a tuition assistance program, many in-house training opportunities and the Gen-Probe E-Learning program, through which employees can take over 100 online courses to develop their skills at their own pace. Gen-Probe also has many social activities for its employees, in-house exercise classes, an on-site gym and car wash.

FINANCIALS: Sales and profits are in thousands of dollars—add 000 to get the full amount. 2006 Note: Financial information for 2006 was not available for all companies at press time.

2006 Sales: $354,764	2006 Profits: $59,498	U.S. Stock Ticker: GPRO
2005 Sales: $305,965	2005 Profits: $60,089	Int'l Ticker: Int'l Exchange:
2004 Sales: $269,707	2004 Profits: $54,575	Employees: 925
2003 Sales: $207,191	2003 Profits: $35,330	Fiscal Year Ends: 12/31
2002 Sales: $155,597	2002 Profits: $13,007	Parent Company:

SALARIES/BENEFITS:

Pension Plan:	ESOP Stock Plan:	Profit Sharing:	Top Exec. Salary: $645,000	Bonus: $470,000
Savings Plan: Y	Stock Purch. Plan: Y		Second Exec. Salary: $363,000	Bonus: $110,000

OTHER THOUGHTS:

Apparent Women Officers or Directors: 3
Hot Spot for Advancement for Women/Minorities: Y

LOCATIONS: ("Y" = Yes)

West:	Southwest:	Midwest:	Southeast:	Northeast:	International:
Y					

Note: Financial information, benefits and other data can change quickly and may vary from those stated here.

GENTA INC

www.genta.com

Industry Group Code: 325412 Ranks within this company's industry group: Sales: 165 Profits: 162

Drugs:		Other:	Clinical:	Computers:	Other:
Discovery:	Y	AgriBio:	Trials/Services:	Hardware:	Specialty Services:
Licensing:		Genomics/Proteomics:	Laboratories:	Software:	Consulting:
Manufacturing:		Tissue Replacement:	Equipment/Supplies:	Arrays:	Blood Collection:
Development:	Y		Research/Development Svcs.:	Database Management:	Drug Delivery:
Generics:			Diagnostics:		Drug Distribution:

TYPES OF BUSINESS:

Anticancer & Related Diseases Drugs
Antisense Drugs

BRANDS/DIVISIONS/AFFILIATES:

Genasense
Ganite

CONTACTS: *Note: Officers with more than one job title may be intentionally listed here more than once.*

Raymond P. Warrell, Jr., CEO
Richard J. Moran, CFO/Sr. VP
Bob D. Brown, VP-Research & Tech.
Thomas N. Julian, VP-Tech. Dev.
Bharat M. Mehta, VP-Mfg. Oper.
Loretta M. Itri, Pres., Pharmaceutical Dev./Chief Medical Officer
Lloyd Sanders, Sr. VP-Commercial Oper.
Raymond P. Warrell, Jr., Chmn.

Phone: 908-286-9800	Fax: 908-464-1701
Toll-Free:	
Address: 200 Connell Dr., Berkeley Heights, NJ 07922 US	

GROWTH PLANS/SPECIAL FEATURES:

Genta, Inc. is a biopharmaceutical company engaged in pharmaceutical research and development of drugs for the treatment of cancer and related diseases. The company's research portfolio consists of two major programs: DNA/RNA medicines and small molecules. The DNA/RNA medicines program includes drugs that are based on using modifications of either DNA or RNA as drugs that can be used to treat disease. The program includes technologies such as antisense, decoys and small interfering or micro RNAs. The lead drug from this program is an investigational antisense compound knows as Genasense. Genasense is designed to block the production of a protein known as Bcl-2, a protein that fortifies cancer cells against treatment, and is being developed primarily as a means of amplifying the cytotoxic effects of other anticancer treatments. The small molecules program includes drugs that are based on gallium-containing compounds. The lead drug from this program is Ganite, which is FDA approved for the treatment of patients with symptomatic cancer-related hypercalcemia that is resistant to hydration. The firm is engaged in developing new formulations of gallium-containing compounds that may be orally absorbed. Genta owns 85 patents and has 87 pending patents applications in the U.S. and foreign countries.

The company offers its employees benefits that include a 401(k) plan; medical, dental and life insurance; and a stock options plan.

FINANCIALS: Sales and profits are in thousands of dollars—add 000 to get the full amount. 2006 Note: Financial information for 2006 was not available for all companies at press time.

2006 Sales: $ 708	2006 Profits: $-56,781	U.S. Stock Ticker: GNTAD
2005 Sales: $26,585	2005 Profits: $-2,203	Int'l Ticker: Int'l Exchange:
2004 Sales: $14,615	2004 Profits: $-32,685	Employees: 55
2003 Sales: $6,659	2003 Profits: $-50,109	Fiscal Year Ends: 12/31
2002 Sales: $3,600	2002 Profits: $-74,500	Parent Company:

SALARIES/BENEFITS:

Pension Plan:	ESOP Stock Plan:	Profit Sharing:	Top Exec. Salary: $460,000	Bonus: $50,000
Savings Plan: Y	Stock Purch. Plan: Y		Second Exec. Salary: $445,200	Bonus: $

OTHER THOUGHTS:

Apparent Women Officers or Directors: 2
Hot Spot for Advancement for Women/Minorities: Y

LOCATIONS: ("Y" = Yes)

West:	Southwest:	Midwest:	Southeast:	Northeast:	International:
				Y	

GENVEC INC

www.genvec.com

Industry Group Code: 325412 Ranks within this company's industry group: Sales: 111   Profits: 103

Drugs:		Other:		Clinical:	Computers:	Other:
Discovery:	Y	AgriBio:		Trials/Services:	Hardware:	Specialty Services:
Licensing:		Genomics/Proteomics:	Y	Laboratories:	Software:	Consulting:
Manufacturing:	Y	Tissue Replacement:		Equipment/Supplies:	Arrays:	Blood Collection:
Development:	Y			Research/Development Svcs.:	Database Management:	Drug Delivery:
Generics:				Diagnostics:		Drug Distribution:

TYPES OF BUSINESS:
Gene-Based Therapeutic Drugs & Vaccines
Cancer Drugs

BRANDS/DIVISIONS/AFFILIATES:
TNFerade
AdPEDF
TherAtoh
Biobypass

CONTACTS: *Note: Officers with more than one job title may be intentionally listed here more than once.*
Paul H. Fischer, CEO
Paul H. Fischer, Pres.
Douglas J. Swirsky, CFO
C. Richter King, Sr. VP-Research
Mark O. Thornton, Sr. VP-Prod. Dev.
Douglas J. Swirsky, Treas./Sec.
Milan Kovacevic, VP-Clinical Oper.
Bryan T. Butman, VP-Vector Oper.
Zola P. Horovitz, Chmn.

Phone: 240-632-0740	Fax: 240-632-0735

Toll-Free: 877-943-6832
Address: 65 W. Watkins Mill Rd., Gaithersburg, MD 20878 US

GROWTH PLANS/SPECIAL FEATURES:
GenVec, Inc. is a biopharmaceutical company that develops gene-based therapeutic drugs and vaccines. The company's lead product candidate, TNFerade biologic, is being developed for use in the treatment of cancer. The drug is an adenovector, or DNA carrier, and is administered directly into tumors. TNFerade is currently the subject of a randomized, controlled Phase II/III pivotal trial for first line-treatment of inoperable, locally-advanced pancreatic cancer. The firm received positive interim data for Phase II trials. TNFerade is also being evaluated for possible use in the treatment of other types of cancer. Other therapeutic programs include AdPEDF, developed for wet age-related macular degeneration, which causes blindness; and TherAtoh, developed for delivering the human atonal gene to trigger the production of therapeutic proteins by cells in the inner ear. The company discontinued patient enrollment in the Biobypass clinical trial for treatment of severe coronary artery disease. GenVec and its collaborators also have multiple vaccines in development, which use the firm's adenovector technology. The company has collaborations with the National Institute of Allergy and Infectious Diseases (NIAID) to develop an HIV vaccine; the U.S. Naval Medical Research Center and the PATH Malaria Vaccine Initiative to develop vaccines for malaria; and with the U.S. Department of Homeland Security and the U.S. Department of Agriculture to develop for food-and-mouth diseases. The firm has access to 272 issued, allowed or pending patents worldwide. In June 2007, GenVec announced that an adenovector-based vaccine for the prevention of HIV entered a Phase I clinical trial at the Vaccine Research Center of the NIAID.

The company offers its employees medical, dental and vision insurance; life, AD&D, short- and long-term disability insurance; a 401(k); an employee assistance program; tuition reimbursement; and an employee stock purchase program. Additional benefits include car washes, dry cleaning services and credit union membership.

FINANCIALS: Sales and profits are in thousands of dollars—add 000 to get the full amount. 2006 Note: Financial information for 2006 was not available for all companies at press time.

2006 Sales: $18,923	2006 Profits: $-19,272	U.S. Stock Ticker: GNVC
2005 Sales: $26,554	2005 Profits: $-13,992	Int'l Ticker: Int'l Exchange:
2004 Sales: $11,853	2004 Profits: $-18,894	Employees: 109
2003 Sales: $10,520	2003 Profits: $-21,261	Fiscal Year Ends: 12/31
2002 Sales: $8,414	2002 Profits: $-25,598	Parent Company:

SALARIES/BENEFITS:
Pension Plan:	ESOP Stock Plan:	Profit Sharing:	Top Exec. Salary: $344,693	Bonus: $100,000
Savings Plan: Y	Stock Purch. Plan: Y		Second Exec. Salary: $216,098	Bonus: $46,000

OTHER THOUGHTS:
Apparent Women Officers or Directors: 1
Hot Spot for Advancement for Women/Minorities:

LOCATIONS: ("Y" = Yes)
West:	Southwest:	Midwest:	Southeast:	Northeast:	International:
				Y	

GENZYME BIOSURGERY   www.genzymebiosurgery.com

Industry Group Code: 339113 Ranks within this company's industry group: Sales: 2   Profits: 3

Drugs:	Other:	Clinical:	Computers:	Other:
Discovery:	AgriBio:	Trials/Services:	Hardware:	Specialty Services:
Licensing:	Genomics/Proteomics:	Laboratories:	Software:	Consulting:
Manufacturing:	Tissue Replacement: Y	Equipment/Supplies: Y	Arrays:	Blood Collection:
Development:		Research/Development Svcs.:	Database Management:	Drug Delivery:
Generics:		Diagnostics:		Drug Distribution:

TYPES OF BUSINESS:

Equipment-Surgery & Orthopedic Products
Burn Treatment Products
Biomaterials & Biotherapeutics

GROWTH PLANS/SPECIAL FEATURES:

Genzyme Biosurgery (GB), a division of Genzyme Corp., develops and markets a portfolio of devices, biomaterials and biotherapeutics primarily for the general surgery market. GB's products are used for osteoarthritis relief, adhesion prevention, hernia repair, cartilage repair, burn treatment and bulk sodium hyaluronate (HA) powder. Synvisc, a treatment for the pain and immobility associated with osteoarthritis of the knee, is an elastic viscous fluid that acts as a shock absorber and lubricant for the knee joint when injected. Genzyme's adhesion prevention products include Seprafilm adhesion barrier, used in abdominal and pelvic surgery; Sepramesh Biosurgical Composite, used in hernia repair surgery; and Seprapack and Sepragel Sinus, which are used after nasal or sinus surgery. Carticel, which is used to treat damaged articular cartilage in the knee, is an injection of a patient's cultured cartilage cells into a knee that has had an inadequate response to a prior arthroscopic or other surgical repair procedure. Epicel, made from cultured epidermal autografts grown from the patient's own skin cells in approximately 16 days, is a permanent skin replacement product for severe burn victims. HyluMed is GB's sterile and medical grade HA powder, which is useful in helping the body adapt to medical implants and to accept medications. GB's research pipeline currently includes early studies for third generation cartilage repair and viscosupplement products, as well as Phase III trials for expanded uses of Synvisc. The company markets its products directly to physicians and hospital administrators throughout the U.S. and Europe. It also uses Genzyme Corp.'s network of distributors to sell certain products in the U.S., Europe, Asia and Latin America.

BRANDS/DIVISIONS/AFFILIATES:

Genzyme Corp.
Synvisc
Seprafilm
Seprapack
Sepragel Sinus
Carticel
Epicel
HyluMed

CONTACTS: Note: Officers with more than one job title may be intentionally listed here more than once.

C. Ann Merrifield, Pres.
Ellen C. Reifsneider, VP-Human Resources
Caren Arnstein, VP-Corp. Comm.
Henri A. Termeer, Chmn./CEO/Pres., Genzyme Corp.

Phone: 617-252-7500	Fax: 617-252-7600
Toll-Free:	
Address: 55 Cambridge Pkwy., Cambridge, MA 02142 US	

FINANCIALS: Sales and profits are in thousands of dollars—add 000 to get the full amount. 2006 Note: Financial information for 2006 was not available for all companies at press time.

2006 Sales: $	2006 Profits: $	U.S. Stock Ticker: Subsidiary
2005 Sales: $	2005 Profits: $	Int'l Ticker: Int'l Exchange:
2004 Sales: $	2004 Profits: $	Employees: 3,800
2003 Sales: $	2003 Profits: $	Fiscal Year Ends: 12/31
2002 Sales: $240,100	2002 Profits: $-177,600	Parent Company: GENZYME CORP

SALARIES/BENEFITS:

Pension Plan:	ESOP Stock Plan:	Profit Sharing:	Top Exec. Salary: $1,300,000	Bonus: $1,770,000
Savings Plan: Y	Stock Purch. Plan: Y		Second Exec. Salary: $625,000	Bonus: $455,000

OTHER THOUGHTS:

Apparent Women Officers or Directors: 3
Hot Spot for Advancement for Women/Minorities: Y

LOCATIONS: ("Y" = Yes)

West:	Southwest:	Midwest:	Southeast:	Northeast:	International:
				Y	

Note: Financial information, benefits and other data can change quickly and may vary from those stated here.

GENZYME CORP   www.genzyme.com

Industry Group Code: 325412   Ranks within this company's industry group: Sales: 22   Profits: 98

Drugs:		Other:		Clinical:		Computers:		Other:	
Discovery:	Y	AgriBio:		Trials/Services:		Hardware:		Specialty Services:	
Licensing:		Genomics/Proteomics:	Y	Laboratories:		Software:		Consulting:	
Manufacturing:	Y	Tissue Replacement:		Equipment/Supplies:	Y	Arrays:		Blood Collection:	
Development:	Y			Research/Development Svcs.:		Database Management:		Drug Delivery:	
Generics:				Diagnostics:	Y			Drug Distribution:	

TYPES OF BUSINESS:
Pharmaceuticals Discovery & Development
Genetic Disease Treatments
Surgical Products
Diagnostic Products
Genetic Testing Services
Oncology Products
Biomaterials
Medical Devices

BRANDS/DIVISIONS/AFFILIATES:
Renagel
Cerezyme
Fabrazyme
Aldurazyme
Thyrogen
Thymoglobulin
Carticel
AnorMED

CONTACTS: *Note: Officers with more than one job title may be intentionally listed here more than once.*
Henri A. Termeer, CEO
Henri A. Termeer, Pres.
Michael S. Wyzga, CFO/Chief Acct. Officer/Exec. VP
Zoltan Csimma, Chief Human Resources Officer/Sr. VP
Alan E. Smith, Chief Scientific Officer/Sr. VP-Research
Thomas J. DesRosier, General Counsel/Sr. VP/Chief Patent Counsel
Mark Bamforth, Sr. VP-Corp. Oper. & Pharmaceuticals
Richard H. Douglas, Sr. VP-Corp. Dev.
Elliott D. Hillback, Jr., Sr. VP-Corp. Affairs
Evan M. Lebson, VP/Treas.
Mara G. Aspinall, Pres., Genetics
John Butler, Pres., Genzyme Renal
Ann Merrifield, Pres., Biosurgery
Donald E. Pogorzelski, Pres., Genzyme Diagnostics
Henri A. Termeer, Chmn.
Sandford D. Smith, Pres., Intl. Group

GROWTH PLANS/SPECIAL FEATURES:
Genzyme Corporation is a major biotech drug manufacturer operating through six major units: renal, therapeutics, transplant, biosurgery, diagnostics/genetics and oncology. The renal unit specializes in treatments for patients suffering from diseases including chronic renal failure. Its main products are Renagel, which reduces phosphorus levels in hemodialysis patients, and Hectorol, a treatment for secondary hyperparathyroidism. The therapeutics unit's products and services are used in the treatment of genetic disorders and other chronic debilitating diseases, including lysosomal storage disorders (LSDs). Its main products include Cerezyme/Ceredase for Type I Gaucher disease, Fabrazyme for Fabry disease, Myozyme for Pompe disease and Thyrogen, a diagnostic agent used in the follow-up treatment of patients with thyroid cancer. The transplant unit's products address pre-transplantation, prevention and treatment of acute rejection in organ transplantation, as well as other autoimmune disorders. Its products include two immunosuppressive polyclonal antibodies, Thymoglobulin and Lymphoglobulin. The biosurgery unit's portfolio of devices, biomaterials and biotherapeutics are used primarily in the orthopedic, general surgery and severe burn treatment markets. Products include Synvisc for the treatment of pain caused by osteoarthritis of the knee; and the Carticel line of products, repairing knee cartilage after trauma. The genetics unit develops, manufactures and distributes in vitro diagnostic products with an emphasis on point-of-care products for the in-hospital and out-of-hospital rapid test segment. Additionally, Genzyme operates an Other division that incorporates the company's diagnostic products, oncology, bulk pharmaceuticals and cardiovascular businesses. In late 2006, Genzyme acquired AnorMED a biotechnology company.

Phone: 617-252-7500   **Fax:** 617-252-7600
Toll-Free:
Address: 500 Kendall St., Cambridge, MA 02142 US

FINANCIALS: Sales and profits are in thousands of dollars—add 000 to get the full amount. 2006 Note: Financial information for 2006 was not available for all companies at press time.

2006 Sales: $3,187,013	2006 Profits: $-16,797	**U.S. Stock Ticker: GENZ**
2005 Sales: $2,734,842	2005 Profits: $441,489	**Int'l Ticker:** Int'l Exchange:
2004 Sales: $2,201,145	2004 Profits: $86,527	Employees: 8,000
2003 Sales: $1,713,900	2003 Profits: $-67,600	Fiscal Year Ends: 12/31
2002 Sales: $1,329,500	2002 Profits: $-13,100	Parent Company:

SALARIES/BENEFITS:
Pension Plan:	ESOP Stock Plan:	Profit Sharing:	Top Exec. Salary: $1,365,000	Bonus: $1,759,500
Savings Plan: Y	Stock Purch. Plan: Y		Second Exec. Salary: $650,000	Bonus: $445,000

OTHER THOUGHTS:
Apparent Women Officers or Directors: 2
Hot Spot for Advancement for Women/Minorities: Y

LOCATIONS: ("Y" = Yes)
West:	Southwest:	Midwest:	Southeast:	Northeast:	International:
Y	Y	Y	Y	Y	Y

GENZYME ONCOLOGY

www.genzymeoncology.com

Industry Group Code: 325412 Ranks within this company's industry group: Sales: Profits:

Drugs:		Other:		Clinical:	Computers:	Other:
Discovery:	Y	AgriBio:		Trials/Services:	Hardware:	Specialty Services:
Licensing:		Genomics/Proteomics:	Y	Laboratories:	Software:	Consulting:
Manufacturing:	Y	Tissue Replacement:		Equipment/Supplies:	Arrays:	Blood Collection:
Development:	Y			Research/Development Svcs.:	Database Management:	Drug Delivery:
Generics:				Diagnostics:		Drug Distribution:

TYPES OF BUSINESS:

Drugs-Gene-Based
Cancer Vaccines
Angiogenesis Inhibitors
Drug Discovery Platforms

BRANDS/DIVISIONS/AFFILIATES:

Genzyme Corp.
SAGE
Tasidotin Hydrochloride
Tumor Endothelial Marker (TEM)
Campath
Clolar
ILEX Oncology, Inc.
DENSPM Melan-AMART

CONTACTS: Note: Officers with more than one job title may be intentionally listed here more than once.

Mark J. Enyedy, Pres.
Katherine W. Klinger, Sr. VP-R&D
Frederic J. Vinick, Sr.VP-Drug Discovery

Phone: 617-761-8777	Fax: 617-761-8918
Toll-Free:	
Address: 55 Cambridge Pkwy., Cambridge, MA 02142 US	

GROWTH PLANS/SPECIAL FEATURES:

Genzyme Oncology, a division of Genzyme Corp., is dedicated to creating cancer vaccines and angiogenesis inhibitors through the integration of its genomics, gene and cell therapy, small-molecule drug discovery and protein therapeutic capabilities. The company's research consists of two approaches to cancer treatment: Tasidotin Hydrochloride and antiangiogenesis. Tasidotin Hydrochloride is a synthetic dolastatin analog currently being studied in Phase II metastatic melanoma, non-small cell lung cancer and prostate cancer clinical trials. Genzyme Oncology's anti-angiogenesis products include its portfolio of proprietary tumor endothelial markers (TEMs), proteins believed to be involved with tumor angiogenesis, which has grown to include over 400 genes. The company is collaborating with researchers at Johns Hopkins University in its attempt to expand its portfolio of TEMs by applying SAGE technology to endothelial cell samples. SAGE facilitates rapid, accurate analysis of gene expression patterns with the ability to identify previously unknown genes, tumor antigens and anti-angiogenic factors, analyzing the effects of drugs on human tissue and gaining insight into disease pathways. Genzyme Oncology licenses SAGE to pharmaceutical, biotechnology and genomics companies. Genzyme Oncology also plans to pursue small molecules targeting the converse of TEMs, the normal endothelial markers (NEMs). The company's recent acquisition of ILEX Oncology, Inc. gave the firm its marketed cancer drugs Campath and Clolar. Campath is indicated for the treatment of B-cell chronic lymphocytic leukemia, while Clolar recently received approval in the U.S. for the treatment of children with refractory or relapsed acute lymphoblastic leukemia. It is the first new leukemia treatment approved specifically for children in more than a decade. The company is also testing Campath as a treatment for non-Hodgkin's lymphoma and multiple sclerosis. Genzyme Oncology's current research pipeline also includes DENSPM, undergoing a Phase I/II clinical trial for the treatment of hepatocellular carcinoma.

FINANCIALS: Sales and profits are in thousands of dollars—add 000 to get the full amount. 2006 Note: Financial information for 2006 was not available for all companies at press time.

2006 Sales: $	2006 Profits: $	U.S. Stock Ticker: Subsidiary
2005 Sales: $	2005 Profits: $	Int'l Ticker: Int'l Exchange:
2004 Sales: $	2004 Profits: $	Employees: 5,200
2003 Sales: $	2003 Profits: $	Fiscal Year Ends: 12/31
2002 Sales: $9,400	2002 Profits: $-23,700	Parent Company: GENZYME CORP

SALARIES/BENEFITS:

Pension Plan:	ESOP Stock Plan:	Profit Sharing:	Top Exec. Salary: $	Bonus: $
Savings Plan: Y	Stock Purch. Plan: Y		Second Exec. Salary: $	Bonus: $

OTHER THOUGHTS:

Apparent Women Officers or Directors: 1
Hot Spot for Advancement for Women/Minorities:

LOCATIONS: ("Y" = Yes)

West:	Southwest:	Midwest:	Southeast:	Northeast:	International:
				Y	

Note: Financial information, benefits and other data can change quickly and may vary from those stated here.

GERON CORPORATION

www.geron.com

Industry Group Code: 325412 Ranks within this company's industry group: Sales: 146 Profits: 132

Drugs:		Other:		Clinical:	Computers:	Other:
Discovery:	Y	AgriBio:		Trials/Services:	Hardware:	Specialty Services:
Licensing:	Y	Genomics/Proteomics:		Laboratories:	Software:	Consulting:
Manufacturing:		Tissue Replacement:	Y	Equipment/Supplies:	Arrays:	Blood Collection:
Development:	Y			Research/Development Svcs.:	Database Management:	Drug Delivery:
Generics:				Diagnostics:		Drug Distribution:

TYPES OF BUSINESS:
Drug Discovery & Development
Telomerase Technologies
Human Stem Cell Technologies

BRANDS/DIVISIONS/AFFILIATES:
Start Licensing, Inc.

CONTACTS: Note: Officers with more than one job title may be intentionally listed here more than once.
Thomas B. Okarma, CEO
Thomas B. Okarma, Pres.
David L. Greenwood, CFO/Exec. VP
Calvin B. Harley, Chief Scientific Officer
David J. Earp, Chief Patent Counsel
David J. Earp, Sr. VP-Bus. Dev.
David L. Greenwood, Treas./Sec.
Alan B. Colowick, Pres., Oncology
Jane S. Lebkowski, Sr. VP-Regenerative Medicine
Melissa A. Kelly Behrs, Sr. VP-Therapeutic Dev./Oncology
Alexander E. Barkas, Chmn.

Phone: 650-473-7700	Fax: 650-473-7750

Toll-Free:
Address: 230 Constitution Dr., Menlo Park, CA 94025 US

GROWTH PLANS/SPECIAL FEATURES:
Geron Corp. develops biopharmaceuticals for the treatment of cancer and chronic degenerative diseases, including spinal cord injury, heart failure, diabetes and HIV/AIDS. The company's therapies are based on telomerase and human embryonic stem cell technologies. Telomeres enable cell division, protect chromosomes from degradation and act as molecular clocks for cellular aging. The enzyme telomerase restores telomere length, which shortens as cells multiply, extending a cell's ability to replicate. Geron is developing anti-cancer therapies based on telomerase inhibitors, telomerase therapeutic vaccines and telomerase-based oncolytic viruses. The company seeks to use human embryonic stem cells (hESC) as a potential source for manufacturing replacement cells and tissues for organ repair. The firm is now testing six different hESC-derived therapeutic cell types in animal models. The most advanced hESC-derived product, GRNOPC1, is targeted for the treatment of spinal cord injury. Geron's second hESC product, GRNCM1, is a population of cariomyocytes, the contractile cells of the heart, which is intended for the treatment of patients with myocardial disease. The firm owns or licenses over 145 U.S. and 240 foreign patents, with more than 450 pending applications worldwide. Geron has formed Start Licensing, Inc., a joint-venture with Exeter Life Sciences, to manage and license a portfolio of animal reproductive and cloning technologies. In July 2007, the company announced the initiation of a clinical trial of GRN163L, a telomerase inhibitor drug, in patients with advanced non-small cell lung cancer.

The company offers its employees medical, dental and vision insurance; life and AD&D insurance; short- and long-term disability; a 401(k) plan; an employee stock purchase plan; and flexible spending accounts. In addition, the firm offers a paid winter break between Christmas and New Year's Day.

FINANCIALS: Sales and profits are in thousands of dollars—add 000 to get the full amount. 2006 Note: Financial information for 2006 was not available for all companies at press time.

2006 Sales: $3,277	2006 Profits: $-31,365	**U.S. Stock Ticker: GERN**
2005 Sales: $6,158	2005 Profits: $-33,689	Int'l Ticker: Int'l Exchange:
2004 Sales: $1,053	2004 Profits: $-79,558	Employees: 103
2003 Sales: $1,174	2003 Profits: $-29,883	Fiscal Year Ends: 12/31
2002 Sales: $1,200	2002 Profits: $-33,900	Parent Company:

SALARIES/BENEFITS:

Pension Plan:	ESOP Stock Plan:	Profit Sharing:	Top Exec. Salary: $475,000	Bonus: $256,500
Savings Plan: Y	Stock Purch. Plan: Y		Second Exec. Salary: $365,000	Bonus: $147,800

OTHER THOUGHTS:
Apparent Women Officers or Directors: 2
Hot Spot for Advancement for Women/Minorities: Y

LOCATIONS: ("Y" = Yes)

West:	Southwest:	Midwest:	Southeast:	Northeast:	International:
Y					

Note: Financial information, benefits and other data can change quickly and may vary from those stated here.

GILEAD SCIENCES INC

www.gilead.com

Industry Group Code: 325412 Ranks within this company's industry group: Sales: 23   Profits: 202

Drugs:		Other:		Clinical:		Computers:		Other:	
Discovery:	Y	AgriBio:		Trials/Services:		Hardware:		Specialty Services:	
Licensing:	Y	Genomics/Proteomics:		Laboratories:		Software:		Consulting:	
Manufacturing:	Y	Tissue Replacement:		Equipment/Supplies:		Arrays:		Blood Collection:	
Development:	Y			Research/Development Svcs.:	Y	Database Management:		Drug Delivery:	
Generics:				Diagnostics:				Drug Distribution:	

TYPES OF BUSINESS:

Viral & Bacterial Infections Drugs
Respiratory & Cardiopulmonary Diseases Drugs

BRANDS/DIVISIONS/AFFILIATES:

Vistide
Truvada
Emtriva
Atripla
AmBisome
Herpera
Flolan
Myogen, Inc.

CONTACTS: *Note: Officers with more than one job title may be intentionally listed here more than once.*

John C. Martin, CEO
John F. Milligan, COO
John C. Martin, Pres.
John F. Milligan, CFO
Kristen M. Metza, Sr. VP-Human Resources
Norbert W. Bischofberger, Exec. VP-R&D/Chief Scientific Officer
Anthony D. Caracciolo, Sr. VP-Mfg.
Gregg H. Alton, General Counsel/Sr. VP
Anthony D. Caracciolo, Sr. VP-Oper.
Kevin Young, Exec. VP-Commercial Oper.
Michael J. Gerber, Sr. VP-Clinical Research
Richard J. Gorczynski, Sr. VP-Pulmonary Research
William A. Lee, Sr. VP-Research
James M. Denny, Chmn.
Paul Carter, Sr. VP-Int'l Commercial Oper.

Phone: 650-574-3000	Fax: 650-578-9264
Toll-Free: 800-445-3235	
Address: 333 Lakeside Dr., Foster City, CA 94404 US	

GROWTH PLANS/SPECIAL FEATURES:

Gilead Sciences, Inc. is a biopharmaceutical company that discovers, develops and commercializes therapeutics for the treatment of life-threatening diseases such as viral and bacterial infections. The company expanded its efforts to include respiratory and cardiopulmonary diseases. The firm maintains research, development, manufacturing, sales and marketing facilities in the U.S., Europe and Australia. Gilead currently has eight products on the market: Vistide, Truvada and Emtriva, which are oral medicines used as part of a combination therapy to treat HIV; Atripla, an oral formulation for treatment of HIV; Hespera, an oral medication used for treatment of Hepatitis B; AmBisome, an antifugal agent to treat serious invasive fungal infections; Vistide, an antiviral medication for the treatment of cytomegalovirus retinitis in patients with AIDS; and Flolan, an injected medication for the long-term intravenous treatment of primary pulmonary hypertension. The company also derives revenues from licensing agreements for Macugen, a macular degeneration treatment developed by OSI Pharmaceutical, Inc.; DaunoXome, used for treatment of AIDS-related Kaposi's sarcoma sold by Diatos S.A.; and Tamiflu, an influenza medication sold by F. Hoffman-LaRoche. In addition, the firm has collaboration with numerous companies and its own research and development program. In June 2006, Gilead purchased the Canadian subsidiary Raylo Chemicals from Degussa AG for about $145 million, to supply some of the active pharmaceutical agents for Gilead products. In August 2006, the company acquired Corus Pharma, Inc., a company engaged in drug discovery related to respiratory and infectious diseases. In November 2006, the firm purchased Myogen, Inc., a company engaged primarily in drug discovery related to cardiopulmonary diseases and other cardiovascular disorders, for roughly $2.5 billion. In June 2007, FDA approved Letairis, the company's treatment of pulmonary arterial hypertension.

The company offers its employees a 401(k) plan, a stock purchase plan, an employee assistance plan, health benefits and tuition reimbursement.

FINANCIALS: Sales and profits are in thousands of dollars—add 000 to get the full amount. 2006 Note: Financial information for 2006 was not available for all companies at press time.

2006 Sales: $3,026,139	2006 Profits: $-1,189,957	U.S. Stock Ticker: GILD
2005 Sales: $2,028,400	2005 Profits: $813,914	Int'l Ticker: Int'l Exchange:
2004 Sales: $1,324,621	2004 Profits: $449,371	Employees: 2,515
2003 Sales: $867,864	2003 Profits: $-72,003	Fiscal Year Ends: 12/31
2002 Sales: $466,800	2002 Profits: $72,100	Parent Company:

SALARIES/BENEFITS:

Pension Plan:	ESOP Stock Plan:	Profit Sharing:	Top Exec. Salary: $997,917	Bonus: $1,450,000
Savings Plan: Y	Stock Purch. Plan: Y		Second Exec. Salary: $607,500	Bonus: $526,125

OTHER THOUGHTS:

Apparent Women Officers or Directors: 4
Hot Spot for Advancement for Women/Minorities: Y

LOCATIONS: ("Y" = Yes)

West:	Southwest:	Midwest:	Southeast:	Northeast:	International:
Y				Y	Y

Note: Financial information, benefits and other data can change quickly and may vary from those stated here.

GLAXOSMITHKLINE PLC

www.gsk.com

Industry Group Code: 325412 **Ranks within this company's industry group:** Sales: 3   Profits: 3

Drugs:		Other:		Clinical:	Computers:		Other:	
Discovery:	Y	AgriBio:		Trials/Services:	Hardware:		Specialty Services:	
Licensing:		Genomics/Proteomics:		Laboratories:	Software:		Consulting:	
Manufacturing:	Y	Tissue Replacement:		Equipment/Supplies:	Arrays:		Blood Collection:	
Development:	Y			Research/Development Svcs.:	Database Management:		Drug Delivery:	
Generics:				Diagnostics:			Drug Distribution:	

TYPES OF BUSINESS:

Prescription Medications
Asthma Drugs
Respiratory Drugs
Antibiotics
Antivirals
Dermatological Drugs
Over-the-Counter & Nutritional Products

BRANDS/DIVISIONS/AFFILIATES:

Lanoxin
Flovent
Paxil
Domantis, Ltd.
Zantac
Nicorette
PLIVA-Istraivacki Institut
Citrucel

CONTACTS: *Note: Officers with more than one job title may be intentionally listed here more than once.*

Jean-Pierre Garnier, CEO
Julian Heslop, CFO
Daniel Phelan, Sr. VP-Human Resources
Tadataka Yamada, Chmn., R&D
Ford Calhoun, CIO
David Pulman, Pres., Global Mfg. & Supply
Rupert Bondy, General Counsel/Sr. VP
David Stout, Pres., Pharm. Oper.
Jennie Younger, Sr. VP-Corp. Comm. & Partnerships
Marc Dunoyer, Pres., Pharmaceuticals, Japan
Chris Viehbacher, Pres., Pharmaceuticals U.S.
Andrew Witty, Pres., Pharmaceuticals, Europe
John Clarke, Pres., Consumer Healthcare
Christopher Gent, Chmn.
Russell Greig, Pres., Pharmaceuticals, Intl.

Phone: 44-20-8047-5000	Fax: 44-20-8047-7807
Toll-Free: 888-825-5249	
Address: 980 Great West Rd., Brentford, Middlesex, TW8 9GS UK	

GROWTH PLANS/SPECIAL FEATURES:

GlaxoSmithKline (GSK) is a leading research-based pharmaceutical company formed from the merger of Glaxo Wellcome and SmithKline Beecham. Its subsidiaries consist of global drug and health companies engaged in the creation, discovery, development, manufacturing and marketing of pharmaceuticals and other consumer health products. The firm's major markets are based in the U.S., Japan, U.K., Spain and Italy. GSK operates in three industry segments: pharmaceuticals, vaccines and consumer health care. Recently developed medications include Paxil for depression and anxiety; Lamictal for epilepsy and bipolar disorder; Ventolin for bronchitis; and Augmentin for bacterial infections. Additionally, GSK also designs prescription medications for the treatment of heart and circulatory conditions, cancer, HIV/AIDS and malaria. Consumer health care products include over-the-counter medication such as Citrucel and Nicorette; oral care products such as Aquafresh; and nutritional products such as Boost. GSK's vaccines are designed to treat life-threatening illnesses such as hepatitis A, diphtheria, influenza and bacterial meningitis. GSK has also recently developed a new therapeutic vaccine which prevents cancer patients from relapsing after recovery. Research and development operations takes place at 14 sites in five countries on clinical trials and new drug discovery technologies. Its seven centers focus on cardiovascular and urogenital systems; metabolic and viral disorders; microbial or musculoskeletal diseases; neurological and gastrointestinal problems; psychiatric conditions; respiratory diseases and biopharmaceutical products. In 2006 and early 2007, GlaxoSmithKline received approval from the FDA to distribute Tykerb in combination with Xeloda to treat breast cancer, Avandamet, a treatment for type 2 diabetes, and Wellbutrin XL. GSK is also in the process of acquiring Praecis Pharmaceuticals and have recently acquired both Domantis Ltd, a company which focuses on antibody therapies, and CNS, the manufacturer of Breathe Right nasal strips and Fiberchoice, in 2006.

U.S. employees of GSK are offered a competitive benefits package including health care and other insurance, an employee assistance program, dependent care resources and corporate discounts.

FINANCIALS: Sales and profits are in thousands of dollars—add 000 to get the full amount. 2006 Note: Financial information for 2006 was not available for all companies at press time.

2006 Sales: $45,595,800	2006 Profits: $10,793,000	**U.S. Stock Ticker: GSK**
2005 Sales: $37,783,631	2005 Profits: $8,400,952	**Int'l Ticker: GSK** Int'l Exchange: London-LSE
2004 Sales: $34,863,347	2004 Profits: $7,015,930	Employees: 110,000
2003 Sales: $17,251,000	2003 Profits: $1,949,000	Fiscal Year Ends: 12/31
2002 Sales: $31,819,000	2002 Profits: $5,903,000	Parent Company:

SALARIES/BENEFITS:

Pension Plan: Y	ESOP Stock Plan: Y	Profit Sharing:	Top Exec. Salary: $1,523,000	Bonus: $2,250,000
Savings Plan: Y	Stock Purch. Plan:		Second Exec. Salary: $949,136	Bonus: $

OTHER THOUGHTS:

Apparent Women Officers or Directors: 1
Hot Spot for Advancement for Women/Minorities:

LOCATIONS: ("Y" = Yes)

West:	Southwest:	Midwest:	Southeast:	Northeast:	International:
Y	Y	Y	Y	Y	Y

Note: Financial information, benefits and other data can change quickly and may vary from those stated here.

GLYCOGENESYS INC

www.glycogenesys.com

Industry Group Code: 325412 Ranks within this company's industry group: Sales: Profits:

Drugs:	Other:	Clinical:	Computers:	Other:
Discovery:	AgriBio:	Trials/Services:	Hardware:	Specialty Services:
Licensing:	Genomics/Proteomics:	Laboratories:	Software:	Consulting:
Manufacturing:	Tissue Replacement:	Equipment/Supplies:	Arrays:	Blood Collection:
Development: Y		Research/Development Svcs.:	Database Management:	Drug Delivery:
Generics:		Diagnostics:		Drug Distribution:

TYPES OF BUSINESS:

Drugs-Carbohydrate-Based
Drugs-Cancer

BRANDS/DIVISIONS/AFFILIATES:

SafeScience Products, Inc.
International Gene Group, Inc.

GROWTH PLANS/SPECIAL FEATURES:

GlycoGenesys, Inc. is a biotechnology company developing pharmaceutical products based on carbohydrate compounds and related technologies. The company's lead product candidate is GCS-100, a complex carbohydrate intended to fight blood-borne and solid tumor cancers and their metastasis, either on its own or in combination with chemotherapy. Licensed from David Platt and Wayne State University and the Barbara Ann Karmanos Cancer Institute, GCS-100 may be able to block the spread of cancer cells, cut off the blood supply to tumors and trigger programmed cell death. GCS-100LE is currently in Phase I/II testing for multiple myeloma at the Dana-Farber Cancer Institute. The company has two wholly-owned non-operating subsidiaries: International Gene Group, Inc. and SafeScience Products, Inc., which the company now plans to sell or otherwise dispose of.

CONTACTS: Note: Officers with more than one job title may be intentionally listed here more than once.

Bradley J. Carver, CEO
Bradley J. Carver, Pres.
John W. Burns, CFO/Sr. VP/Corp. Sec.
William O. Fabbri, General Counsel
Frederick E. Pierce, II, VP-Bus. Dev.
Bradley J. Carver, Interim Chmn.

Phone: 617-422-0674	Fax: 617-422-0675
Toll-Free: 800-260-0675	
Address: 31 St. James Ave., 8th Fl., Boston, MA 02116 US	

FINANCIALS: Sales and profits are in thousands of dollars—add 000 to get the full amount. 2006 Note: Financial information for 2006 was not available for all companies at press time.

2006 Sales: $	2006 Profits: $	U.S. Stock Ticker: GLGSQ.PK
2005 Sales: $	2005 Profits: $	Int'l Ticker: Int'l Exchange:
2004 Sales: $	2004 Profits: $-10,125	Employees: 16
2003 Sales: $	2003 Profits: $-7,622	Fiscal Year Ends: 12/31
2002 Sales: $	2002 Profits: $-10,114	Parent Company:

SALARIES/BENEFITS:

Pension Plan:	ESOP Stock Plan:	Profit Sharing:	Top Exec. Salary: $228,800	Bonus: $26,590
Savings Plan: Y	Stock Purch. Plan:		Second Exec. Salary: $208,000	Bonus: $20,144

OTHER THOUGHTS:

Apparent Women Officers or Directors:
Hot Spot for Advancement for Women/Minorities:

LOCATIONS: ("Y" = Yes)

West:	Southwest:	Midwest:	Southeast:	Northeast: Y	International:

GTC BIOTHERAPEUTICS INC

www.gtc-bio.com

Industry Group Code: 325414 Ranks within this company's industry group: Sales: 8 Profits: 11

Drugs:		Other:		Clinical:	Computers:		Other:	
Discovery:	Y	AgriBio:	Y	Trials/Services:	Hardware:		Specialty Services:	
Licensing:	Y	Genomics/Proteomics:	Y	Laboratories:	Software:		Consulting:	
Manufacturing:	Y	Tissue Replacement:		Equipment/Supplies:	Arrays:		Blood Collection:	
Development:	Y			Research/Development Svcs.:	Database Management:		Drug Delivery:	
Generics:				Diagnostics:			Drug Distribution:	

TYPES OF BUSINESS:
Recombinant Proteins
Drugs-Anticoagulants
Transgenic Animals

BRANDS/DIVISIONS/AFFILIATES:
Genzyme Transgenics Corp.
Atryn

CONTACTS: Note: Officers with more than one job title may be intentionally listed here more than once.
Geoffrey F. Cox, CEO
Geoffrey F. Cox, Pres.
John B. Green, CFO
Harry M. Meade, Sr. VP-R&D
Daniel S. Woloshen, General Counsel/Sr. VP
Gregory Liposky, Sr. VP-Oper.
Ashley Lawton, VP-Bus. Dev.
Thomas E. Newberry, VP-Corp. Comm.
John B. Green, Sr. VP-Finance/Treas.
Carol A. Ziomek, VP-Dev.
Suzanne Groet, VP-Therapeutic Protein Dev.
Richard A. Scotland, Sr. VP-Regulatory Affairs
Frederick J. Finnegan, VP-Commercial Dev.
Geoffrey F. Cox, Chmn.

Phone: 508-620-9700	Fax: 508-370-3797
Toll-Free:	
Address: 175 Crossing Blvd., Framingham, MA 01702-9322 US	

GROWTH PLANS/SPECIAL FEATURES:
GTC Biotherapeutics, Inc. (GTC) applies transgenic technology to develop recombinant proteins for human therapeutic uses. The company uses transgenic animals that express specific recombinant proteins in their milk. Its technology platform includes the ability to generate animals and provide for animal husbandry, breeding and milking, as well as the ability to purify the milk to a clarified intermediate bulk material that may undergo manufacturing to obtain a clinical grade product. The firm generates transgenic animals through microinjection and nuclear transfer, and it expects to rely primarily on nuclear transfer techniques in new program development work. GTC uses goats in most of its commercial development programs due to the relatively short gestation times and relatively high milk production volume of the animals. The company's leading product is ATryn for patients with hereditary antithrombin deficiency undergoing surgical procedures. Antithrombin is an important protein found in the bloodstream with anticoagulant and anti-inflammatory properties. The drug is currently in Phase 3 of clinical trials in the U.S., while it has been approved for use in the European Union. It is also in Phase 1 clinical trials for disseminated intravascular coagulation, which is an acquired deficiency of antithrombin that occurs in sepsis. In addition to its transgenic technologies, GTC is developing a recombinant human serum albumin, a malaria vaccine, and a CD137 antibody to solid tumors. The firm has development partnerships with companies including Abbott Labs, Bristol-Myers Squibb, Centocor, Elan Pharmaceuticals, Progenics and ImmunoGen. In January 2006, the company received a patent covering the production of therapeutic proteins in the milk of transgenic mammals. In August 2006, the European Commission approved GTC's product ATryn for marketing.

GTC offers its employees tuition reimbursement and health coverage.

FINANCIALS: Sales and profits are in thousands of dollars—add 000 to get the full amount. 2006 Note: Financial information for 2006 was not available for all companies at press time.

2006 Sales: $6,128	2006 Profits: $-33,345	U.S. Stock Ticker: GTCB
2005 Sales: $4,152	2005 Profits: $-30,112	Int'l Ticker: Int'l Exchange:
2004 Sales: $6,626	2004 Profits: $-29,493	Employees: 153
2003 Sales: $9,764	2003 Profits: $-29,537	Fiscal Year Ends: 1/01
2002 Sales: $10,400	2002 Profits: $-24,300	Parent Company:

SALARIES/BENEFITS:

Pension Plan:	ESOP Stock Plan:	Profit Sharing:	Top Exec. Salary: $458,640	Bonus: $157,225
Savings Plan: Y	Stock Purch. Plan: Y		Second Exec. Salary: $294,840	Bonus: $76,688

OTHER THOUGHTS:
Apparent Women Officers or Directors: 2
Hot Spot for Advancement for Women/Minorities: Y

LOCATIONS: ("Y" = Yes)

West:	Southwest:	Midwest:	Southeast:	Northeast:	International:
				Y	

HARVARD BIOSCIENCE INC   www.harvardbioscience.com

Industry Group Code: 334500 Ranks within this company's industry group: Sales: 5   Profits: 5

Drugs:	Other:	Clinical:		Computers:		Other:
Discovery:	AgriBio:	Trials/Services:		Hardware:		Specialty Services:
Licensing:	Genomics/Proteomics:	Laboratories:		Software:		Consulting:
Manufacturing:	Tissue Replacement:	Equipment/Supplies:	Y	Arrays:		Blood Collection:
Development:		Research/Development Svcs.:	Y	Database Management:		Drug Delivery:
Generics:		Diagnostics:				Drug Distribution:

TYPES OF BUSINESS:
Apparatus & Scientific Instruments

BRANDS/DIVISIONS/AFFILIATES:
Asys Hitech GmbH
GE Healthcare
Thermo Fisher Scientific, Inc.
VWR
Harvard Apparatus

CONTACTS: *Note: Officers with more than one job title may be intentionally listed here more than once.*
Chane Graziano, CEO
Susan Luscinski, COO
David Green, Pres.
Bryce Chicoyne, CFO
Mark Norige, Pres., Harvard Apparatus Bus. Unit
David Strack, Pres., Genomic Solutions & Union Biometrica
David Parr, Managing Dir.-Biochrom Bus. Unit
Chane Graziano, Chmn.

Phone: 508-893-8999	Fax: 508-429-5732
Toll-Free: 800-272-2775	
Address: 84 October Hill Rd., Holliston, MA 01746 US	

GROWTH PLANS/SPECIAL FEATURES:

Harvard Bioscience, Inc. is a global developer, manufacturer and marketer of a broad range of specialized products, primarily apparatus and scientific instruments. Products are targeted toward two major application areas: ADMET screening and molecular biology. ADMET screening is used to identify compounds that have toxic side effects or undesirable physiological or pharmacological properties. These pharmacological properties consist of absorption distribution, metabolism and elimination. The company's products in this area include absorption diffusion chambers, 96 well equilibrium dialysis plate, organ testing systems, precision infusion pumps, cell injection systems, ventilators and electroporation products. The molecular biology products are mainly scientific instruments such as spectrophotometers and plate readers that analyze light to detect and quantify a wide range of molecular and cellular processes or apparatus such as gel electrophoresis units. These products can quantify the amount of DNA, RNA or protein in a sample; can use chromatography to separate the amino acids in a sample; and can use the method of gel electrophoresis to separate and purify DNA, RNA and proteins. Harvard Bioscience has manufacturing operations in the U.S., the U.K., Germany and Austria with sales facilities in France and Canada. The company sells its products to thousands of researches in over 100 countries through its direct sales force; 1,100 page catalog; its website, and through distributors, including GE Healthcare, Thermo Fisher Scientific, Inc. and VWR. Sales from the Harvard Apparatus catalog generated roughly 32% of revenue in 2006. Customers include primarily research scientists at pharmaceutical and biotechnology companies, universities and government laboratories, including the U.S. National Institutes of Health. In June 2006, Harvard Bioscience acquired, through Austrian subsidiary Asys Hitech GmbH, select assets of the microplate reader and washer product lines from Anthos Labtec Instruments GmbH, a subsidiary of Beckman Coulter, Inc.

FINANCIALS: Sales and profits are in thousands of dollars—add 000 to get the full amount. 2006 Note: Financial information for 2006 was not available for all companies at press time.

2006 Sales: $76,181	2006 Profits: $-2,341	**U.S. Stock Ticker: HBIO**
2005 Sales: $67,431	2005 Profits: $-31,877	**Int'l Ticker:** Int'l Exchange:
2004 Sales: $64,745	2004 Profits: $-2,329	Employees: 299
2003 Sales: $52,024	2003 Profits: $4,260	Fiscal Year Ends: 12/31
2002 Sales: $57,380	2002 Profits: $ 737	Parent Company:

SALARIES/BENEFITS:

Pension Plan:	ESOP Stock Plan:	Profit Sharing:	Top Exec. Salary: $486,000	Bonus: $282,000
Savings Plan:	Stock Purch. Plan:		Second Exec. Salary: $400,000	Bonus: $233,000

OTHER THOUGHTS:
Apparent Women Officers or Directors: 1
Hot Spot for Advancement for Women/Minorities:

LOCATIONS: ("Y" = Yes)

West:	Southwest:	Midwest:	Southeast:	Northeast:	International:
Y				Y	Y

HEMAGEN DIAGNOSTICS INC

www.hemagen.com

Industry Group Code: 325413 Ranks within this company's industry group: Sales: 24 Profits: 14

Drugs:		Other:		Clinical:		Computers:		Other:	
Discovery:		AgriBio:		Trials/Services:		Hardware:		Specialty Services:	
Licensing:	Y	Genomics/Proteomics:		Laboratories:		Software:		Consulting:	
Manufacturing:	Y	Tissue Replacement:		Equipment/Supplies:	Y	Arrays:		Blood Collection:	
Development:				Research/Development Svcs.:		Database Management:		Drug Delivery:	
Generics:				Diagnostics:	Y			Drug Distribution:	

TYPES OF BUSINESS:

Medical Diagnostics Products
Clinical Chemical Analysis Products
Clinical Chemistry Reagents

BRANDS/DIVISIONS/AFFILIATES:

Hemagen Diagnosticos Comercio
EasyLyte
EasyVet
Analyst
Reagent Applications/RAICHEM
Endochek
Virgo
Reagents Applications, Inc.

CONTACTS: Note: Officers with more than one job title may be intentionally listed here more than once.

William P. Hales, CEO
William P. Hales, Pres.
Catherine M. Davidson, CFO
Catherine M. Davidson, Controller
William P. Hales, Chmn.

Phone: 443-367-5500	Fax: 410-997-7812
Toll-Free: 800-495-2180	
Address: 9033 Red Branch Rd., Columbia, MD 21045 US	

GROWTH PLANS/SPECIAL FEATURES:

Hemagen Diagnostics, Inc. develops, manufactures and markets proprietary medical diagnostic test kits. Hemagen has three different product lines. The Virgo product line of diagnostic test kits is used to aid in the diagnosis of certain autoimmune and infectious diseases, using enzyme-linked immunosorbent assay (ELISA), immunoflourescence, and hemaglutination technology. Hemagen also manufactures and markets a complete line of clinical chemistry reagents through its wholly owned subsidiary Reagents Applications, Inc., under the brand name Raichem. In addition, Hemagen manufactures and sells the Analyst, an FDA-cleared clinical chemistry analyzer used to measure important constituents in human and animal blood, and the Endochek, a clinical chemistry analyzer used to measure important constituents in animal blood. Hemagen currently offers more than 150 test kits that have been cleared by the FDA for sale in the U.S. Brand-name products are sold directly to physicians, veterinarians, laboratories and blood banks within the U.S. Products are also made available through private-label agreements with global distributors of medical supplies. A network of international distributors represents the company's various product lines in overseas markets. In total, more than 1,000 customers use Hemagen products worldwide. In South America, Hemagen's operations are managed by the company's majority-owned Brazilian subsidiary, Hemagen Diagnosticos Comercio, Importacao e Exportacao, Ltd. Hemagen has increased its holding significantly by finalizing the purchase of this Brazilian subsidiary, which enables the company to prioritize expanding its South and Central American client base. Medical devices sold by Hemagen include the EasyLyte (human) and the EasyVet (veterinary) electrolyte analyzers, which are manufactured by Medica Corporation.

FINANCIALS: Sales and profits are in thousands of dollars—add 000 to get the full amount. 2006 Note: Financial information for 2006 was not available for all companies at press time.

2006 Sales: $7,250	2006 Profits: $ 313	U.S. Stock Ticker: HMGN	
2005 Sales: $7,586	2005 Profits: $-1,337	Int'l Ticker: Int'l Exchange:	
2004 Sales: $7,471	2004 Profits: $-3,599	Employees: 46	
2003 Sales: $8,473	2003 Profits: $-1,265	Fiscal Year Ends: 9/30	
2002 Sales: $9,500	2002 Profits: $-1,800	Parent Company:	

SALARIES/BENEFITS:

Pension Plan:	ESOP Stock Plan: Y	Profit Sharing:	Top Exec. Salary: $176,250	Bonus: $
Savings Plan: Y	Stock Purch. Plan:		Second Exec. Salary: $84,000	Bonus: $

OTHER THOUGHTS:

Apparent Women Officers or Directors: 1
Hot Spot for Advancement for Women/Minorities:

LOCATIONS: ("Y" = Yes)

West:	Southwest:	Midwest:	Southeast:	Northeast:	International:
Y				Y	Y

Note: Financial information, benefits and other data can change quickly and may vary from those stated here.

HEMISPHERX BIOPHARMA INC   www.hemispherx.net

Industry Group Code: 325412 Ranks within this company's industry group: Sales: 159   Profits: 105

Drugs:		Other:		Clinical:	Computers:	Other:
Discovery:		AgriBio:		Trials/Services:	Hardware:	Specialty Services:
Licensing:		Genomics/Proteomics:	Y	Laboratories:	Software:	Consulting:
Manufacturing:	Y	Tissue Replacement:		Equipment/Supplies:	Arrays:	Blood Collection:
Development:	Y			Research/Development Svcs.:	Database Management:	Drug Delivery:
Generics:				Diagnostics:		Drug Distribution:

TYPES OF BUSINESS:
RNA-Related Drugs
HIV Treatments
Antivirals

BRANDS/DIVISIONS/AFFILIATES:
Oragen
Ampligen
Alferon N
Alferon LDO

CONTACTS: *Note: Officers with more than one job title may be intentionally listed here more than once.*
William A. Carter, CEO
Anthony Bonelli, COO
Anthony Bonelli, Pres.
Robert E. Peterson, CFO
Mei-June Liao, VP-Regulatory Affairs, Quality Control and R&D
Robert Hansen, VP-Mfg.
David R. Strayer, Medical Dir./Regulatory Affairs
Carol A. Smith, VP-Mfg. Quality Assurance
Carol A. Smith, VP-Process Dev.
William A. Carter, Chmn.

Phone: 215-988-0080	Fax: 215-988-1739
Toll-Free:	
Address: 1 Penn Ctr., 1617 JFK Blvd., 6th Fl., #660, Philadelphia, PA 19103 US	

GROWTH PLANS/SPECIAL FEATURES:
Hemispherx Biopharma, Inc. is a biopharmaceutical company which develops and manufactures drugs for the treatment of viral and immune-based disorders. The company has established a strong platform of laboratory, pre-clinical and clinical data for the purpose of commercializing its portfolio of nucleic acid drugs. The company's proprietary drug technology uses specifically configured RNA. Hemispherx's flagship products include the antiviral/immunotherapeutic drugs Ampligen and Oragen, as well its Alferon N Injection. Ampligen is a synthetic double stranded RNA configured to address a variety of chronic diseases and viral disorders such as HIV and malignant melanoma. Oragen drugs, orally administered relatives of Ampligen, are in the early pre-clinical stages of development. Alferon, in either its injection form (N) or its oral form (LDO), is a purified, natural source alpha interferon (interferon alfa-n3, human leukocyte derived). Alferon N is currently the only natural interferon currently sold in the U.S. and is also approved for sale in Mexico, Germany, Singapore, and Hong Kong. In addition to these flagship drugs and projects, Hemispherx has a patent portfolio of over 120 patents issued worldwide. The company currently maintains strategic alliances with Biovail, Chronix Biomedical, Hollister-Stier Laboratories LLC, Laboratorios Del Dr. Esteve S.A. and the National Institute of Infectious Diseases in Tokyo, Japan. In April 2007, Hemispherx entered into an agreement with Armada Health Care to launch its natural interferon, Alferon N Injection into Armada's Specialty Pharmacy infrastructure. Also in April 2007, Hemispherx announced that Japan's Ministry of Health, Labor and Welfare issued authorization to its National Institute of Infectious Diseases approving their budget to advance studies indicating that an H5N1 influenza vaccine co-administered intranasally with Ampligen protected against mutated strains of the virus, and that the seasonal trivalent influenza vaccine co-administered intranasally with Ampligen maintained efficacy even when challenged with the H5N1 influenza virus.

FINANCIALS: Sales and profits are in thousands of dollars—add 000 to get the full amount. 2006 Note: Financial information for 2006 was not available for all companies at press time.

2006 Sales: $ 933	2006 Profits: $-19,399	**U.S. Stock Ticker: HEB**
2005 Sales: $1,083	2005 Profits: $-13,213	**Int'l Ticker:** Int'l Exchange:
2004 Sales: $1,229	2004 Profits: $-24,140	Employees: 62
2003 Sales: $ 657	2003 Profits: $-14,770	Fiscal Year Ends: 12/31
2002 Sales: $ 300	2002 Profits: $-7,400	Parent Company:

SALARIES/BENEFITS:
Pension Plan:	ESOP Stock Plan:	Profit Sharing:	Top Exec. Salary: $655,686	Bonus: $166,624
Savings Plan: Y	Stock Purch. Plan: Y		Second Exec. Salary: $259,164	Bonus: $64,791

OTHER THOUGHTS:
Apparent Women Officers or Directors: 2
Hot Spot for Advancement for Women/Minorities: Y

LOCATIONS: ("Y" = Yes)
West:	Southwest:	Midwest:	Southeast:	Northeast:	International:
				Y	Y

HESKA CORP
www.heska.com

Industry Group Code: 325412B **Ranks within this company's industry group: Sales: 1   Profits: 1**

Drugs:		Other:		Clinical:		Computers:		Other:	
Discovery:	Y	AgriBio:		Trials/Services:		Hardware:		Specialty Services:	
Licensing:		Genomics/Proteomics:		Laboratories:		Software:		Consulting:	
Manufacturing:	Y	Tissue Replacement:		Equipment/Supplies:	Y	Arrays:		Blood Collection:	
Development:	Y			Research/Development Svcs.:		Database Management:		Drug Delivery:	
Generics:				Diagnostics:	Y			Drug Distribution:	

TYPES OF BUSINESS:
Drugs-Animal Health & Pet Care
Veterinary Diagnostics Products & Services
Animal Dietary Supplements
Veterinary Vaccines

BRANDS/DIVISIONS/AFFILIATES:
Core Companion Animal Health
HESKA Feline UltraNasal FVRCP Vaccine
SPOTCHEM EZ Chemistry Analyzer
HESKA Vet CBC-Diff Hematology Analyzer
i-STAT Portable Clinical Analyzer
HESKA Vet/Ox G2 Digital Monitor
E.R.D.-Healthscreen Urine Test
Diamond Animal Health

CONTACTS: *Note: Officers with more than one job title may be intentionally listed here more than once.*
Robert B. Grieve, CEO
Robert B. Grieve, Pres.
Jason A. Napolitano, CFO/Exec. VP/Treas.
Todd M. Gilson, VP-Mktg.
Mark D. Cicotello, VP-Human Resources
Malcolm A. Hammerton, VP-IT
Michael J. McGinley, VP-Tech. Affairs
Nancy Wisnewski, VP-Prod. Dev.
John R. Flanders, General Counsel/VP/Corp. Sec.
Michael J. McGinley, VP-Oper.
Michael A. Bent, Controller/Principle Acct. Officer
Michael J. McGinley, Gen. Mgr.-Heska Des Moines
G. Lynn Snodgrass, VP-Sales
Nancy Wisnewski, VP-Tech. Customer Service
Donald L. Wassom, Managing Dir.-Heska AG/Dir.-Global Allergy
Robert B. Grieve, Chmn.
Joseph H. Ritter, Exec. VP-Global Bus. Oper.

Phone: 970-493-7272   **Fax:** 970-619-3003
Toll-Free: 800-464-3752
Address: 3760 Rocky Mountain Ave., Loveland, CO 80538 US

GROWTH PLANS/SPECIAL FEATURES:
Heska Corp. focuses on the discovery, development and marketing of animal health care products. The company uses biotechnology to create a broad range of diagnostic, therapeutic and vaccine products for dogs, cats and other animals. The business is divided into two segments: Core Companion Animal Health (CCA) and Other Vaccines, Pharmaceuticals and Products (OVP), previously Diamond Animal Health. The CCA segment includes diagnostic and monitoring instruments and supplies, single-use diagnostic tests, vaccines, pharmaceuticals and nutritional supplements, primarily for canine and feline use. Its line of veterinary diagnostic instruments includes the i-STAT Portable Clinical Analyzer, the HESKA Vet CBC-Diff Hematology Analyzer and the SPOTCHEM EZ Chemistry Analyzer to measure blood chemistry. Heska's diagnostic tests include tests for heartworms, allergy tests and early renal damage detection products. The E.R.D.-Healthscreen Urine Tests can identify dogs and cats at risk for kidney disease before the majority of kidney function is lost. It also sells the HESKA Feline Ultranasal FVRCP Vaccine, a three-way modified live vaccine to prevent disease caused by respiratory viruses. The OVP business manufactures vaccines and pharmaceutical products for cattle, small mammals, horses and fish that are marketed and distributed by third parties. In addition, OVP manufactures certain companion animal health products for marketing and sale by Heska. Other products include heartworm prevention tablets, a fatty acid supplement, vitamins and a chewable thyroid supplement.

FINANCIALS: Sales and profits are in thousands of dollars—add 000 to get the full amount. 2006 Note: Financial information for 2006 was not available for all companies at press time.
2006 Sales: $75,060	2006 Profits: $1,828	**U.S. Stock Ticker:** HSKA
2005 Sales: $69,437	2005 Profits: $ 282	**Int'l Ticker:** Int'l Exchange:
2004 Sales: $67,691	2004 Profits: $-4,815	Employees: 299
2003 Sales: $65,300	2003 Profits: $-3,500	Fiscal Year Ends: 12/31
2002 Sales: $51,300	2002 Profits: $-8,700	Parent Company:

SALARIES/BENEFITS:
Pension Plan:	ESOP Stock Plan:	Profit Sharing:	Top Exec. Salary: $341,000	Bonus: $251,378
Savings Plan:	Stock Purch. Plan:		Second Exec. Salary: $221,500	Bonus: $114,300

OTHER THOUGHTS:
Apparent Women Officers or Directors: 2
Hot Spot for Advancement for Women/Minorities: Y

LOCATIONS: ("Y" = Yes)
West:	Southwest:	Midwest:	Southeast:	Northeast:	International:
Y		Y			Y

Note: Financial information, benefits and other data can change quickly and may vary from those stated here.

HI-TECH PHARMACAL CO INC   www.hitechpharm.com

Industry Group Code: 325416 **Ranks within this company's industry group:** Sales: 8   Profits: 6

Drugs:		Other:		Clinical:	Computers:		Other:	
Discovery:		AgriBio:		Trials/Services:	Hardware:		Specialty Services:	Y
Licensing:		Genomics/Proteomics:		Laboratories:	Software:		Consulting:	
Manufacturing:	Y	Tissue Replacement:		Equipment/Supplies:	Arrays:		Blood Collection:	
Development:	Y			Research/Development Svcs.:	Database Management:		Drug Delivery:	
Generics:	Y			Diagnostics:			Drug Distribution:	

TYPES OF BUSINESS:

Drugs-Generic
Nutritional Products
Over-the-Counter Products
Ophthalmic Products
Manufacturing Contract Services
Inhalation Products
Diabetes Products
Branded Drugs

BRANDS/DIVISIONS/AFFILIATES:

Diabetic Tussin
DiabetiSweet
DiabetDerm
Multi-Betic
DiabetTrim
Naprelan
Tanafed
Brometane DX

CONTACTS: Note: Officers with more than one job title may be intentionally listed here more than once.

David S. Seltzer, CEO
David S. Seltzer, Pres.
William Peters, CFO/VP
Edwin A. Berrios, VP-Mktg. & Sales
Polireddy Dondetti, Sr. Dir.-R&D
James P. Tracy, VP-Info. Systems
Tanya Akimova, Dir.-New Bus. Dev.
Margaret Santorufo, Controller
Gary M. April, Pres., Health Care Prod. Div./Divisional VP-Sale
Joanne Curri, Dir.-Regulatory Affairs
Pudpong Poolsuk, Sr. Dir.-Science
Bernard Seltzer, Chmn.

Phone: 631-789-8228	Fax: 631-789-8429
Toll-Free:	
Address: 369 Bayview Ave., Amityville, NY 11701-2802 US	

GROWTH PLANS/SPECIAL FEATURES:

Hi-Tech Pharmacal Co., Inc. is a manufacturer and marketer of generic and branded prescription, over-the-counter and nutritional products that are sold in liquid and cream forms. A wide range of products are produced for various disease states, including asthma, bronchial disorders, dermatological disorders, allergies, pain, stomach, oral care, neurological disorders and other conditions. The company is also a manufacturer of ophthalmic, optic and inhalation products and provides sterile manufacturing contract services. The company's customers consist of generic distributors, drug wholesalers, chain drug stores, mass merchandise chains and mail-order pharmacies. Business has mainly depended upon the following customers: McKesson Corporation, Walgreens, Cardinal Health, Inc., CVS, AmeriSourceBergen Corporation and Wal-Mart. Among Pharmacal's leading generic products are Albuterol Sulfate Inhalation Solution, Albuterol Sulfate Syrup, Amantadine HCL Syrup, Brometane DX, Lactulose Solution, Valproic Acid, Hydroxyzine Syrup, Poly-Vitamin Drops and Tri-Vitamin Drops. The firm's health care products division develops and markets a line of branded products primarily for people with diabetes, including Diabetic Tussin, a leading sugar-free over-the-counter cough medication; DiabetiSweet, an aspartame-free sugar substitute that can be used for baking and cooking; Multi-betic, a daily multi-vitamin; DiabetDerm, a diabetic skin care line; and DiabetTrim, a weight management product. The company's only prescription brand, Tanafed, was acquired in 2006. In 2006, sales of generic pharmaceuticals represented 78% of total sales, sales of the health care products line of over-the-counter products accounted for 18% of total sales and sales of branded prescription products represented 4% of total sales.

FINANCIALS: Sales and profits are in thousands of dollars—add 000 to get the full amount. 2006 Note: Financial information for 2006 was not available for all companies at press time.

2006 Sales: $78,020	2006 Profits: $11,453	**U.S. Stock Ticker:** HITK
2005 Sales: $67,683	2005 Profits: $8,288	**Int'l Ticker:** Int'l Exchange:
2004 Sales: $56,366	2004 Profits: $6,592	Employees: 246
2003 Sales: $47,400	2003 Profits: $5,700	Fiscal Year Ends: 4/30
2002 Sales: $33,300	2002 Profits: $3,500	Parent Company:

SALARIES/BENEFITS:

Pension Plan:	ESOP Stock Plan: Y	Profit Sharing:	Top Exec. Salary: $382,000	Bonus: $277,000
Savings Plan: Y	Stock Purch. Plan:		Second Exec. Salary: $285,000	Bonus: $

OTHER THOUGHTS:

Apparent Women Officers or Directors: 3
Hot Spot for Advancement for Women/Minorities: Y

LOCATIONS: ("Y" = Yes)

West:	Southwest:	Midwest:	Southeast:	Northeast:	International:
				Y	

HOLLIS-EDEN PHARMACEUTICALS   www.holliseden.com

Industry Group Code: 325412   Ranks within this company's industry group: Sales: 172   Profits: 128

Drugs:		Other:		Clinical:	Computers:	Other:	
Discovery:	Y	AgriBio:		Trials/Services:	Hardware:	Specialty Services:	
Licensing:		Genomics/Proteomics:		Laboratories:	Software:	Consulting:	
Manufacturing:		Tissue Replacement:		Equipment/Supplies:	Arrays:	Blood Collection:	
Development:	Y			Research/Development Svcs.:	Database Management:	Drug Delivery:	
Generics:				Diagnostics:		Drug Distribution:	

TYPES OF BUSINESS:

Drugs-Immune System Regulation
Drugs-Infectious Diseases
Drugs-Hormonal Imbalances
Drugs-Anti-Radiation

BRANDS/DIVISIONS/AFFILIATES:

Neumune
Immunitin

CONTACTS: Note: Officers with more than one job title may be intentionally listed here more than once.

Richard B. Hollis, CEO
Richard B. Hollis, Pres.
Robert L. Marsella, Sr. VP-Mktg.
James M. Frincke, Chief Scientific Officer
Eric J. Loumeau, General Counsel/VP
Robert L. Marsella, Sr. VP-Bus. Dev.
Robert W. Weber, Chief Acct. Officer/VP/Controller
Dwight R. Stickney, VP-Medical Affairs
Christopher L. Reading, Exec. VP-Scientific Dev.
Richard B. Hollis, Chmn.

Phone: 858-587-9333	Fax: 858-558-6470
Toll-Free:	
Address: 4435 Eastgate Mall, Ste. 400, San Diego, CA 92121 US	

GROWTH PLANS/SPECIAL FEATURES:

Hollis-Eden Pharmaceuticals, Inc., a development-stage pharmaceutical company, is engaged in the discovery, development and commercialization of products for the treatment of diseases and disorders against which the body is unable to mount an appropriate immune response. It has focused its initial technology efforts on a series of potent hormones and hormone analogs, labeled immune regulating hormones (IRHs), which it believes are key components of the body's natural regulatory system. The firm seeks to use these compounds as a hormone replacement therapy to reestablish balance to the immune system in situations of immune dysregulation. Preclinical and early clinical studies have indicated that these compounds have the ability to significantly reduce a number of well-known inflammatory mediators while increasing innate and adaptive immunity and reversing bone marrow suppression. The compounds are safe, cost-effective to manufacture and unlikely to produce resistance. The lead compound in this area is Neumune (HE2100), which is being co-developed with the U.S. military. Acute radiation injury causes bone marrow suppression and, consequently, a weakening of the immune system. The company hopes that Neumune will guard against most of the deleterious effects of initial exposure to radiation by protecting the bone marrow. Hollis-Eden is currently pursuing an advanced purchase contract to provide the U.S. government with Neumune for the Strategic National Stockpile, for use by first responders and civilians in the event of attack by nuclear or radiological weapons. Neumune is currently undergoing Phase I trials. The company also has IRH compounds, such as Immunitin, in development for the treatment of HIV and malaria. Immunitin is currently in Phase II trials. Hollis-Eden owns or has obtained a license to numerous U.S. and foreign patents and foreign patent applications.

FINANCIALS: Sales and profits are in thousands of dollars—add 000 to get the full amount. 2006 Note: Financial information for 2006 was not available for all companies at press time.

2006 Sales: $ 444	2006 Profits: $-30,231	U.S. Stock Ticker: HEPH
2005 Sales: $ 56	2005 Profits: $-29,441	Int'l Ticker: Int'l Exchange:
2004 Sales: $ 63	2004 Profits: $-24,757	Employees: 66
2003 Sales: $	2003 Profits: $-25,700	Fiscal Year Ends: 12/31
2002 Sales: $	2002 Profits: $-17,500	Parent Company:

SALARIES/BENEFITS:

Pension Plan:	ESOP Stock Plan:	Profit Sharing:	Top Exec. Salary: $512,000	Bonus: $385,000
Savings Plan: Y	Stock Purch. Plan:		Second Exec. Salary: $330,050	Bonus: $66,010

OTHER THOUGHTS:

Apparent Women Officers or Directors:
Hot Spot for Advancement for Women/Minorities:

LOCATIONS: ("Y" = Yes)

West:	Southwest:	Midwest:	Southeast:	Northeast:	International:
Y					

Note: Financial information, benefits and other data can change quickly and may vary from those stated here.

HOSPIRA INC   www.hospira.com

Industry Group Code: 339113 Ranks within this company's industry group: Sales: Profits:

Drugs:		Other:		Clinical:		Computers:		Other:	
Discovery:	Y	AgriBio:		Trials/Services:		Hardware:	Y	Specialty Services:	
Licensing:	Y	Genomics/Proteomics:		Laboratories:		Software:	Y	Consulting:	
Manufacturing:	Y	Tissue Replacement:		Equipment/Supplies:	Y	Arrays:		Blood Collection:	
Development:				Research/Development Svcs.:		Database Management:		Drug Delivery:	
Generics:				Diagnostics:				Drug Distribution:	

TYPES OF BUSINESS:

Pharmaceutical Development
Generic Pharmaceuticals
Medication Delivery Systems
Anesthetics
Injectable Medications
Diagnostic Imaging Agents
Drug Library Software
Contract Manufacturing

BRANDS/DIVISIONS/AFFILIATES:

Abbott Laboratories
One 2 One
Mayne Pharma Limited

CONTACTS: Note: Officers with more than one job title may be intentionally listed here more than once.

Christopher B. Begley, CEO
Terrence C. Kearney, COO
Thomas E. Werner, CFO
Edward A. Ogunro, Sr. VP-R&D/Chief Science Officer
Brian J. Smith, General Counsel/Sr. VP
Thomas E. Werner, Sr. VP-Finance
David A. Jones, Chmn.

Phone: 224-212-2000	Fax: 224-212-3350
Toll-Free: 877-946-7747	
Address: 275 N. Field Dr., Lake Forest, IL 60045 US	

GROWTH PLANS/SPECIAL FEATURES:

Hospira, Inc. is a global specialty pharmaceutical and medication delivery company. The company's primary operations involve the research and development of generic pharmaceuticals, pharmaceuticals based on proprietary pharmaceuticals whose patents have expired. It was established in 2003 by Abbott Laboratories in order to facilitate the company's planned spin-off in the manufacturing and sales of hospital products, including injectable pharmaceuticals and medication delivery systems. Operating independently since May 2004, Hospira oversees most of Abbott's former hospital products division and select operations formerly managed by Abbott's international segment. The company's activities include the development, manufacture and marketing of specialty injectable pharmaceuticals and medication delivery systems that deliver drugs and intravenous (I.V.) fluids. Additionally, the firm offers contract manufacturing services to proprietary pharmaceutical and biotechnology companies for formulation development, filling and finishing of injectable pharmaceuticals. Hospira's specialty injectable pharmaceutical products primarily consist of generic injectable pharmaceuticals including analgesics, anesthetics, anti-infectives, cardiovascular drugs, oncology drugs and others. Hospira's specialty injectable pharmaceutical products include Precedex, a proprietary sedative that is used in the intensive care setting. The firm's medication delivery systems offer management systems, which include electronic drug delivery pumps, safety software, administration sets and accessories, and related services; and infusion therapy solutions and products that are used to deliver I.V. fluids and medications to patients. Through its One 2 One manufacturing services group, Hospira provides contract manufacturing services for formulation development, filling and finishing of injectable drugs worldwide. The company works with its customers to develop stable injectable forms of their drugs, and Hospira fills and finishes those and other drugs into containers and packaging selected by the customer. The customer then sells the finished products under its own label. In early 2007, Hospira acquired Mayne Pharma Limited, and injectable pharmaceuticals manufacturer based in Australia.

FINANCIALS: Sales and profits are in thousands of dollars—add 000 to get the full amount. 2006 Note: Financial information for 2006 was not available for all companies at press time.

2006 Sales: $2,688,505	2006 Profits: $237,679	**U.S. Stock Ticker: HSP**
2005 Sales: $2,626,696	2005 Profits: $235,638	**Int'l Ticker:** Int'l Exchange:
2004 Sales: $2,645,036	2004 Profits: $301,552	Employees: 13,000
2003 Sales: $2,400,200	2003 Profits: $260,400	Fiscal Year Ends: 12/31
2002 Sales: $2,405,100	2002 Profits: $246,400	Parent Company:

SALARIES/BENEFITS:

Pension Plan: Y	ESOP Stock Plan:	Profit Sharing:	Top Exec. Salary: $766,154	Bonus: $1,038,423
Savings Plan: Y	Stock Purch. Plan:		Second Exec. Salary: $362,615	Bonus: $391,633

OTHER THOUGHTS:

Apparent Women Officers or Directors:
Hot Spot for Advancement for Women/Minorities:

LOCATIONS: ("Y" = Yes)

West:	Southwest:	Midwest:	Southeast:	Northeast:	International:
Y	Y	Y	Y	Y	Y

Note: Financial information, benefits and other data can change quickly and may vary from those stated here.

HUMAN GENOME SCIENCES INC
www.hgsi.com

Industry Group Code: 325412 Ranks within this company's industry group: Sales: 103   Profits: 197

Drugs:		Other:		Clinical:	Computers:	Other:
Discovery:	Y	AgriBio:		Trials/Services:	Hardware:	Specialty Services:
Licensing:		Genomics/Proteomics:	Y	Laboratories:	Software:	Consulting:
Manufacturing:	Y	Tissue Replacement:		Equipment/Supplies:	Arrays:	Blood Collection:
Development:	Y			Research/Development Svcs.:	Database Management:	Drug Delivery:
Generics:				Diagnostics:		Drug Distribution:

TYPES OF BUSINESS:
Oncology, Immunology & Infectious Diseases Drugs

BRANDS/DIVISIONS/AFFILIATES:
LymphoStat-B
Albuferon
Abthrax

CONTACTS: *Note: Officers with more than one job title may be intentionally listed here more than once.*
H. Thomas Watkins, CEO
H. Thomas Watkins, Pres.
Timothy C. Barabe, CFO/Sr. VP
Susan Bateson McKay, Sr. VP-Human Resources
David C. Stump, Exec. VP-R&D
James H. Davis, General Counsel/Exec. VP
Curan M. Simpson, Sr. VP-Oper.
Jerry Parrott, VP-Corp. Comm.
Barry A. Labinger, Chief Commercial Officer/Exec. VP
Argeris N. Karabelas, Chmn.

Phone: 301-309-8504 **Fax:** 301-309-8512
Toll-Free:
Address: 14200 Shady Grove Rd., Rockville, MD 20850 US

GROWTH PLANS/SPECIAL FEATURES:
Human Genome Sciences, Inc. (HGS) is a commercially focused drug development company with three products advancing into late-stage clinical development: Albuferon for chronic hepatitis, LymphoStat-B for systemic lupus erythematosus and ABthrax for anthrax disease. The company also has a pipeline of compounds in earlier stages of clinical development in oncology, immunology and infectious disease, including rheumatoid arthritis and HIV/AIDS. The firm's partners conduct clinical trials of additional drugs to treat cardiovascular, metabolic and central nervous system diseases and advanced a number of products derived from the company's technology to clinical development. The Albuferon collaborator is Novartis and the LymphoStat-B collaborator is GlaxoSmithKline. GlaxoSmithKline entered several small molecule drugs into clinical development including GSK480848, for the treatment of atherosclerosis; GSK462795, for the treatment of bone disease; and GSK649868, for the treatment of sleep disorders. HGS entered into collaboration agreements for the co-development and co-commercialization of its Albuferon and LymphoStat-B products. The company has roughly 560 U.S. patents covering genes and proteins. In February 2007, the company announced that it initiated a Phase III clinical trial of Albuferon in combination with ribavirin for treatment of chronic hepatitis C genotypes 2 and 3; and Phase III clinical trial of LymphoStat-B in patients with active systemic lupus erythematosus.

The company offers its employees medical, dental and vision insurance; flexible spending accounts; life and AD&D insurance; short- and long-term disability; a 401(k) plan; an employee stock purchase plan; and education assistance.

FINANCIALS: Sales and profits are in thousands of dollars—add 000 to get the full amount. 2006 Note: Financial information for 2006 was not available for all companies at press time.

2006 Sales: $25,755	2006 Profits: $-210,327	U.S. Stock Ticker: HGSI
2005 Sales: $19,113	2005 Profits: $-239,439	Int'l Ticker: Int'l Exchange:
2004 Sales: $3,831	2004 Profits: $-242,898	Employees: 880
2003 Sales: $8,168	2003 Profits: $-185,324	Fiscal Year Ends: 12/31
2002 Sales: $3,600	2002 Profits: $-219,700	Parent Company:

SALARIES/BENEFITS:
| Pension Plan: | ESOP Stock Plan: | Profit Sharing: | Top Exec. Salary: $650,000 | Bonus: $650,000 |
| Savings Plan: Y | Stock Purch. Plan: Y | | Second Exec. Salary: $460,000 | Bonus: $391,000 |

OTHER THOUGHTS:
Apparent Women Officers or Directors: 1
Hot Spot for Advancement for Women/Minorities:

LOCATIONS: ("Y" = Yes)
West:	Southwest:	Midwest:	Southeast:	Northeast:	International:
				Y	

Note. Financial information, benefits and other data can change quickly and may vary from those stated here.

HYCOR BIOMEDICAL INC

www.hycorbiomedical.com

Industry Group Code: 325413 Ranks within this company's industry group: Sales: Profits:

Drugs:	Other:	Clinical:		Computers:		Other:	
Discovery:	AgriBio:	Trials/Services:		Hardware:	Y	Specialty Services:	
Licensing:	Genomics/Proteomics:	Laboratories:		Software:		Consulting:	
Manufacturing:	Tissue Replacement:	Equipment/Supplies:	Y	Arrays:		Blood Collection:	
Development:		Research/Development Svcs.:		Database Management:		Drug Delivery:	
Generics:		Diagnostics:	Y			Drug Distribution:	

TYPES OF BUSINESS:

Medical Diagnostics Products
Allergy & Autoimmune Diagnostics
Urinalysis Products

BRANDS/DIVISIONS/AFFILIATES:

Stratagene Corp.
Agilent Technologies, Inc.
Hycor Biomedical, Ltd.
Hycor Biomedical GmbH
KOVA Urinalysis
KOVA-Trol
KOVA Refractrol SP
HY-TEC

CONTACTS: Note: Officers with more than one job title may be intentionally listed here more than once.

Nelson F. Thune, Gen. Mgr.
Cheryl Graham, Dir.-Human Resources
Nelson F. Thune, Sr. VP-Oper.
Vance Mitchell, Dir.-Oper.

Phone:	Fax: 714-933-3222
Toll-Free: 800-382-2527	
Address: 7272 Chapman Ave., Garden Grove, CA 92841 US	

GROWTH PLANS/SPECIAL FEATURES:

Hycor Biomedical, Inc., a subsidiary of Stratagene Corp., researches, develops and manufactures medical diagnostic products for clinical laboratories and specialty physicians, with a particular focus on allergy and autoimmune testing and urinalysis products. It operates two wholly-owned foreign subsidiaries, Hycor Biomedical GmbH, based in Germany, and Hycor Biomedical, Ltd., in the U.K. The company's KOVA family of urinalysis products include KOVA Urinalysis, the market leader in Standardized Microscopic Urinalysis; KOVA-Trol tri-level, lyophilized urine dipstick chemistry control, which is stable for 5-7 days at room temperature and for four months when frozen; KOVA Liqua-Trol, a ready-to-use bi-level liquid control, useable with most brands of urine chemistry dipsticks; and KOVA Refractrol SP, a tri-level control product with a 24-month shelf life designed for use with temperature compensated and non-compensated refractometers. Hycor's allergy diagnostic product line, which includes HY-TEC specific IgE tests, features radio-immunoassays and enzymatic immunoassays that test for reactions to more than 1,000 allergens. The company's autoimmune products division manufactures devices (branded as Autostat II) that test for disorders such as systemic lupus erythematosus and rheumatoid arthritis. Hycor also produces HY-TEC 288, an automated consolidated workstation capable of storing up to 50 allergy and 100 autoimmune samples, conducting 288 tests per run, providing bar-coded sample identification, high precision robotic liquid handling and real-time incubation control. Most recently, Hycor added to its in vitro allergy test menu the Stachybotrys allergen, one of the most common molds implicated in sick building syndrome, a condition with wide-ranging symptoms thought to be caused by the mold infestation of damp areas due to floods or leaks. In June 2007, Agilent Technologies, Inc. acquired Stratagene, Hycor's parent company, for $250 million.

FINANCIALS: Sales and profits are in thousands of dollars—add 000 to get the full amount. 2006 Note: Financial information for 2006 was not available for all companies at press time.

2006 Sales: $	2006 Profits: $	U.S. Stock Ticker: Subsidiary
2005 Sales: $	2005 Profits: $	Int'l Ticker: Int'l Exchange:
2004 Sales: $	2004 Profits: $	Employees: 148
2003 Sales: $19,800	2003 Profits: $ 700	Fiscal Year Ends: 12/31
2002 Sales: $18,600	2002 Profits: $-1,300	Parent Company: STRATAGENE CORP

SALARIES/BENEFITS:

Pension Plan: Y	ESOP Stock Plan:	Profit Sharing:	Top Exec. Salary: $255,131	Bonus: $51,026
Savings Plan: Y	Stock Purch. Plan: Y		Second Exec. Salary: $237,083	Bonus: $47,417

OTHER THOUGHTS:

Apparent Women Officers or Directors: 1
Hot Spot for Advancement for Women/Minorities:

LOCATIONS: ("Y" = Yes)

West:	Southwest:	Midwest:	Southeast:	Northeast:	International:
Y					Y

ICON PLC

www.iconclinical.com

ndustry Group Code: 541710   Ranks within this company's industry group: Sales: 5   Profits: 4

Drugs:		Other:		Clinical:		Computers:		Other:	
Discovery:		AgriBio:		Trials/Services:		Hardware:	Y	Specialty Services:	
Licensing:		Genomics/Proteomics:		Laboratories:		Software:	Y	Consulting:	Y
Manufacturing:		Tissue Replacement:		Equipment/Supplies:		Arrays:		Blood Collection:	
Development:				Research/Development Svcs.:	Y	Database Management:		Drug Delivery:	
Generics:				Diagnostics:				Drug Distribution:	

TYPES OF BUSINESS:

Clinical Trial Research Services
Data Management & Analysis
Laboratory Services
Regulatory Affairs Consulting
Interactive Voice Response Systems
Strategic Consulting & Marketing

BRANDS/DIVISIONS/AFFILIATES:

ICON Central Laboratores
ICON Clinical Research
ICON Contracting Solutions
ICON Development Solutions
ICON Medical Imaging
ICOPhone
ICOnet
ICOlabs

CONTACTS: *Note: Officers with more than one job title may be intentionally listed here more than once.*

Peter Gray, CEO
Malcolm Burgess, COO
Ciaran Murray, CFO
Peter Sowood, Chief Scientific Officer, ICON Clinical Research
Bill Taaffe, Pres., Corp. Dev.
John Hubbard, Pres., ICON Clinical Research, U.S.
Dan Weng, Pres., ICON Clinical Research, ROW
Alan Morgan, Pres., ICON Clinical Research, Europe
Thomas Frey, Pres., ICON Development Solutions
John Climax, Chmn.

Phone: 353-1-291-2000	Fax: 353-1-216-2700
Toll-Free:	
Address: S. County Bus. Park, Leopardstown, Dublin, 18 UK	

GROWTH PLANS/SPECIAL FEATURES:

ICON plc is a clinical research organization that provides clinical research, drug development, regulatory affairs, laboratory services and imaging to the pharmaceutical and biotechnology industries throughout Europe, North America and the Pacific Rim. The company operates in five sectors: ICON Central Laboratories, ICON Clinical Research, ICON Contracting Solutions, ICON Development Solutions and ICON Medical Imaging. ICON Central Laboratories offers services that range from complex microbiology and chemistries to simply safety tests. Current research focuses on clinical chemistry, hematology, toxicology, infectious disease, immunology/serology, flow cytometry, pharmacogenomics and molecular diagnostics. ICON Clinical Research specializes in planning management, execution and analysis of Phase IIb-IV clinical trials. This sector has conducted over 1,000 clinical studies in over 31,000 centers across in the world in areas such as the CNS, oncology, cardiology, transplant dermatology, pediatrics, respiratory and urology. ICON Contracting Solutions offers temporary or permanent personnel to meet an organization's research needs, which can include staff shortages, unpredicted monitoring issues and elevated study initiatives. ICON Development Solutions offers strategy, management and execution of products and early phase clinical planning development, particularly in the areas of regulatory affairs and pharmacokinetics/biopharmaceutics. Lastly, ICON Medical Imaging provides image based product development services to biotechnology, pharmaceutical and medical device industries. ICON also offers technological advancements which include the ICOPhone, a device used to increase accuracy, efficiency and cost savings associated with the conduction of clinical trials; ICOnet, a tool that provides on-line access to CRF pages, status reports and project management documentation; and Electronic Data Capture (EDC), a system that provides electronic data collection and data management for clinical trials. In 2007, ICON Central Laboratories launched ICOlabs, a remote data access tool for pharmaceutical and biotechnology companies.

ICON offers its employees tuition reimbursement, health benefits and an employee assistance program.

FINANCIALS: Sales and profits are in thousands of dollars—add 000 to get the full amount. 2006 Note: Financial information for 2006 was not available for all companies at press time.

2006 Sales: $455,597	2006 Profits: $38,304	**U.S. Stock Ticker: ICLR**
2005 Sales: $326,658	2005 Profits: $13,545	**Int'l Ticker: IJF** Int'l Exchange: Dublin-ISE
2004 Sales: $443,875	2004 Profits: $25,742	Employees: 4,290
2003 Sales: $340,971	2003 Profits: $18,283	Fiscal Year Ends: 5/31
2002 Sales: $156,600	2002 Profits: $14,200	Parent Company:

SALARIES/BENEFITS:

Pension Plan:	ESOP Stock Plan:	Profit Sharing:	Top Exec. Salary: $	Bonus: $
Savings Plan: Y	Stock Purch. Plan:		Second Exec. Salary: $	Bonus: $

OTHER THOUGHTS:

Apparent Women Officers or Directors: 1
Hot Spot for Advancement for Women/Minorities:

LOCATIONS: ("Y" = Yes)

West:	Southwest:	Midwest:	Southeast:	Northeast:	International:
Y		Y	Y	Y	Y

Note: Financial information, benefits and other data can change quickly and may vary from those stated here.

ICOS CORPORATION

www.icos.com

Industry Group Code: 325412 **Ranks within this company's industry group:** Sales: Profits:

Drugs:		Other:	Clinical:	Computers:	Other:
Discovery:	Y	AgriBio:	Trials/Services:	Hardware:	Specialty Services:
Licensing:		Genomics/Proteomics:	Laboratories:	Software:	Consulting:
Manufacturing:	Y	Tissue Replacement:	Equipment/Supplies:	Arrays:	Blood Collection:
Development:	Y		Research/Development Svcs.:	Database Management:	Drug Delivery:
Generics:			Diagnostics:		Drug Distribution:

TYPES OF BUSINESS:

Pharmaceuticals Discovery & Development
Erectile Dysfunction Treatment
Cancer & Inflammatory Drugs

BRANDS/DIVISIONS/AFFILIATES:

Cialis
Lilly ICOS, LLC
AndroGel
Tadalafil
Eli Lilly & Co.

CONTACTS: *Note: Officers with more than one job title may be intentionally listed here more than once.*

Paul N. Clark, CEO
Paul N. Clark, Pres.
Michael A. Stein, CFO/Sr. VP
Gary L. Wilcox, Exec. VP-Oper.
Kevin M. Egan, VP-Bus. Dev.
Paul N. Clark, Chmn.

Phone: 425-485-1900	Fax: 425-489-0356
Toll-Free:	
Address: 22021 20th Ave. SE, Bothell, WA 98021 US	

GROWTH PLANS/SPECIAL FEATURES:

ICOS Corporation, a subsidiary of Eli Lilly & Company, researches and develops new pharmaceuticals, determining points of intervention within various diseases to create specific and efficacious drugs. The company combines its capabilities in molecular, cellular and structural biology, high-throughput drug screening, medicinal chemistry and gene expression profiling to develop products with significant commercial potential. ICOS' approved product, Cialis, treats erectile dysfunction (ED) and is marketed throughout the world through a joint venture established with Eli Lilly, known as Lilly ICOS, which survived ICOS being acquired by Eli Lilly. Cialis is an oral inhibitor of the PDE5 enzyme, a relaxing muscle tissue, and over 112 clinical studies, with over 15,000 test subjects were required to demonstrate its feasibility in the clinical phases. Other product candidates in development include Tadalafil for treating benign prostatic hyperplasia (enlargement of the prostate gland), in Phase II studies. Treatments for cancer, psoriasis and various inflammatory and autoimmune diseases are in preclinical development. In January 2007, the company was acquired by Eli Lilly & Co. for $2.3 billion.

Employees are offered flexible spending accounts, tuition reimbursement, an on-site child care facility, adoption assistance and credit union membership. Additionally, the firm offers on-site laundry, casual dress, team sports, company picnics, mixers and an annual golf tournament.

FINANCIALS: **Sales and profits are in thousands of dollars—add 000 to get the full amount. 2006 Note: Financial information for 2006 was not available for all companies at press time.**

2006 Sales: $	2006 Profits: $	U.S. Stock Ticker: Subsidiary
2005 Sales: $71,410	2005 Profits: $-74,842	Int'l Ticker: Int'l Exchange:
2004 Sales: $74,608	2004 Profits: $-198,248	Employees: 700
2003 Sales: $75,100	2003 Profits: $-125,500	Fiscal Year Ends: 12/31
2002 Sales: $92,900	2002 Profits: $-161,600	Parent Company: ELI LILLY AND COMPANY

SALARIES/BENEFITS:

Pension Plan:	ESOP Stock Plan:	Profit Sharing:	Top Exec. Salary: $900,000	Bonus: $809,800
Savings Plan: Y	Stock Purch. Plan:		Second Exec. Salary: $525,000	Bonus: $226,200

OTHER THOUGHTS:

Apparent Women Officers or Directors:
Hot Spot for Advancement for Women/Minorities:

LOCATIONS: ("Y" = Yes)

West:	Southwest:	Midwest:	Southeast:	Northeast:	International:
Y					

IDERA PHARMACEUTICALS INC   www.iderapharma.com

Industry Group Code: 325412   Ranks within this company's industry group: Sales: 149   Profits: 94

Drugs:		Other:		Clinical:	Computers:	Other:
Discovery:	Y	AgriBio:		Trials/Services:	Hardware:	Specialty Services:
Licensing:	Y	Genomics/Proteomics:	Y	Laboratories:	Software:	Consulting:
Manufacturing:		Tissue Replacement:		Equipment/Supplies:	Arrays:	Blood Collection:
Development:	Y			Research/Development Svcs.:	Database Management:	Drug Delivery:
Generics:				Diagnostics:		Drug Distribution:

TYPES OF BUSINESS:

Drugs-Targeted Immune Therapies
Synthetic DNA
Drugs-Infectious Disease
Drugs-Cancer

BRANDS/DIVISIONS/AFFILIATES:

Hybridon, Inc.
IMO-2055
Amplivax
IMO-2125

CONTACTS: *Note: Officers with more than one job title may be intentionally listed here more than once.*

Sudhir Agrawal, CEO
Robert W. Karr, Pres.
Sudhir Agrawal, Chief Scientific Officer
Steven J. Ritter, Intellectual Property Counsel
David M. Lough, Dir.-Bus. Dev.
Alice Bexon, VP-Clinical Dev.
Timothy M. Sullivan, VP-Dev. Programs
James B. Wyngaarden, Chmn.

Phone: 617-679-5500	Fax: 617-679-5592

Toll-Free:
Address: 345 Vassar St., Cambridge, MA 02139 US

GROWTH PLANS/SPECIAL FEATURES:

Idera Pharmaceuticals, Inc., formerly Hybridon, Inc., is a biotechnology company engaged in the discovery and development of novel therapeutics that modulate immune responses through Toll-like receptors (TLR) for the treatment of multiple diseases, including: cancer, infectious diseases, asthma/allergy, autoimmune diseases and for use in combination with therapeutic and prophylactic vaccines. The company is currently partnering with Novartis International Pharmaceuticals, Ltd. to develop treatments for asthma and allergy indications, as well as with Merck & Co. to research, develop and commercialize Idera's TLR agonists into therapeutic and prophylactic vaccines being developed by Merck for cancer, infectious diseases and Alzheimer's disease. The company has seven drug candidates, including the two programs with its partners described above, two of which have entered Phase I & II stages (viz. IMO-2055, a TLR9 agonist, indicated for use in renal cell carcinoma, in Phase II clinical trials; and IMO-2055+ Chemo, for solid tumors, currently in Phase I trials), as well as one candidate at the preclinical development stage (viz. IMO-2125, a TLR9 agonist, for Hepatitis C). The remaining candidates are in the research phase. Idera has also licensed to The Immune Response Corporation its Amplivax drug, for use in development of a novel vaccine for prevention and treatment of HIV. In 2007, the company announced that Novartis has decided to extend the term of the research phase of their collaboration by one year.

FINANCIALS: Sales and profits are in thousands of dollars—add 000 to get the full amount. 2006 Note: Financial information for 2006 was not available for all companies at press time.

2006 Sales: $2,421	2006 Profits: $-16,525	U.S. Stock Ticker: IDP
2005 Sales: $2,467	2005 Profits: $-13,706	Int'l Ticker: Int'l Exchange:
2004 Sales: $ 942	2004 Profits: $-12,735	Employees: 33
2003 Sales: $ 897	2003 Profits: $-17,211	Fiscal Year Ends: 12/31
2002 Sales: $30,255	2002 Profits: $16,972	Parent Company:

SALARIES/BENEFITS:

Pension Plan:	ESOP Stock Plan:	Profit Sharing:	Top Exec. Salary: $445,000	Bonus: $450,000
Savings Plan: Y	Stock Purch. Plan: Y		Second Exec. Salary: $375,000	Bonus: $250,000

OTHER THOUGHTS:

Apparent Women Officers or Directors: 1
Hot Spot for Advancement for Women/Minorities:

LOCATIONS: ("Y" = Yes)

West:	Southwest:	Midwest:	Southeast:	Northeast:	International:
				Y	

IDEXX LABORATORIES INC

www.idexx.com

Industry Group Code: 325413 Ranks within this company's industry group: Sales: 2   Profits: 1

Drugs:	Other:	Clinical:		Computers:		Other:	
Discovery:	AgriBio:	Trials/Services:		Hardware:		Specialty Services:	
Licensing:	Genomics/Proteomics:	Laboratories:	Y	Software:	Y	Consulting:	Y
Manufacturing:	Tissue Replacement:	Equipment/Supplies:	Y	Arrays:		Blood Collection:	
Development:		Research/Development Svcs.:		Database Management:		Drug Delivery:	
Generics:		Diagnostics:	Y			Drug Distribution:	

TYPES OF BUSINESS:

Veterinary Laboratory Testing & Consulting
Point-of-Care Diagnostic Products
Veterinary Pharmaceuticals
Information Management Software
Food & Water Testing Products

BRANDS/DIVISIONS/AFFILIATES:

Colilert-18
SNAP
Parallux
Colisure
SURPASS
Vita-Tech Canada, Inc.

CONTACTS: Note: Officers with more than one job title may be intentionally listed here more than once.

Jonathan W. Ayers, CEO
Jonathan W. Ayers, Pres.
Merilee Raines, CFO/VP
William C. Wallen, Sr. VP/Chief Scientific Officer
Conan R. Deady, General Counsel/Sec./VP
Irene C. Kerr, VP-Worldwide Oper.
Merilee Raines, Treas.
James Polewaczyck, VP-Rapid Assay & Digital Radiography
Michael Williams, VP-Instrument Diagnostics
Thomas J. Dupree, VP-Companion Animal Group
S. Sam Fratoni, VP
Jonathan W. Ayers, Chmn.
Ali Naqui, VP-Dairy, Water, APAC & Latin America Oper.

Phone: 207-556-0300	Fax: 207-556-0346
Toll-Free: 800-548-6733	
Address: 1 IDEXX Dr., Westbrook, ME 04092-2041 US	

GROWTH PLANS/SPECIAL FEATURES:

IDEXX Laboratories, Inc. develops, manufactures and distributes products and provides services for the veterinary and the food and water testing markets.   The company operates in three business segments: products and services for the veterinary market, referred to as the Companion Animal Group; water quality products; and products for production animal health, referred to as the Production Animal Segment. The company also operated a smaller segment that comprises products for dairy quality, referred to as Dairy.   Its primary business focus is on animal health. IDEXX's companion animal and equine veterinary offerings include in-clinic diagnostic tests and instrumentation, laboratory services, pharmaceuticals and veterinary practice information management software, as well as a range of single-use, hand-held test kits.   The principal single-use tests, sold under the SNAP name, include a feline combination test, the SNAP Combo FIV antibody/FeLV antigen test, which enables veterinarians to test simultaneously for feline leukemia virus and feline immunodeficiency virus; a canine combination test, the SNAP 3Dx, which tests simultaneously for Lyme disease, Ehrlichia canis and heartworm; and a canine heartworm-only test. IDEXX has received FDA approval for its topical equine anti-inflammatory SURPASS.   In addition, it also provides assay kits, software and instrumentation for accurate assessment of infectious disease in production animals, such as cattle, swine and poultry. The company currently offers commercial veterinary laboratory and consulting services in the U.S. through facilities located in Arizona, California, Colorado, Illinois, Maryland, Massachusetts, New Jersey, Oregon and Texas.   The water quality segment's products include Colilert-18 and Colisure tests, which simultaneously detect total coliforms and E. coli in water. IDEXX's two principal products for use in testing for antibiotic residue in milk are the SNAP Beta-lactam test and the Parallux system. In 2007, the company acquired Vita-Tech Canada, Inc., the largest provider of reference laboratory testing services to veterinary offices in Canada.

FINANCIALS: Sales and profits are in thousands of dollars—add 000 to get the full amount. 2006 Note: Financial information for 2006 was not available for all companies at press time.

2006 Sales: $739,117	2006 Profits: $93,678	**U.S. Stock Ticker: IDXX**
2005 Sales: $638,095	2005 Profits: $78,254	**Int'l Ticker:** Int'l Exchange:
2004 Sales: $549,181	2004 Profits: $78,332	Employees: 3,900
2003 Sales: $475,992	2003 Profits: $57,090	Fiscal Year Ends: 12/31
2002 Sales: $412,700	2002 Profits: $45,400	Parent Company:

SALARIES/BENEFITS:

Pension Plan:	ESOP Stock Plan:	Profit Sharing:	Top Exec. Salary: $600,000	Bonus: $650,000
Savings Plan: Y	Stock Purch. Plan: Y		Second Exec. Salary: $350,000	Bonus: $255,000

OTHER THOUGHTS:

Apparent Women Officers or Directors: 2
Hot Spot for Advancement for Women/Minorities: Y

LOCATIONS: ("Y" = Yes)

West:	Southwest:	Midwest:	Southeast:	Northeast:	International:
Y	Y	Y	Y	Y	Y

IDM PHARMA INC

www.idm-biotech.com

Industry Group Code: 325412 Ranks within this company's industry group: Sales: 126 Profits: 113

Drugs:		Other:		Clinical:		Computers:		Other:	
Discovery:		AgriBio:		Trials/Services:		Hardware:		Specialty Services:	
Licensing:	Y	Genomics/Proteomics:	Y	Laboratories:		Software:		Consulting:	
Manufacturing:		Tissue Replacement:		Equipment/Supplies:		Arrays:		Blood Collection:	
Development:	Y			Research/Development Svcs.:		Database Management:		Drug Delivery:	
Generics:				Diagnostics:				Drug Distribution:	

TYPES OF BUSINESS:

Drugs-Cancer
Cancer Cell Destroying Products
Tumor Recurrence Prevention
Drugs-Vaccines

BRANDS/DIVISIONS/AFFILIATES:

Epimmune, Inc.
Immuno-Designed Molecules (IDM) S.A.
Junovan (Mepact)
MAK (Monocyte-derived Activated Killer)
Bexidem
Uvidem
Collidem
EP-2101

CONTACTS: *Note: Officers with more than one job title may be intentionally listed here more than once.*

Timothy P. Walbert, CEO
Timothy P. Walbert, Pres.
Robert De Vaere, CFO/Sr. VP
Bonnie Mills, VP-Clinical Oper.
Herve D. de Lamotte, VP-Finance
Bonnie Mills, General Mgr.-US Oper.
Herve D. de Lamotte, General Mgr.-IDM SA
Sylvie Gregoire, Chmn.

Phone: 949-470-4751	Fax: 949-470-6470

Toll-Free:

Address: 9 Parker, Ste. 100, Irvine, CA 92618-1605 US

GROWTH PLANS/SPECIAL FEATURES:

IDM Pharma Inc., the result of the 2005 combination of Epimmune, Inc. and Immuno-Designed Molecules (IDM) S.A., is a biopharmaceutical company focused on developing products to treat and control cancer while maintaining the patient's quality of life. Subsequent to the combination of the two companies, Epimmune changed its name to IDM Pharma Inc., while IDM S.A. became the company's wholly-owned subsidiary. IDM has five products in clinical development, one having completed Phase III trials, three in Phase II trials and one in Phase I. The lead product candidate, Junovan (mifamurtide for injection, known as Mepact in Europe), is part of a new family of immunotherapeutic agents that are designed to destroy residual cancer cells by activating the body's natural defenses. Junovan activates certain immune cells called macrophages in vivo (inside the body), in order to enhance their ability to destroy cancer cells. Junovan recently completed a Phase III clinical trial for the treatment of osteosarcoma, the most common form of bone cancer. The company's other cytotoxic product involves MAK (Monocyte-derived Activated Killer) cells, which are produced using the patient's own white blood cells by activating them to recognize and destroy tumor cells. IDM has one MAK product in clinical development, Bexidem, which is in Phase II/III development for the treatment of superficial bladder cancer. Products designed to prevent tumor recurrence include EP-2101 (a synthetic vaccine in clinical development for the treatment of non-small cell lung cancer); and two dendritophages (which are cell-based vaccines of dendritic cells, a type of specialized immune cell derived from the patient's own white blood cells), Uvidem and Collidem (for the treatment of melanoma and colorectal cancer, respectively). The company holds 129 issued patents and 75 patent applications. IDM has a major product development partnership with Sanofi-Aventis in cancer and immunotherapy.

FINANCIALS: Sales and profits are in thousands of dollars—add 000 to get the full amount. 2006 Note: Financial information for 2006 was not available for all companies at press time.

2006 Sales: $11,286	2006 Profits: $-23,455	U.S. Stock Ticker: IDMI
2005 Sales: $8,539	2005 Profits: $-39,209	Int'l Ticker: Int'l Exchange:
2004 Sales: $5,805	2004 Profits: $-31,657	Employees: 32
2003 Sales: $6,088	2003 Profits: $-18,432	Fiscal Year Ends: 12/31
2002 Sales: $7,100	2002 Profits: $-6,500	Parent Company:

SALARIES/BENEFITS:

Pension Plan:	ESOP Stock Plan:	Profit Sharing:	Top Exec. Salary: $385,000	Bonus: $50,000
Savings Plan: Y	Stock Purch. Plan: Y		Second Exec. Salary: $240,335	Bonus: $40,000

OTHER THOUGHTS:

Apparent Women Officers or Directors: 1
Hot Spot for Advancement for Women/Minorities:

LOCATIONS: ("Y" = Yes)

West:	Southwest:	Midwest:	Southeast:	Northeast:	International:
Y					Y

ILLUMINA INC

www.illumina.com

Industry Group Code: 334500 Ranks within this company's industry group: Sales: 4   Profits: 4

Drugs:	Other:		Clinical:		Computers:		Other:	
Discovery:	AgriBio:		Trials/Services:		Hardware:		Specialty Services:	Y
Licensing:	Genomics/Proteomics:	Y	Laboratories:		Software:	Y	Consulting:	
Manufacturing:	Tissue Replacement:		Equipment/Supplies:	Y	Arrays:	Y	Blood Collection:	
Development:			Research/Development Svcs.:		Database Management:		Drug Delivery:	
Generics:			Diagnostics:				Drug Distribution:	

TYPES OF BUSINESS:

Instruments-Genetic Variation Measurement
Array Technology
Digital Microbead Technology
Software
Genotyping Services

BRANDS/DIVISIONS/AFFILIATES:

BeadArray
Sentrix Array Matrix
Sentrix BeadChip
Oligator
International HapMap Project
CyVera Corp.
BeadXpress System
Solexa, Inc.

CONTACTS: *Note: Officers with more than one job title may be intentionally listed here more than once.*

Jay T. Flatley, CEO
John R. Stuelpnagel, COO
Jay T. Flatley, Pres.
Christian Henry, CFO/Sr. VP
Christian G. Cabou, General Counsel/Sr. VP
Arthur L. Holden, Sr. VP-Corp. & Market Dev.
Maurissa Bornstein, Mgr.-Public Rel.
Peter J. Fromen, Sr. Dir.-Investor Rel.
John R. Stuelpnagel, Sr. VP/Gen. Mgr.-Microarray Business
John West, Sr. VP/Gen. Mgr.-DNA Sequencing
Tristan Orpin, Sr. VP-Commercial Oper.
William H. Rastetter, Chmn.

Phone: 858-202-4566	Fax: 858-202-4766
Toll-Free: 800-809-4566	
Address: 9885 Towne Centre Dr., San Diego, CA 92121-1975 US	

GROWTH PLANS/SPECIAL FEATURES:

Illumina, Inc. is a leading developer of tools for the large-scale analysis of genetic variation and function. The firm's tools provide genomic information that can be used to improve drugs and therapies, customize diagnoses and treatments and cure disease. Its patented BeadArray technology, which boasts the capability of performing multiple assays simultaneously, is deployed in two formats, both currently marketed under the Sentrix brand: the Array Matrix and the BeadChip. These use fiber optics to achieve a level of array miniaturization that allows experimentation to be easily scaled up. Illumina arranges its arrays in patterns that match the wells of industry-standard microtiter plates, allowing higher throughput than other technologies. The company's other, complementary technology, Oligator, permits parallel synthesis of the millions of pieces of DNA necessary to perform large-scale genetic analysis. Illumina has acquired two technologies as the result of mergers. The first is the VeraCode technology, acquired from CyVera Corporation. This technology is similar to the BeadArray, but is best for lower multiplex projects, and will be utilized in the company's BeadXpress system, launched in 2007. The company also acquired DNA sequencing technology from Solexa in 2007. Illumina additionally provides genotyping services for other companies, as well as software, benchtop and production systems, installation and certain warranty services for its products. Illumina is one of five U.S. research groups, chosen by the National Institutes of Health, participating in the International HapMap Project, a global consortium aimed at creating a detailed map of genetic variation. In January 2007, Illumina completed the acquisition of Solexa, Inc. for approximately $600 million.

Illumina offers its employees flexible spending accounts, an educational assistance program, training opportunities, referral bonuses, an incentive plan, an on-site gym, wellness benefits and many company activities.

FINANCIALS: Sales and profits are in thousands of dollars—add 000 to get the full amount. 2006 Note: Financial information for 2006 was not available for all companies at press time.

2006 Sales: $184,586	2006 Profits: $39,968	U.S. Stock Ticker: ILMN
2005 Sales: $73,501	2005 Profits: $-20,874	Int'l Ticker: Int'l Exchange:
2004 Sales: $50,583	2004 Profits: $-6,225	Employees: 596
2003 Sales: $28,035	2003 Profits: $-27,063	Fiscal Year Ends: 12/31
2002 Sales: $10,000	2002 Profits: $-40,300	Parent Company:

SALARIES/BENEFITS:

Pension Plan:	ESOP Stock Plan:	Profit Sharing:	Top Exec. Salary: $463,462	Bonus: $149,175
Savings Plan: Y	Stock Purch. Plan: Y		Second Exec. Salary: $319,231	Bonus: $58,500

OTHER THOUGHTS:

Apparent Women Officers or Directors: 1
Hot Spot for Advancement for Women/Minorities:

LOCATIONS: ("Y" = Yes)

West:	Southwest:	Midwest:	Southeast:	Northeast:	International:
Y				Y	Y

IMCLONE SYSTEMS INC   www.imclone.com

Industry Group Code: 325412 Ranks within this company's industry group: Sales: 37   Profits: 24

Drugs:		Other:		Clinical:	Computers:	Other:	
Discovery:	Y	AgriBio:		Trials/Services:	Hardware:	Specialty Services:	Y
Licensing:		Genomics/Proteomics:		Laboratories:	Software:	Consulting:	
Manufacturing:	Y	Tissue Replacement:		Equipment/Supplies:	Arrays:	Blood Collection:	
Development:	Y			Research/Development Svcs.:	Database Management:	Drug Delivery:	
Generics:				Diagnostics:		Drug Distribution:	

TYPES OF BUSINESS:

Drugs-Cancer
Diagnostic Products
Vaccines

BRANDS/DIVISIONS/AFFILIATES:

ERBITUX
Cetuximab

CONTACTS: *Note: Officers with more than one job title may be intentionally listed here more than once.*

Michael Howerton, CFO/Sr. VP
Lisa M. Cammy, VP-Human Resources
Phillip Frost, Exec. VP/Chief Scientific Officer
Daniel J. Connor, General Counsel/Sr. VP/Corp. Sec.
Michael P. Bailey, Sr. VP-Commercial Oper.
Margery B. Fischbein, VP-Bus. Dev.
Andrea F. Rabney, VP-Corp. Comm.
Ana I. Stancic, Sr. VP-Finance
Richard P. Crowley, Sr. VP-Biopharmaceutical Oper.
Eric K. Rowinsky, Sr. VP/Chief Medical Officer
Elizabeth Yamashita, VP-Regulatory CMC & Oper.
Margaret Dalesandro, VP-Project Mgmt.
Carl C. Icahn, Chmn.

Phone: 212-645-1405	Fax: 212-645-2054
Toll-Free:	
Address: 180 Varick St., New York, NY 10014 US	

GROWTH PLANS/SPECIAL FEATURES:

ImClone Systems, Inc. is a biopharmaceutical company that is engaged in the research and development of novel cancer treatments through growth factor and angiogenesis inhibitors. The firm's lead product is ERBITUX, a growth factor blocker that has been approved by the FDA for treatment of metastatic colorectal cancer. In 2006, the FDA approved the use of ERBITUX in combination with radiation therapy for the treatment of advanced squamous cell carcinoma in the head and neck. ImClone is currently working in collaboration with its partners, BMS and Merck KgaA in developing phase I and II evaluations of ERBITUX for the treatment of carcinomas of the brain, esophagus, stomach, rectum, cervix, endometrium, ovary, bladder, prostate and breast. In addition to the development and commercialization of ERBITUX, the company has developed several investigational agents that are in various stages of clinical development, and the firm is also developing monoclonal antibodies to inhibit angiogenesis. ImClone's research and development department has also recently begun production of antibody therapeutics that interfere with anti-apoptic signaling and survival mechanisms of cancer cells. In February 2007, ImClone and Bristol-Myers Squibb submitted an application for the use of ERBITUX as a treatment for metastatic colorectal cancer to the Japanese Pharmaceuticals and Medical Devices Agency.

The company's employee benefits package includes tuition reimbursement.

FINANCIALS: Sales and profits are in thousands of dollars—add 000 to get the full amount. 2006 Note: Financial information for 2006 was not available for all companies at press time.

2006 Sales: $677,847	2006 Profits: $370,674	U.S. Stock Ticker: IMCL
2005 Sales: $383,673	2005 Profits: $86,496	Int'l Ticker: Int'l Exchange:
2004 Sales: $388,690	2004 Profits: $113,653	Employees: 993
2003 Sales: $80,830	2003 Profits: $-112,502	Fiscal Year Ends: 12/31
2002 Sales: $60,005	2002 Profits: $-157,949	Parent Company:

SALARIES/BENEFITS:

Pension Plan:	ESOP Stock Plan:	Profit Sharing:	Top Exec. Salary: $486,000	Bonus: $
Savings Plan: Y	Stock Purch. Plan: Y		Second Exec. Salary: $320,077	Bonus: $190,000

OTHER THOUGHTS:

Apparent Women Officers or Directors: 4
Hot Spot for Advancement for Women/Minorities: Y

LOCATIONS: ("Y" = Yes)

West:	Southwest:	Midwest:	Southeast:	Northeast:	International:
				Y	

IMMTECH INTERNATIONAL   www.immtech-international.com

Industry Group Code: 325412 Ranks within this company's industry group: Sales: 145   Profits: 88

Drugs:		Other:		Clinical:		Computers:		Other:	
Discovery:	Y	AgriBio:		Trials/Services:		Hardware:		Specialty Services:	
Licensing:		Genomics/Proteomics:		Laboratories:		Software:		Consulting:	
Manufacturing:		Tissue Replacement:		Equipment/Supplies:		Arrays:		Blood Collection:	
Development:	Y			Research/Development Svcs.:		Database Management:		Drug Delivery:	
Generics:				Diagnostics:				Drug Distribution:	

TYPES OF BUSINESS:
Pharmaceuticals Discovery & Development
Drugs-Antifungal
Drugs-Infectious Diseases
Malaria Treatments

BRANDS/DIVISIONS/AFFILIATES:
pafuramidine maleate
DB289

CONTACTS: *Note: Officers with more than one job title may be intentionally listed here more than once.*
Eric L. Sorkin, CEO
Gary C. Parks, CFO
Lawrence A. Potempa, VP-Research/Chief Scientific Officer
Daniel M. Schmitt, VP-Licensing & Comm. Dev.
Gary C. Parks, Corp. Sec.
Gary C. Parks, Treas.
Cecilia Chan, Exec. Dir.
Carol Ann Olson, Sr. VP-Pharmaceutical Dev./Chief Medical Officer
Norman A. Abood, VP-Discovery Programs
Eric L. Sorkin, Chmn.

Phone: 877-898-8038	Fax: 212-791-2917
Toll-Free:	
Address: 1 North End Ave., New York, NY 10282 US	

GROWTH PLANS/SPECIAL FEATURES:

Immtech International focuses on the discovery and commercialization of therapeutics for the treatment of patients afflicted with opportunistic infectious diseases, cancer or compromised immune systems, such as those infected with HIV. The company focuses on new drugs to treat fungal diseases, based on a technology platform for the design of pharmaceutical compounds referred to as dications, which bind to the surface of infected DNA and prevent it from spreading. Immtech has formed alliances with the World Health Organization, National Center for Infectious Diseases, National Institute of Health, the Kenya Trypanosomiasis Research Institute and the Swiss Tropical Institute. Several research agreements with the University of North Carolina at Chapel Hill, Duke University, Auburn University and Georgia State University are in place, helping Immtech further its goals. By utilizing a proprietary library of aromatic cation compounds, Immtech's intention is to develop and commercialize a pipeline of new oral drugs that treat infectious diseases and other disorders. The company is beginning Phase III pivotal human clinical trials of the drug pafuramidine maleate (DB289) for the treatment of Pneumocystis carinii Pneumonia (PCP) and trypanosomiasis and a Phase II clinical trial of the drug for malaria. In January 2007, Immtech was awarded a patent for a compound that may be useful for treating bovine viral diarrhea virus and the hepatitis C virus. In June 2007 the FDA granted Immtech orphan drug designation for DB289 for the treatment of malaria. Also in June 2007, Immtech announced that the company granted an exclusive license to Par Pharmaceutical Companies, Inc. to commercialize DB298 in the United States for the treatment of PCP in AIDS patients. Immtech and Par may also collaborate to develop DB289 as a preventative therapy for patients at risk of developing PCP, including people living with HIV, cancer and other immunosuppressive conditions.

FINANCIALS: Sales and profits are in thousands of dollars—add 000 to get the full amount. 2006 Note: Financial information for 2006 was not available for all companies at press time.

2006 Sales: $3,575	2006 Profits: $-15,526	U.S. Stock Ticker: IMM
2005 Sales: $5,931	2005 Profits: $-13,433	Int'l Ticker: Int'l Exchange:
2004 Sales: $2,416	2004 Profits: $-12,846	Employees: 26
2003 Sales: $1,608	2003 Profits: $-4,679	Fiscal Year Ends: 3/31
2002 Sales: $3,500	2002 Profits: $-3,300	Parent Company:

SALARIES/BENEFITS:

Pension Plan:	ESOP Stock Plan:	Profit Sharing:	Top Exec. Salary: $263,294	Bonus: $
Savings Plan:	Stock Purch. Plan:		Second Exec. Salary: $201,234	Bonus: $

OTHER THOUGHTS:
Apparent Women Officers or Directors: 2
Hot Spot for Advancement for Women/Minorities: Y

LOCATIONS: ("Y" = Yes)

West:	Southwest:	Midwest:	Southeast:	Northeast:	International:
		Y		Y	Y

MMUCELL CORPORATION

www.immucell.com

industry Group Code: 325412B Ranks within this company's industry group: Sales: 2   Profits: 2

rugs:		Other:		Clinical:		Computers:		Other:	
iscovery:	Y	AgriBio:		Trials/Services:		Hardware:		Specialty Services:	
censing:		Genomics/Proteomics:		Laboratories:		Software:		Consulting:	
anufacturing:	Y	Tissue Replacement:		Equipment/Supplies:	Y	Arrays:		Blood Collection:	
evelopment:	Y			Research/Development Svcs.:		Database Management:		Drug Delivery:	
enerics:				Diagnostics:	Y			Drug Distribution:	

YPES OF BUSINESS:

iagnostic Tests & Products
nimal Health Drugs

BRANDS/DIVISIONS/AFFILIATES:

Wipe Out Dairy Wipes
California Mastitis Test
Mastik
MastOut
irst Defense
t

CONTACTS: *Note: Officers with more than one job title may be intentionally listed here more than once.*

Michael F. Brigham, CEO
Michael F. Brigham, Pres.
oseph H. Crabb, Chief Scientific Officer/VP
Michael F. Brigham, Sec.
Michael F. Brigham, Treas.

hone: 207-878-2770	Fax: 207-878-2117
oll-Free: 800-466-8235	
ddress: 56 Evergreen Dr., Portland, ME 04103 US	

GROWTH PLANS/SPECIAL FEATURES:

ImmuCell Corp. is a biotechnology company serving veterinarians and producers in the dairy and beef industries with products that improve animal health and productivity. The company is engaged in the sale of diagnostic tests and products for therapeutic and preventive use against certain infectious diseases in animals and humans. The firm's leading product, First Defense, is an oral medicine manufactured from cows' colostrum using the proprietary vaccine and milk protein purification technologies. The drug targets the 'calf scours' diseases, which cause diarrhea and dehydration in newborn calves and often leads to serious sickness and even death. ImmuCell also sells three products designed to aid in the management of mastitis (inflammation of the mammary gland) caused by bacterial infections. The company's second leading source of product sales is Wipe Out Dairy Wipes, which consist of pre-moistened, biodegradable towelettes that are impregnated with Nisin, a natural antibacterial peptide, to prepare the teat area of a cow in advance of milking. The firm also offers MASTIK (Mastitis Antibiotic Susceptibility Test Kit), which helps veterinarians and producers select the antibiotic most likely to be effective in the treatment of individual cases of mastitis; California Mastitis Test, which can be performed at cow side for early detection of mastitits; and rjt (Rapid Johne's Test) that can identify cattle with symptomatic Johne's disease. ImmuCell had a product development agreement with Pfizer Inc. for MastOut, a nisin-based treatment for mastitis, in which ImmuCell granted Pfizer a worldwide exclusive license to sell the product. The agreement was terminated in July 2007. In 2006, the company discontinued the manufacture of rpt (Rapid Progesterone Test).

FINANCIALS: Sales and profits are in thousands of dollars—add 000 to get the full amount. 2006 Note: Financial information for 2006 was not available for all companies at press time.

006 Sales: $4,801	2006 Profits: $ 647	U.S. Stock Ticker: ICCC	
005 Sales: $4,983	2005 Profits: $ 708	Int'l Ticker: Int'l Exchange:	
004 Sales: $3,696	2004 Profits: $ 144	Employees: 30	
003 Sales: $3,357	2003 Profits: $ 411	Fiscal Year Ends: 12/31	
002 Sales: $6,200	2002 Profits: $ 900	Parent Company:	

SALARIES/BENEFITS:

ension Plan:	ESOP Stock Plan:	Profit Sharing:	Top Exec. Salary: $149,375	Bonus: $45,000
avings Plan:	Stock Purch. Plan:		Second Exec. Salary: $90,672	Bonus: $22,668

OTHER THOUGHTS:

pparent Women Officers or Directors:
ot Spot for Advancement for Women/Minorities:

LOCATIONS: ("Y" = Yes)

West:	Southwest:	Midwest:	Southeast:	Northeast:	International:
				Y	

IMMUNE RESPONSE CORP (THE)   www.imnr.com

Industry Group Code: 325412 Ranks within this company's industry group: Sales: 160   Profits: 28

Drugs:		Other:		Clinical:	Computers:		Other:	
Discovery:	Y	AgriBio:		Trials/Services:	Hardware:		Specialty Services:	
Licensing:		Genomics/Proteomics:		Laboratories:	Software:		Consulting:	
Manufacturing:		Tissue Replacement:		Equipment/Supplies:	Arrays:		Blood Collection:	
Development:	Y			Research/Development Svcs.:	Database Management:		Drug Delivery:	
Generics:				Diagnostics:			Drug Distribution:	

TYPES OF BUSINESS:

Drugs-Autoimmune Diseases
Drugs-Cancer & HIV
Drugs-Multiple Sclerosis
Drugs-Arthritis

BRANDS/DIVISIONS/AFFILIATES:

Remune
IR103
NeuroVax
Immune Response Corporation (The)

CONTACTS: *Note: Officers with more than one job title may be intentionally listed here more than once.*

Joseph F. O'Neill, CEO
Michael K. Green, COO
Joseph F. O'Neill, Pres.
Michael K. Green, CFO
Richard Bartholomew, VP-R&D
Peter Lowry, VP-Mfg.
Georgia Theofan, VP-Clinical Dev.
Robert E. Knowling, Jr., Chmn.

Phone: 760-431-7080	Fax: 760-431-8636
Toll-Free:	
Address: 5931 Darwin Ct., Carlsbad, CA 92008 US	

GROWTH PLANS/SPECIAL FEATURES:

Orchestra Therapeutics, Inc., formerly The Immune Response Corporation, is an immuno-pharmaceutical firm focused on the discovery and development of novel treatments for autoimmune diseases. The lead immune-based therapeutic product is NeuroVax for the treatment of multiple sclerosis, or MS. In addition to MS, the company has proprietary technology and prior clinical experience for clinical evaluation of TCR peptide-based immune-based therapies for rheumatoid arthritis (RA) and psoriasis. The targeted strategy behind the company's autoimmune therapies is reflected in the name Orchestra. Rather than disrupting the function of the entire immune system, these therapeutic vaccines are designed to elicit a very specific response (akin to correcting one instrument in an orchestra that is out of tune) to help control disease. NeuroVax is an investigational T-Cell Receptor peptide vaccine for the treatment of relapsing-remitting forms of MS, which appears to work by enhancing levels of FOXP3+ Regulatory T-cells within the immune system, which may help control levels of pathogenic T-cells in MS patients. Data from the company's most recent Phase II clinical trial in MS showed that reduced levels of FOXP3 can be restored to normal levels after repeated vaccinations with NeuroVax. The company recently announced the injection of the first patient in a large multi-center Phase II study to assess the safety and efficacy of NeuroVax. The company is in discussions with several academic institutions to conduct preclinical work on therapeutic vaccines to treat psoriasis and RA. Based on findings to be derived from these product development programs, the company plant to initiate Phase I trials in one of these new autoimmune areas in 2008. Orchestra also has two HIV product candidates, REMUNE and IR103. In 2007 the company changed its name to Orchestra in order to better reflect its expanded focus on the treatment of autoimmune diseases.

FINANCIALS: Sales and profits are in thousands of dollars—add 000 to get the full amount. 2006 Note: Financial information for 2006 was not available for all companies at press time.

2006 Sales: $ 932	2006 Profits: $195,290	U.S. Stock Ticker: OCHT
2005 Sales: $ 44	2005 Profits: $-17,313	Int'l Ticker: Int'l Exchange:
2004 Sales: $ 323	2004 Profits: $-29,959	Employees: 39
2003 Sales: $ 66	2003 Profits: $-28,799	Fiscal Year Ends: 12/31
2002 Sales: $ 47	2002 Profits: $-30,835	Parent Company:

SALARIES/BENEFITS:

Pension Plan:	ESOP Stock Plan:	Profit Sharing:	Top Exec. Salary: $283,590	Bonus: $225,000
Savings Plan: Y	Stock Purch. Plan: Y		Second Exec. Salary: $227,596	Bonus: $135,000

OTHER THOUGHTS:

Apparent Women Officers or Directors: 1
Hot Spot for Advancement for Women/Minorities:

LOCATIONS: ("Y" = Yes)

West:	Southwest:	Midwest:	Southeast:	Northeast:	International:
Y				Y	

IMMUNOGEN INC

www.immunogen.com

Industry Group Code: 325412 Ranks within this company's industry group: Sales: 95 Profits: 101

Drugs:		Other:		Clinical:		Computers:		Other:	
Discovery:	Y	AgriBio:	Y	Trials/Services:		Hardware:		Specialty Services:	
Licensing:	Y	Genomics/Proteomics:		Laboratories:		Software:		Consulting:	
Manufacturing:		Tissue Replacement:		Equipment/Supplies:		Arrays:		Blood Collection:	
Development:	Y			Research/Development Svcs.:		Database Management:		Drug Delivery:	
Generics:				Diagnostics:				Drug Distribution:	

TYPES OF BUSINESS:

Anticancer Biopharmaceuticals
Tumor-Activated Drugs

BRANDS/DIVISIONS/AFFILIATES:

HuN901-DM1
HuC242-DM4
AVE9633
Transtuzumab-DM1
SAR3419
AVE1642

CONTACTS: Note: Officers with more than one job title may be intentionally listed here more than once.

Mitchel Sayare, CEO
Mitchel Sayare, Pres.
Daniel Junius, CFO/Sr. VP-Finance
Linda Buono, Sr. Dir.-Human Resources
Thomas Chittenden, Exec. Dir.-Research
Godfrey Amphlett, Exec. Dir.-Process & Prod. Dev.
Thomas Lauzon, Exec. Dir.-Mfg. Oper.
Leland Webster, Exec. Dir.-Bus. Dev.
Carol Hausner, Exec. Dir.-Corp. Comm
Carol Hausner, Exec. Dir.-Investor Rel.
David G. Foster, Corp. Controller
Robert J. Lutz, Exec. Dir.-Preclinical Research
John Lambert, Sr. VP-Pharmaceutical Dev.
Mitchel Sayare, Chmn.

Phone: 617-995-2500	Fax: 617-995-2510

Toll-Free:

Address: 128 Sidney St., Cambridge, MA 02139 US

GROWTH PLANS/SPECIAL FEATURES:

ImmunoGen develops targeted therapeutics for the treatment of cancer. The company's Tumor-Activated Prodrug (TAP) technology uses antibodies to deliver a cytotoxic agent specifically to cancer cells. The TAP technology is designed to enable the creation of potent well-tolerated anticancer products. The firm currently has four products in clinical research: HuN901-DM1, in Phase I and II for treatment of small-cell lung cancer, certain neuroendocrine cancers and certain hematological malignancies; HuC242-DM4, in Phase I for treatment of gastrointestinal cancers; AVE9633, in Phase I for treatment of acute myeloid leukemia; and Transtuzumab-DM1, in Phase I for treatment of HER2-positive metastatic breast cancers. ImmunoGen owns the rights to develop and commercialize huN901-DM1 and huC242-DM4. Products in preclinical research include AVE1642 for treatment of solid tumors and certain hematological malignancies; and SAR3419 for treatment of B-cell malignancies including non-Hodgkin's lymphoma. ImmunoGen's collaborative partners include Amgen, Inc.; Biogen Idec, Inc.; Boehringer Ingelheim International CmbH; Centocor, Inc.; Genetech, Inc.; Millennium Pharmaceuticals, Inc.; the sanofi-aventis Group; and Biotest AG. In June 2007, the firm announced positive clinical findings with three TAP compounds, trastuzumab-DM1, huN901-DM1 and huC242-DM4, for the treatment of solid tumors.

The company offers its employees health and dental coverage; life and AD&D insurance; a 401(k) plan; tuition reimbursement; and stock options.

FINANCIALS: Sales and profits are in thousands of dollars—add 000 to get the full amount. 2006 Note: Financial information for 2006 was not available for all companies at press time.

2006 Sales: $32,088	2006 Profits: $-17,834	U.S. Stock Ticker: IMGN
2005 Sales: $35,718	2005 Profits: $-10,951	Int'l Ticker: Int'l Exchange:
2004 Sales: $25,956	2004 Profits: $-5,917	Employees: 192
2003 Sales: $7,628	2003 Profits: $-19,982	Fiscal Year Ends: 6/30
2002 Sales: $5,900	2002 Profits: $-14,600	Parent Company:

SALARIES/BENEFITS:

Pension Plan:	ESOP Stock Plan:	Profit Sharing:	Top Exec. Salary: $420,264	Bonus: $142,994
Savings Plan: Y	Stock Purch. Plan: Y		Second Exec. Salary: $304,067	Bonus: $85,200

OTHER THOUGHTS:

Apparent Women Officers or Directors: 3
Hot Spot for Advancement for Women/Minorities: Y

LOCATIONS: ("Y" = Yes)

West:	Southwest:	Midwest:	Southeast:	Northeast:	International:
				Y	

Note: Financial information, benefits and other data can change quickly and may vary from those stated here.

IMMUNOMEDICS INC

www.immunomedics.com

Industry Group Code: 325413 Ranks within this company's industry group: Sales: 28   Profits: 23

Drugs:		Other:		Clinical:		Computers:		Other:	
Discovery:	Y	AgriBio:		Trials/Services:		Hardware:		Specialty Services:	
Licensing:	Y	Genomics/Proteomics:		Laboratories:		Software:		Consulting:	
Manufacturing:	Y	Tissue Replacement:		Equipment/Supplies:		Arrays:		Blood Collection:	
Development:	Y			Research/Development Svcs.:		Database Management:		Drug Delivery:	
Generics:				Diagnostics:	Y			Drug Distribution:	

TYPES OF BUSINESS:

Monoclonal Antibody-Based Products
Cancer & Autoimmune Products

BRANDS/DIVISIONS/AFFILIATES:

Epratuzumab
CEA-Scan
LeukoScan
LymphoScan
AFP-Scan
UCB S.A.

CONTACTS: *Note: Officers with more than one job title may be intentionally listed here more than once.*

Cynthia L. Sullivan, CEO
Cynthia L. Sullivan, Pres.
Gerald G. Gorman, CFO
David M. Goldenberg, Chief Strategic Officer
Gerald G. Gorman, VP-Finance
David M. Goldenberg, Chmn.

Phone: 973-605-8200	Fax: 973-605-8282
Toll-Free:	
Address: 300 American Rd., Morris Plains, NJ 07950 US	

GROWTH PLANS/SPECIAL FEATURES:

Immunomedics, Inc. is a biopharmaceutical company focused on the development of monoclonal antibody-based products for the targeted treatment of cancer, autoimmune and other serious diseases. The company developed a number of proprietary technologies that allow it to create humanized antibodies that can be used either alone in unlabeled or 'naked' form, or conjugated with radioactive isotopes, chemotherapeutics or toxins, in each case to create targeted agents. Diagnostic imaging products manufactured by Immunomedics include CEA-Scan, for the detection of colorectal, lung and breast cancers, and LeukoScan, for the detection of bone infections. CEA-Scan has been approved in the U.S., Canada and Europe, and LeukoScan has been approved in Canada, Europe and Australia. The company also offers ImmuSTRIP, a laboratory test kit for the detection of human anti-murine antibodies. Products under development include LymphoScan (epratuzumab) for the detection of non-Hodgkin's lymphoma and AFP-Scan for the diagnosis of primary liver and germ-cell cancers. In addition to diagnostic materials, Immunomedics also has a number of therapeutic product candidates currently going through clinical trials. Most of these target such autoimmune diseases as systemic lupus erythematosus and Sjogren's syndrome, B-cell non-Hodgkins lymphoma and various solid tumors. One of the most prominent of these is Epratuzumab, which is currently on the FDA fast track for the treatment of lupus and Sjogren's syndrome. The firm has a portfolio of 108 U.S. and 250 foreign patents. In May 2006, the company reached a collaboration and license agreement with UCB S.A., a leading global pharmaceutical company, granting UCB exclusive rights to develop, market and sell epratuzumab worldwide for all autoimmune applications.

The company offers its employees medical, dental and vision insurance; flexible spending accounts; life insurance; a 401(k) plan; and tuition reimbursement.

FINANCIALS: Sales and profits are in thousands of dollars—add 000 to get the full amount. 2006 Note: Financial information for 2006 was not available for all companies at press time.

2006 Sales: $4,353	2006 Profits: $-28,764	**U.S. Stock Ticker: IMMU**	
2005 Sales: $3,813	2005 Profits: $-26,758	**Int'l Ticker:** Int'l Exchange:	
2004 Sales: $4,306	2004 Profits: $-22,355	Employees: 106	
2003 Sales: $13,719	2003 Profits: $-7,874	Fiscal Year Ends: 6/30	
2002 Sales: $14,300	2002 Profits: $-3,700	Parent Company:	

SALARIES/BENEFITS:

Pension Plan:	ESOP Stock Plan:	Profit Sharing:	Top Exec. Salary: $520,000	Bonus: $104,000
Savings Plan: Y	Stock Purch. Plan:		Second Exec. Salary: $455,000	Bonus: $80,000

OTHER THOUGHTS:

Apparent Women Officers or Directors: 2
Hot Spot for Advancement for Women/Minorities: Y

LOCATIONS: ("Y" = Yes)

West:	Southwest:	Midwest:	Southeast:	Northeast:	International:
				Y	Y

Note: Financial information, benefits and other data can change quickly and may vary from those stated here.

IMPAX LABORATORIES INC

www.impaxlabs.com

Industry Group Code: 325412A Ranks within this company's industry group: Sales: Profits:

Drugs:		Other:		Clinical:		Computers:		Other:	
Discovery:		AgriBio:		Trials/Services:		Hardware:		Specialty Services:	
Licensing:		Genomics/Proteomics:		Laboratories:		Software:		Consulting:	
Manufacturing:	Y	Tissue Replacement:		Equipment/Supplies:		Arrays:		Blood Collection:	
Development:	Y			Research/Development Svcs.:		Database Management:		Drug Delivery:	Y
Generics:	Y			Diagnostics:				Drug Distribution:	

TYPES OF BUSINESS:

Drug Delivery Systems
Drugs-Generic
Pharmaceutical Discovery & Development
Drugs-Central Nervous System

BRANDS/DIVISIONS/AFFILIATES:

Global Pharmaceuticals
IMPAX Pharmaceuticals

CONTACTS: Note: Officers with more than one job title may be intentionally listed here more than once.

Larry Hsu, CEO
Larry Hsu, Pres.
Arthur A. Koch, Jr., CFO
Charles V. Hildenbrand, Sr. VP-Oper.
Arthur A. Koch, Jr., Sr. VP-Finance
David S. Doll, Exec. VP-Commercial Oper.
Charles Hsaio, Chmn.

Phone: 510-476-2000	Fax: 510-471-3200
Toll-Free:	
Address: 30831 Huntwood Ave., Hayward, CA 94544 US	

GROWTH PLANS/SPECIAL FEATURES:

IMPAX Laboratories, Inc. is a technology-based, specialty pharmaceutical company focused on the development and commercialization of generic and brand-name pharmaceuticals. In the generic market, the firm is primarily focusing on selected controlled-release versions of brand-name drugs. In the brand-name drugs market, the company is developing products for the treatment of central nervous system disorders. The company markets branded products through the IMPAX Pharmaceuticals division and generic products through the Global Pharmaceuticals division. IMPAX has developed several proprietary controlled-release delivery technologies that can be utilized with a variety of oral dosage forms and drug release rates. These include the concentric multiple-particulate delivery system, the timed multiple-action delivery system and the dividable multiple-action delivery system. Generic products include controlled-release versions of Dantrium, Ditropan XL, Prilosec, Claritin, Zyban and Wellbutrin.

IMPAX employees receive a benefits package that includes stock options; flexible spending accounts for dependent care and medical expenses; and a confidential employee assistance program.

FINANCIALS: Sales and profits are in thousands of dollars—add 000 to get the full amount. 2006 Note: Financial information for 2006 was not available for all companies at press time.

2006 Sales: $	2006 Profits: $	U.S. Stock Ticker: IPXL.PK	
2005 Sales: $	2005 Profits: $	Int'l Ticker: Int'l Exchange:	
2004 Sales: $	2004 Profits: $	Employees: 453	
2003 Sales: $58,818	2003 Profits: $-14,207	Fiscal Year Ends: 12/31	
2002 Sales: $24,500	2002 Profits: $-20,000	Parent Company:	

SALARIES/BENEFITS:

Pension Plan:	ESOP Stock Plan: Y	Profit Sharing:	Top Exec. Salary: $266,539	Bonus: $100,000
Savings Plan: Y	Stock Purch. Plan: Y		Second Exec. Salary: $263,100	Bonus: $46,250

OTHER THOUGHTS:

Apparent Women Officers or Directors:
Hot Spot for Advancement for Women/Minorities:

LOCATIONS: ("Y" = Yes)

West:	Southwest:	Midwest:	Southeast:	Northeast:	International:
Y				Y	

Note: Financial information, benefits and other data can change quickly and may vary from those stated here.

IMS HEALTH INC

www.imshealth.com

Industry Group Code: 541910 Ranks within this company's industry group: Sales: 1   Profits: 1

Drugs:	Other:	Clinical:	Computers:		Other:	
Discovery:	AgriBio:	Trials/Services:	Hardware:		Specialty Services:	Y
Licensing:	Genomics/Proteomics:	Laboratories:	Software:	Y	Consulting:	Y
Manufacturing:	Tissue Replacement:	Equipment/Supplies:	Arrays:		Blood Collection:	
Development:		Research/Development Svcs.:	Database Management:	Y	Drug Delivery:	
Generics:		Diagnostics:			Drug Distribution:	

TYPES OF BUSINESS:

Market Research - Pharmaceuticals
Pharmaceutical Sales Tracking
Health Care Databases
Software-Sales Management & Market Research
Physician Profiling
Industry Audits
Prescription Tracking Reporting Services

BRANDS/DIVISIONS/AFFILIATES:

Drug Distribution Data (DDD) Service
ValueMedics

CONTACTS: *Note: Officers with more than one job title may be intentionally listed here more than once.*

David R. Carlucci, CEO
Giles V. J. Pajot, COO
David R. Carlucci, Pres.
Leslye G. Katz, CFO/Sr. VP
Bruce F. Boggs, Sr. VP-Global Mktg. & External Affairs
Robert H. Steinfeld, General Counsel/Sr. VP/Corp. Sec.
Murray L. Aitken, Sr. VP-Corp. Strategy
Stephen Phua, VP/General Mgr.-Asia Pacific
Tatsuyuki Saeki, Pres. & Representative Dir.-Japan
William J. Nelligan, Pres., IMS America
Giles V. J. Pajot, Exec. VP/Pres., Global Bus. Mgmt.
David R. Carlucci, Chmn.
Kevin Knightly, Pres., EMEA

Phone: 203-319-4700	Fax: 203-319-4701
Toll-Free:	
Address: 1499 Post Rd., Fairfield, CT 06824 US	

GROWTH PLANS/SPECIAL FEATURES:

IMS Health, Inc. is one of the world's leading providers of market intelligence to the pharmaceutical and health care industries. IMS provides products designed for sales force effectiveness, as well as products for portfolio optimization and launch/brand management. The firm's services include consultation with clients regarding pharmaceutical market trends and development of customized software applications and data warehouse tools. The firm's sales force effectiveness product offerings have produced as much as half of the firm's worldwide revenue. The firm's drug distribution data (DDD) service falls within the scope sales force effectiveness. The DDD allows clients to track the flow of sales for their products through various channels of distribution, including direct sales by pharmaceutical manufacturers and indirect sales through wholesalers and other sources. The firm's prescription tracking reporting services also work within the sales force effectiveness segment, helping companies effectively market to the drug prescribers. The portfolio optimization services encompass pharmaceutical and prescription audits, which in turn reveal a company's product sales broken down into therapeutic class, package size and dosage form. IMS's launch and brand management services offer consulting services to firms in each stage of a product/brand's lifecycle. The firm also provides services related to managed care and consumer health. Concerning consumer health, IMS is currently focused on providing over-the-counter (OTC) manufacturers insights into consumer purchasing dynamics and promotional impact. In 2007, the firm announced the acquisition of ValueMedics Research LLC, a healthcare research and consulting firm.

FINANCIALS: Sales and profits are in thousands of dollars—add 000 to get the full amount. 2006 Note: Financial information for 2006 was not available for all companies at press time.

2006 Sales: $1,958,588	2006 Profits: $315,511	U.S. Stock Ticker: RX
2005 Sales: $1,754,791	2005 Profits: $284,091	Int'l Ticker: Int'l Exchange:
2004 Sales: $1,569,045	2004 Profits: $285,422	Employees: 7,400
2003 Sales: $1,381,800	2003 Profits: $639,000	Fiscal Year Ends: 12/31
2002 Sales: $1,428,100	2002 Profits: $266,100	Parent Company:

SALARIES/BENEFITS:

Pension Plan:	ESOP Stock Plan:	Profit Sharing:	Top Exec. Salary: $801,250	Bonus: $1,041,625
Savings Plan: Y	Stock Purch. Plan: Y		Second Exec. Salary: $675,780	Bonus: $646,123

OTHER THOUGHTS:

Apparent Women Officers or Directors: 2
Hot Spot for Advancement for Women/Minorities: Y

LOCATIONS: ("Y" = Yes)

West:	Southwest:	Midwest:	Southeast:	Northeast:	International:
Y				Y	Y

INCYTE CORP

www.incyte.com

Industry Group Code: 325412 Ranks within this company's industry group: Sales: 99 Profits: 172

Drugs:		Other:		Clinical:		Computers:		Other:	
Discovery:	Y	AgriBio:		Trials/Services:		Hardware:		Specialty Services:	
Licensing:	Y	Genomics/Proteomics:	Y	Laboratories:		Software:		Consulting:	
Manufacturing:		Tissue Replacement:		Equipment/Supplies:		Arrays:		Blood Collection:	
Development:	Y			Research/Development Svcs.:	Y	Database Management:		Drug Delivery:	
Generics:				Diagnostics:				Drug Distribution:	

TYPES OF BUSINESS:
Drug Discovery & Development
Genetic Information Research
Gene Expression Services

BRANDS/DIVISIONS/AFFILIATES:
Incyte Genomics
BioKnowledge Library (BKL)
CCR5 Antagonists
NCB9471
NCB13739
Dexelvucitabine

CONTACTS: Note: Officers with more than one job title may be intentionally listed here more than once.
Paul A. Friedman, CEO
Paul A. Friedman, Pres.
David C. Hastings, CFO/Exec. VP
Paula J. Swain, Exec. VP-Human Resources
Brian W. Metcalf, Chief Drug Discovery Scientist/Exec. VP
Patricia A. Schreck, General Counsel/Exec. VP
John A. Keller, Chief Bus. Officer/Exec. VP
Richard U. DeSchutter, Chmn.

Phone: 302-498-6700	Fax: 302-425-2750
Toll-Free:	
Address: Rte. 141 & Henry Clay Rd., Bldg. E336, Wilmington, DE 19880 US	

GROWTH PLANS/SPECIAL FEATURES:
Incyte Corp., formerly Incyte Genomics, discovers and develops small molecule drugs for the treatment of HIV, diabetes, oncology and inflammation. The company was historically involved with marketing and selling access to its proprietary genomic information database, the BioKnowledge Library (BKL), however it has since discontinued that service. The company currently has multiple products in its pipeline, all within preclinical, Phase I and Phase II clinical studies. For HIV, Incyte is developing CCR5 Antagonists, which belong to a new class of drugs called HIV Entry Inhibitors. The product works by blocking HIV before the virus enters the cell and begins the replication process. The lead product candidate is INCB9471, and currently is in Phase II clinical trials. Incyte's diabetes program relies on products that block the enzyme 11ßHSD1, which may potentially reduce insulin resistance and restore glycemic control in type 2 diabetes. The product INCB13739 is currently in Phase II clinical study. The cancer study program includes products which block epidermal growth factor receptor pathways, which are normally tightly regulated but become out of control in instances of cancer, especially solid tumors. Incyte currently has one product in Phase II clinical studies for cancer. In inflammation program uses CCR2 antagonists which block macrophage accumulation around multiple sclerosis related lesions. An excess of macrophages leads to loss of muscle control, vision, balance and sensation. The company has one product for inflammation in preclinical study. Incyte also is researching the use and control of Janus-associated Kinase Inhibitors (JAK), which may be useful in treating multiple inflammatory diseases, myeloproliferative disorders, and certain cancers. In April 2006, Incyte was forced to discontinue development of Dexelvucitabine, a previous lead candidate for the treatment of HIV, due to complications arising in patients undergoing the treatment.

FINANCIALS: Sales and profits are in thousands of dollars—add 000 to get the full amount. 2006 Note: Financial information for 2006 was not available for all companies at press time.

2006 Sales: $27,643	2006 Profits: $-74,166	U.S. Stock Ticker: INCY	
2005 Sales: $7,846	2005 Profits: $-103,043	Int'l Ticker:	Int'l Exchange:
2004 Sales: $14,146	2004 Profits: $-164,817	Employees: 186	
2003 Sales: $47,092	2003 Profits: $-166,463	Fiscal Year Ends: 12/31	
2002 Sales: $101,612	2002 Profits: $-136,885	Parent Company:	

SALARIES/BENEFITS:
Pension Plan:	ESOP Stock Plan:	Profit Sharing:	Top Exec. Salary: $543,577	Bonus: $393,120
Savings Plan:	Stock Purch. Plan:		Second Exec. Salary: $350,010	Bonus: $180,797

OTHER THOUGHTS:
Apparent Women Officers or Directors: 2
Hot Spot for Advancement for Women/Minorities: Y

LOCATIONS: ("Y" = Yes)
West:	Southwest:	Midwest:	Southeast:	Northeast:	International:
Y				Y	

Note: Financial information, benefits and other data can change quickly and may vary from those stated here.

# INDEVUS PHARMACEUTICALS INC																		www.indevus.com

Industry Group Code: 325412 Ranks within this company's industry group: Sales: 77   Profits: 155

Drugs:		Other:		Clinical:	Computers:		Other:	
Discovery:		AgriBio:		Trials/Services:	Hardware:		Specialty Services:	
Licensing:	Y	Genomics/Proteomics:		Laboratories:	Software:		Consulting:	
Manufacturing:	Y	Tissue Replacement:		Equipment/Supplies:	Arrays:		Blood Collection:	
Development:	Y			Research/Development Svcs.:	Database Management:		Drug Delivery:	
Generics:				Diagnostics:			Drug Distribution:	

TYPES OF BUSINESS:

Pharmaceuticals Acquisition & Development
Drugs-Urological, Gynecological & Men's Health
Hormone Replacement Products

BRANDS/DIVISIONS/AFFILIATES:

Interneuron Pharmaceuticals
Valera Pharmaceuticals, Inc.
Supprelin LA
PRO 2000
Sanctura
Citicoline
Nebido
Delatestryl

CONTACTS: *Note: Officers with more than one job title may be intentionally listed here more than once.*

Glenn L. Cooper, CEO
Thomas F. Farb, COO
Thomas F. Farb, Pres.
Michael W. Rogers, CFO/Exec. VP/Treas.
John Tucker, Exec. VP/Chief Sales & Mktg. Officer
Bobby W. Sandage, Jr., Exec. VP-R&D/Chief Scientific Officer
Mark S. Butler, Exec. VP/Chief Admin. Officer
Mark S. Butler, General Counsel
Noah D. Beerman, Exec. VP/Chief Bus. Officer
Dale Ritter, Sr. VP-Finance
Kurt Lewis, Sr. VP-Sales & Mktg.
Glenn L. Cooper, Chmn.

Phone: 781-861-8444	Fax: 781-861-3830
Toll-Free:	
Address: 33 Hayden Ave., Lexington, MA 02421 US	

GROWTH PLANS/SPECIAL FEATURES:

Indevus Pharmaceuticals, Inc. acquires, develops and commercializes products that treat a variety of diseases. The firm focuses on product development in the urology, gynecology and men's health markets. Indevus' product line includes Sarafem, licensed to Eli Lilly & Co., as a FDA-approved treatment for premenstrual dysphoric disorder. The company, through a partnership with Espirit, Inc., also markets Sanctura, a muscarinic receptor antagonist designed to treat patients with overactive bladders. The company's product candidates include Pagoclone, which may treat panic and generalized anxiety disorders; Citicoline for ischemic stroke; PRO 2000, a topical microbicide that may prevent the transmission of HIV and other sexually transmitted diseases; IP 751, an anti-inflammatory and analgesic compound; and aminocandin, an anti-fungal compound designed to treat a broad spectrum of systemic infections. Indevus has licensed exclusive U.S. rights to market Nebido, a long-acting injectable testosterone preparation for the treatment of male hypogonadism, from Schering AG. The firm's strategy is to focus on individual compounds with proven results, as opposed to more unproven drug discovery technologies. In 2007, the company launched SUPPRELIN LA subcutaneous implant, a drug intended for the treatment of children with Central Precocious Puberty. Indevus completed a merger with Valera Pharmaceuticals, Inc. in April 2007.

Indevus Pharmaceuticals offers its employees benefits including an assistance plan, stock options and long and short term disability coverage.

FINANCIALS: Sales and profits are in thousands of dollars—add 000 to get the full amount. 2006 Note: Financial information for 2006 was not available for all companies at press time.

2006 Sales: $50,452	2006 Profits: $-50,554	U.S. Stock Ticker: IDEV	
2005 Sales: $33,336	2005 Profits: $-53,218	Int'l Ticker: Int'l Exchange:	
2004 Sales: $18,726	2004 Profits: $-68,212	Employees: 158	
2003 Sales: $5,245	2003 Profits: $-31,812	Fiscal Year Ends: 9/30	
2002 Sales: $4,400	2002 Profits: $-17,600	Parent Company:	

SALARIES/BENEFITS:

Pension Plan:	ESOP Stock Plan: Y	Profit Sharing:	Top Exec. Salary: $477,000	Bonus: $131,509
Savings Plan: Y	Stock Purch. Plan: Y		Second Exec. Salary: $333,900	Bonus: $55,928

OTHER THOUGHTS:

Apparent Women Officers or Directors:
Hot Spot for Advancement for Women/Minorities:

LOCATIONS: ("Y" = Yes)

West:	Southwest:	Midwest:	Southeast:	Northeast:	International:
				Y	

INFINITY PHARMACEUTICALS INC

www.ipi.com

ndustry Group Code: 541710 Ranks within this company's industry group: Sales: 16 Profits: 16

Drugs:		Other:	Clinical:	Computers:	Other:	
Discovery:	Y	AgriBio:	Trials/Services:	Hardware:	Specialty Services:	Y
Licensing:		Genomics/Proteomics:	Laboratories:	Software:	Consulting:	
Manufacturing:		Tissue Replacement:	Equipment/Supplies:	Arrays:	Blood Collection:	
Development:	Y		Research/Development Svcs.:	Database Management:	Drug Delivery:	
Generics:			Diagnostics:		Drug Distribution:	

TYPES OF BUSINESS:

Research & Development Services
Drug Discovery Services
Drugs-Cancer

BRANDS/DIVISIONS/AFFILIATES:

Discovery Partners International, Inc.
MedImmune, Inc.
IPI-504
Novartis Institutes of BioMedical Research

CONTACTS: Note: Officers with more than one job title may be intentionally listed here more than once.

Steven H. Holtzman, CEO
Julian Adams, Pres.
Julian Adams, Chief Scientific Officer
Jeffrey K. Tong, VP-Corp. & Prod. Dev.
Gerald E. Quirk, General Counsel/VP
Adelene Q. Perkins, Chief Bus. Officer/Exec. VP
Steven J. Kafka, VP-Finance
David S. Grayzel, VP-Clinical Dev. & Medical Affairs
Vito J. Palombella, VP-Biology
Steven J. Kafka, VP-Strategic Prod. Planning
James L. Wright, VP-Pharmaceutical Dev.
Steven H. Holtzman, Chmn.

Phone: 617-453-1000	Fax: 617-453-1001

Toll-Free: 877-248-6417
Address: 780 Memorial Dr., Cambridge, MA 02139 US

GROWTH PLANS/SPECIAL FEATURES:

Infinity Pharmaceuticals, Inc. (IPI) discovers, develops and delivers medicines for the treatment of cancer and related conditions. IPI has built a pipeline of product candidates utilizing small molecule drug technologies for multiple cancer indications. Being developed in collaboration with MedImmune, Inc. and the company's lead product candidate, IPI-504 is currently undergoing a Phase I clinical trial in patients with Gleevec-refractory gastrointestinal stromal tumors, as well as a Phase I/II clinical trial in patients with advanced non-small cell lung cancer. IPI-504 inhibits heat shock protein 90, which inhibition is believed to have broad therapeutic potential for patients with solid and hematological tumors. The company's next most advanced program, also being developed in collaboration with MedImmune, is directed against the Hedgehog cell pathway, which normally regulates tissue and organ formation during embryonic development. When abnormally activated during adulthood, the Hedgehog pathway is believed to play a central role in allowing the proliferation and survival of certain cancer-causing cells, and is implicated in many of the most deadly cancers. IPI plans to select a clinical candidate in its Hedgehog pathway inhibitor program during 2007. The company's third largest program, in collaboration with the Novartis Institutes of BioMedical Research, is aimed at identifying small molecule compounds which inhibit the Bc1-2 family of proteins, key regulators of programmed cell death. Cancers with abnormally high levels of Bc1-2 are believed to evade cell death and thus become increasingly resistant to chemotherapy. In September 2006, Discovery Partners International changed its corporate name to Infinity Pharmaceuticals, Inc. after a subsidiary of DPI's merged with and into IPI to become Infinity Discovery, Inc.

Infinity offers its employees a First Time Homebuyer Assistance Program; health and dental insurance; pre-tax Flexible Spending Accounts; Life-Balance support programs; paid sabbatical leave eligibility; and discounted home and auto insurance.

FINANCIALS: Sales and profits are in thousands of dollars—add 000 to get the full amount. 2006 Note: Financial information for 2006 was not available for all companies at press time.

2006 Sales: $18,495	2006 Profits: $-28,448	U.S. Stock Ticker: INFI
2005 Sales: $ 522	2005 Profits: $-36,369	Int'l Ticker: Int'l Exchange:
2004 Sales: $44,268	2004 Profits: $3,903	Employees: 133
2003 Sales: $45,209	2003 Profits: $1,059	Fiscal Year Ends: 12/31
2002 Sales: $41,315	2002 Profits: $-62,113	Parent Company:

SALARIES/BENEFITS:

Pension Plan:	ESOP Stock Plan:	Profit Sharing:	Top Exec. Salary: $450,000	Bonus: $182,000
Savings Plan: Y	Stock Purch. Plan: Y		Second Exec. Salary: $375,000	Bonus: $109,500

OTHER THOUGHTS:

Apparent Women Officers or Directors: 1
Hot Spot for Advancement for Women/Minorities:

LOCATIONS: ("Y" = Yes)

West:	Southwest:	Midwest:	Southeast:	Northeast:	International:
				Y	

Note: Financial information, benefits and other data can change quickly and may vary from those stated here.

INSITE VISION INC

www.insitevision.com

Industry Group Code: 325412A **Ranks within this company's industry group:** Sales: 21   Profits: 11

Drugs:		Other:		Clinical:		Computers:		Other:	
Discovery:	Y	AgriBio:		Trials/Services:		Hardware:		Specialty Services:	
Licensing:		Genomics/Proteomics:	Y	Laboratories:		Software:		Consulting:	
Manufacturing:		Tissue Replacement:		Equipment/Supplies:		Arrays:		Blood Collection:	
Development:	Y			Research/Development Svcs.:		Database Management:		Drug Delivery:	Y
Generics:				Diagnostics:	Y			Drug Distribution:	

TYPES OF BUSINESS:

Drug Delivery Systems
Drugs-Ophthalmic
Eye Disease Diagnostics

BRANDS/DIVISIONS/AFFILIATES:

DuraSite
OcuGene
AzaSite
AzaSite Plus
AzaSite Otic
AzaSite Extra
AquaSite
IV-403

CONTACTS: Note: Officers with more than one job title may be intentionally listed here more than once.

S. Kumar Chandrasekaran, CEO
S. Kumar Chandrasekaran, Pres.
S. Kumar Chandrasekaran, CFO
Sandra C. Heine, VP-Admin.
Lyle M. Bowman, VP-Oper.
Lyle M. Bowman, VP-Dev.
Joyce Strand, Sr. Dir.-Corp. Comm.
Joyce Strand, Sr. Dir.-Investor Rel.
Sandra C. Heine, VP-Finance
David F. Heniges, VP/General Mgr.-Commercial Opportunities
Ronald H. Carlson, VP-Regulatory Affairs & Quality
S. Kumar Chandrasekaran, Chmn.

Phone: 510-865-8800	Fax: 510-865-5700
Toll-Free:	
Address: 965 Atlantic Ave., Alameda, CA 94501 US	

GROWTH PLANS/SPECIAL FEATURES:

InSite Vision, Inc. is an ophthalmic product development company focusing on therapies that treat ocular infections glaucoma and ocular diseases. Products are based or proprietary DuraSite eye drop-based drug delivery technology, a patented delivery system that allows medication to stay in the eye for several hours rather than just a few minutes, reducing the number of doses the patient must take and lowering the potential for adverse side effects. InSite's lead product, AzaSite (a DuraSite formulation of azithromycin, a broad spectrum antibiotic) has been approved by the FDA for the treatment of bacterial conjunctivitis. InSite has entered into an initial agreement with Inspire Pharmaceuticals to commercialize AzaSite in the U.S. and Canada. Three other products in the AzaSite family, AzaSite Plus, AzaSite Extra and AzaSite Otic, are currently in development. AzaSite Plus is under development for the treatment of bacterial infection and inflammation and is currently in Phase I trials. AzaSite Otic is under development for the treatment of bacterial infections (otic) and is in preclinical development, as is AzaSite Extra, for the treatment of ocular infection. The company's first product utilizing the DuraSite technology, AquaSite dry eye treatment, was launched as an over-the-counter medication by CIBA Vision. InSite's collaborative glaucoma genetics program is focused on discovering genes associated with the disease. OcuGene, the company's first genetic tool, is commercially available to those with glaucoma. ISV-205, another collaborative product candidate for potential treatment (topical) for inflammation and analgesia, is in Phase II trials. InSite sold its IV-403, a treatment for ocular infections, to Bausch & Lomb. In 2007, the company signed a licensing agreement with Pfizer, Inc., which will enhance the position and marketability of InSite's AzaSite franchise.

InSite offers employees medical, dental and vision coverage; travel assistance; educational assistance; and a health club program.

FINANCIALS: Sales and profits are in thousands of dollars—add 000 to get the full amount. 2006 Note: Financial information for 2006 was not available for all companies at press time.

2006 Sales: $ 2	2006 Profits: $-16,611	**U.S. Stock Ticker:** ISV	
2005 Sales: $ 4	2005 Profits: $-15,215	**Int'l Ticker:** Int'l Exchange:	
2004 Sales: $ 542	2004 Profits: $-5,514	Employees: 44	
2003 Sales: $ 134	2003 Profits: $-6,751	Fiscal Year Ends: 12/31	
2002 Sales: $ 100	2002 Profits: $-10,900	Parent Company:	

SALARIES/BENEFITS:

Pension Plan:	ESOP Stock Plan:	Profit Sharing:	Top Exec. Salary: $390,000	Bonus: $200,000
Savings Plan: Y	Stock Purch. Plan: Y		Second Exec. Salary: $235,000	Bonus: $

OTHER THOUGHTS:

Apparent Women Officers or Directors: 2
Hot Spot for Advancement for Women/Minorities: Y

LOCATIONS: ("Y" = Yes)

West:	Southwest:	Midwest:	Southeast:	Northeast:	International:
Y					

NSMED INCORPORATED
www.insmed.com

Industry Group Code: 325412 Ranks within this company's industry group: Sales: 158   Profits: 160

Drugs:		Other:		Clinical:	Computers:	Other:
Discovery:	Y	AgriBio:	Y	Trials/Services:	Hardware:	Specialty Services:
Licensing:	Y	Genomics/Proteomics:		Laboratories:	Software:	Consulting:
Manufacturing:	Y	Tissue Replacement:		Equipment/Supplies:	Arrays:	Blood Collection:
Development:	Y			Research/Development Svcs.:	Database Management:	Drug Delivery:
Generics:				Diagnostics:		Drug Distribution:

TYPES OF BUSINESS:
Drugs-Metabolic & Endocrine Diseases
Drugs-Cancer

BRANDS/DIVISIONS/AFFILIATES:
Insmed Pharmaceuticals
IPLEX
rhIGFBP-3
INSM-18
SomatoKine

CONTACTS: *Note: Officers with more than one job title may be intentionally listed here more than once.*
Geoffrey Allan, CEO
Ronald D. Gunn, COO
Geoffrey Allan, Pres.
Kevin P. Tully, CFO
Kevin P. Tully, Treas./Controller
Doug Farrar, VP-Insmed Therapeutic Proteins
Steve Glover, Pres., Follow-on Biologics
Geoffrey Allan, Chmn.

Phone: 804-565-3000	Fax: 804-565-3500

Toll-Free:
Address: 8720 Stony Point Pkwy., Ste. 200, Richmond, VA 23235 US

GROWTH PLANS/SPECIAL FEATURES:
Insmed, Inc. discovers and develops pharmaceutical products for the treatment of metabolic and endocrine diseases. The firm's business strategy includes retaining commercial rights to its products; establishing corporate partnerships in its target markets; outsourcing manufacturing to deploy resources efficiently; and acquiring additional products and technologies. The company has acquired non-exclusive patent rights to the use of an insulin-like growth factor (IGF-I) therapy for the treatment of extreme or severe insulin-resistant diabetes from Fujisawa Pharmaceutical Co., Ltd. Under the terms of the agreement, Insmed will obtain worldwide rights in territories, excluding Japan, where a valid patent claim exists, including the U.S. and Europe. The company's lead product, IPLEX (formerly SomatoKine), approved by the FDA in December 2005, is a composition of recombinant human IGF-1 and IGF binding protein 3. IPLEX is now being studied as a treatment for several serious medical conditions including Myotonic Muscular Dystrophy (MMD), ALS (Lou Gehrig's Disease), HIV- Associated Adipose Redistribution Syndrome (HARS) and retinopathy of prematurity (ROP). The company has two oncology compounds, rhIGFBP-3 and INSM-18, in development. Preclinical models show that one or both treatments interact with the IGF-1 system to reduce tumor growth in models of breast, prostate, lung, colorectal and head and neck cancers. In January 2007, the Italian Ministry of Health requested Insmed to make IPLEX available to physicians in Italy to treat patients with Amyotrophic Lateral Sclerosis (ALS).

FINANCIALS: Sales and profits are in thousands of dollars—add 000 to get the full amount. 2006 Note: Financial information for 2006 was not available for all companies at press time.

2006 Sales: $ 991	2006 Profits: $-56,139	**U.S. Stock Ticker: INSM**
2005 Sales: $ 131	2005 Profits: $-40,929	Int'l Ticker: Int'l Exchange:
2004 Sales: $ 137	2004 Profits: $-27,203	Employees: 157
2003 Sales: $ 150	2003 Profits: $-10,298	Fiscal Year Ends: 12/31
2002 Sales: $2,000	2002 Profits: $-36,400	Parent Company:

SALARIES/BENEFITS:

Pension Plan:	ESOP Stock Plan:	Profit Sharing:	Top Exec. Salary: $396,467	Bonus: $
Savings Plan: Y	Stock Purch. Plan: Y		Second Exec. Salary: $270,866	Bonus: $

OTHER THOUGHTS:
Apparent Women Officers or Directors:
Hot Spot for Advancement for Women/Minorities:

LOCATIONS: ("Y" = Yes)

West:	Southwest:	Midwest:	Southeast:	Northeast:	International:
				Y	

Note: Financial information, benefits and other data can change quickly and may vary from those stated here.

INSPIRE PHARMACEUTICALS INC   www.inspirepharm.com

Industry Group Code: 325412 Ranks within this company's industry group: Sales: 88 Profits: 146

Drugs:		Other:		Clinical:		Computers:		Other:	
Discovery:	Y	AgriBio:		Trials/Services:		Hardware:		Specialty Services:	
Licensing:	Y	Genomics/Proteomics:		Laboratories:		Software:		Consulting:	
Manufacturing:		Tissue Replacement:		Equipment/Supplies:		Arrays:		Blood Collection:	
Development:	Y			Research/Development Svcs.:		Database Management:		Drug Delivery:	
Generics:				Diagnostics:				Drug Distribution:	

TYPES OF BUSINESS:

Drugs-Respiratory & Ocular
Cystic Fibrosis Treatment
Retinal Disease Treatment
Dry Eye Treatment
Allergic Conjunctivitis
Cardiovascular Disease Treatment

BRANDS/DIVISIONS/AFFILIATES:

Restasis
Elestat
AzaSite
Prolacria

CONTACTS: Note: Officers with more than one job title may be intentionally listed here more than once.

Christy L. Shaffer, CEO
Christy L. Shaffer, Pres.
Thomas R. Staab, II, CFO
Jeff W. Sampere, VP-Mktg.
Benjamin R. Yerxa, Chief Scientific Officer
Sean K. Blake, Dir.-IT
W.M. Mulchi, VP-Tech. Oper.
Denise M. Sheehan, VP-Prod. Planning
W.M. Mulchi, VP-Mfg.
Joseph M. Spagnardi, General Counsel/Sr. VP/Sec.
Mary B. Bennett, Exec. VP-Oper.
Donald J. Kellerman, Sr. VP-Dev.
Mary B. Bennett, Exec. VP-Comm.
Jenny R. Kobin, VP-Investor Rel.
Mark A. Siemek, Dir.-Financial Planning & Analysis & SEC Reports
R. Kim Brazzell, Sr. VP-Ophthalmic R&D
Benjamin R. Yerxa, Exec. VP-Strategic Oper.
Joseph K. Schachle, Exec. VP/Chief-Commercial Operations
Gerald W. St. Peter, Sr. VP-Sales & Managed Markets
Kenneth B. Lee, Jr., Chmn.

Phone: 919-941-9777	Fax: 919-941-9797
Toll-Free: 877-800-4536	
Address: 4222 Emperor Blvd., Ste. 200, Durham, NC 27703-8466 US	

GROWTH PLANS/SPECIAL FEATURES:

Inspire Pharmaceuticals, Inc. is a biopharmaceutical company dedicated to discovering, developing and commercializing prescription pharmaceutical products in disease areas with significant commercial markets and unmet medical needs. The firm's focus is in the ophthalmic and respiratory therapeutic areas, in addition to the design and synthesis of P2 receptor related agonists, where it has significant expertise. Inspire's ophthalmic products and product candidates are concentrated in the allergic conjunctivitis, dry eye disease, cystic fibrosis, acute cardiac care, seasonal allergic rhinitis and glaucoma indications. The firm's product portfolio includes Elestat for allergic conjunctivitis, Restasis for dry eye disease and AzaSite for bacterial conjunctivitis. Elestat and Restasis are licensed to Allergan and currently marketed in the U.S. Products in clinical development include Prolacria, diquafosol tetrasodium, for dry eye, which has finished four Phase 3 trials and may be planning one additional; denufosol tetrasodium (INS37217 Respiratory) for cystic fibrosis, currently in Phase 3 trials; Bilastine oral antihistamine for allergic rhinitis, currently in Phase 3 trials; Epinastine nasal spray for allergic rhinitis, currently in Phase 3; and INS115644 for glaucoma, currently in Phase 1. In February 2006, the company signed a development and license agreement with Boehringer Ingelheim International GmbH, granting Inspire certain exclusive rights to develop and market an intranasal dosage form of Epinastine in the U.S. and Canada, for the treatment and prevention of seasonal allergic rhinitis. In August 2006, Inspire terminated its Phase 2 study of for platelet aggregation inhibitor, INS50589 Antiplatelet, due to safety complications. In April 2007, AzaSite was approved by the FDA for the treatment of bacterial conjunctivitis.

Inspire offers its employees benefits including health insurance, an employee assistance program, and vacation packages. Additionally, it offers monthly on-site massage therapists, weekly yoga and Pilates classes, as well as group activities and parties.

FINANCIALS: Sales and profits are in thousands of dollars—add 000 to get the full amount. 2006 Note: Financial information for 2006 was not available for all companies at press time.

2006 Sales: $37,059	2006 Profits: $-42,115	U.S. Stock Ticker: ISPH
2005 Sales: $23,266	2005 Profits: $-31,847	Int'l Ticker: Int'l Exchange:
2004 Sales: $11,068	2004 Profits: $-44,069	Employees: 170
2003 Sales: $5,200	2003 Profits: $-31,395	Fiscal Year Ends: 12/31
2002 Sales: $4,900	2002 Profits: $-24,700	Parent Company:

SALARIES/BENEFITS:

Pension Plan:	ESOP Stock Plan:	Profit Sharing: Y	Top Exec. Salary: $431,553	Bonus: $197,284
Savings Plan: Y	Stock Purch. Plan:		Second Exec. Salary: $271,690	Bonus: $103,574

OTHER THOUGHTS:

Apparent Women Officers or Directors: 11
Hot Spot for Advancement for Women/Minorities: Y

LOCATIONS: ("Y" = Yes)

West:	Southwest:	Midwest:	Southeast:	Northeast:	International:
Y				Y	

INSTITUT STRAUMANN AG

www.straumann.com

ndustry Group Code: 339113 Ranks within this company's industry group: Sales: 7   Profits: 6

Drugs:	Other:	Clinical:	Computers:	Other:
Discovery:	AgriBio:	Trials/Services:	Hardware:	Specialty Services:
Licensing:	Genomics/Proteomics:	Laboratories:	Software:	Consulting:
Manufacturing:	Tissue Replacement: Y	Equipment/Supplies: Y	Arrays:	Blood Collection:
Development:		Research/Development Svcs.:	Database Management:	Drug Delivery:
Generics:		Diagnostics:		Drug Distribution: Y

TYPES OF BUSINESS:

Dental Implants
Dental Tissue Regeneration
Dental Drugs, Manufacturing

BRANDS/DIVISIONS/AFFILIATES:

Straumann Dental Implant System
etkon
Daishin Implant System
Biora AB
Emdogain
SLActive
CARES
DenTech

CONTACTS: Note: Officers with more than one job title may be intentionally listed here more than once.

Gilbert Achermann, CEO
Marco Gadola, CFO
Mark Hill, Head-Corp. Comm.
Marianne Burgi, Head-Market & Product Support
Sandro Matter, Head-Products Div.
Rudolf Maag, Chmn.
Wolfgang Becker, Head-Sales, Europe

Phone: 41-61-965-11-11	Fax: 41-61-965-11-01

Toll-Free:
Address: Peter Merian-Weg 12, Basel, CH-4052 Switzerland

GROWTH PLANS/SPECIAL FEATURES:

Institut Straumann AG is the world's second-largest producer of dental implants and a major provider of dental tissue regeneration products. Based in Switzerland, Straumann maintains subsidiaries and distributors in countries throughout the world, with its most important markets being Germany and the U.S. The company is organized into two segments, the implants division and the biologics division. The implants division produces a range of products for the Straumann Dental Implant System, which includes implants, prosthetic components and instruments. Brand names include the TE implant and the SynOcta prosthetic product lines. The biologics division focuses on biotechnology for the reconstruction of dental bone and soft tissue. The company acquired Biora AB, the Swedish developer of Enamel Matrix Derivative (EMD) protein, as a major steppingstone in the expansion of its biologics activities. Its Emdogain gel for the treatment of periodontitis has firmly positioned the company in the market for periodontal tissue regeneration. Some of the company's most recent developments include SLActive, a new implant surface technology that decreases the healing time after surgery and increases general stability of the dental implants it is employed in; and Straumann Bone Ceramic, a synthetic bone graft substitute that actually augments the patient's jaw bone structure to be able to support advanced dental appliances. A new program, the Computer Aided Restoration Service (or CARES), offers an exceptionally precise way to design custom prostheses. Recent acquisitions of the company include etkon, a company focused on conventional and implant-based tooth restoration using CAD/CAM technology; and Daishin Implant System, the exclusive distributor of Straumann implant products and services in Japan.

FINANCIALS: Sales and profits are in thousands of dollars—add 000 to get the full amount. 2006 Note: Financial information for 2006 was not available for all companies at press time.

2006 Sales: $	2006 Profits: $	**U.S. Stock Ticker:**
2005 Sales: $387,056	2005 Profits: $97,363	**Int'l Ticker: STMN** Int'l Exchange: Zurich-SWX
2004 Sales: $327,082	2004 Profits: $78,210	Employees: 1,340
2003 Sales: $269,400	2003 Profits: $62,900	Fiscal Year Ends: 12/31
2002 Sales: $215,350	2002 Profits: $43,860	Parent Company:

SALARIES/BENEFITS:

Pension Plan:	ESOP Stock Plan:	Profit Sharing:	Top Exec. Salary: $	Bonus: $
Savings Plan:	Stock Purch. Plan:		Second Exec. Salary: $	Bonus: $

OTHER THOUGHTS:

Apparent Women Officers or Directors: 1
Hot Spot for Advancement for Women/Minorities:

LOCATIONS: ("Y" = Yes)

West:	Southwest:	Midwest:	Southeast:	Northeast:	International:
				Y	Y

Note: Financial information, benefits and other data can change quickly and may vary from those stated here.

INTEGRA LIFESCIENCES HOLDINGS CORPwww.integra-ls.com

Industry Group Code: 339113 Ranks within this company's industry group: Sales: 9 Profits: 9

Drugs:		Other:	Clinical:		Computers:		Other:
Discovery:		AgriBio:	Trials/Services:		Hardware:		Specialty Services:
Licensing:		Genomics/Proteomics:	Laboratories:		Software:		Consulting:
Manufacturing:	Y	Tissue Replacement:	Equipment/Supplies:	Y	Arrays:		Blood Collection:
Development:	Y		Research/Development Svcs.:		Database Management:		Drug Delivery:
Generics:			Diagnostics:				Drug Distribution:

TYPES OF BUSINESS:

Medical Equipment Manufacturing
Implants & Biomaterials
Absorbable Medical Products
Tissue Regeneration Technology
Neurosurgery Products
Skin Replacement Products

BRANDS/DIVISIONS/AFFILIATES:

Integra NeuroSciences
Integra Plastic and Reconstructive Surgery
JARIT Surgical Instruments, Inc.
DuraGen Dural Graft Matrix
NeuraGen Nerve Guide
DenLite
LXY Healthcare
Kinetikos Medical, Inc.

CONTACTS: Note: Officers with more than one job title may be intentionally listed here more than once.

Stuart M. Essig, CEO
Gerard S. Carlozzi, COO/Exec. VP
Stuart M. Essig, Pres.
Maureen B. Bellantoni, CFO/Exec. VP
Deborah A. Leonetti, Sr. VP-Mktg.
Wilma J. Davis, Sr. VP-Human Resources
Simon J. Archibald, Chief Scientific Officer/VP-Clinical Affairs
Randy Gottlieb, CIO/VP
Linda Littlejohns, VP-Clinical Dev.
Donald R. Nociolo, Sr. VP-Mfg. Oper.
John B. Henneman, III, Chief Admin. Officer/Exec. VP/Corp. Sec.
Richard D. Gorelick, General Counsel/VP
John Bostjancic, VP-Corp. Dev.
John Bostjancic, VP-Investor Rel.
David B. Holtz, Sr. VP-Finance/Treas.
Howard Jamner, Chmn.-Jarit Surgical Instruments
Judith E. O'Grady, Sr. VP-Quality, Regulatory & Clinical Affairs
Jerry Corbin, VP/Corp. Controller
Robert D. Paltridge, Pres., Extremity Reconstruction
Richard E. Caruso, Chmn.
Zeev Hadass, VP-European Oper.

Phone: 609-275-0500	Fax: 609-275-5363
Toll-Free: 800-654-2873	
Address: 311 C Enterprise Dr., Plainsboro, NJ 08536 US	

GROWTH PLANS/SPECIAL FEATURES:

Integra Lifesciences Holdings Corporation develops, manufactures and markets medical devices, implants and biomaterials for use in neurotrauma, neurosurgery, plastic and reconstructive surgery and general surgery. The company's two primary product lines include neurosurgical/orthopedic implants and medical/surgical implants. The Neuro/Ortho implants group provides dural grafts for the repair of dura mater, dermal regeneration and engineered wound dressing, implants for small bone and joint fixation; repair of peripheral nerves and hydrocephalus management. The MedSurg Equipment product group produces ultrasonic surgery systems for tissue ablation, cranial stabilization and brain retraction systems and instrumentation for use in general, neurosurgical, spinal and plastic and reconstructive surgery and dental procedures. Patented products of the company include the DuraGen Dural Graft Matrix, the NeuraGen Nerve Guide, the INTEGRA Dermal Regeneration Template and the INTEGRA Bilayer Matrix Wound Dressing, which incorporates Integra's proprietary absorbable implant technology. Products principally focus on injuries that involve the brain, cranium, spine and central nervous system and the repair and reconstruction of soft tissue. Integra's products are marketed and sold through its subsidiaries, Integra NeuroSciences, Integra Plastic and Reconstructive Surgery, Jarit and Miltex Surgical Instruments, Inc. and other strategic alliances. Integra's sales teams are located throughout the United States, Canada, Germany, the U.K., the Belenux region and France. In August 2006, the firm acquired Kinetikos Medical, Inc., a developer and manufacturer of orthopedic implants and surgical devices for small bones and joints. In May 2007, Integra announced that it would acquire the assets of the pain management business of Physician Industries, Inc. for $4 million. The company also closed the acquisition of LXY Healthcare, a manufacturer of fiber optic headlight systems for the medical industry, and acquired the DenLite illuminated mirror product line from Welch Allyn, Inc. in 2007.

FINANCIALS: Sales and profits are in thousands of dollars—add 000 to get the full amount. 2006 Note: Financial information for 2006 was not available for all companies at press time.

2006 Sales: $419,297	2006 Profits: $29,407	**U.S. Stock Ticker: IART**
2005 Sales: $277,935	2005 Profits: $37,194	**Int'l Ticker:** Int'l Exchange:
2004 Sales: $229,825	2004 Profits: $17,197	Employees: 1,750
2003 Sales: $166,695	2003 Profits: $26,861	Fiscal Year Ends: 12/31
2002 Sales: $112,600	2002 Profits: $35,300	Parent Company:

SALARIES/BENEFITS:

Pension Plan:	ESOP Stock Plan:	Profit Sharing:	Top Exec. Salary: $500,000	Bonus: $500,000
Savings Plan: Y	Stock Purch. Plan: Y		Second Exec. Salary: $420,000	Bonus: $168,000

OTHER THOUGHTS:

Apparent Women Officers or Directors: 7
Hot Spot for Advancement for Women/Minorities: Y

LOCATIONS: ("Y" = Yes)

West:	Southwest:	Midwest:	Southeast:	Northeast:	International:
Y		Y		Y	Y

INTERMUNE PHARMACEUTICALS INC   www.intermune.com

Industry Group Code: 325412   Ranks within this company's industry group: Sales: 66   Profits: 186

Drugs:		Other:		Clinical:	Computers:	Other:
Discovery:		AgriBio:		Trials/Services:	Hardware:	Specialty Services:
Licensing:	Y	Genomics/Proteomics:		Laboratories:	Software:	Consulting:
Manufacturing:		Tissue Replacement:		Equipment/Supplies:	Arrays:	Blood Collection:
Development:	Y			Research/Development Svcs.:	Database Management:	Drug Delivery:
Generics:				Diagnostics:		Drug Distribution:

TYPES OF BUSINESS:

Drugs-Pulmonology & Hepatology
Infectious Disease & Cancer Treatments
Osteoporosis Treatments

BRANDS/DIVISIONS/AFFILIATES:

Actimmune

CONTACTS: *Note: Officers with more than one job title may be intentionally listed here more than once.*

Daniel G. Welch, CEO
Daniel G. Welch, Pres.
John Hodgman, CFO/Sr. VP
Howard Simon, Sr. VP-Human Resources
Williamson Bradford, Sr. VP-Clinical Science
Robin Steele, General Counsel/Sr. VP/Corp. Sec.
Tom Kassberg, Sr. VP-Commercial Oper.
Tom Kassberg, VP-Corp. Dev.
Lawrence Blatt, Chief Science Officer
Marianne Armstrong, Chief Medical Affairs/Regulatory Officer
Steven Porter, Chief Medical Officer
Cynthia Robinson, Sr. VP-Dev. Oper.
William R. Ringo, Jr., Chmn.

Phone: 415-466-2200	Fax: 415-466-2300
Toll-Free:	
Address: 3280 Bayshore Blvd., Brisbane, CA 94005 US	

GROWTH PLANS/SPECIAL FEATURES:

InterMune, Inc. is a biotechnology company that develops and commercializes innovative products and therapies for pulmonary and hepatic diseases. The company's core product is Actimmune, which is used for the treatment of severe, malignant osteopetrosis and chronic granulomatous disease. Actimmune may also be used as possible treatment for idiopathic pulmonary fibrosis (IPF) and ovarian cancer. The active ingredient in Actimmune, Interferon gamma-1b, consists of a human protein that plays a key role in preventing the formation of excessive scars or fibrotic tissue and also acts as a potent stimulator of the immune system. The company's developing drug pipeline falls into two categories: pulmonology and hepatology. InterMune is currently developing two pulmonary therapies for the treatment of idiopathic pulmonary fibrosis: interferon gamma-1b and pirfenidone. In hepatology, the company is developing PEG-Alfacon-1 as a treatment for chronic hepatitis C virus (HCV) infections and a protease inhibitor program. In October 2006, Intermune signed a collaboration agreement with Roche, which will allow the companies to develop and commercialize Intermune's HCV protease inhibitor program.

InterMune offers employees medical, dental and vision plans; a tuition reimbursement program; flexible spending accounts; a college savings program; and monetary rewards through its Quest for the Best! Employee Referral Program.

FINANCIALS: Sales and profits are in thousands of dollars—add 000 to get the full amount. 2006 Note: Financial information for 2006 was not available for all companies at press time.

2006 Sales: $90,784	2006 Profits: $-107,206	U.S. Stock Ticker: ITMN
2005 Sales: $110,496	2005 Profits: $-5,235	Int'l Ticker: Int'l Exchange:
2004 Sales: $128,680	2004 Profits: $-59,478	Employees: 195
2003 Sales: $144,862	2003 Profits: $-97,001	Fiscal Year Ends: 12/31
2002 Sales: $112,000	2002 Profits: $-114,300	Parent Company:

SALARIES/BENEFITS:

Pension Plan:	ESOP Stock Plan:	Profit Sharing:	Top Exec. Salary: $583,440	Bonus: $
Savings Plan: Y	Stock Purch. Plan: Y		Second Exec. Salary: $303,348	Bonus: $

OTHER THOUGHTS:

Apparent Women Officers or Directors: 3
Hot Spot for Advancement for Women/Minorities: Y

LOCATIONS: ("Y" = Yes)

West:	Southwest:	Midwest:	Southeast:	Northeast:	International:
Y					

INTERNATIONAL ISOTOPES   www.intisoid.com

Industry Group Code: 325000 Ranks within this company's industry group: Sales: 7   Profits: 7

Drugs:		Other:		Clinical:		Computers:		Other:	
Discovery:		AgriBio:		Trials/Services:		Hardware:		Specialty Services:	Y
Licensing:		Genomics/Proteomics:		Laboratories:		Software:		Consulting:	
Manufacturing:	Y	Tissue Replacement:		Equipment/Supplies:	Y	Arrays:		Blood Collection:	
Development:				Research/Development Svcs.:		Database Management:		Drug Delivery:	
Generics:				Diagnostics:				Drug Distribution:	

TYPES OF BUSINESS:

Radioactive Isotopes Manufacturing
Nuclear Medicine Standards Publishing
Assay Analysis Services
Fluorine Gas
Depleted Uranium Oxide Products
Gemstone Processing

BRANDS/DIVISIONS/AFFILIATES:

CONTACTS: Note: Officers with more than one job title may be intentionally listed here more than once.

Steve T. Laflin, CEO
Steve T. Laflin, Pres.
Ralph M. Richart, Chmn.

Phone: 208-524-5300	Fax: 208-524-1411
Toll-Free: 800-699-3108	
Address: 4137 Commerce Cir., Idaho Falls, ID 83401 US	

GROWTH PLANS/SPECIAL FEATURES:

International Isotopes, Inc. (INIS), together with its subsidiary International Isotopes Idaho, Inc., manufactures high purity fluoride gas and depleted uranium oxide products using the Fluorine Extraction Process (FEP). The company's business consists of five segments: nuclear medicine reference and calibration standards (NMR&CS); cobalt products; radiochemical products; fluorine extraction process products; and radiological processing services. The NMR&CS segment consists of the manufacture of sources and standards associated with SPECT (single photon emission computed tomography), patient positioning and calibration or operational testing of dose measuring equipment for the nuclear pharmacy business. These items include: flood sources, dose calibrators, rod sources, flexible and rigid rulers, spot markers, pen point markers and a host of specialty items. INIS is an exclusive manufacturer of these products for RadQual LLC. In the cobalt products segment, INIS manufactures a wide range of teletherapy and experimental irradiator capsules, as well as its own high and medium specific activity bulk cobalt (1mm x 1mm nickel plated pellets) and offers recycling services of expended cobalt sources. In the radiochemical products segment, INIS manufactures and distributes various high quality radiochemicals for medical, industrial or research applications. The list of available isotopes includes cobalt-60 and barium-133. The fluorine products segment consists of seven recently acquired patents for the FEP, which the company plans to use to produce several high purity fluorine products, such as geranium tetrafluoride (GT). The radiological processing services segment concerns a wide array of miscellaneous services, including: radiological engineering and health physics consultant services; Type A package design and certification testing; on site radioactive shipping services; and sealed source wipe test kits, analysis and results reporting. In 2007, INIS was awarded a grant by the National Science Foundation (NSF) for an investigation into the use of GT for production of hydrofluorocarbon refrigerants.

The company offers employees medical and dental insurance and educational reimbursement.

FINANCIALS: Sales and profits are in thousands of dollars—add 000 to get the full amount. 2006 Note: Financial information for 2006 was not available for all companies at press time.

2006 Sales: $4,470	2006 Profits: $-1,037	U.S. Stock Ticker: INIS
2005 Sales: $2,985	2005 Profits: $- 983	Int'l Ticker: Int'l Exchange:
2004 Sales: $2,848	2004 Profits: $- 845	Employees: 25
2003 Sales: $2,100	2003 Profits: $- 600	Fiscal Year Ends: 12/31
2002 Sales: $2,181	2002 Profits: $ 149	Parent Company:

SALARIES/BENEFITS:

Pension Plan:	ESOP Stock Plan:	Profit Sharing:	Top Exec. Salary: $145,080	Bonus: $15,000
Savings Plan:	Stock Purch. Plan:		Second Exec. Salary: $99,567	Bonus: $

OTHER THOUGHTS:

Apparent Women Officers or Directors:
Hot Spot for Advancement for Women/Minorities:

LOCATIONS: ("Y" = Yes)

West:	Southwest:	Midwest:	Southeast:	Northeast:	International:
Y					

INTROGEN THERAPEUTICS INC
www.introgen.com

Industry Group Code: 325412 Ranks within this company's industry group: Sales: 157 Profits: 123

Drugs:		Other:	Clinical:	Computers:	Other:	
Discovery:	Y	AgriBio:	Trials/Services:	Hardware:	Specialty Services:	
Licensing:		Genomics/Proteomics:	Laboratories:	Software:	Consulting:	
Manufacturing:		Tissue Replacement:	Equipment/Supplies:	Arrays:	Blood Collection:	
Development:	Y		Research/Development Svcs.:	Database Management:	Drug Delivery:	Y
Generics:			Diagnostics:		Drug Distribution:	

TYPES OF BUSINESS:
Drugs-Cancer
Gene Therapy Products

BRANDS/DIVISIONS/AFFILIATES:
Advexin
Adenoviral Delivery System

CONTACTS: *Note: Officers with more than one job title may be intentionally listed here more than once.*
David G. Nance, CEO
David G. Nance, Pres.
James W. Albrecht, Jr., CFO
Robert E. Sobol, Sr. VP-Medical & Scientific Affairs
Kerstin B. Menander, VP-Clinical Dev.
Peter M. Clarke, VP-Prod. & Tech. Processes
J. David Enloe, Jr., Sr. VP-Oper.
C. Channing Burke, Dir.-Corp. Comm.
Lou Zumstein, Associate VP-Research
Sunil Chada, Associate VP-Clinical Research
David L. Parker, VP-Intellectual Property
John N. Kapoor, Chmn.
Max W. Talbott, Sr. VP-Worldwide Commercial Dev.

Phone: 512-708-9310	Fax: 512-708-9311
Toll-Free:	
Address: 301 Congress Ave., Ste. 1850, Austin, TX 78701 US	

GROWTH PLANS/SPECIAL FEATURES:
Introgen Therapeutics is a leading developer of gene therapy products for the treatment of cancer and other diseases. Introgen's drug discovery and development programs have uncovered innovative approaches using therapeutic genes to treat cancer by directly addressing the genetic abnormalities associated with the disease. Its molecular therapies also precipitate cell death (without harming healthy cells), conduct cell cycle control and enhance the production of normal cancer-fighting proteins. The firm's lead product candidate, Advexin, combines the naturally occurring p53 tumor suppressor gene with Introgen's clinically proven adenoviral delivery system. The critical importance of the p53 gene in controlling tumor growth indicates that Advexin is applicable to multiple cancers. Introgen is in pivotal Phase III clinical studies of Advexin in head and neck cancer and has completed a Phase II clinical trial in non-small cell lung cancer. The company's research department is looking into the applications on Li-Fraumeni Syndrome, a rare genetic disorder characterized by a mutation of the p53 gene. Introgen's other candidates are INGN 241, a cancer suppressor, which is in Phase III trials; INGN 225, a tumor vaccine, which is in phase II trials; INGN 401, a systemic nanoparticle therapy that is in Phase II trials; and NGN 234, a topical formulation for oral cancers, which is in Phase II clinical development. In sum, the company controls an intellectual property portfolio that includes over 300 pending and issued patents for molecular therapy technologies and adenovirus production. According to a recent study, the populations that respond best to Advexin molecular treatment are those who react adversely toward traditional cancer treatments such as radiation or chemotherapy. The company's Advexin product celebrated good news in 2007 when it received result from Phase II trials indicating an increased survival rate in patients with head and neck cancer taking Advexin.

FINANCIALS: Sales and profits are in thousands of dollars—add 000 to get the full amount. 2006 Note: Financial information for 2006 was not available for all companies at press time.

2006 Sales: $1,151	2006 Profits: $-28,801	**U.S. Stock Ticker: INGN**
2005 Sales: $1,867	2005 Profits: $-26,103	**Int'l Ticker:** Int'l Exchange:
2004 Sales: $1,808	2004 Profits: $-24,387	Employees: 74
2003 Sales: $ 304	2003 Profits: $-19,326	Fiscal Year Ends: 12/31
2002 Sales: $1,200	2002 Profits: $-26,100	Parent Company:

SALARIES/BENEFITS:
Pension Plan:	ESOP Stock Plan:	Profit Sharing:	Top Exec. Salary: $538,750	Bonus: $
Savings Plan: Y	Stock Purch. Plan: Y		Second Exec. Salary: $325,000	Bonus: $

OTHER THOUGHTS:
Apparent Women Officers or Directors: 1
Hot Spot for Advancement for Women/Minorities:

LOCATIONS: ("Y" = Yes)
West:	Southwest:	Midwest:	Southeast:	Northeast:	International:
	Y				

Note: Financial information, benefits and other data can change quickly and may vary from those stated here.

INVITROGEN CORPORATION

www.invitrogen.com

Industry Group Code: 325413 Ranks within this company's industry group: Sales: 1   Profits: 27

Drugs:	Other:		Clinical:		Computers:		Other:	
Discovery:	AgriBio:		Trials/Services:		Hardware:		Specialty Services:	
Licensing:	Genomics/Proteomics:	Y	Laboratories:		Software:		Consulting:	
Manufacturing:	Tissue Replacement:		Equipment/Supplies:	Y	Arrays:	Y	Blood Collection:	
Development:			Research/Development Svcs.:	Y	Database Management:		Drug Delivery:	
Generics:			Diagnostics:				Drug Distribution:	

TYPES OF BUSINESS:

Equipment-Gene Cloning Kits
Microarrays
Reagents
RNA & DNA Libraries
Qdot Nanocrystals

BRANDS/DIVISIONS/AFFILIATES:

Cell Culture Systems (CCS)
BioPixels
Quantum Dot Corporation
BioDiscovery
iGene
Sentigen Holding Corp.
BioReliance
Grand Island Biological Company (GIBCO)

CONTACTS: Note: Officers with more than one job title may be intentionally listed here more than once.

Gregory T. Lucier, CEO
David F. Hoffmeister, CFO/Sr. VP
Bernd Brust, Sr. VP-Global Sales
Peter Leddy, Sr. VP-Human Resources
Claude D. Benchimol, Sr. VP-R&D
Karen S. Gibson, CIO/Sr. VP
John A. Cottingham, General Counsel/Sr. VP/Corp. Sec.
Nicolas M. Barthelemy, Sr. VP-Global Oper.
John D. Thompson, Sr. VP-Strategy & Corp. Dev.
Amanda Clardy, VP-Investor Rel.
Kelli Richard, VP-Finance/Chief Acct. Officer
David Onions, Chief Medical Officer
Kip Miller, Sr. VP-BioDiscovery
Kornelikia Zgonc, VP-Functional Excellence
Nicolas M. Barthelemy, Sr. VP-Cell Culture Systems
Gregory T. Lucier, Chmn.

Phone: 760-603-7200	Fax: 760-603-7229
Toll-Free: 800-955-6288	
Address: 1600 Faraday Ave., Carlsbad, CA 92008 US	

GROWTH PLANS/SPECIAL FEATURES:

Invitrogen develops, manufactures and markets research tool kits and cell culture media to the life sciences research, drug discovery, diagnostics and manufacturing industries. The company's line of over 700 research kits and reagents simplify and improve gene cloning, gene expression and gene analysis techniques for functional genomics and drug discovery processes. Invitrogen markets products such as gel-based separation technologies, antibodies and protoarrays. The company's TOPO TA Cloning Kits reduce the time required for ligation reaction from 12 hours to five minutes, while increasing the success rate from 50% to over 90%. The company divides its product and services into two broad segments: BioDiscovery and Cell Culture Systems (CCS). The BioDiscovery segment includes Invitrogen's functional genomics, cell biology and drug discovery product lines. This encompasses products for cloning and manipulation of DNA; the examination and regulation of RNA; and the capturing, separation and analyzing of proteins. CCS reagents and research kits simplify the process of gene acquisition, cloning, expression and analysis techniques. CCS markets its products under Grand Island Biological Company GIBCO, which includes items such as sera, growth factors and cell and tissue culture media. CCS also creates stable cell lines and optimizes production processes used for the production of therapeutic drugs through its PD-Direct Service. In addition, Invitrogen also offers iGene, a searchable database that allows users to find targeted antibodies, assays, proteins, xPCR and RNAi products. The company's acquisitions of Quantum Dot Corporation and the BioPixels business unit of BioCrystal, Ltd. have added significant semiconductor nanotechnology products and services to Invitrogen's portfolio. In 2006, Invitrogen completed the acquisition of Sentigen Holding Corp., which adds assay development through a novel screening approach of protein coupled receptors to the company's platform technologies. In April 2007, Invitrogen completed the sale of its BioReliance business unit to Avista Capital Partners for $210 million.

FINANCIALS: Sales and profits are in thousands of dollars—add 000 to get the full amount. 2006 Note: Financial information for 2006 was not available for all companies at press time.

2006 Sales: $1,263,485	2006 Profits: $-191,049	**U.S. Stock Ticker: IVGN**
2005 Sales: $1,198,452	2005 Profits: $132,046	**Int'l Ticker:** Int'l Exchange:
2004 Sales: $1,023,851	2004 Profits: $88,825	Employees: 4,835
2003 Sales: $777,738	2003 Profits: $60,130	Fiscal Year Ends: 12/31
2002 Sales: $648,600	2002 Profits: $47,700	Parent Company:

SALARIES/BENEFITS:

Pension Plan:	ESOP Stock Plan:	Profit Sharing:	Top Exec. Salary: $904,615	Bonus: $
Savings Plan: Y	Stock Purch. Plan: Y		Second Exec. Salary: $439,423	Bonus: $

OTHER THOUGHTS:

Apparent Women Officers or Directors: 3
Hot Spot for Advancement for Women/Minorities: Y

LOCATIONS: ("Y" = Yes)

West:	Southwest:	Midwest:	Southeast:	Northeast:	International:
Y		Y		Y	Y

Note: Financial information, benefits and other data can change quickly and may vary from those stated here.

IOMED INC

www.iomed.com

Industry Group Code: 325412A Ranks within this company's industry group: Sales: 11   Profits: 7

Drugs:		Other:		Clinical:	Computers:		Other:	
Discovery:		AgriBio:		Trials/Services:	Hardware:		Specialty Services:	
Licensing:		Genomics/Proteomics:		Laboratories:	Software:		Consulting:	
Manufacturing:	Y	Tissue Replacement:		Equipment/Supplies:	Arrays:		Blood Collection:	
Development:	Y			Research/Development Svcs.:	Database Management:		Drug Delivery:	Y
Generics:				Diagnostics:			Drug Distribution:	

TYPES OF BUSINESS:

Drug Delivery Systems
Transdermal Drug Delivery
Local Inflammation Therapies
Local Anesthetic Delivery
Ophthalmic Therapies

BRANDS/DIVISIONS/AFFILIATES:

Phoresor
Companion80
Iontocaine
OcuPhor
TransQ
IOGEL
Hybresis System

CONTACTS:
Note: Officers with more than one job title may be intentionally listed here more than once.
Robert J. Lollini, CEO
Robert J. Lollini, Pres.
Brian L. Mower, CFO
Jessica Barrett, Dir.-Sales & Mtg.
James D. Isaacson, Dir.-Developmental Eng.
James R. Cronkrite, Dir.-Mfg.
Betsy Truax, Consultant-Investor Rel.
James Isaacson, Dir.-Design
Curtis Jensen, Dir.-Quality & Regulatory Affairs
Peter J. Wardle, Chmn.

Phone: 801-975-1191	**Fax:** 801-972-9072
Toll-Free: 800-621-3347	
Address: 2441 S. 3850 W., Ste. A, Salt Lake City, UT 84120 US	

GROWTH PLANS/SPECIAL FEATURES:

IOMED, Inc. develops, manufactures and markets drug transport systems, which are primarily used to treat acute local inflammation in sports and occupational medicine and physical therapy applications. IOMED markets products used to deliver corticosteroids for local inflammation via iontophoresis, a needle-free process that delivers ionized, water-soluble drugs through the skin by applying an electrical current. The firm's patented Phoresor system is designed for clinical use and includes a microprocessor to control drug dosages and single-use transdermal patch kits. These products enable the site-specific, non-invasive administration of soluble salts or other drugs to the body as an alternative to hypodermic injection, avoiding the pain of needle insertion and minimizing both the infiltration of carrier fluids and damage to traumatized tissue caused by needles. Application-specific patch kits including IOGEL and TransQ have been designed to meet the demands of physical therapists and sports medicine professionals in the administration of drugs, primarily the corticosteroid dexamethasone, for inflammation. IOMED also markets the Companion80 self-contained iontophoresis mobile system that patients can use while resuming daily activities. The company previously offered the dermal anesthesia Iontocaine, which is FDA approved, but has discontinued marketing the product because of low historic revenues. IOMED has also discontinued the development of its products for ophthalmic diseases, including OcuPhor, until the firm is able to gain substantial support of a collaborative partner for the product. In July 2007, IOMED agreed to be acquired by and merged with ReAble Therapeutics, Inc. In February 2007, the company announced the introduction of its new Hybresis System, a wireless and miniaturized controller to be used with iontophoresis patches.

IOMED offers its employees educational assistance, leaves of absence, use of an on-site recreation room and other benefits.

FINANCIALS: Sales and profits are in thousands of dollars—add 000 to get the full amount. 2006 Note: Financial information for 2006 was not available for all companies at press time.

2006 Sales: $10,843	2006 Profits: $ 600	**U.S. Stock Ticker: IOX**
2005 Sales: $11,426	2005 Profits: $ 425	**Int'l Ticker:** Int'l Exchange:
2004 Sales: $12,189	2004 Profits: $ 954	Employees: 48
2003 Sales: $11,935	2003 Profits: $ 785	Fiscal Year Ends: 6/30
2002 Sales: $11,800	2002 Profits: $-2,400	Parent Company:

SALARIES/BENEFITS:

Pension Plan:	ESOP Stock Plan:	Profit Sharing:	Top Exec. Salary: $281,000	Bonus: $
Savings Plan: Y	Stock Purch. Plan:		Second Exec. Salary: $144,000	Bonus: $

OTHER THOUGHTS:

Apparent Women Officers or Directors: 2
Hot Spot for Advancement for Women/Minorities: Y

LOCATIONS: ("Y" = Yes)

West:	Southwest:	Midwest:	Southeast:	Northeast:	International:
Y					

Note: Financial information, benefits and other data can change quickly and may vary from those stated here.

ISIS PHARMACEUTICALS INC

www.isispharm.com

Industry Group Code: 325412 Ranks within this company's industry group: Sales: 107 Profits: 150

Drugs:		Other:		Clinical:		Computers:		Other:	
Discovery:	Y	AgriBio:		Trials/Services:		Hardware:		Specialty Services:	Y
Licensing:	Y	Genomics/Proteomics:	Y	Laboratories:		Software:		Consulting:	
Manufacturing:	Y	Tissue Replacement:		Equipment/Supplies:	Y	Arrays:		Blood Collection:	
Development:	Y			Research/Development Svcs.:		Database Management:		Drug Delivery:	
Generics:				Diagnostics:	Y			Drug Distribution:	

TYPES OF BUSINESS:

Drugs-Antisense Technology
Antisense Technology
Biosensors
Small-Molecule Drugs

BRANDS/DIVISIONS/AFFILIATES:

Vitravene
Ibis Technologies
Alicaforsen
TIGER
IBIS T-5000 Universal Pathogen Sensor
Isis Biosciences, Inc.

CONTACTS: Note: Officers with more than one job title may be intentionally listed here more than once.

Stanley T. Crooke, CEO
B. Lynne Parshall, CFO/Exec. VP/Sec.
C. Frank Bennett, Sr. VP-Research
Richard K. Brown, VP-Bus. Dev.
Michael J. Treble, VP/Pres., Ibis Biosciences, Inc.
Mark K. Wedel, VP-Dev./Chief Medical Officer
David J. Ecker, VP/Chief Scientific Officer-Ibis
Jeffrey M. Jonas, Exec. VP-Clinical Dev.
Stanley T. Crooke, Chmn.

Phone: 760-931-9200	Fax: 760-603-2700
Toll-Free:	
Address: 1896 Rutherford Rd., Carlsbad, CA 92008-7208 US	

GROWTH PLANS/SPECIAL FEATURES:

Isis Pharmaceuticals, Inc. (Isis) is a pioneer in the area of RNA-based antisense technology, which involves direct application and interaction of gene sequence information to combat various diseases. The company successfully developed Vitravene, the first antisense drug to achieve marketing clearance, as a treatment for cytomegalovirus retinitis for patients living with AIDS. Isis subsequently licensed Vitravene to Novartis Ophthalmics. In addition to these drugs, the company has four compounds in the pre-clinical development stages and is partnered with various pharmaceutical companies for the development of 12 others. Isis is the owner or exclusive licensee of approximately 1,500 RNA-based drug discovery and development patents. The company's Ibis Technologies division focuses on technologies to support the creation of biosensors to identify biological agents and the development of small-molecule antibacterial and antiviral drugs that bind to RNA. Ibis has developed a biosensor system, entitled TIGER (Triangulation Identification for Genetic Evaluation of Risks), designed to identify infectious organisms. Ibis Biosciences, Inc., a wholly-owned subsidiary of Isis, has developed and commercialized the Ibis T5000 Biosensor System for rapid identification and characterization of organisms. The company transitioned from research and development to commercial production in 2006.

FINANCIALS: Sales and profits are in thousands of dollars—add 000 to get the full amount. 2006 Note: Financial information for 2006 was not available for all companies at press time.

2006 Sales: $24,532	2006 Profits: $-45,903	U.S. Stock Ticker: ISIS
2005 Sales: $40,133	2005 Profits: $-72,401	Int'l Ticker: Int'l Exchange:
2004 Sales: $42,624	2004 Profits: $-142,864	Employees: 274
2003 Sales: $49,990	2003 Profits: $-94,996	Fiscal Year Ends: 12/31
2002 Sales: $58,320	2002 Profits: $-56,413	Parent Company:

SALARIES/BENEFITS:

Pension Plan:	ESOP Stock Plan:	Profit Sharing:	Top Exec. Salary: $582,218	Bonus: $423,855
Savings Plan: Y	Stock Purch. Plan: Y		Second Exec. Salary: $458,898	Bonus: $292,318

OTHER THOUGHTS:

Apparent Women Officers or Directors: 1
Hot Spot for Advancement for Women/Minorities:

LOCATIONS: ("Y" = Yes)

West:	Southwest:	Midwest:	Southeast:	Northeast:	International:
Y					

JAZZ PHARMACEUTICALS   www.jazzpharmaceuticals.com

Industry Group Code: 325412   Ranks within this company's industry group: Sales: 82   Profits: 164

Drugs:		Other:	Clinical:	Computers:	Other:
Discovery:	Y	AgriBio:	Trials/Services:	Hardware:	Specialty Services:
Licensing:	Y	Genomics/Proteomics:	Laboratories:	Software:	Consulting:
Manufacturing:		Tissue Replacement:	Equipment/Supplies:	Arrays:	Blood Collection:
Development:	Y		Research/Development Svcs.:	Database Management:	Drug Delivery:
Generics:			Diagnostics:		Drug Distribution:

TYPES OF BUSINESS:
Pharmaceuticals Discovery & Development
Neurological & Psychiatric Therapeutics

BRANDS/DIVISIONS/AFFILIATES:
Xyrem
Antizol

CONTACTS: *Note: Officers with more than one job title may be intentionally listed here more than once.*
Samuel R. Saks, CEO
Robert M. Myers, Pres.
Matthew K. Fust, CFO
Edwin W. Luker, VP-Sales
Lynn Hughes, VP-Human Resources
Mark G. Eller, VP-Research
Michael DesJardin, VP-Prod. Dev.
Nandan Oza, VP-Mfg.
Carol A. Gamble, General Counsel/Sr. VP/Corp. Sec.
Janne L. T. Wissel, Sr. VP-Dev.
Robert M. Myers, Chief Bus. Officer
Felissa H. Cagan, VP-Intellectual Property
Philip Perera, Chief Medical Officer
Julie Anne Smith, VP-Mktg. & New Prod. Planning
Bruce C. Cozadd, Chmn.
Nandan Oza, VP-Supply Chain

Phone: 650-496-3777	Fax: 650-496-3781
Toll-Free:	
Address: 3180 Porter Dr., Palo Alto, CA 94304 US	

GROWTH PLANS/SPECIAL FEATURES:
Jazz Pharmaceuticals, Inc. is a company with expertise in the development and commercialization of products that address issues in neurology and psychiatry. The company applies formulations and drug delivery technologies to known drug compounds to improve the efficacy and reduce adverse side effects of existing therapies. Jazz's general strategy is to target niche products that might not interest larger pharmaceutical companies. Also central to Jazz's business plan is the aggressive establishment of an innovative product portfolio through in-licensing, direct acquisition and collaboration. At present the company has been approved to market two drugs: Xyrem, an oral compound for the cataplexy suffered by narcoleptics, and Antizol, an antidote for ethylene glycol, such as antifreeze, and methanol poisoning. Antizol is the only antidote FDA-approved for the treatment of ethylene glycol and methanol poisoning. Through the National Organization of Rare Disorders, Inc. (NORD), a non-profit organization dedicated to the identification, treatment and cure of rare "orphan diseases", Jazz also provides Patient Assistance Programs on Xyrem for U.S. patients who are uninsured or underinsured. In March 2007, Jazz launched its Jazz Pharmaceuticals Fellowship Training Grant in Sleep Medicine, which will award four $50,000 one-year grants at the beginning of the 2008-2009 academic year for research focused in the field of sleep medicine. Jazz announced its initial public offering of 6,000,000 shares of common stock in May 2007, and began trading on the NASDAQ Global Market in June 2007.

Among companies of 101 to 500 employees in the Greater Bay Area, Jazz Pharmaceuticals was placed 8th for the 2007 Best Places to Work by the San Francisco Business Times, Silicon Valley/San Jose Business Journal and East Bay Business Times.

FINANCIALS: Sales and profits are in thousands of dollars—add 000 to get the full amount. 2006 Note: Financial information for 2006 was not available for all companies at press time.

2006 Sales: $44,856	2006 Profits: $-59,391	**U.S. Stock Ticker: JAZZ**
2005 Sales: $21,442	2005 Profits: $-85,156	**Int'l Ticker:** Int'l Exchange:
2004 Sales: $	2004 Profits: $	Employees: 185
2003 Sales: $	2003 Profits: $	Fiscal Year Ends: 12/31
2002 Sales: $	2002 Profits: $	Parent Company:

SALARIES/BENEFITS:

Pension Plan:	ESOP Stock Plan:	Profit Sharing:	Top Exec. Salary: $	Bonus: $
Savings Plan:	Stock Purch. Plan:		Second Exec. Salary: $	Bonus: $

OTHER THOUGHTS:
Apparent Women Officers or Directors: 5
Hot Spot for Advancement for Women/Minorities: Y

LOCATIONS: ("Y" = Yes)

West:	Southwest:	Midwest:	Southeast:	Northeast:	International:
Y					

JOHNSON & JOHNSON

www.jnj.com

Industry Group Code: 325412 Ranks within this company's industry group: Sales: 1 Profits: 2

Drugs:		Other:		Clinical:		Computers:		Other:	
Discovery:	Y	AgriBio:		Trials/Services:		Hardware:		Specialty Services:	
Licensing:		Genomics/Proteomics:		Laboratories:		Software:		Consulting:	
Manufacturing:	Y	Tissue Replacement:		Equipment/Supplies:	Y	Arrays:		Blood Collection:	
Development:	Y			Research/Development Svcs.:		Database Management:		Drug Delivery:	
Generics:				Diagnostics:	Y			Drug Distribution:	

TYPES OF BUSINESS:

Personal Health Care & Hygiene Products
Sterilization Products
Surgical Products
Pharmaceuticals
Skin Care Products
Baby Care Products
Contact Lenses
Medical Equipment

BRANDS/DIVISIONS/AFFILIATES:

Risperdal
Mylanta
Band-Aid
Tylenol
Monistat
McNeil Nutritionals, LLC
Hand Innovations
Conor Medsystems, Inc.

CONTACTS: Note: Officers with more than one job title may be intentionally listed here more than once.

William C. Weldon, CEO
James T. Lenehan, Pres.
Dominic J. Caruso, CFO
Kaye Foster-Check, VP-Human Resources
Per A. Peterson, Chmn., R&D Pharmaceuticals Group
Russell C. Deyo, General Counsel/Chief Compliance Officer
Dominic J. Caruso, VP-Finance
Colleen A. Goggins, Worldwide Chmn., Consumer & Personal Care
Christine A. Poon, Worldwide Chmn., Medicine & Nutritionals
Joseph C. Scodari, Chmn., Pharmaceuticals Group
Michael J. Dormer, Chmn., Medical Devices
William C. Weldon, Chmn.

Phone: 732-524-0400	Fax: 732-524-3300
Toll-Free:	
Address: One Johnson & Johnson Plaza, New Brunswick, NJ 08933 US	

GROWTH PLANS/SPECIAL FEATURES:

Johnson & Johnson, founded in 1886, is one of the world's most comprehensive and well-known manufacturers of health care products. The firm owns more than 250 companies in over 90 countries and markets its products in almost every country in the world. Johnson & Johnson's worldwide operations are divided into three segments: consumer, pharmaceutical, and medical devices and diagnostics. The company's principal consumer goods are personal care and hygiene products, including nonprescription drugs, adult skin and hair care, baby care, oral care, first aid and sanitary protection products. Major consumer brands include Mylanta, Band-Aid, Tylenol, Aveeno and Monistat. The pharmaceutical segment covers a wide spectrum of health fields, including antifungal, anti-infective, cardiovascular, dermatology, immunology, pain management, psychotropic and women's health. Among its pharmaceutical products are Risperdal, an antipsychotic used to treat schizophrenia, and Remicade for the treatment of Crohn's disease and rheumatoid arthritis. In the medical devices and diagnostics segment, Johnson & Johnson makes a number of products including suture and mechanical wound closure products, surgical instruments, disposable contact lenses, joint replacement products and intravenous catheters. Johnson & Johnson is pursuing nanotechnology applications in the biomedical fields primarily through research and funding agreements with other biotech companies, including Cordis. In early 2006, the company acquired Hand Innovations, Animas Corporation and Guidant Corporation. The $24 billion Guidant purchase will contribute an extensive line of life-saving technology for cardiac and vascular patients. In December 2006, Johnson & Johnson acquired Pfizer, Inc.'s Consumer Healthcare unit for $16.6 billion. In early 2007, Johnson & Johnson acquired Conor Medsystems, Inc., a cardiovascular device developer.

Johnson & Johnson offers its employees comprehensive heath and wellness. Some locations offer on-site child care centers and Nurture Space programs through which new mothers get counseling on how to return to work while breastfeeding.

FINANCIALS: Sales and profits are in thousands of dollars—add 000 to get the full amount. 2006 Note: Financial information for 2006 was not available for all companies at press time.

2006 Sales: $53,324,000	2006 Profits: $11,053,000	**U.S. Stock Ticker:** JNJ
2005 Sales: $50,514,000	2005 Profits: $10,060,000	**Int'l Ticker:** Int'l Exchange:
2004 Sales: $47,348,000	2004 Profits: $8,180,000	Employees: 122,200
2003 Sales: $41,862,000	2003 Profits: $7,197,000	Fiscal Year Ends: 12/31
2002 Sales: $36,298,000	2002 Profits: $6,597,000	Parent Company:

SALARIES/BENEFITS:

Pension Plan: Y	ESOP Stock Plan:	Profit Sharing:	Top Exec. Salary: $1,659,231	Bonus: $7,461,440
Savings Plan: Y	Stock Purch. Plan:		Second Exec. Salary: $1,023,846	Bonus: $2,574,880

OTHER THOUGHTS:

Apparent Women Officers or Directors: 3
Hot Spot for Advancement for Women/Minorities: Y

LOCATIONS: ("Y" = Yes)

West:	Southwest:	Midwest:	Southeast:	Northeast:	International:
Y	Y	Y	Y	Y	Y

KENDLE INTERNATIONAL INC

www.kendle.com

Industry Group Code: 325412 Ranks within this company's industry group: Sales: 45 Profits: 49

Drugs:	Other:	Clinical:		Computers:		Other:	
Discovery:	AgriBio:	Trials/Services:	Y	Hardware:		Specialty Services:	Y
Licensing:	Genomics/Proteomics:	Laboratories:		Software:	Y	Consulting:	Y
Manufacturing:	Tissue Replacement:	Equipment/Supplies:		Arrays:		Blood Collection:	
Development:		Research/Development Svcs.:	Y	Database Management:		Drug Delivery:	
Generics:		Diagnostics:				Drug Distribution:	

TYPES OF BUSINESS:

Pharmaceutical Development-Clinical Trials
Statistical Analysis
Technical Writing
Regulatory Assistance
Consulting Services
Clinical Trial Software
Clinical Data Management
e-Learning

BRANDS/DIVISIONS/AFFILIATES:

eKendleCollege
TrialWare
TrialWeb
TrialBase
TrialView
TriaLine
eKendleCollege
Latin American CRO Int'l Clinical Research, Ltd.

CONTACTS: Note: Officers with more than one job title may be intentionally listed here more than once.

Candace Kendle, CEO
Christopher Bergen, COO
Christopher Bergen, Pres.
Karl Brenkert, III, CFO/Sr. VP
Simon Higginbotham, Chief Mktg. Officer/VP
Karen L. Crone, VP-Global Human Resources
Gary Wedig, CIO
Anthony Forcellini, VP-Strategic Dev.
Anthony Forcellini, Treas.
Dennis Hurley, VP-Global Clinical Dev. Latin America
Martha Feller, VP-Global Clinical Dev. North America
Dieter Seitz-Tutter, VP-Global Clinical Dev. Europe
Cynthia L. Verst, VP-Late Phase
Candace Kendle, Chmn.

Phone: 513-381-5550	Fax: 513-381-5870
Toll-Free: 800-733-1572	
Address: 1200 Carew Tower, 441 Vine St., Cincinnati, OH 45202 US	

GROWTH PLANS/SPECIAL FEATURES:

Kendle International, Inc. is a leading global clinical research organization that provides a broad range of Phase I-IV global clinical development services to the biopharmaceutical industry. The company augments the research and development activities of biopharmaceutical companies by offering value-added clinical research services and proprietary information technology designed to reduce drug development time and expense. The firm operates in two segments: Early Stage, which handles all Phase I testing services; and Late Stage, which handles all Phase II-IV services. Kendle's services include clinical trial management, clinical data management, statistical analysis, technical writing and regulatory consulting and representation. It runs a state-of-the-art clinical pharmacology unit in the Netherlands, where it offers services for drugs undergoing clinical trials. The company's therapeutic expertise covers fields such as cardiovascular, dermatology, hematology, oncology, respiratory and women's health. Through its health care communications division, the firm provides organizational, meeting management and publication services to various professional associations and pharmaceutical companies. The firm's proprietary TrialWare product line includes a database management system, TrialBase; an interactive voice response patient randomization system, TriaLine; a validated medical imaging system, TrialView; a global project management system, TrialWatch; an Internet based collaborative tool, TrialWeb; and a late phase technology system, Trial4. Additionally, the company operates eKendleCollege, an online e-learning division that runs seminars and training programs, focusing on the organization of clinical trials. In 2006, Kendle acquired Latin America CRO International Clinical Research Limited, a late stage clinical research company with operations in Argentina, Brazil, Chile and Colombia. Later that year, the company acquired the late phase service business of Charles River Laboratories International, Inc.

Kendle offers its employees flexible work schedules, business-casual dress and continuing education through its corporate university and tuition reimbursement program, as well as the option to telecommute.

FINANCIALS: Sales and profits are in thousands of dollars—add 000 to get the full amount. 2006 Note: Financial information for 2006 was not available for all companies at press time.

2006 Sales: $373,936	2006 Profits: $8,530	U.S. Stock Ticker: KNDL
2005 Sales: $250,639	2005 Profits: $10,674	Int'l Ticker: Int'l Exchange:
2004 Sales: $215,868	2004 Profits: $3,572	Employees: 3,050
2003 Sales: $209,657	2003 Profits: $-1,690	Fiscal Year Ends: 12/31
2002 Sales: $214,000	2002 Profits: $-54,800	Parent Company:

SALARIES/BENEFITS:

Pension Plan:	ESOP Stock Plan:	Profit Sharing: Y	Top Exec. Salary: $344,578	Bonus: $50,170
Savings Plan: Y	Stock Purch. Plan: Y		Second Exec. Salary: $298,619	Bonus: $38,647

OTHER THOUGHTS:

Apparent Women Officers or Directors: 6
Hot Spot for Advancement for Women/Minorities: Y

LOCATIONS: ("Y" = Yes)

West:	Southwest:	Midwest:	Southeast:	Northeast:	International:
		Y		Y	Y

KERYX BIOPHARMACEUTICALS INC   www.keryx.com

Industry Group Code: 325412 Ranks within this company's industry group: Sales: 169   Profits: 171

Drugs:		Other:		Clinical:	Computers:		Other:	
Discovery:		AgriBio:		Trials/Services:	Hardware:		Specialty Services:	
Licensing:	Y	Genomics/Proteomics:		Laboratories:	Software:		Consulting:	
Manufacturing:		Tissue Replacement:		Equipment/Supplies:	Arrays:		Blood Collection:	
Development:	Y			Research/Development Svcs.:	Database Management:		Drug Delivery:	
Generics:				Diagnostics:			Drug Distribution:	

TYPES OF BUSINESS:

Pharmaceuticals Development
Drugs-Cancer
Drugs-Diabetes

BRANDS/DIVISIONS/AFFILIATES:

ACCESS Oncology, Inc.
Keryx Biomedical Technologies Ltd.
KRX-0401
Sulonex
Zerenex
Accumin
Neryx Biopharmaceuticals, Inc.
Accumin Diagnostics, Inc.

CONTACTS: *Note: Officers with more than one job title may be intentionally listed here more than once.*

Michael S. Weiss, CEO
I. Craig Henderson, Pres.
Ronald C. Renaud, Jr., CFO/Treas.
Beth F. Levine, General Counsel/Sr. VP/Corp. Sec.
Ronald Renaud, Jr., VP-Investor Rel.
Mark Stier, Chief Acct. Officer
Michael S. Weiss, Chmn.

Phone: 212-531-5965	Fax: 212-531-5961
Toll-Free:	
Address: 750 Lexington Ave., 20th Fl., New York, NY 10022 US	

GROWTH PLANS/SPECIAL FEATURES:

Keryx Biopharmaceuticals, Inc. is a biopharmaceutical company focused on the development and commercialization of pharmaceutical products for the treatment of life-threatening diseases including diabetes and cancer. The company operates in three segments: diagnostics, services and products. The diagnostics business sells diagnostic products for the direct measurement of total, intact urinary albumin. The services business provides clinical trial management and site recruitment services to other biotechnology and pharmaceutical companies. The products business focuses on the acquisition, development and commercialization of medically important, novel pharmaceutical products for the treatment of life-threatening diseases, including diabetes and cancer. Sulonex (sulodexide), previously KRX-101, a treatment for the life-threatening kidney disease caused by diabetes (diabetic nephropathy) is in Phase III and IV trials with the FDA. Zerenex, a treatment for elevated phosphate levels found in the late stages of renal disease, as well as two cancer treatments, are currently in Phase II trials. Keryx is working on several other drug products as well, including three drugs to be used in the field of oncology that are in Phase II development and two drugs, one in the area of oncology and the other focused on neurology, that are in pre-clinical trials. The company has not yet received approval for the sale of any of its drug candidates. Keryx owns three subsidiaries located in the U.S. and two in Israel: ACCESS Oncology, Inc.; Neryx Biopharmaceuticals, Inc.; Accumin Diagnostics, Inc.; Keryx (Israel) Ltd.; and Keryx Biomedical Technologies Ltd., which also operates in Israel.

FINANCIALS: Sales and profits are in thousands of dollars—add 000 to get the full amount. 2006 Note: Financial information for 2006 was not available for all companies at press time.

2006 Sales: $ 534	2006 Profits: $-73,764	U.S. Stock Ticker: KERX	
2005 Sales: $ 574	2005 Profits: $-26,895	Int'l Ticker: Int'l Exchange:	
2004 Sales: $ 809	2004 Profits: $-32,943	Employees: 40	
2003 Sales: $	2003 Profits: $-9,108	Fiscal Year Ends: 12/31	
2002 Sales: $	2002 Profits: $-11,783	Parent Company:	

SALARIES/BENEFITS:

Pension Plan:	ESOP Stock Plan:	Profit Sharing:	Top Exec. Salary: $375,000	Bonus: $
Savings Plan:	Stock Purch. Plan:		Second Exec. Salary: $300,000	Bonus: $

OTHER THOUGHTS:

Apparent Women Officers or Directors: 1
Hot Spot for Advancement for Women/Minorities:

LOCATIONS: ("Y" = Yes)

West:	Southwest:	Midwest:	Southeast:	Northeast:	International:
				Y	

KING PHARMACEUTICALS INC
www.kingpharm.com

Industry Group Code: 325412 Ranks within this company's industry group: Sales: 28 Profits: 25

Drugs:		Other:		Clinical:		Computers:		Other:	
Discovery:		AgriBio:		Trials/Services:		Hardware:		Specialty Services:	
Licensing:	Y	Genomics/Proteomics:		Laboratories:		Software:		Consulting:	
Manufacturing:	Y	Tissue Replacement:		Equipment/Supplies:		Arrays:		Blood Collection:	
Development:	Y			Research/Development Svcs.:	Y	Database Management:		Drug Delivery:	
Generics:				Diagnostics:				Drug Distribution:	

TYPES OF BUSINESS:

Pharmaceuticals Acquisition & Manufacturing
Prescription Pharmaceuticals - Diversified

BRANDS/DIVISIONS/AFFILIATES:

Bicillin
Cytomel
Skelaxin
Nordette
Levoxyl
Altace
Thrombin-JMI
EpiPen

CONTACTS: Note: Officers with more than one job title may be intentionally listed here more than once.

Brian A. Markison, CEO
Brian A. Markison, Pres.
Joseph Squicciarino, CFO
Eric J. Bruce, CTO
James W. Elrod, General Counsel
James E. Green, Exec. VP-Corp. Affairs
Stephen J. Andrzejewski, Chief Commercial Officer
Brian A. Markison, Chmn.

Phone: 423-989-8000	Fax: 423-274-8677
Toll-Free: 800-776-3637	
Address: 501 5th St., Bristol, TN 37620 US	

GROWTH PLANS/SPECIAL FEATURES:

King Pharmaceuticals, Inc. (KPI) manufactures branded pharmaceutical products in four therapeutic areas: cardiology/metabolic, neurosciences, hospital acute care products and other. The company in-licenses and acquires branded products from larger companies, by seeking out novel pharmaceuticals with the potential for enhancement through company focused promotion and marketing. Key products in the company's branded pharmaceutical segment include Altace, Corzide, Corgard, Levoxyl and Cytomel for cardiovascular/metabolic therapy; Skelaxin, Avinze and Sonata for neural therapy and Thrombin-JMI, Bicillin, Synercid and Intal for hospital/acute care treatments. Notable products of KPI include Levoxyl, a thyroid hormone replacement or supplemental therapy for hypothyroidism; Avinza, an extended release formulation of morphine for severe pain and Thrombin-JMI, which controls minor bleeding during surgery. The company also performs a variety of research and development, manufacturing, packaging, quality control, business development and regulatory management activities. Subsidiaries of the firm include King Pharmaceuticals Research and Development; Monarch Pharmaceuticals, Inc.; Meridian Medical Technologies, Inc.; Parkedale Pharmaceuticals, Inc.; King Pharmaceuticals of Nevada, Inc.; and Monarch Pharmaceuticals Ireland Limited. Meridian Medical Technologies consists of the company's auto-injector business while another segment, Royalties, develops and licenses the third party products, Adenoscan and Adenocard, which function as imaging agents and as a converter of irregular heart rhythms to normal sinus rhythms, respectively. In recent news, KPI announced the sale of its Rochester, Michigan manufacturing facility to KHP Pharmaceuticals in mid-2007. In February 2007, the company acquired the rights to Avinza, a once-daily treatment for moderate to severe in patients that require continuous opioid therapy, in the U.S. and Canada. The company also obtained an exclusive license from Vascular Solutions, Inc. to market Thrombi-Pad and Thrombi-Gel hemostats, which are designed for use outside of catheterization and electrophysiology laboratories.

FINANCIALS: Sales and profits are in thousands of dollars—add 000 to get the full amount. 2006 Note: Financial information for 2006 was not available for all companies at press time.

2006 Sales: $1,988,500	2006 Profits: $288,949	U.S. Stock Ticker: KG
2005 Sales: $1,772,881	2005 Profits: $117,833	Int'l Ticker: Int'l Exchange:
2004 Sales: $1,304,364	2004 Profits: $-160,288	Employees: 2,806
2003 Sales: $1,492,789	2003 Profits: $91,954	Fiscal Year Ends: 12/31
2002 Sales: $1,179,500	2002 Profits: $255,100	Parent Company:

SALARIES/BENEFITS:

Pension Plan:	ESOP Stock Plan:	Profit Sharing:	Top Exec. Salary: $825,000	Bonus: $
Savings Plan: Y	Stock Purch. Plan:		Second Exec. Salary: $466,033	Bonus: $

OTHER THOUGHTS:

Apparent Women Officers or Directors:
Hot Spot for Advancement for Women/Minorities:

LOCATIONS: ("Y" = Yes)

West:	Southwest:	Midwest:	Southeast:	Northeast:	International:
		Y	Y	Y	

Note: Financial information, benefits and other data can change quickly and may vary from those stated here.

KOS PHARMACEUTICALS INC   www.kospharm.com

Industry Group Code: 325412 Ranks within this company's industry group: Sales: Profits:

Drugs:		Other:	Clinical:	Computers:	Other:	
Discovery:		AgriBio:	Trials/Services:	Hardware:	Specialty Services:	
Licensing:		Genomics/Proteomics:	Laboratories:	Software:	Consulting:	
Manufacturing:	Y	Tissue Replacement:	Equipment/Supplies:	Arrays:	Blood Collection:	
Development:	Y		Research/Development Svcs.:	Database Management:	Drug Delivery:	Y
Generics:			Diagnostics:		Drug Distribution:	

TYPES OF BUSINESS:

Drugs-Cardiovascular & Respiratory Diseases
Respiratory Drug Delivery Systems

BRANDS/DIVISIONS/AFFILIATES:

Niaspan
Advicor
Azmacort
Flutiform

CONTACTS: Note: Officers with more than one job title may be intentionally listed here more than once.

Adrian Adams, CEO
Adrian Adams, Pres.
Christopher Kiritsy, CFO/Exec. VP
Aaron Berg, VP-Mktg.
Susan Taylor, Sr. VP-Human Resources
Ralf Rosskamp, Exec. VP-R&D
Daiva Bajorunas, VP-Product Realization
Perry Genova, VP-Biomedical Eng.
Juan Rodriguez, Sr. VP-Corp. Admin.
Andrew I. Koven, General Counsel/Corp. Sec./Exec. VP
Richard King, Exec. VP-Commercial Oper.
Sundar Kodiyalam, Sr. VP-Corp. Dev. & Licensing
Juan Rodriguez, Controller
Mark McGovern, Exec. VP/Chief Medical Officer
Marvin Blanford, Sr. VP-Drug Regulatory, Safety & Compliance
Akwete Lex Adjei, VP-Aerosol R&D
Daniel M. Bell, Chmn.

Phone: 609-495-0500	Fax: 609-495-0920
Toll-Free:	
Address: 1 Cedar Brook Dr., Cranbury, NJ 08512-3618 US	

GROWTH PLANS/SPECIAL FEATURES:

Kos Pharmaceuticals, Inc. (Kos), a subsidiary of Abbot Laboratories, develops and produces proprietary prescription products for the treatment of chronic cardiovascular and respiratory diseases. The company currently has three pharmaceuticals on the market. Niaspan, a treatment for cholesterol disorders, is the first once-daily form of niacin approved by the FDA. Advicor, which combines Niaspan with lovastatin, is the first and only FDA-approved dual-component therapy for the treatment of multiple lipic disorders. Kos also offers Azmacort, an inhaled asthma medication recently acquired from Aventis. In addition, the firm has a number of projects in various stages of development. Kos' heat shock protein 90, or Hsp90 inhibitor, KOS-953, is in Phase I and II clinical trials, primarily for multiple myeloma and HER2 positive breast cancer KOS-953 is the company's proprietary formulation of 17-AAG, a geldanamycin analog. In addition, intravenous and oral formulations of the second-generation Hsp90 inhibitor KOS-1022, are currently in Phase I clinical trials. These compounds have a novel mechanism of action targeting multiple pathways involved in cancer cell growth and survival. KOS is also developing KOS-862 in Phase I clinical trials in breast cancer. KOS-862 is an epothilone with a mechanism of action similar to taxanes, one of the most successful classes of anti-tumor agents introduced in the last decade. Its follow-on epothilone, KOS-1584, is in Phase clinical trials. The epothilone program is partnered with Hoffmann-La Roche, Inc. and F. Hoffmann-La Roche Ltd. collectively Roche, through a global development and commercialization agreement. In November 2006, the company agreed to be acquired by Abbott Laboratories for $3.7 billion.

Kos provides its employees with tuition reimbursement credit union access and an employee assistance program.

FINANCIALS: Sales and profits are in thousands of dollars—add 000 to get the full amount. 2006 Note: Financial information for 2006 was not available for all companies at press time.

2006 Sales: $	2006 Profits: $	U.S. Stock Ticker: Subsidiary
2005 Sales: $751,700	2005 Profits: $118,103	Int'l Ticker: Int'l Exchange:
2004 Sales: $497,104	2004 Profits: $142,319	Employees: 1,361
2003 Sales: $293,907	2003 Profits: $59,414	Fiscal Year Ends: 12/31
2002 Sales: $172,700	2002 Profits: $-20,800	Parent Company: ABBOT LABORATORIES

SALARIES/BENEFITS:

Pension Plan:	ESOP Stock Plan: Y	Profit Sharing:	Top Exec. Salary: $425,000	Bonus: $105,000
Savings Plan: Y	Stock Purch. Plan: Y		Second Exec. Salary: $350,000	Bonus: $75,000

OTHER THOUGHTS:

Apparent Women Officers or Directors: 3
Hot Spot for Advancement for Women/Minorities: Y

LOCATIONS: ("Y" = Yes)

West:	Southwest:	Midwest:	Southeast:	Northeast:	International:
			Y	Y	

KOSAN BIOSCIENCES INC

www.kosan.com

Industry Group Code: 325412 Ranks within this company's industry group: Sales: 120 Profits: 126

Drugs:		Other:		Clinical:		Computers:		Other:	
Discovery:	Y	AgriBio:		Trials/Services:		Hardware:		Specialty Services:	
Licensing:	Y	Genomics/Proteomics:	Y	Laboratories:		Software:		Consulting:	
Manufacturing:		Tissue Replacement:		Equipment/Supplies:		Arrays:		Blood Collection:	
Development:	Y			Research/Development Svcs.:		Database Management:		Drug Delivery:	
Generics:				Diagnostics:				Drug Distribution:	

TYPES OF BUSINESS:

Drugs-Polyketide Manipulation
Cancer Treatments
Infectious Disease Treatments
Gastrointestinal Motility Treatments
Hsp90 inhibitors

BRANDS/DIVISIONS/AFFILIATES:

Sloan-Kettering Institute for Cancer Research
Roche

CONTACTS: *Note: Officers with more than one job title may be intentionally listed here more than once.*

Robert G. Johnson, Jr., CEO
Gary S. Titus, CFO/Sr. VP
Pieter Timmermans, Sr. VP-Drug Discovery & Preclinical Research
Peter J. Licari, Sr. VP-Mfg.
Margaret A. Horn, General Counsel/Sr. VP-Legal
Peter J. Licari, Sr. VP-Oper.
Margaret A. Horn, Sr. VP-Corp. Dev.

Phone: 510-732-8400	Fax: 510-732-8401
Toll-Free:	
Address: 3832 Bay Center Pl., Hayward, CA 94545 US	

GROWTH PLANS/SPECIAL FEATURES:

Kosan Biosciences, Inc. uses its proprietary technologies to develop anticancer drug candidates from an important class of natural compounds known as polyketides. Polyketides are produced in small quantities by microorganisms. They are used in a variety of therapeutic products, including antibiotics, anticancer drugs, cholesterol treatments and immunosuppressants. Unfortunately, the highly useful compounds are complex and hard to synthesize artificially. Kosan has solved this problem through genetic manipulation. The four components of Kosan's technology platform are polyketide gene alteration, chemo-biosynthesis, heterologous over-expression and combinatorial biosynthesis. Using this platform, the firm makes improved versions of known polyketide pharmaceutical products and modifies existing products to serve new therapeutic areas. Currently, the company is focusing its development efforts on treatments for cancer, infectious diseases and gastrointestinal motility disorders. Kosan has two research and development agreements with the National Cancer Institute to develop geldanamycin analogs, called Hsp90 inhibitors, which sensitize cancer cells to anticancer agents. These products are in early- to mid-clinical trials. The firm has also collaborated with the Sloan-Kettering Institute for Cancer Research and Roche to develop epothilone compounds for cancer treatment, which are also in early to mid clinical trials.

Kosan offers its employees life insurance and commuter checks, as well as annual company social events.

FINANCIALS: Sales and profits are in thousands of dollars—add 000 to get the full amount. 2006 Note: Financial information for 2006 was not available for all companies at press time.

2006 Sales: $13,506	2006 Profits: $-29,469	U.S. Stock Ticker: KOSN
2005 Sales: $13,410	2005 Profits: $-29,637	Int'l Ticker: Int'l Exchange:
2004 Sales: $22,892	2004 Profits: $-22,126	Employees: 82
2003 Sales: $31,389	2003 Profits: $-9,668	Fiscal Year Ends: 12/31
2002 Sales: $9,600	2002 Profits: $-20,900	Parent Company:

SALARIES/BENEFITS:

Pension Plan:	ESOP Stock Plan:	Profit Sharing:	Top Exec. Salary: $388,333	Bonus: $
Savings Plan:	Stock Purch. Plan: Y		Second Exec. Salary: $300,000	Bonus: $25,000

OTHER THOUGHTS:

Apparent Women Officers or Directors: 1
Hot Spot for Advancement for Women/Minorities:

LOCATIONS: ("Y" = Yes)

West:	Southwest:	Midwest:	Southeast:	Northeast:	International:
Y					

KV PHARMACEUTICAL CO
www.kvpharmaceutical.com

Industry Group Code: 325412A Ranks within this company's industry group: Sales: 2 Profits: 3

Drugs:		Other:		Clinical:		Computers:		Other:	
Discovery:	Y	AgriBio:		Trials/Services:		Hardware:		Specialty Services:	
Licensing:		Genomics/Proteomics:		Laboratories:		Software:		Consulting:	
Manufacturing:	Y	Tissue Replacement:		Equipment/Supplies:		Arrays:		Blood Collection:	
Development:	Y			Research/Development Svcs.:		Database Management:		Drug Delivery:	Y
Generics:	Y			Diagnostics:				Drug Distribution:	

TYPES OF BUSINESS:

Pharmaceutical Products
Drug Delivery & Formulation Technologies
Specialty Ingredients
Taste Masking Systems
Branded Prescription Pharmaceuticals

BRANDS/DIVISIONS/AFFILIATES:

ETHEX Corporation
Ther-Rx Corporation
Particle Dynamics, Inc.
Particle and Coating Technologies, Inc.
Liquette
Gynazole-1
FlavorTech
Micromask

CONTACTS: Note: Officers with more than one job title may be intentionally listed here more than once.

Marc S. Hermelin, CEO
Gerald P. Mitchell, CFO/VP
Eric D. Moyermann, Pres., Pharmaceutical Manufacturing
David S. Hermelin, VP-Corp. Strategy & Oper. Analysis
Catherine Biffignani, VP-Corp. Comm.
Catherine Biffignani, VP-Investor Rel.
Richard H. Chibnall, VP-Finance
Raymond F. Chiostri, Chmn./CEO-Particle Dynamics, Inc.
Patricia K. McCullough, CEO-ETHEX Corp.
Philip J. Vogt, Pres., ETHEX Corp.
Paul T. Brady, Pres., Particle Dynamics, Inc.

Phone: 314-645-6600	Fax: 314-644-2419
Toll-Free:	
Address: 2503 S. Hanley Rd., St. Louis, MO 63144 US	

GROWTH PLANS/SPECIAL FEATURES:

KV Pharmaceutical Co. is a pharmaceutical company that develops, acquires, manufactures and markets technologically branded and generic prescription pharmaceutical products. The company develops a wide variety of drug delivery and formulation technologies, which are primarily focused in four areas: SITE RELEASE bioadhesives; tastemasking; oral controlled release; and oral quick dissolving tablets. The firm incorporates these technologies in the products it markets to control and improve the absorption and utilization of active pharmaceutical compounds. KV manufactures and markets these specialty pharmaceutical products through three wholly-owned subsidiaries: ETHEX Corporation, which targets generic and non-branded market segments; Ther-Rx Corporation, featuring branded product lines; and Particle Dynamics, Inc., a specialty ingredient products company. KV's operating units currently feature some 15 drug delivery technologies. The company's site-specific treatments, including SITE RELEASE, isolate drugs to a specific area of the body to increase their effectiveness. Taste-masking systems and quick-dissolving tablets are aimed at improving the taste of drugs and are marketed under names including Liquette, FlavorTech and Micromask. ETHEX, the company's major subsidiary, offers more than 100 products in four major categories: cardiovascular, women's health, pain management and respiratory/cough/cold. Ther-Rx markets branded prescription pharmaceutical products currently representing six brands in the cardiovascular, women's health and oral hematinic categories with a portfolio of 10 distinctive products. Particle Dynamics develops and markets technically advanced, value-added specialty ingredient products to the pharmaceutical, nutritional and personal care industries. The segment's vitamin-making technology is sold under the following three formulas: Destab, Descote and MicroMask. In June 2007, the company acquired Particle and Coating Technologies, Inc., a privately held St. Louis-based company.

KV offers its employees health, dental, life, accidental death and dismemberment insurance; an employee stock option plan; a 401(k) plan; a profit sharing plan; short- and long-term disability; educational assistance; and an employee assistance program.

FINANCIALS: Sales and profits are in thousands of dollars—add 000 to get the full amount. 2006 Note: Financial information for 2006 was not available for all companies at press time.

2006 Sales: $367,618	2006 Profits: $15,787	U.S. Stock Ticker: KV.A
2005 Sales: $303,493	2005 Profits: $33,269	Int'l Ticker: Int'l Exchange:
2004 Sales: $283,900	2004 Profits: $45,800	Employees: 1,145
2003 Sales: $245,000	2003 Profits: $28,100	Fiscal Year Ends: 3/31
2002 Sales: $204,100	2002 Profits: $31,500	Parent Company:

SALARIES/BENEFITS:

Pension Plan:	ESOP Stock Plan:	Profit Sharing: Y	Top Exec. Salary: $1,393,290	Bonus: $2,061,440
Savings Plan: Y	Stock Purch. Plan: Y		Second Exec. Salary: $370,555	Bonus: $

OTHER THOUGHTS:

Apparent Women Officers or Directors: 2
Hot Spot for Advancement for Women/Minorities: Y

LOCATIONS: ("Y" = Yes)

West:	Southwest:	Midwest:	Southeast:	Northeast:	International:
		Y			

Note: Financial information, benefits and other data can change quickly and may vary from those stated here.

LA JOLLA PHARMACEUTICAL
www.ljpc.com

Industry Group Code: 325412 Ranks within this company's industry group: Sales: Profits: 143

Drugs:		Other:	Clinical:	Computers:	Other:
Discovery:	Y	AgriBio:	Trials/Services:	Hardware:	Specialty Services:
Licensing:		Genomics/Proteomics:	Laboratories:	Software:	Consulting:
Manufacturing:		Tissue Replacement:	Equipment/Supplies:	Arrays:	Blood Collection:
Development:	Y		Research/Development Svcs.:	Database Management:	Drug Delivery:
Generics:			Diagnostics:		Drug Distribution:

TYPES OF BUSINESS:
Drugs-Autoimmune Diseases
Lupus Treatments
Small-Molecule Therapeutics

BRANDS/DIVISIONS/AFFILIATES:
Toleragens
Tolerance Technology
Riquent
La Jolla Limited

CONTACTS: Note: Officers with more than one job title may be intentionally listed here more than once.
Deirdre Gillespie, CEO
Deirdre Gillespie, Pres.
Niv E. Caviar, CFO/Exec. VP
Matthew D. Linnik, Chief Scientific Officer/Exec. VP
Paul C. Jenn, VP-Product Dev.
Luke Seikkula, VP-Mfg.
Gail A. Sloan, Corp. Sec.
Josefina Elchico, VP-Quality Oper.
Niv E. Caviar, Chief Bus. Officer
Gail A. Sloan, VP-Finance
Michael Tansey, Chief Medical Officer/Exec. VP
Lisa Hulle, VP-Regulatory Affairs
Craig R. Smith, Chmn.

Phone: 858-452-6600   Fax: 858-626-2851
Toll-Free:
Address: 6455 Nancy Ridge Dr., San Diego, CA 92121-2249 US

GROWTH PLANS/SPECIAL FEATURES:
La Jolla Pharmaceutical Company (LJPC) is a biopharmaceutical company that is focused on the research and development of therapeutic products for the treatment of certain life-threatening antibody-mediated diseases. LJPC currently owns 111 issued patents and has 65 pending patent applications for its various technologies and drug candidates. The firm's lead product, Riquent, is specifically designed to treat lupus renal disease by preventing or delaying renal flares. The firm's proprietary Tolerance Technology program is intended to treat the underlying cause of these diseases rather than only the symptoms, which prevents the severe negative side effects that are usually experienced with standard treatments that suppress the immune system. In addition to the development of Riquent, the company plans to apply its technology to treat other life-threatening conditions such as myasthenia, a form of muscular paralysis; and Rh hemolytic disease in newborns to prevent the destruction of fetal red blood cells. In addition, LJPC is also developing an SSAO inflammation program, which involves the use of novel, orally-active small molecules for the treatment of autoimmune diseases and acute and chronic inflammatory disorders. La Jolla Limited, based in England, develops Riquent throughout Europe. In February 2007, the firm implemented several enhancements to strengthen its Phase III study after consultations with the U.S. FDA. Future enhancements will focus on higher doses, an increased sample size, analyses of a broader patient population and the combination of Phase II clinical pharmacology studies with Phase III studies.

LJPC provides employees with a tuition reimbursement program, flexible spending accounts, 11 paid holidays and short and long-term disability.

FINANCIALS: Sales and profits are in thousands of dollars—add 000 to get the full amount. 2006 Note: Financial information for 2006 was not available for all companies at press time.
2006 Sales: $	2006 Profits: $-39,445	U.S. Stock Ticker: LJPC
2005 Sales: $	2005 Profits: $-27,363	Int'l Ticker: Int'l Exchange:
2004 Sales: $	2004 Profits: $-40,544	Employees: 84
2003 Sales: $	2003 Profits: $-38,838	Fiscal Year Ends: 12/31
2002 Sales: $	2002 Profits: $-13,799	Parent Company:

SALARIES/BENEFITS:
Pension Plan:	ESOP Stock Plan: Y	Profit Sharing:	Top Exec. Salary: $305,482	Bonus: $
Savings Plan: Y	Stock Purch. Plan:		Second Exec. Salary: $292,789	Bonus: $50,000

OTHER THOUGHTS:
Apparent Women Officers or Directors: 5
Hot Spot for Advancement for Women/Minorities: Y

LOCATIONS: ("Y" = Yes)
West:	Southwest:	Midwest:	Southeast:	Northeast:	International:
Y					Y

Note: Financial information, benefits and other data can change quickly and may vary from those stated here.

LANNETT COMPANY INC

www.lannett.com

Industry Group Code: 325416 Ranks within this company's industry group: Sales: 9   Profits: 7

Drugs:		Other:		Clinical:		Computers:		Other:	
Discovery:		AgriBio:		Trials/Services:		Hardware:		Specialty Services:	
Licensing:		Genomics/Proteomics:		Laboratories:		Software:		Consulting:	
Manufacturing:	Y	Tissue Replacement:		Equipment/Supplies:		Arrays:		Blood Collection:	
Development:	Y			Research/Development Svcs.:		Database Management:		Drug Delivery:	
Generics:				Diagnostics:				Drug Distribution:	Y

TYPES OF BUSINESS:

Drugs-Generic
Drug Delivery System Development

BRANDS/DIVISIONS/AFFILIATES:

Acetazolamide
Butalbital
Digoxin
Diphenoxylate
Levothyroxine Sodium
Primidone
Hydromorphone
Phentermine

CONTACTS: *Note: Officers with more than one job title may be intentionally listed here more than once.*

Arthur P. Bedrosian, CEO
Arthur P. Bedrosian, Pres.
Brian Kearns, CFO
Kevin Smith, VP-Mktg. & Sales
Bernard Sandiford, VP-Oper.
Brian Kearns, Treas.
William Farber, Chmn.
William Schreck, VP-Logistics

Phone: 215-333-9000	Fax: 215-333-9004
Toll-Free: 800-325-9994	
Address: 9000 State Rd., Philadelphia, PA 19136 US	

GROWTH PLANS/SPECIAL FEATURES:

Lannett Company, Inc. develops, manufactures, markets and distributes pharmaceutical products sold under generic names, marketing them primarily to drug wholesalers, retail drug chains, repackagers, distributors and government agencies.   The company's extensive range of products includes: Acetazolamide for glaucoma; Butalbital with aspirin, caffeine and codeine for migraines; Clindamycin, an antibiotic; Digoxin for congestive heart failure; Dicyclomine, a treatment for irritable bowels; Diphenoxylate with atropine sulfate for diarrhea; Levothyroxine Sodium for thyroid deficiency; Methocarbamol, which is a muscle relaxer; Phentermine, a pill for weight-loss; Phenylpropanolamine, a treatment for incontinence; Primidone for epilepsy; Terbutaline, a medication for bronchospasms; and Unithroid, a thyroid medication.   All of the products currently manufactured and sold by the company are prescription products.   Unithroid, Butalbital, Digoxin, Primidone and Levothyroxine are the company's key products; they contributed to more than 80% of Lannett's 2006 net sales (down from 97% in 2004 and 93% in 2005).   Lannett currently manufactures only oral, solid dosage forms (tablets and capsules), but it is pursuing partnerships and research contracts for the development of liquid and injectable products.   Recently, the company announced the FDA's approval for its supplemental Abbreviated New Drug Application for Probenecid Tablets, which are indicated for the treatment of hyperuricemia, which is associated with gout and gouty arthritis. In 2007, the company received approval for Danazol (a gonadotropin inhibitor, a sex hormone) 50mg and 100mg capsules, as well as approval to market Baclofen 10mg tablets for the alleviation of signs and symptoms of spasticity resulting from multiple sclerosis, as well as the commencement of marketing for Meloxicam for arthritis. The company also acquired a privately-owned manufacturer/supplier of bulk active pharmaceutical ingredients.

FINANCIALS: Sales and profits are in thousands of dollars—add 000 to get the full amount. 2006 Note: Financial information for 2006 was not available for all companies at press time.

2006 Sales: $64,060	2006 Profits: $4,969	U.S. Stock Ticker: LCI
2005 Sales: $44,902	2005 Profits: $-32,780	Int'l Ticker: Int'l Exchange:
2004 Sales: $63,781	2004 Profits: $13,215	Employees: 193
2003 Sales: $42,487	2003 Profits: $11,667	Fiscal Year Ends: 6/30
2002 Sales: $25,126	2002 Profits: $7,196	Parent Company:

SALARIES/BENEFITS:

Pension Plan:	ESOP Stock Plan:	Profit Sharing:	Top Exec. Salary: $278,641	Bonus: $92,970
Savings Plan: Y	Stock Purch. Plan: Y		Second Exec. Salary: $193,572	Bonus: $20,712

OTHER THOUGHTS:

Apparent Women Officers or Directors:
Hot Spot for Advancement for Women/Minorities:

LOCATIONS: ("Y" = Yes)

West:	Southwest:	Midwest:	Southeast:	Northeast:	International:
				Y	

LARGE SCALE BIOLOGY CORP

www.lsbc.com

Industry Group Code: 325412 Ranks within this company's industry group: Sales: Profits:

Drugs:		Other:		Clinical:		Computers:		Other:	
Discovery:		AgriBio:	Y	Trials/Services:		Hardware:		Specialty Services:	
Licensing:		Genomics/Proteomics:	Y	Laboratories:		Software:		Consulting:	
Manufacturing:	Y	Tissue Replacement:	Y	Equipment/Supplies:		Arrays:		Blood Collection:	
Development:	Y			Research/Development Svcs.:	Y	Database Management:		Drug Delivery:	
Generics:				Diagnostics:	Y			Drug Distribution:	

TYPES OF BUSINESS:

Drug Discovery & Development
Proteomics & Genomics
Protein-Based Drugs
Protein Markers

BRANDS/DIVISIONS/AFFILIATES:

GENEWARE
GRAMMR
BAMF

CONTACTS: Note: Officers with more than one job title may be intentionally listed here more than once.

Kevin J. Ryan, CEO
Ronald J. Artale, COO/Sr. VP
Kevin J. Ryan, Pres.
Ronald J. Artale, CFO
Laurence K. Grill, Chief Scientific Officer/Sr. VP-Research
Stephen J. Garger, Exec. VP-Biomanufacturing
Daniel Tuse, VP-Bus. Dev.
Michael D. Centron, VP/Treas.
Greg Pogue, VP/Exec. Program Mgr.
Robert L. Erwin, Chmn.

Phone: 707-446-5501	Fax: 707-446-3917

Toll-Free:

Address: 3333 Vaca Valley Pkwy., Ste. 1000, Vacaville, CA 95688 US

GROWTH PLANS/SPECIAL FEATURES:

Large Scale Biology Corp. (LSBC) is a biotechnology-based drug company developing therapeutic products based on its unique technologies. Its areas of expertise lie in plant- and virus-based proteins and vaccines. Some notable products include aprotinin, a protease inhibitor used in medical, research and manufacturing applications; alpha-galactosidase A, for the treatment of Fabry disease; vaccines for human and animal health care, such as antiviral and anticancer applications; and GRAMMR, a technology used in shuffling gene sequences for use in developing hosts with particular traits. LSBC's technologies are designed to analyze both genes and proteins in an automated fashion. From the company's inception, it has focused on providing research and development for customers through its GENEWARE, genomics, proteomics and bioinformatics platforms, but in recent years LSBC's focus has shifted to developing and manufacturing its own products. The company has an exclusive license with Cincinnati Children's Hospital for the development of human lysosomal acid lipase for treating atherosclerosis, a circulatory disease. In November 2006, the company achieved a benchmark in its efforts to develop and apply novel methods for improving proteins through molecular evolution. LSBC's internally developed, patented methods for gene shuffling and lead selection comprise an integrated process through which protein molecules with improved function can be developed and evaluated.

FINANCIALS: Sales and profits are in thousands of dollars—add 000 to get the full amount. 2006 Note: Financial information for 2006 was not available for all companies at press time.

2006 Sales: $		2006 Profits: $		U.S. Stock Ticker: LSBC.PK	
2005 Sales: $		2005 Profits: $		Int'l Ticker: Int'l Exchange:	
2004 Sales: $1,800		2004 Profits: $-17,500		Employees: 78	
2003 Sales: $3,600		2003 Profits: $-25,300		Fiscal Year Ends: 12/31	
2002 Sales: $2,600		2002 Profits: $-33,200		Parent Company:	

SALARIES/BENEFITS:

Pension Plan:	ESOP Stock Plan:	Profit Sharing:	Top Exec. Salary: $242,500	Bonus: $24,000
Savings Plan: Y	Stock Purch. Plan: Y		Second Exec. Salary: $214,200	Bonus: $

OTHER THOUGHTS:

Apparent Women Officers or Directors:
Hot Spot for Advancement for Women/Minorities:

LOCATIONS: ("Y" = Yes)

West:	Southwest:	Midwest:	Southeast:	Northeast:	International:
Y		Y			

Note: Financial information, benefits and other data can change quickly and may vary from those stated here.

LEINER HEALTH PRODUCTS INC

www.leiner.com

Industry Group Code: 325411   Ranks within this company's industry group: Sales: 2   Profits: 6

Drugs:		Other:		Clinical:	Computers:		Other:	
Discovery:		AgriBio:		Trials/Services:	Hardware:		Specialty Services:	Y
Licensing:		Genomics/Proteomics:		Laboratories:	Software:		Consulting:	
Manufacturing:	Y	Tissue Replacement:		Equipment/Supplies:	Arrays:		Blood Collection:	
Development:				Research/Development Svcs.:	Database Management:		Drug Delivery:	
Generics:	Y			Diagnostics:			Drug Distribution:	

TYPES OF BUSINESS:

Vitamins, Mineral and Nutritional Supplements
Over-the-Counter Pharmaceuticals
Contract Manufacturing

BRANDS/DIVISIONS/AFFILIATES:

Your Life
Pharmacist Formula

CONTACTS: Note: Officers with more than one job title may be intentionally listed here more than once.

Robert M. Kaminski, CEO
Robert K. Reynolds, COO
Robert K. Reynolds, Pres.
Kevin McDonnell, CFO/Exec. VP
Kevin J. Lanigan, Exec. VP/Corp. General Manager
Charles F. Baird, Chmn.
Robert J. la Ferriere, Sr. Exec. VP-Supply Chain

Phone: 310-835-8400	Fax: 310-952-7760
Toll-Free:	
Address: 901 E. 233rd St., Carson, CA 90745 US	

GROWTH PLANS/SPECIAL FEATURES:

Leiner Health Products, Inc. (LHP) is a supplier of store brand vitamins, mineral and nutritional supplements (VMS) products and store brand over the counter (OTC) pharmaceuticals. In addition, the firm offers contract manufacturing services to consumer products and pharmaceutical companies. Most of the company's VMS and OTC products are manufactured for its customers to sell as their own store brands. LHP also sells VMS products under its own brand name, Your Life, and OTC products under Pharmacist Formula. The VMS products include more than 225 products in roughly 1,550 stock keeping units in the U.S. sold in tablet and capsule forms and in varying sizes with different potencies, flavors and coatings. Vitamin products include national brand equivalents that compare to the dietary ingredients in Caltrate, Centrum and One-A-Day vitamin products such as vitamin C and E and folic acid; minerals such as calcium; supplements such as chondroitin glucosamine, fish oil, Co-Q10; and herbal products. VMS products accounted for roughly 61% of net sales in 2007. In addition, Your Life brand is the largest VMS brand provider to the U.S. military. The OTC pharmaceuticals principally consist of analgesics, cough, cold and allergy medications and digestive aids. LHP's U.S. operations consist of four operating subsidiaries, principally Leiner Health Products LLC. The Canadian operations consist of Vita Health. The company's products are sold in all 50 states to more than 54,000 outlets, including nine of the ten largest food retailers. In April 2007, LHP's partner, Dr. Reddy Laboratories terminated the OTC distribution agreement, the famotidine supply agreement and the supply agreement between it and the company.

The company offers its employees medical, dental and vision insurance; life and AD&D insurance; short- and long-term disability; business travel accident insurance; an employee assistance program; a 401(k) plan; education assistance and a college savings program.

FINANCIALS: Sales and profits are in thousands of dollars—add 000 to get the full amount. 2006 Note: Financial information for 2006 was not available for all companies at press time.

2006 Sales: $669,561	2006 Profits: $-3,768	U.S. Stock Ticker: Private
2005 Sales: $684,901	2005 Profits: $-47,912	Int'l Ticker: Int'l Exchange:
2004 Sales: $661,045	2004 Profits: $39,056	Employees: 1,995
2003 Sales: $574,281	2003 Profits: $145,730	Fiscal Year Ends: 3/31
2002 Sales: $	2002 Profits: $	Parent Company:

SALARIES/BENEFITS:

Pension Plan:	ESOP Stock Plan:	Profit Sharing:	Top Exec. Salary: $475,000	Bonus: $565,250
Savings Plan: Y	Stock Purch. Plan:		Second Exec. Salary: $325,000	Bonus: $886,750

OTHER THOUGHTS:

Apparent Women Officers or Directors:
Hot Spot for Advancement for Women/Minorities:

LOCATIONS: ("Y" = Yes)

West:	Southwest:	Midwest:	Southeast:	Northeast:	International:
Y				Y	Y

LEXICON PHARMACEUTICALS INC   www.lexgen.com

Industry Group Code: 325412 Ranks within this company's industry group: Sales: 70   Profits: 159

Drugs:		Other:		Clinical:		Computers:		Other:	
Discovery:	Y	AgriBio:		Trials/Services:		Hardware:		Specialty Services:	
Licensing:		Genomics/Proteomics:	Y	Laboratories:		Software:		Consulting:	
Manufacturing:		Tissue Replacement:		Equipment/Supplies:		Arrays:		Blood Collection:	
Development:				Research/Development Svcs.:	Y	Database Management:	Y	Drug Delivery:	
Generics:				Diagnostics:				Drug Distribution:	

TYPES OF BUSINESS:
Drug Discovery & Development
Genetic Databases
Research & Development-Genetics

BRANDS/DIVISIONS/AFFILIATES:
OmniBank
OmniBank E-biology
Genome5000
LexVision

CONTACTS: Note: Officers with more than one job title may be intentionally listed here more than once.
Arthur T. Sands, CEO
Arthur T. Sands, Pres.
Julia P. Gregory, CFO/Exec. VP
Walter F. Colbert, Sr. VP-Human Resources & Corp. Svcs.
Brian P. Zambrowicz, Chief Scientific Officer/Exec. VP
Jeffrey L. Wade, General Counsel/Exec. VP
Tamar D. Howson, Exec. VP-Bus. Dev.
Lance K. Ishimoto, Sr. VP-Intellectual Property
Alan J. Main, Exec. VP-Pharmaceutical Research
James R. Piggott, Sr. VP-Pharmaceutical Biology
Philip M. Brown, VP-Clinical Dev.
Samuel L. Barker, Chmn.

Phone: 281-863-3000	Fax: 281-863-8088
Toll-Free:	
Address: 8800 Technology Forest Pl., The Woodlands, TX 77381-1160 US	

GROWTH PLANS/SPECIAL FEATURES:
Lexicon Pharmaceuticals, Inc., formerly Lexicon Genetics Incorporated, is a biopharmaceutical company that focuses on the discovery and development of new medical treatments for human diseases. Though none of its drug programs have yet to advance into clinical development, Lexicon currently has more than 60 targets in drug discovery programs. Potential drug targets are determined through the Genome5000 program, an experiment that employs knockout technology to disrupt the function of genes in mice in order to determine the proper pharmaceutical treatment for the physiological and behavioral functions of each gene. Potential small molecule, antibody and protein drugs are used for in vivo-validated drug targets and are pushed through preclinical development studies for regulatory filings. Discovery efforts are focused in six therapeutic areas: diabetes and obesity; cardiovascular disease; psychiatric and neurological disorders; cancer; immune system disorders; and ophthalmic disease. The firm also works through strategic collaborations and alliances with companies such as Bristol-Myers Squibb to develop small-molecule drugs in the neuroscience field. Lexicon also works with Genentech, Inc. in order to discover the functions of secreted proteins and potential antibody targets and Takeda Pharmaceutical Company Limited in the field of high blood pressure medication. Lexicon receives fees and royalty payments from pharmaceutical and biotechnology companies for access to its technologies and discoveries. The company maintains OmniBank, where more than 270,000 embryonic cell clones are cataloged and stored, and LexVision, a database which catalogues the gene trapping and targeting technologies discovered through its mice experimentations. In 2006, Lexicon entered into an agreement with Cetek Corporation to access Cetek's library of natural products for use in Lexicon's drug discovery process.

The company offers its employees relocation assistance, educational assistance plans, flexible spending accounts and life insurance.

FINANCIALS: Sales and profits are in thousands of dollars—add 000 to get the full amount. 2006 Note: Financial information for 2006 was not available for all companies at press time.

2006 Sales: $72,798	2006 Profits: $-54,311	U.S. Stock Ticker: LEXG
2005 Sales: $75,680	2005 Profits: $-36,315	Int'l Ticker: Int'l Exchange:
2004 Sales: $61,740	2004 Profits: $-47,172	Employees: 585
2003 Sales: $42,838	2003 Profits: $-64,198	Fiscal Year Ends: 12/31
2002 Sales: $35,200	2002 Profits: $-59,700	Parent Company:

SALARIES/BENEFITS:
Pension Plan:	ESOP Stock Plan: Y	Profit Sharing:	Top Exec. Salary: $473,000	Bonus: $300,000
Savings Plan: Y	Stock Purch. Plan:		Second Exec. Salary: $329,000	Bonus: $80,000

OTHER THOUGHTS:
Apparent Women Officers or Directors: 3
Hot Spot for Advancement for Women/Minorities: Y

LOCATIONS: ("Y" = Yes)
West:	Southwest:	Midwest:	Southeast:	Northeast:	International:
	Y			Y	

Note: Financial information, benefits and other data can change quickly and may vary from those stated here.

LIFE SCIENCES RESEARCH

www.lsrinc.net

Industry Group Code: 541710   Ranks within this company's industry group: Sales: 8   Profits: 12

Drugs:	Other:	Clinical:		Computers:		Other:
Discovery:	AgriBio:	Trials/Services:	Y	Hardware:		Specialty Services:
Licensing:	Genomics/Proteomics:	Laboratories:	Y	Software:		Consulting:
Manufacturing:	Tissue Replacement:	Equipment/Supplies:		Arrays:		Blood Collection:
Development:		Research/Development Svcs.:	Y	Database Management:		Drug Delivery:
Generics:		Diagnostics:				Drug Distribution:

TYPES OF BUSINESS:

Contract Research
Drug Development Services
Non-Clinical Safety Testing

BRANDS/DIVISIONS/AFFILIATES:

Huntingdon Life Sciences Group
Princeton Research Center
Huntingdon and Eye

CONTACTS: *Note: Officers with more than one job title may be intentionally listed here more than once.*

Andrew H. Baker, CEO
Brian Cass, Pres.
Richard Michaelson, CFO
Julian Griffiths, Dir.-Oper.
Richard Michaelson, Corp. Sec.
Andrew H. Baker, Chmn.

Phone: 732-649-9961	**Fax:** 732-649-0021
Toll-Free:	
Address: Mettlers Rd., P.O. Box 2360, East Millstone, NJ 08875 US	

GROWTH PLANS/SPECIAL FEATURES:

Life Sciences Research, Inc. (LSR), formerly Huntingdon Life Sciences Group, is a global contract research organization (CRO) providing product development services to the pharmaceutical, agrochemical and biotechnology industries. It utilizes leading technology and capability to support its clients in non-clinical safety testing of new compounds in early stage development and assessment. The purpose of LSR's work is to identify risks to humans, animals or the environment resulting from the use or manufacture of a wide range of chemicals which are essential components of LSR's clients' products. The company's services are designed to meet the regulatory requirements of governments around the world. LSR operates research facilities in the U.S. (the Princeton Research Center, New Jersey) and the U.K. (Huntingdon and Eye, England). Life Sciences Research's services address safety concerns surrounding a diverse range of products, spanning such areas as agricultural herbicides and other pesticides, medical devices, veterinary medicines, and specialty chemicals used in the manufacture of pharmaceutical intermediates and manufactured foodstuffs and products. The company changed its name and relocated to the U.S., in order to escape pressure from the animal rights activist group Stop Huntingdon Animal Cruelty (SHAC) and others.

FINANCIALS: Sales and profits are in thousands of dollars—add 000 to get the full amount. 2006 Note: Financial information for 2006 was not available for all companies at press time.

2006 Sales: $192,217	2006 Profits: $-14,872	**U.S. Stock Ticker: LSR**
2005 Sales: $172,013	2005 Profits: $1,491	**Int'l Ticker:** Int'l Exchange:
2004 Sales: $157,551	2004 Profits: $17,594	Employees: 1,407
2003 Sales: $132,434	2003 Profits: $3,728	Fiscal Year Ends: 12/31
2002 Sales: $115,742	2002 Profits: $2,697	Parent Company:

SALARIES/BENEFITS:

Pension Plan:	ESOP Stock Plan:	Profit Sharing:	Top Exec. Salary: $608,256	Bonus: $152,064
Savings Plan:	Stock Purch. Plan:		Second Exec. Salary: $608,256	Bonus: $152,064

OTHER THOUGHTS:

Apparent Women Officers or Directors:
Hot Spot for Advancement for Women/Minorities:

LOCATIONS: ("Y" = Yes)

West:	Southwest:	Midwest:	Southeast:	Northeast:	International:
				Y	Y

LIFECELL CORPORATION   www.lifecell.com

Industry Group Code: 325414 Ranks within this company's industry group: Sales: 5   Profits: 5

Drugs:		Other:		Clinical:		Computers:		Other:	
Discovery:	Y	AgriBio:		Trials/Services:		Hardware:		Specialty Services:	Y
Licensing:		Genomics/Proteomics:		Laboratories:		Software:		Consulting:	
Manufacturing:	Y	Tissue Replacement:	Y	Equipment/Supplies:		Arrays:		Blood Collection:	
Development:	Y			Research/Development Svcs.:	Y	Database Management:		Drug Delivery:	
Generics:				Diagnostics:				Drug Distribution:	

TYPES OF BUSINESS:

Tissue Replacement Products
Skin Replacement Technology
Bone Grafting Technology
Regenerative Medicine

GROWTH PLANS/SPECIAL FEATURES:

LifeCell Corporation specializes in regenerative medicine, developing and manufacturing products geared toward the repair, replacement and preservation of human tissues. The company has developed and patented several proprietary technologies, including a method for producing an acellular tissue matrix, a method for cell preservation through signal transduction and a method for freeze-drying biological cells and tissues without damage. Products are used in reconstructive, orthopedic and urogynecologic surgical procedures to repair soft tissue defects. The company's products include AlloDerm, for plastic reconstructive, general surgical, burn and periodontal procedures; Cymetra, a particulate form of AlloDerm suitable for injection; GraftJacket, for orthopedic applications and lower extremity wounds; AlloCraft DBM, for bone grafting procedures; and Repliform, for urogynecologic surgical procedures. The firm markets AlloDerm in the United States for plastic reconstructive, general surgical and burn applications and Cymetra to hospital-based surgeons through its direct sales and marketing organization. BioHorizons Implant Systems, Inc. is LifeCell's exclusive distributor in the U.S. and certain international markets of AlloDerm and AlloDerm GBR for use in periodontal applications. Boston Scientific Corp. is the exclusive worldwide sales and marketing agent of Repliform for use in urogynecology. Wright Medical Group is the exclusive distributor in the U.S. and certain international markets for GraftJacket. Stryker Corp. is the exclusive distributor in the U.S. for AlloCraft DBM.

The company offers its employees a 401(k); medical, dental and vision insurance; life and AD&D insurance; short- and long-term disability; an employee assistance program; and a college savings plan.

BRANDS/DIVISIONS/AFFILIATES:

AlloDerm
Cymetra
Repliform
Graft Jacket
AlloCraft DBM

CONTACTS: Note: Officers with more than one job title may be intentionally listed here more than once.

Paul G. Thomas, CEO
Paul G. Thomas, Pres.
Steven T. Sobieski, CFO
Steven T. Sobieski, Sr. VP-Admin.
Steven T. Sobieski, VP-Finance
Bruce Lamb, Sr. VP-Dev. & Regulatory Affairs
Lisa N. Colleran, Sr. VP-Commercial Oper.
Paul G. Thomas, Chmn.

Phone: 908-947-1100	Fax: 908-947-1200
Toll-Free:	
Address: 1 Millennium Way, Branchburg, NJ 08876 US	

FINANCIALS: Sales and profits are in thousands of dollars—add 000 to get the full amount. 2006 Note: Financial information for 2006 was not available for all companies at press time.

2006 Sales: $141,680	2006 Profits: $20,469	U.S. Stock Ticker: LIFC
2005 Sales: $94,398	2005 Profits: $12,044	Int'l Ticker: Int'l Exchange:
2004 Sales: $61,127	2004 Profits: $7,184	Employees: 335
2003 Sales: $40,249	2003 Profits: $18,672	Fiscal Year Ends: 12/31
2002 Sales: $34,400	2002 Profits: $1,400	Parent Company:

SALARIES/BENEFITS:

Pension Plan:	ESOP Stock Plan:	Profit Sharing:	Top Exec. Salary: $477,750	Bonus: $390,417
Savings Plan: Y	Stock Purch. Plan:		Second Exec. Salary: $272,225	Bonus: $130,913

OTHER THOUGHTS:

Apparent Women Officers or Directors: 1
Hot Spot for Advancement for Women/Minorities:

LOCATIONS: ("Y" = Yes)

West:	Southwest:	Midwest:	Southeast:	Northeast:	International:
				Y	

LIFECORE BIOMEDICAL INC

www.lifecore.com

Industry Group Code: 339113 Ranks within this company's industry group: Sales: 12   Profits: 15

Drugs:	Other:	Clinical:	Computers:	Other:
Discovery:	AgriBio:	Trials/Services:	Hardware:	Specialty Services:
Licensing:	Genomics/Proteomics:	Laboratories:	Software:	Consulting:
Manufacturing: Y	Tissue Replacement: Y	Equipment/Supplies: Y	Arrays:	Blood Collection:
Development:		Research/Development Svcs.:	Database Management:	Drug Delivery:
Generics:		Diagnostics:		Drug Distribution:

TYPES OF BUSINESS:

Biomaterials & Medical Device Manufacturing
Bone Regeneration Products
Surgical Devices
Dental Implants

BRANDS/DIVISIONS/AFFILIATES:

Support Plus
Lifecore Prima Implant System

CONTACTS: *Note: Officers with more than one job title may be intentionally listed here more than once.*

Dennis J. Allingham, CEO
Dennis J. Allingham, Pres.
David M. Noel, CFO
Kipling Thacker, VP-New Bus. Dev.
David M. Noel, VP-Finance
Andre P. Decarie, VP-Mktg. & Sales, Oral Restorative Div.
Larry D. Hiebert, VP/General Mgr.-Hyaluronan Div.

Phone: 952-368-4300	Fax: 952-368-3411
Toll-Free:	
Address: 3515 Lyman Blvd., Chaska, MN 55318 US	

GROWTH PLANS/SPECIAL FEATURES:

Lifecore Biomedical, Inc. develops and manufactures biomaterials and medical devices with applications in various surgical markets. The company operates two divisions, the hyaluronan division and the oral restorative division. Its hyaluronan division is principally involved in the development and manufacture of products utilizing hyaluronan, a naturally occurring polysaccharide that is widely distributed in the extracellar matrix of connective tissues in both animals and humans. This division sells primarily to three medical segments: ophthalmic, orthopedic and veterinary. Lifecore also supplies hyaluronan to customers pursuing other medical applications, such as wound care, aesthetic surgery, medical device coatings, tissue engineering, drug delivery and pharmaceuticals. Lifecore's oral restorative division develops and markets precision surgical and prosthetic devices for the restoration of damaged or deteriorating dentition and associated support tissues. The company's dental implants are permanently implanted in the jaw for tooth replacement therapy as long-term support for crowns, bridges and dentures. It also offers bone regenerative products for the repair of bone defects resulting from periodontal disease and tooth loss. Additionally, the oral restorative division provides professional support services to its dental surgery clients through comprehensive education curricula, as provided in the company's various Support Plus programs and surgical courses. These professional continuing education programs are designed to train restorative clinicians and their auxiliary teams in the principles of tooth replacement therapy and practice management. Recently, the company announced the proprietary Lifecore Prima Implant System.

Lifecore offers its employees an employee assistance program and an on-site fitness center. Lifecore has been certified by the Minnesota Safety and Health Achievement and Recognition Program, a recognition of a higher standard of safety and health programs that goes beyond required OSHA standards.

FINANCIALS: Sales and profits are in thousands of dollars—add 000 to get the full amount. 2006 Note: Financial information for 2006 was not available for all companies at press time.

2006 Sales: $63,097	2006 Profits: $7,040	U.S. Stock Ticker: LCBM
2005 Sales: $55,695	2005 Profits: $17,511	Int'l Ticker: Int'l Exchange:
2004 Sales: $47,036	2004 Profits: $ 707	Employees: 202
2003 Sales: $42,400	2003 Profits: $- 400	Fiscal Year Ends: 6/30
2002 Sales: $38,800	2002 Profits: $-4,700	Parent Company:

SALARIES/BENEFITS:

Pension Plan:	ESOP Stock Plan:	Profit Sharing:	Top Exec. Salary: $300,000	Bonus: $135,000
Savings Plan: Y	Stock Purch. Plan: Y		Second Exec. Salary: $175,632	Bonus: $29,450

OTHER THOUGHTS:

Apparent Women Officers or Directors:
Hot Spot for Advancement for Women/Minorities:

LOCATIONS: ("Y" = Yes)

West:	Southwest:	Midwest:	Southeast:	Northeast:	International:
		Y			Y

Note: Financial information, benefits and other data can change quickly and may vary from those stated here.

LIGAND PHARMACEUTICALS INC

www.ligand.com

Industry Group Code: 325412 Ranks within this company's industry group: Sales: 61 Profits: 134

Drugs:		Other:		Clinical:	Computers:	Other:
Discovery:	Y	AgriBio:	Y	Trials/Services:	Hardware:	Specialty Services:
Licensing:	Y	Genomics/Proteomics:		Laboratories:	Software:	Consulting:
Manufacturing:		Tissue Replacement:		Equipment/Supplies:	Arrays:	Blood Collection:
Development:	Y			Research/Development Svcs.:	Database Management:	Drug Delivery:
Generics:				Diagnostics:		Drug Distribution:

TYPES OF BUSINESS:

Drugs-Diversified
Small-Molecule Drugs

BRANDS/DIVISIONS/AFFILIATES:

Ontak
Targretin
AVINZA
Panretin

CONTACTS: Note: Officers with more than one job title may be intentionally listed here more than once.

John L. Higgins, CEO
John L. Higgins, Pres.
John P. Sharp, CFO
Audrey Warfield-Graham, Dir.-Human Resources
Martin D. Meglasson, VP-Discovery Research
Charles Berkman, General Counsel/VP/Corp. Sec.
Syed Kazmi, VP-Bus. Dev. & Strategic Planning
Erika Luib, Investor Rel.
John P. Sharp, VP-Finance
Zofia E. Dziewanowska, VP-Clinical R&D
John W. Kozarich, Chmn.

Phone: 858-550-7500	Fax: 858-550-7506
Toll-Free:	
Address: 10275 Science Center Dr., San Diego, CA 92121-1117 US	

GROWTH PLANS/SPECIAL FEATURES:

Ligand Pharmaceuticals develops new drugs that address a wide variety of medical needs, including cancer, skin diseases, hormone-related diseases, osteoporosis, metabolic disorders and cardiovascular and inflammatory diseases. The company has recently sold off all of its commercial business so that it can better focus on the research and development of new drugs. The firm uses intracellular receptor (IR) technology, which was developed as a result of Ligand's work in the field of gene transcription. The company utilizes its IR technology in three programs: Thrombopoietin (TPO) Program, Selective Androgen Receptor Modulators (SARM) Program and the Selective Glucocorticoid Receptor Modulators (SGRM) Program. The TPO program focuses on drugs that mimic the activity of thrombopoietin, which is used in the treatment and prevention of Thrombocytopenia. These drugs have indication for cancer, hepatitis C and other blood cell formation disorders. The TPO product is called LGD-4665 and is Phase I clinical study. The SARM program consists of orally active, non-steroidal molecules that are able to selectively antagonize androgen receptors in different tissues. This technology has uses in the treatment of osteoporosis, sexual dysfunction, frailty, prostate cancer, acne and other disorders. The SARM product LGD-3303 is in pre-clinical testing. The SGRM program currently does not have any drug candidates, but it may prove useful in the treatment of cancer, inflammation and other indications. In October 2006, Ligand sold its oncology product line, including the rights to ONTAK, Targretin capsules, Targretin gel and Panretin gel, to Eisai Co., LTC (Tokyo) and Eisai Inc. (New Jersey) for approximately $205 million. In January 2007, the company announced that the company would be restructuring, resulting in a decrease of about 76% of company positions and the closure of its U.K. subsidiary. In February 2007, the company sold its AVINZA product line to King Pharmaceuticals, Inc. for approximately $280.4 million.

FINANCIALS: Sales and profits are in thousands of dollars—add 000 to get the full amount. 2006 Note: Financial information for 2006 was not available for all companies at press time.

2006 Sales: $140,960	2006 Profits: $-31,743	U.S. Stock Ticker: LGND
2005 Sales: $123,010	2005 Profits: $-36,399	Int'l Ticker: Int'l Exchange:
2004 Sales: $112,112	2004 Profits: $-45,141	Employees: 122
2003 Sales: $55,324	2003 Profits: $-96,471	Fiscal Year Ends: 12/31
2002 Sales: $96,600	2002 Profits: $-32,600	Parent Company:

SALARIES/BENEFITS:

Pension Plan:	ESOP Stock Plan:	Profit Sharing:	Top Exec. Salary: $471,000	Bonus: $295,716
Savings Plan: Y	Stock Purch. Plan: Y		Second Exec. Salary: $401,250	Bonus: $

OTHER THOUGHTS:

Apparent Women Officers or Directors: 2
Hot Spot for Advancement for Women/Minorities: Y

LOCATIONS: ("Y" = Yes)

West:	Southwest:	Midwest:	Southeast:	Northeast:	International:
Y					

Note: Financial information, benefits and other data can change quickly and may vary from those stated here.

LONZA GROUP

www.lonza.com

Industry Group Code: 325000 Ranks within this company's industry group: Sales: 5   Profits: 6

Drugs:		Other:		Clinical:		Computers:		Other:	
Discovery:		AgriBio:	Y	Trials/Services:		Hardware:		Specialty Services:	Y
Licensing:		Genomics/Proteomics:		Laboratories:		Software:		Consulting:	
Manufacturing:	Y	Tissue Replacement:		Equipment/Supplies:	Y	Arrays:		Blood Collection:	
Development:				Research/Development Svcs.:	Y	Database Management:		Drug Delivery:	
Generics:				Diagnostics:				Drug Distribution:	

TYPES OF BUSINESS:

Chemicals Manufacturing
Fine Chemicals & Pharmaceutical Intermediates
Biocides
Polymers & Additives
Active Drug Ingredients
Monoclonal Antibody Drugs
Materials Research

BRANDS/DIVISIONS/AFFILIATES:

Alusuisse-Lonza Group
Lonza Biotec
Lonza Biologics
Cambrex Corporation
Lonza Custom Manufacturing

CONTACTS: *Note: Officers with more than one job title may be intentionally listed here more than once.*

Stefan Borgas, CEO
Toralf Haag, CFO
Marcela Cechova, Head-Global Human Resources
Laura Gerber, Head-Investor Rel.
Rosario Valido, Head-Polymer Intermediates
Lukas Utiger, Head-Organic Fine & Performance Chemicals
Stephan Kutzer, Head-Biopharmaceuticals
Uwe H. Bohlke, Head-Exclusive Synthesis
Rolf Soiron, Chmn.

Phone: 41-61-316-8111	Fax: 41-61-316-9111
Toll-Free:	
Address: Muenchensteinerstrasse 38, Basel, 4002 Switzerland	

GROWTH PLANS/SPECIAL FEATURES:

The Swiss-based Lonza Group brings together a globa portfolio of companies engaged in the production and supply of active chemical ingredients, intermediates and biotechnology solutions for a range of clients in the pharmaceutical and agrochemical industries. The group structure was established in 1999 through the de-merger of its core business units from Swiss aluminum and industria conglomerate Alusuisse-Lonza Group. Lonza Group is focused on chemical production and materials research for pharmaceutical applications, and its exclusive synthesis unit is a leading contract manufacturer of the advanced intermediates and active ingredients used in drugs. Lonza Group's biotechnology companies, including Lonza Biotec and Lonza Biologics, are engaged in activities including the production of fine chemical intermediates through microbia fermentation and mammalian cell culture research. The group creates products and solutions for the nutrition, hygiene and personal care and wood and water treatment markets. Other group activities include manufacturing of biocides, polymers and additives, as well as generation and distribution of hydroelectric power to industrial customers. Its polymer intermediates division provides unsaturated polyester resins, compounds and composites especially used in the electronics, transportation and construction market sectors. Lonza Group is in a partnership with Swiss drug company NovImmune to manufacture clinical-grade monoclonal antibody drugs, potentially accessing the whole pipeline of NovImmune's own products and those licensed to third-party partners. NovImmune has been successful in developing fully human therapeutic monoclonal antibodies based on transgenic mouse technology (hu-mice) and antibody display technology. Lonza Group business units will oversee production through to market commercialization In February 2007, Lonza acquired the research bio-products and microbial biopharmaceutical business of the Cambrex Corporation.

FINANCIALS: Sales and profits are in thousands of dollars—add 000 to get the full amount. 2006 Note: Financial information for 2006 was not available for all companies at press time.

2006 Sales: $2,371,810	2006 Profits: $180,690	U.S. Stock Ticker:
2005 Sales: $2,069,430	2005 Profits: $154,300	Int'l Ticker: LONN Int'l Exchange: Zurich-SWX
2004 Sales: $1,927,800	2004 Profits: $121,900	Employees: 5,668
2003 Sales: $1,804,800	2003 Profits: $73,300	Fiscal Year Ends: 12/31
2002 Sales: $1,827,700	2002 Profits: $159,300	Parent Company:

SALARIES/BENEFITS:

Pension Plan:	ESOP Stock Plan:	Profit Sharing:	Top Exec. Salary: $	Bonus: $
Savings Plan:	Stock Purch. Plan:		Second Exec. Salary: $	Bonus: $

OTHER THOUGHTS:

Apparent Women Officers or Directors: 3
Hot Spot for Advancement for Women/Minorities: Y

LOCATIONS: ("Y" = Yes)

West:	Southwest:	Midwest:	Southeast:	Northeast:	International:
		Y		Y	Y

LORUS THERAPEUTICS INC

www.lorusthera.com

Industry Group Code: 325412 Ranks within this company's industry group: Sales: Profits: 84

Drugs:		Other:		Clinical:	Computers:		Other:
Discovery:	Y	AgriBio:		Trials/Services:	Hardware:		Specialty Services:
Licensing:		Genomics/Proteomics:		Laboratories:	Software:		Consulting:
Manufacturing:	Y	Tissue Replacement:		Equipment/Supplies:	Arrays:		Blood Collection:
Development:	Y			Research/Development Svcs.:	Database Management:		Drug Delivery:
Generics:				Diagnostics:			Drug Distribution:

TYPES OF BUSINESS:

Drugs-Cancer
Antisense Compounds
Low-Molecular-Weight Compounds

BRANDS/DIVISIONS/AFFILIATES:

Imutec Pharma
GeneSense Technologies, Inc.
Virulizin

CONTACTS: Note: Officers with more than one job title may be intentionally listed here more than once.

Aiping H. Young, CEO
Aiping H. Young, Pres.
Elizabeth Williams, Acting CFO
Saeid Babaei, Associate Dir.-Corp. Affairs
Elizabeth Williams, Dir.-Finance
Graham Strachan, Chmn.

Phone: 416-798-1200	Fax: 416-798-2200
Toll-Free:	
Address: 2 Meridian Rd., Toronto, ON M9W 4Z7 Canada	

GROWTH PLANS/SPECIAL FEATURES:

Lorus Therapeutics, Inc. is a biopharmaceutical company focused on the research, development and commercialization of drug products and technologies for the treatment of cancer. The company is engaged in basic drug discovery research as well as the development of novel technologies and clinical drug candidates. The firm's lead candidate, Virulizin, is an immunotherapy product that finished expanded Phase III clinical trials as a treatment for pancreatic cancer. The drug is also being investigated with regard to indications for malignant melanoma, Kaposi's sarcoma, lung cancer, breast cancer and uterine, ovarian and cervical cancers. Virulizin has only been approved for sale in Mexico. Lorus is also engaged in advanced stages of development for two antisense cancer drugs, GTI-2040 and GTI-2501. These compounds use a highly selective and targeted approach to inhibiting the formation of disease-causing proteins and have demonstrated encouraging results as potential cancer treatments. GTI-2040 is being developed by Lorus, along with the U.S. National Cancer Institute. The drug is in expanded Phase II clinical trials in association with chemotherapy regimens. GTI-2501 is undergoing Phase II clinical trials in Canada for the treatment of hormone-refractory prostate cancer and expanded Phase II testing in the U.S. for renal cell carcinoma, or kidney cancer. Lorus is also actively pursuing research programs for small-molecule compounds, including the out-licensed development and commercialization of the clotrimazole analog NC 381 by Cyclacel, Ltd. The company also has a collaboration with the University of Toronto, based on Lorus' recent discovery, following three years of research, of novel low-molecular-weight compounds with anticancer and antibacterial activity. In July 2007, the company finished its corporate reorganization, which resulted in $8.5 million cash for the newly reorganized company.

FINANCIALS: Sales and profits are in thousands of dollars—add 000 to get the full amount. 2006 Note: Financial information for 2006 was not available for all companies at press time.

2006 Sales: $	2006 Profits: $-14,900	U.S. Stock Ticker: LRP
2005 Sales: $ 100	2005 Profits: $-17,600	Int'l Ticker: LOR Int'l Exchange: Toronto-TSX
2004 Sales: $ 400	2004 Profits: $-22,300	Employees: 52
2003 Sales: $	2003 Profits: $-12,000	Fiscal Year Ends: 5/31
2002 Sales: $	2002 Profits: $-8,800	Parent Company:

SALARIES/BENEFITS:

Pension Plan:	ESOP Stock Plan:	Profit Sharing:	Top Exec. Salary: $327,861	Bonus: $50,299
Savings Plan:	Stock Purch. Plan:		Second Exec. Salary: $246,460	Bonus: $30,370

OTHER THOUGHTS:

Apparent Women Officers or Directors: 2
Hot Spot for Advancement for Women/Minorities: Y

LOCATIONS: ("Y" = Yes)

West:	Southwest:	Midwest:	Southeast:	Northeast:	International:
					Y

Note: Financial information, benefits and other data can change quickly and may vary from those stated here.

LUMINEX CORPORATION

www.luminexcorp.com

Industry Group Code: 325413 Ranks within this company's industry group: Sales: 17   Profits: 13

Drugs:	Other:	Clinical:		Computers:	Other:
Discovery:	AgriBio:	Trials/Services:		Hardware:	Specialty Services:
Licensing:	Genomics/Proteomics:	Laboratories:		Software:	Consulting:
Manufacturing:	Tissue Replacement:	Equipment/Supplies:	Y	Arrays:	Blood Collection:
Development:		Research/Development Svcs.:		Database Management:	Drug Delivery:
Generics:		Diagnostics:	Y		Drug Distribution:

TYPES OF BUSINESS:

Medical Diagnostics
Bioassays
Software
xMAP Testing

BRANDS/DIVISIONS/AFFILIATES:

xMAP
Luminex 100 IS System
Luminex HTS System
Luminex 200 System
xPONENT
MagPlex Magnetic Microspheres
Tm

CONTACTS: Note: Officers with more than one job title may be intentionally listed here more than once.

Patrick J. Balthrop, CEO
Patrick J. Balthrop, Pres.
Harriss T. Currie, CFO
Gregory J. Gosch, VP-Mktg. & Sales
John C. Carrano, VP-R&D
David S. Reiter, General Counsel/VP
Russell W. Bradley, VP-Bus. Dev. & Strategic Planning
Harriss T. Currie, VP-Finance/Treas.
James W. Jacobson, Chief Scientific Officer/VP
Randel S. Marfin, VP-Luminex Bioscience Group
David S. Reiter, Corp. Sec.
Oliver H. Meek, VP-Quality Assurance/Regulatory Affairs
G. Walter Loewenbaum, II, Chmn.

Phone: 512-219-8020	Fax: 512-219-5195
Toll-Free: 888-219-8020	
Address: 12212 Technology Blvd., Austin, TX 78727 US	

GROWTH PLANS/SPECIAL FEATURES:

Luminex Corporation manufactures and markets biological testing technologies for the life sciences industry. The firm's technologies enable biological tests (called bioassays) to detect biochemicals, proteins and genes in samples for the purpose of protein expression profiling, genomic research, genetic disease, immunodiagnostics and biodefense/environmental testing. The company's xMAP system can perform up to 100 bioassays on a single drop of fluid. The xMAP system makes use of microspheres, lasers, digital signal processing and traditional chemistry in order to run various diagnostic tests. The xMAP system is an industry-leading technology because of the rapid and precise results it produces from a relatively small sample. The firm offers a variety of products featuring its xMAP technology to various partners in several industries. Products include the Luminex 100 IS System, an analyzer based on the principles of flow chemistry; Luminex HTS System, a high throughput screening system which performs thousands of bioassays daily; the Luminex 200 System, used by clinical and research laboratory professionals; and various microspheres, all of which are designed for use in the Luminex Systems. Partners include companies within the research/drug discovery fields, such as Radix BioSolutions and Cayman Chemical Company, as well as companies within clinical diagnostics fields, such as Inverness Medical and Bayer HealthCare. In 2006, Luminex and partner Exiquon A/S unveiled a new line of microRNA products. Also in 2006, the firm received a contract from the Department of Homeland Security to develop a system detect weaponized biological pathogens in the air. In early 2007, the firm released two new technologies, the xPONENT software and MagPlex magnetic microspheres, both of which will enhance ease-of-use and automation capabilities in the xMAP technologies. The firm additionally acquired Tm, a molecular diagnostics firm, for $44 million.

FINANCIALS: Sales and profits are in thousands of dollars—add 000 to get the full amount. 2006 Note: Financial information for 2006 was not available for all companies at press time.

2006 Sales: $52,989	2006 Profits: $1,507	U.S. Stock Ticker: LMNX	
2005 Sales: $42,313	2005 Profits: $-2,666	Int'l Ticker:	Int'l Exchange:
2004 Sales: $35,880	2004 Profits: $-3,605	Employees: 303	
2003 Sales: $26,292	2003 Profits: $-4,209	Fiscal Year Ends: 12/31	
2002 Sales: $13,000	2002 Profits: $-24,900	Parent Company:	

SALARIES/BENEFITS:

Pension Plan:	ESOP Stock Plan:	Profit Sharing:	Top Exec. Salary: $400,000	Bonus: $400,000
Savings Plan: Y	Stock Purch. Plan:		Second Exec. Salary: $225,500	Bonus: $128,071

OTHER THOUGHTS:

Apparent Women Officers or Directors:
Hot Spot for Advancement for Women/Minorities:

LOCATIONS: ("Y" = Yes)

West:	Southwest:	Midwest:	Southeast:	Northeast:	International:
	Y				Y

Note: Financial information, benefits and other data can change quickly and may vary from those stated here.

MACROCHEM CORPORATION

www.macrochem.com

Industry Group Code: 325412A Ranks within this company's industry group: Sales: Profits: 5

Drugs:		Other:		Clinical:	Computers:		Other:	
Discovery:	Y	AgriBio:		Trials/Services:	Hardware:		Specialty Services:	
Licensing:		Genomics/Proteomics:		Laboratories:	Software:		Consulting:	
Manufacturing:	Y	Tissue Replacement:		Equipment/Supplies:	Arrays:		Blood Collection:	
Development:	Y			Research/Development Svcs.:	Database Management:		Drug Delivery:	Y
Generics:				Diagnostics:			Drug Distribution:	

TYPES OF BUSINESS:

Drug Delivery Technology

BRANDS/DIVISIONS/AFFILIATES:

SEPA
MacroDerm
DermaPass
Opterone
EcoNail

CONTACTS: Note: Officers with more than one job title may be intentionally listed here more than once.

Robert J. DeLuccia, CEO
Robert J. DeLuccia, Pres.
Bernard R. Patriacca, CFO/VP
Glenn E. Deegan, General Counsel/Corp. Sec./VP
John L. Zabriskie, Chmn.

Phone: 781-862-4003	Fax: 781-862-4338
Toll-Free:	
Address: 110 Hartwell Ave., Lexington, MA 02421 US	

GROWTH PLANS/SPECIAL FEATURES:

MacroChem is a specialty pharmaceutical company that develops and aims to commercialize its products. Its current portfolio of product candidates is based on its proprietary drug delivery technologies: SEPA, MacroDerm and DermaPass. SEPA, which stands for soft enhancement of percutaneous absorption, enhances the efficiency and rate of diffusion of drugs into and through the skin. The company's MacroDerm drug delivery technology encompasses a family of low to moderate molecular weight polymers that impede dermal drug or chemical penetration. A potential use of this product is to prevent the skin's absorbing insect repellent. The DermaPass family of transdermal absorption enhancers features a different drug delivery profile than SEPA, which may have a wider range of active pharmaceutical ingredients. MacroChem's lead product candidate is EcoNail, a topically applied, SEPA-based econazole lacquer for the treatment of onychomycosis, a condition commonly known as nail fungus. A Phase II trial of EcoNail completed patient enrollment in 2007. The company's other clinical-stage product candidate is Opterone, a topically applied, SEPA-based testosterone cream designed to treat male hypogonadism. Male hypogonadism is a condition in which men have levels of circulating testosterone below the normal range; its symptoms include low energy levels, decreased sexual performance, loss of sex drive, increased body fat or loss of muscle mass. MacroChem is seeking a partner to advance the development of Opterone in a Phase II trial. In recent news, the company acquired an exclusive option to acquire the rights to pexiganan, a small peptide anti-infective for topical treatment of patients with mild diabetic foot infection, from Genaera Corporation.

FINANCIALS: Sales and profits are in thousands of dollars—add 000 to get the full amount. 2006 Note: Financial information for 2006 was not available for all companies at press time.

2006 Sales: $	2006 Profits: $1,187	U.S. Stock Ticker: MACM.OB
2005 Sales: $	2005 Profits: $-6,091	Int'l Ticker: Int'l Exchange:
2004 Sales: $	2004 Profits: $-8,275	Employees: 6
2003 Sales: $1,010	2003 Profits: $-5,662	Fiscal Year Ends: 12/31
2002 Sales: $ 100	2002 Profits: $-7,500	Parent Company:

SALARIES/BENEFITS:

Pension Plan:	ESOP Stock Plan:	Profit Sharing:	Top Exec. Salary: $276,000	Bonus: $50,000
Savings Plan: Y	Stock Purch. Plan:		Second Exec. Salary: $191,666	Bonus: $

OTHER THOUGHTS:

Apparent Women Officers or Directors:
Hot Spot for Advancement for Women/Minorities:

LOCATIONS: ("Y" = Yes)

West:	Southwest:	Midwest:	Southeast:	Northeast:	International:
				Y	

MALLINCKRODT INC
www.mallinckrodt.com

Industry Group Code: 325413 Ranks within this company's industry group: Sales: Profits:

Drugs:		Other:	Clinical:		Computers:		Other:
Discovery:		AgriBio:	Trials/Services:		Hardware:		Specialty Services:
Licensing:		Genomics/Proteomics:	Laboratories:		Software:		Consulting:
Manufacturing:	Y	Tissue Replacement:	Equipment/Supplies:	Y	Arrays:		Blood Collection:
Development:	Y		Research/Development Svcs.:		Database Management:		Drug Delivery:
Generics:	Y		Diagnostics:	Y			Drug Distribution:

TYPES OF BUSINESS:
Imaging Agents & Radiopharmaceuticals
Diagnostics Products
Bulk Analgesic Pharmaceuticals
Generic Pharmaceuticals
Active Pharmaceutical Ingredients
Medical Devices
Respiratory Products

BRANDS/DIVISIONS/AFFILIATES:
Puritan-Bennett
Shiley
DAR
Nellcor
OptiMARK
OxiFirst
NeutroSpec
Covidien Ltd.

CONTACTS: *Note: Officers with more than one job title may be intentionally listed here more than once.*
Douglas A. McKinney, CFO
Lisa Britt, VP-Human Resources
Douglas A. McKinney, VP-Finance
Scott Drake, Pres., Respiratory Div.
Michael J. Collins, Pres., Pharmaceuticals Div.
Steven Hanley, Pres., Imaging Div.

Phone: 314-654-2000	Fax: 314-654-5381
Toll-Free: 800-325-8888	
Address: 675 McDonnell Blvd., Hazelwood, MO 63042 US	

GROWTH PLANS/SPECIAL FEATURES:
Mallinckrodt, Inc., a subsidiary of Covidien Ltd. (formerly Tyco Healthcare Group), develops and manufactures a wide range of medical products and devices, primarily used by hospitals for diagnostic and treatment purposes. It is a world leader in bulk analgesic pharmaceuticals and a leader in generic dosage pharmaceuticals. The company's imaging group produces a full line of imaging agents and radiopharmaceuticals, including ultrasound and MRI contrast agents, x-ray contrast media, catheters and injection systems. Through its respiratory segment, the firm offers a variety of products, including anesthesia, airway management and temperature management devices; medical gases; oxygen therapy and asthma management products; sleep diagnostics and therapy devices; and blood analysis products. Its products are sold under the Shiley, DAR, Nellcor and Puritan-Bennett brands. Mallinckrodt's pharmaceuticals division is focused on providing pain relief and addiction therapy, with a product line that includes codeine, morphine, methadone, naltrexone (for alcohol addiction) and methylphenidate (for attention deficit/hyperactivity disorder). It also supplies raw materials and active pharmaceutical ingredients, makes generic drugs and provides contract manufacturing and other services. Its Brand Pharmaceuticals business acquires and markets other companies' under-promoted products. The company is the developer of the OxiFirst fetal oxygen monitoring system, a new technology that enables obstetricians to monitor fetal oxygenation during labor and delivery. The firm's imaging division has developed OptiMARK, the first and only MRI contrast agent that is FDA-approved for administration by power injection.

FINANCIALS: Sales and profits are in thousands of dollars—add 000 to get the full amount. 2006 Note: Financial information for 2006 was not available for all companies at press time.

2006 Sales: $	2006 Profits: $	**U.S. Stock Ticker: Subsidiary**
2005 Sales: $	2005 Profits: $	**Int'l Ticker:** Int'l Exchange:
2004 Sales: $	2004 Profits: $	Employees: 11,000
2003 Sales: $	2003 Profits: $	Fiscal Year Ends: 9/30
2002 Sales: $3,000,000	2002 Profits: $	Parent Company: COVIDIEN LTD

SALARIES/BENEFITS:

Pension Plan:	ESOP Stock Plan:	Profit Sharing:	Top Exec. Salary: $732,840	Bonus: $4,072,400
Savings Plan:	Stock Purch. Plan:		Second Exec. Salary: $322,793	Bonus: $1,095,600

OTHER THOUGHTS:
Apparent Women Officers or Directors: 1
Hot Spot for Advancement for Women/Minorities:

LOCATIONS: ("Y" = Yes)

West:	Southwest:	Midwest:	Southeast:	Northeast:	International:
Y		Y		Y	Y

MANHATTAN PHARMACEUTICALS INC

www.manhattanpharma.com

Industry Group Code: 325412 Ranks within this company's industry group: Sales: Profits: 74

Drugs:		Other:		Clinical:	Computers:		Other:	
Discovery:	Y	AgriBio:		Trials/Services:	Hardware:		Specialty Services:	
Licensing:		Genomics/Proteomics:		Laboratories:	Software:		Consulting:	
Manufacturing:		Tissue Replacement:		Equipment/Supplies:	Arrays:		Blood Collection:	
Development:	Y			Research/Development Svcs.:	Database Management:		Drug Delivery:	Y
Generics:				Diagnostics:			Drug Distribution:	

TYPES OF BUSINESS:

Pharmaceuticals Discovery & Development
Drug Delivery Systems

BRANDS/DIVISIONS/AFFILIATES:

Oleoyl-esterone
Propofol
PTH (1-34)
Altoderm
Altolyn
Hedrin

CONTACTS: *Note: Officers with more than one job title may be intentionally listed here more than once.*

Douglas Abel, CEO
Douglas Abel, Pres.
Michael G. McGuinness, CFO
Alan G. Harris, Chief Medical Officer
Michael G. McGuinness, Corp. Sec.

Phone: 212-582-3950	Fax: 212-582-3957
Toll-Free:	
Address: 810 7th Ave, 4th Fl., New York, NY 10019 US	

GROWTH PLANS/SPECIAL FEATURES:

Manhattan Pharmaceuticals, Inc. is a clinical-stage pharmaceutical company developing novel, high-value drug candidates primarily in the areas of endocrine/metabolic disease and dermatologic/immunologic disorders. With a pipeline consisting of six clinical-stage product candidates, the company is developing potential therapeutics for large, underserved patient populations seeking superior treatments for conditions including common obesity, morbid obesity, psoriasis and atopic dermatitis (eczema). Product candidates in the pipeline include: Oral Oleoyl-Estrone; Topical PTH (1-34); Altoderm; Altolyn; Hedrin; and Propofol Lingual Spray. Oleoyl-esterone (OE) is under Phase II development for the treatment of obesity and morbid obesity and is an orally administered hormone attached to a fatty-acid that has been shown to cause significant weight loss in preclinical animal studies regardless of dietary modifications. PTH (1-34) is a peptide that regulates epidermal cell growth and differentiation currently under Phase I development as a topical treatment for psoriasis and additional hyperproliferative skin disorders. Altoderm is a topical formulation of cromolyn sodium, a non steroidal, anti-inflammatory agent that has a long history of use in treating allergic inflammatory diseases such as asthma and allergic rhinoconjunctivitis. Altoderm is currently in the preclinical stage in the U.S. and is being studied in a second, ongoing, Phase III clinical trial in Europe. Altolyn is a novel tablet formulation of cromolyn sodium that has been formulated using site specific drug delivery technology. Altolyn is still in the preclinical development stage. Hedrin is a novel, convenient, non-insecticide treatment for head lice and is in preclinical development in the U.S. and is on the market in Europe and the U.K. Propofol Lingual Spray, a proprietary lingual spray technology used to deliver propofol for pre-procedural sedation prior to diagnostic, therapeutic or endoscopic procedures, is currently in Phase I trials. Hedrin was acquired in June 2007, licensed by Manhattan from Thornton & Ross Limited.

FINANCIALS: Sales and profits are in thousands of dollars—add 000 to get the full amount. 2006 Note: Financial information for 2006 was not available for all companies at press time.

2006 Sales: $	2006 Profits: $-9,695	**U.S. Stock Ticker: MHA**	
2005 Sales: $	2005 Profits: $-19,141	**Int'l Ticker:** Int'l Exchange:	
2004 Sales: $	2004 Profits: $-5,896	Employees: 8	
2003 Sales: $	2003 Profits: $-5,961	Fiscal Year Ends: 12/31	
2002 Sales: $ 500	2002 Profits: $-1,547	Parent Company:	

SALARIES/BENEFITS:

Pension Plan:	ESOP Stock Plan:	Profit Sharing:	Top Exec. Salary: $325,000	Bonus: $233,333
Savings Plan: Y	Stock Purch. Plan:		Second Exec. Salary: $252,083	Bonus: $107,500

OTHER THOUGHTS:

Apparent Women Officers or Directors:
Hot Spot for Advancement for Women/Minorities:

LOCATIONS: ("Y" = Yes)

West:	Southwest:	Midwest:	Southeast:	Northeast:	International:
				Y	

Note: Financial information, benefits and other data can change quickly and may vary from those stated here.

MARTEK BIOSCIENCES CORP   www.martekbio.com

Industry Group Code: 325411 Ranks within this company's industry group: Sales: 4   Profits: 3

Drugs:		Other:		Clinical:		Computers:		Other:	
Discovery:	Y	AgriBio:		Trials/Services:		Hardware:		Specialty Services:	
Licensing:		Genomics/Proteomics:		Laboratories:		Software:		Consulting:	
Manufacturing:		Tissue Replacement:		Equipment/Supplies:	Y	Arrays:		Blood Collection:	
Development:	Y			Research/Development Svcs.:		Database Management:		Drug Delivery:	
Generics:				Diagnostics:				Drug Distribution:	

TYPES OF BUSINESS:

Microalgae-Based Product Manufacturing
Drugs-Nutritional Oils
Dietary Supplements
Fluorescent Products
Fermentation Processes

BRANDS/DIVISIONS/AFFILIATES:

Martek Biosciences Kingtree
ARASCO
PBXL Dyes
DHASCO
SureLight
CryptoLight
SensiLight
FermPro Manufacturing, LP

CONTACTS: Note: Officers with more than one job title may be intentionally listed here more than once.

Steve Dubin, CEO
Peter A. Nitze, COO/Exec. VP
David Abramson, Pres.
Peter L. Buzy, CFO
Joe Buron, Sr. VP-Sales & Mktg.
James H. Flatt, Sr. VP-R&D
Barney Easterling, Sr. VP-Mfg.
Peter L. Buzy, Exec. VP-Admin.
David M. Feitel, General Counsel/Sr. VP/Corp. Sec.
David Abramson, Sr. VP-Bus. Dev.
Peter L. Buzy, Exec. VP-Finance/Treas.
Tim Fealey, Sr. VP/Chief Innovation Officer
Henry Linsert, Jr., Chmn.

Phone: 410-740-0081	Fax: 410-740-2985
Toll-Free:	
Address: 6480 Dobbin Rd., Columbia, MD 21045 US	

GROWTH PLANS/SPECIAL FEATURES:

Martek Biosciences commercially develops products that promote health and wellness through its portfolio of nutritional supplements and advanced diagnostic aids. The company's nutritional products group currently manufactures and sells two patented nutritional fatty acids, DHA (docosahexaenoic acid), under the DHASCO and life'sDHA brand names, and ARA (arachidonic acid), under the ARASCO brand name. life'sDHA is a vegetarian source of omega-3 fatty acids and is used in infant formula, perinatal products, foods and beverages and dietary supplements while ARA is an omega-6 fatty acid that is primarily used in infant formula. These fatty acids assist in the development of the optic and central nervous systems in newborns and also promote adult mental and cardiovascular health, which lowers the risk for the development of cardiovascular disease, Alzheimer's disease and dementia. Martek is additionally investigating the potential benefits of DHA in breast cancer, cystic fibrosis and other visual and neurological disorders. The company's licensees and clients for DHA and ARA include Abbott Laboratories, Nestle, Abbott Laboratories and Wyeth Nutritionals, which produces enhanced infant formula for Wal-Mart under the private label, Parent's Choice with LIPIDS. In addition to nutritional products, Martek provides contract manufacturing services, which customizes production processes for the synthesis of enzymes, specialty chemicals, vitamins and agricultural specialty products. The firm has also developed a range of fluorescent markers from microalgae, which attaches antibodies to compounds of interest in order to mark the compound for further investigations during biological screening processes. The company's fluorescent products group markets these superfluor dyes under the brand names, SureLight, CryptoLight and SensiLight. These products have been used to accelerate drug screening assays and also have additional applications in flow cytometry and microarray imaging. Martek recently introduced Sprintissimo Drinkable Yogurt with life'sDHA, Fujisan Sushi with life'sDHA and plans to launch Horizon Organic Milk Plus DHA with life'sDHA.

FINANCIALS: Sales and profits are in thousands of dollars—add 000 to get the full amount. 2006 Note: Financial information for 2006 was not available for all companies at press time.

2006 Sales: $270,654	2006 Profits: $17,811	**U.S. Stock Ticker: MATK**
2005 Sales: $217,852	2005 Profits: $15,284	**Int'l Ticker:** Int'l Exchange:
2004 Sales: $184,493	2004 Profits: $47,048	Employees: 506
2003 Sales: $114,737	2003 Profits: $15,992	Fiscal Year Ends: 10/31
2002 Sales: $45,500	2002 Profits: $-24,200	Parent Company:

SALARIES/BENEFITS:

Pension Plan:	ESOP Stock Plan:	Profit Sharing:	Top Exec. Salary: $483,333	Bonus: $220,000
Savings Plan: Y	Stock Purch. Plan:		Second Exec. Salary: $475,000	Bonus: $190,000

OTHER THOUGHTS:

Apparent Women Officers or Directors:
Hot Spot for Advancement for Women/Minorities:

LOCATIONS: ("Y" = Yes)

West:	Southwest:	Midwest:	Southeast:	Northeast:	International:
Y		Y		Y	

Note: Financial information, benefits and other data can change quickly and may vary from those stated here.

MATRITECH INC
www.matritech.com

Industry Group Code: 325413 Ranks within this company's industry group: Sales: 23 Profits: 17

Drugs:	Other:	Clinical:	Computers:	Other:
Discovery:	AgriBio:	Trials/Services:	Hardware:	Specialty Services:
Licensing:	Genomics/Proteomics:	Laboratories:	Software:	Consulting:
Manufacturing:	Tissue Replacement:	Equipment/Supplies:	Arrays:	Blood Collection:
Development:		Research/Development Svcs.:	Database Management:	Drug Delivery:
Generics:		Diagnostics: Y		Drug Distribution:

TYPES OF BUSINESS:
Medical Diagnostics Products
Point-of-Care Diagnostic Kits
Cancer Diagnostics

BRANDS/DIVISIONS/AFFILIATES:
NMP22 Test Kit
NMP22 BladderChek
NMP179
Matritech GmbH

CONTACTS: Note: Officers with more than one job title may be intentionally listed here more than once.
Stephen D. Chubb, CEO
David L. Corbet, COO
David L. Corbet, Pres.
Richard A. Sandberg, CFO
John E. Quigley, VP-Mktg.
Gary Fagan, VP-R&D
Patricia Randall, General Counsel/VP
Richard A. Sandberg, VP-Finance
Melodie R. Domurad, VP-Clinical & Regulatory Affairs
David G. Kolasinski, VP-Sales
Stephen D. Chubb, Chmn.
Franz Maier, Pres., Matritech GmbH, Europe

Phone: 617-928-0820	Fax: 617-928-0821

Toll-Free:
Address: 330 Nevada St., Newton, MA 02460 US

GROWTH PLANS/SPECIAL FEATURES:
Matritech, Inc. develops, manufactures and markets innovative cancer diagnostic products based on its proprietary nuclear matrix protein (NMP) technology. The nuclear matrix, a three-dimensional protein framework within the nucleus of cells, plays a fundamental role in determining cell type. The company has demonstrated that there are differences in the types and amounts of NMPs found in cancerous and normal tissue and believes the detection of such differences provides important diagnostic information about cellular abnormalities, including cancer. Matritech has developed minimally invasive cancer diagnostic tests for bladder and cervical cancer and is developing additional tests for breast, colon and prostate cancer. The firm's the first product based on NMP technology was NMP22 Test Kit for bladder cancer. The kit has been approved for sale in the U.S., Germany, Japan, China and several other countries worldwide. The company has also developed a point-of-care format known as NMP22 BladderChek, which tests for cancer based on a single urine sample. Results from these tests are available on a same-day basis, and are performed mainly by urologists. The NMP was a technology licensed from MIT in the late 1980's, and the firm's patent portfolio has grown to over 15 since. Matritech has granted Sysmex Corp., a Japanese diagnostic company, an exclusive license for the worldwide use of its NMP179 cervical cancer detection technology. Sysmex is working to develop an automated Pap smear using the firm's cervical cancer biomarker. The firm's subsidiary Matritech GmbH distributes Matritech products in Germany, as well as allergy tests and other diagnostics. The NMP22 Test Kit and the NMP22 BladderChek have been approved by the FDA on four separate occasions and account for about 91% of the company's sales. In November 2006, the Matritech signed an agreement with Innovations, Inc. to manufacture and sell the NMP22 BladderChek system over-the-counter.

FINANCIALS: Sales and profits are in thousands of dollars—add 000 to get the full amount. 2006 Note: Financial information for 2006 was not available for all companies at press time.

2006 Sales: $12,195	2006 Profits: $-11,935	U.S. Stock Ticker: MZT
2005 Sales: $10,415	2005 Profits: $-7,865	Int'l Ticker: Int'l Exchange:
2004 Sales: $7,483	2004 Profits: $-11,123	Employees: 75
2003 Sales: $4,400	2003 Profits: $-7,900	Fiscal Year Ends: 12/31
2002 Sales: $3,300	2002 Profits: $-8,300	Parent Company:

SALARIES/BENEFITS:
Pension Plan:	ESOP Stock Plan:	Profit Sharing:	Top Exec. Salary: $298,815	Bonus: $
Savings Plan:	Stock Purch. Plan: Y		Second Exec. Salary: $242,176	Bonus: $

OTHER THOUGHTS:
Apparent Women Officers or Directors: 2
Hot Spot for Advancement for Women/Minorities: Y

LOCATIONS: ("Y" = Yes)
West:	Southwest:	Midwest:	Southeast:	Northeast:	International:
				Y	Y

Note: Financial information, benefits and other data can change quickly and may vary from those stated here.

MAXYGEN INC

www.maxygen.com

Industry Group Code: 325412 Ranks within this company's industry group: Sales: 106   Profits: 92

Drugs:	Other:	Clinical:	Computers:	Other:
Discovery:	AgriBio:	Trials/Services:	Hardware:	Specialty Services:
Licensing:	Genomics/Proteomics: Y	Laboratories:	Software:	Consulting:
Manufacturing: Y	Tissue Replacement:	Equipment/Supplies:	Arrays:	Blood Collection:
Development: Y		Research/Development Svcs.:	Database Management:	Drug Delivery:
Generics:		Diagnostics:		Drug Distribution:

TYPES OF BUSINESS:

Drug Discovery & Development
Improved & Novel Pharmaceuticals
Research Services
Chemicals
Research & Development-Molecular Evolution

BRANDS/DIVISIONS/AFFILIATES:

MolecularBreeding
DNAShuffling
MaxyScan
Codexis, Inc.
Avidia
Maxy-G34
Maxy-alpha
InterMune

CONTACTS: Note: Officers with more than one job title may be intentionally listed here more than once.

Russell Howard, CEO
Elliot Goldstein, COO
Lawrence Briscoe, CFO/Sr. VP
Stuart Pollard, Chief Scientific Officer/Sr. VP
Michael Rabson, General Counsel/Sr. VP
Grant Yonehiro, Sr. VP-Oper.
Grant Yonehiro, Sr. VP-Global Bus. Dev.
Santosh Vetticaden, Chief Medical Officer
Isaac Stein, Chmn.

Phone: 650-298-5300	Fax: 650-364-2715
Toll-Free:	
Address: 301 Galveston Dr., Redwood City, CA 94063 US	

GROWTH PLANS/SPECIAL FEATURES:

Maxygen, Inc. is a biotechnology company that works on the discovery and development of improved protein pharmaceuticals for the treatment of diseases and serious medical conditions. Technologies developed by Maxygen include MolecularBreeding, a process that mimics the natural events of evolution using DNAShuffling, a recombination process that generates a diverse library of novel DNA sequences. MolecularBreeding allows the company to rapidly move from product concept to IND-ready drug candidate, lowering costs associated with research and expanding the potential for discovery. The company's MaxyScan screening systems than selects individual proteins with desired characteristics from gene variants within the library for additional experimentation. Maxygen currently has four product candidates in different clinical study phases: Maxy-G34, designed to treat neutropenia; Maxy-alpha, for treatment of hepatitis C; Maxy VII, treatment of uncontrollable bleeding; and Maxy-Gamma, designed to treat pulmonary fibrosis. In addition to developing products, Maxygen invests in HIV vaccines and Codexis, Inc, a biotechnology company that improves the manufacturing process of small molecular pharmaceutical products. Maxygen has also entered into several strategic collaborations with Roche to co-develop its MAXY-CI product candidates for treatment of uncontrolled bleeding. Maxygen is working with InterMune in order to develop and commercialize interferon gamma products. In October 2006, Maxygen purchased Avidia from Amgen for $17.8 million. This newly acquired company will be used for the development of a new class of therapeutic peptides for treatment in the fields of autoimmunity, inflammation and oncology.

Maxygen offers its employees comprehensive medical insurance, service awards, concierge service, a credit union, annual Costco memberships, health club membership and tuition reimbursement. The company also pays 75% of employee mass transit costs and provides on-site bicycle lockers.

FINANCIALS: Sales and profits are in thousands of dollars—add 000 to get the full amount. 2006 Note: Financial information for 2006 was not available for all companies at press time.

2006 Sales: $25,021	2006 Profits: $-16,482	**U.S. Stock Ticker: MAXY**
2005 Sales: $14,501	2005 Profits: $-18,436	**Int'l Ticker:** Int'l Exchange:
2004 Sales: $16,275	2004 Profits: $9,342	Employees: 151
2003 Sales: $30,528	2003 Profits: $-44,964	Fiscal Year Ends: 12/31
2002 Sales: $41,800	2002 Profits: $-33,900	Parent Company:

SALARIES/BENEFITS:

Pension Plan:	ESOP Stock Plan: Y	Profit Sharing:	Top Exec. Salary: $455,000	Bonus: $182,000
Savings Plan: Y	Stock Purch. Plan: Y		Second Exec. Salary: $366,261	Bonus: $181,130

OTHER THOUGHTS:

Apparent Women Officers or Directors: 1
Hot Spot for Advancement for Women/Minorities:

LOCATIONS: ("Y" = Yes)

West:	Southwest:	Midwest:	Southeast:	Northeast:	International:
Y					Y

MDS INC

www.mdsintl.com

ndustry Group Code: 621511 **Ranks within this company's industry group:** Sales: 2   Profits: 2

Drugs:		Other:		Clinical:		Computers:		Other:	
Discovery:	Y	AgriBio:		Trials/Services:	Y	Hardware:		Specialty Services:	Y
Licensing:		Genomics/Proteomics:	Y	Laboratories:	Y	Software:		Consulting:	Y
Manufacturing:		Tissue Replacement:		Equipment/Supplies:	Y	Arrays:		Blood Collection:	
Development:	Y			Research/Development Svcs.:	Y	Database Management:		Drug Delivery:	
Generics:				Diagnostics:	Y			Drug Distribution:	Y

TYPES OF BUSINESS:
Drug Discovery & Development Services
Analytic Instruments & Technology
Imaging Agents
Medical & Surgical Supplies
Irradiation Systems
Health Care Product Distribution
Venture Capital

BRANDS/DIVISIONS/AFFILIATES:
MDS Nordion
MDS Sciex
MDS Diagnostic Services
MDS Pharma Services
MDS Capital Corp.
MDS Analytical Technologies
Molecular Devices Corporation

CONTACTS: *Note: Officers with more than one job title may be intentionally listed here more than once.*
Stephen P. DeFalco, CEO
Stephen P. DeFalco, Pres.
Douglas S. Prince, CFO
James M. Reid, Exec. VP-Global Human Resources
Thomas E. Gernon, CIO
Ken Horton, General Counsel
Kennith L. Horton, Exec. VP-Corp. Dev.
Sharon Mathers, Sr. VP-External Comm.
Sharon Mathers, Sr. VP-Investor Rel.
Douglas S. Prince, Exec. VP-Finance
Steve West, Pres., MDS Nordion
David Spaight, Pres., MDS Pharma Svcs.
Andrew W. Boorn, Pres., MDS Analytical Technologies
John Mayberry, Chmn.

Phone: 416-213-4082   **Fax:** 416-675-0688
Toll-Free: 888-637-7222
Address: 2700 Matheson Blvd. E., Mississauga, ON L4W 4V9 Canada

GROWTH PLANS/SPECIAL FEATURES:
MDS, Inc. is a global biotechnology firm operating in health, life sciences and venture capital areas, with operations in 28 countries around the world, including locations in North and South America, Europe, Asia and Africa. The life sciences segment includes MDS Pharma Services, MDS Analytical Technologies and MDS Nordion, providing drug discovery and development services; analytical instruments and technology solutions; and radioisotopes, radiation and related technologies. Customers include pharmaceutical and biotechnology companies, hospitals and health care professionals. MDS Pharma Services provides contract research services to pharmaceutical manufacturers and biotechnology companies, focusing particularly on pre-clinical and early clinical drug development. MDS Nordion's cobalt-60 irradiation systems are used to sterilize over much of the world's disposable medical supplies, to irradiate blood for blood transfusions and to destroy harmful bacteria on produce. MDS Analytical Technologies, which is the technological section of the company, has key markets in the US, Western Europe and Japan. MDS also owns 47% of MDS Capital Corp., which manages funds totaling over $1 billion. In November 2006, the company sold its 50% interest in Source Medical to its partner, Cardinal Health, Inc., for $79 million. In February 2007, MDS sold its Canadian laboratories service to Borealis Infrastructure Management Inc. for $1.3 billion. In March 2007, MDS acquired Molecular Devices Corporation for $615 million, and created a new segment called MDS Analytical Technologies, which is a merger between MDS Sciex and the newly acquired company.

MDS offers its employees a subsidized full-service cafeteria, an employee managed fitness center, health and wellness resources and many other benefits.

FINANCIALS: Sales and profits are in thousands of dollars—add 000 to get the full amount. 2006 Note: Financial information for 2006 was not available for all companies at press time.
2006 Sales: $1,017,200	2006 Profits: $123,100	**U.S. Stock Ticker: MDZ**
2005 Sales: $1,296,908	2005 Profits: $26,999	**Int'l Ticker: MDS** Int'l Exchange: Toronto-TSX
2004 Sales: $1,447,000	2004 Profits: $42,000	Employees: 8,800
2003 Sales: $1,364,000	2003 Profits: $36,000	Fiscal Year Ends: 10/31
2002 Sales: $1,150,000	2002 Profits: $67,000	Parent Company:

SALARIES/BENEFITS:
Pension Plan: Y	ESOP Stock Plan:	Profit Sharing:	Top Exec. Salary: $	Bonus: $
Savings Plan:	Stock Purch. Plan: Y		Second Exec. Salary: $	Bonus: $

OTHER THOUGHTS:
Apparent Women Officers or Directors: 1
Hot Spot for Advancement for Women/Minorities:

LOCATIONS: ("Y" = Yes)
West:	Southwest:	Midwest:	Southeast:	Northeast:	International:
Y	Y	Y	Y	Y	Y

MEDAREX INC
www.medarex.com

Industry Group Code: 325412 Ranks within this company's industry group: Sales: 78   Profits: 195

Drugs:		Other:		Clinical:	Computers:	Other:
Discovery:	Y	AgriBio:		Trials/Services:	Hardware:	Specialty Services:
Licensing:	Y	Genomics/Proteomics:	Y	Laboratories:	Software:	Consulting:
Manufacturing:		Tissue Replacement:		Equipment/Supplies:	Arrays:	Blood Collection:
Development:	Y			Research/Development Svcs.:	Database Management:	Drug Delivery:
Generics:				Diagnostics:		Drug Distribution:

TYPES OF BUSINESS:
Drugs-Human Monoclonal Antibodies
Transgenic Mouse Technology
Drugs-Cancer
Drugs-Autoimmune Disease

BRANDS/DIVISIONS/AFFILIATES:
UltiMAb Human Antibody Development System
HuMAb-Mouse
KM-Mouse
Genmab

CONTACTS: *Note: Officers with more than one job title may be intentionally listed here more than once.*
Howard H. Pien, CEO
Howard H. Pien, Pres.
Christian S. Schade, CFO
Nils Lonberg, Sr. VP/Dir.-Scientific
Geoffrey M. Nichol, Sr. VP-Product Dev.
Christian S. Schade, Sr. VP-Admin.
W. Bradford Middlekauff, General Counsel/Sr. VP/Corp. Sec.
Ronald A. Pepin, Sr. VP-Bus. Dev.
Jean Mantuano, Corp. Comm.
Christian S. Schade, Sr. VP-Finance
Irwin Lerner, Chmn.

Phone: 609-430-2880	Fax: 609-430-2850
Toll-Free:	
Address: 707 State Rd., Princeton, NJ 08540-1437 US	

GROWTH PLANS/SPECIAL FEATURES:
Medarex, Inc. is a leading biopharmaceutical company devoted to the production of human monoclonal antibody-based therapeutics to fight various diseases. The firm's UltiMAb Human Antibody Development System uses transgenic mice, specifically Medarex's HuMAb-Mouse and the KM-Mouse, which was developed collaboratively with Kirin Brewing Co., Ltd. In this system, mouse-derived antibody gene expression is suppressed and effectively replaced with human antibody gene expression. With UltiMAb, the company can create many types of antibodies that are fully human, with no mouse protein sequences embedded in the genetic code. Consequently, UltiMAb antibodies are less likely to be rejected by patients. In addition, the antibodies are likely to have more favorable safety profiles and the human body may eliminate them less rapidly, potentially reducing the required frequency and amount of dosing. The company has 34 product candidates, developed in-house or through collaborations, that are currently in clinical testing stages, including treatments for melanoma; lymphoma; breast, prostate, kidney and other cancers; rheumatoid arthritis and inflammatory diseases. The UltiMAb technology is licensed to pharmaceutical and biotechnology companies that also pay royalties on commercial sales of their products. Over 45 pharmaceutical and biotechnology companies have collaborative or licensing agreements with Medarex to jointly develop opportunities for new antibodies and to commercialize products. Some of the most recent alliances have been formed with Bristol-Myers Squibb to develop cancer treatments. The company owns an interest in Genmab (roughly 10.8%), a Danish biotechnology company that works with HuMAb-Mouse technology. In May 2007, Medarex and Mitsubishi Pharma, a subsidiary of Mistubishi Chemical Holdings Corp., entered into a collaborative agreement to research and develop a potential treatment for autoimmune disorders.

FINANCIALS: Sales and profits are in thousands of dollars—add 000 to get the full amount. 2006 Note: Financial information for 2006 was not available for all companies at press time.

2006 Sales: $48,646	2006 Profits: $-181,701	**U.S. Stock Ticker: MEDX**
2005 Sales: $51,455	2005 Profits: $-148,012	**Int'l Ticker:** Int'l Exchange:
2004 Sales: $12,474	2004 Profits: $-186,392	Employees: 492
2003 Sales: $11,200	2003 Profits: $-129,200	Fiscal Year Ends: 12/31
2002 Sales: $39,500	2002 Profits: $-157,500	Parent Company:

SALARIES/BENEFITS:

Pension Plan:	ESOP Stock Plan:	Profit Sharing:	Top Exec. Salary: $837,900	Bonus: $670,320
Savings Plan: Y	Stock Purch. Plan:		Second Exec. Salary: $485,100	Bonus: $266,805

OTHER THOUGHTS:
Apparent Women Officers or Directors:
Hot Spot for Advancement for Women/Minorities:

LOCATIONS: ("Y" = Yes)

West:	Southwest:	Midwest:	Southeast:	Northeast:	International:
Y				Y	

MEDICINES CO (THE)   www.themedicinescompany.com

Industry Group Code: 325412 Ranks within this company's industry group: Sales: 53 Profits: 35

Drugs:		Other:		Clinical:		Computers:		Other:	
Discovery:		AgriBio:		Trials/Services:		Hardware:		Specialty Services:	
Licensing:	Y	Genomics/Proteomics:		Laboratories:		Software:		Consulting:	
Manufacturing:	Y	Tissue Replacement:		Equipment/Supplies:		Arrays:		Blood Collection:	
Development:	Y			Research/Development Svcs.:		Database Management:		Drug Delivery:	
Generics:				Diagnostics:				Drug Distribution:	

TYPES OF BUSINESS:

Pharmaceuticals Acquisition & Development
Acute Care Hospital Products
Anticoagulants

BRANDS/DIVISIONS/AFFILIATES:

Angiomax
Cleviprex (Clevidine)
Cangrelor
Angiox

CONTACTS: Note: Officers with more than one job title may be intentionally listed here more than once.

Clive A. Meanwell, CEO
John P. Kelley, COO
John P. Kelley, Pres.
Glenn Sblendorio, CFO/Exec. VP
Catharine Newberry, Sr. VP/Chief Human Strategy Officer
Paul M. Antinori, General Counsel/VP
Clive A. Meanwell, Chmn.

Phone: 973-656-1616	Fax: 973-656-0746
Toll-Free:	
Address: 8 Campus Dr., Parsippany, NJ 07054 US	

GROWTH PLANS/SPECIAL FEATURES:

The Medicines Company (TMC) is a pharmaceutical company specializing in acute care hospital products. The firm acquires, develops and commercializes pharmaceutical products in late stages of development. TMC's first product, Angiomax, is an intravenous direct thrombin inhibitor approved for use as an anticoagulant in patients undergoing coronary angioplasty. The firm is currently developing two additional late-stage pharmaceutical products as potential acute care hospital products. The first of these, Cleviprex (clevidipine), is an intravenous drug intended for the control of blood pressure in intensive care patients who require rapid and precise control of blood pressure. Currently Cleviprex is in a clinical trial program comprised of five Phase III trials. The company's second potential product, cangrelor, is an intravenous antiplatelet agent that prevents platelet activation and aggregation, which we believe has potential advantages in the treatment of vascular disease. Cangrelor is currently in Phase III trials, for both safety and effectiveness in preventing ischemic events in patients who require coronary angioplasty. These trials are being conducted simultaneously, in the hope that they will serve as the basis for a submission for approval in the U.S. and the European Union. Its revenue to date has been generated almost entirely from sales of Angiomax in the U.S. Angiomax is currently being sold through third-party distributors in Canada, Israel and New Zealand. TMC has recently received approval from the European Commission to market Angiomax in all 25 member states of the European Union. The drug, known in Europe as Angiox, is marketed in Europe by Nycomed Group and Grupo Ferrer, TMC's European distribution partners. In 2007, the company announced favorable findings from its Phase III trial of Cleviprex, as compared with current intravenous antihypertensive agents in controlling perioperative hypertension (a potentially harmful elevation in blood pressure just before, during and/or after surgery).

FINANCIALS: Sales and profits are in thousands of dollars—add 000 to get the full amount. 2006 Note: Financial information for 2006 was not available for all companies at press time.

2006 Sales: $213,952	2006 Profits: $63,726	**U.S. Stock Ticker: MDCO**
2005 Sales: $150,207	2005 Profits: $-7,753	**Int'l Ticker:** Int'l Exchange:
2004 Sales: $144,251	2004 Profits: $16,999	Employees: 289
2003 Sales: $85,591	2003 Profits: $-16,870	Fiscal Year Ends: 12/31
2002 Sales: $38,301	2002 Profits: $-45,831	Parent Company:

SALARIES/BENEFITS:

Pension Plan:	ESOP Stock Plan:	Profit Sharing:	Top Exec. Salary: $516,000	Bonus: $387,000
Savings Plan: Y	Stock Purch. Plan: Y		Second Exec. Salary: $400,000	Bonus: $250,000

OTHER THOUGHTS:

Apparent Women Officers or Directors: 1
Hot Spot for Advancement for Women/Minorities:

LOCATIONS: ("Y" = Yes)

West:	Southwest:	Midwest:	Southeast:	Northeast:	International:
				Y	Y

Note: Financial information, benefits and other data can change quickly and may vary from those stated here.

MEDICIS PHARMACEUTICAL CORP   www.medicis.com

Industry Group Code: 325412 Ranks within this company's industry group: Sales: 46   Profits: 173

Drugs:		Other:		Clinical:	Computers:		Other:	
Discovery:		AgriBio:		Trials/Services:	Hardware:		Specialty Services:	
Licensing:		Genomics/Proteomics:		Laboratories:	Software:		Consulting:	
Manufacturing:	Y	Tissue Replacement:		Equipment/Supplies:	Arrays:		Blood Collection:	
Development:	Y			Research/Development Svcs.:	Database Management:		Drug Delivery:	
Generics:				Diagnostics:			Drug Distribution:	

TYPES OF BUSINESS:
Dermatological, Aesthetic & Podiatric Conditions Drugs
Acne Treatment
Topical Creams
Wrinkle Treatment

BRANDS/DIVISIONS/AFFILIATES:
OMNICEF
RESTYLANE
SOLODYN
TRIAZ
VANOS
ZIANA
PERLANE

CONTACTS: Note: Officers with more than one job title may be intentionally listed here more than once.
Jonah Shacknai, CEO
Mark A. Prygocki, Sr., CFO/Exec. VP
Richard J. Havens, Exec. VP-Sales & Mktg.
Mitchell S. Wortzman, Chief Scientific Officer/Exec. VP
Joseph P. Cooper, Exec. VP-Prod. Dev.
Jason Hanson, General Counsel/Exec. VP
Joseph P. Cooper, Exec. VP-Corp. Dev.
Mark A. Prygocki, Sr., Treas.
Jonah Shacknai, Chmn.

Phone: 602-808-8800	Fax: 602-808-0822
Toll-Free:	
Address: 8125 N. Hayden Rd., Scottsdale, AZ 85258 US	

GROWTH PLANS/SPECIAL FEATURES:
Medicis Pharmaceutical Corp. is an independent specialty pharmaceutical company that develops and markets products for treatment of dermatological, aesthetic and podiatric conditions. The company offers a broad range of products addressing various conditions or aesthetic improvement, including facial wrinkles; acne; fungal infections; rosacea; hyperpigmentation; photoaging; psoriasis; skin and skin-structure infections; seborrheic dermatitis; and cosmesis (improvement in the texture and appearance of skin). Part of the firm's products include OMNICEF, a patented oral cephalosporin for skin and skin-structure infections; RESTYLANE, an injectable gel for treatment of moderate to severe facial wrinkles and folds such as nasolabial folds; SOLODYN, a once daily dosage in the treatment of inflammatory lesions of non-nodular moderate to severe acne vulgaris in patients 12 or older; TRIAZ, a topical patented gel and cleanser and patent-pending pad treatments for acne; VANOS, a high potency topical corticosteroid indicated for the relief of inflammatory and pruritic manifestations of corticosteroid responsive dermatoses in patients 12 years of age or older; and ZIANA, a once daily topical gel treatment for acne vulgaris in patients 12 or older. Medicis customers include wholesale pharmaceutical distributors such as Cardinal Health, Inc. and McKesson Corp. and other major drug chains. In May 2007, Medicis announced that FDA approved the dermal filler PERLANE for implantation into deep dermis to superficial subcutis for the correction of moderate to severe facial folds and wrinkles.

Medicis offers its employees a 401(k) plan; medical, dental and life insurance; short- and long-term disability; educational assistance; and flexible spending accounts.

FINANCIALS: Sales and profits are in thousands of dollars—add 000 to get the full amount. 2006 Note: Financial information for 2006 was not available for all companies at press time.

2006 Sales: $349,242	2006 Profits: $-75,849	U.S. Stock Ticker: MRX
2005 Sales: $376,899	2005 Profits: $64,990	Int'l Ticker: Int'l Exchange:
2004 Sales: $303,722	2004 Profits: $30,840	Employees: 391
2003 Sales: $247,500	2003 Profits: $51,300	Fiscal Year Ends: 12/31
2002 Sales: $212,800	2002 Profits: $50,000	Parent Company:

SALARIES/BENEFITS:

Pension Plan:	ESOP Stock Plan:	Profit Sharing:	Top Exec. Salary: $1,020,000	Bonus: $895,050
Savings Plan: Y	Stock Purch. Plan:		Second Exec. Salary: $496,000	Bonus: $362,700

OTHER THOUGHTS:
Apparent Women Officers or Directors: 1
Hot Spot for Advancement for Women/Minorities:

LOCATIONS: ("Y" = Yes)

West:	Southwest:	Midwest:	Southeast:	Northeast:	International:
	Y				

MEDIMMUNE INC

www.medimmune.com

Industry Group Code: 325412 Ranks within this company's industry group: Sales: 30 Profits: 37

Drugs:		Other:		Clinical:	Computers:	Other:
Discovery:		AgriBio:		Trials/Services:	Hardware:	Specialty Services:
Licensing:		Genomics/Proteomics:		Laboratories:	Software:	Consulting:
Manufacturing:	Y	Tissue Replacement:		Equipment/Supplies:	Arrays:	Blood Collection:
Development:	Y			Research/Development Svcs.:	Database Management:	Drug Delivery:
Generics:				Diagnostics:		Drug Distribution:

TYPES OF BUSINESS:

Pharmaceuticals Development & Manufacturing
Drugs-Cancer, Infectious & Immune Diseases

BRANDS/DIVISIONS/AFFILIATES:

MedImmune Ventures, Inc.
Ethyol
Synagis
FluMist
CytoGam
NeuTrexin

CONTACTS: Note: Officers with more than one job title may be intentionally listed here more than once.

David M. Mott, CEO
David M. Mott, Pres.
Peter Greenleaf, Sr. VP-Mktg. & Sales
Pamela J. Lupien, Sr. VP-Human Resources
James F. Young, Pres., R&D
William C. Bertrand, Jr., General Counsel/Sr. VP-Legal Affairs
Bernardus N. M. Machielse, Exec. VP-Oper.
Ed Mathers, Sr. VP-Corp. Dev. & Venture
Gail Folena-Wasserman, Sr. VP-Dev.
Edward M. Connor, Exec. VP/Chief Medical Officer
Peter A. Kiener, Sr. VP-Research
Linda J. Peters, Sr. VP-Regulatory Affairs
Wayne T. Hockmeyer, Chmn.

Phone: 301-398-0000	Fax: 301-527-4200

Toll-Free:

Address: 1 MedImmune Way, Gaithersburg, MD 20878 US

GROWTH PLANS/SPECIAL FEATURES:

MedImmune, Inc. is a biotechnology company engaged in the development and production of pharmaceuticals for infectious diseases, cancer and inflammatory disease. The company markets four products: Synagis, Ethyol, FluMist, and Cytogram. Synagis is a treatment for respiratory syncytial virus (RSV). The drug was the first monoclonal antibody approved for an infectious disease and has become a primary pediatric product for the prevention of RSV, a major cause of viral pneumonia and brochiolitis in infants and children. It is marketed in 49 countries by MedImmune and Abbott Laboratories. Ethyol is marketed as a treatment for the reduction of toxicities from cancer chemotherapy and radiotherapy in 60 countries worldwide, mainly through affiliates of Schering-Plough. FluMist is a nasally administered flu vaccine for influenzas A and B. CytoGam is a preventative treatment for the cytomegalovirus disease arising from organ transplant complications. The company also sells NeuTrexin, a lipid-soluble analog of methotrexate, which is a treatment for moderate-to-severe Pheumocystis carinii pneumonia in immunocompromised patients. Regarding its product candidate pipeline, MedImmune focuses its research and development efforts in the therapeutic areas of infectious disease, inflammatory diseases and cancer. The company has eight candidates in the infectious disease category, five candidates in the immunology category and 11 candidates in the oncology category. In May 2006, MedImmune was awarded a $170 million, five-year contract from U.S. Health and Human Services (HHS) to develop cell-based, seasonal and pandemic influenza vaccine technology. In April 2007, the firm agreed to be purchased by AstraZeneca PLC for $15.6 billion.

FINANCIALS: Sales and profits are in thousands of dollars—add 000 to get the full amount. 2006 Note: Financial information for 2006 was not available for all companies at press time.

2006 Sales: $1,276,800	2006 Profits: $48,700	**U.S. Stock Ticker: MEDI**
2005 Sales: $1,243,900	2005 Profits: $-16,600	**Int'l Ticker:** Int'l Exchange:
2004 Sales: $1,141,100	2004 Profits: $-3,800	Employees: 2,538
2003 Sales: $1,054,334	2003 Profits: $183,204	Fiscal Year Ends: 12/31
2002 Sales: $847,700	2002 Profits: $-1,098,000	Parent Company:

SALARIES/BENEFITS:

Pension Plan:	ESOP Stock Plan:	Profit Sharing:	Top Exec. Salary: $991,667	Bonus: $1,300,000
Savings Plan: Y	Stock Purch. Plan: Y		Second Exec. Salary: $570,833	Bonus: $480,000

OTHER THOUGHTS:

Apparent Women Officers or Directors: 4
Hot Spot for Advancement for Women/Minorities: Y

LOCATIONS: ("Y" = Yes)

West:	Southwest:	Midwest:	Southeast:	Northeast:	International:
Y				Y	Y

Note: Financial information, benefits and other data can change quickly and may vary from those stated here.

MEDTOX SCIENTIFIC INC

www.medtox.com

Industry Group Code: 621511 **Ranks within this company's industry group:** Sales: 3 Profits: 3

Drugs:	Other:	Clinical:		Computers:		Other:	
Discovery:	AgriBio:	Trials/Services:	Y	Hardware:		Specialty Services:	Y
Licensing:	Genomics/Proteomics:	Laboratories:	Y	Software:		Consulting:	
Manufacturing:	Tissue Replacement:	Equipment/Supplies:	Y	Arrays:		Blood Collection:	
Development:		Research/Development Svcs.:		Database Management:		Drug Delivery:	
Generics:		Diagnostics:	Y			Drug Distribution:	

TYPES OF BUSINESS:

Diagnostic Device Manufacturing
Forensic & Clinical Lab Services
Diagnostic Drug Screening Devices
Specialty Courier Services
Drug Diagnostics Services

BRANDS/DIVISIONS/AFFILIATES:

MEDTOX Diagnostics, Inc.
MEDTOX Laboratories, Inc.
PROFILE-II Test System
VERDICT-II
ClearCourse
Sure-Screen
Drug Abuse Recognition System (DARS)
CMS Courier

CONTACTS: Note: Officers with more than one job title may be intentionally listed here more than once.

Richard J. Braun, CEO
Richard J. Braun, Pres.
Kevin J. Wiersma, CFO
James A. Schoonover, Chief Mktg. Officer/VP-Sales & Mktg.
Susan E. Puskas, VP-Human Resources
Susan E. Puskas, VP-Quality & Regulatory Affairs
B. Mitchell Owens, VP/COO-MEDTOX Diagnostics, Inc.
Kevin J. Wiersma, VP/COO-Laboratory Div./CFO-MEDTOX Scientific, Inc.
Richard J. Braun, Chmn.

Phone: 651-636-7466	Fax: 651-636-5351
Toll-Free: 800-832-3244	
Address: 402 W. County Rd. D, St. Paul, MN 55112 US	

GROWTH PLANS/SPECIAL FEATURES:

MEDTOX Scientific, Inc. manufactures and distributes diagnostic devices and provides forensic and toxicological laboratory services. Subsidiary MEDTOX Laboratories, Inc. provides laboratory drug testing services for corporations, medical facilities and the federal government. It also offers specialty services including emergency clinical toxicology; medical diagnostics; therapeutic drug monitoring; clinical testing for the pharmaceutical industry; heavy metal, trace element and solvent analysis; and logistics, data and program management services. The firm's other subsidiary, MEDTOX Diagnostics, Inc., based in Burlington, North Carolina, manufactures fast and inexpensive drug screening devices and assays for the corporate, health care, criminal justice, temporary service and drug rehabilitation markets. The PROFILE-II and VERDICT-II products are used for detecting marijuana, cocaine, opiates, amphetamines and phencyclidine (PCP) in human urine, primarily for workplaces, hospitals and government agencies. The PROFILE-II ER system also tests for tricyclic antidepressants, barbiturates, methadone and benzodiazepines, and it is used in hospital and clinical applications. Among the company's other products are the EZ-SCREEN test, which are utilized in agricultural diagnostics to detect mycotoxins and antibiotic residues. The company also offers its ClearCourse comprehensive drug testing program, which consists of: Drug Abuse Recognition System (DARS), which includes deviceless drug abuse recognition training; SureScreen drug testing devices; WEBTOX online data management; and MedTox' own laboratory assistance for confirmation testing. The company's Consolidated Medical Services (CMS Courier) division is a full-service specialty courier specializing in the transportation of medical specimens and equipment.

MEDTOX offers its employees medical and dental insurance; tuition reimbursement; gym membership discounts; and many other benefits.

FINANCIALS: Sales and profits are in thousands of dollars—add 000 to get the full amount. 2006 Note: Financial information for 2006 was not available for all companies at press time.

2006 Sales: $69,804	2006 Profits: $4,548	**U.S. Stock Ticker:** MTOX
2005 Sales: $63,047	2005 Profits: $3,318	**Int'l Ticker:** Int'l Exchange:
2004 Sales: $56,736	2004 Profits: $1,821	Employees: 460
2003 Sales: $51,473	2003 Profits: $- 308	Fiscal Year Ends: 12/31
2002 Sales: $52,000	2002 Profits: $11,700	Parent Company:

SALARIES/BENEFITS:

Pension Plan:	ESOP Stock Plan:	Profit Sharing:	Top Exec. Salary: $300,000	Bonus: $894,000
Savings Plan: Y	Stock Purch. Plan:		Second Exec. Salary: $192,000	Bonus: $310,000

OTHER THOUGHTS:

Apparent Women Officers or Directors: 1
Hot Spot for Advancement for Women/Minorities:

LOCATIONS: ("Y" = Yes)

West:	Southwest:	Midwest:	Southeast:	Northeast:	International:
		Y		Y	

MERCK & CO INC

www.merck.com

Industry Group Code: 325412 **Ranks within this company's industry group:** Sales: 8   Profits: 7

Drugs:		Other:	Clinical:	Computers:	Other:	
Discovery:	Y	AgriBio:	Trials/Services:	Hardware:	Specialty Services:	Y
Licensing:		Genomics/Proteomics:	Laboratories:	Software:	Consulting:	
Manufacturing:	Y	Tissue Replacement:	Equipment/Supplies:	Arrays:	Blood Collection:	
Development:	Y		Research/Development Svcs.:	Database Management:	Drug Delivery:	
Generics:			Diagnostics:		Drug Distribution:	

TYPES OF BUSINESS:

Pharmaceuticals Development & Manufacturing
Cholesterol Drugs
Hypertension Drugs
Heart Failure Drugs
Allergy & Asthma Drugs
Animal Health Products
Vaccines
Preventative Drugs

BRANDS/DIVISIONS/AFFILIATES:

Merck Institute for Science Education
Sirna Therapeutics, Inc.
Singulair
Propecia
Cozaar
Fosamax
Gardasil
Vioxx

CONTACTS: Note: Officers with more than one job title may be intentionally listed here more than once.

Richard Clark, CEO
Richard Clark, Pres.
Peter Kellogg, CFO/Exec. VP
Mirian Graddick-Weir, Sr. VP-Human Resources
Peter Kim, Pres., Research Laboratories
Chris Scalet, CIO/Sr. VP-Svcs.
Willie Deese, Pres., Mfg. Div.
Kenneth Frazier, General Counsel/Sr. VP
Peter Kellogg, Exec. VP-Bus. Dev. & Licensing Activities
Peter Kellogg, Exec. VP-Investor Rel.
Peter H. Loescher, Pres., Global Human Health
Richard Clark, Chmn.
Per Wold-Olsen, Pres., Human Health, EMEA

Phone: 908-423-1000	**Fax:** 908-735-1253
Toll-Free:	
Address: 1 Merck Dr., Whitehouse Station, NJ 08889-0100 US	

GROWTH PLANS/SPECIAL FEATURES:

Merck & Co., Inc. is a leading research-driven pharmaceutical company that manufactures a broad range of products sold in approximately 150 countries. These products include therapeutic and preventative drugs generally sold by prescription and medications used to control and alleviate disease. As one of the world's largest pharmaceutical companies, Merck's line of big selling medicine includes drugs like Zocor for cholesterol; Fosamax for the prevention of osteoporosis; and Cozaar and Hyzaar for high blood pressure. The company also manufactures Propecia, a popular treatment for male pattern baldness, and Singulair, a seasonal allergy and asthma medicine, both of which are offered in tablet form. Merck's bestselling Vioxx product, an arthritis and pain medication, was discontinued in 2004 due to ongoing litigation. The withdrawal of this drug has serious negative implications for the company. Vioxx's return to the market hinges on an FDA review. In addition to medicines, the company manufactures vaccines such as Rotateq, designed to prevent gastroenteritis in infants and children; and Gardasil, a new product designed to reduce the probabilities of cancer. In May 2006, Merck agreed to acquire Abmaxis, Inc. and GlycoFi, Inc. At the end of December 2006, Merck acquired Sirna Therapeutics, Inc., which develops drugs based on RNA interference, for $1.1 billion. In 2006, five Merck products were approved by the FDA: Gardasil, a vaccine for HPV, which causes cervical cancer and warts; Januvia, which improves blood sugar control for patients with type 2 diabetes; Zostavax, a vaccine to prevent shingles; RotaTeq, a vaccine to prevent rotavirus gastroenteritis in infants; and Zolinza, which treats advanced cutaneous T-cell lymphoma.

Merck offers its employees on-site fitness facilities, day care and summer camp programs, tuition reimbursement and financial planning assistance.

FINANCIALS: Sales and profits are in thousands of dollars—add 000 to get the full amount. 2006 Note: Financial information for 2006 was not available for all companies at press time.

2006 Sales: $22,636,000	2006 Profits: $4,433,800	**U.S. Stock Ticker:** MRK
2005 Sales: $22,011,900	2005 Profits: $4,631,300	**Int'l Ticker:** Int'l Exchange:
2004 Sales: $22,938,600	2004 Profits: $5,813,400	Employees: 60,000
2003 Sales: $22,485,900	2003 Profits: $6,830,900	Fiscal Year Ends: 12/31
2002 Sales: $21,445,800	2002 Profits: $7,149,500	Parent Company:

SALARIES/BENEFITS:

Pension Plan: Y	ESOP Stock Plan:	Profit Sharing:	Top Exec. Salary: $1,183,334	Bonus: $1,800,000
Savings Plan: Y	Stock Purch. Plan:		Second Exec. Salary: $828,130	Bonus: $875,000

OTHER THOUGHTS:

Apparent Women Officers or Directors: 4
Hot Spot for Advancement for Women/Minorities: Y

LOCATIONS: ("Y" = Yes)

West:	Southwest:	Midwest:	Southeast:	Northeast:	International:
Y	Y	Y	Y	Y	Y

MERCK KGAA

www.merck.de

Industry Group Code: 325412 Ranks within this company's industry group: Sales: 17   Profits: 17

Drugs:		Other:		Clinical:		Computers:		Other:	
Discovery:	Y	AgriBio:		Trials/Services:		Hardware:		Specialty Services:	
Licensing:	Y	Genomics/Proteomics:		Laboratories:		Software:		Consulting:	
Manufacturing:	Y	Tissue Replacement:		Equipment/Supplies:		Arrays:		Blood Collection:	
Development:	Y			Research/Development Svcs.:		Database Management:		Drug Delivery:	
Generics:	Y			Diagnostics:				Drug Distribution:	

TYPES OF BUSINESS:

Pharmaceuticals
Over-the-Counter Drugs & Vitamins
Generic Drugs
Chemicals
LCD Components
Reagents & Diagnostics
Nanotechnology Research

BRANDS/DIVISIONS/AFFILIATES:

Serono S.A.
Merck Serono S.A.
Merck Ethicals
EMD Chemicals Inc.
EpiPen

CONTACTS: Note: Officers with more than one job title may be intentionally listed here more than once.

Karl-Ludwig Kley, CEO
Karl-Ludwig Kley, Dir.-Human Resources
Bernd Reckmann, Dir.-Info. Svcs.
Bernd Reckmann, Dir.-Eng.
Michael Romer, Dir.-Legal
Michael Becker, Dir.-Bus. Dev.
Michael Romer, Dir.-Corp. Comm.
Michael Becker, Dir.-Acct.
Bernd Reckmann, Mgr.-Darmstadt & Gernsheim Sites
Elmar Schnee, Dir.-Pharmaceuticals Bus. Sector
Walter W. Zywottek, Dir.-Chemicals Bus. Sector
Meiken Krebs, Mgr.-EMD Chemicals Inc.
Wilhelm Simson, Chmn.
Bernd Reckmann, Dir.-Purchasing

Phone: 49-615-1720	Fax: 49-615-1720-00
Toll-Free:	
Address: Frankfurter St. 250, Darmstadt, 64293 Germany	

GROWTH PLANS/SPECIAL FEATURES:

Merck KGAA is a leading global pharmaceuticals and chemicals company with 176 companies in 56 countries. The firm is headquartered in Germany and it maintains important research sites in France, Spain, Great Britain, the U.S. and Japan.   Merck is involved in two main lines of business: pharmaceuticals and chemicals.   Its pharmaceuticals business manufactures prescription drugs for the treatment of cancer, metabolic disorders, and cardiovascular diseases; as well as vitamins, minerals, food supplements, cold remedies and natural remedies. Merck's Generics division is an important part of its pharmaceutical business, with over 400 products in its portfolio. One of the division's areas of focus is on life saving single-dose drugs such as EpiPen, an epinephrine autoinjector for the treatment of acute allergic emergencies. The division also manufactures inhalation sprays for life-threatening respiratory diseases in premature infants and for treating chronic obstructive bronchitis and asthma in children. The firm's chemical business manufactures components for LCDs in televisions, PC monitors, notebooks and mobile phones. In addition, this segment has a laboratory business, which creates reagents and test kits for research laboratories and environmental science.   The business also creates products and services for the entire drug development and manufacturing chain and manufactures pigments for industrial applications and for the cosmetics industry. In 2006, Merck acquired Serono S.A., which was renamed Merck Serono S.A. This acquisition will be merged with Merck Ethicals division in 2007 and operate as the new Merck Serono division under the company's Pharmaceuticals business. In February 2007, Merck announced that it was partnering with U.S. nanotechnology developer Nano Terra LLC to research ways in which nanotechnology could enhance the precision, control, and physical properties of chemicals produced by Merck.   In May 2007, Mylan Laboratories, Inc. agreed to acquire the generic-drugs unit of Merck for $6.7 billion.

FINANCIALS: Sales and profits are in thousands of dollars—add 000 to get the full amount. 2006 Note: Financial information for 2006 was not available for all companies at press time.

2006 Sales: $8,352,950	2006 Profits: $1,335,880	**U.S. Stock Ticker:**
2005 Sales: $7,697,680	2005 Profits: $898,150	**Int'l Ticker: MRK** Int'l Exchange: Frankfurt-Euronext
2004 Sales: $	2004 Profits: $	Employees:
2003 Sales: $	2003 Profits: $	Fiscal Year Ends: 12/31
2002 Sales: $	2002 Profits: $	Parent Company:

SALARIES/BENEFITS:

Pension Plan:	ESOP Stock Plan:	Profit Sharing:	Top Exec. Salary: $	Bonus: $
Savings Plan:	Stock Purch. Plan:		Second Exec. Salary: $	Bonus: $

OTHER THOUGHTS:

Apparent Women Officers or Directors: 1
Hot Spot for Advancement for Women/Minorities:

LOCATIONS: ("Y" = Yes)

West:	Southwest:	Midwest:	Southeast:	Northeast:	International:
Y		Y		Y	Y

MERCK SERONO SA

www.serono.com

Industry Group Code: 325412 Ranks within this company's industry group: Sales: Profits:

Drugs:		Other:		Clinical:	Computers:	Other:	
Discovery:	Y	AgriBio:		Trials/Services:	Hardware:	Specialty Services:	
Licensing:		Genomics/Proteomics:	Y	Laboratories:	Software:	Consulting:	
Manufacturing:	Y	Tissue Replacement:		Equipment/Supplies:	Arrays:	Blood Collection:	
Development:	Y			Research/Development Svcs.:	Database Management:	Drug Delivery:	Y
Generics:				Diagnostics:		Drug Distribution:	

TYPES OF BUSINESS:

Pharmaceuticals Development
Fertility Drugs
Neurology Drugs
Growth & Metabolism Drugs
Dermatology Drugs
Oncology Research

BRANDS/DIVISIONS/AFFILIATES:

GONAL-f
Ovidrel/Ovitrelle
Luveris
Crinone
Cetrotide
Rebif
Saizen
Merck KGaA

CONTACTS: *Note: Officers with more than one job title may be intentionally listed here more than once.*

Elmer Schnee, CEO/Dir.
Olaf Klinger, CFO
Francois Naef, Chief Admin. Officer
Michael Becker, Chmn.

Phone: 41-22-414-3000	**Fax:** 41-22-731-2179
Toll-Free:	
Address: 9, chemin des Mines, Case postale 54, Geneva, CH-1211 20 Switzerland	

GROWTH PLANS/SPECIAL FEATURES:

Merck Serono SA (SRA), formerly Serono SA, was formed in January 2007 when Merck KGaA acquired Serono in a $13 billion transaction. SRA is a global biotechnology company that focuses on product development in four therapeutic areas: reproductive health, growth and metabolism, multiple sclerosis and dermatology. The company currently markets GONAL-f, Ovitrelle, Luveris, Crinone and Cetrotide for female infertility; Rebif and Rebiject for multiple sclerosis; Saizen for growth hormone deficiency in children and Raptiva for psoriasis. SRA has also launched a web site, fertility.com, which addresses infertility issues and provides support for couples who are seeking or undergoing treatments. The firm markets its products in Europe, Asia, Latin America and North America and is involved in numerous worldwide partnerships with companies such as Amgen and Astrazeneca. It also established the SRA Biotech Center (SBC) for the sole purpose of manufacturing bulk pharmaceutical products under U.S., European, Japanese, Australian and Swiss pharmaceutical regulations. SRA operates in 44 countries and has a large network of research and development personnel at three global research institutes: SRA Pharmaceutical Institute in Geneva, Switzerland; SRA Research Institute in Boston; and an Italian facility dedicated to pharmacological research. Currently, SRA has eight biotechnology products on the market and more than 25 ongoing preclinical development projects. SRA research principally centers on genetics and functional genomics in order to identify new therapeutic proteins in areas such as the autoimmune/inflammatory system and oncology. SRA has also focused in particular on treatments for rheumatoid arthritis, osteoarthritis, lymphomas and leukemia. In 2007, SRA launched easypod, an electric growth hormone injection device used for the administration of Saizen. Since the company's recent acquisition, numerous changes will take place, which include the reorganization of its executive board and the modification of all of its logos and names in its global facilities.

FINANCIALS: Sales and profits are in thousands of dollars—add 000 to get the full amount. 2006 Note: Financial information for 2006 was not available for all companies at press time.

2006 Sales: $	2006 Profits: $	**U.S. Stock Ticker: Subsidiary**
2005 Sales: $2,586,400	2005 Profits: $-105,300	**Int'l Ticker: SEO** Int'l Exchange: Zurich-SWX
2004 Sales: $2,177,900	2004 Profits: $494,200	Employees: 4,900
2003 Sales: $1,858,000	2003 Profits: $390,000	Fiscal Year Ends: 12/31
2002 Sales: $1,546,500	2002 Profits: $320,800	Parent Company: MERCK KGAA

SALARIES/BENEFITS:

Pension Plan: Y	ESOP Stock Plan: Y	Profit Sharing: Y	Top Exec. Salary: $	Bonus: $
Savings Plan:	Stock Purch. Plan:		Second Exec. Salary: $	Bonus: $

OTHER THOUGHTS:

Apparent Women Officers or Directors:
Hot Spot for Advancement for Women/Minorities:

LOCATIONS: ("Y" = Yes)

West:	Southwest:	Midwest:	Southeast:	Northeast:	International:
				Y	Y

MERIDIAN BIOSCIENCE INC   www.meridianbioscience.com

Industry Group Code: 325413 Ranks within this company's industry group: Sales: 11 Profits: 6

Drugs:		Other:		Clinical:		Computers:		Other:	
Discovery:		AgriBio:		Trials/Services:		Hardware:		Specialty Services:	
Licensing:		Genomics/Proteomics:		Laboratories:		Software:		Consulting:	
Manufacturing:	Y	Tissue Replacement:		Equipment/Supplies:	Y	Arrays:		Blood Collection:	
Development:	Y			Research/Development Svcs.:		Database Management:		Drug Delivery:	
Generics:				Diagnostics:	Y			Drug Distribution:	

TYPES OF BUSINESS:

Diagnostic Test Kits
Contract Manufacturing
Bulk Antigens, Antibodies & Reagents

BRANDS/DIVISIONS/AFFILIATES:

Biodesign
OEM Concepts
Viral Antigens
cGMP
ImmunoCard STAT EHEC

CONTACTS: *Note: Officers with more than one job title may be intentionally listed here more than once.*

William J. Motto, CEO
John A. Kraeutler, COO
John A. Kraeutler, Pres.
Melissa A. Lueke, CFO/VP
Todd W. Motto, VP-Sales & Mktg.
Kenneth J. Kozak, VP-R&D
Lawrence J. Baldini, Exec. VP-Info. Systems
Lawrence J. Baldini, Exec. VP-Oper.
Susan D. Rolih, VP-Regulatory Affairs & Quality Systems
William J. Motto, Chmn.
Antonio A. Interno, Sr. VP/Pres./Managing Dir.-Europe

Phone: 513-271-3700	Fax: 513-271-3762
Toll-Free:	
Address: 3471 River Hills Dr., Cincinnati, OH 45244 US	

GROWTH PLANS/SPECIAL FEATURES:

Meridian Biosciences, Inc. is a life science company that develops, manufactures, sells and distributes diagnostic test kits, primarily for certain respiratory, gastrointestinal, viral and parasitic infectious diseases; manufactures and distributes bulk antigens, antibodies and reagents; and contract manufactures proteins and other biologicals. The company operates in three segments: U.S. diagnostics, European diagnostics and life science. The U.S. diagnostics segment focuses on the development, manufacture, sale and distribution of diagnostic test kits, which utilize immunodiagnostic technologies that test samples of blood, urine, stool and other body fluids or tissue for the presence of antigens and antibodies of specific infectious diseases. Products also include transport media that store and preserve specimen samples from patient collection to laboratory testing. The European diagnostics segment focuses on the sale and distribution of diagnostic test kits. Its sales and distribution network consists of direct sales forces in Belgium, France, Holland and Italy; and independent distributors in other European, African and Middle Eastern countries. The life sciences segment focuses on the development, manufacture, sale and distribution of bulk antigens, antibodies and reagents, as well as contract development and manufacturing services. The segment is represented by four product-line brands: Biodesign, which represents monoclonal and polyclonal antibodies and assay reagents; OEM Concepts, which represents contract ascites and antibody production services; Viral Antigens, which represents viral proteins; and cGMP biologics, which represents contract development and manufacturing services for drug and vaccine discovery and development. Meridian's core diagnostic products generated 79% of revenue in 2006. In February 2007, the company received clearance from the FDA to market ImmunoCard STAT EHEC, a test for the diagnosis of E. coli infection.

The company offers its employees health, dental and life insurance; a 401(k) plan; stock options; an employee stock purchase plan; a profit-sharing plan; tuition reimbursement; and an employee assistance plan.

FINANCIALS: Sales and profits are in thousands of dollars—add 000 to get the full amount. 2006 Note: Financial information for 2006 was not available for all companies at press time.

2006 Sales: $108,413	2006 Profits: $18,325	**U.S. Stock Ticker: VIVO**
2005 Sales: $92,965	2005 Profits: $12,565	**Int'l Ticker:** Int'l Exchange:
2004 Sales: $79,606	2004 Profits: $9,185	Employees: 363
2003 Sales: $65,864	2003 Profits: $7,018	Fiscal Year Ends: 9/30
2002 Sales: $59,100	2002 Profits: $5,000	Parent Company:

SALARIES/BENEFITS:

Pension Plan:	ESOP Stock Plan:	Profit Sharing: Y	Top Exec. Salary: $456,155	Bonus: $490,875
Savings Plan: Y	Stock Purch. Plan: Y		Second Exec. Salary: $360,573	Bonus: $388,500

OTHER THOUGHTS:

Apparent Women Officers or Directors: 2
Hot Spot for Advancement for Women/Minorities: Y

LOCATIONS: ("Y" = Yes)

West:	Southwest:	Midwest:	Southeast:	Northeast:	International:
				Y	Y

MGI PHARMA INC
www.mgipharma.com

Industry Group Code: 325412 Ranks within this company's industry group: Sales: 47 Profits: 145

Drugs:		Other:		Clinical:	Computers:	Other:
Discovery:		AgriBio:		Trials/Services:	Hardware:	Specialty Services:
Licensing:	Y	Genomics/Proteomics:		Laboratories:	Software:	Consulting:
Manufacturing:		Tissue Replacement:		Equipment/Supplies:	Arrays:	Blood Collection:
Development:	Y			Research/Development Svcs.:	Database Management:	Drug Delivery:
Generics:				Diagnostics:		Drug Distribution:

TYPES OF BUSINESS:
Pharmaceuticals Acquisition & Development
Drugs-Oncology & Rheumatology

BRANDS/DIVISIONS/AFFILIATES:
Aloxi Injection
Gliadel
Salagen
Hexalen
Aquavan
Dacogen
Saforis

CONTACTS: Note: Officers with more than one job title may be intentionally listed here more than once.
Lonnie Moulder, Jr., CEO
Eric Loukas, COO/Exec. VP
Lonnie Moulder, Jr., Pres.
William F. Spengler, CFO/Sr. VP
Mary L. Hedley, Sr. VP/Chief Scientific Officer
Hugh E. Miller, Chmn.

Phone: 952-346-4700   Fax: 952-346-4800
Toll-Free: 800-562-5580
Address: 5775 W. Old Shakopee Rd., Ste. 100, Bloomington, MN 55437-3174 US

GROWTH PLANS/SPECIAL FEATURES:
MGI PHARMA, INC., along with its subsidiaries, is a biopharmaceutical company focused in oncology and acute care products that acquires, researches, develops and commercializes pharmaceutical products that address the unmet needs of patients. Its goal is to become a leading biopharmaceutical company through application of its core competencies of product research, acquisition, development and commercialization, which it applies toward its portfolio of oncology and acute care products and product candidates. The company acquires intellectual property or product rights from others after they have completed the basic research to discover the compounds that will become the company's product candidates or marketable products. It has facilities in Bloomington, Minnesota; Lexington, Massachusetts; and Baltimore, Maryland. The company's marketable and currently marketed products include Aloxi, an injection for chemotherapy-induced nausea and vomiting; Dacogen, for injection, to treat myelodysplastic syndrome; Gliadel wafers for malignant glioma at time of initial surgery and glioblastoma multiforme; Salagen tablets for symptoms of radiation-induced dry mouth in head and neck cancer patients, as well as dry mouth, plus dry eyes outside the U.S., in Sjogren's syndrome patients; and Hexalen capsules for ovarian cancer. Its product pipeline includes: Aloxi injection and capsules, both having completed Phase III trials; Dacogen for injection, in Phase II & III; Aquavan injection, for minimal to moderate sedation of patients undergoing brief diagnostic or surgical procedures, in Phase III trials; Saforis powder for oral suspension, for oral mucositis, which received an FDA approvable letter in 2006; as well as several others in Phase I and II and two preclinical candidates.

The firm provides employees with flexible spending accounts; seminar and professional membership reimbursement; a confidential employee assistance program; a continuing education program; credit union membership; discount off-site airport parking; and discounted tickets for entertainment and cultural activities.

FINANCIALS: Sales and profits are in thousands of dollars—add 000 to get the full amount. 2006 Note: Financial information for 2006 was not available for all companies at press time.

2006 Sales: $342,788	2006 Profits: $-40,161	U.S. Stock Ticker: MOGN
2005 Sales: $279,362	2005 Profits: $-132,410	Int'l Ticker: Int'l Exchange:
2004 Sales: $195,667	2004 Profits: $-85,723	Employees: 540
2003 Sales: $49,385	2003 Profits: $-61,908	Fiscal Year Ends: 12/31
2002 Sales: $28,200	2002 Profits: $-36,100	Parent Company:

SALARIES/BENEFITS:
Pension Plan: Y	ESOP Stock Plan:	Profit Sharing:	Top Exec. Salary: $470,417	Bonus: $
Savings Plan: Y	Stock Purch. Plan: Y		Second Exec. Salary: $344,167	Bonus: $

OTHER THOUGHTS:
Apparent Women Officers or Directors: 1
Hot Spot for Advancement for Women/Minorities:

LOCATIONS: ("Y" = Yes)
West:	Southwest:	Midwest:	Southeast:	Northeast:	International:
		Y		Y	

Note: Financial information, benefits and other data can change quickly and may vary from those stated here.

MIGENIX INC

www.mbiotech.com

Industry Group Code: 325412 **Ranks within this company's industry group:** Sales: 168 Profits: 77

Drugs:		Other:	Clinical:	Computers:	Other:
Discovery:	Y	AgriBio:	Trials/Services:	Hardware:	Specialty Services:
Licensing:		Genomics/Proteomics:	Laboratories:	Software:	Consulting:
Manufacturing:		Tissue Replacement:	Equipment/Supplies:	Arrays:	Blood Collection:
Development:	Y		Research/Development Svcs.:	Database Management:	Drug Delivery:
Generics:			Diagnostics:		Drug Distribution:

TYPES OF BUSINESS:

Pharmaceuticals Discovery & Development
Pre-Clinical Research & Development
Acne Treatment
Antivirals

BRANDS/DIVISIONS/AFFILIATES:

Micrologix Biotech, Inc.
Omigard
Celgosivir

CONTACTS: Note: Officers with more than one job title may be intentionally listed here more than once.

James DeMesa, CEO
James DeMesa, Pres.
Arthur J. Ayers, CFO
Jacob J. Clement, Sr. VP-Science/Chief Science Officer
Jacob J. Clement, Sr. VP-Tech.
AnnKatrin Petersen-Japelli, VP-Clinical Dev.
K. David Campagnari, VP-Oper.
William D. Milligan, Sr. VP-Corp. Dev./Chief Bus. Officer
Arthur J. Ayers, VP-Finance
Neil Howell, VP-Research
C. Robert Cory, VP-Bus. Dev.
David Scott, Chmn.

Phone: 604-221-9666	Fax: 604-221-9688
Toll-Free: 800-665-1968	
Address: BC Research Bldg., 3650 Wesbrook Mall, Vancouver, BC V6S 2L2 Canada	

GROWTH PLANS/SPECIAL FEATURES:

Migenix Inc., formerly Micrologix Biotech, Inc., is a biopharmaceutical company focusing on the discovery and development of anti-infective and anti-degenerative drugs. The company's drug candidates are based on the development of improved derivatives of naturally occurring peptide compounds found in the defense systems of virtually all life forms. Currently, the company has no products approved for marketing and is involved only in the aspects of research and development of pharmaceutical drugs. It has five product candidates in human clinical development, multiple product opportunities in pre-clinical development and several early-stage technologies in various stages of research and evaluation. Currently, Omigard is a leading antimicrobial drug product for the reduction of life-threatening bacterial and fungal bloodstream infections related to the use of central venous catheters in seriously ill patients. The drug is currently in Phase III of clinical trials and the company has partnered with Cadence Pharmaceuticals in the development of the drug. Migenix has also completed a Phase IIb clinical trial for CLS-001, a drug candidate for the treatment of rosacea and acne. MX 4509 is a sodium salt compound designed to treat Alzheimer's disease. The firm has acquired the global rights to Celgosivir, or MBI 3253, a Phase II clinical-stage compound. The drug is an orally administered antiviral agent that can inhibit the replication of a broad range of enveloped viruses, including hepatitis C virus or HCV, by itself and in combination with Ribavirin or Interferon-a, two of the most common HCV therapies.

FINANCIALS: Sales and profits are in thousands of dollars—add 000 to get the full amount. 2006 Note: Financial information for 2006 was not available for all companies at press time.

2006 Sales: $ 538	2006 Profits: $-10,720	**U.S. Stock Ticker: MGIFF.PK**
2005 Sales: $2,200	2005 Profits: $-9,463	**Int'l Ticker: MGI** Int'l Exchange: Toronto-TSX
2004 Sales: $2,667	2004 Profits: $-11,215	Employees: 48
2003 Sales: $7,746	2003 Profits: $-11,215	Fiscal Year Ends: 4/30
2002 Sales: $	2002 Profits: $	Parent Company:

SALARIES/BENEFITS:

Pension Plan:	ESOP Stock Plan:	Profit Sharing:	Top Exec. Salary: $354,818	Bonus: $63,416
Savings Plan:	Stock Purch. Plan:		Second Exec. Salary: $313,101	Bonus: $26,632

OTHER THOUGHTS:

Apparent Women Officers or Directors: 1
Hot Spot for Advancement for Women/Minorities:

LOCATIONS: ("Y" = Yes)

West:	Southwest:	Midwest:	Southeast:	Northeast:	International:
Y					Y

MILLENNIUM PHARMACEUTICALS INC   www.mlnm.com

Industry Group Code: 325412 Ranks within this company's industry group: Sales: 41   Profits: 149

Drugs:		Other:		Clinical:	Computers:	Other:
Discovery:	Y	AgriBio:		Trials/Services:	Hardware:	Specialty Services:
Licensing:		Genomics/Proteomics:	Y	Laboratories:	Software:	Consulting:
Manufacturing:	Y	Tissue Replacement:		Equipment/Supplies:	Arrays:	Blood Collection:
Development:	Y			Research/Development Svcs.:	Database Management:	Drug Delivery:
Generics:				Diagnostics:		Drug Distribution:

TYPES OF BUSINESS:

Pharmaceuticals Discovery & Development
Gene-Based Drug Discovery Platform
Small-Molecule Drugs
Diagnostic Products

BRANDS/DIVISIONS/AFFILIATES:

COR Therapeutics, Inc.
VELCADE
INTEGRILIN
Millennium University

CONTACTS: *Note: Officers with more than one job title may be intentionally listed here more than once.*

Deborah Dunsire, CEO
Kevin Starr, COO
Deborah Dunsire, Pres.
Marsha Fanucci, CFO/Sr. VP
Stephen M. Gansler, Sr. VP-Human Resources
Robert I. Tepper, VP-R&D
Laurie B. Keating, General Counsel
Anna Protopapas, VP-Corp. Dev.
Clare Midgley, VP-Global Corp. Affairs
Christophe Bianchi, Exec. VP-Commercial Oper.
Joseph Bolen, Chief Scientific Officer
Nancy Simonian, Chief Medical Officer
Peter Smith, Sr. VP-Non-Clinical Development Sciences
Kenneth Weg, Chmn.

Phone: 617-679-7000	Fax: 617-374-7788
Toll-Free: 800-390-5663	
Address: 40 Landsdowne St., Cambridge, MA 02139 US	

GROWTH PLANS/SPECIAL FEATURES:

Millennium Pharmaceuticals researches and manufactures small-molecule, biotherapeutic and predictive medicine products. The company integrates large-scale genetics, genomics, high-throughput screening and informatics into a drug discovery platform that accelerates the development of therapeutic and diagnostic products. The firm identifies important genes, determines their functions, validates drug and product development targets, formulates assays based on these targets and identifies product candidates. Millenium's research primarily emphasizes treatments for cancer and inflammation. It focuses on developing small-molecule drugs, typically formulated into pills for oral consumption, as well as proteins and monoclonal antibodies, usually only available as injections. The company licenses its platforms to various pharmaceutical and biotechnology firms in exchange for royalties from products developed with the technology. Millennium's VELCADE was one of the first proteasome inhibitors to earn FDA-approval for multiple myeloma, a cancer of the blood. Sales from the VELCADE product made up about 45% of the firm's 2006 revenue, and the company continues to focus its efforts on the product in other areas. The firm also receives a large share of its revenue from royalties culled from INTEGRILIN, a drug used to treat acute coronary syndrome, marketed by Schering-Plough.

The firm's Millennium University offers employees a variety of in-house workshops and off-site seminars designed to enhance career development.

FINANCIALS: Sales and profits are in thousands of dollars—add 000 to get the full amount. 2006 Note: Financial information for 2006 was not available for all companies at press time.

2006 Sales: $486,830	2006 Profits: $-43,953	U.S. Stock Ticker: MLNM
2005 Sales: $558,308	2005 Profits: $-198,249	Int'l Ticker: Int'l Exchange:
2004 Sales: $448,206	2004 Profits: $-252,297	Employees: 947
2003 Sales: $433,687	2003 Profits: $-483,687	Fiscal Year Ends: 12/31
2002 Sales: $353,000	2002 Profits: $-590,200	Parent Company:

SALARIES/BENEFITS:

Pension Plan:	ESOP Stock Plan:	Profit Sharing:	Top Exec. Salary: $457,619	Bonus: $166,137
Savings Plan: Y	Stock Purch. Plan: Y		Second Exec. Salary: $392,512	Bonus: $157,338

OTHER THOUGHTS:

Apparent Women Officers or Directors: 5
Hot Spot for Advancement for Women/Minorities: Y

LOCATIONS: ("Y" = Yes)

West:	Southwest:	Midwest:	Southeast:	Northeast:	International:
Y				Y	Y

Note: Financial information, benefits and other data can change quickly and may vary from those stated here.

MILLIPORE CORP

www.millipore.com

Industry Group Code: 334500 Ranks within this company's industry group: Sales: 3   Profits: 3

Drugs:	Other:	Clinical:		Computers:		Other:	
Discovery:	AgriBio:	Trials/Services:		Hardware:		Specialty Services:	Y
Licensing:	Genomics/Proteomics:	Laboratories:	Y	Software:		Consulting:	
Manufacturing:	Tissue Replacement:	Equipment/Supplies:	Y	Arrays:		Blood Collection:	
Development:		Research/Development Svcs.:		Database Management:		Drug Delivery:	
Generics:		Diagnostics:				Drug Distribution:	

TYPES OF BUSINESS:

Biotechnology Instruments
Fluid Analysis, Identification & Purification Equipment
Chromatography Technologies

BRANDS/DIVISIONS/AFFILIATES:

Direct-Q 3
Lynx S2S
MicroSafe, B.V.
Newport Bio Systems
Serologicals Corporation

CONTACTS: *Note: Officers with more than one job title may be intentionally listed here more than once.*

Martin D. Madaus, CEO
Martin D. Madaus, Pres.
Kathleen B. Allen, CFO/VP
Bruce Bonnevier, VP-Worldwide Human Resources
Dennis W. Harris, Chief Scientific Officer/VP
Peter C. Kershaw, VP-Worldwide Mfg. Oper.
Jeffrey Rudin, General Counsel/VP/Corp. Sec.
Charles F. Wagner, VP-Strategic & Corp. Dev.
Anthony Mattachione, Corp. Controller
Dominique F. Baly, VP/Pres., Bioscience Div.
Gregory J. Sam, VP-Quality
Jean-Paul Mangeolle, VP/Pres., Bioprocess Div.
Geoffrey Helliwell, Treas.
Martin D. Madaus, Chmn.
Geoffrey F. Ide, VP-Millipore Int'l
Peter C. Kershaw, VP-Global Supply Chain

Phone: 978-715-4321	Fax: 800-645-5439
Toll-Free: 800-645-5476	
Address: 290 Concord Rd., Billerica, MA 01821 US	

GROWTH PLANS/SPECIAL FEATURES:

Millipore Corp. is a multinational bioscience company that provides technologies, tools and services for the development and production of new therapeutic drugs, vaccines and detection tools. The company's products and services are based on technologies such as filtration, chromatography, cell culture supplements, antibodies and cell lines that are offered through its two segments: the Bioscience division and the Bioprocess division. Millipore's Bioscience division offers products and services for drug discovery, gene and protein research and additional research in the fields of molecular biology, cell biology and immunodetection and general laboratory applications for the life sciences research market. The Bioscience division is organized around four specific market segments: biotools for the separation, isolation and purification of biological samples; research reagents such as antibodies, dyes and biochemical reagents; drug discovery reagent for the analysis of drug candidates; and laboratory water purification systems that remove contaminants for critical laboratory analysis. The Bioprocess division offers specialized services in process development, scale-up, production, validation and quality assurance of therapeutics for the biopharmaceutical manufacturing market. Bioprocess provides bio-products and technologies for the manufacturing of biologic drugs in mammalian cell cultures; filtration, purification and chromatography technologies to clarify, concentrate, purify and remove viruses; process monitoring tools for the sampling and testing of drugs and intermediate products and advanced manufacturing systems for use in sterile biomanufacturing environments. The firm operates 11 manufacturing sites in Massachusetts, New Hampshire, France, Ireland and Puerto Rico, and a total of 47 offices worldwide. In April 2006, Millipore acquired Newport Bio Systems, a manufacturer of bioprocess consumable products. In July 2006, Millipore acquired Serologicals Corporation for $1.5 billion. This acquisition greatly expands Millipore's product line and sales force, and creates a firm with combined annual revenues of $1.4 billion.

Millipore offers employees flexible spending accounts, tuition reimbursement, employee assistance programs and adoption assistance.

FINANCIALS: Sales and profits are in thousands of dollars—add 000 to get the full amount. 2006 Note: Financial information for 2006 was not available for all companies at press time.

2006 Sales: $1,255,371	2006 Profits: $96,984	**U.S. Stock Ticker: MIL**
2005 Sales: $991,031	2005 Profits: $80,168	**Int'l Ticker:** Int'l Exchange:
2004 Sales: $883,263	2004 Profits: $105,556	Employees: 6,100
2003 Sales: $799,622	2003 Profits: $100,796	Fiscal Year Ends: 12/31
2002 Sales: $704,300	2002 Profits: $83,700	Parent Company:

SALARIES/BENEFITS:

Pension Plan: Y	ESOP Stock Plan:	Profit Sharing:	Top Exec. Salary: $680,769	Bonus: $
Savings Plan: Y	Stock Purch. Plan:		Second Exec. Salary: $325,633	Bonus: $

OTHER THOUGHTS:

Apparent Women Officers or Directors: 1
Hot Spot for Advancement for Women/Minorities:

LOCATIONS: ("Y" = Yes)

West:	Southwest:	Midwest:	Southeast:	Northeast:	International:
Y	Y	Y	Y	Y	Y

MIRAVANT MEDICAL TECHNOLOGIES

www.miravant.com

Industry Group Code: 325412 Ranks within this company's industry group: Sales: Profits:

Drugs:		Other:		Clinical:		Computers:		Other:	
Discovery:	Y	AgriBio:		Trials/Services:		Hardware:		Specialty Services:	
Licensing:		Genomics/Proteomics:		Laboratories:		Software:		Consulting:	
Manufacturing:		Tissue Replacement:		Equipment/Supplies:	Y	Arrays:		Blood Collection:	
Development:	Y			Research/Development Svcs.:		Database Management:		Drug Delivery:	Y
Generics:				Diagnostics:				Drug Distribution:	

TYPES OF BUSINESS:

Drugs-Photodynamic Therapy
Macular Degeneration Treatments
Drugs-Cardiovascular
Medical Devices
Drug Delivery Systems-Topical

BRANDS/DIVISIONS/AFFILIATES:

PhotoPoint
SnET2
Miravant Cardiovascular, Inc.
Miravant Systems, Inc.
Miravant Pharmaceuticals, Inc.
Photrex

CONTACTS: *Note: Officers with more than one job title may be intentionally listed here more than once.*

Gary S. Kledzik, CEO
David E. Mai, Pres.
John M. Philpott, CFO
Glenn A. Wilson, VP-Pharm.
Gary S. Kledzik, Chmn.

Phone: 805-685-9880	Fax: 805-685-7682
Toll-Free:	
Address: 336 Bollay Dr., Santa Barbara, CA 93117 US	

GROWTH PLANS/SPECIAL FEATURES:

Miravant Medical Technologies is a pharmaceutical company operating through its subsidiaries: Miravant Pharmaceuticals, Inc., Miravant Systems, Inc. and Miravant Cardiovascular, Inc. The subsidiaries develop drugs and medical devices utilizing Miravant's proprietary PhotoPoint photodynamic therapy (PDT) technology. The principle behind PDT is that drugs inside the body can be activated through low power, non-thermal laser light at the target site. The drugs transform light energy into chemical energy whereby they can act upon abnormal cells and blood vessels. Miravant's trademark drug is called Photrex. Photrex began clinical trials in the U.K., Central and Eastern Europe in mid-2005. The drug is a treatment for age-related macular eye degeneration, which functions by destroying abnormal blood vessels behind the eye. Besides ophthalmology, the company's other area of focus, undertaken by the subsidiary Miravant Cardiovascular, is cardiovascular disease. The firm has drugs designed for other applications, such as plaque psoriasis (Phase II), angioplasty, vascular graft intimal hyperplasia (Phase I) and tumor cells, currently in pre-clinical and clinical studies. Miravant is developing a new drug-delivery topical gel to deliver PhotoPoint MV9411 through the skin, for the treatment of psoriasis.

FINANCIALS: Sales and profits are in thousands of dollars—add 000 to get the full amount. 2006 Note: Financial information for 2006 was not available for all companies at press time.

2006 Sales: $	2006 Profits: $	**U.S. Stock Ticker: MRVT**	
2005 Sales: $	2005 Profits: $	**Int'l Ticker:** Int'l Exchange:	
2004 Sales: $	2004 Profits: $	Employees: 56	
2003 Sales: $	2003 Profits: $-7,465	Fiscal Year Ends: 12/31	
2002 Sales: $ 499	2002 Profits: $-15,960	Parent Company:	

SALARIES/BENEFITS:

Pension Plan:	ESOP Stock Plan: Y	Profit Sharing:	Top Exec. Salary: $384,313	Bonus: $391,009
Savings Plan: Y	Stock Purch. Plan:		Second Exec. Salary: $295,625	Bonus: $314,606

OTHER THOUGHTS:

Apparent Women Officers or Directors:
Hot Spot for Advancement for Women/Minorities:

LOCATIONS: ("Y" = Yes)

West:	Southwest:	Midwest:	Southeast:	Northeast:	International:
Y					

MONSANTO CO

www.monsanto.com

Industry Group Code: 115112 Ranks within this company's industry group: Sales: 2 Profits: 1

Drugs:	Other:		Clinical:	Computers:	Other:
Discovery:	AgriBio:	Y	Trials/Services:	Hardware:	Specialty Services:
Licensing:	Genomics/Proteomics:		Laboratories:	Software:	Consulting:
Manufacturing:	Tissue Replacement:		Equipment/Supplies:	Arrays:	Blood Collection:
Development:			Research/Development Svcs.:	Database Management:	Drug Delivery:
Generics:			Diagnostics:		Drug Distribution:

TYPES OF BUSINESS:

Agricultural Biotechnology Products & Chemicals Manufacturing
Herbicides
Seeds
Genetic Products
Lawn & Garden Products

BRANDS/DIVISIONS/AFFILIATES:

Far-Go
Roundup
Harness
Degree
Lasso
Scotts Miracle-Gro Company
Fielder's Choice Direct
Dow AgroSciences, LLC

CONTACTS: Note: Officers with more than one job title may be intentionally listed here more than once.

Hugh Grant, CEO
Hugh Grant, Pres.
Terrell K. Crews, CFO/Exec. VP
Steven C. Mizell, Sr. VP-Human Resources
Robert T. Fraley, CTO/Exec. VP
Mark Leidy, Exec. VP-Mfg.
Janet M. Holloway, Chief of Staff/VP
David F. Snively, General Counsel/Sr. VP
Cheryl P. Morley, Sr. VP-Corp. Strategy
Scarlett Lee Foster, VP-Investor Rel.
Richard B. Clark, Controller/VP
Gerald A. Steiner, Exec. VP-Commercial Acceptance
Carl Casale, Exec. VP-North American Commercial
Steve Padgette, VP-Biotech.
Hugh Grant, Chmn.
Brett D. Begemann, Exec. VP-Int'l Commercial

Phone: 314-694-4296	Fax: 314-694-8394
Toll-Free:	
Address: 800 N. Lindbergh Blvd., St. Louis, MO 63167 US	

GROWTH PLANS/SPECIAL FEATURES:

Monsanto Co. is a global provider of agricultural products for farmers. The company operates in two principal business segments: Seeds and Genomics; and Agricultural Productivity. The Seeds and Genomics segment is responsible for producing seed brands and patenting genetic traits that enable seeds to resist insects, disease, drought and/or weeds. Major seed brands produced by Monsanto include Agroceres, Asgrow, DEKALB, Stoneville, Vistive, Monsoy, Holden's Foundation Seeds, American Seeds, Inc., Seminis, Royal Sluis and Petoseed. The company's genetic trait products include Roundup Ready traits in soybeans, corn, canola and cotton; Bollgard and Bollgard II traits in cotton; and YieldGard Corn Borer and YieldGard Rootworm traits in corn. The Agricultural Productivity segment produces herbicide products and animal agricultural products that improve dairy cow productivity (Posilac bovine somatotropin) and increase the amount of meat on swine (Monsanto Choice Genetics line). The firm's branded herbicides include Roundup, Harness, Degree, Machete, Maverick, Certainty, Outrider and Monitor. Monsanto market its seeds and commercial herbicides through a variety of channels and directly to farmers. The company's Posilac bovine somatotropin and swine genetics products are sold and shipped directly to farmers/producers. Residential herbicides are marketed through the Scotts Miracle-Gro Company. In August 2006, Monsanto agreed to acquire Delta & Pine Land Co. for $1.5 billion in cash. In December 2006, Monsanto's American Seeds, Inc. subsidiary acquired the Landec Corporation's direct marketing and seed sales company, Fielder's Choice Direct. Monsanto and Landec have entered into a five-year agreement under which Monsanto is the exclusive sales and marketing agent for Landec's Intellicoat seed coating.

Monsanto offers its employees health, disability and life insurance; a cash balance pension plan; a 401(k); and reimbursement accounts. Monsanto also has adoption assistance; automobile and home insurance; and an incentive reward program tied to performance of individuals, teams and lines of business.

FINANCIALS: Sales and profits are in thousands of dollars—add 000 to get the full amount. 2006 Note: Financial information for 2006 was not available for all companies at press time.

2006 Sales: $7,344,000	2006 Profits: $689,000	U.S. Stock Ticker: MON
2005 Sales: $6,294,000	2005 Profits: $255,000	Int'l Ticker: Int'l Exchange:
2004 Sales: $5,457,000	2004 Profits: $267,000	Employees: 17,500
2003 Sales: $3,373,000	2003 Profits: $-23,000	Fiscal Year Ends: 12/31
2002 Sales: $4,673,000	2002 Profits: $-1,693,000	Parent Company:

SALARIES/BENEFITS:

Pension Plan: Y	ESOP Stock Plan:	Profit Sharing:	Top Exec. Salary: $1,087,500	Bonus: $1,958,000
Savings Plan: Y	Stock Purch. Plan: Y		Second Exec. Salary: $523,673	Bonus: $650,000

OTHER THOUGHTS:

Apparent Women Officers or Directors: 5
Hot Spot for Advancement for Women/Minorities: Y

LOCATIONS: ("Y" = Yes)

West:	Southwest:	Midwest:	Southeast:	Northeast:	International:
Y	Y	Y	Y	Y	Y

Note: Financial information, benefits and other data can change quickly and may vary from those stated here.

MYLAN LABORATORIES INC

www.mylan.com

Industry Group Code: 325416 Ranks within this company's industry group: Sales: 6   Profits: 3

Drugs:		Other:		Clinical:		Computers:		Other:	
Discovery:	Y	AgriBio:		Trials/Services:		Hardware:		Specialty Services:	
Licensing:		Genomics/Proteomics:		Laboratories:		Software:		Consulting:	
Manufacturing:	Y	Tissue Replacement:		Equipment/Supplies:		Arrays:		Blood Collection:	
Development:	Y			Research/Development Svcs.:		Database Management:		Drug Delivery:	Y
Generics:	Y			Diagnostics:				Drug Distribution:	

TYPES OF BUSINESS:

Drugs-Generic
Generic Pharmaceuticals
Branded Pharmaceuticals

BRANDS/DIVISIONS/AFFILIATES:

UDL Laboratories, Inc.
Mylan Pharmaceuticals, Inc.
Mylan Technologies, Inc.
Matrix Laboratories
Docpharma
nebivolol

CONTACTS: Note: Officers with more than one job title may be intentionally listed here more than once.

Robert J. Coury, CEO
Edward J. Borkowski, CFO
Rajiv Malik, Head-Global Tech. Oper.
Stuart A. Williams, Chief Legal Officer
David F. Mulder, Sr. VP-Corp. & Bus. Dev.
Patrick Fitzgerald, VP-Public Rel.
Patrick Fitzgerald, VP-Investor Rel.
Daniel C. Rizzo, Jr., Corp. Controller/VP
Robert J. Coury, Vice Chmn.
Harry A. Korman, Pres., Mylan Pharmaceuticals/Sr. VP
Carolyn Myers, Pres., Mylan Technologies/VP
Heather Bresch, Head-North American Oper.
Milan Puskar, Chmn.

Phone: 724-514-1800	Fax: 724-514-1870
Toll-Free:	

Address: 1500 Corporate Dr., Ste. 400, Canonsburg, PA 15317 US

GROWTH PLANS/SPECIAL FEATURES:

Mylan Laboratories, Inc. develops, licenses, manufactures, distributes and markets generic, branded and branded generic pharmaceutical products. Branded products are marketed under brand names through marketing programs that are designed to generate physician and consumer loyalty. Generic products are the chemical and therapeutic equivalents of reference brand drugs. Branded generic products are the brands sold by a holder as generic, often through a licensing agreement with a generic company or through a subsidiary, at the same time other generic competition enters the market. The company reports as two segments: the Mylan segment and the Matrix segment. The Mylan segment operates through three primary business units: Mylan Pharmacuticals, Inc.; UDL Laboratories, Inc.; and Mylan Technologies, Inc. Mylan holds one of the lead market positions in new and refilled prescriptions dispensed among all pharmaceutical companies in the U.S. The company's portfolio of generic products includes over 170 products and covers a range of dosage forms including immediate- and extended-release oral tablets and capsules, as well as transdermal patches. Matrix is one of the world's largest manufacturers of API's (active pharmaceutical ingredients). In Europe the segment operates through Docpharma, a wholly-owned subsidiary. In 2006, the company announced an agreement with Forest Laboratories Holdings, Ltd., a wholly owned subsidiary of Forest Laboratories Inc., for the commercialization, development and distribution of Mylan's nebivolol compound in the U.S. and Canada. Recently, the company acquired Matrix Laboratories. In May 2007, the firm agreed to acquire the generic-drugs unit of Merck KGaA for $6.7 billion.

Mylan offers its employees healthcare coverage, life and disability insurance, an educational assistance program, wellness benefits and time-off benefits, among others.

FINANCIALS: Sales and profits are in thousands of dollars—add 000 to get the full amount. 2006 Note: Financial information for 2006 was not available for all companies at press time.

2006 Sales: $1,257,164	2006 Profits: $184,542	**U.S. Stock Ticker:** MYL
2005 Sales: $1,253,374	2005 Profits: $203,592	**Int'l Ticker:** Int'l Exchange:
2004 Sales: $1,374,617	2004 Profits: $334,609	Employees: 2,900
2003 Sales: $1,269,200	2003 Profits: $272,400	Fiscal Year Ends: 3/31
2002 Sales: $1,104,100	2002 Profits: $260,300	Parent Company:

SALARIES/BENEFITS:

Pension Plan:	ESOP Stock Plan:	Profit Sharing:	Top Exec. Salary: $1,500,000	Bonus: $2,737,500
Savings Plan: Y	Stock Purch. Plan:		Second Exec. Salary: $523,591	Bonus: $500,000

OTHER THOUGHTS:

Apparent Women Officers or Directors: 2
Hot Spot for Advancement for Women/Minorities: Y

LOCATIONS: ("Y" = Yes)

West:	Southwest:	Midwest:	Southeast:	Northeast:	International:
	Y	Y		Y	Y

MYRIAD GENETICS INC

www.myriad.com

Industry Group Code: 325412 Ranks within this company's industry group: Sales: 63   Profits: 141

Drugs:		Other:		Clinical:		Computers:		Other:	
Discovery:	Y	AgriBio:		Trials/Services:		Hardware:		Specialty Services:	
Licensing:		Genomics/Proteomics:	Y	Laboratories:		Software:		Consulting:	
Manufacturing:		Tissue Replacement:		Equipment/Supplies:		Arrays:		Blood Collection:	
Development:	Y			Research/Development Svcs.:		Database Management:		Drug Delivery:	
Generics:				Diagnostics:	Y			Drug Distribution:	

TYPES OF BUSINESS:

Pharmaceuticals Discovery & Development
Cancer Treatments
Alzheimer's Treatment
Cancer Diagnostics

BRANDS/DIVISIONS/AFFILIATES:

Myriad Genetic Laboratories, Inc.
Myriad Pharmaceuticals, Inc.
Flurizan
BRACAnalysis
COLARIS
MELARIS
COLARIS AP
Azixa

CONTACTS: Note: Officers with more than one job title may be intentionally listed here more than once.

Peter D. Meldrum, CEO
Peter D. Meldrum, Pres.
Jay M. Moyes, CFO
Jerry Lanchbury, Exec. VP-Research
Richard Marsh, General Counsel/Exec. VP/Corp. Sec.
S. George Simon, VP-Bus. Dev.
William A. Hockett, III, Exec. VP-Corp. Comm.
James S. Evans, VP-Finance
Gregory C. Critchfield, Pres., Myriad Genetic Laboratories, Inc.
Adrian N. Hobden, Pres., Myriad Pharmaceuticals, Inc.
Mark H. Skolnick, Chief Scientific Officer
Mark C. Capone, COO-Myriad Genetic Laboratories, Inc.
John T. Henderson, Chmn.

Phone: 801-584-3600	Fax: 801-584-3640
Toll-Free:	
Address: 320 Wakara Way, Salt Lake City, UT 84108 US	

GROWTH PLANS/SPECIAL FEATURES:

Myriad Genetics, Inc. is a biopharmaceutical company that develops and markets novel therapeutic and molecular diagnostic products. The company develops a number of proprietary technologies that are aimed at understanding the genetic basis of human disease in order to effectively treat diseases. Myriad's molecular diagnostic business is focused on both predictive medicine and personalized medicine. Predictive medicine analyzes genes and mutations to assess an individual's risk of developing diseases in the future while personalized medicine analyzes genes and mutations to assess a patient's risk in disease progression, disease recurrence and drug response. The company currently owns four commercial predictive medicine products: BRACAnalysis for breast and ovarian cancer; COLARIS for colon and uterine cancer; COLARIS AP for polyp-forming syndromes of colon cancer; and MELARIS for melanoma. In addition, Myriad researchers have also made progressive discoveries in the fields of cancer, Alzheimer's disease and infectious diseases. Myriad's major drug development programs include Flurizan for the treatment of Alzheimer's disease; Azixa for the treatment of solid cancer tumors and brain metastases; MPC-2130 for the treatment of hematologic cancers; MPC-0920 for the treatment of thrombosis; and MPI-49839 for the treatment of AIDS. Myriad enters into a variety of strategic partnerships to discover genes and proteins associated with human disease to elucidate protein networks; to screen small molecule libraries against drug target assays; to develop novel drug candidates and to sequence the genome of entire organisms. In April 2006, Myriad entered into a collaboration with Abbot to identify novel therapeutic targets. In June 2007, Myriad also launched TheraGuide 5-FU, a new form of treatment that helps predict whether cancer patients are likely to suffer serious toxic reaction to the drug, 5-Fluorouracil.

Myriad offers employees tax free reimbursement accounts, incentive stock options and medical and dental insurance programs.

FINANCIALS: Sales and profits are in thousands of dollars—add 000 to get the full amount. 2006 Note: Financial information for 2006 was not available for all companies at press time.

2006 Sales: $114,279	2006 Profits: $-38,189	U.S. Stock Ticker: MYGN
2005 Sales: $82,406	2005 Profits: $-39,978	Int'l Ticker: Int'l Exchange:
2004 Sales: $56,648	2004 Profits: $-40,620	Employees: 722
2003 Sales: $64,321	2003 Profits: $-24,825	Fiscal Year Ends: 6/30
2002 Sales: $53,800	2002 Profits: $-14,000	Parent Company:

SALARIES/BENEFITS:

Pension Plan:	ESOP Stock Plan: Y	Profit Sharing:	Top Exec. Salary: $615,475	Bonus: $338,960
Savings Plan: Y	Stock Purch. Plan:		Second Exec. Salary: $424,475	Bonus: $200,710

OTHER THOUGHTS:

Apparent Women Officers or Directors:
Hot Spot for Advancement for Women/Minorities:

LOCATIONS: ("Y" = Yes)

West:	Southwest:	Midwest:	Southeast:	Northeast:	International:
Y					

Note: Financial information, benefits and other data can change quickly and may vary from those stated here.

NABI BIOPHARMACEUTICALS

www.nabi.com

Industry Group Code: 325412 Ranks within this company's industry group: Sales: 67   Profits: 163

Drugs:		Other:		Clinical:		Computers:		Other:	
Discovery:	Y	AgriBio:		Trials/Services:		Hardware:		Specialty Services:	
Licensing:		Genomics/Proteomics:		Laboratories:		Software:		Consulting:	
Manufacturing:	Y	Tissue Replacement:		Equipment/Supplies:	Y	Arrays:		Blood Collection:	
Development:	Y			Research/Development Svcs.:		Database Management:		Drug Delivery:	
Generics:				Diagnostics:				Drug Distribution:	

TYPES OF BUSINESS:

Drugs-Infectious Disease & Autoimmune Disorders
Vaccines
Addiction Treatments
Antibody Plasma Products

BRANDS/DIVISIONS/AFFILIATES:

Nabi-HB
NicVAX
Civacir
Aloprim
ATG-Fresenius S
PhosLo
Nabi Biologics

CONTACTS: *Note: Officers with more than one job title may be intentionally listed here more than once.*

Leslie Hudson, Interim CEO
Leslie Hudson, Interim Pres.
Jordan I. Siegel, CFO
Raafat E. F. Fahim, Sr. VP-R&D
Jordan I. Siegel, Sr. VP-Admin.
Stephan E. Lawton, Sr. General Counsel/Sr. VP
Raafat E. F. Fahim, Sr. VP-Technical & Production Oper.
Keri P. Mattox, Corp. Comm.
Keri P. Mattox, Investor Rel.
Jordan I. Siegel, Sr.VP-Finance/Treas.
Paul Kessler, Sr. VP-Clinical, Medical & Regulatory Affairs
Constantine Alexander, Sec.
Raafat E. F. Fahim, COO/General Mgr.-Nabi Biologics
Geoffrey F. Cox, Chmn.

Phone: 561-989-5800	Fax: 561-989-5801

Toll-Free: 800-635-1766
Address: 5800 Park of Commerce Blvd. NW, Boca Raton, FL 33487 US

GROWTH PLANS/SPECIAL FEATURES:

Nabi Biopharmaceuticals is a research and development biopharmaceutical company that focuses on vaccines and therapies for the treatment and prevention of infectious, autoimmune and nicotine addiction diseases. The company currently markets two biopharmaceutical products: Nabi-HB (Hepatitis B Immune Globulin (Human)) and Aloprim (allopurinol sodium).   Nabi-HB is a human polyclonal antibody product that prevents hepatitis B infection after accidental exposure to the hepatitis B virus. Aloprim treats elevated uric acid levels in the blood caused by leukemia, lymphoma and solid-tumor malignancies. In addition, Nabi markets a number of antibody plasma products, which are used as key raw materials by global healthcare companies for the manufacturing of pharmaceutical and diagnostic products. The firm also has several products in clinical development: Civacir, a polyclonal antibody for the prevention of re-infection of hepatitis C in liver transplant patients; ATG-Fresenius S, an immunosuppressive polyclonal antibody that prevents solid organ rejection after transplants; NicVAX, a vaccine to treat nicotine addition; and various other vaccines and antibody-based therapies.   In mid-2007, Nabi created the Nabi Biologics strategic business unit (SBU), which will advance the company's protein immunological products and development pipeline.   The company also agreed to sell its Aloprim for Injection to an Ireland-based company, Bioniche Teoranta and also sold its PhosLo product to a U.S. subsidiary of Fresenius Medical Care.

Nabi offers its employees a discount prescription program; flexible spending accounts; access to one of three full-service credit unions; an employee assistance program; an employee referral program; and tuition assistance.

FINANCIALS: Sales and profits are in thousands of dollars—add 000 to get the full amount. 2006 Note: Financial information for 2006 was not available for all companies at press time.

2006 Sales: $89,868	2006 Profits: $-58,703	**U.S. Stock Ticker: NABI**
2005 Sales: $94,149	2005 Profits: $-128,449	**Int'l Ticker:** Int'l Exchange:
2004 Sales: $142,183	2004 Profits: $-50,390	Employees: 653
2003 Sales: $176,570	2003 Profits: $-6,066	Fiscal Year Ends: 12/31
2002 Sales: $196,000	2002 Profits: $2,100	Parent Company:

SALARIES/BENEFITS:

Pension Plan: Y	ESOP Stock Plan:	Profit Sharing:	Top Exec. Salary: $487,500	Bonus: $225,000
Savings Plan: Y	Stock Purch. Plan: Y		Second Exec. Salary: $310,546	Bonus: $145,000

OTHER THOUGHTS:

Apparent Women Officers or Directors:
Hot Spot for Advancement for Women/Minorities:

LOCATIONS: ("Y" = Yes)

West:	Southwest:	Midwest:	Southeast:	Northeast:	International:
			Y	Y	

Note: Financial information, benefits and other data can change quickly and may vary from those stated here.

NANOBIO CORPORATION

www.nanobio.com

Industry Group Code: 325412A Ranks within this company's industry group: Sales: Profits:

Drugs:		Other:	Clinical:	Computers:	Other:
Discovery:	Y	AgriBio:	Trials/Services:	Hardware:	Specialty Services:
Licensing:		Genomics/Proteomics:	Laboratories:	Software:	Consulting:
Manufacturing:		Tissue Replacement:	Equipment/Supplies:	Arrays:	Blood Collection:
Development:	Y		Research/Development Svcs.:	Database Management:	Drug Delivery:
Generics:			Diagnostics:		Drug Distribution:

TYPES OF BUSINESS:

Drug Delivery Systems
Nanoemulsion Technology
Cold Sore Treatments
Drugs-Antimicrobial

BRANDS/DIVISIONS/AFFILIATES:

NanoStat
NanoHPX
NanoTxt

CONTACTS: Note: Officers with more than one job title may be intentionally listed here more than once.

Michael J. Nestor, CEO
David Peralta, COO
David Peralta, CFO
James Baker, Chief Scientific Officer
John Coffey, VP-Bus. Dev.
Mary R. Flack, VP-Clinical & Regulatory Affairs
James Baker, Chmn.

Phone: 734-302-4000	Fax: 734-302-9150
Toll-Free:	
Address: 2311 Green Rd., Ste. A, Ann Arbor, MI 48105 US	

GROWTH PLANS/SPECIAL FEATURES:

NanoBio Corporation is a biopharmaceutical company that develops and markets anti-infection products and mucosal vaccines. The company's main technology platform is NanoStat, which implements oil-in-water emulsions that are roughly 150-400 nanometers in size. Nanoemulsion particles can rapidly penetrate the skin through pores and hair shafts to the site of an infection, where it physically kills lipid-containing organisms by penetrating the outer membrane of pathogenic organisms. NanoStat emulsions have the added benefit of being selectively lethal to targeted microbes without affecting any surrounding skin and mucous membranes. NanoBio's current commercial product pipeline consists of five anti-infective pharmaceuticals and two mucosal vaccines. The firm's two leading anti-infective products are NB-001 (Phase III), a treatment for cold sores, and NB-002 (Phase II) for nail fungus, which are both in the clinical stages of development. NanoStat mucosal vaccines are designed to treat influenza, H5N1, hepatitis B, pneumonia, tuberculosis, small pox anthrax and other various viral and bacterial diseases. NanoBio also holds exclusive intellectual property rights for virucidal, fungicidal, sporicidal and bactericidal applications in a wide spectrum of products that include personal care products, medical products and anti-bioterrorism applications. NanoBio also maintains a research team to discover new applications and methods relating to Varicella zoster (a virus that causes shingles), vaginitis, ocular infections, influenza prophylaxis and drug delivery systems. In August 2006, the company received a $30 million equity investment from Perseus, LLC, which is aiding the firm in expanding its clinical product pipeline.

NanoBio offers employees health benefits, flexible spending accounts, an investment 401(k) plan and annual bonuses that are based on an employee performance ratings and the company's overall achievements.

FINANCIALS: Sales and profits are in thousands of dollars—add 000 to get the full amount. 2006 Note: Financial information for 2006 was not available for all companies at press time.

2006 Sales: $	2006 Profits: $	U.S. Stock Ticker: Private	
2005 Sales: $	2005 Profits: $	Int'l Ticker: Int'l Exchange:	
2004 Sales: $	2004 Profits: $	Employees:	
2003 Sales: $	2003 Profits: $	Fiscal Year Ends: 12/31	
2002 Sales: $	2002 Profits: $	Parent Company:	

SALARIES/BENEFITS:

Pension Plan:	ESOP Stock Plan:	Profit Sharing: Y	Top Exec. Salary: $	Bonus: $
Savings Plan: Y	Stock Purch. Plan:		Second Exec. Salary: $	Bonus: $

OTHER THOUGHTS:

Apparent Women Officers or Directors:
Hot Spot for Advancement for Women/Minorities:

LOCATIONS: ("Y" = Yes)

West:	Southwest:	Midwest:	Southeast:	Northeast:	International:
		Y			

NANOGEN INC

www.nanogen.com

Industry Group Code: 325413 Ranks within this company's industry group: Sales: 21 Profits: 25

Drugs:	Other:		Clinical:		Computers:	Other:	
Discovery:	AgriBio:		Trials/Services:		Hardware:	Specialty Services:	
Licensing:	Genomics/Proteomics:	Y	Laboratories:		Software:	Consulting:	
Manufacturing:	Tissue Replacement:		Equipment/Supplies:		Arrays:	Blood Collection:	
Development:			Research/Development Svcs.:		Database Management:	Drug Delivery:	
Generics:			Diagnostics:	Y		Drug Distribution:	

TYPES OF BUSINESS:

Molecular Diagnostic Tests
Genomics Research
PCR Molecular Reagents
Chips-DNA Analysis
Point-of-Care Tests
Microarrays

BRANDS/DIVISIONS/AFFILIATES:

NanoChip 400
NanoChip Molecular Biology Workstation Cartridge
NT-proBNP
NanoChip Electronic Microarray

CONTACTS: Note: Officers with more than one job title may be intentionally listed here more than once.

Howard C. Birndorf, CEO
David Ludvigson, COO
David Ludvigson, Pres.
Robert Saltmarsh, CFO
Robert Proulx, VP-Mktg.
Graham Lidgard, Sr. VP-R&D
William L. Respess, General Counsel/Sr. VP/Corp. Sec.
David Boudreau, VP-Oper.
Merl F. Hoekstra, VP-Bus. Dev.
Robert Bush, VP-Sales
Van N. Schramm, VP-Regulatory Affairs & Quality Assurance
Rod Wilson, Pres./COO-Point-of-Care Diagnostics Div.
Carl T. Foster, VP-Alliance Mgmt.
Howard C. Birndorf, Chmn.

Phone: 858-410-4600	Fax: 858-410-4952

Toll-Free: 877-626-6436
Address: 10398 Pacific Center Ct., San Diego, CA 92121 US

GROWTH PLANS/SPECIAL FEATURES:

Nanogen, Inc. provides advanced diagnostic technologies and products that investigate, predict and diagnose diseases for research labs, hospitals, urgent care centers and universities. The company's product line includes PCR reagents, its patented Nanochip Electronic Microarray platform and point-of-care diagnostic tests. Nanogen's PCR reagents utilize a minor groove binder (MGB) technology that combines to DNA probes to stabilize hybridization reactions. MGB technology allows an increase in melting temperature, which consequently permits the use of shorter probes with a greater discrimination for DNA probe matching and detection of genetic variables associated with gene variations, mutations and expressions associated with genetic disorders and infectious diseases. The NanoChip Electronic Microarray allows rapid movement of negatively charged DNA and RNA molecules, which allows customers to easily run or customize their own assays. Additional instrumentation in the NanoChip product lines includes the Molecular Biology Workstation Cartridge and the NanoChip 400 system, which are detectors that test for multiple gene markers or mutations at a single site. Nanogen's immunoassay products are focused on cardiac-related health conditions such as congestive heart failure. In late 2006, Nanogen was granted patents for applications of its core technology in nanofabrication and nanomanufacturing, which will allow further miniaturization of the company's microarray testing platforms and the development of integrated electronics and photonics, photovoltaics, fuel cells and batteries.

Nanogen provides employees with educational assistance, a 401(k) savings plan, flexible spending accounts and an employee assistance program. In addition, the company hosts bi-weekly socials with free food and beverages.

FINANCIALS: Sales and profits are in thousands of dollars—add 000 to get the full amount. 2006 Note: Financial information for 2006 was not available for all companies at press time.

2006 Sales: $26,852	2006 Profits: $-49,070	**U.S. Stock Ticker: NGEN**
2005 Sales: $12,544	2005 Profits: $-96,494	**Int'l Ticker:** Int'l Exchange:
2004 Sales: $5,374	2004 Profits: $-38,907	Employees: 288
2003 Sales: $6,713	2003 Profits: $-30,596	Fiscal Year Ends: 12/31
2002 Sales: $17,200	2002 Profits: $-22,200	Parent Company:

SALARIES/BENEFITS:

Pension Plan:	ESOP Stock Plan: Y	Profit Sharing:	Top Exec. Salary: $505,000	Bonus: $
Savings Plan: Y	Stock Purch. Plan: Y		Second Exec. Salary: $350,000	Bonus: $

OTHER THOUGHTS:

Apparent Women Officers or Directors:
Hot Spot for Advancement for Women/Minorities:

LOCATIONS: ("Y" = Yes)

West:	Southwest:	Midwest:	Southeast:	Northeast:	International:
Y					Y

Note: Financial information, benefits and other data can change quickly and may vary from those stated here.

NASTECH PHARMACEUTICAL CO INC   www.nastech.com

Industry Group Code: 325412A Ranks within this company's industry group: Sales: 7   Profits: 14

Drugs:		Other:		Clinical:	Computers:		Other:	
Discovery:	Y	AgriBio:		Trials/Services:	Hardware:		Specialty Services:	
Licensing:	Y	Genomics/Proteomics:		Laboratories:	Software:		Consulting:	
Manufacturing:	Y	Tissue Replacement:		Equipment/Supplies:	Arrays:		Blood Collection:	
Development:	Y			Research/Development Svcs.:	Database Management:		Drug Delivery:	Y
Generics:				Diagnostics:			Drug Distribution:	

TYPES OF BUSINESS:

Drug Delivery Systems
Drugs-Nasally Administered
Tight Junction Biology
RNA Interference Technology

BRANDS/DIVISIONS/AFFILIATES:

Calcitonin Nasal Spray

CONTACTS: *Note: Officers with more than one job title may be intentionally listed here more than once.*

Steven C. Quay, CEO
Steven C. Quay, Pres.
Philip C. Ranker, CFO
Timothy M. Duffy, Exec. VP-Mktg.
Paul H. Johnson, Chief Scientific Officer/Sr. VP-R&D
Philip C. Ranker, Corp. Sec.
David E. Wormuth, Sr. VP-Oper.
Timothy M. Duffy, Exec. VP-Bus. Dev.
Ed Bell, Sr. Mgr.-Investor Rel.
Gordon C. Brandt, Exec. VP-Clinical Research & Medical Affairs
Steven C. Quay, Chmn.

Phone: 425-908-3600	Fax: 425-908-3650
Toll-Free:	
Address: 3450 Monte Villa Pkwy., Bothell, WA 98021 US	

GROWTH PLANS/SPECIAL FEATURES:

Nastech Pharmaceutical Company, Inc. is a leader in nasal, tight junction drug delivery technology. Historically, its core technical competency has involved the research, development and manufacture of nasally administered prescription pharmaceuticals. The company investigates the commercial weaknesses of pharmaceutical products currently available in oral, injectable or other dosage forms and determines the advantages of alternative drug delivery systems for those same drugs. For example, nasal delivery can provide faster absorption into the blood stream than oral products, resulting in faster onset of action. Other possible advantages include lower doses, fewer side effects, greater safety and efficacy, greater convenience and lower overall costs. Nastech is also applying its drug delivery technology to a new class of therapeutics based on RNAi, which may have uses in influenza and other respiratory diseases. Tight junction technology targets tissue barriers called tight junctions in an effort to increase the size of drug molecules that can be delivered through the junctions, improve the amount of the drug that reaches the bloodstream and facilitate the direct delivery of the drug to the central nervous system. Also under development is a Calcitonin nasal spray for treating osteoporosis. Par Pharmaceutical, Inc. has licensed the drug and in February 2006 Nastech's manufacturing facility was approved for its commercial production. Nastech has several other products in the development process including an intranasal parathyroid hormone for osteoporosis, which is in Phase II of clinical development in partnership with Procter & Gamble Pharmaceuticals, and an intranasal Peptide YY obesity treatment that is currently in Phase II trials. The company is also developing treatments for diabetes, influenza, inflammatory diseases and anemia. In addition, Nastech's RNA interference program has two drugs in preclinical development.

Nastech offers employees flexible spending accounts for health and dependant care; and financial support for conferences, publications and seminars.

FINANCIALS: Sales and profits are in thousands of dollars—add 000 to get the full amount. 2006 Note: Financial information for 2006 was not available for all companies at press time.

2006 Sales: $28,490	2006 Profits: $-26,877	**U.S. Stock Ticker: NSTK**
2005 Sales: $7,449	2005 Profits: $-32,163	**Int'l Ticker:** Int'l Exchange:
2004 Sales: $1,847	2004 Profits: $-28,609	Employees: 197
2003 Sales: $19,440	2003 Profits: $-2,141	Fiscal Year Ends: 12/31
2002 Sales: $8,900	2002 Profits: $-13,500	Parent Company:

SALARIES/BENEFITS:

Pension Plan:	ESOP Stock Plan:	Profit Sharing:	Top Exec. Salary: $500,000	Bonus: $214,500
Savings Plan: Y	Stock Purch. Plan:		Second Exec. Salary: $275,000	Bonus: $89,078

OTHER THOUGHTS:

Apparent Women Officers or Directors:
Hot Spot for Advancement for Women/Minorities:

LOCATIONS: ("Y" = Yes)

West:	Southwest:	Midwest:	Southeast:	Northeast:	International:
Y				Y	

NBTY INC
www.nbty.com

Industry Group Code: 325411 Ranks within this company's industry group: Sales: 1   Profits: 1

Drugs:		Other:	Clinical:	Computers:	Other:	
Discovery:		AgriBio:	Trials/Services:	Hardware:	Specialty Services:	Y
Licensing:		Genomics/Proteomics:	Laboratories:	Software:	Consulting:	
Manufacturing:	Y	Tissue Replacement:	Equipment/Supplies:	Arrays:	Blood Collection:	
Development:			Research/Development Svcs.:	Database Management:	Drug Delivery:	
Generics:			Diagnostics:		Drug Distribution:	Y

TYPES OF BUSINESS:

Nutritional Supplements
Vitamins
Diet Aids
Sports Nutrition Products

BRANDS/DIVISIONS/AFFILIATES:

Nature's Bounty
Solgar
Ester-C
Sundown
Puritan's Pride
Le Naturiste
Nature's Way
Holland & Barrett

CONTACTS: *Note: Officers with more than one job title may be intentionally listed here more than once.*

Scott Rudolph, CEO
Harvey Kamil, Pres.
Harvey Kamil, CFO
James P. Flaherty, VP-Mktg. & Advertising
William J. Shanahan, VP-Info. Systems
Michael Slade, Sr. VP-Strategic Planning/Sec.
Scott Rudolph, Chmn.

Phone: 631-567-9500	Fax: 631-567-7148
Toll-Free:	
Address: 90 Orville Dr., Bohemia, NY 11716 US	

GROWTH PLANS/SPECIAL FEATURES:

NBTY, Inc. is a vertically integrated manufacturer, marketer and retailer of a broad line of nutritional supplements in the U.S. and throughout the world. The company offers over 22,000 products, including vitamins, minerals, herbs, sports nutrition products, diet aids and other nutritional supplements. The firm markets its products through four channels of distribution: wholesale distribution to mass merchandisers, drug store chains, supermarkets, independent pharmacies and health food stores under brand names such as Nature's Bounty, Solgar, Ester-C and Sundown; Direct Response/ Puritan's Pride, a U.S. nutritional supplement e-commerce/direct response business segment, under the Puritan's Pride brand in catalogs and through the Internet; North American retail operations including 476 Vitamin World and Nutrition warehouse retail stores, which operate throughout the U.S. in 44 states, Guam, Puerto Rico and the Virgin Islands, and 96 Le Naturiste retail stores in Canada; and European retail operations consisting of 549 Holland & Barrett, GNC (UK) and Nature's Way retail stores operating throughout the UK and Ireland and 68 De Tuinen retail stores operating in the Netherlands. In addition, NBTY sells its products on www.puritan.com and www.vitamins.com. The company has roughly 2,840 manufacturing, shipping and packaging associates throughout the U.S. and 28 such associated in British Columbia. In February 2006, the firm completed a $21 million expansion of its softgel facility in New York, whose production, as a result, increased by 53%. In October 2006, NBTY acquired Zila Nutraceutical, Inc. for roughly $37,500 and renamed it The Ester C Co.

FINANCIALS: Sales and profits are in thousands of dollars—add 000 to get the full amount. 2006 Note: Financial information for 2006 was not available for all companies at press time.

2006 Sales: $1,880,222	2006 Profits: $111,785	**U.S. Stock Ticker: NTY**
2005 Sales: $1,737,187	2005 Profits: $78,137	**Int'l Ticker:** Int'l Exchange:
2004 Sales: $1,652,031	2004 Profits: $111,849	Employees: 10,900
2003 Sales: $1,192,548	2003 Profits: $81,585	Fiscal Year Ends: 9/30
2002 Sales: $964,100	2002 Profits: $95,800	Parent Company:

SALARIES/BENEFITS:

Pension Plan:	ESOP Stock Plan:	Profit Sharing:	Top Exec. Salary: $828,976	Bonus: $500,000
Savings Plan:	Stock Purch. Plan:		Second Exec. Salary: $465,548	Bonus: $350,000

OTHER THOUGHTS:

	LOCATIONS: ("Y" = Yes)					
Apparent Women Officers or Directors:	West:	Southwest:	Midwest:	Southeast:	Northeast:	International:
Hot Spot for Advancement for Women/Minorities:	Y	Y	Y	Y	Y	Y

Note: Financial information, benefits and other data can change quickly and may vary from those stated here.

NEKTAR THERAPEUTICS   www.nektar.com

Industry Group Code: 325412A Ranks within this company's industry group: Sales: 3   Profits: 22

Drugs:		Other:		Clinical:		Computers:		Other:	
Discovery:		AgriBio:		Trials/Services:		Hardware:		Specialty Services:	
Licensing:	Y	Genomics/Proteomics:		Laboratories:		Software:		Consulting:	
Manufacturing:		Tissue Replacement:		Equipment/Supplies:		Arrays:		Blood Collection:	
Development:	Y			Research/Development Svcs.:	Y	Database Management:		Drug Delivery:	Y
Generics:				Diagnostics:				Drug Distribution:	

TYPES OF BUSINESS:

Drug Delivery Systems
PEG-Based Delivery Systems
Molecular & Particle Engineering
Equipment-Inhalers

BRANDS/DIVISIONS/AFFILIATES:

Nektar PEGylation Technology
Nektar Pulmonary Technology
Macugen
PEGASYS
Exubera
Neulasta
Somavert
PEG-INTRON

CONTACTS: *Note: Officers with more than one job title may be intentionally listed here more than once.*

Howard W. Robin, CEO
Hoyoung Huh, COO/Head-Pegylation Bus. Unit
Howard W. Robin, Pres.
Louis Drapeau, CFO
Dorian Rinella, VP-Human Resources & Facilities
David Johnston, Sr. VP-R&D
Gil M. Labrucherie, General Counsel/Sec.
Christopher J. Searcy, Sr. VP-Corp. Dev.
Tim Warner, Sr. VP-Investor Rel./Corp. Affairs
Louis Drapeau, Sr. VP-Finance
Nevan Elam, Sr. VP/Head-Pulmonary Bus. Unit
John S. Patton, Chief Scientific Officer
David Tolley, Sr. VP/General Mgr.-Nektar Alabama
Robert B. Chess, Chmn.

Phone: 650-631-3100	**Fax:** 650-631-3150
Toll-Free:	
Address: 150 Industrial Rd., San Carlos, CA 94070 US	

GROWTH PLANS/SPECIAL FEATURES:

Nektar Therapeutics is a biopharmaceutical company focused on developing technology for unmet medical needs. The company focuses on applying its advanced drug delivery technologies to improve therapeutic molecules' efficacy, safety and convenience and to enable the development of new molecules. The company creates differentiated products by applying its platform technologies to established or novel medicine. The firm creates potential breakthrough products by developing products in collaboration with pharmaceutical and biotechnology companies that seek to improve and differentiate their products and by applying its technologies to already approved drugs to create and develop its own differentiated products. The company's leading technology platforms are Pulmonary Technology and PEGylation Technology. Pulmonary Technology makes drugs inhaleable to deliver them to and through the lungs for both systemic and local lung applications. PEGylation Technology is a chemical process designed to enhance the performance of most drug classes with the potential to improve solubility and stability; increase drug half-life; reduce immune responses to an active drug; and improve the efficacy and/or safety of a molecule in certain instances. These technologies are the platforms that form the basis of all of the company's products that have received FDA approval. Leading products include Macugen treatment for macular degeneration; PEGASYS for chronic Hepatitis C; Somavert human growth hormone receptor antagonist; Definity ultrasound contrast agent; Exubera inhaleable insulin; PEG-INTRON for treating Hepatitis C; and Neulasta for the treatment of neutropenia, a condition where the body produces too few white blood cells. Most of these have been developed in collaboration with other pharmaceutical companies, including Amgen, Eyetech, Pfizer, Roche and Schering-Plough. In September 2006, Nektar Therapeutics together with Zelos Therapeutics announced the initiation of a phase one clinical trial of an inhaled powder formulation of Zelos' parathyroid hormone.

FINANCIALS: Sales and profits are in thousands of dollars—add 000 to get the full amount. 2006 Note: Financial information for 2006 was not available for all companies at press time.

2006 Sales: $217,718	2006 Profits: $-154,761	**U.S. Stock Ticker: NKTR**
2005 Sales: $126,279	2005 Profits: $-185,111	**Int'l Ticker:** Int'l Exchange:
2004 Sales: $114,270	2004 Profits: $-101,886	Employees: 793
2003 Sales: $106,257	2003 Profits: $-65,890	Fiscal Year Ends: 12/31
2002 Sales: $94,800	2002 Profits: $-107,500	Parent Company:

SALARIES/BENEFITS:

Pension Plan:	ESOP Stock Plan:	Profit Sharing:	Top Exec. Salary: $418,459	Bonus: $262,718
Savings Plan: Y	Stock Purch. Plan:		Second Exec. Salary: $336,589	Bonus: $119,968

OTHER THOUGHTS:

Apparent Women Officers or Directors: 2
Hot Spot for Advancement for Women/Minorities: Y

LOCATIONS: ("Y" = Yes)

West:	Southwest:	Midwest:	Southeast:	Northeast:	International:
Y			Y		Y

Note: Financial information, benefits and other data can change quickly and may vary from those stated here.

NEOGEN CORPORATION

www.neogen.com

Industry Group Code: 325413 Ranks within this company's industry group: Sales: 15 Profits: 9

Drugs:		Other:		Clinical:		Computers:		Other:	
Discovery:		AgriBio:	Y	Trials/Services:		Hardware:		Specialty Services:	
Licensing:		Genomics/Proteomics:		Laboratories:		Software:		Consulting:	
Manufacturing:	Y	Tissue Replacement:		Equipment/Supplies:	Y	Arrays:		Blood Collection:	
Development:				Research/Development Svcs.:		Database Management:		Drug Delivery:	
Generics:				Diagnostics:	Y			Drug Distribution:	

TYPES OF BUSINESS:

Sanitary & Livestock Diagnostic Products
Food Safety Test Kits
Animal Health Test Kits
Pharmacology Test Kits
Agricultural Test Kits
Veterinary Instruments
Veterinary Pharmaceuticals

BRANDS/DIVISIONS/AFFILIATES:

Agri-Scan
Acumedia
Dr. Frank's
Triple Heat
Triple Cast
Triple Block
Triple Crown
UriCon

CONTACTS: Note: Officers with more than one job title may be intentionally listed here more than once.

James L. Herbert, CEO
Lon M. Bohannon, COO
Lon M. Bohannon, Pres.
Richard R. Current, CFO/VP
Mark A. Mozola, VP-R&D
Kenneth V. Kodilla, VP-Mfg.
Anthony E. Maltese, VP-Corp. Dev.
Edward L. Bradley, VP-Food Safety
Terri A. Morrical, VP-Animal Safety
Paul S. Satoh, VP-Basic & Exploratory Research
Joseph Madden, VP-Scientific Affairs
James L. Herbert, Chmn.

Phone: 517-372-9200	Fax: 517-372-2006

Toll-Free: 800-234-5333
Address: 620 Lesher Pl., Lansing, MI 48912 US

GROWTH PLANS/SPECIAL FEATURES:

Neogen Corporation and its subsidiaries develop, manufacture and market a diverse line of products for food and animal safety. The company's food safety segment consists primarily of diagnostic kits and complementary products (e.g., dehydrated culture media) marketed by company sales personnel in the U.S., Canada, the U.K. and parts of Europe and by distributors elsewhere to food producers and processors to detect dangerous and/or unintended substances in human and animal food, such as food-born pathogens, natural toxins, food allergens, genetic modifications, ruminant by-products, drug residues, pesticide residues and general sanitation concerns. Neogen's animal safety segment is engaged in the development, manufacture and marketing of pharmaceuticals, rodenticides, disinfectants, vaccines, veterinary instruments, topicals and diagnostic products for the worldwide animal safety market. The majority of these consumable products are marketed through a network of national and international distributors, as well as a number of large farm supply retail chains in the U.S. and Canada. The company's USAD-licensed facility in Tampa, Florida produces immunostimulant products for horses and dogs and its unique equine botulism vaccine. Trademarks include Neogen flask; AccuScan; AccuPoint; Acumedia; Agri-Scan; Agri-Screen; Alert; BetaStar; Centrus; D3 Needles; CD&R; Dr. Frank's; ElectroJac; ELISA Tehnologies; Fura-Zone; Gold Nugget; Paddock&Pasture; Triple Heat; Triple Crown; Triple Cast; Triple Block; TopHoof; UriKare; UriCon; and Vita-15, among many others. The firm also provides a 24-hour support line and training programs for the kits. In 2007, the company announced its receipt of official method status from AOAC International for its proprietary 24-hour GeneQuence salmonella diagnostic test system.

FINANCIALS: Sales and profits are in thousands of dollars—add 000 to get the full amount. 2006 Note: Financial information for 2006 was not available for all companies at press time.

2006 Sales: $72,433	2006 Profits: $7,941	**U.S. Stock Ticker:** NEOG
2005 Sales: $62,756	2005 Profits: $5,916	**Int'l Ticker:** Int'l Exchange:
2004 Sales: $55,498	2004 Profits: $5,099	Employees: 393
2003 Sales: $46,488	2003 Profits: $4,787	Fiscal Year Ends: 5/31
2002 Sales: $41,100	2002 Profits: $3,900	Parent Company:

SALARIES/BENEFITS:

Pension Plan:	ESOP Stock Plan:	Profit Sharing:	Top Exec. Salary: $275,000	Bonus: $150,000
Savings Plan: Y	Stock Purch. Plan: Y		Second Exec. Salary: $175,000	Bonus: $65,000

OTHER THOUGHTS:

Apparent Women Officers or Directors: 1
Hot Spot for Advancement for Women/Minorities:

LOCATIONS: ("Y" = Yes)

West:	Southwest:	Midwest:	Southeast:	Northeast:	International:
		Y	Y		Y

NEOPHARM INC

www.neophrm.com

Industry Group Code: 325412A Ranks within this company's industry group: Sales: 20   Profits: 16

Drugs:		Other:		Clinical:	Computers:	Other:	
Discovery:	Y	AgriBio:		Trials/Services:	Hardware:	Specialty Services:	
Licensing:		Genomics/Proteomics:		Laboratories:	Software:	Consulting:	
Manufacturing:	Y	Tissue Replacement:		Equipment/Supplies:	Arrays:	Blood Collection:	
Development:	Y			Research/Development Svcs.:	Database Management:	Drug Delivery:	Y
Generics:				Diagnostics:		Drug Distribution:	

TYPES OF BUSINESS:

Drug Delivery Systems
Liposomal Drug Delivery System
Tumor-Targeting Toxins
Drugs-Cancer

BRANDS/DIVISIONS/AFFILIATES:

NeoLipid
NeoPhectin

CONTACTS: Note: Officers with more than one job title may be intentionally listed here more than once.

Laurence P. Birch, CEO
Laurence P. Birch, Pres.
Jeffrey W. Sherman, Exec. VP/Chief Medical Officer
Timothy P. Walbert, Exec. VP-Commercial Oper.

Phone: 847-887-0800	Fax: 847-887-9281
Toll-Free:	
Address: 1850 Lakeside Dr., Waukegan, IL 60085 US	

GROWTH PLANS/SPECIAL FEATURES:

NeoPharm, Inc. is a biopharmaceutical company engaged in the research, development and commercialization of drugs for the treatment of cancer. The company has built its drug portfolio based on two novel proprietary technology platforms: the NeoLipid liposomal drug delivery system and a tumor-targeting toxin platform. NeoPharm has four drug candidates in various stages of clinical development: IL13-PE38QQR is a tumor-targeting toxin and has completed Phase III trials for the treatment of brain cancer; LE-SN38 is in Phase II trials for the treatment of colorectal cancer and other solid tumors; LEP-ETU is in the planning stages of Phase III trials for the treatment of breast, lung and ovarian cancer and other solid tumors; and LE-DT is in the pre-trial stage for the treatment of breast cancer, lung cancer, prostate cancer and other solid tumors. The company's other product candidates are based on its NeoLipid drug delivery platform. NeoLipid technology combines drugs or other compounds with NeoPharm's proprietary lipids and allows for the creation of a stable liposome. This physical property is especially important during drug storage after the medication has been administered intravenously to a patient. The firm believes this technology may have applications in a variety of other areas in addition to its current product candidates. In addition, the company plans to leverage its NeoLipid technology to develop therapeutic formulations of small molecules and biologic molecules such as RNAi. NeoPharm has developed novel cationic cardiolipin analogues in order to enhance the capability of delivering nucleic acids, including antisense and siRNA. These analogues form the backbone of the technology for efficient delivery. Customized NeoPhectin formulation is based on the entrapment of nucleic acid molecules leading to the ready-to-use formulation, providing a comprehensive solution for a wide variety of drug development challenges.

FINANCIALS: Sales and profits are in thousands of dollars—add 000 to get the full amount. 2006 Note: Financial information for 2006 was not available for all companies at press time.

2006 Sales: $ 11	2006 Profits: $-33,208	**U.S. Stock Ticker: NEOL**
2005 Sales: $ 543	2005 Profits: $-38,724	**Int'l Ticker:** Int'l Exchange:
2004 Sales: $ 157	2004 Profits: $-57,609	Employees: 53
2003 Sales: $	2003 Profits: $-52,791	Fiscal Year Ends: 12/31
2002 Sales: $	2002 Profits: $-36,492	Parent Company:

SALARIES/BENEFITS:

Pension Plan:	ESOP Stock Plan:	Profit Sharing:	Top Exec. Salary: $425,000	Bonus: $
Savings Plan:	Stock Purch. Plan:		Second Exec. Salary: $275,000	Bonus: $

OTHER THOUGHTS:

Apparent Women Officers or Directors:
Hot Spot for Advancement for Women/Minorities:

LOCATIONS: ("Y" = Yes)

West:	Southwest:	Midwest:	Southeast:	Northeast:	International:
		Y			

NEOPROBE CORPORATION

www.neoprobe.com

Industry Group Code: 325413 Ranks within this company's industry group: Sales: 26   Profits: 16

Drugs:	Other:	Clinical:		Computers:		Other:
Discovery:	AgriBio:	Trials/Services:		Hardware:		Specialty Services:
Licensing:	Genomics/Proteomics:	Laboratories:		Software:		Consulting:
Manufacturing:	Tissue Replacement:	Equipment/Supplies:	Y	Arrays:		Blood Collection:
Development:		Research/Development Svcs.:		Database Management:		Drug Delivery:
Generics:		Diagnostics:	Y			Drug Distribution:

TYPES OF BUSINESS:

Supplies-Intraoperative Cancer Diagnosis
Gamma-Guided Surgical Instruments
Blood Flow Measurement Devices
Targeting Agents

BRANDS/DIVISIONS/AFFILIATES:

Cardiosonix, Ltd.
Activated Cellular Therapy
neo2000 Gamma Detection Systems
Lymphoseek
Quantix
CIRA Biosciences, Inc.
RIGScan CR
Cardinal Health

CONTACTS: *Note: Officers with more than one job title may be intentionally listed here more than once.*

David C. Bupp, CEO
David C. Bupp, Pres.
Brent L. Larson, CFO
Douglas Rash, VP-Mktg.
Anthony K. Blair, VP-Mfg. Oper.
Brent L. Larson, Sec.
Brent L. Larson, VP-Finance/Treas.
Rodger A. Brown, VP-Regulatory Affairs & Quality Assurance
Julius R. Krevans, Chmn.

Phone: 614-793-7500	Fax: 614-793-7520
Toll-Free: 800-793-0079	
Address: 425 Metro Pl. N., Ste. 300, Dublin, OH 43017 US	

GROWTH PLANS/SPECIAL FEATURES:

Neoprobe Corp. is focused on developing and commercializing innovative biomedical products that meet the critical intra-operative, diagnostic and therapeutic treatment needs of patients and physicians. Its primary area of focus is improving cancer surgery outcomes by using its market-leading gamma detection devise in combination with radiopharmaceutical agents also referred to as tracing or targeting agents. Neoprobe has recently formed another majority-owned subsidiary, Cira Biosciences, Inc., to explore the development and commercialization of an activated cellular therapy technology that has shown promising early stage patient-specific treatment potential in oncology, viral (HIV/AIDS and hepatitis) and autoimmune diseases. Neoprobe markets its neo2000 Gamma Detection Systems, detector probes and accessories worldwide for surgical procedures called Intra-operative Lymphatic Mapping (ILM) or sentinel lymph node biopsy, a minimally invasive technique for evaluating the potential spread of cancer to lymph node tissues and organs. The company is also exploring other potential applications for the use of its gamma detection devices in cancer surgery. Neoprobe, through its wholly-owned subsidiary, Cardiosonix Ltd., has developed the Quantix line of blood flow measurement products to be used for a variety of clinical needs, including: real-time monitoring, intra-operative quantification, non-invasive diagnostics and the evaluation of blood flow. Neoprobe's investigational activities are currently focused on three particular developmental product initiatives: Lymphoseek, RIGScan CR and Activated Cellular Therapy (ACT). The company believes each of these technology platforms offers exciting and promising treatment opportunities within their respective disease applications. In 2007, the company signed a term sheet with Cardinal Health for the marketing and distribution of Lymphoseek on an exclusive basis in the U.S. through Cardinal's network of over 150 nuclear pharmacies as well as its wholesale distribution operations to in-hospital nuclear pharmacies.

FINANCIALS: Sales and profits are in thousands of dollars—add 000 to get the full amount. 2006 Note: Financial information for 2006 was not available for all companies at press time.

2006 Sales: $6,051	2006 Profits: $-4,741	**U.S. Stock Ticker:** NEOP.OB	
2005 Sales: $5,919	2005 Profits: $-4,929	**Int'l Ticker:** Int'l Exchange:	
2004 Sales: $5,953	2004 Profits: $-3,541	Employees: 22	
2003 Sales: $6,509	2003 Profits: $-1,798	Fiscal Year Ends: 12/31	
2002 Sales: $4,921	2002 Profits: $2,964	Parent Company:	

SALARIES/BENEFITS:

Pension Plan:	ESOP Stock Plan:	Profit Sharing:	Top Exec. Salary: $305,000	Bonus: $20,000
Savings Plan: Y	Stock Purch. Plan:		Second Exec. Salary: $160,000	Bonus: $6,000

OTHER THOUGHTS:

Apparent Women Officers or Directors:
Hot Spot for Advancement for Women/Minorities:

LOCATIONS: ("Y" = Yes)

West:	Southwest:	Midwest:	Southeast:	Northeast:	International:
		Y			

Note: Financial information, benefits and other data can change quickly and may vary from those stated here.

NEOSE TECHNOLOGIES INC

www.neose.com

Industry Group Code: 325412 Ranks within this company's industry group: Sales: 137 Profits: 120

Drugs:	Other:	Clinical:	Computers:	Other:
Discovery:	AgriBio:	Trials/Services:	Hardware:	Specialty Services:
Licensing:	Genomics/Proteomics: Y	Laboratories:	Software:	Consulting:
Manufacturing: Y	Tissue Replacement:	Equipment/Supplies:	Arrays:	Blood Collection:
Development: Y		Research/Development Svcs.:	Database Management:	Drug Delivery:
Generics:		Diagnostics:		Drug Distribution:

TYPES OF BUSINESS:
Drugs-Therapeutic Proteins

BRANDS/DIVISIONS/AFFILIATES:
GlycoAdvance
GlycoConjugation
GlycoPEGylation
GlycoPEG-EPO
GlycoPEG-GCSF

CONTACTS: Note: Officers with more than one job title may be intentionally listed here more than once.
George J. Vergis, CEO
George J. Vergis, Pres.
A. Brian Davis, CFO/Sr. VP
David A. Zopf, Chief Scientific Officer/Exec. VP
Debra J. Poul, General Counsel/Sr. VP/Corp. Sec.
Kathryn J. Gregory, VP-Bus. Dev. & Licensing
Elliot Morales, Jr., VP-Project Management & Analytical Oper.
Shawn DeFrees, VP-Research
Valerie M. Mulligan, Sr. VP-Regulatory Affairs & Quality
Bruce A. Wallin, Sr. VP-Clinical Dev./Chief Medical officer
L. Patrick Gage, Chmn.

Phone: 215-315-9000 **Fax:** 215-315-9100
Toll-Free:
Address: 102 Rock Rd., Horsham, PA 19044 US

GROWTH PLANS/SPECIAL FEATURES:
Neose Technologies, Inc. is a clinical-stage biopharmaceutical company focused on the development of next-generation therapeutic proteins for the stimulation of red and white blood cells. The firm's core enzymatic technologies, GlycoAdvance and GlycoPEGylation, improve the drug properties of therapeutic proteins by building out, and attaching polyethylene glycol to carbohydrate structures on proteins. These modified proteins offer significant advantages, including less frequent dosing and improved efficacy. Neose's proprietary drug development portfolio consists of two therapeutic protein candidates: GlycoPEG-EPO and GlycoPEG-GCSF. GlycoPEG-EPO (NE-180) is a long-acting version of erythropoietin produced in insect cells, which is used to stimulate production of red blood cells. The product is approved for sale in major markets around the world for treatment of chemotherapy-induced anemia and anemia associated with chronic renal failure. It is also being developed for the treatment of anemia in adult cancer patients with non-myeloid malignancies who are receiving chemotherapy and anemia in patients with chronic kidney disease. GlycoPEG-GCSF is a long-acting version of granulocyte colony stimulating factor that is being co-developed with BioGeneriX AG. The drug is prescribed to stimulate production of neutrophils, a type of white blood cell, and is approved for sale for treatment of neutropenia associated with myelosuppressive chemotherapy. Neose has collaborations with other drug production companies, including Novo Nordisk and BioGeneriX. In September 2006, Neose consolidated its operations by selling its pilot manufacturing plant and corporate headquarters and moving to its Horsham, PA facilities. As a result of the consolidation, approximately 25% of employees lost their jobs.

Neose offers employees benefits including medical and dental insurance; flexible spending accounts; and tuition reimbursement.

FINANCIALS: Sales and profits are in thousands of dollars—add 000 to get the full amount. 2006 Note: Financial information for 2006 was not available for all companies at press time.
2006 Sales: $6,184	2006 Profits: $-27,107	**U.S. Stock Ticker: NTEC**
2005 Sales: $6,137	2005 Profits: $-51,839	**Int'l Ticker:** Int'l Exchange:
2004 Sales: $5,070	2004 Profits: $-41,642	Employees: 78
2003 Sales: $1,435	2003 Profits: $-37,681	Fiscal Year Ends: 12/31
2002 Sales: $4,800	2002 Profits: $-26,400	Parent Company:

SALARIES/BENEFITS:
Pension Plan:	ESOP Stock Plan:	Profit Sharing:	Top Exec. Salary: $322,767	Bonus: $262,500
Savings Plan: Y	Stock Purch. Plan:		Second Exec. Salary: $276,205	Bonus: $138,102

OTHER THOUGHTS:
Apparent Women Officers or Directors: 3
Hot Spot for Advancement for Women/Minorities: Y

LOCATIONS: ("Y" = Yes)
West:	Southwest:	Midwest:	Southeast:	Northeast:	International:
				Y	

NEUROBIOLOGICAL TECHNOLOGIES INC   www.ntii.com

Industry Group Code: 325412 Ranks within this company's industry group: Sales: 124   Profits: 121

Drugs:		Other:	Clinical:	Computers:	Other:
Discovery:	Y	AgriBio:	Trials/Services:	Hardware:	Specialty Services:
Licensing:	Y	Genomics/Proteomics:	Laboratories:	Software:	Consulting:
Manufacturing:		Tissue Replacement:	Equipment/Supplies:	Arrays:	Blood Collection:
Development:	Y		Research/Development Svcs.:	Database Management:	Drug Delivery:
Generics:			Diagnostics:		Drug Distribution:

TYPES OF BUSINESS:

Pharmaceuticals Acquisition & Development
Drugs-Neurological Disease

BRANDS/DIVISIONS/AFFILIATES:

Namenda
Memantine
XERECEPT
Viprinex
Celtic Pharmaceutical Holdings LP

CONTACTS: *Note: Officers with more than one job title may be intentionally listed here more than once.*

Paul E. Freiman, CEO
Paul E. Freiman, Pres.
Craig W. Carlson, CFO/VP
Lisa U. Carr, Sr. VP/Chief Medical Officer
David E. Levy, VP-Clinical Development
Karl G. Trass, VP-Regulatory Affairs & Quality Assurance
Warren W. Wasiewski, VP-Clinical Programs
Abraham E. Cohen, Chmn.

Phone: 510-595-6000	Fax: 510-595-6006
Toll-Free:	
Address: 2000 Powell St., Ste. 800, Emeryville, CA 94608 US	

GROWTH PLANS/SPECIAL FEATURES:

Neurobiological Technologies, Inc. (NTI) is a biotechnology company engaged in the business of acquiring and developing central nervous system (CNS) related drug candidates. The company is focused on therapies for neurological conditions that occur in connection with dementia, Alzheimer's disease, ischemic stroke, neuropathic pain and brain cancer. NTI's strategy is to in-license and develop late-stage drug candidates that target major medical needs and that can be rapidly commercialized. The company's first marketable product, Memantine (created with partners Merz Pharmaceuticals GmbH of Frankfurt, Germany and Children's Medical Center Corporation) is an oral treatment for dementia and pain relief. Memantine improves cognitive, psychological, social and motor functions impaired by dementia. It is being marketed by Forest Laboratories, Inc., under the trade name Namenda. NTI is also evaluating XERECEPT for the treatment of peritumoral brain edema (swelling of brain tissue due to tumor), currently in Phase III clinical testing. The company has been awarded orphan drug status for XERECEPT. In addition, Viprinex, a treatment for limiting the damage caused by strokes, is in Phase III clinical trials. Viprinex is licensed from Abbot Laboratories and is currently in Phase III clinical trials for ischemic stroke. NTI sold the worldwide rights to XERCEPT to Celtic Pharmaceutical Holdings L.P. in order to fund the Phase III testing of Viprinex. Celtic and NTI reported the results of a long-term open-label extension study for XERCEPT in June 2007, which indicated that long term therapy with the product in the study's patient focus group appears to be safe and well-tolerated.

FINANCIALS: Sales and profits are in thousands of dollars—add 000 to get the full amount. 2006 Note: Financial information for 2006 was not available for all companies at press time.

2006 Sales: $12,339	2006 Profits: $-27,839	**U.S. Stock Ticker: NTII**
2005 Sales: $3,100	2005 Profits: $-17,322	**Int'l Ticker:** Int'l Exchange:
2004 Sales: $2,786	2004 Profits: $-1,808	Employees: 33
2003 Sales: $1,980	2003 Profits: $-2,686	Fiscal Year Ends: 6/30
2002 Sales: $	2002 Profits: $-4,300	Parent Company:

SALARIES/BENEFITS:

Pension Plan:	ESOP Stock Plan:	Profit Sharing:	Top Exec. Salary: $375,000	Bonus: $425,000
Savings Plan: Y	Stock Purch. Plan: Y		Second Exec. Salary: $275,000	Bonus: $55,000

OTHER THOUGHTS:

Apparent Women Officers or Directors: 1
Hot Spot for Advancement for Women/Minorities:

LOCATIONS: ("Y" = Yes)

West:	Southwest:	Midwest:	Southeast:	Northeast:	International:
Y				Y	

Note: Financial information, benefits and other data can change quickly and may vary from those stated here.

NEUROCRINE BIOSCIENCES INC
www.neurocrine.com

Industry Group Code: 325412 Ranks within this company's industry group: Sales: 86   Profits: 185

Drugs:		Other:		Clinical:	Computers:		Other:	
Discovery:	Y	AgriBio:	Y	Trials/Services:	Hardware:		Specialty Services:	
Licensing:	Y	Genomics/Proteomics:		Laboratories:	Software:		Consulting:	
Manufacturing:		Tissue Replacement:		Equipment/Supplies:	Arrays:		Blood Collection:	
Development:	Y			Research/Development Svcs.:	Database Management:		Drug Delivery:	
Generics:				Diagnostics:			Drug Distribution:	

TYPES OF BUSINESS:
Pharmaceuticals Discovery & Development
Drugs-Neurological & Immune Disorder

BRANDS/DIVISIONS/AFFILIATES:
Indiplon
Urocortin 2
GnRH Antagonist
CRF-R Antagonist

CONTACTS: Note: Officers with more than one job title may be intentionally listed here more than once.
Gary A. Lyons, CEO
Kevin C. Gorman, COO
Gary A. Lyons, Pres.
Tim Coughlin, CFO
Carol Baum, VP-Mktg.
Richard Ranieri, Sr. VP-Human Resources
Dimitri E. Grigoriadis, VP-Research
Bill Wilson, VP-IT
Margaret E. Valeur-Jensen, General Counsel/Sr. VP/Corp. Sec.
Henry Pan, Exec. VP-Clinical Dev.
Haig Bozigian, Sr. VP-Pharmaceutical & Preclinical Dev.
Barbara Finn, VP-Regulatory Affairs & Quality Assurance
Chris O'Brien, Chief Medical Officer
Joseph A. Mollica, Chmn.
D. Bruce Campbell, Sr. VP-Int'l Dev.

Phone: 858-617-7600	Fax: 858-617-7601
Toll-Free:	
Address: 12790 El Camino Real, San Diego, CA 92130 US	

GROWTH PLANS/SPECIAL FEATURES:
Neurocrine Biosciences, Inc. (NBI) is a neuroimmunology company focused on the discovery and development of therapeutics to treat diseases and disorders of the central nervous and immune systems. The company's neuroscience, endocrine and immunology research segments provide a biological understanding of the molecular interactions of the central nervous, immune and endocrine systems. The firm currently has three active research programs, addressing anxiety, depression, gastrointestinal disorders, cachexia, pain, obesity, Parkinson's disease and insomnia. Indiplon, developed in partnership with Pfizer, is NBI's lead drug candidate. The company is working on a small-molecule Gonadotropin-Releasing Hormone (GnRH) receptor antagonist for the treatment of endometriosis. In January 2007, Neurocrine announced positive preliminary results from its second 'proof of concept', safety and efficacy Phase II clinical trial using its GnRH receptor antagonist in patients with endometriosis. In June 2007, Neurocrine resubmitted its New Drug Application (NDA) for indiplon 5 mg and 10 mg capsules for the treatment of insomnia in both adult and elderly patients to the U.S. Food and Drug Administration (FDA). Urocortin 2, which has finished Phase II testing, is being developed for congestive heart failure. The company is also in the process of developing a response to depression and anxiety through its research into the construction of a selective molecular antagonist to the Corticotropin Releasing Factor (CRF) type 1 receptor, which it believes is linked with these disorders.

NBI offers its employees medical and dental coverage; health and dependent care flexible spending accounts and long term care programs; a 401(k); stock options and bonuses based on performance; a 24-hour on-site workout room; a full service café; massage therapy and yoga sessions; and a number of recreational events including summer picnics, holiday parties, Friday afternoon gatherings and an annual bowling tournament.

FINANCIALS: Sales and profits are in thousands of dollars—add 000 to get the full amount. 2006 Note: Financial information for 2006 was not available for all companies at press time.

2006 Sales: $39,234	2006 Profits: $-107,205	**U.S. Stock Ticker: NBIX**
2005 Sales: $123,889	2005 Profits: $-22,191	**Int'l Ticker:** Int'l Exchange:
2004 Sales: $85,176	2004 Profits: $-45,773	Employees: 588
2003 Sales: $139,078	2003 Profits: $-30,256	Fiscal Year Ends: 12/31
2002 Sales: $18,000	2002 Profits: $-94,500	Parent Company:

SALARIES/BENEFITS:
Pension Plan:	ESOP Stock Plan: Y	Profit Sharing:	Top Exec. Salary: $600,000	Bonus: $
Savings Plan: Y	Stock Purch. Plan: Y		Second Exec. Salary: $365,000	Bonus: $

OTHER THOUGHTS:
Apparent Women Officers or Directors: 3
Hot Spot for Advancement for Women/Minorities: Y

LOCATIONS: ("Y" = Yes)
West:	Southwest:	Midwest:	Southeast:	Northeast:	International:
Y					

Note: Financial information, benefits and other data can change quickly and may vary from those stated here.

NEUROGEN CORP
www.neurogen.com

Industry Group Code: 325412 Ranks within this company's industry group: Sales: 131   Profits: 158

Drugs:		Other:	Clinical:	Computers:		Other:
Discovery:	Y	AgriBio:	Trials/Services:	Hardware:		Specialty Services:
Licensing:		Genomics/Proteomics:	Laboratories:	Software:		Consulting:
Manufacturing:		Tissue Replacement:	Equipment/Supplies:	Arrays:		Blood Collection:
Development:	Y		Research/Development Svcs.:	Database Management:	Y	Drug Delivery:
Generics:			Diagnostics:			Drug Distribution:

TYPES OF BUSINESS:
Small-Molecule Drug Development
Drug Discovery Platform
Drugs-Neurological & Psychiatric
Drugs-Pain & Inflammatory
Drugs-Metabolic Disorders
Receptor Biology

BRANDS/DIVISIONS/AFFILIATES:
Accelerated Intelligent Drug Discovery (AIDD)
Focused Compound Library
Virtual Screening
Virtual Library

CONTACTS: Note: Officers with more than one job title may be intentionally listed here more than once.
William H. Koster, CEO
Stephen R. Davis, COO/Exec. VP
William H. Koster, Pres.
Stephen Uden, Exec. VP-R&D
Jeffrey Dill, General Counsel/VP/Corp. Sec.
Tom Pitler, VP-Bus. Dev.
Elaine Dodge, Associate Dir.-Public Rel.
Elaine Dodge, Associate Dir.-Investor Rel.
Alan J. Hutchison, Exec. VP-Discovery Research
James E. Krause, Sr. VP-Biology
Charles J. Manly, VP-Discovery Tech.
Bertrand L. Chenard, Sr. VP-Chemistry & Process Research
Craig Saxton, Chmn.

Phone: 203-488-8201	Fax: 203-481-8683

Toll-Free:
Address: 35 NE Industrial Rd., Branford, CT 06405 US

GROWTH PLANS/SPECIAL FEATURES:
Neurogen Corp. is a small molecule drug discovery and development company targeting new drug candidates to improve the lives of patients suffering from neurological, inflammatory, pain and metabolic disorders. The firm has generated a portfolio of new drug programs through its fully integrated, proprietary drug discovery platform. Its Accelerated Intelligent Drug Discovery (AIDD) system and its expertise in cellular function assays allow the company to rapidly and cost-effectively identify new drug candidates. AIDD is an integrated system, incorporating robotics and computerization that allows scientists to improve on the trial and error approach traditionally associated with discovery and development. The system works in tandem with its focused compound library and trademarked Virtual Screening on its Virtual Library, with several hundred billion virtual compounds. Neurogen, on its own and in partnership with larger drug companies such as Merck and Pfizer, has several small molecule candidates in its product pipeline. These products include a GABA receptor partial agonist in Phase II clinical trials for the treatment of insomnia; vanilloid receptor-1 (VR1) receptor antagonists in Phase II clinical trials for the treatment of pain, which the company has partnered with Merck to develop; a CRF-1 receptor antagonist intended for the treatment of depression or anxiety in preclinical development; and a MCH-1 receptor antagonist intended to treat diabetes and obesity in Phase I clinical development. Neurogen has also in-licensed a drug for Parkinson's disease and Restless Leg Syndrome from Wyeth Pharmaceuticals, which is in Phase I.

FINANCIALS: Sales and profits are in thousands of dollars—add 000 to get the full amount. 2006 Note: Financial information for 2006 was not available for all companies at press time.

2006 Sales: $9,813	2006 Profits: $-53,776	U.S. Stock Ticker: NRGN
2005 Sales: $7,558	2005 Profits: $-37,120	Int'l Ticker: Int'l Exchange:
2004 Sales: $19,180	2004 Profits: $-18,593	Employees: 166
2003 Sales: $6,788	2003 Profits: $-31,576	Fiscal Year Ends: 12/31
2002 Sales: $15,700	2002 Profits: $-23,700	Parent Company:

SALARIES/BENEFITS:
Pension Plan:	ESOP Stock Plan:	Profit Sharing:	Top Exec. Salary: $442,571	Bonus: $179,020
Savings Plan: Y	Stock Purch. Plan:		Second Exec. Salary: $338,350	Bonus: $102,317

OTHER THOUGHTS:
Apparent Women Officers or Directors:
Hot Spot for Advancement for Women/Minorities:

LOCATIONS: ("Y" = Yes)
West:	Southwest:	Midwest:	Southeast:	Northeast:	International:
				Y	

Note: Financial information, benefits and other data can change quickly and may vary from those stated here.

NEXMED INC

www.nexmed.com

Industry Group Code: 325412A Ranks within this company's industry group: Sales: 18   Profits: 8

Drugs:		Other:		Clinical:		Computers:		Other:	
Discovery:		AgriBio:		Trials/Services:		Hardware:		Specialty Services:	
Licensing:		Genomics/Proteomics:	Y	Laboratories:		Software:		Consulting:	
Manufacturing:		Tissue Replacement:	Y	Equipment/Supplies:	Y	Arrays:		Blood Collection:	
Development:	Y			Research/Development Svcs.:		Database Management:		Drug Delivery:	Y
Generics:				Diagnostics:				Drug Distribution:	

TYPES OF BUSINESS:

Drug Delivery Systems
Transdermal Drug Delivery Systems
Sexual Dysfunction Products
Medical Devices

BRANDS/DIVISIONS/AFFILIATES:

NexACT
Befar
Femprox
Alprox-TD

CONTACTS: Note: Officers with more than one job title may be intentionally listed here more than once.

Vivian H. Liu, CEO
Vivian H. Liu, Pres.
Mark Westgate, CFO

Phone: 609-208-9688	**Fax:** 609-208-1868
Toll-Free:	
Address: 350 Corporate Blvd., Robbinsville, NJ 08691 US	

GROWTH PLANS/SPECIAL FEATURES:

NexMed, Inc. is a pharmaceutical and medical technology company with a focus on developing and commercializing therapeutic products based on proprietary delivery systems. The company currently focuses on new and patented topical pharmaceutical products based on a penetration enhancement drug delivery technology called NexACT, which facilitates an active drug's absorption through the skin. It develops topical treatments in forms including cream, gel, patch and tape, and has 12 US patents in connection with its NexACT technology. NexMED's principal product candidates include Alprox-TD and Femprox creams for the treatment of male erectile dysfunction and female sexual arousal disorder, respectively. Alprox-TD has been launched in China under the trademark Befar. Schering AG has the exclusive commercialization rights for Alprox-TD in Europe, Russia, the Middle East, South Africa, Australia and New Zealand. The company is pursuing European marketing approval for Alprox-TD. NexMed, in partnership with Novartis International Pharmesutical Ltd.., is in Phase III trials for developing a topical treatment for nail fungus.

FINANCIALS: Sales and profits are in thousands of dollars—add 000 to get the full amount. 2006 Note: Financial information for 2006 was not available for all companies at press time.

2006 Sales: $1,867	2006 Profits: $-8,043	**U.S. Stock Ticker: NEXM**
2005 Sales: $2,399	2005 Profits: $-15,442	**Int'l Ticker:** Int'l Exchange:
2004 Sales: $ 359	2004 Profits: $-17,024	Employees: 18
2003 Sales: $ 111	2003 Profits: $-17,234	Fiscal Year Ends: 12/31
2002 Sales: $ 148	2002 Profits: $-27,600	Parent Company:

SALARIES/BENEFITS:

Pension Plan:	ESOP Stock Plan:	Profit Sharing:	Top Exec. Salary: $200,000	Bonus: $125,000
Savings Plan: Y	Stock Purch. Plan:		Second Exec. Salary: $160,000	Bonus: $80,000

OTHER THOUGHTS:

Apparent Women Officers or Directors: 1
Hot Spot for Advancement for Women/Minorities:

LOCATIONS: ("Y" = Yes)

West:	Southwest:	Midwest:	Southeast:	Northeast:	International:
				Y	Y

NORTH AMERICAN SCIENTIFIC   www.nasmedical.com

Industry Group Code: 339113 Ranks within this company's industry group: Sales: 10   Profits: 11

Drugs:	Other:	Clinical:		Computers:		Other:
Discovery:	AgriBio:	Trials/Services:		Hardware:		Specialty Services:
Licensing:	Genomics/Proteomics:	Laboratories:		Software:	Y	Consulting:
Manufacturing:	Tissue Replacement:	Equipment/Supplies:	Y	Arrays:		Blood Collection:
Development: Y		Research/Development Svcs.:		Database Management:		Drug Delivery:
Generics:		Diagnostics:	Y			Drug Distribution:

TYPES OF BUSINESS:

Radiation Therapy Products
Radioisotopic Products
Brachytherapy Products
Radiation Therapy Software

BRANDS/DIVISIONS/AFFILIATES:

PEACOCK
BAT
PEREGRINE
BEAK
NOMOS CRANE
AUTOCRANE
CORVUS
MIMiC

CONTACTS: *Note: Officers with more than one job title may be intentionally listed here more than once.*

John B. Rush, CEO
John B. Rush, Pres.
James W. Klingler, CFO/Sr. VP
L. Michael Cutrer, CTO/Exec. VP
David King, Dir.-Legal Affairs/Corp. Sec.
Michael C. Ryan, Pres., NOMOS Radiation Oncology Div.
David Stiles, VP-Sales & Mktg., Brachytherapy Div.
Gary N. Wilmer, Chmn.

Phone: 818-734-8600	Fax: 818-734-5200

Toll-Free: 800-992-6274
Address: 20200 Sunburst St., Chatsworth, CA 91311 US

GROWTH PLANS/SPECIAL FEATURES:

North American Scientific, Inc. (NASI) designs, develops, produces and sells products for radiation therapy treatment, including brachytherapy seeds and treatment planning and delivery technology for intensity modulated radiation therapy (IMRT) and image guided radiation therapy (IGRT). The firm's IMRT treatment planning system is CORVUS, a software-based product that uses images from conventional imaging technologies to create three-dimensional models of the treated area and then tests and rejects beam intensities and angles as it builds a treatment plan. NASI'S multi-leaf collimator, MIMiC, is the first on the market that can deliver radiation from approximately 54 angles. PEACOCK, an IMRT planning and delivery system includes CORVUS as its software-based planning component and MIMiC as its multi-leaf collimator. BAT uses ultrasound images to confirm the location of target organs and tumor sites within the patient's body. BAT reduces error margin from the current standard practice of within two centimeters to within two millimeters. It is predominantly used by hospitals and clinics in treating prostate cancer. PEREGRINE is a radiation dosage calculation software product used to estimate the amount and distribution of the radiation dose in three dimensions that would be absorbed by a patient from a prescribed plan. PEREGRINE can be adjusted for muscle, skin, and bone density, as well as absorption. In addition to NASI's core products, the company also offers accessory products such as BEAK, a device that reduces beam size; NOMOS CRANE, a series of devices used to position patients during radiation therapy; AUTOCRANE, which remotely adjusts the patient's position during treatment; TALON, a device used to position the head during radiation therapy; and the Target Box, a device that provides multiple means of achieving precise patient alignment. In July 2007, the company announced its NOMOS Radiation Oncology division's introduction of BATCAM Multi-probe, a new multi-probe capable ultrasound system.

FINANCIALS: Sales and profits are in thousands of dollars—add 000 to get the full amount. 2006 Note: Financial information for 2006 was not available for all companies at press time.

2006 Sales: $28,988	2006 Profits: $-17,130	**U.S. Stock Ticker:** NASI
2005 Sales: $32,224	2005 Profits: $-55,513	**Int'l Ticker:** Int'l Exchange:
2004 Sales: $24,737	2004 Profits: $-36,307	Employees: 177
2003 Sales: $14,683	2003 Profits: $-9,440	Fiscal Year Ends: 10/31
2002 Sales: $20,800	2002 Profits: $-5,200	Parent Company:

SALARIES/BENEFITS:

Pension Plan:	ESOP Stock Plan:	Profit Sharing:	Top Exec. Salary: $340,700	Bonus: $30,000
Savings Plan: Y	Stock Purch. Plan: Y		Second Exec. Salary: $229,000	Bonus: $20,000

OTHER THOUGHTS:

Apparent Women Officers or Directors:
Hot Spot for Advancement for Women/Minorities:

LOCATIONS: ("Y" = Yes)

West:	Southwest:	Midwest:	Southeast:	Northeast:	International:
Y				Y	Y

NOVARTIS AG

www.novartis.com

Industry Group Code: 325412 Ranks within this company's industry group: Sales: 5   Profits: 4

Drugs:		Other:	Clinical:	Computers:	Other:
Discovery:	Y	AgriBio:	Trials/Services:	Hardware:	Specialty Services:
Licensing:	Y	Genomics/Proteomics:	Laboratories:	Software:	Consulting:
Manufacturing:	Y	Tissue Replacement:	Equipment/Supplies:	Arrays:	Blood Collection:
Development:	Y		Research/Development Svcs.:	Database Management:	Drug Delivery:
Generics:	Y		Diagnostics:		Drug Distribution:

TYPES OF BUSINESS:

Drugs-Diversified
Therapeutic Drug Discovery
Therapeutic Drug Manufacturing
Generic Drugs
Over-the-Counter Drugs
Ophthalmic Products
Nutritional Products
Veterinary Products

BRANDS/DIVISIONS/AFFILIATES:

CIBA Vision
Grand Laboratories, Inc.
ImmTech Biologics, Inc.
Novartis Institute for Biomedical Research, Inc.
Sandoz
Sabex
Gerber Products Co.
Chiron Corp.

CONTACTS: *Note: Officers with more than one job title may be intentionally listed here more than once.*

Daniel Vasella, CEO
Alex Gorsky, COO
Daniel Vasella, Pres.
Raymund Breu, CFO
Jurgen Brokatzky-Geiger, Dir.-Human Resources
Urs Barlocher, Dir.-Legal & Tax Affairs
Ann Bailey, Dir.-Public Affairs & Corp. Comm.
Thomas Ebeling, CEO, Pharmaceuticals
Paul Choffat, CEO, Consumer Health
Mark Fishman, Pres., Novartis Institute for Biomedical Research
Andreas Rummelt, CEO, Sandoz
Daniel Vasella, Chmn.

Phone: 41-61-324-1111	Fax: 41-61-324-8001
Toll-Free:	
Address: Lichtstrasse 35, Basel, CH-4056 Switzerland	

GROWTH PLANS/SPECIAL FEATURES:

Novartis AG (NVS) researches and develops pharmaceuticals as well as a large number of consumer and animal health products. NVS operates in four segments: pharmaceuticals; vaccines and diagnostics; consumer health; and Sandoz Generics. Through its pharmaceuticals division, NVS develops, manufactures and markets prescription medications in a variety of areas, which include treatments for cardiovascular diseases, neurological disorders, respiratory conditions, dermatological conditions and infectious diseases. NVS also provides specialty medications in the fields of oncology, hematology, transplantation, immunology and ophthalmics. The vaccines division produces a large number of vaccinations for influenza, meningitis, rabies and tick-borne encephalitis. In the consumer health sector, NVS manufactures over-the-counter medication such as Maalox and Triaminic, animal health products, medical nutrition products, infant products and CIBA Vision. CIBA Vision is a subsidiary of NVS which researches and develops eye care products, lens care products and ophthalmic surgical products. The animal health division also provides solutions for pest control and offers a number of medications for both pets and farm animals to prevent parasitic and bacterial infections. Another sector of consumer health includes Gerber Products Co., a popular brand providing both infant care and nutrition. Sandoz Generics, is engaged in developing and manufacturing generic pharmaceuticals and currently holds a product list of over 840 compounds in over 5,000 dosage forms. In late 2006, NVS agreed to sell its medical nutrition business to Nestle and also made plans in early 2007 to relocate its vaccines and diagnostics division headquarters to Cambridge, Massachusetts. Recently, the FDA issued its approval for distribution of Tekturna, a high blood pressure medication, and Procleix Tigris System, a blood screening for the West Nile virus. In March 2007, the FDA asked NVS to remove Zelnorm from the U.S. market.

Novartis offers its employees child care, elderly care support, tuition reimbursement and health and fitness programs.

FINANCIALS: Sales and profits are in thousands of dollars—add 000 to get the full amount. 2006 Note: Financial information for 2006 was not available for all companies at press time.

2006 Sales: $36,031,000	2006 Profits: $7,202,000	**U.S. Stock Ticker: NVS**
2005 Sales: $31,005,000	2005 Profits: $6,141,000	**Int'l Ticker: NOVN** Int'l Exchange: Zurich-SWX
2004 Sales: $27,126,000	2004 Profits: $5,601,000	Employees: 100,735
2003 Sales: $24,864,000	2003 Profits: $5,016,000	Fiscal Year Ends: 12/31
2002 Sales: $23,151,000	2002 Profits: $3,546,000	Parent Company:

SALARIES/BENEFITS:

Pension Plan: Y	ESOP Stock Plan:	Profit Sharing:	Top Exec. Salary: $	Bonus: $
Savings Plan:	Stock Purch. Plan:		Second Exec. Salary: $	Bonus: $

OTHER THOUGHTS:

Apparent Women Officers or Directors: 1
Hot Spot for Advancement for Women/Minorities:

LOCATIONS: ("Y" = Yes)

West:	Southwest:	Midwest:	Southeast:	Northeast:	International:
Y		Y	Y	Y	Y

Note: Financial information, benefits and other data can change quickly and may vary from those stated here.

NOVAVAX INC

www.novavax.com

Industry Group Code: 325412A **Ranks within this company's industry group:** Sales: 15 Profits: 12

Drugs:		Other:		Clinical:		Computers:		Other:	
Discovery:	Y	AgriBio:		Trials/Services:		Hardware:		Specialty Services:	
Licensing:		Genomics/Proteomics:		Laboratories:		Software:		Consulting:	
Manufacturing:	Y	Tissue Replacement:		Equipment/Supplies:		Arrays:		Blood Collection:	
Development:	Y			Research/Development Svcs.:	Y	Database Management:		Drug Delivery:	Y
Generics:				Diagnostics:				Drug Distribution:	

TYPES OF BUSINESS:

Drug Delivery Systems
Drugs-Bacterial & Viral Infection
Hormone Replacement Therapies
Contract Research Services

BRANDS/DIVISIONS/AFFILIATES:

Estrasorb
Androsorb
Gynodiol
Novasomes
Sterisomes
VLP Technology
Micellar Nanoparticle Technology
E-selectin Tolerogen

CONTACTS: Note: Officers with more than one job title may be intentionally listed here more than once.

Rahul Singhvi, CEO
Raymond J. Hage, Jr., COO/Sr. VP
Rahul Singhvi, Pres.
James Robinson, VP-Tech. & Quality Oper.
Robert W. Lee, VP-Pharm. Dev.
Penny Heaton, Chief Medical Officer
Gale E. Smith, VP-Vaccine Dev.
Rick A. Bright, VP-Global Influenza Programs
John Lambert, Chmn.

Phone: 240-268-2000	Fax:
Toll-Free:	
Address: 9920 Belward Campus Drive, Rockville, MD 20850 US	

GROWTH PLANS/SPECIAL FEATURES:

Novavax, Inc. is a specialty biopharmaceutical company focused on the development of drug delivery platforms and biologicals. The company's research, development and commercialization efforts are advancing industry standards for vaccine and drug delivery systems. Novavax's flagship product, a micellar nanoparticle (MNP)-based topical emulsion for hormone therapy, is called ESTRASORB. The drug offers a high level of efficacy in treating hot flashes, with additional advantages including ease of use, rapid onset and lower incidence of irritation and nausea. Novavax's MNP technology uses oil and water nanoemulsions (under one micron in diameter) to encapsulate alcohol-soluble drugs and deliver them directly into the bloodstream through topical lotions. This same transdermal technology is being used for future products including ANDROSORB, a testosterone lotion used to treat female sexual dysfunction. This patented delivery platform is envisioned to support a range of drugs and therapeutics, including hormones, antibacterials and antiviral products, as well as central nervous system drugs, anti-inflammatory agents and advanced topical analgesics. Other technologies in development include Novasome and Sterisome, proprietary liposomes that carry encapsulated drugs or are mixed with vaccines as an adjuvant to increase effectiveness. Novavax is also researching virus-like particles, employing non-infectious protein particulate structures to elicit immune system responses. These systems are being explored for potential treatment of infectious diseases including HIV, SARS and influenza.

The company offers its employees comprehensive benefits including a prescription drug plan, flexible spending accounts, paid holidays, personal leave, an employee referral program, stock options and a 401(k) retirement savings plan.

FINANCIALS: Sales and profits are in thousands of dollars—add 000 to get the full amount. 2006 Note: Financial information for 2006 was not available for all companies at press time.

2006 Sales: $4,683	2006 Profits: $-23,068	**U.S. Stock Ticker: NVAX**
2005 Sales: $7,388	2005 Profits: $-11,174	**Int'l Ticker:** Int'l Exchange:
2004 Sales: $8,260	2004 Profits: $-25,920	Employees: 56
2003 Sales: $11,785	2003 Profits: $-17,273	Fiscal Year Ends: 12/31
2002 Sales: $15,000	2002 Profits: $-22,700	Parent Company:

SALARIES/BENEFITS:

Pension Plan:	ESOP Stock Plan:	Profit Sharing:	Top Exec. Salary: $337,510	Bonus: $
Savings Plan: Y	Stock Purch. Plan: Y		Second Exec. Salary: $225,286	Bonus: $

OTHER THOUGHTS:

Apparent Women Officers or Directors: 1
Hot Spot for Advancement for Women/Minorities:

LOCATIONS: ("Y" = Yes)

West:	Southwest:	Midwest:	Southeast:	Northeast:	International:
Y				Y	

Note: Financial information, benefits and other data can change quickly and may vary from those stated here.

NOVEN PHARMACEUTICALS   www.noven.com

Industry Group Code: 325412A Ranks within this company's industry group: Sales: 6   Profits: 2

Drugs:		Other:	Clinical:	Computers:	Other:	
Discovery:	Y	AgriBio:	Trials/Services:	Hardware:	Specialty Services:	
Licensing:		Genomics/Proteomics:	Laboratories:	Software:	Consulting:	
Manufacturing:	Y	Tissue Replacement:	Equipment/Supplies:	Arrays:	Blood Collection:	
Development:	Y		Research/Development Svcs.:	Database Management:	Drug Delivery:	Y
Generics:			Diagnostics:		Drug Distribution:	

TYPES OF BUSINESS:

Drug Delivery Systems
Hormone Replacement Products
Pain Management Products
Central Nervous System Products
Transdermal Drug Delivery Systems

BRANDS/DIVISIONS/AFFILIATES:

Vivelle
Menorest
Daytrana
CombiPatch
Estalis
MethyPatch
Estradot
Novogyne Pharmaceuticals

CONTACTS: Note: Officers with more than one job title may be intentionally listed here more than once.

Robert C. Strauss, CEO
Robert C. Strauss, Pres.
Diane M. Barrett, CFO/VP
W. Neil Jones, VP-Mktg. & Sales
Carolyn Donaldson, VP-Human Resources
Eduardo Abrao, VP-Clinical Dev./Chief Medical Officer
Juan A. Mantelle, CTO/VP
Jeff T. Mihm, General Counsel
Richard Gilbert, VP-Oper.
Paven Handa, Exec. Dir.-Bus. Dev.
Joseph C. Jones, VP-Corp. Affairs
Jeffery P. Eisenberg, Sr. VP-Strategic Alliances
James Harris, VP-Quality Assurance & Control
Robert C. Strauss, Chmn.

Phone: 305-253-5099	Fax: 305-251-1887
Toll-Free:	
Address: 11960 SW 144th St., Miami, FL 33186 US	

GROWTH PLANS/SPECIAL FEATURES:

Noven Pharmaceuticals, Inc. develops and manufactures advanced transdermal drug delivery systems and prescription transdermal products. Its principal commercialized products are transdermal drug delivery systems designed with its DOT Matrix technology for use in hormone replacement therapy. The firm's first product was an estrogen patch for the treatment of menopausal symptoms, marketed under the name Vivelle in the U.S. and Canada and under the name Menorest in Europe and other markets. The company also launched the smallest transdermal estrogen patch ever approved by the U.S. FDA, Vivelle-Dot. Noven markets the product in several foreign countries under the name Estradot. Noven also produces a combination estrogen/progestin transdermal patch for the treatment of menopausal symptoms, marketed under the name CombiPatch in the U.S. and Estalis in Europe and other markets. The firm is developing a range of products, including transdermal delivery systems for the treatment of attention deficit hyperactivity disorder (ADHD). The company markets a daily methylphenidate patch canned Daytrana for the treatment of ADHD called Daytrana through an agreement with Shire. The company also specializes in treatments for central nervous system conditions and pain. Noven has partial ownership of a joint venture formed in partnership with Novartis Pharmaceuticals, called Vivelle Ventures, doing business under the name Novogyne Pharmaceuticals. Novogyne markets Vivelle, Vivelle-Dot and CombiPatch in the U.S. In July 2007, Noven agreed to buy JDS Pharmaceuticals LLC for about $125 million in cash. JDS markets two branded prescription psychiatry products, Lithobid and Pexeva for bipolar disporder and depression, respectively. The company also is advancing a pipeline of other products for the psychiatry and women's health markets.

FINANCIALS: Sales and profits are in thousands of dollars—add 000 to get the full amount. 2006 Note: Financial information for 2006 was not available for all companies at press time.

2006 Sales: $60,689	2006 Profits: $15,988	**U.S. Stock Ticker: NOVN**
2005 Sales: $52,532	2005 Profits: $9,972	**Int'l Ticker:** Int'l Exchange:
2004 Sales: $45,891	2004 Profits: $11,224	Employees: 518
2003 Sales: $43,166	2003 Profits: $11,196	Fiscal Year Ends: 12/31
2002 Sales: $55,400	2002 Profits: $13,900	Parent Company:

SALARIES/BENEFITS:

Pension Plan:	ESOP Stock Plan:	Profit Sharing:	Top Exec. Salary: $587,741	Bonus: $66,121
Savings Plan: Y	Stock Purch. Plan:		Second Exec. Salary: $302,702	Bonus: $24,902

OTHER THOUGHTS:

Apparent Women Officers or Directors: 2
Hot Spot for Advancement for Women/Minorities: Y

LOCATIONS: ("Y" = Yes)

West:	Southwest:	Midwest:	Southeast:	Northeast:	International:
			Y		

NOVO-NORDISK AS

www.novonordisk.com

Industry Group Code: 325412 **Ranks within this company's industry group:** Sales: 19 Profits: 19

Drugs:		Other:		Clinical:	Computers:		Other:	
Discovery:	Y	AgriBio:		Trials/Services:	Hardware:		Specialty Services:	
Licensing:		Genomics/Proteomics:		Laboratories:	Software:		Consulting:	
Manufacturing:	Y	Tissue Replacement:		Equipment/Supplies:	Arrays:		Blood Collection:	
Development:	Y			Research/Development Svcs.:	Database Management:		Drug Delivery:	Y
Generics:				Diagnostics:			Drug Distribution:	

TYPES OF BUSINESS:

Drugs-Diabetes
Hormone Replacement Therapy
Growth Disorder Drugs
Hemophilia Drugs
Insulin Delivery Systems
Educational & Training Services

BRANDS/DIVISIONS/AFFILIATES:

NovoPen
FlexPen
Norditropin SimpleXx
NovoSeven
Activelle
Trisequens
Estrofem
Novo-Nordisk Research US

CONTACTS: *Note: Officers with more than one job title may be intentionally listed here more than once.*

Lars R. Sorensen, CEO
Kare Schultz, COO/Exec. VP
Lars R. Sorensen, Pres.
Jesper Brandgaard, CFO/Exec. VP
Mads K. Thomsen, Chief Science Officer/Exec. VP
Lise Kingo, Exec. VP/Chief of Staff
Sten Scheibye, Chmn.

Phone: 45-4444-8888	Fax: 45-4449-0555

Toll-Free:

Address: 2880 Novo Alle, Basgvaerd, DK-2880 Denmark

GROWTH PLANS/SPECIAL FEATURES:

Novo-Nordisk AS focuses on developing treatments for diabetes, hemostasis management, growth hormone therapy and hormone replacement therapy. With its affiliates, the company has employees in 79 countries, and it markets its products in 179 countries. The firm is a world leader in insulin manufacturing and has the broadest diabetes product line in the world. The NovoPen and FlexPen products are pen-like, multiple-dose injectors that allow patients to easily inject themselves with insulin or hormones. Novo-Nordisk's growth hormone replacement product, Norditropin SimpleXx, is a premixed liquid growth hormone designed to provide the most flexible and accurate dosing. The company's NovoSeven product, a treatment for hemophilia, is a recombinant coagulation factor that enables coagulation to proceed in the absence of natural factors. The firm also manufactures post-menopausal hormone replacement therapy products, including Activelle, Kliogest, Trisequens, Estrofem and Vagifem. In addition, the company offers educational services and training materials for both patients and health care professionals. The company has a licensing agreement with ZymoGenetics for microarrays. Novo-Nordisk also owns and operates a dedicated hemostasis research center in the U.S. The facility, Novo-Nordisk Research US, is the first in the country dedicated to life-threatening bleeding. In February 2007, Novo-Nordisk divested its ownership of Dako A/S, a cancer diagnosis company.

Novo-Nordisk offers its U.S. employees health, life, dental, auto and supplemental insurance, as well as tuition reimbursement.

FINANCIALS: Sales and profits are in thousands of dollars—add 000 to get the full amount. 2006 Note: Financial information for 2006 was not available for all companies at press time.

2006 Sales: $6,913,700	2006 Profits: $1,126,020	**U.S. Stock Ticker:** NVO
2005 Sales: $5,446,472	2005 Profits: $946,073	**Int'l Ticker:** NOVO B Int'l Exchange: Copenhagen-CSE
2004 Sales: $5,324,285	2004 Profits: $859,229	Employees: 22,500
2003 Sales: $4,501,000	2003 Profits: $824,000	Fiscal Year Ends: 12/31
2002 Sales: $3,554,000	2002 Profits: $578,000	Parent Company:

SALARIES/BENEFITS:

Pension Plan:	ESOP Stock Plan:	Profit Sharing:	Top Exec. Salary: $	Bonus: $
Savings Plan: Y	Stock Purch. Plan:		Second Exec. Salary: $	Bonus: $

OTHER THOUGHTS:

Apparent Women Officers or Directors: 2
Hot Spot for Advancement for Women/Minorities: Y

LOCATIONS: ("Y" = Yes)

West:	Southwest:	Midwest:	Southeast:	Northeast:	International:
Y	Y	Y	Y	Y	Y

Note: Financial information, benefits and other data can change quickly and may vary from those stated here.

NOVOZYMES
www.novozymes.com

Industry Group Code: 325414 Ranks within this company's industry group: Sales: Profits: 10

Drugs:	Other:		Clinical:	Computers:	Other:	
Discovery:	AgriBio:		Trials/Services:	Hardware:	Specialty Services:	Y
Licensing:	Genomics/Proteomics:	Y	Laboratories:	Software:	Consulting:	
Manufacturing:	Tissue Replacement:		Equipment/Supplies:	Arrays:	Blood Collection:	
Development:			Research/Development Svcs.:	Database Management:	Drug Delivery:	
Generics:			Diagnostics:		Drug Distribution:	

TYPES OF BUSINESS:
Industrial Enzyme & Microorganism Production
Biopolymers
Pharmaceuticals

BRANDS/DIVISIONS/AFFILIATES:
Novozymes A/S
Ceremix
Delta Biotechnology, Ltd.
Biocon
Neutrase
Attenuzyme
Fungamel
Termamyl

CONTACTS: *Note: Officers with more than one job title may be intentionally listed here more than once.*
Steen Riisgaard, CEO
Steen Riisgaard, Pres.
Benny D. Loft, CFO/Exec. VP
Peder Nielsen, Exec. VP-Mktg. & Sales
Nickie Spile, VP-People & Organization
Per Falholt, Chief Scientific Officer/Exec. VP
Arne W. Schmidt, Exec. VP-Dev., Production & Quality
Anna L. M. Grandjean, VP-Stakeholder Comm. & Sustainability Dev.
Henrik Gurtler, Chmn.

Phone: 45-88-24-99-99	Fax: 45-88-24-99-98
Toll-Free:	
Address: Krogshoejvej 36, Bagsvaerd, 2880 Denmark	

GROWTH PLANS/SPECIAL FEATURES:

Novozymes, a subsidiary of Novozymes A/S, is a biotechnology company that specializes in microbiology, biotechnology and gene technology. The company's core competencies include culture collection; protein design through protein engineering, gene shuffling and molecular modeling; protein chemistry through protein purification, protein sequencing, protein analysis and assay development; pathway engineering; strain development and improvement through promoter selection, manipulation of operons and yield improvement and large scale production that involves lab fermentation and formulation. The firm currently sells over 100 enzymes and microorganisms and manufactures over 700 products that are sold in over 130 countries in the brewing, forest products and oils and fats industries. In the brewing industry, Novozymes offers key auxillary enzymes that are required for the malt brewing process through the brands, Ceremix, Cerezyme Sorghum, Neutrase, Attenuzyme, Fungamyl, Termamyl and Maturex. In the forest products sector, the pulp and paper production processes often add key enzymes such as Pulpzyme HC for bleach boosting; Resinase for pitch control; BAN for starch modification and Aquazym and Nocozym 342 for deinking. Lastly, Novozmes provides enzymatic interesterification and degumming for the production of oil and fats through the brands, Lecitase Ultra and Lipozyme TL IM, and its oil based specialties, Lipozyme TL 100 L and Novozym 435. In addition to its enzymes, the company also markets microorganisms, pharmaceutical proteins and biopolymers. In June 2006, the company signed an agreement to acquire Delta Biotechnology, Ltd. Novozymes opened a new office in Washington D.C. in mid-2007 and also strengthened its presence in India through the acquisition of Biocon, a company that specializes in biopharmaceuticals, contract research, clinical research and enzymes.

Employee benefits at Novozymes' headquarters include new parent leave, child care, a fitness center and holiday cottages. Benefits at the firm's U.S. offices include health club reimbursement, flexible spending accounts, tuition reimbursement and an employee assistance plan.

FINANCIALS: Sales and profits are in thousands of dollars—add 000 to get the full amount. 2006 Note: Financial information for 2006 was not available for all companies at press time.

2006 Sales: $1,251,490	2006 Profits: $167,610	**U.S. Stock Ticker: NVZMY**
2005 Sales: $1,079,840	2005 Profits: $148,025	**Int'l Ticker: NZYM** Int'l Exchange: Copenhagen-CSE
2004 Sales: $1,029,470	2004 Profits: $133,240	Employees:
2003 Sales: $	2003 Profits: $	Fiscal Year Ends: 12/31
2002 Sales: $	2002 Profits: $	Parent Company:

SALARIES/BENEFITS:

Pension Plan: Y	ESOP Stock Plan:	Profit Sharing:	Top Exec. Salary: $	Bonus: $
Savings Plan: Y	Stock Purch. Plan:		Second Exec. Salary: $	Bonus: $

OTHER THOUGHTS:
Apparent Women Officers or Directors: 2
Hot Spot for Advancement for Women/Minorities: Y

LOCATIONS: ("Y" = Yes)

West:	Southwest:	Midwest:	Southeast:	Northeast:	International:
Y				Y	Y

Note: Financial information, benefits and other data can change quickly and may vary from those stated here.

NPS PHARMACEUTICALS INC
www.npsp.com

Industry Group Code: 325412 Ranks within this company's industry group: Sales: 79   Profits: 188

Drugs:		Other:	Clinical:	Computers:	Other:
Discovery:	Y	AgriBio:	Trials/Services:	Hardware:	Specialty Services:
Licensing:	Y	Genomics/Proteomics:	Laboratories:	Software:	Consulting:
Manufacturing:	Y	Tissue Replacement:	Equipment/Supplies:	Arrays:	Blood Collection:
Development:	Y		Research/Development Svcs.:	Database Management:	Drug Delivery:
Generics:			Diagnostics:		Drug Distribution:

TYPES OF BUSINESS:
Small Molecule Drugs & Recombinant Proteins

BRANDS/DIVISIONS/AFFILIATES:
Preotact
Preos
Sensipar
Mimpara

CONTACTS: Note: Officers with more than one job title may be intentionally listed here more than once.
N. Anthony Coles, CEO
Francois Nader, COO/Exec. VP
N. Anthony Coles, Pres.
Gerard J. Michel, CFO
Glenn Melrose, VP-Human Resources
Gregory M. Torre, Sr. VP-Tech. Oper.
Val Antczak, General Counsel/Sr. VP-Legal Affairs
Juergen Lasowski, Sr. VP-Corp. Dev.
Brandi Simpson, Sr. Dir.-Investor Rel.
Alan L. Mueller, VP-Drug Discovery
Gregory M. Torre, Sr. VP-Regulatory Affairs & Quality
Peter G. Tombros, Lead Dir.-Board

Phone: 973-394-8600	Fax: 973-316-6463

Toll-Free:

Address: 300 Interpace Pkwy., 4th Fl., Bldg. B, Parsippany, NJ 07054 US

GROWTH PLANS/SPECIAL FEATURES:
NPS Pharmaceuticals, Inc. is a biopharmaceutical company focused on the development and commercialization of small molecule drugs and recombinant proteins for the treatment of bone and mineral disorders; gastrointestinal disorders; and central nervous system disorders. The company's products include Preos, for the treatment of post-menopausal osteoporosis, which is approved for marketing in Europe under the name Preotact and is awaiting FDA approval for commercialization in the U.S.; the FDA approved cinacalcet HCl for the treatment of hyperparathyroidism developed by the firm's licensees, Amgen and Kirin Brewery, and marketed under the trademark Sensipar in the U.S. and Mimpara in Europe; and teduglutide, for the treatment of the short bowel syndrome, which is in Phase III clinical trial. NPS Pharmaceuticals has collaborative research, development of license agreements with several companies including Amgen, AstraZeneca, GlaxoSmithKline, Janssen, Kirin and Nycomed. The company has been issued roughly 196 patents in the U.S. and granted approximately 890 patents in other countries. In July 2007, the firm sold its Salt Lake City, Utah facility to the University of Utah for roughly $21 million and the Mississauga, Ontario pilot manufacturing and laboratory facility to Transglobe Property Management Services, Ltd for roughly $4 million. Later that month, NPS Pharmaceuticals announced the sale of its royalty entitlement from the European sales of Preotact for roughly $50 million up front and another $25 million if target sales are met.

FINANCIALS: Sales and profits are in thousands of dollars—add 000 to get the full amount. 2006 Note: Financial information for 2006 was not available for all companies at press time.

2006 Sales: $48,502	2006 Profits: $-112,668	**U.S. Stock Ticker:** NPSP
2005 Sales: $12,825	2005 Profits: $-169,723	**Int'l Ticker:** Int'l Exchange:
2004 Sales: $14,237	2004 Profits: $-168,251	Employees: 196
2003 Sales: $9,919	2003 Profits: $-170,395	Fiscal Year Ends: 12/31
2002 Sales: $2,200	2002 Profits: $-86,800	Parent Company:

SALARIES/BENEFITS:

Pension Plan:	ESOP Stock Plan:	Profit Sharing:	Top Exec. Salary: $480,962	Bonus: $245,291
Savings Plan:	Stock Purch. Plan:		Second Exec. Salary: $307,678	Bonus: $92,342

OTHER THOUGHTS:
Apparent Women Officers or Directors: 1
Hot Spot for Advancement for Women/Minorities:

LOCATIONS: ("Y" = Yes)

West:	Southwest:	Midwest:	Southeast:	Northeast:	International:
				Y	Y

NUTRITION 21 INC

www.nutrition21.com

Industry Group Code: 325411 Ranks within this company's industry group: Sales: 7   Profits: 9

Drugs:	Other:	Clinical:	Computers:	Other:
Discovery:	AgriBio:	Trials/Services:	Hardware:	Specialty Services:
Licensing:	Genomics/Proteomics:	Laboratories:	Software:	Consulting:
Manufacturing: Y	Tissue Replacement:	Equipment/Supplies:	Arrays:	Blood Collection:
Development: Y		Research/Development Svcs.:	Database Management:	Drug Delivery:
Generics:		Diagnostics:		Drug Distribution:

TYPES OF BUSINESS:

Dietary Supplements
Cardiovascular & Diabetes Treatments

BRANDS/DIVISIONS/AFFILIATES:

Chromax
Diachrome
Selenomax
Iceland Health, Inc.
Iceland Health Maximum Strength Omega-3
Iceland Health Joint Relief

CONTACTS: Note: Officers with more than one job title may be intentionally listed here more than once.

Paul S. Intlekofer, CEO
Paul S. Intlekofer, Pres.
Alan J. Kirschbaum, CFO
Dean DiMaria, Sr. VP-Mktg. & Sales
James Kormorowski, VP-Scientific Affairs
James Kormorowski, VP-Tech. Svcs.
John Gutfreund, Chmn.

Phone: 914-701-4500	Fax: 914-696-0860
Toll-Free:	
Address: 4 Manhattanville Rd., Purchase, NY 10577 US	

GROWTH PLANS/SPECIAL FEATURES:

Nutrition 21, Inc. specializes in chromium-based therapeutic nutrition supplements for disease-specific conditions, including diabetes, obesity, metabolic syndrome, cardiovascular disease and depression. Chromax chromium picolinate is used as a daily mineral supplement and as an ingredient in sports and animal nutrition products, weight loss products, baked goods and beverages. Nutrition 21 plans to extend the market value of the product by introducing it to consumers interested in heightened metabolic health, and to a market of women aged 35-55. The company's Chromax product is marketed at a variety of major retailers including CVS, Walgreens and Wal-Mart. The firm's branded product Diachrome is a combination of chromium picolinate and biotin, and is patented as a nutrition supplement for people with diabetes. Through an alliance with XLHealth, Diachrome is being tested for efficacy and will be marketed to patients. Nutrition 21 also produces Selenomax, which is a high selenium yeast based nutritional supplement. According to clinical studies, selenium is an important trace mineral that is involved with the immune system and thyroid function. The firm continues to look forward to scientific studies that can demonstrate the connections between chromium picolinate and the control of diabetes. In August 2006, the company acquired Iceland Health, Inc. Iceland Health is a supplier of omega-3 fatty acids and imports Icelandic fish oils. This acquisition marks the beginning of Nutrition 21's entrance into the omega-3 market, with the products Iceland Health Maximum Strength Omega-3 and Iceland Health Joint Relief.

FINANCIALS: Sales and profits are in thousands of dollars—add 000 to get the full amount. 2006 Note: Financial information for 2006 was not available for all companies at press time.

2006 Sales: $10,664	2006 Profits: $-10,317	**U.S. Stock Ticker: NXXI**
2005 Sales: $10,711	2005 Profits: $-7,044	**Int'l Ticker:** Int'l Exchange:
2004 Sales: $10,232	2004 Profits: $-5,901	Employees: 30
2003 Sales: $10,615	2003 Profits: $-10,506	Fiscal Year Ends: 6/30
2002 Sales: $14,700	2002 Profits: $-6,000	Parent Company:

SALARIES/BENEFITS:

Pension Plan:	ESOP Stock Plan:	Profit Sharing:	Top Exec. Salary: $273,125	Bonus: $
Savings Plan: Y	Stock Purch. Plan:		Second Exec. Salary: $230,908	Bonus: $

OTHER THOUGHTS:

Apparent Women Officers or Directors:
Hot Spot for Advancement for Women/Minorities:

LOCATIONS: ("Y" = Yes)

West:	Southwest:	Midwest:	Southeast:	Northeast:	International:
				Y	

NUVELO INC

www.nuvelo.com

Industry Group Code: 325412 Ranks within this company's industry group: Sales: 144 Profits: 192

Drugs:		Other:	Clinical:	Computers:	Other:
Discovery:	Y	AgriBio:	Trials/Services:	Hardware:	Specialty Services:
Licensing:		Genomics/Proteomics:	Laboratories:	Software:	Consulting:
Manufacturing:	Y	Tissue Replacement:	Equipment/Supplies:	Arrays:	Blood Collection:
Development:	Y		Research/Development Svcs.:	Database Management:	Drug Delivery:
Generics:			Diagnostics:		Drug Distribution:

TYPES OF BUSINESS:

Drugs-Cardiovascular & Cancer Treatments
Anticoagulants

BRANDS/DIVISIONS/AFFILIATES:

alfimeprase
RNAPc2
NU206
NU172
Kirin Brewery Company, Ltd.

CONTACTS: *Note: Officers with more than one job title may be intentionally listed here more than once.*

Ted W. Love, CEO
H. Ward Wolff, CFO
Jill M. Pergande, VP-Human Resources
Michael D. Levy, Exec. VP-R&D
Ralph T. Zitnik, VP-Dev.
Lee Bendekgey, General Counsel/Sr. VP
Shelly D. Guyer, VP-Bus. Dev.
Shelly D. Guyer, VP-Investor Rel.
H. Ward Wolff, VP-Finance
Brian S. Kersten, VP-Regulatory Affairs & Quality Assurance
Walter Funk, VP-Research
Ted W. Love, Chmn.

Phone: 650-517-8000	Fax: 650-517-8001

Toll-Free:

Address: 201 Industrial Rd., Ste. 310, San Carlos, CA 94070-6211 US

GROWTH PLANS/SPECIAL FEATURES:

Nuvelo, Inc. is a biopharmaceutical company focused on acute cardiovascular indications and cancer. The firm has three drug candidates currently in clinical trials, as well as one preclinical candidate. The leading drug candidate, alfimeprase, is a thrombolytic agent, or blood clot dissolver. It's currently in two Phase III clinical trials for the treatment of acute peripheral arterial occlusion and catheter occlusion. Nuvelo's second drug candidate, recombinant nematode anticoagulant protein c2, or rNAPc2, is a recombinant version of a naturally occurring protein that has anticoagulant properties. RNAPc2 is currently undergoing Phase II clinical trial for use in treating acute coronary syndromes. The third drug candidate in the cardiovascular portfolio is preclinical NU172, a direct thrombin inhibitor for use as a short-acting anticoagulant during medical procedures. The company has a collaboration agreement with the pharmaceutical division of Kirin Brewery Company, Ltd., for the development and commercialization of NU206, under which Phase I trials commenced in early 2007. NU206 will be tested as a supportive cancer therapy, specifically to treat radiation- and chemotherapy-induced mucositis in the gastrointestinal tract. Another drug candidate was ARC183, a thrombin inhibitor, which completed its Phase 1 clinical trial in 2005 for use as an anticoagulant in coronary artery bypass graft surgery. However, this test showed that the amount of the drug required to achieve the desired effects resulted in a sub-optimal dosage profile. Thus ARC183 was dropped in favor of other thrombin inhibitors. Beyond these candidates, Nuvelo maintains an ongoing discovery program focused on proprietary human gene-encoding proteins and related developmental research. In 2007, Nuvelo received fast track status from the FDA for RNAPc2 in both first- and second-line treatment of metastatic colorectal cancer.

Nuvelo employee benefits include an educational assistance program and company activities including an annual Halloween festival, Thanksgiving feast and winter party.

FINANCIALS: Sales and profits are in thousands of dollars—add 000 to get the full amount. 2006 Note: Financial information for 2006 was not available for all companies at press time.

2006 Sales: $3,888	2006 Profits: $-130,553	**U.S. Stock Ticker: NUVO**
2005 Sales: $ 545	2005 Profits: $-71,611	**Int'l Ticker:** Int'l Exchange:
2004 Sales: $ 195	2004 Profits: $-52,489	Employees: 146
2003 Sales: $1,024	2003 Profits: $-50,187	Fiscal Year Ends: 12/31
2002 Sales: $26,433	2002 Profits: $-44,978	Parent Company:

SALARIES/BENEFITS:

Pension Plan:	ESOP Stock Plan:	Profit Sharing:	Top Exec. Salary: $640,417	Bonus: $289,248
Savings Plan: Y	Stock Purch. Plan: Y		Second Exec. Salary: $398,333	Bonus: $184,828

OTHER THOUGHTS:

Apparent Women Officers or Directors: 2
Hot Spot for Advancement for Women/Minorities: Y

LOCATIONS: ("Y" = Yes)

West:	Southwest:	Midwest:	Southeast:	Northeast:	International:
Y					

NYCOMED

www.nycomed.com

Industry Group Code: 325412 **Ranks within this company's industry group:** Sales: Profits:

Drugs:		Other:	Clinical:	Computers:	Other:
Discovery:	Y	AgriBio:	Trials/Services:	Hardware:	Specialty Services:
Licensing:	Y	Genomics/Proteomics:	Laboratories:	Software:	Consulting:
Manufacturing:	Y	Tissue Replacement:	Equipment/Supplies:	Arrays:	Blood Collection:
Development:	Y		Research/Development Svcs.:	Database Management:	Drug Delivery:
Generics:			Diagnostics:		Drug Distribution:

TYPES OF BUSINESS:

Pharmaceuticals

BRANDS/DIVISIONS/AFFILIATES:

Alvesco
Curosurf
Angiox
Preotact
Matrifen
Beriplast
Riopan
Neosaldina

CONTACTS: *Note: Officers with more than one job title may be intentionally listed here more than once.*

Hakan Bjorklund, CEO
Runar Bjorklund, CFO
Dick Soderberg, Exec. VP-Mktg.
Alfred Goll, Exec. VP-Human Resources
Anders Ullman, Exec. VP-R&D
Thomas Redemann, Dir.-Global Contract Mfg.
Barthold Piening, Exec. VP-Oper.
Kerstin Valinder, Exec. VP-Bus. Dev.
Christoffer Jensen, VP-Comm.
Christian B. Seidelin, VP-Finance
Christian Kanzelmeyer, VP-Licensing
Susanne Hof, Head-External Comm.
Charles Depasse, Exec. VP-Integration
Otto Schwarz, Exec. VP-Commercial Oper.
Poul Haukrog Moller, VP-Int'l Mktg.
Juliane Bruggemann, VP-Supply Chain

Phone: 45-46-77-11-11	Fax: 45-46-75-66-40
Toll-Free:	
Address: Langebjerg 1, Roskilde, 4000 Denmark	

GROWTH PLANS/SPECIAL FEATURES:

Nycomed is a European-based pharmaceutical company engaged in the research, licensing, manufacturing and marketing a wide array of products with application in many therapeutic areas, particularly cardiology, gastroenterology, osteoporosis, respiratory, pain and tissue management. Products include Alvesco, Curosurf and OMNARIS/SOMNAIR for respiratory problems; Angiox and Ebrantil i.v. for cardiology problems; Beriplast, and TachoSil for tissue management; CalciChew and Preotact for treatment of osteoporosis; Matrifen, Neosaldina and Xefo Rapid for pain management; and Pantoprazole and Riopan for gastroenterology problems. The company is privately owned and has production sites in Austria, Brazil, Ireland, Mexico, Estonia, Denmark, Germany, Finland, India, Norway, Belgium, Poland and the U.S. The firm's research and development program is led out of Konstanz, Germany with additional sites in Denmark, India and the U.S. In December 2006, Nycomed acquired Altana Pharma AG, a German pharmaceutical company, for roughly $6.5 billion. Altana Pharma researches therapeutic drugs for the treatment of gastrointestinal and respiratory diseases. Since the closing of the transaction, Altana Pharma has focused on its specialty chemicals business. The respiratory product Alvesco, which is marketed in 26 countries, was also acquired in the sale.

FINANCIALS: Sales and profits are in thousands of dollars—add 000 to get the full amount. 2006 Note: Financial information for 2006 was not available for all companies at press time.

2006 Sales: $	2006 Profits: $	**U.S. Stock Ticker: Private**
2005 Sales: $	2005 Profits: $	**Int'l Ticker:** Int'l Exchange:
2004 Sales: $	2004 Profits: $	Employees: 12,000
2003 Sales: $	2003 Profits: $	Fiscal Year Ends: 12/31
2002 Sales: $	2002 Profits: $	Parent Company:

SALARIES/BENEFITS:

Pension Plan:	ESOP Stock Plan:	Profit Sharing:	Top Exec. Salary: $	Bonus: $
Savings Plan:	Stock Purch. Plan:		Second Exec. Salary: $	Bonus: $

OTHER THOUGHTS:

Apparent Women Officers or Directors: 4
Hot Spot for Advancement for Women/Minorities: Y

LOCATIONS: ("Y" = Yes)

West:	Southwest:	Midwest:	Southeast:	Northeast:	International:
					Y

Note: Financial information, benefits and other data can change quickly and may vary from those stated here.

ONYX PHARMACEUTICALS INC   www.onyx-pharm.com

Industry Group Code: 325412 Ranks within this company's industry group: Sales: 177   Profits: 181

Drugs:		Other:	Clinical:	Computers:	Other:
Discovery:	Y	AgriBio:	Trials/Services:	Hardware:	Specialty Services:
Licensing:		Genomics/Proteomics:	Laboratories:	Software:	Consulting:
Manufacturing:		Tissue Replacement:	Equipment/Supplies:	Arrays:	Blood Collection:
Development:	Y		Research/Development Svcs.:	Database Management:	Drug Delivery:
Generics:			Diagnostics:		Drug Distribution:

TYPES OF BUSINESS:

Pharmaceuticals Discovery & Development
Small-Molecule Drugs
Cancer Treatments

BRANDS/DIVISIONS/AFFILIATES:

Nexavar

CONTACTS: Note: Officers with more than one job title may be intentionally listed here more than once.

Hollings C. Renton, CEO
Hollings C. Renton, Pres.
Gregory W. Schafer, CFO/VP
Edward F. Kenney, Chief Commercial Officer/Exec. VP
Kathleen Stafford, VP-Human Resources
Jeffrey D. Bloss, VP-Clinical Dev.
Gregory J. Giotta, Chief Legal Counsel/VP
Jeanne Y. Jew, VP-Corp. & Commercial Dev.
Julianna R. Wood, VP-Corp. Comm.
Julianna R. Wood, VP-Investor Rel.
Randy A. Kelley, VP-Sales
Laura A. Brege, Chief Bus. Officer/Exec. VP
Henry J. Fuchs, Chief Medical Officer/Exec. VP
Patricia A. Oto, VP-Regulatory Affairs
Hollings C. Renton, Chmn.

Phone: 510-597-6500	Fax: 510-597-6600

Toll-Free:

Address: 2100 Powell St., Emeryville, CA 94608 US

GROWTH PLANS/SPECIAL FEATURES:

Onyx Pharmaceuticals, Inc. is engaged in the discovery, development and commercialization of innovative products that target oncological molecular mechanisms. The company's flagship drug, Nexavar, is the result of a partnership with Bayer Corporation and is aimed at blocking inappropriate growth signals in tumor cells by inhibiting the responsible active enzymes that induce cancer cell growth. Nexavar is intended to inhibit both tumor cell proliferation and angiogenesis and is also approved by the U.S. FDA to treat patients with advanced kidney cancer. In July 2006, Nexavar was approved by the European Commission for the treatment of advanced renal cell carcinoma in patients that have failed to respond or are unsuited for other forms of therapy. Onyx and Bayer are also in the process of conducting Phase III clinical studies of Nexavar as a treatment for advanced primary liver cancer, metastatic melanoma and non-small cell lung cancer. Nexavar is also being tested in Phase II and Phase I clinical trials in combination with standard chemotherapeutic and other anticancer agents for the treatment of lung, breast, and other cancers. In a previous collaboration with Warner-Lambert Company, Onyx developed compounds that modulate the activity of key enzymes that regulate the process of single cell replication in the cell cycle. While the research phase of the cell cycle kinase inhibitor program terminated in 2001, Pfizer has continued the development of the most advanced compound in the program, PD 332991, which will allow Onyx to receive milestone payments and royalties upon commercialization of the product. In June 2007, Bayer and Onyx submitted a new European marketing authorization application to the European Medicines Agency to approve the marketing of Nexacar for the treatment of hepatocellular carcinoma.

Onyx offers employees flexible spending accounts, tuition reimbursement, stock options, membership in a credit union and an employee assistance program.

FINANCIALS: Sales and profits are in thousands of dollars—add 000 to get the full amount. 2006 Note: Financial information for 2006 was not available for all companies at press time.

2006 Sales: $ 250	2006 Profits: $-92,681	**U.S. Stock Ticker:** ONXX
2005 Sales: $1,000	2005 Profits: $-95,174	**Int'l Ticker:** Int'l Exchange:
2004 Sales: $ 500	2004 Profits: $-46,756	Employees: 125
2003 Sales: $	2003 Profits: $-44,969	Fiscal Year Ends: 12/31
2002 Sales: $2,700	2002 Profits: $-45,800	Parent Company:

SALARIES/BENEFITS:

Pension Plan:	ESOP Stock Plan: Y	Profit Sharing:	Top Exec. Salary: $525,000	Bonus: $
Savings Plan: Y	Stock Purch. Plan: Y		Second Exec. Salary: $390,000	Bonus: $

OTHER THOUGHTS:

Apparent Women Officers or Directors: 4
Hot Spot for Advancement for Women/Minorities: Y

LOCATIONS: ("Y" = Yes)

West:	Southwest:	Midwest:	Southeast:	Northeast:	International:
Y					

ORASURE TECHNOLOGIES INC   www.orasure.com

Industry Group Code: 325413 Ranks within this company's industry group: Sales: 16   Profits: 10

Drugs:	Other:	Clinical:		Computers:		Other:	
Discovery:	AgriBio:	Trials/Services:		Hardware:		Specialty Services:	
Licensing:	Genomics/Proteomics:	Laboratories:		Software:		Consulting:	
Manufacturing:	Tissue Replacement:	Equipment/Supplies:	Y	Arrays:		Blood Collection:	
Development:		Research/Development Svcs.:	Y	Database Management:		Drug Delivery:	
Generics:		Diagnostics:	Y			Drug Distribution:	

TYPES OF BUSINESS:
Medical Devices Manufacturing
Oral Fluid Collection Devices
Cryosurgical Products

BRANDS/DIVISIONS/AFFILIATES:
OraQuick
OraSure
Histofreezer
Interceptor

CONTACTS: Note: Officers with more than one job title may be intentionally listed here more than once.
Douglas A. Michels, CEO
Ronald H. Spair, COO
Douglas A. Michels, Pres.
Ronald H. Spair, CFO
Joseph E. Zack, Exec. VP-Sales & Mktg.
Stephen Lee, Chief Science Officer/Exec. VP
Jack E. Jerrett, General Counsel/Sr. VP/Sec.
P. Michael Formica, Exec. VP-Oper.
Mark K. Luna, Sr. VP-Finance/Controller
Kenneth A. Adach, VP-Consumer Mktg.
Douglas Watson, Chmn.

Phone: 610-882-1820	Fax: 610-882-1830
Toll-Free:	
Address: 220 E. First St., Bethlehem, PA 18015 US	

GROWTH PLANS/SPECIAL FEATURES:

OraSure Technologies, Inc. develops, manufactures, markets and sells oral fluid specimen collection devices using the proprietary oral fluid technologies, as well as other diagnostic products including immunoassays and other in vitro diagnostic tests that are used on other specimen types, and other medical devices. The company's diagnostic products include tests that are processed in a laboratory and tests that are performed on a rapid basis at the point of care. The firm's principal platform technologies are the OraSure/Intercept oral fluid collection devices used in conjunction with screening and confirmatory tests for HIV-1 infection and other analytes; the OraQuick rapid test platform for testing oral fluid, whole blood and plasma samples for the presence of various antibodies or analytes; and the Histofreezer cryosurgical system for removal of warts and other benign skin lesions by physicians. OraSure Technologies' products are sold in the U.S. and internationally to various clinical laboratories, hospitals, clinics, community-based organizations and other public health organizations, distributors, government agencies, physicians' offices and commercial and industrial entities. International sales generated 17% of revenue in 2006. In July 2007, the company signed an agreement for the supply of OraQuick HIV-1/2 antibody test with the Supply Chain Management System.

The company offers its employees health and dental coverage; life insurance; short- and long-term disability; flexible spending accounts; a 401(k) plan; and a stock award plan.

FINANCIALS: Sales and profits are in thousands of dollars—add 000 to get the full amount. 2006 Note: Financial information for 2006 was not available for all companies at press time.

2006 Sales: $68,155	2006 Profits: $5,268	**U.S. Stock Ticker: OSUR**
2005 Sales: $69,366	2005 Profits: $27,448	**Int'l Ticker:** Int'l Exchange:
2004 Sales: $54,008	2004 Profits: $- 560	Employees: 250
2003 Sales: $40,451	2003 Profits: $-1,136	Fiscal Year Ends: 12/31
2002 Sales: $32,000	2002 Profits: $-3,300	Parent Company:

SALARIES/BENEFITS:

Pension Plan:	ESOP Stock Plan:	Profit Sharing:	Top Exec. Salary: $433,146	Bonus: $108,400
Savings Plan: Y	Stock Purch. Plan:		Second Exec. Salary: $320,220	Bonus: $87,500

OTHER THOUGHTS:
Apparent Women Officers or Directors:
Hot Spot for Advancement for Women/Minorities:

LOCATIONS: ("Y" = Yes)

West:	Southwest:	Midwest:	Southeast:	Northeast:	International:
				Y	Y

Note: Financial information, benefits and other data can change quickly and may vary from those stated here.

ORCHID CELLMARK INC   www.orchid.com

Industry Group Code: 621511 Ranks within this company's industry group: Sales: 4   Profits: 4

Drugs:	Other:		Clinical:		Computers:		Other:	
Discovery:	AgriBio:		Trials/Services:		Hardware:		Specialty Services:	Y
Licensing:	Genomics/Proteomics:	Y	Laboratories:		Software:		Consulting:	
Manufacturing:	Tissue Replacement:		Equipment/Supplies:		Arrays:		Blood Collection:	
Development:			Research/Development Svcs.:		Database Management:		Drug Delivery:	
Generics:			Diagnostics:	Y			Drug Distribution:	

TYPES OF BUSINESS:

Research-Bioinformatics
Genomics Services
Diagnostic Products & Services
Genetic Databases
DNA Testing
Forensic Testing

BRANDS/DIVISIONS/AFFILIATES:

Orchid Cellmark
Orchid Europe
Orchid GeneScreen

CONTACTS: *Note: Officers with more than one job title may be intentionally listed here more than once.*

Thomas A. Bologna, CEO
Thomas A. Bologna, Pres.
John Deighan, CFO/Corp. Controller
Wiliam Lutz, VP-IT
Warren Meltzer, General Counsel
Nicholas Conti, VP-Corp. Dev.
Mary Bashore, Dir.-Corp. Comm.
Mary Bashore, Dir.-Investor Rel.
Mark D. Stolorow, Exec. Dir.-Forensic Science
Bruce Basarab, VP-North American Sales & Mktg.
George Poste, Chmn.

Phone: 609-750-2200	Fax: 609-750-6400
Toll-Free:	
Address: 4390 US Rte. 1, Princeton, NJ 08540 US	

GROWTH PLANS/SPECIAL FEATURES:

Orchid Cellmark, Inc. (ORCH), formerly Orchid BioSciences, Inc., provides identity genomics services for the forensic and public health markets as well as paternity DNA testing and animal DNA testing. Forensic DNA testing is primarily used to establish and maintain DNA profile databases of individuals arrested or convicted of crimes, to analyze and compare evidence from crime scenes or to determine if a man has fathered a particular child in paternity cases. In agricultural applications, DNA testing services are available for selective trait breeding and traceability applications. Technologies utilized by ORCH include short tandem repeats (STRs) for forensic and paternal testing and single nucleotide polymorphisms (SNPs) for DNA agricultural applications. Agricultural projects currently pursued by the company include scrapie genotyping, a U.K. government project designed to reduce the animal disease, scrapie, on sheep farms. In addition to casework testing, the company provides DNA identification profiles of individuals for national, state and local criminal DNA databases in the U.S. and U.K. ORCH's services have been selected by major police departments in the U.S., including New York City, Chicago, Phoenix and Houston, and London's Metropolitan Police Force, or Scotland Yard, in the U.K. The company has conducted DNA testing for such notable cases as those of O.J. Simpson, Jon Benet Ramsey, Danielle Van Dam, the Unabomber and the Green River murderer and has employed its SNP technology to identify a large number of previously unidentified World Trade Center victims. In late 2006, ORCH was awarded new DNA contracts with a combined total of over $3.5 million by Kent, London and Sussex police forces.

The company offers its employees full medical benefits, credit union membership, reward and recognition programs, employee assistance programs and employee appreciation activities.

FINANCIALS: Sales and profits are in thousands of dollars—add 000 to get the full amount. 2006 Note: Financial information for 2006 was not available for all companies at press time.

2006 Sales: $56,854	2006 Profits: $-11,271	**U.S. Stock Ticker:** ORCH
2005 Sales: $61,609	2005 Profits: $-9,439	**Int'l Ticker:** Int'l Exchange:
2004 Sales: $62,499	2004 Profits: $-8,812	Employees: 403
2003 Sales: $50,627	2003 Profits: $-23,568	Fiscal Year Ends: 12/31
2002 Sales: $50,400	2002 Profits: $-80,100	Parent Company:

SALARIES/BENEFITS:

Pension Plan:	ESOP Stock Plan:	Profit Sharing:	Top Exec. Salary: $368,750	Bonus: $192,500
Savings Plan: Y	Stock Purch. Plan:		Second Exec. Salary: $157,244	Bonus: $

OTHER THOUGHTS:

Apparent Women Officers or Directors: 1
Hot Spot for Advancement for Women/Minorities:

LOCATIONS: ("Y" = Yes)

West:	Southwest:	Midwest:	Southeast:	Northeast:	International:
	Y	Y	Y	Y	Y

ORGANOGENESIS INC   www.organogenesis.com

Industry Group Code: 325414 Ranks within this company's industry group: Sales: 3 Profits: 2

Drugs:	Other:	Clinical:	Computers:	Other:
Discovery:	AgriBio:	Trials/Services:	Hardware:	Specialty Services:
Licensing:	Genomics/Proteomics:	Laboratories:	Software:	Consulting:
Manufacturing:	Tissue Replacement: Y	Equipment/Supplies: Y	Arrays:	Blood Collection:
Development:		Research/Development Svcs.:	Database Management:	Drug Delivery:
Generics:		Diagnostics:		Drug Distribution:

TYPES OF BUSINESS:

Tissue Replacement Products
Wound Dressing Products

BRANDS/DIVISIONS/AFFILIATES:

Apligraf
FortaDerm Antimicrobial
FortaPerm
FortaGen
CuffPatch
Revitix
TestSkin & TestSkin II
BioSTAR

CONTACTS: Note: Officers with more than one job title may be intentionally listed here more than once.

Geoff MacKay, CEO
Gary S. Gillheeney, Sr., COO/Exec. VP
Geoff MacKay, Pres.
Gary S. Gillheeney, Sr., CFO/Exec. VP
Santino Costanzo, VP-Sales, Bio-active Wound Healing
Houda Damaha, Dir.-Human Resources
Vincent Ronfard, Chief Scientist Officer
Phillip Nolan, VP-Mfg. Oper.
Richard Shaw, VP-Finance
Dario Eklund, VP-Bio-Surgery & Bio-Aesthetics
Patrick Bilbo, VP-Clinical & Regulatory Affairs
Susan Chapman, Dir.-Production
Shannon Banks, Dir.-Project Mgmt.

Phone: 781-575-0775	Fax: 781-575-0440
Toll-Free:	
Address: 150 Dan Rd., Canton, MA 02021 US	

GROWTH PLANS/SPECIAL FEATURES:

Organogenesis, Inc. is a tissue-engineering firm that designs, develops and manufactures medical products containing living cells or natural connective tissue. Organogenesis specializes in bio-active wound healing, bio-aesthetics and bio-surgery. Its bio-active wound healing products include: Apligraf, designed for the treatment of venous leg ulcers due to poor blood circulation and for diabetic foot ulcers without tendon, muscle, capsule or bone exposure; Fortaderm Antimicrobial, a collagen based antimicrobial wound dressing that helps wounds heal by providing the dermal scaffolding or structure to help the body's own cells migrate to facilitate closure of the wound; and VCT01, currently in late stage development, the company's next generation of bio-engineered skin substitute, which is self assembled, bi-layered bio engineered skin. Bio-aesthetics products include: Revitix, which utilizes the technology and living cells that make up Apligraf to rejuvenate skin; and TestSkin, along with TestSkin II, models of human skin that mimic the key properties of actual skin, used for simulating in a lab the reaction of skin when exposed to various products/drugs. Bio-surgery products include: CuffPatch for rotator cuff surgery, which repairs the tendons that connect the upper arm bones to the shoulder blade; FortaGen, used for tissue repair in cases of vaginal prolapse (a condition caused by the protrusion (herniation) of a pelvic organ into the vaginal area); FortaPerm, used as a sling to raise a dropped bladder neck, which causes stress urinary incontinence, the leakage of urine as a result of increased pressure (i.e., coughing, laughing, sneezing) upon the bladder, back into a position from which it is able to shut when pressure is applied; and BioSTAR, an implant technology for the treatment of cardiac sources of migraine headaches, strokes and other potential brain attacks, namely a common heart defect called a patent foramen ovale (PFO).

FINANCIALS: Sales and profits are in thousands of dollars—add 000 to get the full amount. 2006 Note: Financial information for 2006 was not available for all companies at press time.

2006 Sales: $	2006 Profits: $	U.S. Stock Ticker: Private
2005 Sales: $	2005 Profits: $	Int'l Ticker: Int'l Exchange:
2004 Sales: $	2004 Profits: $	Employees: 182
2003 Sales: $	2003 Profits: $	Fiscal Year Ends: 12/31
2002 Sales: $	2002 Profits: $	Parent Company:

SALARIES/BENEFITS:

Pension Plan:	ESOP Stock Plan:	Profit Sharing:	Top Exec. Salary: $277,420	Bonus: $56,000
Savings Plan: Y	Stock Purch. Plan:		Second Exec. Salary: $229,836	Bonus: $30,000

OTHER THOUGHTS:

Apparent Women Officers or Directors: 2
Hot Spot for Advancement for Women/Minorities: Y

LOCATIONS: ("Y" = Yes)

West:	Southwest:	Midwest:	Southeast:	Northeast:	International:
				Y	Y

OSCIENT PHARMACEUTICALS INC   www.oscient.com

Industry Group Code: 325412 Ranks within this company's industry group: Sales: 81   Profits: 174

Drugs:		Other:	Clinical:	Computers:	Other:
Discovery:	Y	AgriBio:	Trials/Services:	Hardware:	Specialty Services:
Licensing:		Genomics/Proteomics:	Laboratories:	Software:	Consulting:
Manufacturing:		Tissue Replacement:	Equipment/Supplies:	Arrays:	Blood Collection:
Development:			Research/Development Svcs.:	Database Management:	Drug Delivery:
Generics:			Diagnostics:		Drug Distribution:

TYPES OF BUSINESS:

Pharmaceuticals Discovery & Development
Pharmaceuticals Commercialization
Antibiotics

BRANDS/DIVISIONS/AFFILIATES:

Genome Therapeutics Corporation
Genesoft Pharmaceuticals
FACTIVE
Ramoplanin
ANTARA
Menarini Group
Abbott Canada

CONTACTS: *Note: Officers with more than one job title may be intentionally listed here more than once.*

Steven Rauscher, CEO
Steven Rauscher, Pres.
Philippe M. Maitre, CFO/Sr. VP
Thomas Chen, VP-Mktg.
Joseph Pane, VP-Human Resources
Inder Kaul, VP-Clinical Dev., Medical & Regulatory Affairs
Robert Spadafora, VP-Legal Affairs
Nick Colangelo, Exec. VP-Oper.
Nick Colangelo, Exec. VP-Corp. Dev.
Christopher J. M. Taylor, VP-Corp. Comm.
Christopher J. M. Taylor, VP-Investor Rel.
Glenn Tillotson, VP-Science Strategic Rel.
Elenie Chadbourne, VP-Drug Safety & Pharmacovigilance
David K. Stone, Chmn.

Phone: 781-398-2300	Fax: 781-893-9535
Toll-Free:	
Address: 1000 Winter St., Ste. 2200, Waltham, MA 02451 US	

GROWTH PLANS/SPECIAL FEATURES:

Oscient Pharmaceuticals, formed through the 2004 merger of Genome Therapeutics Corporation and Genesoft Pharmaceuticals, is engaged in the development and commercialization of new therapeutics to address unmet medical needs. The company's lead product, FACTIVE (gemifloxacin mesylate), has been approved in tablet formulation by the FDA for two indications: community-acquired pneumonia of mild to moderate severity and acute bacterial exacerbations of chronic bronchitis. Oscient also develops the drug ANTARA (fenofibrate), which is indicated for the adjunct treatment of hypercholesterolemia (high blood cholesterol) and hypertriglyceridemia (high triglycerides) in combination with diet. The company is also engaged in advanced clinical development of a novel antibiotic candidate, Ramoplanin, for the treatment of Clostridium difficile-associated disease. Oscient is actively developing its sales and marketing infrastructure in conjunction with the commercialization of FACTIVE, as well as building new capacities to function more effectively in the biopharmaceutical industry. It has an agreement with LG Life Sciences, a partner in its FACTIVE development, to outsource commercial interests to a manufacturing facility in South Korea. In January 2007, Oscient granted the commercialization rights to FACTIVE tablets in Europe to Menarini Group, a leading European pharmaceutical company based in Italy. In March 2007, Abbott Canada, the Canadian affiliate of Abbott, launched FACTIVE tablets in Canada for the five-day treatment of acute bacterial exacerbations of chronic bronchitis (AECB). Oscient granted the commercialization rights for FACTIVE to Abbott Canada in August of 2006.

Oscient offers its employees medical, dental, vision, and life insurance; a 401(k) plan; short and long-term disability; an employee stock purchase plan; qualified tuition reimbursement; flexible work schedules; travel assistance; discounted fitness center access; adoption assistance; subsidized cafeteria service; and a range of bonus and incentive plans.

FINANCIALS: Sales and profits are in thousands of dollars—add 000 to get the full amount. 2006 Note: Financial information for 2006 was not available for all companies at press time.

2006 Sales: $46,152	2006 Profits: $-78,477	**U.S. Stock Ticker: OSCI**
2005 Sales: $23,609	2005 Profits: $-88,593	**Int'l Ticker:** Int'l Exchange:
2004 Sales: $6,613	2004 Profits: $-93,271	Employees: 336
2003 Sales: $7,009	2003 Profits: $-29,789	Fiscal Year Ends: 12/31
2002 Sales: $23,000	2002 Profits: $-34,000	Parent Company:

SALARIES/BENEFITS:

Pension Plan:	ESOP Stock Plan:	Profit Sharing:	Top Exec. Salary: $432,115	Bonus: $325,282
Savings Plan: Y	Stock Purch. Plan: Y		Second Exec. Salary: $338,654	Bonus: $206,136

OTHER THOUGHTS:

Apparent Women Officers or Directors: 1
Hot Spot for Advancement for Women/Minorities:

LOCATIONS: ("Y" = Yes)

West:	Southwest:	Midwest:	Southeast:	Northeast:	International:
				Y	

Note: Financial information, benefits and other data can change quickly and may vary from those stated here.

OSI PHARMACEUTICALS INC

www.osip.com

Industry Group Code: 325412 Ranks within this company's industry group: Sales: 44 Profits: 201

Drugs:		Other:		Clinical:		Computers:		Other:	
Discovery:	Y	AgriBio:		Trials/Services:		Hardware:		Specialty Services:	
Licensing:		Genomics/Proteomics:		Laboratories:		Software:		Consulting:	
Manufacturing:	Y	Tissue Replacement:		Equipment/Supplies:		Arrays:		Blood Collection:	
Development:	Y			Research/Development Svcs.:		Database Management:		Drug Delivery:	
Generics:				Diagnostics:				Drug Distribution:	

TYPES OF BUSINESS:

Drugs-Cancer
Drugs-Small-Molecule
Drugs-Diabetes
Drugs-Macular Degeneration

BRANDS/DIVISIONS/AFFILIATES:

Tarceva
Novantrone
Gelclair
OSI Prosidion
Macugen
OSI Eyetech
OSI Oncology
Eyetech Pharmaceuticals, Inc.

CONTACTS: Note: Officers with more than one job title may be intentionally listed here more than once.

Colin Goddard, CEO
Michael G. Atieh, CFO/Exec. VP
Linda E. Amper, VP-Human Resources
Neil Gibson, Chief Scientific Officer/VP
Robert L. Simon, Exec. VP-Pharmaceutical Dev. & Mfg.
Linda E. Amper, VP-Bus. Admin.
Barbara A. Wood, General Counsel/Sec.
Michael G. Atieh, Treas.
Gabriel Leung, Pres., OSI Oncology
Anker Lundemose, Exec. VP/Pres., OSI Prosidion
Paul G. Chaney, Exec. VP/Pres., OSI Eyetech
Robert A. Ingram, Chmn.

Phone: 631-962-2000	Fax: 631-752-3880
Toll-Free:	
Address: 41 Pinelawn Rd., Melville, NY 11747 US	

GROWTH PLANS/SPECIAL FEATURES:

OSI Pharmaceuticals is a biotechnology company that discovers, develops and commercializes molecular targeted therapies addressing major unmet medical needs in oncology ophthalmology and diabetes. The firm has four marketed products: Novantrone, Macugen, Gelclair and Tarceva. Novantrone reduces existing pain and delays cancer-related pain progression for prostate cancer. Gelclair, an oral gel, coats the inside of the mouth and forms a protective barrier to soothe oral lesions, a common side effect of chemotherapy. Tarceva is a small-molecule inhibitor of the epidermal growth factor receptor, the protein product of which is over-expressed in many solid tumor cancers. The drug is an oral, once-a-day drug with approved indication for non-small cell lung cancer. Tarceva is also in trials for indications of solid tumors and kidney cancer. The company currently has the oncology product erlotinib tablets in Phase III clinical trials. The company's OSI Prosidion business unit is focused on diabetes and obesity and is developing PSN9301, a DP-IV inhibitor, currently in Phase II trials. In April 2006, the company submitted a New Drug Application in Japan covering the use of Tarceva for the treatment of advanced or recurrent non-small cell lung cancer. In November 2006, the firm decided to divest its eye disease business, including Macugen, a product for the treatment of AMD (age-related macular degeneration). Macugen is the result of the company's acquisition of Eyetech Pharmaceuticals, Inc. In January 2007, the Prosidion segment outsourced its glucokinase activator, or GKA, program to Eli Lily & Co. for an upfront fee of $25 million. In July 2007, OSI announced an agreement to divest its anti-platelet derived growth factor program to Ophthotech Corporation.

OSI offers its employees stock options, tuition reimbursement, flexible spending plans, an employee assistance program and employee functions and activities.

FINANCIALS: Sales and profits are in thousands of dollars—add 000 to get the full amount. 2006 Note: Financial information for 2006 was not available for all companies at press time.

2006 Sales: $375,696	2006 Profits: $-582,184	U.S. Stock Ticker: OSIP	
2005 Sales: $174,194	2005 Profits: $-157,123	Int'l Ticker:	Int'l Exchange:
2004 Sales: $42,800	2004 Profits: $-260,371	Employees: 611	
2003 Sales: $32,369	2003 Profits: $-181,357	Fiscal Year Ends: 12/31	
2002 Sales: $21,800	2002 Profits: $-218,500	Parent Company:	

SALARIES/BENEFITS:

Pension Plan:	ESOP Stock Plan:	Profit Sharing:	Top Exec. Salary: $611,538	Bonus: $
Savings Plan: Y	Stock Purch. Plan: Y		Second Exec. Salary: $411,309	Bonus: $205,000

OTHER THOUGHTS:

Apparent Women Officers or Directors: 2
Hot Spot for Advancement for Women/Minorities: Y

LOCATIONS: ("Y" = Yes)

West:	Southwest:	Midwest:	Southeast:	Northeast:	International:
Y				Y	Y

OXIGENE INC
www.oxigene.com

Industry Group Code: 325412 Ranks within this company's industry group: Sales: Profits: 87

Drugs:		Other:	Clinical:	Computers:	Other:
Discovery:		AgriBio:	Trials/Services:	Hardware:	Specialty Services:
Licensing:	Y	Genomics/Proteomics:	Laboratories:	Software:	Consulting:
Manufacturing:		Tissue Replacement:	Equipment/Supplies:	Arrays:	Blood Collection:
Development:	Y		Research/Development Svcs.:	Database Management:	Drug Delivery:
Generics:			Diagnostics:		Drug Distribution:

TYPES OF BUSINESS:
Pharmaceuticals Acquisition & Development
Drugs-Cancer
Anti-Inflammatory Agents
Ocular Disease Treatments

BRANDS/DIVISIONS/AFFILIATES:
Zybrestat

CONTACTS: Note: Officers with more than one job title may be intentionally listed here more than once.
Richard Chin, CEO
Richard Chin, Pres.
James B. Murphy, CFO/VP
Dai Chaplin, Chief Scientific Officer/Head-R&D
John A Kollins, Chief Bus. Officer
Peter Harris, Chief Medical Officer
Joel Citron, Chmn.

Phone: 781-547-5900	Fax: 781-547-6800
Toll-Free:	
Address: 230 3rd Ave., Waltham, MA 02451 US	

GROWTH PLANS/SPECIAL FEATURES:
OXiGENE, Inc. is an international biopharmaceutical company engaged principally in the research and development of products for the treatment of cancer and certain ocular diseases. The company in-licenses complementary compounds from academic institutions in order to lead them through early-stage clinical trials, as well as to negotiate contracts with pharmaceutical companies to develop, market and manufacture resulting commercial drugs and clinical products. OXiGENE's primary drug development programs are based on a series of natural products called Combretastatins, which were originally isolated from the African bush willow tree (Combretum caffrum) by researchers at Arizona State University (ASU). ASU has granted the company a worldwide license for the use of the Combretastatins. The company has developed two primary technologies based on Combretastatins: vascular disrupting agents (VDAs) and ortho-quinone prodrugs (OQPs). The company's VDA compund, Zybrestat (CA4P), eliminates tumors by destroying the abnormal blood vessels that enable the tumor to grow. The product is currently in clinical testing for anaplastic thyroid cancer, Phase III; platinum resistant ovarian cancer, entering Phase III; solid tumors, such as lung cancer, in combination with Avastin, Genentech's anti-angiogenic agent, entering Phase II; and myopic macular degeneration, Phase II. The FDA has granted Zybrestat fast-track status for treatment of anaplastic thyroid cancer. OXiGENE's first OQP candidate is OXi4503 for use against solid tumors. The compound not only shuts down blood flow, but can also be metabolized into a compound which could assist with killing the remaining tumor cells at the periphery of the tumor by direct cytotoxic activity against tumor cells. It is currently in Phase I clinical trials.

OXiGENE offers employees healthcare, flexible spending accounts and other benefits.

FINANCIALS: Sales and profits are in thousands of dollars—add 000 to get the full amount. 2006 Note: Financial information for 2006 was not available for all companies at press time.

2006 Sales: $	2006 Profits: $-15,457	U.S. Stock Ticker: OXGN
2005 Sales: $ 1	2005 Profits: $-11,909	Int'l Ticker: Int'l Exchange:
2004 Sales: $ 7	2004 Profits: $-10,024	Employees: 22
2003 Sales: $ 30	2003 Profits: $-8,368	Fiscal Year Ends: 12/31
2002 Sales: $ 335	2002 Profits: $-11,013	Parent Company:

SALARIES/BENEFITS:

Pension Plan:	ESOP Stock Plan:	Profit Sharing:	Top Exec. Salary: $325,000	Bonus: $81,250
Savings Plan: Y	Stock Purch. Plan:		Second Exec. Salary: $220,000	Bonus: $100,000

OTHER THOUGHTS:
Apparent Women Officers or Directors:
Hot Spot for Advancement for Women/Minorities:

LOCATIONS: ("Y" = Yes)

West:	Southwest:	Midwest:	Southeast:	Northeast:	International:
				Y	Y

Note: Financial information, benefits and other data can change quickly and may vary from those stated here.

OXIS INTERNATIONAL INC

www.oxis.com

Industry Group Code: 325412 Ranks within this company's industry group: Sales: 139   Profits: 63

Drugs:		Other:		Clinical:		Computers:		Other:	
Discovery:	Y	AgriBio:		Trials/Services:		Hardware:		Specialty Services:	
Licensing:		Genomics/Proteomics:		Laboratories:		Software:		Consulting:	
Manufacturing:	Y	Tissue Replacement:		Equipment/Supplies:	Y	Arrays:		Blood Collection:	
Development:	Y			Research/Development Svcs.:	Y	Database Management:		Drug Delivery:	
Generics:				Diagnostics:	Y			Drug Distribution:	

TYPES OF BUSINESS:

Drugs-Oxidative Stress-Related Diseases
Small-Molecule Drugs
Diagnostic Products
Specialty Chemicals
Research Services
Veterinary Diagnostics
Anti-Oxidants
Nutraceuticals

BRANDS/DIVISIONS/AFFILIATES:

BioCheck, Inc.
OxisResearch
OXIS Therapeutics, Inc.
Animal Health Profiling
L-ERGO
ERGOLD
Bioxytech Assay Kits

CONTACTS: Note: Officers with more than one job title may be intentionally listed here more than once.

Marvin Hausman, CEO
Marvin Hausman, Pres.
Randy Moeckli, Sr. Dir.-Sale & Mktg.
Marvin Hausman, Chmn.

Phone: 650-212-2568	Fax: 650-573-1969
Toll-Free: 800-547-3686	
Address: 323 Vintage Park Dr., Ste. B, Foster City, CA 94404-1136 US	

GROWTH PLANS/SPECIAL FEATURES:

OXIS International, Inc. develops diagnostic and therapeutic products as well as new technologies to treat oxidative stress diseases, which are disorders associated with damage from free radicals and reactive oxygen species. The firm has three central divisions: OXIS Therapeutics (OT), OxisResearch (OR) and its Ergothioneine (ERGO) division. OT carries out OXIS's pharmaceutical and nutraceutical discovery business and is focused on new drugs to treat diseases associated with tissue damage from free radicals and reactive oxygen species. The division's lead drug candidate, BXT-51072, has completed a Phase IIA clinical trial for treatment of mild to moderate ulcerative colitis. The division's other products in development include GPx mimics, lipid soluble antioxidants, super oxide dismutase and ergothioneine analogs. The OR division rests primarily on its Bioxytech Assay Kit product line, which simplifies the testing of oxidative, antioxidant, nitrosative and inflammatory biomarkers. OR markets research data, commercial diagnostic assays and fine chemicals to research and clinical laboratories, as well as other customers. The company's Ergothioneine division focuses on the production, development and marketing of its patented synthetic L-Ergothioneine antioxidant amino acid. The division's trademarks for the products developed from L-Ergothioneine are L-ERGO and ERGOLD. In addition to these three divisions, the company owns a majority stake in BioCheck, Inc., with which it has an agreement for an option to purchase the remaining 49%. BioCheck offers over 40 enzyme immunoassays in addition to the research and manufacturing operations that accompany the immunoassay products.

FINANCIALS: Sales and profits are in thousands of dollars—add 000 to get the full amount. 2006 Note: Financial information for 2006 was not available for all companies at press time.

2006 Sales: $5,776	2006 Profits: $-4,940	**U.S. Stock Ticker: OXIS**
2005 Sales: $2,497	2005 Profits: $-3,109	**Int'l Ticker:** Int'l Exchange:
2004 Sales: $2,364	2004 Profits: $-2,698	Employees: 6
2003 Sales: $2,740	2003 Profits: $- 791	Fiscal Year Ends: 12/31
2002 Sales: $2,050	2002 Profits: $- 822	Parent Company:

SALARIES/BENEFITS:

Pension Plan:	ESOP Stock Plan:	Profit Sharing:	Top Exec. Salary: $209,000	Bonus: $5,000
Savings Plan: Y	Stock Purch. Plan: Y		Second Exec. Salary: $179,000	Bonus: $

OTHER THOUGHTS:

Apparent Women Officers or Directors: 1
Hot Spot for Advancement for Women/Minorities:

LOCATIONS: ("Y" = Yes)

West:	Southwest:	Midwest:	Southeast:	Northeast:	International:
Y					

PACIFIC BIOMETRICS INC

www.pacbio.com

Industry Group Code: 541710 **Ranks within this company's industry group:** Sales: 19 Profits: 8

Drugs:	Other:	Clinical:		Computers:		Other:	
Discovery:	AgriBio:	Trials/Services:	Y	Hardware:		Specialty Services:	
Licensing:	Genomics/Proteomics:	Laboratories:	Y	Software:		Consulting:	Y
Manufacturing:	Tissue Replacement:	Equipment/Supplies:		Arrays:		Blood Collection:	
Development:		Research/Development Svcs.:	Y	Database Management:		Drug Delivery:	
Generics:		Diagnostics:	Y			Drug Distribution:	

TYPES OF BUSINESS:

Clinical Trials
Laboratory Services
Contract Research & Development
Diagnostic Tests
DNA Amplification Systems

BRANDS/DIVISIONS/AFFILIATES:

PBI Technology, Inc.
Logarithmic Isothermal DNA Amplification
SalivaSac

CONTACTS: *Note: Officers with more than one job title may be intentionally listed here more than once.*

Ronald R. Helm, CEO
Michael Carrosino, CFO
Elizabeth T. Leary, Chief Scientific Officer
Tonya Aggoune, Mgr.-Client & Info. Svcs.
Michael Murphy, Sr. VP-Laboratory Oper.
Janice O'Connor, Dir.-Bus. Dev.
John Jensen, Controller
Mario Ehlers, Chief Medical Officer
Timothy Carlson, Dir.-Laboratory Svcs.
Kenneth Waters, Dir.-Strategic Planning
Kristin Walsh, Mgr.-Quality Assurance
Ronald R. Helm, Chmn.

Phone: 206-298-0068	Fax: 206-298-9838
Toll-Free: 800-767-9151	
Address: 220 W. Harrison St., Seattle, WA 98119 US	

GROWTH PLANS/SPECIAL FEATURES:

Pacific Biometrics, Inc. (PBI) provides specialty reference laboratory and clinical research services to pharmaceutical, biotechnology and laboratory manufacturers. Services are offered through two main sectors: Clinical Trial Support and Diagnostic Product Development. In Clinical Trial Support, tailored databases are customized for each clinical research study protocol while a data analyst facilitates adapted data management plans for all clients. Diagnostic Product Development develops and improves reagents and point-of-care devices for diagnostic companies. PBI specializes in laboratory services for lipids and cardiovascular risk; diabetes; and bone and cartilage metabolism. Services offered include the measurement of cardiovascular disease markers through lipoprotein components, cholesterol, triglycerides, phospholipids and apolipoproteins; testing for diabetes markers such as glucose, HbA1c, microalbumin and non-esterified fatty acids; measurements of hormone and biochemical markers such as pyridonoline, procollagens, osteocalcin; and bone specific-phosphatase cartilage oligomeric matrix protein for osteoporosis, bone and cartilage metabolism. PBI's wholly-owned subsidiary, PBI Technology, Inc., develops and commercializes molecular diagnostic technologies, non-invasive diagnostic devices and early-stage drug candidates. In the past years, PBI has purchased DNA amplification and cell viability technologies from Saigene Corporation. Through this transaction, PBI acquired the Logarithmic Isothermal DNA Amplification (LIDA) technology platform, which rapidly replicates DNA without expensive lab equipment and lengthy procedural protocol. The company also acquired Saigene's cell viability platform, a molecular method that can rapidly distinguish live cells from dead cells. In July 2006, the company was awarded a contract with a multinational pharmaceutical company to provide laboratory data for a Phase IIIb cardiosvascular trial. PBI Technology, Inc. also signed a licensing agreement with Evergreen Innovation Partners to license certain patents for its proprietary SalivaSac technology, which consists of a device that collects the sterile filtrate of saliva in order to determine blood glucose levels.

Pacific offers its employees a stock purchase plan and flexible spending accounts.

FINANCIALS: Sales and profits are in thousands of dollars—add 000 to get the full amount. 2006 Note: Financial information for 2006 was not available for all companies at press time.

2006 Sales: $10,750	2006 Profits: $ 179	**U.S. Stock Ticker: PBME.OB**
2005 Sales: $3,230	2005 Profits: $-2,993	**Int'l Ticker:** Int'l Exchange:
2004 Sales: $4,801	2004 Profits: $-1,884	Employees: 25
2003 Sales: $5,764	2003 Profits: $1,457	Fiscal Year Ends: 6/30
2002 Sales: $4,400	2002 Profits: $ 900	Parent Company:

SALARIES/BENEFITS:

Pension Plan:	ESOP Stock Plan:	Profit Sharing:	Top Exec. Salary: $239,994	Bonus: $15,000
Savings Plan: Y	Stock Purch. Plan: Y		Second Exec. Salary: $181,775	Bonus: $

OTHER THOUGHTS:

Apparent Women Officers or Directors: 4
Hot Spot for Advancement for Women/Minorities: Y

LOCATIONS: ("Y" = Yes)

West:	Southwest:	Midwest:	Southeast:	Northeast:	International:
Y					

PAIN THERAPEUTICS INC   www.paintrials.com

Industry Group Code: 325412 Ranks within this company's industry group: Sales: 76   Profits: 51

Drugs:		Other:		Clinical:		Computers:		Other:	
Discovery:	Y	AgriBio:		Trials/Services:		Hardware:		Specialty Services:	
Licensing:	Y	Genomics/Proteomics:		Laboratories:		Software:		Consulting:	
Manufacturing:		Tissue Replacement:		Equipment/Supplies:		Arrays:		Blood Collection:	
Development:	Y			Research/Development Svcs.:		Database Management:		Drug Delivery:	Y
Generics:				Diagnostics:				Drug Distribution:	

TYPES OF BUSINESS:
Drugs, Opioids
Abuse-Resistant Drug Delivery

BRANDS/DIVISIONS/AFFILIATES:
Oxytrex
Remoxy

CONTACTS: Note: Officers with more than one job title may be intentionally listed here more than once.
Remi Barbier, CEO
Nadav Friedmann, COO
Remi Barbier, Pres.
Peter S. Roddy, CFO
Grant L. Schoenhard, Chief Scientific Officer
Michael Zamloot, Sr. VP-Tech. Oper.
Roger Fu, Pharmaceutical Dev.
Christi Waarich, Sr. Mgr.-Investor Rel.
Peter Butera, VP-Clinical Oper.
Nadav Friedmann, Chief Medical Officer
Michael Marsman, VP-Regulatory Affairs
Remi Barbier, Chmn.

Phone: 650-624-8200	Fax: 650-624-8222

Toll-Free:

Address: 416 Browning Way, S. San Francisco, CA 94080 US

GROWTH PLANS/SPECIAL FEATURES:
Pain Therapeutics, Inc. is a biopharmaceutical company that develops novel drugs for severe pain, particularly opioids, and oncology. The firm has two proprietary drugs in Phase III clinical trials: Oxytrex and Remoxy. There are two drugs in Phase I clinical studies, including PTI-202, an abuse-resistant opioid pain killer, and a monoclonal antibody for metastatic melanoma, a rare but deadly form of skin cancer. The company is also working on a factor IX replacement for hemophilia, which is in pre-clinical study. Oxytrex, the company's lead drug candidate, a novel oral opioid, is a small-molecule drug in development to treat severe chronic pain including low-back, arthritic and cancer pain. Animal testing indicates that the drug may not cause the addiction, tolerance or physical dependence associated with opioid analgesics. Pain Therapeutics believes Oxytrex will serve as an effective substitute for oxycodone, a narcotic drug widely used to treat chronic pain with sales exceeding $2 billion per year. Remoxy is an abuse-deterrent, long-acting time-release version of oxycodone, designed to foil abusers who attempt to use the drug recreationally and to prevent accidental overdose. The drug is currently in Phase III trials. Pain Therapeutics has a collaboration agreement with King Pharmaceuticals to develop and commercialize Remoxy, as well as other abuse-resistant opioid painkillers. In March 2007, the company announced that it has licensed technology to treat hemophilia from the Stanford University School of Medicine.

FINANCIALS: Sales and profits are in thousands of dollars—add 000 to get the full amount. 2006 Note: Financial information for 2006 was not available for all companies at press time.
2006 Sales: $53,918	2006 Profits: $6,188	U.S. Stock Ticker: PTIE
2005 Sales: $5,080	2005 Profits: $-30,670	Int'l Ticker: Int'l Exchange:
2004 Sales: $	2004 Profits: $-37,776	Employees: 41
2003 Sales: $	2003 Profits: $-21,617	Fiscal Year Ends: 12/31
2002 Sales: $	2002 Profits: $-15,925	Parent Company:

SALARIES/BENEFITS:
Pension Plan:	ESOP Stock Plan:	Profit Sharing:	Top Exec. Salary: $502,470	Bonus: $400,000
Savings Plan: Y	Stock Purch. Plan: Y		Second Exec. Salary: $386,185	Bonus: $300,000

OTHER THOUGHTS:
Apparent Women Officers or Directors: 1
Hot Spot for Advancement for Women/Minorities:

LOCATIONS: ("Y" = Yes)
West:	Southwest:	Midwest:	Southeast:	Northeast:	International:
Y					

PALATIN TECHNOLOGIES INC

www.palatin.com

Industry Group Code: 325412 Ranks within this company's industry group: Sales: 109   Profits: 124

Drugs:		Other:		Clinical:		Computers:		Other:	
Discovery:	Y	AgriBio:		Trials/Services:		Hardware:		Specialty Services:	
Licensing:		Genomics/Proteomics:	Y	Laboratories:		Software:		Consulting:	
Manufacturing:		Tissue Replacement:		Equipment/Supplies:		Arrays:		Blood Collection:	
Development:	Y			Research/Development Svcs.:		Database Management:		Drug Delivery:	
Generics:				Diagnostics:	Y			Drug Distribution:	

TYPES OF BUSINESS:

Drugs-Diversified
Sexual Dysfunction Drugs
Inflammation Drugs
Peptide Technology
Diagnostic Imaging Products
Obesity Treatments

BRANDS/DIVISIONS/AFFILIATES:

NeutroSpec
MIDAS
Bremelanotide (PT-141)

CONTACTS: *Note: Officers with more than one job title may be intentionally listed here more than once.*

Carl Spana, CEO
Carl Spana, Pres.
Stephen T. Wills, CFO
Trevor Hallam, Exec. VP-R&D
Stephen T. Wills, Exec. VP-Oper.
Shubh Sharma, VP/Chief Scientific Officer
John K. A. Prendergast, Chmn.

Phone: 609-495-2200	Fax: 609-495-2201
Toll-Free:	
Address: 4-C Cedar Brook Dr., Cranbury, NJ 08512 US	

GROWTH PLANS/SPECIAL FEATURES:

Palatin Technologies, Inc. is a development-stage biopharmaceutical company committed to the discovery, development and commercialization of novel therapeutics. Its primary focus is discovering and developing melanocortin (MC)-based therapeutics. The MC family of receptors has been identified with a variety of conditions and diseases, including sexual dysfunction; obesity; cachexia, which is extreme wasting, generally secondary to a chronic disease; and inflammation. Palatin's patented MIDAS (metal ion-induced distinctive array of structures) platform for drug design and discovery allows the company to synthesize pharmaceuticals that mimic the activity of peptides. MIDAS can generate both receptor antagonists and agonists, to either block or promote metabolic responses. The company is engaged in research and development using this technology to diagnose infections and treat cancer, sexual dysfunction, cachexia, congestive heart failure, obesity and inflammation. Its Bremelanotide (formerly PT-141) product is a nasally administered peptide for the treatment of sexual dysfunction in both genders. Palatin has completed Phase I safety studies and Phase IIa and IIb efficacy studies in male subjects, along with a Phase I safety study and Phase IIa efficacy study in female patients with Female Sexual Dysfunction (FSD). Palatin is partnered with king Pharmaceuticals for the development of this product. NeutroSpec is a radio-labeled monoclonal antibody that binds to white blood cells that collect at sites of infection, thus enabling the infection to be easily and rapidly imaged and detected with a gamma camera. The marketing, sales and distribution of this product have been voluntarily, and indefinitely, suspended by request of the FDA. The company also has products for congestive heart failure, obesity and cachexia in its preclinical pipeline. The company presented the results of clinical testing of Bremelanotide on pre-menopausal women in 2006.

The company provides employee benefits including dental, medical and maternity insurance; education assistance; and tuition reimbursement.

FINANCIALS: Sales and profits are in thousands of dollars—add 000 to get the full amount. 2006 Note: Financial information for 2006 was not available for all companies at press time.

2006 Sales: $19,749	2006 Profits: $-28,959	**U.S. Stock Ticker: PTN**
2005 Sales: $17,957	2005 Profits: $-14,358	**Int'l Ticker:** Int'l Exchange:
2004 Sales: $2,315	2004 Profits: $-26,318	Employees: 85
2003 Sales: $1,270	2003 Profits: $-20,565	Fiscal Year Ends: 6/30
2002 Sales: $ 300	2002 Profits: $-16,100	Parent Company:

SALARIES/BENEFITS:

Pension Plan:	ESOP Stock Plan:	Profit Sharing:	Top Exec. Salary: $350,000	Bonus: $100,000
Savings Plan: Y	Stock Purch. Plan:		Second Exec. Salary: $285,000	Bonus: $75,000

OTHER THOUGHTS:

Apparent Women Officers or Directors:
Hot Spot for Advancement for Women/Minorities:

LOCATIONS: ("Y" = Yes)

West:	Southwest:	Midwest:	Southeast:	Northeast:	International:
				Y	

Note: Financial information, benefits and other data can change quickly and may vary from those stated here.

PANACOS PHARMACEUTICALS INC   www.panacos.com

Industry Group Code: 325414 Ranks within this company's industry group: Sales: Profits:

Drugs:		Other:	Clinical:	Computers:	Other:
Discovery:	Y	AgriBio:	Trials/Services:	Hardware:	Specialty Services:
Licensing:		Genomics/Proteomics:	Laboratories:	Software:	Consulting:
Manufacturing:		Tissue Replacement:	Equipment/Supplies:	Arrays:	Blood Collection:
Development:	Y		Research/Development Svcs.:	Database Management:	Drug Delivery:
Generics:			Diagnostics:		Drug Distribution:

TYPES OF BUSINESS:

Drug Discovery & Development
Antivirals

BRANDS/DIVISIONS/AFFILIATES:

V. I. Pharmaceuticals
Vitex
Bevirimat
PA-1050040
PA-457
PA-040

CONTACTS: Note: Officers with more than one job title may be intentionally listed here more than once.

Alan W. Dunton, CEO
Graham P. Allaway, COO
Alan W. Dunton, Pres.
Peyton J. Marshall, CFO/Exec. VP
Scott McCallister, Chief Medical Officer
David E. Martin, Sr. VP-Drug Dev.
John D. Richards, VP-Mfg. Oper.
Fredrick Schmid, Sr. VP-Bus. Dev.
Fredrick Schmid, Sr. VP-Commercial Oper.
Tom Latengan, VP-Regulatory Affairs
Jeremy Hayward-Surry, Chmn.

Phone: 617-926-1551	Fax: 617-923-2245
Toll-Free:	
Address: 134 Coolidge Ave., Watertown, MA 02472 US	

GROWTH PLANS/SPECIAL FEATURES:

Panacos Pharmaceuticals, Inc., formerly V.I. Pharmaceuticals, which did business as Vitex, discovers and develops small-molecule oral drugs for the treatment of HIV and other major human viral diseases. The company focuses on novel targets in the viral life cycle, including virus maturation and virus fusion. Panacos' lead product candidate is bevirimat, formerly PA-457, is a once-daily oral HIV drug candidate in Phase II clinical testing. The product is one of the first in a new class of drugs that works by maturation inhibition. Maturation inhibition occurs at the end of the virus life cycle as newly formed HIV matures into infectious virus particles. Bevirimat blocks a key step in the processing of a viral core protein called capsid, so that following bevirimat treatment, virus particles released from cells are immature and non-infectious. In addition, the company has a research and development program designed to produce second- and third-generation maturation inhibition products, including PA-1050040, which is a second-generation product in Phase I clinical trials. Panacos also is researching products that inhibit the fusion of the HIV virus with human cells, an event that happens early in the HIV virus life cycle. In February 2007, the company commenced Phase I clinical study of its second-generation maturation inhibitor, PA-040. That study was completed in July 2007. In July 2007, Panacos also began Phase I clinical testing of bevirimat in healthy volunteers to evaluate the effectiveness of two novel liquid formulations of the drug.

FINANCIALS: Sales and profits are in thousands of dollars—add 000 to get the full amount. 2006 Note: Financial information for 2006 was not available for all companies at press time.

2006 Sales: $ 300	2006 Profits: $-38,100	U.S. Stock Ticker: PANC
2005 Sales: $1,000	2005 Profits: $-59,100	Int'l Ticker: Int'l Exchange:
2004 Sales: $1,700	2004 Profits: $-18,200	Employees: 32
2003 Sales: $ 716	2003 Profits: $-22,353	Fiscal Year Ends: 12/31
2002 Sales: $4,225	2002 Profits: $-20,040	Parent Company:

SALARIES/BENEFITS:

Pension Plan:	ESOP Stock Plan:	Profit Sharing:	Top Exec. Salary: $249,737	Bonus: $81,813
Savings Plan: Y	Stock Purch. Plan: Y		Second Exec. Salary: $277,173	Bonus: $84,892

OTHER THOUGHTS:

Apparent Women Officers or Directors:
Hot Spot for Advancement for Women/Minorities:

LOCATIONS: ("Y" = Yes)

West:	Southwest:	Midwest:	Southeast:	Northeast:	International:
				Y	

Note: Financial information, benefits and other data can change quickly and may vary from those stated here.

PAR PHARMACEUTICAL COMPANIES INCwww.parpharm.com

Industry Group Code: 325416 Ranks within this company's industry group: Sales: Profits:

Drugs:		Other:		Clinical:	Computers:	Other:
Discovery:	Y	AgriBio:		Trials/Services:	Hardware:	Specialty Services:
Licensing:		Genomics/Proteomics:		Laboratories:	Software:	Consulting:
Manufacturing:	Y	Tissue Replacement:		Equipment/Supplies:	Arrays:	Blood Collection:
Development:	Y			Research/Development Svcs.:	Database Management:	Drug Delivery:
Generics:	Y			Diagnostics:		Drug Distribution:

TYPES OF BUSINESS:
Drugs-Generic & Branded
Pharmaceutical Intermediates

BRANDS/DIVISIONS/AFFILIATES:
Pharmaceutical Resources, Inc.
Par Pharmaceutical, Inc.
Megace ES
Kali Laboratories, Inc.
FineTech Laboratories Ltd.
Optimer Pharmaceutials, Inc.

CONTACTS: *Note: Officers with more than one job title may be intentionally listed here more than once.*
Patrick G. LePore, CEO
Gerard A. Martino, COO/Exec. VP
Patrick G. LePore, Pres.
Veronica A. Lubatkin, CFO/Exec. VP
Thomas Haughey, General Counsel/Sec./Exec. VP
Paul V. Campanelli, Pres., Generic Products Div.
John A. MacPhee, Pres., Branded Products Div.
John D. Abernathy, Chmn.

Phone: 201-802-4000	Fax: 201-802-4600
Toll-Free:	
Address: 300 Tice Blvd., Woodcliff Lake, NJ 07677 US	

GROWTH PLANS/SPECIAL FEATURES:
Par Pharmaceutical Companies, Inc. (formerly Pharmaceutical Resources, Inc.) develops, manufactures and markets branded and generic pharmaceuticals through its principal subsidiary, Par Pharmaceutical, Inc. Products include treatments for central nervous system disorders, cardiovascular drugs, analgesics, anti-inflammatory products, anti-bacterials, anti-diabetics, antihistamines, anti-virals, cholesterol-lowering drugs and ovulation stimulants. Subsequent to the shipment of the company's first branded product, Megace ES, in late 2005, Par now operates in two segments: generic pharmaceuticals and branded pharmaceuticals. In the generic segment, the company's product line comprises generic prescription drugs consisting of 213 products representing various dosage strengths for 92 separate drugs. These are manufactured principally in the solid oral dosage form (tablet, caplet and two-piece hard shell capsule). Among these are generic versions of Advil, Daypro, Glucophage, Zantac, Clomid, Prozac, Halcion and Prilosec. Par's only product in the branded segment is Megace ES, which was given FDA approval in June 2005 for the treatment of anorexia, cachexia, or an unexplained, significant weight loss in patients with a diagnosis of AIDS. In 2006, Optimer Pharmaceutials, Inc. announced its Phase IIA clinical studies have indicated that PAR-101, another brand product, appears to be efficacious in the treatment of Clostridium difficile-associated diarrhea. With the recent acquisition of Kali Laboratories, Inc. Par more than doubled the size of its research and development capabilities. The company recently received FDA approval to market generic versions of Isotopin SR and Ultracet, for hypertension and pain respectively. The firm recently divested former subsidiary FineTech Laboratories Ltd. In 2007, Par announced the commencement of shipping of metoprolol succinate extended release 100mg and 200mg tablets, as well as the shipment of generic Zantac syrup.

PAR offers employees a 529 college savings plan, career growth opportunities, annual incentive programs and a flexible spending plan, along with health, dental and life insurance.

FINANCIALS: Sales and profits are in thousands of dollars—add 000 to get the full amount. 2006 Note: Financial information for 2006 was not available for all companies at press time.

2006 Sales: $	2006 Profits: $	U.S. Stock Ticker: PRX
2005 Sales: $433,194	2005 Profits: $-8,250	Int'l Ticker: Int'l Exchange:
2004 Sales: $689,107	2004 Profits: $29,246	Employees: 766
2003 Sales: $646,023	2003 Profits: $122,533	Fiscal Year Ends: 9/30
2002 Sales: $381,600	2002 Profits: $79,500	Parent Company:

SALARIES/BENEFITS:

Pension Plan:	ESOP Stock Plan:	Profit Sharing:	Top Exec. Salary: $620,400	Bonus: $395,505
Savings Plan: Y	Stock Purch. Plan: Y		Second Exec. Salary: $410,597	Bonus: $354,375

OTHER THOUGHTS:
Apparent Women Officers or Directors: 1
Hot Spot for Advancement for Women/Minorities:

LOCATIONS: ("Y" = Yes)

West:	Southwest:	Midwest:	Southeast:	Northeast:	International:
				Y	

Note: Financial information, benefits and other data can change quickly and may vary from those stated here.

PAREXEL INTERNATIONAL   www.parexel.com

Industry Group Code: 541710 Ranks within this company's industry group: Sales: 4   Profits: 6

Drugs:	Other:	Clinical:		Computers:		Other:	
Discovery:	AgriBio:	Trials/Services:	Y	Hardware:	Y	Specialty Services:	Y
Licensing:	Genomics/Proteomics:	Laboratories:		Software:	Y	Consulting:	Y
Manufacturing:	Tissue Replacement:	Equipment/Supplies:		Arrays:		Blood Collection:	
Development:		Research/Development Svcs.:	Y	Database Management:	Y	Drug Delivery:	
Generics:		Diagnostics:				Drug Distribution:	

TYPES OF BUSINESS:

Clinical Trial & Data Management
Biostatistical Analysis & Report Production
Medical Marketing Services
Diagnostic Services
Medical Publishing
Web-Based Solutions
Medical Software Solutions
Consulting Services

BRANDS/DIVISIONS/AFFILIATES:

PAREXEL Consulting and Medical Marketing Services
Perceptive Informatics, Inc.
Droit & Pharmacie
Barnett Educational Services
Cambridge Medical Publications
Journal of International Medical Research (The)
Good Clinical Practice

CONTACTS: *Note: Officers with more than one job title may be intentionally listed here more than once.*

Josef H. von Rickenbach, CEO
Carl A. Spalding, COO
Carl A. Spalding, Pres.
James F. Winschel, Jr., CFO/Sr. VP
Michael E. Woehler, Exec. VP/Pres., Clinical Research Svcs.
Ulf Schneider, Chief Admin. Officer/Sr. VP
Susan H. Alexander, General Counsel/Sr. VP/Corp. Sec.
Mark A. Goldberg, Pres., Perceptive Informatics, Inc.
Kurt A. Brykman, Pres., PAREXEL Consulting
Todd A. Joron, Corp. VP/General Mgr.-Perceptive Informatics, Inc.
Jennifer Baird, Dir.-Public Rel., PAREXEL Int'l
Josef H. von Rickenbach, Chmn.

Phone: 781-487-9900	Fax: 781-487-0525
Toll-Free:	
Address: 200 West St., Waltham, MA 02451-1163 US	

GROWTH PLANS/SPECIAL FEATURES:

PAREXEL International Corporation is a biopharmaceutical services company that provides clinical research, medical marketing, consulting and informatics and advanced technology products and services to pharmaceutical, biotechnology and medical device industries worldwide. PAREXEL operates through five business units: clinical research services (CRS), PAREXEL Consulting and Medical Marketing Services (PCMS) and Perceptive Informatics, Inc., an information technology subsidiary. CRS, the company's core business, provides clinical trial management and biostatistics, data management and clinical pharmacology, as well as related medical advisory and investigator site services. PCMS provides technical expertise for regulatory affairs, industry training, publishing, product development, management consulting, registration, commercialization consulting, market development, health policy consulting and strategic reimbursement services. Perceptive Informatics provides technology designed to improve clients' product development processes such as medical imaging services, interactive voice response systems, clinical trials management systems, web-based portals and systems integration. PAREXEL's publications divisions are Droit & Pharmacie, Barnett Educational Services and Cambridge Medical Publications. In June 2006, the firm entered into a joint venture with Synchron Research Services Private Ltd., to form PAREXEL International Synchron Private Limited located in Bangalore, India to conduct clinical trial business operations. In November 2006, PAREXEL acquired California Clinical Trials Medical Group, Inc. and Behavioral and Medical Research, LLC for $65 million. PAREXEL also recently acquired a minority equity interest in the clinical pharmacology business of Synchron Research in Ahmedabad, India; and opened an office in Mexico City to provide clinical research and consulting services.

PAREXEL offers its employees paid time off, career development in the form on the job training and rewards for excellence in performance.

FINANCIALS: Sales and profits are in thousands of dollars—add 000 to get the full amount. 2006 Note: Financial information for 2006 was not available for all companies at press time.

2006 Sales: $614,947	2006 Profits: $23,544	**U.S. Stock Ticker: PRXL**
2005 Sales: $544,726	2005 Profits: $-35,177	**Int'l Ticker:** Int'l Exchange:
2004 Sales: $540,983	2004 Profits: $13,791	Employees: 5,600
2003 Sales: $615,838	2003 Profits: $10,662	Fiscal Year Ends: 6/30
2002 Sales: $564,900	2002 Profits: $13,200	Parent Company:

SALARIES/BENEFITS:

Pension Plan:	ESOP Stock Plan:	Profit Sharing:	Top Exec. Salary: $465,750	Bonus: $286,157
Savings Plan:	Stock Purch. Plan:		Second Exec. Salary: $317,500	Bonus: $217,408

OTHER THOUGHTS:

Apparent Women Officers or Directors: 2
Hot Spot for Advancement for Women/Minorities: Y

LOCATIONS: ("Y" = Yes)

West:	Southwest:	Midwest:	Southeast:	Northeast:	International:
Y		Y	Y	Y	Y

Note: Financial information, benefits and other data can change quickly and may vary from those stated here.

PDL BIOPHARMA

www.pdl.com

Industry Group Code: 325412 Ranks within this company's industry group: Sales: 43   Profits: 191

Drugs:		Other:	Clinical:	Computers:	Other:
Discovery:	Y	AgriBio:	Trials/Services:	Hardware:	Specialty Services:
Licensing:	Y	Genomics/Proteomics:	Laboratories:	Software:	Consulting:
Manufacturing:	Y	Tissue Replacement:	Equipment/Supplies:	Arrays:	Blood Collection:
Development:	Y		Research/Development Svcs.:	Database Management:	Drug Delivery:
Generics:			Diagnostics:		Drug Distribution:

TYPES OF BUSINESS:

Pharmaceuticals Development
Drugs-Kidney Transplant Rejection
Humanized Monoclonal Antibodies
Oncology Drugs
Asthma Drugs
Autoimmune Disease Drugs

BRANDS/DIVISIONS/AFFILIATES:

Protein Design Labs, Inc.
Nuvion
Zenapax
Cardene IV
Retavase

CONTACTS: Note: Officers with more than one job title may be intentionally listed here more than once.

Mark McDade, CEO
Andrew Guggenhime, CFO/Sr. VP
David Iwanicki, VP-Sales & Sales Oper.
Richard Murray, Exe. VP/Chief Scientific Officer
Behrooz Najafi, VP-IT
Robert M. Savel, II, Sr. VP-Tech. Oper.
Eric Emery, VP-Mfg.
Cynthia Shumate, VP-Legal Affairs/Sec./Chief Compliance Officer
Jeanmarie Guenot, VP-Corp. & Bus. Dev.
Mark McCamish, Sr. VP/Chief Medical Officer
Jaisim Shah, VP-Bus. Affairs & Mktg.
Laurie Torres, VP-Corp. Svcs.
L. Patrick Gage, Chmn.

Phone: 510-574-1400	Fax: 510-574-1500

Toll-Free:

Address: 34801 Campus Dr., Fremont, CA 94555 US

GROWTH PLANS/SPECIAL FEATURES:

PDL BioPharma, formerly Protein Design Labs, Inc., is a biopharmaceutical company focused on discovering, developing and commercializing innovative therapies for severe or life threatening illnesses. Currently the company markets and sells products in the acute-care hospital setting in the U.S. and Canada. PDL also receives royalties and other revenues through licensing agreements with numerous biotechnology and pharmaceutical companies based on its proprietary antibody-based platform. These licensing agreements have contributed to the development by its licensees of nine marketed products and cover several antibodies in clinical development. PDL currently has several investigational compounds in clinical development for severe or life-threatening diseases and it has entered into or intends to enter into collaborations with other pharmaceutical or biotechnology companies for the joint development, manufacture and commercialization of certain of these compounds. The company's research platform is focused on the discovery and development of antibodies for the treatment of cancer and autoimmune diseases. PDL's Zenapax was developed for the prevention of kidney transplant rejection. The product is currently licensed to Hoffman-La Roche, Inc. for marketing in the U.S., Europe and other countries. Of the firm's own marketed products, its Cardene IV drug is a calcium channel blocker delivered intravenously for the treatment of hypertension. Retavase, a drug acquired from ESP Pharma, is designed for the management of acute myocardial infarction or heart attacks. In the product pipeline, the company is developing Nuvion, a monoclonal antibody directed at the CD3 antigen on activated T-cells, which advances certain autoimmune diseases. The drug is meant as a treatment for severe ulcerative colitis and is in Phase II clinical testing. The firm's name change at the start of 2006 marked the company's merger with its subsidiary ESP Pharma.

FINANCIALS: Sales and profits are in thousands of dollars—add 000 to get the full amount. 2006 Note: Financial information for 2006 was not available for all companies at press time.

2006 Sales: $414,770	2006 Profits: $-130,020	**U.S. Stock Ticker:** PDLI	
2005 Sales: $280,569	2005 Profits: $-166,577	**Int'l Ticker:** Int'l Exchange:	
2004 Sales: $96,024	2004 Profits: $-53,241	Employees: 1,100	
2003 Sales: $66,686	2003 Profits: $-129,814	Fiscal Year Ends: 12/31	
2002 Sales: $46,400	2002 Profits: $-14,600	Parent Company:	

SALARIES/BENEFITS:

Pension Plan:	ESOP Stock Plan:	Profit Sharing:	Top Exec. Salary: $600,000	Bonus: $300,000
Savings Plan: Y	Stock Purch. Plan: Y		Second Exec. Salary: $375,000	Bonus: $116,250

OTHER THOUGHTS:

Apparent Women Officers or Directors: 4
Hot Spot for Advancement for Women/Minorities: Y

LOCATIONS: ("Y" = Yes)

West:	Southwest:	Midwest:	Southeast:	Northeast:	International:
Y		Y		Y	Y

Note: Financial information, benefits and other data can change quickly and may vary from those stated here.

PENWEST PHARMACEUTICALS CO

www.penw.com

Industry Group Code: 325412A Ranks within this company's industry group: Sales: 16   Profits: 15

Drugs:		Other:		Clinical:		Computers:		Other:	
Discovery:	Y	AgriBio:		Trials/Services:		Hardware:		Specialty Services:	
Licensing:	Y	Genomics/Proteomics:		Laboratories:		Software:		Consulting:	
Manufacturing:	Y	Tissue Replacement:		Equipment/Supplies:		Arrays:		Blood Collection:	
Development:	Y			Research/Development Svcs.:		Database Management:		Drug Delivery:	Y
Generics:				Diagnostics:				Drug Distribution:	

TYPES OF BUSINESS:

Drug Delivery Systems
Nervous System Disorders Drugs

BRANDS/DIVISIONS/AFFILIATES:

TIMERx
Geminex
SyncroDose
Opan ER
Nalbuphine ER
Torsemide ER

CONTACTS: *Note: Officers with more than one job title may be intentionally listed here more than once.*

Jennifer L. Good, CEO
Jennifer L. Good, Pres.
Benjamin Palleiko, CFO
Paul Hayes, Sr. VP-Strategic Mktg.
Anand R. Baichwal, Chief Scientific Officer/Sr. VP-Licensing
Mehrdad Abedin, CIO/VP-IT
Benjamin Palleiko, Sr. VP-Corp. Dev.
Thomas R. Sciascia, Chief Medical Officer/Sr. VP
Amale Hawi, Sr. VP-Pharmaceutical Dev.
Paul E. Freiman, Chmn.

Phone: 203-796-3700	Fax: 203-794-1393
Toll-Free:	
Address: 39 Old Ridgebury Rd., Ste. 11, Danbury, CT 06810 US	

GROWTH PLANS/SPECIAL FEATURES:

Penwest Pharmaceuticals Co. develops pharmaceutical products based on proprietary drug delivery technologies with a focus on products that address disorders of the nervous system.   Opana ER, which FDA approved for marketing in the U.S. in 2006, is an extended release formulation of oxymorphone hydrochloride that the company developed with Endo Pharmaceuticals, Inc. using the TIMERx drug delivery technology.   The drug is an oral extended release opioid analgesic approved for twice-a-day dosing in patients with moderate to severe pain requiring continuous, around-the-clock opioid treatment for an extended period of time. The firm currently develops product candidates designed for the treatment of edema resulting from congestive heart failure.   Products in clinical development include nalbuphine ER, a controlled release formulation of nalbuphine hydrochloride, for the treatment of pain; and torsemide ER, a controlled release formulation of torsemide, a loop diuretic for the treatment of chronic edema, a condition involving excess fluid accumulation resulting from congestive heart failure.   Both drugs are still in Phase I clinical trials. Penwest currently has four proprietary drug delivery technologies: TIMERx, a controlled-release technology; Geminex, a technology enabling drug release at two different rates; SyncroDose, a technology enabling controlled release at the appropriate site in the body; and the gastroretentive system, a technology enabling drug delivery to the upper gastrointestinal tract. The company owns a total of 34 U.S. and 207 foreign patents. In July 2007, the firm entered into a research, development, commercialization and license agreement with Edison Pharmaceuticals, Inc. for the treatment of neurological disorders resulting from defects in cellular energy metabolism.

The company offers its employee medical, dental, vision and prescription drug insurance; disability benefits; life and accident insurance; flexible spending accounts; education reimbursement; a 401(k) plan; an employee stock purchase program; and employee stock options.

FINANCIALS: Sales and profits are in thousands of dollars—add 000 to get the full amount. 2006 Note: Financial information for 2006 was not available for all companies at press time.

2006 Sales: $3,499	2006 Profits: $-31,312	**U.S. Stock Ticker: PPCO**
2005 Sales: $6,213	2005 Profits: $-22,898	**Int'l Ticker:** Int'l Exchange:
2004 Sales: $5,108	2004 Profits: $-23,785	Employees: 75
2003 Sales: $4,678	2003 Profits: $-15,935	Fiscal Year Ends: 12/31
2002 Sales: $42,000	2002 Profits: $-17,100	Parent Company:

SALARIES/BENEFITS:

Pension Plan:	ESOP Stock Plan:	Profit Sharing:	Top Exec. Salary: $342,012	Bonus: $100,000
Savings Plan: Y	Stock Purch. Plan: Y		Second Exec. Salary: $279,262	Bonus: $85,500

OTHER THOUGHTS:

Apparent Women Officers or Directors: 2
Hot Spot for Advancement for Women/Minorities: Y

LOCATIONS: ("Y" = Yes)

West:	Southwest:	Midwest:	Southeast:	Northeast:	International:
				Y	

PERBIO SCIENCE AB   www.perbio.com

Industry Group Code: 325413 Ranks within this company's industry group: Sales: Profits:

Drugs:	Other:	Clinical:		Computers:		Other:
Discovery:	AgriBio:	Trials/Services:		Hardware:		Specialty Services:
Licensing:	Genomics/Proteomics:	Laboratories:		Software:		Consulting:
Manufacturing:	Tissue Replacement:	Equipment/Supplies:	Y	Arrays:		Blood Collection:
Development:		Research/Development Svcs.:		Database Management:		Drug Delivery:
Generics:		Diagnostics:	Y			Drug Distribution:

TYPES OF BUSINESS:
Research Diagnostics Products

BRANDS/DIVISIONS/AFFILIATES:
Pierce
HyClone
Endogen
Dharmacon
Cellomics
Thermo Fisher Scientific, Inc.

CONTACTS: *Note: Officers with more than one job title may be intentionally listed here more than once.*
Ragnar Lindqvist, Sec.
Mats Fischier, Chmn.

Phone: 46-42-26-90-90	**Fax:** 46-42-26-90-98
Toll-Free:	
Address: Knutpunkten 34, Helsingborg, 252 78 Sweden	

GROWTH PLANS/SPECIAL FEATURES:
Perbio Science AB, a subsidiary of Thermo Fisher Scientific, Inc., focuses on protein production and research. The firm specializes in products, systems and services for the study and production of proteins, cell cultures and RNA, primarily for academic and research institutes, pharmaceutical companies and diagnostics companies. The company operates in three segments: bioresearch, cell culture and medical devices. The bioresearch segment develops products used for research in protein chemistry and molecular biology. Products include kits, reagents and services for identifying, quantifying, purifying and modifying proteins, amidites and nucleotides for nucleic acid-based drugs and diagnostics. The cell culture segment offers proper nutrition for animal cells, process liquids for protein purification, sterile liquid handling systems and disposable, sterile liquid processing systems. Products include animal sera such as fetal bovine serum, cosmic calf serum and equine serum; and cell culture media in powder and liquid forms. The medical devices segment provides voice prosthesis that helps laryngectomy patients regain their speech ability; devices for laryngectomized and tracheostomized patients; and devices for the treatment of excessive menstrual bleeding and obesity. Leading brands include Pierce, a provider of research products for use in protein chemistry, sample handling, immunology and chromatography; Endogen, a provider of kits for the detection of infections, multiplex assays testing up to 16 types of antibodies and antibodies to cytokines and kinases; HyClone, a provider of cell-culture products and bioprocessing systems; Dharmacon, a provider of RNA products and technologies; and Cellomics, a provider of cellular measuring tools and screening instruments. Perbio has operating companies in the United Kingdom, France, Germany, Switzerland, the Netherlands and Belgium.

FINANCIALS: Sales and profits are in thousands of dollars—add 000 to get the full amount. 2006 Note: Financial information for 2006 was not available for all companies at press time.

2006 Sales: $	2006 Profits: $	**U.S. Stock Ticker: Subsidiary**
2005 Sales: $	2005 Profits: $	**Int'l Ticker:** Int'l Exchange:
2004 Sales: $	2004 Profits: $	Employees: 1,266
2003 Sales: $	2003 Profits: $	Fiscal Year Ends: 12/31
2002 Sales: $248,600	2002 Profits: $26,300	Parent Company: THERMO FISHER SCIENTIFIC INC

SALARIES/BENEFITS:

Pension Plan:	ESOP Stock Plan:	Profit Sharing:	Top Exec. Salary: $	Bonus: $
Savings Plan:	Stock Purch. Plan:		Second Exec. Salary: $	Bonus: $

OTHER THOUGHTS:
Apparent Women Officers or Directors:
Hot Spot for Advancement for Women/Minorities:

LOCATIONS: ("Y" = Yes)

West:	Southwest:	Midwest:	Southeast:	Northeast:	International: Y

PEREGRINE PHARMACEUTICALS INC www.peregrineinc.com

Industry Group Code: 325412 Ranks within this company's industry group: Sales: 147 Profits: 99

Drugs:		Other:		Clinical:		Computers:		Other:	
Discovery:	Y	AgriBio:		Trials/Services:		Hardware:		Specialty Services:	
Licensing:		Genomics/Proteomics:		Laboratories:		Software:		Consulting:	
Manufacturing:	Y	Tissue Replacement:		Equipment/Supplies:		Arrays:		Blood Collection:	
Development:	Y			Research/Development Svcs.:	Y	Database Management:		Drug Delivery:	
Generics:				Diagnostics:				Drug Distribution:	

TYPES OF BUSINESS:
Cancer & Hepatitis C Drugs
Contract Manufacturing Services

BRANDS/DIVISIONS/AFFILIATES:
Avid Bioservices, Inc.
Cotara
Peregrine Beijing Pharmaceuticals Technology Dev.

CONTACTS: Note: Officers with more than one job title may be intentionally listed here more than once.
Steven W. King, CEO
Steven W. King, Pres.
Paul J. Lytle, CFO
Richard Richieri, Sr. VP-Mfg. & BioProcess Dev.
F. David King, VP-Bus. Dev.
Joseph Shan, Exec. Dir.-Clinical & Regulatory Affairs
Shelley Fussey, VP-Intellectual Property
John Quick, Head-Quality Systems
Thomas A. Waltz, Chmn.
Mary J. Boyd, Head-Bus. Dev., Asia & Europe

Phone: 714-508-6000	Fax: 714-838-5817
Toll-Free:	
Address: 14272 Franklin Ave., Tustin, CA 92780 US	

GROWTH PLANS/SPECIAL FEATURES:
Peregrine Pharmaceuticals, Inc. is a biopharmaceutical company with a portfolio of clinical stage and preclinical product candidates using monoclonal antibodies for the treatment of cancer and viral diseases. The company's products fall under two technology platforms: anti-phosphatidylserine (anti-PS) and tumor necrosis therapy (TNT); TNT carries antibodies to tumors through dying (necrotic) cells that form the nucleus in almost all varieties of tumors. The anti-PS immunotherapeutics are monoclonal antibodies that target and bind to components of cells normally found only on the inner surface of the cell membrane. The TNT technology uses monoclonal antibodies that target and bind DNA and associated histone proteins accessible in dead and dying cells found at the core of solid tumors. The firm's lead anti-PS product, bavituximab, is in separate clinical trials for the treatment of solid cancers and hepatitis C virus infection. Anti-PS therapy targets phospholipid on the external walls and penetrates those cells, leaving the healthy cells undisturbed. Under the firm's TNT technology, the lead product candidate is Cotara, for treatment of brain cancer. Peregrine Pharmaceuticals operates a wholly-owned contract manufacturing subsidiary, Avid Bioservices, Inc. Avid provides several functions for Peregrine such as the manufacturing of all clinical supplies, commercial scale-up of products in clinical trials and assisting with the advancement of new clinical candidates. Avid also provides contract manufacturing services for outside biotechnology and biopharmaceutical companies on a fee-for-service basis. In January 2007, Peregrine Pharmaceuticals announced that it had established a wholly foreign-owned subsidiary in Beijing China. The new company's name is Peregrine Beijing Pharmaceuticals Technology Development, Ltd.

FINANCIALS: Sales and profits are in thousands of dollars—add 000 to get the full amount. 2006 Note: Financial information for 2006 was not available for all companies at press time.

2006 Sales: $3,193	2006 Profits: $-17,061	U.S. Stock Ticker: PPHM
2005 Sales: $4,959	2005 Profits: $-15,452	Int'l Ticker: Int'l Exchange:
2004 Sales: $3,314	2004 Profits: $-14,345	Employees: 116
2003 Sales: $3,921	2003 Profits: $-11,559	Fiscal Year Ends: 4/30
2002 Sales: $3,800	2002 Profits: $-11,700	Parent Company:

SALARIES/BENEFITS:

Pension Plan:	ESOP Stock Plan:	Profit Sharing:	Top Exec. Salary: $303,750	Bonus: $5,809
Savings Plan: Y	Stock Purch. Plan:		Second Exec. Salary: $241,188	Bonus: $4,613

OTHER THOUGHTS:
Apparent Women Officers or Directors: 2
Hot Spot for Advancement for Women/Minorities: Y

LOCATIONS: ("Y" = Yes)

West:	Southwest:	Midwest:	Southeast:	Northeast:	International:
Y					Y

Note: Financial information, benefits and other data can change quickly and may vary from those stated here.

PERRIGO CO   www.perrigo.com

Industry Group Code: 325416 Ranks within this company's industry group: Sales: 4   Profits: 5

Drugs:		Other:		Clinical:	Computers:		Other:	
Discovery:		AgriBio:		Trials/Services:	Hardware:		Specialty Services:	Y
Licensing:		Genomics/Proteomics:		Laboratories:	Software:		Consulting:	
Manufacturing:	Y	Tissue Replacement:		Equipment/Supplies:	Arrays:		Blood Collection:	
Development:	Y			Research/Development Svcs.:	Database Management:		Drug Delivery:	
Generics:	Y			Diagnostics:			Drug Distribution:	

TYPES OF BUSINESS:

Generic Prescription Drugs
Over-the-Counter Pharmaceuticals
Nutritional Products
Active Pharmaceutical Ingredients
Consumer Products

BRANDS/DIVISIONS/AFFILIATES:

Careline
Neca
Natural Formula
Perrigo New York, Inc.
Perrigo Israel Pharmaceuticals, Ltd.
Quimica Y Famarcia S.A. de C.V.
Wrafton Laboratories, Ltd.
Qualis, Inc.

CONTACTS: Note: Officers with more than one job title may be intentionally listed here more than once.

Joseph C. Papa, CEO
Joseph C. Papa, Pres.
Judy L. Brown, CFO/Exec. VP
Todd W. Kingma, General Counsel/Exec. VP/Sec.
John T. Hendrickson, Exec. VP-Global Oper.
Moshe Arkin, General Manager-Perrigo Global Generics & API
David T. Gibbons, Chmn.
Refael Lebel, Exec VP/General Manager-Perrigo Israel
John T. Hendrickson, Exec. VP-Supply Chain

Phone: 269-673-8451	Fax: 269-673-9128
Toll-Free:	
Address: 515 Eastern Ave., Allegan, MI 49010 US	

GROWTH PLANS/SPECIAL FEATURES:

Perrigo Co. is a global healthcare supplier and the world's largest manufacturer of over-the-counter pharmaceutical and nutritional products for the store brand market. The company also develops and manufactures generic prescription drugs, active pharmaceutical ingredients (API) and consumer products. The firm operates through three segments: consumer healthcare, prescription pharmaceuticals and API. The consumer healthcare segment makes a broad line of products including analgesics, cough/cold/allergy/sinus, gastrointestinal, smoking cessation, first aid, vitamin and nutritional supplement products. The pharmaceuticals segment's primary activity is the development, manufacture and sale of generic prescription drug products, generally for the U.S. market. The company currently markets roughly 150 generic prescription products to approximately 100 customers. The API segment develops, manufactures and markets API for the drug industry and branded pharmaceutical companies. In addition, Perrigo's operations also include the Israel consumer products segment, which consist of cosmetics, toiletries and detergents generally sold under brands such as Careline, Neca and Natural Formula; and the Israel pharmaceutical and diagnostic products segment, which includes the marketing and manufacturing of branded prescription drugs under long-term exclusive licenses and the importation of pharmaceutical, diagnostics and other medical products into Israel based on exclusive agreement with the manufacturers. Perrigo operates through several wholly-owned subsidiaries. In the U.S., these subsidiaries consist primarily of L. Perrigo Co.; Perrigo Co. of South Carolina; and Perrigo New York, Inc. Outside the U.S., the subsidiaries consist primarily of Perrigo Israel Pharmaceuticals, Ltd.; Chemagis, Ltd.; Quimica Y Farmacia S.A. de C.V.; Wrafton Laboratories, Ltd.; and Perrigo U.K., Ltd. Wal-Mart accounted for roughly 22% of net sales in 2006. In May 2007, Perrigo announced that FDA approved the marketing of the firm's over-the-counter coated nicotine polacrilex gum. In July 2007, the company acquired Qualis, Inc., a private manufacturer of store brand pediculicide products, for roughly $12 million.

FINANCIALS: Sales and profits are in thousands of dollars—add 000 to get the full amount. 2006 Note: Financial information for 2006 was not available for all companies at press time.

2006 Sales: $1,366,821	2006 Profits: $71,400	U.S. Stock Ticker: PRGO
2005 Sales: $1,024,098	2005 Profits: $-325,983	Int'l Ticker: Int'l Exchange:
2004 Sales: $898,204	2004 Profits: $80,567	Employees: 5,969
2003 Sales: $826,000	2003 Profits: $54,000	Fiscal Year Ends: 6/30
2002 Sales: $826,300	2002 Profits: $50,200	Parent Company:

SALARIES/BENEFITS:

Pension Plan:	ESOP Stock Plan:	Profit Sharing:	Top Exec. Salary: $750,000	Bonus: $487,500
Savings Plan:	Stock Purch. Plan:		Second Exec. Salary: $400,000	Bonus: $275,000

OTHER THOUGHTS:

Apparent Women Officers or Directors: 1
Hot Spot for Advancement for Women/Minorities:

LOCATIONS: ("Y" = Yes)

West:	Southwest:	Midwest:	Southeast:	Northeast:	International:
				Y	Y

Note: Financial information, benefits and other data can change quickly and may vary from those stated here.

PFIZER INC

www.pfizer.com

Industry Group Code: 325412 Ranks within this company's industry group: Sales: 2 Profits: 1

Drugs:		Other:		Clinical:		Computers:		Other:	
Discovery:	Y	AgriBio:		Trials/Services:		Hardware:		Specialty Services:	
Licensing:		Genomics/Proteomics:		Laboratories:		Software:		Consulting:	
Manufacturing:	Y	Tissue Replacement:		Equipment/Supplies:		Arrays:		Blood Collection:	
Development:	Y			Research/Development Svcs.:		Database Management:		Drug Delivery:	
Generics:				Diagnostics:				Drug Distribution:	

TYPES OF BUSINESS:

Drugs-Diversified
Prescription Pharmaceuticals
Veterinary Pharmaceuticals
Capsule Manufacturing

BRANDS/DIVISIONS/AFFILIATES:

Celebrex
Viagra
Zoloft
Zyrtec
Aricept
Lipitor
Relpax
Pharmacia Corporation

CONTACTS: *Note: Officers with more than one job title may be intentionally listed here more than once.*

Jeffrey B. Kindler, CEO
Alan G. Levin, CFO/Sr. VP
Mary McLeod, Leader-Worldwide Talent Dev. & Human Resources
John L. LaMattina, Sr. VP/Pres., Pfizer Global R&D
Greg Vahle, Sr. VP-Global Bus. Svcs.
Jonathan White, Sr. VP-Tech. & Bus. Innovation
Nick Saccomano, Sr. VP-Exploratory Science & Tech.
Joseph M. Feczko, Sr. VP/Chief Medical Officer
Nat Ricciardi, Pres., Pfizer Global Mfg.
Allen Waxman, Sr. VP/General Counsel
Ian Read, Sr. VP/Pres., Worldwide Pharmaceutical Oper.
Ed Harrigan, Sr. VP-Worldwide Bus. Dev.
Rich Bagger, Sr. VP-Worldwide Public Affairs & Policy
Amal Naj, Sr. VP-Worldwide Investor Dev. & Strategy
Bill Roche, VP-Finance, Strategic Mgmt. & Functional Support
David Shedlarz, Vice Chmn.
Andreas Fibig, Sr. VP/General Mgr.-Powers Bus. Unit
Oliver Brandicourt, Sr. VP/General Mgr.-Pratt Bus. Unit
Susan Silbermann, Sr. VP-Worldwide Commercial Dev.
Jeffrey B. Kindler, Chmn.
Jean Michel Halfon, Area Pres., Latin America, Africa, ME & Canada

Phone: 212-573-2323	Fax: 212-573-7851
Toll-Free:	
Address: 235 E. 42nd St., New York, NY 10017-5755 US	

GROWTH PLANS/SPECIAL FEATURES:

Pfizer, Inc. is a research-based global drug company that discovers, develops, manufactures and markets prescription medicines for humans and animals. The company operates in two business segments: Human Health and Animal Health. The Human Health segment includes treatments for cardiovascular and metabolic diseases (including Lipitor and Norvasc), central nervous system disorders (including Zoloft, Xanax and Neurontin), arthritis and pain (Celebrex), infectious and respiratory diseases (including Spiriva and Exubera), cancer (including Ellence, Sutent and Camptosar), eye disease, endocrine disorders (Genotropin) and allergies (notably Zyrtec), among others. The company's Viagra is the leading treatment for erectile dysfunction and one of the world's most recognized pharmaceutical brands. The Animal Health Segment discovers, develops and sells products for the prevention and treatment of diseases in livestock and companion animals. Its products include parasiticides (including Revolution), anti-inflammatories (Rimadyl), vaccines, antibiotics and related medicines. Pfizer's staff includes 12,000 medical researchers who support substantial research and development projects and investments, totaling $7.6 billion in 2006. Currently, the company holds about 169 new molecular entities, 73 product-line extensions and more than 400 compounds in discovery research. In a major development, Pfizer acquired Pharmacia Corporation, making it the largest pharmaceutical company in the U.S., Europe, Japan and Latin America. In June 2006, Johnson & Johnson agreed to acquire Pfizer's consumer health division, maker of Benadryl, Listerine and many other over-the-counter health items, for $16.6 billion dollars. In early 2007, Pfizer announced it would eliminate 10,000 positions (10% of its workforce) over the next two years. This includes more than 20% of its European sales force and 2,200 U.S. sales representatives. Additionally, the firm plans to close up to five of its research laboratories.

Pfizer offers its employees free prescription drugs, tuition reimbursement, adoption assistance, a referral program, educational loans and four-year college scholarships for children of employees.

FINANCIALS: Sales and profits are in thousands of dollars—add 000 to get the full amount. 2006 Note: Financial information for 2006 was not available for all companies at press time.

2006 Sales: $48,371,000	2006 Profits: $19,337,000	U.S. Stock Ticker: PFE
2005 Sales: $47,405,000	2005 Profits: $8,085,000	Int'l Ticker: Int'l Exchange:
2004 Sales: $48,988,000	2004 Profits: $11,361,000	Employees: 106,000
2003 Sales: $45,188,000	2003 Profits: $3,910,000	Fiscal Year Ends: 12/31
2002 Sales: $32,373,000	2002 Profits: $9,126,000	Parent Company:

SALARIES/BENEFITS:

Pension Plan: Y	ESOP Stock Plan:	Profit Sharing:	Top Exec. Salary: $2,270,500	Bonus: $
Savings Plan: Y	Stock Purch. Plan:		Second Exec. Salary: $1,220,300	Bonus: $1,383,000

OTHER THOUGHTS:

Apparent Women Officers or Directors: 25
Hot Spot for Advancement for Women/Minorities: Y

LOCATIONS: ("Y" = Yes)

West:	Southwest:	Midwest:	Southeast:	Northeast:	International:
Y	Y	Y	Y	Y	Y

Note: Financial information, benefits and other data can change quickly and may vary from those stated here.

PHARMACEUTICAL PRODUCT DEVELOPMENT INC
www.ppdi.com

Industry Group Code: 541710 **Ranks within this company's industry group:** Sales: 2 Profits: 1

Drugs:		Other:		Clinical:		Computers:		Other:	
Discovery:		AgriBio:		Trials/Services:		Hardware:		Specialty Services:	Y
Licensing:		Genomics/Proteomics:		Laboratories:		Software:	Y	Consulting:	Y
Manufacturing:		Tissue Replacement:		Equipment/Supplies:		Arrays:		Blood Collection:	
Development:				Research/Development Svcs.:	Y	Database Management:		Drug Delivery:	
Generics:				Diagnostics:				Drug Distribution:	

TYPES OF BUSINESS:
Contract Research
Drug Discovery & Development Services
Clinical Data Consulting Services
Medical Marketing & Information Support Services
Drug Development Software
Medical Device Development

BRANDS/DIVISIONS/AFFILIATES:
PPD Discovery
PPD Development
CSS Informatics
PPD Medical Communications
PPD Virtual

CONTACTS: *Note: Officers with more than one job title may be intentionally listed here more than once.*
Fredric N. Eshelman, CEO
Linda Baddour, CFO
Paul S. Covington, Exec. VP-Dev.
William Richardson, Sr. VP-Global Bus. Dev.
Linda Baddour, Treas./Asst. Corp. Sec.
Frederic N. Eshelman, Vice Chmn.
Mark Roseman, VP-Clinical Oper.
Andy Strayer, Sr. VP-Clinical Oper.-Americas & Asia
Randy Marchbanks, VP-Bus. Dev.-Americas
Ernest Mario, Chmn.
Sue Stansfield, Sr. VP-Clinical Oper. & Project Mgmt., Europe

Phone: 910-251-0081	Fax: 910-762-5820

Toll-Free:
Address: 3151 S. 17th St., Wilmington, NC 28412 US

GROWTH PLANS/SPECIAL FEATURES:
Pharmaceutical Product Development, Inc. (PPD) provides drug discovery and development services to pharmaceutical and biotechnology companies as well as academic and government organizations. PPD's services are primarily divided into two company segments: PPD Discovery and PPD Development. Through the combined services of these segments, PPD helps pharmaceutical companies through all stages of clinical testing. The stages of testing can be specifically divided into preclinical, phase I, phase II-IIIb and post-approval. In the preclinical stages of drug testing, PPD provides information concerning the pharmaceutical composition of a new drug, its safety, its formulaic design and how it will be administered to children and adults. During phase I of testing, PPD conducts healthy volunteer clinics, provides data management services and guides companies/laboratories through regulatory affairs. In phase II and III tests, PPD oversees the later stages of product development and government approval, providing project management and clinical monitoring. In the post-approval stage of a drug's development, PPD provides technology and marketing services aimed to maximize the new drug's lifecycle. PPD has experience conducting research and drug development in several areas, including antiviral studies, cardiovascular diseases, critical care studies, endocrine/metabolic studies, hematology/oncology studies, immunology studies and ophthalmology studies. The firm additionally conducts regional, national and global studies and research projects through offices in 28 countries worldwide. In 2006, PPD signed an agreement with PDL BioPharma, Inc., whereby the firm will perform molecular profiling to discover biomarkers. Additionally, CSS Informatics, the clinical and safety data management division of PPD, entered into an agreement with i-clinics, Ltd., a software provider of data acquisition and management systems. In early 2007, PPD opened a new facility in Lanarkshire, Scotland.

FINANCIALS: Sales and profits are in thousands of dollars—add 000 to get the full amount. 2006 Note: Financial information for 2006 was not available for all companies at press time.

2006 Sales: $1,247,682	2006 Profits: $156,652	**U.S. Stock Ticker:** PPDI
2005 Sales: $1,037,090	2005 Profits: $119,897	**Int'l Ticker:** Int'l Exchange:
2004 Sales: $841,256	2004 Profits: $91,684	Employees: 9,150
2003 Sales: $726,983	2003 Profits: $46,310	Fiscal Year Ends: 12/31
2002 Sales: $562,600	2002 Profits: $39,900	Parent Company:

SALARIES/BENEFITS:
Pension Plan:	ESOP Stock Plan:	Profit Sharing:	Top Exec. Salary: $688,733	Bonus: $475,000
Savings Plan: Y	Stock Purch. Plan:		Second Exec. Salary: $348,189	Bonus: $175,000

OTHER THOUGHTS:
Apparent Women Officers or Directors: 2
Hot Spot for Advancement for Women/Minorities: Y

LOCATIONS: ("Y" = Yes)
West:	Southwest:	Midwest:	Southeast:	Northeast:	International:
Y		Y	Y	Y	Y

Note: Financial information, benefits and other data can change quickly and may vary from those stated here.

PHARMACOPEIA DRUG DISCOVERY www.pharmacopeia.com

Industry Group Code: 541710 Ranks within this company's industry group: Sales: 17   Profits: 15

Drugs:	Other:	Clinical:	Computers:	Other:
Discovery:	AgriBio:	Trials/Services:	Hardware:	Specialty Services:
Licensing:	Genomics/Proteomics:	Laboratories:	Software: Y	Consulting:
Manufacturing:	Tissue Replacement:	Equipment/Supplies:	Arrays:	Blood Collection:
Development:		Research/Development Svcs.: Y	Database Management:	Drug Delivery:
Generics:		Diagnostics:		Drug Distribution:

TYPES OF BUSINESS:

Drug Discovery Technology
Molecular Combinational Chemistry
Molecular Modeling & Simulation Software
Chemical Databases

BRANDS/DIVISIONS/AFFILIATES:

ECLiPS

CONTACTS: *Note: Officers with more than one job title may be intentionally listed here more than once.*

Leslie Browne, CEO
Leslie Browne, Pres.
Brian M. Posner, CFO
David M. Floyd, Chief Scientific Officer/Exec. VP
Stephen Costalas, General Counsel/Corp. Sec./Exec. VP
Simon Tomlinson, Sr. VP-Bus. Dev.
Brian Posner, Treas./Exec. VP
Maria Webb, VP-Preclinical Research, Biological & Pharm.
Rene Belder, VP-Clinical & Regulatory Affairs
Joseph Mollica, Chmn.

Phone: 609-452-3600	Fax: 609-452-3672
Toll-Free:	
Address: 3000 Eastpark Blvd., Cranbury, NJ 08512 US	

GROWTH PLANS/SPECIAL FEATURES:

Pharmacopeia Drug Discovery, Inc. is engaged in the design, development and marketing of products and services that are intended to improve and accelerate drug discovery and chemical development. The company serves a large number of patients suffering from unmet medical needs, often focusing on conditions such as rheumatoid arthritis, chronic obstructive pulmonary disease, oncology and metabolic diseases. Currently, Pharmacopeia has ongoing programs for each of those conditions, all of which are now in clinical Phase One. To investigate potential chemical compounds for development, the company uses a proprietary chemical encoding/decoding process called ECLiPS technology. It allows chemists to perform thousands of reactions at a time on polymer beads. In addition, Pharmacopeia utilizes an ultra-high-throughput screening platform that enables the screening of hundred of thousands to millions of compounds per week. Pharmacopeia has also developed proprietary software to support its drug discovery activities. Its customers typically fund the research and provide for significant milestone payments when chemicals have passed through the lead discovery and optimization phases and are ready for development. The revenue is concentrated mainly in the company's two largest collaborators, Schering-Plough and N.V. Organon. In early 2006, the company entered into a drug discovery and development alliance with GlaxoSmithKline. It also licensed certain rights to novel therapeutic candidates from Bristol-Meyers Squibb, specifically those related to the treatment of cardiovascular disease. Also in 2006, Pharmacopeia formed an alliance with Cephalon, Inc. to discover and develop new drugs, which afforded the firm an up-front program access fee of $15 million. In 2007, Schering-Plough initiated Phase One clinical trials of a drug developed in partnership with Pharmacopeia to potentially treat metabolic diseases.

Pharmacopeia offers its employees health, vision, dental and prescription drug coverage benefits; flexible spending accounts; group life insurance; short and long term disability coverage; and education reimbursement.

FINANCIALS: Sales and profits are in thousands of dollars—add 000 to get the full amount. 2006 Note: Financial information for 2006 was not available for all companies at press time.

2006 Sales: $16,936	2006 Profits: $-27,764	U.S. Stock Ticker: PCOP
2005 Sales: $20,403	2005 Profits: $-17,138	Int'l Ticker: Int'l Exchange:
2004 Sales: $24,359	2004 Profits: $-17,420	Employees: 150
2003 Sales: $29,503	2003 Profits: $-2,848	Fiscal Year Ends: 12/31
2002 Sales: $29,304	2002 Profits: $-2,142	Parent Company:

SALARIES/BENEFITS:

Pension Plan:	ESOP Stock Plan:	Profit Sharing:	Top Exec. Salary: $379,167	Bonus: $82,000
Savings Plan: Y	Stock Purch. Plan: Y		Second Exec. Salary: $285,542	Bonus: $90,000

OTHER THOUGHTS:

Apparent Women Officers or Directors: 3
Hot Spot for Advancement for Women/Minorities: Y

LOCATIONS: ("Y" = Yes)

West:	Southwest:	Midwest:	Southeast:	Northeast:	International:
				Y	

PHARMACYCLICS INC   www.pcyc.com

Industry Group Code: 325412 Ranks within this company's industry group: Sales: 180   Profits: 147

Drugs:	Other:	Clinical:	Computers:	Other:
Discovery:	AgriBio:	Trials/Services:	Hardware:	Specialty Services:
Licensing:	Genomics/Proteomics:	Laboratories:	Software:	Consulting:
Manufacturing:	Tissue Replacement:	Equipment/Supplies:	Arrays:	Blood Collection:
Development: Y		Research/Development Svcs.:	Database Management:	Drug Delivery:
Generics:		Diagnostics:		Drug Distribution:

TYPES OF BUSINESS:

Pharmaceuticals Development
Drugs-Cancer & Cardiovascular Disease
Texaphyrins

BRANDS/DIVISIONS/AFFILIATES:

Xcytrin
PCI-24781
Bruton's Tyrosine Kinase Inhibitors
Alimta
Taxotere

CONTACTS: Note: Officers with more than one job title may be intentionally listed here more than once.

Richard A. Miller, CEO
Richard A. Miller, Pres.
Leiv Lea, CFO
David Loury, VP-Preclinical Sciences
Hugo Madden, VP-Tech. Dev.
Leiv Lea, VP-Admin.
Michael Inouye, Sr. VP-Corp. & Commercial Dev.
Leiv Lea, VP-Finance
Markus F. Renschler, VP-Clinical Dev.
Gregory Hemmi, VP-Chemical Oper.
Richard M. Levy, Chmn.

Phone: 408-774-0330	Fax: 408-774-0340
Toll-Free:	
Address: 995 E. Arques Ave., Sunnyvale, CA 94085-4521 US	

GROWTH PLANS/SPECIAL FEATURES:

Pharmacyclics, Inc. is a pharmaceutical company developing innovative treatments for cancer and other diseases. The company is working on a range of clinical and laboratory programs and is pursuing the development of novel drugs aimed at specific pathways. The company's most advanced product candidate is Xcytrin (motexafin gadolinium), a rationally designed small molecule, the first in a class of drugs that concentrates in cancer cells and disrupts cellular metabolism by a unique mechanism of action. Xcytrin is in late-stage human clinical trials for lung cancer and other oncology indications. The company is also developing several other product candidates derived from structure-based research. A series of histone deacetylase (HDAC) inhibitors appear to inhibit tumor growth by interfering with cancer cell DNA transcription. Pharmacyclics' lead HDAC inhibitor is PCI-24781, which is in Phase I clinical trials. The company has also developed a small molecule inhibitor of Factor VIIa. Factor VIIa inhibitors appear to inhibit tumor growth and angiogenesis when complexed with intrinsic tissue factor. Pharmacyclics, through its Bruton's Tyrosine Kinase Inhibitors program, has developed drugs that can inhibit an enzyme required for early B-cells to mature into fully functioning cells. Tyrosine kinase inhibitors prevent B-cell maturation and are undergoing investigation for the treatment of lymphomas and autoimmune disorders. Pharmacyclics has research collaboration agreements with the National Cancer Institute and the University of Texas. The firm holds 76 U.S. patents and has 21 other pending U.S. patent applications. In 2007, the company announced interim results from two ongoing Phase II trials supporting the use of Xcytrin Injection, in combination with Alimta (pemetrexed) and in combination with Taxotere (docetaxel) as a second-line treatment for patients with non-small cell lung cancer (NSCLC) who failed at least one platinum-based chemotherapy regimen. In the trials, 85% and 87% of patients, respectively, have achieved stabilization of their tumors.

FINANCIALS: Sales and profits are in thousands of dollars—add 000 to get the full amount. 2006 Note: Financial information for 2006 was not available for all companies at press time.

2006 Sales: $ 181	2006 Profits: $-42,158	**U.S. Stock Ticker:** PCYC
2005 Sales: $	2005 Profits: $-31,048	**Int'l Ticker:** Int'l Exchange:
2004 Sales: $	2004 Profits: $-29,165	Employees: 114
2003 Sales: $	2003 Profits: $-28,298	Fiscal Year Ends: 6/30
2002 Sales: $	2002 Profits: $-36,600	Parent Company:

SALARIES/BENEFITS:

Pension Plan:	ESOP Stock Plan:	Profit Sharing:	Top Exec. Salary: $421,688	Bonus: $
Savings Plan: Y	Stock Purch. Plan: Y		Second Exec. Salary: $296,549	Bonus: $

OTHER THOUGHTS:

Apparent Women Officers or Directors:
Hot Spot for Advancement for Women/Minorities:

LOCATIONS: ("Y" = Yes)

West:	Southwest:	Midwest:	Southeast:	Northeast:	International:
Y					

PHARMANET DEVELOPMENT GROUP INC
www.pharmanet.com

Industry Group Code: 541710 Ranks within this company's industry group: Sales: 7 Profits: 5

Drugs:	Other:	Clinical:		Computers:		Other:	
Discovery:	AgriBio:	Trials/Services:	Y	Hardware:		Specialty Services:	
Licensing:	Genomics/Proteomics:	Laboratories:	Y	Software:	Y	Consulting:	
Manufacturing:	Tissue Replacement:	Equipment/Supplies:		Arrays:		Blood Collection:	
Development:		Research/Development Svcs.:	Y	Database Management:		Drug Delivery:	
Generics:		Diagnostics:				Drug Distribution:	

TYPES OF BUSINESS:
Contract Research
Clinical Trial Services
Bioanalytical Laboratory Services
Clinical Trial Software

BRANDS/DIVISIONS/AFFILIATES:
SFBC International, Inc.
PharmaNet, Inc.
Anapharm, Inc.

CONTACTS: *Note: Officers with more than one job title may be intentionally listed here more than once.*
Jeffry P. McMullen, CEO
Jeffrey P. McMullen, Pres.
John P. Hamill, CFO/Exec. VP
Mark Di Ianni, Exec. VP-Strategic Initiatives
Marc LeBel, Exec. VP-Laboratories
Johane Boucher-Champagne, Exec. VP-Early Clinical Dev./Pres. & CEO-Anapharm
Mark Di Ianni, Pres., Early Stage Dev.
Thomas J. Newman, Exec. VP-Late Stage Dev./CEO-PharmaNet, Inc.
Jack Levine, Chmn.

Phone: 609-951-6800	Fax: 609-514-0390
Toll-Free:	
Address: 504 Carnegie Ctr., Princeton, NJ 08540 US	

GROWTH PLANS/SPECIAL FEATURES:
PharmaNet Development Group Inc. (PharmaNet), formerly SFBC International, Inc., is a global drug development services company, which provides a broad range of both early and late stage clinical drug development services to branded pharmaceutical, biotechnology and generic drug and medical device companies around the world. PharmaNet offers a full suite of services to its clients as well, including early clinical pharmacology research; biostatistics and data management; scientific and medical writing; and regulatory audit planning. The company reports in two operating segments: early stage and late stage clinical development. In early stage clinical development services, it specializes primarily in the areas of Phase I clinical trials and bioanalytical laboratory services. The company operates two early stage clinical trial facilities located in Quebec City and Montreal in Canada. PharmaNet has developed and currently maintains extensive databases of available individuals who have indicated an interest in participating in future early stage clinical trials. PharmaNet provides late stage clinical development, including Phase II through Phase IV clinical trials, and related services through a network of 31 offices. It uses a full line of proprietary software products specifically designed to support clinical development activities. These web-based products, which the company believes complies with both FDA and international guidelines and regulations governing the conduct of clinical trials, facilitate the collection, management and reporting of clinical trial information. The firm has acquired a number of subsidiaries, which include PharmaNet, Inc., a late stage development subsidiary, and Anapharm, Inc., which is involved with early stage clinical trials and bioanalytical research. In May 2006, PharmaNet announced that it plans to cease all operations located in Florida, including its Miami and Ft. Myers facilities.

FINANCIALS: Sales and profits are in thousands of dollars—add 000 to get the full amount. 2006 Note: Financial information for 2006 was not available for all companies at press time.

2006 Sales: $406,955	2006 Profits: $-36,025	**U.S. Stock Ticker: PDGI**
2005 Sales: $361,506	2005 Profits: $4,779	Int'l Ticker: Int'l Exchange:
2004 Sales: $111,894	2004 Profits: $19,659	Employees: 2,089
2003 Sales: $93,784	2003 Profits: $11,582	Fiscal Year Ends: 12/31
2002 Sales: $64,740	2002 Profits: $7,868	Parent Company:

SALARIES/BENEFITS:

Pension Plan:	ESOP Stock Plan:	Profit Sharing:	Top Exec. Salary: $663,531	Bonus: $975,000
Savings Plan: Y	Stock Purch. Plan:		Second Exec. Salary: $475,000	Bonus: $214,965

OTHER THOUGHTS:
Apparent Women Officers or Directors: 1
Hot Spot for Advancement for Women/Minorities:

LOCATIONS: ("Y" = Yes)

West:	Southwest:	Midwest:	Southeast:	Northeast:	International:
Y		Y	Y	Y	Y

PHARMION CORP

www.pharmion.com

Industry Group Code: 325412 Ranks within this company's industry group: Sales: 51   Profits: 179

Drugs:		Other:		Clinical:		Computers:		Other:	
Discovery:		AgriBio:		Trials/Services:		Hardware:		Specialty Services:	
Licensing:	Y	Genomics/Proteomics:		Laboratories:		Software:		Consulting:	
Manufacturing:		Tissue Replacement:		Equipment/Supplies:		Arrays:		Blood Collection:	
Development:	Y			Research/Development Svcs.:		Database Management:		Drug Delivery:	
Generics:				Diagnostics:				Drug Distribution:	Y

TYPES OF BUSINESS:

Hematology & Oncology Therapeutics

BRANDS/DIVISIONS/AFFILIATES:

Vidaza
Innohep
Refludan
Thalidomide Pharmion
Cabrellis Pharmaceuticals

CONTACTS: *Note: Officers with more than one job title may be intentionally listed here more than once.*

Patrick J. Mahaffy, CEO
Patrick J. Mahaffy, Pres.
Erle T. Mast, CFO/Exec. VP
Pam Herriott, VP-Human Resources
Andrew R. Allen, Chief Medical Officer/Exec. VP
Jeffrey P. Davis, VP-IT
Joe Como, VP-Global Mfg.
Steven DuPont, General Counsel/VP
Barrie L. Alioth, VP-Strategic Planning
Jay T. Backstrom, VP-Global Medical & Safety
Gillian C. Ivers-Read, Exec. VP-Dev. Oper.
Michael Cosgrave, Chief Commercial Officer

Phone: 720-564-9100	Fax: 720-564-9191

Toll-Free: 866-742-7646

Address: 2525 28th St., Ste. 200, Boulder, CO 80301 US

GROWTH PLANS/SPECIAL FEATURES:

Pharmion Corp. is a global pharmaceutical company that acquires, develops and commercializes products for the treatment of hematology and oncology patients. The company's primary commercial products include Vidaza, marketed and sold as an approved treatment for myelodysplastic syndromes in the U.S., Switzerland, Israel and the Philippines; and Thaliodomide Pharmion 50mg, a therapy used for the treatment of multiple myeloma and certain other forms of cancer, approved in Australia, New Zealand, Turkey, Israel, South Korea and Thailand. Together, these two drugs generated roughly 92% of total net sales in 2006. Other products include Innohep, a low molecular weight heparin sold in the U.S.; and Refludan, an anti-thrombin agent sold in Europe and other countries outside the U.S. and Canada. The company has its own research, regulatory, development and sales and marketing organization in the U.S., the E.U. and Australia. The firm also owns a distribution network to reach the hematology and oncology markets in several additional countries throughout Europe, the Middle East and Asia. In January 2006, the company obtained commercialization rights from MethylGene, Inc. for MethylGene's histone deacetylase (HDAC) inhibitors in North America, Europe, the Middle East and certain other markets. The deal includes MGCD0103, MethylGene's lead HDAC inhibitor, as well as MethylGene's pipeline of second-generation HDAC inhibitor compounds for oncology indications. In November 2006, the firm acquired Cabrellis Pharmaceuticals, a clinical-stage private oncology company dedicated to the development of amrubicin for treatment of small cell lung cancer in North America and the E.U.

FINANCIALS: Sales and profits are in thousands of dollars—add 000 to get the full amount. 2006 Note: Financial information for 2006 was not available for all companies at press time.

2006 Sales: $238,646	2006 Profits: $-91,012	**U.S. Stock Ticker: PHRM**
2005 Sales: $221,244	2005 Profits: $2,269	**Int'l Ticker:** Int'l Exchange:
2004 Sales: $130,171	2004 Profits: $-17,537	Employees: 417
2003 Sales: $25,539	2003 Profits: $-50,059	Fiscal Year Ends: 12/31
2002 Sales: $4,735	2002 Profits: $-34,697	Parent Company:

SALARIES/BENEFITS:

Pension Plan:	ESOP Stock Plan:	Profit Sharing:	Top Exec. Salary: $901,538	Bonus: $
Savings Plan:	Stock Purch. Plan:		Second Exec. Salary: $475,000	Bonus: $267,000

OTHER THOUGHTS:

Apparent Women Officers or Directors: 2
Hot Spot for Advancement for Women/Minorities: Y

LOCATIONS: ("Y" = Yes)

West:	Southwest:	Midwest:	Southeast:	Northeast:	International:
Y					Y

PHARMOS CORP
www.pharmoscorp.com

Industry Group Code: 325412 Ranks within this company's industry group: Sales: Profits: 138

Drugs:		Other:	Clinical:	Computers:	Other:	
Discovery:	Y	AgriBio:	Trials/Services:	Hardware:	Specialty Services:	
Licensing:		Genomics/Proteomics:	Laboratories:	Software:	Consulting:	
Manufacturing:		Tissue Replacement:	Equipment/Supplies:	Arrays:	Blood Collection:	
Development:	Y		Research/Development Svcs.:	Database Management:	Drug Delivery:	Y
Generics:			Diagnostics:		Drug Distribution:	

TYPES OF BUSINESS:
Neurological Diseases Drugs
Drug Delivery Systems
Synthetic Cannabinoids
Cannabinoids

BRANDS/DIVISIONS/AFFILIATES:
Dextofisopam
Dexanibinol
NanoEmulsion
Cannabinor
Vela Pharmaceuticals, Inc.

CONTACTS: Note: Officers with more than one job title may be intentionally listed here more than once.
Elkan R. Gamzu, CEO
Alan L. Rubino, COO
Alan L. Rubino, Pres.
S. Colin Neill, CFO/Sr. VP
S. Colin Neill, Sec./Treas.
Alon Michal, VP-Finance/General Manager
Iris Alroy, Sr. VP-Discovery
Haim Aviv, Chmn.

Phone: 732-452-9556	Fax: 732-452-9557
Toll-Free:	
Address: 99 Wood Ave. S., Ste. 311, Iselin, NJ 08830 US	

GROWTH PLANS/SPECIAL FEATURES:
Pharmos Corp. is a biopharmaceutical company that discovers and develops novel therapeutics focusing on specific diseases of the nervous system including disorders of the brain-gut axis (GI/IBS), pain/inflammation and autoimmune disorders. The company's lead product, Dextofisopam, positively completed a Phase IIa trial, and has initiated a Phase IIb trial, treating irritable bowel syndrome. Pharmos' discovery efforts are focused primarily on CB2-selective compounds, which are small molecule cannabinoid receptor agonists that bind preferentially to CB2 receptors found primarily in peripheral immune cells and peripheral nervous system. Cannabinator, the lead product candidate, completed on the two Phase IIa test in pain indications with the intravenous formulation and is scheduled to complete the second Phase IIa test and initiate a Phase I test with the oral formulation during 2007. Other compounds from Pharmos' proprietary synthetic cannabinoid library are in pre-clinical studies targeting nociceptive, neuropathic and visceral pain, as well as autoimmune diseases, including multiple sclerosis, rheumatoid arthritis and inflammatory bowel disease. The company also has a family of synthetic cannbinoid compounds that do not bind to cannabinoid receptors. The firm seeks to partner Dexanabinol, the lead compound from this family, as a preventive agent against cognitive impairment following cardiac surgery. Other products include NanoEmulsion, a drug delivery system that completed a second Phase I trial that confirmed its safety and tolerability of a more optimized NE diclofenac cream formulation in 2006 and commenced a Phase IIa trial in June 2007; VP103, which completed Phase IIa clinical development as a treatment for neropathic pain and female hypoactive sexual desire disorder; and tianeptine, a potential follow-up product to Dextofisopam, which completed late-preclinical development for the treatment of irritable bowel syndrome. In October 2006, the firm acquired Vela Pharmaceuticals, Inc., which specializes in the development of medicines related to diseases of the nervous system disorders.

FINANCIALS: Sales and profits are in thousands of dollars—add 000 to get the full amount. 2006 Note: Financial information for 2006 was not available for all companies at press time.

2006 Sales: $	2006 Profits: $-35,137	U.S. Stock Ticker: PARS
2005 Sales: $	2005 Profits: $-2,930	Int'l Ticker: Int'l Exchange:
2004 Sales: $	2004 Profits: $-21,968	Employees: 54
2003 Sales: $	2003 Profits: $-18,485	Fiscal Year Ends: 12/31
2002 Sales: $	2002 Profits: $-17,100	Parent Company:

SALARIES/BENEFITS:
Pension Plan:	ESOP Stock Plan:	Profit Sharing:	Top Exec. Salary: $298,284	Bonus: $
Savings Plan:	Stock Purch. Plan:		Second Exec. Salary: $249,063	Bonus: $

OTHER THOUGHTS:
Apparent Women Officers or Directors: 1
Hot Spot for Advancement for Women/Minorities:

LOCATIONS: ("Y" = Yes)
West:	Southwest:	Midwest:	Southeast:	Northeast:	International:
				Y	Y

Note: Financial information, benefits and other data can change quickly and may vary from those stated here.

POLYDEX PHARMACEUTICALS

www.polydex.com

Industry Group Code: 325414 Ranks within this company's industry group: Sales: 9   Profits: 7

Drugs:		Other:		Clinical:		Computers:		Other:	
Discovery:		AgriBio:		Trials/Services:		Hardware:		Specialty Services:	
Licensing:		Genomics/Proteomics:		Laboratories:		Software:		Consulting:	
Manufacturing:	Y	Tissue Replacement:		Equipment/Supplies:		Arrays:		Blood Collection:	
Development:	Y			Research/Development Svcs.:	Y	Database Management:		Drug Delivery:	Y
Generics:				Diagnostics:				Drug Distribution:	Y

TYPES OF BUSINESS:

Drugs-Raw Ingredients
Contraceptive Development
HIV & STD Preventatives

BRANDS/DIVISIONS/AFFILIATES:

Ushercell
Dextran Products, Ltd.
Dextran
CONRAD
Organichem Corporation
Pantheon, Inc.
BCY, Inc.

CONTACTS: *Note: Officers with more than one job title may be intentionally listed here more than once.*

George G. Usher, CEO
Sharon L. Wardlaw, COO
George G. Usher, Pres.
John A. Luce, CFO
Sharon L. Wardlaw, Corp. Sec.
Sharon L. Wardlaw, Treas.
George G. Usher, Chmn.

Phone: 416-755-2441	Fax: 242-328-6919
Toll-Free:	
Address: 421 Comstock Rd., Toronto, ON M1L 2H5 Canada	

GROWTH PLANS/SPECIAL FEATURES:

Polydex Pharmaceuticals, Ltd. develops and manufactures dextran-based powder and liquid products in bulk for use in drug delivery and in therapeutic compounds through its subsidiary, Dextran Products, Ltd. Dextran is the generic name applied to certain synthetic compounds formed by bacterial growth on sugar from sugar cane and is used as a volume expander for blood plasma, a moisture retainer in cosmetics and a vaccine adjuvant. Product derivatives include dextran sulphate for DNA hybridization, ulcer treatment and blood coagulation, among other applications; and DEAE (diethlyaminoethyl) dextran, a treatment for obesity, hyperlipemia, hypercholestremia and a vaccine adjuvant. The firm also offers an iron-infused dextran in various strengths. Polydex has one product in testing phases, Ushercell, a high-molecular-weight cellulose sulphate compound intended as a contraceptive anti-microbial agent to protect against transmission of certain sexual diseases. Two Phase II contraceptive clinical studies on the drug are in progress. Polydex is part of a research and development agreement with the University of British Columbia, several Canadian hospitals and a company owned by an affiliate whereby Polydex supplies funding and equipment for the research of a low molecular weight dextran believed to be capable of being used as a treatment for Cystic Fibrosis. In return, the company is allowed exclusive rights to manufacture, market and sell any products developed from the research. Recently, the CONRAD program of the Eastern Virginia Medical School, a Polydex strategic partner, announced that it will begin two Phase III HIV prevention trails using Ushercell. Polydex maintains strategic partnerships with Organichem Corporation, which manufactures Ushercell base materials; Pantheon, Inc., which processes the drug into a gel; and BCY, Inc. and CONRAD, which collaborate with the company in conducting clinical trials. In addition to its dextran-based products, the firm also owns a patented method for the production of cellulose sulphate.

FINANCIALS: Sales and profits are in thousands of dollars—add 000 to get the full amount. 2006 Note: Financial information for 2006 was not available for all companies at press time.

2006 Sales: $5,265	2006 Profits: $-1,489	U.S. Stock Ticker: POLXF
2005 Sales: $6,372	2005 Profits: $1,139	Int'l Ticker: Int'l Exchange:
2004 Sales: $14,100	2004 Profits: $ 100	Employees: 24
2003 Sales: $12,800	2003 Profits: $- 700	Fiscal Year Ends: 1/31
2002 Sales: $12,200	2002 Profits: $- 200	Parent Company:

SALARIES/BENEFITS:

Pension Plan:	ESOP Stock Plan:	Profit Sharing:	Top Exec. Salary: $256,788	Bonus: $85,596
Savings Plan:	Stock Purch. Plan:		Second Exec. Salary: $150,000	Bonus: $

OTHER THOUGHTS:

Apparent Women Officers or Directors: 1
Hot Spot for Advancement for Women/Minorities:

LOCATIONS: ("Y" = Yes)

West:	Southwest:	Midwest:	Southeast:	Northeast:	International:
		Y			Y

Note: Financial information, benefits and other data can change quickly and may vary from those stated here.

PONIARD PHARMACEUTICALS INC   www.poniard.com

Industry Group Code: 325412 Ranks within this company's industry group: Sales: Profits: 112

Drugs:		Other:	Clinical:	Computers:	Other:
Discovery:		AgriBio:	Trials/Services:	Hardware:	Specialty Services:
Licensing:	Y	Genomics/Proteomics:	Laboratories:	Software:	Consulting:
Manufacturing:		Tissue Replacement:	Equipment/Supplies:	Arrays:	Blood Collection:
Development:	Y		Research/Development Svcs.:	Database Management:	Drug Delivery:
Generics:			Diagnostics:		Drug Distribution:

TYPES OF BUSINESS:
Drugs-Cancer & Heart Disease

BRANDS/DIVISIONS/AFFILIATES:
NeoRx Corporation
Picoplatin

CONTACTS: *Note: Officers with more than one job title may be intentionally listed here more than once.*
Jerry McMahon, CEO
Ronald A. Martell, COO
Ronald A. Martell, Pres.
Caroline M. Loewy, CFO
Anna L. Wight, VP-Legal
Cheni Kwok, VP-Bus. Dev.
Julie Rathbun, Corp. Comm.
David A. Karlin, Sr. VP-Clinical Dev. & Regulatory Affairs
Jerry McMahon, Chmn.

Phone: 650-583-3774	Fax: 650-583-3789
Toll-Free:	
Address: 7000 Shoreline Ct., Ste. 270, S. San Francisco, CA 94080 US	

GROWTH PLANS/SPECIAL FEATURES:
Poniard Pharmaceuticals, Inc., formerly NeoRx Corporation, develops therapeutic biopharmaceuticals for the treatment of cancer, cardiovascular diseases and inflammatory diseases. In particular, the company is currently focused on platinum based cancer treatment agents.   Poniards lead product candidate is Picoplatin, a platinum chemotherapeutic agent candidate to be used in place of traditional platinum cancer treatment agents. Typical platinum chemotherapeutic agents provide mixed results, depending on the nature of the cancer. Potential benefits of Picoplatin are that is broadly applicable for the treatment of solid tumors; the compound may have less severe side effects than other platinum treatments; and it may be possible to use this product as treatment for platinum-sensitive, -resistant and -refractory disease. Picoplatin has received orphan drug designation from the FDA for the treatment of small cell lung cancer (SCLC). The product is currently in Phase II clinical trials for the treatment of SCLC, and an international Phase III SPEAR (Study of Picoplatin Efficacy After Relapse) trial is currently being held. Picoplatin is also being assessed in Phase I and Phase II trials for patients with stage IV metastatic colorectal cancer and patients with hormone refractory prostate cancer. The company is also initiating a Phase I trial of an oral version of the drug.

FINANCIALS: Sales and profits are in thousands of dollars—add 000 to get the full amount. 2006 Note: Financial information for 2006 was not available for all companies at press time.

2006 Sales: $	2006 Profits: $-23,294	**U.S. Stock Ticker: PARD**
2005 Sales: $ 15	2005 Profits: $-20,997	**Int'l Ticker:** Int'l Exchange:
2004 Sales: $1,015	2004 Profits: $-19,371	Employees: 32
2003 Sales: $10,531	2003 Profits: $-5,059	Fiscal Year Ends: 12/31
2002 Sales: $11,100	2002 Profits: $-23,100	Parent Company:

SALARIES/BENEFITS:

Pension Plan:	ESOP Stock Plan:	Profit Sharing:	Top Exec. Salary: $400,977	Bonus: $180,389
Savings Plan: Y	Stock Purch. Plan:		Second Exec. Salary: $270,404	Bonus: $74,080

OTHER THOUGHTS:
Apparent Women Officers or Directors: 3
Hot Spot for Advancement for Women/Minorities: Y

LOCATIONS: ("Y" = Yes)

West:	Southwest:	Midwest:	Southeast:	Northeast:	International:
Y	Y				

Note: Financial information, benefits and other data can change quickly and may vary from those stated here.

POZEN INC
www.pozen.com

Industry Group Code: 325412 Ranks within this company's industry group: Sales: 119   Profits: 104

Drugs:		Other:	Clinical:	Computers:	Other:
Discovery:	Y	AgriBio:	Trials/Services:	Hardware:	Specialty Services:
Licensing:		Genomics/Proteomics:	Laboratories:	Software:	Consulting:
Manufacturing:		Tissue Replacement:	Equipment/Supplies:	Arrays:	Blood Collection:
Development:	Y		Research/Development Svcs.:	Database Management:	Drug Delivery:
Generics:			Diagnostics:		Drug Distribution:

TYPES OF BUSINESS:
Drugs, Analgesic
Migraine Therapy

BRANDS/DIVISIONS/AFFILIATES:
Trexima

CONTACTS: *Note: Officers with more than one job title may be intentionally listed here more than once.*
John R. Plachetka, CEO
John R. Plachetka, Pres.
William L. Hodges, CFO/Sr. VP-Finance
Marshall E. Reese, Exec. VP-Product Dev.
William L. Hodges, VP-Admin.
Gilda M. Thomas, General Counsel/Sr. VP
John Barnhardt, VP-Finance & Admin./Principal Acct. Officer
John R. Plachetka, Chmn.

Phone: 919-913-1030	Fax: 919-913-1039
Toll-Free:	
Address: 1414 Raleigh Rd., Chapel Hill, NC 27517 US	

GROWTH PLANS/SPECIAL FEATURES:
Pozen, Inc. is a pharmaceutical company focused primarily on products which can provide improved efficacy, safety or patient convenience in the treatment of migraine, acute and chronic pain and other pain related indications. The firm is currently developing Trexima in collaboration with GlaxoSmithKline. Trexima is GSK's proposed brand name for the combination of sumatriptan succinate, formulated with GSK's RT Technology and naproxen sodium in a single tablet designed for the acute treatment of migraine. Trexima incorporates Pozen's MT 400 technology, which refers to its proprietary combinations of a triptan and a non-steroidal anti-inflammatory drug (NSAID). Under this technology, Pozen seeks to develop product candidates that provide acute migraine therapy by combining the activity of two drugs that act by different mechanisms to reduce the pain and associated symptoms of migraine. The firm recently completed its second Phase III trial on Trexima and (after an early 2007 revision of its response to the FDA, following its initial letter and the FDA's initial response) it expects an action letter for the NDA from the FDA within 2007. Pozen is also developing product candidates that combine a type of acid inhibitor, a proton pump inhibitor (PPI), with an NSAID. These product candidates are intended to provide management of pain and inflammation. The company anticipates it may continue to incur operating losses over the next several years as it completes development and seeks regulatory approval for its product candidates, develops other candidates and acquires and develop product portfolios in other areas. These results may vary, depending on several factors, including: the progress of Trexima, as well as its PA and PN candidates; the acquisition and/or in-licensing and development of other therapeutic product candidates; and costs related to the pending class action lawsuit relating to the approvability of MT 100 and MT 300.

FINANCIALS: Sales and profits are in thousands of dollars—add 000 to get the full amount. 2006 Note: Financial information for 2006 was not available for all companies at press time.

2006 Sales: $13,517	2006 Profits: $-19,310	**U.S. Stock Ticker: POZN**
2005 Sales: $28,647	2005 Profits: $1,959	**Int'l Ticker:** Int'l Exchange:
2004 Sales: $23,088	2004 Profits: $-5,261	Employees: 35
2003 Sales: $3,717	2003 Profits: $-14,863	Fiscal Year Ends: 12/31
2002 Sales: $	2002 Profits: $-24,555	Parent Company:

SALARIES/BENEFITS:

Pension Plan:	ESOP Stock Plan:	Profit Sharing:	Top Exec. Salary: $464,815	Bonus: $470,250
Savings Plan: Y	Stock Purch. Plan:		Second Exec. Salary: $304,331	Bonus: $113,012

OTHER THOUGHTS:
Apparent Women Officers or Directors: 1
Hot Spot for Advancement for Women/Minorities:

LOCATIONS: ("Y" = Yes)

West:	Southwest:	Midwest:	Southeast:	Northeast:	International:
				Y	

PRA INTERNATIONAL   www.prainternational.com

Industry Group Code: 541710 Ranks within this company's industry group: Sales: Profits:

Drugs:	Other:	Clinical:		Computers:		Other:
Discovery:	AgriBio:	Trials/Services:	Y	Hardware:		Specialty Services:
Licensing:	Genomics/Proteomics:	Laboratories:		Software:		Consulting:
Manufacturing:	Tissue Replacement:	Equipment/Supplies:		Arrays:		Blood Collection:
Development:		Research/Development Svcs.:	Y	Database Management:	Y	Drug Delivery:
Generics:		Diagnostics:				Drug Distribution:

TYPES OF BUSINESS:
Clinical Research & Testing Services
Clinical Development Services
Clinical Trials
Data Management Services

BRANDS/DIVISIONS/AFFILIATES:
Pharmaceutical Research Associates, Inc.
PRA E-TMF
Sterling Synergy Systems Private Ltd.
Flex DMA
Pharma Bio-Research

CONTACTS: *Note: Officers with more than one job title may be intentionally listed here more than once.*
Terrance J. Bieker, CEO
Colin Shannon, COO
Colin Shannon, Pres.
Linda Baddour, CFO/Exec. VP
Monika M. Pietrek, Exec. VP-Scientific & Medical Affairs
Kent Thoelke, VP-Global Product Dev. Svcs.
David Dockhorn, Exec. VP-Global Clinical Oper.
Bruce Teplitzky, Exec. VP-Bus. Dev.
Steve Powell, VP-Data Mgmt.
William M. Walsh, III, Sr. VP-Corp. Dev.
Melvin Booth, Chmn.

Phone: 703-464-6300	**Fax:** 703-464-6301
Toll-Free:	
Address: 12120 Sunset Hills Rd., Ste. 600, Reston, VA 20190 US	

GROWTH PLANS/SPECIAL FEATURES:
PRA International is a contract research organization (CRO) that provides clinical drug development services to pharmaceutical and biotechnology companies around the world. CROs typically assist companies in developing drug compounds, biologics, drug delivery devices and the attainment of certain regulatory approvals necessary to market these technologies. In addition, PRA provides a broad array of services in clinical development programs, the creation of drug development and regulatory strategy plans; the utilization of bioanalytical laboratory testing and the development of integrated global clinical databases. Bioanalytical sample testing includes the use of LC-MS Machines, the Packard Robotic System, the HPLL System and the Immuno-Analysis suite. PRA also provides data management services such as electronic data capture, data monitoring and database development. The company supports its data services through the use of its proprietary PRA E-TMF enterprise-wide electronic document management system, which provides clients with faster document handling and archival services, and Flex DMA, which allows direct entry of clinical trial data by investigational sites. Clinical trials in the U.S. are largely centered around PRA's facilities in Kansas. With a comprehensive suite of services and a worldwide network of facilities, PRA currently reports a client roster of more than 295 companies across a range of sectors in the pharmaceutical/biotechnology industries and has conducted over 2,300 clinical trial projects since 1999. The firm operates offices in North America, South America, Europe, Australia and Africa. In July 2006, PRA completed the acquisition of Pharma Bio-Research, an early phase clinical development and bioanalytical laboratory company in Zuidlaren, The Netherlands.

FINANCIALS: Sales and profits are in thousands of dollars—add 000 to get the full amount. 2006 Note: Financial information for 2006 was not available for all companies at press time.

2006 Sales: $338,166	2006 Profits: $26,845	**U.S. Stock Ticker:** PRAI
2005 Sales: $326,244	2005 Profits: $32,223	**Int'l Ticker:** Int'l Exchange:
2004 Sales: $307,644	2004 Profits: $20,749	Employees: 2,700
2003 Sales: $289,997	2003 Profits: $13,247	Fiscal Year Ends: 12/31
2002 Sales: $201,000	2002 Profits: $5,600	Parent Company:

SALARIES/BENEFITS:

Pension Plan:	ESOP Stock Plan:	Profit Sharing:	Top Exec. Salary: $419,583	Bonus: $
Savings Plan: Y	Stock Purch. Plan: Y		Second Exec. Salary: $293,917	Bonus: $

OTHER THOUGHTS:

	LOCATIONS: ("Y" = Yes)					
Apparent Women Officers or Directors:	West:	Southwest:	Midwest:	Southeast:	Northeast:	International:
Hot Spot for Advancement for Women/Minorities:	Y	Y	Y		Y	Y

Note: Financial information, benefits and other data can change quickly and may vary from those stated here.

PROGENICS PHARMACEUTICALS   www.progenics.com

Industry Group Code: 325412 Ranks within this company's industry group: Sales: 72   Profits: 108

Drugs:		Other:	Clinical:	Computers:	Other:
Discovery:	Y	AgriBio:	Trials/Services:	Hardware:	Specialty Services:
Licensing:		Genomics/Proteomics:	Laboratories:	Software:	Consulting:
Manufacturing:	Y	Tissue Replacement:	Equipment/Supplies:	Arrays:	Blood Collection:
Development:	Y		Research/Development Svcs.:	Database Management:	Drug Delivery:
Generics:			Diagnostics:		Drug Distribution:

TYPES OF BUSINESS:

Pharmaceuticals Development
Drugs-HIV
Drugs-Cancer
Small-Molecule Drugs
Viral Entry Inhibitors
Vaccines

BRANDS/DIVISIONS/AFFILIATES:

MNTX
PSMA Development Company LLC
GMK
PRO 140
ProVax

CONTACTS: *Note: Officers with more than one job title may be intentionally listed here more than once.*

Paul J. Maddon, CEO
Robert A. McKinney, CFO
William C. Olson, VP-R&D
Thomas A. Boyd, Sr. VP-Prod. Dev.
Nitya G. Ray, VP-Mfg.
Mark R. Baker, General Counsel/Sr. VP
Walter M. Capone, VP-Oper.
Richard W. Krawiec, VP-Corp. Affairs
Robert A. McKinney, VP-Finance/Treas.
Paul J. Maddon, Chief Science Officer
Robert J. Isreal, Sr. VP-Medical Affairs
Alton B. Kremer, VP-Clinical Research
Paul F. Jacobson, Co-Chmn.
Kurt W. Briner, Co-Chmn.

Phone: 914-789-2800	Fax: 914-789-2817
Toll-Free:	
Address: 777 Old Saw Mill River Rd., Tarrytown, NY 10591 US	

GROWTH PLANS/SPECIAL FEATURES:

Progenics Pharmaceuticals, Inc. develops and commercializes therapeutic products to treat patients with debilitating conditions and life-threatening diseases. The firm's principal programs are directed toward symptom management and supportive care in the treatment of HIV infection and cancer. Progenics has four product candidates in clinical development and several others in preclinical development. The firm's product candidate in the area of symptom management and supportive care is methylnaltrexone, or MNTX, which is designed to reverse the side effects of opioid pain medications while maintaining pain relief. MNTX is currently in Phase III clinical testing. Progenics is also developing MNTX for the management of post-operative bowel dysfunction, in collaboration with Wyeth Pharmaceuticals, a division of Wyeth. In May 2007, the FDA accepted a New Drug Application (NDA) for this drug. In the area of HIV infection, Progenics is developing viral entry inhibitors, which are molecules designed to inhibit the virus' ability to enter certain types of immune system cells. The company has completed Phase I clinical trials of its PRO 140 viral entry inhibitor, a potential treatment for HIV infection. The company is also conducting research on ProVax, which is being developed as a candidate HIV vaccine. ProVax is currently in pre-clinical trials. In addition, Progenics is developing immunotherapies for prostate cancer, including monoclonal antibodies directed against prostate specific membrane antigen (PSMA), a protein found on the surface of prostate cancer cells. The firm's PSMA programs are conducted through PSMA Development Company LLC, a joint venture with Cytogen Corporation. The company's cancer vaccine, GMK, is in Phase III clinical trials for the treatment of malignant melanoma.

Progenics offers employees medical, dental and life insurance; a 401(k) plan; an employee stock options plan; an employee stock purchase plan; flexible spending accounts; short and long-term disability; tuition reimbursement; and an employee assistance plan.

FINANCIALS: Sales and profits are in thousands of dollars—add 000 to get the full amount. 2006 Note: Financial information for 2006 was not available for all companies at press time.

2006 Sales: $69,906	2006 Profits: $-21,618	U.S. Stock Ticker: PGNX
2005 Sales: $9,486	2005 Profits: $-69,429	Int'l Ticker: Int'l Exchange:
2004 Sales: $9,576	2004 Profits: $-42,018	Employees: 191
2003 Sales: $7,461	2003 Profits: $-30,985	Fiscal Year Ends: 12/31
2002 Sales: $10,100	2002 Profits: $-20,800	Parent Company:

SALARIES/BENEFITS:

Pension Plan:	ESOP Stock Plan: Y	Profit Sharing:	Top Exec. Salary: $565,000	Bonus: $401,577
Savings Plan: Y	Stock Purch. Plan: Y		Second Exec. Salary: $340,000	Bonus: $125,000

OTHER THOUGHTS:

Apparent Women Officers or Directors: 1
Hot Spot for Advancement for Women/Minorities:

LOCATIONS: ("Y" = Yes)

West:	Southwest:	Midwest:	Southeast:	Northeast:	International:
				Y	

PROMEGA CORP

www.promega.com

Industry Group Code: 325413 Ranks within this company's industry group: Sales: Profits:

Drugs:	Other:	Clinical:		Computers:	Other:	
Discovery:	AgriBio:	Trials/Services:		Hardware:	Specialty Services:	Y
Licensing:	Genomics/Proteomics:	Laboratories:		Software:	Consulting:	
Manufacturing:	Tissue Replacement:	Equipment/Supplies:	Y	Arrays:	Blood Collection:	
Development:		Research/Development Svcs.:		Database Management:	Drug Delivery:	
Generics:		Diagnostics:			Drug Distribution:	

TYPES OF BUSINESS:
Life Sciences Solutions & Technical Support
Specialty Biochemicals

BRANDS/DIVISIONS/AFFILIATES:
Promega Biosciences, Inc.
Terso Solutions

CONTACTS: Note: Officers with more than one job title may be intentionally listed here more than once.
William A. Linton, CEO
William A. Linton, Pres.
Laura Francis, CFO
Randy Dimond, CTO/VP
Laura Francis, VP-Finance
William A. Linton, Chmn.

Phone: 608-274-4330	Fax: 608-277-2516
Toll-Free: 800-356-9526	
Address: 2800 Woods Hollow Rd., Madison, WI 53711 US	

GROWTH PLANS/SPECIAL FEATURES:
Promega Corp. provides solutions and technical support to the life sciences industry. The company's over 1,450 products aid scientists in life science research, especially genomics, proteomics and cellular analysis. The firm's products are comprised of kits and reagents, as well as integrated solutions for life sciences research and drug discovery. Reagents include DNA and RNA purification; genotype analysis; protein expression and analysis; and DNA sequencing. Integrated solutions provide customers with personal automation tools that combine instrumentation and reagents into one system. Promega's Terso Solutions subsidiary focuses on leveraging RFID technology to solve customers' needs for complete inventory management solutions. Subsidiary Promega Biosciences, Inc. based in California, is a leading manufacturer of specialty biochemicals, with expertise in nucleic acid chemistry, lipids, bioluminescence and fluorescence/labeling dyes. The company has U.S. and foreign patents in areas including nucleic acid purification, human identification, and bioluminescence, coupled with in vitro transcription and translation and cell biology. The firm has branches in 11 countries and sells its products directly and through more than 50 global distributors.

The company offers its employees health and dental insurance; a 401(k) plan; flexible spending accounts; life insurance; short- and long-term disability; long-term care insurance; tuition assistance; and an employee assistance program. Additional perks include on-campus Farmer's Market, convenience services and wellness programs, which include on-site fitness classes, exercise facilities, yoga classes and massage therapy.

FINANCIALS: Sales and profits are in thousands of dollars—add 000 to get the full amount. 2006 Note: Financial information for 2006 was not available for all companies at press time.

2006 Sales: $	2006 Profits: $	U.S. Stock Ticker: Private
2005 Sales: $	2005 Profits: $	Int'l Ticker: Int'l Exchange:
2004 Sales: $169,500	2004 Profits: $	Employees: 755
2003 Sales: $155,800	2003 Profits: $	Fiscal Year Ends: 12/31
2002 Sales: $112,700	2002 Profits: $	Parent Company:

SALARIES/BENEFITS:
Pension Plan:	ESOP Stock Plan:	Profit Sharing:	Top Exec. Salary: $	Bonus: $
Savings Plan: Y	Stock Purch. Plan:		Second Exec. Salary: $	Bonus: $

OTHER THOUGHTS:
Apparent Women Officers or Directors: 1
Hot Spot for Advancement for Women/Minorities:

LOCATIONS: ("Y" = Yes)
West:	Southwest:	Midwest:	Southeast:	Northeast:	International:
Y		Y		Y	Y

Note: Financial information, benefits and other data can change quickly and may vary from those stated here.

PROTEIN POLYMER TECHNOLOGIES   www.ppti.com

Industry Group Code: 325411 Ranks within this company's industry group: Sales: 9   Profits: 7

Drugs:		Other:		Clinical:		Computers:		Other:	
Discovery:		AgriBio:		Trials/Services:		Hardware:		Specialty Services:	
Licensing:		Genomics/Proteomics:		Laboratories:		Software:		Consulting:	
Manufacturing:	Y	Tissue Replacement:	Y	Equipment/Supplies:	Y	Arrays:		Blood Collection:	
Development:	Y			Research/Development Svcs.:		Database Management:		Drug Delivery:	Y
Generics:				Diagnostics:				Drug Distribution:	

TYPES OF BUSINESS:
Medical Products Development
Medical Products-Bodily Repair
Drug Delivery Devices
Protein Polymers

BRANDS/DIVISIONS/AFFILIATES:
Spine Wave, Inc.
NuCore

CONTACTS: *Note: Officers with more than one job title may be intentionally listed here more than once.*
William N. Plamondon III, CEO
Joseph Cappello, VP-R&D
Joseph Cappello, CTO
John E. Flowers, VP-Oper.
John E. Flowers, VP-Planning
Erin Davis, Dir.-Comm
Erin Davis, Dir.-Investor Rel.
Franco A. Ferrari, VP-Laboratory Oper. & Polymer Prod.
William N. Plamondon III, Chmn.

Phone: 858-558-6064	Fax: 858-558-6477

Toll-Free:
Address: 10655 Sorrento Valley Rd., San Diego, CA 92121 US

GROWTH PLANS/SPECIAL FEATURES:
Protein Polymer Technologies, Inc. (PPT) is a development-stage company that uses its proprietary protein-based biomaterials technology to research, produce and clinically test medical products that aid in the natural process of bodily repair. Its product focus includes surgical adhesives and sealants, soft-tissue augmentation products, wound-healing matrices, drug delivery devices and surgical adhesion barriers. PPT uses biomaterials that can mimic the natural properties and functions of proteins and peptides, and interact with a cell's enzymatic reactions to spur protein formation. The company has developed surgical tissue sealants which combine the biocompatibility of fibrin glues, without the risks associated with use of blood-derived products, with high strength and fast setting times. The product adheres well to tissue to seal gas and fluid leaks and comes in resorbable and non-resorbable formulations. This product is in preclinical studies. PPT has developed protein polymers for use on dermal wounds, particularly chronic wounds such as decubitous ulcers, which assist the tissue to heal in such a way that it regenerates as functional tissue instead of scar tissue. The company's urethral bulking agent for the treatment of female stress urinary incontinence is currently in clinical trials. The FDA approved the company's Investigational Device Exemption (IDE), which allows the testing of the safety and effectiveness of the incontinence product in women over the age of 40 who have become incontinent due to the shifting of their bladder or the weakening of the muscle at its base. Another application for the firm's bodily repair technology is spine applications. PPT used the company's patented tissue adhesive technology to create Spine Wave's NuCore intervertebral disc repair material. PPT is a technical partner of Spine Wave, Inc., and the company manufactured the NuCore material for Spine Wave's clinical trials.

FINANCIALS: Sales and profits are in thousands of dollars—add 000 to get the full amount. 2006 Note: Financial information for 2006 was not available for all companies at press time.

2006 Sales: $ 605	2006 Profits: $-7,878	**U.S. Stock Ticker: PPTI.OB**
2005 Sales: $ 867	2005 Profits: $-5,822	**Int'l Ticker:** Int'l Exchange:
2004 Sales: $ 453	2004 Profits: $-3,565	Employees: 18
2003 Sales: $1,617	2003 Profits: $-2,193	Fiscal Year Ends: 12/31
2002 Sales: $3,000	2002 Profits: $-1,000	Parent Company:

SALARIES/BENEFITS:
Pension Plan:	ESOP Stock Plan:	Profit Sharing:	Top Exec. Salary: $300,000	Bonus: $
Savings Plan: Y	Stock Purch. Plan:		Second Exec. Salary: $280,000	Bonus: $

OTHER THOUGHTS:
	LOCATIONS: ("Y" = Yes)					
	West:	Southwest:	Midwest:	Southeast:	Northeast:	International:
Apparent Women Officers or Directors:	Y					
Hot Spot for Advancement for Women/Minorities:						

Note: Financial information, benefits and other data can change quickly and may vary from those stated here.

QIAGEN NV

www.qiagen.com

Industry Group Code: 325413 Ranks within this company's industry group: Sales: 3   Profits: 3

Drugs:	Other:		Clinical:		Computers:		Other:	
Discovery:	AgriBio:		Trials/Services:		Hardware:		Specialty Services:	Y
Licensing:	Genomics/Proteomics:	Y	Laboratories:		Software:		Consulting:	
Manufacturing:	Tissue Replacement:		Equipment/Supplies:	Y	Arrays:		Blood Collection:	
Development:			Research/Development Svcs.:		Database Management:		Drug Delivery:	
Generics:			Diagnostics:	Y			Drug Distribution:	

TYPES OF BUSINESS:

Supplies-Plasmid Purification
Genomics Analysis Products
Diagnostic Products

BRANDS/DIVISIONS/AFFILIATES:

Qiagen, GmbH
Tianwei Times
Nextal Biotechnology, Inc.
artus GmbH
Shenzhen PG Biotech Co. Ltd.
Gentra Systems, Inc.

CONTACTS: *Note: Officers with more than one job title may be intentionally listed here more than once.*

Peer Schatz, CEO
Peer Schatz, Pres.
Roland Sackers, CFO
Bernd Uder, Sr. VP-Global Sales
Gerhard Sohn, VP-Global Human Resources
Joachim Schorr, Sr. VP-Global Research & Dev.
Douglas Liu, VP-Global Oper.
Ulrich Schriek, VP-Corp. Bus. Dev.
Solveigh Mahler, Dir.-Investor Rel. & Public Relations
Michael Collasius, VP-Automated Systems
Thomas Schweins, VP-Mktg. & Strategy
Detlev H. Riesner, Chmn.

Phone: 31-77-320-8400	Fax: 31-77-320-8409
Toll-Free:	
Address: Spoorstraat 50, Venlo, 5911 KJ The Netherlands	

GROWTH PLANS/SPECIAL FEATURES:

Qiagen NV provides technologies and products for the separation and purification of nucleic acids, used in genomics, molecular diagnostics and genetic vaccination and gene therapy. Through its many subsidiaries, Qiagen offers over 320 products in over 40 countries worldwide. The firm's products and services find application in animal and veterinary research, biomedical research, biosecurity and biodefense, epigenetics, genetic identity and forensics, gene expression analysis, gene silencing, influenza research, molecular diagnostics, pharmacogenomics, plant research and protein science fields. Qiagen recently spun off its synthetic DNA business into a new, privately owned company called Operon Biotechnologies, Inc. Qiagen also owns Tianwei Times, Nextal Biotechnology, Inc., artus GmbH and Shenzhen PG Biotech Co. Ltd. Tianwei, based in Beijing, develops, manufactures and supplies nucleic acid sample preparation consumables in China. Nextal provides proprietary sample preparation tools that make protein crystallization more accessible. artus GmbH is a leader in PCR molecular diagnostic systems. Shenzhen PG Biotech develops, manufactures and supplies PCR based molecular diagnostic kits in China. In May 2006, Qiagen entered upon an agreement to acquire Gentra Systems, Inc., a Minnesota based company that develops, manufactures and supplies non-solid phase nucleic acid purification products. The company entered into a distribution agreement in May 2007 with Whatman, a leader in life sciences and enabling technologies worldwide. The arrangement allows Qiagen to provide Whatman FTA and FTA-based kits to its target markets.expression analysis, gene silencing, influenza research, molecular diagnostics, pharmacogenomics, plant research and protein science fields.

Qiagen's German subsidiary, Qiagen GmbH, was recently voted one of the top ten companies to work for by the Corporate Research Foundation and Geva Institute in Munich, Germany. Qiagen offers its employees a Performance Enhancement System, various sabbatical programs and performance oriented compensation programs, in-house corporate childcare, company sponsored fitness and medical facilities, and recreational and health facilities and programs.

FINANCIALS: Sales and profits are in thousands of dollars—add 000 to get the full amount. 2006 Note: Financial information for 2006 was not available for all companies at press time.

2006 Sales: $465,778	2006 Profits: $70,539	**U.S. Stock Ticker:** QGEN
2005 Sales: $298,395	2005 Profits: $62,225	**Int'l Ticker: QIA** Int'l Exchange: Frankfurt-Euronext
2004 Sales: $380,629	2004 Profits: $48,705	Employees: 1,954
2003 Sales: $351,404	2003 Profits: $42,850	Fiscal Year Ends: 12/31
2002 Sales: $298,600	2002 Profits: $23,100	Parent Company:

SALARIES/BENEFITS:

Pension Plan:	ESOP Stock Plan:	Profit Sharing: Y	Top Exec. Salary: $	Bonus: $
Savings Plan:	Stock Purch. Plan: Y		Second Exec. Salary: $	Bonus: $

OTHER THOUGHTS:

Apparent Women Officers or Directors: 1
Hot Spot for Advancement for Women/Minorities:

LOCATIONS: ("Y" = Yes)

West:	Southwest:	Midwest:	Southeast:	Northeast:	International:
Y				Y	Y

QLT INC

www.qltinc.com

Industry Group Code: 325412 Ranks within this company's industry group: Sales: 57   Profits: 183

Drugs:		Other:		Clinical:		Computers:		Other:	
Discovery:	Y	AgriBio:		Trials/Services:		Hardware:		Specialty Services:	
Licensing:	Y	Genomics/Proteomics:		Laboratories:		Software:		Consulting:	
Manufacturing:	Y	Tissue Replacement:		Equipment/Supplies:		Arrays:		Blood Collection:	
Development:	Y			Research/Development Svcs.:		Database Management:		Drug Delivery:	Y
Generics:				Diagnostics:				Drug Distribution:	

TYPES OF BUSINESS:

Drugs-Photodynamic
Cancer Treatments
Eye Disease Treatments
Drug Delivery-Controlled & Extended Release

BRANDS/DIVISIONS/AFFILIATES:

ACZONE
Visudyne
Eligard
Atrigel
QLT USA, Inc.

CONTACTS: *Note: Officers with more than one job title may be intentionally listed here more than once.*

Robert L. Butchofsky, CEO
Robert L. Butchofsky, Pres.
Cameron Nelson, CFO
Daniel Wattier, VP-Mktg. & Sales
Linda Lupini, Sr. VP-Human Resources & Organizational Dev.
Peter O'Callaghan, General Counsel
Alexander R. Lussow, VP-Bus. Dev.
Therese Hayes, VP-Corp. Comm.
Therese Hayes, VP-Investor Rel.
Cameron Nelson, VP-Finance
Alain Curaudeau, Sr. VP-Portfolio & Project Mgmt.
C. Boyd Clarke, Chmn.
Michael Duncan, Pres., QLT USA, Inc.

Phone: 604-707-7000	**Fax:** 604-707-7001

Toll-Free: 800-663-5486

Address: 887 Great Northern Way, Vancouver, BC V5T 4T5 Canada

GROWTH PLANS/SPECIAL FEATURES:

QLT, Inc. is a global biopharmaceutical company that focuses on the discovery, development and commercialization of new therapies. Its research and development efforts are focused on pharmaceutical products in the fields of ophthalmology and dermatology. In addition, the company utilizes two unique technology platforms, photodynamic therapy and Atrigel, to create products such as Visudyne and Eligard. The firm utilizes photodynamic therapy to stifle and kill abnormally growing tissue. The treatment involves, first, the administering of the drug to target proteins, and second, the activation of the drug through externally provided light. The firm's second-generation photosensitizer, Visudyne, is one of the few approved treatments for wet age-related macular degeneration (AMD), the leading cause of blindness in people over the age of 50. Co-developed and marketed by Novartis Opthalmics, Visudyne is also approved for treatment of subfoveal choroidal neovascularization (CNV) in more than 75 countries. Atrigel is a system consisting of biodegradable polymers, which are injected into tissue sites, containing pharmaceuticals that are released on a controlled basis. The company's Eligard drug, developed and marketed by QLT USA and a treatment for prostate cancer, utilizes the Atrigel system in its delivery. The firm's most advanced proprietary dermatological product, Aczone, is approved in the U.S. and Canada, but it is not yet marketed. The firm's third and final drug delivery system employs a solvent microparticle system to mix a dissolved drug and the microparticle suspension of the drug for prolonged drug delivery. The Aczone drug uses the technology to provide delivery of dapsone, for the treatment of acne primarily in teenagers. To focus its business on the research and development of proprietary products in its core therapeutic areas, in December 2006 the company sold the generic dermatology business, dental businesses and the manufacturing facility of QLT USA to Tolmar, Inc.

FINANCIALS: Sales and profits are in thousands of dollars—add 000 to get the full amount. 2006 Note: Financial information for 2006 was not available for all companies at press time.

2006 Sales: $175,100	2006 Profits: $-101,600	**U.S. Stock Ticker:** QLTI
2005 Sales: $241,973	2005 Profits: $-325,412	**Int'l Ticker: QLT** Int'l Exchange: Toronto-TSX
2004 Sales: $186,072	2004 Profits: $-165,709	Employees: 254
2003 Sales: $146,750	2003 Profits: $44,817	Fiscal Year Ends: 12/31
2002 Sales: $110,500	2002 Profits: $13,600	Parent Company:

SALARIES/BENEFITS:

Pension Plan:	ESOP Stock Plan:	Profit Sharing:	Top Exec. Salary: $481,481	Bonus: $
Savings Plan: Y	Stock Purch. Plan:		Second Exec. Salary: $335,797	Bonus: $26,455

OTHER THOUGHTS:

Apparent Women Officers or Directors: 2
Hot Spot for Advancement for Women/Minorities: Y

LOCATIONS: ("Y" = Yes)

West:	Southwest:	Midwest:	Southeast:	Northeast:	International:
Y					Y

Note: Financial information, benefits and other data can change quickly and may vary from those stated here.

QUEST DIAGNOSTICS INC   www.questdiagnostics.com

Industry Group Code: 621511 Ranks within this company's industry group: Sales: 1   Profits: 1

Drugs:	Other:	Clinical:		Computers:		Other:	
Discovery:	AgriBio:	Trials/Services:		Hardware:		Specialty Services:	
Licensing:	Genomics/Proteomics:	Laboratories:	Y	Software:		Consulting:	
Manufacturing:	Tissue Replacement:	Equipment/Supplies:		Arrays:		Blood Collection:	
Development:		Research/Development Svcs.:		Database Management:		Drug Delivery:	
Generics:		Diagnostics:	Y			Drug Distribution:	

TYPES OF BUSINESS:

Services-Testing & Diagnostics
Clinical Laboratory Testing
Clinical Trials Testing
Esoteric Testing Laboratories

BRANDS/DIVISIONS/AFFILIATES:

Cardio CRP
HEPTIMAX
Focus Diagnostics, Inc.
CF Complete
CellSearch
Bio-Intact PTH
Leumeta
LabOne

CONTACTS: *Note: Officers with more than one job title may be intentionally listed here more than once.*

Surya N. Mohapatra, CEO
Surya N. Mohapatra, Pres.
Robert A. Hagemann, CFO/Sr. VP
Robert E. Peters, VP-Mktg. & Sales
Michael E. Prevoznik, General Counsel/VP-Legal & Compliance
Gary Samuels, VP-Corp. Comm. & Media Rel.
Laura Park, VP-Investor Rel.
David M. Zewe, Sr. VP-Diagnostic Testing Oper.
Surya N. Mohapatra, Chmn.

Phone: 201-393-5000	Fax: 201-729-8920
Toll-Free: 800-222-0446	
Address: 1290 Wall Street W., Lyndhurst, NJ 07071 US	

GROWTH PLANS/SPECIAL FEATURES:

Quest Diagnostics, Inc. is one of the largest clinical laboratory testing companies in the U.S., offering a broad array of diagnostic testing and related services to the health care industry. The firm's operations consist of routine, esoteric and clinical trials testing. Quest operates through its national network of 2,000 patient service centers, principal laboratories in more than 35 major metropolitan areas along with approximately 150 rapid-response laboratories, as well as esoteric testing laboratories on both coasts. Routine tests measure various important bodily health parameters such as the functions of the kidney, heart, liver, thyroid and other organs. Tests in this category include blood cholesterol level tests, complete blood cell counts, pap smears, HIV-related tests, urinalyses, pregnancy and prenatal tests, and substance-abuse tests. Esoteric tests require more sophisticated equipment and technology, professional attention and highly skilled personnel. The firm's tests in this field include Cardio CRP and HEPTIMAX. Quest's two esoteric testing laboratories, which operate as Quest Diagnostics Nichols Institute located in San Juan Capistrano, California, are among the leading esoteric clinical testing laboratories in the world. Esoteric tests involve a number of medical fields including endocrinology, genetics, immunology, microbiology, oncology, serology and special chemistry. Clinical trial testing primarily involves assessing the safety and efficacy of new drugs to meet FDA requirements, with services including Bio-Intact PTH. In recent news, Quest received FDA clearance to market its Plexus HerpeSelect 1 and 2 IgG test kit. Additionally, the company acquired diagnostic testing company AmeriPath, Inc. for approximately $2 billion. In July 2007, Quest introduced a new diagnostic testing technique to help physicians diagnose genetic metabolic disorders such as phenylketonuria (PKU) and homocystinuria.

Quest offers employees educational assistance, adoption assistance, free lab testing, annual development training and credit union access.

FINANCIALS: Sales and profits are in thousands of dollars—add 000 to get the full amount. 2006 Note: Financial information for 2006 was not available for all companies at press time.

2006 Sales: $6,268,659	2006 Profits: $586,421	**U.S. Stock Ticker:** DGX
2005 Sales: $5,456,726	2005 Profits: $546,277	**Int'l Ticker:** Int'l Exchange:
2004 Sales: $5,066,986	2004 Profits: $499,195	Employees: 41,000
2003 Sales: $4,737,958	2003 Profits: $436,717	Fiscal Year Ends: 12/31
2002 Sales: $4,108,100	2002 Profits: $322,200	Parent Company:

SALARIES/BENEFITS:

Pension Plan:	ESOP Stock Plan:	Profit Sharing: Y	Top Exec. Salary: $1,023,000	Bonus: $100,000
Savings Plan: Y	Stock Purch. Plan: Y		Second Exec. Salary: $481,415	Bonus: $

OTHER THOUGHTS:

Apparent Women Officers or Directors: 3
Hot Spot for Advancement for Women/Minorities: Y

LOCATIONS: ("Y" = Yes)

West:	Southwest:	Midwest:	Southeast:	Northeast:	International:
Y	Y	Y	Y	Y	Y

Note: Financial information, benefits and other data can change quickly and may vary from those stated here.

QUESTCOR PHARMACEUTICALS   www.questcor.com

Industry Group Code: 325412 Ranks within this company's industry group: Sales: 122   Profits: 76

Drugs:		Other:		Clinical:		Computers:		Other:	
Discovery:	Y	AgriBio:		Trials/Services:		Hardware:		Specialty Services:	
Licensing:		Genomics/Proteomics:		Laboratories:		Software:		Consulting:	
Manufacturing:		Tissue Replacement:		Equipment/Supplies:		Arrays:		Blood Collection:	
Development:	Y			Research/Development Svcs.:		Database Management:		Drug Delivery:	
Generics:				Diagnostics:				Drug Distribution:	Y

TYPES OF BUSINESS:

Drugs-Neurology
Multiple Sclerosis Treatment

BRANDS/DIVISIONS/AFFILIATES:

HP Acthar Gel
Doral

CONTACTS: *Note: Officers with more than one job title may be intentionally listed here more than once.*

Don M. Baily, Interim Pres.
George M. Stuart, CFO
Stephen L. Cartt, Exec. VP-Corp. Dev.
Eric Liebler, Sr. VP-Comm.
George M. Stuart, Sr. VP-Finance
Steven Halladay, Sr. VP-Clinical & Regulatory Affairs
Eric Liebler, Sr. VP-Strategic Planning
David J. Medieros, VP-Pharmaceutical Oper.
Albert Hansen, Chmn.

Phone: 510-400-0700	Fax: 510-400-0799
Toll-Free:	
Address: 3260 Whipple Rd., Union City, CA 94587-1217 US	

GROWTH PLANS/SPECIAL FEATURES:

Questcor Pharmaceuticals, Inc. is an integrated specialty pharmaceutical company that focuses on novel therapeutics for the treatment of diseases and disorders of the central nervous system (CNS). The company currently owns and markets two commercial products: H.P. Acthar Gel (Acthar) and Doral. Acthar is a natural source, highly purified preparation of the adrenal corticotropin hormone, which is specially formulated to provide prolonged release after intramuscular or subcutaneous injection. The product is indicated for use in acute exacerbations of MS and is prescribed for patients that have MS and experience painful, episodic flares. The company is also in the process of getting Acthar approved for infantile spasms, a condition for which no other drug has been approved. Doral (quazepam) is a non-narcotic, selective benzodiazepine receptor agonist that is indicated for the treatment of insomnia, characterized by difficulty in falling asleep, frequent nocturnal awakenings and/or early morning awakenings. Sleep disturbance and insomnia are very common side effects of many neurological diseases and disorders such as MS, Epilepsy, Parkinson's disease and Alzheimer's disease. The company is also developing QSC-001, in conjunction with Eurand, which is an orally dissolving tablet for severe pain. Eurand is providing the flavor mechanism. In May 2006, Questcor acquired the product Doral from MedPointe for $2.5 Million and a possible future milestone payment of $1.5 million.

FINANCIALS: Sales and profits are in thousands of dollars—add 000 to get the full amount. 2006 Note: Financial information for 2006 was not available for all companies at press time.

2006 Sales: $12,788	2006 Profits: $-10,109	U.S. Stock Ticker: QSC
2005 Sales: $14,162	2005 Profits: $7,392	Int'l Ticker: Int'l Exchange:
2004 Sales: $18,404	2004 Profits: $- 832	Employees: 70
2003 Sales: $14,100	2003 Profits: $-3,800	Fiscal Year Ends: 12/31
2002 Sales: $13,800	2002 Profits: $-2,800	Parent Company:

SALARIES/BENEFITS:

Pension Plan:	ESOP Stock Plan:	Profit Sharing:	Top Exec. Salary: $315,000	Bonus: $118,125
Savings Plan:	Stock Purch. Plan: Y		Second Exec. Salary: $257,000	Bonus: $61,680

OTHER THOUGHTS:

Apparent Women Officers or Directors:
Hot Spot for Advancement for Women/Minorities:

LOCATIONS: ("Y" = Yes)

West:	Southwest:	Midwest:	Southeast:	Northeast:	International:
Y					

Note: Financial information, benefits and other data can change quickly and may vary from those stated here.

QUINTILES TRANSNATIONAL CORP   www.quintiles.com

Industry Group Code: 541710 Ranks within this company's industry group: Sales: Profits:

Drugs:		Other:	Clinical:		Computers:		Other:	
Discovery:		AgriBio:	Trials/Services:	Y	Hardware:		Specialty Services:	Y
Licensing:	Y	Genomics/Proteomics:	Laboratories:	Y	Software:		Consulting:	Y
Manufacturing:		Tissue Replacement:	Equipment/Supplies:		Arrays:		Blood Collection:	
Development:			Research/Development Svcs.:	Y	Database Management:		Drug Delivery:	
Generics:			Diagnostics:				Drug Distribution:	

TYPES OF BUSINESS:

Contract Research
Pharmaceutical, Biotech & Medical Device Research
Consulting & Training Services
Sales & Marketing Services

BRANDS/DIVISIONS/AFFILIATES:

PharmaBio Development
Innovex
Pharma Services Holding, Inc.
Medical Action Communication
Q.E.D. Communications

CONTACTS: Note: Officers with more than one job title may be intentionally listed here more than once.

Dennis Gillings, CEO
John Ratliff, COO
Mike Troullis, Acting CFO
Stephen DeCherny, Pres., Global Clinical Research Org.
William Deam, CIO
Oppel Greef, Pres., Clinical Tech. Svcs.
Ron Wooten, Exec. VP-Corp. Dev.
Hywel Evans, Pres., Quintiles Global Commercialization
Derek Winstanly, Exec. VP-Strategic Bus. Partnerships
Dennis Gillings, Chmn.

Phone: 919-998-2000	Fax: 919-998-9113
Toll-Free:	
Address: 4709 Creekstone Dr., Ste. 200, Durham, NC 27703 US	

GROWTH PLANS/SPECIAL FEATURES:

Quintiles Transnational Corp. provides full-service contract research, sales and marketing services to the global pharmaceutical, biotechnology and medical device industries. The company is one of the world's top contract research organizations (CROs), and it provides a broad range of contract services to speed the process from development to peak sales of a new drug or medical device. Quintiles operates through offices in 50 countries, organized in three primary business segments: the product development group, the commercialization group and the PharmaBio Development group. The product development group provides a full range of drug development services from strategic planning and preclinical services to regulatory submission and approval. The commercial services group, which operates under the Innovex brand, engages in sales force deployment and strategic marketing services as well as consulting services and training for its customers. Within the group, Medical Action Communication uses proven science and marketing technique to advertise products to a potential audience; while Q.E.D. Communications works with product managers in tailoring promotional programs, sales training and testing. The PharmaBio development group works with the other service groups to enter into strategic transactions that it believes will position the company to explore new opportunities and areas for potential growth. PharmaBio also acquires the rights to market pharmaceutical products. In early 2007, Quintiles partnered with Onmark, a group purchasing organization for medical community-based oncology practices, to increase its number of patient recruits in the U.S.

Quintiles Transnational offers its employees a comprehensive benefits package including on-the-job training, recreational activities and community support activities.

FINANCIALS: Sales and profits are in thousands of dollars—add 000 to get the full amount. 2006 Note: Financial information for 2006 was not available for all companies at press time.

2006 Sales: $	2006 Profits: $	U.S. Stock Ticker: Private	
2005 Sales: $2,398,583	2005 Profits: $ 648	Int'l Ticker: Int'l Exchange:	
2004 Sales: $1,956,254	2004 Profits: $-7,427	Employees: 16,000	
2003 Sales: $2,046,000	2003 Profits: $-7,430,000	Fiscal Year Ends: 12/31	
2002 Sales: $1,992,400	2002 Profits: $127,400	Parent Company:	

SALARIES/BENEFITS:

Pension Plan:	ESOP Stock Plan:	Profit Sharing:	Top Exec. Salary: $706,061	Bonus: $825,000
Savings Plan: Y	Stock Purch. Plan:		Second Exec. Salary: $471,224	Bonus: $709,000

OTHER THOUGHTS:

Apparent Women Officers or Directors:
Hot Spot for Advancement for Women/Minorities:

LOCATIONS: ("Y" = Yes)

West:	Southwest:	Midwest:	Southeast:	Northeast:	International:
Y	Y	Y	Y	Y	Y

Note: Financial information, benefits and other data can change quickly and may vary from those stated here.

RANBAXY LABORATORIES LIMITED   www.ranbaxy.com

Industry Group Code: 325416 Ranks within this company's industry group: Sales: 3   Profits: 4

Drugs:		Other:		Clinical:		Computers:		Other:	
Discovery:		AgriBio:		Trials/Services:		Hardware:		Specialty Services:	
Licensing:	Y	Genomics/Proteomics:		Laboratories:		Software:		Consulting:	
Manufacturing:	Y	Tissue Replacement:		Equipment/Supplies:		Arrays:		Blood Collection:	
Development:	Y			Research/Development Svcs.:		Database Management:		Drug Delivery:	Y
Generics:	Y			Diagnostics:	Y			Drug Distribution:	

TYPES OF BUSINESS:

Generic Pharmaceuticals
Active Pharmaceutical Ingredients
Drugs-Anti-Retroviral & HIV/AIDS
Drug Delivery Systems
Veterinary Pharmaceuticals
Specialty Chemicals
Diagnostics

BRANDS/DIVISIONS/AFFILIATES:

Ranbaxy Pharmaceuticals, Inc.
Ohm Laboratories, Inc.
Revital
Pepfiz
Gesdyp
Garlic Pearls
The Rainbow Coalition
Terapia

CONTACTS: Note: Officers with more than one job title may be intentionally listed here more than once.

Malvinder Mohan Singh, CEO
Atul Sobti, COO
Pushpinder Bindra, Pres.
Ram S. Ramasunder, CFO
Bhagwat Yagnik, VP-Global Human Resources
Himadri Sen, Pres., R&D
Pushpinder Bindra, CTO
Dipak Chattaraj, Pres., Corp. Dev.
Ramesh L. Adige, Exec. Dir.-Corp. Affairs & Global Corp. Comm.
Anurag Kalra, Sr. Mgr.-Investor Rel.
Satish Chawla, VP-Global Internal Audit
Pradip Bhatnagar, Sr. VP-New Drug Discover Research
Jay Deshmukh, Sr. VP-Intellectual Property
Peter Burema, Pres., Global Pharmaceutical Bus.
Harpal Singh, Chmn.

Phone: 91-124-413-5000	Fax: 91-124-4106490
Toll-Free:	
Address: Plot 90, Sector 32, Gurgaon, Haryana 122001 India	

GROWTH PLANS/SPECIAL FEATURES:

Ranbaxy Laboratories Limited is a leading Indian manufacturer and marketer of generic pharmaceuticals, branded generics, over-the-counter medications and active pharmaceutical ingredients. It has manufacturing operations in 11 countries, a ground presence in 49 countries and its products are available in over 125 nations worldwide. Ranbaxy operates in the U.S. through Ranbaxy Pharmaceuticals, Inc.; Ranbaxy Laboratories, Inc.; and Ohm Laboratories, Inc. With the purchase of RPG (Aventis) SA, Ranbaxy established itself as one of the leading generic drug companies in France and continues to pursue European and international expansion. The company's research and development activities focus on select infectious diseases, metabolic diseases, inflammatory/respiratory disease and oncology. It conducts major research collaborations with GlaxoSmithKline. The company's API (active pharmaceutical ingredient) business has over 50 products in its portfolio and supplies APIs to other generic manufacturing businesses worldwide. The company also has a global consumer healthcare segment, which includes the brands Revital, Pepfiz, Gesdyp and Garlic Pearls. This segment also uses a communication platform called The Rainbow Coalition, a program that targets both physicians and consumers, to assist in providing knowledge about the company's products to the public. Ranbaxy has recently acquired a number of businesses worldwide. In March 2006, the company announced that it has acquired a 96.7% stake in Romanian pharmaceutical company Terapia, for $324 million. Additionally, it acquired Ethimed NV, a generics company in Belgium, also in March 2006. In April 2006, the company acquired the unbranded generics business of GlaxoSmithKline in Italy. In July 2006, Ranbaxy acquired the Mudogen generic business of GlaxoSmithKline in Spain. In May 2007, the company finalized the acquisition of Be-Tabs Pharmaceuticals (Pty) Limited in South Africa for $70 million. Ranbaxy offers employees group life insurance and medical insurance.

FINANCIALS: Sales and profits are in thousands of dollars—add 000 to get the full amount. 2006 Note: Financial information for 2006 was not available for all companies at press time.

2006 Sales: $1,405,500	2006 Profits: $116,800	**U.S. Stock Ticker: RBXLF.PK**
2005 Sales: $1,182,500	2005 Profits: $56,500	**Int'l Ticker: 500359** Int'l Exchange: Bombay-BSE
2004 Sales: $1,265,200	2004 Profits: $160,000	Employees: 11,000
2003 Sales: $	2003 Profits: $	Fiscal Year Ends: 12/31
2002 Sales: $	2002 Profits: $	Parent Company:

SALARIES/BENEFITS:

Pension Plan: Y	ESOP Stock Plan:	Profit Sharing:	Top Exec. Salary: $	Bonus: $
Savings Plan:	Stock Purch. Plan:		Second Exec. Salary: $	Bonus: $

OTHER THOUGHTS:

Apparent Women Officers or Directors:
Hot Spot for Advancement for Women/Minorities:

LOCATIONS: ("Y" = Yes)

West:	Southwest:	Midwest:	Southeast:	Northeast:	International:
			Y	Y	Y

Note: Financial information, benefits and other data can change quickly and may vary from those stated here.

REGENERON PHARMACEUTICALS INC www.regeneron.com

Industry Group Code: 325412 Ranks within this company's industry group: Sales: 73   Profits: 184

Drugs:		Other:		Clinical:	Computers:	Other:
Discovery:	Y	AgriBio:		Trials/Services:	Hardware:	Specialty Services:
Licensing:		Genomics/Proteomics:	Y	Laboratories:	Software:	Consulting:
Manufacturing:	Y	Tissue Replacement:		Equipment/Supplies:	Arrays:	Blood Collection:
Development:	Y			Research/Development Svcs.:	Database Management:	Drug Delivery:
Generics:				Diagnostics:		Drug Distribution:

TYPES OF BUSINESS:
Drugs-Diversified
Protein-Based Drugs
Small-Molecule Drugs
Genetics & Transgenic Mouse Technologies

BRANDS/DIVISIONS/AFFILIATES:
VelocImmune
VelociGene
VelociMouse
VEGF Trap
VEGF Trap-Eye
IL-1 Trap

CONTACTS: *Note: Officers with more than one job title may be intentionally listed here more than once.*
Leonard S. Schleifer, CEO
Leonard S. Schleifer, Pres.
Murray A. Goldberg, CFO
George D. Yancopoulos, Exec. VP/Chief Scientific Officer
Randall G. Rupp, Sr. VP-Mfg. & Process Sciences
Murray A. Goldberg, Sr. VP-Admin.
Stuart A. Kolinski, General Counsel/Sr. VP/Sec.
Murray A. Goldberg, Sr. VP-Finance/Treas./Asst. Sec.
George D. Yancopoulos, Pres., Regeneron Research Laboratories
Neil Stahl, Sr. VP-R & D Sciences
P. Roy Vagelos, Chmn.

Phone: 914-345-7400	Fax: 914-347-2847
Toll-Free:	
Address: 777 Old Saw Mill River Rd., Tarrytown, NY 10591-6707 US	

GROWTH PLANS/SPECIAL FEATURES:
Regeneron Pharmaceuticals, Inc. is a biopharmaceutical company that discovers, develops and intends to commercialize pharmaceutical drugs for the treatment of serious medical conditions. The firm is currently focusing on the three clinical development programs: VEGF Trap in oncology; VEGF Trap eye formulation (VEGF Trap-Eye) in eye diseases using intraocular delivery; and IL-1 Trap in various systemic inflammatory indications. The VEGF Trap oncology development program is being developed jointly with the sanofi-aventis Group through a 2003 agreement. Regeneron's preclinical research programs are in the areas of oncology and angiogenesis, ophthalmology, metabolic and related diseases, muscle diseases and disorders, inflammation and immune diseases, bone and cartilage, pain and cardiovascular diseases. The company expects that its next generation of product candidates will be based on its proprietary technologies for developing Traps and Human Monoclonal Antibodies. Since its inception the company has not generated any sales or profits from the commercialization of any of its product candidates. Regeneron's proprietary technologies include VelociGene, VelociMouse and VelocImmune, among others. The VelociGene technology allows precise DNA manipulation and gene staining, helping to identify where a particular gene is active in the body. VelociMouse technology allows for the direct and immediate generation of genetically altered mice from ES cells, avoiding the lengthy process involved in generating and breeding knock-out mice from chimeras. VelocImmune is a novel mouse technology platform for producing fully human monoclonal antibodies. In 2007, the company announced its entry into a licensing agreement with AstraZeneca for Regeneron's VelocImmune technology in its internal research programs to discover human monoclonal antibodies.

FINANCIALS: Sales and profits are in thousands of dollars—add 000 to get the full amount. 2006 Note: Financial information for 2006 was not available for all companies at press time.

2006 Sales: $63,447	2006 Profits: $-102,337	U.S. Stock Ticker: REGN
2005 Sales: $66,193	2005 Profits: $-95,446	Int'l Ticker: Int'l Exchange:
2004 Sales: $174,017	2004 Profits: $41,699	Employees: 573
2003 Sales: $57,497	2003 Profits: $-107,458	Fiscal Year Ends: 12/31
2002 Sales: $22,000	2002 Profits: $-124,400	Parent Company:

SALARIES/BENEFITS:
Pension Plan:	ESOP Stock Plan:	Profit Sharing:	Top Exec. Salary: $685,000	Bonus: $420,000
Savings Plan: Y	Stock Purch. Plan:		Second Exec. Salary: $544,300	Bonus: $335,000

OTHER THOUGHTS:
Apparent Women Officers or Directors:
Hot Spot for Advancement for Women/Minorities:

LOCATIONS: ("Y" = Yes)
West:	Southwest:	Midwest:	Southeast:	Northeast:	International:
				Y	

Note: Financial information, benefits and other data can change quickly and may vary from those stated here.

REPLIGEN CORPORATION

www.repligen.com

Industry Group Code: 325412 Ranks within this company's industry group: Sales: 121 Profits: 56

Drugs:		Other:	Clinical:	Computers:	Other:	
Discovery:	Y	AgriBio:	Trials/Services:	Hardware:	Specialty Services:	
Licensing:		Genomics/Proteomics:	Laboratories:	Software:	Consulting:	
Manufacturing:	Y	Tissue Replacement:	Equipment/Supplies:	Arrays:	Blood Collection:	
Development:	Y		Research/Development Svcs.:	Database Management:	Drug Delivery:	
Generics:			Diagnostics:		Drug Distribution:	

TYPES OF BUSINESS:

Neuropsychiatric Drugs
Autoimmune Disorder Treatments

BRANDS/DIVISIONS/AFFILIATES:

SecreFlo
Protein A
Uridine
Secretin
RG2417

CONTACTS: *Note: Officers with more than one job title may be intentionally listed here more than once.*

Walter C. Herlihy, CEO
Walter C. Herlihy, Pres.
Daniel Muehl, CFO
James R. Rusche, Sr. VP-R&D
Daniel Muehl, Sec.
Daniel P. Witt, VP-Oper.
Laura Whitehouse Pew, VP-Market Dev.
Paul Schimmel, Co-Chmn.
Alexander Rich, Co-Chmn.

Phone: 781-250-0111	Fax: 781-250-0115

Toll-Free: 800-622-2259

Address: 41 Seyton St., Bldg. 1, Ste. 100, Waltham, MA 02453
US

GROWTH PLANS/SPECIAL FEATURES:

Repligen Corp. develops therapeutics for the treatment of diseases of the central nervous system. The company also manufactures protein A, which is used in the production of many therapeutic monoclonal antibodies. The firm currently sells two products: Protein A; and SecreFlo, a synthetic form of the hormone secretin, which is used as an aid in the diagnosis of certain diseases of the pancreas. Repligen has products in the development stage for neuropsychiatric disorders: secretin, a hormone produced in the small intestine that regulates the function of the pancreas as part of the process of digestion, evaluated by the company for improvement of MRI imaging of the pancreas; uridine, a biological compound essential for the synthesis of DNA and RNA, being tested by the firm under an oral formulation; RG2417, for the treatment of bipolar disorder; and transcription enhancers for Friedreich's ataxia. Repligen sells its Protein A products primarily to value-added resellers including GEHC and Applied Biosystems, Inc., as well as through distributors and certain foreign markets; and SecreFlo to gastroenterologists in the U.S. The company owns or has exclusive rights to more than 15 issued U.S. patents and corresponding foreign equivalents.

The company offers its employees benefits that include health and dental insurance; life and long-term disability coverage; a 401(k) plan; and equity participation in a publicly traded company.

FINANCIALS: Sales and profits are in thousands of dollars—add 000 to get the full amount. 2006 Note: Financial information for 2006 was not available for all companies at press time.

2006 Sales: $12,911	2006 Profits: $ 697	**U.S. Stock Ticker: RGEN**
2005 Sales: $9,360	2005 Profits: $-2,984	**Int'l Ticker:** Int'l Exchange:
2004 Sales: $6,914	2004 Profits: $-9,551	Employees: 45
2003 Sales: $7,771	2003 Profits: $-4,537	Fiscal Year Ends: 3/31
2002 Sales: $4,300	2002 Profits: $-4,500	Parent Company:

SALARIES/BENEFITS:

Pension Plan:	ESOP Stock Plan:	Profit Sharing:	Top Exec. Salary: $334,000	Bonus: $82,551
Savings Plan: Y	Stock Purch. Plan:		Second Exec. Salary: $227,000	Bonus: $38,272

OTHER THOUGHTS:

Apparent Women Officers or Directors: 1
Hot Spot for Advancement for Women/Minorities:

LOCATIONS: ("Y" = Yes)

West:	Southwest:	Midwest:	Southeast:	Northeast:	International:
				Y	

REPROS THERAPEUTICS INC

www.reprosrx.com

Industry Group Code: 325412 Ranks within this company's industry group: Sales: 167   Profits: 83

Drugs:		Other:		Clinical:	Computers:		Other:	
Discovery:	Y	AgriBio:		Trials/Services:	Hardware:		Specialty Services:	
Licensing:		Genomics/Proteomics:		Laboratories:	Software:		Consulting:	
Manufacturing:		Tissue Replacement:		Equipment/Supplies:	Arrays:		Blood Collection:	
Development:	Y			Research/Development Svcs.:	Database Management:		Drug Delivery:	
Generics:				Diagnostics:			Drug Distribution:	

TYPES OF BUSINESS:

Drugs-Fertility & Sexual Dysfunction
Reproductive System Disorder Treatment
Hormonal Disorder Treatment

BRANDS/DIVISIONS/AFFILIATES:

Proellex
Androxal
VASOMAX
Bimexes
ERxin

CONTACTS: *Note: Officers with more than one job title may be intentionally listed here more than once.*

Joseph S. Podolski, CEO
Joseph S. Podolski, Pres.
Louis Ploth, CFO
Ronald Wiehle, VP-R&D
Louis Ploth, Corp. Sec.
Louis Ploth, VP-Bus. Dev.
Andre van As, Sr. VP- Clinical & Regulatory Affairs
Andre van As, Chief Medical Officer
Daniel F. Cain, Chmn.

Phone: 281-719-3400	Fax: 281-719-3446
Toll-Free:	
Address: 2408 Timberloch Pl., Ste. B-7, The Woodlands, TX 77380 US	

GROWTH PLANS/SPECIAL FEATURES:

Repros Therapeutics, Inc., formerly Zonagen, Inc., develops products for the treatment of hormonal and reproductive system disorders. Proellex, its leading product candidate, is an orally active small molecule compound being developed for two indications: the treatment of uterine fibroids and the treatment of endometriosis. The National Uterine Fibroid Foundation estimates that as many as 80% of all women in the United States have uterine fibroids, and one in four of these women have symptoms severe enough to require treatment. According to The Endometriosis Association, endometriosis affects 5.5 million women in the United States and Canada. As a new chemical entity, Proellex is required to undergo the full regulatory approval process, including a two-year carcinogenicity study, which the company began in mid-2006. For the treatment of uterine fibroids, Proellex is undergoing Phase II clinical trials, which are expected to be completed by mid-2007. For the treatment of endometriosis, Proellex is undergoing a Phase I/II trial. The company's other product candidate, Androxal, is an orally available small molecule compound being developed for the treatment of testosterone deficiency in men. Androxal targets the symptoms of secondary hypogonadism (a male condition associated with aging and the subsequent decline of the male hormone testosterone), and is currently in Phase III testing in the U.S. The company has a third, phentolamine-based product, VASOMAX, which has been marketed in Mexico and Brazil for the treatment of male erectile dysfunction (ED). VASOMAX is not currently marketed in the U.S. because of the FDA's partial clinical hold on phentolamine-based products. Repros met with the Ministry of Health in Mexico during the first quarter 2006 regarding its second generation phentolamine-based products for the treatment of ED: Bimexes, an oral therapy for men with mild to moderate ED, and ERxin, an injectable therapy for the treatment of severe ED.

FINANCIALS: Sales and profits are in thousands of dollars—add 000 to get the full amount. 2006 Note: Financial information for 2006 was not available for all companies at press time.

2006 Sales: $ 596	2006 Profits: $-14,195	**U.S. Stock Ticker: RPRX**
2005 Sales: $ 634	2005 Profits: $-7,391	**Int'l Ticker:** Int'l Exchange:
2004 Sales: $ 257	2004 Profits: $-3,697	Employees: 8
2003 Sales: $1,015	2003 Profits: $-3,329	Fiscal Year Ends: 12/31
2002 Sales: $4,500	2002 Profits: $-3,900	Parent Company:

SALARIES/BENEFITS:

Pension Plan: Y	ESOP Stock Plan:	Profit Sharing:	Top Exec. Salary: $330,750	Bonus: $98,398
Savings Plan:	Stock Purch. Plan: Y		Second Exec. Salary: $209,475	Bonus: $52,369

OTHER THOUGHTS:

Apparent Women Officers or Directors:
Hot Spot for Advancement for Women/Minorities:

LOCATIONS: ("Y" = Yes)

West:	Southwest:	Midwest:	Southeast:	Northeast:	International:
	Y				

Note: Financial information, benefits and other data can change quickly and may vary from those stated here.

RIGEL PHARMACEUTICALS INC   www.rigel.com

Industry Group Code: 325412 Ranks within this company's industry group: Sales: 92   Profits: 140

Drugs:		Other:	Clinical:	Computers:	Other:
Discovery:	Y	AgriBio:	Trials/Services:	Hardware:	Specialty Services:
Licensing:	Y	Genomics/Proteomics:	Laboratories:	Software:	Consulting:
Manufacturing:		Tissue Replacement:	Equipment/Supplies:	Arrays:	Blood Collection:
Development:	Y		Research/Development Svcs.:	Database Management:	Drug Delivery:
Generics:			Diagnostics:		Drug Distribution:

TYPES OF BUSINESS:
Biopharmaceuticals Development
Small-Molecule Drugs
Drugs-Cancer & Inflammatory Diseases
Drugs-Viral Diseases

GROWTH PLANS/SPECIAL FEATURES:

Rigel Pharmaceuticals, Inc. is a biotechnology company focused on developing novel small-molecule drugs to meet medical needs in the fields of inflammatory diseases, cancer and viral diseases. The firm's lead candidate for the treatment of rheumatoid arthritis, R788, functions by inhibiting IgG receptor signaling in macrophages and B-cells and is in Phase II of the development process. The company's cancer treatment candidate, R763, is a specific inhibitor of Aurora kinase, shown to block proliferation of trigger apoptosis in several tumor cell lines. Rigel is in partnership with Merck Serono for the development of R763, which is currently in Phase I clinical trials. R343, a SYK kinase inhibitor and the company's asthma and allergy treatment candidate, functions by inhibiting the IgE receptor signaling in respiratory tract mast cells. The company holds a partnership with Pfizer for the development of R343, which is currently in preclinical development. The firm has several other drugs in the preclinical development stages, including drugs for the treatment of hepatitis C and the prevention of transplant rejection. Rigel has partnerships with pharmaceutical companies including Merck Serono, Pfizer, Johnson & Johnson, Merck and Novartis for the treatment of tumor growth, asthma, transplant rejection, autoimmune disease and chronic bronchitis. In October 2006, Rigel chose the compound R348 as its next candidate to advance into preclinical development. The drug has the potential to become a therapy for autoimmune diseases.

Rigel offers employees education reimbursement, subsidized lunches, an employee assistance program and a wellness benefits program. In addition, the company organizes summer picnics, weekly happy hours, corporate softball games and other activities.

BRANDS/DIVISIONS/AFFILIATES:
R788
R763
Merck Serono
R343
R348

CONTACTS: Note: Officers with more than one job title may be intentionally listed here more than once.
James M. Gower, CEO
Raul R. Rodriguez, COO/Exec. VP
Ryan Maynard, CFO/Sr. VP
Donald G. Payan, Chief Scientific Officer/Exec. VP
Dolly Vance, General Counsel/Sr. VP-Intellectual Property
Robin Cooper, Sr. VP-Pharmaceutical Sciences
Elliott B. Grossbard, Sr. VP-Medical Dev.
James M. Gower, Chmn.

Phone: 650-624-1100	Fax: 650-624-1101

Toll-Free:

Address: 1180 Veterans Blvd., South San Francisco, CA 94080 US

FINANCIALS: Sales and profits are in thousands of dollars—add 000 to get the full amount. 2006 Note: Financial information for 2006 was not available for all companies at press time.

2006 Sales: $33,473	2006 Profits: $-37,637	**U.S. Stock Ticker: RIGL**
2005 Sales: $16,526	2005 Profits: $-45,256	**Int'l Ticker:** Int'l Exchange:
2004 Sales: $4,733	2004 Profits: $-56,255	Employees: 152
2003 Sales: $11,055	2003 Profits: $-41,197	Fiscal Year Ends: 12/31
2002 Sales: $15,788	2002 Profits: $-37,030	Parent Company:

SALARIES/BENEFITS:

Pension Plan:	ESOP Stock Plan:	Profit Sharing:	Top Exec. Salary: $455,000	Bonus: $150,150
Savings Plan: Y	Stock Purch. Plan: Y		Second Exec. Salary: $395,000	Bonus: $130,350

OTHER THOUGHTS:
Apparent Women Officers or Directors: 1
Hot Spot for Advancement for Women/Minorities:

LOCATIONS: ("Y" = Yes)

West:	Southwest:	Midwest:	Southeast:	Northeast:	International:
Y					

Note: Financial information, benefits and other data can change quickly and may vary from those stated here.

ROCHE HOLDING LTD

www.roche.com

Industry Group Code: 325412 Ranks within this company's industry group: Sales: 6   Profits: 5

Drugs:		Other:		Clinical:		Computers:		Other:	
Discovery:	Y	AgriBio:		Trials/Services:		Hardware:		Specialty Services:	
Licensing:		Genomics/Proteomics:		Laboratories:		Software:		Consulting:	
Manufacturing:	Y	Tissue Replacement:		Equipment/Supplies:		Arrays:		Blood Collection:	
Development:	Y			Research/Development Svcs.:		Database Management:		Drug Delivery:	
Generics:				Diagnostics:	Y			Drug Distribution:	

TYPES OF BUSINESS:

Pharmaceuticals Manufacturing
Antibiotics
Diagnostics
Cancer Drugs
Virology Products
HIV/AIDS Treatments
Transplant Drugs

BRANDS/DIVISIONS/AFFILIATES:

Genentech
F. Hoffmann-La Roche, Ltd.
Aleve
Rennie
Tamiflu
Herceptin
Chugai Pharmaceuticals
Bioveris Corporation

CONTACTS: *Note: Officers with more than one job title may be intentionally listed here more than once.*

Franz B. Humer, CEO
Erich Hunziker, CFO
Pascal Soriot, Head-Strategic Mktg.
Gottlieb Keller, VP-Human Resources
Jonathan Knowles, Dir.-Global Research
Eduard E. Holdener, Dir.-Global Pharma Dev./Chief Medical Officer
Pascal Soriot, Head-Commercial Oper.
Rolf D. Schlapfer, Head-Global Corp. Comm.
William M. Burns, CEO-Roche Pharmaceuticals
Severin Schwan, CEO-Roche Diagnostics
Osamu Nagayama, Pres./CEO-Chugai
Peter Hug, Head-Pharma Partnering
Franz B. Humer, Chmn.

Phone: 41-61-688-1111	Fax: 41-61-691-9391
Toll-Free:	
Address: Grenzacherstrasse 124, Basel, 4070 Switzerland	

GROWTH PLANS/SPECIAL FEATURES:

Roche Holding, Ltd. is one of the world's largest health care companies, occupying an industry-leading position in the global diagnostics market and ranking as one of the top producers of pharmaceuticals, with particular market penetration in the areas of cancer drugs, virology and transplantation medicine. Group operations currently extend to some 150 countries, with additional alliances and research and development agreements with corporate and institutional partners furthering Roche's collective reach. Among the company's related corporate interests are majority ownership holdings in Genentech and Japanese pharmaceutical firm Chugai. Roche's products include the cancer drugs Avastin, Bondronat, Xeloda, Herceptin and Mabthera/Rituxan; the antibiotic Rocephin; the HIV/AIDS treatments Viracept, Fortovase and Fuzeon; and Tamiflu, which is used to prevent and treat influenza. Roche has invested heavily in diagnostics, both through internal resource development and through selective acquisitions. Roche companies control proprietary diagnostic technologies across a range of areas, including advanced DNA tests, leading consumer diabetes monitoring devices and applied sciences methodologies for laboratory research. As part of the recent mobilization of Tamiflu, the company recently came to an agreement with Gilead, which will coordinate the commercialization and manufacture of drug, important in the case of a flu pandemic. In early 2007, Roche granted GlaxoSmithKline Consumer Healthcare the non-prescription rights to the anti-obesity drug orlistat; these global rights exclude the U.S. and Japan. In April 2007, the firm agreed to acquire BioVeris Corp., a diagnostic company located in the U.S., for $600 million. The acquisition will strengthen Roche's position in the immunochemistry market.

FINANCIALS: Sales and profits are in thousands of dollars—add 000 to get the full amount. 2006 Note: Financial information for 2006 was not available for all companies at press time.

2006 Sales: $34,851,500	2006 Profits: $7,116,030	**U.S. Stock Ticker: RHHBY**
2005 Sales: $27,385,668	2005 Profits: $5,189,777	**Int'l Ticker: RO** Int'l Exchange: Zurich-SWX
2004 Sales: $22,767,021	2004 Profits: $5,446,567	Employees: 68,218
2003 Sales: $25,132,100	2003 Profits: $2,470,500	Fiscal Year Ends: 12/31
2002 Sales: $21,422,800	2002 Profits: $-2,901,500	Parent Company:

SALARIES/BENEFITS:

Pension Plan:	ESOP Stock Plan:	Profit Sharing:	Top Exec. Salary: $	Bonus: $
Savings Plan:	Stock Purch. Plan:		Second Exec. Salary: $	Bonus: $

OTHER THOUGHTS:

Apparent Women Officers or Directors: 2
Hot Spot for Advancement for Women/Minorities: Y

LOCATIONS: ("Y" = Yes)

West:	Southwest:	Midwest:	Southeast:	Northeast:	International:
Y	Y	Y	Y	Y	Y

Note: Financial information, benefits and other data can change quickly and may vary from those stated here.

ROSETTA INPHARMATICS LLC

www.rii.com

Industry Group Code: 541710 Ranks within this company's industry group: Sales: 20   Profits: 13

Drugs:	Other:	Clinical:	Computers:	Other:
Discovery:	AgriBio:	Trials/Services:	Hardware:	Specialty Services:
Licensing:	Genomics/Proteomics:	Laboratories:	Software: Y	Consulting:
Manufacturing:	Tissue Replacement:	Equipment/Supplies:	Arrays: Y	Blood Collection:
Development:		Research/Development Svcs.: Y	Database Management:	Drug Delivery:
Generics:		Diagnostics:		Drug Distribution:

TYPES OF BUSINESS:

Biotechnology Research
Bioinformatics Software
Gene Expression Research
DNA Microarrays

BRANDS/DIVISIONS/AFFILIATES:

Merck and Co.
Rosetta Biosoftware
Rosetta Resolver
Rosetta Syllego
Elucidator

CONTACTS: Note: Officers with more than one job title may be intentionally listed here more than once.

Douglas E. Bassett, Jr., Co-Site Head
Stephen H. Friend, Pres.
Peter S. Linsley, Exec. Dir.-Research
Deborah A. Kessler, Exec. Dir.-Oper.
Alan B. Sachs, Exec. Dir.-Molecular Profiling
Eric Schadt, Sr. Scientific Dir.-Research Genetics

Phone: 206-802-7000	Fax: 206-802-6501
Toll-Free:	
Address: 401 Terry Ave. N., Seattle, WA 98109 US	

GROWTH PLANS/SPECIAL FEATURES:

Rosetta Inpharmatics, a wholly-owned subsidiary of Merck and Co., uses gene expression research and DNA microarray technologies to support Merck's drug discovery efforts and to enhance drug development activities. Through its Rosetta Biosoftware business unit, the company continues the commercial release of the Rosetta Resolver gene expression data analysis system, an enterprise-level bioinformatics software package launched in 2000 and currently licensed to many of the world's leading academic research institutions and life sciences corporations. The firm has recently released two new software packages: the Rosetta Syllego, for genetic data management; and the Elucidator, for protein expression analysis. The company's expertise has led to new applications of gene expression technologies in areas including molecular toxicology, biomarker discovery and disease classification. Rosetta Inpharmatics is committed to using genomic research and data analysis to enable more accurate selection of drug targets and more efficient drug development, while also developing new tools to extend the analysis of gene expression data generated by DNA microarrays. These developments not only support the pharmaceutical industry, but also bring new capabilities to other sectors, including agrochemical research and biotechnology.

Rosetta Inpharmatics has a lively work atmosphere, which includes clubs for rock climbers and for theater goers. The company also has a post-doctorate program to facilitate the transition of fellows to independent scientific investigators in a rigorous research setting.

FINANCIALS: Sales and profits are in thousands of dollars—add 000 to get the full amount. 2006 Note: Financial information for 2006 was not available for all companies at press time.

2006 Sales: $	2006 Profits: $	**U.S. Stock Ticker:** Subsidiary
2005 Sales: $	2005 Profits: $	**Int'l Ticker:** Int'l Exchange:
2004 Sales: $	2004 Profits: $	Employees:
2003 Sales: $	2003 Profits: $	Fiscal Year Ends: 12/31
2002 Sales: $	2002 Profits: $	Parent Company: MERCK & CO INC

SALARIES/BENEFITS:

Pension Plan: Y	ESOP Stock Plan:	Profit Sharing:	Top Exec. Salary: $	Bonus: $
Savings Plan: Y	Stock Purch. Plan:		Second Exec. Salary: $	Bonus: $

OTHER THOUGHTS:

Apparent Women Officers or Directors: 1
Hot Spot for Advancement for Women/Minorities:

LOCATIONS: ("Y" = Yes)

West:	Southwest:	Midwest:	Southeast:	Northeast:	International:
Y					

SALIX PHARMACEUTICALS

www.salix.com

Industry Group Code: 325412 Ranks within this company's industry group: Sales: 54   Profits: 42

Drugs:	Other:	Clinical:	Computers:	Other:
Discovery:	AgriBio:	Trials/Services:	Hardware:	Specialty Services:
Licensing:	Genomics/Proteomics:	Laboratories:	Software:	Consulting:
Manufacturing: Y	Tissue Replacement:	Equipment/Supplies:	Arrays:	Blood Collection:
Development: Y		Research/Development Svcs.:	Database Management:	Drug Delivery:
Generics:		Diagnostics:		Drug Distribution:

TYPES OF BUSINESS:

Pharmaceuticals Development & Manufacturing
Drugs-Gastroenterology

BRANDS/DIVISIONS/AFFILIATES:

Colazal
Azasan
Proctocort
Anusol-HC
OsmoPrep
Xifaxan
Visicol
DIACOL

CONTACTS: *Note: Officers with more than one job title may be intentionally listed here more than once.*

Carolyn J. Logan, CEO
Carolyn J. Logan, Pres.
Adam C. Derbyshire, CFO
William P. Forbes, VP-R&D
Adam C. Derbyshire, Sr. VP-Admin.
William P. Forbes, Chief Dev. Officer
Adam C. Derbyshire, Sr. VP-Finance
John F. Chappell, Chmn.

Phone: 919-862-1000	**Fax:** 919-862-1095
Toll-Free: 888-802-9956	
Address: 1700 Perimeter Park Dr., Morrisville, NC 27560 US	

GROWTH PLANS/SPECIAL FEATURES:

Salix Pharmaceuticals is a specialty pharmaceutical company dedicated to acquiring, developing and commercializing prescription drugs used in the treatment of a variety of gastrointestinal diseases, which affect the digestive tract. The company seeks to identify late-stage or approved proprietary therapeutics for in-licensing, which subsequently advances the new drugs through regulatory procedures and final product development stages. Salix's first two products are currently in commercial distribution, supported by an in-house, 150-member sales and marketing team that is aggressively targeting the gastroenterology community. Colazal (balsalazide disodium) treats ulcerative colitis, while Azasan (azathioprine tablets), originally intended to suppress immune response in organ transplant recipients, is marketed by Salix as a treatment for Crohn's disease and ulcerative colitis. The company also sells Xifaxam (rifaximin), a gastrointestinal-specific oral antibiotic; Visicol, a product indicated for cleansing of the bowel as a preparation for colonoscopy; OsmoPrep tablets; MoviPrep oral solution; Anusol-HC rectal suppositories; and Proctocort, which is available in a cream form that is indicated for the relief of the inflammatory and pruritic manifestations of corticosteroid-responsive dermatoses, and in a suppository form, which is indicated for use in inflamed hemorrhoids and postirradiation proclitis. In early 2007, Salix also added Pepcid Oral Suspension and Diuril Oral Suspension to its line of products by acquiring the rights to them from Merck & Co., Inc. The primary product candidates Salix is developing are: balsalazide disodium tablets, which the company intends to sell for the treatment of ulcerative colitis; a patented, granulated formula of mesalamine, which it intends to sell for the treatment of ulcerative colitis; rifaximin for various additional potential indications; and Sanvar IR, which it intends to sell as a treatment of acute esophagal variceal bleeding. In 2007, the company licensed the exclusive rights to market DIACOL 1500 mg tablets to Dr. Falk Pharma GmbH of Germany.

FINANCIALS: Sales and profits are in thousands of dollars—add 000 to get the full amount. 2006 Note: Financial information for 2006 was not available for all companies at press time.

2006 Sales: $208,533	2006 Profits: $31,510	**U.S. Stock Ticker:** SLXP
2005 Sales: $154,903	2005 Profits: $-60,585	**Int'l Ticker:** Int'l Exchange:
2004 Sales: $105,496	2004 Profits: $6,839	Employees: 240
2003 Sales: $55,807	2003 Profits: $-20,101	Fiscal Year Ends: 12/31
2002 Sales: $33,456	2002 Profits: $-24,742	Parent Company:

SALARIES/BENEFITS:

Pension Plan: Y	ESOP Stock Plan:	Profit Sharing:	Top Exec. Salary: $615,000	Bonus: $280,000
Savings Plan: Y	Stock Purch. Plan:		Second Exec. Salary: $325,000	Bonus: $147,000

OTHER THOUGHTS:

Apparent Women Officers or Directors: 1
Hot Spot for Advancement for Women/Minorities:

LOCATIONS: ("Y" = Yes)

West:	Southwest:	Midwest:	Southeast:	Northeast:	International:
				Y	

SANGAMO BIOSCIENCES INC

www.sangamo.com

Industry Group Code: 541710 Ranks within this company's industry group: Sales: 6 Profits: 17

Drugs:		Other:		Clinical:		Computers:		Other:	
Discovery:	Y	AgriBio:		Trials/Services:		Hardware:		Specialty Services:	
Licensing:		Genomics/Proteomics:	Y	Laboratories:		Software:		Consulting:	
Manufacturing:		Tissue Replacement:	Y	Equipment/Supplies:		Arrays:		Blood Collection:	
Development:	Y			Research/Development Svcs.:	Y	Database Management:		Drug Delivery:	
Generics:				Diagnostics:				Drug Distribution:	

TYPES OF BUSINESS:
Drug Research & Development
Gene Expression Regulation Therapies
Transcription Factor Technology

BRANDS/DIVISIONS/AFFILIATES:
Universal GeneTools
ZFP
ZFP TF
ZFN
City of Hope
Juvenile Diabetes Research Foundation

CONTACTS: Note: Officers with more than one job title may be intentionally listed here more than once.
Edward O. Lanphier, CEO
Edward O. Lanphier, Pres.
Carl Pabo, Chief Scientific Officer/Sr. VP
Edward J. Rebar, Sr. Dir.-Tech.
Greg S. Zante, VP-Admin.
David Ichikawa, Sr. VP-Bus. Dev.
Greg S. Zante, VP-Finance
Sean Brennan, Sr. Dir.-Intellectual Property
Dale Ando, Chief Medical Officer/VP-Therapeutic Dev.
Eric T. Rhodes, Sr. Dir.-Commercial Dev.
Phillip Gregory, VP-Research

Phone: 510-970-6000	Fax: 510-236-8951
Toll-Free:	
Address: 501 Canal Blvd., Ste. A100, Richmond, CA 94804 US	

GROWTH PLANS/SPECIAL FEATURES:
Sangamo BioSciences, Inc. is a biotechnology company that researches, develops and commercializes novel transcription factors for gene regulation and modification. The company's proprietary technology platform is based on engineering a naturally occurring class of proteins called zinc finger DNA-binding proteins (ZFPs). ZFPs are designed to target any gene within the genome of any organism. Engineered ZFPs can be used to manufacture ZFP transcription factors; ZFP TFs, which are proteins that bind to DNA in order to turn genes on or off; and ZFNs, or zinc finger nucleases, which cut genomic DNA at pre-selected sequence locations. The firm intends to establish ZPF TFs for commercial applications in small-molecule drugs, pharmaceutical discovery, human monoclonal antibody development, DNA diagnostics and agricultural and industrial biotechnologies. Sangamo has also developed a line of ZFP Therapeutic products that can regulate or modify targets at the DNA level, which have previously proven to be intractable in conventional methods of drug discovery. The company has entered into a technological partnership with Charles River Laboratories, Inc. to apply ZFP TF technology towards the creation of a gene-altered rat model for use in the development of new drugs and therapies for cancer. In recent news, Sangamo and City of Hope have announced a license agreement and research collaboration to develop treatments for brain cancer. The Juvenile Diabetes Research Foundation has also agreed to fund human clinical studies of a new ZFP Therapeutic for diabetic neuropathy. In 2006, Sangamo entered into a definite agreement to acquire the angiogenesis program from Edwards Lifesciences Corporation.

Sangamo offers its employees health insurance, life insurance and a 401(k).

FINANCIALS: Sales and profits are in thousands of dollars—add 000 to get the full amount. 2006 Note: Financial information for 2006 was not available for all companies at press time.

2006 Sales: $7,885	2006 Profits: $-17,864	U.S. Stock Ticker: SGMO	
2005 Sales: $2,484	2005 Profits: $-13,293	Int'l Ticker: Int'l Exchange:	
2004 Sales: $1,315	2004 Profits: $-13,818	Employees: 75	
2003 Sales: $2,579	2003 Profits: $-10,433	Fiscal Year Ends: 12/31	
2002 Sales: $4,300	2002 Profits: $-29,800	Parent Company:	

SALARIES/BENEFITS:
Pension Plan:	ESOP Stock Plan: Y	Profit Sharing:	Top Exec. Salary: $418,000	Bonus: $230,000
Savings Plan: Y	Stock Purch. Plan:		Second Exec. Salary: $315,000	Bonus: $90,000

OTHER THOUGHTS:
Apparent Women Officers or Directors: 2
Hot Spot for Advancement for Women/Minorities: Y

LOCATIONS: ("Y" = Yes)
West:	Southwest:	Midwest:	Southeast:	Northeast:	International:
Y					

SANKYO CO LTD

www.sankyo.co.jp

Industry Group Code: 325412 Ranks within this company's industry group: Sales: Profits:

Drugs:		Other:		Clinical:		Computers:		Other:	
Discovery:		AgriBio:		Trials/Services:		Hardware:		Specialty Services:	
Licensing:		Genomics/Proteomics:		Laboratories:		Software:		Consulting:	
Manufacturing:	Y	Tissue Replacement:		Equipment/Supplies:		Arrays:		Blood Collection:	
Development:	Y			Research/Development Svcs.:		Database Management:		Drug Delivery:	
Generics:				Diagnostics:			Y	Drug Distribution:	

TYPES OF BUSINESS:

Pharmaceuticals Manufacturing
Medical Devices
Diagnostics
Veterinary Drugs

BRANDS/DIVISIONS/AFFILIATES:

Mevalotin
Olmetec
Calblock
Lulu
Shin-Sankyo Ichoyaku
Regain
Daiichi Pharmaceutical Co.
Daiichi Sankyo

CONTACTS: Note: Officers with more than one job title may be intentionally listed here more than once.

Yasuhiro Ikegami, Pres.
Koichi Hirai, VP-Human Resources
Koichi Hirai, VP-Admin.
Shigemichi Kondo, VP-Corp. Comm.
Takashi Yoshino, Dir.-Admin. & Human Resources
Tetsuo Takato, Chmn.

Phone: 81-3-5255-7111	Fax: 81-3-5255-7035
Toll-Free:	
Address: 3-5-1 Nihonbashi-honcho, Chuo-ku, Tokyo, 103-8426 Japan	

GROWTH PLANS/SPECIAL FEATURES:

Sankyo Co., Ltd. manufactures, markets and imports pharmaceuticals, medical devices and diagnostics and veterinary drugs. The firm focuses on cardiovascular diseases, carbohydrate metabolic diseases, bone and joint diseases, immunological and allergic diseases, cancer and infectious diseases. It operates through 23 branch offices, 65 satellite offices, five research and development centers and four manufacturing factories. Sankyo also works with 40 group affiliate companies. The company focus on cardiovascular diseases includes drugs for arteriosclerosis, heart disease, kidney disease, hypertension and hyperlipemia. Branded company products include Mevalotin, a cholesterol-lowering drug that serves to suppress cholesterol synthesis in the liver; Olmetec, a angiotensin II receptor antagonist designed to relieve hypertension; Calblock, a calcium-channel antagonist for the treatment of high blood pressure; and a variety of self-medication alternatives such as Lulu, Shin-Sankyo Ichoyaku and Regain. The firm is researching drugs for bone and joint diseases, including osteoporosis, rheumatoid arthritis and osteoarthritis; immunological and allergic diseases; cancer; and infectious diseases including bacterial and fungal diseases. The firm has laboratories in Europe, South America and the U.S. The firm recently acquired Daiichi Pharmaceuticals, for about $7.7 billion, and following a full integration of the companies by the year 2007, the company will go by the name Daiichi Sankyo.

FINANCIALS: Sales and profits are in thousands of dollars—add 000 to get the full amount. 2006 Note: Financial information for 2006 was not available for all companies at press time.

2006 Sales: $	2006 Profits: $	U.S. Stock Ticker: Subsidiary
2005 Sales: $	2005 Profits: $	Int'l Ticker: Int'l Exchange:
2004 Sales: $5,645,000	2004 Profits: $410,900	Employees: 5,401
2003 Sales: $4,755,500	2003 Profits: $282,400	Fiscal Year Ends: 3/31
2002 Sales: $4,138,100	2002 Profits: $292,500	Parent Company: DAIICHI SANKYO CO LTD

SALARIES/BENEFITS:

Pension Plan:	ESOP Stock Plan:	Profit Sharing:	Top Exec. Salary: $	Bonus: $
Savings Plan:	Stock Purch. Plan:		Second Exec. Salary: $	Bonus: $

OTHER THOUGHTS:

Apparent Women Officers or Directors:
Hot Spot for Advancement for Women/Minorities:

LOCATIONS: ("Y" = Yes)

West:	Southwest:	Midwest:	Southeast:	Northeast:	International:
Y					Y

SANOFI-AVENTIS

www.sanofi-synthelabo.fr

Industry Group Code: 325412 Ranks within this company's industry group: Sales: 4   Profits: 6

Drugs:		Other:		Clinical:	Computers:		Other:	
Discovery:	Y	AgriBio:		Trials/Services:	Hardware:		Specialty Services:	
Licensing:		Genomics/Proteomics:		Laboratories:	Software:		Consulting:	
Manufacturing:	Y	Tissue Replacement:		Equipment/Supplies:	Arrays:		Blood Collection:	
Development:	Y			Research/Development Svcs.:	Database Management:		Drug Delivery:	
Generics:				Diagnostics:			Drug Distribution:	

TYPES OF BUSINESS:

Pharmaceuticals Development & Manufacturing
Over-the-Counter Drugs
Cardiovascular Drugs
CNS Drugs
Oncology Drugs
Diabetes Drugs
Generics
Vaccines

BRANDS/DIVISIONS/AFFILIATES:

Aprovel
Plavix
Allegra
Depakine
Stilnox
Sanofi-Pasteur
Eloxatin
Sanofi-Synthelabo

CONTACTS: *Note: Officers with more than one job title may be intentionally listed here more than once.*

Gerard Le Fur, CEO
Jean-Claude Leroy, CFO/Exec. VP
Heinz Werner Meier, Sr. VP-Human Resources
Donna Vitter, General Counsel
Oliver Jacquesson, VP-Bus. Dev.
Michael Labie, Sr. VP-Corp. Comm.
Mark Cluzel, Sr. VP-Science & Medical Affairs
Hanspeter Spek, Exec. VP-Pharm. Oper.
Gilles Lhernould, VP-Industrial Oper.
David Williams, Sr. VP-Vaccines
Jean-Francois Dehecq, Chmn.
Greg Irace, Sr. VP-Pharm. Oper. USA

Phone: 33-1-53-77-4000	**Fax:** 33-1-53-77-4622

Toll-Free:

Address: 174 Ave. de France, Paris, 75013 France

GROWTH PLANS/SPECIAL FEATURES:

Sanofi-Aventis (SNY) is an international pharmaceutical group engaged in the research, development, manufacturing and marketing of primarily prescription pharmaceutical products. The company was created by the $67-billion merger of Sanofi-Synthelabo and Aventis in 2004, creating one of the largest pharmaceutical companies in the world. SNY conducts research and produces major pharmaceutical products in seven major therapeutic areas: cardiovascular diseases, thrombosis, metabolic disorders, oncology, central nervous system (CNS) disorders and vaccines. The firm's cardiovascular medications include the blood pressure medication, Aprovel and the anti-clotting agent, Plavix. In the field of oncology, SNY manufactures products which include Eloxatin, a treatment for colon-rectal cancer, and Taxotere as medication for breast cancer patients. CNS medications include Stilnox, the world's leading prescription insomnia medication, and Depakine, a treatment for epilepsy. Products in the internal medicine sector include the antihistamine, Allegra, as well as Xatral, a treatment for enlarged prostrates. SNY's subsidiary company, Sanofi-Pasteur, produces more than 20 different vaccines that are distributed in over 150 countries. SNY operates on 300 sites in 100 countries and its research and development sector markets at least 15 to 20 compounds per year. The research department aims to explore new molecular and physiopathological approaches in order to develop better formulations of its pharmaceutical products. In late 2006, SNY sold all of its interests in Rhodia to BNP Paribas. The FDA also approved new indications for formerly developed drugs, which allows patients with acute ST-segment elevation myocardial infarction to use Plavix and head and neck cancer patients to take the prescribed drug, Taxotere.

FINANCIALS: Sales and profits are in thousands of dollars—add 000 to get the full amount. 2006 Note: Financial information for 2006 was not available for all companies at press time.

2006 Sales: $38,722,100	2006 Profits: $6,003,540	**U.S. Stock Ticker: SNY**	
2005 Sales: $37,272,700	2005 Profits: $3,538,800	**Int'l Ticker: SAN** Int'l Exchange: Paris-Euronext	
2004 Sales: $20,377,000	2004 Profits: $-4,890,000	Employees: 100,298	
2003 Sales: $10,118,000	2003 Profits: $2,610,000	Fiscal Year Ends: 12/31	
2002 Sales: $7,823,000	2002 Profits: $1,847,000	Parent Company:	

SALARIES/BENEFITS:

Pension Plan:	ESOP Stock Plan:	Profit Sharing:	Top Exec. Salary: $	Bonus: $
Savings Plan:	Stock Purch. Plan: Y		Second Exec. Salary: $	Bonus: $

OTHER THOUGHTS:

Apparent Women Officers or Directors: 3
Hot Spot for Advancement for Women/Minorities: Y

LOCATIONS: ("Y" = Yes)

West:	Southwest:	Midwest:	Southeast:	Northeast:	International:
Y	Y	Y	Y	Y	Y

SAVIENT PHARMACEUTICALS INC   www.savientpharma.com

Industry Group Code: 325412 Ranks within this company's industry group: Sales: 80   Profits: 36

Drugs:		Other:		Clinical:	Computers:		Other:	
Discovery:	Y	AgriBio:		Trials/Services:	Hardware:		Specialty Services:	
Licensing:		Genomics/Proteomics:		Laboratories:	Software:		Consulting:	
Manufacturing:	Y	Tissue Replacement:		Equipment/Supplies:	Arrays:		Blood Collection:	
Development:	Y			Research/Development Svcs.:	Database Management:		Drug Delivery:	
Generics:				Diagnostics:			Drug Distribution:	

TYPES OF BUSINESS:

Pharmaceuticals Discovery & Development
Weight Gain Products
Hormone Therapy
Contraception
Cancer treatment

BRANDS/DIVISIONS/AFFILIATES:

Oxandrin
Puricase
Soltamax
Prosatide
Mircette
Oxandrolone

CONTACTS: *Note: Officers with more than one job title may be intentionally listed here more than once.*

Christopher Clement, CEO
Christopher Clement, Pres.
Brian J. Hayden, CFO/Sr. VP
Zeb Horowitz, Sr. VP/Chief Medical Officer
Paul Hamelin, VP-Commercial Oper.
Philip Yachmetz, Chief Bus. Officer/Exec. VP
Brian J. Hayden, Treas.
Robert Lamm, Sr. VP-Quality & Regulatory Affairs
Stephen Jaeger, Chmn.

Phone: 732-418-9300	Fax: 732-418-0570
Toll-Free:	
Address: 1 Tower Ctr., 14th Fl., E. Brunswick, NJ 08816 US	

GROWTH PLANS/SPECIAL FEATURES:

Savient Pharmaceuticals, Inc. is engaged in the development, manufacture and marketing of both genetically engineered and niche-focused, specialty pharmaceutical products. Through a combination of internal research and development, acquisitions, collaborative relationships and licensing arrangements, the company has developed a number of therapeutics. Currently, the primary product marketed worldwide by Savient is Oxandrin, an oral anabolic agent primarily used to promote weight gain. Oxandrin is the company's only product available in the U.S., and it is used primary for AIDS patients suffering from weigh loss, although the firm is exploring the drug's applications on frail, elderly patients. It also distributes a generic version of the same drug, oxandrolone. Savient is currently conducting clinical studies for the drug Puricase, a genetically engineered enzyme conjugate being studied for the elimination of excess uric acid in individuals suffering from severe gout. The drug is currently engaged in Phase III development. Recently, the company made the decision to end its development of Prosatide; to sell the licensing rights to the oral contraception, Mircette; and to terminate its research and development phase of Soltamax, an oral breast cancer treatment. In 2006, continuing its trend of divestiture to concentrate company resources on Oxandrin, the company completed the sale of its injectable testosterone product Delatestryl. The company further restructured its commercial operations in 2006 and 2007 such that it currently operates within one Specialty Pharmaceutical segment, which includes sales of Oxandrin and oxandrolone and the research and development of Puricase. The elimination of its 19 person Oxandrin field sales force in early 2007 was part of this restructuring. The company recently sold its Rosemont Pharmaceuticals and BTG Israel subsidiaries, reporting them as discontinued operations.

Savient offers employees: flexible spending accounts for health and dependant care; a variety of insurance options; and a summer hours program.

FINANCIALS: Sales and profits are in thousands of dollars—add 000 to get the full amount. 2006 Note: Financial information for 2006 was not available for all companies at press time.

2006 Sales: $47,514	2006 Profits: $60,325	U.S. Stock Ticker: SVNT
2005 Sales: $49,495	2005 Profits: $5,968	Int'l Ticker: Int'l Exchange:
2004 Sales: $62,353	2004 Profits: $-27,515	Employees: 67
2003 Sales: $101,103	2003 Profits: $12,454	Fiscal Year Ends: 12/31
2002 Sales: $103,000	2002 Profits: $9,700	Parent Company:

SALARIES/BENEFITS:

Pension Plan:	ESOP Stock Plan:	Profit Sharing:	Top Exec. Salary: $432,600	Bonus: $285,516
Savings Plan: Y	Stock Purch. Plan: Y		Second Exec. Salary: $343,505	Bonus: $198,375

OTHER THOUGHTS:

Apparent Women Officers or Directors:
Hot Spot for Advancement for Women/Minorities:

LOCATIONS: ("Y" = Yes)

West:	Southwest:	Midwest:	Southeast:	Northeast:	International:
				Y	Y

Note: Financial information, benefits and other data can change quickly and may vary from those stated here.

SCHERING-PLOUGH CORP

www.sch-plough.com

Industry Group Code: 325412 Ranks within this company's industry group: Sales: 14 Profits: 18

Drugs:		Other:	Clinical:	Computers:	Other:
Discovery:	Y	AgriBio:	Trials/Services:	Hardware:	Specialty Services:
Licensing:		Genomics/Proteomics:	Laboratories:	Software:	Consulting:
Manufacturing:	Y	Tissue Replacement:	Equipment/Supplies:	Arrays:	Blood Collection:
Development:	Y		Research/Development Svcs.:	Database Management:	Drug Delivery:
Generics:			Diagnostics:		Drug Distribution:

TYPES OF BUSINESS:

Drugs-Diversified
Anti-Infective & Anti-Cancer Drugs
Dermatologicals
Cardiovascular Drugs
Animal Health Products
Over-the-Counter Drugs
Foot & Sun Care Products
Genomics Research

BRANDS/DIVISIONS/AFFILIATES:

NASONEX
CLARINEX
AVELOX
LEVITRA
Afrin
Tinactin
Akzo Nobel N.V.
Organon Biosciences N.V.

CONTACTS: *Note: Officers with more than one job title may be intentionally listed here more than once.*

Fred Hassan, CEO
Robert J. Bertolini, CFO/Exec. VP
C. Ron Cheeley, Sr. VP-Global Human Resources
Thomas P. Koestler, Exec. VP/Pres., Schering-Plough Research Institute
Thomas Sabatino, Jr., Exec. VP/General Counsel
Carrie S. Cox, Exec. VP/Pres., Global Pharmaceuticals
Lori Queisser, Sr. VP-Global Compliance & Bus. Practices
Raul E. Kohan, Group Dir.-Global Specialty Oper.
Brent Saunders, Sr. VP/Pres., Consumer Health Care
Fred Hassan, Chmn.

Phone: 908-298-4000	Fax: 908-298-7653
Toll-Free:	
Address: 2000 Galloping Hill Rd., Kenilworth, NJ 07033-0530 US	

GROWTH PLANS/SPECIAL FEATURES:

Schering-Plough Corporation (SGP) develops, manufactures and markets global health care products which are divided into three sectors: prescription pharmaceuticals, consumer products and animal health products. The SGP prescription pharmaceuticals segment produces both primary and specialty care advanced drug therapies. In the primary sector, SGP manufactures allergy/respiratory medication such as NASONEX and CLARINEX; antibiotics such as AVELOX and LEVITRA for male erectile dysfunction disorder. In the specialty care sector, notable products produced by this company include anti-inflammatory, anti-viral and antifungal medications as well as treatments in the fields of oncology and coronary care. SGP's consumer products division markets a number of over-the-counter drugs, foot care and sun care products. These products include Dr. Scholl's foot care products; Lotrimin and Tinactin antifungal cream; Afrin nasal decongestant spray; Correctol laxative tablets and Coppertone sun care products. SGP Animal Health manufactures a broad range of pharmaceuticals and biological products for pets and livestock. Animal health products include antibiotics such as Nuflor; Banamine, an anti-inflammatory drug; and a broad range of vaccines, parasiticides, sutures, bandages and nutritional products. The company's research sector, Schering-Plough Research Institute, has pioneered many new advances in biotechnology and immunology in support of drug discovery and the enhancement of existing prescription products. SGP currently markets its products in the U.S., Canada, Europe, Latin America and Asia. In late 2007, this company will acquire Organon Biosciences N.V. as well as the human and health care sectors of Akzo Nobel N.V. In 2006, SGP gained approval from both the FDA and the European Union to market NOFAXIL, a treatment for Oropharyngeal Candidiasis. SGP also began restructuring its current facilities in Latin American and the expansion of its discovery operations in Cambridge, Massachusetts.

FINANCIALS: Sales and profits are in thousands of dollars—add 000 to get the full amount. 2006 Note: Financial information for 2006 was not available for all companies at press time.

2006 Sales: $10,594,000	2006 Profits: $1,143,000	U.S. Stock Ticker: SGP
2005 Sales: $9,508,000	2005 Profits: $269,000	Int'l Ticker: Int'l Exchange:
2004 Sales: $8,272,000	2004 Profits: $-947,000	Employees: 33,500
2003 Sales: $8,334,000	2003 Profits: $-92,000	Fiscal Year Ends: 12/31
2002 Sales: $10,180,000	2002 Profits: $1,974,000	Parent Company:

SALARIES/BENEFITS:

Pension Plan: Y	ESOP Stock Plan:	Profit Sharing:	Top Exec. Salary: $1,556,250	Bonus: $3,861,600
Savings Plan:	Stock Purch. Plan:		Second Exec. Salary: $937,500	Bonus: $1,448,300

OTHER THOUGHTS:

Apparent Women Officers or Directors: 2
Hot Spot for Advancement for Women/Minorities: Y

LOCATIONS: ("Y" = Yes)

West:	Southwest:	Midwest:	Southeast:	Northeast:	International:
Y	Y	Y	Y	Y	Y

SCHIFF NUTRITION INTERNATIONAL INC
www.schiffnutrition.com

Industry Group Code: 325411 **Ranks within this company's industry group:** Sales: 5   Profits: 4

Drugs:	Other:	Clinical:	Computers:	Other:
Discovery:	AgriBio:	Trials/Services:	Hardware:	Specialty Services:
Licensing:	Genomics/Proteomics:	Laboratories:	Software:	Consulting:
Manufacturing:	Tissue Replacement:	Equipment/Supplies:	Arrays:	Blood Collection:
Development:		Research/Development Svcs.:	Database Management:	Drug Delivery:
Generics:		Diagnostics:		Drug Distribution:

TYPES OF BUSINESS:
Vitamins/Nutrition Manufacturing & Specialty Retailing
Sports Nutrition Products
Weight Management Products
Nutrition Bars

GROWTH PLANS/SPECIAL FEATURES:
Schiff Nutrition International, Inc., formerly Weider Nutrition International, Inc., develops, manufactures, markets, distributes and sells vitamins, nutritional supplements, weight management and sports nutrition products in the form of capsules, tablets and nutritional bars. Schiff sells and distributes its products through mass volume retailers, health food stores and distributors, drug stores, supermarkets, health clubs and gyms. Its leading domestic product brands are Schiff, Tiger's Milk and Fi-Bar. Schiff brand vitamin products include multivitamins, such as Single Day; individual vitamins, such as Vitamin B and Vitamin C; minerals, such as Calcium; specialty formulas for men and women, such as Prostate Health and Folic Acid; and other specialty formulas, such as Melatonin Plus, Niacin and Lutein. Manufactured private label products are sold to key retailers for distribution under their store brand names. Private label products include vitamins and minerals; specialty supplements, such as joint care products; Vitamin B; and Calcium.

Schiff offers its employees comprehensive benefits, including a 401(k) and stock options.

BRANDS/DIVISIONS/AFFILIATES:
Tiger's Milk
Schiff
Fi-Bar
Move Free
Move Free Advanced
Weider Global Nutrition, LLC
Weider Nutrition International, Inc.
Single Day

CONTACTS: *Note: Officers with more than one job title may be intentionally listed here more than once.*
Bruce J. Wood, CEO
Bruce J. Wood, Pres.
Joseph W. Baty, CFO/Exec. VP
Daniel A. Thomson, General Counsel/Sr. VP
Tom Elitharp, Exec. VP-Oper.
Daniel A. Thomson, Sr. VP-Bus. Dev.
Daniel A. Thomson, Sr. VP/Corp. Sec.
Eric Weider, Chmn.

Phone: 801-975-5000	Fax: 801-972-2223
Toll-Free:	
Address: 2002 S. 5070 W., Salt Lake City, UT 84104-4726 US	

FINANCIALS: Sales and profits are in thousands of dollars—add 000 to get the full amount. 2006 Note: Financial information for 2006 was not available for all companies at press time.

2006 Sales: $178,372	2006 Profits: $15,839	**U.S. Stock Ticker:** WNI
2005 Sales: $173,095	2005 Profits: $6,569	**Int'l Ticker:** Int'l Exchange:
2004 Sales: $168,127	2004 Profits: $8,887	Employees: 382
2003 Sales: $240,900	2003 Profits: $-7,500	Fiscal Year Ends: 5/31
2002 Sales: $311,100	2002 Profits: $-7,500	Parent Company:

SALARIES/BENEFITS:
Pension Plan:	ESOP Stock Plan:	Profit Sharing:	Top Exec. Salary: $474,000	Bonus: $426,600
Savings Plan: Y	Stock Purch. Plan: Y		Second Exec. Salary: $256,000	Bonus: $166,400

OTHER THOUGHTS:
Apparent Women Officers or Directors:
Hot Spot for Advancement for Women/Minorities:

LOCATIONS: ("Y" = Yes)
West:	Southwest:	Midwest:	Southeast:	Northeast:	International:
Y					Y

Note: Financial information, benefits and other data can change quickly and may vary from those stated here.

SCICLONE PHARMACEUTICALS

www.sciclone.com

Industry Group Code: 325412 Ranks within this company's industry group: Sales: 94   Profits: 55

Drugs:		Other:		Clinical:		Computers:		Other:	
Discovery:		AgriBio:		Trials/Services:		Hardware:		Specialty Services:	
Licensing:	Y	Genomics/Proteomics:		Laboratories:		Software:		Consulting:	
Manufacturing:	Y	Tissue Replacement:		Equipment/Supplies:		Arrays:		Blood Collection:	
Development:	Y			Research/Development Svcs.:		Database Management:		Drug Delivery:	
Generics:				Diagnostics:				Drug Distribution:	

TYPES OF BUSINESS:

Pharmaceuticals Acquisition & Development
Immune System Enhancers
Hepatitis Therapies
Drug Manufacturing

BRANDS/DIVISIONS/AFFILIATES:

SciClone Pharmaceuticals International, Ltd.
ZADAXIN
SCV-07

CONTACTS: Note: Officers with more than one job title may be intentionally listed here more than once.

Friedhelm Blobel, CEO
Friedhelm Blobel, Pres.
Richard A. Waldron, CFO
Randy J. McBeath, VP-Mktg.
Cynthia W. Tuthill, VP-Scientific Affairs
Sriram Vemuri, VP-Product Dev.
Sriram Vemuri, VP-Mfg.
Israel Rios, Chief Medical Officer
Dean Woodman, Chmn.
Hans P. Schmid, Managing Dir.-SciClone Pharmaceuticals Int'l Ltd.

Phone: 650-358-3456	Fax: 650-358-3469
Toll-Free:	
Address: 901 Mariner's Island Blvd., Ste. 205, San Mateo, CA 94404 US	

GROWTH PLANS/SPECIAL FEATURES:

SciClone Pharmaceuticals, Inc. develops and commercializes pharmaceutical and biological therapeutic compounds that are acquired or in-licensed at the stage of late pre-clinical or early clinical development. The firm's lead product ZADAXIN is currently being evaluated in two Phase III hepatitis C virus (HCV) clinical trials in the U.S. The company is awaiting results from the second of two FDA Phase III clinical trials for treatment of HCV (the first trial did not prove the statistical clinical benefit that the company anticipated). ZADAXIN is also being evaluated in other late-stage clinical trials for the treatment of hepatitis B virus and certain cancers. The drug is a pure synthetic preparation of thymosin alpha 1, a natural substance that circulates in the body and is instrumental in the immune response to viral infections and certain cancers. The drug was given orphan status by the FDA for the treatment of stage IIb-IV malignant melanoma. ZADAXIN is approved for sale in over 34 countries internationally, primarily in Asia, the Middle East and Latin America, and 93% of the company's sales of ZADAXIN are in China, with revenues of $30 million in 2006. SciClone's other proprietary drug development candidate is SCV-07, which is in Phase II testing for the treatment of viral and other infectious diseases. SCV-07 is a synthetic dipeptide that has demonstrated immunomodulatory activity by increasing T-cell differentiation and function, biological processes that are necessary for the body to fight off infection. The company acquired exclusive worldwide rights, outside of Russia, to SCV-07 from Verta, Ltd., a biotechnology company located in St. Petersburg, Russia. Subsidiary SciClone Pharmaceuticals International, Ltd markets the firm's drugs.

SciClone offers employees medical, dental, vision and life insurance; short and long-term disability coverage; flexible spending accounts; a 401(k) plan; and an employee stock purchase program.

FINANCIALS: Sales and profits are in thousands of dollars—add 000 to get the full amount. 2006 Note: Financial information for 2006 was not available for all companies at press time.

2006 Sales: $32,662	2006 Profits: $ 727	**U.S. Stock Ticker: SCLN**
2005 Sales: $28,334	2005 Profits: $-7,713	**Int'l Ticker:** Int'l Exchange:
2004 Sales: $24,396	2004 Profits: $-13,278	Employees: 166
2003 Sales: $32,538	2003 Profits: $-5,275	Fiscal Year Ends: 12/31
2002 Sales: $17,800	2002 Profits: $-10,000	Parent Company:

SALARIES/BENEFITS:

Pension Plan:	ESOP Stock Plan:	Profit Sharing:	Top Exec. Salary: $359,897	Bonus: $
Savings Plan: Y	Stock Purch. Plan: Y		Second Exec. Salary: $325,686	Bonus: $95,750

OTHER THOUGHTS:

Apparent Women Officers or Directors: 1
Hot Spot for Advancement for Women/Minorities:

LOCATIONS: ("Y" = Yes)

West:	Southwest:	Midwest:	Southeast:	Northeast:	International:
Y					Y

SCIELE PHARMA INC   www.horizonpharm.com

Industry Group Code: 325412 Ranks within this company's industry group: Sales: 49   Profits: 38

Drugs:		Other:		Clinical:	Computers:	Other:	
Discovery:		AgriBio:		Trials/Services:	Hardware:	Specialty Services:	Y
Licensing:	Y	Genomics/Proteomics:		Laboratories:	Software:	Consulting:	
Manufacturing:		Tissue Replacement:		Equipment/Supplies:	Arrays:	Blood Collection:	
Development:				Research/Development Svcs.:	Database Management:	Drug Delivery:	
Generics:				Diagnostics:		Drug Distribution:	

TYPES OF BUSINESS:
Drugs-Acquisition & Licensing
Prescription Drug Sales & Marketing

GROWTH PLANS/SPECIAL FEATURES:

Sciele Pharma Inc., formerly First Horizon Pharmaceutical Corp., markets and sells brand-name prescription products. The firm focuses on cardiology and women's health. Sciele acquires or licenses pharmaceutical products that are not actively marketed but have high sales growth potential, are promotion-sensitive and complement Sciele's existing products. It enlists third-party manufacturers to manufacture its products. The company promotes its portfolio of 17 branded prescription products, ten of which account for approximately 95% of its total sales, through a nationwide sales and marketing force. The company targets high-prescribing physicians in cardiology, obstetrics/gynecology, pediatrics and gastroenterology. Sciele's products include Sular for hypertension; Fortamet as an adjunct to diet and exercise in order to lower blood glucose in Type 2 Diabetes patients; Altoprev for cholesterol reduction and coronary heart disease; triglide for primary hypercholesterolemia and mixed dyslipidemia; nitrolingual pumpspray for acute relief of an attack or prophylaxis of angina pectoris due to coronary artery disease; Prenate Elite, a prenatal vitamin; OptiNate, a prenatal vitamin; and Ponstel capsules, a primary dysmenorrheal in patients over the age of 14. Sciele has a 15-year exclusive licensing agreement with SkyePharma to market and distribute a cardiovascular product, Fenofibrate. In June 2007, the company acquired pediatric specialty pharmaceutical comapny Alliant Pharmaceuticals, Inc. for approximately $109.75 million. In July 2007, the company signed an agreement to market hypertension and attention deficit hyperactivity disorder (ADHD) medicine CLONICEL, and the company acquired a head lice asphyxiation product developed by Summers Laboratories, Inc.

BRANDS/DIVISIONS/AFFILIATES:
First Horizon Pharmaceutical Corp.
Prenatal Elite
Robinul
Tanafed
Optinate
Fortamet
Altoprev
Glycopyrrolate

CONTACTS: *Note: Officers with more than one job title may be intentionally listed here more than once.*
Patrick Fourteau, CEO
Patrick Fourteau, Pres.
Darrell Borne, CFO/Exec. VP
Sam F. Gibbons, VP-Sales
Alan T. Roberts, VP-Scientific Affairs
Leslie Zacks, VP-Legal
Michael Mavrogordato, VP-Global Bus. Dev.
Edward Schutter, Exec. VP/Chief Comm. Officer
Darrell Borne, Treas.
Larry M. Dillaha, Exec. VP/Chief Medical Officer
John N. Kapoor, Chmn.

Phone: 770-442-9707	Fax: 770-442-9594
Toll-Free:	
Address: 5 Concourse Pkwy., Ste. 1800, Atlanta, GA 30328 US	

FINANCIALS: Sales and profits are in thousands of dollars—add 000 to get the full amount. 2006 Note: Financial information for 2006 was not available for all companies at press time.

2006 Sales: $293,181	2006 Profits: $45,244	**U.S. Stock Ticker:** SCRX
2005 Sales: $216,358	2005 Profits: $39,209	**Int'l Ticker:** Int'l Exchange:
2004 Sales: $151,967	2004 Profits: $26,554	Employees: 782
2003 Sales: $95,305	2003 Profits: $-1,738	Fiscal Year Ends: 12/31
2002 Sales: $115,200	2002 Profits: $6,100	Parent Company:

SALARIES/BENEFITS:

Pension Plan:	ESOP Stock Plan: Y	Profit Sharing:	Top Exec. Salary: $285,000	Bonus: $273,743
Savings Plan: Y	Stock Purch. Plan:		Second Exec. Salary: $200,000	Bonus: $135,600

OTHER THOUGHTS:
Apparent Women Officers or Directors:
Hot Spot for Advancement for Women/Minorities:

LOCATIONS: ("Y" = Yes)

West:	Southwest:	Midwest:	Southeast:	Northeast:	International:
			Y	Y	

Note: Financial information, benefits and other data can change quickly and may vary from those stated here.

SCIOS INC
www.sciosinc.com

Industry Group Code: 325412 Ranks within this company's industry group: Sales: Profits:

Drugs:		Other:		Clinical:		Computers:		Other:	
Discovery:	Y	AgriBio:		Trials/Services:		Hardware:		Specialty Services:	
Licensing:		Genomics/Proteomics:		Laboratories:		Software:		Consulting:	
Manufacturing:	Y	Tissue Replacement:		Equipment/Supplies:		Arrays:		Blood Collection:	
Development:	Y			Research/Development Svcs.:	Y	Database Management:	Y	Drug Delivery:	
Generics:				Diagnostics:				Drug Distribution:	

TYPES OF BUSINESS:
Cardiovascular & Inflammatory Diseases Drugs

BRANDS/DIVISIONS/AFFILIATES:
Natrecor
ADHERE

CONTACTS: *Note: Officers with more than one job title may be intentionally listed here more than once.*
James R. Mitchell, Pres.
Tao Fu, VP-Corp. Planning & Dev.
Jim Barr, VP-Finance

Phone: 650-564-5000	Fax: 650-564-7070
Toll-Free:	
Address: 1900 Charleston Rd., Mountain View, CA 94039 US	

GROWTH PLANS/SPECIAL FEATURES:

Scios, Inc., a subsidiary of Johnson & Johnson, is a biopharmaceutical company developing treatments for cardiovascular and inflammatory diseases. The firm's technology platform fuses classical medicinal chemistry with the most recent advances in disease-based gene array, bioinformatics and computational chemistry. The company's lead product, Natrecor, is an intravenous cardiovascular drug approved to treat acute congestive heart failure (CHF). Natrecor is human B-type natriuretic peptide (hBNP) that has been manufactured from E. coli using recombinant DNA technology. Natrecor is intravenously admitted to the bloodstream, where the hBNP attaches to the particulate guanylate cyclase receptor of vascular smooth muscle and endothelial cells, which in turn assists in relaxing the muscles that surround the blood vessels and allows the blood to flow more smoothly without causing the heartbeat to speed up or become irregular. The firm's current research and development program focuses on the discovery of new therapeutics for other cardiovascular diseases. The firm has previously done research in the field of protein kinases, particularly p38 MAP, which controls the production of growth factors and the creation of molecules produced by the immune system that causes inflammation. It also has previously worked in the field of small-molecule TGF-beta inhibitors, which are involved with diseases such as cancer, CHF and live cirrhosis. Scios also has a program call ADHERE, Acute Decompensated Heart Failure National Registry, which is an observational registry that lists data on heart failure patient treatment and associated outcomes. In March 2007, Scios released the results from a Natrecor study, which provides renal and mortality safety data on the drug in patients with advanced chronic decompensated heart failure.

The company offers its employees medical and dental coverage; life, disability, long-term care and accident insurance; health and wellness services; and savings and pension plans.

FINANCIALS: Sales and profits are in thousands of dollars—add 000 to get the full amount. 2006 Note: Financial information for 2006 was not available for all companies at press time.

2006 Sales: $	2006 Profits: $	**U.S. Stock Ticker: Subsidiary**
2005 Sales: $	2005 Profits: $	**Int'l Ticker:** Int'l Exchange:
2004 Sales: $	2004 Profits: $	Employees: 509
2003 Sales: $	2003 Profits: $	Fiscal Year Ends: 12/31
2002 Sales: $107,300	2002 Profits: $-88,100	Parent Company: JOHNSON & JOHNSON

SALARIES/BENEFITS:

Pension Plan: Y	ESOP Stock Plan:	Profit Sharing:	Top Exec. Salary: $460,000	Bonus: $500,000
Savings Plan: Y	Stock Purch. Plan:		Second Exec. Salary: $267,750	Bonus: $180,000

OTHER THOUGHTS:
Apparent Women Officers or Directors:
Hot Spot for Advancement for Women/Minorities:

LOCATIONS: ("Y" = Yes)

West:	Southwest:	Midwest:	Southeast:	Northeast:	International:
Y					

SEATTLE GENETICS

www.seattlegenetics.com

Industry Group Code: 325412 **Ranks within this company's industry group:** Sales: 128   Profits: 139

Drugs:		Other:	Clinical:	Computers:	Other:
Discovery:	Y	AgriBio:	Trials/Services:	Hardware:	Specialty Services:
Licensing:		Genomics/Proteomics:	Laboratories:	Software:	Consulting:
Manufacturing:		Tissue Replacement:	Equipment/Supplies:	Arrays:	Blood Collection:
Development:	Y		Research/Development Svcs.:	Database Management:	Drug Delivery:
Generics:			Diagnostics:		Drug Distribution:

TYPES OF BUSINESS:

Biopharmaceuticals Development
Cancer Treatments
Monoclonal Antibodies

BRANDS/DIVISIONS/AFFILIATES:

SGN-40
SGN-30
SGN-33
SGN-35
SGN-70
SGN-75
Antibody-Drug Conjugates (ADCs)

CONTACTS: *Note: Officers with more than one job title may be intentionally listed here more than once.*

Clay B. Siegall, CEO
Clay B. Siegall, Pres.
Todd Simpson, CFO
Pamela A. Trail, Chief Scientific Officer
Morris Z. Rosenberg, Sr. VP-Dev.
Thomas C. Reynolds, Chief Medical Officer
Iqbal S. Grewal, VP-Preclincal Therapeutics
Eric L. Dobmeier, Chief Bus. Officer
Peter D. Senter, VP-Chemistry
Felix J. Baker, Chmn.

Phone: 425-527-4000	**Fax:** 425-527-4001
Toll-Free:	
Address: 21823 30th Dr. SE, Bothell, WA 98021 US	

GROWTH PLANS/SPECIAL FEATURES:

Seattle Genetics develops monoclonal antibody (mAb)-based therapies for the treatment of cancer and autoimmune diseases. Its strategy is to advance its portfolio of product candidates in diseases with unmet medical need and significant market potential. The company has an exclusive, worldwide license agreement with Genentech to develop and commercialize its lead product candidate SGN-40. Its research and development activities focus on mAb-based therapies for human cancers including lung, renal cell, Hodgkin's disease, non-Hodgkin's lymphoma, multiple myeloma and melanoma. Products in the developmental stage include: SGN-30, in Phase II testing for systemic anaplastic large cell lymphoma (ALCL), cutaneous ALCL and Hodgkin's disease; SGN-40, which is in Phase I testing for multiple myeloma and non-Hodgkin's lymphoma, in Phase I/II for chronic lymphocytic leukemia and in a preclinical stage for Hodgkin's disease, Waldenstrom's macroglobulineamia, bladder and renal cancer; SGN-33, which is in Phase I for acute myeloid leukemia; SGN-35, which is in Phase I for myelodysplastic syndromes and an investigational new drug (IND) for hematologic malignancies; SGN-70, which is an IND for hematologic malignancies, renal cancer and immunologic diseases; and SGN-75, which is a preclinical candidate to treat renal cancer, hematologic malignancies and immunologic diseases. These product candidates represent applications of Seattle Genetics' primary platform technologies: genetically engineered monoclonal antibodies and antibody-drug conjugates (ADC). Each of these platforms is designed to support therapies that are able to identify and kill cancer cells while limiting damage to normal tissue. Seattle Genetics currently has license agreements for its proprietary ADC technology with Genentech, Protein Design Labs, Agensys and CuraGen. In 2007, the FDA granted Orphan Drug designations to the SGN-33 and SGN-35 programs.

FINANCIALS: Sales and profits are in thousands of dollars—add 000 to get the full amount. 2006 Note: Financial information for 2006 was not available for all companies at press time.

2006 Sales: $10,005	2006 Profits: $-36,015	**U.S. Stock Ticker: SGEN**
2005 Sales: $9,757	2005 Profits: $-29,433	**Int'l Ticker:** Int'l Exchange:
2004 Sales: $6,701	2004 Profits: $-35,439	Employees: 151
2003 Sales: $5,074	2003 Profits: $-22,086	Fiscal Year Ends: 12/31
2002 Sales: $1,684	2002 Profits: $-23,161	Parent Company:

SALARIES/BENEFITS:

Pension Plan:	ESOP Stock Plan:	Profit Sharing:	Top Exec. Salary: $435,000	Bonus: $206,388
Savings Plan: Y	Stock Purch. Plan: Y		Second Exec. Salary: $275,473	Bonus: $103,086

OTHER THOUGHTS:

Apparent Women Officers or Directors: 1
Hot Spot for Advancement for Women/Minorities:

LOCATIONS: ("Y" = Yes)

West:	Southwest:	Midwest:	Southeast:	Northeast:	International:
Y					

Note: Financial information, benefits and other data can change quickly and may vary from those stated here.

SENETEK PLC

www.senetekplc.com

Industry Group Code: 325412 Ranks within this company's industry group: Sales: 132   Profits: 54

Drugs:		Other:		Clinical:	Computers:		Other:	
Discovery:		AgriBio:		Trials/Services:	Hardware:		Specialty Services:	
Licensing:	Y	Genomics/Proteomics:		Laboratories:	Software:		Consulting:	
Manufacturing:	Y	Tissue Replacement:		Equipment/Supplies:	Arrays:		Blood Collection:	
Development:	Y			Research/Development Svcs.:	Database Management:		Drug Delivery:	
Generics:				Diagnostics:			Drug Distribution:	

TYPES OF BUSINESS:

Drugs-Sexual Dysfunction
Anti-Aging Products

BRANDS/DIVISIONS/AFFILIATES:

Invicorp
Kinetin
Zeatin
PRK 124

CONTACTS: *Note: Officers with more than one job title may be intentionally listed here more than once.*

Frank J. Massino, CEO
William O'Kelly, CFO
Wade H. Nichols, General Counsel
Wade H. Nichols, Exec. VP-Corp. Dev.
Frank J. Massino, Chmn.

Phone: 707-226-3900	Fax: 707-259-6241
Toll-Free:	
Address: 831 Latour Ct., Ste. A, Napa, CA 94558 US	

GROWTH PLANS/SPECIAL FEATURES:

Senetek plc develops and markets proprietary anti-aging products and products for the treatment of sexual dysfunction. In the anti-aging market, the company offers Kinetin, an antioxidant skin cream that has been shown in clinical trials to reduce fine lines and blotchiness and increase the amount of moisture retained by the skin. Senetek markets an entire line of Kinetin products, as well as Zeatin, an analog of Kinetin that is indicated for dermatological applications including treatment of psoriasis. Kinetin products are currently marketed worldwide under a variety of names, including Almay and The Body Shop, by Valeant Pharmaceuticals International, Obagi Medical Products, Med Beauty AG and many others. For erectile dysfunction, the company markets Invicorp, a self-administered injection that delivers a combination therapy of two specific drugs, vasoactive intestinal polypeptide and phentolamine mesylate. Invicorp is meant to be a second-line treatment, after oral therapies have failed. The drug is licensed to Ardana for distribution in Europe. Senetek also sponsors research on the treatment of conditions related to aging, with the agreement that it will have rights to whatever products result from the research. The firm recently announced the completion of pre-clinical studies and the beginning of clinical trials for Zeatin as a topical anti-aging skin treatment. Senetek holds 130 patents and 97 trademarks. In 2006, the company announced successful results in its initial clinical trial of PRK 124, a new cytokinin anti-aging compound. Also, in the same year, Senetek announced the licensing for the manufacture and marketing in North America of its Invicorp product to Plethora Solutions; and the company sold its patents, trademarks and automated manufacturing equipment for its proprietary disposable autoinjector for self-administration of parenteral drugs, including epinephrine for emergency treatment of anaphylactic shock from peanut and other allergies.

FINANCIALS: Sales and profits are in thousands of dollars—add 000 to get the full amount. 2006 Note: Financial information for 2006 was not available for all companies at press time.

2006 Sales: $8,431	2006 Profits: $1,883	**U.S. Stock Ticker:** SNTK
2005 Sales: $5,871	2005 Profits: $-1,739	**Int'l Ticker:** Int'l Exchange:
2004 Sales: $7,550	2004 Profits: $ 566	Employees: 9
2003 Sales: $8,226	2003 Profits: $-4,994	Fiscal Year Ends: 12/31
2002 Sales: $9,400	2002 Profits: $1,500	Parent Company:

SALARIES/BENEFITS:

Pension Plan:	ESOP Stock Plan:	Profit Sharing:	Top Exec. Salary: $319,000	Bonus: $100,000
Savings Plan:	Stock Purch. Plan:		Second Exec. Salary: $120,961	Bonus: $

OTHER THOUGHTS:

Apparent Women Officers or Directors:
Hot Spot for Advancement for Women/Minorities:

LOCATIONS: ("Y" = Yes)

West:	Southwest:	Midwest:	Southeast:	Northeast:	International:
Y					Y

Note: Financial information, benefits and other data can change quickly and may vary from those stated here.

SEPRACOR INC

www.sepracor.com

Industry Group Code: 325412 Ranks within this company's industry group: Sales: 31 Profits: 29

Drugs:		Other:		Clinical:	Computers:	Other:
Discovery:	Y	AgriBio:		Trials/Services:	Hardware:	Specialty Services:
Licensing:	Y	Genomics/Proteomics:		Laboratories:	Software:	Consulting:
Manufacturing:	Y	Tissue Replacement:		Equipment/Supplies:	Arrays:	Blood Collection:
Development:	Y			Research/Development Svcs.:	Database Management:	Drug Delivery:
Generics:				Diagnostics:		Drug Distribution:

TYPES OF BUSINESS:

Pharmaceuticals Discovery & Development
Respiratory Treatments
Central Nervous System Disorder Treatments

BRANDS/DIVISIONS/AFFILIATES:

XOPENEX
LUNESTA
ALLEGRA
CLARINEX
XUSA/XYZAL
BROVANA

CONTACTS: Note: Officers with more than one job title may be intentionally listed here more than once.

Adrian Adams, CEO
William J. O'Shea, COO
Adrian Adams, Pres.
David P. Southwell, CFO
Mark H. N. Corrigan, Exec. VP-R&D
Robert F. Scumaci, Exec. VP-Tech. Oper.
Robert F. Scumaci, Exec. VP-Admin.
Andrew I. Koven, General Counsel/Exec. VP/Corp. Sec.
David P. Southwell, Exec. VP-Corp. Planning & Dev.
Robert F. Scumaci, Exec. VP-Finance
David P. Southwell, Exec. VP-Licensing
W. James O'Shea, Vice Chmn.
Timothy J. Barberich, Chmn.

Phone: 508-481-6700	**Fax:** 508-357-7499
Toll-Free: 800-245-5961	
Address: 84 Waterford Dr., Marlborough, MA 01752 US	

GROWTH PLANS/SPECIAL FEATURES:

Sepracor, Inc. is a research-based pharmaceutical company whose goal is to discover, develop and market products that are directed toward serving unmet medical needs, particularly in the treatment of respiratory and central nervous system disorders. The company also develops and markets improved versions of widely prescribed drugs. These versions, known as improved chemical entities, feature enhancements such as reduced side effects, increased therapeutic efficacy, improved dosage forms and, in some cases, additional indications. Serpacor's lead product is XOPENEX, an inhalation solution used in nebulizers for patients with asthma or chronic obstructive pulmonary disease (COPD). BROVANA (arformoterol tartrate) was approved in October 2006 as a long term, twice-daily treatment for COPD. The product is administered by nebulizer. The company also markets LUNESTA (eszopiclone) for the treatment of insomnia in adults. Sepracor has entered into a license agreement with Eisai Co. Ltd. for the development and manufacture of LUNESTA in Japan, which is scheduled to finish clinical trials in Japan by 2010. In addition, the company has designed treatments for asthma, depression and restless legs syndrome. Sepracor markets its own and other companies' products through its sales force, co-promotion agreements and out-licensing partnerships. Due to the firm's patents relating to the chemicals desloratadine, fexofenadine and levocetirizine, Sepracor has out-licensing agreements with Schering-Plough for CLARINEX, Aventis for ALLEGRA and UCB Farchim SA for its XUSAL/XYZAL products, all of which are allergy medications. In April 2007, the company commercially launched its newly approved prescription drug, BROVANA.

Sepracor offers its employees tuition reimbursement, adoption reimbursement, back-up childcare and a comprehensive health plan.

FINANCIALS: Sales and profits are in thousands of dollars—add 000 to get the full amount. 2006 Note: Financial information for 2006 was not available for all companies at press time.

2006 Sales: $1,196,534	2006 Profits: $184,562	**U.S. Stock Ticker:** SEPR
2005 Sales: $820,928	2005 Profits: $3,927	**Int'l Ticker:** Int'l Exchange:
2004 Sales: $380,877	2004 Profits: $-296,910	Employees: 2,470
2003 Sales: $344,040	2003 Profits: $-135,936	Fiscal Year Ends: 12/31
2002 Sales: $239,000	2002 Profits: $-276,500	Parent Company:

SALARIES/BENEFITS:

Pension Plan:	ESOP Stock Plan:	Profit Sharing:	Top Exec. Salary: $875,000	Bonus: $525,000
Savings Plan: Y	Stock Purch. Plan: Y		Second Exec. Salary: $525,000	Bonus: $236,250

OTHER THOUGHTS:

Apparent Women Officers or Directors:
Hot Spot for Advancement for Women/Minorities:

LOCATIONS: ("Y" = Yes)

West:	Southwest:	Midwest:	Southeast:	Northeast:	International:
				Y	

Note: Financial information, benefits and other data can change quickly and may vary from those stated here.

SEQUENOM INC

www.sequenom.com

Industry Group Code: 325413 Ranks within this company's industry group: Sales: 19 Profits: 19

Drugs:	Other:		Clinical:		Computers:		Other:	
Discovery:	AgriBio:		Trials/Services:		Hardware:		Specialty Services:	
Licensing:	Genomics/Proteomics:	Y	Laboratories:		Software:	Y	Consulting:	
Manufacturing:	Tissue Replacement:		Equipment/Supplies:		Arrays:	Y	Blood Collection:	
Development:			Research/Development Svcs.:		Database Management:	Y	Drug Delivery:	
Generics:			Diagnostics:				Drug Distribution:	

TYPES OF BUSINESS:

Chips-DNA Arrays
Genotype Analysis Software
Cell Research Database

BRANDS/DIVISIONS/AFFILIATES:

MassARRAY
MassARRAY SNP Genotyping
MassARRAY QGE
MassARRAY EpiTYPER
MassARRAY SNP Discovery
Oligonucleotide Quality Control
Axiom Biotechnologies
iPLEX

CONTACTS: *Note: Officers with more than one job title may be intentionally listed here more than once.*

Harry Stylli, CEO
Michael Monko, Sr. VP-Sales & Mktg.
Elizabeth Dragon, Sr. VP-R&D
Charles R. Cantor, Chief Scientific Officer
Clarke Neumann, General Counsel/VP
Larry Myres, VP-Oper.
Clarke Neumann, Dir.-Investor Rel.
John Sharp, VP-Finance
Jeffrey Otto, Dir.-Genetic Svcs.
Steve Owings, VP-Commercial Dev., Prenatal Diagnostics
Harry Stylli, Chmn.

Phone: 858-202-9000	Fax: 858-202-9001
Toll-Free:	
Address: 3595 John Hopkins Ct., San Diego, CA 92121-1331 US	

GROWTH PLANS/SPECIAL FEATURES:

Sequonom, Inc. is an industrial genomics firm that provides products and services for biomedical research, molecular medicine and agricultural and non-invasive prenatal testing markets. The company's main source of revenue is derived from MassARRAY, a hardware and software application with consumable chips and reagents that analyzes high performance nucleic acids and quantitatively measures genetic target material and variations. The MassARRAY system offers cost-effective methods for numerous types of DNA analysis applications, which can range from SNP genotyping and allelotyping; quantitative gene expression analysis; and quantitative methylation marker analysis. New platform products introduced in 2006 include the MassARRAY SNP Genotyping, an analyzer of genetic variations in order to determine impacts on human health; MassARRAY QGE to accurately measure levels of gene expression; MassARRAY EpiTYPER to assess methylation ratios of multiple samples; MassARRAY SNP Discovery; and Oligonucleotide Quality Control, which can also be used to identify genetic markers and to analyze up to several thousand single nucleotide polymorphisms (SNPs) within samples. Customers of MassARRAY include clinical research laboratories, biotechnology companies, universities and government agencies in North America, Europe, India, Israel, Japan, Korea, New Zealand and Turkey. In addition to biotechnological uses, MassARRAY is also utilized to provide county-of-origin verification, age verification and national ID programs for traceability content of crops. This may also include genetic analysis of crops for enhanced traits, such as nutritional quality, disease resistance and crop yields. Livestock-focused service providers have also started using MassARRAY for genotyping solutions. In recent news, health organizations such as M.D. Anderson's Kleberg Center have utilized the MassARRAY system for epigenomic studies and genotyping. International research institutions have also reported that with the use of this technology, significant advancements were made in understanding cancer, diabetes and drug-resistant malaria.

SEQUENOM offers employees health benefits, college savings plans, personal accident expense plans and cancer indemnity plans.

FINANCIALS: Sales and profits are in thousands of dollars—add 000 to get the full amount. 2006 Note: Financial information for 2006 was not available for all companies at press time.

2006 Sales: $28,496	2006 Profits: $-17,577	**U.S. Stock Ticker:** SQNM	
2005 Sales: $19,421	2005 Profits: $-26,537	**Int'l Ticker:** Int'l Exchange:	
2004 Sales: $22,449	2004 Profits: $-34,625	Employees: 123	
2003 Sales: $30,252	2003 Profits: $-36,681	Fiscal Year Ends: 12/31	
2002 Sales: $30,900	2002 Profits: $-205,600	Parent Company:	

SALARIES/BENEFITS:

Pension Plan:	ESOP Stock Plan:	Profit Sharing:	Top Exec. Salary: $324,000	Bonus: $
Savings Plan: Y	Stock Purch. Plan:		Second Exec. Salary: $277,308	Bonus: $

OTHER THOUGHTS:

Apparent Women Officers or Directors: 1
Hot Spot for Advancement for Women/Minorities:

LOCATIONS: ("Y" = Yes)

West:	Southwest:	Midwest:	Southeast:	Northeast:	International:
Y				Y	Y

SERACARE LIFE SCIENCES INC

www.seracare.com

Industry Group Code: 325414 **Ranks within this company's industry group:** Sales: Profits:

Drugs:		Other:		Clinical:		Computers:		Other:	
Discovery:		AgriBio:		Trials/Services:		Hardware:		Specialty Services:	Y
Licensing:		Genomics/Proteomics:		Laboratories:		Software:		Consulting:	
Manufacturing:	Y	Tissue Replacement:		Equipment/Supplies:		Arrays:		Blood Collection:	Y
Development:				Research/Development Svcs.:	Y	Database Management:		Drug Delivery:	
Generics:				Diagnostics:	Y			Drug Distribution:	

TYPES OF BUSINESS:

Diagnostics Products
Blood & Plasma Collection & Processing
Research Support Services
Bovine Serum
Biological Product Database

BRANDS/DIVISIONS/AFFILIATES:

Boston Biomedica, Inc.
SeraCare Global Repository
SeraCare BioServices
SeraCare BioProcessing
BBI Diagnostics
SeraCare Diagnostic Products

CONTACTS: Note: Officers with more than one job title may be intentionally listed here more than once.

Susan L. N. Vogt, CEO
Susan L. N. Vogt, Pres.
Gregory A. Gould, CFO
Bill Smutny, VP-Mktg. & Sales
Kathleen W. Benjamin, VP-Human Resources
Mark Manak, Chief Scientific Officer
Richard D'Allessandro, CIO
Ron Dilling, VP-Oper.
Jeff R. Livingstone, VP-Bus. Dev.
Jeffrey Hirthy, VP-Finance Oper.
Kathi Shea, VP-BioServices Oper.
Barry D. Plost, Chmn.

Phone: 508-580-1900	Fax: 508-580-2202
Toll-Free: 800-676-1881	
Address: 375 West Street, West Bridgewater, MA 02379 US	

GROWTH PLANS/SPECIAL FEATURES:

SeraCare Life Sciences, Inc. develops, manufactures and sells a broad range of biological based materials and services; their products are the requisite components for the manufacture of diagnostic tests, commercial bioproduction of therapeutic drugs and additional research applications in various medical and scientific fields. The company's offerings include plasma-based therapeutic products, diagnostic products and reagents, cell culture products, specialty plasmas and in vitro stabilizers; in addition, it offers clinically annotated DNA, RNA, serum, and tissue specimens ethically collected from consenting donors with a variety of patient conditions. SeraCare's business is divided into two main segments: biopharmaceutical and diagnostics. The biopharmaceutical segment is further organized into three divisions: SeraCare BioProcessing, which provides human and animal products for use by pharmaceutical developers and therapeutic drug manufacturers; SeraCare BioServices, which provides research support services to government and commercial clients; and the SeraCare Global Repository, a collection of over 600,000 biological specimens of normal and diseased human serum, plasma, DNA, RNA and tissue from 120,000 donors. The diagnostics segment is divided into SeraCare Diagnostic Products and BBI Diagnostics, both of which provide a wide range of products to manufacturers of diagnostics.

FINANCIALS: Sales and profits are in thousands of dollars—add 000 to get the full amount. 2006 Note: Financial information for 2006 was not available for all companies at press time.

2006 Sales: $	2006 Profits: $	U.S. Stock Ticker: SRLSE.PK	
2005 Sales: $	2005 Profits: $	Int'l Ticker: Int'l Exchange:	
2004 Sales: $28,441	2004 Profits: $4,155	Employees: 242	
2003 Sales: $23,202	2003 Profits: $2,616	Fiscal Year Ends: 9/30	
2002 Sales: $25,307	2002 Profits: $3,624	Parent Company:	

SALARIES/BENEFITS:

Pension Plan:	ESOP Stock Plan:	Profit Sharing:	Top Exec. Salary: $191,666	Bonus: $100,000
Savings Plan:	Stock Purch. Plan:		Second Exec. Salary: $152,249	Bonus: $

OTHER THOUGHTS:

Apparent Women Officers or Directors: 3
Hot Spot for Advancement for Women/Minorities: Y

LOCATIONS: ("Y" = Yes)

West:	Southwest:	Midwest:	Southeast:	Northeast:	International:
				Y	

SHIRE PLC

www.shire.com

Industry Group Code: 325412 Ranks within this company's industry group: Sales: Profits:

Drugs:		Other:		Clinical:	Computers:		Other:	
Discovery:	Y	AgriBio:		Trials/Services:	Hardware:		Specialty Services:	
Licensing:	Y	Genomics/Proteomics:		Laboratories:	Software:		Consulting:	
Manufacturing:	Y	Tissue Replacement:		Equipment/Supplies:	Arrays:		Blood Collection:	
Development:	Y			Research/Development Svcs.:	Database Management:		Drug Delivery:	Y
Generics:				Diagnostics:			Drug Distribution:	

TYPES OF BUSINESS:

Drugs-Diversified
Drug Delivery Technology
Small-Molecule Drugs

BRANDS/DIVISIONS/AFFILIATES:

Adderall
Daytrana
Carbatrol
Pentasa
Colazide
Replagal
Elaprase
Vyvanse

CONTACTS: Note: Officers with more than one job title may be intentionally listed here more than once.

Matthew Emmens, CEO
Angus Russell, CFO
Eliseo Salinas, Chief Scientific Officer/Exec. VP-Global R&D
Anita Graham, Chief Admin. Officer/Exec. VP-Corp. Bus. Svcs.
Tatjana May, General Counsel/Exec. VP-Global Legal Affairs/Sec.
Barbara Deptula, Exec. VP-Bus. Dev.
Clea Rosenfeld, VP-Investor Rel.
Joseph Rus, Exec. VP-Market Alliance & New Market Dev.
Mike Cola, Pres., Specialty Pharmaceuticals
David D. Pendergast, Pres., Human Genetic Therapies
Eric Rojas, Dir.-Investor Rel. North America
Matthew Emmens, Chmn.

Phone: 44-1256-894-000	Fax: 44-1256-894-708

Toll-Free:

Address: Hampshire Int'l Business Park, Chineham, Basingstoke, Hampshire RG24 8EP UK

GROWTH PLANS/SPECIAL FEATURES:

Shire plc is an international specialty pharmaceutical company with a strategic focus on four therapeutic areas: attention deficit and hyperactivity disorder (ADHD), human genetic therapies (HGT), gastrointestinal and renal diseases. Shire's main product for the treatment of ADHD is Adderall XR, designed to provide a day-long treatment with one morning dose. Daytrana, a transdermal delivery system for the once daily treatment of ADHD, is the first and only patch medication approved by the FDA for the pediatric treatment of ADHD. Carbatrol is the company's epilepsy treatment product. For the treatment of ulcerative colitis, Shire markets Pentasa and Colazide. Shire's HGT products include Replagal for the treatment of Fabry disease, a rare genetic disorder resulting from the deficiency of an enzyme involved in the breakdown of fats, and Elaprase for the treatment of Hunter syndrome, a rare genetic disorder interfering with the processing of certain waste substances in the body. For renal disease Shire markets Fosrenol, a phosphate binder for use in end-stage renal failure patients receiving dialysis. For the treatment of myeloproliferative disorders, a group of diseases in which one or more blood cell types are overproduced, Shire markets Agrylin in the United States and Xagrid in Europe. Other products marketed by Shire include Reminyl, CalciChew, Lodine, Solaraze and Vaniqa. The company additionally receives royalties on antiviral products for HIV and Hepatitis B based on certain of its patents. In February 2007, the firm agreed to acquire New River Pharmaceuticals for roughly $2.6 billion. Shire's newest ADHD medication, Vyvanse, was made available in the U.S. in July 2007.

Shire offers its employees business travel accident insurance, flexible spending account arrangements, an employee assistance program, a wellness/work-life program, a health and fitness subsidy, wellness rooms, adoption assistance, parenting leave and educational assistance.

FINANCIALS: Sales and profits are in thousands of dollars—add 000 to get the full amount. 2006 Note: Financial information for 2006 was not available for all companies at press time.

2006 Sales: $	2006 Profits: $	**U.S. Stock Ticker:** SHPGY
2005 Sales: $1,599,300	2005 Profits: $-410,800	**Int'l Ticker:** SHP.L Int'l Exchange: London-LSE
2004 Sales: $1,363,200	2004 Profits: $269,000	Employees: 1,833
2003 Sales: $1,237,101	2003 Profits: $276,051	Fiscal Year Ends: 12/31
2002 Sales: $1,037,300	2002 Profits: $250,600	Parent Company:

SALARIES/BENEFITS:

Pension Plan:	ESOP Stock Plan:	Profit Sharing:	Top Exec. Salary: $1,105,000	Bonus: $1,985,000
Savings Plan: Y	Stock Purch. Plan: Y		Second Exec. Salary: $701,000	Bonus: $971,000

OTHER THOUGHTS:

Apparent Women Officers or Directors: 4
Hot Spot for Advancement for Women/Minorities: Y

LOCATIONS: ("Y" = Yes)

West:	Southwest:	Midwest:	Southeast:	Northeast:	International:
				Y	Y

Note: Financial information, benefits and other data can change quickly and may vary from those stated here.

SHIRE-BIOCHEM INC

www.shire.com

Industry Group Code: 325412 Ranks within this company's industry group: Sales: Profits:

Drugs:		Other:	Clinical:	Computers:	Other:
Discovery:	Y	AgriBio:	Trials/Services:	Hardware:	Specialty Services:
Licensing:		Genomics/Proteomics:	Laboratories:	Software:	Consulting:
Manufacturing:	Y	Tissue Replacement:	Equipment/Supplies:	Arrays:	Blood Collection:
Development:	Y		Research/Development Svcs.:	Database Management:	Drug Delivery:
Generics:			Diagnostics:		Drug Distribution:

TYPES OF BUSINESS:

Pharmaceuticals Discovery & Development
Influenza Vaccine
Drugs-HIV

BRANDS/DIVISIONS/AFFILIATES:

Shire Pharmaceuticals Group
Biochem Pharma
Shire PLC
3TC
Aderall XR
Agrylin
Alertec
Hectorol

CONTACTS: *Note: Officers with more than one job title may be intentionally listed here more than once.*

Joseph Rus, CEO
Joseph Rus, Pres.
Claude Perron, VP/Gen. Mgr.

Phone: 514-787-2300	Fax: 514-787-2427
Toll-Free:	

Address: 2250 Alfred-Nobel Blvd., Ste. 500, Ville Saint-Laurent, QC H4S 2C9 Canada

GROWTH PLANS/SPECIAL FEATURES:

Shire BioChem, Inc. is a Canadian company that focuses on research, development and commercialization of innovative products for the prevention and treatment of human diseases. The company is also in charge of marketing products in Canada. The firm is a subsidiary of Shire PLC, formerly Shire Pharmaceuticals Group, a global specialty pharmaceutical company with a focus on developing and marketing products in the areas of central nervous system, gastrointestinal and renal diseases. The parent company has operations in pharmaceutical markets in the U.S., Canada, the U.K., France, Italy, Spain and Germany, as well as a specialist drug delivery unit in the U.S. Shire PLC's world headquarters are in Basingstroke, UK, with its U.S. headquarters are in Wayne, PA. Shire BioChem currently engages in researching cancer and infectious diseases such as HIV and influenza. The firm's 3TC is the cornerstone of the therapies that are now standard treatments for HIV and AIDS. 3TC is available worldwide and is marketed in partnership with GlaxcoSmithKline. Among other products Shire BioChem currently has on the market in Canada are Aderall XR, for ADHD; Agrylin, treatment of essential thrombocythaemia; Alertec; Amatine; Diastat; Estrace; Fosrenol, reduces serum phosphate in patients that suffer from end-stage renal disease; Hectorol; and Zanaflex.

FINANCIALS: Sales and profits are in thousands of dollars—add 000 to get the full amount. 2006 Note: Financial information for 2006 was not available for all companies at press time.

2006 Sales: $	2006 Profits: $	**U.S. Stock Ticker: Subsidiary**
2005 Sales: $	2005 Profits: $	**Int'l Ticker:** Int'l Exchange:
2004 Sales: $	2004 Profits: $	Employees: 500
2003 Sales: $	2003 Profits: $	Fiscal Year Ends: 12/31
2002 Sales: $	2002 Profits: $	Parent Company: SHIRE PLC

SALARIES/BENEFITS:

Pension Plan:	ESOP Stock Plan:	Profit Sharing:	Top Exec. Salary: $	Bonus: $
Savings Plan: Y	Stock Purch. Plan:		Second Exec. Salary: $	Bonus: $

OTHER THOUGHTS:

Apparent Women Officers or Directors: 2
Hot Spot for Advancement for Women/Minorities: Y

LOCATIONS: ("Y" = Yes)

West:	Southwest:	Midwest:	Southeast:	Northeast:	International:
					Y

SICOR INC

www.sicor.com

Industry Group Code: 325412 Ranks within this company's industry group: Sales: Profits:

Drugs:		Other:		Clinical:		Computers:		Other:	
Discovery:		AgriBio:		Trials/Services:		Hardware:		Specialty Services:	
Licensing:	Y	Genomics/Proteomics:		Laboratories:	Y	Software:		Consulting:	
Manufacturing:	Y	Tissue Replacement:		Equipment/Supplies:		Arrays:		Blood Collection:	
Development:	Y			Research/Development Svcs.:	Y	Database Management:		Drug Delivery:	
Generics:				Diagnostics:				Drug Distribution:	

TYPES OF BUSINESS:

Drugs-Diversified
Bulk Pharmaceutical Ingredients
Contract Manufacturing & Research
Injectable Pharmaceuticals

BRANDS/DIVISIONS/AFFILIATES:

Teva Pharmaceuticals Industries, Ltd.
Sicor Biotech UAB
Sicor Pharmeceuticals, Inc.
Sicor S.p.A.

CONTACTS: Note: Officers with more than one job title may be intentionally listed here more than once.

Marvin Samson, CEO
Marvin Samson, Pres.
David C. Dreyer, Interim CFO
Welsley N. Fach, General Counsel
Donald E. Panoz, Chmn.

Phone: 949-455-4700	Fax:
Toll-Free:	
Address: 17 Hughes, Irvine, CA 92618 US	

GROWTH PLANS/SPECIAL FEATURES:

Sicor, Inc., which is a wholly owned subsidiary of TEVA Pharmaceuticals USA, uses its internal research and development capabilities, together with its operational manufacturing and regulatory experience, to develop a wide range of pharmaceutical products. Sicor markets nearly 20 oncolytic and 30 specialty injectable products. It concentrates on products and technologies that face significant barriers to entering the worldwide market. The company operates through its three subsidiaries. Sicor Pharmaceuticals, Inc. focuses primarily on injectable products used in oncology and anesthesiology. Sicor S.p.A., based in Italy, produces specialty bulk drug substances. Finally, Sicor Biotech UAB, based in Lithuania, provides manufacturing, research and development laboratories. Through its own sales force and its marketing partners, the company also markets over 125 finished dosage products in the U.S., Europe, Latin America and elsewhere. Sicor has concentrated its expansion toward securing marketing approval for new drugs and marketing them first in developing nations, then in Europe and finally in the U.S. The company recently received ANDA (Abbreviated New Drug Application) approval for a number of drugs, pushing its total number to more than 40.

FINANCIALS: Sales and profits are in thousands of dollars—add 000 to get the full amount. 2006 Note: Financial information for 2006 was not available for all companies at press time.

2006 Sales: $	2006 Profits: $	U.S. Stock Ticker: Subsidiary
2005 Sales: $	2005 Profits: $	Int'l Ticker: Int'l Exchange:
2004 Sales: $	2004 Profits: $	Employees: 1,905
2003 Sales: $	2003 Profits: $	Fiscal Year Ends:
2002 Sales: $456,000	2002 Profits: $128,300	Parent Company: TEVA PHARMACEUTICAL INDUSTRIES

SALARIES/BENEFITS:

Pension Plan:	ESOP Stock Plan:	Profit Sharing: Y	Top Exec. Salary: $500,000	Bonus: $463,417
Savings Plan: Y	Stock Purch. Plan: Y		Second Exec. Salary: $368,125	Bonus: $158,036

OTHER THOUGHTS:

Apparent Women Officers or Directors:
Hot Spot for Advancement for Women/Minorities:

LOCATIONS: ("Y" = Yes)

West:	Southwest:	Midwest:	Southeast:	Northeast:	International:
Y					Y

Note: Financial information, benefits and other data can change quickly and may vary from those stated here.

SIGA TECHNOLOGIES INC   www.siga.com

Industry Group Code: 325412 **Ranks within this company's industry group:** Sales: 133   Profits: 75

Drugs:		Other:	Clinical:	Computers:	Other:	
Discovery:	Y	AgriBio:	Trials/Services:	Hardware:	Specialty Services:	
Licensing:		Genomics/Proteomics:	Laboratories:	Software:	Consulting:	
Manufacturing:		Tissue Replacement:	Equipment/Supplies:	Arrays:	Blood Collection:	
Development:	Y		Research/Development Svcs.:	Database Management:	Drug Delivery:	Y
Generics:			Diagnostics:		Drug Distribution:	

TYPES OF BUSINESS:

Drugs-Infectious Diseases
Vaccines
Antibiotics
Biothreat Rapid-Response Therapeutics
Mucosal Drug Delivery

BRANDS/DIVISIONS/AFFILIATES:

SIGA-246
ST-294
ST-193

CONTACTS: *Note: Officers with more than one job title may be intentionally listed here more than once.*

Eric A. Rose, CEO
Thomas N. Konatich, CFO
Dennis E. Hruby, Chief Scientific Officer
Thomas N. Konatich, Treas.
Eric A. Rose, Chmn.

Phone: 212-672-9100	Fax: 212-697-3130
Toll-Free:	
Address: 420 Lexington Ave., Ste. 408, New York, NY 10170 US	

GROWTH PLANS/SPECIAL FEATURES:

SIGA Technologies is a development-stage biotechnology company focused on the discovery, development and commercialization of vaccines, antibiotics and anti-infectives for serious infectious diseases. The major focus of the company's activities is on products for use in defense against biological warfare agents such as Smallpox and Arenaviruses (hemorrhagic fevers). Its lead product, SIGA-246, is an orally administered anti-viral drug that targets the smallpox virus. In December 2005, the FDA accepted SIGA's investigational new drug (IND) application for SIGA-246 and granted the program Fast Track status. In December 2006, the FDA granted Orphan Drug designation to SIGA-246 for the prevention and treatment of smallpox. SIGA's antiviral programs are designed to prevent or limit the replication of the viral pathogen. Its anti-infectives programs are aimed at the increasingly serious problem of drug resistance. The company is also developing a technology for the mucosal delivery of vaccines which may allow the vaccines to activate the immune system at the mucus lined surfaces of the body (the mouth, the nose, the lungs and the gastrointestinal and urogenital tracts), the sites of entry of most infectious agents. As a result of the success of SIGA's efforts to develop products for use against agents of biological warfare, it has not spent significant resources to further the development of its anti-infective and vaccine technologies. Further biological warfare defense products in the company's portfolio include ST-294, an anti-arenavirus drug candidate. The firm's other technology platforms include bacterial commensal vector vaccine delivery systems, surface protein expression systems, bio-threat rapid-response therapeutics and bacterial bio-shielding. The company has collaborations with organizations including the National Institutes of Health and TransTech Pharma. In 2007, the company announced the successful results of a proof of concept guinea pig trial of its lead Lassa fever virus drug, ST-193.

FINANCIALS: Sales and profits are in thousands of dollars—add 000 to get the full amount. 2006 Note: Financial information for 2006 was not available for all companies at press time.

2006 Sales: $7,258	2006 Profits: $-9,899	**U.S. Stock Ticker: SIGA**
2005 Sales: $8,477	2005 Profits: $-2,288	**Int'l Ticker:** Int'l Exchange:
2004 Sales: $1,839	2004 Profits: $-9,373	Employees: 44
2003 Sales: $ 732	2003 Profits: $-5,277	Fiscal Year Ends: 12/31
2002 Sales: $ 300	2002 Profits: $-3,300	Parent Company:

SALARIES/BENEFITS:

Pension Plan: Y	ESOP Stock Plan:	Profit Sharing:	Top Exec. Salary: $233,333	Bonus: $117,500
Savings Plan: Y	Stock Purch. Plan:		Second Exec. Salary: $229,166	Bonus: $174,500

OTHER THOUGHTS:

Apparent Women Officers or Directors:
Hot Spot for Advancement for Women/Minorities:

LOCATIONS: ("Y" = Yes)

West:	Southwest:	Midwest:	Southeast:	Northeast:	International:
Y				Y	

SIGMA ALDRICH CORP

www.sigmaaldrich.com

Industry Group Code: 325000 Ranks within this company's industry group: Sales: 6 Profits: 5

Drugs:	Other:	Clinical:		Computers:		Other:	
Discovery:	AgriBio:	Trials/Services:		Hardware:		Specialty Services:	
Licensing:	Genomics/Proteomics:	Laboratories:		Software:		Consulting:	
Manufacturing:	Tissue Replacement:	Equipment/Supplies:	Y	Arrays:		Blood Collection:	
Development:		Research/Development Svcs.:		Database Management:		Drug Delivery:	
Generics:		Diagnostics:	Y			Drug Distribution:	

TYPES OF BUSINESS:

Chemicals Manufacturing
Biotechnology Equipment
Pharmaceutical Ingredients
Fine Chemicals
Chromatography Products

BRANDS/DIVISIONS/AFFILIATES:

Research Essentials
Research Specialties
Research Biotech
SAFC
Iropharm
Advanced Separation Technologies, Inc.
Epichem Group, Ltd.

CONTACTS: *Note: Officers with more than one job title may be intentionally listed here more than once.*

Jai Nagarkatti, CEO
Jai Nagarkatti, Pres.
Mike Hogan, CFO
Doug Rau, VP-Human Resources
Carl Turza, CIO
Mike Hogan, Chief Admin. Officer/Sec.
Karen Miller, Controller
Giles Cottier, Pres., Research Essentials
Dave Julien, Pres., Research Specialties
Kirk Richter, Treas.
Steven Walton, VP-Quality & Safety
David R. Harvey, Chmn.

Phone: 314-771-5765	**Fax:** 314-771-5757

Toll-Free: 800-521-8956
Address: 3050 Spruce St., St. Louis, MO 63103 US

GROWTH PLANS/SPECIAL FEATURES:

Sigma Aldrich Corp. is a life science and high technology company that develops, manufactures, purchases and distributes a broad range of biochemicals and organic chemicals. The company offers roughly 100,000 chemicals (including 45,000 chemicals manufactured in-house) and 30,000 equipment products used for scientific and genomic research; biotechnology; pharmaceutical development; the diagnosis of disease; and pharmaceutical and high technology manufacturing. Sigma Aldrich operates in four segments: Research Essentials, which sells biological buffers, cell culture reagents, biochemicals, chemicals, solvents and other reagents and kits; Research Specialties, which provides organic chemicals, biochemicals, analytical reagents, chromatography consumables, reference materials and high-purity products; Research Biotech, which supplies immunochemical, molecular biology, cell signaling and neuroscience biochemicals and kits used in biotechnology, genomic, proteomic and other life science research applications; and SAFC (Fine Chemicals), which offers large-scale organic chemicals and biochemicals used in development and production by pharmaceutical, biotechnology, industrial and diagnosis companies. The company operates in 35 countries, selling its products in over 165 countries and servicing over 1 million customers. Customers include commercial laboratories; pharmaceutical and industrial companies; universities; diagnostics, chemical and biotechnology companies and hospitals; non-profit organizations; and governmental institutions. In 2006, Sigma Aldrich acquired Honeywell International's Iropharm unit, a custom chemical synthesis business located in Arklow, Ireland, increasing SAFC's API and pharmaceutical intermediates manufacturing capabilities; and Advanced Separation Technologies, Inc., a chiral chromatography business in NJ. In February 2007, Sigma Aldrich acquired Epichem Group, Ltd., a U.K.-based company, for roughly $60 million.

FINANCIALS: Sales and profits are in thousands of dollars—add 000 to get the full amount. 2006 Note: Financial information for 2006 was not available for all companies at press time.

2006 Sales: $1,797,500	2006 Profits: $276,800	**U.S. Stock Ticker:** SIAL
2005 Sales: $1,666,500	2005 Profits: $258,300	**Int'l Ticker:** Int'l Exchange:
2004 Sales: $1,409,200	2004 Profits: $232,900	Employees: 7,299
2003 Sales: $1,298,146	2003 Profits: $193,102	Fiscal Year Ends: 12/31
2002 Sales: $1,207,000	2002 Profits: $130,700	Parent Company:

SALARIES/BENEFITS:

Pension Plan:	ESOP Stock Plan:	Profit Sharing:	Top Exec. Salary: $600,000	Bonus: $416,874
Savings Plan:	Stock Purch. Plan:		Second Exec. Salary: $430,000	Bonus: $222,955

OTHER THOUGHTS:

Apparent Women Officers or Directors: 1
Hot Spot for Advancement for Women/Minorities:

LOCATIONS: ("Y" = Yes)

West:	Southwest:	Midwest:	Southeast:	Northeast:	International:
		Y			Y

Note: Financial information, benefits and other data can change quickly and may vary from those stated here.

SIMCERE PHARMACEUTICAL GROUP   www.simcere.com

Industry Group Code: 325412 Ranks within this company's industry group: Sales: 62   Profits: 43

Drugs:		Other:		Clinical:	Computers:		Other:	
Discovery:	Y	AgriBio:		Trials/Services:	Hardware:		Specialty Services:	
Licensing:		Genomics/Proteomics:		Laboratories:	Software:		Consulting:	
Manufacturing:	Y	Tissue Replacement:		Equipment/Supplies:	Arrays:		Blood Collection:	
Development:	Y			Research/Development Svcs.:	Database Management:		Drug Delivery:	
Generics:	Y			Diagnostics:			Drug Distribution:	

TYPES OF BUSINESS:

Branded Generic Pharmaceuticals
Drug Research

BRANDS/DIVISIONS/AFFILIATES:

Endu
Bicun
Zailin
Yingtaiqing
Anqi
Biqi
Simcere Medgenn Bio-Pharmaceutical Co., Ltd.

CONTACTS: Note: Officers with more than one job title may be intentionally listed here more than once.

Jinsheng Ren, CEO
Frank Zhigang Zhao, CFO
Yat Ming Chu, VP-Mktg. & Sales
Xiaojin Yin, VP-R&D
Jindong Zhou, VP-Mfg.
Haibo Qian, Sec.
Eric Wang Lam Cheung, VP-Investor Rel.
Jinsheng Ren, Chmn.

Phone: 86-25-8556-6666	Fax: 85-25-8547-1729
Toll-Free:	
Address: No. 699-18 Xuan Wu Ave., Xuan Wu District, Nanjing, Jiangsu Province, 210042 China	

GROWTH PLANS/SPECIAL FEATURES:

Simcere Pharmaceutical Group is a Chinese manufacturer and supplier of branded generic pharmaceuticals. The company manufactures and sells 35 pharmaceutical products and is the exclusive distributor of three additional pharmaceutical products marketed under the firm's brand name. The firm's products are used for treatment of a wide range of diseases such as cancer; cerebrovascular and cardiovascular diseases; infections; arthritis; diarrhea; allergies; respiratory conditions; and urinary conditions. Simcere Pharmaceutical's products include Bicun, an anti-stroke medication and first synthetic free radical scavenger sold in China; Zailin, a generic amoxicillin granule antibiotic; Endu, an anticancer medication and first recombinant human endostatin injection in China; Yingtaiqing, a generic diclofenac sodium sustained-release capsule for inflammation and pain relief; Anqi, a generic amoxicillin with clavulanate potassium antibiotic; and Biqi, an OTC generic smectite powder for diarrhea. In addition, the Chinese FDA has approved the manufacture and sale of 100 other company products. Simcere Pharmaceutical has three manufacturing bases, two nationwide sales and marketing subsidiaries and one research and development center. In February 2006, the company established a joint laboratory for drug discovery with Tsinghua University. In September 2006, the company acquired 80% of Shandong Simcere Medgenn Bio-Pharmaceutical Co., Ltd. for roughly $26 million and launched Endu, also known as Endostar, its patented anti-cancer drug, the first recombinant human endostatin to be successfully commercialized. In April 2007, the firm started to trade on the New York Stock Exchange. In June 2007, Simcere Pharmaceutical entered into an agreement to acquire an additional 10% interest in Shandong Simcere Medgenn Bio-Pharmaceutical Co., Ltd. for roughly $3.5 million.

The company offers its employees benefits that include medical insurance and a housing plan.

FINANCIALS: Sales and profits are in thousands of dollars—add 000 to get the full amount. 2006 Note: Financial information for 2006 was not available for all companies at press time.

2006 Sales: $121,800	2006 Profits: $22,100	U.S. Stock Ticker: SCR
2005 Sales: $91,300	2005 Profits: $12,700	Int'l Ticker: Int'l Exchange:
2004 Sales: $68,100	2004 Profits: $5,600	Employees: 1,838
2003 Sales: $	2003 Profits: $	Fiscal Year Ends:
2002 Sales: $	2002 Profits: $	Parent Company:

SALARIES/BENEFITS:

Pension Plan:	ESOP Stock Plan:	Profit Sharing:	Top Exec. Salary: $	Bonus: $
Savings Plan:	Stock Purch. Plan:		Second Exec. Salary: $	Bonus: $

OTHER THOUGHTS:

Apparent Women Officers or Directors:
Hot Spot for Advancement for Women/Minorities:

LOCATIONS: ("Y" = Yes)

West:	Southwest:	Midwest:	Southeast:	Northeast:	International:
					Y

SKYEPHARMA PLC
www.skyepharma.com

Industry Group Code: 325412A Ranks within this company's industry group: Sales: Profits:

Drugs:		Other:	Clinical:	Computers:	Other:	
Discovery:	Y	AgriBio:	Trials/Services:	Hardware:	Specialty Services:	
Licensing:	Y	Genomics/Proteomics:	Laboratories:	Software:	Consulting:	
Manufacturing:	Y	Tissue Replacement:	Equipment/Supplies:	Arrays:	Blood Collection:	
Development:	Y		Research/Development Svcs.:	Database Management:	Drug Delivery:	Y
Generics:			Diagnostics:		Drug Distribution:	

TYPES OF BUSINESS:

Drug Delivery Systems
Generic Drugs

BRANDS/DIVISIONS/AFFILIATES:

Solaraze
Xatra
Madopar DR
Diclofenac
Cordicant-Uno
Corumo
Foradil Certihaler
Pulmicoart

CONTACTS: Note: Officers with more than one job title may be intentionally listed here more than once.

Frank Condella, CEO
Ken Cunningham, COO
Peter Laing, Dir.-Corp. Comm.
Peter Grant, Dir.-Finance
Argeris Karabelas, Chmn.

Phone: 44-20-7491-1777	**Fax:** 44-20-7491-3338

Toll-Free:

Address: 105 Piccadilly, London, W1J 7NJ UK

GROWTH PLANS/SPECIAL FEATURES:

SkyePharma PLC is a worldwide provider of drug delivery technologies. Its drug delivery products and new drugs are manufactured under the SkyePharma name, as well as under contract to global pharmaceutical companies, including GlaxoSmithKline and Abbott Laboratories. SkyePharma's four platform technologies consist of oral, topical, inhalation and enhanced solubility systems. Dedicated research facilities support these four areas with primary operations in Switzerland, as well as in Lyon, France. The company additionally develops generic and super-generic drugs (drugs that are difficult to replicate, or that are improved versions of off-patent originals). SkyePharma products are divided into oral, inhalable, topical and soluble. The firm's products include oral treatments Xatra for genito-urinary diseases, Madopar DR for Parkinson's disease, Diclofenac for arthritis, Cordicant-Uno for hypertension, and Coruno for angina. Inhalation products in development include Foradil Certihaler, Pulmicort, Formoterol HFA, QAB 149 and Flutiform for asthma; marketed injectable drugs DepoCyt for oncology uses and DepoDur for acute pain. The topical drug under development is Solaraze for Actinic Keratosis; and the soluble drug Triglide is for cardiovascular diseases. SkyePharma, Inc., a U.S.-based subsidiary, develops sustained-release therapeutic products based on DepoFoam, a proprietary injectable drug delivery technology. DepoFoam formulations are capable of releasing drugs over a few days or over several weeks. In March 2007, the company sold its injectable business segment to Paul Capital Partners and Paul Capital Healthcare.

FINANCIALS: Sales and profits are in thousands of dollars—add 000 to get the full amount. 2006 Note: Financial information for 2006 was not available for all companies at press time.

2006 Sales: $	2006 Profits: $	**U.S. Stock Ticker:**
2005 Sales: $105,500	2005 Profits: $-87,500	**Int'l Ticker: SKP** Int'l Exchange: London-LSE
2004 Sales: $119,200	2004 Profits: $-46,600	Employees: 420
2003 Sales: $94,800	2003 Profits: $-77,100	Fiscal Year Ends: 12/31
2002 Sales: $112,100	2002 Profits: $1,800	Parent Company:

SALARIES/BENEFITS:

Pension Plan: Y	ESOP Stock Plan:	Profit Sharing:	Top Exec. Salary: $685,860	Bonus: $288,190
Savings Plan:	Stock Purch. Plan: Y		Second Exec. Salary: $590,870	Bonus: $231,840

OTHER THOUGHTS:

Apparent Women Officers or Directors:
Hot Spot for Advancement for Women/Minorities:

LOCATIONS: ("Y" = Yes)

West:	Southwest:	Midwest:	Southeast:	Northeast:	International:
					Y

SONUS PHARMACEUTICALS   www.sonuspharma.com

Industry Group Code: 325412A Ranks within this company's industry group: Sales: 9   Profits: 13

Drugs:		Other:	Clinical:	Computers:	Other:	
Discovery:	Y	AgriBio:	Trials/Services:	Hardware:	Specialty Services:	
Licensing:		Genomics/Proteomics:	Laboratories:	Software:	Consulting:	
Manufacturing:		Tissue Replacement:	Equipment/Supplies:	Arrays:	Blood Collection:	
Development:	Y		Research/Development Svcs.:	Database Management:	Drug Delivery:	Y
Generics:			Diagnostics:		Drug Distribution:	

TYPES OF BUSINESS:

Drug Delivery Systems
Drugs-Cancer

BRANDS/DIVISIONS/AFFILIATES:

TOCOSOL
TOCOSOL Paclitaxel
TOCOSOL Camptothecin

CONTACTS: Note: Officers with more than one job title may be intentionally listed here more than once.

Michael A. Martino, CEO
Michael A. Martino, Pres.
Alan Fuhrman, CFO/Sr. VP
Tom D'Orazio, VP-Strategic Mktg.
Ingrid Rasch, VP-Human Resources
Lynn Gold, VP-R&D
Neile A. Grayson, VP-Corp. Dev. & Strategic Planning
Pamela Dull, Mgr.-Investor Rel.
Craig Eudy, Controller
Dean R. Kessler, VP-Preclinical Dev.
Elaine Waller, VP-Regulatory Affairs & Quality Assurance
Wayne Rebich, VP-Financial Planning & Bus. Processes
Richard Daifuku, Acting Chief Medical Officer

Phone: 425-487-9500	Fax: 425-489-0626
Toll-Free:	
Address: 22026 20th Ave. SE, Bothell, WA 98021 US	

GROWTH PLANS/SPECIAL FEATURES:

Sonus Pharmaceuticals, Inc. is focused on the development of drugs that may offer improved effectiveness, safety, tolerability and administration for the treatment of cancer and related therapies. The company bases all new development off of its proprietary TOCOSOL technology platform, which has been designed to address the formulation challenges of therapeutic drugs. The technology uses vitamin E and vitamin E derivatives to solubilize, stabilize and formulate drugs with the goal of enhancing their delivery, safety and efficacy. The firm's leading oncology candidate, TOCOSOL Paclitaxel, is a new formulation of paclitaxel, one of the world's most widely prescribed anti-cancer drugs, approved in the U.S. for the treatment of breast, ovarian and non-small cell lung cancers and Kaposi's sarcoma. TOCOSOL Paclitaxel is a ready-to-use, injectable paclitaxel emulsion formulation. TOCOSOL Paclitaxel is in Phase III clinical trials. Its second oncology drug candidate, TOCOSOL Camptothecin Injectable Emulsion, is a novel camptothecin derivative discovered and formulated with Sonus' proprietary TOCOSOL technology, which is currently in Phase I clinical Trials. The company was awarded a patent for TOCOSOL Camptothecin in June 2007. Also in 2007, the company announced that its TOCOSOL Paclitaxel product demonstrated less neuropathy (nerve damage) in preclinical study compared to approved taxane products: Taxotere, Taxol and ABRAXANE.

Sonus Pharmaceuticals offers employees flexible reimbursement accounts; employee assistance programs; tuition reimbursement; and reimbursement of professional and health club dues.

FINANCIALS: Sales and profits are in thousands of dollars—add 000 to get the full amount. 2006 Note: Financial information for 2006 was not available for all companies at press time.

2006 Sales: $22,392	2006 Profits: $-23,551	**U.S. Stock Ticker: SNUS**
2005 Sales: $8,254	2005 Profits: $-21,097	**Int'l Ticker:** Int'l Exchange:
2004 Sales: $	2004 Profits: $-16,311	Employees: 61
2003 Sales: $ 25	2003 Profits: $-10,467	Fiscal Year Ends: 12/31
2002 Sales: $ 25	2002 Profits: $-11,636	Parent Company:

SALARIES/BENEFITS:

Pension Plan:	ESOP Stock Plan:	Profit Sharing:	Top Exec. Salary: $374,040	Bonus: $
Savings Plan: Y	Stock Purch. Plan: Y		Second Exec. Salary: $245,699	Bonus: $

OTHER THOUGHTS:

Apparent Women Officers or Directors: 4
Hot Spot for Advancement for Women/Minorities: Y

LOCATIONS: ("Y" = Yes)

West:	Southwest:	Midwest:	Southeast:	Northeast:	International:
Y					

SPECIALTY LABORATORIES INC   www.specialtylabs.com

Industry Group Code: 621511 Ranks within this company's industry group: Sales: Profits:

Drugs:	Other:	Clinical:		Computers:		Other:	
Discovery:	AgriBio:	Trials/Services:		Hardware:		Specialty Services:	
Licensing:	Genomics/Proteomics:	Laboratories:	Y	Software:		Consulting:	
Manufacturing:	Tissue Replacement:	Equipment/Supplies:		Arrays:		Blood Collection:	
Development:		Research/Development Svcs.:		Database Management:		Drug Delivery:	
Generics:		Diagnostics:	Y			Drug Distribution:	

TYPES OF BUSINESS:
Clinical Laboratory Tests
Assays

BRANDS/DIVISIONS/AFFILIATES:
AmeriPath, Inc.
DataPassport
DataPassport MD

CONTACTS: *Note: Officers with more than one job title may be intentionally listed here more than once.*

Phone: 661-799-6543	Fax: 661-799-6634
Toll-Free: 800-421-7110	
Address: 27027 Tourney Rd., Valencia, CA 91355 US	

GROWTH PLANS/SPECIAL FEATURES:

Specialty Laboratories, Inc. (SL), a subsidiary of AmeriPath, Inc., is a research-based clinical laboratory, predominantly focused on developing and performing esoteric clinical laboratory tests, referred to as assays. The firm offers a comprehensive menu of more than 2,500 assays, many of which it developed through internal research and development efforts and which are used to diagnose, evaluate and monitor patients in the areas of endocrinology; genetics; infectious diseases; neurology; pediatrics; urology; allergy and immunology; cardiology and coagulation; hepatology; microbiology; oncology; rheumatology; women's health; dermatopathology; gastroenterology; nephrology; pathology; and toxicology. Assays include procedures in the areas of molecular diagnostics, protein chemistry, cellular immunology and advanced microbiology. Commonly ordered assays include viral and bacterial detection and drug therapy monitoring assays; autoimmune panels; and complex cancer evaluations. In addition, SL owns proprietary information technology that accelerates and automates test ordering and results reporting with customers. Current information technology products include DataPassport client interface module, designed to take advantage of Internet-based technologies; and DataPassportMD, a web-based laboratory order entry and resulting system. The company's primary customers are hospitals, independent clinical laboratories and physicians.

The company offers its employees health and vision plans; a 401(k) plan; short- and long-term disability plans; a health and dependent care account; an employee assistance program; and an education reimbursement program.

FINANCIALS: Sales and profits are in thousands of dollars—add 000 to get the full amount. 2006 Note: Financial information for 2006 was not available for all companies at press time.

2006 Sales: $	2006 Profits: $	U.S. Stock Ticker: Subsidiary
2005 Sales: $	2005 Profits: $	Int'l Ticker: Int'l Exchange:
2004 Sales: $134,803	2004 Profits: $-12,950	Employees: 689
2003 Sales: $119,653	2003 Profits: $-6,361	Fiscal Year Ends: 12/31
2002 Sales: $140,200	2002 Profits: $-13,400	Parent Company: AMERIPATH INC

SALARIES/BENEFITS:

Pension Plan:	ESOP Stock Plan:	Profit Sharing:	Top Exec. Salary: $422,908	Bonus: $
Savings Plan: Y	Stock Purch. Plan:		Second Exec. Salary: $262,000	Bonus: $

OTHER THOUGHTS:
Apparent Women Officers or Directors: 3
Hot Spot for Advancement for Women/Minorities: Y

LOCATIONS: ("Y" = Yes)

West:	Southwest:	Midwest:	Southeast:	Northeast:	International:
Y					

Note: Financial information, benefits and other data can change quickly and may vary from those stated here.

SPECTRAL DIAGNOSTICS INC

www.spectraldx.com

Industry Group Code: 325413 Ranks within this company's industry group: Sales: 25 Profits:

Drugs:	Other:	Clinical:		Computers:		Other:
Discovery:	AgriBio:	Trials/Services:		Hardware:		Specialty Services:
Licensing:	Genomics/Proteomics:	Laboratories:		Software:		Consulting:
Manufacturing:	Tissue Replacement:	Equipment/Supplies:	Y	Arrays:		Blood Collection:
Development:		Research/Development Svcs.:		Database Management:		Drug Delivery:
Generics:		Diagnostics:	Y			Drug Distribution:

TYPES OF BUSINESS:

Medical Diagnostics Products
Cardiac Diagnostics
Sepsis Diagnostics
Antibodies
Plasma Separation Membranes

BRANDS/DIVISIONS/AFFILIATES:

CardiacSTATus
Endotoxin Activity Assays
IDx, Inc
Technology Partnerships Canada
Industry Canada

CONTACTS: Note: Officers with more than one job title may be intentionally listed here more than once.

Paul M. Walker, CEO
Paul M. Walker, Pres.
Tony Businskas, CFO
Mark Pride, VP-Mktg. & Sales
Nisar A. Shaikh, Sr. VP-Science Affairs
Robert Verhagen, VP-Bus. Dev.

Phone: 416-626-3233	Fax: 416-626-7383
Toll-Free: 888-426-4264	
Address: 135 The West Mall, Toronto, ON M9C 1C2 Canada	

GROWTH PLANS/SPECIAL FEATURES:

Spectral Diagnostics, Inc. is a medical diagnostics company that focuses on developing, manufacturing and selling rapid assays that direct cost-effective patient management. Its activities are primarily involved with diagnostics solutions for sepsis and West Nile Virus. Spectral makes Endotoxin Activity Assays (EAA), rapid, whole blood tests used to identify patients who are at risk for sepsis (infection). With the development of the EAA test as a rapid indicator of endotoxin, physicians are assisted in stratifying patients into those who are at low risk of severe sepsis and those who require directed anti-sepsis or anti-infection therapy. Spectral also has an in house antibody generation and assay development team. Over 90% of the antibodies used the commercial kits are generated in house. In addition, Spectral supplies membranes for plasma separation and collection, eliminating the need for centrifuges due to capillary action between separating and collecting membranes. The company's latest product, its RapidWN West Nile Virus rapid diagnostic test, diagnoses patients potentially infected with West Nile Virus in less than an hour. The product received marketing approval for the product from the FDA in December 2006. The company also produces high performance polyclonal and monoclonal antibodies, and recombinant proteins. Spectral has successfully engineered and produced over 30 antigens in bacteria and has created several patented novel molecules.

FINANCIALS: Sales and profits are in thousands of dollars—add 000 to get the full amount. 2006 Note: Financial information for 2006 was not available for all companies at press time.

2006 Sales: $6,200	2006 Profits: $	U.S. Stock Ticker:
2005 Sales: $8,200	2005 Profits: $	Int'l Ticker: SDI Int'l Exchange: Toronto-TSX
2004 Sales: $9,000	2004 Profits: $-1,600	Employees: 70
2003 Sales: $9,800	2003 Profits: $-3,000	Fiscal Year Ends: 3/31
2002 Sales: $8,300	2002 Profits: $3,400	Parent Company:

SALARIES/BENEFITS:

Pension Plan:	ESOP Stock Plan:	Profit Sharing:	Top Exec. Salary: $	Bonus: $
Savings Plan: Y	Stock Purch. Plan: Y		Second Exec. Salary: $	Bonus: $

OTHER THOUGHTS:

Apparent Women Officers or Directors:
Hot Spot for Advancement for Women/Minorities:

LOCATIONS: ("Y" = Yes)

West:	Southwest:	Midwest:	Southeast:	Northeast:	International:
Y		Y		Y	Y

SPECTRUM PHARMACEUTICALS INC
www.spectrumpharm.com
Industry Group Code: 325412 Ranks within this company's industry group: Sales: 140   Profits: 111

Drugs:		Other:		Clinical:		Computers:		Other:	
Discovery:	Y	AgriBio:		Trials/Services:		Hardware:		Specialty Services:	
Licensing:	Y	Genomics/Proteomics:		Laboratories:		Software:		Consulting:	
Manufacturing:	Y	Tissue Replacement:		Equipment/Supplies:		Arrays:		Blood Collection:	
Development:	Y			Research/Development Svcs.:		Database Management:		Drug Delivery:	
Generics:	Y			Diagnostics:				Drug Distribution:	

TYPES OF BUSINESS:
Oncology & Urology Drugs
Cancer Treatments
Generic Drugs

BRANDS/DIVISIONS/AFFILIATES:
Satraplatin
Levofolonic
Sumatriptan
EOquin
Ortataxel
Ozarelix

CONTACTS: *Note: Officers with more than one job title may be intentionally listed here more than once.*
Rajesh Shrotriya, CEO
Rajesh C. Shrotriya, Pres.
Luigi Lenaz, Chief Scientific Officer
William Pedranti, General Counsel/VP
Russell L. Skibsted, Chief Business Officer/Sr. VP
Shyam Kumaria, VP-Finance
Chuck O. Chuckwumerije, VP-Global Quality Oper.
Daniel Pertschuk, VP-Medical Affairs
Ashok Gore, Sr. VP-Regulatory Compliance/Pharmaceutical Oper.
George Uy, VP-Mktg., Oncology & Urology
Rajesh Shrotriya, Chmn.

Phone: 949-788-6700	Fax: 949-788-6706
Toll-Free:	
Address: 157 Technology Dr., Irvine, CA 92618 US	

GROWTH PLANS/SPECIAL FEATURES:
Spectrum Pharmaceuticals, Inc. is a biopharmaceutical company that acquires and advances a diversified portfolio of drug candidates, with a focus on oncology, urology and other critical health challenges. The firm addresses both proprietary drug development as well as generic drug production. The company currently has ten drugs in development, including five in late stage clinical development. Products include Satraplatin, for which the firm collaborated with GBC Biotech AG, for the treatment of hormone refractory prostrate cancer; levofolonic acid, for which the company submitted a new drug application (NDA) with FDA, for treatment of colorectal cancer; sumatriptan injection, a generic form of GSK's Imitrex injection, for which the firm submitted an abbreviated NDA, for the treatment of migraines; and EOquin, which completed a Phase II clinical trial, for the treatment of non-invasive bladder cancer. Spectrum has various drugs in preclinical development for the treatment of such diseases as hyperphosphatemia in end-stage renal disease and chemotherapy induced neuropathy. The company received approval for three drugs: ciprofloxacin tablets, carboplatin liquid injection and fluconazole tablets. In April 2006, Spectrum acquired Targent's entire oncology drug asset portfolio, including levofolinic acid, which the company intends as a treatment for osteogenic sarcoma and colorectal cancer. In January 2007, the company initiated a Phase IIb clinical trial for ozarelix in patients with benign prostrate hypertrophy. In May 2007, the firm announced the initiation of a Phase III clinical trial for EOquin, for patients with non-invasive bladder cancer. In July 2007, Spectrum acquired from Indena S.p.A. worldwide rights to ortataxel, a Phase II third generation taxane classified as a new chemical entity that demonstrated clinical activity in taxane-refractory tumors. That same month, the company withdrew its Satraplatin NDA for accelerated approval from the FDA. The firm plans on resubmitting the drug with additional survival analysis of the SPARC trial.

FINANCIALS: Sales and profits are in thousands of dollars—add 000 to get the full amount. 2006 Note: Financial information for 2006 was not available for all companies at press time.

2006 Sales: $5,673	2006 Profits: $-23,284	U.S. Stock Ticker: SPPI
2005 Sales: $ 577	2005 Profits: $-18,642	Int'l Ticker: Int'l Exchange:
2004 Sales: $ 258	2004 Profits: $-12,286	Employees: 50
2003 Sales: $1,000	2003 Profits: $-10,390	Fiscal Year Ends: 12/31
2002 Sales: $2,400	2002 Profits: $-17,600	Parent Company:

SALARIES/BENEFITS:
Pension Plan:	ESOP Stock Plan:	Profit Sharing:	Top Exec. Salary: $500,000	Bonus: $250,000
Savings Plan: Y	Stock Purch. Plan:		Second Exec. Salary: $350,000	Bonus: $100,000

OTHER THOUGHTS:
Apparent Women Officers or Directors:
Hot Spot for Advancement for Women/Minorities:

LOCATIONS: ("Y" = Yes)
West:	Southwest:	Midwest:	Southeast:	Northeast:	International:
Y					

STEMCELLS INC

www.stemcellsinc.com

Industry Group Code: 325414 Ranks within this company's industry group: Sales: 11 Profits: 8

Drugs:		Other:		Clinical:	Computers:		Other:	
Discovery:	Y	AgriBio:		Trials/Services:	Hardware:		Specialty Services:	
Licensing:		Genomics/Proteomics:	Y	Laboratories:	Software:		Consulting:	
Manufacturing:		Tissue Replacement:	Y	Equipment/Supplies:	Arrays:		Blood Collection:	
Development:	Y			Research/Development Svcs.:	Database Management:		Drug Delivery:	
Generics:				Diagnostics:			Drug Distribution:	

TYPES OF BUSINESS:

Cell-Based Therapeutics

BRANDS/DIVISIONS/AFFILIATES:

HuCNS-SC

CONTACTS: Note: Officers with more than one job title may be intentionally listed here more than once.

Martin McGlynn, CEO
Ann Tsukamoto, COO
Martin McGlynn, Pres.
Rodney Young, CFO
Rodney Young, VP-Admin.
Ken Stratton, General Counsel
Rodney Young, VP-Finance
Stephen Huhn, VP/Head-Neural Program
Elizabeth Leininger, VP-Regulatory Affairs & Quality Assurance
Maria Millian, VP/Head-Liver Program
Nobuko Uchida, VP-Stem Cell Biology
John J. Schwartz, Chmn.

Phone: 650-475-3100	Fax: 650-475-3101
Toll-Free:	
Address: 3155 Porter Dr., Palo Alto, CA 94304 US	

GROWTH PLANS/SPECIAL FEATURES:

StemCells, Inc. is focused on the discovery and development of cell-based therapeutics to treat damage to, or degeneration of, major organ systems including the central nervous system, liver and pancreas. The company seeks to identify and purify rare stem and progenitor cells; expand and bank them as transplantable cells; and then develop therapeutic agents. The firm uses cells derived from donated fetal or adult tissue sources. StemCells is currently conducting a Phase I clinical trial to evaluate the safety and efficacy of its lead product, the human neural stem cell (HuCNS-SC), as a treatment for infantile and late infantile neuronal ceroid lipofuscinosis, two forms of a group of disorders often referred to as Batten disease. In addition, the company identified a population of cells derived from liver tissue, the human liver engrafting cells (hLEC), which when transplanted into a mouse model of liver degeneration show long-term engraftment, differentiate into human hepatocytes, secrete human hepatic proteins and form structural elements of the liver. Based on this data, the firm is developing the hLEC for potential therapeutic applications to liver diseases. Liver stem cells may be useful in the treatment of diseases such as hepatitis, liver failure, blood-clotting disorder, cirrhosis of the liver and liver cancer. StemCells has also identified a candidate stem cell of the pancreas and this program is at the research stage. The company's research and development programs focus on isolating and developing stem and progenitor cells that can serve as a basis for protecting or replacing diseased or injured cells. The firm owns 48 issued U.S. patents and 130 foreign patents. In June 2007, StemCells announced that the Phase I clinical trial of the HuCNS-SC product candidate would proceed to the high-dose cohort after no safety issues were identified that would preclude advancing the trial to the next dose level.

FINANCIALS: Sales and profits are in thousands of dollars—add 000 to get the full amount. 2006 Note: Financial information for 2006 was not available for all companies at press time.

2006 Sales: $ 93		2006 Profits: $-18,948		**U.S. Stock Ticker: STEM**	
2005 Sales: $ 206		2005 Profits: $-11,738		**Int'l Ticker:** Int'l Exchange:	
2004 Sales: $ 141		2004 Profits: $-15,330		Employees: 46	
2003 Sales: $ 273		2003 Profits: $-12,291		Fiscal Year Ends: 12/31	
2002 Sales: $ 400		2002 Profits: $-10,400		Parent Company:	

SALARIES/BENEFITS:

Pension Plan:	ESOP Stock Plan:	Profit Sharing:	Top Exec. Salary: $357,115	Bonus: $90,720
Savings Plan:	Stock Purch. Plan:		Second Exec. Salary: $270,192	Bonus: $50,000

OTHER THOUGHTS:

Apparent Women Officers or Directors: 4
Hot Spot for Advancement for Women/Minorities: Y

LOCATIONS: ("Y" = Yes)

West:	Southwest:	Midwest:	Southeast:	Northeast:	International:
Y					

STIEFEL LABORATORIES INC   www.stiefel.com

Industry Group Code: 325412 Ranks within this company's industry group: Sales: Profits:

Drugs:		Other:	Clinical:	Computers:	Other:	
Discovery:		AgriBio:	Trials/Services:	Hardware:	Specialty Services:	
Licensing:		Genomics/Proteomics:	Laboratories:	Software:	Consulting:	
Manufacturing:	Y	Tissue Replacement:	Equipment/Supplies:	Arrays:	Blood Collection:	
Development:	Y		Research/Development Svcs.:	Database Management:	Drug Delivery:	Y
Generics:			Diagnostics:		Drug Distribution:	

TYPES OF BUSINESS:
Dermatological & Skin Care Products

BRANDS/DIVISIONS/AFFILIATES:
Evoclin
Duac
Physiogel
DermaVite
Oilatum
LactiCare
Olux

CONTACTS: *Note: Officers with more than one job title may be intentionally listed here more than once.*
Charles W. Stiefel, CEO
Jeffrey S. Thompson, COO
Charles W. Stiefel, Pres.
Teresita L. Brunken, CFO
William Eaglstein, VP-R&D Pipeline Dev. & Scientific Affairs
Amy E. Button, Media Contact
Larry Staubach, VP-Global Medical Affairs & Pharmacovigilance
Charles W. Stiefel, Chmn.

Phone: 305-443-3800	Fax: 305-443-3467
Toll-Free:	
Address: 255 Alhambra Cir., Coral Gables, FL 33134 US	

GROWTH PLANS/SPECIAL FEATURES:
Stiefel Laboratories, Inc. is a specialized pharmaceutical company that focuses on the advancement of dermatology and skin care. The company's products treat a wide range of dermatological ailments including acne, psoriasis, fungal infections, eczema, dry skin, oily skin, rosacea, seborrhea, pruritus and sun damaged and aging skin. The firm's products include Evoclin and Duac, acne skin care products; Physiogel, a skin and hair product; Olux, used for scalp treatment; and LactiCare emollient moisturizers for dry skin. In addition, Stiefel Laboratories offer biopsy skin punch equipment, DermaVite dietary supplement and Oilatum mild unscented soap. The company has subsidiaries in more than 30 countries and its products are available in more than 100 countries. In December 2006, Stiefel Laboratories acquired Connectics Corp., a specialty pharmaceutical company that develops and commercializes products focused exclusively on the treatment of dermatological conditions, for roughly $640 million.

FINANCIALS: Sales and profits are in thousands of dollars—add 000 to get the full amount. 2006 Note: Financial information for 2006 was not available for all companies at press time.

2006 Sales: $	2006 Profits: $	U.S. Stock Ticker: Private
2005 Sales: $184,264	2005 Profits: $33,958	Int'l Ticker: Int'l Exchange:
2004 Sales: $144,355	2004 Profits: $19,015	Employees: 394
2003 Sales: $75,331	2003 Profits: $-4,100	Fiscal Year Ends:
2002 Sales: $52,800	2002 Profits: $-16,600	Parent Company:

SALARIES/BENEFITS:
Pension Plan:	ESOP Stock Plan:	Profit Sharing:	Top Exec. Salary: $530,000	Bonus: $325,000
Savings Plan:	Stock Purch. Plan:		Second Exec. Salary: $393,083	Bonus: $190,000

OTHER THOUGHTS:
Apparent Women Officers or Directors: 2
Hot Spot for Advancement for Women/Minorities: Y

LOCATIONS: ("Y" = Yes)
West:	Southwest:	Midwest:	Southeast:	Northeast:	International:
Y			Y	Y	Y

STRATAGENE CORP

www.stratagene.com

Industry Group Code: 325413 Ranks within this company's industry group: Sales: 13   Profits: 15

Drugs:	Other:		Clinical:		Computers:		Other:	
Discovery:	AgriBio:		Trials/Services:		Hardware:	Y	Specialty Services:	
Licensing:	Genomics/Proteomics:	Y	Laboratories:		Software:	Y	Consulting:	
Manufacturing:	Tissue Replacement:		Equipment/Supplies:	Y	Arrays:	Y	Blood Collection:	
Development:			Research/Development Svcs.:		Database Management:	Y	Drug Delivery:	
Generics:			Diagnostics:	Y			Drug Distribution:	

TYPES OF BUSINESS:

Laboratory Instrument Manufacturing
Gene-Sequencing Software
Clinical Diagnostic Equipment

BRANDS/DIVISIONS/AFFILIATES:

KOVA Microscopic Urinalysis System
PicoFuge
StrataCooler
Absolutely RNA
GeneMorph II
ArrayAssist
PathwayAssist
Hycor

CONTACTS: *Note: Officers with more than one job title may be intentionally listed here more than once.*

Joseph A. Sorge, CEO
Steve Martin, CFO
Ronni L. Sherman, General Counsel
Nelson F. Thune, Sr. VP-Oper.
Nelson F. Thune, Gen. Mgr., Hycor Biomedical Inc.
Joseph A. Sorge, Chmn.
John R. Pouk, VP-Int'l Oper.

Phone: 858-535-5400	Fax: 858-535-0071
Toll-Free: 800-424-5444	
Address: 1011 N. Torrey Pines Rd., La Jolla, CA 92037 US	

GROWTH PLANS/SPECIAL FEATURES:

Stratagene Corp., a subsidiary of Agilent Technologies, Inc., develops, manufactures and markets lab instruments and materials for the molecular biology, genomics, proteomics, drug discovery and toxicology fields. The firm's products are divided between the Research Supplies and Clinical Diagnostics segments. Within Research Supplies, the company's gene analysis products, which help researchers to study gene activity, genetic code sequences and gene dosage, account for about a third of total company revenue. Also within the segment are protein analysis and gene discovery products. Protein analysis tools make the pathways within and between cells visible to create, purify and measure proteins; while gene discovery products assist researchers in DNA or RNA discovery through purification and amplification. In the Clinical Diagnostics segment, Stratagene offers urinalysis through the KOVA Microscopic Urinalysis System, which provides laboratories tools such as plastic containers, tubes and pipettes and microscopic slides for conducting bio-chemical urinalysis. Stratagene offers an allergy diagnostic product line, with a complete offering of radioimmunoassay and enzymatic immunoassay procedures for specific allergies, as well as traditional screening tests. The firm's autoimmune diagnostic product line includes tests used to diagnose and monitor autoimmune disorders such as rheumatoid arthritis and systemic lupus erythematosus. Stratagene markets products in over 60 countries worldwide, with manufacturing facilities in California, Texas and Edinburgh, Scotland. The firm's goal is to capture at least 20% of the molecular diagnostic market by 2010. In June 2007, the firm was acquired by Agilent Technologies, Inc. for $245.5 million. Stratagene will run as a separate division of Agilent.

StrataGene offers employees an on-site workout facility and credit union membership.

FINANCIALS: Sales and profits are in thousands of dollars—add 000 to get the full amount. 2006 Note: Financial information for 2006 was not available for all companies at press time.

2006 Sales: $95,557	2006 Profits: $ 59	**U.S. Stock Ticker: Subsidiary**
2005 Sales: $130,285	2005 Profits: $7,788	**Int'l Ticker:** Int'l Exchange:
2004 Sales: $84,813	2004 Profits: $7,438	Employees: 453
2003 Sales: $69,703	2003 Profits: $3,252	Fiscal Year Ends: 12/31
2002 Sales: $64,000	2002 Profits: $ 100	Parent Company: AGILENT TECHNOLOGIES

SALARIES/BENEFITS:

Pension Plan:	ESOP Stock Plan:	Profit Sharing:	Top Exec. Salary: $467,608	Bonus: $139,933
Savings Plan: Y	Stock Purch. Plan: Y		Second Exec. Salary: $328,884	Bonus: $98,514

OTHER THOUGHTS:

Apparent Women Officers or Directors: 1
Hot Spot for Advancement for Women/Minorities:

LOCATIONS: ("Y" = Yes)

West:	Southwest:	Midwest:	Southeast:	Northeast:	International:
Y	Y				Y

SUPERGEN INC   www.supergen.com

Industry Group Code: 325412 Ranks within this company's industry group: Sales: 87   Profits: 93

Drugs:		Other:		Clinical:	Computers:	Other:
Discovery:		AgriBio:		Trials/Services:	Hardware:	Specialty Services:
Licensing:	Y	Genomics/Proteomics:		Laboratories:	Software:	Consulting:
Manufacturing:		Tissue Replacement:		Equipment/Supplies:	Arrays:	Blood Collection:
Development:	Y			Research/Development Svcs.:	Database Management:	Drug Delivery:
Generics:				Diagnostics:		Drug Distribution:

TYPES OF BUSINESS:

Pharmaceuticals Acquisition & Development
Oncology Drugs

BRANDS/DIVISIONS/AFFILIATES:

Nipent
Mitomycin
Orathecin
Surface Safe

CONTACTS: *Note: Officers with more than one job title may be intentionally listed here more than once.*

James S. Manuso, CEO
James S. Manuso, Pres.
Michael Molkentin, CFO
David Bearss, Chief Scientist
Sanjeev Redkar, VP-Mfg. & Pre-Clinical Dev.
Michael Molkentin, Corp. Sec.
Timothy L. Enns, Sr. VP-Bus. Dev.
Timothy L. Enns, Sr. VP-Corp. Comm.
Mary M. Vegh, Mgr.-Investor Rel.
Audrey F. Jakubowski, Chief Regulatory & Quality Officer
Gregory Berk, Chief Medical Officer
Wayne Davis, Sr. VP-Clinical Research
Michael McCullar, VP-Drug Discovery Oper.
James S. Manuso, Chmn.

Phone: 925-560-0100	Fax: 925-560-0101

Toll-Free:

Address: 4140 Dublin Blvd., Ste. 200, Dublin, CA 94568 US

GROWTH PLANS/SPECIAL FEATURES:

SuperGen Inc. develops and commercializes pharmaceutical products intended to treat life-threatening diseases, particularly cancer. The company's strategy is to identify, acquire and develop pharmaceutical products that are in the later stages of development. SuperGen possesses a large portfolio of proprietary oncological drugs that treat a variety of solid tumors and hematological malignancies. Dacogen is a therapeutic product that decreases the amount of methylation, the replacement of hydrogen with methyl, at certain DNA sites to assist patients with myeldysplastic syndrome. The drug is currently under review for marketing in Europe, through the company's subsidiary EuroGen Pharmaceuticals, and is marketed in the U.S. through MGI Pharma. Products in preclinical development include MP470 tyrosine kinase inhibitor/rad 51 supressor, which is just enterering Phase I clinical trials and MP529 aurora A kinase inhibitor in preclinical study. The company also has multiple other candidates in its pipeline that have not yet reached preclinical stages. In January 2006, the firm acquired the assets of Montigen Pharmaceuticals, which includes its research and development team, a proprietary drug discovery platform known as Climb, and late-stage pre-clinical compounds targeting aurora-A kinase and members of the tyrosine kinase receptor family. In June 2006, SuperGen sold its oncology products Nipent, which is a chemotherapy agent used in the treatment of hairy cell leukemia, and SurfaceSafe, which is a patented two-step application kit used to clean and inactivate anti-cancer drugs on chemotherapy work surfaces, for $34 million. Also in 2006, the product Orathecin, an oral chemotherapy compound classified as a topoisomerase I inhibitor that causes single-strand breaks in the DNA of rapidly dividing tumor cells, was withdrawn from the regulatory review process.

The firm provides employees with benefits such as dental and vision coverage, as well as a voluntary cancer and accident insurance plan.

FINANCIALS: Sales and profits are in thousands of dollars—add 000 to get the full amount. 2006 Note: Financial information for 2006 was not available for all companies at press time.

2006 Sales: $38,083	2006 Profits: $-16,487	**U.S. Stock Ticker:** SUPG
2005 Sales: $30,169	2005 Profits: $-14,482	**Int'l Ticker:** Int'l Exchange:
2004 Sales: $31,993	2004 Profits: $-46,860	Employees: 89
2003 Sales: $11,494	2003 Profits: $-53,470	Fiscal Year Ends: 12/31
2002 Sales: $15,300	2002 Profits: $-49,500	Parent Company:

SALARIES/BENEFITS:

Pension Plan:	ESOP Stock Plan:	Profit Sharing:	Top Exec. Salary: $459,242	Bonus: $350,000
Savings Plan: Y	Stock Purch. Plan: Y		Second Exec. Salary: $397,098	Bonus: $113,516

OTHER THOUGHTS:

Apparent Women Officers or Directors: 2
Hot Spot for Advancement for Women/Minorities: Y

LOCATIONS: ("Y" = Yes)

West:	Southwest:	Midwest:	Southeast:	Northeast:	International:
Y					

SYNBIOTICS CORP

www.synbiotics.com

Industry Group Code: 325412B Ranks within this company's industry group: Sales: Profits:

Drugs:	Other:	Clinical:	Computers:	Other:
Discovery:	AgriBio:	Trials/Services:	Hardware:	Specialty Services: Y
Licensing:	Genomics/Proteomics:	Laboratories:	Software: Y	Consulting:
Manufacturing: Y	Tissue Replacement:	Equipment/Supplies: Y	Arrays:	Blood Collection:
Development: Y		Research/Development Svcs.:	Database Management:	Drug Delivery:
Generics:		Diagnostics: Y		Drug Distribution:

TYPES OF BUSINESS:

Veterinary Products Manufacturing
Animal Health Diagnostics
Animal Pregnancy Tests
Breeding Services
Flock Management Software

BRANDS/DIVISIONS/AFFILIATES:

Synbiotics Europe
Fungassay
Ovassay
ViraCHEK
WITNESS
Profile
Tuberculin PPD
Flu Detect

CONTACTS: Note: Officers with more than one job title may be intentionally listed here more than once.

Paul R. Hays, CEO
Paul R. Hays, Pres.
Keith A. Butler, CFO/VP
B. Kent Luther, VP-Mktg. & Sales
Mark Mellencamp, VP-R&D
Clifford J. Frank, VP-Strategic Projects

Phone:	Fax: 816-464-3521
Toll-Free: 800-228-4305	
Address: 12200 NW Ambassador Dr., Ste. 101, Kansas City, MO 64163 US	

GROWTH PLANS/SPECIAL FEATURES:

Synbiotics Corp. develops, manufactures and markets animal health products and services to veterinarians and breeders. The firm produces products for poultry, cows, swine, dogs, horses, cats and non-human primates, and is able to treat animal coagulation, dermatophytes, gastrointestinal parasites, tuberculosis, heartworm, parvovirus and arthritis. Synbiotics' brand names include Fungassay for dermatophyte infections; Ovassay for identifying gastrointestinal parasites; Tuberculin PPD (purified protein derivative), an intradermal test for mammalian tuberculosis in cattle, goats, swine and non-human primates; ViraCHEK, a coronavirus identifier; and WITNESS, the company's highly protected heartworm identification kit. Synbiotics also markets domestic animal pregnancy tests and the Profile flock management software. The company offers canine reproduction services that include a referral network and a freezing center that allows breeders to conduct long-distance breeding. In May 2007, the canine freezing center relocated to Kansas City, MO to a building which utilizes state-of-the-art cryogenic storage technology. The firm's subsidiary, Synbiotics Europe, based in Lyon, France, offers the same products and services to the European Union. Synbiotics manufactures most of its products at its facilities located in San Diego, California and Lyon, France. However, in August 2006 the company began to relocate its San Diego operations to a new location in Kansas City, MO. In November 2006, the company's product Flu Detect, a rapid antigen diagnostic for the detection of avian flu, was approved by the USDA.

FINANCIALS: Sales and profits are in thousands of dollars—add 000 to get the full amount. 2006 Note: Financial information for 2006 was not available for all companies at press time.

2006 Sales: $	2006 Profits: $	U.S. Stock Ticker: SYNB.PK
2005 Sales: $	2005 Profits: $	Int'l Ticker: Int'l Exchange:
2004 Sales: $19,219	2004 Profits: $- 647	Employees: 93
2003 Sales: $	2003 Profits: $	Fiscal Year Ends: 12/31
2002 Sales: $21,671	2002 Profits: $-14,401	Parent Company:

SALARIES/BENEFITS:

Pension Plan:	ESOP Stock Plan:	Profit Sharing:	Top Exec. Salary: $250,000	Bonus: $20,000
Savings Plan:	Stock Purch. Plan:		Second Exec. Salary: $188,700	Bonus: $

OTHER THOUGHTS:

Apparent Women Officers or Directors:
Hot Spot for Advancement for Women/Minorities:

LOCATIONS: ("Y" = Yes)

West:	Southwest:	Midwest:	Southeast:	Northeast:	International:
Y				Y	Y

SYNGENTA AG
www.syngenta.com

Industry Group Code: 115112 Ranks within this company's industry group: Sales: 1   Profits: 2

Drugs:	Other:		Clinical:	Computers:	Other:
Discovery:	AgriBio:	Y	Trials/Services:	Hardware:	Specialty Services:
Licensing:	Genomics/Proteomics:		Laboratories:	Software:	Consulting:
Manufacturing:	Tissue Replacement:		Equipment/Supplies:	Arrays:	Blood Collection:
Development:			Research/Development Svcs.:	Database Management:	Drug Delivery:
Generics:			Diagnostics:		Drug Distribution:

TYPES OF BUSINESS:
Agricultural Biotechnology Products & Chemicals Manufacturing
Crop Protection Products
Seeds

BRANDS/DIVISIONS/AFFILIATES:
Dual Gold
Bicept Magnum
Acanto
Score
Syngenta Biotechnology, Inc.
Conrad Fafard, Inc.
Emergent Genetics Vegetable A/S
Fischer Group

CONTACTS: *Note: Officers with more than one job title may be intentionally listed here more than once.*
Michael Pragnell, CEO
Domenico Scala, CFO
Bruce Bissell, VP-Human Resources
David Lawrence, VP-R&D
David Lawrence, VP-Tech.
Christoph Mader, VP-Legal & Taxes/Sec.
Mark Peacock, VP-Global Oper.
John Atkin, Head-Bus. Dev.
John Atkin, COO-Crop Protection
Michael Mack, COO-Seeds
Martin Taylor, Chmn.

Phone: 41-61-697-1111	Fax: 41-61-323-2424
Toll-Free:	
Address: Schwarzwaldalle 215, Basel, 4058 Switzerland	

GROWTH PLANS/SPECIAL FEATURES:
Syngenta AG is one of the world's largest agrochemical companies and a leading worldwide supplier of conventional and bioengineered crop protection and seeds. Products designed for crop protection include herbicides, fungicides and insecticides. Additionally, the firm produces seeds for field crops, vegetables and flowers. Its leading marketed products include the following: Dual Gold and Bicept Magnum herbicides; Acanto, Score and Amistar fungicides; and Proclaim, an insecticide. The seeds that Syngenta markets are for field crops, such as corn, soybeans, sugar beets, sunflowers and oilseed rape (canola); fruits and vegetables, including tomatoes, lettuce, melons, squash, cabbages, peppers, beans and radishes; and a wide variety of garden plants such as begonias, lavender, sage and many other seasonal flowers and herbs, some of which can also be purchased as plugs or full-grown plants. The company spends a significant amount of resources on research and development, with major laboratories located in Basel and Stein, Switzerland; Jealott's Hill, in the U.K.; and Syngenta Biotechnology, Inc., in North Carolina. Spending on research and development efforts regularly exceed $800 million per year. Syngenta's plant science division is engaged in collaborations with several companies and universities, including Wageningen Agricultural University in the Netherlands. The research and development activities at Syngenta are currently devoted to advances in genomics, crop protection, health assessment, environmental issues, genetic mapping and traits development. The company's gene technology has become so refined that single genes can be isolated from a type of plant material and transferred to the DNA of another. This process allows manipulation of such traits as nutrient composition, appearance, and even the specifics of the taste of a certain crop. In 2006, the firm acquired both Conrad Fafard, Inc. and Emergent Genetics Vegetable A/S. In March 2007, the firm acquired the Fischer Group, a vegetative flower company, for roughly $67 million.

FINANCIALS: Sales and profits are in thousands of dollars—add 000 to get the full amount. 2006 Note: Financial information for 2006 was not available for all companies at press time.

2006 Sales: $8,046,000	2006 Profits: $634,000	**U.S. Stock Ticker: SYT**
2005 Sales: $8,104,000	2005 Profits: $622,000	**Int'l Ticker: SYNN** Int'l Exchange: Zurich-SWX
2004 Sales: $7,269,000	2004 Profits: $428,000	Employees: 19,000
2003 Sales: $6,578,000	2003 Profits: $268,000	Fiscal Year Ends: 12/31
2002 Sales: $6,197,000	2002 Profits: $-27,000	Parent Company:

SALARIES/BENEFITS:

Pension Plan: Y	ESOP Stock Plan:	Profit Sharing:	Top Exec. Salary: $	Bonus: $
Savings Plan:	Stock Purch. Plan:		Second Exec. Salary: $	Bonus: $

OTHER THOUGHTS:
Apparent Women Officers or Directors: 1
Hot Spot for Advancement for Women/Minorities:

LOCATIONS: ("Y" = Yes)

West:	Southwest:	Midwest:	Southeast:	Northeast:	International:
Y	Y	Y	Y	Y	Y

SYNOVIS LIFE TECHNOLOGIES INC   www.synovislife.com

Industry Group Code: 339113 Ranks within this company's industry group: Sales: 11   Profits: 10

Drugs:	Other:	Clinical:		Computers:		Other:
Discovery:	AgriBio:	Trials/Services:		Hardware:		Specialty Services:
Licensing:	Genomics/Proteomics:	Laboratories:		Software:		Consulting:
Manufacturing:	Tissue Replacement:	Equipment/Supplies:	Y	Arrays:		Blood Collection:
Development:		Research/Development Svcs.:	Y	Database Management:		Drug Delivery:
Generics:		Diagnostics:				Drug Distribution:

TYPES OF BUSINESS:

Surgical & Interventional Treatment Products
Implantable Biomaterials

BRANDS/DIVISIONS/AFFILIATES:

Synovis Surgical Innovations
Synovis Micro Companies Alliance
Synovis Interventional Solutions
Microvascular Anastomotis Coupler
Peri-Strips
Veritas
Tissue-Guard
4Closure Surgical Fascia Closure System

CONTACTS: *Note: Officers with more than one job title may be intentionally listed here more than once.*

Richard W. Kramp, CEO
Richard W. Kramp, Pres.
Brett A. Reynolds, CFO
B. Nicholas Oray, VP-R&D
Brett A. Reynolds, Corp. Sec.
Brett A. Reynolds, VP-Finance
Michael K. Campbell, Pres., Micro Companies Alliance, Inc.
Mary L. Frick, VP-Regulatory Affairs & Quality Assurance
Mary L. Frick, VP-Clinical Affairs
Timothy M. Scanlan, Chmn.

Phone: 651-796-7300	Fax: 651-642-9018
Toll-Free: 800-255-4018	
Address: 2575 University Ave. W., St. Paul, MN 55114 US	

GROWTH PLANS/SPECIAL FEATURES:

Synovis Life Technologies, Inc. is a diversified medical device company engaged in developing, manufacturing and bringing to market products for the surgical and interventional treatment of disease. The company operates in two business segments: the surgical business and the interventional business. The surgical business develops, through subsidiaries Synovis Innovations and Synovis Micro Companies Alliance, manufactures, markets and sells implantable biomaterial products, devices for microsurgery and surgical tools. Biometrical products include Peri-Strips, a biomaterial stapling buttress used as reinforcement at the surgical stable line to reduce the risk of potentially fatal leaks; Tissue-Guard, used to repair and replace damaged tissue in cardiac, vascular, thoracic, abdominal and neuron surgeries; and Veritas, for use in pelvic floor reconstruction, stress urinary incontinence treatment, vaginal and rectal prolapse repair, hernia repair as well as soft tissue repair. A device for microsurgery includes the Microvascular Anastomotic Coupler, which enables microsurgeons in numerous surgical specialties, including plastic and reconscructive. The interventional business, through subsidiary Synovis Interventional Solutions, develops, engineers, prototypes and manufactures coils, helices, stylets, guidewires and other complex micro-wire, polymer and micro-machined metal components used in minimally invasive devices for cardiac rhythm management, neurostimulation, vascular and other procedures. In addition, the interventional business designs and develops proprietary technology platforms that can be adapted for customers. In April 2007, Synovis acquired 4Closure Surgical Fascia Closure System, a device and operating method for closure of puncture in the fascia, a layer of connective tissue on the inner surface of the chest or abdominal wall, from Fascia Closure Systems, LLC.

FINANCIALS: Sales and profits are in thousands of dollars—add 000 to get the full amount. 2006 Note: Financial information for 2006 was not available for all companies at press time.

2006 Sales: $55,835	2006 Profits: $-1,481	**U.S. Stock Ticker: SYNO**
2005 Sales: $60,256	2005 Profits: $ 883	**Int'l Ticker:** Int'l Exchange:
2004 Sales: $55,044	2004 Profits: $1,278	Employees: 390
2003 Sales: $57,989	2003 Profits: $4,973	Fiscal Year Ends: 10/31
2002 Sales: $39,962	2002 Profits: $3,041	Parent Company:

SALARIES/BENEFITS:

Pension Plan:	ESOP Stock Plan:	Profit Sharing:	Top Exec. Salary: $375,000	Bonus: $15,000
Savings Plan: Y	Stock Purch. Plan:		Second Exec. Salary: $225,000	Bonus: $4,500

OTHER THOUGHTS:

Apparent Women Officers or Directors: 2
Hot Spot for Advancement for Women/Minorities: Y

LOCATIONS: ("Y" = Yes)

West:	Southwest:	Midwest:	Southeast:	Northeast:	International:
		Y			

TAKEDA PHARMACEUTICAL COMPANY LTD www.takeda.com

Industry Group Code: 325412 Ranks within this company's industry group: Sales: 15   Profits: 11

Drugs:		Other:		Clinical:	Computers:	Other:	
Discovery:	Y	AgriBio:		Trials/Services:	Hardware:	Specialty Services:	
Licensing:	Y	Genomics/Proteomics:		Laboratories:	Software:	Consulting:	
Manufacturing:	Y	Tissue Replacement:		Equipment/Supplies:	Arrays:	Blood Collection:	
Development:	Y			Research/Development Svcs.:	Database Management:	Drug Delivery:	
Generics:				Diagnostics:		Drug Distribution:	

TYPES OF BUSINESS:

Pharmaceuticals Discovery & Development
Over-the-Counter Drugs
Vitamins
Chemicals
Agricultural & Food Products

BRANDS/DIVISIONS/AFFILIATES:

Takeda America Holdings, Inc.
Takeda Research Investment, Inc.
TAP Pharmaceutical Products Inc.
Takeda Europe Research and Development Center Ltd.
Takeda Ireland Limited
Boie-Takeda Chemicals, Inc.
Paradigm Therapeutics Ltd.
Prevacid

CONTACTS: *Note: Officers with more than one job title may be intentionally listed here more than once.*

Yasuchika Hasegawa, Pres.
Tsutomu Miura, Gen. Mgr.-Ethical Prod. Mktg. Dept.
Tsudoi Miyoshi, Gen. Mgr.-Human Resources Dept.
Hiroshi Shinha, Gen. Mgr.-Legal Dept.
Yasuhiko Yamanaka, Gen. Mgr.-Corp. Strategy & Planning Dept.
Toyoji Yoshida, Gen. Mgr.-Corp. Comm. Dept.
Hiroaki Ogata, Gen. Mgr.-Bus. Dev. & Global Licensing Dept.
Hiroshi Sakiyama, Gen. Mgr.-Tokyo Branch
Teruo Sakurada, Gen. Mgr.-Osaka Branch
Hiroshi Ohtsuki, Pres., Consumer Healthcare Company
Kunio Takeda, Chmn.
Naohisa Takeda, Gen. Mgr.-Dept. of Europe & Asia

Phone: 81-6-6204-2111	Fax: 81-6-6204-2880
Toll-Free:	
Address: 1-1, Doshomachi 4-chome, Chuo-ku,, Osaka, 540-8645 Japan	

GROWTH PLANS/SPECIAL FEATURES:

Takeda Pharmaceutical Company Ltd., based in Japan, is an international research-based company focused on pharmaceuticals.   One of the largest pharmaceutical companies in Japan, it operates three research centers and an international marketing network that includes 13 overseas bases in the U.S., Europe and Asia.   Takeda discovers, develops, manufactures and markets pharmaceutical products in two categories: ethical and consumer health care drugs.   Ethical drugs, as the firm denominates them, constitute about 81.5% of company sales.   This segment includes the anti-prostatic cancer agent leuprolide acetate, marketed as Lupron Depot, Enantone, Prostap and Leuplin; the anti-peptic ulcer agent lansoprazole, marketed as Prevacid, Ogast, Takepron and other names; the anti-hypertensive agent candesartan cilexetil, marketed as Blopress, Kenzen and Amias; and the anti-diabetic agent pioglitazone hydrochloride, marketed as Actos.   The company's consumer health care division focuses on the over-the-counter drug market.   Takeda's main consumer brands include Alinamin, a vitamin B1 derivative; Benza, a cold remedy; and Hicee, a vitamin C preparation.   Outside Japan,   Takeda's   subsidiaries   include   development, production, and sales and marketing companies, as well as holding companies in the U.S. and Europe.   Overseas markets account for 44.3% of total sales. The company also operates subsidiaries with the agro, food, urethane chemicals and bulk vitamin businesses.   Within research and development, Takeda focuses on the life-style related diseases, oncology, urologic diseases, central nervous system   diseases   and   gastroenterology   life   cycle management.   The company pursues alliances with other pharmaceutical manufacturers, biotechnology companies, universities and other research institutions to efficiently introduce key technologies.   Takeda's plans are to increase sales of its four international strategic products (Leuprorelin, Lansoprazole, Candesartan and Pioglitazone); to expand its overseas operations by reinforcing infrastructures, and to restructure its non-pharmaceutical businesses.   In March 2007, Takeda announced its intentions to acquire Paradigm Therapeutics Ltd.

FINANCIALS: Sales and profits are in thousands of dollars—add 000 to get the full amount. 2006 Note: Financial information for 2006 was not available for all companies at press time.

2006 Sales: $10,360,744	2006 Profits: $2,677,342	**U.S. Stock Ticker: TKPHF**
2005 Sales: $10,441,300	2005 Profits: $2,579,600	**Int'l Ticker: 4502** Int'l Exchange: Tokyo-TSE
2004 Sales: $10,284,200	2004 Profits: $2,700,300	Employees: 15,069
2003 Sales: $8,728,500	2003 Profits: $2,267,600	Fiscal Year Ends: 3/31
2002 Sales: $7,577,100	2002 Profits: $1,776,600	Parent Company:

SALARIES/BENEFITS:

Pension Plan:	ESOP Stock Plan:	Profit Sharing:	Top Exec. Salary: $	Bonus: $
Savings Plan:	Stock Purch. Plan:		Second Exec. Salary: $	Bonus: $

OTHER THOUGHTS:

		LOCATIONS: ("Y" = Yes)					
		West:	Southwest:	Midwest:	Southeast:	Northeast:	International:
Apparent Women Officers or Directors:		Y		Y			Y
Hot Spot for Advancement for Women/Minorities:							

Note: Financial information, benefits and other data can change quickly and may vary from those stated here.

TANOX INC
www.tanox.com

Industry Group Code: 325412 Ranks within this company's industry group: Sales: 75   Profits: 61

Drugs:		Other:	Clinical:	Computers:	Other:
Discovery:	Y	AgriBio:	Trials/Services:	Hardware:	Specialty Services:
Licensing:		Genomics/Proteomics:	Laboratories:	Software:	Consulting:
Manufacturing:	Y	Tissue Replacement:	Equipment/Supplies:	Arrays:	Blood Collection:
Development:	Y		Research/Development Svcs.:	Database Management:	Drug Delivery:
Generics:			Diagnostics:		Drug Distribution:

TYPES OF BUSINESS:
Drugs-Immunologic & Infectious Disease
Monoclonal Antibodies
Asthma & Allergy Treatments

BRANDS/DIVISIONS/AFFILIATES:
Xolair
Ibalizumab
TNX-650
TNX-355

CONTACTS: Note: Officers with more than one job title may be intentionally listed here more than once.
Danong Chen, CEO
Edward Hu, COO/Sr. VP
Danong Chen, Pres.
Hugo Santos, VP-Human Resources
Zhengbin Yao, VP-Research
Katie-Pat Bowman, General Counsel/VP/Corp. Sec.
Gregory P. Guidroz, Corp. Comm.
Gregory P. Guidroz, Investor Rel.
Gregory P. Guidroz, VP-Finance
Brian Kim, VP-Quality
Nancy T. Chang, Chmn.

Phone: 713-578-4000	Fax: 713-578-5000
Toll-Free:	
Address: 10555 Stella Link, Houston, TX 77025-5631 US	

GROWTH PLANS/SPECIAL FEATURES:
Tanox, Inc. discovers and develops therapeutic monoclonal antibodies to address the areas of immune-mediated diseases, infectious diseases, inflammation and oncology. The company's first product, Xolair, was developed in collaboration with Genentech, Inc. and Novartis Pharma, A.G. (Novartis), and is the exclusive trademark of Novartis. Xolair is currently labeled for treatment of adults and adolescents with moderate-to-severe persistent asthma who have a positive skin test or reactivity to a perennial aeroallergen and whose symptoms are inadequately controlled with inhaled corticosteroids. Xolair is an anti-immunoglobulin E, or anti-IgE, antibody that has been shown to decrease the incidence of asthma exacerbations in these patients. Safety and efficacy have not been established in other allergic conditions. Tanox is also assessing another product, Ibalizumab (TNX-355), as a treatment for HIV infection. The drug is a humanized, non-immunosuppressive monoclonal antibody that works by blocking the ability of HIV to penetrate CD4 cells. The product is currently in Phase 2 clinical trials. Ibalizumab is the result of a licensing agreement between Tanox and Biogen Idec. The firm has focused work in genetically engineered monoclonal antibodies, man-made antibodies that target a specific antigen. The lead product is TNX-650, which is in Phase I clinical study for the treatment of Hodgkin's lymphoma. In 2006, the company began Phase I trials of TNX-650 for the treatment of asthma. In January 2007, Tanox agreed to be acquired by Genentech, Inc. for approximately $919 million. Once the agreement is finalized, the company expects to be merged with a subsidiary of Genentech.

Tanox offers its employees a comprehensive benefits package that includes medical, dental and vision coverage; flexible spending accounts; paid time off; and a college savings plan.

FINANCIALS: Sales and profits are in thousands of dollars—add 000 to get the full amount. 2006 Note: Financial information for 2006 was not available for all companies at press time.

2006 Sales: $56,137	2006 Profits: $-2,568	**U.S. Stock Ticker: TNOX**
2005 Sales: $44,687	2005 Profits: $-19,424	**Int'l Ticker:** Int'l Exchange:
2004 Sales: $20,506	2004 Profits: $-10,290	Employees: 197
2003 Sales: $18,487	2003 Profits: $-4,638	Fiscal Year Ends: 12/31
2002 Sales: $600	2002 Profits: $-26,000	Parent Company:

SALARIES/BENEFITS:
Pension Plan:	ESOP Stock Plan:	Profit Sharing:	Top Exec. Salary: $463,202	Bonus: $147,666
Savings Plan: Y	Stock Purch. Plan:		Second Exec. Salary: $205,583	Bonus: $44,018

OTHER THOUGHTS:
Apparent Women Officers or Directors: 3
Hot Spot for Advancement for Women/Minorities: Y

LOCATIONS: ("Y" = Yes)
West:	Southwest:	Midwest:	Southeast:	Northeast:	International:
Y	Y				

TAPESTRY PHARMACEUTICALS INCwww.tapestrypharma.com

Industry Group Code: 325412 Ranks within this company's industry group: Sales: Profits: 96

Drugs:		Other:	Clinical:	Computers:	Other:
Discovery:	Y	AgriBio:	Trials/Services:	Hardware:	Specialty Services:
Licensing:	Y	Genomics/Proteomics:	Laboratories:	Software:	Consulting:
Manufacturing:		Tissue Replacement:	Equipment/Supplies:	Arrays:	Blood Collection:
Development:	Y		Research/Development Svcs.:	Database Management:	Drug Delivery:
Generics:			Diagnostics:		Drug Distribution:

TYPES OF BUSINESS:

Cancer Therapeutics

BRANDS/DIVISIONS/AFFILIATES:

NaPro BioTherapeutics, Inc.
TPI 287

CONTACTS: *Note: Officers with more than one job title may be intentionally listed here more than once.*

Leonard P. Shaykin, CEO
Martin Batt, COO/Sr. VP
Donald H. Picker, Pres.
Gordon Link, Jr., CFO/Sr. VP
James D. McChesney, Chief Scientific Officer
Gilles H. Tapolsky, VP-Prod. Dev.
Kai P. Larson, General Counsel/VP
Sandra Silberman, Chief Medical Officer
David L. Emerson, VP-Cancer Biology
Lawrence Helson, VP-Bio Research
Brenda P. Fielding, Dir.-Regulatory Affairs
Leonard P. Shaykin, Chmn.

Phone: 303-516-8500	Fax: 303-530-1296

Toll-Free:

Address: 4840 Pearl E. Cir., Ste. 300W, Boulder, CO 80301 US

GROWTH PLANS/SPECIAL FEATURES:

Tapestry Pharmaceuticals, Inc., formerly NaPro BioTherapeutics, Inc., is a biopharmaceutical company focused on the development of proprietary therapies for the treatment of cancer. The company is also engaged in evaluating new therapeutic agents and related technologies. All the firm's products and technologies are in the early stages of development. Tapestry Pharmaceuticals has two products in the most advanced stages of development: TPI 287, for the treatment of prostrate, non-small cell lung and other cancers; and TIP 287 oral formulation, for the treatment of prostrate, non-small cell lung and other cancer. In January 2006, the company discontinued its research activities relating to the Huntington's disease program and its development activities on TPI 284, its peptide linked cytotoxic compound. In May 2007, the firm announced that it was developing an oral formulation of TPI 287 and that it opened enrollment in a Phase II clinical trial for TPI 287 in patients with hormone refractory prostrate cancer. Tapestry Pharmaceuticals also plans to commence an additional Phase II trial in patients with glioblastoma multiforme, a primary cancer of the central nervous system; and a third Phase II trial in patients with cancer of the pancreas.

FINANCIALS: Sales and profits are in thousands of dollars—add 000 to get the full amount. 2006 Note: Financial information for 2006 was not available for all companies at press time.

2006 Sales: $	2006 Profits: $-16,652	**U.S. Stock Ticker: TPPH**
2005 Sales: $	2005 Profits: $-17,538	**Int'l Ticker:** Int'l Exchange:
2004 Sales: $	2004 Profits: $-24,174	Employees: 38
2003 Sales: $	2003 Profits: $38,128	Fiscal Year Ends: 12/31
2002 Sales: $34,200	2002 Profits: $-8,700	Parent Company:

SALARIES/BENEFITS:

Pension Plan:	ESOP Stock Plan:	Profit Sharing:	Top Exec. Salary: $370,000	Bonus: $350,000
Savings Plan:	Stock Purch. Plan:		Second Exec. Salary: $270,000	Bonus: $140,000

OTHER THOUGHTS:

Apparent Women Officers or Directors: 2
Hot Spot for Advancement for Women/Minorities: Y

LOCATIONS: ("Y" = Yes)

West:	Southwest:	Midwest:	Southeast:	Northeast:	International:
Y				Y	

Note: Financial information, benefits and other data can change quickly and may vary from those stated here.

TARGETED GENETICS CORP

www.targen.com

Industry Group Code: 325412 Ranks within this company's industry group: Sales: 129   Profits: 136

Drugs:		Other:		Clinical:	Computers:		Other:	
Discovery:	Y	AgriBio:		Trials/Services:	Hardware:		Specialty Services:	
Licensing:		Genomics/Proteomics:	Y	Laboratories:	Software:		Consulting:	
Manufacturing:		Tissue Replacement:		Equipment/Supplies:	Arrays:		Blood Collection:	
Development:	Y			Research/Development Svcs.:	Database Management:		Drug Delivery:	
Generics:				Diagnostics:			Drug Distribution:	

TYPES OF BUSINESS:

Gene Therapeutics

BRANDS/DIVISIONS/AFFILIATES:

Mydicar
tgAAC94
tgAAC09
Celladon Corp.

CONTACTS: *Note: Officers with more than one job title may be intentionally listed here more than once.*

H. Stewart Parker, CEO
H. Stewart Parker, Pres.
David J. Poston, CFO
Barrie J. Carter, Chief Scientific Officer/Exec. VP
B.G. Susan Robinson, VP-Bus. Dev.
David J. Poston, VP-Finance/Treas.
Pervin Anklesaria, VP-Therapeutic Dev.
Richard W. Peluso, VP-Process Dev.
Jeremy Curnock Cook, Chmn.

Phone: 206-623-7612	Fax: 206-223-0288
Toll-Free:	
Address: 1100 Olive Way, Ste. 100, Seattle, WA 98101 US	

GROWTH PLANS/SPECIAL FEATURES:

Targeted Genetics Corp. is a clinical-stage biotechnology company that develops gene therapeutics. The company's gene therapeutics consist of a delivery vehicle, called a vector, and genetic material. The role of the vector is to carry the genetic material into a target cell. Once delivered into the cell, the gene can express or direct production of the specific proteins encoded by the gene. The firm develops and manufactures adeno-associated viral (AAV) vectors, or AAV-based gene therapeutics. Targeted Genetics has two products that are in clinical trials for treatment of inflammatory arthritis and HIV/AIDS: tgAAC94, an AAV delivery of TNF-alpha antagonist; and tgAAC09 & HVDDT, an AAV delivery of HIV antigen. The firm's two preclinical product candidates, both in development with collaboration partners, are aimed at congestive heart failure and Huntington's disease. The company has ceased development of tgAAVCF, a product candidate for treating cystic fibrosis. In May 2007, Targeted Genetics and Celladon Corp. announced the commencement of a Phase I clinical trial of Mydicar in patients with cardiomyopathy and symptoms of heart failure. The drug uses an AAV vector to deliver the SERCA2a gene to heart muscle tissue. In July 2007, the company received a patent related to its AAV technology platform. Later that month, the firm announced that further trials of its tgAAC94 for treatment of inflammatory arthritis were placed on hold due to a serious adverse event and subsequent death that occurred to one of the patients enrolled in the program.

The company offers its employees medical, vision and dental insurance; life and AD&D insurance; short- and long-term disability; an employee assistance program; a 401(k) plan; and stock options.

FINANCIALS: Sales and profits are in thousands of dollars—add 000 to get the full amount. 2006 Note: Financial information for 2006 was not available for all companies at press time.

2006 Sales: $9,864	2006 Profits: $-33,990	**U.S. Stock Ticker:** TGEN	
2005 Sales: $6,874	2005 Profits: $-19,198	**Int'l Ticker:** Int'l Exchange:	
2004 Sales: $9,652	2004 Profits: $-14,257	Employees: 70	
2003 Sales: $14,073	2003 Profits: $-14,833	Fiscal Year Ends: 12/31	
2002 Sales: $19,300	2002 Profits: $-23,800	Parent Company:	

SALARIES/BENEFITS:

Pension Plan:	ESOP Stock Plan:	Profit Sharing:	Top Exec. Salary: $418,000	Bonus: $62,700
Savings Plan: Y	Stock Purch. Plan: Y		Second Exec. Salary: $278,000	Bonus: $41,700

OTHER THOUGHTS:

Apparent Women Officers or Directors: 1
Hot Spot for Advancement for Women/Minorities:

LOCATIONS: ("Y" = Yes)

West:	Southwest:	Midwest:	Southeast:	Northeast:	International:
Y					

TARO PHARMACEUTICAL INDUSTRIES   www.taro.com

Industry Group Code: 325416 Ranks within this company's industry group: Sales: Profits:

Drugs:		Other:	Clinical:	Computers:	Other:
Discovery:	Y	AgriBio:	Trials/Services:	Hardware:	Specialty Services:
Licensing:	Y	Genomics/Proteomics:	Laboratories:	Software:	Consulting:
Manufacturing:	Y	Tissue Replacement:	Equipment/Supplies:	Arrays:	Blood Collection:
Development:	Y		Research/Development Svcs.:	Database Management:	Drug Delivery:
Generics:	Y		Diagnostics:		Drug Distribution:

TYPES OF BUSINESS:

Drugs-Generic & Proprietary
Over-the-Counter Analgesics
Vitamins
Anti-Cancer Drugs
Dermatological Drugs

BRANDS/DIVISIONS/AFFILIATES:

Taro Pharmaceuticals U.S.A., Inc.
Kusch Manual
RxDesktop
LUSTRA
Ovide
Clotrimazole/Betamethasomne
Phenytoin
Terconazole

CONTACTS: *Note: Officers with more than one job title may be intentionally listed here more than once.*

Thomas E. McClary, CFO/VP
Zahava Rafalowicz, VP-Mktg. & Sales
Inbal Rothman, VP-Human Resources & Community Affairs
Avraham Yacobi, Sr. VP-R&D
Tzvi Tal, VP-IT
Roman Kaplan, VP-Tech. Oper. & Pharmaceuticals
Noam Shamir, VP-Industrial Eng.
Rebecca A. Roof, Interim Chief Admin. & Restructuring Officer
Tal Levitt, Corp. Sec.
Tal Levitt, Sr. VP-Corp. Affairs/Treas.
Hannah Bayer, VP-Finance/Chief Acct. Mgr.
Hagai Reingold, VP-API Div.
Mariana Bacalu, VP-Quality Affairs
Daniel Saks, VP-Corp. Affairs
Barrie Levitt, Chmn.
Noam Shamir, VP-Supply Chain

Phone: 972-9-971-1800	Fax: 972-9-955-7443

Toll-Free:

Address: Italy House, Euro Park, Yakum, 60972 Israel

GROWTH PLANS/SPECIAL FEATURES:

Taro Pharmaceutical Industries is a multinational pharmaceutical company that discovers, develops, manufactures and markets a wide range of both proprietary and generic health care products, from over-the-counter and prescription analgesics to vitamins. Its products address illnesses varying from the common cold to cancer, and are mainly used in dermatology, cardiology, neurology and pediatrics. The firm primarily operates through three entities: Taro Pharmaceutical Industries (Israel) and its subsidiaries Taro Pharmaceuticals, Inc. (Canada) and Taro U.S.A. Taro produces more than 200 pharmaceutical products, including topical preparations such as creams, ointments, gels and solutions; oral medications such as tablets, capsules, powders and liquids; and sterile products such as injectables, ophthalmic drops and powders. Top products manufactured by Taro include LUSTRA, a treatment for dyschromia, or discolored skin; Ovide, a lotion for the treatment of head lice; Warfarin, sodium tablets for certain heart conditions; Clotrimazole and Betamethasomne Dipropionate Creams, for the treatment of the topical effects of fungus, which are the generic equivalent of Lotrisone; Phenytoin Oral Suspension, for the treatment of epilepsy; and Terconazole Vaginal Cream, for treating yeast infections. Taro also funds the Kusch Manual, a dermatology diagnosis manual for physicians, and RxDesktop, a program for health care professionals that contains conversation tools, dosage calculators and other information. The firm has six subsidiaries for distribution and manufacturing in the U.S., Canada, Israel, the U.K. and Ireland. Currently, the firm is in clinical trials of T2000, a non-sedating barbiturate compound. In May 2007, the company agreed to be acquired by Sun Pharmaceutical Industries, Ltd. In July 2007, the FDA approved Taro's Abbreviated New Drug Application for Terbinafine Hydrochloride Cream 1%, the bioequivalant to Lamisil.

FINANCIALS: Sales and profits are in thousands of dollars—add 000 to get the full amount. 2006 Note: Financial information for 2006 was not available for all companies at press time.

2006 Sales: $	2006 Profits: $	U.S. Stock Ticker: TAROF.PK	
2005 Sales: $	2005 Profits: $	Int'l Ticker: Int'l Exchange:	
2004 Sales: $284,100	2004 Profits: $11,100	Employees: 1,400	
2003 Sales: $315,500	2003 Profits: $61,200	Fiscal Year Ends: 12/31	
2002 Sales: $211,600	2002 Profits: $44,600	Parent Company:	

SALARIES/BENEFITS:

Pension Plan: Y	ESOP Stock Plan:	Profit Sharing:	Top Exec. Salary: $	Bonus: $
Savings Plan:	Stock Purch. Plan: Y		Second Exec. Salary: $	Bonus: $

OTHER THOUGHTS:

Apparent Women Officers or Directors: 4
Hot Spot for Advancement for Women/Minorities: Y

LOCATIONS: ("Y" = Yes)

West:	Southwest:	Midwest:	Southeast:	Northeast:	International:
				Y	Y

Note: Financial information, benefits and other data can change quickly and may vary from those stated here.

TECHNE CORP

www.techne-corp.com

Industry Group Code: 325413 Ranks within this company's industry group: Sales: 7   Profits: 2

Drugs:	Other:	Clinical:		Computers:	Other:
Discovery:	AgriBio:	Trials/Services:		Hardware:	Specialty Services:
Licensing:	Genomics/Proteomics:	Laboratories:		Software:	Consulting:
Manufacturing:	Tissue Replacement:	Equipment/Supplies:	Y	Arrays:	Blood Collection:
Development:		Research/Development Svcs.:		Database Management:	Drug Delivery:
Generics:		Diagnostics:	Y		Drug Distribution:

TYPES OF BUSINESS:

Biotechnology Products
Reagents, Antibodies & Assay Kits
Hematology Products

BRANDS/DIVISIONS/AFFILIATES:

Research and Diagnostic Systems, Inc.
R&D Systems Europe, Ltd.
R&D Systems GmbH
Whole Blood Flow Cytometry Control
Whole Blood Glucose/Hemoglobin Control
Fortron Bio Science, Inc.
BiosPacific, Inc.

CONTACTS: *Note: Officers with more than one job title may be intentionally listed here more than once.*

Thomas E. Oland, CEO
Thomas E. Oland, Pres.
Gergory J. Melson, CFO
Lea Simoane, Dir.-Human Resources
Richard A. Krzyzek, VP-Research
Gregory J. Melson, VP-Finance
Marcel Veronneau, VP-Hematology Oper.
Thomas E. Oland, Treas.
Roger C. Lucas, Vice Chmn.
Thomas E. Oland, Chmn.

Phone: 612-379-8854	Fax: 612-379-6580
Toll-Free: 800-343-7475	
Address: 614 McKinley Pl. NE, Minneapolis, MN 55413-2610 US	

GROWTH PLANS/SPECIAL FEATURES:

Techne Corp. is a holding company that operates via two subsidiaries: Research and Diagnostic Systems, Inc. (R&D Systems) and R&D Systems Europe, Ltd. (R&D Europe). R&D Systems manufactures biological products in two major segments hematology controls, which are used in clinical and hospital laboratories to monitor the accuracy of blood analysis instruments; and biotechnology products, which including purified proteins and antibodies that are sold exclusively to the research market and assay kits that are sold to the research and clinical diagnostic markets. R&D Europe distributes biotechnology products throughout Europe and also operates a sales office in France and its German sales subsidiary, R&D Systems GmbH. In recent years, R&D Systems has also expanded its product portfolio to include enzymes and intracellular cell signaling reagents such as proteases, kinases and phosphatases for diseases such as cancer, Alzheimer's ,arthritic, autoimmunity, diabetes, hypertension, obesity, AIDS and SARS. Techne also produces controls and calibrators for a variety of medical brands such as Abbott Diagnostics, Beckman Coulter, Bayer Technicon and Sysmex. In the hematology sector, the company's Whole Blood Flow Cytometry Control is used to identify and quantify white blood cells by their surface antigens while linearity and reportable range controls assess the linearity of hematology analyzers for white blood cells, red blood ells, platelets and reticulocytes. R&D Systems is currently engaged in ongoing research and development in all of its major product lines and is particularly focused on the release of new cytokines, antibodies and cytokine assay kits for the coming year.

FINANCIALS: Sales and profits are in thousands of dollars—add 000 to get the full amount. 2006 Note: Financial information for 2006 was not available for all companies at press time.

2006 Sales: $202,617	2006 Profits: $73,351	**U.S. Stock Ticker: TECH**
2005 Sales: $178,700	2005 Profits: $66,100	**Int'l Ticker:** Int'l Exchange:
2004 Sales: $161,257	2004 Profits: $52,928	Employees: 577
2003 Sales: $145,000	2003 Profits: $45,400	Fiscal Year Ends: 6/30
2002 Sales: $130,900	2002 Profits: $27,100	Parent Company:

SALARIES/BENEFITS:

Pension Plan:	ESOP Stock Plan:	Profit Sharing: Y	Top Exec. Salary: $254,000	Bonus: $
Savings Plan: Y	Stock Purch. Plan: Y		Second Exec. Salary: $251,400	Bonus: $45,755

OTHER THOUGHTS:

Apparent Women Officers or Directors: 2
Hot Spot for Advancement for Women/Minorities: Y

LOCATIONS: ("Y" = Yes)

West:	Southwest:	Midwest:	Southeast:	Northeast:	International:
		Y			Y

Note: Financial information, benefits and other data can change quickly and may vary from those stated here.

TELIK INC

www.telik.com

ndustry Group Code: 325412 **Ranks within this company's industry group:** Sales: Profits: 175

Drugs:		Other:	Clinical:	Computers:		Other:
Discovery:	Y	AgriBio:	Trials/Services:	Hardware:		Specialty Services:
Licensing:		Genomics/Proteomics:	Laboratories:	Software:	Y	Consulting:
Manufacturing:		Tissue Replacement:	Equipment/Supplies:	Arrays:		Blood Collection:
Development:	Y		Research/Development Svcs.:	Database Management:		Drug Delivery:
Generics:			Diagnostics:			Drug Distribution:

TYPES OF BUSINESS:

Drugs-Cancer
Small-Molecule Drugs

BRANDS/DIVISIONS/AFFILIATES:

Target-Related Affinity Profiling (TRAP)
TELCYTA
TELINTRA

CONTACTS: Note: Officers with more than one job title may be intentionally listed here more than once.

Michael M. Wick, CEO
Cynthia M. Butitta, COO
Michael M. Wick, Pres.
Cynthia M. Butitta, CFO
Gail L. Brown, Sr. VP/Chief Medical Officer
William P. Kaplan, General Counsel/VP/Corp. Sec.
Marc L. Steuer, Sr. VP-Bus. Dev.
Michael M. Wick, Chmn.

Phone: 650-845-7700	Fax: 650-845-7800
Toll-Free:	
Address: 3165 Porter Dr., Palo Alto, CA 94304 US	

GROWTH PLANS/SPECIAL FEATURES:

Telik, Inc. discovers, develops and commercializes small-molecule pharmaceuticals, mainly for the treatment of specific cancers. The company's proprietary Target-Related Affinity Profiling (TRAP) chemoinformatics technology is used by Telik and its partners to rapidly identify promising chemicals for development.   TRAP can select a small sample from its large compound library and effectively identify small, biologically active molecules. The company currently has two product candidates in clinical trials. TELCYTA is a novel tumor-activated compound that is currently in Phase III clinical trials for use against ovarian and non-small-cell lung cancers and in Phase II of development for colorectal and breast cancers. The drug product candidate is designed to be activated in cancer cells through binding to the GST P1-1 protein, which is elevated in many human cancers, even more so in patients who have already been treated with other standard chemotherapy drugs. Once bound to the protein inside a cancer cell, a chemical reaction occurs that releases TELCYTA and causes programmed cancer death.   TELINTRA, a small-molecule bone marrow stimulant, is in Phase II trials for myelodysplastic syndrome (MDS), a form of pre-leukemia. The product is used in the treatment of blood disorders with low blood cell levels, such as neutropenia or anemia. The company has also recently identified a third product candidate called TLK58747, which has shown significant anti-tumor activity in human breast, pancreatic and colon tumors while in preclinical study. In February 2007, the company announced a corporate restructuring that will result in the loss of approximately 25% of its workforce.

FINANCIALS: Sales and profits are in thousands of dollars—add 000 to get the full amount. 2006 Note: Financial information for 2006 was not available for all companies at press time.

2006 Sales: $	2006 Profits: $-79,624	U.S. Stock Ticker: TELK
2005 Sales: $ 19	2005 Profits: $-75,542	Int'l Ticker: Int'l Exchange:
2004 Sales: $ 163	2004 Profits: $-69,817	Employees: 118
2003 Sales: $ 436	2003 Profits: $-50,642	Fiscal Year Ends: 12/31
2002 Sales: $1,300	2002 Profits: $-34,800	Parent Company:

SALARIES/BENEFITS:

Pension Plan:	ESOP Stock Plan:	Profit Sharing:	Top Exec. Salary: $494,000	Bonus: $
Savings Plan:	Stock Purch. Plan: Y		Second Exec. Salary: $344,000	Bonus: $

OTHER THOUGHTS:

Apparent Women Officers or Directors: 2
Hot Spot for Advancement for Women/Minorities: Y

LOCATIONS: ("Y" = Yes)

West:	Southwest:	Midwest:	Southeast:	Northeast:	International:
Y					

TEVA PHARMACEUTICAL INDUSTRIES www.tevapharm.com

Industry Group Code: 325416 Ranks within this company's industry group: Sales: 1   Profits: 1

Drugs:	Other:	Clinical:	Computers:	Other:
Discovery:	AgriBio:	Trials/Services:	Hardware:	Specialty Services:
Licensing:	Genomics/Proteomics:	Laboratories:	Software:	Consulting:
Manufacturing: Y	Tissue Replacement:	Equipment/Supplies:	Arrays:	Blood Collection:
Development: Y		Research/Development Svcs.: Y	Database Management:	Drug Delivery:
Generics: Y		Diagnostics:		Drug Distribution:

TYPES OF BUSINESS:

Drugs-Generic
Active Pharmaceutical Ingredients

BRANDS/DIVISIONS/AFFILIATES:

Teva Pharmaceuticals USA
Pharmachemie BV
Copaxone
Ivax Corporation

CONTACTS: Note: Officers with more than one job title may be intentionally listed here more than once.

Shlomo Yanai, CEO
Shlomo Yanai, Pres.
Dan S. Suesskind, CFO
Judith Vardi, VP-Israel Pharmaceutical Sales
Bruria Sofrin, VP-Human Resources
Ben-Zion Weiner, Chief R&D Officer
Shmuel Ben Zvi, VP-IT/Planning & Economics
Rodney Kasan, CTO/VP
Uzi Karniel, General Counsel/Corp. Sec.
Itzhak Krinsky, VP-Bus. Dev.
Doron Blachar, VP-Finance
George S. Barrett, VP-N. America/CEO/Pres., Teva North America
Amir Elstein, VP-Global Specialty Pharmaceutical Prod.
Gerard Van Odlijk, VP/CEO/Pres., Teva Pharmaceuticals Europe B.V.
Jacob Winter, VP-Global Generic Resources
Eli Hurvitz, Chmn.
Chaim Hurvitz, VP-Int'l

Phone: 972-3-926-7267	Fax: 972-3-923-4050
Toll-Free:	
Address: 5 Basel St., Petach Tikva, 49131 Israel	

GROWTH PLANS/SPECIAL FEATURES:

Teva Pharmaceutical Industries, Ltd., based in Israel, is a global pharmaceutical company that produces, distributes and sells pharmaceutical products internationally. Teva has additional manufacturing and marketing facilities in North America and Europe. Eighty percent of the firm's sales are in North America and Europe. The firm focuses on human pharmaceuticals (HP) and active pharmaceutical ingredients (API). The HP segment produces generic drugs in all major therapeutics in a variety of dosage forms, including capsules, tablets, creams, ointments and liquids. The API segment, which accounts for 90% of total sales, distributes ingredients to manufacturers worldwide, in addition to supporting its own pharmaceutical products. Teva also manufactures innovative drugs in niche markets through its research and development efforts. Subsidiary Teva Pharmaceuticals USA is a dominant figure in the manufacture and distribution of generic drugs in the U.S. In addition, subsidiary Pharmachemie BV has the largest market share in generics in the Netherlands. The firm's two chief products are Copaxone, a branded treatment for MS, currently marketed and sold in 42 countries including the U.S., and Azilect, a new treatment for Parkinson's marketed in Israel, the U.S., Canada, Europe and Turkey. The firm's R&D activities focus on the development of innovative molecules for treatment of multiple sclerosis; other neurological and neurodegenerative diseases (Alzheimer's, Parkinson's and ALS); auto-immune and inflammatory diseases (Lupus and psoriasis); and diseases in the field of oncology (Hermato-oncological applications and Glioma). In January 2006, Teva acquired Ivax Corporation, a U.S.-based multinational generics company. The results of this purchase and the subsequent merger boosted Teva's operations to cover over 50 markets and expanded operations to 44 plants, 15 research centers and 18 API sites.

FINANCIALS: Sales and profits are in thousands of dollars—add 000 to get the full amount. 2006 Note: Financial information for 2006 was not available for all companies at press time.

2006 Sales: $8,408,000	2006 Profits: $546,000	**U.S. Stock Ticker: TEVA**
2005 Sales: $5,250,000	2005 Profits: $1,072,000	**Int'l Ticker: TEVA** Int'l Exchange: Tel Aviv-TASE
2004 Sales: $4,799,000	2004 Profits: $-331,800	Employees: 26,670
2003 Sales: $3,276,400	2003 Profits: $691,000	Fiscal Year Ends: 12/31
2002 Sales: $2,518,600	2002 Profits: $410,300	Parent Company:

SALARIES/BENEFITS:

Pension Plan: Y	ESOP Stock Plan:	Profit Sharing:	Top Exec. Salary: $	Bonus: $
Savings Plan: Y	Stock Purch. Plan: Y		Second Exec. Salary: $	Bonus: $

OTHER THOUGHTS:

Apparent Women Officers or Directors: 6
Hot Spot for Advancement for Women/Minorities: Y

LOCATIONS: ("Y" = Yes)

West:	Southwest:	Midwest:	Southeast:	Northeast:	International:
Y	Y	Y	Y	Y	Y

THERAGENICS CORP

www.theragenics.com

Industry Group Code: 339113 Ranks within this company's industry group: Sales: Profits:

Drugs:	Other:	Clinical:		Computers:		Other:
Discovery:	AgriBio:	Trials/Services:		Hardware:		Specialty Services:
Licensing:	Genomics/Proteomics:	Laboratories:		Software:		Consulting:
Manufacturing:	Tissue Replacement:	Equipment/Supplies:	Y	Arrays:		Blood Collection:
Development:		Research/Development Svcs.:	Y	Database Management:		Drug Delivery:
Generics:		Diagnostics:				Drug Distribution:

TYPES OF BUSINESS:

Medical Devices
Surgical Products

BRANDS/DIVISIONS/AFFILIATES:

TheraSeed
CE Mark
CP Medical Corp.
Galt Medical Corp.

CONTACTS: Note: Officers with more than one job title may be intentionally listed here more than once.

M. Christine Jacobs, CEO
M. Christine Jacobs, Pres.
Frank J. Tarallo, CFO
Bruce W. Smith, Exec. VP-Strategy & Bus. Dev.
Frank J. Tarallo, Treas.
James R. Eddings, Pres., Galt Medical
Patrick J. Ferguson, Pres., CP Medical
R. Michael O'Bannon, Exec. VP-Organizational Dev.
M. Christine Jacobs, Chmn.

Phone: 770-271-0233	Fax: 770-831-5294
Toll-Free:	
Address: 5203 Bristol Industrial Way, Buford, GA 30518 US	

GROWTH PLANS/SPECIAL FEATURES:

Theragenics Corp. is a medical device company serving the cancer treatment and surgical markets. The company operates in two segments: the brachytherapy seed business and the surgical products business. The brachytherapy business segment produces, markets and sells TheraSeed, the firm's premier palladium-103 prostrate cancer treatment device; I-seed, its iodone-125 based prostrate cancer treatment device; and other related products and services. Theragenics is the world's largest producer of palladium-103, the radioactive isotope that supplies the therapeutic radiation for its TheraSeed device. TheraSeed, marketed in Europe with the brand name CE Mark, is an implant the size of a grain of rice that is used primarily in treating localized prostate cancer with a one-time, minimally invasive procedure. The implant emits radiation within the immediate prostate area, killing the tumor while sparing surrounding organs from significant radiation exposure. Physicians, hospitals and other healthcare providers, primarily located in the United States, utilize the TheraSeed device. The majority of TheraSeed sales are channeled through one third-party distributor. The surgical products business segment consists of wound closure and vascular access products. Wound closure include sutures, needles and other surgical products with applications in, among other areas, urology, veterinary, cardiology, orthopedics, plastic surgery and dental. Vascular access includes introducers and guidewires used in the interventional radiology, interventional cardiology and vascular surgery markets. Theragenics' subsidiaries, CP Medical Corp. and Galt Medical Corp., accounted for roughly 36% of revenue in 2006. In August 2006, Theragenics acquired Galt Medical Corp., a private manufacturer of disposable medical devices utilized for vascular access, for roughly $32.7 million.

The company offers its employees medical, dental and life insurance; short- and long-term disability protection; a 401(k) plan; an employee stock purchase plan; and an on-site wellness center.

FINANCIALS: Sales and profits are in thousands of dollars—add 000 to get the full amount. 2006 Note: Financial information for 2006 was not available for all companies at press time.

2006 Sales: $54,096	2006 Profits: $6,865	U.S. Stock Ticker: TGX
2005 Sales: $44,270	2005 Profits: $-29,006	Int'l Ticker: Int'l Exchange:
2004 Sales: $33,338	2004 Profits: $-4,310	Employees: 315
2003 Sales: $35,600	2003 Profits: $- 300	Fiscal Year Ends: 12/31
2002 Sales: $41,900	2002 Profits: $5,600	Parent Company:

SALARIES/BENEFITS:

Pension Plan:	ESOP Stock Plan:	Profit Sharing:	Top Exec. Salary: $493,000	Bonus: $270,000
Savings Plan: Y	Stock Purch. Plan: Y		Second Exec. Salary: $255,000	Bonus: $127,100

OTHER THOUGHTS:

Apparent Women Officers or Directors: 2
Hot Spot for Advancement for Women/Minorities: Y

LOCATIONS: ("Y" = Yes)

West:	Southwest:	Midwest:	Southeast:	Northeast:	International:
Y	Y		Y		

THERMO FISHER SCIENTIFIC INC   www.thermofisher.com

Industry Group Code: 421450 Ranks within this company's industry group: Sales: 1   Profits: 1

Drugs:		Other:		Clinical:		Computers:		Other:	
Discovery:		AgriBio:		Trials/Services:	Y	Hardware:		Specialty Services:	Y
Licensing:		Genomics/Proteomics:		Laboratories:		Software:		Consulting:	
Manufacturing:	Y	Tissue Replacement:		Equipment/Supplies:	Y	Arrays:		Blood Collection:	
Development:				Research/Development Svcs.:	Y	Database Management:		Drug Delivery:	
Generics:				Diagnostics:	Y			Drug Distribution:	

TYPES OF BUSINESS:

Laboratory Equipment & Supplies Distribution
Contract Manufacturing
Equipment Callibration & Repair
Clinical Trial Services
Laboratory Workstations
Clinical Consumables
Diagnostic Reagents
Custom Chemical Synthesis

BRANDS/DIVISIONS/AFFILIATES:

Fisher Scientific International
Thermo Electron Group
Qualigens Fine Chemicals

CONTACTS: *Note: Officers with more than one job title may be intentionally listed here more than once.*

Marijin E. Dekkers, CEO
Marijin E. Dekkers, Pres.
Peter M. Wilver, CFO
Stephen G. Sheehan, Sr. VP-Human Resources
Seth H. Hoogasian, General Counsel/Sec./Sr. VP
Kenneth J. Apicerno, VP-Investor Rel./Treas.
Peter E. Hornstra, VP/Chief Acct. Officer
Marc N. Casper, Exec. VP
Guy Broadbent, Sr. VP
Alan J. Malus, Sr. VP
Fredric T. Walder, Sr. VP-Customer Excellence
Joseph R. Massaro, Sr. VP-Global Bus. Svcs.

Phone: 781-622-1000	Fax: 781-622-1207
Toll-Free: 800-678-5599	
Address: 81 Wyman St., Waltham, MA 02454 US	

GROWTH PLANS/SPECIAL FEATURES:

Thermo Fisher Scientific Inc., formerly Fisher Scientific International, is a distributor of products and services principally to the scientific-research and clinical laboratory markets. It serves over 350,000 customers in biotechnology and pharmaceutical companies; colleges and universities; medical-research institutions; hospitals; reference, quality-control, process-control and research and development labs in various industries; and government agencies. Thermo Fisher offers an array of products and services, from biochemicals, cell-culture media and proprietary RNAi technology to rapid diagnostic tests, safety products and other consumable supplies. The company's services include pharmaceutical services for Phase III and IV clinical trials, laboratory instrument calibration and repair, contract manufacturing, custom chemical synthesis, combinatorial chemistry, specialized packaging and supply chain management. Thermo Fisher's laboratory workstations segment engages in the manufacture and sale of laboratory furniture and fume hoods and also provides laboratory design services. The company was formed by the 2006 acquisition of Fisher Scientific International by Thermo Electron Group for $10.6 billion. In July 2007, Thermo Fisher announced plans to acquire Qualigens Fine Chemicals, a division of GlaxoSmithKine Pharmaceuticals Ltd., based in India.

Thermo Fisher offers its employees health and dental insurance and a 401(k) plan.

FINANCIALS: Sales and profits are in thousands of dollars—add 000 to get the full amount. 2006 Note: Financial information for 2006 was not available for all companies at press time.

2006 Sales: $3,791,600	2006 Profits: $168,900	**U.S. Stock Ticker:** TMO
2005 Sales: $2,633,000	2005 Profits: $223,200	**Int'l Ticker:** Int'l Exchange:
2004 Sales: $2,206,000	2004 Profits: $361,800	Employees:
2003 Sales: $1,899,400	2003 Profits: $200,000	Fiscal Year Ends: 12/31
2002 Sales: $1,849,400	2002 Profits: $309,700	Parent Company: THERMO ELECTRON GROUP

SALARIES/BENEFITS:

Pension Plan:	ESOP Stock Plan:	Profit Sharing:	Top Exec. Salary: $	Bonus: $
Savings Plan: Y	Stock Purch. Plan:		Second Exec. Salary: $	Bonus: $

OTHER THOUGHTS:

Apparent Women Officers or Directors:
Hot Spot for Advancement for Women/Minorities:

LOCATIONS: ("Y" = Yes)

West:	Southwest:	Midwest:	Southeast:	Northeast:	International:
Y		Y	Y	Y	Y

THIRD WAVE TECHNOLOGIES INC   www.twt.com

Industry Group Code: 325413 Ranks within this company's industry group: Sales: 20   Profits: 20

Drugs:		Other:		Clinical:		Computers:		Other:	
Discovery:		AgriBio:		Trials/Services:		Hardware:		Specialty Services:	
Licensing:		Genomics/Proteomics:	Y	Laboratories:		Software:		Consulting:	
Manufacturing:		Tissue Replacement:		Equipment/Supplies:	Y	Arrays:		Blood Collection:	
Development:				Research/Development Svcs.:		Database Management:		Drug Delivery:	
Generics:				Diagnostics:	Y			Drug Distribution:	

TYPES OF BUSINESS:
Genetic Analysis Products
Assays

BRANDS/DIVISIONS/AFFILIATES:
Invader
Cleavase
Invader InPlex
Invader Plus
Universal Invader Program

CONTACTS: Note: Officers with more than one job title may be intentionally listed here more than once.
Kevin T. Conroy, CEO
Kevin T. Conroy, Pres.
Maneesh Arora, CFO/Sr. VP
John Bellano, VP-Sales
Jorge Garces, VP-R&D
Cindy Ahn, General Counsel/VP/Corp. Sec.
Gregory Hamilton, VP-Oper.
Gregory Hamilton, VP-Finance
Ivan D. Trifunovich, Sr. VP
David Thompson, Chmn.

Phone: 608-273-8933	Fax: 608-273-8618
Toll-Free: 888-898-2357	
Address: 502 S. Rosa Rd., Madison, WI 53719 US	

GROWTH PLANS/SPECIAL FEATURES:
Third Wave Technologies, Inc. is a leading developer, manufacturer and marketer of genetic analysis products for clinical diagnostics and studies. Its patented genetic analysis platform, Invader, offers several advantages over conventional genetic analysis technologies. Invader relies upon the company's proprietary Cleavase enzyme for testing rather than using a complex copying technique known as a polymerase chain reaction (PCR). Available in ready-to-use formats, Third Wave's products are compatible with existing automation processes and detection platforms. These advantages make the Invader platform a convenient solution for genetic analysis, with applications ranging from disease discovery to patient care, including large-scale disease association studies, drug response marker profiling and molecular diagnostics. The company's Invader Plus product relies on a combination of the Invader platform with traditional PCR amplification. Developed in collaboration with 3M, the Invader InPlex product combines Invader chemistry with 3M's microfluidic technology, which helps to improve speed, efficiency and ease of use; in addition to utilizing existing laboratory equipment. In addition to Invader products, Third Wave offers assays for cardiovascular disease and deep-vein thrombosis detection, as well as tests relating to animal and plant genetics. The company's headquarters are located in Madison, Wisconsin. In November 2006, Third Wave announced the launch of its Universal Invader Program, which enables customers to customize their own Invader based assays.

FINANCIALS: Sales and profits are in thousands of dollars—add 000 to get the full amount. 2006 Note: Financial information for 2006 was not available for all companies at press time.

2006 Sales: $28,027	2006 Profits: $-18,887	**U.S. Stock Ticker: TWTI**
2005 Sales: $23,906	2005 Profits: $-22,346	**Int'l Ticker:** Int'l Exchange:
2004 Sales: $46,493	2004 Profits: $-1,942	Employees: 162
2003 Sales: $36,320	2003 Profits: $-8,116	Fiscal Year Ends: 12/31
2002 Sales: $30,900	2002 Profits: $-40,900	Parent Company:

SALARIES/BENEFITS:

Pension Plan:	ESOP Stock Plan:	Profit Sharing:	Top Exec. Salary: $375,000	Bonus: $282,769
Savings Plan:	Stock Purch. Plan:		Second Exec. Salary: $268,125	Bonus: $159,000

OTHER THOUGHTS:
Apparent Women Officers or Directors: 1
Hot Spot for Advancement for Women/Minorities:

LOCATIONS: ("Y" = Yes)

West:	Southwest:	Midwest:	Southeast:	Northeast:	International:
		Y			

TITAN PHARMACEUTICALS

www.titanpharm.com

Industry Group Code: 325412 Ranks within this company's industry group: Sales: 188   Profits: 90

Drugs:		Other:		Clinical:		Computers:		Other:	
Discovery:	Y	AgriBio:		Trials/Services:		Hardware:		Specialty Services:	
Licensing:		Genomics/Proteomics:		Laboratories:		Software:		Consulting:	
Manufacturing:	Y	Tissue Replacement:		Equipment/Supplies:		Arrays:		Blood Collection:	
Development:	Y			Research/Development Svcs.:		Database Management:		Drug Delivery:	Y
Generics:				Diagnostics:				Drug Distribution:	

TYPES OF BUSINESS:

Central Nervous System, Cardiovascular & Bone Diseases Therapeutics
Drug Delivery Systems

BRANDS/DIVISIONS/AFFILIATES:

Spheramine
Probuphine
Iloperidone
Gallium Maltolate
ProNeura

CONTACTS: *Note: Officers with more than one job title may be intentionally listed here more than once.*

Louis R. Bucalo, CEO
Sunil Bhonsle, COO/Exec. VP
Louis R. Bucalo, Pres.
Robert E. Farrell, CFO/Exec. VP
Louis R. Bucalo, Chmn.

Phone: 650-244-4990	Fax: 650-244-0715
Toll-Free:	
Address: 400 Oyster Point Blvd., Ste. 505, S. San Francisco, CA 94080 US	

GROWTH PLANS/SPECIAL FEATURES:

Titan Pharmaceuticals, Inc. is a biopharmaceutical company that develops proprietary therapeutics for the treatment of central nervous system disorders, cardiovascular disease, bone disease and other disorders. The company is focused on clinical development of probuphine, for the treatment of opioid dependence; iloperidone, in collaboration with Vanda Pharmaceutical, Inc., for the treatment of schizophrenia and related psychotic disorders; spheramine, in collaboration with Bayer Schering Pharma AG, for the treatment of advanced Parkinson's disease; DITPA, for the treatment of cardiovascular disease; and gallium maltolate, for the treatment of bone related diseases, chronic bacterial infections and cancer. The firm also offers a long-term drug delivery system that can be used for up to a year and is placed subcutaneously in the patient, called ProNeura; and its proprietary cell-coated microcarrier technology, which enables the development of cell-based therapies for minimally invasive, site-specific delivery to the central nervous system of therapeutic factors where they are needed.   The company utilizes grants from government agencies to fund development of its product candidates. In June 2007, the firm announced completion of enrollment in a Phase IIb clinical study of spheramine in the treatment of advanced Parkinson's disease.

FINANCIALS: Sales and profits are in thousands of dollars—add 000 to get the full amount. 2006 Note: Financial information for 2006 was not available for all companies at press time.

2006 Sales: $ 32	2006 Profits: $-15,737	U.S. Stock Ticker: TTP	
2005 Sales: $ 89	2005 Profits: $-22,462	Int'l Ticker: Int'l Exchange:	
2004 Sales: $ 31	2004 Profits: $-26,004	Employees: 38	
2003 Sales: $ 89	2003 Profits: $-29,889	Fiscal Year Ends: 12/31	
2002 Sales: $2,900	2002 Profits: $-28,200	Parent Company:	

SALARIES/BENEFITS:

Pension Plan:	ESOP Stock Plan:	Profit Sharing:	Top Exec. Salary: $366,325	Bonus: $
Savings Plan:	Stock Purch. Plan:		Second Exec. Salary: $279,208	Bonus: $

OTHER THOUGHTS:

Apparent Women Officers or Directors:
Hot Spot for Advancement for Women/Minorities:

LOCATIONS: ("Y" = Yes)

West:	Southwest:	Midwest:	Southeast:	Northeast:	International:
Y					

Note: Financial information, benefits and other data can change quickly and may vary from those stated here.

TORREYPINES THERAPEUTICS INC
www.torreypinestherapeutics.com

Industry Group Code: 325412 Ranks within this company's industry group: Sales: 130   Profits: 116

Drugs:		Other:		Clinical:		Computers:		Other:	
Discovery:	Y	AgriBio:		Trials/Services:		Hardware:		Specialty Services:	
Licensing:	Y	Genomics/Proteomics:		Laboratories:		Software:		Consulting:	
Manufacturing:		Tissue Replacement:		Equipment/Supplies:		Arrays:		Blood Collection:	
Development:	Y			Research/Development Svcs.:		Database Management:		Drug Delivery:	
Generics:				Diagnostics:				Drug Distribution:	

TYPES OF BUSINESS:
Pharmaceutical Acquisition & Development
Biopharmaceuticals
Drug Development
Central Nervous System Disorder Treatment Products
Alzheimer's Disease Treatment Products
CIAS Treatment Products
Migraine Treatment Products

BRANDS/DIVISIONS/AFFILIATES:
AXONYX, Inc.
tezampanel
phenserine
posiphen
bisnorcymserine
Eisai Co., Ltd.

CONTACTS: Note: Officers with more than one job title may be intentionally listed here more than once.
Neil Kurtz, CEO
Evelyn Graham, COO
Neil Kurtz, Pres.
Craig Johnson, CFO/VP-Finance
Steven Wagner, Chief Scientific Officer
Paul Schneider, General Counsel/VP
Sue Mellbery, VP-Project Mgmt.
Michael Murphy, Sr. VP/Chief Medical Officer

Phone: 858-623-5665	Fax: 858-623-5666
Toll-Free:	
Address: 11085 N Torrey Pines Rd., Ste. 300, La Jolla, CA 92037 US	

GROWTH PLANS/SPECIAL FEATURES:
TorreyPines Therapeutics, Inc., formerly AXONYX, Inc., is a biopharmaceutical company engaged in the discovery, development, and commercialization of novel small molecules to treat diseases and disorders of the central nervous system. AXONYX, Inc. became TorreyPines Therapeutics, Inc. following a merger with TorreyPines in October 2006. The company's therapeutic focus is in two areas: chronic pain, including migraine and neuropathic pain; and cognitive disorders, including cognitive impairment associated with schizophrenia and Alzheimer's disease. For chronic pain, TorreyPines has two product candidates in clinical trials: tezampanel, currently in a Phase IIb trial for the abortive treatment of migraines; and NGX426, in a Phase I clinical trial for chronic pain treatment. Tezampanel does not constrict blood vessels, as current migraine treatment options do, but instead is believed to work by selectively blocking the transmission of pain signals to the brain. For cognitive disorders, NGX267 is the company's lead product candidate for the treatment of Cognitive Impairment Associated with Schizophrenia (CIAS), and is currently in a third Phase I clinical trial. The company expects that NGX267 would be used as adjunctive therapy to current antipsychotic therapy to treat schizophrenia. Another product candidate for the treatment of CIAS in preclinical development is NGX292. Both NGX267 and NGX292 may also be evaluated by TorreyPines for the potential treatment of Alzheimer's disease. TorreyPines has an additional four product candidates in development focused on the treatment of Alzheimer's disease: phenserine, for which the company has completed Phase III clinical trials and is currently pursuing out-licensing opportunities; posiphen, which has completed Phase I clinical trials; and bisnorcymserine and NGX555, both of which are in preclinical development. TorreyPines is also conducting two drug discovery programs in collaboration with Eisai Co., Ltd. which are focused on discovering and validating small molecules and novel molecular targets for Alzheimer's disease.

FINANCIALS: Sales and profits are in thousands of dollars—add 000 to get the full amount. 2006 Note: Financial information for 2006 was not available for all companies at press time.

2006 Sales: $9,850	2006 Profits: $-25,377	**U.S. Stock Ticker: TPTX**
2005 Sales: $7,967	2005 Profits: $-11,542	**Int'l Ticker:** Int'l Exchange:
2004 Sales: $3,551	2004 Profits: $-10,356	Employees: 43
2003 Sales: $1,000	2003 Profits: $-8,106	Fiscal Year Ends: 12/31
2002 Sales: $	2002 Profits: $-6,300	Parent Company:

SALARIES/BENEFITS:

Pension Plan:	ESOP Stock Plan:	Profit Sharing:	Top Exec. Salary: $354,272	Bonus: $135,364
Savings Plan:	Stock Purch. Plan:		Second Exec. Salary: $311,850	Bonus: $58,747

OTHER THOUGHTS:
Apparent Women Officers or Directors: 2
Hot Spot for Advancement for Women/Minorities: Y

LOCATIONS: ("Y" = Yes)

West:	Southwest:	Midwest:	Southeast:	Northeast:	International:
Y					Y

Note: Financial information, benefits and other data can change quickly and may vary from those stated here.

TOYOBO CO LTD

www.toyobo.co.jp

Industry Group Code: 313000 **Ranks within this company's industry group:** Sales: 1   Profits: 1

Drugs:		Other:	Clinical:	Computers:	Other:
Discovery:	Y	AgriBio:	Trials/Services:	Hardware:	Specialty Services:
Licensing:		Genomics/Proteomics:	Laboratories:	Software:	Consulting:
Manufacturing:	Y	Tissue Replacement:	Equipment/Supplies:	Arrays:	Blood Collection:
Development:			Research/Development Svcs.:	Database Management:	Drug Delivery:
Generics:			Diagnostics:		Drug Distribution:

TYPES OF BUSINESS:

Textile & Fiber Manufacturing
Advanced Materials
Biomedical Products
Industrial Textiles
Plastics & Films
Engineering
Logistics Services
Real Estate

BRANDS/DIVISIONS/AFFILIATES:

Zylon
Dyneema

CONTACTS: Note: Officers with more than one job title may be intentionally listed here more than once.

Ryuzou Sakamoto, COO
Ryuzou Sakakmoto, Pres.
Fumishige Imamura, Sr. Managing Dir.-Credit, Audit & Finance
Kazuyuki Yabuki, Sr. Gen. Mgr.-Corp. R&D
Kenji Hayashi, Sr. Managing Dir.-Labor & Admin.
Kenji Hayashi, Sr. Managing Dir.-Law
Shigeaki Kogamo, Sr. General Mgr.-Corp. Planning
Fumishige Imamura, Dir.-Audit Dept., Finance, Acct., Control Dept.
Hiroyuki Kagawa, Dir.-Fibers & Textiles
Kanji Aono, Sr. Managing Dir.-Films
Fumiaki Miyoshi, General Mgr.-High Functional Prod.
Yoshihisa Kawamura, Dir.-Bioscience & Medical
Junji Tsumura, Chmn.

Phone: 81-6-6348-3111	Fax: 81-6-6348-3206
Toll-Free:	
Address: 2-8, Dojima Hama 2-chome, Kit-ku, Osaka, 530-8230 Japan	

GROWTH PLANS/SPECIAL FEATURES:

Toyobo Co. Ltd., founded as a textile business in 1882, has expanded into domains that utilize its core technologies of polymerization, modification, processing and bioscience. Currently, the company operates in four segments: fibers and textiles, plastics, bio and medical and other business. The fibers and textiles segment produces a range of fabric materials, including polyester, nylon, spandex, polynostic, cotton and wool. The company has also developed super-fibers, which combine the strength of metal with the lightness of fibers. These include Zylon, a fiber with high thermal resistance and Dyneema, a super-strong fiber that is lighter than water. Toyobo's bioscience and medical segment is a leading manufacturer of antibody drugs, and was the first company in Japan to receive approval for the production of biopharmaceuticals using genetically engineered animal cells. The plastics segment produces polyester film for soft packaging and optical polyester film for IT equipment, as well as various plastics and resins. The company also operates several other businesses, including engineering, real estate, information processing services and logistics services. Toyobo has recently revised its business structure and reduced assets, especially in unprofitable businesses, such as common fabrics and textiles for the clothing sector. This is part of a managing initiative to focus more on businesses that the company sees as more profitable, stable and in tune with today's world and market.   Toyobo intends to aggressively expand its businesses in Japan and abroad and start working on management reform, focusing on enhancing, integrating and evolving core technologies in order to accelerate the creation of new products and businesses.

FINANCIALS: Sales and profits are in thousands of dollars—add 000 to get the full amount. 2006 Note: Financial information for 2006 was not available for all companies at press time.

2006 Sales: $3,418,200	2006 Profits: $107,100	**U.S. Stock Ticker: TYOBY**
2005 Sales: $3,660,500	2005 Profits: $113,500	**Int'l Ticker: 3101** Int'l Exchange: Tokyo-TSE
2004 Sales: $3,531,400	2004 Profits: $82,900	Employees: 3,273
2003 Sales: $3,140,500	2003 Profits: $-58,100	Fiscal Year Ends: 3/31
2002 Sales: $2,888,000	2002 Profits: $100,700	Parent Company:

SALARIES/BENEFITS:

Pension Plan:	ESOP Stock Plan:	Profit Sharing:	Top Exec. Salary: $	Bonus: $
Savings Plan:	Stock Purch. Plan:		Second Exec. Salary: $	Bonus: $

OTHER THOUGHTS:

Apparent Women Officers or Directors:
Hot Spot for Advancement for Women/Minorities:

LOCATIONS: ("Y" = Yes)

West:	Southwest:	Midwest:	Southeast:	Northeast:	International:
				Y	Y

Note: Financial information, benefits and other data can change quickly and may vary from those stated here.

TRIMERIS INC
www.trimeris.com

Industry Group Code: 325412 Ranks within this company's industry group: Sales: 89 Profits: 50

Drugs:		Other:	Clinical:	Computers:	Other:	
Discovery:	Y	AgriBio:	Trials/Services:	Hardware:	Specialty Services:	Y
Licensing:		Genomics/Proteomics:	Laboratories:	Software:	Consulting:	
Manufacturing:		Tissue Replacement:	Equipment/Supplies:	Arrays:	Blood Collection:	
Development:	Y		Research/Development Svcs.:	Database Management:	Drug Delivery:	
Generics:			Diagnostics:		Drug Distribution:	

TYPES OF BUSINESS:
Antivirals
HIV Drugs

BRANDS/DIVISIONS/AFFILIATES:
FUZEON

CONTACTS:
Note: Officers with more than one job title may be intentionally listed here more than once.

E. Lawrence Hill, COO
E. Lawrence Hill, Acting Pres.
George W. Koszalka, Exec. VP-Scientific Oper.
Neil Graham, Sr. VP-Clinical Dev. & Medical Affairs
Robert R. Bonczek, General Counsel
Walter Capone, VP-Commercial Oper.
Neil Graham, Chief Medical Officer
Jeffrey M. Lipton, Chmn.

Phone: 919-419-6050	Fax: 919-419-1816

Toll-Free:

Address: 3500 Paramount Pkwy., Morrisville, NC 27560 US

GROWTH PLANS/SPECIAL FEATURES:

Trimeris, Inc. is a biopharmaceutical company primarily engaged in the discovery, development and commercialization of a class of antiviral drug treatments called fusion inhibitors. Fusion inhibitors impair viral fusion, a complex process by which viruses attach to, penetrate and infect host cells. The firm focuses on Fuzeon, an HIV fusion inhibitor that was developed in collaboration with F. Hoffman-La Roche, Ltd. When used in combination with other anti-HIV drugs, FUZEON has been shown to reduce the amount of HIV in the blood and increase the number of T-cells. The drug is approved for marketing by FDA in combination with other anti-HIV drugs for the treatment of HIV-1 infection in treatment-experienced patients with evidence of HIV-1 replication despite ongoing anti-HIV therapy; and by the European Agency for the Evaluation of Medicinal Products in exceptional circumstances. The company has partnerships with Array BioPharma, ChemBridge Research Laboratories, Inc., F. Hoffman-La Roche, Ltd. and Tranzyme, Inc. Trimeris' research and development program focuses primarily on treating viral diseases by identifying mechanisms for blocking viral entry. The company owns or has exclusive licenses to 41 issued U.S. patents. In July 2007, the firm initiated a new trial designed to evaluate the efficacy and safety of Fuzeon in combination with an investigational integrase inhibitor.

The company offers its employees health, dental and life insurance; short- and long-term disability; a 401(k) plan; flexible spending accounts; an employee stock purchase plan; and an educational assistance program.

FINANCIALS:
Sales and profits are in thousands of dollars—add 000 to get the full amount. 2006 Note: Financial information for 2006 was not available for all companies at press time.

2006 Sales: $36,980	2006 Profits: $7,384	**U.S. Stock Ticker: TRMS**
2005 Sales: $19,059	2005 Profits: $-8,106	**Int'l Ticker:** Int'l Exchange:
2004 Sales: $6,708	2004 Profits: $-40,088	Employees: 64
2003 Sales: $3,719	2003 Profits: $-65,703	Fiscal Year Ends: 12/31
2002 Sales: $1,133	2002 Profits: $-75,678	Parent Company:

SALARIES/BENEFITS:

Pension Plan:	ESOP Stock Plan:	Profit Sharing:	Top Exec. Salary: $487,008	Bonus: $133,900
Savings Plan: Y	Stock Purch. Plan: Y		Second Exec. Salary: $472,008	Bonus: $190,000

OTHER THOUGHTS:
Apparent Women Officers or Directors:
Hot Spot for Advancement for Women/Minorities:

LOCATIONS: ("Y" = Yes)

West:	Southwest:	Midwest:	Southeast:	Northeast:	International:
				Y	

Note: Financial information, benefits and other data can change quickly and may vary from those stated here.

TRINITY BIOTECH PLC

www.trinitybiotech.com

Industry Group Code: 325413 Ranks within this company's industry group: Sales: 10 Profits: 12

Drugs:	Other:	Clinical:		Computers:		Other:	
Discovery:	AgriBio:	Trials/Services:		Hardware:		Specialty Services:	
Licensing:	Genomics/Proteomics:	Laboratories:		Software:		Consulting:	
Manufacturing:	Tissue Replacement:	Equipment/Supplies:	Y	Arrays:		Blood Collection:	
Development:		Research/Development Svcs.:		Database Management:		Drug Delivery:	
Generics:		Diagnostics:	Y			Drug Distribution:	

TYPES OF BUSINESS:

Medical Diagnostics Products
Immunoassay Technology

BRANDS/DIVISIONS/AFFILIATES:

AMAX
Bartels
Capillus
Destiny
UniGold
Recombigen
Captia
MarDx

CONTACTS: Note: Officers with more than one job title may be intentionally listed here more than once.

Ronan O'Caoimh, CEO
James Walsh, COO
Brendan K. Farrell, Pres.
Rory Nealon, CFO
Ronan O'Caoimh, Chmn.

Phone: 353-1276-9800	**Fax:** 353-1276-9888
Toll-Free:	
Address: 1 Southern Cross, IDA Business Park, Bray, Wicklow Ireland	

GROWTH PLANS/SPECIAL FEATURES:

Trinity Biotech plc is engaged in developing, manufacturing and marketing diagnostic products that use immunoassay technologies. Trinity has five manufacturing sites in Ireland, Germany, Sweden and the U.S., and sells its products to point-of-care and clinical laboratories in 80 countries worldwide. The company's product offerings are used for testing and diagnosis in autoimmune, infectious and sexually transmitted diseases (STDs), diabetes and disorders of the blood, liver and intestine. Trinity's products can be divided into four lines, which include haemostasis, point-of-care, infectious diseases and clinical chemistry. The haemostasis product line includes test kits and instrumentation for the detection of blood disorders. It includes the products Biopool, Amax and Destiny. Point-of-care products primarily include tests for the presence of HIV antibodies, including the products UniGold, Capillus and Recombigen. The infectious disease product line tests for a wide range of issues, such as autoimmune diseases, hormonal imbalances, STDs, intestinal infections and many others. Products include Bartels, Captia and MarDx. Trinity's clinical chemistry products have proven performance in the diagnosis of many disease states from liver and kidney disease to G6PDH deficiency which is an indicator of haemolytic anaemia. Since its inception in 1992, Trinity has expanded through a series of acquisitions. In June 2006, the company acquired the haemostasis product line, which includes haemostasis test kits and automated instruments, from bioMerieux Inc. for approximately $44.4 million. In October 2006, Trinity acquired the French distribution business of Laboratories Nephrotek SARL for approximately $1.175 million.

FINANCIALS: Sales and profits are in thousands of dollars—add 000 to get the full amount. 2006 Note: Financial information for 2006 was not available for all companies at press time.

2006 Sales: $118,674	2006 Profits: $3,276	**U.S. Stock Ticker: TRIB**
2005 Sales: $98,560	2005 Profits: $5,280	**Int'l Ticker: TWU** Int'l Exchange: Dublin-ISE
2004 Sales: $80,008	2004 Profits: $5,714	Employees: 826
2003 Sales: $65,700	2003 Profits: $5,800	Fiscal Year Ends: 12/31
2002 Sales: $52,000	2002 Profits: $5,000	Parent Company:

SALARIES/BENEFITS:

Pension Plan:	ESOP Stock Plan:	Profit Sharing:	Top Exec. Salary: $552,000	Bonus: $246,000
Savings Plan:	Stock Purch. Plan:		Second Exec. Salary: $419,000	Bonus: $157,000

OTHER THOUGHTS:

Apparent Women Officers or Directors: 1
Hot Spot for Advancement for Women/Minorities: Y

LOCATIONS: ("Y" = Yes)

West:	Southwest:	Midwest:	Southeast:	Northeast:	International:
Y		Y		Y	Y

TRIPOS INC

www.tripos.com

Industry Group Code: 511212 Ranks within this company's industry group: Sales: 5 Profits: 4

Drugs:		Other:		Clinical:	Computers:		Other:	
Discovery:	Y	AgriBio:	Y	Trials/Services:	Hardware:		Specialty Services:	Y
Licensing:		Genomics/Proteomics:		Laboratories:	Software:	Y	Consulting:	Y
Manufacturing:		Tissue Replacement:		Equipment/Supplies:	Arrays:		Blood Collection:	
Development:				Research/Development Svcs.:	Database Management:		Drug Delivery:	
Generics:				Diagnostics:			Drug Distribution:	

TYPES OF BUSINESS:

Biotech Software, Research Products & Services
Clinical Software
Consulting Services
Chemical Compound Libraries
Integrated Data Systems

BRANDS/DIVISIONS/AFFILIATES:

SYBYL
Benchware
LeadDiscovery
LeadQuest
LeadHopping
Discovery Chemistry
Tripos Discovery Research Centre
Optive Research, Inc.

CONTACTS: Note: Officers with more than one job title may be intentionally listed here more than once.

John P. McAlister III, CEO
John P. McAlister III, Pres.
Richard D. Cramer III, Chief Scientific Officer/Sr. VP-Science
Mary P. Woodward, Sr. VP-Strategic Dev.
Philip Small, VP-High Throughput Chemistry
Janet Heenan, VP-European Human Resources
Ralph S. Lobdell, Chmn.

Phone: 314-647-1099	**Fax:** 314-647-9241

Toll-Free: 800-323-2960
Address: 1699 S. Hanley Rd., St. Louis, MO 63144-2913 US

GROWTH PLANS/SPECIAL FEATURES:

Tripos, Inc., through its many subsidiaries worldwide, is a leader in discovery services, informatics and products for life science organizations. Its products and services are used primarily in the preclinical phases of new pharmaceutical development, the equivalent pre-approval phase of agrochemical product development and the product discovery phases of chemical research. The company's discovery informatics products include SYBYL, a virtual discovery laboratory and molecular modeling environment; Benchware, a suite of applications used to support synthesis and testing decisions, manipulate research information and facilitate experimental collaboration; and various operating platform technologies. Additionally, Tripos produces custom IT solutions designed to reduce the amount of time involved in drug discovery. Tripos Discovery Research Centre's U.K. laboratories offer research products and services such as compound design and synthesis, molecular analysis and off-the-shelf libraries of compounds. The company's LeadDiscovery and Discovery Chemistry programs are designed to facilitate and accelerate the drug discovery process, offering computational design tools, chemical discovery initiation points and patent infringement prevention technologies. The company's discovery research business is underpinned by LeadQuest, a library of over 80,000 drug-relevant, synthetically feasible compounds. Tripos has long-standing relationships with numerous leading pharmaceutical and biotechnology companies. Major customers and collaborators include Bristol-Myers Squibb, Servier, Wyeth Pharmaceuticals, Bayer AG, European Molecular Biology Laboratory, Cara Therapeutics and the German Cancer Research Centre. In March 2007, Tripos sold its discovery informatics business, the company's principal operating unit, to Vector Capital Corporation for $26.2 million, which has since turned that segment into a newly private company that will continue to develop its existing portfolio. In June 2007, the company sold its discovery research business, including all related products and services, to Commonwealth Biotechnologies. The company intends to repay existing debt and then dissolve and liquidate the business. The initial liquidating distribution is expected to be made in late 2007.

FINANCIALS: Sales and profits are in thousands of dollars—add 000 to get the full amount. 2006 Note: Financial information for 2006 was not available for all companies at press time.

2006 Sales: $27,384	2006 Profits: $-38,593	**U.S. Stock Ticker:** TRPS.OB
2005 Sales: $27,981	2005 Profits: $-4,288	**Int'l Ticker:** Int'l Exchange:
2004 Sales: $20,478	2004 Profits: $ 232	Employees: 327
2003 Sales: $54,148	2003 Profits: $2,100	Fiscal Year Ends: 12/31
2002 Sales: $51,100	2002 Profits: $ 900	Parent Company:

SALARIES/BENEFITS:

Pension Plan:	ESOP Stock Plan:	Profit Sharing:	Top Exec. Salary: $380,000	Bonus: $
Savings Plan: Y	Stock Purch. Plan: Y		Second Exec. Salary: $240,000	Bonus: $

OTHER THOUGHTS:

Apparent Women Officers or Directors: 3
Hot Spot for Advancement for Women/Minorities: Y

LOCATIONS: ("Y" = Yes)

West:	Southwest:	Midwest:	Southeast:	Northeast:	International:
		Y			

Note: Financial information, benefits and other data can change quickly and may vary from those stated here.

UCB SA

www.ucb-group.com

Industry Group Code: 325412 Ranks within this company's industry group: Sales: 25   Profits: 23

Drugs:		Other:	Clinical:		Computers:		Other:
Discovery:	Y	AgriBio:	Trials/Services:		Hardware:		Specialty Services:
Licensing:		Genomics/Proteomics:	Laboratories:		Software:		Consulting:
Manufacturing:	Y	Tissue Replacement:	Equipment/Supplies:	Y	Arrays:		Blood Collection:
Development:	Y		Research/Development Svcs.:		Database Management:		Drug Delivery:
Generics:			Diagnostics:				Drug Distribution:

TYPES OF BUSINESS:

Pharmaceuticals Development
Industrial Chemical Products
Allergy & Respiratory Treatments
Central Nervous System Disorder Treatments

BRANDS/DIVISIONS/AFFILIATES:

UCB Pharma
UCB Surface Specialties
Keppra
Zyzal
Xyzal
Zyrtec
Lortab
Metadate

CONTACTS: *Note: Officers with more than one job title may be intentionally listed here more than once.*

Roch Doliveux, CEO
Detlef Thielgen, CFO/Exec. VP
Jean-Pierre Pradier, Exec. VP-Human Resources
Melanie Lee, Exec. VP-R&D
Bob Trainor, General Counsel/Exec. VP
Bill Robinson, Exec. VP-Global Oper.
Antje Witte, VP-Comm.
Antje Witte, VP-Investor Rel.
Francois Meurgey, Sr. VP-Commercialization
Baron Jacobs, Chmn.

Phone: 32-2-559-99-99	Fax: 32-2-599-99-00
Toll-Free:	
Address: Allee de la Recherche, 60, Brussels, B-1070 Belgium	

GROWTH PLANS/SPECIAL FEATURES:

UCB SA is a Belgian biopharmaceutical firm focusing on severe diseases in the fields of the central nervous system, including epilepsy; inflammation, including allergy; and oncology. The firm's main products include antiepileptics, antiallergics, cerebral function regulators, painkillers, blood pressure treatments and tranquillizers, among others. Keppra, the firm's lead product, is available in 46 countries for the treatment of epilepsy. UCB Pharma's anti-allergics include Xyzal, an anti-allergic designed to treat and prevent persistent rhinitis in children, characterized by severe and long-lasting allergic symptoms with a tendency to evolve towards allergic asthma; and Zyrtec, the only anti-allergic drug approved for use in children from the age of six months and the highest-selling antihistamine in the world. Other products include Nootropil, a cerebral function regulator used to treat adults and the elderly; Atarax, a non-benzodiazepinic tranquillizer, which has given its name to a therapeutic class, the ataraxics; Lortab, an analgesic approved in the USA for the relief of moderate to moderately severe pain; Tussionex, a 12-hour cough suppressant approved in the USA; Metadate CD/Equasym XL used in the treatment of Attention Deficit Hyperactivity Disorder (ADHD); and BUP-4, a once daily treatment for urinary incontinence. In January 2007, UCB acquired more than 88% of Schwarz Pharma AG's shares. In March 2007, the company sold its stake in Cytec Industries Inc.

FINANCIALS: Sales and profits are in thousands of dollars—add 000 to get the full amount. 2006 Note: Financial information for 2006 was not available for all companies at press time.

2006 Sales: $2,987,500	2006 Profits: $501,100	**U.S. Stock Ticker:**
2005 Sales: $2,602,500	2005 Profits: $961,760	**Int'l Ticker: UCB** Int'l Exchange: Brussels-Euronext
2004 Sales: $2,132,440	2004 Profits: $419,100	Employees: 11,403
2003 Sales: $4,176,800	2003 Profits: $426,400	Fiscal Year Ends: 12/31
2002 Sales: $3,063,900	2002 Profits: $346,600	Parent Company:

SALARIES/BENEFITS:

Pension Plan:	ESOP Stock Plan:	Profit Sharing:	Top Exec. Salary: $	Bonus: $
Savings Plan:	Stock Purch. Plan:		Second Exec. Salary: $	Bonus: $

OTHER THOUGHTS:

Apparent Women Officers or Directors: 1
Hot Spot for Advancement for Women/Minorities:

LOCATIONS: ("Y" = Yes)

West:	Southwest:	Midwest:	Southeast:	Northeast:	International:
			Y	Y	Y

UNIGENE LABORATORIES

www.unigene.com

Industry Group Code: 325412 Ranks within this company's industry group: Sales: 138   Profits: 79

Drugs:		Other:	Clinical:	Computers:	Other:	
Discovery:	Y	AgriBio:	Trials/Services:	Hardware:	Specialty Services:	Y
Licensing:	Y	Genomics/Proteomics:	Laboratories:	Software:	Consulting:	
Manufacturing:	Y	Tissue Replacement:	Equipment/Supplies:	Arrays:	Blood Collection:	
Development:	Y		Research/Development Svcs.:	Database Management:	Drug Delivery:	Y
Generics:			Diagnostics:		Drug Distribution:	

TYPES OF BUSINESS:

Peptides Research & Production
Drug Delivery Systems

BRANDS/DIVISIONS/AFFILIATES:

Fortical
Forcaltonin

CONTACTS: *Note: Officers with more than one job title may be intentionally listed here more than once.*

Warren Levy, CEO
Warren P. Levy, Pres.
Nozer M. Mehta, VP-Biological R&D
James Gilligan, VP-Product Dev.
Paul Shields, VP-Mfg. Oper.
Ronald Levy, Sec./Exec. VP
William Steinhauer, VP-Finance
Jay Levy, Treas.
Jay Levy, Chmn.

Phone: 973-882-0860	Fax: 973-882-8277

Toll-Free:

Address: 110 Little Falls Rd., Fairfield, NJ 07004 US

GROWTH PLANS/SPECIAL FEATURES:

Unigene Laboratories, Inc. is a biopharmaceutical company that focuses on the research, production and delivery of small proteins, or peptides, for medical use. The company has a patented manufacturing technology for producing many peptides, as well as patented oral and nasal delivery technologies that have been shown to deliver medically useful amounts to various peptides into the bloodstream. The firm's primary focus is on the development of calcitonin and other peptide products for the treatment of osteoporosis and other indications. Unigene Laboratories licensed worldwide rights to its manufacturing and delivery technologies for oral parathyroid hormone to GlaxoSmithKline. In the U.S., the company licensed its nasal calcitonin product, trademarked Fortical, to Upsher-Smith Laboratories, Inc. Fortical, aimed at treating osteoporosis, was the company's first drug approval in the U.S. The firm licensed worldwide rights to its patented manufacturing technology for the production of calcitonin to Novartis Pharma AG. In addition to Fortical, Unigene Laboratories has an injectable calcitonin product, Forcaltonin, that is approved for sale in the E.U. for osteoporosis indications. The company holds nine U.S. patents. The firm derives revenue solely from domestic sources. In August 2007, Unigene Laboratories announced the commencement of a clinical study in humans in the U.S. with its proprietary formulation of oral calcitonin for the treatment of osteoporosis.

FINANCIALS: Sales and profits are in thousands of dollars—add 000 to get the full amount. 2006 Note: Financial information for 2006 was not available for all companies at press time.

2006 Sales: $6,059	2006 Profits: $-11,784	**U.S. Stock Ticker:** UGNE
2005 Sales: $14,276	2005 Profits: $- 496	**Int'l Ticker:** Int'l Exchange:
2004 Sales: $8,400	2004 Profits: $-5,900	Employees: 91
2003 Sales: $6,024	2003 Profits: $-7,398	Fiscal Year Ends: 12/31
2002 Sales: $2,658	2002 Profits: $-6,337	Parent Company:

SALARIES/BENEFITS:

Pension Plan:	ESOP Stock Plan:	Profit Sharing:	Top Exec. Salary: $246,591	Bonus: $19,800
Savings Plan:	Stock Purch. Plan:		Second Exec. Salary: $228,614	Bonus: $18,900

OTHER THOUGHTS:

Apparent Women Officers or Directors:
Hot Spot for Advancement for Women/Minorities:

LOCATIONS: ("Y" = Yes)

West:	Southwest:	Midwest:	Southeast:	Northeast:	International:
				Y	

UNITED THERAPEUTICS CORP   www.unither.com

Industry Group Code: 325412 Ranks within this company's industry group: Sales: 59   Profits: 32

Drugs:		Other:		Clinical:		Computers:		Other:	
Discovery:	Y	AgriBio:		Trials/Services:		Hardware:		Specialty Services:	
Licensing:		Genomics/Proteomics:		Laboratories:		Software:		Consulting:	
Manufacturing:	Y	Tissue Replacement:		Equipment/Supplies:	Y	Arrays:		Blood Collection:	
Development:	Y			Research/Development Svcs.:		Database Management:		Drug Delivery:	
Generics:				Diagnostics:				Drug Distribution:	

TYPES OF BUSINESS:

Cardiovascular, Cancer & Infectious Diseases Therapeutics
Dietary Supplements
Telecardiology Products

BRANDS/DIVISIONS/AFFILIATES:

Unither Pharma, Inc.
Medicomp, Inc.
HeartBar
Remodulin
OvaRex

CONTACTS: Note: Officers with more than one job title may be intentionally listed here more than once.

Martine Rothblatt, CEO
Roger Jeffs, COO
Roger Jeffs, Pres.
John Ferrari, CFO
Shola Oyewole, CIO
Paul A. Mahon, General Counsel
Paul A. Mahon, Exec. VP-Strategic Planning
John Ferrari, Treas.
Dan Balda, Pres./COO-Medicomp, Inc.
Peter C. Gonze, COO-Unither Pharmaceuticals, Inc.
David Walsh, Exec. VP/COO-Production
Martine Rothblatt, Chmn.

Phone: 301-608-9292	Fax: 301-608-9291
Toll-Free:	
Address: 1110 Spring St., Silver Spring, MD 20910 US	

GROWTH PLANS/SPECIAL FEATURES:

United Therapeutics Corp. (UTC) is a biotechnology company focused on the development and commercialization of therapeutic products for patients with chronic and life-threatening diseases that is active in three therapeutic areas: cardiovascular, cancer and infectious diseases. The company's key therapeutic platforms include: prostacyclin analogs, which are stable synthetic forms of prostacyclin, a molecule produced by the body that affects blood-vessel health and function; immunotherapeutic monoclonal antibodies, which are antibodies that activate patients' immune systems to treat cancer, including OvaRex, which is being developed for the treatment of ovarian cancer; and glycobiology antiviral agents, which are a class of small molecules that have shows preclinical indications of efficacy against a broad range of viruses. UTC's Remodulin has been approved by the FDA for the treatment of pulmonary arterial hypertension in patients with New York Heart Association moderate to severe symptoms to diminish symptoms associated with exercise, and in other countries for similar use. One of the firm's subsidiaries, Unither Pharma, Inc., markets the HeartBar line of products, which are arginine-enriched dietary supplements. Arginine is one of the 20 amino acids necessary for life and is critical for maintaining circulatory function. Subsidiary Medicomp, Inc. manufactures and markets a variety of telecardiology services, including cardiac Holter monitoring, event monitoring and analysis and pacemaker monitoring. Medicomp's services are delivered through its proprietary, miniaturized, digital Decipher Holter recorder/analyzer and its CardioPAL family of event monitors. In March 2007, UTC signed an agreement with Mochida Pharmaceutical Co., Ltd. for the exclusive distribution of Remodulin in Japan.

The company offers its employees medical, vision, dental and prescription insurance; life and AD&D insurance; short- and long-term disability; flexible spending accounts; a 401(k) plan; educational assistance; and an employee assistance program.

FINANCIALS: Sales and profits are in thousands of dollars—add 000 to get the full amount. 2006 Note: Financial information for 2006 was not available for all companies at press time.

2006 Sales: $159,632	2006 Profits: $73,965	U.S. Stock Ticker: UTHR
2005 Sales: $115,915	2005 Profits: $65,016	Int'l Ticker: Int'l Exchange:
2004 Sales: $73,590	2004 Profits: $15,449	Employees: 285
2003 Sales: $53,341	2003 Profits: $-9,969	Fiscal Year Ends: 12/31
2002 Sales: $30,100	2002 Profits: $-23,700	Parent Company:

SALARIES/BENEFITS:

Pension Plan:	ESOP Stock Plan:	Profit Sharing:	Top Exec. Salary: $725,000	Bonus: $300,000
Savings Plan: Y	Stock Purch. Plan:		Second Exec. Salary: $650,000	Bonus: $210,000

OTHER THOUGHTS:

Apparent Women Officers or Directors: 1
Hot Spot for Advancement for Women/Minorities:

LOCATIONS: ("Y" = Yes)

West:	Southwest:	Midwest:	Southeast:	Northeast:	International:
			Y	Y	Y

Note: Financial information, benefits and other data can change quickly and may vary from those stated here.

UNITED-GUARDIAN INC

www.u-g.com

Industry Group Code: 325412 Ranks within this company's industry group: Sales: 125 Profits: 53

Drugs:		Other:		Clinical:		Computers:		Other:	
Discovery:	Y	AgriBio:		Trials/Services:		Hardware:		Specialty Services:	
Licensing:		Genomics/Proteomics:		Laboratories:		Software:		Consulting:	
Manufacturing:	Y	Tissue Replacement:		Equipment/Supplies:		Arrays:		Blood Collection:	
Development:	Y			Research/Development Svcs.:	Y	Database Management:		Drug Delivery:	
Generics:				Diagnostics:				Drug Distribution:	

TYPES OF BUSINESS:

Cosmetic Ingredients & Pharmaceuticals
Personal & Healthcare Products
Organic Chemicals
Test Solutions
Indicators
Dyes
Reagents

BRANDS/DIVISIONS/AFFILIATES:

Guardian Laboratories
Eastern Chemical Corp.
Paragon Organic Chemicals, Inc.
Lubrajel
Recidin
Clorpactin
Klensoft
Confetti Dermal Delivery Flakes

CONTACTS: *Note: Officers with more than one job title may be intentionally listed here more than once.*

Alfred R. Globus, CEO
Kenneth H. Globus, Pres.
Kenneth H. Globus, CFO
Joseph J. Vernice, VP/Manager-R&D
Joseph J. Vernice, Dir.-Tech. Svcs.
Robert S. Rubinger, Dir.-Prod. Dev.
Kenneth H. Globus, General Counsel
Robert S. Rubinger, Treas./Exec. VP/Sec.
Charles W. Castanza, Sr. VP/Dir.-Plant Oper.
Derek Hampson, VP/Manager-Eastern Chemical Corp. Subsidiary
Peter A. Hiltunen, VP/Production Manager
Cecile M. Brophy, Treas./Controller
Alfred R. Globus, Chmn.

Phone: 631-273-0900	Fax: 631-273-0858
Toll-Free: 800-645-5566	
Address: 230 Marcus Blvd., Hauppauge, NY 11788 US	

GROWTH PLANS/SPECIAL FEATURES:

United-Guardian, Inc., through Guardian Laboratories, develops, manufactures and markets cosmetic ingredients; personal and healthcare products; pharmaceuticals; and specialty industrial products. The company also distributes, through Eastern Chemical Corp., an extensive line of fine organic chemicals, test solutions, indicators, dyes and reagents. Guardian Laboratories' two largest marketed product lines are the Lubrajel line of cosmetic ingredients, nondrying water-based moisturizing and lubricating gels that have applications in the cosmetic industry primarily as a moisturizer and in the medical field primarily as a lubricant, which accounted for roughly 70% of revenue in 2006; and Rencidin Irrigation, a urological prescription drug, used primarily to prevent the formation of and to dissolve calcifications in catheters implanted in the urinary bladder, that accounted for roughly 17% of revenue in 2006. Other products include Clorpactin, a microbicidal product used primarily in urology and surgery as an antiseptic for treating a wide range of localized infections in the urinary bladder, the peritoneum, the abdominal cavity, the eye, ear, nose, throat and sinuses; Klensoft, a surfactant that can be used in shampoos, shower gels, makeup removers and other cosmetic formulations; and Confetti Dermal Delivery Flakes, a product line that incorporates various functional oil-soluble ingredients into colorful flakes that can be added to, and suspended in, various water-based products. Eastern Chemical Corp. does not carry out any chemical manufacturing, but it does contract with several custom chemical manufacturers and also will package-to-order for those customers that require it. Paragon Organic Chemicals, Inc. is a wholly-owned subsidiary of United-Guardian that functions as a purchasing arm for Eastern Chemical Corp. It has no assets or sales of its own.

FINANCIALS: Sales and profits are in thousands of dollars—add 000 to get the full amount. 2006 Note: Financial information for 2006 was not available for all companies at press time.

2006 Sales: $12,195	2006 Profits: $2,737	**U.S. Stock Ticker: UG**	
2005 Sales: $12,135	2005 Profits: $2,617	**Int'l Ticker:** Int'l Exchange:	
2004 Sales: $11,123	2004 Profits: $2,475	Employees: 43	
2003 Sales: $11,157	2003 Profits: $2,471	Fiscal Year Ends: 12/31	
2002 Sales: $9,100	2002 Profits: $1,400	Parent Company:	

SALARIES/BENEFITS:

Pension Plan:	ESOP Stock Plan:	Profit Sharing:	Top Exec. Salary: $218,248	Bonus: $53,000
Savings Plan:	Stock Purch. Plan:		Second Exec. Salary: $143,675	Bonus: $12,750

OTHER THOUGHTS:

Apparent Women Officers or Directors: 1
Hot Spot for Advancement for Women/Minorities:

LOCATIONS: ("Y" = Yes)

West:	Southwest:	Midwest:	Southeast:	Northeast:	International:
				Y	

URIGEN PHARMACEUTICALS INC   www.urigen.com

Industry Group Code: 325412 Ranks within this company's industry group: Sales: 33   Profits: 161

Drugs:		Other:	Clinical:	Computers:	Other:
Discovery:		AgriBio:	Trials/Services:	Hardware:	Specialty Services:
Licensing:	Y	Genomics/Proteomics:	Laboratories:	Software:	Consulting:
Manufacturing:		Tissue Replacement:	Equipment/Supplies:	Arrays:	Blood Collection:
Development:	Y		Research/Development Svcs.:	Database Management:	Drug Delivery:
Generics:			Diagnostics:		Drug Distribution:

TYPES OF BUSINESS:
Drugs-Urological

BRANDS/DIVISIONS/AFFILIATES:
Valentis, Inc.
Urigen N.A., Inc.

CONTACTS: Note: Officers with more than one job title may be intentionally listed here more than once.
William J. Garner, CEO
Terry M. Nida, COO
William J. Garner, Pres.
Martin E. Shmagin, CFO
Amie E. Franklin, Mgr.-Clinical, Regulatory & Intellectual Property
Tracy Taylor, Chmn.

Phone: 650-259-0239	Fax: 650-259-0901
Toll-Free:	
Address: 875 Mahler Rd., Ste. 235, Burlingame, CA 94010 US	

GROWTH PLANS/SPECIAL FEATURES:
Urigen Pharmaceuticals, Inc., formerly Valentis, Inc., historically developed cardiovascular therapeutics, however it has changed its focus to the development and commercialization of products for the treatment and diagnosis of urological disorders. The company has two products that are currently in clinical trials. URG101 is a proprietary combination therapy of components approved by global regulatory authorities that is locally delivered to the bladder for rapid relief of pain and urgency. It is in Phase II clinical trials for Painful Bladder syndrome, but the company is also developing indications for radiation cystitis and dyspareunia, or painful intercourse. URG301 is in Phase II trials for both overactive bladder and urethritis. The company is also researching an indication in acute urethral discomfort. In October 2006, the company announced that its product, URG101, was unable to meet the primary endpoint in one of its Phase II trials for chronic pelvic pain, but the company intends to continue with the development of the product. In early July 2007, Valentis entered into a reverse merger with Urigen N.A., Inc., which resulted in Urigen N.A. merging with a subsidiary of Valentis, turning it into a wholly-owned subsidiary of the company. The company (Valentis) changed its name to Urigen Pharmaceuticals, Inc. in late July 2007.

FINANCIALS: Sales and profits are in thousands of dollars—add 000 to get the full amount. 2006 Note: Financial information for 2006 was not available for all companies at press time.

2006 Sales: $ 727	2006 Profits: $-15,337	**U.S. Stock Ticker: URGP.OB**
2005 Sales: $2,177	2005 Profits: $-11,083	**Int'l Ticker:** Int'l Exchange:
2004 Sales: $7,478	2004 Profits: $-6,484	Employees: 8
2003 Sales: $4,000	2003 Profits: $-14,900	Fiscal Year Ends: 6/30
2002 Sales: $3,800	2002 Profits: $-33,100	Parent Company:

SALARIES/BENEFITS:
Pension Plan:	ESOP Stock Plan:	Profit Sharing:	Top Exec. Salary: $44,767	Bonus: $
Savings Plan:	Stock Purch. Plan:		Second Exec. Salary: $37,984	Bonus: $

OTHER THOUGHTS:
Apparent Women Officers or Directors: 1
Hot Spot for Advancement for Women/Minorities:

LOCATIONS: ("Y" = Yes)
West:	Southwest:	Midwest:	Southeast:	Northeast:	International:
Y					

VALEANT PHARMACEUTICALS INTERNATIONAL
www.valeant.com

Industry Group Code: 325412 Ranks within this company's industry group: Sales: 164   Profits: 86

Drugs:		Other:	Clinical:	Computers:	Other:
Discovery:	Y	AgriBio:	Trials/Services:	Hardware:	Specialty Services:
Licensing:	Y	Genomics/Proteomics:	Laboratories:	Software:	Consulting:
Manufacturing:	Y	Tissue Replacement:	Equipment/Supplies:	Arrays:	Blood Collection:
Development:	Y		Research/Development Svcs.:	Database Management:	Drug Delivery:
Generics:			Diagnostics:		Drug Distribution:

TYPES OF BUSINESS:
Prescription & Non-Prescription Pharmaceuticals
Neurology Drugs
Dermatology Drugs
Infectious Diseases Drugs

BRANDS/DIVISIONS/AFFILIATES:
Efudex/Efudix
Dermatix
Oxsoralen-Ultra
Infergen
Virazole
Kinerase
Mestinon
Librax

CONTACTS: Note: Officers with more than one job title may be intentionally listed here more than once.
Timothy C. Tyson, CEO
Timothy C. Tyson, Pres.
Peter J. Blott, CFO/Exec. VP
Geoffrey M. Glass, CIO/Sr. VP
Wesley P. Wheeler, Pres., Global Prod. Dev.
Eileen C. Pruette, General Counsel/Exec. VP
Jeff Misakian, VP-Investor Rel.
Martin N. Mercer, Exec. VP-Int'l
Wesley P. Wheeler, Pres., North America
Robert A. Ingram, Chmn.
Charles J. Bramlage, Pres., Europe

Phone: 949-461-6000	**Fax:** 949-461-6609
Toll-Free: 800-548-5100	
Address: 1 Enterprise, Aliso Viejo, CA 92656 US	

GROWTH PLANS/SPECIAL FEATURES:
Valeant Pharmaceuticals International is a global pharmaceutical company that develops, manufactures and markets a broad spectrum of prescription and non-prescription pharmaceuticals. The company focuses principally in the therapeutic areas of neurology, dermatology and infectious diseases. The firm's products also treat, among others, neuromuscular disorders, cancer, cardiovascular disease, diabetes and psychiatric disorders. Valeant's products are sold through three pharmaceutical segments: North America; International, composed of the Latin America, Asia and Australasia regions; and EMEA, Europe, Middle East and Africa. The company's specialty pharmaceuticals product portfolio comprises roughly 370 branded products with approximately 2,200 stock keeping units. Products in the neurology field include Mestinon, Librax, Migranal, Tasmar and Zelapar; in the dermatology, Efudex, Kinerase, Oxsoralen-Ultra and Dermatrix; and addressing infectious diseases, Infergen and Virazole. Valeant's research and development program focuses on preclinical and clinical development of identified molecules. The company is developing product candidates, including two clinical stage programs: taribavirin and retigabine. The firm's marketing and promotion efforts focus on the Promoted Products, which consists of products sold in more than 100 markets around the world with primary focus on the U.S., Mexico, Poland, Canada, Germany, Spain, Italy, the U.K. and France. In January 2006, Valeant acquired the U.S. and Canadian rights to the hepatitis C treatment drug Infergen from Intermune, Inc. In October 2006, the company signed a distribution agreement with Intendis GmbH for rights to certain dermatological products in the U.K., which include the distribution rights to Finacea, a topical treatment for rosacea. In December 2006, the firm sold its HIV and cancer development programs and certain discovery and preclinical assets to Ardea Biosciences. In January 2007, Valeant licensed the development and commercialization rights to the hepatitis B compound pradefovir to Schering-Plough.

FINANCIALS: Sales and profits are in thousands of dollars—add 000 to get the full amount. 2006 Note: Financial information for 2006 was not available for all companies at press time.

2006 Sales: $907,238	2006 Profits: $-56,565	**U.S. Stock Ticker:** VRX
2005 Sales: $823,886	2005 Profits: $-188,143	**Int'l Ticker:** Int'l Exchange:
2004 Sales: $684,251	2004 Profits: $-154,653	Employees: 3,443
2003 Sales: $685,953	2003 Profits: $-55,640	Fiscal Year Ends: 12/31
2002 Sales: $737,100	2002 Profits: $-134,900	Parent Company:

SALARIES/BENEFITS:

Pension Plan:	ESOP Stock Plan:	Profit Sharing:	Top Exec. Salary: $860,000	Bonus: $645,000
Savings Plan:	Stock Purch. Plan:		Second Exec. Salary: $434,718	Bonus: $256,481

OTHER THOUGHTS:
Apparent Women Officers or Directors: 3
Hot Spot for Advancement for Women/Minorities: Y

LOCATIONS: ("Y" = Yes)

West:	Southwest:	Midwest:	Southeast:	Northeast:	International:
Y					Y

Note: Financial information, benefits and other data can change quickly and may vary from those stated here.

VASOGEN INC

www.vasogen.com

Industry Group Code: 325412 Ranks within this company's industry group: Sales: Profits: 167

Drugs:		Other:	Clinical:	Computers:	Other:
Discovery:	Y	AgriBio:	Trials/Services:	Hardware:	Specialty Services:
Licensing:		Genomics/Proteomics:	Laboratories:	Software:	Consulting:
Manufacturing:		Tissue Replacement:	Equipment/Supplies:	Arrays:	Blood Collection:
Development:	Y		Research/Development Svcs.:	Database Management:	Drug Delivery:
Generics:			Diagnostics:		Drug Distribution:

TYPES OF BUSINESS:

Biopharmaceuticals Development
Immune Modulation Therapies

BRANDS/DIVISIONS/AFFILIATES:

Celacade

CONTACTS: Note: Officers with more than one job title may be intentionally listed here more than once.

Christopher J. Waddick, CEO
Christopher J. Waddick, Pres.
Graham D. Neil, CFO
Catherine Bouchard, VP-Human Resources
Anthony E. Bolton, Chief Science Officer
Anne E. Goodbody, VP-Drug Dev.
Jacqueline Le Saux, VP-Corp. & Legal Affairs
Glenn Neumann, Investor Rel. Contact
Graham D. Neil, VP-Finance
Susan F. Langlois, VP-Regulatory Affairs & Quality Assurance
Eldon R. Smith, Sr. VP-Scientific Affairs/Chief Medical Officer
Michael Shannon, VP-Medical Affairs
Terrance H. Gregg, Chmn.

Phone: 905-817-2000	Fax: 905-569-9231
Toll-Free:	
Address: 2505 Meadowvale Blvd., Mississauga, ON L5N5S2 Canada	

GROWTH PLANS/SPECIAL FEATURES:

Vasogen, Inc. researches, develops and commercializes immune modulation therapies targeting chronic inflammation caused by neurological cardiovascular disorders. The firm has two leading drug therapies, Celacade and VP025. Celacade, currently in Phase III clinical studies, is an immune modulation therapy designed to treat chronic heart failure and peripheral arterial disease. Celacade is a monthly treatment that targets chronic inflammation underlying cardiovascular disease by activating the immune system's anti-inflammatory response to cells undergoing apoptosis, a natural disintegration of cells. During a brief outpatient procedure, a small sample of a patient's blood is drawn into a Celacade single-use disposable cartridge, exposed to controlled oxidative stress utilizing the company's proprietary Celacade device, and then re-administered intramuscularly. VP025 is in Phase I clinical trials and is being developed to treat neuro-inflammatory diseases, including Alzheimer's disease, Parkinson's disease and ALS (Lou Gehrig's disease), that lead to the death of nerve cells and eventual loss of function. VP025 interacts with macrophages and other cells of the immune system, eliciting an anti-inflammatory response. The product has been shown to significantly reduce key measures of inflammation and cell death in the brain and to improve physiological measurements that correlate with memory and learning activity. VP025 has completed Phase I of clinical study. In April 2007, Vasogen announced a collaboration with Grupo Ferrer Internacional, S.A. to commercialize Celacade in The European Union and certain Latin American countries.

FINANCIALS: Sales and profits are in thousands of dollars—add 000 to get the full amount. 2006 Note: Financial information for 2006 was not available for all companies at press time.

2006 Sales: $	2006 Profits: $-62,319	U.S. Stock Ticker: VSGN
2005 Sales: $	2005 Profits: $-79,900	Int'l Ticker: VAS Int'l Exchange: Toronto-TSX
2004 Sales: $	2004 Profits: $-62,800	Employees: 149
2003 Sales: $	2003 Profits: $-24,600	Fiscal Year Ends: 11/30
2002 Sales: $	2002 Profits: $-12,662	Parent Company:

SALARIES/BENEFITS:

Pension Plan:	ESOP Stock Plan:	Profit Sharing:	Top Exec. Salary: $330,086	Bonus: $105,902
Savings Plan:	Stock Purch. Plan:		Second Exec. Salary: $267,431	Bonus: $81,244

OTHER THOUGHTS:

Apparent Women Officers or Directors: 4
Hot Spot for Advancement for Women/Minorities: Y

LOCATIONS: ("Y" = Yes)

West:	Southwest:	Midwest:	Southeast:	Northeast:	International:
					Y

VAXGEN INC

www.vaxgen.com

Industry Group Code: 325412 Ranks within this company's industry group: Sales: Profits:

Drugs:		Other:		Clinical:	Computers:		Other:	
Discovery:	Y	AgriBio:		Trials/Services:	Hardware:		Specialty Services:	
Licensing:		Genomics/Proteomics:		Laboratories:	Software:		Consulting:	
Manufacturing:	Y	Tissue Replacement:		Equipment/Supplies:	Arrays:		Blood Collection:	
Development:	Y			Research/Development Svcs.:	Database Management:		Drug Delivery:	
Generics:				Diagnostics:			Drug Distribution:	

TYPES OF BUSINESS:

Biologic Products

BRANDS/DIVISIONS/AFFILIATES:

Chemo-Sero-Therapeutic Research Institute
Kaketsuken

CONTACTS: *Note: Officers with more than one job title may be intentionally listed here more than once.*

James P. Panek, CEO
James P. Panek, Pres.
Matthew J. Pfeffer, CFO
Matthew J. Pfeffer, Sr. VP-Admin.
Piers Whitehead, VP-Corp. & Bus. Dev.
Lance Ignon, VP-Corp. Affairs
Matthew J. Pfeffer, VP-Finance
Mark Gurwith, Sr. VP-Medical Affairs/Chief Medical Officer
Randall L-W. Caudill, Chmn.

Phone: 650-624-2400	**Fax:** 650-624-1001
Toll-Free:	
Address: 349 Oyster Point Blvd., S. San Francisco, CA 94080 US	

GROWTH PLANS/SPECIAL FEATURES:

VaxGen, Inc. is a biopharmaceutical company that develops, manufactures and commercializes biologic products for the prevention and treatment of human infectious diseases. The company develops vaccines against the inhalation anthrax and smallpox. In collaboration with the Chemo-Sero-Therapeutic Research Institute (Kaketsuken) situated in Kumamoto, Japan, the firm worked on making available an attenuated smallpox vaccine, LC16m8, for use in the U.S. and potentially other markets. The vaccine is already licensed for use in Japan. VaxGen operates a manufacturing facility in California for the commercial-scale production of its anthrax vaccine candidate. The facility can also be used to make a wide range of other biopharmaceutical products. In December 2006, the U.S. government terminated its contract with the company for the supply of recombinant anthrax vaccine. In June 2007, the firm announced that it terminated its collaboration with Kaketsuken to co-develop a next-generation attenuated smallpox vaccine for use in the U.S. and elsewhere due to the lack of commitment from the U.S. government to fund the development of purchase the product.

FINANCIALS: Sales and profits are in thousands of dollars—add 000 to get the full amount. 2006 Note: Financial information for 2006 was not available for all companies at press time.

2006 Sales: $	2006 Profits: $	**U.S. Stock Ticker:** VXGN
2005 Sales: $	2005 Profits: $	**Int'l Ticker:** Int'l Exchange:
2004 Sales: $	2004 Profits: $	Employees: 133
2003 Sales: $14,300	2003 Profits: $-25,500	Fiscal Year Ends: 12/31
2002 Sales: $1,600	2002 Profits: $-31,700	Parent Company:

SALARIES/BENEFITS:

Pension Plan:	ESOP Stock Plan:	Profit Sharing:	Top Exec. Salary: $348,075	Bonus: $115,000
Savings Plan:	Stock Purch. Plan:		Second Exec. Salary: $339,121	Bonus: $200,000

OTHER THOUGHTS:

Apparent Women Officers or Directors:
Hot Spot for Advancement for Women/Minorities:

LOCATIONS: ("Y" = Yes)

West:	Southwest:	Midwest:	Southeast:	Northeast:	International:
Y					

Note: Financial information, benefits and other data can change quickly and may vary from those stated here.

VENTRIA BIOSCIENCE

www.ventria.com

Industry Group Code: 325411   Ranks within this company's industry group: Sales: Profits:

Drugs:	Other:		Clinical:	Computers:	Other:	
Discovery:	AgriBio:	Y	Trials/Services:	Hardware:	Specialty Services:	
Licensing:	Genomics/Proteomics:		Laboratories:	Software:	Consulting:	
Manufacturing: Y	Tissue Replacement:		Equipment/Supplies:	Arrays:	Blood Collection:	
Development: Y			Research/Development Svcs.:	Database Management:	Drug Delivery:	
Generics:			Diagnostics:		Drug Distribution:	

TYPES OF BUSINESS:

Medical Supplies, Manufacturing
Lactoferrin Manufacturing
Lysozyme Manufacturing

BRANDS/DIVISIONS/AFFILIATES:

ExpressPro
ExpressTec
ExpressMab
BioShare

CONTACTS: *Note: Officers with more than one job title may be intentionally listed here more than once.*

Scott E. Deeter, CEO
Scott E. Deeter, Pres.
Terry D. Carlone, General Counsel
Randy Semadeni, VP-Bus. Dev.
Randy Semadeni, VP-Finance
Delia R. Bethell, VP-Clinical Dev.
Victor Hicks, VP/Commercial Dir., InVitria Division
Somen Nandi, Dir.-Molecular Breeding
Greg Unruh, VP/Gen. Mgr.

Phone: 916-921-6148	Fax: 916-921-5611
Toll-Free:	
Address: 4110 N. Freeway Blvd., Sacramento, CA 95834 US	

GROWTH PLANS/SPECIAL FEATURES:

Ventria Bioscience is a biotech company focused on human nutrition and human therapeutics. The company manufactures two products, lactoferrin and lysozyme. Lactoferrin is a globular multifunctional protein with antimicrobial activity. Lactoferrin protein has applications in the alleviation of fungal infections gastrointestinal health, dietary management of acute diarrhea and the treatment of topical infections and inflammations. Lysozyme is an enzyme that attacks the cell walls of many bacterias, and has anti-bacterial, anti-viral and anti-fungal properties. Lysozyme enzyme has applications in the alleviation of fungal infections gastrointestinal health, dietary management of acute diarrhea and the treatment of topical infections and inflammations. The materials are generally found in human breast milk as well as tears, nasogastric secretions, saliva and bronchial secretions. Ventria's ExpressTec platform uses self-pollinating crops, specifically rice and barley, as the production host for these products. Ventria uses the ExpressTec production system as a basis for forming more specialized systems which produce specific molecules. These platforms include ExpressPro for production of proteins, ExpressTide for production of peptides and ExpressMab for production of monoclonal antibodies. The company has a program for researchers called BioShare, in which researchers submit requests for materials and are granted access to a free supply of recombinant proteins and peptides.

FINANCIALS: Sales and profits are in thousands of dollars—add 000 to get the full amount. 2006 Note: Financial information for 2006 was not available for all companies at press time.

2006 Sales: $	2006 Profits: $	U.S. Stock Ticker: Private
2005 Sales: $	2005 Profits: $	Int'l Ticker: Int'l Exchange:
2004 Sales: $	2004 Profits: $	Employees:
2003 Sales: $	2003 Profits: $	Fiscal Year Ends:
2002 Sales: $	2002 Profits: $	Parent Company:

SALARIES/BENEFITS:

Pension Plan:	ESOP Stock Plan:	Profit Sharing:	Top Exec. Salary: $	Bonus: $
Savings Plan:	Stock Purch. Plan:		Second Exec. Salary: $	Bonus: $

OTHER THOUGHTS:

Apparent Women Officers or Directors: 1
Hot Spot for Advancement for Women/Minorities:

LOCATIONS: ("Y" = Yes)

West:	Southwest:	Midwest:	Southeast:	Northeast:	International:
Y					

VERNALIS PLC

www.vernalis.com

Industry Group Code: 325412 Ranks within this company's industry group: Sales: 93   Profits: 177

Drugs:		Other:		Clinical:	Computers:		Other:	
Discovery:	Y	AgriBio:		Trials/Services:	Hardware:		Specialty Services:	Y
Licensing:	Y	Genomics/Proteomics:		Laboratories:	Software:		Consulting:	
Manufacturing:		Tissue Replacement:		Equipment/Supplies:	Arrays:		Blood Collection:	
Development:	Y			Research/Development Svcs.:	Database Management:		Drug Delivery:	
Generics:				Diagnostics:			Drug Distribution:	Y

TYPES OF BUSINESS:

Drugs-Neurology & Acute Pain
Drugs-Parkinson's Disease
Drugs-Migraine
Drugs-Obesity
Drugs-Oncology

BRANDS/DIVISIONS/AFFILIATES:

British Biotech plc
Vernalis Group plc
RiboTargets
Frova
Apokyn
Apokyn Circle of Care

CONTACTS: *Note: Officers with more than one job title may be intentionally listed here more than once.*

Simon Sturge, CEO
Anthony Wier, CFO
Joseph Canny, VP-Mktg.
John Slater, General Counsel
John Hutchison, Dir.-Dev.
Julia Wilson, Head-Corp. Comm
Jean Hubble, VP-Medical Affairs, US
Rick A. Henson, VP-US Sales
Sylvia McBrinn, Sr. VP-US Oper.
Peter Fellner, Chmn.

Phone: 440-118-977-3133	Fax: 440-118-989-9300

Toll-Free:
Address: Oakdene Ct., 613 Reading Rd., Winnersh, RG41 5UA UK

GROWTH PLANS/SPECIAL FEATURES:

Vernalis, formed from the merger of British Biotech plc, Vernalis Group plc and RiboTargets, discovers, researches and develops new compounds for central nervous system disorders.   The company's main two developmental franchises are for pain, including the treatment of migraines, and for neurological disorders, such as Parkinson's disease. Vernalis is also working on products for inflammation, obesity, and cancer.   The firm actively seeks partnerships with pharmaceutical companies to complete the development and marketing of those compounds.   Vernalis' leading drug, Frova, is intended for the oral treatment of migraines and menstrual-associated migraines.   Frova has the longest half-life in the blood, 26 hours, of pain relievers in its class and has been approved worldwide, including in the U.S., Germany and France.   Its newest marketed product Apokyn, acquired from Mylan, treats immobilization due to Parkinson's disease and is the only such drug marketed in the U.S.   The company also has the Apokyn Circle of Care program, which is comprised of a nurse call center, home healthcare visits and a pilot nurse clinical liaison program. The firm has also created V10153, a dual-acting, thrombolytic and antithrombotic agent, with the hopes to treat acute myocardial infarction and other thrombotic diseases including stroke, peripheral arterial occlusion, pulmonary embolism and deep vein thrombosis.   Vernalis is investigating the potential use of certain receptor antagonists to treat obesity at the level of the nervous system, and is working with Hsp90 in an attempt to inhibit cancer growth.   In January 2006, the company launched its U.S. sales force, which focuses on the sale of its two drugs, Apokyn and Frova, to specialty neurological markets.   In May 2007, Vernalis announced that it has formed an oncology drug discovery collaboration with Servier Research Group, which will utilize Vernalis' proprietary drug discovery platform.

FINANCIALS: Sales and profits are in thousands of dollars—add 000 to get the full amount. 2006 Note: Financial information for 2006 was not available for all companies at press time.

2006 Sales: $33,040	2006 Profits: $-85,880	**U.S. Stock Ticker: VNLSY.PK**
2005 Sales: $24,300	2005 Profits: $-56,500	**Int'l Ticker: VER** Int'l Exchange: London-LSE
2004 Sales: $29,100	2004 Profits: $-56,000	Employees: 192
2003 Sales: $15,400	2003 Profits: $-61,000	Fiscal Year Ends: 12/31
2002 Sales: $2,100	2002 Profits: $-24,700	Parent Company:

SALARIES/BENEFITS:

Pension Plan:	ESOP Stock Plan:	Profit Sharing:	Top Exec. Salary: $467,464	Bonus: $232,365
Savings Plan:	Stock Purch. Plan:		Second Exec. Salary: $344,557	Bonus: $106,648

OTHER THOUGHTS:

Apparent Women Officers or Directors: 2
Hot Spot for Advancement for Women/Minorities: Y

LOCATIONS: ("Y" = Yes)

West:	Southwest:	Midwest:	Southeast:	Northeast:	International:
				Y	Y

Note: Financial information, benefits and other data can change quickly and may vary from those stated here.

VERTEX PHARMACEUTICALS INC   www.vpharm.com

Industry Group Code: 325412 Ranks within this company's industry group: Sales: 52   Profits: 196

Drugs:		Other:	Clinical:	Computers:	Other:
Discovery:	Y	AgriBio:	Trials/Services:	Hardware:	Specialty Services:
Licensing:	Y	Genomics/Proteomics:	Laboratories:	Software:	Consulting:
Manufacturing:		Tissue Replacement:	Equipment/Supplies:	Arrays:	Blood Collection:
Development:	Y		Research/Development Svcs.:	Database Management:	Drug Delivery:
Generics:			Diagnostics:		Drug Distribution:

TYPES OF BUSINESS:
Small Molecule Drugs

BRANDS/DIVISIONS/AFFILIATES:
Telaprevir
Lexiva
Telzir
MK-0457
VX-702
VX-770
VX-883
VX-409

CONTACTS: *Note: Officers with more than one job title may be intentionally listed here more than once.*
Joshua Boger, CEO
Joshua Boger, Pres.
Ian F. Smith, CFO/Exec. VP
Peter Mueller, Chief Scientific Officer
Kenneth S. Boger, General Counsel/Sr. VP
Kurt C. Graves, Head-Strategic Dev.
Peter Mueller, Exec. VP-Drug Innovation & Realization
Richard C. Garrison, Sr. VP-Catalyst
Kurt C. Graves, Exec. VP/Chief Commercial Officer
Amit K. Sachdev, Sr. VP-Public Policy & Gov't Affairs
Charles A. Sanders, Chmn.

Phone: 617-444-6100	Fax: 617-444-6680
Toll-Free:	
Address: 130 Waverly St., Cambridge, MA 02139 US	

GROWTH PLANS/SPECIAL FEATURES:
Vertex Pharmaceuticals, Inc. discovers, develops and commercializes small molecule drugs for the treatment of serious diseases. The company concentrates most of its drug development resources on four drug candidates: telaprevir for the treatment of hepatitis C virus infection; VX-702 for the treatment of rheumatoid arthritis and other inflammatory diseases; VX-770 for the treatment of cystic fibrosis; and VX-883 for the treatment of bacterial infection. The firm's lead product is telaprevir, an oral hepatitis C protease inhibitor, which is in Phase IIb clinical trials. The FDA granted fast track designation to telaprevir. Vertex's pipeline includes several drug candidates that are being developed by its collaborators. The most advanced of these drug candidates is MK-0457, an Aurora kinase inhibitor developed by Merck & Co., Inc. for the treatment of cancer. Other collaborations include Cystic Fibrosis Foundation Therapeutics, Inc., with which it is developing VX-770; GlaxoSmithKline, for the development and commercialization of VX-409; Kissei Pharmaceutical Co., Ltd., for the development and commercialization of VX-702. Forsamprenavir calcium, a company discovered compound for the treatment of HIV infection, is marketed by the firm's collaborator GlaxoSmithKlein plc as Lexiva in the U.S. and Telzir in Europe. In June 2006, Vertex entered into a collaboration agreement with Janssen Pharmaceutica, N.V., a Johnson & Johnson company relating to telaprevir, under which Vertex retains exclusive commercial rights to the drug in North American and will lead the clinical development program. Janssen will be responsible for the commercialization of telaprevir, including the manufacture of its own commercial supply of the drug for the Janssen territories, which include territories outside of North American and the Far East.

FINANCIALS: Sales and profits are in thousands of dollars—add 000 to get the full amount. 2006 Note: Financial information for 2006 was not available for all companies at press time.

2006 Sales: $216,356	2006 Profits: $-206,891	U.S. Stock Ticker: VRTX
2005 Sales: $160,890	2005 Profits: $-203,417	Int'l Ticker: Int'l Exchange:
2004 Sales: $102,717	2004 Profits: $-166,247	Employees: 962
2003 Sales: $69,141	2003 Profits: $-196,767	Fiscal Year Ends: 12/31
2002 Sales: $161,100	2002 Profits: $-108,600	Parent Company:

SALARIES/BENEFITS:

Pension Plan:	ESOP Stock Plan:	Profit Sharing:	Top Exec. Salary: $593,921	Bonus: $705,600
Savings Plan:	Stock Purch. Plan:		Second Exec. Salary: $449,729	Bonus: $354,707

OTHER THOUGHTS:
Apparent Women Officers or Directors: 2
Hot Spot for Advancement for Women/Minorities: Y

LOCATIONS: ("Y" = Yes)

West:	Southwest:	Midwest:	Southeast:	Northeast:	International:
Y				Y	Y

VIACELL INC
www.viacellinc.com

Industry Group Code: 325414 Ranks within this company's industry group: Sales: 10 Profits: 12

Drugs:	Other:		Clinical:	Computers:	Other:	
Discovery:	AgriBio:		Trials/Services:	Hardware:	Specialty Services:	Y
Licensing:	Genomics/Proteomics:		Laboratories:	Software:	Consulting:	
Manufacturing:	Tissue Replacement:	Y	Equipment/Supplies:	Arrays:	Blood Collection:	
Development:			Research/Development Svcs.:	Database Management:	Drug Delivery:	
Generics:			Diagnostics:		Drug Distribution:	

TYPES OF BUSINESS:
Human Cells-Based Medicine
Stem Cell Research

BRANDS/DIVISIONS/AFFILIATES:
ViaCord
ViaCyte
Mothers Work, Inc.

CONTACTS: Note: Officers with more than one job title may be intentionally listed here more than once.
Marc D. Beer, CEO
Mark D. Beer, Pres.
John F. Thero, CFO/Sr. VP
Morey Kraus, CTO
Jim Corbett, Pres., ViaCell Reproductive Health
Mary T. Thistle, Sr. VP-Bus. Dev.-ViaCell Reproductive Health
Vaughn M. Kailian, Chmn.

Phone: 617-914-3400 **Fax:** 617-577-9018
Toll-Free:
Address: 245 1st St., 15th Fl., Cambridge, MA 02142 US

GROWTH PLANS/SPECIAL FEATURES:
ViaCell, Inc. is a biotechnology company dedicated to enabling the widespread application of human cells as medicine. The company has a reproductive health business that generates revenue from sales of ViaCord, a service offering through which expectant families can preserve the baby's umbilical cord blood for possible future medical use. Stem cells from umbilical cord blood are a treatment option for over 40 diseases, including certain blood cancers and genetic diseases. The company's lead drug candidate, ViaCyte, is being studied for its potential to broaden reproductive choices for women through the cryo-preservation of human unfertilized eggs. The firm's research and development efforts focus on investigating the potential for new therapeutic uses of umbilical cord blood-derived and adult stem cells and on technology for expanding populations of these cells. The areas of concentration include cancer, cardiac disease and diabetes. ViaCell has processing and storage facilities in Hebron, Kentucky and an additional research and development operation in Singapore. In August 2006, the company entered into a data license and marketing services agreement with Mothers Work, Inc., the world's largest designer and retailer of maternity apparel. Under the term of the agreement, Mothers Work granted the firm an exclusive license within the field of preserving stem cell from cord blood and other sources to market directly to those Mothers Work customers who affirmatively agreed to permit disclosure of their data and information. In early 2007, ViaCell decided not to advance CB001, comprised of stem cells isolated from umbilical cord blood and expanded using the company's proprietary selective amplification technology, in future clinical trials.

The company offers its employees health, dental and vision insurance; life and AD&D coverage; disability insurance; a 401(k) plan; an educational assistance program; and fitness, health & wellness benefits.

FINANCIALS: Sales and profits are in thousands of dollars—add 000 to get the full amount. 2006 Note: Financial information for 2006 was not available for all companies at press time.
2006 Sales: $54,426	2006 Profits: $-21,330	**U.S. Stock Ticker:** VIAC
2005 Sales: $44,443	2005 Profits: $-14,677	**Int'l Ticker:** Int'l Exchange:
2004 Sales: $38,274	2004 Profits: $-21,097	Employees: 254
2003 Sales: $31,880	2003 Profits: $-55,468	Fiscal Year Ends: 12/31
2002 Sales: $	2002 Profits: $	Parent Company:

SALARIES/BENEFITS:
Pension Plan:	ESOP Stock Plan:	Profit Sharing:	Top Exec. Salary: $363,462	Bonus: $137,592
Savings Plan: Y	Stock Purch. Plan:		Second Exec. Salary: $284,278	Bonus: $78,650

OTHER THOUGHTS:
Apparent Women Officers or Directors: 1
Hot Spot for Advancement for Women/Minorities:

LOCATIONS: ("Y" = Yes)
West:	Southwest:	Midwest:	Southeast:	Northeast:	International:
		Y		Y	Y

Note: Financial information, benefits and other data can change quickly and may vary from those stated here.

VICAL INC

www.vical.com

Industry Group Code: 325412 **Ranks within this company's industry group:** Sales: 118   Profits: 110

Drugs:		Other:		Clinical:		Computers:		Other:	
Discovery:	Y	AgriBio:		Trials/Services:		Hardware:		Specialty Services:	
Licensing:	Y	Genomics/Proteomics:		Laboratories:		Software:		Consulting:	
Manufacturing:		Tissue Replacement:		Equipment/Supplies:		Arrays:		Blood Collection:	
Development:	Y			Research/Development Svcs.:		Database Management:		Drug Delivery:	
Generics:				Diagnostics:				Drug Distribution:	

TYPES OF BUSINESS:

Drug Delivery Systems
Cancer, Infectious Disease & Metabolic Disease Drugs & Vaccines

BRANDS/DIVISIONS/AFFILIATES:

Allovectin-7
Vaxfectin

CONTACTS: *Note: Officers with more than one job title may be intentionally listed here more than once.*

Vijay B. Samant, CEO
Vijay B. Samant, Pres.
Jill M. Church, CFO/VP
Alain P. Rolland, Sr. VP-Prod. Dev.
Kevin R. Bracken, VP-Mfg.
Jill M. Church, Sec.
Alan Engbring, Exec. Dir.-Investor Rel.
Robert M. Jackman, Sr. VP-Bus. Oper.
Ronald B. Moss, VP-Clinical Dev.
Larry R. Smith, VP-Vaccine Research
R. Gordon Douglas, Chmn.

Phone: 858-646-1100	Fax: 858-646-1150
Toll-Free:	
Address: 10390 Pacific Ctr. Ct., San Diego, CA 92121 US	

GROWTH PLANS/SPECIAL FEATURES:

Vical, Inc. researches and develops biopharmaceutical products based on its patented DNA delivery technologies for the prevention and treatment of serious or life-threatening diseases. The company's research areas include vaccines for use in high-risk population for infectious disease targets; vaccines for general pediatric, adolescent and adult populations for infectious disease applications; and cancer vaccines or immunotherapies. The firm has four active independent development programs in the areas of infectious disease and cancer, which include a Phase III clinical trial using the Allovectin-7 immunotherapeutic in patients with metastatic melanoma; a Phase II clinical trial using the cytomegalovirus DNA vaccine in hematopoietic cell transplant patients; a Phase I clinical trial of electroporation-enhanced delivery of interleukin-2 DNA that utilizes the company's delivery technology with an initial indication in metastatic melanoma; and a pandemic influenza DNA vaccine candidate using the proprietary Vaxfectin as an adjuvant. Vical has licenses its technologies to companies including Merck & Co., Inc.; the Sanofi-Aventis Groups; and AnGes. The National Institute of Health has clinical stage vaccine programs based on the company's technology in five infectious disease targets: HIV, pandemic influenza, Ebola, West Nile virus and severe acute respiratory syndrome. The firm also has two veterinary vaccine studies, licensed to Merial, Ltd. and Aqua Health, Ltd., for infectious diseases and cancer in household animals. In March 2007, Vical announced that licensee Merial received conditional approved from the U.S. Department of Agriculture to market a therapeutic DNA vaccine designed to treat melanoma in dogs.

FINANCIALS: Sales and profits are in thousands of dollars—add 000 to get the full amount. 2006 Note: Financial information for 2006 was not available for all companies at press time.

2006 Sales: $14,740	2006 Profits: $-23,148	**U.S. Stock Ticker: VICL**
2005 Sales: $12,003	2005 Profits: $-24,357	**Int'l Ticker:** Int'l Exchange:
2004 Sales: $14,545	2004 Profits: $-23,733	Employees: 147
2003 Sales: $8,078	2003 Profits: $-24,449	Fiscal Year Ends: 12/31
2002 Sales: $7,000	2002 Profits: $-27,900	Parent Company:

SALARIES/BENEFITS:

Pension Plan:	ESOP Stock Plan:	Profit Sharing:	Top Exec. Salary: $435,000	Bonus: $220,000
Savings Plan:	Stock Purch. Plan:		Second Exec. Salary: $283,000	Bonus: $50,000

OTHER THOUGHTS:

Apparent Women Officers or Directors: 1
Hot Spot for Advancement for Women/Minorities:

LOCATIONS: ("Y" = Yes)

West:	Southwest:	Midwest:	Southeast:	Northeast:	International:
Y					

VION PHARMACEUTICALS INC

www.vionpharm.com

Industry Group Code: 325412 Ranks within this company's industry group: Sales: 189 Profits: 115

Drugs:		Other:		Clinical:		Computers:		Other:	
Discovery:	Y	AgriBio:		Trials/Services:		Hardware:		Specialty Services:	
Licensing:		Genomics/Proteomics:		Laboratories:		Software:		Consulting:	
Manufacturing:		Tissue Replacement:		Equipment/Supplies:		Arrays:		Blood Collection:	
Development:	Y			Research/Development Svcs.:		Database Management:		Drug Delivery:	Y
Generics:				Diagnostics:				Drug Distribution:	

TYPES OF BUSINESS:

Cancer Treatment Drugs
Drug Delivery Systems

BRANDS/DIVISIONS/AFFILIATES:

Triapine
Tumor Amplified Protein Expression Therapy (TAPET)
Cloretazine

CONTACTS: Note: Officers with more than one job title may be intentionally listed here more than once.

Alan Kessman, CEO
Howard B. Johnson, Pres.
Howard B. Johnson, CFO
Ivan King, VP-R&D
James Tanguay, VP-Chemistry, Mfg. & Control
Meghan Fitzgerald, Chief Bus. Officer
Karen Schmedlin, VP-Finance/Chief Acct. Officer
Aileen Ryan, VP-Regulatory Affairs
Ann Cahill, VP-Clinical Dev.
William R. Miller, Chmn.

Phone: 203-498-4210	Fax: 203-498-4211
Toll-Free:	
Address: 4 Science Park, New Haven, CT 06511 US	

GROWTH PLANS/SPECIAL FEATURES:

Vion Pharmaceuticals, Inc. is pharmaceutical company engaged in the development of products for the treatment of cancer. The company's portfolio of product candidates consists of two small molecule anticancer agents in clinical development and additional small molecules in preclinical development. The firm also developed a drug delivery technology for the treatment of cancer. Vion's lead product candidate, Cloretazine, is an alkylating (DNA-damaging) agent for the treatment of acute myelogenous leukemia (AML). The drug received two fast track designations from FDA for the treatment of relapsed AML and elderly poor-risk AML and an orphan drug designation for the treatment of AML in the U.S. and the E.U. Triapine, the second product candidate in clinical trials, is a small molecule that in preclinical models inhibits the enzyme ribonucleotide reductase and therefore prevents the replication of tumor cells by blocking a critical step in DNA synthesis. Products in preclinical development include VN40541, a cytotoxic compound for the treatment of cancer. The firm's drug delivery technology, TAPET (Tumor Amplified Protein Expression Therapy), uses genetically altered strains of salmonella bacteria as a vehicle for delivering anticancer agents directly to solid tumors. Because the salmonella bacteria accumulate in tumors, the TAPET system will expose the tumor to a greater concentration of the anti-cancer medicine on a continuous basis, without damaging healthy tissue to as great a degree as if the cancer treatment circulated freely through the body. In December 2006, Vion entered into a manufacturing agreement for Cloretazine with Ben Venue Laboratories, a division of Boehringer Ingelheim, under which Ben Venue will manufacture the finished drug product. In May 2007, the firm suspended a Phase III trial for Cloretazine for patients with relapsed AML.

The company offers its employees medical and dental insurance; life, disability and long-term care insurance; a 401(k) plan; an employee stock purchase plan; and an education reimbursement.

FINANCIALS: Sales and profits are in thousands of dollars—add 000 to get the full amount. 2006 Note: Financial information for 2006 was not available for all companies at press time.

2006 Sales: $ 22	2006 Profits: $-25,347	**U.S. Stock Ticker: VION**	
2005 Sales: $ 23	2005 Profits: $-18,041	**Int'l Ticker:** Int'l Exchange:	
2004 Sales: $ 275	2004 Profits: $-16,055	Employees: 39	
2003 Sales: $ 375	2003 Profits: $-11,838	Fiscal Year Ends: 12/31	
2002 Sales: $ 200	2002 Profits: $-12,300	Parent Company:	

SALARIES/BENEFITS:

Pension Plan:	ESOP Stock Plan:	Profit Sharing:	Top Exec. Salary: $445,619	Bonus: $95,000
Savings Plan: Y	Stock Purch. Plan: Y		Second Exec. Salary: $286,000	Bonus: $61,150

OTHER THOUGHTS:

Apparent Women Officers or Directors: 4
Hot Spot for Advancement for Women/Minorities: Y

LOCATIONS: ("Y" = Yes)

West:	Southwest:	Midwest:	Southeast:	Northeast:	International:
				Y	

Note: Financial information, benefits and other data can change quickly and may vary from those stated here.

VIRAGEN INC

www.viragen.com

Industry Group Code: 325412 Ranks within this company's industry group: Sales: 174   Profits: 102

Drugs:		Other:		Clinical:	Computers:	Other:
Discovery:	Y	AgriBio:		Trials/Services:	Hardware:	Specialty Services:
Licensing:	Y	Genomics/Proteomics:	Y	Laboratories:	Software:	Consulting:
Manufacturing:	Y	Tissue Replacement:		Equipment/Supplies:	Arrays:	Blood Collection:
Development:	Y			Research/Development Svcs.:	Database Management:	Drug Delivery:
Generics:				Diagnostics:		Drug Distribution:

TYPES OF BUSINESS:

Drugs-Human Interferons
Avian Transgenics
Human Immune System Research
Protein-Based Drugs
Cancer Vaccines

BRANDS/DIVISIONS/AFFILIATES:

Multiferon
OVA System

CONTACTS: *Note: Officers with more than one job title may be intentionally listed here more than once.*

Charkes A. Rice, CEO
Charles A. Rice, Pres.
Dennis W. Healey, CFO
Dennis W. Healey, Corp. Sec.
Douglas W. Calder, Dir.-Comm.
Patrick Yeramian, Consulting Medical Dir.
Orjan Norberg, Managing Dir.-ViraNative AB
Karen Jervis, VP/Managing Dir.-VSL
Carl N. Singer, Chmn.

Phone: 954-233-8746	Fax: 954-233-1414
Toll-Free:	
Address: 865 SW 78th Ave., Ste. 100, Plantation, FL 33324 US	

GROWTH PLANS/SPECIAL FEATURES:

Viragen, Inc. focuses on the research, development, manufacture and commercialization of innovated technologies and products used to treat infectious diseases and cancers. The company is a pioneer in the science of avian transgenics, using the OVA System, wherefrom it intends to produce high quality proteins and antibodies in the egg whites of transgenic chickens. The company has development and manufacturing operations in Umea, Sweden; research and development activities in Edinburgh, Scotland; and corporate headquarters in Plantation, Florida. Its product portfolio includes three leading candidate-type products. Multiferon is a natural leukocyte-derived multi-subtype interferon alpha, used in the treatment of a number of viral diseases and cancer indications. It is approved for sale in Bulgaria, Chile, Mexico, Philippines and Sweden as a second-line therapy for the treatment of diseases in which patients show an initial response to recombinant alpha interferon followed by treatment failure; it is also approved for sale in Egypt, Hong Kong, Indonesia, Myanmar, and South Africa as a second-line therapy for the treatment of hairy cell leukemia and chronic myelogenous leukemia. VG101 is an antibody to the GD3 antigen, which is over-expressed on malignant melanoma tumors, thereby preventing the body's natural immune system from stopping cancer cell growth and proliferation. VG102 is an antibody to the CD55 antigen, which is over-expressed on nearly all solid, cancerous tumors and which prevents the body's natural immune system from killing cancer cells. In May 2006, the company announced that the U.S. Army Medical Research Institute of Infectious Diseases has agreed to commence a series of in vivo studies to determine the potential of Multiferon as an anti-viral product capable of defending against biowarfare. In June 2007, Viragen announced that it is halting its development of the OVA System in order to more fully focus its resources on advancing the company's anti-cancer efforts.

FINANCIALS: Sales and profits are in thousands of dollars—add 000 to get the full amount. 2006 Note: Financial information for 2006 was not available for all companies at press time.

2006 Sales: $ 391	2006 Profits: $-18,215	**U.S. Stock Ticker: VRAI.PK**
2005 Sales: $ 279	2005 Profits: $-26,207	**Int'l Ticker:** Int'l Exchange:
2004 Sales: $ 266	2004 Profits: $-18,177	Employees: 54
2003 Sales: $ 630	2003 Profits: $-17,348	Fiscal Year Ends: 6/30
2002 Sales: $1,300	2002 Profits: $-11,100	Parent Company:

SALARIES/BENEFITS:

Pension Plan:	ESOP Stock Plan:	Profit Sharing:	Top Exec. Salary: $300,000	Bonus: $
Savings Plan:	Stock Purch. Plan:		Second Exec. Salary: $210,000	Bonus: $

OTHER THOUGHTS:

Apparent Women Officers or Directors: 1
Hot Spot for Advancement for Women/Minorities:

LOCATIONS: ("Y" = Yes)

West:	Southwest:	Midwest:	Southeast:	Northeast:	International:
			Y		Y

VIRBAC CORP   www.virbaccorp.com

Industry Group Code: 325412B   Ranks within this company's industry group: Sales: Profits:

Drugs:		Other:		Clinical:	Computers:	Other:	
Discovery:	Y	AgriBio:		Trials/Services:	Hardware:	Specialty Services:	
Licensing:		Genomics/Proteomics:		Laboratories:	Software:	Consulting:	
Manufacturing:	Y	Tissue Replacement:		Equipment/Supplies:	Arrays:	Blood Collection:	
Development:	Y			Research/Development Svcs.:	Database Management:	Drug Delivery:	
Generics:				Diagnostics:		Drug Distribution:	

TYPES OF BUSINESS:

Drugs-Animal Health & Pet Care
Contract Manufacturing-Animal Health Products

BRANDS/DIVISIONS/AFFILIATES:

PetRelief
Zema
Petrodex

CONTACTS: *Note: Officers with more than one job title may be intentionally listed here more than once.*

Erik R. Martinez, CEO
Eric R. Martinez, Pres.
Eric Maree, Chmn.

Phone: 817-831-5030	**Fax:** 817-831-8327
Toll-Free: 800-338-3659	
Address: 3200 Meacham Blvd., Fort Worth, TX 76137 US	

GROWTH PLANS/SPECIAL FEATURES:

Virbac Corporation develops, manufactures, markets, distributes and sells a variety of pet and companion animals health products; these products focus on dermatological, parasiticidal, dental and livestock. The company has three segments: veterinary, which provides animal health products to veterinary clinics throughout North America; consumer, which sells over-the-counter products for companion animal health to national accounts, distributors and wholesalers; and contract manufacturing, which offers a range of services and specialized expertise in the manufacture of regulated animal health products. Products include flea and tick treatments, canine heartworm preventives, ear cleaners, endocrinology treatments, euthanasia drugs, aquarium water conditioners, gastrointestinal products and specialty chemicals. The company also offers dermatological and oral hygiene products for pets and companion animals, under the PetRelief, Zema and Petrodex brand names. In May 2006, Virbac announced it was granted marketing authorization in India for the SAG2 strain in the form of an oral bait vaccine for stray dogs; the vaccine is a treatment for rabies, which affects more than 50,000 people per year around the world. In October 2006, Virbac Corporation became a wholly-owned subsidiary of Virbac SA after Virbac SA increased its share from 60.09% to 100%. As a result, the parent company delisted Virbac Corporation from the American stock markets. The parent company, Virbac SA, is based in Carros, France, but sells veterinary products in over 100 countries and has 24 foreign subsidiaries.

FINANCIALS: Sales and profits are in thousands of dollars—add 000 to get the full amount. 2006 Note: Financial information for 2006 was not available for all companies at press time.

2006 Sales: $	2006 Profits: $	**U.S. Stock Ticker:** Subsidiary
2005 Sales: $80,778	2005 Profits: $3,873	**Int'l Ticker:** Int'l Exchange:
2004 Sales: $77,115	2004 Profits: $1,471	Employees: 269
2003 Sales: $67,077	2003 Profits: $-5,004	Fiscal Year Ends: 10/31
2002 Sales: $63,800	2002 Profits: $3,400	Parent Company: VIRBAC SA

SALARIES/BENEFITS:

Pension Plan:	ESOP Stock Plan:	Profit Sharing:	Top Exec. Salary: $305,363	Bonus: $
Savings Plan:	Stock Purch. Plan:		Second Exec. Salary: $151,470	Bonus: $

OTHER THOUGHTS:

Apparent Women Officers or Directors:
Hot Spot for Advancement for Women/Minorities:

LOCATIONS: ("Y" = Yes)

West:	Southwest:	Midwest:	Southeast:	Northeast:	International:
	Y	Y			Y

VIROPHARMA INC

www.viropharma.com

Industry Group Code: 325412 Ranks within this company's industry group: Sales: 58   Profits: 34

Drugs:		Other:	Clinical:	Computers:	Other:
Discovery:		AgriBio:	Trials/Services:	Hardware:	Specialty Services:
Licensing:	Y	Genomics/Proteomics:	Laboratories:	Software:	Consulting:
Manufacturing:		Tissue Replacement:	Equipment/Supplies:	Arrays:	Blood Collection:
Development:	Y		Research/Development Svcs.:	Database Management:	Drug Delivery:
Generics:			Diagnostics:		Drug Distribution:

TYPES OF BUSINESS:

Oral Antibiotics
Hepatitis C Drugs
Infectious Diseases Drugs

BRANDS/DIVISIONS/AFFILIATES:

Vanococin HCl Capsules
Maribavir
HCV-796

CONTACTS: *Note: Officers with more than one job title may be intentionally listed here more than once.*

Michel de Rosen, CEO
Vincent J. Milano, COO
Michel de Rosen, Pres.
Vincent J. Milano, CFO/VP
Colin Broom, Chief Scientific Officer/VP
Thomas F. Doyle, General Counsel/VP/Sec.
Daniel B. Soland, VP/Chief Commercial Officer
Michel de Rosen, Chmn.
Robert G. Pietrusko, VP-Global Regulatory Affairs & Quality

Phone: 610-458-7300	Fax: 610-458-7380
Toll-Free:	
Address: 397 Eagleview Blvd., Exton, PA 19341 US	

GROWTH PLANS/SPECIAL FEATURES:

ViroPharma, Inc. is a biopharmaceutical company that develops and commercializes products that address serious infectious diseases, with a focus on products used by physician specialists or in hospital settings. The company's only marketed product is Vancocin HC1 capsules, an oral antibiotic for the treatment of antibiotic-associated pseudomembranous colitis and enterocolitis, including methicilin-resistant strains. The firm is developing maribavir, currently in Phase III for the prevention and treatment of cytomegalovirus diseases and HCV-796, currently in Phase II for the treatment of hepatitis C virus. ViroPharma licensed the U.S. and Canadian rights for a third product candidate, an intranasal formulation of pleconaril, to Schering-Plough for the treatment of picornavirus infections. Customers of Vancocin include wholesalers who then distribute the drug to pharmacies, hospitals and long term care facilities. In April 2006, ViroPharma entered into an agreement with Alpharma, Inc. for the manufacturing of the active pharmaceutical ingredient for Vancocin. In May 2007, the company announced the decision to establish a European subsidiary that would be in charge of developing and commercializing maribavir in Europe. In June 2007, the firm announced that FDA granted fast track designation for HCV-796 for treatment of hepatitis C virus infection. In July 2007, ViroPharma announced that it had initiated a Phase III clinical trial for maribavir for patients undergoing a liver transplant procedure.

The company offers its employees medical and dental benefits; a 401(k) plan; and stock options.

FINANCIALS: Sales and profits are in thousands of dollars—add 000 to get the full amount. 2006 Note: Financial information for 2006 was not available for all companies at press time.

2006 Sales: $167,181	2006 Profits: $66,666	**U.S. Stock Ticker: VPHM**
2005 Sales: $132,417	2005 Profits: $113,705	**Int'l Ticker:** Int'l Exchange:
2004 Sales: $22,389	2004 Profits: $-19,534	Employees: 67
2003 Sales: $1,612	2003 Profits: $-36,942	Fiscal Year Ends: 12/31
2002 Sales: $5,500	2002 Profits: $-15,800	Parent Company:

SALARIES/BENEFITS:

Pension Plan:	ESOP Stock Plan:	Profit Sharing:	Top Exec. Salary: $375,000	Bonus: $171,562
Savings Plan: Y	Stock Purch. Plan: Y		Second Exec. Salary: $317,000	Bonus: $142,277

OTHER THOUGHTS:

Apparent Women Officers or Directors:
Hot Spot for Advancement for Women/Minorities:

LOCATIONS: ("Y" = Yes)

West:	Southwest:	Midwest:	Southeast:	Northeast:	International:
				Y	

Note: Financial information, benefits and other data can change quickly and may vary from those stated here.

VYSIS INC

www.vysis.com

Industry Group Code: 325413 Ranks within this company's industry group: Sales: Profits:

Drugs:		Other:		Clinical:		Computers:		Other:	
Discovery:		AgriBio:		Trials/Services:		Hardware:	Y	Specialty Services:	
Licensing:		Genomics/Proteomics:	Y	Laboratories:		Software:		Consulting:	
Manufacturing:		Tissue Replacement:		Equipment/Supplies:	Y	Arrays:	Y	Blood Collection:	
Development:				Research/Development Svcs.:		Database Management:		Drug Delivery:	
Generics:				Diagnostics:	Y			Drug Distribution:	

TYPES OF BUSINESS:

Medical Diagnostics Products
Reagents
Microarray Technology
DNA Probes
Genomics Workstations
Genetic Diagnostic Products

BRANDS/DIVISIONS/AFFILIATES:

Abbott Laboratories
Fluorescence In Situ Hybridization (FISH)
GenoSensor Microarray System
PathVysion
UroVysion
AneuVysion
LSI Locus Specific Identifier DNA Probes
HYBrite Hybridization System

CONTACTS: *Note: Officers with more than one job title may be intentionally listed here more than once.*

Steven A. Seeling, Div. VP
Susan Zint, Dir.-Human Resources
David Reiners, Mgr.-Mktg. Comm.

Phone: 224-361-7000	**Fax:** 224-361-7138
Toll-Free: 800-553-7042	
Address: 1300 E. Touhy, Des Plaines, IL 60018-3315 US	

GROWTH PLANS/SPECIAL FEATURES:

Vysis, Inc., a subsidiary of Abbott Laboratories, develops and markets genomics-based clinical products for the evaluation and management of cancer, prenatal disorders and other genetic diseases. The firm's technology platforms, which it not only uses for its own research but markets for other firms, are Fluorescence in Situ Hybridization (FISH) and the GenoSensor Microarray System, both designed to detect gene and chromosome copy numbers; comparative genomic hybridization agents; and the VP 2000 Processor, an integrated workstation with a variety of tools for preparing slides and specimens. Vysis's developed products include: PathVysion HER-2 DNA Probe Kit, which enables assessment of the HER-2/NEU breast cancer gene; the UroVysion Bladder Cancer Kit, for detecting specific chromosome abnormalities indicative of bladder cancer reoccurrence; and the AneuVysion Assay, detecting trisomy 13, 18, 21 and sex chromosome anomalies for prenatal screenings. Vysis also produces many products relating to more general DNA analysis, such as CEP Chromosome Enumeration DNA Probes, LSI Locus Specific Identifier DNA Probes and TelVysion Telomere DNA Probes. The SpectraVysion Assay provides distinct identification of the 24 human chromosomes for anomalies that are not identifiable using traditional cytogenetic banding methods. HYBrite Hybridization System provides for FISH procedures through a hands-free, denaturation/hybridization system eliminating hazardous solutions such as formamide and ethanol.

FINANCIALS: Sales and profits are in thousands of dollars—add 000 to get the full amount. 2006 Note: Financial information for 2006 was not available for all companies at press time.

2006 Sales: $	2006 Profits: $	**U.S. Stock Ticker:** Subsidiary
2005 Sales: $	2005 Profits: $	**Int'l Ticker:** Int'l Exchange:
2004 Sales: $	2004 Profits: $	Employees: 133
2003 Sales: $	2003 Profits: $	Fiscal Year Ends: 12/31
2002 Sales: $	2002 Profits: $	Parent Company: ABBOTT LABORATORIES

SALARIES/BENEFITS:

Pension Plan:	ESOP Stock Plan:	Profit Sharing:	Top Exec. Salary: $	Bonus: $
Savings Plan:	Stock Purch. Plan:		Second Exec. Salary: $	Bonus: $

OTHER THOUGHTS:

Apparent Women Officers or Directors: 1
Hot Spot for Advancement for Women/Minorities:

LOCATIONS: ("Y" = Yes)

West:	Southwest:	Midwest:	Southeast:	Northeast:	International:
		Y			Y

WARNER CHILCOTT PLC

www.warnerchilcott.com

Industry Group Code: 325412 Ranks within this company's industry group: Sales: Profits:

Drugs:		Other:	Clinical:	Computers:	Other:
Discovery:		AgriBio:	Trials/Services:	Hardware:	Specialty Services:
Licensing:		Genomics/Proteomics:	Laboratories:	Software:	Consulting:
Manufacturing:	Y	Tissue Replacement:	Equipment/Supplies:	Arrays:	Blood Collection:
Development:	Y		Research/Development Svcs.:	Database Management:	Drug Delivery:
Generics:			Diagnostics:		Drug Distribution:

TYPES OF BUSINESS:

Pharmaceuticals Development & Manufacturing
Contraceptives
Hormone Therapies
Vitamins
Dermatology Treatments

BRANDS/DIVISIONS/AFFILIATES:

Loestrin 24
Estrostep
femhrt
Ovcon
Estrace
Taclonex
Dovonex
Doryx

CONTACTS: Note: Officers with more than one job title may be intentionally listed here more than once.

Roger Boissonneault, CEO
Roger Boissonneault, Pres.
Paul Herendeen, CFO/Exec. VP
Herman Ellman, Sr. VP-Clinical Dev.
Leland H. Cross, Sr. VP-Tech. Oper.
Izumi Hara, General Counsel/Sr. VP/Corp. Sec.
Anthony D.Bruno, Exec. VP-Corp. Dev.
W. Carlton Reichel, Pres., Pharmaceuticals
Alvin Howard, Sr. VP-Regulatory Affairs

Phone: 973-442-3200	Fax: 973-442-3283
Toll-Free: 800-521-8813	
Address: 100 Enterprise Dr., Rockaway, NJ 07866 US	

GROWTH PLANS/SPECIAL FEATURES:

Warner Chilcott (WC) is a specialty pharmaceutical company that focuses on developing, manufacturing and marketing branded prescription pharmaceutical products in women's healthcare and dermatology. Its franchises are comprised of complementary portfolios of established, branded, development stage and new products, including recently launched products Loestrin 24 and Taclonex. Its women's healthcare franchise is anchored by its strong presence in the hormonal contraceptive and hormone therapy categories and its dermatology franchise is built on its established positions in the markets for psoriasis and acne therapies. In April 2006, it launched Loestrin 24 Fe, an oral contraceptive with a novel patented 24-day dosing regimen, with the goal of growing the market share position it achieved with its Ovcon and Estrostep products in the hormonal contraceptive market. The company also has a significant presence in the hormone therapy market, primarily through its products femhrt and Estrace Cream. In dermatology, its psoriasis product Dovonex enjoys the leading position in the U.S. for the non-steroidal topical treatment of psoriasis. WC strengthened and extended its position in the market for psoriasis therapies with the April 2006 launch of Taclonex, the first once-a-day topical psoriasis treatment that combines betamethasone dipropionate, a corticosteroid, with calcipotriene, the active ingredient in Dovonex. The company's product Doryx is the leading branded oral tetracycline in the U.S. for the treatment of acne. Recently, the company launched Dorys delayed-release tablets. In 2007, WC announced collaborations with various entities, including: Foamix; Watson Pharmaceuticals; LEO Pharma; and Paratek.

FINANCIALS: Sales and profits are in thousands of dollars—add 000 to get the full amount. 2006 Note: Financial information for 2006 was not available for all companies at press time.

2006 Sales: $	2006 Profits: $	U.S. Stock Ticker: Private
2005 Sales: $	2005 Profits: $	Int'l Ticker: Int'l Exchange:
2004 Sales: $	2004 Profits: $	Employees: 960
2003 Sales: $432,300	2003 Profits: $96,200	Fiscal Year Ends: 12/31
2002 Sales: $235,200	2002 Profits: $145,100	Parent Company:

SALARIES/BENEFITS:

Pension Plan:	ESOP Stock Plan:	Profit Sharing:	Top Exec. Salary: $	Bonus: $
Savings Plan:	Stock Purch. Plan:		Second Exec. Salary: $	Bonus: $

OTHER THOUGHTS:

Apparent Women Officers or Directors:
Hot Spot for Advancement for Women/Minorities:

LOCATIONS: ("Y" = Yes)

West:	Southwest:	Midwest:	Southeast:	Northeast:	International:
				Y	Y

WATSON PHARMACEUTICALS INC
www.watson.com

Industry Group Code: 325416 Ranks within this company's industry group: Sales: 2 Profits: 9

Drugs:		Other:		Clinical:		Computers:		Other:	
Discovery:	Y	AgriBio:		Trials/Services:		Hardware:		Specialty Services:	
Licensing:		Genomics/Proteomics:		Laboratories:		Software:		Consulting:	
Manufacturing:	Y	Tissue Replacement:		Equipment/Supplies:		Arrays:		Blood Collection:	
Development:	Y			Research/Development Svcs.:		Database Management:		Drug Delivery:	
Generics:	Y			Diagnostics:				Drug Distribution:	Y

TYPES OF BUSINESS:

Generic Pharmaceuticals
Branded Drugs
Urology Drugs
Anti-Hypertensive Drugs
Psychiatric Drugs
Pain Management Drugs
Dermatology Drugs
Nephrology Drugs

BRANDS/DIVISIONS/AFFILIATES:

Watson Laboratories
Watson Pharma
Rugby
Ferrlecit
Trelstar Depot
Oxytrol
Anda
Andrx Corp.

CONTACTS: Note: Officers with more than one job title may be intentionally listed here more than once.

Allen Chao, CEO
Susan Skara, Sr. VP-Human Resources
Charles D. Ebert, Sr. VP-R&D
Thomas R. Giordano, CIO/Sr. VP
David A. Buchen, General Counsel/Sr. VP/Sec.
Edward F. Heimers, Jr., Exec. VP/Pres., Brand Division
David C. Hsia, Sr. VP-Scientific Affairs
Gordon Munro, Sr. VP-Quality Assurance
Thomas R. Russillo, Exec. VP/Pres., U.S. Generics Division
Allen Chao, Chmn.

Phone: 951-493-5300	Fax: 973-355-8301
Toll-Free:	
Address: 311 Bonnie Cir., Corona, CA 92880 US	

GROWTH PLANS/SPECIAL FEATURES:

Watson Pharmaceuticals, Inc. manufactures and distributes over 25 branded and over 150 generic pharmaceutical products. The company offers generic versions of popular brand-name pharmaceuticals including Zyban, Wellbutrin, Lorcet, Vicodin, Percocet, Ocycontin, Nicorette, Lortab, Triphasil and Demulen. The firm markets its generic products through a network of 25 sales and marketing professionals under the Watson Laboratories and Watson Pharma labels. Over-the-counter products are sold under the Rugby label or private label. Generic products accounted for roughly 77% of net sales in 2006. Watson Pharmaceuticals' brand business segment develops, manufactures, markets, sells and distributes products primarily through two sales and marketing groups: specialty groups and nephrology. The specialty products include urology, anti-hypertensive, psychiatry, pain management and dermatology products and a genital warts treatment. Brand names include Trelstar Depot, Trelstar LA and Oxytrol. The nephrology product line consists of products for the treatment of iron deficiency anemia. The primary product is Ferrlecit, indicated for patients undergoing hemodialysis in conjunction with erythropoietin therapy. The company markets its brand products through 333 sales professionals and offers trademarked off-patent products to physicians, hospitals and healthcare professionals. Brand products accounted for roughly 19% of total revenue in 2006. The company's distribution business consists of Anda, Anda Pharmaceuticals and Valmed. During 2006, Watson Pharmaceuticals entered into an agreement with Solvay Pharmaceuticals for the promotion of AndroGel to urologists in the U.S. In November 2006, the company acquired Andrx Corp., a pharmaceutical products distributor, for roughly $1.9 billion. In June 2007, the firm received FDA approval forbypropion hydrochloride tablets, the generic equivalent to Wellbutrin XL tablets, for the treatment of major depressive disorder.

The company offers its employees medical, dental and vision insurance; a 401(k) plan; life and AD&D insurance; domestic partner coverage; business travel accident insurance; short- and long-term disability; pet insurance; and tuition reimbursement.

FINANCIALS: Sales and profits are in thousands of dollars—add 000 to get the full amount. 2006 Note: Financial information for 2006 was not available for all companies at press time.

2006 Sales: $1,979,244	2006 Profits: $-445,005	U.S. Stock Ticker: WPI
2005 Sales: $1,646,203	2005 Profits: $138,557	Int'l Ticker: Int'l Exchange:
2004 Sales: $1,640,551	2004 Profits: $150,018	Employees: 5,830
2003 Sales: $1,457,722	2003 Profits: $202,864	Fiscal Year Ends: 12/31
2002 Sales: $1,223,200	2002 Profits: $175,800	Parent Company:

SALARIES/BENEFITS:

Pension Plan:	ESOP Stock Plan:	Profit Sharing:	Top Exec. Salary: $937,692	Bonus: $993,000
Savings Plan: Y	Stock Purch. Plan:		Second Exec. Salary: $420,038	Bonus: $

OTHER THOUGHTS:

Apparent Women Officers or Directors: 2
Hot Spot for Advancement for Women/Minorities: Y

LOCATIONS: ("Y" = Yes)

West:	Southwest:	Midwest:	Southeast:	Northeast:	International:
Y		Y	Y	Y	Y

Note: Financial information, benefits and other data can change quickly and may vary from those stated here.

WHATMAN PLC

www.whatman.co.uk

Industry Group Code: 334500 Ranks within this company's industry group: Sales: Profits:

Drugs:	Other:	Clinical:		Computers:		Other:	
Discovery:	AgriBio:	Trials/Services:		Hardware:		Specialty Services:	
Licensing:	Genomics/Proteomics:	Laboratories:		Software:		Consulting:	
Manufacturing:	Tissue Replacement:	Equipment/Supplies:	Y	Arrays:		Blood Collection:	
Development:		Research/Development Svcs.:		Database Management:		Drug Delivery:	
Generics:		Diagnostics:	Y			Drug Distribution:	

TYPES OF BUSINESS:

Equipment-Filtration Systems
Diagnostic Supplies

BRANDS/DIVISIONS/AFFILIATES:

Multiwell
UNIFILTER
Schleicher & Schuell GmbH
FAST Slides
Protagen AG UNIclone
Hangzhou Whatman-Xinhua Filter Paper Co. Ltd.

CONTACTS: *Note: Officers with more than one job title may be intentionally listed here more than once.*

Keran May, CEO
Ian Bonnar, COO
Chris Rickard, Dir.-Finance
John Simmonds, Corp. Sec.
Bob Thain, Chmn.

Phone: 440-208-326-1740	Fax: 440-208-326-1741
Toll-Free:	
Address: 27 Great West Rd., Brentford, Middlesex TW89BW UK	

GROWTH PLANS/SPECIAL FEATURES:

Whatman plc manufactures and distributes separation and filtration products used in laboratories, health care facilities and bioscience research. The firm offers over 100 different products for filtration including student kits, DNA purification and Ph testers. The firm's subsidiary companies are expansions to different countries and manufacturing plants, and are located in Asia Pacific, Germany, Ireland, the U.S., Japan, Canada and Belgium. Whatman products serve three markets: LabSciences, MedTech and BioScience. The LabSciences unit produces chromatography products used for purification; extraction products, filter papers, filtration devices, membrane filters and specialty products. The MedTech sector makes diagnostic components and medical devices such as purification and sterilization tools used by OEMs. The BioScience unit produces the Multiwell and UNIFILTER lines of filter plates and products used for nucleic acid sample preparation. Schleicher & Schuell GmbH (S&S), a Whatman subsidiary, is a manufacturer of diagnostic and blotting membranes; paper, glass fiber and non-woven pads; surface-modified media; microarray slides; and reagent kits for life science research. In early 2006, with the assistance of S&S technology, the firm announced that two of its most promising technologies, FAST Slides and the Protagen AG UNIclone human protein expression library, have enabled the discovery of novel auto-antigens associated with the autoimmune disease alopecia areata. In July 2007, Whatman, along with Hangzhou Xinhua Paper Industry Co., Ltd., the leading Chinese manufacturer of chemical analysis paper and related paper products, announced their joint venture company, Hangzhou Whatman-Xinhua Filter Paper Co. Ltd. has commenced trading.

FINANCIALS: Sales and profits are in thousands of dollars—add 000 to get the full amount. 2006 Note: Financial information for 2006 was not available for all companies at press time.

2006 Sales: $	2006 Profits: $	**U.S. Stock Ticker: WTPLF.PK**
2005 Sales: $	2005 Profits: $	**Int'l Ticker: WHM** Int'l Exchange: London-LSE
2004 Sales: $159,500	2004 Profits: $-1,000	Employees: 812
2003 Sales: $149,000	2003 Profits: $-3,300	Fiscal Year Ends: 12/31
2002 Sales: $140,000	2002 Profits: $-32,300	Parent Company:

SALARIES/BENEFITS:

Pension Plan:	ESOP Stock Plan:	Profit Sharing:	Top Exec. Salary: $	Bonus: $
Savings Plan:	Stock Purch. Plan:		Second Exec. Salary: $	Bonus: $

OTHER THOUGHTS:

Apparent Women Officers or Directors:
Hot Spot for Advancement for Women/Minorities:

LOCATIONS: ("Y" = Yes)

West:	Southwest:	Midwest:	Southeast:	Northeast:	International:
				Y	Y

Note: Financial information, benefits and other data can change quickly and may vary from those stated here.

WYETH
www.wyeth.com

Industry Group Code: 325412 Ranks within this company's industry group: Sales: 10   Profits: 9

Drugs:		Other:		Clinical:	Computers:	Other:	
Discovery:	Y	AgriBio:		Trials/Services:	Hardware:	Specialty Services:	
Licensing:		Genomics/Proteomics:	Y	Laboratories:	Software:	Consulting:	
Manufacturing:	Y	Tissue Replacement:		Equipment/Supplies:	Arrays:	Blood Collection:	
Development:	Y			Research/Development Svcs.:	Database Management:	Drug Delivery:	
Generics:				Diagnostics:		Drug Distribution:	

TYPES OF BUSINESS:
Drugs-Diversified
Wholesale Pharmaceuticals
Animal Health Care Products
Biologicals
Vaccines
Over-the-Counter Drugs
Women's Health Care Products
Nutritional Supplements

BRANDS/DIVISIONS/AFFILIATES:
Chap Stick
Premarin
Dimetapp
Advil
Robitussin
Preparation H
Centrum
Wyeth K.K.

CONTACTS:
Note: Officers with more than one job title may be intentionally listed here more than once.

Robert Essner, CEO
Bernard Poussot, COO
Bernard Poussot, Pres.
Rene Lewin, VP-Human Resources
Jeffrey Keisling, CIO
Lawrence Stein, General Counsel/Sr. VP
Thomas Hofstaetter, Sr. VP-Bus. Dev.
Marilyn Rhudy, VP-Public Affairs
Justin Victoria, VP-Investor Rel.
John Kelly, VP-Finance Oper.
James Pohlman, VP-Corp. Strategic Initiatives
Joseph Mahady, Sr. VP
Robert Ruffolo, Sr. VP
Robert E. Landry, Jr., Treas.
Robert Essner, Chmn.

Phone: 973-660-5000	Fax: 973-660-7026

Toll-Free:
Address: 5 Giralda Farms, Madison, NJ 07940-0874 US

GROWTH PLANS/SPECIAL FEATURES:
Wyeth is a global leader in pharmaceuticals, consumer health care products and animal health care products. The firm discovers, develops, manufactures, distributes and sells a diversified line of products arising from three divisions: pharmaceuticals, consumer health care and animal care. The pharmaceuticals segment is itself divided into women's health care, neuroscience, vaccines and infectious disease, musculoskeletal, internal medicine, hemophilia and immunology and oncology. The division sells branded and generic pharmaceuticals, biological and nutraceutical products as well as animal biological products and pharmaceuticals. Its branded products include Premarin, Prempro, Premphase, Triphasil, Ativan, Effexor, Altace, Inderal, Zoton, Protonix and Enbrel. The consumer health care segment's products include analgesics, cough/cold/allergy remedies, nutritional supplements, lip balm and hemorrhoidal, antacid, asthma and other relief items sold over-the-counter. The segment's well-known over-the-counter products include Advil, cold medicines Robitussin and Dimetapp and nutritional supplement Centrum, as well as Chap Stick, Caltrate, Preparation H and Solgar. The company's animal health care products include vaccines, pharmaceuticals, endectocides (dewormers that control both internal and external parasites) and growth implants under the brand names LymeVax, Duramune and Fel-O-Vax. The company recently upped its stake in Wyeth/Takeda Pharmaceutical Company Limited joint venture Wyeth K.K. to 80%, after the purchase of an additional 10 percent stake in 2006.

FINANCIALS:
Sales and profits are in thousands of dollars—add 000 to get the full amount. 2006 Note: Financial information for 2006 was not available for all companies at press time.

2006 Sales: $20,350,655	2006 Profits: $4,196,706	U.S. Stock Ticker: WYE
2005 Sales: $18,755,790	2005 Profits: $3,656,298	Int'l Ticker: Int'l Exchange:
2004 Sales: $17,358,028	2004 Profits: $1,233,997	Employees: 50,060
2003 Sales: $15,850,600	2003 Profits: $2,051,600	Fiscal Year Ends: 12/31
2002 Sales: $14,584,000	2002 Profits: $4,447,200	Parent Company:

SALARIES/BENEFITS:
| Pension Plan: Y | ESOP Stock Plan: | Profit Sharing: | Top Exec. Salary: $1,590,000 | Bonus: $2,700,000 |
| Savings Plan: Y | Stock Purch. Plan: | | Second Exec. Salary: $840,000 | Bonus: $1,260,000 |

OTHER THOUGHTS:
Apparent Women Officers or Directors: 1
Hot Spot for Advancement for Women/Minorities:

LOCATIONS: ("Y" = Yes)
West:	Southwest:	Midwest:	Southeast:	Northeast:	International:
Y	Y	Y	Y	Y	Y

Note: Financial information, benefits and other data can change quickly and may vary from those stated here.

XECHEM INTERNATIONAL

www.xechem.com

Industry Group Code: 325412 Ranks within this company's industry group: Sales: 179   Profits: 78

Drugs:		Other:	Clinical:	Computers:	Other:
Discovery:	Y	AgriBio:	Trials/Services:	Hardware:	Specialty Services:
Licensing:		Genomics/Proteomics:	Laboratories:	Software:	Consulting:
Manufacturing:	Y	Tissue Replacement:	Equipment/Supplies:	Arrays:	Blood Collection:
Development:	Y		Research/Development Svcs.: Y	Database Management:	Drug Delivery:
Generics:	Y		Diagnostics:		Drug Distribution:

TYPES OF BUSINESS:

Pharmaceuticals Development & Manufacturing
Drugs-Proprietary
Drugs-Generic
Research & Development Services
Fine Chemicals
Nutraceuticals

BRANDS/DIVISIONS/AFFILIATES:

Xechem, Inc.
GinsengOnce
GinkgoOnce
GarlicOnce
Gugulon
Xechem (India) Pvt., Ltd.
XetaPharm
NICOSAN

CONTACTS: *Note: Officers with more than one job title may be intentionally listed here more than once.*

Robert Swift, Interim CEO
Robert Swift, Interim Pres.
H. Scott English, Internal Dir.-Investor Rel.
Benjamin S. White, III, Dir.-Finance & Accounting
Renuka Misra, Dir.-Natural Products
Robert Swift, Interim Chmn.

Phone: 732-247-3300	Fax: 732-247-4090
Toll-Free:	
Address: 100 Jersey Ave., Bldg. B, Ste. 310, New Brunswick, NJ 08901 US	

GROWTH PLANS/SPECIAL FEATURES:

Xechem International, Inc. is a fully integrated pharmaceutical and nutraceutical company and provider generic and proprietary drugs. It is also a public holding company whose principal subsidiary, Xechem, Inc., is engaged primarily in applying its proprietary extraction, isolation and purification technology to the production and manufacture of paclitaxel (commonly referred to in the scientific literature as Taxol, a registered trademark of Bristol-Myers Squibb Company). Paclitaxel is an anticancer compound used for the treatment of ovarian, breast, small cell lung cancers and AIDS related Kaposi sarcomas. The company has successfully isolated pure paclitaxel and has received a process patent on this technology. The company has submitted to the FDA a Drug Master File (or DMF) for the facility and the bulk paclitaxel product. The company has been issued five U.S. patents on paclitaxel and its second generation analogs from the U.S. Patent and Trademark Office and has over 34 international patents pending. Subsidiaries include: Xechem (India) Pvt., Ltd. in New Delhi, India; XetaPharm, Inc., which develops and markets over-the-counter (OTC) natural health supplements NUTRACEUTICALS such as GingkoOnce, GarlicOnce, GinsengOnce, Gugulon and CoEnzyme Q-10; Xechem Laboratories, Inc.; Xechem U.K. Ltd.; and Xechem Pharmaceuticals Nigeria Ltd. The company utilizes its international network of ethnobotanists, local folklore healers (shamans) and chemists to screen natural products used by folklore healers for their therapeutic value to develop those plants, their extracts and pure compounds which demonstrate promising activity in the primary screen. In 2007, the company announced that subsidiary Xechem Pharmaceuticals Nigeria Ltd. would increase its pilot scale production of NICOSAN, allowing the company to bring NICOSAN to as many as 30,000 patients per month, resulting in net sales of approximately $500,000 per month. NICOSAN targets sickle cell anemia shown to substantially reduce the degree of sickling of the red blood cells of those afflicted.

FINANCIALS: Sales and profits are in thousands of dollars—add 000 to get the full amount. 2006 Note: Financial information for 2006 was not available for all companies at press time.

2006 Sales: $ 202	2006 Profits: $-11,130	**U.S. Stock Ticker:** XKEM
2005 Sales: $ 6	2005 Profits: $-10,039	**Int'l Ticker:** Int'l Exchange:
2004 Sales: $ 168	2004 Profits: $-17,606	Employees: 80
2003 Sales: $ 318	2003 Profits: $-5,497	Fiscal Year Ends: 12/31
2002 Sales: $ 269	2002 Profits: $-3,599	Parent Company:

SALARIES/BENEFITS:

Pension Plan:	ESOP Stock Plan:	Profit Sharing:	Top Exec. Salary: $324,167	Bonus: $
Savings Plan:	Stock Purch. Plan: Y		Second Exec. Salary: $71,500	Bonus: $

OTHER THOUGHTS:

Apparent Women Officers or Directors:
Hot Spot for Advancement for Women/Minorities:

LOCATIONS: ("Y" = Yes)

West:	Southwest:	Midwest:	Southeast:	Northeast:	International:
				Y	Y

Note: Financial information, benefits and other data can change quickly and may vary from those stated here.

XOMA LTD
www.xoma.com

Industry Group Code: 325412   Ranks within this company's industry group: Sales: 97   Profits: 156

Drugs:		Other:		Clinical:	Computers:	Other:
Discovery:	Y	AgriBio:		Trials/Services:	Hardware:	Specialty Services:
Licensing:	Y	Genomics/Proteomics:	Y	Laboratories:	Software:	Consulting:
Manufacturing:	Y	Tissue Replacement:		Equipment/Supplies:	Arrays:	Blood Collection:
Development:	Y			Research/Development Svcs.:	Database Management:	Drug Delivery:
Generics:				Diagnostics:		Drug Distribution:

TYPES OF BUSINESS:
Therapeutic Antibodies

BRANDS/DIVISIONS/AFFILIATES:
Raptiva
Lucentis
Neuprex
Genetech, Inc.
Opebacan
XOMA 052
XOMA 629

CONTACTS: Note: Officers with more than one job title may be intentionally listed here more than once.
John L. Castello, CEO
John L. Castello, Pres.
J. David Boyle II, CFO
Charles C. Wells, VP-Human Resources
Mary Haak-Frendscho, VP-Preclinical R&D
Charles C. Wells, VP-IT
Christopher J. Margolin, General Counsel/VP/Sec.
Robert S. Tenerowicz, VP-Oper.
J. David Boyle II, VP-Finance
Patrick J. Scannon, Chief Biotechnology Officer/Exec. VP
Daniel P. Cafaro, VP-Regulatory Affairs
Arthur Shedden, VP-Clinical Dev.
Calvin L. McGoogan, VP-Quality & Facilities
John L. Castello, Chmn.

Phone: 510-214-7200	Fax: 510-644-2011
Toll-Free:	
Address: 2910 7th St., Berkeley, CA 94710 US	

GROWTH PLANS/SPECIAL FEATURES:

Xoma, Ltd. is a biopharmaceutical company that discovers and develops therapeutic antibodies, primarily directed toward treatments for cancer and immune disorders. The company has interests in two marketed antibody products and is developing other antibody and protein therapeutic products. Raptiva, marketed by Genetech, Inc., is a humanized therapeutic monoclonal antibody developed to treat immune system disorders. The drug is approved in the U.S. for the treatment of moderate-to-severe plaque psoriasis and marketed in over 50 countries. Lucentis, by Genetech is an antibody fragment against vascular endothelial growth factor for the treatment of age-related macular degeneration. The drug was approved by FDA in June 2006 and in the E.U., where it is distributed by Novartis AG, in January 2007. The drug uses the company's bacterial cell expression technology. Neuprex, an injectable formulation of opebacan, a modified recombinant fragment of human bactericidal/permeability-increasing (BPI) protein, is in Phase I/II clinical trials in patients undergoing allogeneic hematopoietic stem cell transplantation. In September 2006, the European Medicines Agency granted an orphan medicinal product designation to Neuprex in meningococcal sepsis, a bacterial infection predominantly affecting young children. Other drugs undergoing clinical trials include HCD122 (formerly CHIR-12.12), a fully human anti-CD40 antagonist antibody intended as a treatment for B-cell mediated diseases; XOMA 052 (formerly XMA005.2), a monoclonal antibody designed to be used as an injectable therapeutic for treating multiple inflammatory indications; and XOMA 629 (a reformulation of XMP.629), a topical anti-bacterial formulation of a BPI-derived peptide under development as a possible treatment for acne. Xoma has collaborations with several companies, including Genetech, Inc.; Takeda; Schering-Plough; Millenium; and Triton.

The company offers its employees medical, dental and vision plans; disability programs; life insurance; a 401(k) plan; an employee stock purchase plan; an employee assistance program; educational assistance; and access to a credit union.

FINANCIALS: Sales and profits are in thousands of dollars—add 000 to get the full amount. 2006 Note: Financial information for 2006 was not available for all companies at press time.

2006 Sales: $29,498	2006 Profits: $-51,841	U.S. Stock Ticker: XOMA
2005 Sales: $18,669	2005 Profits: $2,779	Int'l Ticker: Int'l Exchange:
2004 Sales: $3,665	2004 Profits: $-78,942	Employees: 255
2003 Sales: $24,412	2003 Profits: $-58,653	Fiscal Year Ends: 12/31
2002 Sales: $29,900	2002 Profits: $-33,200	Parent Company:

SALARIES/BENEFITS:

Pension Plan:	ESOP Stock Plan:	Profit Sharing:	Top Exec. Salary: $500,000	Bonus: $144,745
Savings Plan: Y	Stock Purch. Plan: Y		Second Exec. Salary: $360,000	Bonus: $55,224

OTHER THOUGHTS:
Apparent Women Officers or Directors: 1
Hot Spot for Advancement for Women/Minorities:

LOCATIONS: ("Y" = Yes)

West:	Southwest:	Midwest:	Southeast:	Northeast:	International:
Y					

Note: Financial information, benefits and other data can change quickly and may vary from those stated here.

ZILA INC

www.zila.com

Industry Group Code: 325412 **Ranks within this company's industry group:** Sales: 98   Profits: 125

Drugs:		Other:		Clinical:		Computers:		Other:	
Discovery:	Y	AgriBio:		Trials/Services:		Hardware:		Specialty Services:	
Licensing:	Y	Genomics/Proteomics:		Laboratories:		Software:		Consulting:	
Manufacturing:	Y	Tissue Replacement:		Equipment/Supplies:	Y	Arrays:		Blood Collection:	
Development:	Y			Research/Development Svcs.:	Y	Database Management:		Drug Delivery:	
Generics:				Diagnostics:	Y			Drug Distribution:	

TYPES OF BUSINESS:

Cancer Detection Products
Dental Products

BRANDS/DIVISIONS/AFFILIATES:

ViziLite
OraTest
Peridex
Zila Pharmaceuticals, Inc.
Zila, Ltd.
Zila Swab Technoolgies, Inc.
Zila Biotechnology, Inc.
Professional Dental Technologies, Inc.

CONTACTS: Note: Officers with more than one job title may be intentionally listed here more than once.

Frank J. Bellizzi, Pres.
Lawrence A. Gyenes, CFO
Gary Klinefelter, General Counsel/VP
Diane Klein, Treas./VP
David R. Bethune, Chmn.

Phone: 602-266-6700	Fax: 602-234-2264
Toll-Free:	
Address: 5227 N. 7th St., Phoenix, AZ 85014 US	

GROWTH PLANS/SPECIAL FEATURES:

Zila, Inc. is a provider of preventive healthcare technologies and products, focusing on enhanced body defense and the detection of pre-disease states. Zila is a holding company that conducts its operations through two business units: pharmaceuticals and biotechnology. The pharmaceutical business unit includes the ViziLite chemiluminescent disposable light product and the adjunct product ViziLite Plus with T-Blue630 for the illumination and marketing of oral mucosal abnormalities; and Peridex periodontal rinse, an antibacterial oral rinse used between dental visits for the treatment of gingivitis and periodontal disease. The segment includes the Nevada based Zila Pharmaceuticals, Inc.; the U.K. company Zila, Ltd.; and Zila Swab Technologies, Inc. The biotechnology business unit is the company's research, development and licensing division specializing in precancer/cancer detection through the Zila tolonium chloride squamous cell cancer detection technology. The segment is the manager of the OraTest product, an oral cancer diagnostic system. The unit includes Zila Biotechnology, Inc. and Zila Technical, Inc. Zila no long operates a nutraceutical business unit, which manufactured and marketed Ester-C, a patented form of vitamin C sold in 24 countries, and Ester-E, a proprietary enhanced form of vitamin E. In October 2006, the company sold Zila Nutraceuticals, Inc. to NBTY, Inc. for roughly $40.5 million. In November 2006, the firm purchased Professional Dental Technologies, Inc., a private dental products company, for roughly $34 million. The acquisition allowed Zila to transform into a cancer detection company.

The company offers its employees health insurance; flexible spending accounts; retirement and savings benefits; and stock options.

FINANCIALS: Sales and profits are in thousands of dollars—add 000 to get the full amount. 2006 Note: Financial information for 2006 was not available for all companies at press time.

2006 Sales: $28,188	2006 Profits: $-29,346	**U.S. Stock Ticker: ZILA**
2005 Sales: $43,489	2005 Profits: $1,099	**Int'l Ticker:** Int'l Exchange:
2004 Sales: $36,682	2004 Profits: $-4,375	Employees: 116
2003 Sales: $39,210	2003 Profits: $7,246	Fiscal Year Ends: 7/31
2002 Sales: $34,900	2002 Profits: $-12,000	Parent Company:

SALARIES/BENEFITS:

Pension Plan: Y	ESOP Stock Plan:	Profit Sharing:	Top Exec. Salary: $356,688	Bonus: $
Savings Plan: Y	Stock Purch. Plan: Y		Second Exec. Salary: $216,602	Bonus: $

OTHER THOUGHTS:

Apparent Women Officers or Directors: 3
Hot Spot for Advancement for Women/Minorities: Y

LOCATIONS: ("Y" = Yes)

West:	Southwest:	Midwest:	Southeast:	Northeast:	International:
	Y				Y

ZYMOGENETICS INC

www.zymogenetics.com

Industry Group Code: 325412 Ranks within this company's industry group: Sales: 105 Profits: 190

Drugs:		Other:		Clinical:		Computers:		Other:	
Discovery:	Y	AgriBio:		Trials/Services:		Hardware:		Specialty Services:	
Licensing:	Y	Genomics/Proteomics:	Y	Laboratories:		Software:		Consulting:	
Manufacturing:	Y	Tissue Replacement:		Equipment/Supplies:		Arrays:		Blood Collection:	
Development:	Y			Research/Development Svcs.:		Database Management:		Drug Delivery:	
Generics:				Diagnostics:				Drug Distribution:	

TYPES OF BUSINESS:

Therapeutic Proteins
Hemostasis, Inflammatory & Autoimmune Diseases Drugs
Cancer & Viral Infections Drugs

BRANDS/DIVISIONS/AFFILIATES:

rhThrombin
Interleukin-21
PEG-IFN
Atacicept
Merck Serono
Novo Nordisk
Bayer HealthCare

CONTACTS: *Note: Officers with more than one job title may be intentionally listed here more than once.*

Bruce L. A. Carter, CEO
Douglas E. Williams, Pres.
James A. Johnson, CFO/Exec. VP
Michael J. Dwyer, Sr. VP-Sales & Mktg.
Darren R. Hamby, Sr. VP-Human Resources
Douglas E. Williams, Chief Scientific Officer
Vaughn B. Himes, Sr. VP-Tech. Oper.
Suzanne M. Shema, General Counsel/Sr. VP
James A. Johnson, Treas.
Nicole Onetto, Sr. VP/Chief Medical Officer
Bruce L. A. Carter, Chmn.

Phone: 206-442-6600	Fax: 206-442-6608
Toll-Free: 800-775-6686	
Address: 1201 Eastlake Ave. E., Seattle, WA 98102 US	

GROWTH PLANS/SPECIAL FEATURES:

ZymoGenetics, Inc. discovers, develops, manufactures and commercializes therapeutic proteins for the treatment of human diseases. The company's current therapeutic focus is in the areas of hemostasis; inflammatory and autoimmune diseases; cancer; and viral infections. The firm's most advanced product candidate, rhThrombin, which is being developed as a replacement for plasma-derived hemostatic products, completed Phase III testing and is currently under regulatory review in the U.S. Other products include atacicept (formerly known as TACI-Ig), a soluble receptor with potential applications for the treatment of cancer and autoimmune diseases; Interleukin-21, a cytokine with potential applications for the treatment of cancer; and PEG-IFN (formerly known as IL-29), a cytokine with potential applications for the treatment of viral infections. ZymoGenetics collaborates with Merck Serono for atacicept and research, development and commercialization of protein and antibody therapeutics derived from the company's proprietary portfolio of genes and proteins; and Novo Nordisk for Interleukin-21. ZymoGenetics contributed to the discovery or development of six recombinant protein products currently on the market:Novolin, NovoSeven, Regranex, GEM 21S, GlucanGen and Cleactor. The company holds more than 295 unexpired issued or allowed U.S. patents and over 660 U.S. patent application pending. In addition, the firm has more than 650 issued or allowed foreign patents. In December 2006, ZymoGenetics initiated a Phase II clinical trial for atacicept in patients with rheumatoid arthritis. In June 2007, the company entered into a global collaboration with Bayer HealthCare for the development and commercialization of rhThrombin, having received FDA approval for the drug, administered by spray device as an aid to controlling bleeding during surgery, in March 2007.

The company offers its employees medical, dental and vision insurance; a 401(k) plan; stock options; an employee assistance program; short- and long-term disability; life and AD&D insurance; and tuition reimbursement.

FINANCIALS: Sales and profits are in thousands of dollars—add 000 to get the full amount. 2006 Note: Financial information for 2006 was not available for all companies at press time.

2006 Sales: $25,380	2006 Profits: $-130,002	U.S. Stock Ticker: ZGEN
2005 Sales: $42,909	2005 Profits: $-78,027	Int'l Ticker: Int'l Exchange:
2004 Sales: $35,694	2004 Profits: $-88,756	Employees: 298
2003 Sales: $25,957	2003 Profits: $-59,571	Fiscal Year Ends: 12/31
2002 Sales: $52,775	2002 Profits: $-30,416	Parent Company:

SALARIES/BENEFITS:

Pension Plan:	ESOP Stock Plan:	Profit Sharing:	Top Exec. Salary: $591,544	Bonus: $297,415
Savings Plan: Y	Stock Purch. Plan: Y		Second Exec. Salary: $395,833	Bonus: $151,200

OTHER THOUGHTS:

Apparent Women Officers or Directors: 3
Hot Spot for Advancement for Women/Minorities: Y

LOCATIONS: ("Y" = Yes)

West:	Southwest:	Midwest:	Southeast:	Northeast:	International:
Y					

Note: Financial information, benefits and other data can change quickly and may vary from those stated here.

ADDITIONAL INDEXES

CONTENTS:

INDEX OF FIRMS NOTED AS HOT SPOTS FOR ADVANCEMENT FOR WOMEN & MINORITIES

4SC AG
ABBOTT LABORATORIES
ACAMBIS PLC
ACCELRYS INC
ADOLOR CORP
AFFYMETRIX INC
AGILENT TECHNOLOGIES INC
ALCON INC
ALKERMES INC
AMGEN INC
AMYLIN PHARMACEUTICALS INC
ANTIGENICS INC
APPLERA CORPORATION
APPLIED BIOSYSTEMS GROUP
ARENA PHARMACEUTICALS INC
ARIAD PHARMACEUTICALS INC
ASTRAZENECA PLC
AUTOIMMUNE INC
AVIGEN INC
AXCAN PHARMA INC
BARR PHARMACEUTICALS INC
BAUSCH & LOMB INC
BAXTER INTERNATIONAL INC
BIO RAD LABORATORIES INC
BIOANALYTICAL SYSTEMS INC
BIOMARIN PHARMACEUTICAL INC
BIOMIRA INC
BIOPURE CORPORATION
BIOSITE INC
BIOTECH HOLDINGS LTD
BIOVAIL CORPORATION
BRISTOL MYERS SQUIBB CO
CALIPER LIFE SCIENCES
CAMBREX CORP
CAMBRIDGE ANTIBODY TECHNOLOGY LIMITED
CARACO PHARMACEUTICAL LABORATORIES
CARDIOME PHARMA CORP
CARDIUM THERAPEUTICS INC
CELERA GENOMICS GROUP
CELL GENESYS INC
CELL THERAPEUTICS INC
CEPHEID
CHARLES RIVER LABORATORIES INTERNATIONAL INC
CIPHERGEN BIOSYSTEMS INC
COMMONWEALTH BIOTECHNOLOGIES INC
COMPUGEN LTD
COVIDIEN LTD
CURAGEN CORPORATION
CV THERAPEUTICS INC
CYPRESS BIOSCIENCE INC
DENDREON CORPORATION
DENDRITE INTERNATIONAL INC
DIGENE CORPORATION

DISCOVERY LABORATORIES INC
DUPONT AGRICULTURE & NUTRITION
DURECT CORP
DYAX CORP
E I DU PONT DE NEMOURS & CO (DUPONT)
ELAN CORP PLC
ELI LILLY AND COMPANY
ELSEVIER MDL
ENCORIUM GROUP INC
ENCYSIVE PHARMACEUTICALS INC
ENTREMED INC
ENZO BIOCHEM INC
EPIX PHARMACEUTICALS INC
ERESEARCH TECHNOLOGY INC
EXELIXIS INC
FLAMEL TECHNOLOGIES SA
FORBES MEDI-TECH INC
FOREST LABORATORIES INC
GENELABS TECHNOLOGIES INC
GENENTECH INC
GENEREX BIOTECHNOLOGY
GEN-PROBE INC
GENTA INC
GENZYME BIOSURGERY
GENZYME CORP
GERON CORPORATION
GILEAD SCIENCES INC
GTC BIOTHERAPEUTICS INC
HEMISPHERX BIOPHARMA INC
HESKA CORP
HI-TECH PHARMACAL CO INC
IDEXX LABORATORIES INC
IMCLONE SYSTEMS INC
IMMTECH INTERNATIONAL
IMMUNOGEN INC
IMMUNOMEDICS INC
IMS HEALTH INC
INCYTE CORP
INSITE VISION INC
INSPIRE PHARMACEUTICALS INC
INTEGRA LIFESCIENCES HOLDINGS CORP
INTERMUNE PHARMACEUTICALS INC
INVITROGEN CORPORATION
IOMED INC
JAZZ PHARMACEUTICALS
JOHNSON & JOHNSON
KENDLE INTERNATIONAL INC
KOS PHARMACEUTICALS INC
KV PHARMACEUTICAL CO
LA JOLLA PHARMACEUTICAL
LEXICON PHARMACEUTICALS INC
LIGAND PHARMACEUTICALS INC
LONZA GROUP
LORUS THERAPEUTICS INC
MATRITECH INC
MEDIMMUNE INC
MERCK & CO INC
MERIDIAN BIOSCIENCE INC

MILLENNIUM PHARMACEUTICALS INC
MONSANTO CO
MYLAN LABORATORIES INC
NEKTAR THERAPEUTICS
NEOSE TECHNOLOGIES INC
NEUROCRINE BIOSCIENCES INC
NOVEN PHARMACEUTICALS
NOVO-NORDISK AS
NOVOZYMES
NUVELO INC
NYCOMED
ONYX PHARMACEUTICALS INC
ORGANOGENESIS INC
OSI PHARMACEUTICALS INC
PACIFIC BIOMETRICS INC
PAREXEL INTERNATIONAL
PDL BIOPHARMA
PENWEST PHARMACEUTICALS CO
PEREGRINE PHARMACEUTICALS INC
PFIZER INC
PHARMACEUTICAL PRODUCT DEVELOPMENT INC
PHARMACOPEIA DRUG DISCOVERY
PHARMION CORP
PONIARD PHARMACEUTICALS INC
QLT INC
QUEST DIAGNOSTICS INC
ROCHE HOLDING LTD
SANGAMO BIOSCIENCES INC
SANOFI-AVENTIS
SCHERING-PLOUGH CORP
SERACARE LIFE SCIENCES INC
SHIRE PLC
SHIRE-BIOCHEM INC
SONUS PHARMACEUTICALS
SPECIALTY LABORATORIES INC
STEMCELLS INC
STIEFEL LABORATORIES INC
SUPERGEN INC
SYNOVIS LIFE TECHNOLOGIES INC
TANOX INC
TAPESTRY PHARMACEUTICALS INC
TARO PHARMACEUTICAL INDUSTRIES
TECHNE CORP
TELIK INC
TEVA PHARMACEUTICAL INDUSTRIES
THERAGENICS CORP
TORREYPINES THERAPEUTICS INC
TRINITY BIOTECH PLC
TRIPOS INC
VALEANT PHARMACEUTICALS INTERNATIONAL
VASOGEN INC
VERNALIS PLC
VERTEX PHARMACEUTICALS INC
VION PHARMACEUTICALS INC
WATSON PHARMACEUTICALS INC
ZILA INC
ZYMOGENETICS INC

INDEX OF SUBSIDIARIES, BRAND NAMES AND AFFILIATIONS

Brand or subsidiary, followed by the name of the related corporation

3gAllergy; **DIAGNOSTIC PRODUCTS CORPORATION**
3-Nitro; **ALPHARMA INC**
3TC; **SHIRE-BIOCHEM INC**
4Closure Surgical Fascia Closure System; **SYNOVIS LIFE TECHNOLOGIES INC**
4SCan Technology; **4SC AG**
A. Aarons, Inc.; **BRADLEY PHARMACEUTICALS INC**
Abbot Laboratories; **EXPERIMENTAL & APPLIED SCIENCES INC**
Abbott Canada; **OSCIENT PHARMACEUTICALS INC**
Abbott Laboratories; **HOSPIRA INC**
Abbott Laboratories; **VYSIS INC**
Abbott Molecular; **APPLERA CORPORATION**
ABELCET; **ENZON PHARMACEUTICALS INC**
Abelcet; **CEPHALON INC**
Abraxane; **ABRAXIS BIOSCIENCE INC**
Abraxis BioScience; **ABRAXIS BIOSCIENCE INC**
Abraxis Pharmaceutical Products; **ABRAXIS BIOSCIENCE INC**
Abreva; **AVANIR PHARMACEUTICALS**
Absolutely RNA; **STRATAGENE CORP**
Absorbable Polymer Technologies, Inc.; **DURECT CORP**
Abthrax; **CAMBRIDGE ANTIBODY TECHNOLOGY LIMITED**
Abthrax; **HUMAN GENOME SCIENCES INC**
AC Vaccine; **AVAX TECHNOLOGIES INC**
ACAM2000; **ACAMBIS PLC**
ACAM-FLU-A; **ACAMBIS PLC**
Acanto; **SYNGENTA AG**
Accelerated Intelligent Drug Discovery (AIDD); **NEUROGEN CORP**
ACCESS Oncology, Inc.; **KERYX BIOPHARMACEUTICALS INC**
Accord; **ACCELRYS INC**
Accretropin; **CANGENE CORP**
Accumin; **KERYX BIOPHARMACEUTICALS INC**
Accumin Diagnostics, Inc.; **KERYX BIOPHARMACEUTICALS INC**
AccuTrac; **COMMONWEALTH BIOTECHNOLOGIES INC**
Acemannan Hydrogel; **CARRINGTON LABORATORIES INC**
Acetazolamide; **LANNETT COMPANY INC**
Aciphex/Pariet; **EISAI CO LTD**
AcrySof; **ALCON INC**
ACTCellerate; **ADVANCED CELL TECHNOLOGY INC**
Actelion Pharmaceuticals U.S., Inc.; **ACTELION LTD**
Actelion-1; **ACTELION LTD**
Actigenics; **CEPHEID**

Actimmune; **INTERMUNE PHARMACEUTICALS INC**
Actiq; **CEPHALON INC**
Activase; **GENENTECH INC**
Activated Cellular Therapy; **NEOPROBE CORPORATION**
Activated Checkpoint Therapy; **ARQULE INC**
Activelle; **NOVO-NORDISK AS**
AcuForm; **DEPOMED INC**
Acumedia; **NEOGEN CORPORATION**
Acyclovir; **EMISPHERE TECHNOLOGIES INC**
ACZONE; **QLT INC**
ADAGEN; **ENZON PHARMACEUTICALS INC**
AdapT; **CEL-SCI CORPORATION**
Adderall; **SHIRE PLC**
Adenoviral Delivery System; **INTROGEN THERAPEUTICS INC**
Aderall XR; **SHIRE-BIOCHEM INC**
ADHERE; **SCIOS INC**
Adhibit; **ANGIOTECH PHARMACEUTICALS**
AdPEDF; **GENVEC INC**
Advanced Inhalation Research (AIR); **ALKERMES INC**
Advanced Separation Technologies, Inc.; **SIGMA ALDRICH CORP**
AdvantEdge; **EXPERIMENTAL & APPLIED SCIENCES INC**
AdvantEdge; **ABBOTT LABORATORIES**
Advate; **BAXTER INTERNATIONAL INC**
Advexin; **INTROGEN THERAPEUTICS INC**
Advicor; **KOS PHARMACEUTICALS INC**
Advil; **WYETH**
AEOL 10150; **AEOLUS PHARMACEUTICALS INC**
AEOL 11207; **AEOLUS PHARMACEUTICALS INC**
Aerobid; **FOREST LABORATORIES INC**
AeroChamber Plus; **FOREST LABORATORIES INC**
Aerosurf; **DISCOVERY LABORATORIES INC**
AERx; **ARADIGM CORPORATION**
AERx Insulin Diabetes Management System; **ARADIGM CORPORATION**
AFP-Scan; **IMMUNOMEDICS INC**
Afrin; **SCHERING-PLOUGH CORP**
AG-707; **ANTIGENICS INC**
Agencourt Personal Genomics; **APPLIED BIOSYSTEMS GROUP**
AGI-1067; **ATHEROGENICS INC**
AGI-1096; **ATHEROGENICS INC**
Agilent Technologies, Inc.; **HYCOR BIOMEDICAL INC**
Agrinomica and Renessen, LLC; **EXELIXIS INC**
Agrinomics, LLC; **EXELIXIS PLANT SCIENCES INC**
Agri-Scan; **NEOGEN CORPORATION**
Agroproducts Corey, S.A. de C.V.; **DUPONT AGRICULTURE & NUTRITION**
Agrylin; **SHIRE-BIOCHEM INC**
AIR Epinephrine; **ALKERMES INC**
AIR Insulin; **ALKERMES INC**
Akorn New Jersey, Inc.; **AKORN INC**

INDEX OF SUBSIDIARIES, BRAND NAMES AND AFFILIATIONS, CONT.

INDEX OF SUBSIDIARIES, BRAND NAMES AND AFFILIATIONS, CONT.

INDEX OF SUBSIDIARIES, BRAND NAMES AND AFFILIATIONS, CONT.

INDEX OF SUBSIDIARIES, BRAND NAMES AND AFFILIATIONS, CONT.

California Planting Cotton Seed Distributors; **BAYER CORP**
Caliper Technologies; **CALIPER LIFE SCIENCES**
Cambrex Bio Science Walkersville, Inc.; **CAMBREX CORP**
Cambrex Corporation; **LONZA GROUP**
Cambridge Antibody Technology Group; **ASTRAZENECA PLC**
Cambridge Antibody Technology Group plc; **CAMBRIDGE ANTIBODY TECHNOLOGY LIMITED**
Cambridge Medical Publications; **PAREXEL INTERNATIONAL**
Campath; **GENZYME ONCOLOGY**
Camptosar; **ENZON PHARMACEUTICALS INC**
CANASA; **AXCAN PHARMA INC**
Cangene Corporation; **CHESAPEAKE BIOLOGICAL LAB**
CANGENUS; **CANGENE CORP**
CANGENUS; **CANGENE CORP**
Cangrelor; **MEDICINES CO (THE)**
Cannabinor; **PHARMOS CORP**
CA-p (MN); **DENDREON CORPORATION**
Caphosol; **CYTOGEN CORPORATION**
Capillus; **TRINITY BIOTECH PLC**
CapityalBio Corporation; **AVIVA BIOSCIENCES CORP**
Captia; **TRINITY BIOTECH PLC**
Caraloe, Inc.; **CARRINGTON LABORATORIES INC**
Carbatrol; **SHIRE PLC**
Cardene IV; **PDL BIOPHARMA**
CardiacSTATus; **SPECTRAL DIAGNOSTICS INC**
Cardinal Health; **NEOPROBE CORPORATION**
Cardio CRP; **QUEST DIAGNOSTICS INC**
Cardiolite; **BRISTOL MYERS SQUIBB CO**
CardioPass; **CARDIOTECH INTERNATIONAL**
Cardiosonix, Ltd.; **NEOPROBE CORPORATION**
CardioTech International, Ltd.; **CARDIOTECH INTERNATIONAL**
Cardium Biologics; **CARDIUM THERAPEUTICS INC**
Cardizem LA; **BIOVAIL CORPORATION**
Careline; **PERRIGO CO**
CARES; **INSTITUT STRAUMANN AG**
Cargill Hybrid Seeds; **DOW AGROSCIENCES LLC**
CarraSmart; **CARRINGTON LABORATORIES INC**
Carticel; **GENZYME BIOSURGERY**
Carticel; **GENZYME CORP**
CAT-354; **CAMBRIDGE ANTIBODY TECHNOLOGY LIMITED**
Catalyst; **ACCELRYS INC**
Catheter & Disposables Technology, Inc.; **CARDIOTECH INTERNATIONAL**
CCR5 Antagonists; **INCYTE CORP**
CE Mark; **THERAGENICS CORP**
CEA; **DENDREON CORPORATION**
CEA-Scan; **IMMUNOMEDICS INC**

Cehmische Fabrik WIBARCO GmBH; **BASF AG**
Cel-1000; **CEL-SCI CORPORATION**
Celacade; **VASOGEN INC**
Celebrex; **PFIZER INC**
Celera Diagnostics; **CELERA GENOMICS GROUP**
Celera Diagnostics; **APPLERA CORPORATION**
Celera Discovery System; **CELERA GENOMICS GROUP**
Celera Discovery System; **APPLERA CORPORATION**
Celera Genomics; **APPLERA CORPORATION**
Celexa; **FOREST LABORATORIES INC**
Celgosivir; **MIGENIX INC**
Cell Culture Systems (CCS); **INVITROGEN CORPORATION**
Cell Therapeutics Europe S.r.l.; **CELL THERAPEUTICS INC**
Celladon Corp.; **TARGETED GENETICS CORP**
CellBeads; **BIOCOMPATIBLES INTERNATIONAL PLC**
Cellegy Canada, Inc.; **CELLEGY PHARMACEUTICALS**
CellMed; **BIOCOMPATIBLES INTERNATIONAL PLC**
Cellomics; **PERBIO SCIENCE AB**
Cellophane; **E I DU PONT DE NEMOURS & CO (DUPONT)**
CellSearch; **QUEST DIAGNOSTICS INC**
Celsis Development Services; **CELSIS INTERNATIONAL PLC**
Celsis Laboratory Group; **CELSIS INTERNATIONAL PLC**
Celtic Pharmaceutical Holdings LP; **NEUROBIOLOGICAL TECHNOLOGIES INC**
Cenestin; **BARR PHARMACEUTICALS INC**
Centricity; **GE HEALTHCARE**
Centrum; **WYETH**
Ceremix; **NOVOZYMES**
Cerezyme; **GENZYME CORP**
Cetherin; **ALSERES PHARMACEUTICALS INC**
CETi; **AVANT IMMUNOTHERAPEUTICS**
Cetilistat; **ALIZYME PLC**
Cetrorelix; **AETERNA ZENTARIS INC**
Cetrotide; **AETERNA ZENTARIS INC**
Cetrotide; **MERCK SERONO SA**
Cetuximab; **IMCLONE SYSTEMS INC**
CF Complete; **QUEST DIAGNOSTICS INC**
CG0070; **CELL GENESYS INC**
CG5757; **CELL GENESYS INC**
cGMP; **MERIDIAN BIOSCIENCE INC**
Chap Stick; **WYETH**
Chattem (U.K.), Ltd.; **CHATTEM INC**
Cheminformatics Programs; **ACCELRYS INC**
Chemo-Sero-Therapeutic Research Institute; **VAXGEN INC**
ChimeriVax-Dengue; **ACAMBIS PLC**
ChimeriVax-JE; **ACAMBIS PLC**

INDEX OF SUBSIDIARIES, BRAND NAMES AND AFFILIATIONS, CONT.

INDEX OF SUBSIDIARIES, BRAND NAMES AND AFFILIATIONS, CONT.

INDEX OF SUBSIDIARIES, BRAND NAMES AND AFFILIATIONS, CONT.

INDEX OF SUBSIDIARIES, BRAND NAMES AND AFFILIATIONS, CONT.

INDEX OF SUBSIDIARIES, BRAND NAMES AND AFFILIATIONS, CONT.

INDEX OF SUBSIDIARIES, BRAND NAMES AND AFFILIATIONS, CONT.

Hangzhou Whatman-Xinhua Filter Paper Co. Ltd.;
WHATMAN PLC
Hansa Chemie International; **BASF AG**
Haptoguard, Inc.; **ALTEON INC**
Harmony; **ADVANCED BIONICS CORPORATION**
Harness; **MONSANTO CO**
Harvard Apparatus; **HARVARD BIOSCIENCE INC**
HaveItAll; **BIO RAD LABORATORIES INC**
HCV-796; **VIROPHARMA INC**
HeartBar; **UNITED THERAPEUTICS CORP**
Hectorol; **SHIRE-BIOCHEM INC**
Hedgehog; **CURIS INC**
Hedrin; **MANHATTAN PHARMACEUTICALS INC**
Helinx; **CERUS CORPORATION**
Hemagen Diagnosticos Comercio; **HEMAGEN
DIAGNOSTICS INC**
Hemopure; **BIOPURE CORPORATION**
HepaGam B; **CANGENE CORP**
Heparin; **EMISPHERE TECHNOLOGIES INC**
HepArrest; **COMMONWEALTH
BIOTECHNOLOGIES INC**
HepaSphere SAP; **BIOSPHERE MEDICAL INC**
HEPTIMAX; **QUEST DIAGNOSTICS INC**
Herceptin; **GENENTECH INC**
Herceptin; **ROCHE HOLDING LTD**
Herculex; **DOW AGROSCIENCES LLC**
hERGexpress; **AVIVA BIOSCIENCES CORP**
Herpera; **GILEAD SCIENCES INC**
HESKA Feline UltraNasal FVRCP Vaccine; **HESKA
CORP**
HESKA Vet CBC-Diff Hematology Analyzer; **HESKA
CORP**
HESKA Vet/Ox G2 Digital Monitor; **HESKA CORP**
HetaCool; **BIOTIME INC**
Hexalen; **MGI PHARMA INC**
Hextend; **BIOTIME INC**
HiRes; **ADVANCED BIONICS CORPORATION**
HiResolution Bionic Ear System; **ADVANCED
BIONICS CORPORATION**
Histofreezer; **ORASURE TECHNOLOGIES INC**
Histostat; **ALPHARMA INC**
Hoffman-La Roche; **GENE LOGIC INC**
Hoffman-La Roche, Inc.; **EMISPHERE
TECHNOLOGIES INC**
Holland & Barrett; **NBTY INC**
HP Acthar Gel; **QUESTCOR PHARMACEUTICALS**
HuC242-DM4; **IMMUNOGEN INC**
HuCNS-SC; **STEMCELLS INC**
HuMAb-Mouse; **MEDAREX INC**
Humalog; **ELI LILLY AND COMPANY**
HUMIRA; **CAMBRIDGE ANTIBODY
TECHNOLOGY LIMITED**
HUMIRA; **ABBOTT LABORATORIES**
HuN901-DM1; **IMMUNOGEN INC**
Huntingdon and Eye; **LIFE SCIENCES RESEARCH**

Huntingdon Life Sciences Group; **LIFE SCIENCES
RESEARCH**
Hybresis System; **IOMED INC**
Hybrid Capture; **DIGENE CORPORATION**
Hybridon, Inc.; **IDERA PHARMACEUTICALS INC**
HYBrite Hybridization System; **VYSIS INC**
HyClone; **PERBIO SCIENCE AB**
Hycor; **STRATAGENE CORP**
Hycor Biomedical GmbH; **HYCOR BIOMEDICAL INC**
Hycor Biomedical, Ltd.; **HYCOR BIOMEDICAL INC**
Hydromorphone; **LANNETT COMPANY INC**
HydroThane; **CARDIOTECH INTERNATIONAL**
HyluMed; **GENZYME BIOSURGERY**
HY-TEC; **HYCOR BIOMEDICAL INC**
Hyvisc; **ANIKA THERAPEUTICS INC**
Ibalizumab; **TANOX INC**
IBIS T-5000 Universal Pathogen Sensor; **ISIS
PHARMACEUTICALS INC**
Ibis Technologies; **ISIS PHARMACEUTICALS INC**
Iceland Health Joint Relief; **NUTRITION 21 INC**
Iceland Health Maximum Strength Omega-3;
NUTRITION 21 INC
Iceland Health, Inc.; **NUTRITION 21 INC**
ICOlabs; **ICON PLC**
ICON Central Laboratores; **ICON PLC**
ICON Clinical Research; **ICON PLC**
ICON Contracting Solutions; **ICON PLC**
ICON Development Solutions; **ICON PLC**
ICON Medical Imaging; **ICON PLC**
ICOnet; **ICON PLC**
ICOPhone; **ICON PLC**
Icos Corp.; **ELI LILLY AND COMPANY**
Icy Hot; **CHATTEM INC**
IDx, Inc; **SPECTRAL DIAGNOSTICS INC**
iGene; **INVITROGEN CORPORATION**
IL-1 Trap; **REGENERON PHARMACEUTICALS INC**
ILEX Oncology, Inc.; **GENZYME ONCOLOGY**
Illomedin; **BAYER SCHERING PHARMA AG**
Iloperidone; **TITAN PHARMACEUTICALS**
ImmTech Biologics, Inc.; **NOVARTIS AG**
IMMULITE; **DIAGNOSTIC PRODUCTS
CORPORATION**
IMMULITE 1000; **DIAGNOSTIC PRODUCTS
CORPORATION**
IMMULITE 2500; **DIAGNOSTIC PRODUCTS
CORPORATION**
IMMULITE Turbo; **DIAGNOSTIC PRODUCTS
CORPORATION**
Immune Response Corporation (The); **IMMUNE
RESPONSE CORP (THE)**
Immune Tolerance; **BIOMARIN PHARMACEUTICAL
INC**
Immunitin; **HOLLIS-EDEN PHARMACEUTICALS**
ImmunoCard STAT EHEC; **MERIDIAN BIOSCIENCE
INC**

INDEX OF SUBSIDIARIES, BRAND NAMES AND AFFILIATIONS, CONT.

INDEX OF SUBSIDIARIES, BRAND NAMES AND AFFILIATIONS, CONT.

INDEX OF SUBSIDIARIES, BRAND NAMES AND AFFILIATIONS, CONT.

Madopar DR; **SKYEPHARMA PLC**
MagPlex Magnetic Microspheres; **LUMINEX CORPORATION**
MAK (Monocyte-derived Activated Killer); **IDM PHARMA INC**
Mallinckrodt Pharmaceuticals; **COVIDIEN LTD**
Manapol; **CARRINGTON LABORATORIES INC**
MarDx; **TRINITY BIOTECH PLC**
Maribavir; **VIROPHARMA INC**
Martek Biosciences Kingtree; **MARTEK BIOSCIENCES CORP**
MassARRAY; **SEQUENOM INC**
MassARRAY EpiTYPER; **SEQUENOM INC**
MassARRAY QGE; **SEQUENOM INC**
MassARRAY SNP Discovery; **SEQUENOM INC**
MassARRAY SNP Genotyping; **SEQUENOM INC**
Mastik; **IMMUCELL CORPORATION**
MastOut; **IMMUCELL CORPORATION**
Material Studio 4.1; **ACCELRYS INC**
Matrifen; **NYCOMED**
Matritech GmbH; **MATRITECH INC**
Matrix Laboratories; **MYLAN LABORATORIES INC**
Maxim Total Knee System; **BIOMET INC**
Maxipime; **ELAN CORP PLC**
Maxy-alpha; **MAXYGEN INC**
Maxy-G34; **MAXYGEN INC**
MaxyScan; **MAXYGEN INC**
Mayne Pharma Limited; **HOSPIRA INC**
McNeil Nutritionals, LLC; **JOHNSON & JOHNSON**
MDL Available Chemicals Directory; **ELSEVIER MDL**
MDL Isentris; **ELSEVIER MDL**
MDS Analytical Technologies; **MDS INC**
MDS Capital Corp.; **MDS INC**
MDS Diagnostic Services; **MDS INC**
MDS Nordion; **MDS INC**
MDS Pharma Services; **MDS INC**
MDS Sciex; **MDS INC**
Mead Johnson Nutritionals; **BRISTOL MYERS SQUIBB CO**
Medical Action Communication; **QUINTILES TRANSNATIONAL CORP**
Medicomp, Inc.; **UNITED THERAPEUTICS CORP**
MediGene AG; **BRADLEY PHARMACEUTICALS INC**
MedImmune Ventures, Inc.; **MEDIMMUNE INC**
MedImmune, Inc.; **INFINITY PHARMACEUTICALS INC**
MedImmune, Inc.; **CERUS CORPORATION**
Medion Pharma Co., Ltd.; **CUBIST PHARMACEUTICALS INC**
Medisorb; **ALKERMES INC**
MEDTOX Diagnostics, Inc.; **MEDTOX SCIENTIFIC INC**
MEDTOX Laboratories, Inc.; **MEDTOX SCIENTIFIC INC**

Medtronic; **BIOCOMPATIBLES INTERNATIONAL PLC**
Medusa; **FLAMEL TECHNOLOGIES SA**
Megace ES; **PAR PHARMACEUTICAL COMPANIES INC**
Megan Egg; **AVANT IMMUNOTHERAPEUTICS**
Megan Vac 1; **AVANT IMMUNOTHERAPEUTICS**
MEK; **ARRAY BIOPHARMA INC**
MEKK Technology; **ATHEROGENICS INC**
MELARIS; **MYRIAD GENETICS INC**
Memantine; **NEUROBIOLOGICAL TECHNOLOGIES INC**
Menarini Group; **OSCIENT PHARMACEUTICALS INC**
Menorest; **NOVEN PHARMACEUTICALS**
Merck & Co., Inc.; **ARENA PHARMACEUTICALS INC**
Merck and Co.; **ROSETTA INPHARMATICS LLC**
Merck Ethicals; **MERCK KGAA**
Merck Institute for Science Education; **MERCK & CO INC**
Merck KGaA; **MERCK SERONO SA**
Merck Serono; **AETERNA ZENTARIS INC**
Merck Serono; **RIGEL PHARMACEUTICALS INC**
Merck Serono; **ZYMOGENETICS INC**
Merck Serono S.A.; **MERCK KGAA**
MessageAMP; **APPLIED BIOSYSTEMS GROUP**
Mestinon; **VALEANT PHARMACEUTICALS INTERNATIONAL**
Metadate; **UCB SA**
Metformin XL; **FLAMEL TECHNOLOGIES SA**
MethyPatch; **NOVEN PHARMACEUTICALS**
Mevalotin; **SANKYO CO LTD**
Micellar Nanoparticle Technology; **NOVAVAX INC**
Microdur; **DURECT CORP**
Micrologix Biotech, Inc.; **MIGENIX INC**
Micromask; **KV PHARMACEUTICAL CO**
Micropump; **FLAMEL TECHNOLOGIES SA**
MicroSafe, B.V.; **MILLIPORE CORP**
Microvascular Anastomotis Coupler; **SYNOVIS LIFE TECHNOLOGIES INC**
MIDAS; **PALATIN TECHNOLOGIES INC**
Miikana Therapeutics, Inc.; **ENTREMED INC**
Millennium University; **MILLENNIUM PHARMACEUTICALS INC**
Milnacipran; **CYPRESS BIOSCIENCE INC**
MIMiC; **NORTH AMERICAN SCIENTIFIC**
Mimotopes Pty Ltd; **COMMONWEALTH BIOTECHNOLOGIES INC**
Mimpara; **NPS PHARMACEUTICALS INC**
MiniOpticon; **BIO RAD LABORATORIES INC**
Mio Relax; **BENTLEY PHARMACEUTICALS INC**
Miravant Cardiovascular, Inc.; **MIRAVANT MEDICAL TECHNOLOGIES**
Miravant Pharmaceuticals, Inc.; **MIRAVANT MEDICAL TECHNOLOGIES**

INDEX OF SUBSIDIARIES, BRAND NAMES AND AFFILIATIONS, CONT.

INDEX OF SUBSIDIARIES, BRAND NAMES AND AFFILIATIONS, CONT.

INDEX OF SUBSIDIARIES, BRAND NAMES AND AFFILIATIONS, CONT.

INDEX OF SUBSIDIARIES, BRAND NAMES AND AFFILIATIONS, CONT.

INDEX OF SUBSIDIARIES, BRAND NAMES AND AFFILIATIONS, CONT.

INDEX OF SUBSIDIARIES, BRAND NAMES AND AFFILIATIONS, CONT.

INDEX OF SUBSIDIARIES, BRAND NAMES AND AFFILIATIONS, CONT.

SeraCare Diagnostic Products; **SERACARE LIFE SCIENCES INC**
SeraCare Global Repository; **SERACARE LIFE SCIENCES INC**
Serologicals Corporation; **MILLIPORE CORP**
Serono S.A.; **MERCK KGAA**
Sesonique; **BARR PHARMACEUTICALS INC**
SevoFlo; **ABBOTT LABORATORIES**
SFBC International, Inc.; **PHARMANET DEVELOPMENT GROUP INC**
SGN-30; **SEATTLE GENETICS**
SGN-33; **SEATTLE GENETICS**
SGN-35; **SEATTLE GENETICS**
SGN-40; **SEATTLE GENETICS**
SGN-70; **SEATTLE GENETICS**
SGN-75; **SEATTLE GENETICS**
Shearform; **BIOVAIL CORPORATION**
ShellGel; **ANIKA THERAPEUTICS INC**
Shenzhen PG Biotech Co. Ltd.; **QIAGEN NV**
Shigella; **AVANT IMMUNOTHERAPEUTICS**
Shiley; **MALLINCKRODT INC**
Shin-Sankyo Ichoyaku; **SANKYO CO LTD**
Shionogi; **AETERNA ZENTARIS INC**
Shire Pharmaceuticals Group; **SHIRE-BIOCHEM INC**
Shire PLC; **SHIRE-BIOCHEM INC**
Sico, Inc.; **AKZO NOBEL NV**
Sicor Biotech UAB; **SICOR INC**
Sicor Pharmeceuticals, Inc.; **SICOR INC**
Sicor S.p.A.; **SICOR INC**
Siegfried, Ltd.; **CELGENE CORP**
SIGA-246; **SIGA TECHNOLOGIES INC**
Signet Laboratories, Inc.; **COVANCE INC**
Silicon Genetics; **AGILENT TECHNOLOGIES INC**
Simcere Medgenn Bio-Pharmaceutical Co., Ltd.; **SIMCERE PHARMACEUTICAL GROUP**
Similac; **ABBOTT LABORATORIES**
Single Day; **SCHIFF NUTRITION INTERNATIONAL INC**
SingleCell; **DIVERSA CORPORATION**
Singulair; **MERCK & CO INC**
Sirius Laboratories, Inc.; **DUSA PHARMACEUTICALS INC**
Sirna Therapeutics, Inc.; **MERCK & CO INC**
Skelaxin; **KING PHARMACEUTICALS INC**
SLActive; **INSTITUT STRAUMANN AG**
Sloan-Kettering Institute for Cancer Research; **KOSAN BIOSCIENCES INC**
Smart Cycler; **CEPHEID**
Smart Cycler II; **CEPHEID**
SmartSpec Plus; **BIO RAD LABORATORIES INC**
SNAP; **IDEXX LABORATORIES INC**
SnET2; **MIRAVANT MEDICAL TECHNOLOGIES**
SofLens; **BAUSCH & LOMB INC**
Solae Company (The); **DUPONT AGRICULTURE & NUTRITION**
Solaraze; **SKYEPHARMA PLC**
Solaraze; **BRADLEY PHARMACEUTICALS INC**
Solexa, Inc.; **ILLUMINA INC**
Solgar; **NBTY INC**
Soliris; **ALEXION PHARMACEUTICALS INC**
SOLODYN; **MEDICIS PHARMACEUTICAL CORP**
Soltamax; **SAVIENT PHARMACEUTICALS INC**
Soltamox; **CYTOGEN CORPORATION**
SomatoKine; **INSMED INCORPORATED**
Somavert; **NEKTAR THERAPEUTICS**
Sonazoid for Injection; **DAIICHI SANKYO CO LTD**
Sorona; **E I DU PONT DE NEMOURS & CO (DUPONT)**
Spasfon; **CEPHALON INC**
Spezyme Fred,; **GENENCOR INTERNATIONAL INC**
Spheramine; **TITAN PHARMACEUTICALS**
Spine Wave, Inc.; **PROTEIN POLYMER TECHNOLOGIES**
Spirulina Pacifica; **CYANOTECH CORPORATION**
SPOTCHEM EZ Chemistry Analyzer; **HESKA CORP**
Squalamine lactate; **GENAERA CORPORATION**
ST-193; **SIGA TECHNOLOGIES INC**
ST-294; **SIGA TECHNOLOGIES INC**
Staarvisc-II; **ANIKA THERAPEUTICS INC**
STARGEN; **GENENCOR INTERNATIONAL INC**
Start Licensing, Inc.; **GERON CORPORATION**
STEALTH; **ALZA CORP**
Stem Cell Summit; **BURRILL & COMPANY**
Sterisomes; **NOVAVAX INC**
Sterling Synergy Systems Private Ltd.; **PRA INTERNATIONAL**
Stilnox; **SANOFI-AVENTIS**
STIMUVAX; **BIOMIRA INC**
StrataCooler; **STRATAGENE CORP**
Stratagene Corp.; **HYCOR BIOMEDICAL INC**
Straumann Dental Implant System; **INSTITUT STRAUMANN AG**
Striant; **COLUMBIA LABORATORIES INC**
Study Tracker; **COVANCE INC**
Sucanon; **BIOTECH HOLDINGS LTD**
SUDCA; **AXCAN PHARMA INC**
Sulonex; **KERYX BIOPHARMACEUTICALS INC**
Sumatriptan; **SPECTRUM PHARMACEUTICALS INC**
Sun Pharmaceutical Industries; **CARACO PHARMACEUTICAL LABORATORIES**
Sundex, LLC; **CHATTEM INC**
Sundown; **NBTY INC**
SuperGen, Inc.; **AVI BIOPHARMA INC**
Support Plus; **LIFECORE BIOMEDICAL INC**
Supprelin LA; **INDEVUS PHARMACEUTICALS INC**
Supro; **DUPONT AGRICULTURE & NUTRITION**
SureLight; **MARTEK BIOSCIENCES CORP**
Sure-Screen; **MEDTOX SCIENTIFIC INC**
Surface Safe; **SUPERGEN INC**
Surfaxin; **DISCOVERY LABORATORIES INC**
SURPASS; **IDEXX LABORATORIES INC**
SYBYL; **TRIPOS INC**

INDEX OF SUBSIDIARIES, BRAND NAMES AND AFFILIATIONS, CONT.

SYMLIN; **AMYLIN PHARMACEUTICALS INC**
Synagis; **MEDIMMUNE INC**
Synbiotics Europe; **SYNBIOTICS CORP**
SyncroDose; **PENWEST PHARMACEUTICALS CO**
Synera; **ENDO PHARMACEUTICALS HOLDINGS INC**
Syneture; **COVIDIEN LTD**
Syngenta Biotechnology, Inc.; **SYNGENTA AG**
Synovis Interventional Solutions; **SYNOVIS LIFE TECHNOLOGIES INC**
Synovis Micro Companies Alliance; **SYNOVIS LIFE TECHNOLOGIES INC**
Synovis Surgical Innovations; **SYNOVIS LIFE TECHNOLOGIES INC**
Synvisc; **GENZYME BIOSURGERY**
Taclonex; **WARNER CHILCOTT PLC**
Tadalafil; **ICOS CORPORATION**
Takeda America Holdings, Inc.; **TAKEDA PHARMACEUTICAL COMPANY LTD**
Takeda Europe Research and Development Center Ltd.; **TAKEDA PHARMACEUTICAL COMPANY LTD**
Takeda Ireland Limited; **TAKEDA PHARMACEUTICAL COMPANY LTD**
Takeda Research Investment, Inc.; **TAKEDA PHARMACEUTICAL COMPANY LTD**
Tamiflu; **ROCHE HOLDING LTD**
Tanafed; **SCIELE PHARMA INC**
Tanafed; **HI-TECH PHARMACAL CO INC**
TAP Pharmaceutical Products Inc.; **TAKEDA PHARMACEUTICAL COMPANY LTD**
Tarceva; **OSI PHARMACEUTICALS INC**
Target-Related Affinity Profiling (TRAP); **TELIK INC**
Targretin; **LIGAND PHARMACEUTICALS INC**
Taro Pharmaceuticals U.S.A., Inc.; **TARO PHARMACEUTICAL INDUSTRIES**
Tasidotin Hydrochloride; **GENZYME ONCOLOGY**
TAXOL; **BRISTOL MYERS SQUIBB CO**
Taxotere; **PHARMACYCLICS INC**
TAXUS; **ANGIOTECH PHARMACEUTICALS**
TBC3711; **ENCYSIVE PHARMACEUTICALS INC**
TBC4746; **ENCYSIVE PHARMACEUTICALS INC**
Tecadenoson; **CV THERAPEUTICS INC**
Technology Partnerships Canada; **SPECTRAL DIAGNOSTICS INC**
Teflon; **E I DU PONT DE NEMOURS & CO (DUPONT)**
Tegretol; **CARACO PHARMACEUTICAL LABORATORIES**
Telaprevir; **VERTEX PHARMACEUTICALS INC**
TELCYTA; **TELIK INC**
TELINTRA; **TELIK INC**
Telzir; **VERTEX PHARMACEUTICALS INC**
TempRx; **BIOSPHERE MEDICAL INC**
Terapia; **RANBAXY LABORATORIES LIMITED**
Terconazole; **TARO PHARMACEUTICAL INDUSTRIES**

Termamyl; **NOVOZYMES**
Terso Solutions; **PROMEGA CORP**
Testogel; **BAYER SCHERING PHARMA AG**
TestSkin & TestSkin II; **ORGANOGENESIS INC**
Teva Pharmaceuticals Industries, Ltd.; **SICOR INC**
Teva Pharmaceuticals USA; **TEVA PHARMACEUTICAL INDUSTRIES**
tezampanel; **TORREYPINES THERAPEUTICS INC**
tgAAC09; **TARGETED GENETICS CORP**
tgAAC94; **TARGETED GENETICS CORP**
Thalidomide Pharmion; **PHARMION CORP**
Thalomid; **CELGENE CORP**
The Flood Company; **AKZO NOBEL NV**
The Rainbow Coalition; **RANBAXY LABORATORIES LIMITED**
Thelin; **ENCYSIVE PHARMACEUTICALS INC**
TheraPei Pharmaceuticals, Inc.; **FORBES MEDI-TECH INC**
Therapore; **AVANT IMMUNOTHERAPEUTICS**
TheraSeed; **THERAGENICS CORP**
TherAtoh; **GENVEC INC**
Thermo Electron Group; **THERMO FISHER SCIENTIFIC INC**
Thermo Fisher Scientific, Inc.; **HARVARD BIOSCIENCE INC**
Thermo Fisher Scientific, Inc.; **PERBIO SCIENCE AB**
Ther-Rx Corporation; **KV PHARMACEUTICAL CO**
Thrombin-JMI; **KING PHARMACEUTICALS INC**
Thymoglobulin; **GENZYME CORP**
Thyrogen; **GENZYME CORP**
Tianwei Times; **QIAGEN NV**
Tiazac; **FOREST LABORATORIES INC**
TIGER; **ISIS PHARMACEUTICALS INC**
Tiger's Milk; **SCHIFF NUTRITION INTERNATIONAL INC**
TIGRIS; **GEN-PROBE INC**
TILLING; **ARCADIA BIOSCIENCES**
TIMERx; **PENWEST PHARMACEUTICALS CO**
Tinactin; **SCHERING-PLOUGH CORP**
Tissue Factor VIIa; **BIOCRYST PHARMACEUTICALS INC**
Tissue Repair Cells; **AASTROM BIOSCIENCES INC**
Tissue Repair Co.; **CARDIUM THERAPEUTICS INC**
Tissue-Guard; **SYNOVIS LIFE TECHNOLOGIES INC**
Tm; **LUMINEX CORPORATION**
TNFerade; **GENVEC INC**
TNKase; **GENENTECH INC**
TNX-355; **TANOX INC**
TNX-650; **TANOX INC**
TOBI; **CHIRON CORP**
TOCOSOL; **SONUS PHARMACEUTICALS**
TOCOSOL Camptothecin; **SONUS PHARMACEUTICALS**
TOCOSOL Paclitaxel; **SONUS PHARMACEUTICALS**
Toleragens; **LA JOLLA PHARMACEUTICAL**

INDEX OF SUBSIDIARIES, BRAND NAMES AND AFFILIATIONS, CONT.

INDEX OF SUBSIDIARIES, BRAND NAMES AND AFFILIATIONS, CONT.

VetScan HM2; **ABAXIS INC**
VetScan VS2; **ABAXIS INC**
ViaCord; **VIACELL INC**
ViaCyte; **VIACELL INC**
Viagra; **PFIZER INC**
Vidaza; **PHARMION CORP**
VIG; **CANGENE CORP**
Vioxx; **MERCK & CO INC**
Viprinex; **NEUROBIOLOGICAL TECHNOLOGIES INC**
ViraCHEK; **SYNBIOTICS CORP**
Viral Antigens; **MERIDIAN BIOSCIENCE INC**
Virazole; **VALEANT PHARMACEUTICALS INTERNATIONAL**
Virgo; **HEMAGEN DIAGNOSTICS INC**
Virtual Library; **NEUROGEN CORP**
Virtual Screening; **NEUROGEN CORP**
Virulizin; **LORUS THERAPEUTICS INC**
Visicol; **SALIX PHARMACEUTICALS**
Vistide; **GILEAD SCIENCES INC**
Visudyne; **QLT INC**
Vitagel Surgical Hemostat; **ANGIOTECH PHARMACEUTICALS**
Vita-Tech Canada, Inc.; **IDEXX LABORATORIES INC**
Vitex; **VI TECHNOLOGIES INC**
Vitravene; **ISIS PHARMACEUTICALS INC**
Vivelle; **NOVEN PHARMACEUTICALS**
Vivitrol; **ALKERMES INC**
Vivitrol; **CEPHALON INC**
Vivola; **FORBES MEDI-TECH INC**
ViziLite; **ZILA INC**
VLP Technology; **NOVAVAX INC**
v-protectant; **ATHEROGENICS INC**
VWR; **HARVARD BIOSCIENCE INC**
VX-409; **VERTEX PHARMACEUTICALS INC**
VX-702; **VERTEX PHARMACEUTICALS INC**
VX-770; **VERTEX PHARMACEUTICALS INC**
VX-883; **VERTEX PHARMACEUTICALS INC**
Vyvanse; **SHIRE PLC**
Walter Lorenz Surgical, Inc.; **BIOMET INC**
Watson Laboratories; **WATSON PHARMACEUTICALS INC**
Watson Pharma; **WATSON PHARMACEUTICALS INC**
Watson Pharmaceuticals, Inc.; **DEPOMED INC**
Webforce; **DENDRITE INTERNATIONAL INC**
WebPhage; **DYAX CORP**
Weider Global Nutrition, LLC; **SCHIFF NUTRITION INTERNATIONAL INC**
Weider Nutrition International, Inc.; **SCHIFF NUTRITION INTERNATIONAL INC**
Wellbutrin XL; **BIOVAIL CORPORATION**
Whole Blood Flow Cytometry Control; **TECHNE CORP**
Whole Blood Glucose/Hemoglobin Control; **TECHNE CORP**
WI Harper Group; **AVIVA BIOSCIENCES CORP**

WideStrike; **DOW AGROSCIENCES LLC**
William C. Conner Research Center; **ALCON INC**
WinRho SDF; **CANGENE CORP**
Wintershall AG; **BASF AG**
Wipe Out Dairy Wipes; **IMMUCELL CORPORATION**
WITNESS; **SYNBIOTICS CORP**
Wrafton Laboratories, Ltd.; **PERRIGO CO**
Wyeth K.K.; **WYETH**
Xatra; **SKYEPHARMA PLC**
Xcytrin; **PHARMACYCLICS INC**
Xechem (India) Pvt., Ltd.; **XECHEM INTERNATIONAL**
Xechem, Inc.; **XECHEM INTERNATIONAL**
Xenerex; **AVANIR PHARMACEUTICALS**
Xenogen Corp.; **CALIPER LIFE SCIENCES**
XERECEPT; **NEUROBIOLOGICAL TECHNOLOGIES INC**
XetaPharm; **XECHEM INTERNATIONAL**
Xifaxan; **SALIX PHARMACEUTICALS**
Xigris; **ELI LILLY AND COMPANY**
xMAP; **LUMINEX CORPORATION**
Xolair; **TANOX INC**
XOMA 052; **XOMA LTD**
XOMA 629; **XOMA LTD**
XOPENEX; **SEPRACOR INC**
xPONENT; **LUMINEX CORPORATION**
XUSA/XYZAL; **SEPRACOR INC**
XYOTAX; **CELL THERAPEUTICS INC**
Xyrem; **JAZZ PHARMACEUTICALS**
Xyzal; **UCB SA**
Yamanouchi Pharmaceutical Co., Ltd.; **ASTELLAS PHARMA INC**
Yasmin; **BAYER SCHERING PHARMA AG**
Yingtaiqing; **SIMCERE PHARMACEUTICAL GROUP**
Yokogawa Analytical Systems; **AGILENT TECHNOLOGIES INC**
Your Life; **LEINER HEALTH PRODUCTS INC**
ZADAXIN; **SCICLONE PHARMACEUTICALS**
Zailin; **SIMCERE PHARMACEUTICAL GROUP**
Zantac; **GLAXOSMITHKLINE PLC**
Zavesca; **ACTELION LTD**
Zeatin; **SENETEK PLC**
Zema; **VIRBAC CORP**
Zenapax; **PDL BIOPHARMA**
Zeniva; **AVANIR PHARMACEUTICALS**
Zenyth Therapeutics; **CSL LIMITED**
Zerenex; **KERYX BIOPHARMACEUTICALS INC**
Zero Order Release System; **BIOVAIL CORPORATION**
ZFN; **SANGAMO BIOSCIENCES INC**
ZFP; **SANGAMO BIOSCIENCES INC**
ZFP TF; **SANGAMO BIOSCIENCES INC**
ZIANA; **MEDICIS PHARMACEUTICAL CORP**
Zila Biotechnology, Inc.; **ZILA INC**
Zila Pharmaceuticals, Inc.; **ZILA INC**
Zila Swab Technoolgies, Inc.; **ZILA INC**

INDEX OF SUBSIDIARIES, BRAND NAMES AND AFFILIATIONS, CONT.

Zila, Ltd.; **ZILA INC**
ZLB Behring; **CSL LIMITED**
ZLB Plasma Services; **CSL LIMITED**
Zoamix; **ALPHARMA INC**
Zoloft; **PFIZER INC**
Zomig; **ASTRAZENECA PLC**
Zonegran; **EISAI CO LTD**
Zovirax; **BIOVAIL CORPORATION**
Zybrestat; **OXIGENE INC**
Zydone; **ENDO PHARMACEUTICALS HOLDINGS INC**
Zylon; **TOYOBO CO LTD**
Zymark Corp.; **CALIPER LIFE SCIENCES**
Zyprexa; **ELI LILLY AND COMPANY**
Zyrtec; **PFIZER INC**
Zyrtec; **UCB SA**
Zyzal; **UCB SA**

Plunkett Research, Ltd. 2008 Catalog

Market research, analysis, trends, statistics & companies in 29 industry sectors

▶ **Printed Almanacs with CD-ROM Database**

▶ **Instant Online Access**

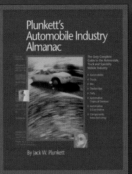

▶ Doing market research?

▶ Working a case study?

▶ Looking for business prospects?

▶ Writing a business plan?

▶ Looking for a job?

▶ Training employees?

▶ Supporting a sales department?

▶ Planning a business strategy?

Plunkett Research, Ltd.

▶ When you need accurate, timely, innovative information

▶ 713.932.0000 www.plunkettresearch.com

Get market research, industry analysis and corporate profiles in one resource!

▸ Each exciting industry title contains all of the information you need to fully understand the trends, technologies, finances and leading companies of a specific industry.

▸ Whatever field you are researching, from InfoTech to Energy to Retailing, you'll find all of the answers you need in one value-packed, industry-specific resource.

▸ **Take a tour at www.plunkettresearch.com**

Get reference titles packed with up to 700 pages of in-depth information

Here's what you'll find inside each Plunkett Research industry book:

▸ Glossary
▸ Industry Overview and Trends
▸ Segment by Segment Analysis
▸ Statistics, Charts, Tables
▸ Technologies
▸ Industry Contacts and Associations
▸ Finances
▸ Profiles of Hundreds of Leading Companies
▸ Thorough Indexes and Cross-Indexes
▸ CD-ROM Database

▸ Doing market research?
▸ Looking for business prospects?
▸ Writing a business plan?
▸ Planning a business strategy?
▸ Looking for a job?

Plunkett's Outsourcing & Offshoring Industry Almanac 2008

Get the complete picture of the booming world of outsourcing: trends, statistics, glossary, contacts and profiles of leading companies from the U.S. to India to China.

Plunkett's Wireless, Wi-Fi, RFID & Cellular Industry Almanac 2008

Few sectors offer more potential for growth and change than the wireless industry. See the trends and the leading firms in RFID inventory management, cell phones and much more.

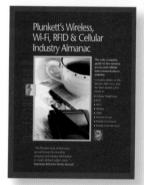

Plunkett's Renewable, Alternative & Hydrogen Energy Industry Almanac 2008

Billions of dollars are pouring into alternative energy production, research and development on a global basis. Learn about the trends, the terms, the leaders and the statistics in safer nuclear, cleaner coal, fuel cells and much more.

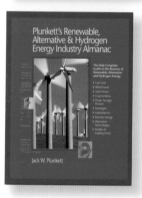

It's easy to place your order

Printed Almanacs with CD-ROM Database
▸ Use the order form
▸ Order online at www.plunkettresearch.com
▸ By phone at 713.932.0000
▸ By fax at 713.932.7080
▸ By mail at:
P.O. Drawer 541737 Houston, TX 77254

Plunkett Research Online
▸ Call for access pricing and online demo:
713.932.0000

"A critically important reference book" *Midwest Review of Books*

"A uniquely accessible book; highly recommended for business and career collections." *Library Journal*

Quantity

Advertising, Branding & Marketing
__ Plunkett's Advertising & Branding Industry Almanac 2008
ISBN 978-1-59392-109-5 Apr 2008 $299.99 451 pages

Airlines, Hotel & Travel
__ Plunkett's Airline, Hotel & Travel Industry Almanac 2008
ISBN 978-1-59392-093-7 Sept 2007 $299.99 463 pages

Apparel & Textiles
__ Plunkett's Apparel & Textiles Industry Almanac 2008
ISBN 978-1-59392-110-1 Apr 2008 $299.99 445 pages

Automobile
__ Plunkett's Automobile Industry Almanac 2008
ISBN 978-1-59392-094-4 Oct 2007 $299.99 538 pages

Chemicals, Coatings & Plastics
__ Plunkett's Chemicals, Coatings & Plastics Industry Almanac 2008
ISBN 978-1-59392-091-3 Jul 2007 $299.99 491 pages

Computers, E-Commerce & Internet
__ Plunkett's E-Commerce & Internet Business Almanac 2008
ISBN 978-1-59392-105-7 Mar 2008 $299.99 568 pages
__ Plunkett's InfoTech Industry Almanac 2008
ISBN 978-1-59392-104-0 Feb 2008 $299.99 650 pages

Consulting, Outsourcing & Offshoring
__ Plunkett's Consulting Industry Almanac 2008
ISBN 978-1-59392-113-2 May 2008 $299.99 372 pages
__ Plunkett's Outsourcing & Offshoring Industry Almanac 2008
ISBN 978-1-59392-088-3 Jun 2007 $299.99 439 pages

Energy
__ Plunkett's Energy Industry Almanac 2008
ISBN 978-1-59392-099-9 Dec 2007 $299.99 660 pages
__ Plunkett's Renewable, Alternative &
Hydrogen Energy Industry Almanac 2008
ISBN 978-1-59392-100-2 Dec 2007 $299.99 350 pages

Engineering, Research & Nanotechnology
__ Plunkett's Engineering & Research Industry Almanac 2008
ISBN 978-1-59392-111-8 May 2008 $299.99 532 pages
__ Plunkett's Nanotechnology & MEMS Industry Almanac 2008
ISBN 978-1-59392-114-9 Jun 2008 $299.99 400 pages

Entertainment & Media
__ Plunkett's Entertainment & Media Industry Almanac 2008
ISBN 978-1-59392-103-3 Jan 2008 $299.99 555 pages

Financial Services, Banking, Insurance, Investments & Mortgages
__ Plunkett's Banking, Mortgages & Credit Industry Almanac 2008
ISBN 978-1-59392-098-2 Nov 2007 $299.99 458 pages
__ Plunkett's Insurance Industry Almanac 2008
ISBN 978-1-59392-097-5 Nov 2007 $299.99 442 pages
__ Plunkett's Investment & Securities Industry Almanac 2008
ISBN 978-1-59392-102-6 Jan 2008 $299.99 452 pages

Food & Beverage
__ Plunkett's Food Industry Almanac 2008
ISBN 978-1-59392-106-4 Mar 2008 $299.99 541 pages

Health Care & Biotech
__ Plunkett's Biotech & Genetics Industry Almanac 2008
ISBN 978-1-59392-087-6 Sept 2007 $299.99 560 pages
__ Plunkett's Health Care Industry Almanac 2008
ISBN 978-1-59392-096-8 Oct 2007 $299.99 715 pages

Quantity

Job Seeker & Careers
__ The Almanac of American Employers 2008
ISBN 978-1-59392-095-1 Oct 2007 $279.99 748 pages
__ Plunkett's Companion to The Almanac
of American Employers 2008
ISBN 978-1-59392-107-1 Mar 2008 $279.99 689 pages

Middle Market Companies
__ Plunkett's Almanac of Middle Market Companies 2008
ISBN 978-1-59392-092-0 Jul 2007 $299.99 636 pages

Real Estate & Construction
__ Plunkett's Real Estate & Construction Industry Almanac 2008
ISBN 978-1-59392-112-5 May 2008 $299.99 520 pages

Retail
__ Plunkett's Retail Industry Almanac 2008
ISBN 978-1-59392-101-9 Dec 2007 $299.99 630 pages

Sports
__ Plunkett's Sports Industry Almanac 2008
ISBN 978-1-59392-089-0 Jul 2007 $299.99 456 pages

Quantity

Telecommunications & Wireless
__ Plunkett's Telecommunications Industry Almanac 2008
ISBN 978-1-59392-086-9 Aug 2007 $299.99 622 pages
__ Plunkett's Wireless, Wi-Fi, RFID & Cellular Industry Almanac 2008
ISBN 978-1-59392-090-6 Jul 2007 $299.99 441 pages

Transportation, Supply Chain & Logistics
__ Plunkett's Transportation, Supply Chain & Logistics Industry Almanac 2008
ISBN 978-1-59392-108-8 Mar 2008 $299.99 650 pages

Call today for Plunkett Online pricing and subscriptions

Plunkett Research Order Form for Printed Almanacs with CD-ROM Database

STANDING ORDER ☐ Check here for a standing order and receive a 10% discount on future editions of the book(s) you have ordered, which will be shipped to you automatically. You may cancel your standing order at any time.

METHOD OF PAYMENT (Our Federal ID number is 74-2440918)

Purchase order number (if any): _____

☐ Check enclosed ☐ Bill me ☐ Credit card (check one) ☐ VISA ☐ MasterCard ☐ American Express

Credit Card No. _____ Expiration Date: _____

SHIP TO: Name _____

Title/Department _____

Organization_____

Address _____

City_____ State _____ Zip _____

Telephone _____ Fax _____

E-mail_____

Subtotal	$ _____
Shipping/handling add $9.50 per book US ground	$ _____
Next day air in the US, add $28.50 per book	$ _____
Canada, ground add $11.00 per book	$ _____
FedEx outside USA, add $58.50 per book	$ _____
In Texas add 8.25% sales tax if you are not tax exempt	$ _____
TOTAL	$ _____

Plunkett Research, Ltd.
P. O. Drawer 541737 • Houston, Texas 77254 USA
Phone: 713.932.0000 • FAX: 713.932.7080
www.plunkettresearch.com

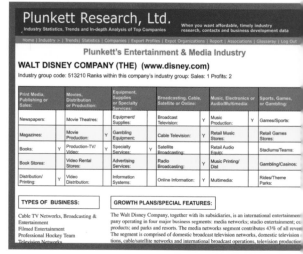
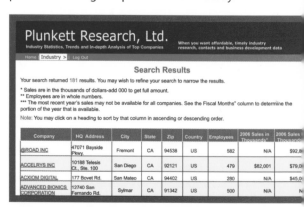